International Conference on Particle Physics and Astrophysics 2015 (ICPPA-2015)

Journal of Physics: Conference Series Volume 675

Moscow, Russia
5 – 10 October 2015

Editors:

Mikhail Skorohvatov
Mikhail Danilov
Arkady Galper
Anatoly Petrukhin
Valery Dmitrenko

ISBN: 978-1-5108-2082-1
ISSN: 1742-6588

Printed from e-media with permission by:

Curran Associates, Inc.
57 Morehouse Lane
Red Hook, NY 12571

Some format issues inherent in the e-media version may also appear in this print version.

Printed by Curran Associates, Inc. (2016)

For permission requests, please contact the Institute of Physics at the address below.

Institute of Physics
Dirac House, Temple Back
Bristol BS1 6BE UK

Phone: 44 1 17 929 7481
Fax: 44 1 17 920 0979

techtracking@iop.org

Additional copies of this publication are available from:

Curran Associates, Inc.
57 Morehouse Lane
Red Hook, NY 12571 USA
Phone: 845-758-0400
Fax: 845-758-2634
Email: curran@proceedings.com
Web: www.proceedings.com

Table of contents

Volume 675, 2016
International Conference on Particle Physics and Astrophysics (ICPPA-2015)
5–10 October 2015, Moscow, Russia
Editors- Mikhail Skorohvatov, Mikhail Danilov, Arkady Galper, Anatoly Petrukhin and Valery Dmitrenko

Preface

Particle physics, 2016

Nuclear physics, 2016

Methods of experimental physics, 2016

042050

Journal of Physics: Conference Series **675** (2016) 001001 doi:10.1088/1742-6596/675/0/001001

International conference on particle physics and astrophysics

The International Conference on Particle Physics and Astrophysics (ICPPA-2015) was held in Moscow, Russia, from October 5 to 10, 2015. The conference is organized by Center of Fundamental Research and Particle Physics of National Research Nuclear University "MEPhI". The aim of the Conference is to promote contacts between scientists and development of new ideas in fundamental research. We bring together experts and young scientists working on experimental and theoretical aspects of nuclear, particle, astroparticle physics and cosmology.

The conference covers a wide range of topics such as accelerator physics, (astro) particle physics, cosmic rays, cosmology and methods of experimental physics - detectors and instruments. These directions are unified by development of the Standard Model (SM) which is evidently not complete. There are deviations from the Standard Model – neutrino oscillations, the dark matter existence. Together with strong interactions, they are main subjects of the Conference.

New results from LHC collider as well as its future upgrade are discussed with the Higgs as the main point for discussion. Substantial development of experimental tools for astrophysical observations and new results from cosmic ray experiments is one of the main subjects of the conference. Various aspects of strong interaction are discussed. Among them: Charmonium and Bottomonium states, Flavor physics at Super B factories, Exotic Nuclei in Astrophysics.

Another subject for discussion is the neutrino physics, promising and unique way to get new knowledge. In this content, several talks on BOREXINO experiment where new results in neutrino oscillations are presented.

Special session is devoted to PAMELA experiment - 9 years in orbit and to the future GAMMA–400 gamma-ray telescope with following main scientific goals: indirect dark matter origin study by the gamma-ray astronomy methods, discrete astrophysical sources observations, diffuse background γ-emission analysis, investigation of high energy γ-emission from GRBs and other astrophysical objects including Sun, explore of high energy e−e+ fluxes, research of high energy light nuclei fluxes.

Several talks concern the cosmic ray physics with high energy muons in large volume detectors – IceCube, NEVOD are among them.

Also γ-quanta and charged particles generation in astrophysical sources and planets magnetospheres were discussed in the frameworks of Cosmic Rays subsection.

Cosmological problems such as an origin and features of the dark matter, space reionization, smallness of the cosmological constant and black hole origin are discussed at the theoretical section of the conference.

Special workshops are devoted to NEVOD experiment, Pamela Experiment, SiPM and Electronics.

International Advisory Committee

S Rubin, NRNU MEPhI, Moscow, Russia – chair man
A Galper, NRNU MEPhI, Moscow, Russia – vice chair man
A Petrukhin, NRNU MEPhI, Moscow, Russia - vice chair man
M Skorokhvatov, NRC Kurchatov Institute, Moscow, Russia - vice chair man
P Berghaus, DESY, D-15735 Zeuthen
M Danilov, ITEP NRC Kurchatov Institute, Moscow, Russia

V Dmitrenko, NRNU MEPhI, Moscow, Russia
J Fabris, UFES, Brasil
M Khlopov, APC, Paris, France
R Konoplich, NYU, New York, USA
Yu Kudenko, INR RAS, Russia
P Picozza University Tor Vergata Rome, Italy
C Spiering, DESY, D-15735 Zeuthen, Germany
Y Suvorov, University of California Los Ahgeles (UCLA), Los Angeles, USA
E Troitskaya, SINP MSU, Moscow, Russia
A Vacchi INFN, Triest, Italy

Local Organizing Committee (NRNU MEPhI, Moscow, Russia)

I Arkhangelskaja – Scientific secretary
P Buzhan
E Chasovikov
S Lazarev
O Murashova
R Pritula
L Savushkina
A Shustov
E Soldatov
P Teterin

Journal of Physics: Conference Series **675** (2016) 001002 doi:10.1088/1742-6596/675/0/001002

International conference on particle physics and astrophysics

1. Scientific Editors

Section 1: Particle physics
Mikhail Skorohvatov – NRC Kurchatov Institute, Moscow, Russia

Section 2: Nuclear hpysics
Mikhail Danilov – ITEP NRC Kurchatov Institute, Moscow, Russia

Section 3: Cosmic rays
Arkady Galper – NRNU MEPhI, Moscow, Russia
Anatoly Petrukhin – NRNU MEPhI, Moscow, Russia

Section 4: Methods of experimental physics
Valery Dmitrenko – NRNU MEPhI, Moscow, Russia

2. Publishing Editors

Irene Arkhangelskaja – NRNU MEPhI, Moscow, Russia
Pavel Buzhan – NRNU MEPhI, Moscow, Russia

The organizers of the conference would like to thank for the support from National Research Nuclear University MEPhI in the framework of the Russian Academic Excellence Project (contract No. 02.a03.21.0005, 27.08.2013).

5100

RUSSIAN ACADEMIC
EXCELLENCE PROJECT

Journal of Physics: Conference Series **675** (2016) 001004 doi:10.1088/1742-6596/675/0/001004

Peer review statement

All papers published in this volume of *Journal of Physics: Conference Series* have been peer reviewed through processes administered by the proceedings Editors. Reviews were conducted by expert referees to the professional and scientific standards expected of a proceedings journal published by IOP Publishing.

The development of cross disciplinary studies in cosmology and particle physics on the platform of a Scientific-Educational complex of Virtual Institute of Astroparticle physics (VIA)

M Yu Khlopov[1,2,3]

[1] National Research Nuclear University MEPhI (Moscow Engineering Physics Institute), Kashirskoe highway 31, Moscow, 115409, Russia
[2] Centre for Cosmoparticle Physics "Cosmion" 115409 Moscow, Russia
[3] APC laboratory 10, rue Alice Domon et Léonie Duquet
75205 Paris Cedex 13, France

E-mail: khlopov@apc.univ-paris7.fr

Abstract. The overview of the main problems and methods of cosmoparticle physics, studying fundamental relationship of micro- and macro-worlds, is accompanied by the demonstration of the facility of Virtual Institute of Astroparticle physics (VIA) activity as a well proven platform for scientific collaborative work and education at distance.

1. Introduction

Cosmoparticle physics [1, 2, 3, 4, 5] studies the fundamental basis and mutual relationship between micro-and macro-worlds in the proper combination of physical, astrophysical and cosmological signatures. It originated from the modern trend of particle physics to go beyond the Standard model and from the development of the modern cosmology, implying new physics to explain inflation, baryosynthesis and dark matter and energy - its basic elements, proved by observations but having no grounds in the physics of the Standard model. Here we concentrate on some aspects of this relationship, which arise in the problem of cosmological Dark Matter (DM).

Extensions of the standard model imply new symmetries and new particle states. The respective symmetry breaking induces new fundamental physical scales in particle theory. In particle theory Noether's theorem relates the exact symmetry to conservation of respective charge. If the symmetry is strict, the charge is strictly conserved. The lightest particle, bearing this charge, is stable. It gives rise to the fundamental relationship between dark matter candidates and particle symmetry beyond the Standard model.

If the symmetry is broken, the mechanism of the symmetry breaking implies restoration of the symmetry at high temperatures and densities. Such high temperatures and densities should have naturally arisen at the early stages of cosmological evolution. It makes Big Bang Universe natural laboratory of particle physics, not only due to possibility of creation of hypothetical particles in the early Universe, but also owing to reflection of the hierarchy of particle symmetry breaking in cosmological phase transitions.

2. Cosmoparticle physics of dark matter

According to the modern cosmology, the dark matter corresponds to $\sim$ 25% of the total cosmological density. This form of matter (see e.g. [5, 6] for review and reference) should be stable and saturate the measured dark matter density. Formation of the Large Scale Structure of the Universe from small initial density fluctuations is one of the most important reasons for the *nonbaryonic* nature of the dark matter that is decoupled from matter and radiation at least before the beginning of matter dominated stage and provides the effective growth of these fluctuations before recombination. The easiest way to satisfy these conditions is to assume that the dark matter candidates from the physics beyond the Standard model are Weakly Interacting massive particles (WIMPs) (see [6] for a review). However, problems of direct and indirect dark matter searches may imply more sophisticated solutions for this problem.

On the other hand, the initial density fluctuations, coming from the very early Universe are also originated from physics beyond the Standard model. It links the primordial seeds of galaxy formation to effects of particle theory at very high energies.

2.1. Dark matter and physics of very early Universe

2.1.1. Charge symmetric case. WIMPs The simplest primordial form of new physics is the gas of new stable massive particles, originated from early Universe. For particles with the mass m, at high temperature $T > m$ the equilibrium condition, $n \cdot \sigma v \cdot t > 1$ is valid, if their annihilation cross section $\sigma > 1/(mm_{Pl})$ is sufficiently large to establish the equilibrium. At $T < m$ such particles go out of equilibrium and their relative concentration freezes out. This is the main idea of calculation of primordial abundance for Weakly Interacting Massive Particles (WIMPs).

If ordinary particles are among the products of WIMP annihilation, even their small fraction can annihilate in the Galaxy causing significant effect in cosmic rays and gamma background. This effect, first revealed in [7] and then proved for even subdominant fraction of annihilating dark matter in [8], is now in the basis of indirect dark matter searches in cosmic rays [9].

The process of WIMP annihilation to ordinary particles, considered in $t$channel, determines their scattering cross section on ordinary particles and thus relates the primordial abundance of WIMPs to their scattering rate in the ordinary matter. Forming nonluminous massive halo of our Galaxy, WIMPs can penetrate the terrestrial matter and scatter on nuclei in underground detectors. The strategy of direct WIMP searches implies detection of recoil nuclei from this scattering.

Weak interaction of dark matter particles with matter can be directly detected by recoil caused by these particles in the matter. This idea was first proposed for massive neutrinos in the diploma work by Zeldovich's pupil V.Shvartsman in 1969 and was published in [11]. Direct search of weakly interactive massive particles (WIMPs) is now under way (see [12] for review and references) giving controversial results, what may find a nontrivial explanation by dark atoms.

The process inverse to annihilation of WIMPs corresponds to their production in collisions of ordinary particles. It should lead to effects of missing mass and energy-momentum, being the challenge for experimental search for production of dark matter candidates at accelerators, e.g. at the LHC.

2.1.2. Charge asymmetry of dark matter The fact that particles are not absolutely stable means that the corresponding charge is not strictly conserved and generation particle charge asymmetry is possible, as it is assumed for ordinary baryonic matter. At sufficiently strong particle annihilation cross section excessive particles (antiparticles) can dominate in the relic density, leaving exponentially small admixture of their antiparticles (particles) in the same way as primordial excessive baryons dominate over antibaryons in baryon asymmetric Universe. In this case *Asymmetric dark matter* doesn't lead to significant effect of particle annihilation in the modern Universe and can be searched for either directly in underground detectors or indirectly by effects of decay or condensation and structural transformations of e.g. neutron stars (see [10] for recent review and references).

2.2. Dark atoms

2.2.1. Charged stable relics. *Dark atoms* A possibility that new stable charged leptons and quarks can exist was proposed in Glashow's tera-helium solution for dark matter [13]. This solution came from the assumed million times heavier partners of quarks and charged leptons, among which tera-E-electron and tera-U quark were sufficiently stable to form neutral [(UUU)EE] tera-helium "atoms" that behaved like WIMPs. The main problem for this solution was to suppress the abundance of positively charged species bound with ordinary electrons, which behave as anomalous isotopes of hydrogen or helium. This problem turned to be unresolvable [14], since the model [13] predicted stable tera-electrons E^- with charge -1. As soon as primordial helium is formed in the Standard Big Bang Nucleosynthesis (SBBN) it captures all the free E^- in positively charged $(HeE)^+$ ion, preventing any further suppression of positively charged species. Therefore, in order to avoid anomalous isotopes overproduction, stable particles with charge -1 (and corresponding antiparticles) should be absent, so that stable negatively charged particles should have charge -2 only.

2.2.2. Charged dark atoms' constituents Elementary particle frames for heavy stable -2 charged species are provided by: (a) stable "antibaryons" $\bar{U}\bar{U}\bar{U}$ formed by anti-U quark of fourth generation (b) AC-leptons, predicted in the extension of standard model, based on the approach of almost-commutative geometry; (c) Technileptons and anti-technibaryons in the framework of walking technicolor models (WTC). (see [6] for review and references). Since all these models also predict corresponding +2 charge antiparticles, cosmological scenario should provide mechanism of their suppression, what can naturally take place in the asymmetric case, corresponding to excess of -2 charge species, O^{--}. Then their positively charged antiparticles can effectively annihilate in the early Universe.

If new stable species belong to non-trivial representations of electroweak SU(2) group, sphaleron transitions at high temperatures can provide the relationship between baryon asymmetry and excess of -2 charge stable species, as it was demonstrated in the case of WTC in [15].

2.2.3. Dark atoms cosmology A cosmological model of O-helium dark matter can explain the puzzles of direct dark-matter searches. The proposed explanation is based on a specific mechanism for low-energy binding of OHe with nuclei. Within the uncertainty of nuclear physics parameters there exists a range for which the OHe binding energy with sodium is in the interval 2-4 keV. The annual modulation in the radiative capture of OHe to this bound state leads to the corresponding energy release observed as an ionization signal in DAMA/NaI and DAMA/LIBRA. An interesting feature of this explanation is the conclusion that the ionization signal may be absent in detectors containing light (e.g. ^3He) or heavy (e.g. Xe) elements. Therefore a test of the results of the DAMA/NaI and DAMA/LIBRA experiments by other experimental groups can become a very non-trivial task.

The advantages of the OHe Warmer-than-Cold-dark-matter scenario is that it is minimally related to the parameters of new physics and is dominantly based on the effects of known atomic and nuclear physics. These effects need proper quantitative description of OHe interaction with matter and its development confronted with the experimental data will provide the complete test of the composite dark matter model. The model can explain the observed excess of low and high energy positrons in the Galaxy for the fixed values of O^{--} masses around 1 TeV. It challenges search for stable double charged particles at accelerators and cosmic rays as direct experimental probe for charged constituents of dark atoms of dark matter.

3. VIA platform for cosmoparticle physics

The development of cosmoparticle physics implies cross disciplinary studies and Virtual Institute of Astroparticle Physics (VIA) [16] was organized as a platform for this development. The activity of the Institute takes place on its website (see [17] for the latest review) in a form of regular weekly videoconferences with VIA lectures, covering all the theoretical and experimental activities in astroparticle physics and related topics. The library of records of these lectures, talks and their presentations is accomplished by multi-lingual Forum.

One of the interesting forms of VIA activity is the educational work at distance. For the last six years M.Khlopov's MEPhI course "Introduction to cosmoparticle physics" is given in the form of VIA videoconferences and the records of these lectures and their presentations are put in the corresponding directory of the Forum [17]. Having attended the VIA course of lectures in order to be admitted to exam students should put on Forum a post with their small thesis. Professor's comments and proposed corrections are put in a Post reply so that students should continuously present on Forum improved versions of work until it is accepted as satisfactory. Then they are admitted to pass their exam. The record of videoconference with their oral exam is also put in the corresponding directory of Forum. Such procedure provides completely transparent way of evaluation of students' knowledge. Since 2014 the second part of this course is used for an application of VIA system as a possible supplementary tool for Massive Online Open Courses (MOOC) activity [17]. In the context of MOOC VIA facility can be used for individual online work with advanced students.

The experience gained in MEPhI in the use of VIA facility makes natural appearance of VIA Session at the present conference.

Acknowledgements

I thank the Organizers for the unique possibility to use VIA system for discussion at distance the problems of dark matter. This work was performed within the framework of the Center FRPP supported by MEPhI Academic Excellence Project (contract 02.03.21.0005, 27.08.2013), supported by the Ministry of Education and Science of the Russian Federation, Project No. 3.472.2014/K and the work on the forms of dark matter by grant RFBR 14-22-03048.

References

[1] Sakharov A D 1989 *USSR Academy of Sciences journal* **4** 39
[2] Khlopov M Yu 2001 *Herald of the Russian Academy of Sciences* **71** 1133
[3] Khlopov M Yu 1999 *Cosmoparticle physics* (New York-London-Hong Kong-Singapore:World Scientific)
[4] Khlopov M Yu 2011 *Basics of Cosmoparticle physics* (Moscow:URSS Publishing)
[5] Khlopov M 2012 *Fundamentals of Cosmic Particle physics* (Cambridge:CISP-SPRINGER)
[6] Khlopov M Yu 2013 *Int. J. Mod. Phys.* A **28** 1330042
[7] Zeldovich Ya B *et al.* 1980 *Sov. J. Nucl. Phys.* **31** 664
[8] Fargion D *et al.* 1995 *Phys. Rev.* D **52** 1828
[9] Jungman G, Kamionkowski M and Griest K 1996 *Phys. Rept.* **267** 195
[10] Petraki K and Volkas R R 2013 *Int. J. Mod. Phys.* A **28** 1330028
[11] Shvartsman V F *et al.* 1982 *JETP Lett.* **36** 277
[12] Bernabei R *et al.* 2013 *Int. J. Mod. Phys.* A **28** 1330022
[13] Glashow S L 2005 *arXiv*:0504287
[14] Fargion D and Khlopov M 2005 *Gravitation and Cosmology* **19** 219 (*arXiv*:0507087)
[15] Khlopov M Y and Kouvaris C 2008 *Phys. Rev.* D **77** 065002
[16] Khlopov M Y 2008 *arXiv*:0801.0376
[17] Khlopov M Y 2014 *Bled Workshops in Physics* **15** 285

The Canfranc Underground Laboratory: a multidisciplinary underground facility

A Ianni
Canfranc Underground Laboratory, Paseo de los Ayerbe, s/n E-22880, Canfranc Estaciòn (Huesca), Spain

E-mail: aianni@lsc-canfranc.es

Abstract. General characteristics of deep underground laboratories are described. The Canfranc Underground Laboratory is reviewed. Infrastructures and experiments at the Canfranc Underground Laboratory are presented. Future possibilities for the Canfranc Underground Laboratory are discussed.

1. Introduction

Research activities in deep underground sites started in 1965 with the first measurement of horizontal muons induced by atmospheric neutrinos inside mines in India [1] and South Africa [2]. These two locations were very deep, 2700m and 3200m, respectively. In 1966 the Baksan Neutrino Observatory started operating in the Caucasus [3]. This research facility has been the first fully dedicated to underground science with horizontal access. In 1968 the first solar neutrino was detected in the Homestake mine, USA [4]. In the Kamioka mine, Japan, a new underground facility with horizontal access started operating in 1983 with the Kamiokande water Cherenkov detector [5]. In 1987 the Gran Sasso Laboratory, Italy, became the largest underground facility for Astroparticle Physics [6]. Since than a number of other underground sites have been built, the most active ones being: SNOLab in Canada [7], Canfranc Laboratory (LSC) in Spain [8], Modane (LSM) in France [9], Boulby in the UK [10], SURF in USA [11], INO in India [12] and CJPL in China [13]. These so-called Deep Underground Laboratories (DULs) have a number of fundamental features which we briefly review. Access: it can be vertical or horizontal. Horizontal access facilitates operations underground, transport of equipments. Muon flux: it depends on depth and at present, to make an example, roughly changes from 10^{-3} $m^{-2}s^{-1}$ at LSC and LNGS to 10^{-6} $m^{-2}s^{-1}$ at SNOLab and CJPG. Radiogenic neutrons flux: it does not depend on depth for DULs but on local geology and material used for lining the underground cavity; the flux is of order 10^{-2} $m^{-2}s^{-1}$ and the energy spectrum below 10 MeV. Cosmogenic neutrons flux: it depends on depth; the flux is of order ten times or more smaller than for radiogenic neutrons and the energy range is very broad. Gamma background: it does not depend on depth but on local environment; it is of order 10^4 $m^{-2}s^{-1}$. Radon: it depends on local environment, in DULs is of order 50-100 Bq/m^3 due to forced ventilation; it can have seasonal variations; in Boulby, a salt mine, it is of order < 1 Bq/m^3.

The research carried out in DULs is mainly on Astroparticle Physics, namely neutrino physics (atmospheric neutrinos, solar neutrinos, neutrinos from the Earth, neutrinos supernovae, neutrinos from beams, neutrinos from power reactors, neutrinos from artificial sources) and direct dark matter observation. Due to the low background environment rare processes are searched for in DULs: neutrinoless double beta decay is the most important one.

In recent years DULs are also used to carry out geophysics research on local and teleseismic scales. In 1990s it became known that life extends deep underground in extreme environments. A large number of microorganisms live in subsurface. The interest in learning about life in underground has grown and DULs are getting involved.

From the above considerations it turns out that DULs, at the present time, are multidisciplinary infrastructures for research.

In this paper we review in details the Canfranc facility (LSC) in Spain. In section 2 we give a general description of the LSC. In sections 3, 4 and 5 we discuss research activities at the LSC. In Section 6 we draw our conclusions.

2. The Canfranc Underground Laboratory

The Canfranc Laboratory (LSC) is located at about 1100 m above sea level in the Spanish Pyrenees. In 1985, A. Morales and collaborators from the University of Zaragoza started using the abandoned train tunnel at Canfranc to carry out Astroparticle Physics experiments. At that time only old service cavities were used. At present, this space is known with the name LAB780, taking into account the distance to the Spanish entrance to the train tunnel. In 1994, a 8-km road tunnel was excavated to connect Spain and France, the so-called Somport tunnel. During the excavation an experimental hall of 118 m^2 was built at about 2500 m away from the Spanish entrance on a side of the train tunnel. This cavity is known as LAB2500. In 2006, a larger excavation was started with the idea to make at Canfranc an international underground facility for Astroparticle Physics. This excavation is divided into two halls, named A and B. The new laboratory is known as LAB2400. In figure 1 we show the underground space at Canfranc. The large Hall A 40x15x12(h) m^3 was built in the direction to CERN, thinking about the possibility to receive a neutrino beam.

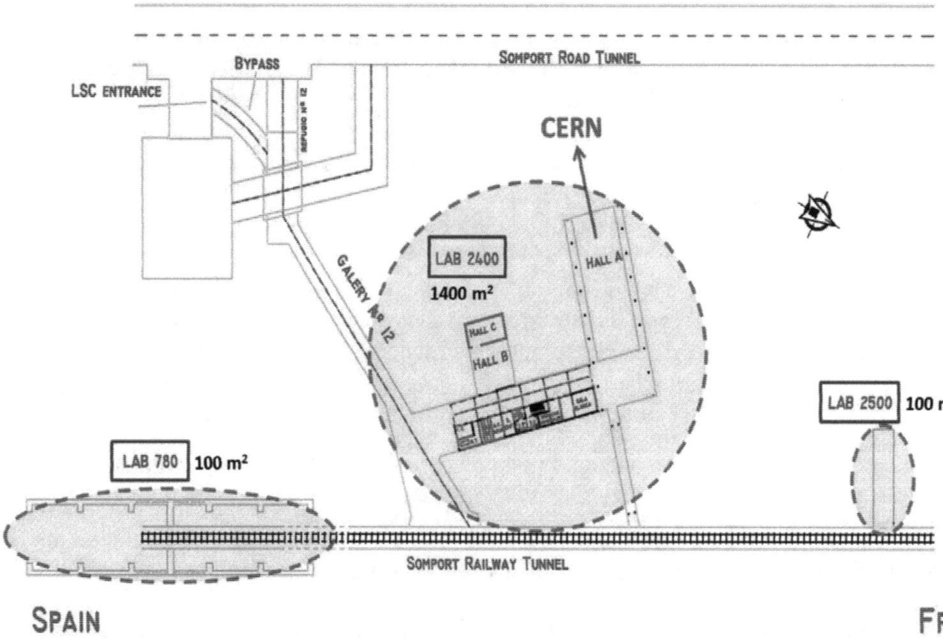

Figure 1. Layout of underground space at Canfranc Laboratory.

The total underground space available for research activities at Canfranc is about 10000 m^3 which corresponds to a total surface of 1600 m^2. In the LAB2400, Hall B has dimensions 15x10x8(h) m^3. The protocol to access the underground area establishes that users enter from the road tunnel and exit from the train tunnel.

2.1. Service facilities at LSC
LSC is equipped with a number of facilities fundamental for all underground laboratories:

- Screening facility equipped with seven high purity germanium (HpGe) γ-spectrometers (p-type), one SAGe well, one proportional α/β-counter, an alpha spectrometer (from 2016), a NaI(Tl) scintillator 3" x 3", four AlphaGuard detectors for radon monitoring, a Rn detector with mBq/m^3 sensitivity (from 2016);
- Underground Clean Room: ISO7 and ISO6 in a sector;
- Radon Abatement system (from 2016): to produce Rn-free air;
- A Chemistry Laboratory on surface;
- A Workshop on surface and underground;
- A computing facility.

In table 1 we report the background rates of some HpGe at LSC in the low energy range.

Table 1. Background rate in counts/day of some of the HpGe at LSC in the lower energy bin.

Detector	100-2700 keV
GeOroel	148±1
GeTobazo	436±2
GeLatuca	314±2
GeAspe	433±1

The chemistry laboratory at LSC can produce electroformed copper which has a lower radioactivity contamination than high quality commercial copper.

2.3. Monitoring
LSC has two monitoring programs: a first one for radon and underground water; a second for convergence measurements. In collaboration with the University of Zaragoza a number of measurements on radon background and water quality parameters are carried out on a regular basis. In addition, a unique feature of LSC, to comply safety requirements for operating tunnels, data are collected to detect signs of instability underground. The conventional method of convergence is implemented together with a set of 18 optic fibers equipped with temperature and humidity sensors. Data has been collected for more than four years on a number of locations underground.

2. The Canfranc Underground Laboratory
The main research program at LSC is the direct detection of dark matter and neutrinoless double beta decay. In addition support activities for projects performed in other laboratories are carried out by means of the facilities reported in the previous Section. In table 2 we summarize all projects which have been done or are in progress at LSC.

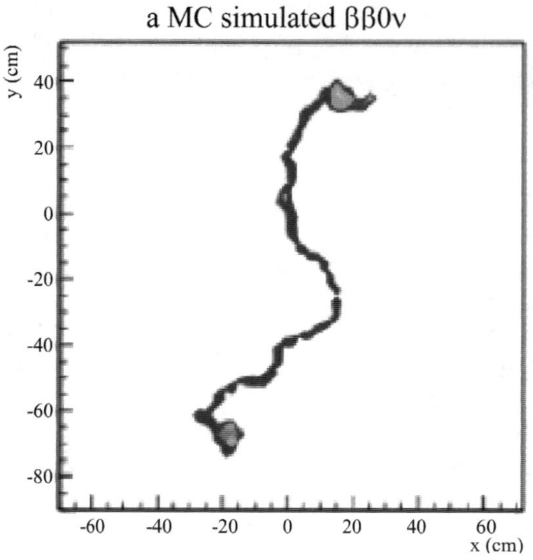

Figure 2. Monte Carlo simulation of a double beta decay event in NEXT.

Table 2. Experiments and experimental activities at LSC.

Project	Status
ROSEBUD (DM[a] R&D)	Stopped
ANAIS(DM[a])	Running
ArDM (DM[a])	Running
NEXT-NEW (DBD[b])	Demonstrator
BiPo (DBD[b] R&D)	Running
BiPo (DBD[b] R&D)	Running
SuperKGd (R&D)	

[a]DM=Dark Matter.
[b]DBD=Double Beta Decay.
See text for details.

ANAIS [14] will make use in 2016 of 112 kg of high purity NaI(Tl) to search for the annual modulation observed in DAMA/LIBRA. Eventually ANAIS will use 250 kg of NaI(Tl). ArDM [15] makes use of 2 tons of liquid argon in a two-phase TPC to search for WIMPs recoil. At present it has been operated in single phase. The two-phase mode will start in 2016.

NEXT [16] aims to search for neutrinoless double beta decay with 100 kg of Xe enriched at 90% in ^{136}Xe. NEXT exploits a TPC in high pressure Xe (15 bars) and two readout planes: the cathode with photomultipliers to detect the energy deposited in the Xe gas; the anode with SiPMs to determine the topology of the event by means of drifting ionization electrons. NEXT will be able to distinguish signals from double electrons (DBD) with respect to signals with single electrons (background). This feature improves the background subtraction for DBD search. At present, NEXT is commissioning a 10 kg demonstrator named NEXT-NEW. The demonstrator has 50 cm drift field, 12 phtomultipliers at the cathode and 1800 SiPMs at the anode. The inner TPC vessel surface is coated with TPB. In figure 2 we show a Monte Carlo simulation of a double beta decay event with two blobs at the end of the tracks. On the contrary, a background event induced by a β or γ will produce just one blob at the end of the track.

BiPo is a set-up to measure ^{214}Bi and ^{208}Tl contamination at μBq/kg level on planar geometry. The method of tagging searches for Bi-Po space and time $\beta-\alpha$ correlated coincidences in the ^{238}U and ^{232}Th chains. The set-up has an active surface of 3.6 m^2 equipped with 24 sectors, each with two photomultipliers and scintillators. At present BiPo is used to determine the contamination of ^{82}Se foils for the SuperNEMO project [17].

SuperKGd is an experimental activity which aims to select a low radioactivity salt with gadolinium, $Gd_2(SO_4)_3$ for SuperKamiokande. SuperKGd makes use of the screening facility at LSC.

3. Nuclear Astrophysics

At LSC there is interest in developing a facility for nuclear astrophysics. This effort is known as the Canfranc Underground Nuclear Astrophysics (CUNA) project. A feasibility study to make a new excavation has been performed with this purpose. CUNA aims to measure the most relevant cross sections in the s-process, namely:

$$^{22}Ne(\alpha,n)^{25}Mg \text{ and } ^{13}C(\alpha,n)^{16}O \qquad (1)$$

The measurement is performed by tagging the neutron produced. Therefore, the neutron background is limiting the sensitivity of the project. For this reason the CUNA collaboration has carried out measurements in Hall A at LSC to determine the neutron background from the underground environment at Canfranc. Six large, 60 cm in length and 2.54 cm in diameter, ^{3}He counters at 20 atm have been used, each with a different thickness polyethylene shielding to measure the energy spectrum of incoming neutrons. The counters exploit the well known reaction: $^3\text{He} + n \rightarrow {}^3\text{H} + p$ with Q=764 keV. In figure 2 we show the result of a calibration with a ^{252}Cf neutron source [17]. The peak at 764 keV is clearly visible. The result of the measurement gave: $\phi_n = (3.47\pm0.3)\text{x}10^{-6}$ cm^{-2}s^{-1}.

In spite of other projects planned in DULs on nuclear astrophysics as LUNA-MV at LNGS, the physics case of CUNA remains valid. However, the future of CUNA depends on the possibility to make a large international collaboration which can support the cost of the machine and detectors to be installed underground.

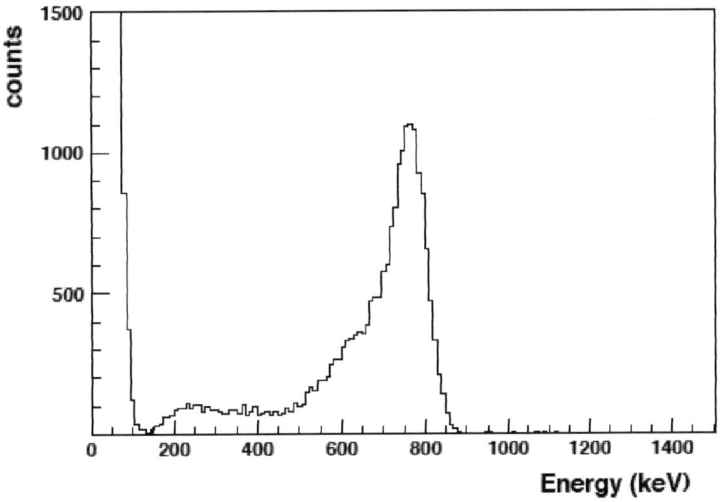

Figure 2. Calibration of ^{3}He counter performed with a ^{252}Cf source.

4. Geophysics

LSC is equipped with a geodynamic facility which aims to study local and teleseismic events. The facility consists of: 1) a broadband seismometer and accelerometer: 2) two 70 m long laser strainmeters; 3) two GPS stations on surface in the surrounding of the underground site. The two lasers are installed in LAB780 and in one by-pass gallery (gal 16) between the road and train tunnels, respectively. They are aligned at 90^0 one with respect to the other. The instrument has a nominal sensitivity of $\Delta L/L < 10^{-12}$ and a bandwidth ranging from zero to 200 Hz. A displacement of $\Delta L/L=10^{-9}$ corresponds to $\Delta L=0.07\mu m$. The LSC site is exceptionally low noise and the strainmeters has studied non-linear ocean load tides at 120 km distance in the Gulf of Biscay. The observed amplitude and phase are well in agreement with predictions. Hydrological signatures due to the load of rain water and from discharge of the close by river Aragon have been studied with the strainmeters.

5. Life in extreme environemts

Since 1990s it has been known that life extends in the subsurface and in extreme environments. Several questions need to be answered about this finding [19]: what is the extent and diversity of deep life? What are the limiting factors for life in underground? How does the deep biosphere interact with geosphere and atmosphere? What are the types of energy sources in subsurface environments? DULs offer an opportunity to carry out this research. At present scientists are trying to identify a network of sites where perform sampling according to a standard protocol [19]. LSC offers the train tunnel and other underground locations for this purpose. In this context, at present, the GOLLUM project has been proposed to the LSC Scientific Committee. GOLLUM aims to characterize microbial communities underground by extraction of DNA in rock samples (metagenomics). For a review on this subject see [20].

6. Conclusions

In conclusion LSC is the second larger underground laboratory in Europe after LNGS. It is a multidisciplinary facility equipped with a number of Services to support scientists and underground installations. At present, the research activity at LSC goes from dark matter and neutrinoless double beta decay, to geophysics and biology in the subsurface. At LSC new space is available to carry out research. The infrastructure used by ROSEBUD [21], a hut $3x3x4.8m^3$ inside a Faraday cage and acoustically isolated, is open for new projects. This hut is suitable for rare events searches such as dark matter, chameleons underground, small set-ups for double beta decay. In Hall A about 100 m^2 are at disposal for new research activities.

Acknowledgments
The author wishes to acknowledge the organizers of the International Conference on Particle Physics and Astrophysics for the invitation to report about the Canfranc Underground Laboratory.

References

[1] Achar I C V *et al.* 1965 *Phys. Lett.* **18** 196
[2] Reines F *et al.* 1965 *Phys. Lett.* **15** 429
[3] Kuzminov V V 2012 *Eurp. Phys. J.* **127** 114
[4] Davis R *et al.* 1964 *Phys. Rev. Lett.* **12** 302
[5] Suzuki Y 2012 *Eurp. Phys. J.* **127** 111
[6] Votano L 2012 *Eurp. Phys. J.* **127** 109
[7] Smith N J T 2012 *Eurp. Phys. J.* **127** 108
[8] Bettini A 2012 *Eurp. Phys. J.* **127** 112
[9] Piquemal F 2012 *Eurp. Phys. J.* **127** 110
[10] URL *http://www.boulby.stfc.ac.uk/Boulby/*
[11] Lesko K T 2012 *Eurp. Phys. J.* **127** 107
[12] Mondal N K 2012 *Eurp. Phys. J.* **127** 106

Journal of Physics: Conference Series **675** (2016) 012002 doi:10.1088/1742-6596/675/1/012002

[13] Chen H 2012 *Eurp. Phys. J.* **127** 105
[14] Cuesta C (ANAIS collaboration) 2015 Background analysis and status of the ANAIS dark matter project *Low Radiation Techniques workshop, 18-20 March 2015*
Amaré J *et al.* (ANAIS collaboration) Production and relevance of cosmogenic radionuclides in NaI(Tl) crystals *arXiv*:1505.06102
[15] ArDM collaboration 2015 *arXiv*:1505.02443
[16] NEXT collaboration 2014 *Adv. High Energy Phys.* 907067
[17] Arnold R *et al.* (SuperNEMO collaboration) 2010 *Eur. Phys. J.* C **70** 927
[18] Jordan D *et al.* 2013 *Astroparticle Physics* **42** 1
[19] Kieft T L *et al.* 2015 Workshop to develop deep-life continental scientific drilling projects *Sci. Dri.* **19** 43
[20] 2015 *DULIA-bio Workshop, LCS, 13-14 October 2015*
[21] Coron N *et al.* (ROSEBUD collaboration) 2013 *Astropart. Phys.* **47** 31

Journal of Physics: Conference Series **675** (2016) 012003 doi:10.1088/1742-6596/675/1/012003

The inverse beta decay: a study of cross section

A Oralbaev[1], **M Skorokhvatov**[1,2] **and O Titov**[1]

[1] NRC "Kurchatov Institute", 123182, Moscow, Russia

[2] National Research Nuclear University MEPhI (Moscow Engineering Physics Institute), Kashirskoe highway 31, Moscow, 115409, Russia

E-mail: `oralbaev_ay@nrcki.ru`, `skorokhvatov_md@nrcki.ru`, `titov-o@mail.ru`

Abstract. Knowledge of antineutrino interaction cross-sections is an important and necessary ingredient in many measurements. With the advent of new precision experiments, the demands on better understanding of neutrino interactions is becoming even greater. The purpose of this report is to survey our current knowledge of the inverse beta decay cross-sections and to do a comparison the theoretical analysis with experimental data.

The inverse beta decay (IBD) reaction

$$\bar{\nu}_e + p \rightarrow n + e^+ \tag{1}$$

has played an important role in the first observation of free antineutrinos and up to now the reaction (1) is used in the most of reactor antineutrino experiments. It also provides the detection of geo-neutrinos and the search for sterile neutrino.

The energy threshold of IBD, $E = 1.806$ MeV, is low enough to observe antineutrino from different sources. The neutrino energy $E_{\bar{\nu}}$ and the observable positron energy E_e are related by the formula $E_{\bar{\nu}} \rightarrow E_e + 1.3$ MeV, neglecting the small neutron recoil.

Compared to other neutrino interactions, the reaction (1) has relatively large cross section and its "naive" theoretical description at the low energy ($<$ 10 MeV) is rather simple. The vector and the axial weak formfactors, g_V and g_A respectively, make the main contribution to the cross section. Neglecting all the other formfactors one derives the cross section:

$$\sigma_0(E_\nu) = \frac{1}{\pi}(G_V^2 + 3G_A^2)\overline{p}\,\overline{E}, \tag{2}$$

where E and p are the positron energy and momentum, the "effective" beta decay constants G_V and G_A were defined through the Fermi constant, G_F, and the Cabibbo angle, V_{ud}, by means of the relation:

$$(G_V^2 + 3G_A^2) = (G_F V_{ud})^2(g_V^2 + 3g_A^2). \tag{3}$$

Under the fixed constants the formula (2) provides the accuracy of few percent for prediction of the IBD cross section. In the early 1980s, the experimental technique had been developed, and the problem of more precise calculations emerged. In order to provide accurate calculation of the cross section, one must take into account the following corrections:

(i) "Outer" and "inner" radiative corrections δ_{rad} and Δ_{rad} [1, 2, 3, 4];

(ii) Nucleon recoil and weak magnetism δ_{rec+WM} [2, 3];

(iii) Threshold correction δ_{thr} [1].

Thus, the overall IBD cross section takes form:

$$\sigma = \sigma_0(1+\delta_{rad})(1+\Delta_{rad})(1+\delta_{rec+WM})(1+\delta_{thr}). \tag{4}$$

Since the middle of the 80s these corrections were repeatedly calculated using different methods. In the most of works the coinciding results were received that gives us confidence to consider these calculations very reliable. Figures 1,2 present the correction functions (the inner radiative correction $\Delta_{rad} = 0.0238(4)$ [5] is omitted because it does not depend on energy). It is evident that corrections vary within several percent depending on energy. It is necessary to pay attention to the threshold correction which is important for the description of IBD cross section round the energy of 1.8 MeV. Thus it is possible to draw a conclusion that applying corrections, its precision at relevant energies for antineutrino observation ($<$ 10 MeV) is better than 0.2 %. Of course, again it is fair at the fixed values of beta decay constants and to determine the total precision of IBD cross section we have to consider the experimental data on the combination $(G_V^2+3G_A^2)$ of beta decay constants.

There is a direct way to measure $(G_V^2+3G_A^2)$ from experimental study of free neutron lifetime τ. The neutron lifetime has been measured in the two sets of experiments — with neutron beams by decay-in-flight method and with storage of ultracold neutrons by a neutron confinement method. Unfortunately, there is a significant discrepancy between the measurements of τ in the beam and the storage experiments. Given disagreement [6] is essential to having a reliable $(G_V^2+3G_A^2)$ value. The beam measurements yield the value 888.0 ± 2.1 s for the lifetime, higher than the average of 879.6 ± 0.8 s for storage experiments [6]. From these data the numerical values of factors $(G_V^2+3G_A^2)$ are:

$$\left(G_V^2+3G_A^2\right)_{beam} = 7.516(18)\cdot 10^{-10}\ \mathrm{GeV}^{-4}; \tag{5}$$

$$\left(G_V^2+3G_A^2\right)_{storage} = 7.59(8)\cdot 10^{-10}\ \mathrm{GeV}^{-4}; \tag{6}$$

This discrepancy of 1% is large enough that is the dominant uncertainty now in the prediction of the IBD cross section.

It should be noted that there is also another way (but not direct) of definition of the constants, using data from nuclear superallowed transitions and from angular correlation coefficient measurements in neutron β-decay (measuring G_V^2 and a ratio $\lambda = G_A/G_V$, respectively). Using the recommended values for $G_V^2 = 1.29106(57)\cdot 10^{-10}\,\mathrm{GeV}^{-4}$ [7] and $\lambda = -1.2748(13)$ [8], the factor $(G_V^2+3G_A^2)$ is equal to $\left(G_V^2(1+3\lambda^2)\right)_{corr.} = 7.583(15)\cdot 10^{-10}\mathrm{GeV}^{-4}$, that is close to data from storage experiments.

The IBD cross section has been measured in the reactor-based short-baseline experiments. Nuclear reactors are an intense source of 0-10 MeV antineutrinos, resulting from many fission products of nuclear fuel. Despite many studies let us consider the two most reliable measurements with the largest statistics that were carried out in the BUGEY-4 [9] and the Daya Bay [10] experiments. The measured cross section averaged over energy spectrum of antineutrinos (in units of cm^2 per fission):

$$\sigma_{exp}\ \text{(Bugea-4)} = (5.75 \pm 0.08)\cdot 10^{-43}\ \mathrm{cm}^2/\text{fission} \tag{7}$$

$$\sigma_{exp}\ \text{(Daya Bay)} = (5.92 \pm 0.14)\cdot 10^{-43}\ \mathrm{cm}^2/\text{fission} \tag{8}$$

The energy spectrum of reactor antineutrinos depends on the composition of the main fuel isotopes: ^{235}U, ^{238}U, ^{239}Pu and ^{241}Pu. Corresponding contributions of fissile isotopes to antineutrino flux in considered experiments were:

$$^{235}\mathrm{U} :^{238}\mathrm{U} :^{239}\mathrm{Pu} :^{241}\mathrm{Pu}\ \text{(Bugey-4)} \rightarrow 0.538 : 0.078 : 0.328 : 0.056 \tag{9}$$

$$^{235}\mathrm{U} :^{238}\mathrm{U} :^{239}\mathrm{Pu} :^{241}\mathrm{Pu}\ \text{(Daya Bay)} \rightarrow 0.586 : 0.076 : 0.288 : 0.050 \tag{10}$$

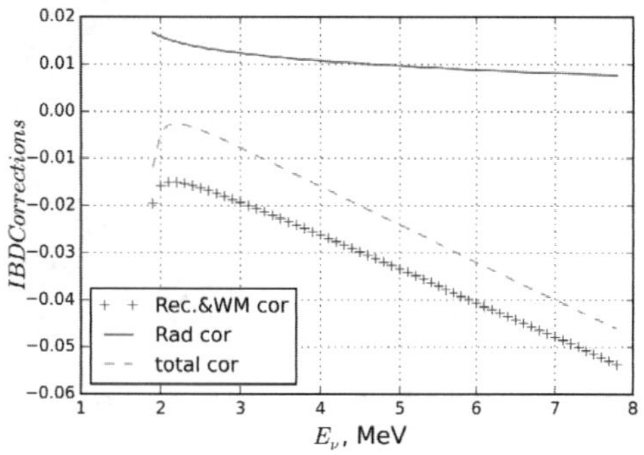

Figure 1. IBD corrections

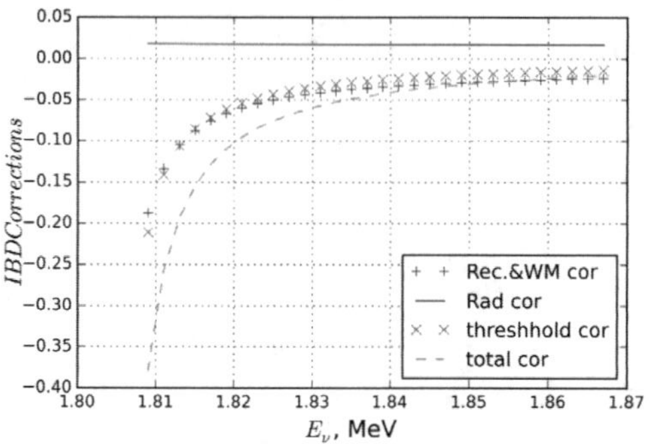

Figure 2. IBD correction (near threshold)

The measured IBD cross sections are consistent for both experiments after correcting for the difference of fission fractions.

The expected cross section σ_{expec} can be calculated by means of the integration of the formula (4) under the antineutrino spectra. The energy spectrum of reactor antineutrinos was predicted in 80-s, using conversion procedure based on the beta spectra of fuel isotopes, measured at ILL [11]. The value of σ_{expec} for this case is in agreement with measurements within 3% uncertainty: $\sigma_{meas}/\sigma_{expec} \approx 0.99 \pm 0.03$. Recently improved treatments of reactor antineutrinos flux and spectrum were presented [12] to be higher than the measurements: $\sigma_{meas}/\sigma_{expec} \approx 0.94 \pm 0.03$. Sometimes this discrepancy is considered as a manifestation of new physics. But the most likely explanation is a difficulty of fissile antineutrino modeling and an underestimation of contributions of unknown fission fragments to a generation of the antineutrino energy spectrum.

In summary, we have reviewed the accuracy of the IBD cross section. The expected IBD cross section value has uncertainty of about 1 percent connected with the current uncertainty in the values of the beta decay constants combination. The contributions of some corrections are rather small and they are calculated reliably. Exact comparison with experimental data of reactor-based experiments is complicated mainly due to incomplete antineutrino energy spectrum knowledge. Thus, it is desirable to conduct measurements with an intense source of well-known spectrum.

Acknowledgments

The work was partially supported by the Russian Foundation for Basic Research under grant 16-02-00616 and by MEPhI Academic Excellence Project (contract 02.a03.21.0005, 27.08.2013).

References

[1] Vogel P 1984 *Phys. Rev.* D **29** 1918
Fayans S A 1985 *Sov. J. Nucl. Phys.* **42** 590
[2] Raha U, Myhrer F and Kubodera K 2012 *Phys. Rev.* C **85** 045502 (*arXiv*:1112.2007)
[3] Ivanov A N *et al.* 2013 *Phys. Rev.* C **88** 055501 (*arXiv*:1306.1995)
[4] Vogel P and Beacom J F 1999 *Phys. Rev.* D **60** 53003
[5] Marciano W J and Sirlin A 2006 *Phys. Rev. Lett.* **96** 032002
[6] Bowman J D *et al.* 2014 *arXiv*:1410.5311 [nucl-ex]
[7] Hardy J C and Towner I S 2014 *arXiv*:1411.5987 [nucl-ex]
[8] Mund D *et al.* 2013 *Phys. Rev. Lett.* **110** 172502 (*arXiv*:1204.0013)
[9] Declais Y *et al.* 1994 *Phys. Lett.* B **338** 383
[10] An F P *et al.* [Daya Bay Collaboration] 2015 *arXiv*:1508.04233 [hep-ex]
[11] Schreckenbach K *et al.* 1985 *Phys. Lett.* B **160** 325
Von Feilitzsch F, Hahn A A and Schreckenbach K 1982 *Phys. Lett.* B **118** 162

Hahn A A *et al.* 1989 *Phys. Lett.* B **218** 365
Vogel P *et al.* 1981 *Phys. Rev.* C **24** 1543
[12] Mueller T A *et al.* 2011 *Phys. Rev.* C **83** 054615
Huber P 2011 *Phys. Rev.* C **84** 024617

Dark matter distribution and annihilation at the Galactic center

V I Dokuchaev[1,2] and Yu N Eroshenko[1]

[1] Institute for Nuclear Research of the Russian Academy of Sciences, 60-letiya Oktyabrya prospekt 7a, Moscow, 117312, Russia

[2] National Research Nuclear University MEPhI (Moscow Engineering Physics Institute), Kashirskoe highway 31, Moscow, 115409, Russia

E-mail: dokuchaev@inr.ac.ru, eroshenko@inr.ac.ru

Abstract. We describe a promising method for measuring the total dark matter mass near a supermassive black hole at the Galactic center based on observations of nonrelativistic precession of the orbits of fast S0 stars. An analytical expression for the precession angle has been obtained under the assumption of a power-law profile of the dark matter density. The awaited weighing of the dark matter at the Galactic center provides the strong constraints on the annihilation signal from the neuralino dark matter particle candidate. The mass of the dark matter necessary for the explanation of the observed excess of gamma-radiation owing to the annihilation of the dark matter particles has been calculated with allowance for the Sommerfeld effect.

In recent years significant advances have been achieved in the observations of stars gravitationally bound with the supermassive black hole SgrA* at the center of the Galaxy. Several so-called S0 stars, which move at very high velocities ($> 10^3$ km/s) in almost elliptic orbits around a very compact supermassive object, are observed in the infrared range [1, 2, 3, 4]. According to the measured parameters of the Kepler orbits of S0 stars, the mass of supermassive black hole SgrA* is $M_{\rm BH} = (4.1 \pm 0.4) \times 10^6 M_\odot$ [1, 2, 3, 4]. Independent and currently most accurate values of the mass $M_{\rm BH}$ and spin (Kerr parameter) a of the SgrA* black hole are determined from the observations of quasi-periodic oscillations with the average periods of 11.5 and 19 min [6, 7]. They are $M_{\rm BH} = (4.2 \pm 0.2) \times 10^6 M_\odot$ and $a = 0.65 \pm 0.05$ [8, 9].

At the center of the Galaxy, in addition to the supermassive black hole SgrA*, there are invisible sources of mass such as compact gas clouds, dim stars and their remnants, and a distributed mass in the form of the diffuse dark matter. This additional mass would result in the deviation of the total Newtonian gravitational potential from the point mass potential of the black hole $U = -GM_{\rm BH}/r$. As a result, the orbits of S0 stars gravitationally bound with the black hole would be unclosed and precess. The openness of the orbit of the most studied S0-2 star will be measured in the next one or two years. Thus, the total mass of the dark matter within the orbit of this star with a characteristic radius of 0.005 pc will be determined.

The existence of fast S0 stars provides a unique possibility of reconstructing the gravitational potential and measuring the mass distribution at the center of the Galaxy by fitting their orbits. We discuss and develop a method for studying the distribution of the dark matter at the center of the Galaxy by measuring the precession angle of orbits of S0 stars. Nowadays it is known that the distributed mass within the orbit of the S0-2 star is no more than $3-4$ % of the mass of the supermassive black hole SgrA*. It is noteworthy that the expected measurement of the nonrelativistic precession of the orbit of the S0-2 star will allow either improving the indicated bound on the distributed dark mass by two or three orders of magnitude or determining this dark mass.

The analytical formulas are derived general for the precession of S0 star orbits with a power law density profile of the dark matter distribution in the Galactic center. These formulas make it possible to easily determine the additional distributed mass from the measured precession angle. An additional independent method for determining the distribution of the dark matter is the search for a possible annihilation signal from the center of the Galaxy. We calculated [13] (see figure 1 and 2) the mass of the SUSY neutralino dark matter necessary for the explanation of the excess of gamma-radiation from the center of the Galaxy detected recently by the Fermi-LAT space gamma-telescope [10, 11].

In particular, we determined the dependence of the additional mass both on the profile of the central peak of the dark matter density and on the annihilation cross section of dark matter particles taking into account the Sommerfeld enhancement effect (see figure 1).

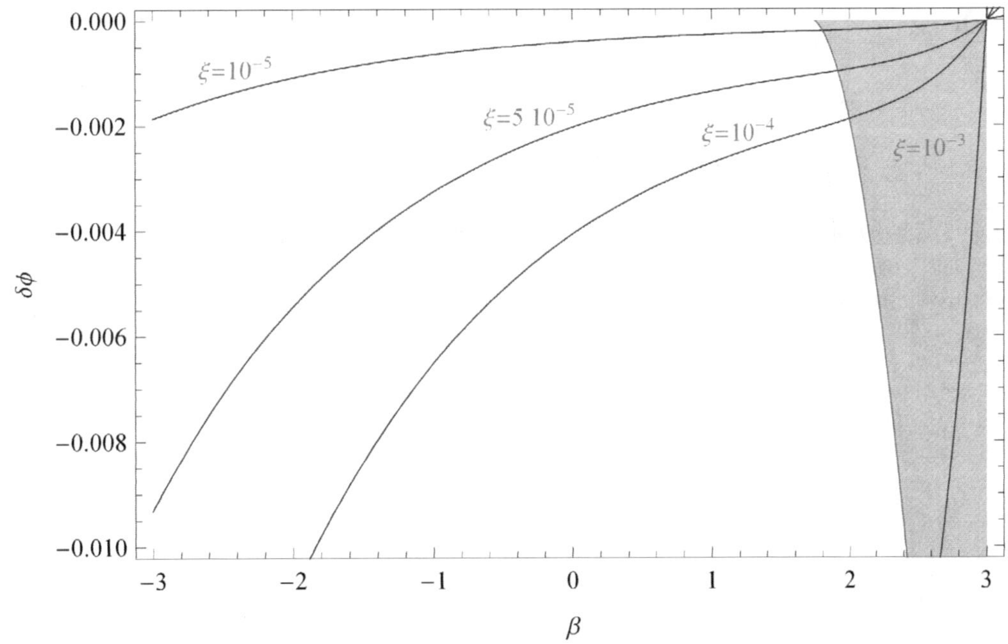

Figure 1. Shift angle of the apsis of the orbit of the star in one turn $\delta\phi$ versus the exponent of the power-law spectrum of the dark matter for realistic values of the mass fraction of the dark matter ξ within the orbit of the S0-2 star. The indicated region is excluded by the constraints caused by the annihilation of dark matter particles if the dark matter makes the main contribution to the mass fraction of the dark matter ξ.

We consider the power-law density profile of matter responsible for the correction δU to the potential of the black hole $\rho(r) = \rho_h(r/r_h)^{-\beta}$, where ρ_h, r_h, and β are the parameters. The corresponding total mass of the dark matter within the sphere with the radius r is

$$M_{\rm DM}(r) = \frac{4\pi\rho_h r_h^{\beta}}{3-\beta}\left[r^{3-\beta} - R_{\rm min}^{3-\beta}\right], \tag{1}$$

where $R_{\rm min}$ is the minimum radius to which the density profile expands. We define the mass

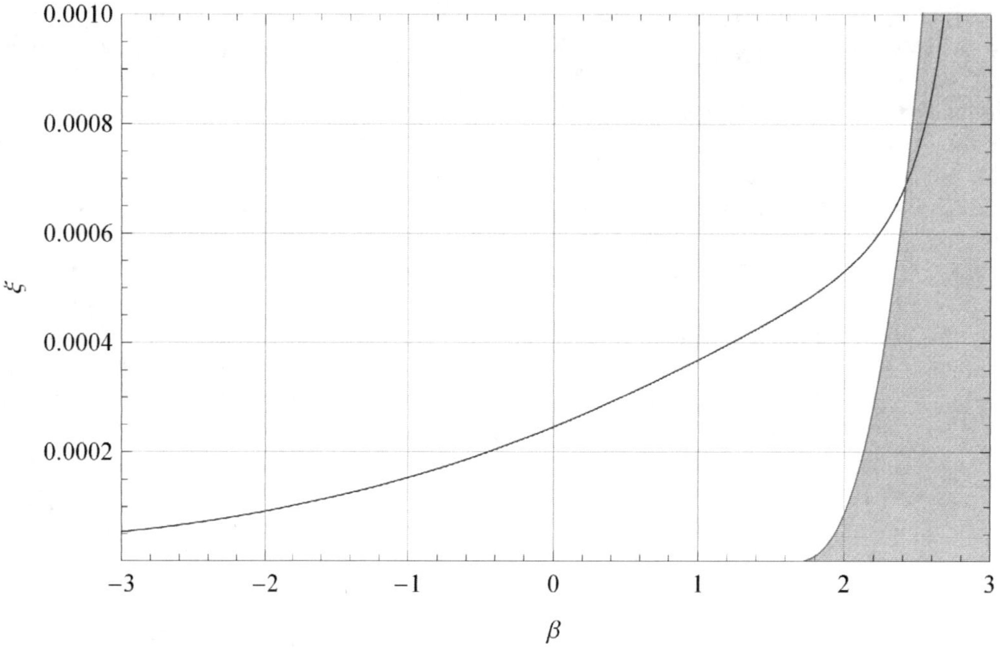

Figure 2. Mass fraction of the dark matter ξ versus the exponent β in the density profile at the precession angle $\delta\phi = 0.01$. The indicated region is excluded by the constraints caused by the annihilation of dark matter particles.

fraction of the dark matter within the orbit of the S0 star $\xi = [M_{\rm DM}(r_a) - M_{\rm DM}(r_p)]/M_{\rm BH}$, which is significant for the subsequent analysis.

In the presence of a small correction δU to the Newtonian potential of the black hole, the precession angle of the orbit of a probe particle (S0-2 star) in one turn is (see [12], section 15, Problem 3)

$$\delta\phi = \frac{\partial}{\partial L}\left(\frac{2m}{L}\int\limits_0^\pi r^2(\phi)\delta U d\phi\right). \tag{2}$$

Here, integration is performed with the trajectory of the particle in the form of an unperturbed elliptic orbit $r(\phi) = p(1 + e\cos\phi)^{-1}$, where e is the eccentricity of the ellipse, $p = L^2/(GM_{\rm BH}m) = a(1-e^2)$ is the parameter of the orbit, a is the major semiaxis, and L is the conserved angular momentum of the star with the mass m. The observed parameters of the Kepler orbit of the S0-2 star: the eccentricity $e = 0.898 \pm 0.0034$, the radius of the pericenter $r_p = a(1-e) = 0.585$ mpc, and the radius of the apocenter $r_a = a(1+e) = 9.42$ mpc. We note that, in the case of relativistic precession, the orbit would rotate in the direction of the rotation of the star, but Newtonian precession (2) occurs in the opposite direction, i.e., $\delta\phi < 0$.

The calculation [13] of the precession angle of the orbit of the star in the time of one turn around the black hole $\delta\phi$ gives an expression with one hypergeometric function ${}_2F_1(a,b;c;z)$:

$$\delta\phi = -\frac{4\pi^2\rho_h r_h^\beta p^{3-\beta}}{(1-e)^{4-\beta}M_{\rm BH}}\,{}_2F_1\left(4-\beta,\frac{3}{2};3;-\frac{2e}{1-e}\right), \tag{3}$$

where $p = L^2/(GM_{\rm BH}m) = a(1-e^2)$ is the parameter of the elliptic orbit, e is the eccentricity of the orbital ellipse of the S0-2 star, a is the major semiaxis, and L is the conserved angular

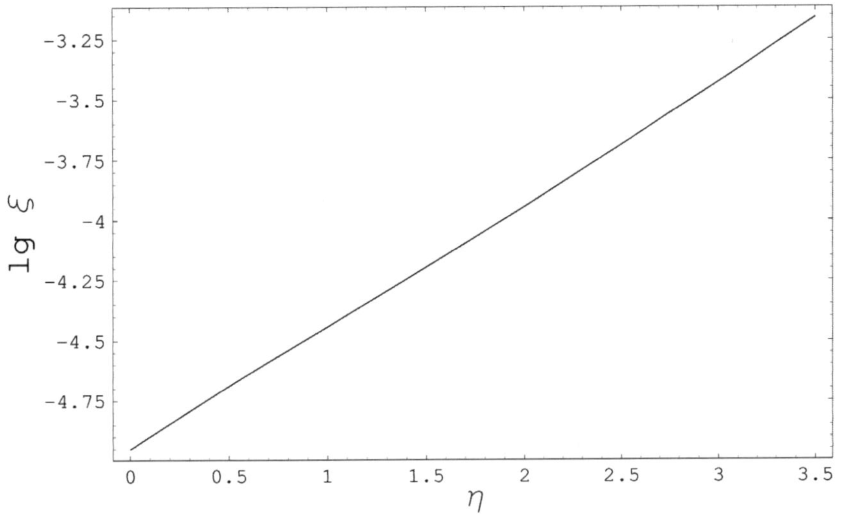

Figure 3. Mass fraction of the dark matter ξ required by the observed excess of gamma radiation in the case of the annihilation of the dark matter particles with the cross section due to the Sommerfeld enhancement effect $\langle\sigma v\rangle \propto v^{-\eta}$ versus an index η.

momentum of the star with the mass m.

The currently existing observation accuracy is still insufficient for the measurement of the precession angle of fast S0 stars and the distributed invisible mass. However, there is a high probability of reaching in the near future the accuracy required either for the measurement of the precession angle or for the determination of a strong bound, which in turn will make it possible to impose stringent dynamic constraints on the additional dark mass.

Acknowledgments

This work was supported in part by the Division of Physical Sciences of the Russian Academy of Sciences (project no. 0031-2015-0002), by the Russian Foundation for Basic Research grant no. 13-02-00257 and by the Leading Scientific Schools grant no. NSh-3110.2014.2.

References

[1] Ghez A M *et al.* 2008 *Astrophys. J.* **689** 1044
[2] Gillessen S *et al.* 2009 *Astrophys. J.* **692** 1075
[3] Gillessen S *et al.* 2009 *Astrophys. J.* **707** L114
[4] Meyer L *et al.* 2012 *Science* **338** 84
[5] Shapiro S L and Teukolsky S A 1985 *Astrophys. J. Lett.* **292** L41
[6] Aschenbach B *et al.* 2004 *Astron. Astrophys.* **417** 71
[7] Genzel R *et al.* 2003 *Nature* **425** 934
[8] Dokuchaev V I 2014 *Gen. Relativ. Gravit.* **46** 1832
[9] Dokuchaev V I and Eroshenko Yu N 2015 to appear in *Phys. Usp.* **58** (8)
[10] Daylan T *et al.* *arXiv*:1402.6703 [astro-ph.HE]
[11] Fields B D, Shapiro S L and Shelton J 2014 *Phys. Rev. Lett.* **113** 151302
[12] Landau L D and Lifshitz E M 1988 *Course of Theoretical Physics, Vol. 1: Mechanics* (New York: Pergamon Press), chapter 3
[13] Dokuchaev V I and Eroshenko Yu N 2015 *JETP Lett.* **101** 777

International Conference on Particle Physics and Astrophysics (ICPPA-2015) IOP Publishing
Journal of Physics: Conference Series **675** (2016) 012005 doi:10.1088/1742-6596/675/1/012005

Perspectives of DSNB neutrino researches in modern detectors

L Borodikhina[1,2]

[1] National Research Nuclear University MEPhI (Moscow Engineering Physics Institute), Kashirskoe highway 31, Moscow, 115409, Russia

[2] National Research Center Kurchatov Institute, Akademika Kurchatova pl. 1, Moscow, 123182, Russia

E-mail: liudmila.borodikhina@lngs.infn.it

Abstract. Studies of diffuse supernova neutrino background (DSNB) by modern underground detectors are reviewed. DSNB neutrino fluxes, their spectra and current experimental limits on their flux are discussed. Currently the best upper limit on DSNB neutrino flux is 2.9 $cm^{-2}s^{-1}$. Also posibilities to improve upper limits on future detectors and perspectivies of DSNB neutrino detection are discussed.

1. Diffuse Supernova Neutrinos

The Diffuse Supernova Neutrino Background (DSNB) is the flux of neutrinos and antineutrinos, which were emitted by all core-collapse supernovae in the Universe. Total rate and spectral shape of DSNB neutrinos could give new information about supernova core-collapse rate and neutrino emission from supernova.

1.1. Spectrum

The DSNB flux spectrum at the Earth can be written as[1]:

$$\frac{d\varphi}{dE_\nu} = \int_0^{z_{max}} dz \frac{N_{SN}(E(z+1))R_{SN}}{H_0\sqrt{(z+1)^3\Omega_m + \Omega_\Lambda}} \tag{1}$$

where N_{Sn} – neutrino emission spectrum from each individual supernova, R_{SN} – cosmic supernova rate (it is a function of redshift and star formation rate), H_0 – Hubble constant, $\Omega_m = 0.3, \Omega_\Lambda = 0.7$ – the relative densities of the matter and dark matter in the Universe.

1.2. Astrophysical inputs

DSNB predictions are dependent from astrophysical inputs: the cosmic supernova rate, the star formation rate(SFR), the neutrino emission spectra from core-collapse supernova.

Star formation rate is defined by the mass of the formed star in a volume unit. The SFR function can be written in the form[2]:

$$R_{SN} = 10^{-4} yr^{-1} Mpc^{-3} \begin{cases} (1+z)^{3.28}, z < 1; \\ (1+z)^{-0.26}, 1 < z < 4.5; \\ (1+z)^{-7.8}, 4.5 < z; \end{cases} \tag{2}$$

The energy spectrum for neutrino of each flavour is expected to be thermal near the surface of the supernova and can be parametrized as:

$$N_{SN} \simeq \frac{(1+\beta_\nu)^{1+\beta_\nu} L_\nu}{\Gamma(1+\beta_\nu) E_{0\nu}^2} \frac{E}{E_{0\nu}}^{2} e^{(1+\beta_\nu)\frac{E}{E_{0\nu}}}, \tag{3}$$

where E — neutrino energy, L_ν — expected luminosity of individual flavour of ν, $E_{0\nu}$ — average energy, β_ν — pinch factor $\sim 2-5$ describing the shape of the spectrum.

Currently, there are several MC calculations performed by different groups, predicting the neutrino spectral shape: Lawrence-Livermore National Laboratory group(LL)[3], Keil, Raffelt, Janka (KRJ) group[4] and Thompson, Burrows, Pinto (TBP) group[5]. General difference between models is in time of bursts simulation.

Neutrino energy spectrum is thermal and neutrinos average energy is $\sim 10-20$ MeV, ν_e and $\bar{\nu_e}$ will have colder spectra with respect to ν_μ and ν_τ.

Studies of DSNB neutrinos were performed on several detectors but diffuse neutrinos from supernovae were not detected yet, nevertheless some upper limits for diffuse neutrino flux were obtained.

2. Super-Kamiokande

Super-Kamiokande detector is a 50 kton water Cherenkov detector located in the Kamioka mine in Japan. More information about the detector can be found in [6].

Super-Kamiokande detects electron antineutrinos via inverse beta decay reaction:

$$\bar{\nu_e} + p \longrightarrow n + e^+ \tag{4}$$

In 2012 upper limit for DSNB electron antineutrino was set on. Limit is $2.9\bar{\nu_e}\ cm^{-2}s^{-1}$in $E > 16$ MeV (17.3 MeV E_ν) energy range[6].

In the paper [7] SK collaboration lowered neutrino energy threshold from 17.3 MeV to 13.3 MeV by including neutron tagging. In further researches the collaboration expects that the energy threshold will be lowered to 10 MeV.

3. KamLAND

The KamLAND is liquid scintillator detector located in underground laboratory Kamioka Observatory in Japan.

Results of searches for extraterrestrial electron antineutrino sources were published in 2012. As a result of these researches, limit for diffuse supernova electron antineutrino flux was obtained and its value is 139 $cm^{-2}s^{-1}$ in $8.3 < E < 31.8$ MeV energy range[8].

4. SNO

SNO is one kton the water-Cherenkov detector, which used heavy water as a target. It is located in the Inco, Ltd. Creighton nickel mine near Sudbury, Ontario, Canada at a depth of 6010 m water equivalent.

The diffuse supernova electron neutrinos and antineutrinos were detected in this detector through the reactions: $\nu + d \longrightarrow \nu + n + p$, $\nu_e + d \longrightarrow p + p + e^-$, $\bar{\nu_e} + d \longrightarrow n + n + e^+$. Upper limit on the supernova neutrino flux was obtained in $22.9 < E < 36.9$ MeV energy range and this limit is 70 $cm^{-2}s^{-1}$ [9], $\bar{\nu_e}$ from unknown sources flux limit is $2 \cdot 10^6\ cm^{-2}s^{-1}$ for energy range $4 < E < 8$ MeV and for energies up to 14.8 MeV limit is $3.4 \cdot 10^4\ cm^{-2}s^{-1}$[10].

5. Borexino

Borexino is a large volume liquid organic scintillator detector located in the underground national laboratory Gran-Sasso, Italy.

DSNB electron antineutrinos detect in Borexino via inverse beta decay reaction.

Borexino collaboration has published the limit for $\bar{\nu_e}$ from unknown sources in the energy range $1.8 < E < 17.8$ MeV [11]. This limit value is 760 $cm^{-2}s^{-1}$.

Summary of all detectors results for all antineutrino extraterrestrial sources are shown in figure 1.

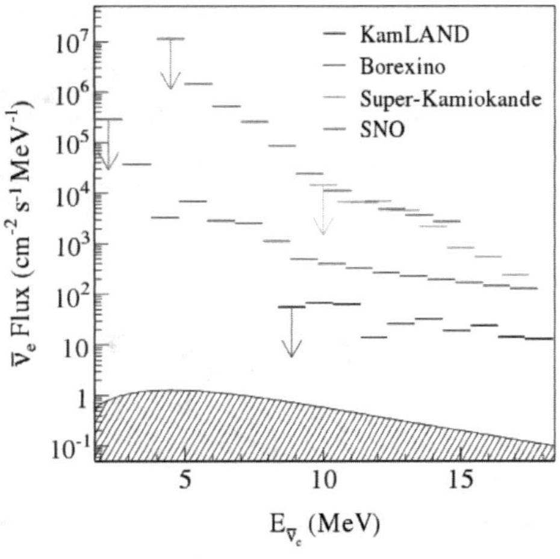

Figure 1. Model independent upper limits at 90% flux from different detectors

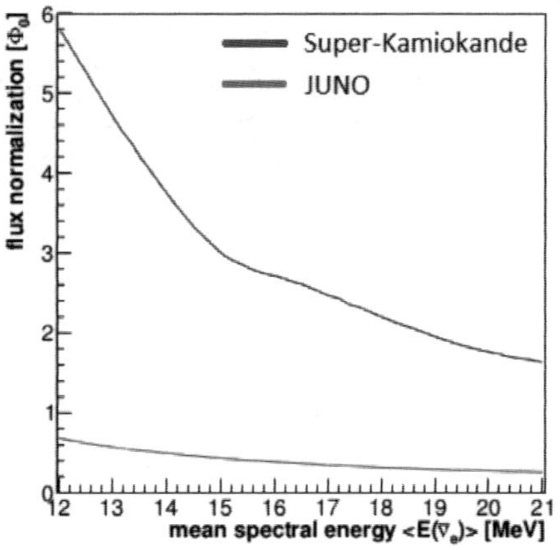

Figure 2. Predicted upper limit contour (90% C.L.) if JUNO finds for absence DSNB signal

6. Prospects for future measurements

The Jiangmen Underground Neutrino Observatory (JUNO) will be a 20 kton liquid scintillator underground detector, being constructed in Guangdong, China.

Diffuse supernova neutrinos will be detected by inverse beta-decay reactions.

Expected DSNB rate is 1.5 - 2.6 events per year. Expected upper limit for diffuse electron antineutrino flux in absence of signal is shown on the figure 2 together with Super-Kamiokande results for comparison.

The collaboration expects upper limit value will be 0.2 $cm^{-2}s^{-1}$ after 10 years in energy range above 17 MeV[12].

7. Conclusion

Modern results limit neutrino fluxes in energy range from 8.3 MeV to 40 MeV. Since Borexino detector has the least energy threshold, this low-background detector has an unique chance to set the upper limit for DSNB neutrinos in the low-energy area.

Acknowledgments

This work was performed within the framework of the Center of FRPP supported by MEPhI Academic Excellence Project (contract No. 02.a03.21.0005, 27.08.2013) and this work was supported by RFBR grant 14-22-03031

References

[1] Beacom J F 2010 The diffuse supernova neutrino background *Ann. Rev. Nucl. Part. Sci.* **60** 439-62
[2] Lunardini C 2010 Diffuse supernova neutrinos at underground laboratories *arXiv*:1007.3252 [astro-ph.CO]
[3] Ando S and Sato K 2004 Relic neutrino background from cosmological supernovae *New J. Phys.* **6** 170
[4] Keil M Th, Raffelt G G and Janka H 2003 Monte Carlo study of supernova neutrino spectra formation *Astrophys. J.* **590** 971-91
[5] Thompson T A, Burrows A and Pinto P A 2003 *Astrophys. J.* **592** 434
[6] The Super-Kamiokande Collaboration 2012 Supernova relic neutrino search at super-Kamiokande *Phys. Rev.* D **85** 052007
[7] The Super-Kamiokande Collaboration 2015 Supernova relic neutrino search with neutron tagging at Super-Kamiokande-IV *Astropart. Phys.* **60** 41
[8] The KamLAND Collaboration 2012 Search for extraterrestrial antineutrino sources with the KamLAND detector *Astrophys. J.* **745** 193
[9] The SNO Collaboration 2006 Search for neutrinos from the solar hep reaction and the diffuse supernova neutrino background with the Sudbury Neutrino Observatory *Astrophys. J.* **653** 1545-51
[10] The SNO Collaboration 2004 Electron antineutrino search at the Sudbury Neutrino Observatory *Phys. Rev.* D **70** 093014
[11] Borexino Collaboration 2011 Study of solar and other unknown anti-neutrino fluxes with Borexino at LNGS *Phys. Lett.* B **696** 191-6
[12] Fengpeng An *et al.* 2015 Neutrino Physics with JUNO *arXiv*:1507.05613 [physics.ins-det]

Low energy neutrinos from gamma-ray bursts: experimental search status

M A Toropova[1,2]

[1] National Research Nuclear University MEPhI (Moscow Engineering Physics Institute), Kashirskoe highway 31, Moscow, 115409, Russia

[2] National Research Center "Kurchatov Institute", Akademika Kurchatova square 1, Moscow, 123182, Russia

E-mail: toropovam@yandex.ru

Abstract. Gamma-ray bursts (GRBs) are the most energetic known events in the Universe. Though gamma-ray telescopes observe about one GRB event per day, the nature of this phenomenon is not yet totally understood. Many theoretical models predict emission of neutrinos of all types in a wide energy range. In this talk we review experimental searches of GRB neutrinos in MeV energy range. The searches of this kind had been performed by several experiments: SuperKamiokande, SNO and KamLAND. Also the similar study is now in progress in Borexino collaboration.

1. Introduction

Gamma-ray bursts are one of the most spectacular and bright objects in the Universe which produce great amount of energy in a short period of time. This phenomenon has been staying the relevant problem in modern astrophysics for several decades. GRBs were explored for the first time in the 60s of previous century, but still their nature is not well-known, there is no full information about GRB engine and all processes occurring in its center. According to the most popular theoretical models we can observe GRBs due to merging of two massive objects (two neutron stars or neutron star and black hole) [1] or core collapse of massive, rapidly rotating star [2]. These models predict that significant part of produced energy is taken away by neutrinos of all flavors. Thus, search for neutrinos in a wide energy range correlated in time with GRBs can help to understand better this kind of events.

Neutrinos of all types were previously searched for in several experiments. IceCube [3], ANTARES [4], AMANDA [5], Baikal [6] and SuperKamiokande [7] collaborations have searched for neutrinos and antineutrinos in the TeV and PeV energy range. Search for MeV neutrinos was performed by SuperKamiokande [7], SNO [8] and KamLAND [9] and is now in progress in Borexino collaboration.

The average rate of observed GRBs is about 1 event per day. GRBs are detected by several satellites, such as Swift, Fermi, Integral and more. There are two admitted types of GRBs concerning their duration. Short GRBs have duration less than 2 seconds and are presumed to be connected with merging of two objects, duration of long bursts is from 2 to 1000 seconds, in theoretical models they are commonly connected with core collapse of massive star.

In this report we review all performed experimental searches for low energy neutrinos (MeV energy range) from GRBs.

2. Review of experimental searches for neutrinos emitted in GRBs

As it was said before searches for MeV neutrinos and antineutrinos were done by SuperKamiokande, SNO and KamLAND neutrino experiments. These detectors have different construction and properties and have opportunity to detect neutrinos via different reactions. Moreover, methods applied for analysis are also different. Detailed descriptions of analysis performed and results obtained by all collaborations are presented below.

2.1. SuperKamiokande

SuperKamiokande is a 50 kiloton water Cherenkov detector located in the Kamioka Mine near the city of Hida, Gifu Prefecture, Japan. The main goals of this experiment are the search for proton decay, studies of solar and atmospheric neutrinos, and watching for supernovaes.

Search for neutrinos from GRBs was performed by SuperKamiokande in 2002 [7]. Detector is able to register neutrino in the wide energy range, analysis for low energy (7-80 MeV) and high energy (200 MeV-200 GeV) was made independently. In this work all GRBs registered from April 1996 to May 2000 were considered, the BATSE online catalog [10] was used as GRBs data base. SuperKamiokande collaboration made an assumption that neutrinos in GRBs had been ejected simultaneously with light. Three time windows were chosen for analysis: ± 10, ± 100 and ± 1000 seconds centered on the GRB time. Besides, 24-hour period before the moment of GRB registration was splitted into 24 time windows to search for neutrino signal excess within these windows. It was made in order to search for neutrinos that may precede GRB's photons as it was observed in 1987 during supernovae explosion. There was no evidence of neutrinos correlated with GRBs in this analysis. The fluence upper limits were set for neutrinos and antineutrinos with 90% confidence level. Since there is no one definite model of GRBs and therefore neutrino spectrum is not known, limits on neutrino fluence were expressed in terms of "Green's function" fluence for mono-energetic neutrinos.

2.2. SNO

SNO detector contained 1 kiloton of pure heavy water. It was the experiment located in Creighton Mine in Sudbury, Ontario, Canada and was designed to detect neutrinos from the Sun.

Analysis of correlations between GRBs and neutrino signal in detector was performed by SNO in 2014 [8], they have used Swift satellite [11] data as the source of information about GRBs. Available energy for SNO detector to register neutrinos is in range from 5 to 13 MeV. Two time windows were chosen. The first window is 3 hours prior to GRB time (in assumption that neutrinos produced before the photons, the same as for SN 1987a) and the second window is 3 hours after the moment of GRB registration (in order to detect neutrinos from GRB's afterglows and additionally to consider delay between neutrinos and photons due to non-zero neutrino mass). The excess of signal above the background was not showed up. The same as for SuperKamiokande results, the fluence upper limits were obtained for all neutrinos and antineutrinos (in terms of "Green's function" fluence for mono-energetic neutrinos).

2.3. KamLAND

KamLAND is liquid scintillator neutrino detector located in the underground laboratory near Toyama, Japan. KamLAND experiment is performing researches of neutrino oscillations, reactor, solar and geoneutrinos.

KamLAND collaboration has published results of the search for electron antineutrino associated with GRBs in 2015 [9]. Electron antineutrinos in this analysis were detected via inverse beta-decay reaction. GRB events detected by Swift, Fermi, Integral, AGILE and more satellites were used. Analysis was independently made for short and long GRBs. Data taking period consists of two periods: from August 2002 to September 2011 (antineutrino events with energies from 7.5 to 100 MeV were searched in first period) and from September 2011 to June 2013 (lower limit for energy range was shifted to 0.9 MeV). Time windows for correlation search are different for each GRB and depend on GRB duration, also in estimation of time windows durations difference between times of registration of light and neutrinos was considered. No time coincidences of neutrino signal and GRB were found. The fluence upper limit for electron antineutrino was estimated (in terms of "Green's function" fluence for mono-energetic neutrinos).

3. Results comparison

Let us compare results from experiments described above.

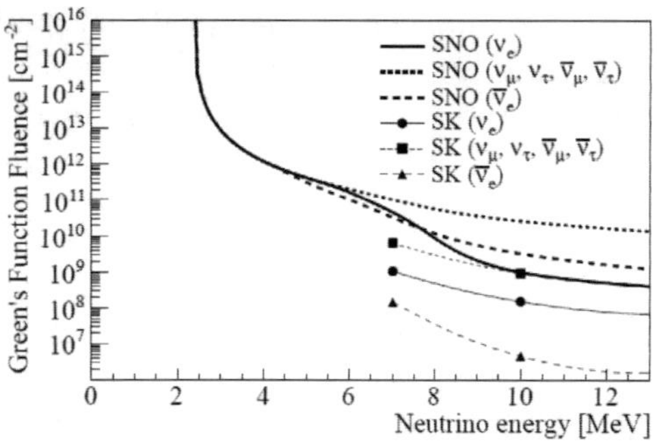

Figure 1. Fluence upper limits versus neutrino energy, results of SuperKamiokande and SNO collaborations [8]. Lines connecting points are guide to the eyes only.

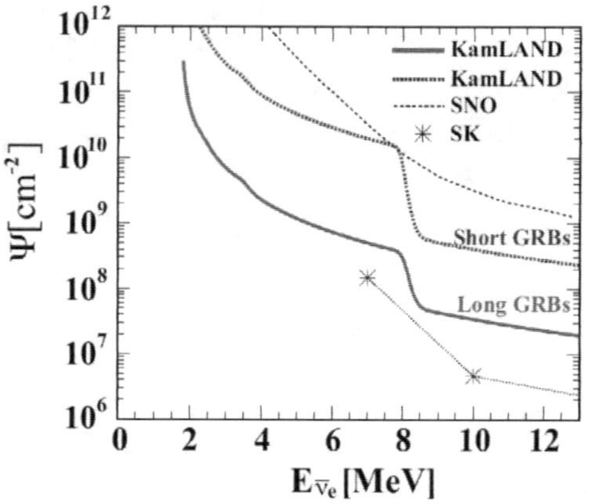

Figure 2. Fluence upper limits versus electron antineutrino energy, results of SuperKamiokande, SNO and KamLAND collaborations [9]. Lines connecting points are guide to the eyes only.

From figure 1 can be seen that the strongest limits for all neutrino types for energy above 7 MeV were obtained by SuperKamiokande. The best result for energy lower then 7 MeV for all neutrino types except electron antineutrinos was obtained by SNO. And as it can be seen in figure 2 for the electron antineutrinos with energy 2-7 MeV the strongest limit is the KamLAND's one.

4. Borexino

Borexino is liquid scintillator detector located at the Laboratori Nazionali del Gran Sasso (LNGS), near the town of L'Aquila, Italy. The main aim of Borexino is to measure fluxes of solar neutrinos, also Borexino allows to detect atmospheric and geoneutrinos, and neutrinos from different astrophysical sources.

Analysis of GRBs and neutrinos coincidences is now in progress in Borexino collaboration. IceCube data base [12] is used as catalog of GRBs in period from December 2007 to February 2015. There are two independent DAQ systems in Borexino, both of them are used in analysis. Fluence upper limits for neutrinos and antineutrinos are going to be obtained in energy region from 2 to 20 MeV in order to compare with another results.

5. Conclusion

Experimental searches for low energy neutrinos from GRBs have a long history and though no proof of existence of such kind of neutrino was found, modern neutrino detectors help to learn more about GRB phenomenon by setting limits on fluences for all types of neutrinos.

Acknowledgments
The work was supported by RFBR grant 14-22-03031 and partially supported by MEPhI Academic Excellence Project (contract No. 02.a03.21.0005, 27.08.2013).

References
[1] Meszaros P 2006 *Rep. Prog. Phys* **69** 2259
[2] Lee W H and Ramirez-Ruiz E 2006 *Astrophys. J.* **641** 961
[3] Abbasi R U *et al.* (IceCube Collaboration) 2010 *Astrophys. J.* **710** 346
[4] Aarstem M G *et al.* (IceCube Collaboration) 2015 *Ap. J.* **805** L5
[5] Achterberg A *et al.* (IceCube Collaboration) 2007 *Astrophys. J.* **664** 397
[6] Avrorin A *et al.* (Baikal Collaboration) 2009 *arXiv*:0910.4327
[7] Fukuda S *et al.* (Super-Kamiokande Collaboration) 2002 *Astrophys. J.* **578** 317
[8] Aharmim B *et al.* (SNO Collaboration) 2014 *Astropart. Phys.* **55** 1
[9] Asakura K *et al.* (KamLAND Collaboration) 2015 *Ap. J.* **806** 87
[10] NASA 2000 *http://gammaray.msfc.nasa.gov/batse/grb/catalog/current/*
[11] NASA 2006 *http://swift.gsfc.nasa.gov/docs/swift/archive/grb_table.html*
[12] IceCube Collaboration 2015 *http://grbweb.icecube.wisc.edu/help.php*

Journal of Physics: Conference Series **675** (2016) 012007 doi:10.1088/1742-6596/675/1/012007

CPT breaking and electric charge non-conservation

V A Novikov

National Research Nuclear University MEPhI (Moscow Engineering Physics Institute), Kashirskoe highway 31, Moscow, 115409, Russia

E-mail: novikov@itep.ru

Abstract. We demonstrate that CPT-violation due to the e^+e^- mass difference breaks an electric current conservation and generates a photon mass. Cosmological bounds on the photon mass lead to the bound for e^+e^- mass difference that is 10-15 orders of magnitude stronger than any direct experimental bounds.

My talk is based on a paper written in collaboration with A D Dolgov[1].

There is a good old tradition to parametrize CPT violation by attributing different masses to particle and antiparticle (see PDG [2]). This tradition is traced to the theory of $K - \bar{K}$-mesons oscillation, that is equivalent to a non-hermitian Quantum Mechanics (QM) with two degrees of freedom. Diagonal elements of 2×2 Hamiltonian matrix represent masses for particle and antiparticle. Their unequality breaks CPT-symmetry. Such strategy has no explicit loop-holes and is still used for parametrization of CPT-symmetry violation in D and B meson oscillations.

Quantum Field Theory deals with infinite number d.o.f. The very straightforward generalization of CPT-conserving QFT to CPT-violating QFT was made by Barenboim *et al.* (2001) [3]. They represented complex scalar field as an infinite sum over modes and attributed different masses for particle and antiparticle

$$\phi(x) = \sum_{\mathbf{q}} \left\{ a(\mathbf{q}) \frac{1}{\sqrt{2E}} e^{-i(Et - \mathbf{qx})} + b^+(\mathbf{q}) \frac{1}{\sqrt{2\tilde{E}}} e^{i(\tilde{E}t - \mathbf{qx})} \right\} , \tag{1}$$

where(m, E) and $(\tilde{m}, \tilde{E})$ are masses and energies for particle and antiparticle with momentum $\mathbf{q}$ respectively. Here $a(\mathbf{q})$ and $b(\mathbf{q})$ and $a^+(\mathbf{q})$ and $b^+(\mathbf{q})$ are the creation and annihilaton operators for particles and antiparticles respectively. They obey the standard Bose commutator relations. In this formalism one can calculate v.e.v. of field operators, i.e. the Wightman functions:

$$< \phi(x), \phi(y)^+ >= D^+(x - y; m) \, , \tag{2}$$

$$< \phi(x)^+, \phi(y) >= D^-(x - y; \tilde{m}) \, . \tag{3}$$

They are given by the standard Lorentz-invariant Pauli-Jordan functions but with different masses.

Greenberg (2002) [4] noticed that such theory is nonlocal and acausal. The commutator of two fields is equal to the difference $D^+(x - y; m) - D^-(x - y; \tilde{m})$ and does not vanish for space-like separation! In this sense the theory is not a Lorentz-invariant one. Moreover the Feynman propagator is explicitly non-invariant:

$$D_F(q) = \frac{1}{(2E(\mathbf{q}))} \frac{1}{(q_0 - E(\mathbf{q}))} - \frac{1}{(2\tilde{E}(\mathbf{q}))} \frac{1}{(q_0 + \tilde{E}(\mathbf{q}))} \tag{4}$$

and can be rewritten in invariant form only if $m = \tilde{m}$.These arguments demonstrate that any local fields theory that violates CPT symmetry necessarily violates Lorentz invariance.

We have shown recently that the theories with different masses for particle and antiparticle break also a conservation law for a local electric current [6]. In the theory where photon interacts with non-conserving electric current nothing protects photon from being massive and at the first loop one expects non-zero m_γ:

$$m_\gamma^2 = C\,\frac{\alpha}{\pi}\,\Delta m^2\,. \tag{5}$$

The coefficient C can be calculated for any particilar convention about QFT with different masses for particle and antiparticle. In paper [3] we argued that there is no reasonable model for local QFT where $m \neq \tilde{m}$ and no reliable theoretical frameworks for calculations of C. Still even with uncertain coefficient C the relation (5) is extremely interesting. Indeed an upper bound on the photon mass produces a bound on the mass difference for electron and positron. As it is follows from equation (5):

$$\Delta m_e < 20\, m_\gamma/\sqrt{C}\,, \tag{6}$$

where we have to substitute for m_γ the upper limit on the photon mass.

These limits and discussion of their validity are presented in the review [7]. The Earth based experiments give for the Compton wave length of the photon $\lambda_C > 8\cdot 10^7$ cm, i.e. $m_\gamma < 3\cdot 10^{-13}$ eV, and respectively $\Delta m < 6\cdot 10^{-12}$ eV, nine orders of magnitude stronger than PDG bound. From the measurement of the magnetic field of the Jupiter it follows that the Compton wave length of photon is larger than $5{\cdot}10^{10}$ cm or $m_\gamma < 4{\cdot}10^{-16}$ eV, and respectively $\Delta m < 8{\cdot}10^{-15}$ eV. The strongest solar system bound is obtained from the analysis of the solar wind extended up to the Pluto orbit [8]: $\lambda_C > 2\cdot 10^{13}$ cm, i.e. $m_\gamma < 10^{-18}$ eV. This is an "official" limit present by the Particle Data Group [2]. The corresponding bound on the electron-positron mass difference is $\Delta m < 2\cdot 10^{-17}$ eV, which is almost 14 orders of magnitude stronger than the direct bound on Δm.

The strongest existing bound follows from the the existence of the large scale magnetic fields in galaxies [9]: $\lambda_C > 10^{22}$ cm and $m_\gamma < 2\cdot 10^{-27}$ eV. Correspondingly $\Delta m < 4\cdot 10^{-26}$ eV, which is 23 orders of magnitude stronger than the direct limit on the electron-positron mass difference.

It is instructive to present a sample of actual calculations of m_γ. We stress again that there is no one example of a local Lorentz invariant Field Theory with non-zero Δm. Here we simply start with 'a la Barenboim-Greenberg decomposition for an electron-positron spinor field operator $\Psi(x)$.

$$\Psi(x) = \sum_{\mathbf{p}} \left\{ a(\mathbf{p})\frac{u(p)e^{-ipx}}{\sqrt{2\omega(p)}} + b^+(\mathbf{p})\frac{u(-\mathbf{p})e^{i\tilde{p}x}}{\sqrt{2\tilde{\omega}(p)}} \right\} \tag{7}$$

$$\{a(\mathbf{p}), a^+(\mathbf{p}')\} = \delta_{\mathbf{p},\mathbf{p}'}\ , \quad \text{etc.} \tag{8}$$

The first term in this decomposition annihilates electron with mass m, while the second term creates positron with mass $\tilde{m}$. Creation and annihilation operators obey the standard anti-commutation relations. We also assume the validity of the usual local product of field operators for the electric current

$$j_\mu(x) = \bar{\Psi}(x)\gamma_\mu\Psi(x)\ \ .$$

Because of the electron-positron mass difference this current is not conserved, $\partial j(x) \neq 0$.

Electron-positron pair contribute into the photon polarization operator.

$$\Pi_{\mu\nu} = (ie^2)\int \frac{d^D p}{(2\pi)^D}\mathrm{Tr}\frac{1}{\hat{p} - m_1}\gamma_\nu\frac{1}{\hat{p}-\hat{q}-m_2}\gamma_\mu = \tilde{g}_{\mu\nu}\Pi_T(q^2) + g_{\mu\nu}\Pi_L(q^2), \tag{9}$$

where $\tilde{g}_{\mu\nu} = g_{\mu\nu} - q_\mu q_\nu / q^2$. This divergent integral has to be regularized and we choose covariant dimensional regularization. For non-conserving currents a longitudinal function $\Pi_L(q^2)$ has to be generated and nonzero $\Pi_L(0) \neq 0$ corresponds to non-zero photon mass

$$q_\mu \Pi_{\mu\nu} = q_\nu \Pi_L(q^2) = ie^2 \int \frac{d^D p}{(2\pi)^D} \mathrm{Tr}\left[\frac{1}{\hat{p} - m_1}\right] \gamma_\nu \left[\frac{1}{\hat{p} - \hat{q} - m_L}\right] \hat{q} \ . \tag{10}$$

In the standard case of equal masses the $\hat{q}$ is a difference of two inverse propagators and we reproduce the standard Ward identity. In our case $\hat{q}$ is a difference of inverse propagators plus mass difference:

$$\hat{q} = (\hat{p} - m_1) - (\hat{p} - \hat{q} - m_2) + (m_1 - m_2) \ . \tag{11}$$

Substituting this formula into (10) we get

$$\begin{aligned} q_\nu \Pi_L(q^2) &= ie^2(m_1 - m_2) \int \frac{d^D p}{(2\pi)^D} \frac{\mathrm{Tr}\left[(\hat{p} + m_1)\gamma_\nu (\hat{p} - \hat{q} + m_2\right]}{(p^2 - m_1^2)[(p-q)^2 - m_2^2]} = \\ &= 4e^2(m_1 - m_2) q_\nu \int_0^1 dx \int \frac{d^D p}{(2\pi)^D} \frac{m_2 x - m_1(1-x)}{[p^2 + \Delta^2]^2} \ , \end{aligned} \tag{12}$$

where $\Delta^2 = m_1^2(1-x) + m_2^2 x - q^2 x(1-x)$. For $q^2 = 0$ the integral is trivial and one gets that

$$m_\gamma^2 = \Pi_L(0) = \frac{\alpha}{2\pi} \left[m_2 - m_1\right]^2 \left[\ln \frac{\Lambda^2}{m^2} - \frac{5}{3}\right] \ , \tag{13}$$

where Λ is a cut-off. Thus the photon mass is divergent and has to be renormalized. Formally it can be an arbitrary number, even zero. But if loop calculations have any physical sense for such theories this number has to be proportional to fine coupling constant and disappear for equal mass, i.e. we arrive to equation (5).

Acknowledgements
I am grateful to A D Dolgov for long discussions. This work was performed within the framework of the Center FRPP supported by MEPhI Academic Excellence Project (contract 02.03.21.0005, 27.08.2013).

References

[1] Dolgov A D and Novikov V A 2014 *Phys.Lett.* B **732** 244
[2] Beringer J *et al.* (Particle Data Group) 2012 *Phys. Rev.* D **86** 010001
[3] Barenboim G *et al.* 2002 *JHEP* **0210** 001
Barenboim G, Borissov L and Lykken J D 2002 *Phys. Lett.* B **534** 106
[4] Greenberg O W 2002 *Phys. Rev. Lett.* **89** 231602
[5] Chaichian M *et al.* 2011 *Phys.Lett.* B **699** 177
[6] Dolgov A D and Novikov V A 2012 *JETP Lett.* **95** 594
Novikov V 2012 *Nuovo Cim.* **C035N1** 243
[7] Goldhaber A S and Nieto M M 2010 *Rev. Mod. Phys.* **82** 939
[8] Ryutov D D 2010 *Plasma Phys. Controlled Fusion* **49** 429
[9] Chibisov G V 1976 *Sov. Phys. Usp.* **119** 624 (original Russian text: *Uspek. Fiz. Nauk* 1976 **119** 551)

Journal of Physics: Conference Series **675** (2016) 012008 doi:10.1088/1742-6596/675/1/012008

Online calibration of neutrino liquid scintillator detectors above 10 MeV

A S Chepurnov[1], M B Gromov[1,2,4] and A F Shamarin[3]

[1]Lomonosov Moscow State University Skobeltsyn Institute of Nuclear Physics, 119234 Moscow, Russia
[2]Lomonosov Moscow State University, Faculty of Physics, 119234 Moscow, Russia
[3]OOO Marathon, 119234 Moscow, Russia

E-mail: `gromov@physics.msu.ru`, `aschepurnov@gmail.com`

Abstract. Online calibration of neutrino liquid scintillator detector at energies above 10 MeV is very important for study of such rare process as supernova and for correct calculation of backgrounds if spectral properties is the focus of researches. The traditional procedure implies the usage of radioactive sources with well-known spectral properties but such approach is limited by available radioactive sources, upper possible energies (~10-11 MeV) and dangerous for ultra low background environment of modern detectors. The approach we propose is based on simulation of events with controllable UV double LED pulser. The LED's main wavelength fits the scintillator excitation wavelength. This technique allows to simulate physical events within the detector in very wide energy range from a few hundred keV to about 50 MeV. Additional studies like pile-up analysis can be performed due to double-LEDs scheme which generates two delayed signals with different adjustable amplitudes. The delay time is also adjustable parameter.

The large liquid scintillator neutrino detectors have been created in recent decades. To exploit them for study of various rare processes that take place in the energy range from 10 to 100 MeV and for correct account of backgrounds in the same energies online calibration is strongly necessary. But the traditional method based on the usage of radioactive sources with well-known spectral properties cannot be applied. There are several main problems in this case. First of all, the source must be small. It means the dimensions are not much more than the accuracy of the spatial reconstruction of events in the detector. Therefore only the sources which are based on natural radioactivity can be used. But all of them including complex sources like AmBe have energies of radiation below $10 - 11$ MeV. Finally the radioactive sources are very dangerous for ultra low background environment of modern detectors. The way out is to use a powerful light pulses with adjustable amplitudes.

During recent years the different light sources have been actively applied in large volume liquid scintillator detectors for technical purposes. For example, the condition of photomultipliers can be tested with a laser. Another example is a definition of the spatial position of a radioactive source in the detectors during a calibration campaign. It can be done by using a few cameras attached to the inner surface of central tank and a small light source (laser or LED) placed in immediate vicinity from the radioactive one. These instances of works have been performed in the Borexino experiment [1, 2].

[4] To whom any correspondence should be addressed.

The concept of the online calibration above 10 MeV by light sources implies an assumption that the energy of simulated signal is directly proportional to the number of emitted photons. There are three light sources: a laser, laser diode and LED. The laser has two main advantages. The first one is a power of radiation. For example, one laser could be used to calibrate simultaneously plenty PMTs by dividing the signal with a splitter. The high power provides a sufficient number of photons if the laser is exploited for reconstruction of the radioactive source position. The second advantage is a duration of pulses. It should be about 10 – 200 ns like a typical duration of real events. On the other hand the main disadvantage of the laser is the high cost of components of the calibration system if it's planned to control some signal parameters like amplitude, duration, rate of pulses and delayed time between them. It's especially a huge problem in case of UV lasers. The last one is useful when the task is to simulate real events which produce the primary radiation that is invisible for PMTs. Thus the UV laser doesn't suit for calibration. Moreover the systems with any type of laser diode are expensive as well.

Another solution is to use a LED as a source of light. LEDs are cheap. Since the beginning of the 1990s the new types of LEDs, in particular blue and UV, have appeared. But it was impossible to make the calibration system due to the power of LEDs was insufficient and there were no blue and UV LED drivers. The situation has changed over time.

A standard blue or UV LED has a power about 0.1–1.0 mW. It's easy to understand whether this value is high or not. Let P denotes the power of the LED, $E = 50\,\mathrm{MeV}$ – the maximum energy of simulated event, $n_{\mathrm{eff}} = 10$ – the attenuation of the pulse in the system, $\tau = 20\,\mathrm{ns}$ – the signal duration, m – the number of photons per pulse and $\lambda = 250\,\mathrm{nm}$ – the wavelength. Then

$$P = \frac{E \cdot n_{\mathrm{eff}}}{\tau} \approx 4\,\mathrm{mW} \tag{1}$$

and

$$m \approx \frac{2\pi\lambda n_{\mathrm{eff}} E}{\hbar c} \approx 4 \cdot 10^9 \tag{2}$$

It's important to note that the assessment is conservative and rough. On the other hand, it is implied that the LED worked in continuous mode. If the pulses are rare (i.e. the pulsed mode) the maximum diode current may be higher by an order. But the power of a few mW is preferred for the calibration above 10 MeV.

The scheme of easily controllable, reliable, compact and very cheap driver for nanosecond LED pulser was proposed by J.S.Kapustinsky and his colleagues in 1985 [3]. It has become relatively popular since that time. For instance, the pulsers are based on this scheme were applied for time and amplitude calibration in the Baikal experiment NT-200 [4] and in the ANTARES experiment [5] located in the Mediterranean Sea. The scheme was called "Kapustinsky's driver". Lubsandorzhiev and Vyatchin adapted it for InGaN/GaN ultraviolet and blue LEDs produced by Nichia Chemical [6]. So that the way to construct a cheap system for online calibration with LED pulser has been opened.

The common simplified scheme of the proposed system is presented in figure 1. Borexino is indicated since the system is expected to be implemented in this detector and for completeness of the illustration. There are three main ideas which formed the basis of the method.

1. **Adjustable input of the LED pulser.** It means that the amplitude, rate and delayed time of the simulated signals are adjustable parameters. They are set remotely by operator. The duration of pulses is specified during the design. On hardware level a generator produces starting pulses that define two parameters – the rate and delayed time, and a control circuit manages the supply voltage of the LED driver that influences to the amplitude.
2. **Double-LEDs scheme.** The LED driver has two diode so that it's possible to generate two close peak. The delayed time between them can be changed up to 0. It's useful for simplified pile-up analysis and for simulation of inverse β - decay events.

Journal of Physics: Conference Series **675** (2016) 012008 doi:10.1088/1742-6596/675/1/012008

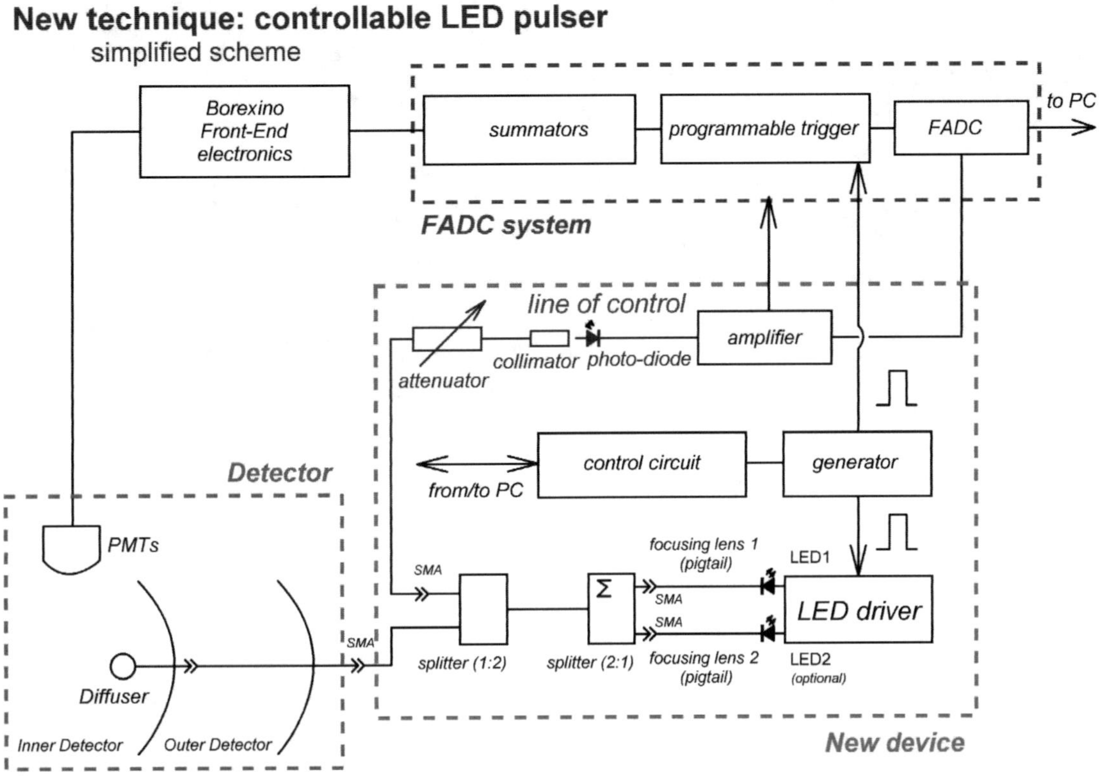

Figure 1. The simplified scheme of the proposed system for online calibration of Borexino.

3. **Controllable output of the LED pulser (line of control or power meter).** It allows to measure the pulse energy. One uses the method of relative measurements. For calibration of the line of control a few test set of the pulser signals are written and compared with the other calibration data collected during the cycle of measurements with radioactive sources. The energy scale up to 10 MeV is built. It must be linear. Then the scale is extrapolated implementing the main assumption of the proposed procedure that the energy of simulated signal is directly proportional to the number of emitted photons. The structure of the power meter is very simple. This is an optic system that consists of an attenuator, focusing lens/collimator, photodiode with a preamplifier and amplifier. A trigger and ADC are absent due to the line of control naturally integrates into the DAQ of the detector.

The current design of the calibration system implies the use of UV LEDs with wavelength 265 ± 5 nm and the maximum pulse power of 6 mW at 200 mA. The parameters of the diodes were fitted to properties of the Borexino scintillator (PC+PPO). The LED driver is slightly modified Lubsandorzhiev's LED driver. The first prototype that was made a year ago is shown in figure 2. A typical oscillogram obtained on a test stand is shown in figure 3. The test stand is a black box with one PMT (ETE 9823B) and the prototype of the device.

It's worth noting that the off-axis calibration can be performed with the new system. It depends on the source insertion mechanism. The sketch of the off-axis calibration in case of Borexino is presented in figure 4. One detail should be taken into account: a small diffuser must be placed at the end of the fiber to reduce anisotropy of radiation.

The new calibration system is quite universal. Therefore it's planned to apply it in other

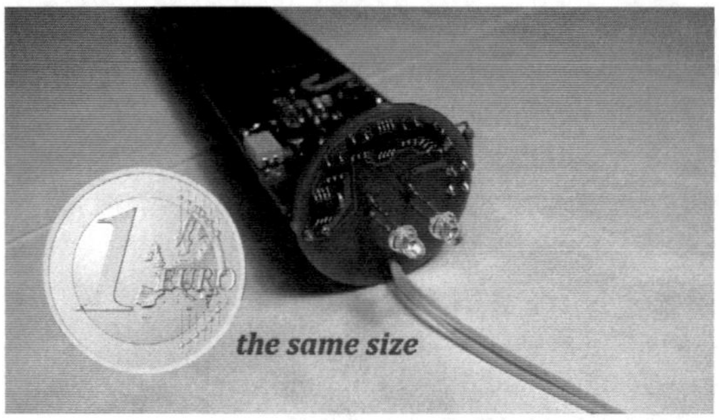

Figure 2. A working prototype of the LED driver.

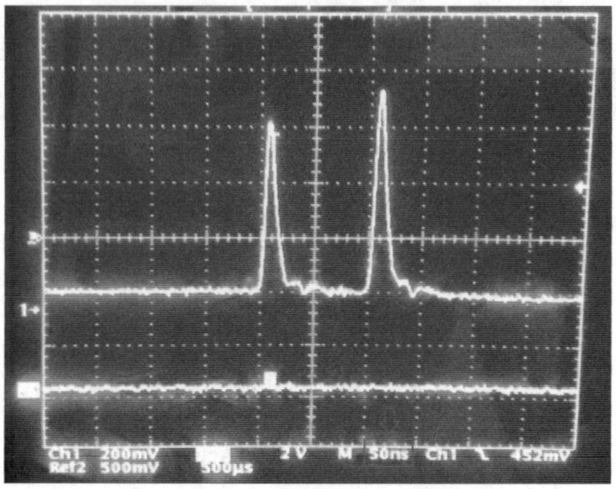

Figure 3. The example of simulated signals, which were obtained with test stand based on the ETE 9823B PMT.

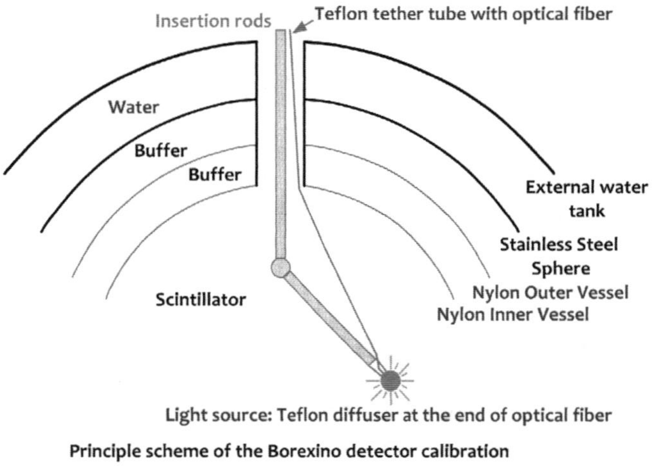

Figure 4. The illustration of the principle scheme of the Borexino online calibration.

experiments. The system will be useful for hyper projects like JUNO [7] and Hyper -Kamiokande [8]. Also the further research and technological development of the method will be based on the iDREAM [9] and Borexino experiments.

Acknowledgements

The research is made possible by funding from RFBR: Grants 14-22-0301ofi_m and 14-02-31381mol_a. We are grateful to Machulin I.N., Lukyanchenko G.A. and Marcov D.S. for help in studies and useful discussions.

References

[1] Agostini M *et al.* (Borexino Collaboration) 2009 *Nucl. Instrum. Meth.* A **600** 568–93 (*arXiv:* 0806.2400v1)
[2] Back H *et al.* (Borexino Collaboration) 2012 *JINST* **7** P10018 (*arXiv:*1207.4816v2)
[3] Kapustinsky J S *et al.* 1985 *Nucl. Instrum. Meth.* A **241** 612–13
[4] Belolaptikov I A *et al.* 1997 *Astropart. Phys.* **7** 263–82
[5] McMillan J E *et al.* 2001 *Proc. of 27th ICRC* (Hamburg) p 1287

[6] Lubsandorzhiev B K and Vyatchin Y E 2006 *JINST* **1** T06001 (*arXiv*:physics/0410281v1)
[7] Adam T *et al.* (JUNO Collaboration) 2015 *arXiv*:1508.07166v2
[8] Abe K *et al.* (Hyper-Kamiokande working group) 2011 *arXiv*:1109.3262v1
[9] Gromov M B *et al.* (iDREAM Collaboration) 2015 The industrial iDREAM detector for neutrino monitoring of the operating modes of nuclear reactors *Moscow University Physics Bulletin* **3**

Journal of Physics: Conference Series **675** (2016) 012009 doi:10.1088/1742-6596/675/1/012009

Electromagnetic modulation of monochromatic neutrino beams

A L Barabanov[1,2], **O A Titov**[2]

[1] NRC "Kurchatov Institute" , 123182, Moscow, Russia

[2] Moscow Institute of Physics and Technology, 141700, Dolgoprudny, Moscow Region, Russia

E-mail: `titov-o@mail.ru`

Abstract. We discuss the possibility to produce a modulated monochromatic neutrino beam. Monochromatic neutrinos can be obtained in electron capture by nuclei of atoms or ions. Hydrogen-like ions are of particular interest. It is shown that monochromatic neutrino beam from such hydrogen-like ions with nuclei of non-zero spin can be modulated because of different probabilities of electron capture from hyperfine states. Modulation arises by means of inducing of electromagnetic transitions between the hyperfine states. Requirements for the hydrogen-like ions with necessary properties are discussed. A list of the appropriate nuclei for such ions is presented.

1. Introduction

"Clean" neutrino beams (with neutrinos of single flavour) with known intensity and energy spectrum are of great interest for studying the neutrino properties and the details of neutrino-involving processes. A comprehensive list of applications for such beams is given, *e.g.*, in [1]. In [2] it was proposed to use accelerated β^+- and β^-- decaying nuclei (or radioactive ions), held in a storage ring, as a source for such neutrino beams (β-beams). A useful modification has been advanced in [3, 4]: one can use the ions with electron-capturing nuclei as a neutrino source. The advantage of the electron capture (EC) is that in the ion rest frame the neutrino energy E_ν^0 is completely definite. Thus, in the laboratory frame at $\gamma \gg 1$, one can obtain a high-energy neutrino beam, which is collimated and monochromatic (EC-beam).

We discuss another possible improvement: EC-beam could be modulated, or, in other words, transformed into a sequence of neutrino pulses. This can increase the efficiency of neutrino detection. A modulated beam causes a modulated signal in the detector which can be separated from constant or randomly fluctuating background noise (the similar advantage is pointed out in [5] for neutrinos produced by interaction of a modulated high-energy electron beam with a target).

The capture of an electron e^- by the nucleus in a H-like ion (with emission of an electron neutrino ν_e) is similar to the muon capture by the nucleus in a muonic atom (with emission of a muonic neutrino ν_μ). In both cases, a lepton of spin 1/2 is captured from the $|1s\rangle$ - state. In more exact terms, if the nucleus has a non-zero spin I, then the capture occurs from the hyperfine state with definite total angular momentum $F = I \pm 1/2$. In general, the decay probability highly depends on the value of F. This dependence for muon capture was first found out in

1950s and it was called "the hyperfine effect" [6, 7]. This effect was repeatedly observed (see, *e.g.*, [8] and references therein).

The hyperfine effect is most pronounced in pure Gamow–Teller transitions, *i.e.*

$$I^{\pi} \to I'^{\pi}, \qquad I' = I \pm 1, \tag{1}$$

where I' is the daughter nucleus spin. In the Gamow–Teller approximation, the total angular momentum consists of the daughter nucleus spin I' and the neutrino spin $1/2$. Therefore, due to the total angular momentum conservation the following condition must be satisfied:

$$F = I' \pm 1/2. \tag{2}$$

This means that the mother nucleus can capture e^- or μ^- from only one of the hyperfine states: if $I' = I - 1$, the capture occurs from the state $F = I - 1/2$, while in the case of $I' = I + 1$ the capture occurs from the state $F = I + 1/2$.

It is clear that, inducing transitions between the states with "allowed" and "forbidden" F values, one can impact the capture rate, and, therefore, modulate the corresponding neutrino emission. The idea to use electromagnetic radiation to induce transitions between hyperfine states and change the capture probability was first stated for muonic atoms in [9]. In 1990s, the hyperfine effect for the H-like ions was rediscovered in [10] (evidently, due to the prospects for experimental studies of H-like ions with heavy β^+-radioactive nuclei). The authors of [10] also noted the possibility to impact the EC rate by using electromagnetic radiation.

Our purpose was to select the appropriate β^+-radioactive nuclei (in fact, H-like ions with such nuclei), that could be used as sources of modulated monochromatic neutrino beams. This report briefly summarizes our results.

2. Possible source isotopes for modulated beams

The nuclei of source ions have to satisfy several requirements. The mother nuclei must have non-zero spin. The spins of the initial and the final nuclear states have to differ by 1, their parities have to be the same. To provide high monochromaticity of the neutrino beam, β^+ decay must be highly suppressed and the nuclear transition through electron capture must go predominantly to only one final state (we admitted the branching larger than 98% of total decay probability). In order to produce a beam of significant intensity, the nuclei have to be rather short-living (we set the upper limit on the half-lives $T_{1/2} < 10^6$ s $\simeq 11.6$ d).

Table 1 shows the nuclear transitions that satisfy the requirements stated above.

3. Hyperfine structure and modulation

In the point-like nucleus approximation, the distance between two hyperfine levels is determined by the squared modulus $|\psi_{g.s.}(0)|^2$ of the electron wavefunction in the ion, taken at the origin (at the nucleus), and by the nuclear spin I and magnetic moment μ (see, *e.g.*, [11]). For ions with high Z, there are significant contributions (of the order of 5%) into the hyperfine splitting from finite-size nuclear corrections and radiative corrections.

Note that the sign of the nuclear magnetic moment μ determines the sequence of hyperfine levels in a H-like ion: when $\mu > 0$, the state with angular momentum $F = I - 1/2$ is lower, while in the case of $\mu < 0$ the lower state has $F = I + 1/2$. In Gamow–Teller transitions in H-like ions, EC decay from the lower state could be either allowed or forbidden. Accordingly, we define two types of H-like ions: A type and F type, depending on whether the EC decay is allowed or forbidden for the lower state.The type of an ion is determined, first, by the sign of the nuclear magnetic moment μ and, second, by the relation between the initial and the final nuclear spins I and I'.

Journal of Physics: Conference Series **675** (2016) 012009 doi:10.1088/1742-6596/675/1/012009

Table 1. The pairs of mother and daughter nuclei, related by Gamow–Teller transition following the electron capture. Here ${}^A_ZX$ is the mother nucleus (of spin and parity I^π and half-life $T_{1/2}$), undergoing the Gamow–Teller transition into one state of the daughter nucleus ${}_{Z-1}^{A}X'$ (of spin and parity I'^π and of excitation energy E') with probability $P \geq 98$ %. Q_{EC} is the energy release in the transition. The data are taken from [12].

${}^A_ZX$	I^π	$T_{1/2}$	${}_{Z-1}^{A}X'$	I'^π	E', keV	Q_{EC}, keV	P, %
${}^{71}_{32}$Ge	$1/2^-$	11.4 d	${}^{71}_{31}$Ga	$3/2^-$	0	232.6	100
${}^{107}_{48}$Cd	$5/2^+$	6.5 h	${}^{107}_{47}$Ag*	$7/2^+$	93.1	1323.2	99.7
${}^{118m}_{51}$Sb	8^-	5.0 h	${}^{118}_{50}$Sn*	7^-	2574.8	1332	98.3
${}^{131}_{55}$Cs	$5/2^+$	9.7 d	${}^{131}_{54}$Xe	$3/2^+$	0	354.8	100
${}^{135}_{57}$La	$5/2^+$	19.5 h	${}^{135}_{56}$Ba	$3/2^+$	0	1207	98.1
${}^{163}_{68}$Er	$5/2^-$	75 m	${}^{163}_{67}$Ho	$7/2^-$	0	1211	99.9
${}^{165}_{68}$Er	$5/2^-$	10.4 h	${}^{165}_{67}$Ho	$7/2^-$	0	378	100

The modulation methods for neutrino beams, produced by H-like ions of A and F types, are different. Indeed, for A-type ions, the transition from the lower to the upper state, induced by a resonant electromagnetic field, is accompanied by an interruption of the neutrino emission. For F-type ions, on the contrary, such transition will initiate the neutrino emission. In both cases, the reverse transition from the upper level to the lower one could be either spontaneous or induced by an external resonant electromagnetic wave.

To our opinion, H-like ions of F type are of special interest. Such ions decay predominantly or exclusively from the upper state of the hyperfine structure. Hence, their lower state is practically stable. This circumstance may be used to accumulate ions in a storage ring. Then, by electromagnetic transitions to the upper state (arranged in π-pulses), one may form neutrino pulses.

4. Results and discussion

Table 2 gives the following parameters of H-like ions: the hyperfine splitting $|\Delta_{HF}|$ of the ground state , the resonant wavelength λ_{HF}, the lifetime $\tau_{HF} = 1/w_{HF}$ of the upper hyperfine state with respect to spontaneous emission. For the ion with the nucleus ^{118m}Sb we give values of τ_{HS} both for positive (+) and for negative (−) magnetic moment μ.

All the selected nuclei have rather large half-lives (for a nucleus, accelerated to Lorentz factor γ, lifetime increases by γ times). This means that the total number of radioactive ions in a storage ring should be large, in order to get the neutrino beam of significant intensity. Of course, the required beam intensity, as well as its energy (and, thus, the factor γ), depends on the experimental problem.

5. Conclusion

It turns out that the list of appropriate nuclei is rather short. However, the H-like ions with these nuclei noticeably differ from each other both by the lifetime with respect to the electron

Table 2. The main properties of the H-like ions that could be used as sources of monochromatic and modulated neutrino beams. Here ${}^A_ZX$ is the mother nucleus, which undergoes the Gamow–Teller transition $I^\pi \to I'^\pi$ into the final state of the daughter nucleus, μ is the magnetic moment of the mother nucleus in units of nuclear magneton μ_N, A or F is the type of the H-like ion, $|\Delta_{HF}|$ is the absolute value of hyperfine splitting of the H-like ion ground state, λ_{HF} is the resonant wavelength of hyperfine transition, τ_{HF} is the lifetime of the upper hyperfine state with respect to spontaneous emission.

${}^A_ZX$	$I^\pi \to I'^\pi$	μ/μ_N	Type	$\|\Delta_{HF}\|$, eV	λ_{HF}, μm	τ_{HF}, s
${}^{71}_{32}$Ge	$1/2^- \to 3/2^-$	+0.55	F	0.041	30.2	1024
${}^{107}_{48}$Cd	$5/2^+ \to 7/2^+$	−0.615	A	0.105	11.8	26.3
${}^{118m}_{51}$Sb	$8^- \to 7^-$	2.32		0.433	2.86	0.46(+), 0.41(−)
${}^{131}_{55}$Cs	$5/2^+ \to 3/2^+$	+3.54	A	0.973	1.27	0.046
${}^{135}_{57}$La	$5/2^+ \to 3/2^+$	+3.70	A	1.162	1.06	0.027
${}^{163}_{68}$Er	$5/2^- \to 7/2^-$	+0.56	F	0.346	3.58	1.03
${}^{165}_{68}$Er	$5/2^- \to 7/2^-$	+0.64	F	0.399	3.10	0.67

capture as well as by the hyperfine structure of the ground states. The latter is important for the modulation method. Given a specific problem, one can choose a suitable sort of H-like ions from our list and use it as a source of modulated monochromatic neutrino beams.

References

[1] Espinoza C, Lazauskas R and Volpe C 2012 *Phys. Rev.* D **86** 113016
[2] Zucchelli P 2002 *Phys. Lett.* B **532** 166
[3] Sato J 2005 *Phys. Rev. Lett.* **95** 131804
[4] Bernabeu J *et al.* 2005 *JHEP* **12** 14
[5] Ginzburg I F 2015 *Nucl. Instrum. Meth.* A **772** 89
[6] Bernstein J *et al.* 1958 *Phys. Rev.* **111** 313
[7] Primakoff H 1959 *Rev. Mod. Phys.* **31** 802
[8] Wiaux V *et al.* 2002 *Phys. Rev.* C **65** 025503
[9] Batkin I S, Smirnov Yu G and Churakova T A 1977 *Sov. J. Nucl. Phys.* **26** 16
[10] Folan L M and Tsifrinovich V I 1995 *Phys. Rev. Lett.* **74** 499
[11] Shabaev V M *et al.* 1997 *Phys. Rev.* A **56** 252
[12] Live chart of nuclides *https://www-nds.iaea.org*

Journal of Physics: Conference Series **675** (2016) 012010 doi:10.1088/1742-6596/675/1/012010

Termosyphon cryogenic system for RED-100 detector

V Sosnovtsev, I Tolstukhin, A Shakirov* and R Shafigullin
National Research Nuclear University MEPhI (Moscow Engineering Physics Institute), Kashirskoe highway 31, Moscow, 115409, Russia

E-mail: *Shalexey91@mail.ru

Abstract. A cryogenic system based on a two-phase closed tubular thermosyphon with 12 mm diameter copper tube is developed. It was used for thermal stabilization of the liquid xenon emission detector RED-100. The nitrogen refrigerant cooled down with a free-boiling liquid nitrogen bath has been used. It was shown that the system supports the RED100 operation at temperature 166 K with accuracy ±1K.

1. Introduction

1.1. About RED 100

The RED 100 emission detector based on liquid xenon is being created by the efforts of the RED (Russian Emission Detectors) collaboration for detecting and investigating the effect of coherent neutrino scattering off heavy nuclei. The principle of registration was discussed in detail in the paper [1]. Such a detector can register single electrons produced in massive active medium from rare ionizing interactions with low-energy signature.

1.2. Challenges for cryostatting system

The RED 100 detector contains 200 kg of liquid xenon in a 100 kg titanium cryostat. The total mass of the detector including xenon, cryostat, PMTs and internal light-collecting structure is about 500 kg. The challenge for the cryogenic system is to condense xenon and to maintain a constant temperature in the liquid at about 166K with an accuracy of1K during long time required for the experiment. The additional challenges are:

1) Minimizing the mass of peripheral devices near the detector.
2) Deposition of the detector inside a massive defence or in a deep well for providing low-background experiment.

The last requirement means that the detector should be removed to a distance of several meters from the data acquisition system, gas purifying system and cryostatting systems for the detector. The described thermosyphon technology meets the specified requirements.

2. Thermosyphon principle of operation

The thermosyphon or gravity-assisted heat pipe consists of three basic sections shown in figure 1: a cooling section (condenser) located above a heating section (evaporator) and a passive adiabatic section connecting the two active sections [2].

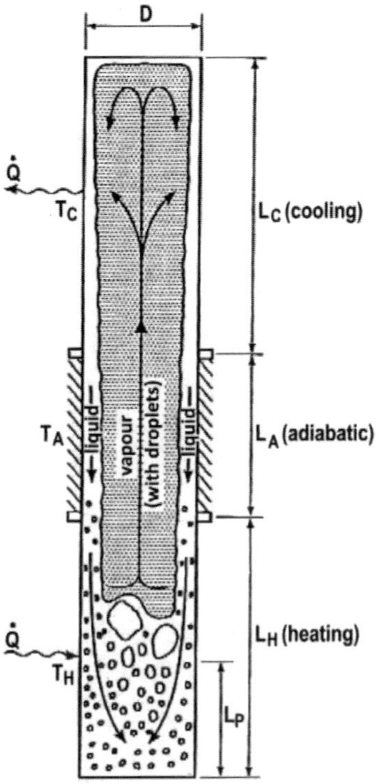

Figure 1. The operating principle of a two-phase closed tubular thermosyphon (heat pipe).

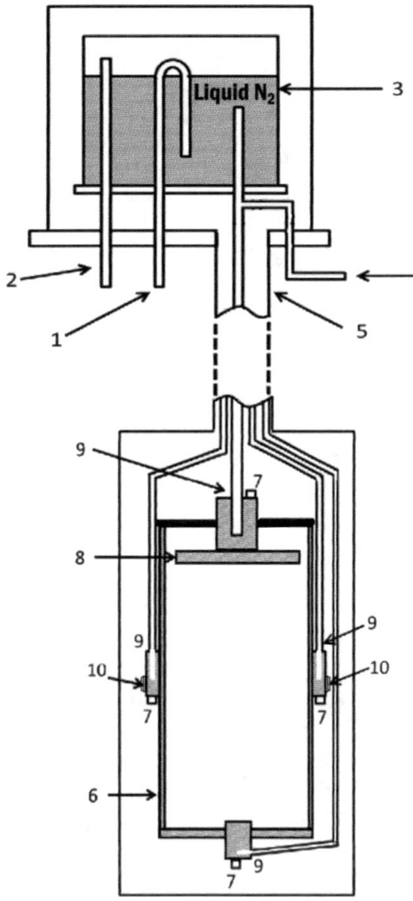

Figure 2. Schematic drawing of the experimental setup: 1 – liquid nitrogen inlet; 2 – ventilation outlet; 3 – liquid nitrogen tank;4 –nitrogen fluid supply from charging system; 5 – vacuum jacket; 6 – copper roll; 7 – thermal resistors Pt-100; 8-copper condenser; 9 – evaporators; 10 – heaters.

The condenser of LC length is filled with gas, which is depositing heat energy Q into a cooling machine operating in a good thermal contact with this section. The condensate generated inside the condenser falls down through the adiabatic section into the evaporator, where the liquid is boiling and absorbing the heat Q. The generated vapor is rising up into the condenser, returning the heat to the cooling machine, condensing in the liquid phase, then, the heat transfer cycle repeats. Since the operation of the thermosyphon relies upon the gravitational force, the evaporator must be located below the condenser. The schematic drawing of the experimental setup is shown in figure 2.

3. Thermal stabilization system

Two types of thermosyphon was considered. The first one is a bellow flex hose thermosyphon consist of standard high-vacuum stainless steel bellow hose with 1/2" inside diameter with the same cooling and heating sections as reported in [3]. The other one is a closed tubular thermosyphon consist of 12 mm diameter copper pipe. A comparison of the operation for these two types is shown in figure 3.

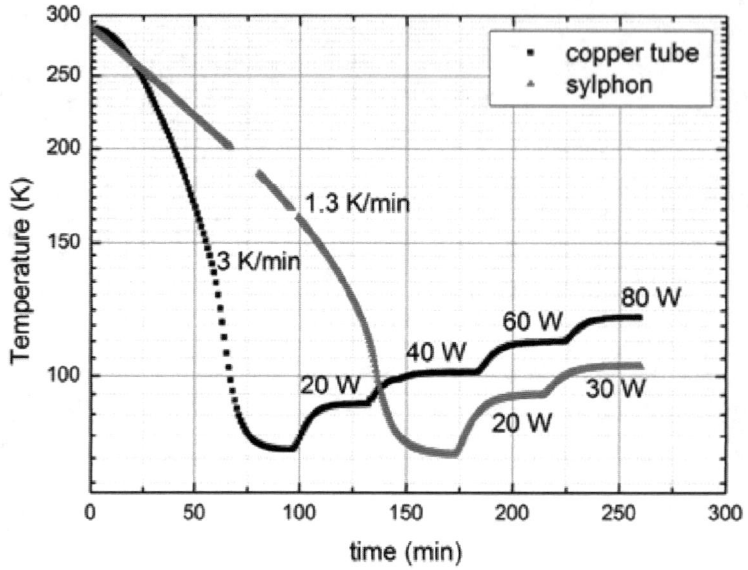

Figure 3. Temperature-time dependences both for copper pipe (black line) and for bellow flex hose thermosyphon (red line). The error bars are within data points.

Figure 3 shows the temperature of the cold head under different heat load conditions. At the beginning of the test, the thermosyphon is vacuumed at room temperature. The cooling power is applied to the cold head via the evaporator as soon as the thermosyphon is charged with nitrogen fluid. When the lowest temperature 80 K is achieved, the additional electric power is applied with the electrical heater (15 Ohm, 200 W resistor) installed on the cold head. The temperatures rise with the increase of the heat load. The steady states are observed at different heat powers.

From this comparison we conclude that the thermosyphon based on copper tube has an advantage, because it has larger cooling capacity.

In order to maintain a constant temperature it was used an automatic control system which regulates the amount of nitrogen in the thermosyphons. The algorithm of the automatic control system is as follows:

1) Turn on the thermosyphon. The injection valve which providing nitrogen from an external gas source into the heat pipe is opened for a time Δt_1;
2) The temperature of the heat exchanger reaches To-ΔT;
3) Turn off the thermosyphon. The relief valve which connect heat pipe with atmosphere is opened for a time Δt_2;
4) The heat exchanger temperature begins to rise and reaches To + ΔT due to the external heat;
5) The injection valve is opened again for the period Δt_1 and the cooling cycle is repeated. Constants Δt_1, Δt_2 are selected empirically for each thermosyphon individually.

4. System launching

The result of the cryotest for copper pipe is shown in figure 4. The red line corresponds to the side of the detector, and the black line – to the bottom. The graph has two plateaus, corresponding to different temperatures stabilize. It was shown that we can maintain a constant temperature of the detector at 166 K with an accuracy of 1 K.

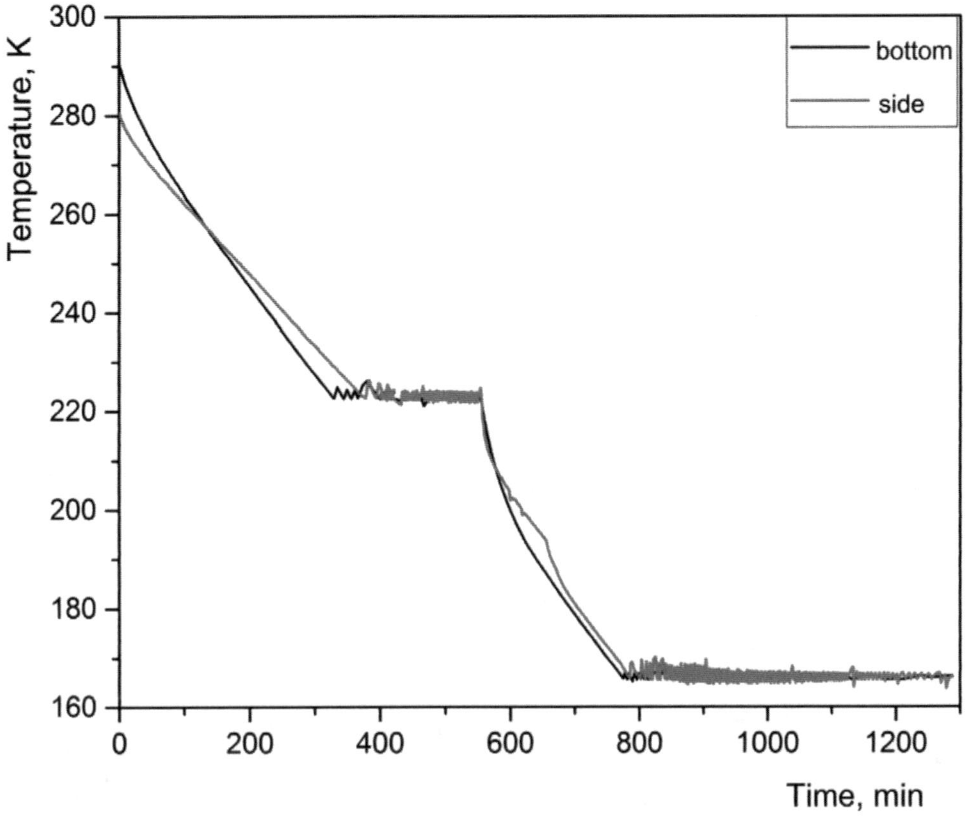

Figure 4. Dynamic temperature stabilization using the copper pipe thermosyphons for bottom and side of the detector. The error bars are within data points.

5. Conclusion

It was shown that the cryogenic system based on closed two-phase thermosyphon using nitrogen as an active medium can provide thermal stabilization of emission detector based on liquid xenon at required temperature with an accuracy of 1K. Such a system was investigated and successfully tested.

Acknowledgements

This research is supported by the Ministry of Education and Science of the Russian Federation under the Contract of No. 11.G34.31.0049 from October 19, 2011 and within the framework of the Centre for Nanostructured Electronics supported by MEPhI Academic Excellence Project (contract № 02.a03.21.0005, 27.08.2013).

References

[1] Bolozdynya A 2010 *Emission detectors* (Singapore: World Scientific)
[2] Lock G S H1992 *The Tubular Thermosyphon: Variation on a theme* (Oxford: Oxford University Press)
[3] Bolozdynya A I*et al.* 2015 *Int. J. of Heat and Mass Tr.* **80** 159-62

Digital trigger system for the RED-100 detector based on the unit in VME standard

D Yu Akimov[1,2], V A Belov[1,2], A I Bolozdynya[2], Yu V Efremenko[2,3], V A Kaplin[2] and P P Naumov[2,*]

[1] SSC RF Institute for Theoretical and Experimental Physics of National Research Centre “Kurchatov Institute”, 25 Bolshaya Cheremushkinskaya, Moscow, 117218, Russian Federation

[2] National Research Nuclear University MEPhI (Moscow Engineering Physics Institute), 31 Kashirskoe highway, Moscow, 115409, Russian Federation

[3] University of Tennessee,1408 Circle Dr, Knoxville, TN 37996-1200, USA

E-mail: [*]ddr727@yandex.ru

Abstract. The system for forming a trigger for the RED-100 liquid xenon detector has been developed. The trigger can be generated for all types of events required to calibrate the detector and data acquisition, including events with one ionization electron. The system has an event detection mechanism where each event is assigned with the timestamp and event type. The trigger system is required in the systems searching for rare events to keep only the necessary information from the ADC array. The characteristics and implementation of the trigger system that provides high efficiency operation even at low-energy events have been described.

1. Introduction

International scientific collaboration COHERENT [1] with the participation of Laboratory for experimental nuclear physics MEPhI created RED-100 detector which is designed to search and study the effect of coherent elastic neutrino-nucleus scattering (CENNS) [2, 3]. Spallation Neutron Source (SNS), Oak Ridge National Laboratory (USA) is considered as a source of neutrinos for measurement. Because of the smallness of the neutrino energy from source expected recoil energy is very low (<10 keV).

The RED-100 experimental setup is a two-phase liquid xenon time projection chamber [3]. The principle of operation of such detectors is illustrated in figure 1. Particles interacting with dense noble liquefied gas medium produces excitation and ionization. The excitation creates prompt scintillation (S1 signal). Under the influence of the applied electric field ionization electrons drift along applied electric field towards the surface and then enter into the vapor phase where produce proportional scintillation or electroluminescense (S2 signal). The time difference between S1 and S2 signals defines Z position of the event, while the X-Y position is determined from the pattern of S2 light in the photodetecting system. The 3D event reconstruction allows to define a fiducial volume eliminating potentially background events originated in the vicinity of the detector walls. Measurement of the ratio of S2 to S1 signal amplitudes gives an ability to discriminate between electron recoils and nuclear recoils providing an efficient background rejection.

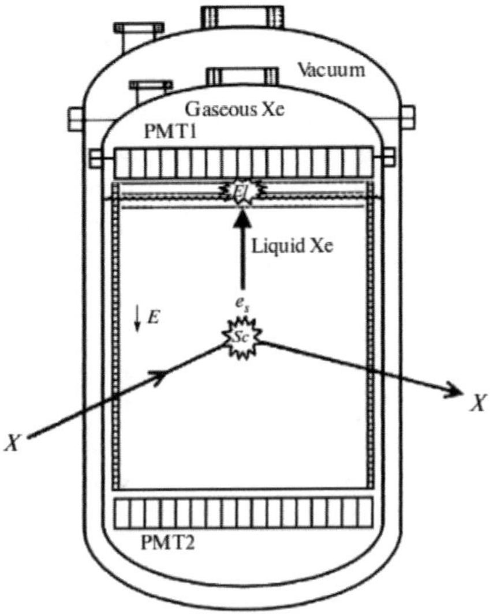

Figure 1. RED-100 detector scheme and operation principle.

Detectors of this type are widely used in measurements of low-energy recoil to search for dark matter [5]. A distinctive feature of emission detectors is the significant (to a few hundred photons per electron) ionization yield of electroluminescence. This allows you to achieve very high sensitivity, up to the registration of single electron ionization (SE) [4]. With such low-energy electroluminescence signal is divided into its component signals of single photoelectrons (SPE), distributed along the entire length of the electroluminescent signal, which in this case is significant 2 microseconds.

In the RED-100 detector light signals are registered by two arrays of 19 PMTs each located above and below the active volume (see figure 1). The energy range of ionization from one electron is up to 100 keV, which corresponds to the total number of photons detectable from 10^2 to $5 * 10^5$. This puts high requirements on the trigger generation scheme (both the lower threshold and the dynamic range).

2. Statement of the problem

Usually simple discrimination of signals in the overall channel after smoothing filter provides the trigger on scintillation and luminescence. In the experiment on SNS expected useful signal is less than 10 SE, which requires a low threshold of the trigger to be lowered to the level of several SE. Amplitude resolution of single-photon pulses PMT is quite large (FWHM ≈ 100%), which in the case of analog signals summation lead to high errors in estimating the amount of SPE. The effect of spontaneous emission SE [4,6] creates a background signal greater than useful signal by 2-3 orders [2]. The use of discriminators for each channel, and then the analog summation significantly improves the amplitude resolution, but can be still insufficient. An additional problem is that the scintillation signals can provide the same amount of that SPE as SE, thus have much higher energy and have to be excluded by trigger. To fulfill all requirements for a trigger it was decided to use discriminators in each channel and the digital circuit (Field-Programmable Gate Array - FPGA), receiving signals and precisely counting the incoming SPE.

To implement the required functionality it was decided to use a CAEN V1495 digital module with custom firmware and 3 CAEN V895 discriminator modules. The use of FPGA unit in V1495 simplifies modification and enhances capabilities of the trigger system. Firmware Development program was carried out with the use of computer-aided design (CAD) Quartus 9.1.

3. Implementation of the trigger system

The signals from each PMT come to individual discriminators configured to activation on the SPE, and then go straight into the digital module V1495. The internal structure of the digital module can be divided into four parts: counting units, local trigger units, coincidence unit and event registration unit. Local trigger units form the trigger signal when the input signals according to certain conditions. These signals are sent to a coincidence unit, which determines what types of events should match for generating the main trigger. Event registration unit saves in the internal buffer memory information on the main trigger activation, such as a timestamp and the event type, and generates a signal that starts recording event in an array of ADCs. Upon request, the trigger system transmits information about the registered events from its buffer to the data acquisition system DAQ.

SPE counting was implemented by summing the number of pulses in a specified-width time window. The resulting sum is compared with the lower and upper thresholds for this type of event. There are two counting units for an independent pulse counting from upper and lower arrays of the PMT. The width of the time window and the thresholds are set in the registers of V1495 unit by the DAQ system. Functional diagram of the counting unit is shown in figure 2.

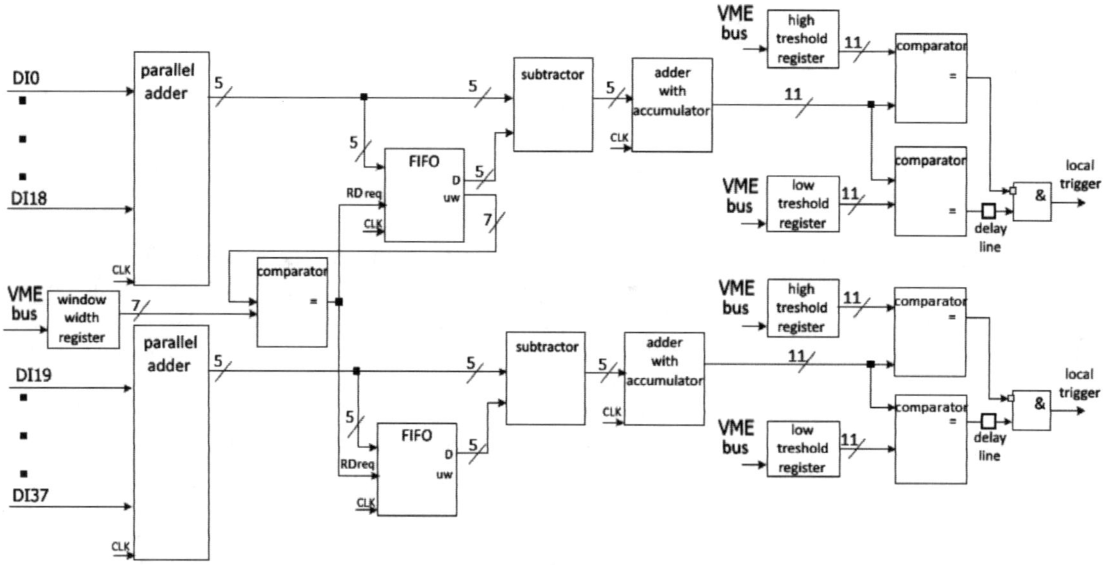

Figure 2. Functional diagram of the counting unit and local trigger units.

The circuit is clocked by the main generator of frequency 100 MHz. The signals from the discriminator (DI) fed to a parallel adder, which every 10 ns summarizes the values of all 19 inputs. Each sum is fed to the subtractor and FIFO, which promotes incoming value through all FIFO words. Thus, the delay time equal 10 ns*(capacity of FIFO). The value of a parallel adder is fed to the subtractor, where it is subtracted from the delayed value from the FIFO. The adder with accumulator summarizes the values obtained from the subtractor, which may be either positive or negative, and each clock cycle stores in the accumulator result being the current count of pulses in the time window.

Adjusting the width of the window is implemented as follows. The trigger system provides window width register in which the DAQ computer records the required value. Each clock cycle FIFO gives the number of used words (uw). This number is compared with that which is specified in the window width register, and in case of equality signal to read FIFO (RDreq) is issued.

The coincidence unit (figure 3) is implemented on the AND, OR, NOT logic elements.
Bits of the channel enable register are set to 1 when the corresponding channels are enabled and for disabled channels bits are set to 0. The output of each local trigger and the output of the corresponding bit are fed to an OR gate with one NOT input. When enabled channels coincide, scheme & (AND)

generates signal to indicate a match. This signal enters the latch circuit for generating the output signal duration of 1 μs. The generated trigger signal passes through the circuit specifying the minimum interval between triggers, and if the interval is not violated, is issued to the ADC output array. The generation of the trigger can be disabled by sending signal to the external input inhibit. But event will be registered.

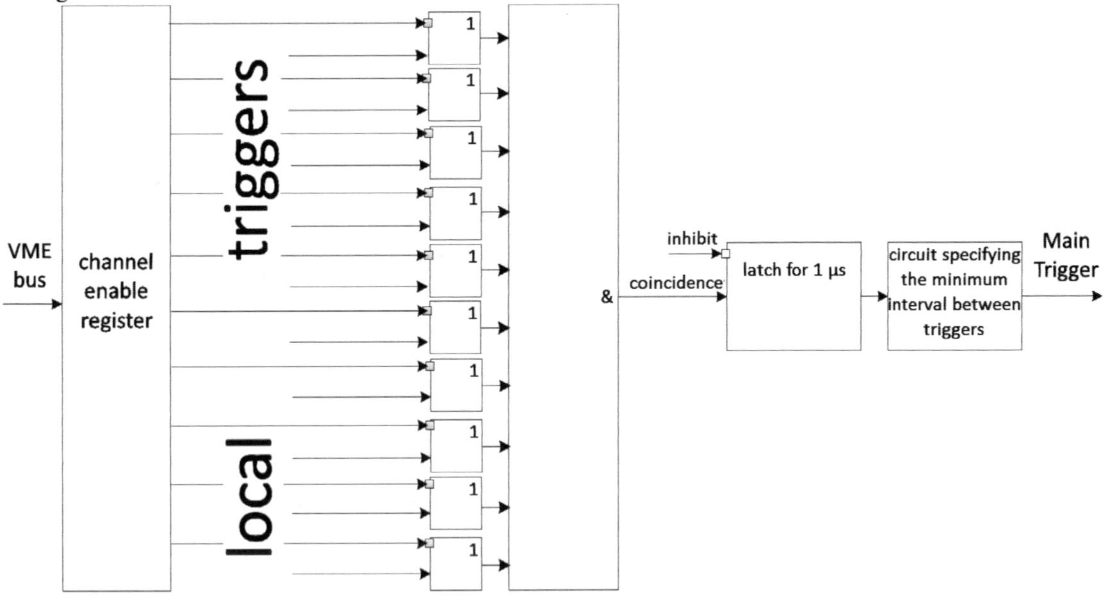

Figure 3. Functional diagram of the coincidence unit.

Bits of the channel enable register are set to 1 when the corresponding channels are enabled and for disabled channels bits are set to 0. The output of each local trigger and the output of the corresponding bit are fed to an OR gate with one NOT input. When enabled channels coincide, scheme & (AND) generates signal to indicate a match. This signal enters the latch circuit for generating the output signal duration of 1 ms. The generated trigger signal passes through the circuit specifying the minimum interval between triggers, and if the interval is not violated, is issued to the ADC output array. The generation of the trigger can be disabled by sending signal to the external input inhibit. But event will be registered.

Event registration unit is a circuit detecting the last arrived local trigger before the coincidence. The type of event is determined by number of channel from a local trigger arrived. The unit is provided with FIFO and time counter (24 bit width). Counter time must be tied to the DAQ time because of its overflow. FIFO depth is 4096 words (32 bit word width). Bits 31-28 are reserved, 27-24 - type of event code, 23-0 - time stamp of event in microseconds. The possibility not to generate the main trigger signal when FIFO overflows, or overwrite the oldest events was implemented.

4. Test results

Testing the SPE counting unit was carried out using arbitrary waveform generator, which are configured to generate pulses of known quantity. The maximum operating frequency of the system is 100 MHz, the maximum trigger delay after the appearance of the event is 10 ns. The circuit, part of the basis of the firmware for the on-board FPGA of CAEN V1495, occupied 50% of all its resources.

To test the system in the conditions close to the real, simplified model of one channel as part of installation (a photomultiplier in a dark box, the LED lighting it, and the discriminator) was assembled. The pulse of amplitude small enough so that the PMT can register single photoelectrons and 1 ms width was send from the signal generator to the LED. The average number of photoelectrons was varied by changing the amplitude of the supplied signal. The discriminator threshold was set at

0.5 SPE amplitude. The signal was fed to a discriminator FlashADC and was recorded in a file together with the sign of the trigger, the trigger was taken from the primary generator. Waveforms was analyzed further using a processing program that finds and counts the number of pulses in each event. Then the histogram distribution of the number of events by the number of pulses in the event for those events where there is a sign of the trigger, and for all events was drawn. The ratio of the histograms shows the efficiency shown in figure 4.

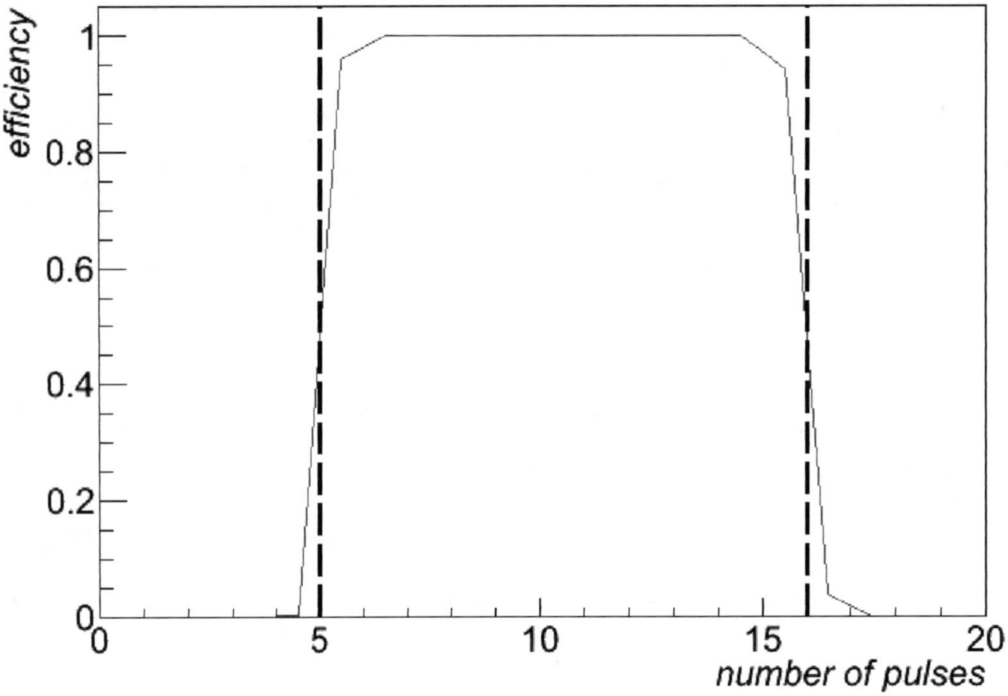

Figure 4. The histograms for triggered events and for all events.

It is easy to see that the trigger is generated within strictly defined between the lower and upper thresholds (5 SPE and the SPE 16, respectively), while the efficiency is 100%. Operation of the trigger system was also tested in several data sets during a session of measurements carried out by the RED-1 detector (prototype of the RED-100 detector) in December 2013. By results of test defects are not found, the module operates normally.

Acknowledgements
This study was supported by RF Government under contracts of NRNU MEPhI with the Ministry of Education and Science of №11.G34.31.0049 from October 19, 2011 and within the framework of the Centre for Nanostructured Electronics supported by MEPhI Academic Excellence Project (contract № 02.a03.21.0005, 27.08.2013).

References
[1] Akimov D Yu *et al.* *arXiv:*1310.0125
[2] Akimov D Yu *et al.* 2013 *J. Instrum.* **8** P10023
[3] Chepel V and Araújo H 2013 *J. Instrum.* **8** R04001
[4] Akerib D S *et al.* 2014 *Phys. Rev. Lett.* **112** 9 091303
[5] Santos E *et al.* 2011 *J. High Energy Phys.* **1112** 115
[6] Akimov D Yu *et al.* 2012 *Instrum. Exp. Tech.* **55** 423

Journal of Physics: Conference Series **675** (2016) 012012 doi:10.1088/1742-6596/675/1/012012

Understanding the detector behavior through Montecarlo and calibration studies in view of the SOX measurement

A Caminata[1], M Agostini[2], K Altenmüller[2], S Appel[2], G Bellini[3], J Benziger[4], N Berton[5], D Bick[6], G Bonfini[7], D Bravo[8], B Caccianiga[3], F Calaprice[9], P Cavalcante[7], A Chepurnov[10], K Choi[11], M Cribier[5], D D'Angelo[3], S Davini[12], A Derbin[13], L Di Noto[1], I Drachnev[12], M Durero[5], A Empl[14], A Etenko[15], S Farinon[1], V Fischer[5], K Fomenko[16], D Franco[17], F Gabriele[7], J Gaffiot[5], C Galbiati[9], C Ghiano[1], M Giammarchi[3], M Goeger-Neff[2], A Goretti[9], M Gromov[10], C Hagner[6], T Houdy[5], E Hungerford[14], Aldo Ianni[7], Andrea Ianni[9], N Jonquères[18], K Jedrzejczak[19], M Kaiser[6], V Kobychev[20], D Korablev[16], G Korga[7], V Kornoukhov[30], D Kryn[17], T Lachenmaier[21], T Lasserre[5], M Laubenstein[7], B Lehnert[22], J Link[8], E Litvinovich[15,23], F Lombardi[7], P Lombardi[3], L Ludhova[3], G Lukyanchenko[15,23], I Machulin[15,23], S Manecki[8], W Maneschg[24], S Marcocci[12], J Maricic[11], G Mention[5], E Meroni[3], M Meyer[6], L Miramonti[3], M Misiaszek[19,7], M Montuschi[25], P Mosteiro[9], V Muratova[13], R Musenich[1], B Neumair[2], L Oberauer[2], M Obolensky[17], F Ortica[26], M Pallavicini[1], L Papp[2], L Perasso[1], A Pocar[27], G Ranucci[3], A Razeto[7], A Re[3], A Romani[26], R Roncin[7,17], N Rossi[7], S Schönert[2], L Scola[5], D Semenov[13], H Simgen[24], M Skorokhvatov[15,23], O Smirnov[16], A Sotnikov[16], S Sukhotin[15], Y Suvorov[28,15], R Tartaglia[7], G Testera[1], J Thurn[22], M Toropova[15], E Unzhakov[13], C Veyssière[5], A Vishneva[16], M Vivier[5], R B Vogelaar[8], F von Feilitzsch[2], H Wang[28], S Weinz[29], J Winter[29], M Wojcik[19], M Wurm[29], Z Yokley[8], O Zaimidoroga[16], S Zavatarelli[1], K Zuber[22] and G Zuzel[19]

[1] Dipartimento di Fisica, Università degli Studi e INFN, 16146 Genova, Italy
[2] Physik-Department and Excellence Cluster Universe, Technische Universität München, 85748 Garching, Germany
[3] Dipartimento di Fisica, Università degli Studi e INFN, 20133 Milano, Italy
[4] Chemical Engineering Department, Princeton University, Princeton, NJ 08544, USA
[5] Commissariat à l'Énergie Atomique et aux Énergies Alternatives, Centre de Saclay, IRFU, 91191 Gif-sur-Yvette, France
[6] Institut für Experimentalphysik, Universität, 22761 Hamburg, Germany
[7] INFN Laboratori Nazionali del Gran Sasso, 67010 Assergi (AQ), Italy
[8] Physics Department, Virginia Polytechnic Institute and State University, Blacksburg, VA 24061, USA
[9] Physics Department, Princeton University, Princeton, NJ 08544, USA
[10] Lomonosov Moscow State University Skobeltsyn Institute of Nuclear Physics, 119234 Moscow, Russia
[11] Department of Physics and Astronomy, University of Hawai'i, Honolulu, HI 96822, USA

[12] Gran Sasso Science Institute (INFN), 67100 L'Aquila, Italy
[13] St. Petersburg Nuclear Physics Institute NRC Kurchatov Institute, 188350 Gatchina, Russia
[14] Department of Physics, University of Houston, Houston, TX 77204, USA
[15] NRC Kurchatov Institute, 123182 Moscow, Russia
[16] Joint Institute for Nuclear Research, 141980 Dubna, Russia
[17] AstroParticule et Cosmologie, Université Paris Diderot, CNRS/IN2P3, CEA/IRFU, Observatoire de Paris, Sorbonne Paris Cité, 75205 Paris Cedex 13, France
[18] Commissariat à l'Énergie Atomique et aux Énergies Alternatives, Centre de Saclay, DEN/DM2S/ SEMT/BCCR, 91191 Gif-sur-Yvette, France
[19] M. Smoluchowski Institute of Physics, Jagiellonian University, 30059 Krakow, Poland
[20] Kiev Institute for Nuclear Research, 06380 Kiev, Ukraine
[21] Kepler Center for Astro and Particle Physics, Universität Tübingen, 72076 Tübingen, Germany
[22] Department of Physics, Technische Universität Dresden, 01062 Dresden, Germany
[23] National Research Nuclear University MEPhI (Moscow Engineering Physics Institute), Kashirskoe highway 31, Moscow, 115409, Russia
[24] Max-Planck-Institut für Kernphysik, 69117 Heidelberg, Germany
[25] Dipartimento di Fisica e Scienze della Terra Università degli Studi di Ferrara e INFN, Via Saragat 1-44122, Ferrara, Italy
[26] Dipartimento di Chimica, Biologia e Biotecnologie, Universita' degli Studi e INFN, 06123 Perugia, Italy
[27] Amherst Center for Fundamental Interactions and Physics Department, University of Massachusetts, Amherst, MA 01003, USA
[28] Physics and Astronomy Department, University of California Los Angeles (UCLA), Los Angeles, California 90095, USA
[29] Institute of Physics and Excellence Cluster PRISMA, Johannes Gutenberg Universität Mainz, 55099 Mainz, Germany
[30] Institute for Theoretical and Experimental Physics, 117218 Moscow, Russia

E-mail: `alessio.caminata@ge.infn.it`

Abstract. Borexino is an unsegmented neutrino detector operating at LNGS in central Italy. The experiment has shown its performances through its unprecedented accomplishments in the solar and geoneutrino detection. These performances make it an ideal tool to accomplish a state-of-the-art experiment able to test the existence of sterile neutrinos (SOX experiment). For both the solar and the SOX analysis, a good understanding of the detector response is fundamental. Consequently, calibration campaigns with radioactive sources have been performed over the years. The calibration data are of extreme importance to develop an accurate Monte Carlo code. This code is used in all the neutrino analyses. The Borexino-SOX calibration techniques and program and the advances on the detector simulation code in view of the start of the SOX data taking are presented.

1. Introduction

Although the collected neutrino experimental data well fit into the three-flavor oscillation model, several short-baseline neutrino experiments have reported anomalies which significantly deviates from the three active neutrino pictures ([1, 2, 3, 4]). The SOX project aims to use the Borexino detector to investigate the existence of sterile neutrinos in the Δm^2_{14} region of $\sim 1\ eV^2$ [5]. The first part of the project consists in deploying a 150 kCi $^{144}Ce-^{144}Pr\ \overline{\nu}_e$ source in a dedicated pit located 8.25 m below the detector's center. Antineutrinos are detected in Borexino by means of inverse beta decay (IBD, 1.8 MeV threshold) on protons. IBD events are clearly tagged using the space-time coincidence between the prompt e^+ signal and the subsequent neutron capture ($\tau = 254\ \mu s$ [6]) event. Consequently, accidental background is almost negligible. The $^{144}Ce -^{144} Pr$ source has been identified as a suitable $\overline{\nu}_e$ emitter having a long enough half-life

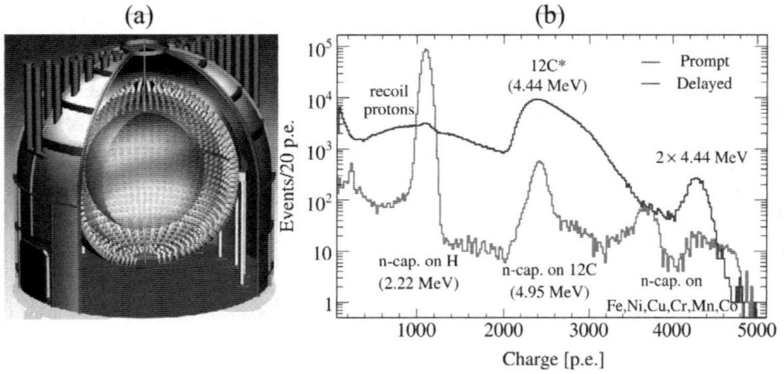

Figure 1. a) schematic view of the Borexino detector. b) Energy spectrum (variable photoelectrons) of the ^{241}Am^{9}Be source. The spectrum is subdivided in prompt (blue) and delayed (red) signals. c) Picture of the AmBe neutron source.

to allow the source production and the transportation to LNGS. The ^{144}Ce source β-decays to ^{144}Pr ($t_{1/2} = 296$ days) which rapidly ($t_{1/2} = 17$ min) β-decays to ^{144}Nd emitting $\overline{\nu}_e$ above IBD threshold (the endpoint of the ^{144}Pr decay is about 3 MeV).

One of the key requirement for the success of the SOX project is a good knowledge of the detector's response. The large dimensions of the detector require a careful mapping of its energy response in different positions within the scintillating volume. Since the energy of the incoming antineutrino is reconstructed from the positron energy, the knowledge of the energy scale is mandatory. Borexino has been calibrated several times in the past (2008 and 2009 calibration campaigns [7]). These data are extremely useful in view of the SOX experiment. Since the SOX and geoneutrino signals have the same features, calibration data acquired for the geoneutrino analysis can be used also for SOX. Nevertheless, a new calibration campaign is foreseen before the arrival of the SOX antineutrino source. These calibration data are of extreme importance in understanding the detector behavior and to increase the reliability of the SOX Monte Carlo simulation code. Since the SOX analysis, the cut efficiency and the sensitivity studies are performed analyzing the output of the simulations, having an accurate simulation code is crucial for a proper data analysis.

2. Detector calibration with gamma and neutron sources

Calibration sources are deployed into the desired location within the scintillator using a series of interconnecting hollow rods, assembled into an arm that can be bent up to 90° once inside the detector. The system is deployed through a pipe connecting the scintillator to the top of the detector (figure 1). To determine the source reference position, seven consumer grade digital cameras are fixed to the stainless steal sphere. The true location of a source is determined in the following way: a laser-illuminated diffuser ball, attached close to the source, is flashed while the CCD cameras take pictures simultaneously. During the 2008 and 2009 calibration campaigns, several sources of different types (α, β, γ, neutron) were inserted into Borexino to test the response to different particle types and to cover the energy region of interest for Borexino (which is larger than the SOX one). A detailed description of all the inserted sources can be found in [7]. Gamma sources cover the energy range between 0.1 MeV and 1.5 MeV. They are placed not only in the detector's center, but also in several positions within the scintillating volume to create a map of the energy response. The sources were dissolved in water and sealed in a quartz vial. A well understanding of the detector response to the gamma sources is necessary for SOX since the energy deposit of the prompt event is due to positron interaction with matter and to the signal due to the two 511 keV annihilation gammas. The $^{241}Am^{9}Be$ neutron source calibration was performed for Borexino ^{8}B neutrino and geoneutrino analyses. However, those data are of fundamental importance for SOX, since the delayed event is due to neutron capture. Neutrons are produced associated with de-excitation γ rays with a total energy of 4.44 MeV. These γ rays, together with the recoil protons from neutron scattering in the medium, are responsible for a prompt scintillator signal (figure 1). Afterwards, neutrons thermalize in the hydrogen-rich organic scintillator and are captured either on protons or carbon nuclei, emitting characteristic 2.22 MeV and 4.95 MeV γ rays (figure 1). These γ rays produce a delayed signal according to the neutron capture time of $\sim$254 μs.

3. Monte Carlo simulation code and the physics modeling the detector response

Particles depositing energy in the scintillator produce scintillation and Cherenkov light which propagates inside the detector and it is detected by the 2212 PMTs. For each event, the number of photon measured and the charge collected by each PMT and the time of arrival of each photon are measured. From these quantities, the energy of the event, its position and, in some cases, the particle type can be extracted. A precise knowledge of the detector response to different particle types and energies is fundamental for properly connecting physical and measurable quantities. The method of evaluation of the detector response function is based on a Monte Carlo simulation that models and predicts the expected shapes of the signal and background. It is an *ab initio* simulation of all the processes influencing the energy deposits in the materials building the detector. The scintillation and Cherenkov light emission, light propagation and detection processes are fully simulated as well as the read-out electronics. The Monte Carlo code produces a set of raw data with the same format of the real raw data, allowing an identical data processing [8]. Previously measured parameters are used as input values of the Monte Carlo code. The energy deposit and all physical processes of the light (from propagation to detection in the PMTs) are simulated using the standard GEANT4 [9] package while a custom C++ code simulates the electronic response. In view of the Borexino phase II solar neutrino analysis and the SOX measurement, the Monte Carlo code has been reviewed. Particular attention have been devoted in developing generators for sterile neutrino signal as well as in improving the reliability of the light collection and the detector's response for events far away from the center.

4. Conclusions

A complete knowledge of the detector response is fundamental for the forthcoming SOX analysis. Calibration campaigns with radioactive sources have been performed over the years. A new one is foreseen in the first months of 2016. The Borexino-SOX Monte Carlo code have been refined to increase the reliability of the simulations especially for events that interacts far from the detector's center. The new calibration data will be extremely useful for a final tuning of the parameters used in the detector simulation.

References

[1] Aguilar A *et al.* 2001 *Phys. Rev.* D **64** 112007
[2] Aguilar A *et al.* 2013 *Phys. Rev. Lett.* **110** 161801
[3] Mention G *et al.* 2011 *Phys. Rev.* D **83** 073006
[4] Giunti C and Laveder M 2011 *Phys. Rev.* C **83** 065504
[5] Bellini G *et al.* 2013 *JHEP* **38** 1308
[6] Agostini M *et al.* 2015 *Phys. Rev.* D **92** 031101
[7] Bellini G *et al.* 2012 *JINST* **7** 10018
[8] Bellini G *et al.* 2014 *Phys. Rev.* D **89** 112007
[9] Geant4 Collaboration, Physics Reference Manual

International Conference on Particle Physics and Astrophysics (ICPPA-2015) IOP Publishing
Journal of Physics: Conference Series **675** (2016) 012013 doi:10.1088/1742-6596/675/1/012013

Visible, invisible and trapped ghosts as sources of wormholes and black universes

S V Bolokhov[1], K A Bronnikov[1,2,3,*], P A Korolyov[1] and M V Skvortsova[2]

[1] Institute of Gravitation and Cosmology, PFUR, 6 Miklukho-Maklaya st., Moscow 117198, Russia
[2] VNIIMS, 46 Ozyornaya st, Moscow 119361, Russia
[3] National Research Nuclear University MEPhI (Moscow Engineering Physics Institute), Kashirskoe highway 31, Moscow 115409, Russia

E-mail: *kb20@yandex.ru

Abstract. We construct explicit examples of globally regular static, spherically symmetric solutions in general relativity with scalar and electromagnetic fields, describing traversable wormholes with flat and AdS asymptotics and regular black holes, in particular, black universes. (A black universe is a regular black hole with an expanding, asymptotically isotropic space-time beyond the horizon.) Such objects exist in the presence of scalar fields with negative kinetic energy ("phantoms", or "ghosts"), which are not observed under usual physical conditions. To account for that, we consider what we call "trapped ghosts" (scalars whose kinetic energy is only negative in a strong-field region of space-time) and "invisible ghosts", i.e., phantom scalar fields sufficiently rapidly decaying in the weak-field region. The resulting configurations contain different numbers of Killing horizons, from zero to four.

The so-called exotic matter, which violates the weak and null energy conditions (WEC and NEC), is well known to be a necessary ingredient for the construction of wormholes in general relativity and some of its extensions (see, e.g., [1–4] for reviews), although such matter is not observed under usual physical conditions, and its possible existence meets serious theoretical objections mostly related to quantum phenomena (see [5] and references therein). On the other hand, phantom fields naturally appear in some models of string theory [6], supergravities [7] and theories in more than 11 dimensions [8]. One more viewpoint of interest is [9] that a phantom field, being a source of gravity, can exist without its own dynamics and therefore escape the related problems with quantum particle creation.

An important, though not very confident, support from modern cosmological observations allowing for the ratio of pressure to energy density $w = p/\rho < -1$ (see, e.g., [10] and references therein) is one of the reasons for the recent interest in possible phenomena with exotic matter including wormhole construction and properties.

It has also been found that in the presence of exotic matter, say, in the form of a phantom scalar field, not only wormholes are possible but also different types of regular black holes, including the so-called black universes. The latter look from their static regions as "ordinary" black holes in general relativity, but instead of a singularity beyond the horizon, there is an expanding universe which can at large times become isotropic, in particular, de Sitter [11, 12].

In the present note, we briefly describe different extensions of the scalar field solutions of [11] to configurations containing electric or magnetic fields in the framework of general relativity. One of the motivations for their inclusion is that by modern observations there can exist a global magnetic field up to 10^{-15} Gauss, causing correlated orientations of sources remote from each other [13], and some authors admit a possible primordial nature of such a magnetic field.

The most straightforward extension [14] assumes the action

$$S = \frac{1}{2}\int \sqrt{-g}d^4x\Big[R + \varepsilon(\partial\phi)^2 - 2V(\phi) - F_{\mu\nu}F^{\mu\nu}\Big], \tag{1}$$

where R is the scalar curvature, $g = \det(g_{\mu\nu})$, and $F_{\mu\nu}$ is the electromagnetic field tensor, and $\varepsilon = \pm 1$ distinguishes usual, canonical ($\varepsilon = +1$) and phantom ($\varepsilon = -1$) scalar fields. We consider static, spherically symmetric space-times with the general metric

$$ds^2 = A(u)dt^2 - \frac{du^2}{A(u)} - r^2(u)(d\theta^2 + \sin^2\theta d\varphi^2). \tag{2}$$

written in the so-called quasiglobal gauge $g_{00}g_{11} = -1$ convenient for describing Killing horizons which occur at regular zeros of the function $A(u)$ [2]. Assume that the metric is asymptotically flat as $u \to \infty$, then a *wormhole* is a geometry in which $u \in \mathbb{R}$, $A > 0$ at all u, and the metric is asymptotically flat or AdS as $u \to -\infty$. If again $u \in \mathbb{R}$ but at large negative u the function $A(u) \sim -r^2(u) \to -\infty$ (a de Sitter asymptotic), we are dealing with a *black universe*.

In both wormhole and black universe metrics, the area function $r(u)$ has a regular minimum at some $u = u_0$. The WEC and NEC are necessarily violated at such minima due to the Einstein equations: indeed, one of them reads (see equation (4) below) $2Ar''/r = -(T^t_t - T^u_u)$, where T^ν_μ is the stress-energy tensor (SET) and the prime stands for d/du. In terms of the density ρ and radial pressure p_r, the condition $r'' > 0$ near $u = u_0$ implies $\rho + p_r < 0$, which manifests NEC violation. This is true both for an R-region ($A > 0$) where $u = u_0$ is a throat and for a T-region ($A < 0$) where $u = u_0$ is a bouncing point in a Kantowski-Sachs cosmology [2, 15].

The electromagnetic fields can be radial electric (Coulomb) and magnetic (monopole) ones. There is, however, no need to introduce specific electric or magnetic charges (or monopoles): in both wormholes and black universes a radial field can exist without sources due the space-time geometry. In the wormhole case it perfectly conforms to Wheeler's idea of a "charge without charge", in a black universe the situation is basically the same but looks more involved.

The electromagnetic field equations are solved in a general form, leading to the SET $T^\nu_\mu[e] = q^2 r^{-4}(u)\,\mathrm{diag}(1, 1, -1, -1)$, where $q^2 = q_e^2 + q_m^2$ and the constants q_e and q_m are the effective electric and magnetic charges. The remaining independent Einstein equations read

$$(A'r^2)' = -2r^2V + 2q^2/r^2, \tag{3}$$

$$r''/r = -\varepsilon\phi'^2, \tag{4}$$

$$A(r^2)'' - r^2A'' = 2 - 4q^2/r^2. \tag{5}$$

Being interested in wormhole and black universe solutions, we take $\varepsilon = -1$ and, using the inverse problem method, assume for the area function $r = \sqrt{u^2 + b^2} \equiv b\sqrt{x^2 + 1}$, $b = \text{const} > 0$ (the length scale). Then equation (5) is integrated giving for $B(x) \equiv A(x)/r^2(x)$

$$B(x) = B_0 + \frac{1 + q^2 + px}{1 + x^2} + \left(p + \frac{2q^2x}{1 + x^2}\right)\arctan x + q^2\arctan^2 x, \tag{6}$$

where p and B_0 are integration constants. The metric is thus completely known while ϕ and V are now easily found from the field equations. Assuming that our system is asymptotically flat at large x, so that $A \to 1$, $r \to \infty$ and $B \to 0$ as $x \to \infty$, the constants are related by

$$B_0 = -\pi p/2 - \pi^2q^2/4, \qquad p = 3m - \pi q^2. \tag{7}$$

The whole solution is parametrized by two constants m (Schwarzschild mass at $x \to \infty$) and q, the charge. It turns out that even if we restrict ourselves to solutions asymptotically flat as $x \to \infty$ and $m > 0$, there are as many as 10 classes of globally regular space-times which differ in the type of the second asymptotic $x \to -\infty$ (M — Minkowski, dS — de Sitter, AdS — anti-de Sitter) and the number and kinds of horizons, see table 1 (where $n = 1$ refers to simple horizons, $n = 2$ to extremal ones, R- and T-regions are disposed from left to right along the x axis). Two examples of the corresponding Carter-Penrose diagrams are shown in figure 1: one of them, with two horizons, cannot be drawn on a single plane due to region overlapping while the other, with three horizons, occupies the whole plane except the grey triangles. Some of non-asymptotically flat solutions contain four horizons. We refer to [14] for a detailed description of the global causal structures and Carter-Penrose diagrams for different branches of the general solution.

Table 1. Types of asymptotically flat solutions with $m > 0$.

Configuration type, asymptotics $(x \to +\infty)$ — $(x \to -\infty)$	Horizons: number, order n	disposition of R- and T-regions
M – M wormhole	none	R
M – M extremal black hole	1 hor., $n = 2$	RR
M – M black hole	2 hor., $n = 1$ (both)	RTR
M – dS black universe	1 hor., $n = 1$	TR
M – dS black universe	2 hor., $n = 2$ and $n = 1$	TTR
M – dS black universe	2 hor., $n = 1$ and $n = 2$	TRR
M – dS black universe	3 hor., $n = 1$ (each)	TRTR
M – AdS black hole	2 hor., $n = 1$ (both)	RTR
M – AdS extremal black hole	1 hor., $n = 2$	RR
M – AdS wormhole	none	R

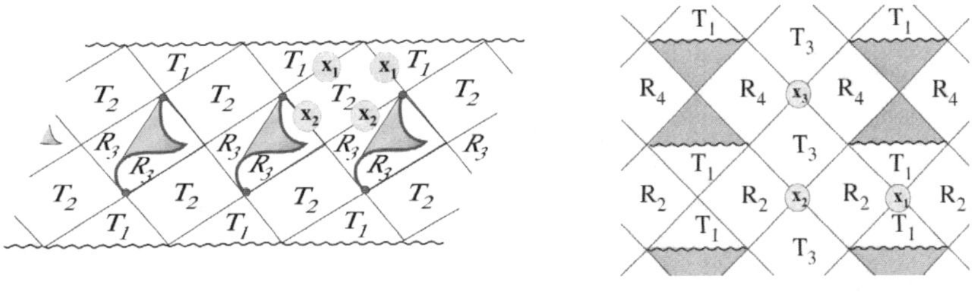

Figure 1. Carter-Penrose diagrams of two space-times presented in table 1, 5th line (left panel) and 7th line (right panel). The letters R and T mark R- and T-regions, x_n mark the horizons, and both are ordered left to right along the x axis.

Since exotic matter or phantom fields are so far not detected, it makes sense to try to avoid their emergence in asymptotic weak-field regions. To do so, we considered a special kind of fields, named "trapped ghosts" [16–18], which have phantom properties only in some restricted strong-field region and become canonical outside it. This happens if we substitute in (1), instead of ε, a function $h(\phi)$ changing its sign at some ϕ value. Choosing, instead of $r = \sqrt{u^2 + b^2}$, the function [17, 18]

$$r(u) = a\frac{x^2 + 1}{\sqrt{x^2 + n}}, \qquad a = \text{const} > 0, \qquad n = \text{const} > 2. \tag{8}$$

we obtain examples of such a behaviour. Indeed, with (8), $r'' > 0$ (corresponding to a ghost field) at $x^2 < n(2n-1)/(n-2)$ and $r'' < 0$ at larger $|x|$, as required. As before, all other quantities are found from the field equations, and the qualitative features of solutions are the same as described above, including the ten classes of regular solutions presented in table 1.

A weak point of such models is the transition surface from normal to phantom fields. It turns out, for example, that if we consider spherically symmetric perturbations of solutions with such a field, the corresponding effective potential has a singularity which should in general lead to a violent instability [15].

Another approach is to consider what may be called an "invisible ghost", i.e., a phantom field that decays sufficiently rapidly in the weak-field region. In [15] this idea was realized with the same ansatz (8) on $r(u)$ but with a sigma-model-like combination of two scalar fields, a phantom one, $\psi(u)$, and a canonical one, $\phi(u)$, where ψ decays at large u much more rapidly than ϕ. Since the function $A(u)$ is, as before, found from equation (5) with known $r(u)$, the whole geometry remains the same, changes only the field content of the system.

A configuration with an "invisible ghost" can also be obtained with a single field $\phi(u)$, but then one should choose a function $r(u)$ like $r(u) = a(1 + x^8)^{1/8}$, closer than before approaching $r = u$ at large u and preserving the condition $r'' > 0$ to ensure the phantom nature of ϕ. The resulting geometries should be qualitatively the same as those discussed in [14, 18]. Indeed, since $r(x) \approx |u|$ at both infinities, the causal structure and the corresponding Carter-Penrose diagrams are completely determined by zeros of $B(x)$ and its asymptotic behaviour.

Summarizing, we have obtained numerous examples of wormhole, regular black hole and black universe solutions to the Einstein equations with scalar and electromagnetic fields as sources, where the scalar fields are phantom in nature at least in the strong field region.

Acknowledgments

The work of KB was performed within the framework of the Center FRPP supported by MEPhI Academic Excellence Project (contract No. 02.a03.21.0005, 27.08.2013).

References

[1] Visser M 1995 *Lorentzian Wormholes: from Einstein to Hawking* (Woodbury: AIP)
[2] Bronnikov K A and Rubin S G 2012 *Black Holes, Cosmology and Extra Dimansions* (Singapore: World Scientific)
[3] Lobo F S N 2013 *EPJ Web Conf.* **58** 01006 (*arXiv*: 1212.1006)
[4] Bronnikov K A and Skvortsova M V 2014 *Grav. Cosmol.* **20** 171 (*arXiv*: 1404.5750)
[5] Bronnikov K A and Starobinsky A A 2007 *JETP Lett.* **85** 1 (*arXiv*: gr-qc/0612032) (Original Russian text: *Pis'ma v ZhETF* **85** 3)
[6] Sen A 2002 *JHEP* **0204** 048; **0207** 065
[7] Nilles H P 1984 *Phys. Rep.* **110** 1
[8] Khviengia N, Khviengia Z, Lu H and Pope C N 1998 *Class. Quantum Grav.* **15** 759
[9] Ellis H G 2015 *Int. J. Mod. Phys. D* **24** 1550069
[10] Ade P A R *et al* (Planck Collaboration) 2015 Planck 2015 results. XIII. Cosmological parameters *arXiv*: 1502.01589
[11] Bronnikov K A and Fabris J C 2006 *Phys. Rev. Lett.* **96** 251101 (*arXiv*: gr-qc/0511109)
[12] Bronnikov K A, Melnikov V N and Dehnen H 2007 *Gen. Rel. Grav.* **39** 973 (*arXiv*: gr-qc/0611022)
[13] Poltis R and Stojkovic D 2010 *Phys. Rev. Lett.* **105** 161301 (*arXiv*: 1004.2704)
[14] Bolokhov S V, Bronnikov K A and Skvortsova M V 2012 *Class. Quantum Grav.* **29** 245006 (*arXiv*: 1208.4619)
[15] Bronnikov K A and Korolyov P A 2015 *Grav. Cosmol.* **21** 157 (*arXiv*: 1503.02956)
[16] Bronnikov K A and Sushkov S V 2010 *Class. Quantum Grav.* **27** 095022 (*arXiv*: 1001.3511)
[17] Bronnikov K A and Donskoy E V 2011 *Grav. Cosmol.* **17** 176 (*arXiv*: 1110.6030)
[18] Bronnikov K A, Donskoy E V and Korolyov P A 2013 *Vestnik RUDN* **2** 139

Estimation of atmospheric neutrinos background in Borexino

V S Atroshchenko[1] **and E A Litvinovich**[1,2]

[1] National Research Centre Kurchatov Institute, Moscow, 123182, Russia
[2] National Research Nuclear University MEPhI (Moscow Engineering Physics Institute), Kashirskoe highway 31, Moscow, 115409, Russia

E-mail: victor.atroshchenko@lngs.infn.it

Abstract. Atmospheric neutrinos are produced in interactions of cosmic rays with atomic nuclei in the Earth's atmosphere. Although their flux is too low for studying in Borexino, atmospheric neutrinos act as a background for other processes. This paper presents the theoretical expected yield of atmospheric neutrinos in Borexino for three neutrino detection reactions: νp-ES, νe-ES and inverse β-decay, as well as the status of Monte-Carlo simulation for $\nu^{12}C$ interaction channels. Calculations were performed based on the only currently known detailed model of atmospheric neutrinos flux at very low energies.

1. Atmospheric neutrinos

Atmospheric neutrinos are produced in the interaction of cosmic rays with atomic nuclei in the Earth's atmosphere. Muon neutrinos are produced in decays of pions (kaons) and muons, whereas electron neutrinos are produced only in the decay of muons. Thus, a flux ratio of muon neutrinos to electron neutrinos is about 2-to-1 (it is true with energies less than 1 GeV). Spectrum of atmospheric neutrinos is associated with the spectrum of cosmic rays, and extends up to ultrahigh energies.

FLUKA atmospheric neutrinos flux at low energy from [1] was used. According to the article, the overall absolute flux uncertainty does not exceed 25%. Flux spectra shown in figure 1. Smooth curves obtained by quadratic spline interpolation.

The most numerous particles in the Borexino scintillator are electrons, protons and carbon-12 nuclei (see table 1). Neutrino interactions with these particles were considered.

2. Neutrino proton elastic scattering reaction channel

The differential cross section of neutrino on free proton elastic scattering (νp-ES) in terms of proton kinetic energy T_p is [2]

$$\frac{d\sigma}{dT_p} = \frac{G_F^2 m_p^3}{4\pi E_\nu^2}\left[A \mp B\frac{(s-u)}{m_p^2} + C\frac{(s-u)^2}{m_p^4}\right] \tag{1}$$

In this formula the minus (plus) is for (anti)neutrinos, s and u are Mandelstam variables; and A, B, C are functions of Q^2 - 4-momentum transfer, they contain vector form-factors F_1 and F_2, and axial form-factor in dipole form

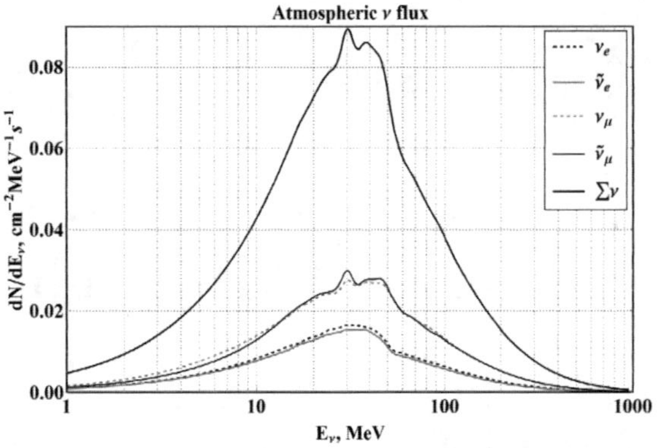

Figure 1. Atmospheric neutrino flux spectra.

Table 1. Particles in Borexino scintillator (number per 100 tons).

e^-	$3.30531 \cdot 10^{31}$
p	$6.00013 \cdot 10^{30}$
^{12}C	$4.45652 \cdot 10^{30}$
^{13}C	$5.00226 \cdot 10^{28}$
^{2}H	$9.00155 \cdot 10^{26}$
^{16}O	$8.52935 \cdot 10^{26}$
^{14}N	$8.51840 \cdot 10^{26}$
^{15}N	$3.12919 \cdot 10^{24}$
^{18}O	$1.70994 \cdot 10^{24}$
^{17}O	$3.24888 \cdot 10^{23}$

$$F_A = \frac{g_A}{\left(1 + Q^2/M_A^2\right)^2}, \tag{2}$$

where $g_A = F_A(Q^2 = 0) = -1.267$ is axial-vector coupling constant, and $M_A = 1.03$ GeV is the axial mass.

Recoil protons kinetic energy distribution in the detector is related with the differential cross-section $d\sigma/dT_p$ and neutrino spectrum dN/dE_ν as

$$\frac{dN}{dT_p} = N_p \int\limits_{E_\nu^{\min}=\frac{1}{2}\left(T_p+\sqrt{T_p^2+2T_p m_p}\right)}^{\infty} dE_\nu \frac{dN_\nu\left(E_\nu\right)}{dE_\nu} \frac{d\sigma\left(E_\nu, T_p\right)}{dT_p}, \tag{3}$$

where N_p is the number of free protons in the Borexino detector (see table 1).

Because of the effect of scintillation quenching, the light yield for protons is suppressed. Visible energy for proton E_p^{vis} is [3]

$$E_p^{vis}(T_p) = \int\limits_0^{T_p} \frac{dE}{1 + kB\frac{dE}{dx}(E)} \tag{4}$$

where dE/dx is stopping power of scintillator material with respect to protons, kB is Birks parameter, $kB = 0.0115$ cm/MeV for protons in the Borexino scintillator. Function dE/dx is computed using the software SRIM [4].

Borexino detector energy resolution $\delta E^{vis} = 5\%/\sqrt{E^{vis}\,[\text{MeV}]}$ is also taken into account.

Calculated events number is in table 2. Events energy distribution shown in figure 2.

3. Inverse beta-decay reaction channel

Inverse beta-decay (IBD) $\tilde{\nu}_e + p \to n + e^+$ is the threshold reaction, possible if

$$E_\nu > E_{thr} = \frac{\left(m_n + m_e\right)^2 - m_p^2}{2m_p} = 1.806 \text{ MeV} \tag{5}$$

Detection of this reaction is based on the method of time coincidences. Prompt event is the positron annihilation in scintillator, and delayed event is the neutron capture mostly by

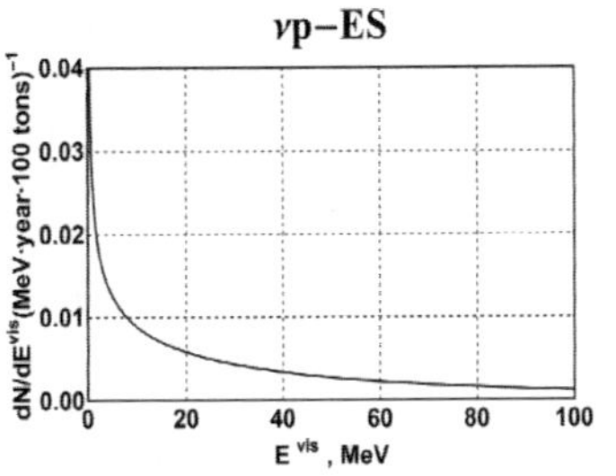

Figure 2. Visible energy distribution of recoil protons in νp-ES for the atmospheric neutrinos in Borexino.

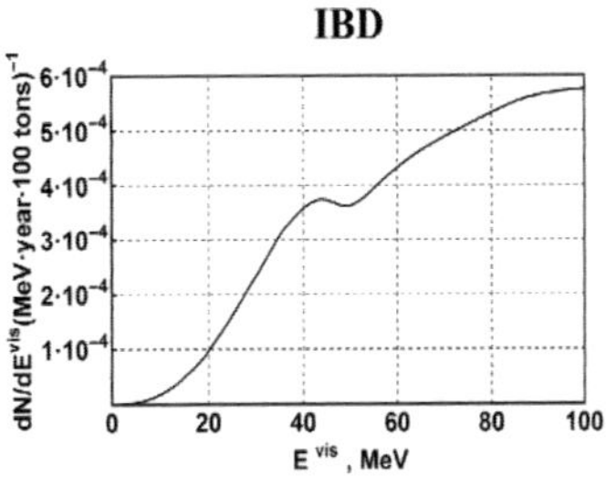

Figure 3. Visible energy distribution of positrons in IBD for the atmospheric neutrinos in Borexino.

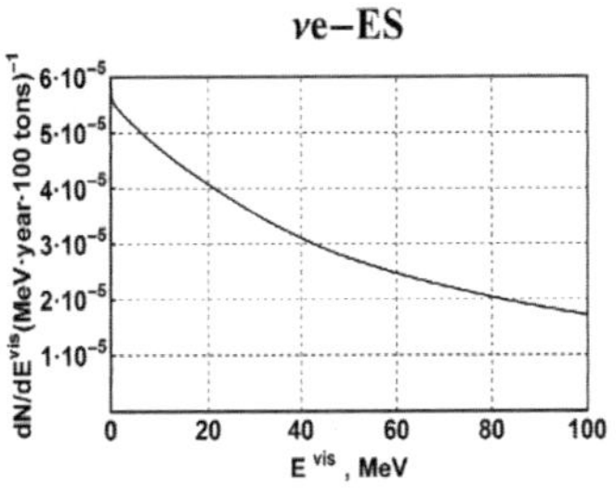

Figure 4. Visible energy distribution of electrons in νe-ES for the atmospheric neutrinos in Borexino.

hydrogen (mean neutron capture time in Borexino scintillator is $(254.5 \pm 1.8)\mu s$ [5]), producing deuterium and photon with energy 2.22 MeV.

The exact inverse beta decay cross-section (including radiative corrections) from [6] is used. Calculated prompt events number is in table 2. Energy distribution shown in figure 3.

4. Neutrino electron elastic scattering reaction channel

According to the Standard Model, (anti)neutrino electron elastic scattering (νe-ES) differential cross-section in terms of electron kinetic energy T_e is

$$\frac{d\sigma}{dT_e}(E_\nu, T_e) = \frac{2G_F^2 m_e}{\pi E_\nu^2}\left[g_L^2 E_\nu^2 + g_R^2 (E_\nu - T_e)^2 - g_L g_R m_e T_e\right] \tag{6}$$

Coupling constants $g_R = \sin^2\theta_W$ for all neutrino types, $g_L = \sin^2\theta_W + 1/2$ for ν_e, $g_L = \sin^2\theta_W - 1/2$ for $\nu_{\mu,\tau}$, $g_R \leftrightarrow g_L$ for antineutrino.

Calculations are analogous to those for νp-ES with $kB(e) = kB(p) = 0.0115$ cm/MeV and dE/dx calculated using ESTAR [7]. Calculated events number is in table 2. Events energy distribution shown in figure 4.

5. Neutrino carbon-12 reactions channels

Because of the large number of possible $\nu^{12}C$ reactions channels, GENIE Neutrino Monte-Carlo Generator (ver 2.8.6) [8] was used. Sample of more than 600 thousands neutrino interaction events was obtained, and normalized by integrating flux times cross-sections, implemented in GENIE.

A special module was developed for Geant4 based Borexino Monte-Carlo generator (BxMC) to read and track final state particles from GENIE neutrino interactions. Unfortunately, nuclear deexcitation is not implemented in GENIE, and nuclear remnants, tagged by GENIE as "low energy hadronic blob", were assumed as ground state nuclei.

Obtained visible events number and energy distributions shown in table 2 and figure 5.

While reactions with pions and muons as final state particles can be easily tagged because of their high deposited energy, reactions producing, for example, low energy protons (and neutrinos) and (pseudo)stable nuclei are hard for tagging. There are also many reactions, that can imitate IBD reaction with protons/electrons/nuclei as prompt events and knocked out neutrons as delayed events. Some residual unstable nuclei have long lifetime ($\tau(^9C, ^8He, ^9Li) \sim 0.2$ s, $\tau(^8Li, ^8B, ^6He) \sim 1.2$ s, $\tau(^{10}C) \sim 27.8$ s, $\tau(^{11}C) \sim 29.3$ min), which means they can be recognized with high efficiency, but with applying long time cut, up to 3 hours for ^{11}C.

It all requires a detailed consideration, which is in progress.

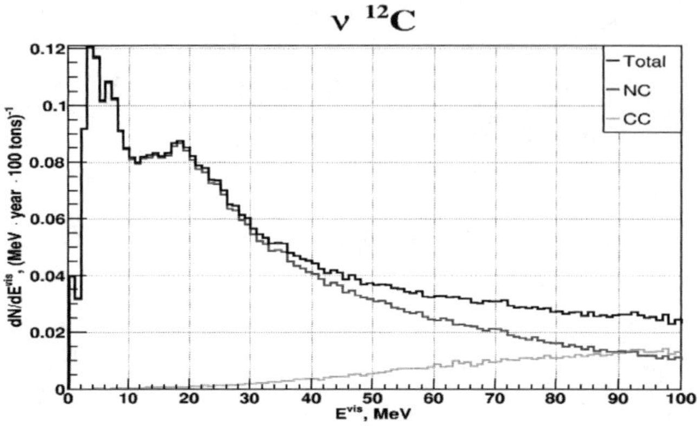

Figure 5. Neutrino carbon-12 interaction reactions yield for the atmospheric neutrinos in the Borexino detector. Visible energy distributions of events for: (blue) neutral current reactions, (green) charged current reactions, (black) all reactions in sum.

Table 2. Reactions yield for atmospheric neutrinos in the Borexino detector (events per year per 100 tons).

	νp-ES	IBD	νe-ES	$\nu^{12}C$-CC	$\nu^{12}C$-NC	$\nu^{12}C$ total
$0.25 \leq E^{vis} \leq 15$ MeV	0.181443	0.000205	0.000725	0.005962	1.282274	1.288701
$1 \leq E^{vis} \leq 100$ MeV	0.397184	0.033788	0.002981	0.619615	4.156067	4.775682

6. Conclusions

Table 2 shows expected events rate number for atmospheric neutrinos interaction reactions in the Borexino detector in two visible energy ranges for two Borexino DAQ systems. Visible energy events distributions are shown in figures 2 to 5. Detection efficiency is not taken into account.

Results of the work show that atmospheric neutrinos should be thought as an important source of background for some kinds of processes in Borexino with few expected events per year.

Acknowledgements
This work was supported by RFBR grant 14-22-03031 and partially supported by MEPhI Academic Excellence Project (contract № 02.a03.21.0005, 27.08.2013).

References

[1] Battistoni G *et al.* 2005 *Astropart. Phys.* **23** 526
[2] Ahrens L A *et al. Phys. Rev.* D **35** 785
[3] Birks J B 1964 *The theory and practice of scintillation counting* (New York: Macmillan)
[4] Ziegler J F, Ziegler M D and Biersack J P 2010 *Nucl. Instrum. Meth.* B **268** 1818
[5] Bellini G *et al.* 2011 *JINST* **6** 5005
[6] Strumia A and Vissani F 2003 *Phys. Lett.* B **564** 42
[7] Berger M J *et al.* 2005 ESTAR, PSTAR, and ASTAR: Computer Programs for Calculating Stopping-Power and Range Tables for Electrons, Protons, and Helium Ions (version 1.2.3) URL `http://physics.nist.gov/Star`
[8] Andreopoulos C *et al.* 2010 *Nucl. Instrum. Meth.* A **614** 87

Development of scanning technique for sTGC detectors production quality control

P E Teterin[1], K A Filippov[1], S P Konovalov[2], G Mikenberg[3], A Romaniouk[1], D Shchukin[1], M Shoa[3], V Smakhtin[3], S Yu Smirnov[1], V V Sosnovtsev[1], V O Tikhomirov[1,2], V I Tsekhosh[2] and K A Vorobev[1]
[1] National Research Nuclear University MEPhI (Moscow Engineering Physics Institute), Kashirskoe highway 31, Moscow, 115409, Russia
[2] P.N. Lebedev Physical Institute, Leninsky prospekt 53, Moscow, 119991, Russia
[3] Weizmann Institute of Science, 234 Herzl St., Rehovot 7610001, Israel

E-mail: vladimir.smakhtin@weizmann.ac.il

Abstract. The innermost station of the ATLAS muon end-cap system will be replaced during the forthcoming Phase-I upgrade of the ATLAS detector. The small-strip Thin Gap Chambers (sTGC) are supposed to operate at harsh radiation conditions of super LHC. X-ray scanning technique for production quality control is proposed to ensure long-term reliability of these chambers. It allows to reveal different types of technological defects critical for sTGC chambers operation.

1. Introduction

The Large Hadron Collider (LHC) was designed to operate with two proton beams of 7 TeV energy in each direction with maximum luminosity of 10^{34} cm^{-2}s^{-1}. However in order to increase statistics of rare events, an upgrade of the LHC machine (super-LHC) is foreseen. Average luminosity of LHC will reach a level of 5×10^{34} cm^{-2}s^{-1}after upgrade. In order to operate in these conditions, all LHC experiments develop now corresponding upgrade programs. One of them is the upgrade of the ATLAS muon spectrometer. In the first phase of the ATLAS upgrade (2018-2019) a new muon end-cap system New Small Wheel (NSW) will be installed [1, 2]. It is based on two technologies: small-strip Thin Gap chambers (sTGC) and Micromegas chambers. The sTGC is mainly devoted to the Level-1 trigger function and also provides track measurement with a precision of about 100 μm for each chamber plane. The trigger detectors should provide bunch-crossing identification, which requires good time resolution and track angle measurements with an accuracy better than 1 mrad [1].

A general concept of the sTGC chamber is based on the technology of Multi-Wire Proportional Chambers, which was developed in 1983 [3]. The sTGC chamber consists of a layer anode wires of 50 μm made from gold plated tungsten spaced by1.8 mm, which is sandwiched between two cathode graphite-sprayed G10 planes at a distance of 1.4 mm from the wire plane. Behind the graphite there are precision strips on one side (which run perpendicular to the wires) and large pads on the other side [2]. Chambers are filled with the n-pentane+CO_2 (45/55%) gas mixture and operate with a voltage of 2.9 kV applied to the anode wires.

sTGC detectors will work in a harsh radiation environment and will be practically inaccessible during run periods. This implies a requirement of very strict quality control procedures during chamber production. One of the most important tests, which insures robust operation under large ionizing particle fluxes, is the one with different types of radioactive sources. It is very important to carry out these tests at early stages of production in order to be able to introduce corrections in the production process and to eliminate any technological defects before the chambers are fully assembled. Various technological defects, such as missing wires, wires lost tension, cathode strip and gap thickness defects, dust particles inside the gap and on the wires, gluing defects, etc. result in gas gain non-uniformity and may cause discharges or wrong interpretation of the measurement results.

Very often techniques used for multi-wire chamber qualification under irradiation require front-end electronics to be installed on a chamber and radioactive source [1, 4]. In this paper the technique based on the X-ray scanner with direct measurement of the chamber current is described.

2. Experimental set-up and test results

The X-ray scanner is an X-Y CNC machine designed to scan the entire sTGC area with X-ray tube and to measure the current from anode wires shown in figure 1.

Figure 1. X-ray scanner prototype.

Few different types of scans are foreseen for the sTGC quality control. One of them is a search for hot spots and global gas gain uniformity measurement. These types of tests are performed with a wide X-ray beam of 3-30 cm in diameter depending on the goal. Another type of the scans are precision scans with the beam of about 1 mm and less, which allow to localize and study in details observed defects without equipment of the detector with a complex readout system.

The X-Y scan machine can operate in both directions in a continuous movement mode with a speed, which can be set in the range of 0.5 -100 mm/s or in step mode with steps 0.5 mm or larger. The scan area is 2220x1410 mm and covers the maximum sTGC chamber size. The positioning accuracy is less than 1 mm at the maximal distance and it is much below 1 mm at smaller distances. The scanner is equipped with X-ray tube Amptek MINI-X with working parameters 40-50 kV and 75-80μA. Tests were performed with different sTGC prototypes. CAEN N1471H high voltage power supply was used for the test. It allows to measure the current with a precision up to 50 pA. Both n-pentane+CO_2 (45+55%) and CO_2 were used as working mixtures. A special graphical user interface and visualization tool described in [5] was used to perform the scans and to analyze the results.

2.1. Current stabilization during irradiation

One of the critical parameters, which has an impact on the scanning technique, is the time needed to obtain correct current measurements after turning on X-ray irradiation or changing irradiation area position. Figure 2 shows the current response of the sTGC filled with CO_2 after irradiation start. One sees that the current stabilizes after 0.3 seconds after turning the X-ray tube on. This time includes X-ray tube reaction and all transition processes inside the chamber. This time can be considered as a limit, which should be taken into account when programming scanner operation modes.

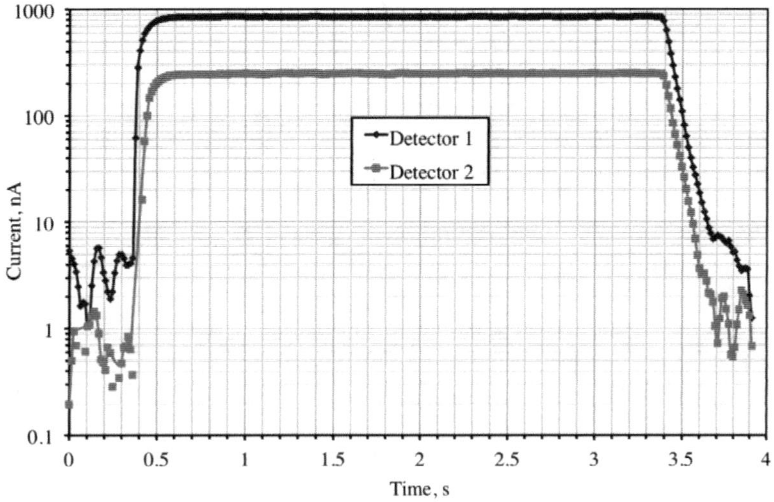

Figure2. The current response of the sTGC prototypes on irradiation with X-ray tube when it was turned on and off. The chamber was filled with CO_2 gas. Lines connecting points are guide to the eyes only.

2.2. Scans of defective areas

For the identification of the defective areas a whole area of the sTGC quadruplet was scanned with X-ray beam of 3 cm in diameter with step of 2.5 cm. The scanning results are presented in these proceedings [6]. Quite many defective areas were found. Detailed analysis of these areas requires more precise scan with a thin beam. For this purpose, a collimator with a slit of 1 mm was used. This collimator reduces the chamber current, however, it is enough to obtain the results with good accuracy.

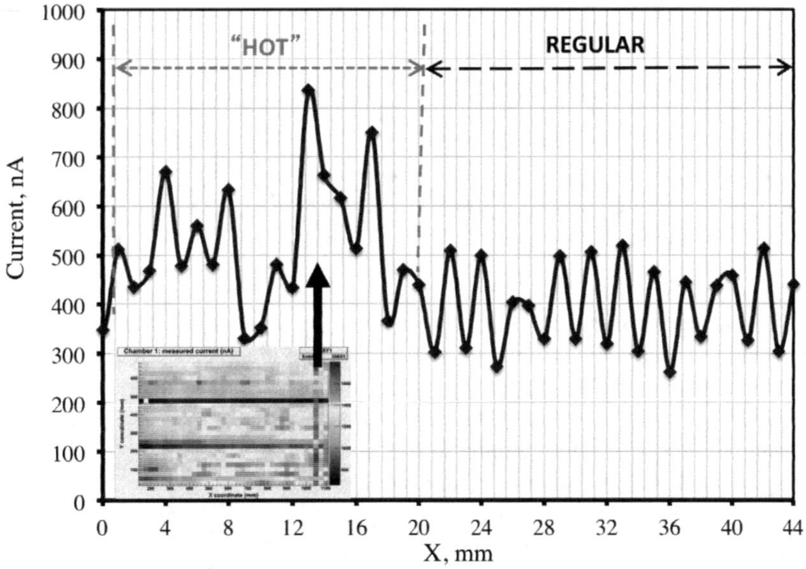

Figure3. Current profile of a defective vertical area in the sTGC chamber. Lines connecting points are guide to the eyes only. Inset: current map of the first chamber with marked defective region.

In figure 3 the results of the detailed scan of the defective area corresponding to one of the anode wires in the first chamber of sTGC quadruplet are shown. X-ray tube operated with the following parameters: 40 keV and 75 μA. In the figure insert the current map of the chamber is shown. Wire on tested chambers are oriented in vertical direction and strips – in horizontal direction. The vertical red area corresponds to the wire defect. For detailed studies,1 mm wide X-ray beam was moved across the wire with a step of 1 mm. The current profile corresponding to a read area on the map is presented in figure 3 as main plot. One sees that significant current modulation corresponds to the position when X-ray beam covers the wires. These modulations tell that a significant part of the gas ionization comes from interaction of X-rays with the wires. Since the distance between the wires is 1.8 mm and the beam width and the step is 1 mm, some beating of the current modulations is observed. It is clearly seen that one wire has a gas gain by a factor of 1.5 larger than the adjacent ones, which tells about its displacement from the anode plane.

Another type of the anode defect was found in the second chamber of the sTGC quadruplet. A "cold" vertical area shown in the insert of figure 4 corresponds to the case of low gas gain. This area was scanned with the thin X-ray beam and the results are shown on the main plot in this figure. The current value close to the dark current was observed in the area between 8 and 10 mm. This area corresponds to missing wires or most likely to the absence of HV on a group of wires.

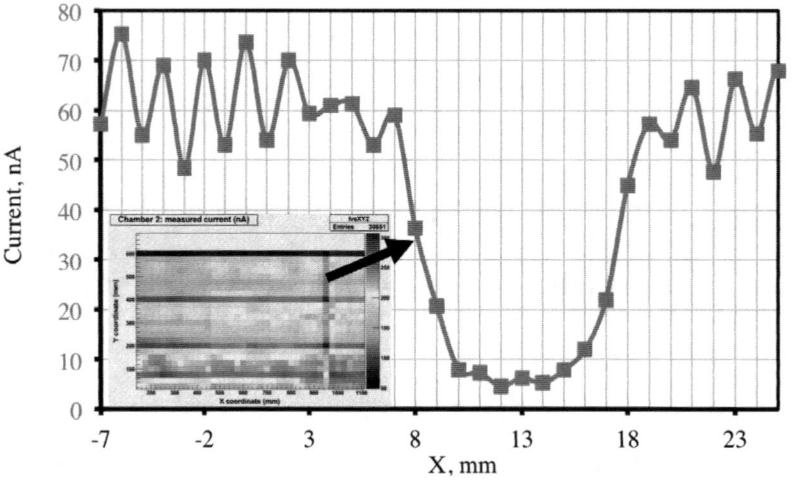

Figure 4. Detailed scan of the defective vertical area in the second chamber of sTGC quadruplet. The current drop between x=8 mm and x=18 mm reflects the fact of either absence of HV on the group of wires or missing wires. Lines connecting points are guide to the eyes only. Insert: current map of the second chamber with marked defective region.

Horizontal rows on the current maps reflect orientation of the cathode strips. In the insert in figure 5 the "cold" horizontal area manifests a strip problem. A scan of this area across the strips was performed with 2 mm X-ray beam and with a step of 2 mm. The distance between strips is small with respect to the size of the beam and one should not expect significant current modulation. Reduction of the gas gain in area between 5 and 10 mm manifests some issue related to the cathode strips. As the coordinate of a particle is calculated by weighting the signals from several strips, this defect will lead to a significant degradation of the coordinate accuracy in this region.

2.3. Precise scan mode

All the experiments described above were carried out in a step-scanning mode. A continuous scanning mode can give more precise information about defect locations and a mechanical layout of chamber components.

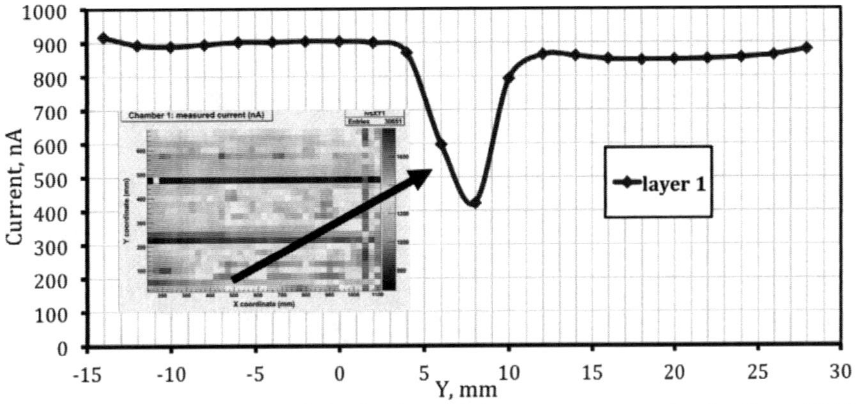

Figure 5. Current profile of the horizontal low current area in the first sTGC chamber. Lines connecting points are guide to the eyes only. Inset: current map of the first chamber with marked defective region.

The HV power supply (CAEN N1471H) used in the tests allows to perform current measurements with a frequency of 1 Hz or less. This actually defines the maximum speed of X-ray beam movement. The movement should not be faster than a half of the collimator width times 1 Hz. Detailed step mode is very time-consuming and from that point of view continuous operation is much more preferable. To compare the continuous measurement method with the step mode, a scan with 1 mm collimator along the wire was performed. This type of scan allows to obtain information concerning the strip structure of the chamber. The chamber with two different strip sizes of 3 and 4 mm was used. The results are presented in figure 6. Contrary to the anode wire effect, the X-ray interactions with Cu-strips reduce X-ray flux and hence the total ionization of the working gas [7]. The peaks in the plot correspond to the place between the strips where Cu is absent and deeps to centers of Cu-strips. One sees that position and amplitude of continuous and step modes of the scan are in good agreement.

The plot was compared with exact strip layout. Change of the cathode strip structure is clearly seen in the figure.

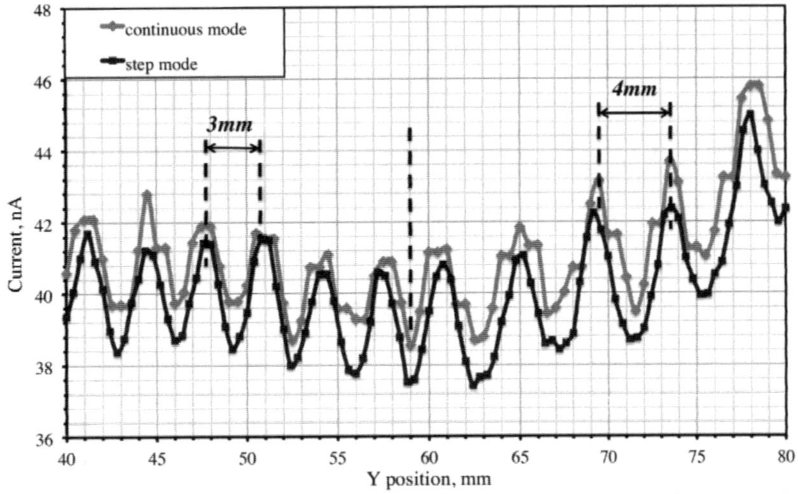

Figure 6. Comparison of current profiles obtained with a continuous and a step scan modes. Continuous mode parameters: collimator 1 mm; scan speed 0.5 mm/sec; 1 measurement/s. Step mode parameters: collimator 1 mm; scan step 0.4 mm; measurement time 5 sec. Lines connecting points are guide to the eyes only.

3. Conclusion

X-ray scanning technique offers a possibility to measure gas gain uniformity and to find hot spots in sTGC chambers. It also allows to perform detailed studies of found defects and to obtain information related to strips and wires, which are not accessible after chamber assembly. Further development of the technique will be important to finalize the sTGC quality control procedure.

Acknowledgements

This work was carried out in the Center of FRPP and was supported by MEPhI Academic Excellence Project (contract No. 02.a03.21.0005, 27.08.2013). We gratefully acknowledge the financial support from the Ministry of Education and Science of Russian Federation (R&D project RFMEFI61014X0005) and grant No.14-22-03053 from the Russian Foundation for Basic Research.

References

[1] ATLAS Collaboration 2013 *New Small Wheel Technical Design Report* CERN-LHCC-2013-006 ATLAS-TDR-020
[2] Stelzer B 2014 *Nucl. Phys.* B *Proc. Suppl.* **00** 1–6
[3] Majewski S *et al.* 1983 *Nucl. Instr. Meth.* A **217** 265
[4] Tanaka S *et al.* 2004 *IEEE Trans. on Nucl. Science* **51** 934
[5] Tikhomirov V et *al.* (*These Proceedings*)
[6] Filippov K *et al.* (*These Proceedings*)
[7] Tsuno S, Kobayashi T and Ye B 2002 *Nucl. Instrum. Meth.* A **482** 667–73

RED-100 detector for the first observation of the elastic coherent neutrino scattering off xenon nuclei

D Yu Akimov [1,2], A K Berdnikova[2], V A Belov[1,2], A I Bolozdynya [2], A A Burenkov[1,2], Yu V Efremenko [2,3], Yu V Gusakov [2,4], A V Etenko [2,5], V A Kaplin[2], A V Khromov [2,6], A M Konovalov [1,2], A G Kovalenko[1,2], E S Kozlova[2], A V Kumpan [2], T D Krakhmalova [2], Yu A Melikyan [2], P P Naumov [2], D G Rudik[1,2], R R Shafigullin [2], A V Shakirov [2], G E Simakov [1,2], V V Sosnovtsev [2], V N Stekhanov [1,2], A A Tobolkin [2] and I A Tolstukhin [2]

[1] SSC RF Institute for Theoretical and Experimental Physics of National Research Centre "Kurchatov Institute", 25 Bolshaya Cheremushkinskaya, Moscow, 117218, Russian Federation

[2] National Research Nuclear University MEPhI (Moscow Engineering Physics Institute), 31 Kashirskoe highway, Moscow, 115409, Russian Federation

[3] University of Tennessee, 1408 Circle Dr, Knoxville, TN 37996-1200, USA

[4] Joint Institute for Nuclear Research, 6 Joliot-Curie, Dubna, Moscow Region, 141980, Russian Federation

[5] National Research Centre "Kurchatov Institute", 1 Akademika Kurchatova Sq, 123182, Moscow, Russian Federation

E-mail: avkhromov@mephi.ru

Abstract. The RED-100 (Russian Emission Detector) is being constructed for the experiment to search for elastic coherent neutrino scattering off atomic nuclei. This fundamental process was predicted several decades ago by the Standard Model of electroweak interactions but has not been discovered yet. The RED-100 is a two-phase emission xenon detector containing ~200 kg of the liquid Xe (~ 100 kg of that is in a fiducial volume). One of the possible sites to carry out the experiment is the SNS (Spallation Neutron Source) facility at Oak Ridge National Laboratory, USA. SNS is the world's most intense pulsed source of neutrinos and unique place to study neutrino properties. The energy spectrum of neutrinos produced at the SNS extends up to ~ 50 MeV and satisfies coherence condition. These neutrinos give kinetic energies of Xe recoils up to a few tens of keV where the response of nuclear recoils is well-known from neutron calibrations of dark matter detectors. The detector will be deployed in the basement under the experimental hall at a distance of ~30 meters from the SNS target. The expected signal and background (neutron and gamma) are estimated for this specific location. The detector details, current status and future plans are provided.

1. Introduction

Coherent elastic neutrino-nucleus scattering (CENNS) was theoretically predicted more than 40 years ago [1]. This is a neutral current process with a cross-section approximately proportional to N^2 (where N is the number of neutrons in the nucleus) due to the coherence enhancement factor. In order to

[6] To whom any correspondence should be addressed

satisfy the coherence condition, neutrinos with energies less than 50 MeV are mostly preferred. The clear understanding of coherent neutrino scattering behavior can be employed for explaining several fundamental physics models. For example, CENNS reaction plays important role in supernova core-collapse processes [2]. As a probe of the Standard Model, any measured deviation from prediction could be an indication of new physics [3]. Moreover, the precise knowledge of coherent neutrino scattering cross-sections in the energy range of solar and atmospheric neutrinos is vitally important for the sensitivity of the next generation direct dark matter detection experiments where neutrinos are an irreducible background [4].

The international collaboration COHERENT has recently been established with the main goal to discover coherent neutrino scattering at SNS (Spallation Neutron Source), Oak Ridge National Laboratory, USA. SNS is a proton accelerator that provides the most intense pulsed neutron beams in the world. Although SNS primarily is constructed for the neutron production, neutrinos with energies up to 50 MeV are created as by-products of stopped pions decaying in the spallation liquid mercury target. The pulsed structure of the beam, high neutrino flux and appropriate energy spectrum make SNS the most attractive site for the first CENNS observation. Although the rate of events for medium-A nuclei is relatively high compared to the conventional neutrino detection processes like elastic neutrino-electron scattering or inverse beta decay reactions, CENNS still remains unobserved due to low energy of nuclear recoils resulted from the neutrino interaction. The energy of the recoils lies in the range of a few keVs or less. Modern low-background detectors for dark matter search have quite well-known response in this energy region that hopefully allows to apply a similar technique for coherent neutrino scattering detection.

It was decided that the best way to unambiguously discover CENNS and investigate N^2 dependence is to use three detector systems (CsI[Na] scintillation crystals, p-type point-contact germanium detectors, and a two-phase liquid xenon time projection chamber (TPC)). In the framework of this joint effort, we propose to use the RED-100 detector as one of suggested technologies. The RED100 is a two-phase liquid xenon time projection chamber that now is fully assembled and commissioning at NRNU MEPhI. It is supposed that the detector will be shipped to USA for installation at the experimental site at SNS in the short-term future.

2. Two-phase emission detector technology

The two-phase principle of particle detection was invented about 40 years ago [5]. Particles interacting with dense noble liquefied gas medium produce excitation and ionization. The excitation creates prompt scintillation (S1 signal). Under the influence of the applied electric field ionization electrons drift along the field towards the surface and then enter the gas phase where produce proportional scintillation or electroluminescense (S2 signal). The time difference between S1 and S2 signals defines the Z position of the event, while the X-Y position is determined from the pattern of S2 light in the photodetecting system. The 3D event reconstruction allows to define a fiducial volume eliminating potential background events originated in the vicinity of the detector walls. Measurement of the ratio of S2 to S1 signals gives an ability to discriminate between electron recoils and nuclear recoils providing an efficient background rejection.

3. RED-100 detector design

In addition to the TPC enclosed in a vacuum cryostat, the RED-100 experimental setup (photo is shown in figure 1b) consists of several ancillary systems. These systems are cryogenics; xenon purification; light collection, electronics and data acquisition; interface breakout-out-box.

3.1. Cryostat and internal structure

The RED-100 cryostat consists of two cylindrical vessels made of VT1-00 titanium. The outer vessel (168 cm in height and 64 cm in diameter) is utilized to hold a vacuum to insulate the inner cold vessel (108 cm in height and 50 in diameter) from external heat load. The cold vessel upper flange is attached to the outer vessel upper dome flange via three hangers thermally isolated in plastic and three flexible bellows. The flanges of both cans are designed for Helicoflex sealing. The detector drawing is shown in figure 1. The inner vessel contains a field cage and ~200 kg of liquid xenon. The field cage is a dodecagonal structure formed by PTFE light reflection panels and enclosing the detector active volume viewed from above and below by two arrays of 19 PMTs each.

Five electrodes: cathode, anode, gate and two shield grids define an electric field inside the detector and are embedded in the PTFE structure. There are 20 copper field-shaping rings between the cathode and anode grids. The field-shaping rings are connected to each other with a 1 GΩ resistor ensuring a uniform electric field. All grids are fabricated in the form of two stainless steel rings pressing a stainless steel mesh (90 % open area) with screws. The distance between the cathode and anode grids is 43 cm. The liquid level is set halfway between the anode and gate grids, which are separated by 2 cm. The top (6 cm to the anode) and bottom (5 cm to the cathode) shield grids prevent the high electric field influence from the adjacent electrodes at the PMT photocathode. The field cage is joined to both PMT arrays with six Ti straps. Every PMT is placed inside a Cu holder that is inserted in a copper 10 cm thick PMT support disc. All copper details are M1 OFHC grade. Gaps between PMT windows in the array are covered with PTFE reflectors for improving light collection.

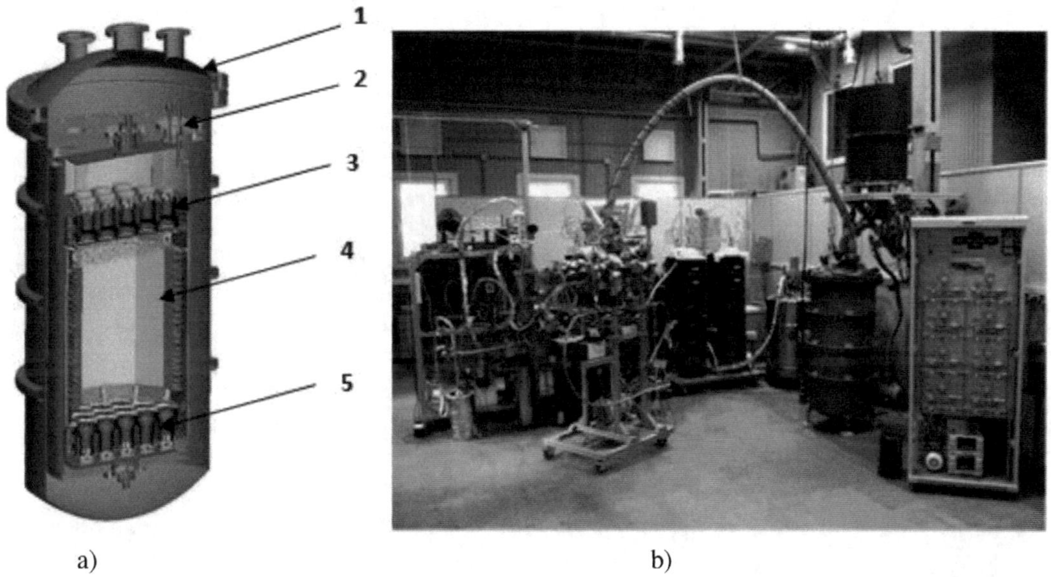

Figure 1. a) Rendering of RED-100 detector: 1) Outer Ti vessel 2) Inner Ti vessel 3)Top PMT array 4) Drift cage with PTFE reflectors 5) Bottom PMT array; b) RED-100 detector at NRNU MEPhI during technical run

3.2. Break-Out-Box

In order to provide a compact detector location inside a passive shielding and minimize amount of radioactive materials near the detector, so called Break-Out-Box (BOB) was constructed. BOB is placed on a mobile platform and connected to the detector inner vessel through three flexible 4 m in length stainless steel bellows. The first bellow is packed with PMT high voltage/signal and instrumentation cables. The other conduit contains three grid high voltage cables. The last comprises a xenon heat exchanger designed in the form of two flexible bellow hoses (for incoming and outcoming flow) passed inside each other. Decoupler boxes for PMT cables, HV bandwidth filters and a commercial ceramic vacuum feedthrough for the cathode HV (up to 50 kV) are instrumented in the BOB structure. The BOB is connected to a gas purification system with two 4 m long flexible hoses.

3.3. Cryogenics
To keep the RED-100 detector at LXe temperatures we developed a cryogenic system based on thermosyphon technology [6]. This system is economical, possesses high cooling power and is capable of maintaining stable temperature throughout the detector operation. A thermosyphon (gravity assisted heat pipe) is a closed tube filled with gaseous nitrogen and consists of three sections: 1) at the top, condenser immersed in a LN_2 Dewar; 2) at the bottom, evaporator attached to the detector inner vessel and 3) passive adiabatic copper pipe, connecting two active sections. In RED-100 we use four thermosyphons (adiabatic section length varies from 2 to 3.5 m) with total cooling power up to 400 W. The most powerful thermosyphon (16 mm in diameter tube) through evaporator section is attached to a massive cooper disc inside the detector inner vessel. This disc plays a role of a cold head serving as a heat sink for cables and providing cooling power for xenon condensation. Other thermosyphons (12 mm in diameter tube) are attached to the bottom and copper thermal shield surrounding the detector inner vessel to maintain a temperature gradient along the vertical direction. The temperature is monitored with ADAM 6015 module utilizing the signal from Pt-100 temperature sensors. This thermosyphon configuration showed stable performance with temperature accuracy better than 1 K. The themosyphons operate in a dynamic regime by adjusting amount of N_2 in the tube. The bottom thermosyphon cold head is equipped with three 100 W heaters assisting LXe recovery.

3.4. Xenon purification system
The operation of LXe emission detectors requires a very low level of electronegative and molecular impurities. In order to achieve reasonable electron lifetime (several meter drift lengths) we constructed a purification system. The purification is accomplished by continuously circulating xenon gas with a double-diaphragm pump through a SAES MonoTorr hot getter. To evaporate LXe (getter works only with gas), liquid is sucked out through a thin tube from the detector bottom by the circulation pump. The heat exchanger is used to mitigate heat load from the condensation of returning gas. Gaseous xenon is stored in 50 liter Luxfer aluminum cylinders placed on weigh scales.

3.5. Light collection, electronics and data acquisition
Since our detector is a light detection device we use a low radioactive 3-inch Hamamatsu R11410-20 photomultiplier tube specially designed for LXe low-background experiments. All PMTs were characterized in terms of single photoelectron response and dark count rate before the implementation in the internal structure. The voltage divider boards are made of Cirlex and supply a positive bias voltage to the PMT with a total resistance of 18.5 MΩ. The performed tests demonstrated that these PMTs satisfy the detector requirements, however, an unusual behavior was noticed for several PMT units [7]. All PMTs were installed in top and bottom arrays according to their quantum efficiency and noise characteristics. To exclude PMT performance deterioration from a bright light, a special circuit was designed to switch off the tubes in case of a cosmic muon crossing the TPC [8]. The electronic system amplifies signals from PMTs and sends it to registration without shaping. We use CAEN V1730 FlashADC fast digitizers with 500 MSmps speed and 14-bit resolution to record waveforms of signals. Waveforms are stored in computer for offline processing. The trigger system is based on specific digital module built on CAEN V1495 board and allows to use a flexible selection criteria with a threshold as low as a single ionization electron [9]. Testing of DAQ system is underway now.

4. Prospects for CENNS observation with RED-100 at SNS
A 1 GeV proton beam hits the liquid mercury target in 700 ns wide bursts with a 60 Hz frequency. The proton interaction produces π^+-mesons which are stopped in the target and decay at rest. These pion decays yield a 30 MeV monochromatic ν_μ flux (prompt neutrinos), followed within 2.2 μs interval by $\bar{\nu}_\mu$ and ν_e with a few tens of MeV energy from μ^+ decays (delayed neutrinos). During a background measurement campaign several possible deployment sites at SNS were studied. The background measurements indicated that the beam-related neutron count in the basement

hallway is four orders of magnitude lower than that in the experimental floor. Moreover, the basement location is protected from cosmic rays by 8 m.w.e. of overburden. The RED-100 response has been simulated taking into consideration the expected neutrino flux at this particular site. The computational model is based on Geant4 and NEST package, describing interactions of charged particles in noble gas based detectors. According to this model, an optimal passive shielding configuration is 10-15 cm of lead and 15 cm of water. Backgrounds from the radioactivity of internal components, external gammas and neutrons from the SNS and induced by cosmic muons were all considered. Approximately 1,700 CENNS events per year are expected for delayed neutrinos under the condition $S1 \geq 2$ ph.e. This allows to apply 1 μs timing cut to eliminate beam-coincident fast neutron backgrounds. The background simulation shows that the expected background is an order of magnitude less than the signal from coherent neutrino scattering. A 5σ measurement of CENNS is expected within a few months of operation.

5. Conclusion

The process of coherent elastic neutrino-nucleus scattering can be discovered with the two-phase liquid xenon RED-100 detector. During a background measurement campaign, it was found out that the best site for deploying the detector is the basement of SNS. For this particular place the detector expected signal and background simulations were performed, showing that RED-100 has a great potential for the first CENNS observation. The detector design and details are described. The RED-100 is under commissioning now at NRNU MEPhI and then will be transferred to SNS.

Acknowledgements

This study was supported by RF Government under contracts of NRNU MEPhI with the Ministry of Education and Science of №11.G34.31.0049 from October 19, 2011 and within the framework of the Centre for Nanostructured Electronics supported by MEPhI Academic Excellence Project (contract № 02.a03.21.0005, 27.08.2013.

References

[1] Freedman D Z 1974 Coherent effects of a weak neutral current *Phys. Rev.* D **9** 1389

[2] Freedman D Z, Schramm D N and Tubbs D L 1977 The weak neutral current and its effect in stellar collapse *Ann. Rev. Nucl. Part. Sci.* **27** 167-207

[3] Barranco J, Miranda O G and Rashba T I 2007 Low energy neutrino experiments sensitivity to physics beyond the standard model *Phys. Rev.* D **76** 073008

[4] Billard J, Figuero-Feliciano E and Strigari L 2014 Implication of neutrino backgrounds on the reach of next generation dark matter direct detection experiments *Phys. Rev.* D **89** 023524

[5] Dolgoshein B A, Lebedenko V N and Rodionov B U 1970 New method of registration ionizing-particle tracks in condensed matter *JETP Lett.* **11** 351-3

[6] Bolozdynya A I *et al.* 2015 The two-phase closed tubular cryogenic thermosiphon *IJHMT* **80** 159-62.

[7] Akimov D Yu *et al.* 2015 Observation of light emission from Hamamatsu R11410-20 photomultiplier tubes *Nucl. Instrum. Meth.* A **794** 0102.

[8] Akimov D Yu *et al.* 2014 A controllable voltage divider for Hamamatsu R11410-20 photomultipliers for use in the RED100 emission detector *Instrum. Exp. Tech.* **57**(5) 108-12.

[9] Naumov P P *et al.* 2014 Digital trigger system for RED-100 (in Russian) *Nucl. Phys. and Engineering* **5**(9) 824-8 (Original Russian title: *Yadernaya Fizika i Enginiring*).

International Conference on Particle Physics and Astrophysics (ICPPA-2015) IOP Publishing
Journal of Physics: Conference Series **675** (2016) 012017 doi:10.1088/1742-6596/675/1/012017

On the temperature evolution of multicomponent dark matter with Coulomb-like interaction

K M Belotsky[1], E A Esipova and A A Kirillov

National Research Nuclear University MEPhI (Moscow Engineering Physics Institute), Kashirskoe highway 31, Moscow, 115409, Russia

E-mail: k-belotsky@yandex.ru, esipovaea@gmail.com, kirillov-aa@yandex.ru

Abstract. We consider model of multicomponent dark matter consisting of WIMP- and SIMP-components, possessing Coulomb-like interaction. The latter, in its term, implies an existence of thermal background of the interaction carriers. We evaluate the temperature evolution of different components for early epoch. It is shown that it can be very different depending on model parameters.

Many attempts of solution of dark matter (DM) problem, undertaken now, are connected with self-interacting multicomponent DM models. They allow to reach the better agreement with observation data [1–6]. We take here a "toy" model where there are three components of DM — SIMP-component, WIMP-component[2], y-background, where the latter is the thermal background of carries of Coulomb-like interaction (y-interaction) which is possessed by WIMPs and SIMPs. The model of OHe with 4th generation fermions can be served as example of similar model [7–10].

Effect of recombination plays a crucial role in such models, since this process can drastically affect dynamics of dark matter (see [11] and references therein). Effectiveness of recombination strongly depends on temperature of the particles of question. Besides that the knowledge of temperature is important even without recombination which also determines cold-like or warm-like DM dynamics. Here we study, in framework of the chosen simple model, how sensitive to variation of a few model parameters (masses of particles $m_{\rm w}$, $m_{\rm s}$ and y-interaction constant (α_y) the temperature evolution of the considered system can be.

The temperature evolution of WIMP- and SIMP-components can be described by the system of equations[3]

$$\begin{cases} \dfrac{dT_{\rm w}}{dT} = -\dfrac{2}{3T}\left\{\dfrac{1}{H}\left[\left\langle\overline{\Delta E}\sigma v\right\rangle_{\rm ws} n_{\rm s} + \left\langle\overline{\Delta E}\sigma v\right\rangle_{\rm wy} n_y\right] - 3T_{\rm w}\right\} \\ \dfrac{dT_{\rm s}}{dT} = -\dfrac{2}{3T}\left\{\dfrac{1}{H}\left[\left\langle\overline{\Delta E}\sigma v\right\rangle_{\rm sw} n_{\rm w} + \left\langle\overline{\Delta E}\sigma v\right\rangle_{\rm sy} n_y + \left\langle\overline{\Delta E}\sigma v\right\rangle_{\rm sp} n_{\rm p}\right] - 3T_{\rm s}\right\} \end{cases} \quad (1)$$

where T is the temperature of CMB, H is the Hubble rate, $n_{\rm w,s,y}$ and $T_{\rm w,s,y}$ are the number density and temperature of WIMP-, SIMP-gases and y-background respectively, $n_{\rm p}$ is the

[1] To whom any correspondence should be addressed.
[2] SIMP — strongly interacting massive particle, WIMP — weakly interacting massive particle.
[3] Derivation of analogous equations can be found in [11, 12]

number density of protons (nucleons), $\overline{\Delta E}$ is the kinematically averaged transferred energy, v is the relative velocity of interacting particles, σ is the interaction cross section, brackets $\langle\ \rangle$ mean averaging over the thermal distribution.

We consider the case when WIMPs and SIMPs are oppositely y-charged and $n_{\mathrm{w}}(T) = n_{\mathrm{s}}(T) \approx \frac{\Omega_{\mathrm{DM}}\rho_{\mathrm{mod}}}{m_{\mathrm{s}}}\left(\frac{T}{T_0}\right)^3$, where $\Omega_{\mathrm{DM}} \approx 0.25$ is relative density of DM, $\rho_{\mathrm{mod}} \approx 5.5\ \mathrm{keV/cm^3}$ and $T_0 \approx 2.73\mathrm{K}$ are the modern total density and CMB temperature respectively.

y-Background temperature is given by $T_y = \kappa^{\frac{1}{3}}T$, where κ is relative contribution of y-background to the entropy.

Energy exchange between WIMP- and SIMP-components can be described by [13]

$$\langle\overline{\Delta E}\sigma v\rangle_{\mathrm{ws}} = -\langle\overline{\Delta E}\sigma v\rangle_{\mathrm{sw}} = 2\pi\alpha_y^2 L \int \left(\frac{1}{T_{\mathrm{w}}} - \frac{1}{T_{\mathrm{s}}}\right) \frac{v^2(\vec{v}_{\mathrm{w}}\vec{v}_{\mathrm{s}}) - (\vec{v}\vec{v}_{\mathrm{w}})(\vec{v}\vec{v}_{\mathrm{s}})}{v^3} f_{\mathrm{w}} f_{\mathrm{s}} d^3v_{\mathrm{w}} d^3v_{\mathrm{s}} =$$
$$= 4\sqrt{2\pi}\alpha_y^2 L \sqrt{\frac{m_{\mathrm{w}}m_{\mathrm{s}}}{(m_{\mathrm{w}}T_{\mathrm{s}} + m_{\mathrm{s}}T_{\mathrm{w}})^3}}(T_{\mathrm{s}} - T_{\mathrm{w}}) \qquad (2)$$

where $L \approx \ln\left(\frac{R_{\mathrm{D}}T}{\alpha_y}\right)$ is the Coulomb logarithm with R_{D} being the Debye radius, $v_{\mathrm{w,s}}$ is the velocity of WIMPs and SIMPs, $f_{\mathrm{w,s}}$ is the Maxwell distribution for them.

We take into account SIMP-nucleon interaction for which we get (see Appendix)

$$\langle\overline{\Delta E}\sigma v\rangle_{\mathrm{sp}} = \frac{8\sqrt{2}}{\sqrt{\pi}}\frac{\mu^2}{T_{\mathrm{s}}T_{\mathrm{p}}}\sigma_{\mathrm{sp}}\left(\frac{T_{\mathrm{s}}}{m_{\mathrm{s}}}\frac{T_{\mathrm{p}}}{m_{\mathrm{p}}}\right)^{\frac{3}{2}}\sqrt{\frac{m_{\mathrm{s}}}{T_{\mathrm{s}}} + \frac{m_{\mathrm{p}}}{T_{\mathrm{p}}}} \cdot (T_{\mathrm{p}} - T_{\mathrm{s}}) \qquad (3)$$

where $\sigma_{\mathrm{sp}} \sim 1\ \mathrm{fm}^2$, $\mu = \frac{m_{\mathrm{s}}m_{\mathrm{p}}}{m_{\mathrm{s}}+m_{\mathrm{p}}}$, m_{p} is the proton mass. For WIMPs- (SIMPs-) y-interaction we have [11]

$$\langle\overline{\Delta E}\sigma v\rangle_{\mathrm{w(s)y}} = \frac{32\pi{\alpha_y}^2}{{m_{\mathrm{w(s)}}}^3}T_y(T_y - T_{\mathrm{w(s)}}) \qquad (4)$$

The results of solution of the system (1) for some different sets of parameter values are presented on the series of plots on figure 1.

They illustrate a diversity of physical situations in the considered multicomponent DM evolution, when different chains of mutual component thermal decouplings happen.

Acknowledgments
This work was performed within the framework of the Center of Fundamental Researches and Particle Physics supported by MEPhI Academic Excellence Project (contract no. 02.a03.21.0005, 27.08.2013), grant of RFBR no. 14-22-03048 and in part by grants of RFBR no. 14-22-03031 and of Ministry of Education and Science of the Russian Federation, no. 3.472.2014/K.

Appendix

Average transferred energy between SIMP and proton in co-moving reference frame (r.f.) is expressed in terms of energy and momentum in r.f. where SIMP is at the rest (l.s.) as

$$\overline{\Delta E}_{\mathrm{sp}}^{\mathrm{r.f.}} = \overline{E}_{\mathrm{s2}}^{\mathrm{r.f.}} - E_{\mathrm{s1}}^{\mathrm{r.f.}} = \frac{\overline{E}_{\mathrm{s2}}^{\mathrm{l.s.}} - v_{\mathrm{s1}}\overline{p}_{\mathrm{s2}}^{\mathrm{l.s.}}\cos\theta_{\vec{v}\vec{v}_{\mathrm{s1}}}}{\sqrt{1 - v_{\mathrm{s1}}^2}} - E_{\mathrm{s1}}^{\mathrm{r.f.}} \qquad (5)$$

where the indices "1" and "2" correspond to initial and final states respectively. Here we used that averaging of momentum vector $\vec{p}_{s2}^{\,\mathrm{l.s.}}$ gives average longitudinal component along $\vec{v}$, $\overline{p}_{\mathrm{s2}}^{\mathrm{l.s.}}$.

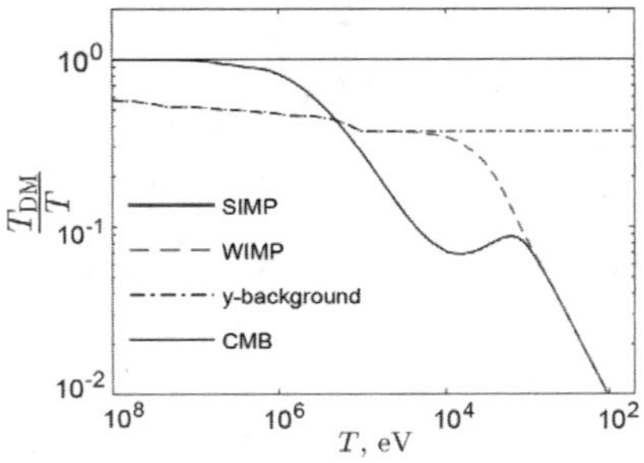

(a) $m_w = 10$ GeV, $m_s = 1000$ TeV, $\alpha_y = 1/100$

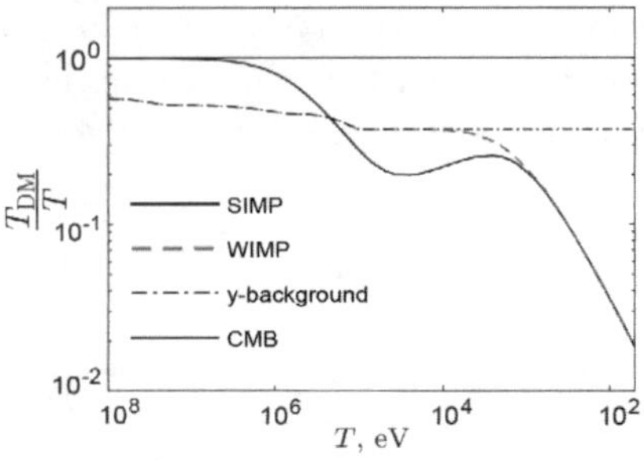

(b) $m_w = 10$ GeV, $m_s = 1000$ TeV, $\alpha_y = 1/30$

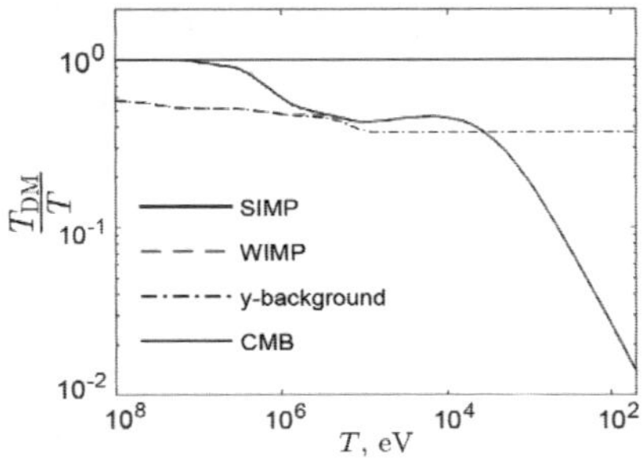

(c) $m_w = 10$ GeV, $m_s = 1$ TeV, $\alpha_y = 1/100$

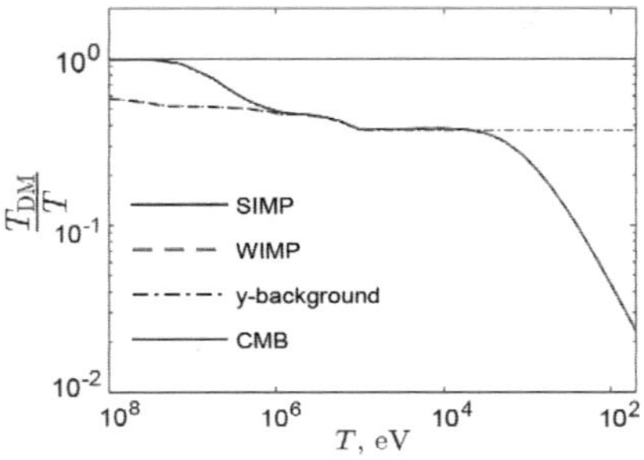

(d) $m_w = 10$ GeV, $m_s = 1$ TeV, $\alpha_y = 1/30$

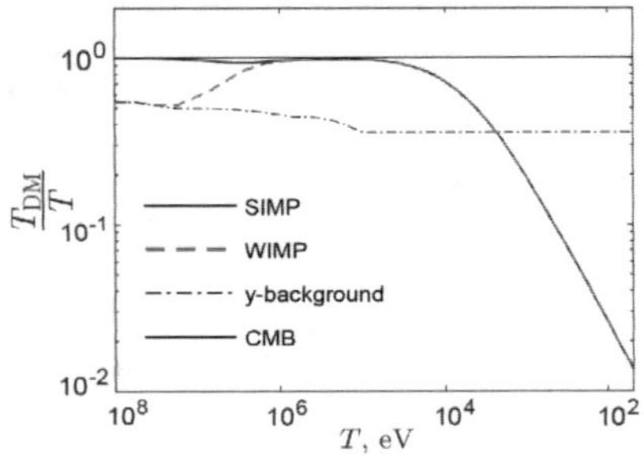

(e) $m_w = 100$ GeV, $m_s = 1$ TeV, $\alpha_y = 1/100$

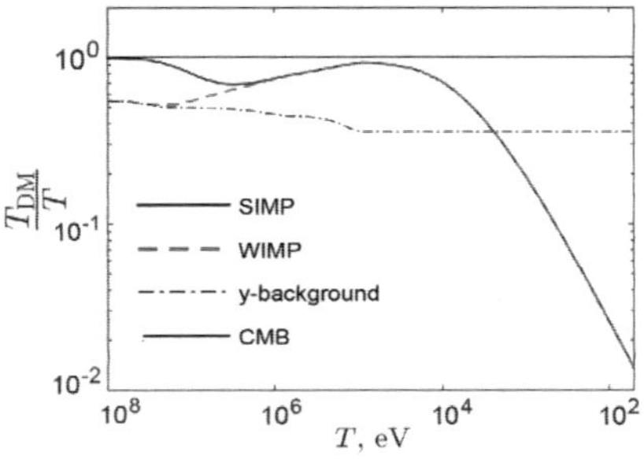

(f) $m_w = 100$ GeV, $m_s = 1$ TeV, $\alpha_y = 1/30$

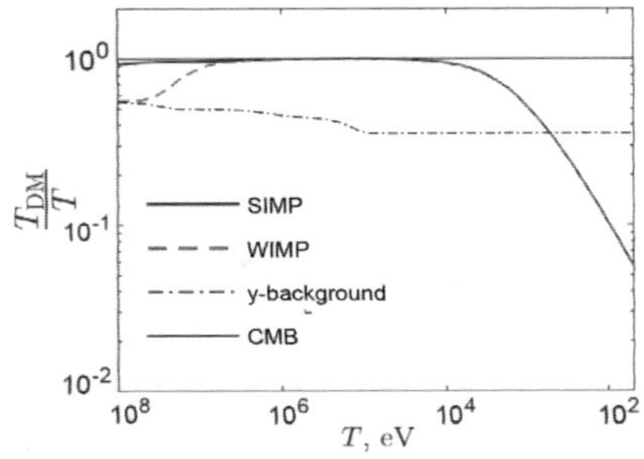

(g) $m_w = 100$ GeV, $m_s = 100$ GeV, $\alpha_y = 1/100$

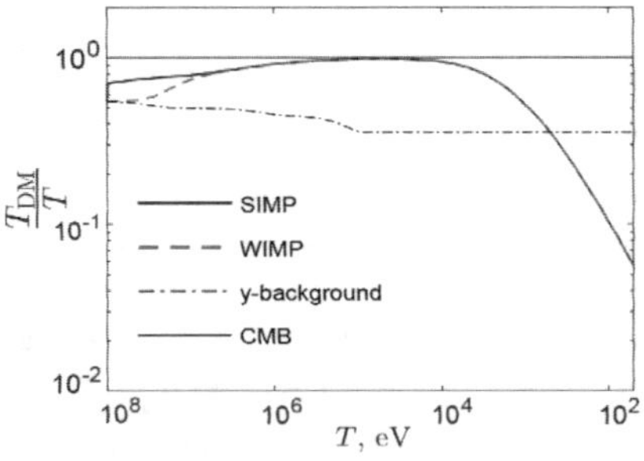

(h) $m_w = 100$ GeV, $m_s = 100$ GeV, $\alpha_y = 1/30$

Figure 1. The evolution of the temperature of DM components T_{DM} in units of CMB one T is shown for different sets of model parameters.

In non-relativistic approximation we have

$$\overline{E}_{s_2}^{\text{l.s.}} \approx m_s + \frac{\mu^2 v^2}{m_s}, \quad \overline{p}_{s_2}^{\text{l.s.}} \approx \mu v. \tag{6}$$

Then

$$\left\langle \overline{\Delta E}\, \sigma v \right\rangle_{\text{sp}} = \sigma_{\text{sp}} \int \left(\frac{\mu^2 v^2}{m_s} - \mu v v_{s_1} \cos\theta_{\vec{v}\vec{v}_{s_1}} \right) v f_s f_p d^3 v_s d^3 v_p \tag{7}$$

and we obtain equation (3).

References

[1] Aoki M *et al.* 2012 *Phys. Rev.* D **86** 076015
[2] Kajiyama Y, Okada H and Toma T 2013 *Phys. Rev.* D **88** 015029
[3] Feldmani D *et al.* 2010 *Phys. Rev.* D **81** 095017
[4] Belotsky K, Kirillov A and Khlopov M 2013 *Physics of Atomic Nuclei* **76** 469–475
[5] Belotsky K, Kirillov A and Khlopov M 2014 *Grav. & Cosmol.* **20** 47–54
[6] Belotsky K *et al.* 2014 *Advances in High Energy Physics* 214258
[7] Khlopov M 2006 *JETP Letters* **83** 1–4
[8] Khlopov M and Kouvaris C 2008 *Phys. Rev.* D **77** 065002
[9] Khlopov M and Kouvaris C 2008 *Phys. Rev.* D **78** 065040
[10] Cudell J, Khlopov M and Wallemacq Q 2014 *Mod. Phys. Lett. A* **29** 1440006
[11] Belotsky K, Esipova E and Kirillov A 2015 *arXiv*:1506.03094
[12] Belotsky K and Kirillov A 2015 *JCAP* **01** 041 (*arXiv*:1409.8601)
[13] Rumer Yu B and Ryvkin M Sh 2001 *Thermodynamics, Statistic Physics and Kinetics* (Novosibirsk University Press)

Visualization tool for X-ray scanner for sTGC detector production quality control

V O Tikhomirov[1,2,4], K A Filippov[1], S P Konovalov[2], G Mikenberg[3], A Romaniouk[1], D Shchukin[1], M Shoa[3], V Smakhtin[3], S Yu Smirnov[1], V V Sosnovtsev[1], P E Teterin[1], V I Tsekhosh[2] and K A Vorobev[1]

[1] National Research Nuclear University MEPhI (Moscow Engineering Physics Institute), Kashirskoe highway 31, Moscow, 115409, Russia

[2] P.N. Lebedev Physical Institute of the Russian Academy of Sciences, Leninsky prospect 53, Moscow, 119991, Russia

[3] Weizmann Institute of Science, 234 Herzl St., Rehovot 7610001, Israel

E-mail: `Vladimir.Tikhomirov@cern.ch`

Abstract. The ATLAS experiment at the Large Hadron Collider has an ambitious program of the detector upgrade to meet an expected rise of accelerator luminosity. The first large system which supposed to be installed in 2019 is the New Small Wheel (NSW) for ATLAS muon spectrometer. In order to ensure high quality and reliability of NSW chambers an X-ray scanning technique is being developed. One of the main components of the X-ray scanner is a special software visualization tool which would allow a fast and clear representation of scanning results and an identification of possible chamber defects.

1. Introduction

ATLAS [1] is one of the largest experiment at the CERN Large Hadron Collider (LHC). To meet an expected rise of LHC luminosity, ATLAS is going to upgrade several detector components during next LHC Long Shutdown period. One of the systems which supposed to be installed in 2019 is the New Small Wheel (NSW) for ATLAS muon spectrometer [2]. To test the quality of muon "small-strip Thin Gap Chambers" (sTGC) at production sites (Canada, Chili, China, Israel and, possibly, Russia) an approach based on scanning of the chambers by X-ray was proposed. A special X-ray scanner was developed by Moscow Institutes involved in the NSW project (MEPhI, LPI and SINP MSU). The first scanner prototype was produced and now is being tested at MEPhI. Description of the scanner and corresponding methodical results can be found in reports [3,4] presented at this Conference.

For X-ray scanner operating, drive and control of scanning process and for visual presentation of results a dedicated software package has to be developed. This package will be an indivisible component of the delivered scanner equipment. An overview of this software tool is presented in this paper with the emphasis on visualization and presentation possibilities of the results obtained during scanning process.

[4] To whom any correspondence should be addressed.

2. X-ray scanner software

The general task of the scanner software development is to integrate into one package the hardware control (scanning process itself, X-ray tube control, high voltage power supply and multimeter) with fast and visual presentation of results obtained during scanning process. The program is based on the ROOT package [5] in Graphical User Interface (GUI) and in analysis/presentation parts. For the hardware control a low-level C/C++ coding is provided.

General view of GUI is shown in figure 1. The program can work under Linux, Windows and Mac operating systems. In the hardware control part the program is still under development. The part for visualization and presentation of the data is in the final stage of development and is described in more details below. Some real results obtained by manual scan of sTGC chambers with a pentane gas mixture are shown below.

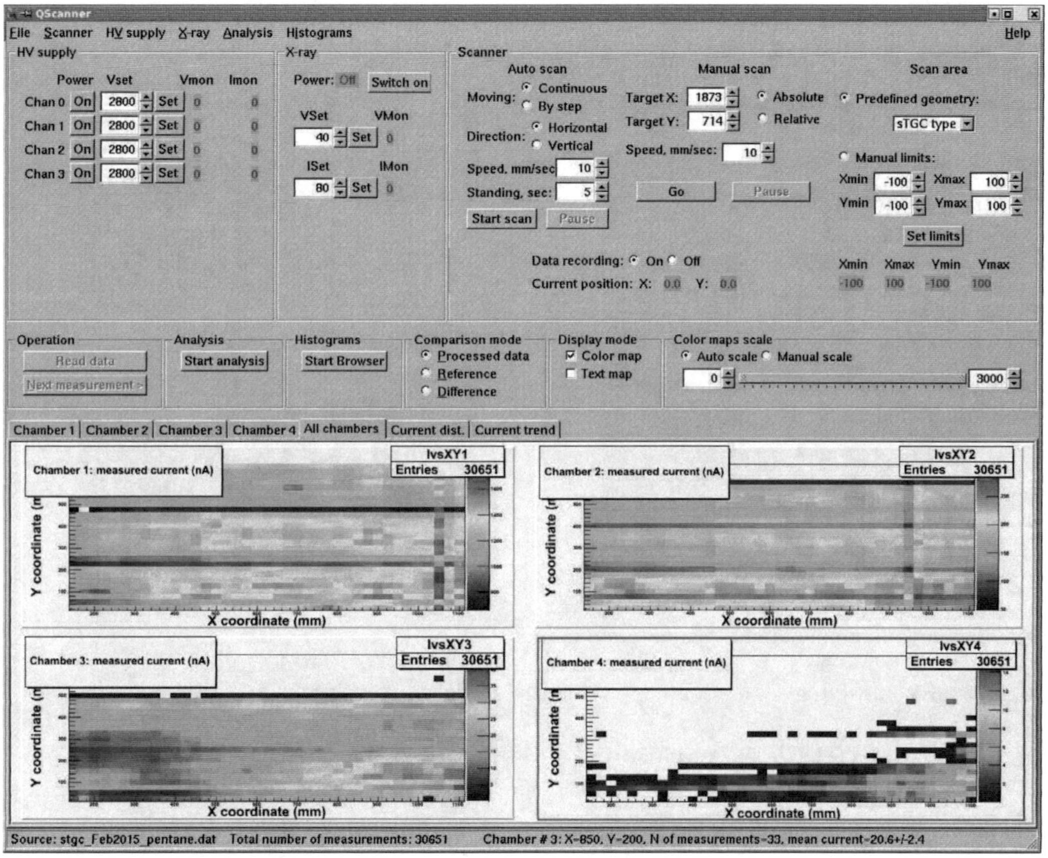

Figure 1. General view of the program GUI. The upper part is dedicated to drive the scanner hardware: high voltage power supply, X-ray tube and stepper motors. The lower part is for results presentation.

3. Visualization tool

During the scanning process the sTGC detectors connected to HV power supply are irradiated by X-ray tube and the current from sTGC anode wires is measured. Any possible local defects in the chambers being tested (non-uniformity in chamber geometry, anode wire defects, loss of gas gain etc) lead to rise or drop in measured chamber current. The easiest and most evident way for quick detection of such current variation is to present the results of current measurement as a two-dimensional map – in numbers or in colors – in XY plane of the irradiated sTGC. As the sTGC is a quadruplet of four identical planes glued together, the map has to be presented for each chamber in the quadruplet. On the lower part of figure 1 an example of such a maps for four chambers is shown. Each channel of two-dimensional ROOT histogram corresponds to X and Y coordinates of sTGC chamber irradiated by X-ray tube and the color presents the measured chamber current at this point – according to adjustable color palette on the right side of histograms.

A user can also mark "Text" check-box to show value of the current in numbers directly on the map field and also on the status bar in the bottom of the GUI. For more detailed view of a particular chamber area one can do double-left mouse click on the map field to open new window with zoomed view (figure 2). Horizontal and vertical sliders can then be used to move to any other map position.

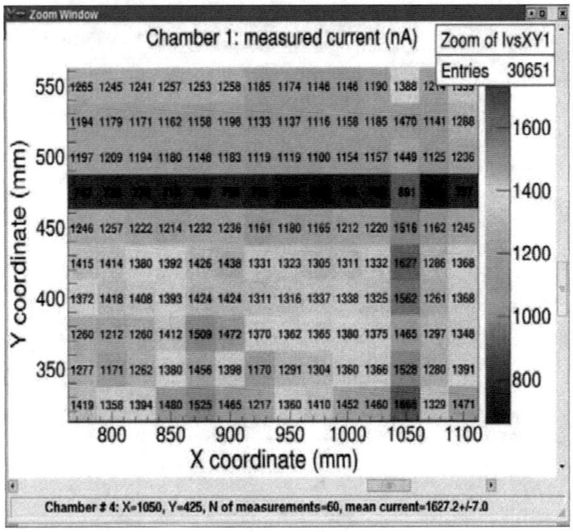

Figure 2. Zoom window with color/text map of currents measured in first sTGC chamber.

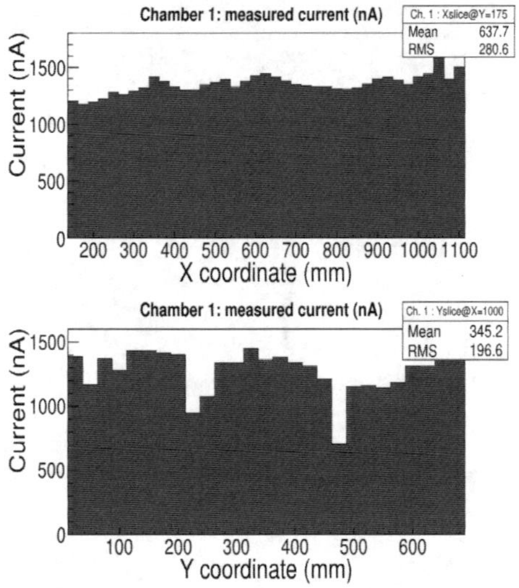

Figure 3. Slices of measured current along X (top) and Y (bottom) axis.

Middle button mouse click on the map opens a new window with two slices – histograms of the current values along X and Y axis (figure 3). It can be useful for quick detection of possible problems in the direction along and across the anode wires orientation.

Since the information about scanning process and the results are kept as ROOT histograms, a user can directly display these histograms via standard ROOT browser. The most important ones (current distributions in the chambers and current trends in time) can also be presented directly in the main GUI window if the corresponding tab button is selected (figures 4 and 5).

4. Examples of usage

Two examples of visualization tool usage are demonstrated in figures 6 and 7. Visible horizontal strip structures are due to chambers design and are not interesting for analysis. The examples also show evident problems in vertical direction which corresponds to the direction of anode wires, at the coordinate with X≈1050 mm in first chamber of the sTGC quadruplet and X≈950 mm in the second chamber. Such fast fault diagnosis will help in sTGC quality control at the chambers production sites.

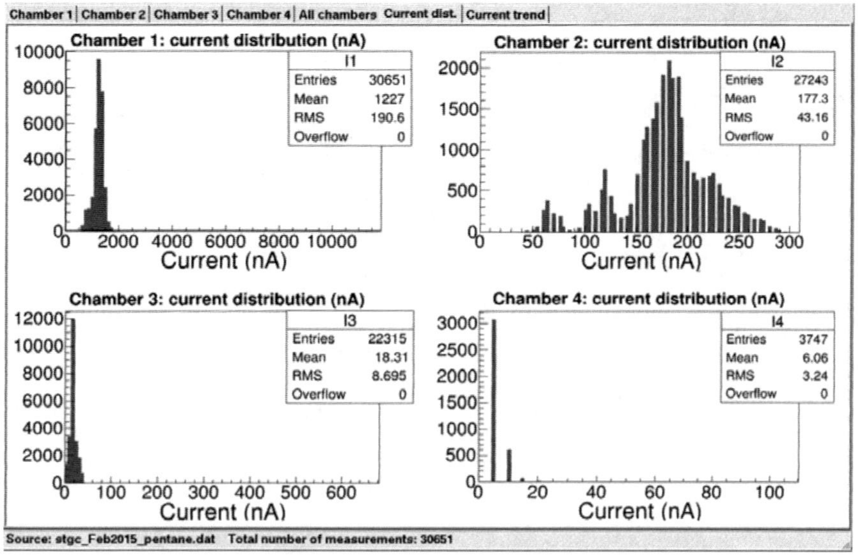

Figure 4. Measured current distributions for four detectors of sTGC quadruplet.

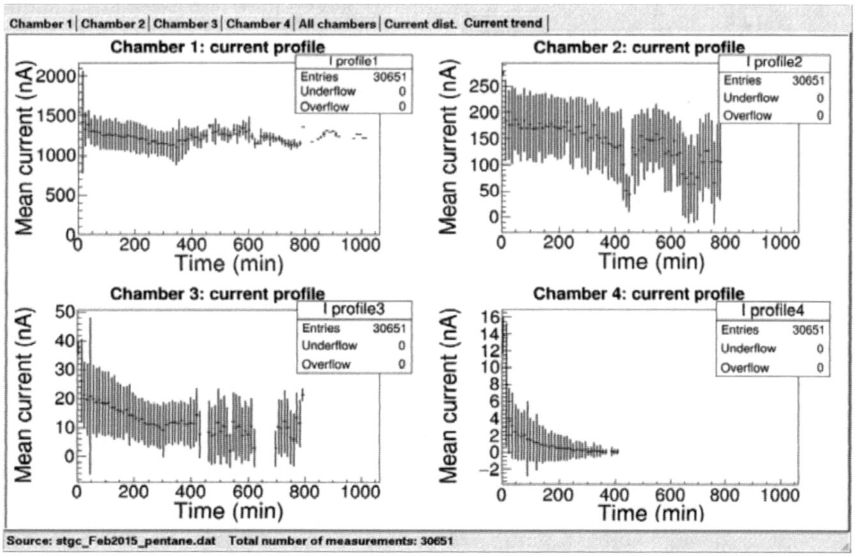

Figure 5. Time evaluation of measured current in four detectors of sTGC quadruplet.

5. Conclusion

A dedicated software tool for operation of X-ray scanner and visualization of the results obtained during scanning is under development. Visualization software provides wide possibilities for results presentation and analysis. It has a key role for identification of possible defects in muon chambers being tested. By the end of 2015 first scanners with corresponding software have to be delivered to ATLAS muon chambers production sites.

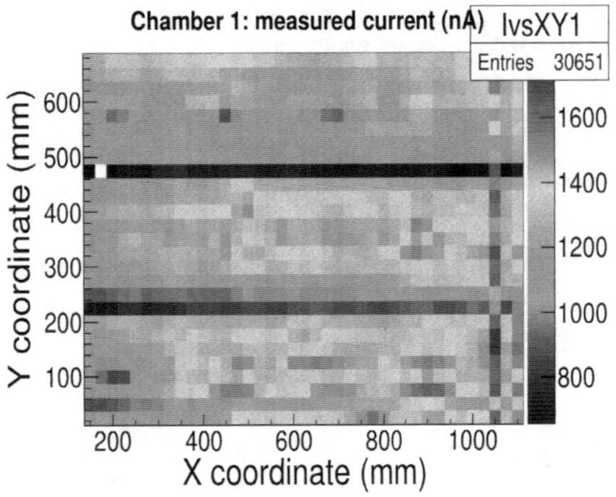

Figure 6. "Hot" area in the chamber at X coordinate ≈1050 mm: possible problem with anode wire offset.

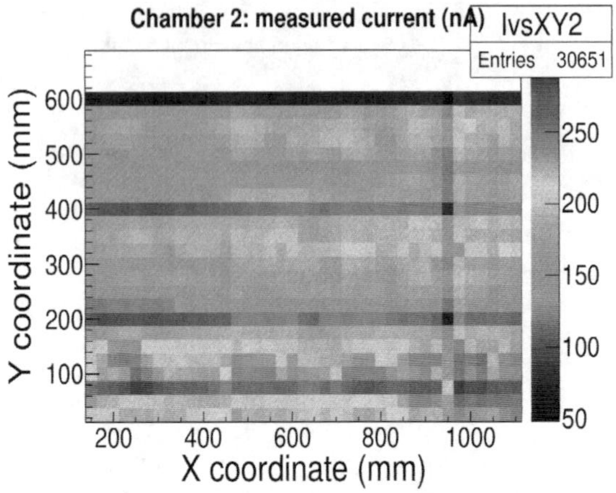

Figure 7. "Cold" area in the chamber at X coordinate ≈950 mm: possible missing anode wire or drop in gas gain.

Acknowledgments

This work was performed within the framework of the Center of FRPP supported by MEPhI Academic Excellence Project (contract number 02.a03.21.0005, 27.08.2013). We gratefully acknowledge the financial support from the Ministry of Education and Science of Russian Federation (R&D Project RFMEFI61014X0005) and grant No.14-22-03053 from the Russian Foundation for Basic Research.

References

[1] The ATLAS Collaboration 2008 *The ATLAS Experiment at the CERN Large Hadron Collider JINST* **3** S08003
[2] The ATLAS Collaboration 2013 *New Small Wheel Technical Design Report* CERN-LHCC-2013-006; ATLAS-TDR-020
[3] Filippov K A *et al.* 2015 Gas mixtures for quality control of the sTGC chambers. *J. Phys.: Conf. Series* (In this Proceedings)
[4] Teterin P E *et al.* 2015 Development of scanning technique for sTGC detector production quality control. *J. Phys.: Conf. Series* (In this Proceedings)
[5] Brun R and Rademakers F 1997 *Nucl. Instrum. Meth.* A **389** 81–86

Excitation spectra of solitary waves in scalar field models with polynomial self-interaction

V A Gani[1,2], V Lensky[1,2,3], M A Lizunova[1,2] and E V Mrozovskaya[1]

[1]National Research Nuclear University MEPhI (Moscow Engineering Physics Institute), Kashirskoe highway 31, Moscow, 115409, Russia
[2]National Research Centre Kurchatov Institute, Institute for Theoretical and Experimental Physics, Bolshaya Cheremushkinskaya str. 25, Moscow, 117218, Russia
[3]Institut für Kernphysik & PRISMA Cluster of Excellence, Johannes Gutenberg Universität Mainz, D-55128 Mainz, Germany

E-mail: `vakhid.gani@gmail.com`

Abstract. We study excitations of solitary waves – the kinks – in scalar models with degree eight polynomial self-interaction in (1 + 1) dimensions. We perform numerical studies of scattering of two kinks with an exponential asymptotic off each other and analyse the occurring resonance phenomena. We connect these phenomena to the energy exchange between the translational and the vibrational modes of the colliding kinks. We also point out that the interaction of two kinks with power-law asymptotic can lead to a long-range interaction between the two kinks.

1. Introduction

Field-theoretical models with polynomial self-interaction are widespread in modern theoretical physics, being applied to various phenomena from cosmology to condensed matter physics. In the latter, the most known example is perhaps the phenomenological Ginzburg-Landau model, where the complex scalar field of Cooper pairs has a polynomial self-interaction of the fourth degree. The φ^4 model with two degenerated minima is commonly used as a model example of spontaneous symmetry breaking, whereas models with more sophisticated potentials of higher polynomial degrees are applied, e.g., to model several consecutive phase transitions [1, 2].

We further consider field models with a single real field and polynomial potentials that are even functions of field strength φ and have degenerate minima. Such models admit topological solutions — the kinks — that interpolate between two adjacent minima of the potential (the vacua of the model). The kinks of the φ^4 and φ^6 models are well studied, yielding many interesting results concerning, e.g., excitations of the kinks [3, 4] and scattering of two kinks off each other [5, 6]. At the same time, similar features of models with self-interaction of higher degrees, used in studies of many important phenomena [7], have only recently started to be explored systematically [1, 8].

Our work studies excitations of the φ^8 kinks and their relation to resonance phenomena in kink-antikink collisions. We connect the occurrence of so-called escape windows to the presence of vibrational excitations.

2. A kink of the φ^8 model and its excitation spectrum

The φ^8 model that we consider is described by the following Lagrangian in $(1+1)$ space-time dimensions, with $\varphi = \varphi(t,x)$ being a real scalar field:

$$\mathcal{L} = \frac{1}{2}\left(\frac{\partial\varphi}{\partial t}\right)^2 - \frac{1}{2}\left(\frac{\partial\varphi}{\partial x}\right)^2 - V(\varphi), \tag{1}$$

where the potential $V(\varphi)$ is a degree eight polynomial with two, three, or four degenerate minima φ_i yielding $V(\varphi_i) = 0$. This Lagrangian develops topological BPS-saturated solutions $\varphi_{i_1 i_2}(x)$, each of them connecting two adjacent minima, φ_{i_1} and φ_{i_2} [9]. To explore the excitation spectra of the kinks, one starts along the standard procedure used to study the linear Lyapunov stability: the kink is perturbed by a small deviation,

$$\delta\varphi(t,x) = \psi(x)\cos(\omega t), \tag{2}$$

and the spectrum of ω^2 coincides to that of the time-independent Schrödinger-like problem with the potential

$$U(x) = \left.\frac{d^2V}{d\varphi^2}\right|_{\varphi_\mathrm{k}(x)}. \tag{3}$$

$U(x)$ is a potential well with different or identical asymptotic values at $x \to \pm\infty$ depending on the symmetry of the initial kink. The BPS kinks are classically stable, hence ω^2 cannot be negative; apart from that, it is easy to show that there always is an eigenmode with $\omega^2 = 0$ — the translational (zero) mode. There can also be one or more discrete (vibrational) modes with $\omega^2 > 0$, lying below the continuum part of the spectrum.

We further consider a variant of the φ^8 that has four degenerate vacua, corresponding to the self-interaction potential

$$V(\varphi) = (\varphi^2 - a^2)^2(\varphi^2 - b^2)^2 \tag{4}$$

that has four degenerate minima $\pm a$, $\pm b$. Following [1], we set $a = (\sqrt{3}-1)/2$, $b = (\sqrt{3}+1)/2$. Equation (4) admits three topological sectors, $(-b,-a)$, $(-a,a)$, and (a,b), each with its kink and the corresponding antikink. We constrain our study to the sector $(-a,a)$, with the kink profile given by [1, 10]:

$$e^{\mu x} = \frac{a+\varphi}{a-\varphi}\left(\frac{b-\varphi}{b+\varphi}\right)^{a/b}, \tag{5}$$

where $\mu = 2\sqrt{2}a(b^2-a^2)$. The antikink of this sector is obtained from equation (5) by substituting $x \to -x$.

Our numerical study of the excitation spectrum of this kink confirms the existence of the zero mode (our result is $\omega_0^2 = -2\times 10^{-8}$, the deviation of which from zero can serve as an uncertainty estimate), and also finds a vibrational mode corresponding to $\omega_1^2 = 2.70491$. The latter excitation manifests itself in resonance phenomena in kink-antikink collsions, as explained below.

3. Kink-antikink collisions in the φ^8 model

A vibrational mode in the excitation spectrum of a kink (antikink) leads under certain conditions to a part of the kink's kinetic energy being transformed into small oscillations of the kink's profile — the resonant energy exchange between the translational and the vibrational modes. This gives rise to a host of exciting phenomena in kink-antikink collsions [5, 6, 11, 12].

In order to explore how this mechanism works in the φ^8 model, we performed a numerical study of kink-antikink scattering in the topological sector described in the previous section, with the self-interaction given by equation (4). The initial configuration was taken as the sum of the kink (5) and the corresponding antikink, located at $t = 0$, respectively, at $x = \pm x_0$, $x_0 = 12.5$, and moving towards each other with the respecitve initial velocities $\pm v_{\rm in}$.

We found that there are several different scattering regimes, depending on the value of $v_{\rm in}$. In particular, there is a critical value $v_{\rm in} = v_{\rm cr} \simeq 0.3160$ such that the kink and the antikink bounce off each other at all $v_{\rm in} \geq v_{\rm cr}$, see figure 1a. This process is only approximately elastic, i.e., $v_{\rm f} < v_{\rm in}$, which can also be seen in figure 1. For example, $v_{\rm f} = 0.23$ at $v_{\rm in} = 0.40$, $v_{\rm f} = 0.36$ at $v_{\rm in} = 0.50$, and $v_{\rm f} = 0.47$ at $v_{\rm in} = 0.60$. At $v_{\rm in} < v_{\rm cr}$, on the other hand, the two kinks capture each other and form a long-lived quasi-bound state, the bion, see figure 1b. This, however, does not happen universally at all $v_{\rm in} < v_{\rm cr}$: we found narrow intervals of $v_{\rm in}$ — the escape windows — where the kink and the antikink escape after two, three, or more bounces (see figure 1c, which shows an escape after two bounces). The phenomenon of escape windows is explained by the resonance energy exchange between the translational and the vibrational modes of the kink (antikink). The resonance condition

$$\omega_R T_{12} = 2\pi n + \delta, \tag{6}$$

where T_{12} is the time between the two bounces, δ is a constant, and n is an integer number, resulted in the empirical value of the resonance frequency $\omega_R = 1.618$ being very close to the frequency of the kink's vibrational mode, $\omega_1 = \sqrt{\omega_1^2} = 1.644$.

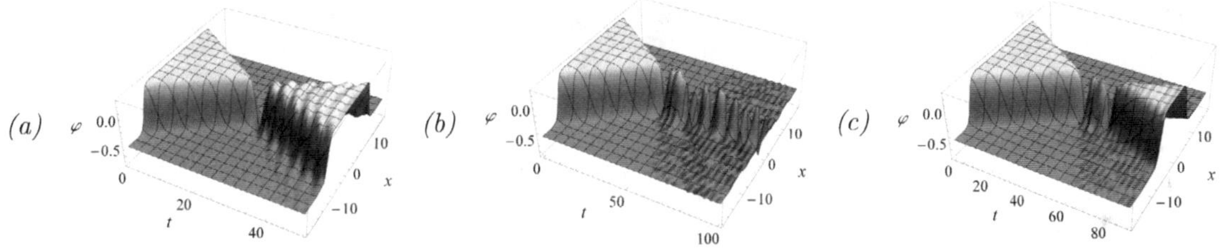

Figure 1. a) Escape at $v_{\rm in} > v_{\rm cr}$; b) Formation of a bion at $v_{\rm in} < v_{\rm cr}$; c) Escape window: escape after two bounces at $v_{\rm in} < v_{\rm cr}$.

4. Conclusion

In conclusion, we performed a first study of excitation spectra of the kinks occurring in the φ^8 field model. We connected the occurrence of a vibrational excitation of a single kink to the resonant energy exchange between this excitation and the translational excitation of the kink, leading to escape windows in the kink-antikink collisions.

We would like to mention a case of special interest that can be realized in various field models. If the model has a vacuum where the second derivative of the self interaction potential turns to zero, the kink that approaches this vacuum does so under a power-law asymptotic, in distinction to the situation considered previously where the asymptotic is exponential. Such kinks can occur in the φ^8 model or models with higher polynomial degrees under certain choice of the potential parameters [1, 10]. Qualitatively, such a power-law asymptotic leads to a long-range interaction between the kink and the corresponding antikink. In particular, it would be interesting to study how well a kink-antikink system with long-range interaction can be treated using the collective coordinate method; we plan to perform such a study of the corresponding sector of the φ^8 model in the future.

Acknowledgments

The authors are very grateful to Prof. A. E. Kudryavtsev for his interest to their work and for enlightening discussions. This work was supported in part by the Russian Federation Government under grant No. NSh-3830.2014.2. V. A. Gani acknowledges support of the Ministry of Education and Science of the Russian Federation, Project No. 3.472.2014/K. M. A. Lizunova thanks the ITEP support grant for junior researchers and gratefully acknowledges financial support from the Dynasty Foundation.

This work was performed within the framework of the Center of Fundamental Research and Particle Physics supported by MEPhI Academic Excellence Project (contract No. 02.a03.21.0005, 27.08.2013).

References

[1] Khare A, Christov T C and Saxena A 2014 Successive phase transitions and kink solutions in ϕ^8, ϕ^{10}, and ϕ^{12} field theories *Phys. Rev.* E **90** 023208

[2] Pavlov S V and Akimov M L 1999 Phenomenological theory of isomorphous phase transitions *Crystall. Rep.* **44** 297-99

[3] Saadatmand D *et al.* 2015 The effect of the ϕ^4 kink's internal mode during scattering on $\mathcal{PT}$-symmetric defect *JETP Lett.* **101** 497

[4] Saadatmand D *et al.* 2015 Kink scattering from a parity-time-symmetric defect in the ϕ^4 model *Commun. Nonlinear Sci. Numer. Simulat.* **29** 267

[5] Dorey P *et al.* 2011 Kink-antikink collisions in the ϕ^6 model *Phys. Rev. Lett.* **107** 091602

[6] Gani V A, Kudryavtsev A E and Lizunova M A 2014 Kink interactions in the (1+1)-dimensional φ^6 model *Phys. Rev.* D **89** 125009

[7] Greenwood E *et al.* 2009 Electroweak vacua, collider phenomenology, and possible connection with dark energy *Phys. Rev.* D **79** 103003

[8] Gani V A, Lensky V and Lizunova M A 2015 Kink excitation spectra in the (1+1)-dimensional φ^8 model *JHEP* **08** 147

[9] Manton N and Sutcliffe P 2004 *Topological Solitons* (Cambridge: Cambridge University Press)

[10] Lohe M A 1979 Soliton structures in $P(\varphi)_2$ *Phys. Rev.* D **20** 3120

[11] Belova T I and Kudryavtsev A E 1997 Solitons and their interactions in classical field theory *Phys. Usp.* **40** 359-86

[12] Gani V A and Kudryavtsev A E 1999 Kink-antikink interactions in the double sine-Gordon equation and the problem of resonance frequencies *Phys. Rev.* E **60** 3305

Scalar triplet on a domain wall

V A Gani[1,2], M A Lizunova[1,2] and R V Radomskiy[1]

[1]National Research Nuclear University MEPhI (Moscow Engineering Physics Institute), Kashirskoe highway 31, Moscow, 115409, Russia

[2]National Research Centre Kurchatov Institute, Institute for Theoretical and Experimental Physics, Bolshaya Cheremushkinskaya str. 25, Moscow, 117218, Russia

E-mail: vakhid.gani@gmail.com

Abstract. We consider a model with a real scalar field with polynomial self-interaction of the fourth degree and a coupled scalar triplet. We demonstrate that there is an exact analytic solution in the form of a domain wall with a localised configuration of the scalar triplet coupled to the wall. We study some properties of this solution.

1. Introduction

Non-Abelian moduli fields are widely discussed in the literature, in particular, in the context of non-Abelian strings [1]-[5].A number of theories allows strings with non-Abelian moduli both in supersymmetric and in non-supersymmetric models.

Recently several simple models were suggested that allow the existence of topological defects (Abrikosov-Nielsen-Olesen string or domain walls) which produce non-Abelian degrees of freedom on their world sheets [2, 6].

We consider a model with a global non-Abelian symmetry G which allows topological defects (strings, vertices, domain walls) to exist. These defects break the global symmetry G to a subgroup H (also global). Then, in addition to translational modes, there will be $\dim G - \dim H$ orientational modes. These moduli fields are governed by a sigma-model with G/H internal space. As an example, see the recent publication [2] that introduced a simple model supporting domain walls. A domain wall breaks the global symmetry, and, as a result, there appear two orientational moduli in addition to a translational modulus. Besides that, in [6] a spin-orbit interaction in the bulk was added that entangles the translational and orientational degrees of freedom.

Most of the main results of [6] have been obtained numerically. We, in turn, investigate the same model mostly analytically. We obtained an exact static solution that can be seen as a domain wall with an attached lump of the χ field and examined some of its properties. We also study the small excitation spectrum of the bare domain wall, with the field $\chi \equiv 0$, and confirm that this domain wall is unstable when the model parameters are chosen as in [6].

2. The model and the exact solution

We work with a real scalar field φ and a real triplet of scalars χ^i in (3+1)-dimensional space-time. The space and time coordinates are (x, y, z) and t, respectively. The dynamics of the system is governed by the Lagrangian

$$\mathcal{L} = \frac{1}{2}\partial_\mu\varphi\partial^\mu\varphi + \frac{1}{2}\partial_\mu\chi^i\partial^\mu\chi^i - \lambda(\varphi^2 - v^2)^2 - \gamma\left[(\varphi^2 - \mu^2)\chi^i\chi^i + \beta(\chi^i\chi^i)^2\right], \quad (1)$$

where $\mu^2 < v^2$, β, λ, and γ are positive constants, and i runs from 1 to 3. The fields φ and χ^i interact due to the term $-\gamma(\varphi^2 - \mu^2)\chi^i\chi^i$ in the Lagrangian.

We further consider static domain-wall-like configurations. If the wall is orthogonal to the z axis, the fields depend only on z,

$$\varphi = \varphi(z), \quad \chi^i = \chi^i(z), \quad (2)$$

and the equations of motion for the fields $\varphi(z)$ and $\chi^i(z)$ that follow from the Lagrangian (1) become

$$\varphi'' = 4\lambda\varphi(\varphi^2 - v^2) + 2\gamma(\chi^j\chi^j)\varphi, \quad \chi^{i\prime\prime} = 2\gamma(\varphi^2 - \mu^2)\chi^i + 4\beta\gamma(\chi^j\chi^j)\chi^i. \quad (3)$$

Here the prime stands for the derivative with respect to z. Since the considered model (and, in particular, the system (3)) allows arbitrary orientation of the field χ^i in the internal space, we choose

$$\chi^i = \chi(z)\begin{pmatrix} 0 \\ 0 \\ 1 \end{pmatrix}. \quad (4)$$

Then the system (3) takes the form:

$$\varphi'' = 4\lambda\varphi(\varphi^2 - v^2) + 2\gamma\chi^2\varphi, \quad \chi'' = 2\gamma\chi(\varphi^2 - \mu^2) + 4\beta\gamma\chi^3. \quad (5)$$

To guess the form of the analytic solution, one has to look at the Lagrangian (1). If $\gamma = 0$ then the fields do not interact, and the free field $\varphi(z)$ is described by the Lagrangian of the $\lambda\varphi^4$ model, with one of the static solutions of the equation of motion being the well-known φ^4 kink [7]:

$$\varphi(z) = \varphi_{\mathrm{K}}(z) = v\tanh\left(\frac{m_\varphi}{2}z\right), \quad (6)$$

where $m_\varphi = \sqrt{8\lambda v^2}$. Now, we assume that when we turn on the interaction between the fields in (1), $\varphi(z)$ holds its general form (6) with the possible scale dilation [8]:

$$\varphi = v\tanh\alpha z. \quad (7)$$

When this ansatz is substituted into the first equation of the system (5), we obtain for $\chi(z)$:

$$\chi = \frac{A}{\cosh\alpha z}. \quad (8)$$

Here the constant A is related to α and the other model parameters by

$$A^2 = \frac{2\lambda v^2 - \alpha^2}{\gamma}. \quad (9)$$

Further substituting (7) and (8) in the second equation of the system (5) and equating coefficients in front of $\cosh^{-1}\alpha z$ and $\cosh^{-3}\alpha z$, we obtain two more relations between the constants:

$$\alpha^2 = 2\gamma(v^2 - \mu^2), \quad \alpha^2 = \gamma(v^2 - 2\beta A^2). \quad (10)$$

Equations (7) and (8), along with the constraints on the model parameters, are the static solution of the system (5). Let us now discuss some features of this solution.

- From equation (9) it follows that $A = 0$ at $\alpha = \sqrt{2\lambda v^2}$, which corresponds to the "bare" kink (6). Besides that, the limit $\gamma \to 0$ can be taken only simultaneously with $\alpha \to \sqrt{2\lambda v^2}$.
- The spatial scales of the fields φ and χ are the same, both of the order of α^{-1}.
- The asymptotics of the fields at $z \to -\infty$ are

$$\varphi \approx -v + 2v\, e^{2\alpha z}, \quad \chi \approx 2A\, e^{\alpha z}. \tag{11}$$

Notice that these expressions coincide with those obtained in [6]. The asymptotics of the fields at $z \to +\infty$ are

$$\varphi \approx v - 2v\, e^{-2\alpha z}, \quad \chi \approx 2A\, e^{-\alpha z}. \tag{12}$$

- On the (φ, χ) plane the solution (7), (8) is a semi-ellipse with semi-axes v and A (we assume that $A > 0$). The ellipse becomes a line segment at $\alpha = \sqrt{2\lambda v^2}$.
- The energy of the solution (7), (8) or, more strictly, the tension of the corresponding plane structure in the three-dimensional space (x, y, z), is given by

$$E = \int_{-\infty}^{+\infty} \left\{ \frac{1}{2}\left(\frac{d\varphi}{dz}\right)^2 + \frac{1}{2}\left(\frac{d\chi}{dz}\right)^2 + \lambda(\varphi^2 - v^2)^2 + \gamma\left[(\varphi^2 - \mu^2)\chi^2 + \beta\chi^4\right] \right\} dz, \tag{13}$$

and is equal to

$$E_{\text{sol}} = \frac{4A^4\beta\gamma + 2v^2(\alpha^2 + 2\lambda v^2) + A^2(\alpha^2 + 2\gamma(v^2 - 3\mu^2))}{3\alpha}. \tag{14}$$

When $A = 0$ (and $\alpha^2 = 2\lambda v^2$), this gives just the energy of the kink (6): $E_{\text{k}} = \dfrac{4\alpha v^2}{3} = \dfrac{2m_\varphi v^2}{3}$.

3. A remark on the linear stability of the solution $\varphi_{\text{k}}(z)$, $\chi \equiv 0$

It is easy to see that $\varphi = \varphi_{\text{k}}(z)$, $\chi \equiv 0$ satisfies the equations of motion following from the Lagrangian (1). Now we will investigate the linear stability of this configuration. To do this, we insert $\varphi = \varphi_{\text{k}}(z)$, $\chi = 0 + \delta\chi(t, z)$, where $\delta\chi(t, z) = \eta(z)\cos\omega t$, in the equations of motion, and get a time-independent Schrödinger-like spectral problem with the spectral parameter ω^2:

$$-8\lambda v^2 \eta_{\zeta\zeta} + 2\gamma\left[v^2 \tanh^2\left(\frac{\zeta}{2}\right) - \mu^2\right]\eta = \omega^2\eta, \tag{15}$$

here $\zeta = m_\varphi z$, and $\eta = \eta(\zeta)$. The eigenvalues of ω are:

$$\omega_n^2 = m_\varphi^2\, \varepsilon_n, \quad \varepsilon_n = \left(1 - \frac{\mu^2}{v^2}\right)\frac{\gamma}{4\lambda} - \frac{1}{16}\left(-1 - 2n + \sqrt{1 + \frac{4\gamma}{\lambda}}\right)^2. \tag{16}$$

The discrete part of the spectrum is defined by $n < s = \dfrac{1}{2}\left(-1 + \sqrt{1 + \dfrac{4\gamma}{\lambda}}\right)$, with n starting from $n = 0$. In particular, with the choice of the model parameters as in [6], we obtain:

$$s \simeq 2.37228, \tag{17}$$

i.e. there are three levels in the discrete part of the spectrum:

$$\epsilon_0 = -0.0119297, \quad \epsilon_1 = 0.924211, \quad \epsilon_2 = 1.36035. \tag{18}$$

The presence of the negative value $\epsilon_0 < 0$ means $\omega_0^2 < 0$, i.e. the configuration $\varphi = \varphi_{\text{k}}(z)$, $\chi \equiv 0$ is unstable. Notice that our value of ϵ_0 differs from that reported in [6].

4. Conclusion

We have considered a model with a real scalar field and a coupled scalar triplet with the Lagrangian (1). We found an exact analytic solution in the form of a domain wall with the coupled scalar triplet localized on the wall. We discussed some properties of this solution and also analysed the linear stability of the bare domain wall, and analytically showed that this bare domain wall solution is unstable.

Acknowledgments

This work was supported in part by the Russian Federation Government under grant No. NSh-3830.2014.2. V. A. Gani acknowledges support of the Ministry of Education and Science of the Russian Federation, Project No. 3.472.2014/K. M. A. Lizunova thanks the ITEP support grant for junior researchers and gratefully acknowledges financial support from the Dynasty Foundation.

This work was performed within the framework of the Center of Fundamental Research and Particle Physics supported by MEPhI Academic Excellence Project (contract No. 02.a03.21.0005, 27.08.2013).

References

[1] Shifman M and Yung A 2013 Abrikosov-Nielsen-Olesen string with non-Abelian moduli and spin-orbit Interactions *Phys. Rev. Lett.* **110** 201602

[2] Shifman M 2013 Simple models with non-Abelian moduli on topological defects *Phys. Rev.* D **87** 025025

[3] Monin S, Shifman M and Yung A 2015 Non-Abelian string of a finite length *Phys. Rev.* D **92** 025011

[4] Monin S and Shifman M 2014 Degeneracy between Abelian and non-Abelian strings *Int. J. Mod. Phys.* A **29** 1450105

[5] Gorsky A and Mikhailov V 2007 Non-Abelian strings in a dense matter *Phys. Rev.* D **76** 105008

[6] Kurianovych E and Shifman M 2014 Non-Abelian moduli on domain walls *Int. J. Mod. Phys.* A **29** 1450193

[7] Belova T I and Kudryavtsev A E 1997 Solitons and their interactions in classical field theory *Phys. Usp.* **40** 359-86

[8] Lensky V A, Gani V A and Kudryavtsev A E 2001 On $U(1)$-charged domain walls *J. Exp. Theor. Phys.* **93** 677-84

Studies of the possibility to use Gas Pixel Detector as a fast trigger tracking device

N Sinev[1], G Bashindzhagyan[2], N Korotkova[2], A Romaniouk[3] and V Tikhomirov[3,4]

[1] University of Oregon, Eugene, OR 97403-1274, USA
[2] M.V.Lomonosov Moscow State University, Leninskie gory, GSP-1, Moscow 119991, Russia
[3] National Research Nuclear University MEPhI (Moscow Engineering Physics Institute), Kashirskoe highway 31, Moscow, 115409, Russia
[4] P. N. Lebedev Physical Institute of the Russian Academy of Sciences, Leninskij Prospekt 53, Moscow, 119991, Russia

E-mail: sinev@slac.stanford.edu, george.bashindzhagyan@cern.ch

Abstract. Gas Pixel Detector (GPD) technology offers new possibilities, which make them very attractive for application in existing and future accelerator experiments and beyond. GPDs combine advantages of silicon and gaseous detectors. They can be produced radiation hard and with low power consumption using relatively cheap technology. Low capacitance of the individual pixel channel allows us to obtain a large signal to noise ratio. Using a time projection method for GPD readout one obtains 3D track image with precise coordinate (31 μm) and angular information (0.40°). This feature would allow us to achieve performance of one GPD layer equal to a few layers of silicon detectors. Implementation of a fast readout and data processing at the front-end level allows one to reconstruct a track segment in less than 1 μs, and to use this information for the first level trigger generation. The relevant algorithms of data acquisition and analysis are described and the results of simulations are presented in this paper.

1. Introduction

Gas pixel detectors (GPD) have been proposed and developed to combine the advantages of a gas chamber and a small pixel readout techniques which is required for future accelerator experiments [1-5]. One GPD consists of cathode and anode planes with 10 - 16 mm drift gap between them filled with gas. The schematic view of GPD structure is shown on figure 1.

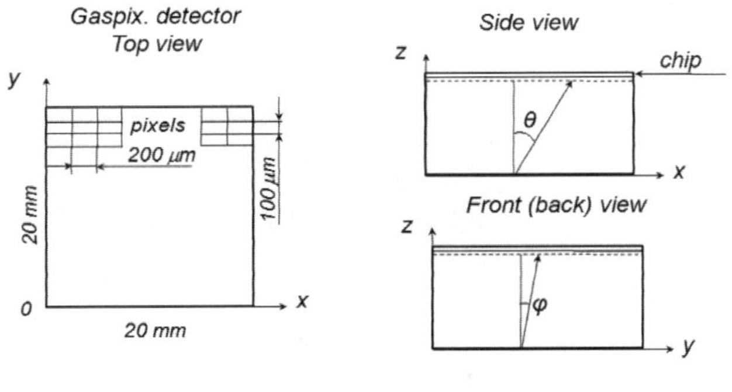

Figure 1. The schematic view of GPD structure.

Anode plane is a large electronic chip with pixel sensors and integrated electronics. There is a mesh in front of the sensors. About 50 μm space between the sensors and mesh is used for gas amplification.

The goal of presented work is to develop readout concept and analysis algorithms allowing to determine a particle trajectory in less than 1 μs and to use this information for a fast trigger generation.

Relatively large 100×200 μm^2 pixels have been selected for simulations to decrease the number of channels. Non square pixel size is aimed to achieve better angular resolution in the plane perpendicular beam axis and to improve an accuracy of p_T measurements in magnetic field. A 20×20 mm^2 chip size and 10 mm gap have been assumed. The other parameters were: number of pixels on X axis – 100, number of pixels on Y axis – 200, voltage between cathode and anode – 4000 V, gas amplification ≥ 1000, electron drift time from cathode to mesh – 200 ns, transverse electron diffusion during 200 ns – 300 μm, readout cycle – 25 ns.

Binary readout has been selected to speed up data acquisition and analysis. Gas and electronic amplifications ensure effective registration of every electron crossing the mesh.

2. Basic approaches to the time-projection method

Electrons generated by an ionizing particle in a gas drift along Z axis transversely to the chip surface (XY plane) independently of the particle trajectory. Electron drift velocity is constant (it does not depend on Z coordinate), because the electric field is uniform between the cathode and mesh. Figure 2 shows XZ projection of the particle track. The Z axis in this figure has two

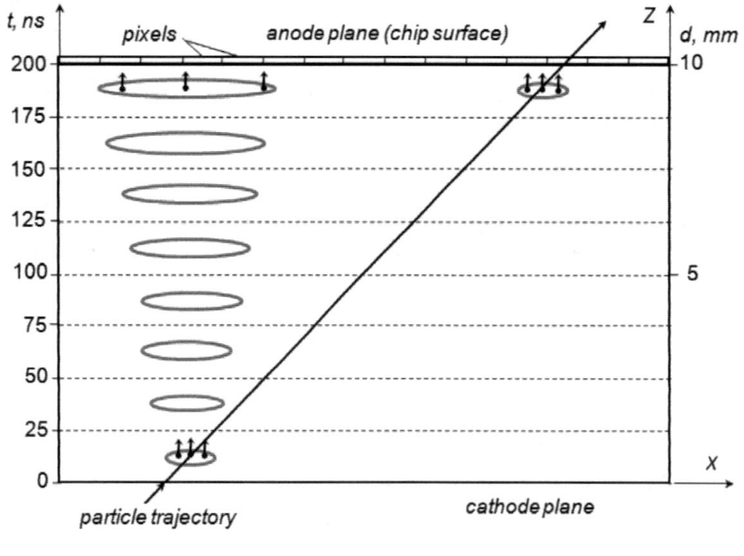

Figure 2. Electrons collecting on the GPD chip plane.

scales: one is the distance from an electron generation point to the chip surface, 0 - 10 mm, and the second one is the corresponding drift time 0 - 200 ns. The track angle to Z axis in this projection we call θ.

Let's assume that the particle crosses the detector at the time $t_0 = 0$. We can neglect the time difference between the particle entering and leaving detector, because it is in the range 0.03 – 0.05 ns, which is very small compared to readout time intervals of 25 ns. It is clear, that the electrons generated close to cathode, will drift to anode in the maximum time, and arrive there at t = 200 ns. But for electrons generated very close to the grid, the drift time will be practically zero. Thus during the 200 ns period all the electrons generated by the particle will be gathered and registered on the chip.

Taking into account that every readout cycle is equal 25 ns, all the available information related to the particle trajectory will be received in 8 cycles. The time scale in figure 2 is divided into 8 parts of 25 ns each, which corresponds to 1/8 of 10 mm or 1.25 mm in distance from cathode to anode. In such a way, the location of fired pixels in each of 8 time intervals reflects the particle coordinates in space (in XY plane) on each of 8 distances from chip surface. In the time interval t = 0 - 25 ns, the average drift time 25 / 2 = 12.5 ns corresponds to the particle-to-chip distance 1.25 / 2 = 0.625 mm. In other words, the average values of X and Y coordinates of the pixels fired during this interval (0 - 25 ns) represent the position of the particle in the detector at the third coordinate Z = 0.625 mm. Subsequently, the location of pixels fired in each of 8 readout cycles give us 8 measurements of 3D particle position within the detector volume, and makes it possible to reconstruct the particle trajectory.

3. Front-end electronics and readout algorithm

The process of transverse diffusion spreads the electrons around the particle trajectory during their up to 200 ns drift through the gas volume. Therefore up to 50 pixels can be fired during 25 ns readout cycle. It allows that this improves the accuracy of the measurements with relatively large pixels, but creates a real problem to transmit and analyze the coordinate of every fired pixel online. Therefore another method of data acquisition and analysis has been used. Every pixel has an integrated fast preamplifier, discriminator and flip-flop (FF) with two fixed current outputs connected to common X and Y lines going along the particular X_i and Y_i rows correspondingly (see figure 3). If a pixel is fired during a cycle, FF turns ON and current flows along the line. If a few (up to 15) pixels have been fired in the same row in the same readout circle, the sum of their currents appears in the corresponding line as shown in figure 4. Because all the pixels generate equal currents the sum is linearly proportional to the number or fired pixels. The end of each line is connected to a 4-bit flash ADC integrated in the X and Y processing blocks PBX and PBY positioned on the two edges of the chip. At the end of each cycle a special command is

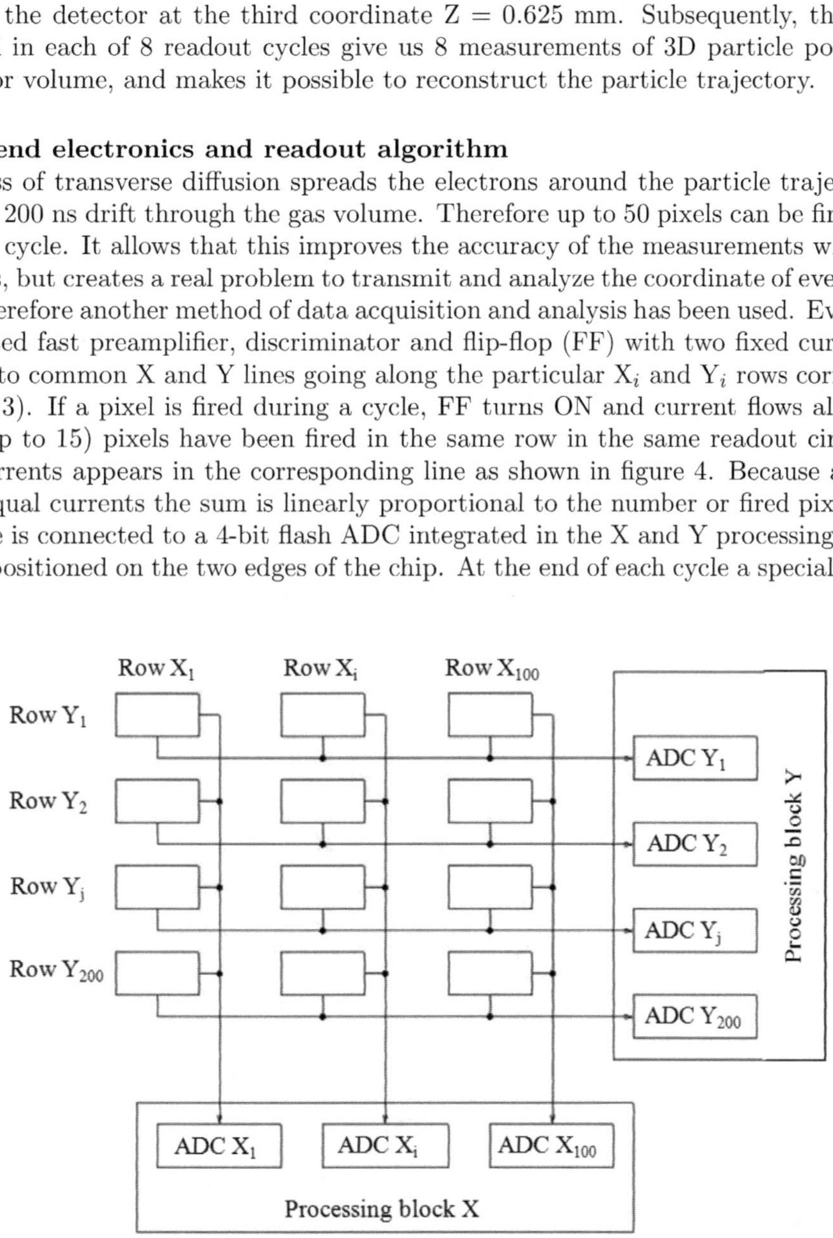

Figure 3. The block diagram of the readout algorithm.

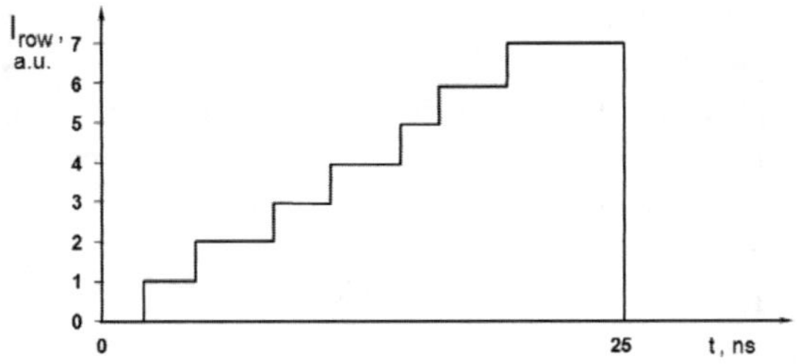

Figure 4. An example of the current in the row line if 7 pixels are fired during one 25 ns cycle.

generated and all the ADCs convert the current into the number of fired pixels in this row. The same command returns all the FFs into starting position preparing the channel to next readout cycle. This method allows us transmit the information from the row to the ADC using only one wire.

4. Reconstruction algorithm

If particle crossed the detector perpendicularly to the chip surface, both angles θ and ϕ are zero and the mean position of fired pixels cluster in XY coordinates is the same for all 8 cycles. Only the number of fired pixels increases from 1-st to 8-th cycle because of transverse diffusion. If one or both angles are not zero the cluster moves along axis X or Y or both of them reflecting particle trajectory within the detector. Figure 5 illustrates a typical cluster structure in XY plane and its projections on both axes for the 3-rd of 8 cycles. To calculate particle position in row number units we use center of mass approach. In each projection the row number R^i_X or R^j_Y is multiplied by the number of fired pixels in this row, n^i_X or n^j_Y, and the sum of the products is divided by the total number of fired pixels in the cluster N_P: $X=\Sigma R^i_X n^i_X/N_P$, $Y=\Sigma R^j_Y n^j_Y/N_P$.

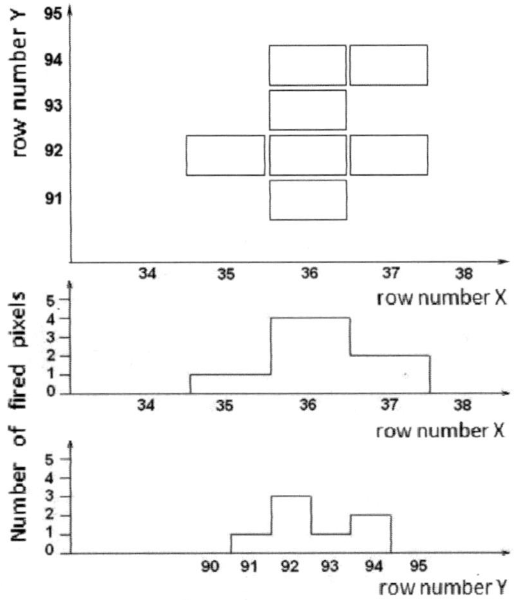

Figure 5. The example of the cluster structure in XY plane and its projections on X and Y axes.

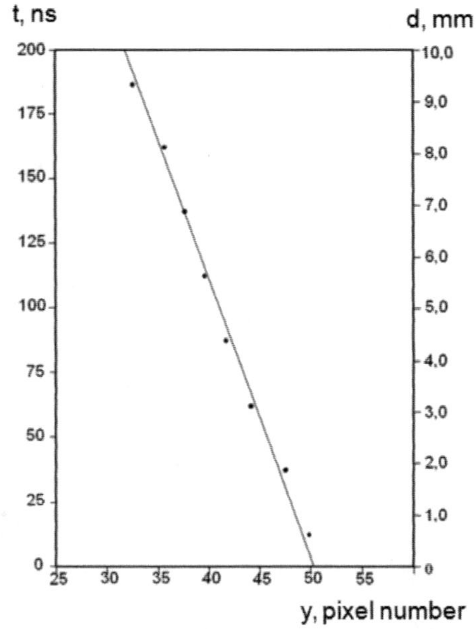

Figure 6. The example of the partice trajectory reconstruction.

This procedure is performed for each of 8 cycles. And the cycle number indicates the corresponding Z coordinate. Thus we can create two plots representing the XZ and YZ projections of the particle trajectory with 8 sets of found coordinates shown as dots in each plot. An example of one simulated event is presented in figure 6, where a straight line is drawn through 8 points using the least-squares method. From there we can determine the particle trajectory, i.e. XY coordinates where the particle crossed the cathode and anode planes as well as both angles: θ in the XZ plane and ϕ in YZ plane.

5. Expected resolution

To estimate the expected resolution dedicated MC simulation has been done. Figure 7 shows the example of simulated distributions for the 20 GeV pions crossing the detector with $\theta = 20°$ and $\phi = 3°$. It is clearly seen that the average values of the reconstructed angles ($\theta = 20.06°$ and $\phi = 3.00°$) is very close to the initial ones, and the r.m.s is about 0.4°. For the reconstructed entry points r.m.s. is 0.26 of pixel size for X and 0.31 for Y coordinates which corresponds to 52 and 31 μm respectively.

Further simulations will be performed with improved algorithm and better resolution is expected.

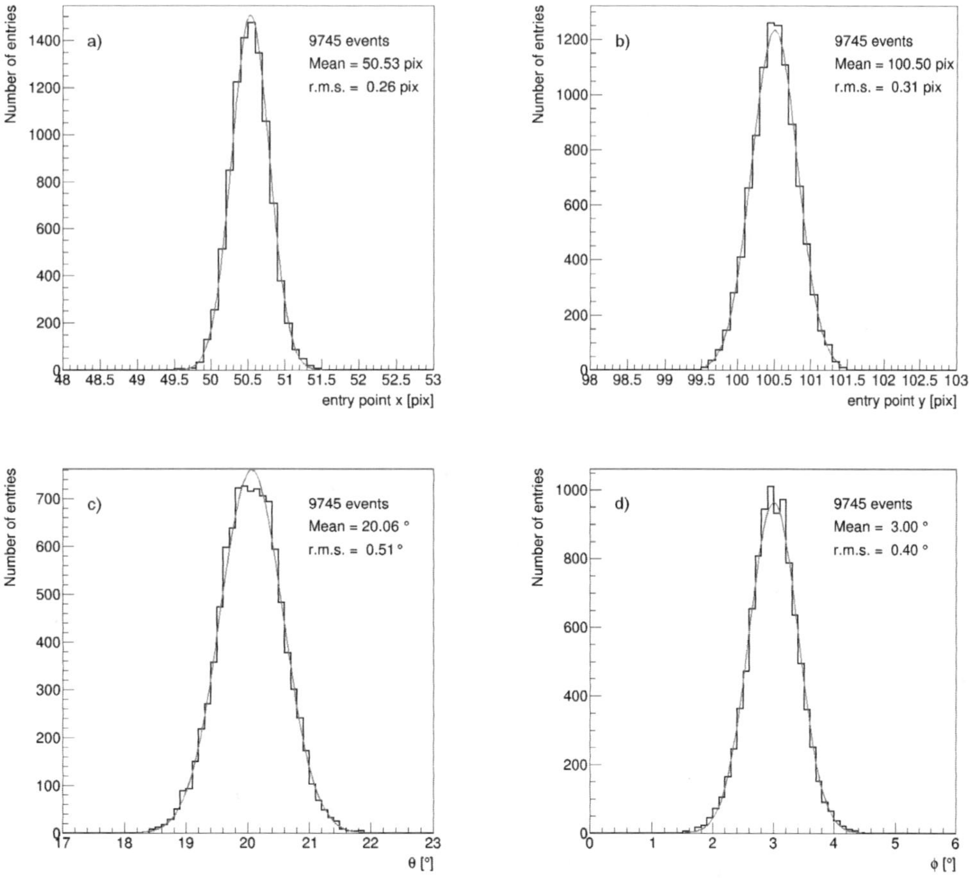

Figure 7. The distributions of the reconstructed track entry points coordinates: **a**) X; **b**) Y; and angles: **c**) θ; **d**) ϕ for 20 GeV pions crossing the GPD with $\theta = 20°$ and $\phi = 3.00°$.

6. Conclusion

The results of the simulations show that particle trajectory can be reconstructed with angular resolution 0.40° and space resolution 31 μm using only one layer of GPD. The proposed algorithms allow using online data analysis for a fast trigger generation.

Acknowledgments

This work was partially supported by Russian Foundation for Basic Research (grant No. 14-22-0353) and by MEPhI Academic Excellence Project (contract No. 02.a03.21.0005, 27.08.2013).

References

[1] Ballazzini R *et al.* 2004 *Nucl. Instrum. Meth.* A **535** 477
[2] Chefdeville M *et al.* 2006 *Nucl. Instrum. Meth.* A **556** 490
[3] Blanco Carballo V *et al.* 2010 *JINST* **5** 02002
[4] Blanco Carballo V *et al.* 2011 *Nucl. Instrum. Meth.* A **629** 118
[5] Hartjes F *et al.* 2013 *Nucl. Instrum. Meth.* A **706** 59

Characterization of the low-background Hamamatsu R11410-20 cryogenic PMTs for the RED100 detector

D Yu Akimov[1, 2], A I Bolozdynya[1], Yu V Efremenko[1, 3], V A Kaplin[1], A V Khromov[1], Yu A Melikyan[1,4,*] and V V Sosnovtsev[1]
[1] National Research Nuclear University MEPhI (Moscow Engineering Physics Institute), Kashirskoe highway 31, Moscow, 115409, Russia
[2] SSC RF Institute for Theoretical and Experimental Physics, 25 Bolshaya Cheremushkinskaya, Moscow, 117218, Russia
[3] University of Tennessee, 1408 Circle Park Drive, Knoxville, Tennessee, U.S.A.

E-mail: *velyour@yandex.ru

Abstract. The RED100 two-phase liquid xenon emission detector for neutrino coherent scattering experiments is equipped with 38 Hamamatsu R11410-20 photomultiplier tubes capable to operate at cryogenic temperatures and made of low background materials. A dedicated characterization procedure has been carried out for each PMT unit to be installed into the detector. The results presented here include single photoelectron analysis, gain curves for a wide range of the bias voltage values, data on dark count rate for 34 PMT samples. Peculiar noise characteristics of selected PMT units are analysed and discussed.

1. Introduction

The process of coherent elastic neutrino-nucleus scattering (CEvNS) was predicted by Standard model but has never been observed [1]. A broad experimental program aimed on study of this effect has been developed, utilizing different substances as working media, including the germanium detector technology [2], low-background CsI technology [3] and xenon emission detector technology, which was chosen by the RED collaboration [4] as the most relevant way to observe CEvNS.

For this purpose the RED100 two-phase cryogenic emission detector filled with ~200 kg of xenon has been developed and constructed at MEPhI to be irradiated by a pulsed neutrino beam at the Spallation Neutron Source (Oak Ridge National Laboratory, USA). An act of coherent neutrino scattering off Xe nucleus in the detector's volume leads to several keV of kinetic energy transfer to the Xe nucleus, causing a scintillation flash in the point of interaction and ionization electrons emission. Ionization electrons can be pulled up to a thin (~1 cm) Xe gas phase in the upper part of the detector with the help of a strong electric field (~500 V/cm) to be further accelerated there and generate an electroluminescent flash.

Light from the both signals is registered by two photodetecting arrays of 19 Hamamatsu R11410-20 photomultiplier tubes each. The tubes have been specially developed for their operation in low-background liquid xenon environments, including the optimization of their radiopurity, pressure-resistance and single photon response parameters. The main results of the these parameters check for

[4] To whom any correspondence should be addressed.

first 34 PMTs installed in the RED100 detector are presented, and the observed deviations in the dark count rate of several PMT units are discussed.

2. Single photon response

Both scintillation and luminescent signals of interest in RED100 are rather faint (up to several hundreds of photoelectrons detected by all PMTs), that's why the knowledge of the PMTs' single photon detection capabilities is important. The LED-induced single photoelectron spectra have been measured for each PMT. The example spectra obtained for one of the PMTs for different bias voltages values are shown in figure 1.

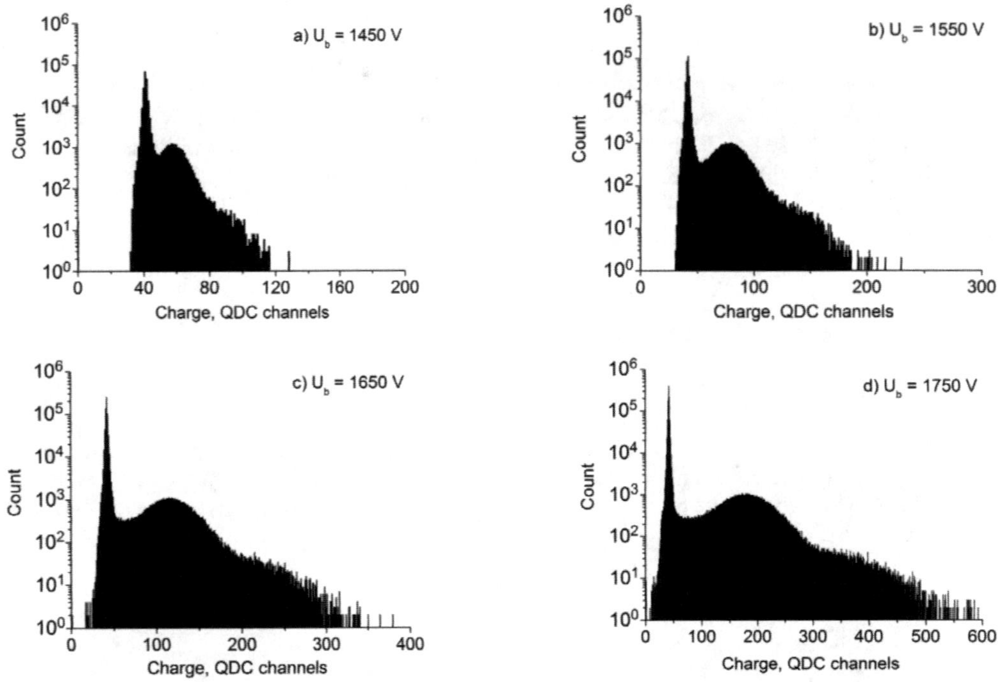

Figure 1. Typical LED-induced single photoelectron spectra, measured with one of the tested PMTs at different bias voltage values.

The "peak-to-valley" ratio is an important parameter of single photoelectron response of a PMT. It is calculated as the single photoelectron peak's maximum count divided by the minimum count in the "valley" between the peak and the pedestal. For the bias voltage values greater than 1500 V no significant difference could be seen in the peak-to-valley ratios of the obtained spectra, which are equal to ~3.7. For optimal bias voltage values, the smallest obtained peak-to-valley ratio for all the tested PMTs is 1.9, while 4.0 is the biggest one.

Once the single photoelectron spectra have been measured for different voltage values, it became possible to plot the distribution of dependencies of single photoelectron mean amplitude on the bias voltage values over all tested PMTs. The resulting plot is shown in figure 2.

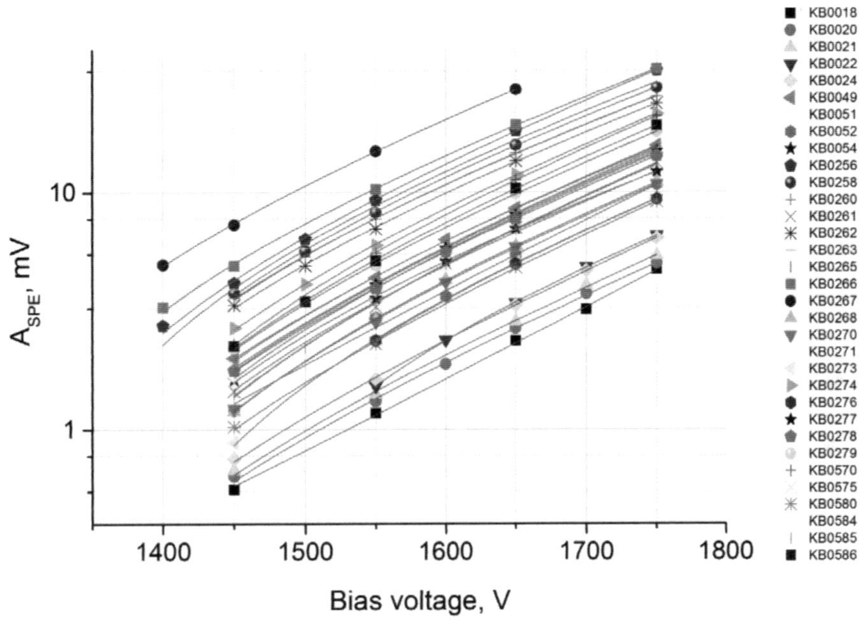

Figure 2. Dependencies of single photoelectron signals mean amplitude (A_{SPE}) on the bias voltage values over all the tested PMTs. Error bars lie within the data points size. Data points are connected using exponential fits (red solid curves).

This plot could help in gain-matching of all the PMTs used in the experiment, and also in setting the discriminator threshold at the level of 1/3 of single photoelectron amplitude while measuring the dark count rate values.

3. The dark count rate

Dark counts in a PMT are mainly caused by thermionic emission from the photocathode's surface [5], thereby having single photoelectron amplitude. Dark signals with greater amplitudes could be generated in a PMT in the processes of ion feedback, described in details for the Hamamatsu R11410-20 PMTs by reference [6]. The other way for large amplitude dark signals generation in a PMT is concluded in a charged particle penetration through the volume of a PMT window, which could lead to Cherenkov light emission. Combination of quartz window and VUV-sensitive photocathode in R11410-20 leads to very large amplitude (hundreds of photoelectrons) dark signals generation in case of atmospheric muons passing through the PMT's window [7].

Figure 3 represents the typical Hamamatsu R11410-20 PMT integral dark count rate amplitude spectrum measured by a scaler at different levels of discriminator's thresholds, confirming the predominance of dark pulses with single photoelectron amplitude.

Distribution of the gain-matched PMTs' dark count rates at discriminator threshold level of 1/3 of single photoelectron amplitude is shown in figure 4. The measurements were done in a light-tight metallic box (2 mm metal thickness) at room temperature after the PMT storage in darkness for at least 16 hours.

As one can see, the dark count rate values for the tested PMTs vary from several hundred hertz to ten kilohertz for almost all the tubes. Data points for the two tubes labeled KB0018 and KB0054 are not shown since dark count rate for these tubes is higher than 100 kHz. Additional studies, described in details here [8], have shown strong evidence of light emission from these PMTs' internal structure, causing these high values of "dark" count rates.

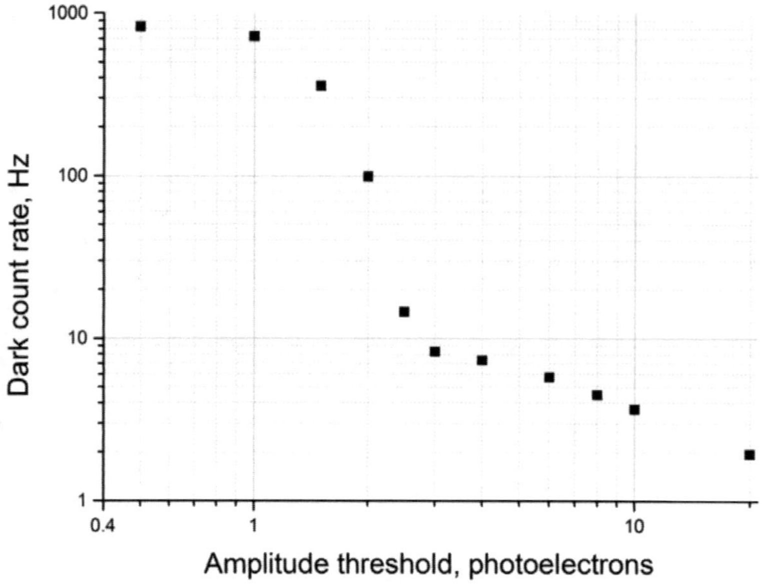

Figure 3. Distribution of dark count rate values measured at different amplitude thresholds. Error bars lie within the data points size.

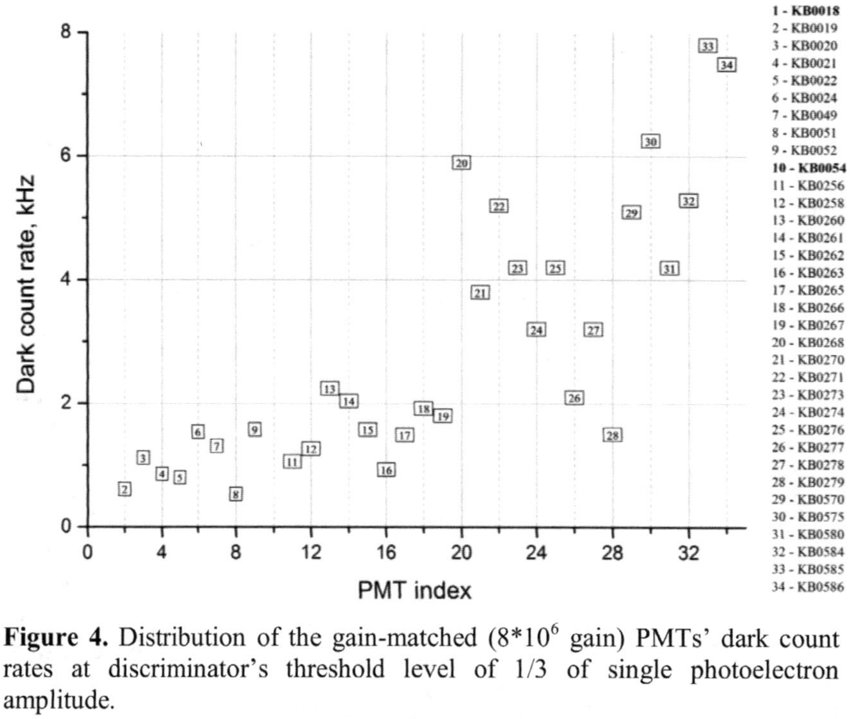

Figure 4. Distribution of the gain-matched ($8*10^6$ gain) PMTs' dark count rates at discriminator's threshold level of 1/3 of single photoelectron amplitude.

The light emission phenomenon is probably caused by specific constructive features of R11410-20 PMTs, concluded in the usage of non-widespread materials, determined by radiopurity reasons: quartz internal insulation, ceramic stem and Kovar metal body [9].

4. Conclusion

The study of the Hamamatsu R11410-20 PMTs parameters showed the propriety of the PMT choice for its use as a basic element of the RED100 photodetecting system due to an excellent single photon detection capability and acceptable noise parameters (for most part of the tested PMTs). It was shown that a predominant part of dark pulses in Hamamatsu R11410-20 is characterized by single photoelectron amplitude, which is consistent with the assumption of thermionic nature of major part of dark pulses in a PMT. At the same time, certain anomalies in the noise characteristics of several PMT units are described. These anomalies are associated with light emission from the PMTs' internal structure, which could be caused by a number of specific constructive features of these PMTs. The tested PMTs could be used in the RED100 detector after preliminary selection.

Acknowledgments

This study was supported by RF Government under contracts of NRNU MEPhI with the Ministry of Education and Science of №11.G34.31.0049 from October 19, 2011 and within the framework of the Centre for Nanostructured Electronics supported by NRNU MEPhI Academic Excellence Project (contract № 02.a03.21.0005, 27.08.2013).

References

[1] Gaponov Yu V and Tikhonov V N 1977 *Sov. J. Nucl. Phys.* **26** (3) 314 (original Russian text: *Yad. Fiz.* **26** 594)
[2] Wong H T 2011 *J. Phys.: Conf. Ser.* **309** 012024
[3] Mikhailik V B and Kraus H 2006 *J. Phys.* D **39** 1181
[4] Akimov D Yu *et al.* 2013 *JINST* **8** P10023
[5] Burle Industries Inc. 1980 *Photomultiplier handbook* USA
[6] Akimov D Yu *et al.* 2015 *Instrum. exp. tech.* **3** 097101 (original Russian text: *Pribory i Tekhnika Eksperimenta* 2015 **3** 097101)
[7] Melikyan Yu A Talk at the International conference on new photodetectors on July 7, 2015, Moscow, Russia
[8] Akimov D Yu *et al.*2015 *Nucl. Instrum. Meth.* A **794** 0102
[9] Yuji Hotta Talk at the DM2014 conference, February 28, 2014

International Conference on Particle Physics and Astrophysics (ICPPA-2015) IOP Publishing
Journal of Physics: Conference Series **675** (2016) 012023 doi:10.1088/1742-6596/675/1/012023

High-energy cosmic antiparticle excess vs. isotropic gamma-ray background problem in decaying dark matter Universe

V V Alekseev[1], K M Belotsky[2], Yu V Bogomolov[1], R I Budaev[2], O A Dunaeva[1], A A Kirillov[1,2], A V Kuznetsov[1], M N Laletin[2], A D Lukyanov[1], V V Malakhov[2], A G Mayorov[1,2], M A Mayorova[2], A F Mosichkin[1], A A Okrugin[1], S A Rodenko[2] and A M Shitova[1]

[1] Yaroslavl State P.G. Demidov University, Sovietskaya st. 14, Yaroslavl, 150000, Russia
[2] National Research Nuclear University MEPhI (Moscow Engineering Physics Institute), Kashirskoe highway 31, Moscow, 115409, Russia

E-mail: maxim.n.laletin@gmail.com

Abstract. We are going to show that any conventional decaying dark matter model, providing an explanation of cosmic antiparticle excess observed by PAMELA and AMS-02, inevitably faces the contradiction with isotropic diffuse gamma-ray background, measured by FERMI/LAT.

1. Introduction

Charged cosmic particles seems to be one of the most essential source of information about our Galaxy, it's structure and composition. In spite of the evident role, which charged cosmic rays play in astrophysics, they also provide a unique tool to search for new physics. An indeed sensitive "probe" to reveal the presence of some exotic objects inside our Galaxy (like dark matter particles, microscopic black holes, Q-balls and etc.) is the abundance of high-energy charged antiparticles (HECA), since their production rate in ordinary astrophysical phenomena rapidly falls to zero with the growth of energy.

The unceasing experimental attempts of studying the antiparticle cosmic rays component had lead to the discovery of high-energy positron excess in PAMELA experiment [1]. The same apparatus has in addition observed the hints of high-energy antiproton excess [2]. Recently AMS-02 experiment (which has also detected the increase of positron fraction [3]and measured it more accurately) confirmed the PAMELA's guess on antiprotons [4]. Thus we have an unexplained surplus over secondaries in both antiproton and positron spectrum, which insinuates the existence of unknown primary sources of energetic antiparticles.

Though a few articles (see for instance [5, 6]) claim that there's no excess at all, but only secondaries, arising due to the uncertainties in cosmic ray propagation model (especially in antiprotons), theoretical endeavours in understanding the nature of HECA source can be mainly divided in two groups. Authors belonging to the first group (e.g. [7, 8]) state that well-known astrophysical objects like old supernova remnants and pulsars are responsible for the aforementioned effects. The second ones (e.g. [9–11]) turn to a new and undiscovered essense (mostly dark matter) to explain the experimental results.

In this work we are going to consider a lax, though demonstrative, proof that any reasonable dark matter model (with conventional isotropical distribution of unstable or annihilating component) producing enough charged antiparticles to explain PAMELA and AMS-02 results is strictly constrained by FERMI/LAT data on diffuse isotropic gamma-ray background [12].

2. Cosmic rays from dark matter decays

Let us consider a common phenomenological dark matter model, where a fraction of dark matter ξ consist of metastable particles X with mass M_X and lifetime τ (in fact an independent parameter of our model defining the fluxes of HECA is ξ/τ, so one can easily increase the amount of metastable dark matter by increasing it's lifetime or vice versa), decaying to various SM particles (for our purposes it is enough to introduce only e^-e^+, $\mu^-\mu^+$, $\tau^-\tau^+$ and $u\bar{u}$ channels since any other rational decay mode like $b\bar{b}$, ZZ or W^-W^+, providing charged antiparticles couldn't significantly improve the result) with different branching ratios. Our aim is to fit the data on charged cosmic particles (especially antiparticles), taking into account the effects of their propagation in the Galaxy (diffusion, re-acceleration, energy losses and etc.). To calculate the propagation of charged particles (including "background" fluxes, such as primary electrons and protons of common astrophysical origin and secondary positrons, electrons and antiprotons from spallation of cosmic rays on interstellar media) and gamma rays brought by these charged particles, traveling through magnetic halo of the Galaxy (inversed compton scattering (ICS) photons and bremmstrahlung), we used GALPROP [13]. In our analysis we assumed that the dark matter density profile is Navarro-Frenk-White (NFW):

$$\rho_{\rm NFW}(r) = \frac{\rho_0}{(r/r_s)\left[1 + r/r_s\right]^2} \tag{1}$$

with scale radius $r_s = 24$ kpc and the factor ρ_0 was chosen to reproduce the local dark matter density $\rho_\odot = 0.39$ GeV/cm^3 at $R_\odot = 8.5$ kpc [14]. The chosen values of propagation parameters corresponds to MED model [15]. The primary (injection) spectra of cosmic rays was simulated using Pythia 8.1 [16].

Besides charged particles and antiparticles decaying dark matter component also produce gamma rays (and neutrinos which, however, have nothing to do with the aim of our analysis) and thus contribute to the isotropic diffuse gamma-ray background. This contribution includes prompt radiation (i.e. final-state radiation and gamma rays from meson decays in case of τ-lepton and hadronic modes), coming to us directly from the point where a dark matter decay took place, and radiation induced by "roaming" charged particles, born in these decays (the last component of gamma rays, as we've mentioned, was calculated using GALPROP). The prompt radiation in its turn can be classified by its origin as Galactic and extragalactic radiation. Gamma-ray flux from decays in the Galactic halo can be calculated as

$$F_\gamma^{\rm G}(E) = \frac{\xi}{4\pi M_X \tau \Delta\Omega} \sum_i {\rm Br}_i \frac{dN_\gamma^i}{dE} \int_{\Delta\Omega} d\Omega \int_{\rm line-of-sight} ds\, \rho_{\rm DM}[r(s,b,l)], \tag{2}$$

where $\Delta\Omega$ is the observable angular region of the sky (in our analysis it corresponds to the range of Galactic coordinates $20^\circ \le b \le 90^\circ$ and $0^\circ \le l < 360^\circ$, where FERMI/LAT measures IGRB), $\rho_{\rm DM}$ is the dark matter density distribution (NFW) as a function of galactocentric radius (which can be expressed as $r(s,b,l) = \sqrt{8.5^2 + s^2 - 2\cdot 8.5 s\cos(b)\cos(l)}$), ${\rm Br}_i$ denotes the branching ratio of i-th decay channel and dN_γ^i/dE is the energy spectrum of primary photons, emitted in that channel.

The extragalactic contribution to IGRB from dark matter decays can be estimated as

$$F_{\gamma}^{\mathrm{E}}(E) = \frac{c\xi\Omega_{\mathrm{DM}}\rho_{\mathrm{crit}}}{4\pi M_X \tau} \int_0^{\min[E_{\max}/E,1100]} \frac{dz}{H_0\sqrt{\Omega_\Lambda + \Omega_m (1+z)^3}} \sum_i \mathrm{Br}_i \frac{dN_\gamma^i}{dE} [(1+z)E], \quad (3)$$

where c is the speed of light, $\rho_{\mathrm{crit}} = 4.9$ keV/cm^3 is the critical density of the Universe, $E_{\max} = M_X/2$ denotes the maximal energy of the photon (and $z = 1100$ corresponds to the beginning of recombination epoch), $\Omega_m = 0.315$, $\Omega_{\mathrm{DM}} = 0.265$ and $\Omega_\Lambda = 0.685$ denotes matter, dark matter and dark energy cosmological densities respectively, $H_0 = 67.3$ km s^{-1} Mpc^{-1} is the modern Hubble expansion rate.

3. The problem of gamma-ray overabundance

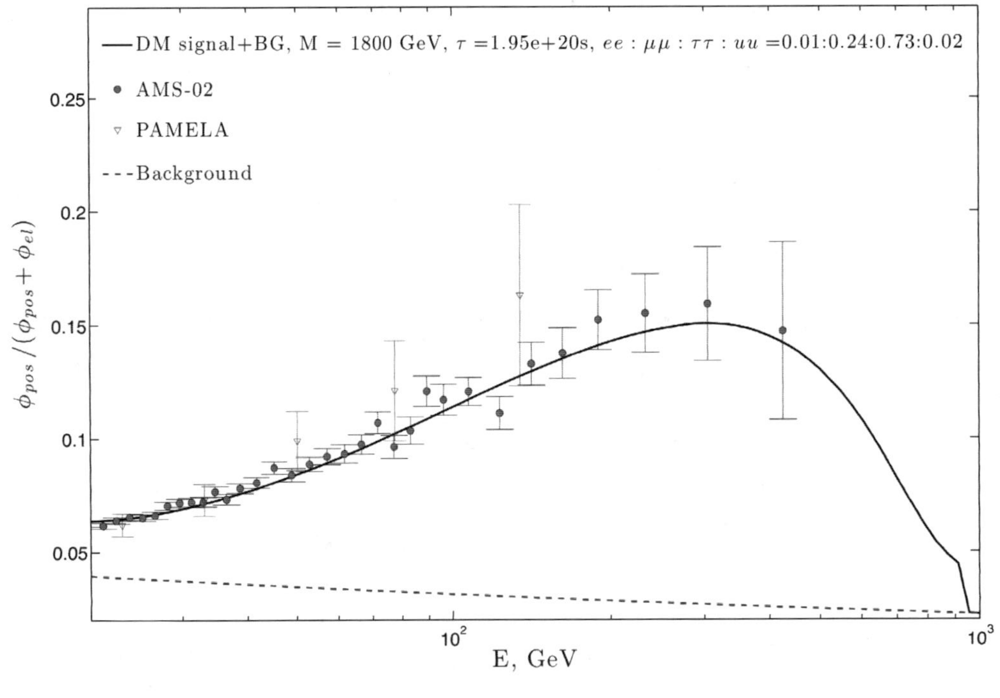

Figure 1. Predicted fraction of positrons in cosmic rays. The values of model parameters corresponds to those listed in the *top left* part of the plot. We compare our results to data in the energy range above 20 GeV, where the effects of solar modulation are insignificant.

As it was noted in [17], any conventional decaying (or annihilating) dark matter model pretending to explain experimental data on both high-energy cosmic antiproton and positron abundance inevitably suffers from diffuse gamma rays "pollution". Here we are going to show a special case, where this problem arise, though it appears to be a global issue (see section 4 for details). Suppose the metastable fraction of dark matter ($\xi \sim 10^{-6}$) has a mass $M_X \sim 1800$ GeV and lifetime $\tau \sim 10^{20}$ s. The exact values of its lifetime and branching ratios were found to adequately reproduce the observed fluxes of HECA. In figures 1-3 we present the predicted positron fraction, antiproton flux and diffuse gamma-ray flux respectively, compared to the corresponding experimental data.

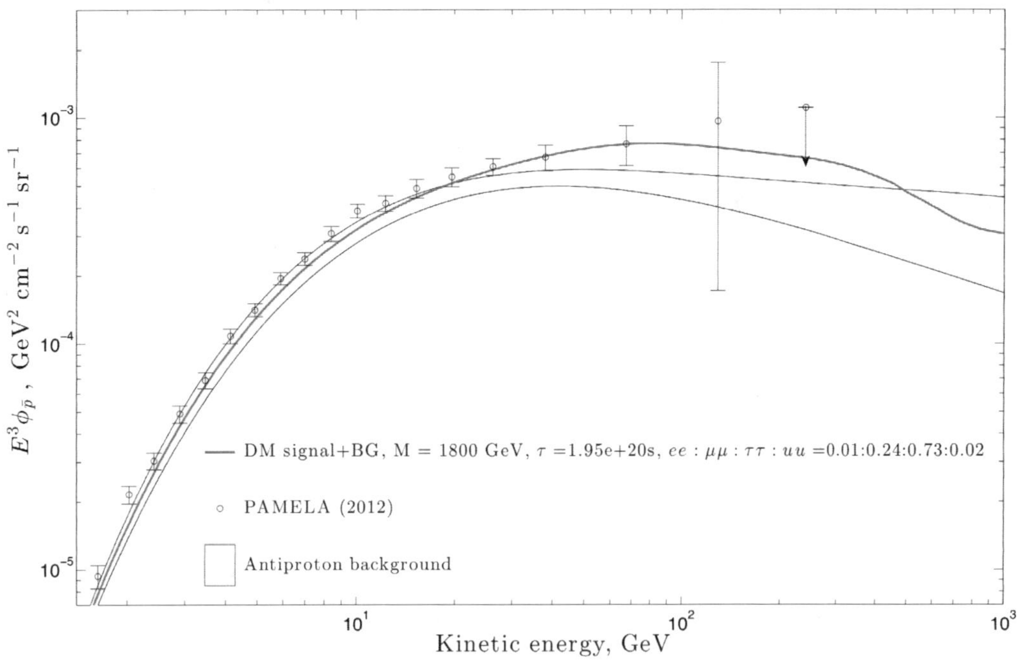

Figure 2. Predicted antiproton flux from dark matter decays, compared to the PAMELA data. The yellow strip corresponds to the flux of secondary and tertiary antiprotons, scattered due to the uncertainties of propagation parameters.

So, as one can clearly see, an attempt to explain the excess of high-energy positrons and antiprotons clashes with IGRB observation (note that in our analysis we do not try to reproduce the IGRB spectrum, but our intention is just not to exceed it). Though we haven't provided a mathematically accurate fit of the data yet, in this case it is quite obvious that a "fine tuning" of the model parameters won't relieve the considering problem.

4. Conclusion

In this work we have demonstrated, how gamma rays accompanying dark matter decays to charged particles complicates the satisfactory explanation of positron and antiproton anomaly in cosmic rays. Despite the fact that we've considered only some special case a number of reasons let us suppose that this problem covers any reasonable dark matter model, assuming isotropical distribution of particles and a set of propagation parameters, consistent with B/C ratio. First of all, variation of mass parameter can not change the situation substantially, because lower values of mass ($\lesssim$ 1 TeV) harden the explanation of high-energy positrons and antiprotons, while greater values of mass ($>$ 1800 GeV) are strictly constrained by IGRB data. Since positron flux is almost insensitive to the choice of propagation model (MIN, MED or MAX) and since the drastic contribution into the gamma-ray spectrum comes mostly from leptonic modes variation of propagation parameters do not help either. The difference between annihilations and decays (as well as the difference between various dark matter distribution profiles) also doesn't affect the gamma-ray spectrum, if the model provides a statistically acceptable fit of (at least) positron data. All these features though should be taken into account while doing the global statistical

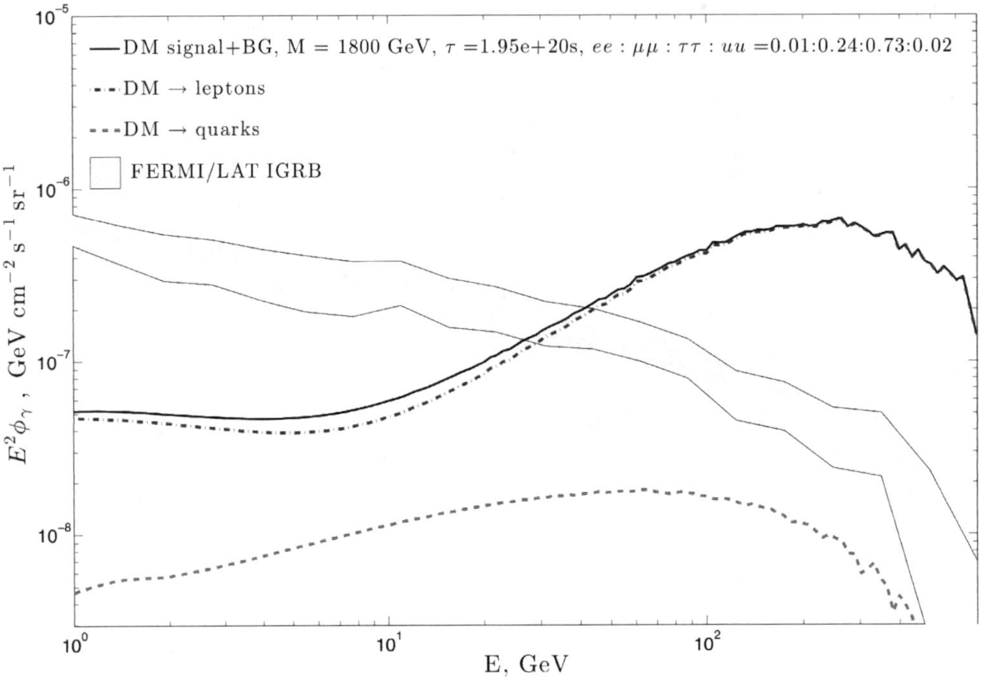

Figure 3. Predicted diffuse gamma-ray flux from dark matter decays, compared to the FERMI/LAT data. The contributions from leptonic (blue dot-dashed) and hadronic (red dashed) channels are shown separately. The flux is averaged over the solid angle corresponding to the range of Galactic coordinates $20° \leq b \leq 90°$ and $0° \leq l < 360°$, in which FERMI/LAT measures IGRB. The uncertainties of IGRB are related to the arbitrary choice of foreground model.

analysis, which will be the next step of our research. Also we propose a possible solution to the discussed problem, requiring a reconsideration of conventional isotropic distribution of decaying dark matter [18].

Acknowledgments

The research is financed by the grant of the Russian Science Foundation (Project No. 15-12-10039). The work by MEPhI group (K M Belotsky, R I Budaev, M N Laletin, V V Malakhov, M A Mayorova and S A Rodenko) on the calculation of cosmic ray fluxes from dark matter decays was performed within the framework of the Center of FRPP supported by MEPhI Academic Excellence Project (contract No. 02.a03.21.0005, 27.08.2013).

References

[1] Adriani O *et al.* (PAMELA Collaboration) 2013 *Phys. Rev. Lett.* **111** 081102
[2] Adriani O *et al.* (PAMELA Collaboration) 2012 *JETP Lett.* **96** 621
[3] Accardo L *et al.* (AMS Collaboration) 2014 *Phys. Rev. Lett.* **113** 121101
[4] AMS Days at CERN and Latest Results: *http://www.ams02.org/2015/04/ams-days-at-cern-and-latest-results-from-the-ams-experiment-onthe- international-space-station*
[5] Giesen G *et al.* 2015 AMS-02 antiprotons, at last! Secondary astrophysical component and immediate implications for Dark Matter *arXiv*:1504.04276

[6] Kappl R, Reinert A and Winkler M W 2015 AMS-02 Antiprotons Reloaded *arXiv*:1506.04145
[7] Kohri K *et al.* 2015 Can we explain AMS-02 antiproton and positron excesses simultaneously by nearby supernovae without pulsars nor dark matter? *arXiv*:1505.01236
[8] Feng J and Zhang H H 2015 Pulsar interpretation of the lepton spectrums measured by AMS-02 *arXiv*:1504.03312
[9] Belotsky K *et al.* 2014 *Adv. High Energy Phys.* **214258** 10
[10] Geng C Q *et al.* 2015 *Phys. Rev.* D **91** 095006
[11] Basak T *et al.* 2015 Explaining AMS-02 positron excess and muon anomalous magnetic moment in dark left-right gauge model *arXiv*:1501.06193
[12] Ackermann M *et al.* (Fermi LAT Collaboration) 2015 *Astrophys. J.* **86** 799
[13] The GALPROP code for cosmic-ray transport and diffuse emission production: *http://galprop.stanford.edu*
[14] Ibarra A, Tran D and Weniger C 2013 *Int. J. Mod. Phys.* **A28** 1330040
[15] Maurin D *et al.* 2001 *Astrophys. J.* **555** 585
[16] Sjostrand T, Mrenna S and Skands P 2008 *Comput. Phys. Commun.* **178** 852
[17] Alekseev V *et al.* 2015 *Physics Procedia* **74C** 48
[18] Alekseev V *et al.* 2015 On a possible solution to gamma-ray overabundance arising in dark matter explanation of cosmic antiparticle excess (*these proceedings*)

Gas mixtures for quality control of the sTGC chambers

K A Filippov[1], S P Konovalov[2], G Mikenberg[3], A Romaniouk[1], D Shchukin[1], M Shoa[3], V Smakhtin[3], S Yu Smirnov[1], V V Sosnovtsev[1], P E Teterin[1], V O Tikhomirov[1,2], V I Tsekhosh[2] and K A Vorobev[1]
[1] National Research Nuclear University MEPhI (Moscow Engineering Physics Institute), Kashirskoe highway 31, Moscow, 115409, Russia
[2] P.N. Lebedev Physical Institute of the Russian Academy of Science, Leninsky prospect 53, Moscow, 119991, Russia
[3] Weizmann Institute of Science, 234 Herzl St., Rehovot 7610001, Israel

E-mail: vladimir.smakhtin@weizmann.ac.il

Abstract. sTGC chambers are designed to operate at super–LHC conditions and will be installed in place during Phase-I upgrade of the ATLAS muon spectrometer. These chambers will provide precise coordinate measurements of the charged particle tracks and level 1 trigger for high p_T muons. It is critical for the ATLAS detector to ensure a robust operation of these chambers during entire sLHC period. A quality control procedure based on X-ray scanner is being developed. Choice of the active gas for these tests is a very important issue. On one hand it should allow to find different types of chamber production defects, on the other hand one has to be sure that found problems are essential for the detector operation in future. Studies of the operation of the sTGC chamber prototype under X-ray irradiation with two gas mixtures (n-pentane/CO_2 and CO_2) were performed. The prototype was irradiated by X-rays with energy up to 50 keV. Particular attention was paid to the study of the "hot" chamber regions.

1. Introduction

The small-strip Thin Gap Chambers (sTGC) were specially designed for conditions of super LHC when luminosity may reach values up to $7{*}10^{34}$ cm^{-2}s^{-1}. The ATLAS detector is expected to run about 10 years under these conditions and will accumulate total integral luminosity of 3000 fb^{-1} [1]. The main features of sTGC technology are:

- Very good spatial resolution (around 100 μm).
- Ability to operate at high loads (20 kHz/cm²).
- High resistance to radiation damage.
- Trigger and coordinate measurement capabilities.

In order to ensure long time operation stability in high radiation environment a very strict quality control procedures must be applied. One of them is a direct operation test under high radiation. For this a scanning technique with using X-ray tube is proposed. This technique is based on irradiation of the chamber with a flux of gamma-ray photons with the energy 40-50 keV. The X-ray tube is installed on the moving carriage of the specially designed scanner, which can work in different modes and make irradiation of entire surface of the chambers (2.2x1.4 m^2). Irradiation tests of TGC detectors were performed in the past using ^{60}Co and ^{90}Sr radioactive sources and have shown very high efficiency. These quality control procedures allowed to find operation defects and helped to make a decision whether to cure them using chamber training under HV and irradiation, or to remove defective wires [2]. However these tests were carried out after the chambers were produced and assembled. Tests with X-ray scanner will complement these procedures allowing to find out defects at early stages of the production process. In addition precision scans with different type of collimators can be performed to check technical alignment of the chamber components [3]. In this paper results of studies of different gas mixtures for the X-ray scanning technique are presented.

2. Test set-up.

The tests were carried out at CERN with the first prototypes of sTGC chambers assembled in quadruplet (figure 1). The first chamber had dimensions 1232 mm wide and 723 mm high. The total gas volume of the quadruplet was about 9 litres.

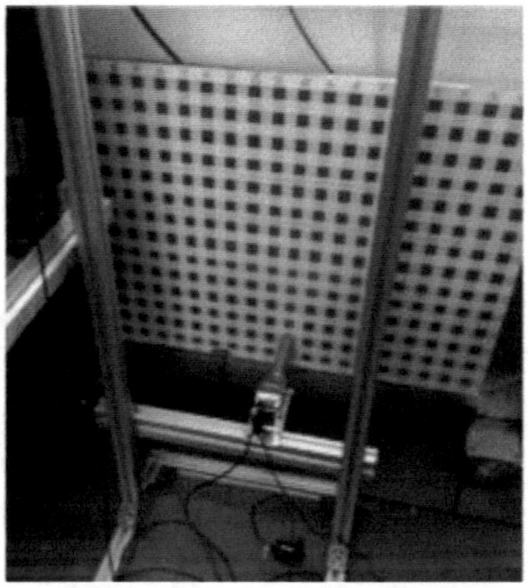

Figure 1. A prototype of sTGC quadruplet used for the tests with X-ray tube equipped with collimator (at the bottom of the chamber).

Two types of gas mixtures were used for the tests: CO_2 – 100% and n-pentane/CO_2 – 45%/55%. The first one is a safe non-flammable gas, which allows to operate chambers at voltages close to the nominal. The second gas mixture is the one used during sTGC operation at LHC. Amptek X-ray tube [4] was used for the tests. Scans were performed in manual mode with 25 mm steps using the collimator of 30 mm in diameter and 15 cm long head fixed on the tube (figure 1). The X-ray tube was running at 40 kV with a current of 80 μA. For CO_2 gas mixture the voltage on chamber was set to 2.9 kV and for n-pentane/CO_2 to 2.8 kV.

The main goal of the tests was to measure gas gain uniformity across the chambers and to study sensitivity of different gases to the sTGC chamber defects.

3. Test results

Figure 2 shows the current in two chambers as a function of applied voltage in logarithmic scale for two gas mixtures. The current is shown in arbitrary units. Because of X-ray attenuation in the first chamber the currents from the second chamber are scaled to overlap the curves. One sees that even at the highest voltages the gas gain is linear for CO_2 mixture. For n-pentane/CO_2 mixture gas gain saturation is observed above 2800 V. At this voltage (the nominal operating voltage) the slope of the curves is smaller than for CO_2 gas. This means that if the chamber operates with this mixture one should expect lower sensitivity to the electric field uniformity across the chamber.

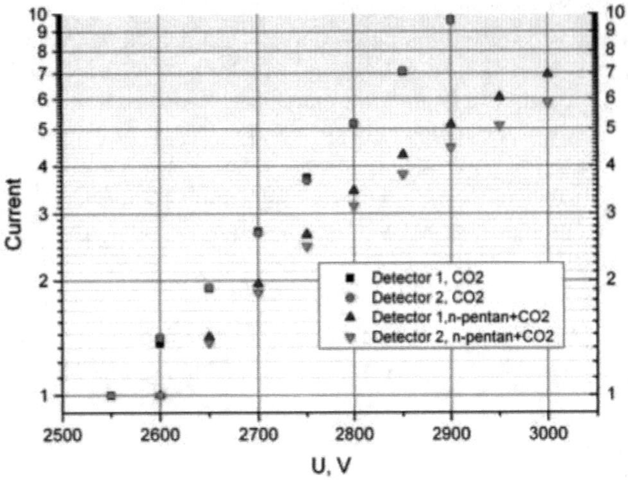

Figure 2. Current (arbitrary units) as a function of applied voltage in two chambers for two gas mixtures.

Map of the currents after the scan of entire chamber operating with CO_2 gas is shown in figure 3. This map was obtained using a special visualisation program developed for the X-ray scanner [5]. One sees a presence of some number of the hot spots where current reaches a level of 5-10 μA. During the tests the chambers were mounted in such a way that the strips were oriented in horizontal direction and wires – in vertical directions. A clear wire defect causing higher gas gain is observed in the right part of the chamber. Low current areas in horizontal direction correspond to the wire supports (see description of the sTGC design in [1]). There is also a significant reduction of gas gain at the bottom left corner of the chamber which has still unknown origin.

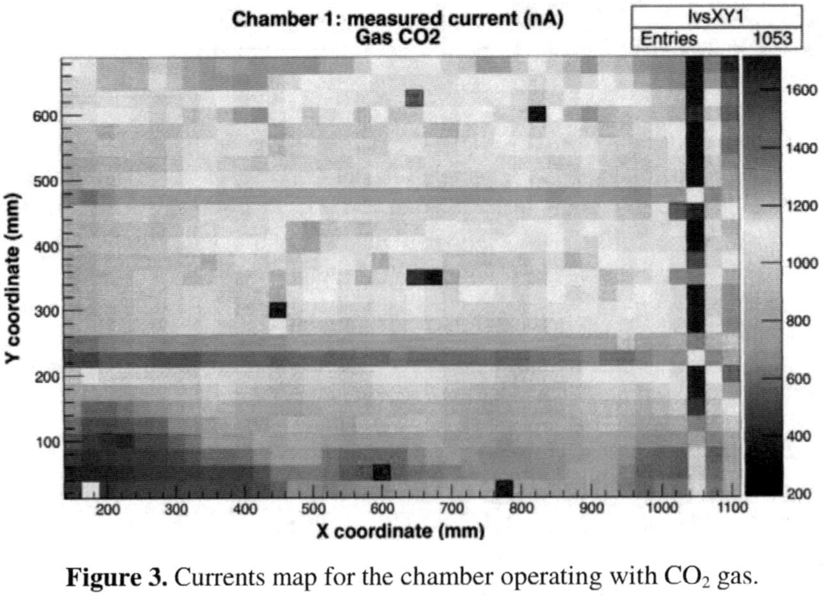

Figure 3. Currents map for the chamber operating with CO_2 gas.

Figure 4 shows similar current map for the first chamber filled with the n-pentane/CO_2 mixture. No hot spots except one were found in that case. The discharge in this spot was quickly cured after training under irradiation. This spot is not shown on the map because scan was done after the problem had been fixed. Less contrast but similar behaviour of the gas gain across the chamber was observed.

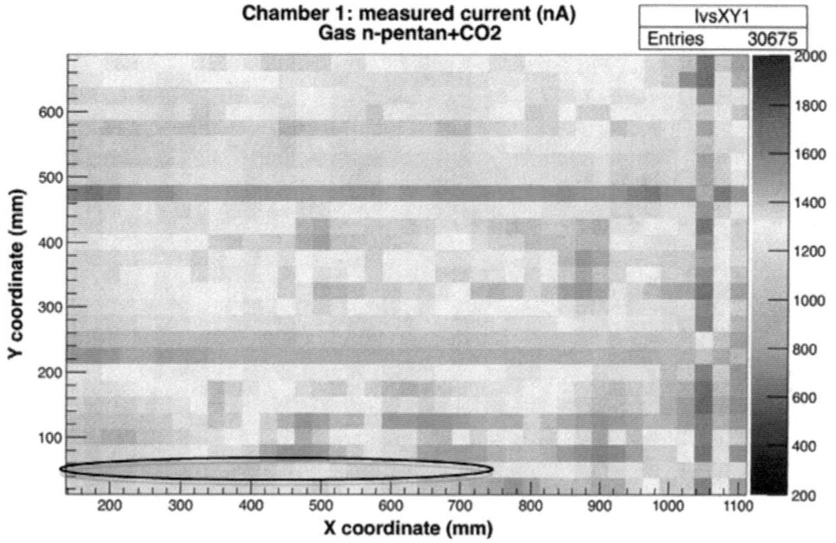

Figure 4. Currents map for the chamber operating with n-pentane/CO_2 gas mixture.

Wire supports, an area with decreased gas gain, as well as defect associated with strips (in black ellipse on the picture) are well visible.

For the second chamber the results are shown in figures 5 and 6. Since the vertical size of this chamber was smaller than that of the first one, no current above Y=620 mm was observed. Similar to the first chamber no hot spots were registered in n-pentane/CO_2 gas mixture.

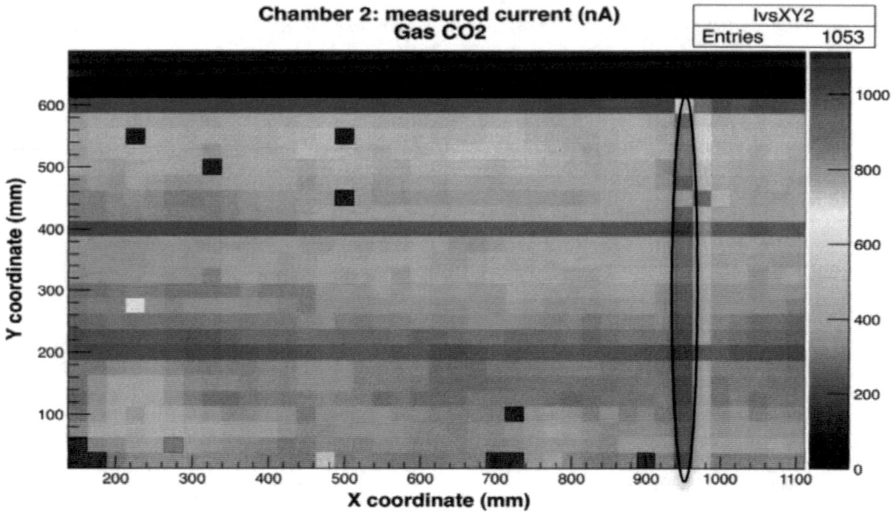

Figure 5. Currents map for the second chamber operating with CO_2 gas.

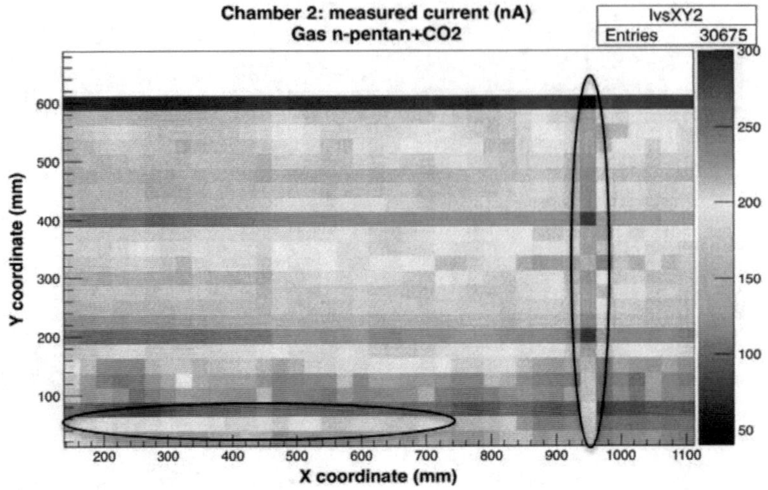

Figure 6. Currents map for the second chamber operating with n-pentane/CO_2 gas mixture.

Many hot spots found in CO_2 gas are related to a surface current on insulators supporting the wires and holding the cathode planes. They are not observed in n-pentane/CO_2 gas mixture because of its excellent quenching properties. Quite a few hot spots in two chambers had a self-sustaining discharge. Figure 7 shows evolution of the current in CO_2 gas mixture during X-ray irradiation of one of the spots with continuous discharge in chamber 2 (coordinates X=900 mm and Y = 25 mm). At the time of about 100 sec the X-ray irradiation was turned on. The current reached the level of about 250 nA and shortly after jumped up to 2500 nA. It did not change significantly during 10 min of irradiation and remained at the level of 1100 nA even after X-ray irradiation had been turned off. A few steps in current after 1400 sec correspond to the moments of lowering down HV on the chamber. The current went down to zero when the voltage had been set to 2700 V.

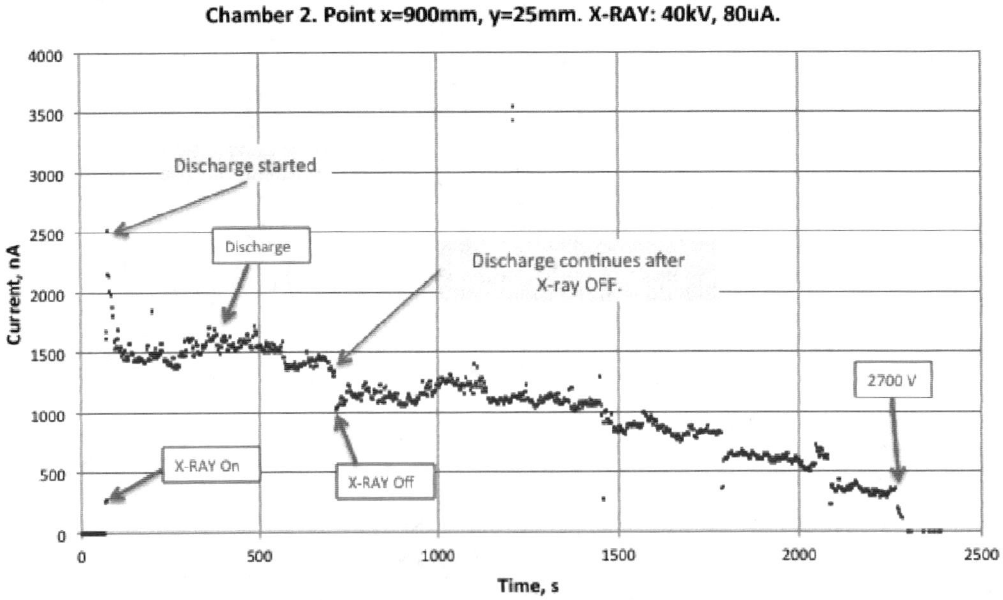

Figure 7. Time evolution of the current in the chamber filled with CO_2 during X-ray irradiation of one of the spots with continuous discharge.

Most likely the reason for this behavior is the Malter effect [6] happening in CO_2 gas mixture. This effect takes place when a thin layer of insulator covers the cathode. Positive ions are accumulated on the surface of this insulator and create a strong electric field being able to provoke electron emission. No discharges were observed in these points in the n-pentane/CO_2 mixture. Absence of the Malter effect in this mixture might be explained by a modification of a surface conductivity by n-pentane gas.

The uniformity of the gas gain measured with two gas mixtures is compared in figure 8. Plots in the left show the current profile measured with CO_2 gas in slices of 25 mm wide along X and Y directions for the crossing point with coordinates X=1000 and Y=175 in the first chamber. Right side plots show similar results for n-pentane/CO_2 mixture. For this mixture the gas gain variation is at least by factor of two less. Significant difference in the gas gain variation at smaller Y coordinate for the tests with CO_2 gas most likely indicate a presence of a mechanical deformation of the chamber.

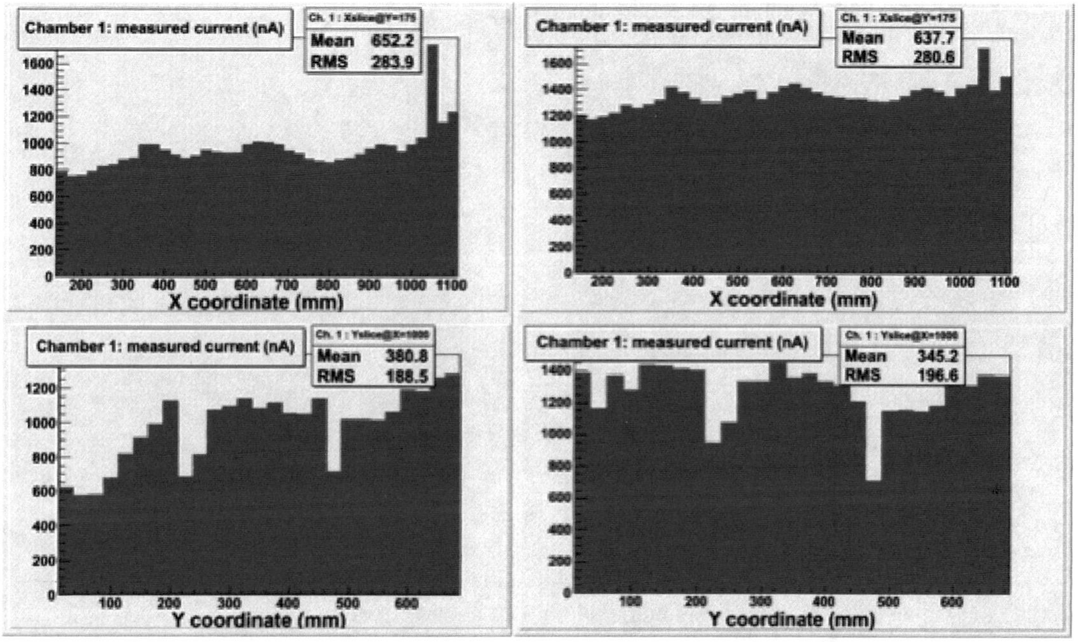

(a) (b)

Figure 8. The current profiles along X and Y directions for CO_2 (a) and n-pentane/CO_2 (b) gas mixtures at the point with coordinates X=1000 and Y=175 in the first chamber.

4. Conclusion

The tests of sTGC chambers irradiated with X-ray source carried out with different gas mixtures demonstrated that the chamber defects are more pronounced when the chamber is filled with CO_2 gas. Quite a few hot spots can be associated with wire or cathode supports. Self-sustaining discharges were found in many places indicating some problem with the cathode surface of the chambers. However, the chambers filled with n-pentane/CO_2 gas mixture showed much higher operation stability and lower sensitivity to the electric field non-uniformity. No areas with self-sustaining discharge or hot spots except one were found for this mixture. This particular spot was cured after a few minutes of radiation at different voltages.

Acknowledgements

This work was carried out in the Center of FRPP and was supported by MEPhI Academic Excellence Project (contract No. 02.a03.21.0005, 27.08.2013). We gratefully acknowledge the financial support from the Ministry of Education and Science of Russian Federation (R&D project

RFMEFI61014X0005) and grant No.14-22-03053 from the Russian Foundation for Basic Research.

References

[1] The ATLAS Collaboration 2013 *New Small Wheel Technical Design Report* CERN-LHCC-2013-006 ATLAS-88 TDR-020
[2] Etzion E *et al.* 2004 The certification of ATLAS thin gap chambers produced in Israel and China *IEEE/NSS Conf. 2004*
[3] Teterin P E *et al.* 2015 Development of scanning technique for sTGC detector production quality control *J. Phys.: Conf. Series* (In this Proceedings)
[4] URL: (*http://www.amptek.com/products/mini-x-ray-tube/*)
[5] Tikhomirov V O *et al.* 2015 Visualization tool for X-ray scanner for sTGC detector production quality control *J. Phys.: Conf. Series* (In this Proceedings)
[6] Malter L 1936 Thin film emission *Phys. Rev* **50** (1) 48-58

Journal of Physics: Conference Series **675** (2016) 012025 doi:10.1088/1742-6596/675/1/012025

Test of the electric charge conservation law with Borexino detector

A Vishneva[1,29,*], M Agostini[2], S Appel[2], G Bellini[3], J Benziger[4], D Bick[5], G Bonfini[6], D Bravo[7], B Caccianiga[3], F Calaprice[8], A Caminata[9], P Cavalcante[6], A Chepurnov[10], D D'Angelo[3], S Davini[11], A Derbin[12], L Di Noto[9], I Drachnev[11], A Etenko[13], K Fomenko[1], D Franco[14], F Gabriele[6], C Galbiati[8], C Ghiano[9], M Giammarchi[3], M Goeger-Neff[2], A Goretti[8], M Gromov[10], C Hagner[5], E Hungerford[15], Aldo Ianni[6], Andrea Ianni[8], K Jedrzejczak[17], M Kaiser[5], V Kobychev[18], D Korablev[1], G Korga[6], D Kryn[14], M Laubenstein[6], B Lehnert[19], E Litvinovich[13,20], F Lombardi[6], P Lombardi[3], L Ludhova[3], G Lukyanchenko[13,20], I Machulin[13,20], S Manecki[7], W Maneschg[22], S Marcocci[11], E Meroni[3], M Meyer[5], L Miramonti[3], M Misiaszek[17,6], M Montuschi[23], P Mosteiro[8], V Muratova[12], B Neumair[2], L Oberauer[2], M Obolensky[14], F Ortica[24], M Pallavicini[9], L Papp[2], L Perasso[9], A Pocar[26], G Ranucci[3], A Razeto[6], A Re[3], A Romani[24], R Roncin[6,14], N Rossi[6], S Schönert[2], D Semenov[12], H Simgen[22], M Skorokhvatov[13,20], O Smirnov[1] A Sotnikov[1], S Sukhotin[13], Y Suvorov[27,13], R Tartaglia[6], G Testera[9], J Thurn[19], M Toropova[13], E Unzhakov[12], R B Vogelaar[7], F von Feilitzsch [2], H Wang[27], S Weinz[28], J Winter[28], M Wojcik[17], M Wurm[28], Z Yokley[7], O Zaimidoroga[1], S Zavatarelli[9], K Zuber[19] and G Zuzel[17] (Borexino Collaboration)

[1] Joint Institute for Nuclear Research, 141980 Dubna, Russia
[2] Physik-Department and Excellence Cluster Universe, Technische Universität München, 85748 Garching, Germany
[3] Dipartimento di Fisica, Università degli Studi e INFN, 20133 Milano, Italy
[4] Chemical Engineering Department, Princeton University, Princeton, NJ 08544, USA
[5] Institut für Experimentalphysik, Universität, 22761 Hamburg, Germany
[6] INFN Laboratori Nazionali del Gran Sasso, 67010 Assergi (AQ), Italy
[7] Physics Department, Virginia Polytechnic Institute and State University, Blacksburg, VA 24061, USA
[8] Physics Department, Princeton University, Princeton, NJ 08544, USA
[9] Dipartimento di Fisica, Università degli Studi e INFN, Genova 16146, Italy
[10] Lomonosov Moscow State University Skobeltsyn Institute of Nuclear Physics, 119234 Moscow, Russia
[11] Gran Sasso Science Institute (INFN), 67100 L'Aquila, Italy
[12] St. Petersburg Nuclear Physics Institute NRC Kurchatov Institute, 188350 Gatchina, Russia
[13] NRC Kurchatov Institute, 123182 Moscow, Russia

[29] Presenter. To whom any correspondence should be addressed.

[14] AstroParticule et Cosmologie, Université Paris Diderot, CNRS/IN2P3, CEA/IRFU, Observatoire de Paris, Sorbonne Paris Cité, 75205 Paris Cedex 13, France
[15] Department of Physics, University of Houston, Houston, TX 77204, USA
[16] Institute for Theoretical and Experimental Physics, 117218 Moscow, Russia
[17] M. Smoluchowski Institute of Physics, Jagiellonian University, 30059 Krakow, Poland
[18] Kiev Institute for Nuclear Research, 06380 Kiev, Ukraine
[19] Department of Physics, Technische Universität Dresden, 01062 Dresden, Germany
[20] National Research Nuclear University MEPhI (Moscow Engineering Physics Institute), Kashirskoe highway 31, Moscow 115409, Russia
[21] Kepler Center for Astro and Particle Physics, Universität Tübingen, 72076 Tübingen, Germany
[22] Max-Planck-Institut für Kernphysik, 69117 Heidelberg, Germany
[23] Dipartimento di Fisica e Scienze della Terra Università degli Studi di Ferrara e INFN, Via Saragat 1-44122, Ferrara, Italy
[24] Dipartimento di Chimica, Università e INFN, 06123 Perugia, Italy
[25] Physics Department, Queen's University, Kingston ON K7L 3N6, Canada
[26] Amherst Center for Fundamental Interactions and Physics Department, University of Massachusetts, Amherst, MA 01003, USA
[27] Physics and Astronomy Department, University of California Los Angeles (UCLA), Los Angeles, California 90095, USA
[28] Institute of Physics and Excellence Cluster PRISMA, Johannes Gutenberg-Universität Mainz, 55099 Mainz, Germany

E-mail: *the2cherry2orchard@gmail.com

Abstract. The new limit on the electron lifetime is obtained from data of the Borexino experiment. The expected signal from the $e \to \gamma\nu$ decay mode is a 256 keV photon detected in liquid scintillator. Because of the extremely low radioactive background level in the Borexino detector it was possible to improve the previous measurement by two orders of magnitude.

1. Introduction

The electric charge conservation law is a fundamental physical principle. There are no hints for violation of this law neither in theory within the Standard model nor in any experiment. Since the electric charge non-conservation (CNC) is admitted in exotic theories such as extra-dimensional theories [1], investigation of such processes is an evident way to search for physics beyond the Standard model.

The most frequently searched for CNC processes are decays of the electron into neutral particles. Two decay modes are usually accounted for experimentally:

$$e \to \gamma\nu, \tag{1}$$

where a monoenergetic 256 keV photon is searched for, and

$$e \to \nu\nu\nu, \tag{2}$$

where only effects due to the electron disappearance would be observed. However, the impossibility of occurrence of such processes is presented in [2], where it is shown that such decays would be followed by a huge amount of low-energy bremsstrahlung photons. For the process (1) it would mean the absence of 256 keV photon while the electron disappearance is more model-independent and the corresponding atomic effects in the case (2) would remain the same. Thus one can see that observing the 256 keV photon from the electron decay would mean not only CNC but also going beyond the Standard model.

2. Overview of experiments

Study of the electron stability has long experimental history. The list of the experiments in which the electron decay was being searched for is presented in table 1. There are also plans for providing analogous studies at present and future experiments [18, 19, 20].

Table 1. Experimental tests for the electron stability.

year	material	limit for $e \to \gamma\nu$	limit for $e \to \nu\nu\nu$	CL	reference
1959	NaI	10^{19}	10^{17}	68%	[3]
1965	NaI	4×10^{22}	2×10^{21}	68%	[4]
1975	Ge	—	5.3×10^{21}	68%	[5]
1979	NaI	3.5×10^{23}	—	68%	[6]
1983	Ge	3×10^{23}	2×10^{22}	68%	[7]
1986	Ge	1.5×10^{25}	—	68%	[8]
1993	Ge	1.63×10^{25}	—	68%	[9]
1995	Ge	2.1×10^{25}	2.6×10^{23}	90%	[10]
1996	Xe	2×10^{25}	1.5×10^{23}	68%	[11]
1999	NaI	—	$(1.5 - 2.4) \times 10^{23}$	90%	[12]
1999	NaI	—	2.4×10^{24}	90%	[13]
2000	Xe	2×10^{26}	—	90%	[14]
2002	PXE	4.6×10^{26}	—	90%	[15]
2007	Ge	1.93×10^{26}	—	90%	[16]
2012	NaI	—	1.2×10^{24}	90%	[17]

2.1. NaI detectors

Experiments based on NaI detectors were the first to provide the limits on the electron stability [3, 4].The expected signal for the mode (2) is a photon with maximal energy of 33.2 keV emitted while filling the vacancy caused by the electron disappearance from the iodine K-shell. The decay to a photon and a neutrino is investigated by searching for the 256 keV photon. Various coincidence techniques are also applied in such detectors. First was the search of simultaneous 256 keV and 33.2 keV photons occurence [4]. Another approach based on the electron capture by a nucleus without the consequent atomic number change was considered recently in [17]. Simultaneous observation of the 33.2 keV photons and the nucleus deexcitation (417.9 keV) would mean the electron disappearance.

2.2. Ge detectors

The electron stability is widely studied using germanium detectors. The main advantage of such detectors is good energy resolution (about 1 keV). In addition, the background level in the region of interest is lower than that in NaI detectors. The expected signal is a photon of energy 11.1 keV for the mode (2) and a 256 keV photon for the mode (1), respectively.

2.3. Liquid scintillators

The strongest limits on the electron lifetime with respect to the decay mode (1) during the last fifteen years have been obtained with liquid scintillation detectors. Their main advantages are large mass and a possibility of purification from radioactive contaminations. The first one was

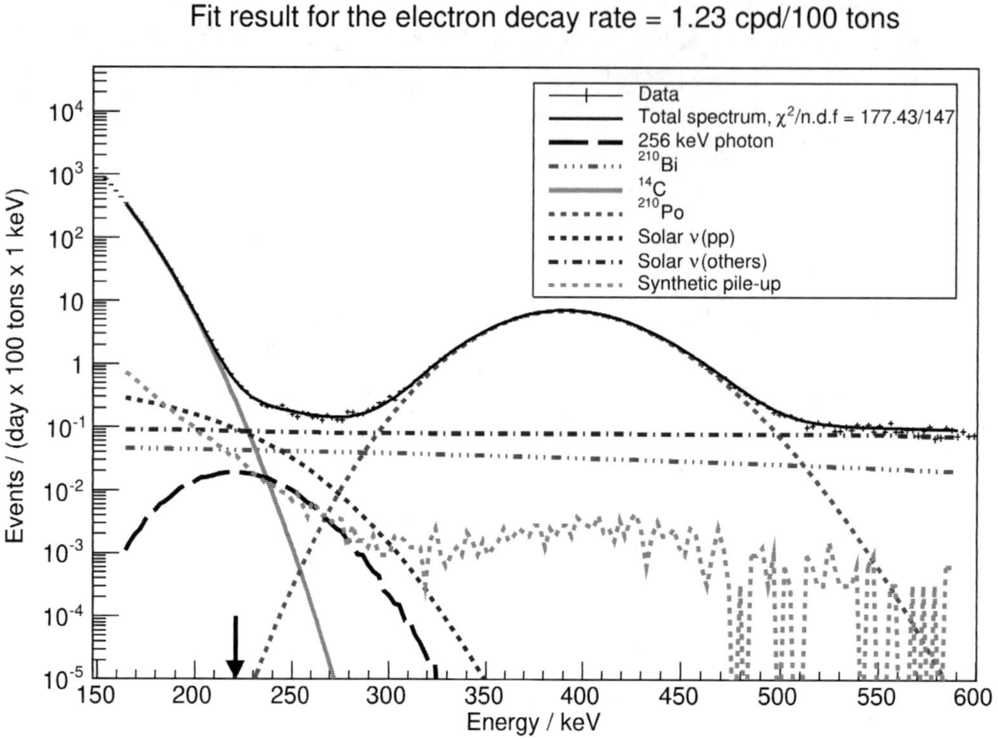

Figure 1. Borexino spectrum composition.

DAMA/LXe experiment [11, 14]. This detector contains 6.5 kg ($\sim$ 2 litres) of liquid xenon. This apparatus has rather low energy threshold and is sensitive to both electron decay modes.

The second one is the prototype of the Borexino detector, CTF-II [21]. Its main goal was to test the purification techniques developed for Borexino. During the tests various scintillators were used. CTF-II was filled with 4 tons of PXE (phenylxylylethane) which has less ionization quenching in comparison with PC (pseudocumene) used in Borexino. Large mass and extremely low background level made it possible to obtain a stronger limit of 4.6×10^{26} years (90% confidence level) on the electron lifetime in shorter exposure time. This result remained the best until the same study was performed in Borexino.

3. The Borexino detector

Borexino is a large volume scintillation detector located deep underground in the Laboratori Nazionali del Gran Sasso [22]. Its active media contains 278 tons of organic liquid scintillator, namely, pseudocumene (1,2,4-trimethylbenzene) with admixture of PPO (2,5-diphenyloxazole) at a concentration of 1.5 g/l. Borexino has extremely low background level in the region of interest, namely 0.15 day^{-1}ton^{-1}keV^{-1}. The energy threshold is above 50 keV so Borexino is not sensitive to the disappearance mode. By comparing sensitivity of CTF and Borexino the expected electron lifetime limit is estimated, which exceeds the previous one at two orders of magnitude.

4. The electron decay search

4.1. Analysis approach

The data set used in the analysis were acquired from January 2012 to May 2013 (Borexino Phase 2). This data set was obtained after the purification campaign [23] which reduced in particular the contamination of ^{85}Kr and ^{210}Bi which give a significant contribution in the low-energy region.

The same 408 days data set is successfully used in the measurement of solar pp-neutrino flux [24]. In this analysis the same energy range (150–600 keV) and parameters used in the fitting procedure are considered. The only difference is the addition of the 256 keV photon line in the fitting function. The sample of spectral fit is shown on Fig. 4.1. One can see that the sought-for peak (marked by arrow) is shifted to the lower energies due to ionization quenching.

4.2. Constraint on the pp-neutrino event rate

The 256 keV photon occurrence is strongly correlated with pp-neutrino event rate. Therefore treating the pp-neutrino rate as a free parameter in the fit leads to non-physical values at the limit. Indeed, the 256 keV photon event rate corresponding to the 90% confidence level ($\simeq$ 12 cpd/100 tons) corresponds to zero pp-neutrino rate which is not consistent with observations by radiochemical experiments [25]. As far as the latter ones are not sensitive to the electron decay it is reasonable to use their results to constrain the pp-neutrino event rate. This constraint gives the limit on the event rate of 1.23 cpd/100 tons. The lifetime limit is expressed as $\tau \geqslant \epsilon \mathrm{N}_e \mathrm{T}/\mathrm{S}_{\mathrm{lim}}$, where N_e is the total number of electrons in the detector, ϵ is the fraction of electrons survived after the fiducial volume cut, T is the exposure time, and $\mathrm{S}_{\mathrm{lim}}$ is the event rate limit. It gives the electron lifetime of $\tau \geqslant 7.2 \times 10^{28}$ years.

4.3. Systematic errors study and final results

The main sources of the systematic errors in this study are the following. The most important is the precision of the scintillator light yield measurement ($\sim$ 1%). It strongly influences the peak position and therefore affects the sensitivity. Another source of systematic errors is the fiducial mass measurement precision, which gives negligible effect. Choice of the energy estimator can also affect the result. In the present study two variables are used as energy estimators, namely, number of PMTs hit in the time intervals of 230 ns and 400 ns. After having accounted for all these effects the lifetime limit has become weaker and the final result for the electron lifetime limit is $\tau_{e \to \gamma\nu} \geqslant 6.6 \times 10^{28}$ years at the 90% confidence level. This study is described in more details in [26].

Acknowledgments

Russian colleagues from MEPhI acknowledge partial support from MEPhI Academic Excellence Project (contract No. 02.a03.21.0005, 27.08.2013).

References

[1] Dubovsky S L, Rubakov V A and Tinyakov P G 2000 *JHEP* **0008** 041
[2] Okun L B 1989 *Sov. Phys. Usp.* **32** 543
[3] Feinberg G and Goldhaber M 1959 *Proc. Natl. Acad. Aci. US* **45** 1301
[4] Moe M K and Reines F 1965 *Phys. Rev.* **140** 992
[5] Steinberg R I *et al.* 1975 *Phys. Rev.* D **12** 2582
[6] Koval'chuk E L, Pomanskiĭ A A and Smol'nikov A A 1979 *JETP Lett.* **29** 145
[7] Belotti E *et al.* 1983 *Phys. Lett.* B **124** 435
[8] Avignone III F T, Brodzinski R L, Hensley W K, Miley H S and Reeves J H 1986 *Phys. Rev.* D **34** 97
[9] Balysh A *et al.* 1993 *Phys. Lett.* B **298** 278
[10] Aharonov Y *et al.* 1995 *Phys. Lett.* B **353** 168
[11] Belli P *et al.* 1996 *Astropart. Phys.* **5** 217-9

[12] Belli P *et al.* 1999 *Phys. Rev.* C **60** 065501
[13] Belli P *et al.* 1999 *Phys. Lett.* B **460** 236
[14] Belli P *et al.* 2000 *Phys. Rev.* D **61** 117301
[15] Back H O *et al.* 2002 *Phys. Lett.* B **525** 29
[16] Klapdor-Kleingrothaus H V, Krivosheina I V and Titkova I V 2007 *Phys. Lett.* B **644** 109
[17] Bernabei R *et al.* 2012 *Eur. Phys. J.* C **72** 1920
[18] Aalseth C E *et al.* 2004 *Phys. Atom. Nucl.* **67** 2002
[19] Ardito R *et al.* 2005 *arXiv*:0501010
[20] Wurm M *et al.* 2012 *Astropart. Phys.* **35** 685-732
[21] Alimonti G *et al.* 1998 *Nucl. Instrum. Meth.* A **406** 411
[22] Alimonti G *et al.* 2009 *Nucl. Instrum. Meth. A* **600** 568-93
[23] Bellini G *et al.* 2014 *Phys. Rev.* D **89** 112007
[24] Bellini G *et al.* 2014 *Nature* **512** 383
[25] Abdurashitov J N *et al.* 2009 *Phys. Rev.* C **80** 015807
[26] Agostini M *et al.* 2015 *arXiv*:1509.01223

On a possible solution to gamma-ray overabundance arising in dark matter explanation of cosmic antiparticle excess

V V Alekseev[1], K M Belotsky[2], Yu V Bogomolov[1], R I Budaev[2], O A Dunaeva[1], A A Kirillov[1,2], A V Kuznetsov[1], M N Laletin[2], A D Lukyanov[1], V V Malakhov[2], A G Mayorov[1,2], M A Mayorova[2], A F Mosichkin[1], A A Okrugin[1], S A Rodenko[2] and A M Shitova[1]

[1] Yaroslavl State P.G. Demidov University, 150000 Sovietskaya st. 14, Yaroslavl, Russia
[2] National Research Nuclear University MEPhI (Moscow Engineering Physics Institute), Kashirskoe highway 31, Moscow, 115409, Russia

E-mail: maxim.n.laletin@gmail.com

Abstract. As we are going to show, some self-interacting dark matter models may provide an intriguing solution to the "cosmic antiparticle excess vs. isotropic diffuse gamma-ray background" problem (an overproduction of diffuse gamma rays, arising in any reasonable decaying or annihilating dark matter model explaining high-energy charged cosmic antiparticles anomalous abundance).

1. Introduction

The existence of dark matter – an unknown substance, which compose almost 26% of our Universe [1], but only perceptible by virtue of its gravitational effects – can hardly surprise scientific community nowadays, however its nature and origin still remain a mystery. Though an enormous amount of hypothesis were proposed to explain this phenomenon a mostly conventional conception of dark matter is a quite massive particle with vanishingly weak interaction with ordinary matter (except the strongly interactive massive particle (SIMP) models, which are however strongly disfavored by experimental searches [2]) and rather rarefied local distribution. Observational facts force us to think that dark matter behaves like an ideal gas, though there's a not only tempting, but a pretty reasonable (from the particle physics point of view) idea, that dark matter should posess some new fundamental interaction. Probably the first thing that comes to mind is the possibility of dark matter annihilations or decays, rarely taking place due to this new interaction. Such an obsessive idea has been actively studied for many decades and is commonly used to explain a broad variety of astrophysical effects, the hints of which we possibly observe (like gamma-ray bursts, emission lines in gamma-ray spectrum of the Galactic center, anomalous abundance of high-energy antiparticles in cosmic rays and etc.), or, on the other hand, to impose constraints on various Standard Model extensions, predicting candidates for dark matter.

Another interesting concept is that dark matter particles can scatter on each other rather than annihilate. It was shown that a new QCD-like strong self-interaction is able of solving the so called "cusp-core" problem [3] and a new wave of interest towards self-interacting dark matter models was caused by the recent observation of Abell 3827 galaxy cluster [4]. However, in order to stay consistent with cosmological and astrophysical observations one should permit only a small fraction of dark matter (e.g. a different species of dark matter) to interact with each other noticeably, while the major part of dark matter reveals the properties of collisionless gas. Except for the strong self-interaction model, which we mentioned, it would be an indeed beautiful theory, in which dark matter particles are charged under an unknown $U(1)$ interaction (or "dark" electromagnetism). Besides suitable conditions for a plenty of indirect (see e.g. [5]) or even direct (see e.g. [6]) dark matter effects it was noted in [7] that this model can also provide a solution for the "cusp-core" problem. Furthermore, the authors of the considering article proposed an exceptionally new idea, that self-interacting dark matter can cool down due to "dark radiation" emission and form a disk-like structure, possibly coinciding with the Galactic plane. Though this idea may seem unrealistic or superfluous at first sight in this work we are going to show, that if self-interacting dark matter also experience decays or annihilation to Standard Model quarks and leptons, such approach can solve the "charged antiparticle excess vs. diffuse gamma-ray background" problem [8, 9].

2. Charged antiparticles from dark matter decays and diffuse gamma-ray overabundance

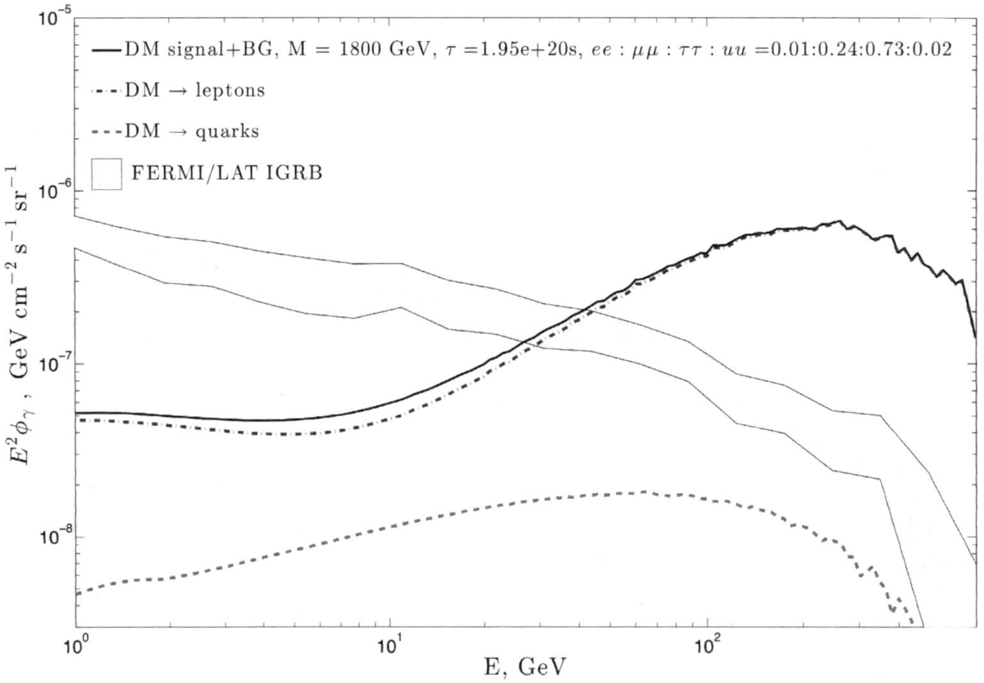

Figure 1. Diffuse gamma-ray flux predicted by the average decaying dark matter model, explaining positron and antiproton excess in cosmic rays (see [9]). The flux is averaged over the solid angle corresponding to the range of Galactic coordinates $20° \leq b \leq 90°$ and $0° \leq l < 360°$, in which FERMI/LAT measures IGRB. The uncertainties of IGRB are related to the arbitrary choice of foreground model.

Let us briefly remind here the basics of "charged antiparticle excess vs. diffuse gamma-ray background" problem (see [9] for a detailed review). Any rational dark matter model (assuming isotropical distribution of metastable or annihilating fraction) capable of high-energy cosmic antiparticle excess explanation (seen by PAMELA and AMS-02 in positrons [10, 11] and antiprotons [12, 13]) predicts a tremendous amount of high-energy gamma rays as well and thus contradicts with FERMI/LAT data on isotropic diffuse gamma-ray background (IGRB) [14].

One should note that this problem can not be solved neither by model parameter tuning nor propagation parameters variation. Though we've considered only decaying dark matter we strongly suppose that the same problem strikes also annihilating dark matter scenario.

3. Dark matter disk model (DDDM) and the solution of gamma-ray overproduction problem

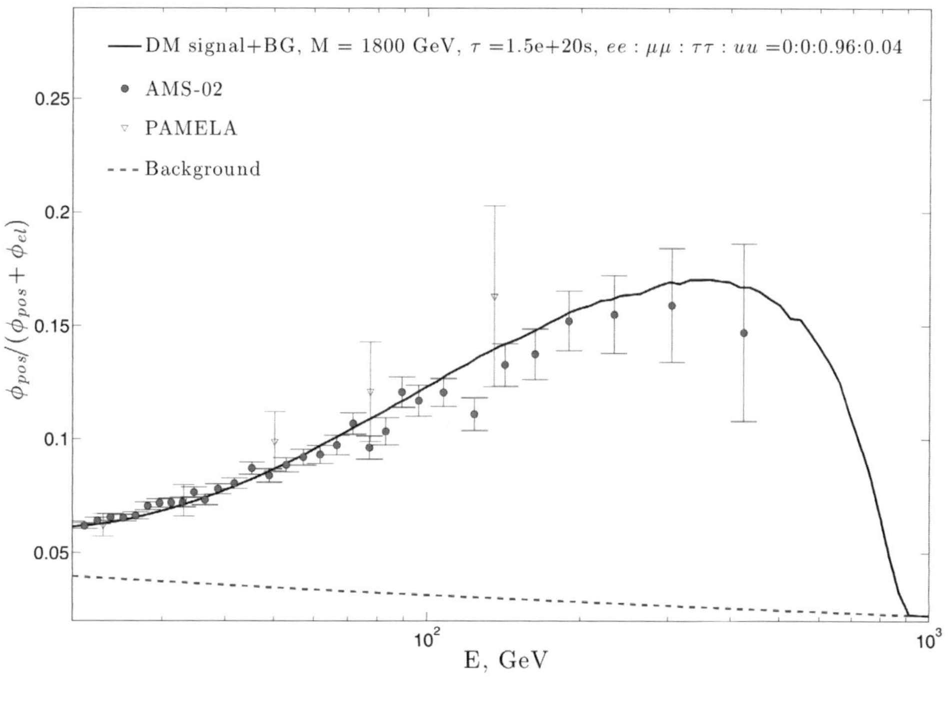

Figure 2. Fraction of positrons in cosmic rays, predicted by the double-disk decaying dark matter model and compared to AMS-02 and PAMELA data. The values of model parameters corresponds to those listed in the *top left* part of the plot.

If decaying or annihilating component of dark matter posess long-range self-interaction, which allows it to cool down due to energy exchange with "dark radiation" bath, then a dark matter disk formation seems possible. Since charged particles (and antiparticles) responsible for anomalous effects in cosmic rays observations can reach the Solar system only from the sources located in the magnetic halo, while gamma rays come to us from the whole large dark matter halo, one can easily see that squeezing decaying dark matter into a disk, following the shape and size of baryonic disk, can significantly reduce the amount of isotropic diffuse gamma rays, while preserving almost the same flux of charged antiparticles. In figures 2-4 we show our results for the same dark matter model, mentioned in section 2, but placed in the thin disk.

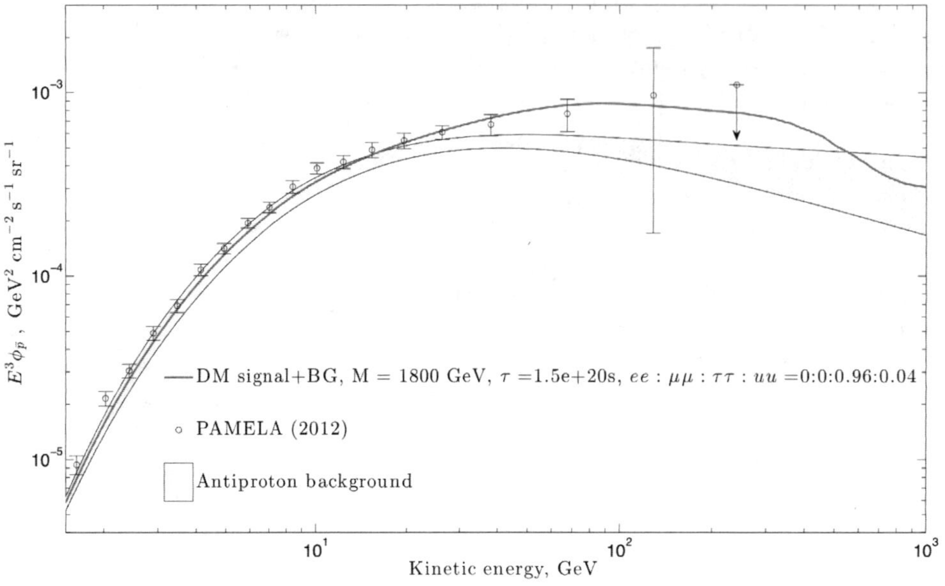

Figure 3. Antiproton flux, predicted by the double-disk decaying dark matter model and compared to the PAMELA data.

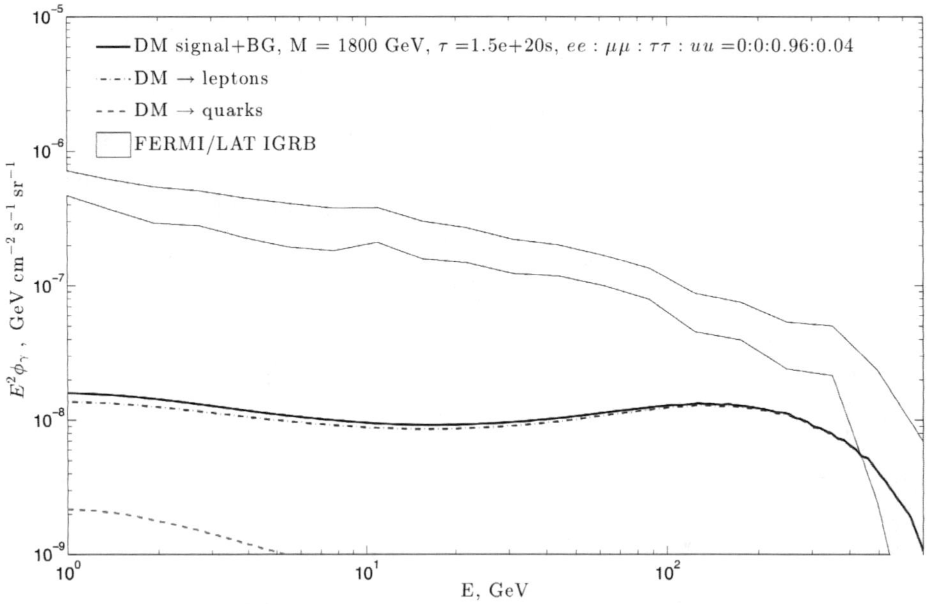

Figure 4. Diffuse gamma-ray flux, predicted by the double-disk decaying dark matter model and compared to the FERMI/LAT data. The separate contributions from leptonic (blue dot-dashed) and hadronic (red dashed) decaying modes are also presented.

4. Conclusion

In this work we have applied the double-disk dark matter concept to the decaying dark matter model and worked out a possible solution to the "cosmic antiparticle excess vs. isotropic diffuse gamma-ray background" problem. However our approach is mostly phenomenological and a more detailed theoretical study of decaying (or annihilating) double-disk dark matter model is needed. Furthermore, we should accurately verify the consistency of our model with cosmological and astrophysical observations (first of all, we should check the predicted gamma-ray flux from the Galactic center and the Oort limit on the local amount of matter). Also we believe that a small anisotropy in high-energy cosmic positron flux may appear in our model. All these questions should be the aim of our future research.

Acknowledgments

The research is financed by the grant of the Russian Science Foundation (Project No. 15-12-10039). The work by MEPhI group (K M Belotsky, R I Budaev, M N Laletin, V V Malakhov, M A Mayorova and S A Rodenko) on the calculation of cosmic ray fluxes from dark matter decays was performed within the framework of the Center of FRPP supported by MEPhI Academic Excellence Project (contract No. 02.a03.21.0005, 27.08.2013).

References

[1] Adam R *et al.* (Planck Collaboration) 2015 *arXiv:1502.01582*
[2] Ericksek A L *et al.* 2007 *Phys. Rev.* D **76** 042007
[3] Spergel D N and Steinhardt P J 2000 *Phys. Rev. Lett.* **84** 3760
[4] Harvey D *et al.* 2015 *Science* **347** 6229 1462
[5] Belotsky K, Khlopov M and Kirillov A 2014 *Adv. High Energy Phys.* **2014** 651247
[6] Cudell J R and Wallemacq Q 2015 *JCAP* **1502** 011
[7] Fan J *et al.* 2013 *Phys. Dark Univ.* **2** 139
[8] Alekseev V *et al.* 2015 *Physics Procedia* **74C** 48
[9] Alekseev V *et al.* 2015 High-energy cosmic antiparticle excess vs. isotropic gamma-ray background problem in decaying dark matter Universe (*these proceedings*)
[10] Adriani O *et al.* (PAMELA Collaboration) 2013 *Phys. Rev. Lett.* **111** 081102
[11] Accardo L *et al.* (AMS Collaboration) 2014 *Phys. Rev. Lett.* **113** 121101
[12] Adriani O *et al.* (PAMELA Collaboration) 2012 *JETP Lett.* **96** 621
[13] AMS Days at CERN and Latest Results: *http://www.ams02.org/2015/04/ams-days-at-cern-and-latest-results-from-the-ams-experiment-onthe- international-space-station*
[14] Ackermann M *et al.* (Fermi LAT Collaboration) 2015 *Astrophys. J.* **86** 799

Journal of Physics: Conference Series **675** (2016) 012027 doi:10.1088/1742-6596/675/1/012027

Measurement of Solar pp-neutrino flux with Borexino: results and implications

O Yu Smirnov[1,29], M Agostini[2], S Appel[2], G Bellini[3], J Benziger[4], D Bick[5], G Bonfini[6], D Bravo[7], B Caccianiga[3], F Calaprice[8], A Caminata[9], P Cavalcante[6], A Chepurnov[10], D D'Angelo[3], S Davini[11], A Derbin[12], L Di Noto[9], I Drachnev[11], A Etenko[13], K Fomenko[1], D Franco[14], F Gabriele[6], C Galbiati[8], C Ghiano[9], M Giammarchi[3], M Goeger-Neff[2], A Goretti[8], M Gromov[10], C Hagner[5], E Hungerford[15], Aldo Ianni[6], Andrea Ianni[8], K Jedrzejczak[17], M Kaiser[5], V Kobychev[18], D Korablev[1], G Korga[6], D Kryn[14], M Laubenstein[6], B Lehnert[19], E Litvinovich[13,20], F Lombardi[6], P Lombardi[3], L Ludhova[3], G Lukyanchenko[13,20], I Machulin[13,20], S Manecki[7], W Maneschg[22], S Marcocci[11], E Meroni[3], M Meyer[5], L Miramonti[3], M Misiaszek[17,6], M Montuschi[23], P Mosteiro[8], V Muratova[12], B Neumair[2], L Oberauer[2], M Obolensky[14], F Ortica[24], M Pallavicini[9], L Papp[2], L Perasso[9], A Pocar[26], G Ranucci[3], A Razeto[6], A Re[3], A Romani[24], R Roncin[6,14], N Rossi[6], S Schönert[2], D Semenov[12], H Simgen[22], M Skorokhvatov[13,20], A Sotnikov[1], S Sukhotin[13], Y Suvorov[27,13], R Tartaglia[6], G Testera[9], J Thurn[19], M Toropova[13], E Unzhakov[12], A Vishneva[1], R B Vogelaar[7], F von Feilitzsch [2], H Wang[27], S Weinz[28], J Winter[28], M Wojcik[17], M Wurm[28], Z Yokley[7], O Zaimidoroga[1], S Zavatarelli[9], K Zuber[19]and G Zuzel[17](Borexino Collaboration)

[1] Joint Institute for Nuclear Research, 141980 Dubna, Russia
[2] Physik-Department and Excellence Cluster Universe, Technische Universität München, 85748 Garching, Germany
[3] Dipartimento di Fisica, Università degli Studi e INFN, 20133 Milano, Italy
[4] Chemical Engineering Department, Princeton University, Princeton, NJ 08544, USA
[5] Institut für Experimentalphysik, Universität, 22761 Hamburg, Germany
[6] INFN Laboratori Nazionali del Gran Sasso, 67010 Assergi (AQ), Italy
[7] Physics Department, Virginia Polytechnic Institute and State University, Blacksburg, VA 24061, USA
[8] Physics Department, Princeton University, Princeton, NJ 08544, USA
[9] Dipartimento di Fisica, Università degli Studi e INFN, Genova 16146, Italy
[10] Lomonosov Moscow State University Skobeltsyn Institute of Nuclear Physics, 119234 Moscow, Russia
[11] Gran Sasso Science Institute (INFN), 67100 L'Aquila, Italy
[12] St. Petersburg Nuclear Physics Institute NRC Kurchatov Institute, 188350 Gatchina, Russia
[13] NRC Kurchatov Institute, 123182 Moscow, Russia
[14] AstroParticule et Cosmologie, Université Paris Diderot, CNRS/IN2P3, CEA/IRFU, Observatoire de Paris, Sorbonne Paris Cité, 75205 Paris Cedex 13, France

[29] Presenter. To whom any correspondence should be addressed.

[15] Department of Physics, University of Houston, Houston, TX 77204, USA
[16] Institute for Theoretical and Experimental Physics, 117218 Moscow, Russia
[17] M. Smoluchowski Institute of Physics, Jagiellonian University, 30059 Krakow, Poland
[18] Kiev Institute for Nuclear Research, 06380 Kiev, Ukraine
[19] Department of Physics, Technische Universität Dresden, 01062 Dresden, Germany
[20] National Research Nuclear University MEPhI (Moscow Engineering Physics Institute), Kashirskoe highway 31, Moscow 115409, Russia
[21] Kepler Center for Astro and Particle Physics, Universität Tübingen, 72076 Tübingen, Germany
[22] Max-Planck-Institut für Kernphysik, 69117 Heidelberg, Germany
[23] Dipartimento di Fisica e Scienze della Terra Università degli Studi di Ferrara e INFN, Via Saragat 1-44122, Ferrara, Italy
[24] Dipartimento di Chimica, Università e INFN, 06123 Perugia, Italy
[25] Physics Department, Queen's University, Kingston ON K7L 3N6, Canada
[26] Amherst Center for Fundamental Interactions and Physics Department, University of Massachusetts, Amherst, MA 01003, USA
[27] Physics and Astronomy Department, University of California Los Angeles (UCLA), Los Angeles, California 90095, USA
[28] Institute of Physics and Excellence Cluster PRISMA, Johannes Gutenberg-Universität Mainz, 55099 Mainz, Germany

E-mail: osmirnov@jinr.ru

Abstract. Measurement of the Solar pp-neutrino flux completed the measurement of Solar neutrino fluxes from the pp-chain of reactions in Borexino experiment. The result is in agreement with the prediction of the Standard Solar Model and the MSW/LMA oscillation scenario. A comparison of the total neutrino flux from the Sun with Solar luminosity in photons provides a test of the stability of the Sun on the 10^5 years time scale, and sets a strong limit on the power production by the unknown energy sources in the Sun.

1. Introduction

The solar photon luminosity $L_\odot = 3.846 \times 10^{26}$ W is measured for a precision of 0.4% [1, 2]. The energy lost by neutrinos adds $L_\nu = 0.023 \cdot L_\odot$ to this value [3]. The solar luminosity constraint on the solar neutrino fluxes can be written as [4]:

$$\frac{L_\odot}{4\pi(1a.u.)^2} = \sum \alpha_i \Phi_i \tag{1}$$

where 1 a.u. is the average Earth-Sun distance, the coefficient α_i is the amount of energy provided to the star by nuclear fusion reactions associated with each of the important solar neutrino fluxes, Φ_i. The numerical values of the α's are determined to an accuracy of 10^{-4} and better.

The estimated uncertainty in the luminosity of the Sun corresponds to less than 3% uncertainty in total solar neutrino flux.

Because of the relation (1) between the solar photon and neutrino luminosity, the measurement of the total neutrino luminosity will provide a test of the stability of the Sun at the time scale of 40000 years [5], the time needed for the radiation born at the center of the Sun to arrive to its surface. Finding a disagreement between $L_\odot$ and L_ν would have significant long term enviromental implications, and in the case of an agreement between the two measurements it would be possible to limit the unknown sources of the solar energy, different from the known thermonuclear fusion of light elements in the pp-chain and CNO-cycle.

The main neutrino sources in the Sun are the pp- and ^{7}Be reactions, providing roughly 91 and 7% of the total neutrino flux respectively. Borexino already measured the ^{7}Be neutrino flux with 5% precision [6], but till recent time the pp-neutrino flux was derived in a differential measurement using solar detectors data.

A number of projects aiming to perform pp-neutrino detection have been put forward in past two decades, but with all the time passed since the proposals, none of them started the operation facing the technical problems with realization. A possibility to use ultrapure liquid organic scintillator as a low energy solar neutrino detector for a first time was discussed in [7, 8]. The authors come to the conclusion that a liquid scintillator detector with an active volume of 10 tons is an appropriate tool to register the solar pp-neutrino if operated at the target level of radiopurity for Borexino and good energy resolution (5% at 200 keV) is achieved.

2. Solar pp-neutrino flux measurement with Borexino detector

The low-energy range, namely 165-590 keV, of the Borexino experimental spectrum has been carefully analyzed with the purpose of the pp-neutrino flux extraction [9]. The data were acquired from January 2012 to May 2013 and correspond to 408 live days of data taking.

The main features of the experimental spectrum, as can be seen in in figure 1, are: a main contribution from the ^{14}C decays at low energies (below 200 keV) and the monoenergetic peak corresponding to 5.3 MeV $\alpha-$particles from ^{210}Po decay. The statistics in the first bins used in the analysis is very high, of the order of 5×10^5events, demanding the development of a very precise model for the studies - the allowed systematic precision at low energy part should be comparable to the statistical fluctuations of 0.14%. The development of such a precise model was a main goal of the analysis.

As one can see from figure 1 the shapes of contributions to the spectrum from ^{14}C and other background components are quite different from the contribution from the expected contribution of the electron recoil spectrum from the Solar pp-neutrino. This fact allows the extraction of the Solar pp-neutrino contribution from the data.

The stability and robustness of the measured pp neutrino interaction rate was verified by performing fits varying initial conditions, including fit energy range, method of pile-up construction, and energy estimator. The distribution of the central values for pp-neutrino interaction rates obtained for all these fit conditions was then used as an estimate of the maximal systematic error (partial correlations between different factors are not excluded).

3. Results and Implications

The solar pp neutrino interaction rate measured by Borexino is $pp = 144 \pm 13(stat) \pm 10(syst)$ cpd/100 t, compatible with the expected rate of $pp_{theor} = 131 \pm 2$ cpd/100 t. The corresponding total solar pp-neutrino flux is $\phi_{pp}(Borex) = (6.6 \pm 0.7) \times 10^{10}$ cm^{-2}s^{-1}, in a good agreement with the combined best fit value of the radiochemical and other solar experiments $\phi_{pp}(other) = (6.14 \pm 0.61) \times 10^{10}$ cm^{-2}s^{-1} [10]. Both are in agreement with the expected value of $6.0 \times (1.000 \pm 0.006) \times 10^{10}$ cm^{-2}s^{-1}.

The survival probability for electron neutrino from pp-reaction is $P_{ee}(Borex) = 0.64 \pm 0.12$.

Taking into account that Borexino and other experiments measurements are independent, the results can be combined:

$$\phi_{pp} = (6.37 \pm 0.46) \times 10^{10} cm^{-2} s^{-1}.$$

The electron neutrino survival probability measured in all solar but Borexino experiment is $P_{ee}(other) = 0.56 \pm 0.08$, combining it with Borexino one we obtain:

$$P_{ee} = 0.60 \pm 0.07,$$

well compatible with theoretical prediction of the MSW/LMA model $0.561^{+0.030}_{-0.042}$.

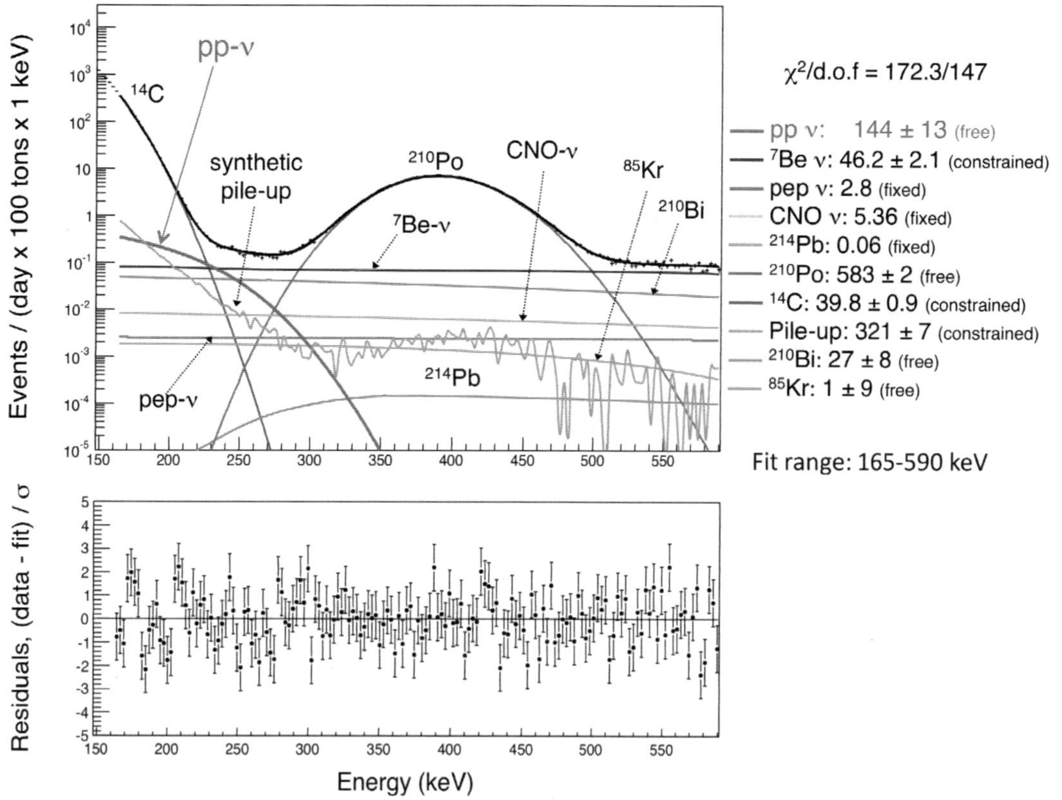

Figure 1. Borexino energy spectrum between 165 and 590 keV (in black). The pp neutrino component is shown in red, the ^{14}C spectrum in dark purple and the synthetic pile-up in light purple. The large green peak is ^{210}Po α-decays. ^{7}Be (dark blue), pep and CNO (light blue) solar neutrinos, and ^{210}Bi (orange) spectra are almost flat in this energy region.

Table 1. The Standard Solar Model predictions (for high metallicity and low metallicity abundances) and current experimental status of the Solar neutrino fluxes measurement. The limits are given for 90% C.L.. The corresponding energy release is calculated in the last column.

Reaction	GS98 [11]	AGS09 [12]	Units $cm^{-2}s^{-1}$	Measurement	MeV/1ν	L $\times 10^{26}$ W·s^{-1}
pp	5.98 ± 0.04	6.03 ± 0.04	$\times 10^{10}$	6.14 ± 0.61 [10] 6.6 ± 0.7 [13] 6.37 ± 0.46	13.10	3.76 ± 0.28
pep	1.44 ± 0.012	1.47 ± 0.012	$\times 10^{8}$	1.6 ± 0.3 [14]	11.92	0.009 ± 0.002
^{7}Be	5.00 ± 0.07	4.56 ± 0.07	$\times 10^{9}$	4.87 ± 0.24 [6]	12.60	0.276 ± 0.014
^{8}B	5.58 ± 0.14	4.59 ± 0.14	$\times 10^{6}$	5.25 ± 0.16 [15]	6.63	1.57 ± 0.05 $\times 10^{-4}$
hep	8.0 ± 2.4	8.3 ± 2.5	$\times 10^{3}$	< 23[16]		
^{13}N	2.96 ± 0.14	2.17 ± 0.14	$\times 10^{8}$	CNO: < 7.4[14]		
^{15}O	2.23 ± 0.15	1.56 ± 0.15	$\times 10^{8}$			
^{17}F	5.52 ± 0.17	3.40 ± 0.16	$\times 10^{6}$			

All available measurements of the solar neutrino fluxes are shown in table 1. The total energy production in the solar reactions observed till now (by detecting corresponding neutrino fluxes) is $(4.04 \pm 0.28) \times 10^{26}$ W·s^{-1} in a good agreement with a total measured $L_{\odot} = 3.846 \times 10^{26}$ W·s^{-1}. There is not much space left for the unknown energy sources, the 90% C.L. lower limit for the total energy production (conservatively assuming zero contribution from the not-observed reactions) is $L_{tot} = 3.68 \times 10^{26}$ W·s^{-1}. If one assumes that such an unknown source exists, its total power with 90% probability can't exceed 0.15×10^{26} W·s^{-1}. In other words no more than 4% of the total energy production in the Sun is left for the unknown energy sources, confirming that the Sun shines due to the thermonuclear fusion reactions.

4. Conclusion

Borexino provided an independent measurement of the Solar pp-neutrino flux, which can be combined with measurements from other solar experiments. A comparison of the neutrino flux from the Sun with Solar luminosity in photons provides a test of the stability of the Sun on the 10^5 years time scale, and allows to set a limit of no more than 4% of the total power production for the unknown energy sources in the Sun at 90% C.L..

Acknowledgments

The Borexino program is made possible by funding from INFN (Italy), NSF (USA), BMBF, DFG, and MPG (Germany), RFBR: Grants 14-22-03031 and 13-02-12140, RFBR-ASPERA-13-02-92440 (Russia), and NCN Poland (UMO-2012/06/M/ST2/00426). We acknowledge the generous support and hospitality of the Laboratori Nazionali del Gran Sasso (LNGS). Russian colleagues from MEPhI acknowledge partial support from MEPhI Academic Excellence Project (contract No. 02.a03.21.0005, 27.08.2013).

References

[1] Chapman G A 1997 *Encyclopedia of Planetary Science, Encyclopedia of Earth Science* (Springer Netherlands) 748
[2] Fröhlich C and Lean J 1998 *Geophysical Research Letters* **25**(23) 4377–80
[3] Bahcall J N 1989 *Neutrino Astrophysics* (Cambridge University Press)
[4] Bahcall J N 2002 *Phys. Rev.* C 025801
[5] Fiorentini G and Ricci B 1999 *Comments on Astrophysics* **1** 49–51
[6] Bellini G *et al.* 2011 *Physical Review Letters* **107**(14) 141302
[7] Smirnov O Ju, Zaimidoroga O A and Derbin A V 2003 *Physics of Atomic Nuclei* **66**(4) 712–23
[8] Derbin A V, Smirnov O.Yu. and Zaimidoroga O A 2004 *Physics of Atomic Nuclei* **67**(11) 2066–72
[9] Bellini G *et al.* 2014 *Nature* **512** 383–6
[10] Ianni A 2014 *Physics of the Dark Universe* **4** 44 – 9
[11] Grevesse N and Sauval A J 1998 *Space Science Reviews* **85**(1-2) 161–74
[12] Asplund M *et al.* 2009 *Annual Review of Astronomy and Astrophysics* **47**(1) 481–522
[13] Bellini G *et al.* 2014 *Phys. Rev.* D **89** 112007
[14] Bellini G *et al.* 2012 *Phys. Rev. Lett.* **108** 051302
[15] Aharmim B *et al.* 2010 *Phys. Rev.* C **81** 055504
[16] Aharmim B *et al.* 2006 *The Astrophysical Journal* **653**(2) 1545

Rotating cylindrical wormholes: a no-go theorem

K A Bronnikov[1,2,3]

[1] Center for Gravitation and Fundam. Metrology, VNIIMS, Ozyornaya 46, Moscow 119361, Russia

[2] Institute of Gravitation and Cosmology, PFUR, Miklukho-Maklaya 6, Moscow 117198, Russia

[3] National Research Nuclear University MEPhI (Moscow Engineering Physics Institute), Kashirskoe highway 31, Moscow 115409, Russia

E-mail: kb20@yandex.ru

Abstract. The existing solutions to the Einstein equations describing rotating cylindrical wormholes are not asymptotically flat and therefore cannot describe wormhole entrances as local objects in our Universe. To overcome this difficulty, flat asymptotic regions are added to wormhole solutions by matching them at some surfaces Σ_- and Σ_+. It is shown, however, that if the wormhole solution is obtained for scalar fields with arbitrary potentials, possibly interacting with an azimuthal electric or magnetic field, then the matter content of one or both thin shells appearing on Σ_- and Σ_+ violate the Null Energy Condition. Thus exotic matter is still necessary for obtaining a twice asymptotically flat wormhole.

Traversable Lorentzian wormholes are widely discussed in gravitational physics since they lead to many effects of interest like time machines or shortcuts between distant parts of space. Large enough wormholes, if any, can lead to observable effects in astronomy [1–3].

In attempts to build realistic wormhole models, the main difficulty is that in general relativity (GR) and some of its extensions a static wormhole geometry requires the presence of "exotic", or phantom matter, that is, matter violating the weak and null energy condition (WEC and NEC), at least near the throat, the narrowest place in a wormhole [4–7]. This result is obtained if the throat is a compact 2D surface with a finite area [6].

Examples of phantom-free wormhole solutions have been obtained in extensions of GR, such as the Einstein-Cartan theory [8], Einstein-Gauss-Bonnet gravity [9], brane worlds [10], etc. We here prefer to consider GR as a theory quite well describing the reality on the macroscopic scale while the extensions more likely concern very large densities and/or curvatures. In GR there are phantom-free wormhole models with axial symmetry, including the Zipoy [11] and superextremal Kerr vacuum solutions as well as solutions with scalar and electromagnetic fields possibly interacting with each other [12, 13]; in all of them, however, a disk that plays the role of a wormhole throat is bounded by a ring singularity whose existence may be though of as an unpleasant price paid for the absence of exotic matter.

The above-mentioned result of [6] does not directly apply to objects like cosmic strings, infinitely stretched along a certain direction, in the simplest case cylindrically symmetric ones. Such systems with and without rotation were discussed, in particular, in [14–17] (see also references therein). It was shown there, with a number of examples, that phantom-free cylindrical wormhole solutions to the Einstein equations are rather easily obtained.

A problem with cylindrical systems is their undesirable asymptotic behavior. To describe wormholes potentially visible to distant observers like ourselves in our very weakly curved Universe, one has to require their flat (or string[4]) asymptotics, which is hard to achieve under cylindrical symmetry.

To overcome this difficulty, it was suggested [15] to build wormholes with flat asymptotic regions on both sides of the throat by matching a wormhole solution to suitable parts of Minkowski space-time, but no successful examples of phantom-free wormholes were so far obtained in this way. In the present paper we further discuss this opportunity and prove a no-go theorem on the conditions under which this cut-and-paste procedure cannot lead to asymptotically flat phantom-free wormholes.

Consider a stationary cylindrically symmetric metric of the form

$$ds^2 = \mathrm{e}^{2\gamma(u)}[dt - E(u)\mathrm{e}^{-2\gamma(u)}\, d\varphi]^2 - \mathrm{e}^{2\alpha(u)}du^2 - \mathrm{e}^{2\mu(u)}dz^2 - \mathrm{e}^{2\beta(u)}d\varphi^2, \tag{1}$$

where u, z and φ are the radial, longitudinal and angular coordinates. This metric is said to describe a wormhole if either (i) the circular radius $r(u) = \mathrm{e}^{\beta(u)}$ has a minimum (called an r-throat) and is large (preferably tends to infinity) far from this minimum or (ii) a similar behavior is observed for the area function $a(u) = \mathrm{e}^{\mu+\beta}$ (its minimum is called an a-throat) [14, 15]. If a wormhole is asymptotically flat at both extremes of the u range, it evidently possesses both kinds of throats.

The vortex gravitational field existing in a space-time with the metric (1) can be characterized by the angular velocity $\omega(u)$ given by [15, 18, 19]

$$\omega = \tfrac{1}{2}(E\mathrm{e}^{-2\gamma})'\mathrm{e}^{\gamma-\beta-\alpha}. \tag{2}$$

under an arbitrary choice of the coordinate u (a prime stands for d/du). Furthermore, the reference frame comoving to a matter distribution in its motion by the angle φ is determined by a zero component T_0^3 of the stress-energy tensor (SET), hence (via the Einstein equations) by the Ricci tensor component $R_0^3 \sim (\omega\mathrm{e}^{2\gamma+\mu})' = 0$, and thus in this reference frame we have

$$\omega = \omega_0\mathrm{e}^{-\mu-2\gamma}, \qquad \omega_0 = \mathrm{const}. \tag{3}$$

It then turns out [15] that the diagonal components of the Ricci (R_μ^ν) and Einstein ($G_\mu^\nu = R_\mu^\nu - \frac{1}{2}\delta_\mu^\nu R$) tensors split into the corresponding components for the static metric (that is, (1) with $E = 0$) plus the ω-dependent addition, so that $G_\mu^\nu = {}_sG_\mu^\nu + {}_\omega G_\mu^\nu$, where ${}_sG_\mu^\nu$ is the static part, and

$${}_\omega G_\mu^\nu = \omega^2\,\mathrm{diag}(-3,\ 1,\ -1,\ 1). \tag{4}$$

Moreover, the tensors ${}_sG_\mu^\nu$ and ${}_\omega G_\mu^\nu$ (each separately) satisfy the conservation law $\nabla_\alpha G_\mu^\alpha = 0$ in this static metric. Hence, by the Einstein equations $G_\mu^\nu = -\varkappa T_\mu^\nu$, the tensor ${}_\omega G_\mu^\nu/\varkappa$ acts as an additional SET with exotic properties (e.g., the effective energy density is $-3\omega^2/\varkappa < 0$), making it easier to satisfy the existence conditions for both r- and a-throats, as confirmed by a number of examples in [15, 17, 19].

In all these examples, however, the wormholes are not asymptotically flat; even more than that, as is clear from equation (3), any solution obtained in this way cannot be asymptotically flat since it would require $\omega \to 0$ along with finite limits of γ and μ, which is incompatible with (3).

The following trick was suggested [15] for obtaining an asymptotically flat configuration: to cut a non-asymptotically flat wormhole configuration at some regular cylinders Σ_+ ($u = u_+$)

[4] Or at least asymptotically flat up to an angular deficit well-known in cosmic string configurations. In what follows, for brevity, we speak of asymptotic flatness though cosmic-string asymptotics could also be mentioned on equal grounds.

and Σ_- ($u = u_-$) on different sides of the throat and to match it there with flat-space regions extending to infinity. Then the junction surfaces comprise thin shells with certain surface SETs, and it remains to check whether or not these SETs satisfy the WEC and NEC.

We will show, however, that with a large class of matter sources of the wormhole solution, it is impossible to obtain the surface SETs on both Σ_+ and Σ_- respecting the NEC.

Indeed, if we choose the harmonic radial coordinate u [20, 21] specified by the relation $\alpha = \beta + \gamma + \mu$, then a certain combination of the Einstein equations takes the form

$$\beta'' - \gamma'' - 4\omega_0^2 e^{2\beta-2\gamma} = \varkappa e^{2\alpha}(T^t_t - T^\varphi_\varphi). \tag{5}$$

Therefore, if the SET of matter satisfies the condition $T^t_t = T^\varphi_\varphi$, equation (5) is easily integrated giving

$$e^{\beta-\gamma} = \frac{1}{2|\omega_0|s(k,u)}, \qquad k = \text{const}, \tag{6}$$

where one more integration constant has been excluded by choosing the origin of the u coordinate, and the function $s(k,u)$ is defined as

$$s(k,u) = \begin{cases} k^{-1}\sinh ku, & k > 0, \quad u \in \mathbb{R}_+; \\ u, & k = 0, \quad u \in \mathbb{R}_+; \\ k^{-1}\sin ku, & k < 0, \quad 0 < u < \pi/|k|. \end{cases} \tag{7}$$

The condition $T^t_t = T^\varphi_\varphi$ holds for a large class of matter Lagrangians in the metric (1), such as, e.g.,

$$L = g^{\mu\nu}\partial_\mu\phi\partial_\nu\phi - 2V(\phi) - P(\phi)F^{\mu\nu}F_{\mu\nu}, \tag{8}$$

with an arbitrary scalar field potential $V(\phi)$ and an arbitrary function $P(\phi)$ characterizing the scalar-electromagnetic interaction, assuming that $\phi = \phi(u)$ and that the Maxwell tensor $F_{\mu\nu}$ describes a stationary azimuthal magnetic field ($F_{21} = -F_{12} \neq 0$) or its electric analogue.

Quite a general observation follows from equation (6): in the case $k < 0$, if the remaining field equations give the metric functions μ and α which are finite and regular on the segment $0 \le u \le \pi/|k|$ (or, equivalently, on any other half-wave of the function $\sin ku$), then the whole configuration is a wormhole, with both r- and a-throats; both $r(u)$ and $a(u)$ tend to infinity as $u \to 0$ and $u \to \pi/|k|$, but these limiting surfaces are singular due to $e^\gamma \to 0$ and $\omega \to \infty$. Configurations with $k \ge 0$ can also be of wormhole nature, as is confirmed by examples of vacuum and scalar-vacuum solutions [15, 17].

Now suppose we have such a wormhole metric and cut it at some regular points $u = u_-$ to the left and $u = u_+$ to the right of both r- and a-throats. Let us try to join it at $u = u_\pm$ to regions of Minkowski space-time with the metric $ds^2_{\rm M} = dt^2 - dx^2 - dz^2 - x^2 d\varphi^2$ outside suitable cylinders $x = x_\pm = $ const. Assuming a rotating reference frame with an angular velocity $\Omega = $ const, we substitute $\varphi \to \varphi + \Omega t$ to obtain

$$-ds^2_{\rm M} = dx^2 + dz^2 + x^2(d\varphi + \Omega dt)^2 - dt^2. \tag{9}$$

The relevant quantities in the notations of (1) are

$$e^{2\gamma} = 1 - \Omega^2 x^2, \qquad e^{2\beta} = \frac{x^2}{1 - \Omega^2 x^2}, \qquad E = \Omega x^2, \qquad \omega = \frac{\Omega}{1 - \Omega^2 x^2}. \tag{10}$$

This metric is stationary and can be matched to an internal metric at $|x| < 1/|\Omega|$, inside the "light cylinder" on which the linear rotational velocity reaches that of light.

Matching of two cylindrically symmetric regions at a surface $\Sigma: \; u = u_0$ means that we identify the two metrics on this surface, so that

$$[\beta] = 0, \qquad [\mu] = 0, \qquad [\gamma] = 0, \qquad [E] = 0, \tag{11}$$

where, as usual, the brackets denote discontinuities: for any $f(u)$, $[f] = f(u_0 + 0) - f(u_0 - 0)$. One should note that in general the metrics on different sides of the surface Σ may be written using different choices of the radial coordinate u, but it does not matter since the quantities involved in all matching conditions used are insensitive to the choice of u.

The next step is to determine the material content of the junction surface Σ according to the Darmois-Israel formalism [22, 23]: in our case of a timelike surface Σ, the surface SET S_a^b is expressed in terms of K_a^b, the extrinsic curvature of Σ, as

$$S_a^b = -\varkappa^{-1}[\tilde{K}_a^b], \qquad \tilde{K}_a^b := K_a^b - \delta_a^b K_c^c, \qquad a, b, c = 0, 2, 3. \tag{12}$$

Assume that the matching conditions (11) on $\Sigma_\pm$, identifying the surfaces $x = x_\pm$ in Minkowski regions and $u = u_\pm$ in the internal region, are fulfilled by fixing the values of $x_\pm$ for given $u_\pm$ and other parameters of the system, including the values of $\Omega = \Omega_\pm$ in each Minkowski region. It is important that we should take $x_+ > 0$ and $x_- < 0$ to adjust the directions of the normal vectors to $\Sigma_\pm$.

Now, the question is whether the surface SETs on $\Sigma_\pm$ can satisfy the WEC whose requirements are

$$S_{00}/g_{00} = \sigma \geq 0, \qquad S_{ab}\xi^a\xi^b \geq 0, \tag{13}$$

where ξ^a is any null vector ($\xi^a\xi_a = 0$) on $\Sigma = \Sigma_\pm$; the second inequality in (13) comprises the NEC as part of the WEC. The conditions (13) are equivalent to

$$[\tilde{K}_{44}/g_{44}] \leq 0, \qquad [K_{ab}\xi^a\xi^b] \leq 0. \tag{14}$$

If we choose two null vectors on Σ in the z and φ directions as

$$\xi^a_{(1)} = (\mathrm{e}^{-\gamma},\ \mathrm{e}^{-\mu},\ 0), \qquad \xi^a_{(2)} = (\mathrm{e}^{-\gamma} + E\mathrm{e}^{-\beta-2\gamma},\ 0,\ \mathrm{e}^{-\beta}), \tag{15}$$

the conditions (14) read [15]

$$[\mathrm{e}^{-\alpha}(\beta' + \mu')] \leq 0, \qquad [\mathrm{e}^{-\alpha}(\mu' - \gamma')] \leq 0, \qquad [\mathrm{e}^{-\alpha}(\beta' - \gamma') + 2\omega] \leq 0. \tag{16}$$

Now we can apply these requirements to our configuration at both junctions. Consider the third condition which contains the function $\beta - \gamma$ given by equation (6). Using it, on Σ_- with $x = x_- < 0$ we obtain

$$\mathrm{e}^{-\alpha(u_-)}\,\frac{(-s' + \operatorname{sign}\omega_0)}{s} + \frac{(1 + \Omega_- x)^2}{|x|(1 - \Omega_-^2 x^2)} \leq 0, \tag{17}$$

and on Σ_+ with $x = x_+ > 0$ we have in a similar way

$$\mathrm{e}^{-\alpha(u_+)}\,\frac{(s' - \operatorname{sign}\omega_0)}{s} + \frac{(1 + \Omega_+ x)^2}{x(1 - \Omega_+^2 x^2)} \leq 0. \tag{18}$$

Here s and $s' = ds/du$ refer to the function $s = s(k, u)$ introduced in (7).

The inequalities (17) and (18) lead to the conclusion that the NEC (hence also the WEC) cannot be satisfied on both Σ_+ and Σ_- simultaneously.

Indeed, assuming $\omega_0 > 0$, the inequality (17) can only hold if $1 - s'(k,u) < 0$ at $u = u_-$. But $s'(k,u) = \{\cosh ku,\ 1,\ \cos|k|u\}$ for $k > 0$, $k = 0$ and $k < 0$, respectively, and only at $k > 0$ we have $1 - s' < 0$. Thus at Σ_- the solution in the internal region should be taken with $k > 0$. On the contrary, (18) can hold only if $1 - s'(k,u) > 0$ at $u = u_+$, which is only possible if $k < 0$. But any particular solution describing the internal region has a fixed value of k, hence the inequalities (17) and (18) are incompatible with each other. Furthermore, if $\omega_0 < 0$, we have in the first term in (17) $-s' - 1 < 0$, making no problem; however, in (18) there is, instead, $s' + 1 > 0$, hence this inequality cannot hold whatever be the parameter k.

We conclude that the NEC is inevitably violated at least on one of the surfaces Σ_+ and Σ_-.

This general result has been obtained for any source of gravity in wormhole solutions such that $T^t_t = T^\varphi_\varphi$ in the metric (1), in particular, those given by (8). Possible solutions with other sources are certainly not ruled out. So the problem of obtaining potentially observable, stationary, nonsingular, phantom-free wormholes in GR remains open.

Acknowledgments

This work was performed within the framework of the Center FRPP supported by MEPhI Academic Excellence Project (contract No. 02.a03.21.0005, 27.08.2013)

References

[1] Doroshkevich A, Hansen J, Novikov I and Shatskiy A 2009 *Int. J. Mod. Phys.* D **18** 1665 (*arXiv*: 0812.0702)
[2] Harko T, Kovacs Z and Lobo F S N 2009 *Phys. Rev.* D **79** 064001 (*arXiv*: 0901.3926)
[3] Kirillov A A and Savelova E P 2013 *Grav. Cosmol.* **19** 92
[4] Morris M, Thorne K S and Yurtsever U 1988 *Phys. Rev. Lett.* **61** 1446
[5] Visser M 1995 *Lorentzian Wormholes: from Einstein to Hawking* (Woodbury: AIP).
[6] Hochberg D and Visser M 1997 *Phys. Rev.* D **56** 4745 (*arXiv*: gr-qc/9704082)
[7] Bronnikov K A and Rubin S G 2012 *Black Holes, Cosmology and Extra Dimensions* (Singapore: World Scientific).
[8] Bronnikov K A and Galiakhmetov A M 2015 Wormholes without exotic matter in Einstein-Cartan theory, to appear in *Grav. Cosmol.* **21** (4) (*arXiv*: 1508.01114)
[9] Maeda H and Nozawa M 2008 *Phys. Rev.* D **78** 024005
[10] Bronnikov K A and Kim S-W 2003 *Phys. Rev.* D **67** 064027 (*arXiv*: gr-qc/0212112)
[11] Zipoy D 1966 *J. Math. Phys.* **7** 1137
[12] Bronnikov K A and Fabris J C 1997 *Class. Quantum Grav.* **14** 831 (*arXiv*: gr-qc/9603037)
[13] Miranda G and Matos T 2015 Exact rotating magnetic traversable wormholes satisfying the energy conditions *arXiv*: 1507.02348
[14] Bronnikov K A and Lemos J P S 2009 *Phys. Rev.* D **79** 104019 (*arXiv*: 0902.2360)
[15] Bronnikov K A, Krechet V G and Lemos J P S 2013 *Phys. Rev.* D **87** 084060 (*arXiv*: 1303.2993)
[16] Bronnikov K A and Skvortsova M V 2014 *Grav. Cosmol.* **20** 171 (*arXiv*: 1404.5750)
[17] Bronnikov K A and Krechet V G 2015 Rotating cylindrical wormholes and energy conditions *arXiv*: 1509.04665
[18] Krechet V G 2007 *Russian Physics Journal* **50**(10) 1021 (Original Russian text: *Izvestiya Vuzov, Fizika* **50**(10) 57)
[19] Krechet V G and Sadovnikov D V 2009 *Grav. Cosmol.* **15** 337 (*arXiv*: 0912.2181)
[20] Bronnikov K A 1979 *J. Phys. A: Math. Gen.* **12** 201
[21] Bronnikov K A and Kovalchuk M A 1979 *Gen. Rel. Grav.* **11** 343
[22] Israel W 1967 *Nuovo Cim.* B **48** 463
[23] Berezin V A, Kuzmin V A and Tkachev I I 1987 *Phys. Rev.* D **36** 2919

International Conference on Particle Physics and Astrophysics (ICPPA-2015) IOP Publishing
Journal of Physics: Conference Series **675** (2016) 012029 doi:10.1088/1742-6596/675/1/012029

Geo-neutrino results with Borexino

R Roncin[1,2,29], M Agostini[3], S Appel[3], G Bellini[4], J Benziger[5], D Bick[6], G Bonfini[1], D Bravo[7], B Caccianiga[4], F Calaprice[8], A Caminata[9], P Cavalcante[1], A Chepurnov[10], D D'Angelo[4], S Davini[11], A Derbin[11], L Di Noto[9], I Drachnev[11], A Etenko[13], K Fomenko[14], D Franco[2], F Gabriele[1], C Galbiati[8], C Ghiano[9], M Giammarchi[4], M Goeger-Neff[3], A Goretti[8], M Gromov[10], C Hagner[6], E Hungerford[15], Aldo Ianni[1], Andrea Ianni[8], K Jedrzejczak[17], M Kaiser[6], V Kobychev[18], D Korablev[14], G Korga[1], D Kryn[2], M Laubenstein[1], B Lehnert[19], E Litvinovich[13,20], F Lombardi[1], P Lombardi[4], L Ludhova[4], G Lukyanchenko[13,20], I Machulin[13,20], S Manecki[7], W Maneschg[22], S Marcocci[11], E Meroni[4], M Meyer[6], L Miramonti[4], M Misiaszek[17,1], M Montuschi[23], P Mosteiro[8], V Muratova[11], B Neumair[3], L Oberauer[3], M Obolensky[2], F Ortica[24], M Pallavicini[9], L Papp[3], L Perasso[9], A Pocar[26], G Ranucci[4], A Razeto[1], A Re[4], A Romani[24], N Rossi[1], S Schönert[3], D Semenov[11], H Simgen[22], M Skorokhvatov[13,20], O Smirnov[14]A Sotnikov[14], S Sukhotin[13], Y Suvorov[27,13], R Tartaglia[1], G Testera[9], J Thurn[19], M Toropova[13], E Unzhakov[11], A Vishneva[14], R B Vogelaar[7], F von Feilitzsch [3], H Wang[27], S Weinz[28], J Winter[28], M Wojcik[17], M Wurm[28], Z Yokley[7], O Zaimidoroga[14], S Zavatarelli[9], K Zuber[19] and G Zuzel[17](Borexino Collaboration)

[1] INFN Laboratori Nazionali del Gran Sasso, 67010 Assergi (AQ), Italy
[2] AstroParticule et Cosmologie, Université Paris Diderot, CNRS/IN2P3, CEA/IRFU, Observatoire de Paris, Sorbonne Paris Cité, 75205 Paris Cedex 13, France
[3] Physik-Department and Excellence Cluster Universe, Technische Universität München, 85748 Garching, Germany
[4] Dipartimento di Fisica, Università degli Studi e INFN, 20133 Milano, Italy
[5] Chemical Engineering Department, Princeton University, Princeton, NJ 08544, USA
[6] Institut für Experimentalphysik, Universität, 22761 Hamburg, Germany
[7] Physics Department, Virginia Polytechnic Institute and State University, Blacksburg, VA 24061, USA
[8] Physics Department, Princeton University, Princeton, NJ 08544, USA
[9] Dipartimento di Fisica, Università degli Studi e INFN, Genova 16146, Italy
[10] Lomonosov Moscow State University Skobeltsyn Institute of Nuclear Physics, 119234 Moscow, Russia
[11] Gran Sasso Science Institute (INFN), 67100 L'Aquila, Italy
[12] St. Petersburg Nuclear Physics Institute NRC Kurchatov Institute, 188350 Gatchina, Russia
[13] NRC Kurchatov Institute, 123182 Moscow, Russia
[14] Joint Institute for Nuclear Research, 141980 Dubna, Russia
[15] Department of Physics, University of Houston, Houston, TX 77204, USA
[16] Institute for Theoretical and Experimental Physics, 117218 Moscow, Russia

[29] Presenter. To whom any correspondence should be addressed.

[17] M. Smoluchowski Institute of Physics, Jagiellonian University, 30059 Krakow, Poland
[18] Kiev Institute for Nuclear Research, 06380 Kiev, Ukraine
[19] Department of Physics, Technische Universität Dresden, 01062 Dresden, Germany
[20] National Research Nuclear University MEPhI (Moscow Engineering Physics Institute), Kashirskoe highway 31, Moscow 115409, Russia
[21] Kepler Center for Astro and Particle Physics, Universität Tübingen, 72076 Tübingen, Germany
[22] Max-Planck-Institut für Kernphysik, 69117 Heidelberg, Germany
[23] Dipartimento di Fisica e Scienze della Terra Università degli Studi di Ferrara e INFN, Via Saragat 1-44122, Ferrara, Italy
[24] Dipartimento di Chimica, Università e INFN, 06123 Perugia, Italy
[25] Physics Department, Queen's University, Kingston ON K7L 3N6, Canada
[26] Amherst Center for Fundamental Interactions and Physics Department, University of Massachusetts, Amherst, MA 01003, USA
[27] Physics and Astronomy Department, University of California Los Angeles (UCLA), Los Angeles, California 90095, USA
[28] Institute of Physics and Excellence Cluster PRISMA, Johannes Gutenberg-Universität Mainz, 55099 Mainz, Germany

E-mail: `romain.roncin@lngs.infn.it`

Abstract. Borexino is a liquid scintillator detector primary designed to observe solar neutrinos. Due to its low background level as well as its position in a nuclear free country, Italy, Borexino is also sensitive to geo-neutrinos. Borexino is leading this interdisciplinary field of neutrino geoscience by studying electron antineutrinos which are emitted from the decay of radioactive isotopes present in the crust and the mantle of the Earth. With 2056 days of data taken between December 2007 and March 2015, Borexino observed 77 antineutrino candidates. If we assume a chondritic Th/U mass ratio of 3.9, the number of geo-neutrino events is found to be $23.7^{+6.5}_{-5.7}\,(\mathrm{stat})\,^{+0.9}_{-0.6}\,(\mathrm{syst})$. With this measurement, Borexino alone is able to reject the null geo-neutrino signal at $5.9\,\sigma$, to claim a geo-neutrino signal from the mantle at 98 % C.L. and to restrict the radiogenic heat production for U and Th between 23 and 36 TW.

1. Introduction

Geo-neutrinos are electron antineutrinos which are produced by the decay of radioactive isotopes present in the crust and the mantle of our planet. Since the chemical composition of the Earth is not yet perfectly known, having a new source of information will help to better understand our planet. The idea of using geo-neutrinos as direct messengers was suggested in 1965 by G. Eder [1] and in 1968 by G. Marx [2] before being reviewed by L.M. Krauss, S.L. Glashow and D.N. Schramm in 1984 [3]. So far, only the KamLAND experiment in Japan [4, 5] and the Borexino experiment in Italy [6, 7, 8] have reported geo-neutrino measurements.

2. Geo-neutrino analysis and results

In Borexino, the detection of geo-neutrinos relies on the signature of the inverse β decay (IBD) reaction $\bar{\nu}_e + p \rightarrow e^+ + n$ where the positron, the "prompt" signal, is followed in time by the neutron capture on hydrogen, the "delayed" signal. The prompt and the delayed signals are correlated in space and time, allowing to accurately identify electron antineutrino signal. With an IBD threshold of 1.806 MeV, only geo-neutrinos coming from the decay of ^{238}U and ^{232}Th chains can be detected.

Despite Italy is a nuclear free country, the dominant background remains electron antineutrinos emitted by abroad nuclear reactors. It is nonetheless possible to estimate the expected number of nuclear reactors events, N_{react}, as follows:

$$N_{\mathrm{react}} = \sum_{r=1}^{R}\sum_{m=1}^{M} \frac{\eta_m}{4\pi L_r^2} P_{rm} \times \int \mathrm{d}E_{\bar{\nu}_e} \sum_{i=1}^{4} \frac{f_i}{E_i} \phi_i(E_{\bar{\nu}_e}) \sigma(E_{\bar{\nu}_e}) P_{ee}(E_{\bar{\nu}_e}, L_r), \quad (1)$$

where r runs over the number of nuclear reactors R considered, m runs over the number of months M considered, η_m stands for the exposure in month m and includes detector efficiency, L_r is the detector-reactor distance, P_{rm} is the effective thermal power of reactor r in month m, i runs over the spectral components of ^{235}U, ^{238}U, ^{239}Pu and ^{241}Pu, f_i is the power fraction of component i, E_i the average energy released per fission of component i, $\phi_i(E_{\bar{\nu}_e})$ the antineutrino energy spectrum per fission of component i, $\sigma(E_{\bar{\nu}_e})$ the IBD cross section and $P_{ee}(E_{\bar{\nu}_e}, L_r)$ the survival probability of the emitted antineutrinos of energy $E_{\bar{\nu}_e}$ created at distance L_r.

Table 1. Estimated background components in terms of number of events taken from [8]. The combined upper limit is obtained by Monte Carlo.

^{9}Li-^{8}He	$0.194^{+0.125}_{-0.089}$
Accidental coincidences	0.221 ± 0.004
Time correlated	$0.035^{+0.029}_{-0.028}$
(α, n) in scintillator	0.165 ± 0.010
(α, n) in buffer	< 0.51
Fast n's (μ in WT)	< 0.01
Fast n's (μ in rock)	< 0.43
Untagged muons	0.12 ± 0.01
Fission in PMTs	0.032 ± 0.003
^{214}Bi-^{214}Po	0.009 ± 0.013
Total	$0.78^{+0.13}_{-0.10}$
	< 0.65 (combined)

Other backgrounds can mimick an IBD reaction in Borexino, like (α, n) background, accidental coincidences and cosmogenic background such as ^{9}Li-^{8}He. In Borexino, the overall background rate is estimated to be a factor 100 lower than the antineutrino one. The estimated background for each components is reported in table 1.

In order to measure the number of geo-neutrinos and antineutrinos from nuclear reactors, we implement an unbinned maximum likelihood fit of the prompt energy spectrum of our antineutrino candidates. We define the log-likelihood function as follows:

$$\begin{aligned} \ln \mathcal{L}(N_{\mathrm{geo}}, N_{\mathrm{react}}, N_{\mathrm{acc}}, N_{\mathrm{LiHe}}, N_{\alpha n}) &= -N_{\mathrm{exp}}(N_{\mathrm{geo}}, N_{\mathrm{react}}, N_{\mathrm{acc}}, N_{\mathrm{LiHe}}, N_{\alpha n}) \\ &+ \sum_{i=1}^{N} \ln\left(f_{\bar{\nu}_e}(E_i, N_{\mathrm{geo}}, N_{\mathrm{react}}) + f_{\mathrm{bg}}(E_i, N_{\mathrm{acc}}, N_{\mathrm{LiHe}}, N_{\alpha n})\right) - \frac{1}{2}\sum_{\mathrm{bg}} \left(\frac{N_{\mathrm{bg}} - (N_{\mathrm{bg}})_{\mathrm{est}}}{(\delta_{\mathrm{bg}})_{\mathrm{est}}}\right)^2, \end{aligned} \quad (2)$$

with:

$$f_{\bar{\nu}_e}(E_i, N_{\mathrm{geo}}, N_{\mathrm{react}}) = f_{\mathrm{geo}}(E_i, N_{\mathrm{geo}}) + f_{\mathrm{react}}(E_i, N_{\mathrm{react}}) \quad (3)$$

$$f_{\mathrm{bg}}(E_i, N_{\mathrm{acc}}, N_{\mathrm{LiHe}}, N_{\alpha n}) = f_{\mathrm{acc}}(E_i, N_{\mathrm{acc}}) + f_{\mathrm{LiHe}}(E_i, N_{\mathrm{LiHe}}) + f_{\alpha n}(E_i, N_{\alpha n}) \quad (4)$$

where N_{exp} corresponds to the expected total number of events and i runs over the $N = 77$ antineutrino candidates. f_{geo}, f_{react}, f_{acc}, f_{LiHe} and $f_{\alpha n}$ are the individual spectra of the geo-neutrinos, the antineutrinos from nuclear reactors, the accidental coincidences, the ^{9}Li-^{8}He events and the (α, n) events. N_{geo} and N_{react} are left as free parameters while the last term

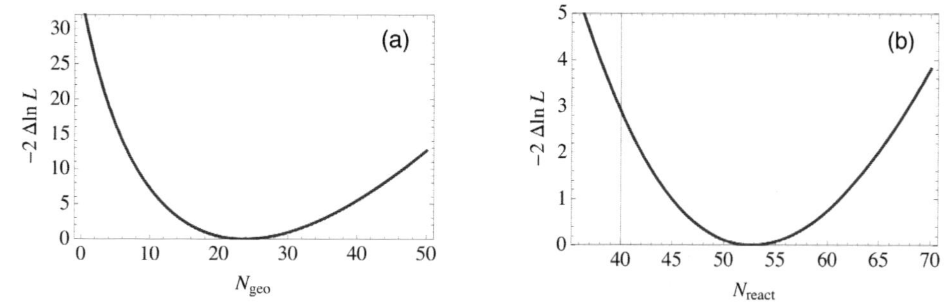

Figure 1. $-2\,\Delta\ln\mathcal{L}$ profiles for N_{geo} (a) and N_{react} (b).

constrains the background components reported in table 1.

If we assume a chondritic Th/U mass ratio of 3.9, our best fit values are $N_{\mathrm{geo}} = 23.7^{+6.5}_{-5.7}\,(\mathrm{stat})\,^{+0.9}_{-0.6}\,(\mathrm{syst})$ and $N_{\mathrm{react}} = 52.7^{+8.5}_{-7.7}\,(\mathrm{stat})\,^{+0.7}_{-0.9}\,(\mathrm{syst})$ events, which is equivalent to $43.5^{+11.8}_{-10.4}\,(\mathrm{stat})\,^{+2.7}_{-2.4}\,(\mathrm{syst})$ and $96.6^{+15.6}_{-14.2}\,(\mathrm{stat})\,^{+4.9}_{-5.0}\,(\mathrm{syst})$ TNU[30] respectively. This result allows to reject the null geo-neutrino signal at $5.9\,\sigma$. Figure 1 shows the $-2\,\Delta\ln\mathcal{L}$ profiles for N_{geo} and N_{react}.

A signal from the mantle can then be assessed by retrieving the crust signal (investigated in [9] and [10]) to the total signal measured in Borexino. Using the geo-neutrino log-likelihood profile and assuming a Gaussian approximation for the crust contribution, one can extract a signal from the mantle equal to $20.9^{+15.1}_{-10.3}$, leading to a 98 % C.L. geo-neutrino signal from the mantle. Finally, a fit where both U and Th spectra are left as free parameters has also been performed, restricting the radiogenic heat production from these isotopes between 23 and 36 TW.

3. Investigation on a possible georeactor

In addition to the standard geo-neutrino analysis, we report an investigation on a possible natural nuclear reactor, called georeactor, standing inside the Earth. We assume this reactor to release a constant power for the whole data taking period. The Monte Carlo spectrum is built such that $^{235}\mathrm{U}/^{238}\mathrm{U}$ has been set to 0.75/0.25 while the Pu contribution is set to 0. The

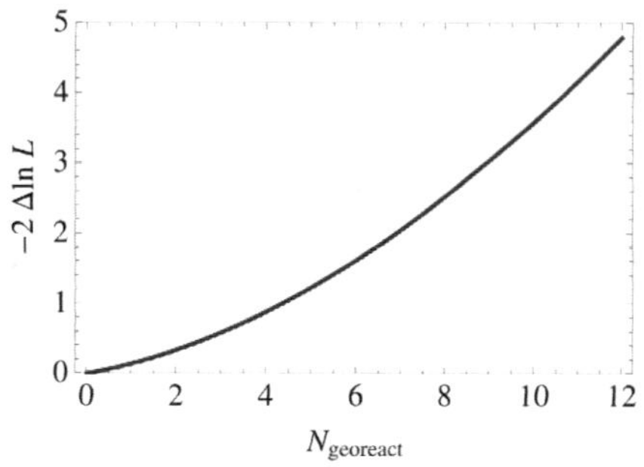

Figure 2. $-2\,\Delta\ln\mathcal{L}$ profile for N_{georeact}.

fit has been done in the energy range above 1510 p.e. in order to get rid of the geo-neutrino spectrum. The background components have been normalized to the [1510, 5000 p.e.] energy

[30] One TNU corresponds to one event detected over one year exposure of 10^{32} target protons at 100 % efficiency.

range of interest and the reactor component has been constrained to the theoretical value and error of 56 $\pm$ 2 (30 $\pm$ 1 in the [1510, 5000 p.e.] energy range of interest).

Figure 2 shows the $-2\,\Delta\ln\mathcal{L}$ profile for N_{georeact}. The best fit value is 0 and the upper limit in terms of number of events is 8.4 (10.5) at 90 % C.L. (95 % C.L.). This limit is usually expressed in terms of TW. On the whole energy range, 1 TW is found to be equal to 4.4 events with an exposure of 5.5×10^{31} proton $\times$ year, oscillation through core and mantle taken into account. It corresponds to 2.5 events in the [1510, 5000 p.e.] energy range of interest, which leads to an upper limit of 3.4 TW (4.2 TW) at 90 % C.L. (95 % C.L.).

4. Conclusion

From 2056 days of data taking, Borexino alone is able to reject the null geo-neutrino signal at $5.9\,\sigma$, to claim a geo-neutrino signal from the mantle at 98 % C.L. and to restrict the radiogenic heat production for U and Th between 23 and 36 TW. With a signal-to-background ratio of the order of 100, Borexino provides a real time spectroscopy of geo-neutrinos. Finally, we have investigated the hypothesis of a georeactor and we have set an upper limit for a 3.4 TW georeactor (4.2 TW) at 90 % C.L. (95 % C.L.).

Acknowledgments

Russian colleagues from MEPhI acknowledge partial support from MEPhI Academic Excellence Project (contract No. 02.a03.21.0005, 27.08.2013).

References

[1] Eder G 1966 *Nucl. Phys.* **78** 657-62
[2] Marx G 1969 *Czech. J. Phys.* B **19** 1471-9
[3] Krauss L M, Glashow S L and Schramm D N 1984 *Nature* **310** 191-8
[4] Araki T *et al.* (KamLAND Collaboration) 2005 *Nature* **436** 499-503
[5] Gando A *et al.* (KamLAND Collaboration) 2013 *Phys. Rev.* D **88** 033001
[6] Bellini G *et al.* (Borexino Collaboration) 2010 *Phys. Lett.* B **687** 299-304
[7] Bellini G *et al.* (Borexino Collaboration) 2013 *Phys. Lett.* B **722** 295-300
[8] Agostini M *et al.* (Borexino Collaboration) 2015 *Phys. Rev.* D **92** 031101
[9] Coltorti M *et al.* 2011 *Geochim. Cosmochim. Acta* **75** 2271
[10] Huang Y *et al.* 2013 *Geochem., Geophys., Geosyst.* **14** 2003-29

The new experiment WAGASCI for water to hydrocarbon neutrino cross section measurement using the J-PARC beam

T Ovsiannikova[1,2], M Antonova[1,2], C Bronner[3], A Blondel[4], A Bonnemaison[5], R Cornat[5], F Cadoux[4], N Chikuma[6], S Cao[7], O Drapier[5], A Izmaylov[1,8], A K Ichikawa[7], O Ferreira[5], Y Favre[7], F Gastaldi[5] , M Gonin[5], T Hayashino[7], F Hosomi[6], Y Hayato[9], T Koga[6], K Kondo[7], K Kin[1]0, Y Kudenko[1,2,11], M Khabibullin[1], A Khotjantsev[1], A Minamino[7], Th A Mueller[5], A Mefodiev[1,11], O Mineev[1], K Nakamura[7], T Nakaya[7], E Noah[4], N Patel[7], B Quilain[7], M Rayner[4], Y Seiya[10], S Suvorov[1,11], N Yershov[1], M Yamamoto[7], K Yamamoto[10] and M Yokoyama[6]

[1] Institute for Nuclear Research of the Russian Academy of Sciences, Russia
[2] National Research Nuclear University MEPhI (Moscow Engineering Physics Institute), Kashirskoe highway 31, Moscow, 115409, Russia
[3] University of Tokyo, Kavli institute for the Physics and Mathematics of the Universe, Japan
[4] Geneva University, Switzerland
[5] LLR Ecole Polytechnique, France
[6] University of Tokyo, Japan
[7] Kyoto University, Japan
[8] Instituto de Fsica Corpuscular (IFIC CSIC), Spain
[9] University of Tokyo, Institute for Cosmic Ray Research, Japan
[10] Osaka City University, Japan
[11] Moscow Institute of Physics and Technology (State University), Russia

E-mail: `ovsyanik@inr.ru`

Abstract. The T2K (Tokai-to-Kamioka) is a long baseline neutrino experiment designed to study various parameters that rule neutrino oscillations, with an intense beam of muon neutrinos. A near detector complex (ND280) is used to constrain non-oscillated flux and hence to predict the expected number of events in the far detector (Super-Kamiokande). The difference in the target material between the far (water) and near (scintillator, hydrocarbon) detectors leads to the main non-canceling systematic uncertainty for the oscillation analysis. In order to reduce this uncertainty a new water grid and scintillator detector, WAGASCI, has been proposed. The detector will be operated at the J-PARC neutrino beam line with the main physics goal to measure the charged current neutrino cross section ratio between water and hydrocarbon with a few percent accuracy. Further physics program may include high-precision measurements of different charged current neutrino interaction channels. The concept of the new detector will be covered together with the actual construction plan.

1. Introduction

The T2K (Tokai-to-Kamioka) is a long-baseline neutrino experiment in Japan with the main goal to study neutrino oscillations. In the T2K oscillation analysis flux and cross section (model) parameters are largely constrained by the near detector (ND280) measurements [1]. The systematics uncertainties on predicted number of signal events for different oscillation channels are presented in table 1 [2]. As can be seen from the table the largest one is a non-canceling uncertainty related to the cross section model and is caused by the difference in the target material between the near (hydrocarbon, CH, as the active target) and far detectors (Super-Kamiokande, Super-K, water Cherenkov detector), and also by the limited acceptance of ND280 (Super-K has 4π coverage). The new water-scintillator detector (WAGASCI, WAter-Grid-SCIintilator-Detector) is proposed to reduce this systematic error with the approach similar to the one previously used to measure the Fe to CH neutrino cross section ratio with the INGRID detector (sandwich iron-scintillator detectors + "proton" module of pure hydrocarbon) of the T2K ND280 complex [3]. The main goals of the proposed detector are:

- measurement of the charge current cross section ratio between water and scintillator targets with 3% accuracy,
- measurement of different charged current neutrino interaction channels with high-precision and large acceptance.

Table 1. Uncertainties for predicted number of signal events for different oscillation modes.

Systematics	$\nu_\mu \to \nu_e$	$\nu_\mu \to \nu_\mu$	$\overline{\nu}_\mu \to \overline{\nu}_\mu$
Flux & XSEC	3.1	2.7	3.4
Non-canceling XSEC	4.7	5.0	10.0
Super-K detector etc.	2.4	3.0	2.1
FSI+SI	2.7	4.0	3.8
Total	6.8	7.7	11.6

The new detector will be located in the ND280 detector pit and hence utilize the same neutrino beam provided by the J-PARC synchrotron. A 1.5° off-axis angle will be used which results in ν flux similar to the one of the ND280 off-axis complex (2.5 degrees), as shown in figure 1. The neutrino energy spectrum is peaked at $\simeq$0.7 GeV.

2. WAGASCI Detector Design

Figure 2 shows the basic design of the WAGASCI. There are two main elements of the detector: the central part is a neutrino interaction target which contains water and hydrocarbon, the target is surrounded by muon range detectors (MRDs). The central part consists of four alternating modules filled with water or hydrocarbon. The key element of the detector is the usage of 3D-grid structure made out of thin scintillator bars of $1000\times25\times3\ mm^3$ each (figure 2). Cells between the active bars are filled with a certain material (water or CH), this design maximizes the fraction of the target material up to 80% level and also provides good particle tracking capabilities allowing to reconstruct tracks emerging at large angles w.r.t. neutrino beam direction. The total size of the target is $1\times1\times2\ m^3$.

The muon identification and muon momenta measurements will be provided using MRDs, which have sandwich structure of alternating iron and scintillator layers. In order to estimate the direction of reconstructed tracks, reduce background from neutrino interactions in MRDs

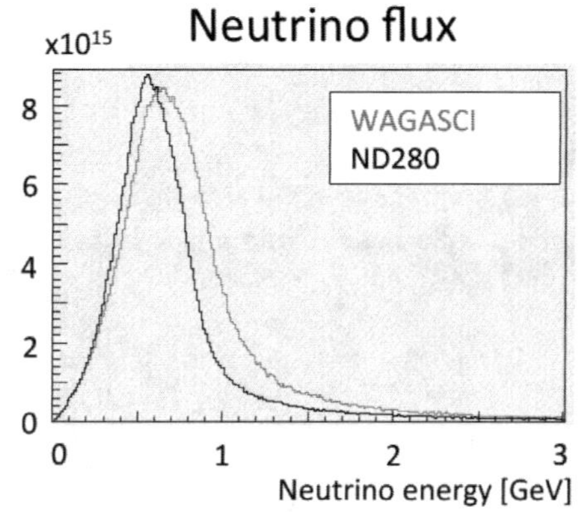

Figure 1. Expected neutrino flux at the site of WAGASCI (red line) and of the T2K off-axis near detector (black line).

and walls of the detector hall, the time of flight information between the target and muon range detectors will be utilized.

Plastic scintillator bars will be used as active elements in both target and surrounding muon range detectors. Light collection is done by utilizing WLS fibers (embedded into the grooves) that transport light to the photosensors, Hamamatsu MPPCs will be used for the latter. For WAGASCI experiment new generation of MPPCs will be used, these sensors have low dark noise rate, wider range of operation over-voltage ($\simeq$4V, allowing to increase the efficiency), and low rate of afterpulses and crorsstalk between the pixels.

It is crucial to obtain high detection efficiency with thin bars of the target area so to fulfill the requirements of the physics program. The performance of plastic scintillator bars was measured with the 600 MeV positron beam. The average light yield was found to be 10-18 p.e. and the detection efficiency was better than 99% for the whole region of scintillator with a threshold set to 1.5 p.e.

The WAGASCI detector will collect data with both polarities of T2K focusing horn system. Operation with the anti-neutrino beam will require reduction of neutrino background which is about 30% in this case. In order to provide charge-based discrimination of daughter particles it is proposed to utilize a Magnetized IroN Detector (MIND) [4] located downstream of the target (as an alternative to the downstream MRD). The MIND has a structure which consists of alternating layers of scintillator detectors and iron plates, the latter are magnetized by normal conducting aluminum coils wound on the surface of each individual plate. The sign of the particle charge is identified by the analysis of the trajectory (track curvature) in the magnetic field. The efficiency to identify correctly the charge is expected to be of 90%, which will allow to decrease the background of neutrino interactions down to 2.9% level.

3. Expected Detector Performance

The expected performance of the WAGASCI detector is studied with Monte Carlo methods. The neutrino flux and neutrino interactions inside the detector are simulated by JNUBEAM [5] and NEUT [6] respectively. Further modeling includes propagation of secondary particles through detector media (by means of GEANT4 [7] toolkit), emission of scintillation light, processes inside WLS fibers and light collection by MPPCs.

A neutrino charged current interaction in the central detector is identified by a track starting from the fiducial volume of the target and entering one of the MRDs, where the MRD is used to

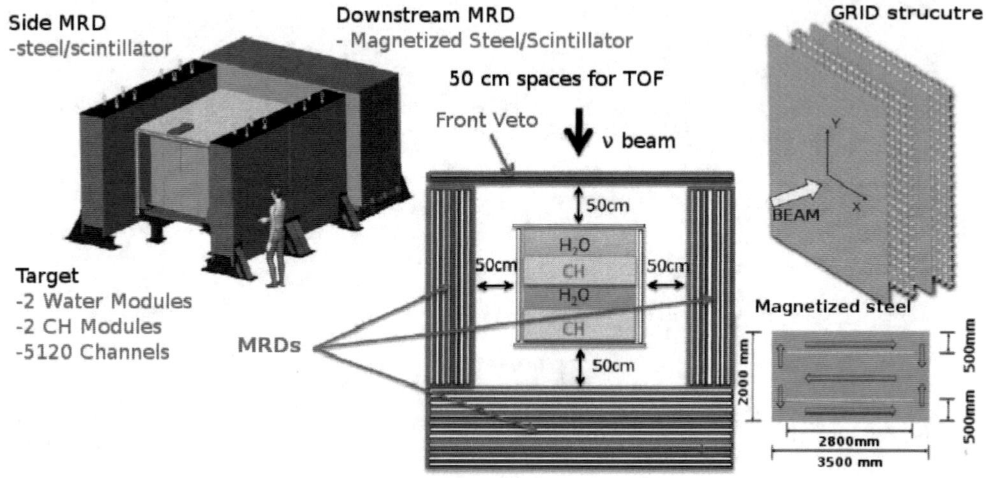

Figure 2. The basic structure of WAGASCI detector including one of the possible designs for MIND plates.

identify a long muon track. Tracks crossing the central part and the MRD are required to stop in the detector. The expected number of detected events from hydrocarbon target and water target with 10^{21} protons on target (POT) are 3×10^3 and 2.5×10^3, respectively. The purity of charged current events for hydrocarbon and water interactions is 91.0% and 75.5%, respectively. Some basic distributions for selected events are shown in figure 3.

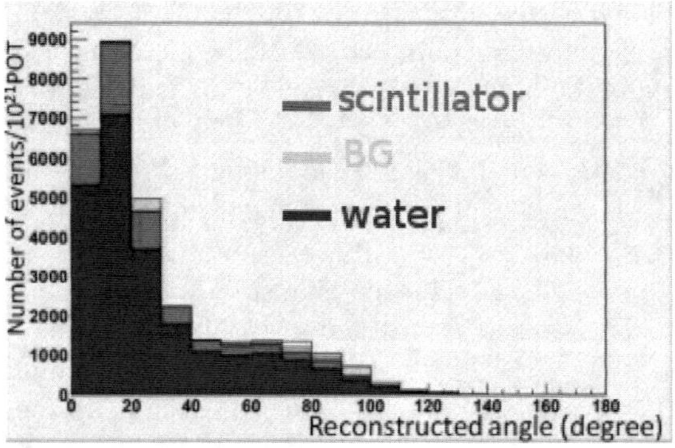

(a)

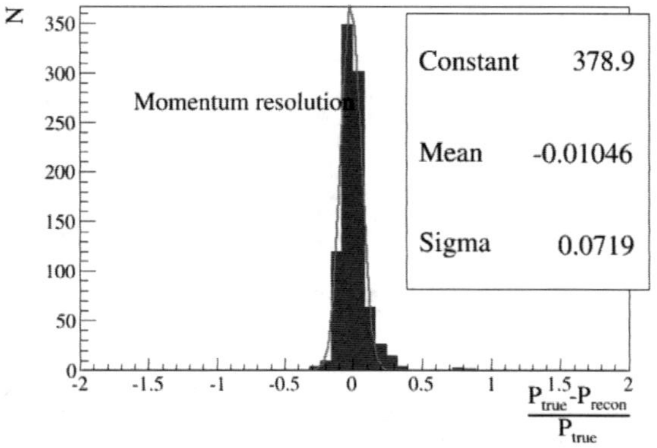

(b)

Figure 3. (a): reconstructed angle of muon candidate for water target events that pass the selection; events split into categories based on true vertex location. (b): $\simeq$7% momentum resolution achieved for tracks stopping in MRDs.

For basic performance tests of the WAGASCI target a test experiment will be carried out. A prototype of the WAGASCI target will be installed in front of the central INGRID module of the T2K near detector to take on-axis neutrino beam data. The total size of WAGASCI prototype is $125 \times 125 \times 46$ cm^3 and the total mass is 1.2 kton. It is expected that the cross section ratio between H_2O and CH will be measured with a total uncertainty of 3 % with 1×10^{20} POT and 5 % with 2×10^{20} POT in the positive and negative (anti-neutrino beam) focusing mode, respectively, although the angular acceptance will be significantly limited w.r.t. to the full WAGASCI. The prototype will be constructed in October-December 2015.

The construction of the full WAGASCI detector will start at the end of 2015 and the physics data taking is scheduled to begin in late fall of 2016.

4. Conclusion

To reduce the systematic uncertainties for the T2K neutrino oscillation analysis and to measure different channels of charged current neutrino interactions the new experiment WAGASCI is proposed. The detector design will allow to measure the cross section ratio between water and hydrocarbon with 3% accuracy. The exact design of the detector and its components is still being finalized. The construction and installation of WAGASCI detector will start at the end of 2015 and the physics operation will start in the fall of 2016.

Acknowledgments

This work was performed within the framework of the Center of Basic Research and Particle Physics of National Research Nuclear University MEPhI partly supported by MEPhI Academic Excellence Project (contract No. 02.a03.21.0005, 27.08.2013) and the Russian Science Foundation Grant No. 14-12-00560.

Reference

[1] Abe K *et al.* (T2K Collaboration) 2011 *Nucl. Instr. Meth.* A **659** 106-35
[2] Abe K *et al.*(T2K Collaboration) 2015 *Prog. Theor. Exp. Phys.* 043C01
[3] Abe K *et al.* (T2K Collaboration) 2014 *Phys. Rev.* D **90** 052010
[4] Asfandiyarov R *et al.* 2014 Proposal for SPS beam time for the baby MIND and TASD neutrino detector prototypes *arXiv:* 1405.6089v1
[5] Abe K *et al.* (T2K Collaboration) 2013 *Phys. Rev.* D **87** 019902
[6] Hayato Y *et al.* 2002 *Nucl. Phys.* B *(Proc. Suppl.)* **112** 171-6
[7] GEANT. Detector Description and Simulation Tool: *www.gean4.cern.ch*

Journal of Physics: Conference Series **675** (2016) 012031 doi:10.1088/1742-6596/675/1/012031

iDREAM: an industrial detector for nuclear reactor monitoring

I V Gribov[1], M B Gromov[2], G A Lukjanchenko[3], G J Novikova[4], B A Obinyakov[3], A Y Oralbaev[3], M D Skorokhvatov[3,5], S V Sukhotin[3], A S Chepurnov[1] and A V Etenko[3]

[1] Skobeltsyn Institute of Nuclear Physics, Moscow State University, Moscow, 119991 Russia
[2] Department of Physics, Moscow State University, Moscow, 119991 Russia
[3] NRC "Kurchatov Institute", 123182, Moscow, Russia
[4] Institute for Nuclear Research, Russian Academy of Sciences, Moscow, 117312 Russia
[5] National Research Nuclear University MEPhI (Moscow Engineering Physics Institute), Kashirskoe highway 31, Moscow, 115409, Russia

E-mail: b.obinyakov@yandex.ru, oralbaev_ay@nrcki.ru, skorokhvatov_md@nrcki.ru, aschepurnov@gmail.com

Abstract. Prototype of industrial reactor antineutrino detector iDREAM is dedicated for an experiment to demonstrate the possibility of remote monitoring of PWR reactor operational modes by neutrino method in real-time in order to avoid undeclared exposure modes for nuclear fuel and unauthorized removal of isotopes. The prototype detector was started up in 2014. To test the detector elements and components of electronics distilled water has been used as a target, which enables the use of Cerenkov radiation from cosmic muons as a physical signal. Also parallel measuring of the long-term stability has been doing for samples of liquid organic scintillator doped with gadolinium and synthesized by different methods.

1. Introduction

Nuclear power belongs to the list of Priority Development Areas in Science, Technology and Engineering in the Russian Federation approved by Russian Government. It is also the fact that Russia is one of few worldwide player on the nuclear power plant world market including Nuclear Power Plant (NPP) R&D&Construction together with complete NPP fuel cycle. In February 2010 the federal target program aimed to bring a new technology platform for the nuclear power industry based on fast reactors have been approved. Rosatom's long-term strategy up to 2050 involves moving to inherently safe nuclear plants using fast reactors with a closed fuel cycle. It is expected that NPP will provide 45-50 % of electricity at that time.

It is important to notice that the development of the neutrino detector for diagnostics of a reactor active zone is one of the announced aims included in to the Federal program "Development the nuclear power complex" within the Federal Targeted Program "The nuclear technologies of the new generation for the time period from 2010 to 2015 and for the period until 2020".

The neutrino method for monitoring the operating modes of nuclear reactors was proposed and tested in the 1980s–1990s at the Kurchatov Institute and it was confirmed experimentally at Rovno NPP[1]. This method is based on the registration and interpretation of inverse beta decay (IBD) products with liquid organic scintillator optimized for neutron capture.

Until recently the detectors of reactor antineutrinos have been designed mainly for research purposes. By the way, a collaboration of Russian institutions (the Skobeltsyn Institute of Nuclear Physics, the Kurchatov Institute, the Institute for Nuclear Research, and the All-Russia Research Institute of Automatics) set as their goal to construct a test prototype of the iDREAM (industrial Detector for REactor Antineutrino Monitoring).

2. Project goal

The original application of iDREAM was to demonstrate industrial version of neutrino method and extend the efficiency of nonproliferation procedures for operating NPP being additional "black-box"-like unattended instrument with secure communication links directly to IAAE. By the way this detector have much wider application areas including research and education. Compact size, low cost and expected high detection efficiency makes the detector a unique research instrument to study perspective NPP reactors including future perspective VVER and fast-neutron reactors. The iDREAM application purpose is to provide a reliable information about active zone condition by measuring antineutrino flux from NPP reactor which cannot be falsified.

3. Monitoring methods

The IBD reaction which is used to detect antineutrinos produces a positron, whose energy depends on the initial antineutrino energy, and a neutron:

$$\bar{\nu}_e + p \rightarrow n + e^+$$

Since the threshold of this reaction is 1.8 MeV, only 25% of the total number of antineutrinos emitted by the reactor (the ones with their energy above the threshold) are detected. The most appropriate target materials are hydrogensaturated transparent plastic or liquid organic scintillators (LOSs). The IBD detection method was originally proposed by Frederick Reines in mid 1950's [2]. The method is based on the detection of two consecutive scintillation flashes (photons of luminescence): the first pulse is due to a positron (the ionization of the medium with subsequent annihilation) while the second pulse is due to neutron capture within the target with the immediate irradiation of gamma which energy and number depends on the capture nuclei. Clear time stamp allows to suppress the background considerably using the delayed coincidence technique.

The number of the detected IBD events is proportional to current power of reactor and directly corresponds generated energy. The calculations and the results of experiments that were conducted earlier show that the detector count rate is about 10^3 events/day at a distance of 10-20 m from the core of a nuclear reactor with an output of, for example, 1.4 GW if the sensitive LOS volume is about 1 m^3 and the efficiency of detection of the IBD products is 60%.

4. Detector construction

The detector (Figure 1) design incorporates two concentric stainless-steel tanks with a common sealed cap. The inner tank is divided by a convex transparent organic glass membrane and is scanned by 16 PMTs. The inner tank volume (1 m^3) below the membrane is filled with a LOS doped with gadolinium and serves as the detector target. The target may be accessed via a vertical pipe that is glued into the membrane center. The upper end of this pipe is outside of the tank. The inner tank volume above the membrane is filled with linear alkylbenzene (LAB), which helps to increase the light collection efficiency and shield the target from the gamma background from PMTs. The outer tank is also divided vertically into two parts for transportation and handling convenience. The circular volume is filled with scintillator liquid and serves as an active shield for the detector target. It also helps to increase the overall efficiency by detecting the gamma quanta that escape the target volume. The circular volume is scanned wby 12 PMTs. A light-protective barrier between the buffer and the circular volumes is located at the bottom side of the cap. The LOS is pumped in and out in the target and in the circular volume via a header using a system of pipes and circular separators at the bottom of the inner and the circular volumes. In order to increase the light collection efficiency, the inner walls of the buffer and the main sensitive volumes are covered with a reflective material (Lumirror E6SR) with a reflection coefficient of up to 90% within the scintillation spectrum of the LOS. Thus, two liquid scintillator types (with and without gadolinium) and a buffer liquid are used in the detector. The total volume of these liquids, which are based on LAB, is roughly 3500 L. Original DAQ system based on industrial 1Gsample/s flash analogue to digital converters (FADCs) is used to process PMT outputs. Use of FADCs allows to introduce in to data analyses pulse shape discrimination technique. Set of summing-discriminating modules of fast electronics together with original flexible programmable trigger based on FPGA have been developed for DAQ system of iDREAM to reduce data stream of raw data. Slow control system based on industrial CAN-bus/CANopen network with a set of industrial sensors and originally developed electronic instruments was developed to control high voltage power supplies, DAQ thresholds, LOS levels and ambient conditions.

5. Developing a liquid organic scintillator

The lack of a stable, inexpensive, nonflammable, and nontoxic LOS was one of the problems that hinder the development of industrial detectors for nuclear reactor monitoring. The LOS instability, which was manifested in the deterioration of its characteristics (light output and transparency), limited the lifetime of such detectors as the RONS and CHOOZ detectors.
The primary requirement for a small–volume ($\sim 1\text{m}^3$) industrial detector consists in maintaining a stable light output and a stable neutron detection efficiency (i.e., preserving a constant concentration of gadolinium for a protracted period of time), while the scintillator transparency is not a crucial parameter, although it may reflect the degree of chemical purity of the LOS components. An original liquid scintillator based on Russian LAB (TU 2414–028–05766480–2006) was developed for the iDREAM instrument. A concentrate containing a base gadolinium substance with a gadolinium concentration that is ten times higher than its resulting concentration in the LOS was proposed for use in order to introduce gadolinium into the LOS. The concentrate with gadolinium was dissolved in the scintillator with the fluor (PPO, 3 g/L) and the shifter (POPOP, 0.03 g/L), which was prepared beforehand [3].

6. Conclusion

To the present time it was developed and adjusted the detector construction, some of modifications are in progress. During the first physical start of the detector with a distilled water all systems were tested and showed an operational suitability. Three LOS samples of scintillator with three different gadolinium compounds are testing for stability.

For the near future it is planning to shift the detector to the underground laboratory specially developed in SINP MSU to study efficiency and background suppression capabilities of the detector. The second long term activity is focused on continuous study of LOS samples stability in order to select the sample for the whole detector. The following step involves the installation of the iDREAM at NPP for final test during reactor campaign. The negotiations about such testing are in progress.

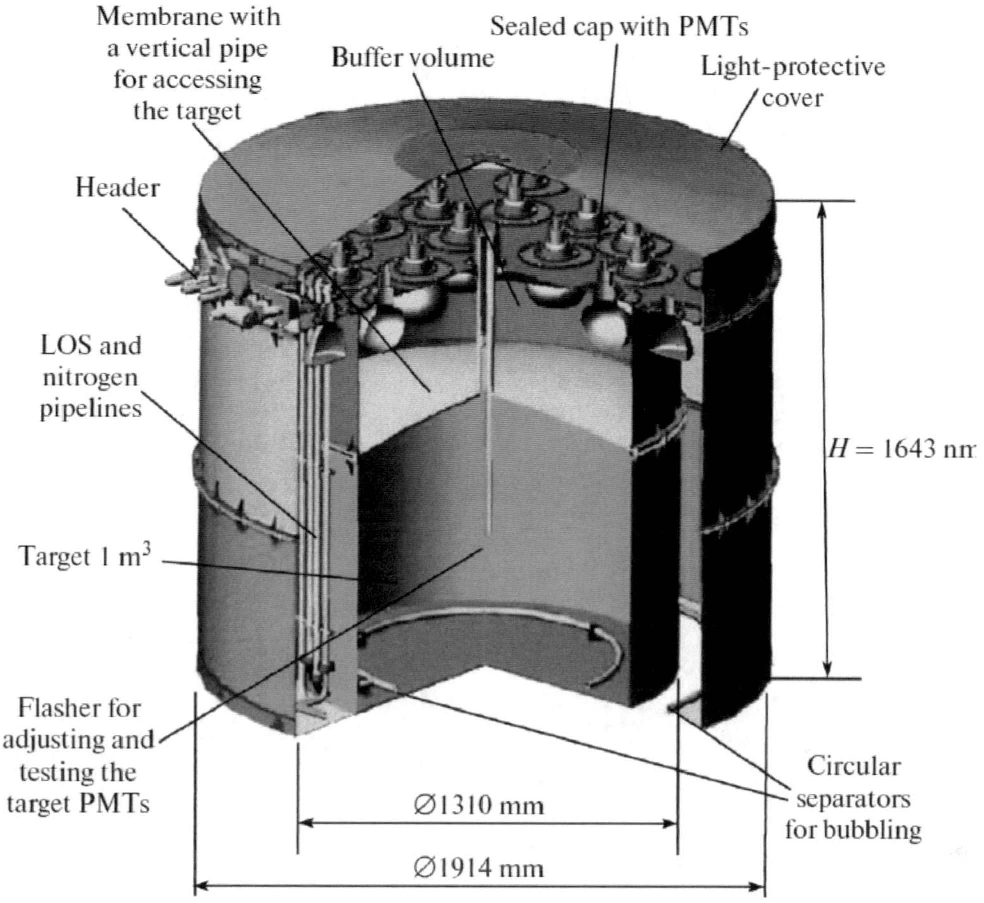

Figure 1. iDREAM Construction

Acknowledgments
This study was supported financially by the Russian Foundation for Basic Research, grants 14-22-0301 ofi_m and 14-02-31381 mol_a.

References
[1] Borovoi A and Mikaelyan L 1978 (in Russian) *At. Energy* **44** 508 (original Russian title: *Atomnaya Energiya*)
Klimov Yu V *et al.* 1990 (in Russian) *Nucl.Phys* **52** 1579 (original Russian title: *Yadernaya Fizika*)
Kopeikin V I, Mikaelyan L A and Sinev V V 1997 *Phys. At. Nucl.* **60** 172
[2] Reines F 1979 *Science* **203** (4375) 11
[3] Novikova G Ya, Bakulina N I and Morgalyuk V P 2014 *Russ. J. Inorg. Chem.* **59** 244
Bakulina N I *et al.* 2014 *Preprint* 1387/2014 (Moscow: Institute for Nuclear Research Press)

International Conference on Particle Physics and Astrophysics (ICPPA-2015) IOP Publishing
Journal of Physics: Conference Series **675** (2016) 012032 doi:10.1088/1742-6596/675/1/012032

The ^{144}Ce source for SOX

M Durero[4], M Vivier[4], M Agostini[24], K Altenmüller[24], S Appel[24], G Bellini[15], J Benziger[19], N Berton[4], D Bick[5], G Bonfini[14], D Bravo[22], B Caccianiga[15], F Calaprice[18], A Caminata[3], P Cavalcante[14], A Chepurnov[25], K Choi[30], M Cribier[4], D D'Angelo[15], S Davini[26], A Derbin[17], L Di Noto[3], I Drachnev[26], A Etenko[12], S Farinon[3], V Fischer[4], K Fomenko[2], D Franco[1], F Gabriele[14], J Gaffiot[4], C Galbiati[18], C Ghiano[3], M Giammarchi[15], M Goeger-Neff[24], A Goretti[18], M Gromov[25], C Hagner[5], T Houdy[4], E Hungerford[27], Aldo Ianni[14], Andrea Ianni[18], N Jonquères[7], K Jedrzejczak[8], M Kaiser[5], V Kobychev[9], D Korablev[2], G Korga[14], V Kornoukhov[10], D Kryn[1], T Lachenmaier[11], T Lasserre[4], M Laubenstein[14], B Lehnert[28], J Link[22], E Litvinovich[12,13], F Lombardi[14], P Lombardi[15], L Ludhova[15], G Lukyanchenko[12,13], I Machulin[12,13], S Manecki[22], W Maneschg[6], S Marcocci[26], J Maricic[30], G Mention[4], E Meroni[15], M Meyer[5], L Miramonti[15], M Misiaszek[8,14], M Montuschi[23], V Muratova[17], R Musenich[3], B Neumair[24], L Oberauer[24], M Obolensky[1], F Ortica[16], M Pallavicini[3], L Papp[24], L Perasso[3], A Pocar[21], G Ranucci[15], A Razeto[14], A Re[15], A Romani[16], R Roncin[14,1], N Rossi[14], S Schönert[24], L Scola[4], D Semenov[17], H Simgen[6], M Skorokhvatov[12,13], O Smirnov[2], A Sotnikov[2], S Sukhotin[12], Y Suvorov[29,12], R Tartaglia[14], G Testera[3], J Thurn[28], M Toropova[12], C Veyssiére[4], E Unzhakov[17], R B Vogelaar[22], F von Feilitzsch[24], H Wang[29], S Weinz[31], J Winter[31], M Wojcik[8], M Wurm[31], Z Yokley[22], O Zaimidoroga[2], S Zavatarelli[3], K Zuber[28] and G Zuzel[8]
Borexino Collaboration

[1]AstroParticule et Cosmologie, Universit Paris Diderot, CNRS/IN2P3, CEA/IRFU, Observatoire de Paris, Sorbonne Paris Cit, 75205 Paris Cedex 13, France
[2]Joint Institute for Nuclear Research, 141980 Dubna, Russia
[3]Dipartimento di Fisica, Università degli Studi e INFN, Genova 16146, Italy
[4]Commissariat à l'Énergie Atomique et aux Énergies Alternatives, Centre de Saclay, IRFU, 91191 Gif-sur-Yvette, France
[5]Institut für Experimentalphysik, Universität, 22761 Hamburg, Germany
[6]Max-Planck-Institut für Kernphysik, 69117 Heidelberg, Germany
[7]Commissariat à l'Énergie Atomique et aux Énergies Alternatives, Centre de Saclay, DEN/DM2S/SEMT/BCCR, 91191 Gif-sur-Yvette, France
[8]M. Smoluchowski Institute of Physics, Jagiellonian University, 30059 Krakow, Poland
[9]Kiev Institute for Nuclear Research, 06380 Kiev, Ukraine
[10]Institute for Theoretical and Experimental Physics, 117218 Moscow, Russia
[11]Kepler Center for Astro and Particle Physics, Universität Tübingen, 72076 Tübingen, Germany.
[12]NRC Kurchatov Institute, 123182 Moscow, Russia

[13]National Research Nuclear University MEPhI (Moscow Engineering Physics Institute), Kashirskoe highway 31, Moscow, 115409, Russia.
[14]INFN Laboratori Nazionali del Gran Sasso, 67010 Assergi (AQ), Italy
[15]Dipartimento di Fisica, Università degli Studi e INFN, 20133 Milano, Italy
[16]Dipartimento di Chimica, Università e INFN, 06123 Perugia, Italy
[17]St. Petersburg Nuclear Physics Institute NRC Kurchatov Institute, 188350 Gatchina, Russia
[18]Physics Department, Princeton University, Princeton, NJ 08544, USA
[19]Chemical Engineering Department, Princeton University, Princeton, NJ 08544, USA
[20]Physics Department, Queen's University, Kingston ON K7L 3N6, Canada
[21]Amherst Center for Fundamental Interactions and Physics Department, University of Massachusetts, Amherst, MA 01003, USA
[22]Physics Department, Virginia Polytechnic Institute and State University, Blacksburg, VA 24061, USA
[23]Dipartimento di Fisica e Scienze della Terra Università degli Studi di Ferrara e INFN, Via Saragat 1-44122, Ferrara, Ital
[24]Physik-Department and Excellence Cluster Universe, Technische Universität München, 85748 Garching, Germany.
[25] Lomonosov Moscow State University Skobeltsyn Institute of Nuclear Physics, 119234 Moscow, Russia.
[26] Gran Sasso Science Institute (INFN), 67100 L'Aquila, Italy.
[27]Department of Physics, University of Houston, Houston, TX 77204, USA.
[28]Department of Physics, Technische Universität Dresden, 01062 Dresden, Germany.
[29]Physics and Astronomy Department, University of California Los Angeles (UCLA), Los Angeles, California 90095, USA.
[30]Department of Physics and Astronomy, University of Hawaii, Honolulu, HI 96822, USA.
[31]Institute of Physics and Excellence Cluster PRISMA, Johannes Gutenberg-Universität Mainz, 55099 Mainz, Germany.

E-mail: mathieu.durero@cea.fr, matthieu.vivier@cea.fr

Abstract. The SOX (Short distance neutrino Oscillations with BoreXino) project aims at testing the light sterile neutrino hypothesis. To do so, two artificials sources of antineutrinos and neutrinos respectively will be consecutively deployed at the Laboratori Nazionali del Gran Sasso (LNGS) in close vicinity to Borexino, a large liquid scintillator detector. This document reports on the $\bar{\nu}_e$ source production and transportation. The source should exhibit a long lifetime and a high decay energy, a requirement fullfilled by the ^{144}Ce-^{144}Pr pair at secular equilibrium. It will be produced at FSUE "Mayak" PA using spent nuclear fuel. It will then be shielded and packed according to international regulation and shipped to LNGS across Europe. Knowledge of the Cerium antineutrino generator (CeANG) parameters is crucial for SOX as it can strongly impact the experiment sensitivity. Several apparatuses are being used or designed to characterize CeANG activity, radioactive emission and content. An overview of the measurements performed so far is presented here.

1. Introduction

The neutrino sector still shows unexplained discrepancies between observations and theoretical predictions. One of them is an electron neutrino deficit in short distance nuclear reactor based experiments, the so-called reactor antineutrino anomaly (RAA) [1]. A similar problem occurs in the source experiments GALLEX and SAGE [2]. This anomaly could be explained by a short distance oscillation between the three known neutrino states and one hypothetical new state with a squared mass difference of $\Delta m^2_{new} \approx 1\text{eV}^2$ with the known neutrinos [3]. Due to constraints on Z boson coupling asserted by LEP [4], this state would be sterile.

The SOX experiment (Short baseline Oscillation in BoreXino) is designed to observe this oscillation, which has an expected length of 1 m order of magnitude, using a big liquid scintillator detector, in addition to the possible interaction rate deficit. Borexino is a spherical liquid scintillator detector, with a fiducial mass of 280 tons, and provides vertex reconstruction with 15 cm precision and 5% energy resolution [5]. With a diameter of 8.5 m, the scintillator vessel allows to probe interaction rate as a function of particle travel distance from the source. $\bar{\nu}_e$ are detected through the inverse beta decay (IBD) process. It has a relatively high cross-section and allows a very good background rejection thanks to its rather unique signature: a time-correlated energy deposit from both the emitted positron and neutron. The study of the flux from a ^{144}Ce-^{144}Pr $\bar{\nu}_e$ generator (CeANG), deployed in a tunnel right under the center of the detector base, is the first phase of the project.

2. CeANG design

2.1. Source selection

The source design of a β decay antineutrino generator must comply with some challenging constraints. First the source lifetime should be long enough to obtain statistically significant results with a 2 years schedule and to make possible the source manufacturing and transportation in a realistic amount of time. Second, the IBD process has a 1.806 MeV threshold so the source $\bar{\nu}_e$ spectrum must significantly extend beyond this threshold. This statement contradicts the previous one as it leads to a nucleus half-life shorter than the day. To circumvent this difficulty, the retained solution is a nuclei couple, where the mother nucleus satisfies the first requirement, and decays with a long half-life in a daughter nucleus having a high energy decay.

Four nuclei couples hold to this description: ^{42}Ar-^{42}K, ^{90}Sr-^{90}Y, ^{106}Ru-^{106}Rh and ^{144}Ce-^{144}Pr. The choice of the best suited couple is based on the following considerations: ease of production at the necessary scale, then available energy for the daughter nucleus decay to optimize the event rate. The ^{144}Ce-^{144}Pr couple is the best candidate, as ^{144}Ce is a relatively abundant fission product from irradiated nuclear fuel and ^{144}Pr exhibits endpoint energy $Q_\beta = 2.997$ MeV well beyond IBD threshold [6]. With a ^{144}Ce half-life of 285 days, the minimal activity for 18 months of SOX data taking is 3.7 PBq.

2.2. Source production

The production of the CeANG requires industrial scale spent nuclear fuel reprocessing. Despite the ~5% cumulative fission yield of ^{144}Ce in commercial reactor fuel, gathering enough material to manufacture the source forces to reprocess tons of spent fuel, depending on the irradiation history and cooling times.

The FSUE "Mayak" PA has demonstrated the ability to handle such radioactive materials and will be in charge of the whole production process. The Cerium extraction begins by the standard PUREX (Plutonium Uranium Redox EXtraction) radiochemical process which allows to separate Uranium and Plutonium from the lighter fission products by liquid-liquid extraction. Then, complex displacement chromatography of rare-earth elements enables to isolate Cerium from the other fission products. The obtained Cerium residue undergoes calcination to form CeO_2 which will be pressed and inserted into a specially designed stainless steel capsule. To satisfy the high activity requirement, the source will contain up to 5 kg of CeO_2. The expected delivery date of the CeANG is end of 2016.

3. Source handling and transport

3.1. Shielding

The ^{144}Pr decay produces γ rays. Among them the most difficult to handle is a 2.2 MeV ray with 0.7% intensity. Given the PBq activity of the source, it requires shielding both for biological safety and to avoid source-induced backgrounds into the detector. The flux must be suppressed by a factor $\gtrsim 10^{12}$.

The CeANG must also fit into the 1 m-width Borexino tunnel, putting constraints on the shielding compactness. The shielding should then have the highest possible density. For these reasons, it is built in tungsten-iron-nickel alloy with $\sim 18\,\mathrm{g/cm^3}$ density. The shielding thickness is 19 cm, leading to a $\phi = 54$ and $H = 60\,\mathrm{cm}$ hollow cylinder in which the stainless steel source capsule will be placed. The total mass of the shielding is 2.4 tons, making it the biggest tungsten piece ever built. It is currently being manufactured at Xiamen Honglu Inc. (China). It will be delivered by the end of the year, for testing the different mechanical interfaces (calorimeter, transportation cask) and for some handling training.

3.2. Transportation

Once the capsule will be inserted in the shielding, the transportation will take place from Mayak to LNGS. The route has been planned and taken care of according to national and international regulations by Areva TN (France). The transportation of the CeANG must use a specific certified cask. Areva TN-MTR model, generally used for commercial nuclear fuel transportation, has been selected because of its capacity to handle the weight of the shielding. A custom basket has been designed to fit the shielding into the cask. The accreditations of the cask and basket as a source container for transport have been obtained, as well as transportation authorizations.

The selected route aims at minimizing border-crossing, which is lengthy and tedious to organize for nuclear material transportation. The cask will be delivered by train at Saint Petersburg, then a dedicated boat will bring it to Le Havre. The final step will be done by truck through France and Italy. The whole transportation is expected to take in any case less than three weeks, allowing for less than 5% activity loss. The delivery should take place at LNGS in December 2016.

4. Measurements for CeANG characterization

4.1. Activity

To guarantee the sensitivity of the SOX experiment over a wide range of sterile ν oscillation parameters, one of the key point is to be able to predict the interaction rate into the detector as a function of the $\bar{\nu}_e$ energy and distance to the source. The source activity is a key information to predict this global interaction rate. In order to measure the activity, two calorimeters are being built within the SOX collaboration. They aim at two independent subpercent activity measurements. The calorimeters are detailed in dedicated talks [7],[8].

4.2. β spectroscopy

To convert the thermal power, which is the raw result of the calorimeter measurements, to an activity, one should know the mean decay energy of the couple ^{144}Ce-^{144}Pr. It depends on the β spectrum shape. This shape also influences directly the $\bar{\nu}_e$ interaction rate prediction as the $\bar{\nu}_e$ spectrum is estimated by converting the β spectrum. To test the parameter space favored by the RAA, the β spectrum shape should be determined at a 0.5% precision. It corresponds to an $\bar{\nu}_e$ interaction rate precision $\leq 1\%$.

The ^{144}Ce and ^{144}Pr spectra contain non-unique forbidden β transitions, for which the theoretical spectral shape is not known at this precision level. Furthermore past measurements of these isotopes show huge discrepancies at the scale of 10 to 15% [9][10]. It is then necessary to make a new β spectrum measurement for the CeANG.

Two β spectrometers are under development for this task, using plastic scintillator. The first have a multiwire-chamber to tag β rays by comparison to γ rays. It is developped by the Technische Universität München and has already been used for ^{238}U fission products spectrometry [11]. The second is designed specifically for SOX by CEA. It aims both at maximizing the light collection and doing fast and repeatable measurements.

FSUE "Mayak" PA has provided a first batch of radioactive samples which are used to check and tune the various apparatuses. Even if the source production process has evolved and the first samples are not representative of the final CeANG, other samples will be sent during the year of production.

Due to the high lifetime difference, ^{144}Ce and ^{144}Pr are in secular equilibrium in source samples. As the ^{144}Pr $\bar{\nu}_e$ spectrum is the only one contributing to the IBD interactions in Borexino, a chemical extraction will be realized to isolate ^{144}Pr. The second spectrometer will then be used to make repeated hour-scale measurements of ^{144}Pr to achieve the best precision on its shape.

4.3. Source impurity assessment

In addition to calorimetry and β spectroscopy, a whole batch of measurements will help to characterize the content of the CeANG, as the production and handling environments could lead to traces of radioactive leftovers in the final source.

The γ spectroscopy is essential to assert the low background level needed by SOX. The CeANG must not show any gamma emitting impurities with an activity higher than 10^{-3} Bq/Bq(β). This measurement is also done on the same source samples provided by FSUE "Mayak" PA using a high purity Germanium crystal detector. The first sample batch has shown to be within this specification.

To check the relative abundance of Cerium and to quantify the presence of long lifetime impurities and possible neutron emitting impurities, we also rely on α spectroscopy and inductive coupled plasma mass spectroscopy (ICP-MS). α spectroscopy test with gridded ionization chamber was particularly focused on ^{244}Cm and ^{241}Am and underlined neutron activity level lesser than 10^{-6} Bq/Bq(β) in agreement with CeANG specifications. Both these nuclei could generate a problematic background as neutrons they emit can mimic an IBD event in the detector. On ICP-MS again no significant impurities were seen on the first tests, and the relative abundance of isotopes 140, 142 and 144 of Cerium were consistent with the sample origin.

5. Conclusion

The SOX project relies heavily on the feasability of the radioactive sources production and their characterisation. The CeANG is by various sides an atypical object which has thus required a thorough process of design and validation for the associated tools. Now these early tasks are giving way to the final production phases of the experiment preparation.

During 2016, several characterization measurements will take place on radioactive samples and be cross checked to ensure a deployment with minimal troubles at the end of the year.

Acknowledgments

Thanks to the European Research Council for support under the Starting Grant No. StG-307184

References

[1] Mention G *et al.* 2011 *Phys. Rev.* D **83** 073006
[2] Giunti C and Laveder M 2011 *Phys. Rev.* C **83** 065504
[3] Giunti C *et al.* 2012 *Phys. Rev.* D **86** 113014
[4] Schael S *et al.* 2006 *Phys. Rept.* **427** 257-454
[5] Alimonti G *et al.* 2009 *Nucl.Instrum.Meth.* A **600** 568 (*arXiv*:0806.2400).
[6] Gaffiot J *et al.* 2015 *Phys. Rev.* D **91** 072005 (*arXiv*:1411.6694v2)
[7] Di Noto L 2015 The high precision calorimetric measurement of the 144Ce activity in the SOX experiment *Int. Conf. on Particle Physics and Astrophysics* October 2015 Moscow
[8] Altenmüller K 2015 Thermal model of the Munich-Genoa calorimeter for SOX *Int. Conf. on Particle Physics and Astrophysics* October 2015 Moscow
[9] Porter F and Day P 1959 *Phys. Rev.* **5** 114
[10] Bosch H *et al.* 1973 *Can. J. Phys.* **51** 2260
[11] Haag N *et al.* 2014 *Phys. Rev. Lett.* **112** 122501

Unified model of primordial black holes and dark matter formation

A V Grobov[1], **A E Dmitriev**[1], **V I Dokuchaev**[1,2] **and S G Rubin**[1]

[1] National Research Nuclear University MEPhI (Moscow Engineering Physics Institute), Kashirskoe highway 31, Moscow, 115409, Russia

[2] Institute for Nuclear Research of the Russian Academy of Sciences, Prospekt 60-letiya Oktyabrya 7a, Moscow, 117312, Russia

E-mail: alexey.grobov@gmail.com, sergeirubin@list.ru, alexdintras@mail.ru, dokuchaev@inr.ac.ru

Abstract. We propose a unified model of primordial black holes and soliton dark matter formation. Dynamic of spherically symmetric clumps of scalar field is considered in Newtonian approximation. The formation of hidden mass of the Universe is discussed. Numerical solution of the system of interacting scalar and gravitational fields is used to obtain the mass of a clumps.

1. Introduction

There are a lot of different ways to form a primordial black holes in early Universe. Some of them based on a phase transition phenomenon occurred during the process of inflation. For instance [1–3] there was elaborated a mechanism of primordial black holes (PBH) formation with wide range of masses. That kind of a mechanism could be suited to match hidden mass of the Universe very well.

Apparently the existence of complex scalar field implies the possibility of scalar clumps formation. Scalar clump is a self-gravitating (quasi)stable soliton configuration of the field. It's existence was predicted by Wheeler and then studied very thoroughly in a number of papers [4–6].

Here we build a toy model and numerically estimate the number of PBHs and scalar clumps with different mass. In frames of that model a lot of bubbles with different value of vacua appears. Some of them could evolve into PBHs, but other stay in form of stable scalar field configuration, such as scalar clumps. Both entities could be used to explain hidden mass of the Universe.

2. Black holes dark matter

Let's start with a mechanism of primordial black holes (PBH) formation based on first order phase transition during inflationary stage of the Universe [7], [8]. Consider complex scalar field $\psi = \chi \exp(i\theta)$ during the inflation process. This field being additional to inflaton is described by the potential:

$$V(\chi, \theta) = \lambda(\chi^2 - \frac{f^2}{2})^2 + \Lambda^4(1 - \cos(\theta)) \tag{1}$$

The process of quantum fluctuations of the field during its classical motion would lead to closed domain walls formation. In the case when domain wall collapses down to its Schwarzschild radius then the black hole forms.

The process of quantum fluctuations at inflation is described in detail in [9–12]. An amplitude of quantum fluctuations obeys the Gauss distribution with the average value $\chi(N)$ equals to the classical component of the field at the e-fold number N

$$dP(\chi_{crit}) = \frac{1}{\sqrt{2\pi(N_u - N)}} \exp\left(-\frac{(\chi_{crit} - \chi(N))^2}{2\delta\chi^2(N_u - N)}\right). \quad (2)$$

Here $\delta\chi = H/(2\pi)$ is the average value of the amplitude of the field fluctuation, H - is a Hubble parameter on inflationary stage and $N_u = 60$ is a duration of the inflationary period.

The example of PBHs spectrum obtained numerically is presented on the figure 1. Since

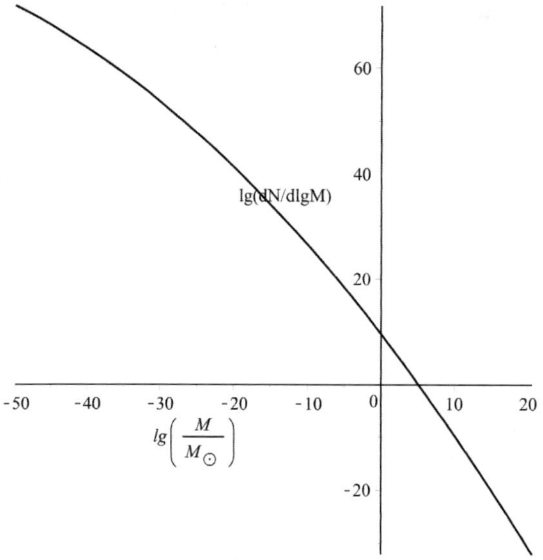

Figure 1. Mass spectrum of PBHs and scalar clumps for the parameter values: $\lambda = 0.0005$, $f = 40$ H, $\chi_0 = 70$ H, $\theta_0 = \frac{14\pi}{15}$, $\Lambda = 0.6$.

the mechanism of PBH formation implies that energy domain wall concentrates inside the Schwarzschild radius r_{Sch} PBHs it should be $r_{Sch} > h$, where h is the width of the wall. If $h > r_{Sch}$ then PBHs can't form. Scalar field configuration formed in that case may be stable or quasi-stable and decay with time. In general one should deal with self-gravitating complex scalar field. Let us simplify the task and consider non-relativistic complex scalar field in frames of Newtonian approximation instead of General Relativity. The system of interacting gravitational and complex scalar field would lead to non-singular scalar configurations, scalar clumps or so-called boson stars [4–6].

Take Einstein action in the form

$$S = S_g + S_m = \frac{M_{pl}^2}{2}\int R\sqrt{-g}d^4x + \int (g_{\mu\nu}\partial_\mu\psi^*\partial_\nu\psi - U(\psi))\sqrt{-g}d^4x. \quad (3)$$

here we take into account only quadratic term in potential $U(\psi) \simeq m^2\psi^*\psi = 2\sqrt{\lambda}f\psi^*\psi$, and metric tensor:

$$ds^2 = g_{\mu\nu}dx^\mu dx^\nu = (1 + 2V)dt^2 - (1 - 2V)(dx^2 + dy^2 + dz^2), \quad (4)$$

here V - is a gravitational potential, $\psi = \phi(x,t)e^{-imt}$ - is a field representation. Following the standard formalism we solve the system of two Poisson equation under assumption of spherical symmetry.

$$E\phi = \frac{1}{2r^2}\frac{\partial}{\partial r}\left(r^2\frac{\partial\phi}{\partial r}\right) - V\phi \tag{5}$$

$$\frac{1}{r^2}\frac{\partial}{\partial r}\left(r^2\frac{\partial V}{\partial r}\right) = \phi\phi^* \tag{6}$$

We use dimensionless variables $mx_{phys}, mt_{phys} \to (x,t)$ and $\sqrt{4\pi G}\phi_{phys} \to (\phi)$ and consider clump in the stationary state.

Scalar clump mass is given by:

$$M = \frac{M_{pl}^2}{m}\int\limits_0^{r_c}\phi^2 4\pi r^2 dr = \frac{M_{pl}^2}{m}I, \tag{7}$$

where m - is a mass of field quanta, r_c - is a characteristic size of a clump, i.e. the distance where gathered main amount of field energy. In general the mass of a clumps depends on the mass of scalar field quanta. Note, that in case of PBH we calculate its mass as energy of domain wall that being collected inside the Schwarzschild radius, but in case of clumps this is not true. Only a part of energy that being delivered by domain wall constitutes the mass of scalar clump. The rest of energy is reemited in space. So the mass distribution of scalar clumps presented on figure 1 should be distorted.

3. Conclusion

Mass of the clump for the parameters we use is about 10 g, but their number is enormous. Such entities along with black holes could easily pretend to be a dark matter, contributing to full energy density of the Universe. Notice that only scalar field that has been recently discovered is the Higgs field, and there is no evidence of scalar clumps existence. So they remain purely theoretical entities by now along with primordial black holes. A mechanism of primordial black hole formation is developed in a number of papers. In turn the model of associated production of scalar clumps and black holes needs to be refined. Since the formation of scalar clump is possible only under a certain parameters and initial value of field, it seems unlikely that all domain walls, that failed to become PBHs, can form scalar clumps. One more thing about scalar clump formation in frames of Newtonian regime. Mass of the clump is given by the formula $M = \frac{M_{pl}^2}{m}I$, and Schwarzschild radius is $r_{Sch} = \frac{2M}{M_{pl}^2} = \frac{2I}{m}$. On the other hand the physical size of the clump $r_{phys} = \frac{r_c}{m}$. If $r_{Sch} > r_{phys}$ or equivalently $2I > r_c$, then we have a contradiction with Newtonian approximation. To sum up the mechanism of numerous scalar clumps production needs further development.

Acknowledgments

This work was performed within the framework of the FRPP Center supported by MEPhI Academic Excellence Project (contract No. 02.a03.21.0005, 27.08.2013). The work of A.V.G., A.E.D. and S.G.R. was supported by The Ministry of education and science of Russian Federation, project 3.472.2014/K. The work of A.V.G. was was supported by Grant of RFBR no. 14-02-31417

References

[1] Grobov A V *et al.* 2011 A Mechanism for Primordial Black Hole Formation *Gravitation and Cosmology* **17** 181-4

[2] Grobov A V, Rubin S G and Shalamova V Yu 2014 *Adv. High Energy Phys.* 319218
[3] Dokuchaev V I, Eroshenko Yu N and Rubin S G *arXiv*:0709.0070
[4] Feinblum D A and McKinley W A 1968 *Phys. Rev.* D **168** 1445
[5] Kaup D J 1968 *Phys. Rev.* D **172** 1331
[6] Ruffini R and Bonazolla S 1969 *Phys. Rev.* D **187** 1767
[7] Khlopov M Yu, Rubin S G and Sakharov A S 2005 *Astropart. Phys.* **23** 265
[8] Rubin S G *arXiv*:astro-ph/051118
[9] Bronnikov K A and Rubin S G *Black Holes, Cosmology and Extra Dimensions* (River Edge, NJ: World Scientific Publishing)
[10] Dokuchaev V I *et al.* 2010 *Astronomy Letters* **36**(11) 773-9
[11] Starobinsky A A 1986 Stochastic de sitter (inflationary) stage in the early universe *Lecture Notes in Physics* vol 246, ed H J de Vega and N Sanchez (Berlin: Springer) 107-26
[12] Rey S J 1987 *Nuclear Physics* B **284** 706-28

Journal of Physics: Conference Series **675** (2016) 012034 doi:10.1088/1742-6596/675/1/012034

Fermion scattering on deformed extra space

A V Grobov, A E Dmitriev and S G Rubin

National Research Nuclear University MEPhI (Moscow Engineering Physics Institute), Kashirskoe highway 31, Moscow, 115409, Russia

E-mail: alexey.grobov@gmail.com, sergeirubin@list.ru, alexdintras@mail.ru

Abstract. Fermions are considered in a multidimensional space with two extra dimensions. Fermion Lagrangian contains coupling to a metric of deformed extra space. We show that the point-like defect on a space with topology of sphere leads to fermion interaction with such defect and calculate cross section of a fermion scattering on such defect.

1. Introduction

The idea of extra dimensions in field theory is proved to be very fruitful [1–3]. We consider a fermion in model with extra space of sphere topology distorted by point-like defect [3]. Let's start with 6-dimensional action for massless fermion:

$$S = \int d^6x\sqrt{|G|}i\overline{\Psi}e_a^A\Gamma_a\nabla_A\Psi, \tag{1}$$

where Ψ - is a six-dimensional Dirac spinor. For simplicity only first four components of the spinor are nonzero.

Consider a manifold $M_4 \times S^2$, where M_4 is ordinary space-time. Then metric could be written in a form of:

$$ds^2 = \eta_{\mu\nu}dx^\mu dx^\nu - r^2(\theta, x)(d\theta^2 + \sin(\theta)^2 d\phi^2). \tag{2}$$

After standard calculations the expression for the 4-dimensional effective action acquires a form:

$$S = \int d^4x(i\overline{\psi}\gamma_\mu\partial_\mu\psi I(x) + i\overline{\psi}\gamma_\mu\psi A_\mu(x)), \tag{3}$$

where

$$I(x) = 2\pi\int r^2\sin(\theta)d\theta,\ A_\mu(x) = \frac{1}{2}\partial_\mu I(x). \tag{4}$$

Thus Lagrangian gets a coupling between fermion and scalar field originated from point-like defect. The size of extra space is supposed to vary depending on 4-dimensional coordinates.

$$r(\theta, x) = a(\theta) + b(\theta)e^{-cr}. \tag{5}$$

a and b functions are defined by numerical solution of multidimensional equations for gravity, c parameter is found from principle of least action.

2. Amplitude and cross-section

To obtain an expression for the cross-section one should write down Feynman rules, considering vector and scalar fields classically. Invariante amplitude in the first order of perturbation theory is:

$$M = i\overline{u_2}\gamma_\mu u_1 p_{1\mu}\chi(q) - \overline{u_2}\gamma_\mu u_1 A_\mu(q), \tag{6}$$

where

$$\chi(q) = \frac{32\pi^2 c\alpha}{(q^2+c^2)^2} + \frac{32\pi^2 c\beta}{(q^2+4c^2)^2},\ A_\mu(q) = \frac{i}{2}\chi(q)q_\mu. \tag{7}$$

Transferred momentum has a form:

$$q_\mu = (0, -\vec{q}),\ q^2 = -4(E^2 - m^2)\sin^2\left(\frac{\theta}{2}\right). \tag{8}$$

Now one can calculate the square of invariante amplitude:

$$\overline{|M|}^2 = 4m^2\chi^2(m^2 + (E^2 - m^2)\sin^2\left(\frac{\theta}{2}\right)). \tag{9}$$

Cross-section of the fermion scattering on classical potential is given by:

$$\frac{d\sigma}{d\Omega} = \frac{1}{16\pi^2}\overline{|M|}^2. \tag{10}$$

Cross-section for electrons is presented on figure 1. In case of non-relativistic electrons cross-section is estimated as 10^{-18} GeV^{-2}.

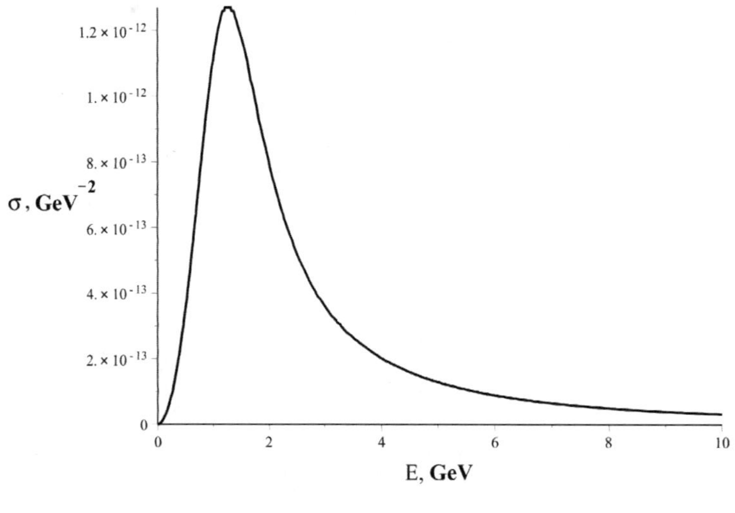

Figure 1. Electron scattering cross-section vs full energy of the electron.

3. Conclusion

Deformed extra space is a promising dark matter candidate. In light of this fermion scattering process may be an observational signature of point-like defect in space. Here we show only the possibility of corresponding process, but of course the study demands further development.

The full study of all possible effect is far beyond the scope of that paper. To simplify the analysis we required that fermion can not propagate in extra space and took into account only four components of 8-component Dirac spinor. So the obtained result should be treated as estimation of more complicated effect. One can see that cross-section is zero for massless fermion, so neutrinos are very unlikely to interact with that extra space. On the other hand nonrelativistic electrons interact with cross-section 10^{-48} cm^2.

Acknowledgments
This work was performed within the framework of the FRPP Center supported by MEPhI Academic Excellence Project (contract No. 02.a03.21.0005, 27.08.2013). The work of A.V.G., A.E.D. and S.G.R. was supported by The Ministry of education and science of Russian Federation, project 3.472.2014/K.

References
[1] Gogberashvili M, Midodashvili P and Singleton D 2007 *JHEP* **0708** 33
[2] Liu Y, Zhang L and Duan Y 2008 *Commun. Theor. Phys.* **49** 1577-9
[3] Gani V A, Dmitriev A E and Rubin S G 2015 *Int. J. Mod. Phys.* D **24** 1545001

The high precision measurement of the ^{144}Ce activity in the SOX experiment

L Di Noto[1,31] on behalf of the SOX collaboration, M Agostini[2], K Althenmüller[2], S Appel[2], G Bellini[3], J Benziger[4], N Berton[5], D Bick[6], G Bonfini[7], D Bravo–Berguño[8], B Caccianiga[3], F Calaprice[9], A Caminata[1], P Cavalcante[7], R Cereseto[1], A Chepurnov[10], K Choi[11], M Cribier[5], D DAngelo[3], S Davini[12], A Derbin[13], I Drachnev[12], M Durero[5], A Etenko[14], S Farinon[1], V Fischer[5], K Fomenko[15], D Franco[16], F Gabriele[7], J Gaffiot[5], C Galbiati[9], C Ghiano[1], M Giammarchi[3], M Göeger-Neff[2], A Goretti[9], M Gromov[10], C Hagner[6], Th Houdy[5], E Hungerford[17], Aldo Ianni[7], Andrea Ianni[9], N Jonquères [18], K Jedrzejczak[19], M Kaiser[6], V Kobychev[20], D Korablev[15], G Korga[7], V Kornoukhov[21], D Kryn[16], T Lachenmaier[22] T Lasserre[5], M Laubenstein[7], T Lehnert[23], J Link[8] E Litvinovich[14,24], F Lombardi[7], P Lombardi[3], L Ludhova[3], G Lukyanchenko[14,24], I Machulin[14,24], S Manecki[8], W Maneschg[25], S Marcocci[12], J Maricic[11], G Mention[5], E Meroni[3], M Meyer[6], L Miramonti[3], M Misiaszek[7,19], M Montuschi[26], P Mosteiro[9], V Muratova [13], R Musenich[1], B Neumair[2], L Oberauer[2], M Obolensky[16], F Ortica[27], M Pallavicini[1], L Papp[2], L Perasso[1], A Pocar[28], G Ranucci[3], A Razeto[7], A Re[3], A Romani[27], R Roncin[7,16], C Rossi[1], N Rossi[7], S Schönert[2], L Scola[5], D Semenov[13] H Simgen[25], M Skorokhvatov[14,24], O Smirnov[15], A Sotnikov[15], S Sukhotin[14], Y Suvorov[14,29], R Tartaglia[7], G Testera[1], J Thurn[23], M Toropova[14], C Veyssière[5], M Vivier[5], E Unzhakov[13], R B Vogelaar[8], F von Feilitzsch[2], H Wang[29], S Weinz[30], J Winter[30], M Wojcik[19], M Wurm [6,30], Z Yokley[8], O Zaimidoroga[15], S Zavatarelli[1], K Zuber[23] and G Zuzel[19]

[1] Dipartimento di Fisica, Università degli Studi e INFN 16146 Genova, Italy
[2] Physik-Department and Excellence Cluster Universe, Technische Universität München, 85748 Garching, Germany
[3] Dipartimento di Fisica, Università degli Studi e INFN, 20133 Milano, Italy
[4] Chemical Engineering Department, Princeton University, Princeton, NJ 08544, USA
[5] Commisariat à l'Énergie Atomique et aux Énergies Alternatives, Centre de Saclay, IRFU, 91191 Gif-sur-Yvette, France
[6] Institut für Experimentalphysik, Universität, 22761 Hamburg, Germany
[7] INFN Laboratori Nazionali del Gran Sasso, 67010 Assergi (AQ), Italy
[8] Physics Department, Virginia Polytechnic Institute and State University, Blacksburg, VA 24061, USA
[9] Physics Department, Princeton University, Princeton, NJ 08544, USA

[31] To whom any correspondence should be addressed.

[10] Lomonosov Moscow State University Skobeltsyn Institute of Nuclear Physics, 119234 Moscow, Russia
[11] Department of Physics and Astronomy, University of Hawaii, Honolulu, HI 96822, USA
[12] Gran Sasso Science Institute (INFN), 67100 L'Aquila, Italy
[13] St. Petersburg Nuclear Physics Institute NRC Kurchatov Institute, 188350 Gatchina, Russia
[14] NRC Kurchatov Institute, 123182 Moscow, Russia
[15] Joint Institute for Nuclear Research, 141980 Dubna, Russia
[16] AstroParticule et Cosmologie, Universite Paris Diderot, CNRS/IN2P3, CEA/IRFU, Observatoire de Paris, Sorbonne Paris Citè, 75205 Paris Cedex 13, France
[17] Department of Physics, University of Houston, Houston, TX 77204, USA
[18] Commisariat à l'Énergie Atomique et aux Énergies Alternatives, Centre de Saclay, DEN/DM2S/SEMT/BCCR, 91191 Gif-sur-Yvette, France
[19] M Smoluchowski Institute of Physics, Jagiellonian University, 30059 Krakow, Poland
[20] Kiev Institute for Nuclear Research, 06380 Kiev, Ukraine
[21] Institute for Theoretical and Experimental Physics, 117218 Moscow, Russia
[22] Kepler Center for Astro and Particle Physics, Universität Tübingen, 72076 Tübingen, Germany
[23] Department of Physics, Technische Universität Dresden, 01062 Dresden, Germany
[24] National Research Nuclear University MEPhI (Moscow Engineering Physics Institute), Kashirskoe highway 31, Moscow, 115409, Russia
[25] Max-Planck-Institut für Kernphysik, 69117 Heidelberg, Germany
[26] Dipartimento di Fisica e Scienze della Terra Università degli Studi di Ferrara e INFN, Via Saragat 1 - 44122, Ferrara, Italy
[27] Dipartimento di Chimica, Università e INFN, 06123 Perugia, Italy
[28] Amherst Center for Fundamental Interactions and Physics Department, University of Massachusetts, Amherst, MA 01003, USA
[29] Physics and Astronomy Department, University of California Los Angeles (UCLA), Los Angeles, California 90095, USA
[30] Institute of Physics and Excellence Cluster PRISMA, Johannes Gutenberg-Universität Mainz, 55099 Mainz, Germany

E-mail: `lea.dinoto@ge.infn.it`

Abstract. In order to perform a resolutive measurement to clarify the neutrino anomalies and to observe possible short distance neutrino oscillations, the SOX (Short distance neutrino Oscillations with BoreXino) experiment is under construction. In the first phase, a 100 kCi ^{144}Ce-^{144}Pr antineutrino source will be placed under the Borexino detector at the Laboratori Nazionali del Gran Sasso (LNGS), in center of Italy, and the rate measurement of the antineutrino events, observed by the very low radioactive background Borexino detector, will be compared with the high precision ($< 1\%$) activity measurement performed by two calorimeters. The source will be embedded in a 19 mm thick tungsten alloy shield and both the calorimeters have been conceived for measuring the thermal heat absorbed by a water flow. In this report the design of the calorimeters will be described in detail and very preliminary results will be also shown.

1. Introduction

Since the neutrino anomalies leave room to the existence of sterile neutrino and to the short distance neutrino oscillations [1–3], the SOX experiment was proposed [4] with the goal of observing the neutrino events emitted by a high activity ^{144}Ce-^{144}Pr artificial source. The source will be placed in a tunnel, 8 m far from the center of Borexino detector and thanks to the very low radioactive background and to the detector properties [5,6], short distance oscillations of eV mass sterile neutrino might be observed and at the same time a disappearance experiment can be performed, if the source activity is also measured with less than 1% precision [4,7].

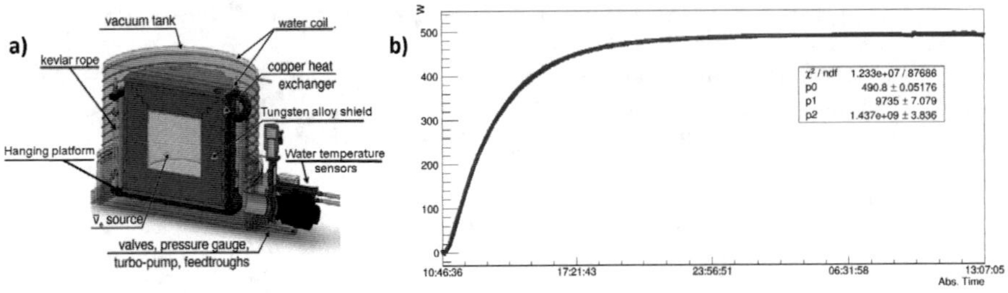

Figure 1. a) Section of the source with the tungsten alloy shield embedded inside the INFN-TUM calorimeter. The main parts of the apparatus are clearly visible. b) Preliminary power measurement performed by using a mock-up electrical source. In the picture the parameter as resulted by fitting the data with the exponential function of equation 2, are reported as well.

As the size of the source embedded in the shield is really close to the available room in the tunnel, the project will be developed in two different phases: the calorimetric measurement of the activity, just before the insertion of the source inside the pit under Borexino (1 week duration) and the second phase, when the source will be installed inside the tunnel and the Borexino data taking will occur (1.5 year duration). Other calorimetric measurements in the between or at the end of the data taking might be also performed in order to increase the accuracy of the final results.

2. Description of the calorimetric measurement

The ^{144}Ce-^{144}Pr artificial source will be produced with around 100 kCi activity at the reactor of the Mayak Production Association in Russia after a chemical extraction of the $^{144}CeO_2$ from exhausted nuclear fuel. The radioactive powder will be sealed inside a double stainless steel container and then embedded in a 19 mm thick safety shield made of a tungsten alloy (see figure 1a). After the production and the transportation to the LNGS, the calorimetric measurement will occur thanks to two different calorimeters, that are under construction and testing by the INFN-TUM and the CEA groups within the SOX collaboration. Both the calorimeters have been conceived to measure the source activity with high precision by knowing the power released by the radiation in the tungsten shield and absorbed by a water flow.

In the INFN-TUM calorimeter the water flows inside a copper heat exchanger, in contact with the tungsten shield (see figure 1a), while in the CEA calorimeter the source with the tungsten is directly immersed in the water inside the calorimeter chamber. In both the systems, neglecting the heat losses P_{lost}, the power P is achieved by measuring with high accuracy the mass flow $\dot{m}$ and the temperatures T_{out} and T_{in} of the water outgoing and entering in the copper heat exchanger, according to the relation:

$$P = \dot{m}c(T_{out} - T_{in}) \tag{1}$$

where c is the specific heat of the water averaged in the T_{in}-T_{out} temperature range. As it will be shown in the following, the calorimeters have been designed in order to minimize P_{lost} that will be estimated during the calibration and testing phase and reduced as much as possible.

In particular in the INFN-TUM calorimeter, for reducing the convection, the copper heat exchanger is kept in vacuum at a pressure lower than $5 \cdot 10^{-5}$ mbar, while the source inside the tungsten, separated through a vacuum flange, is at atmospheric pressure. Secondly for minimizing irradiation, two stages of 10 layers each one of super insulators are placed between the copper and the vacuum tank and at the same time hot water flows inside a pipe built directly on the tank itself (see figure 1a) in order to thermalize the chamber close to the copper temperature. Finally for reducing the conduction, the source inside the copper heat exchanger is placed on an hanging platform, suspended by three kevlar ropes with low heat conductivity. In both the systems a closed water loop is foreseen and in the INFN-TUM calorimeter the water loop was designed in order to accurately control the massflow and the temperature of the water just before the entrance in the vacuum tank. Thanks to a proportional valve controlled by a feedback loop, the water flow, coming from the chiller, can be tunable in the range of 2-15 g/s and stabilized at 0.3% of the nominal value, heavily reducing the oscillations due to the chiller pump. At the same time also the temperature of the water entering inside the calorimeter can be stabilized within 0.1 K in the 13-17 °C range, compensating any environmental temperature fluctuation or chiller instability during all the phases of the measurement. In particular during the calorimetric measurement a massflow of 10 g/s was chosen, giving a difference of temperature of $T_{out} - T_{in} \sim 28$ °C, for a source power of 1200 W correspondent to 100 kCi activity.

On the contrary, during the second phase, since the source will be inserted in the pit inside the copper heat exchanger of the INFN-TUM calorimeter, the water loop will be necessary to cool down the apparatus and the controls feedback will be aimed to stabilize the temperature in the tunnel in order to prevent turbulence effect in the scintillator liquid of the Borexino detector.

3. Preliminary results and conclusions

In the calibration phase, before the source delivery, the set up can be tested with a mock-up source made by a copper cylinder with the same dimensions of the source container, embedded in an Aluminum body that replaces the tungsten shield. Inside the copper cylinder six electrical heaters are placed and they can be turned on with a total maximum power of 1200 W (the same power dissipated by a 100 kCi Ce source), measured very precisely by an amperometer and a voltmeter directly connected with the heaters. The temperature of the copper and of the aluminum part (on the top, on the side and on the bottom) can be monitored and the temperatures of many critical points inside the vacuum chamber (as the platform, the copper jacket and the internal chamber) are acquired as well.

As first test a rough measurement performed with a simple copper coil directly at contact with the aluminum was done at atmospheric pressure with a set power of 497 W. The plot of the measured power calculated by the relation 1 as a function of the time is shown in figure 1b. The data were acquired after the heaters switching on and the fit was done with the exponential function:

$$P(t) = p_0 \left(1 - e^{-\frac{x-p_2}{p_1}}\right) \tag{2}$$

where the time scale is in UNIX time date format. From the fit the measured power at the equilibrium was found around 491 W, less than the set value since the heat losses due to the convection can not be neglected and a first estimation of the time constant $p_1 \simeq 9735$ s $= 2,7$ hours was achieved for the first time.

In conclusion two calorimeters have been built for measuring the source activity with a precision $< 1\%$ and now they are under testing. In the next months many calibrations will be performed with the mock-up source, while the final measurement with the Ce source and the starting of the Borexino data taking is fixed in the autumn 2016.

References

[1] Mention G *et al.* 2011 *Phys. Rev.* D **83** 073006
[2] Aguilar A *et al.* (LSND Collaboration) 2001 *Phys. Rev.* D **64**112007
[3] Aguilar A *et al.* (MiniBooNe Collaboration) 2013 *Phys. Rev. Lett.* **110** 161801
[4] Bellini G *et al.* (Borexino Collaboration) 2013 *JHEP* **08** 038
[5] Bellini G *et al.* (Borexino Collaboration) 2014 *Phys. Rev.* D **89** 112007
[6] Bellini G *et al.* (Borexino Collaboration) 2010 *Phys. Lett.* B **687** 299
[7] Palazzo A 2013 *Mod. Phys. Lett.* A **28** 1330004

International Conference on Particle Physics and Astrophysics (ICPPA-2015) IOP Publishing
Journal of Physics: Conference Series **675** (2016) 012036 doi:10.1088/1742-6596/675/1/012036

Overview and accomplishments of the Borexino experiment

G Ranucci[1,29], M Agostini[2], S Appel[2], G Bellini[1], J Benziger[3], D Bick[4], G Bonfini[5], D Bravo[6], B Caccianiga[1], F Calaprice[7], A Caminata[8], P Cavalcante[5], A Chepurnov[9], D D'Angelo[1], S Davini[10], A Derbin[11], L Di Noto[8], I Drachnev[10], A Etenko[12], K Fomenko[13], D Franco[14], F Gabriele[5], C Galbiati[7], C Ghiano[8], M Giammarchi[1], M Goeger-Neff[2], A Goretti[7], M Gromov[9], C Hagner[4], E Hungerford[15], Aldo Ianni[5], Andrea Ianni[7], K Jedrzejczak[17], M Kaiser[4], V Kobychev[18], D Korablev[13], G Korga[5], D Kryn[14], M Laubenstein[5], B Lehnert[19], E Litvinovich[12,20], F Lombardi[5], P Lombardi[1], L Ludhova[1], G Lukyanchenko[12,20], I Machulin[12,20], S Manecki[6], W Maneschg[22], S Marcocci[10], E Meroni[1], M Meyer[4], L Miramonti[1], M Misiaszek[17,5], M Montuschi[23], P Mosteiro[7], V Muratova[11], B Neumair[2], L Oberauer[2], M Obolensky[14], F Ortica[24], M Pallavicini[8], L Papp[2], L Perasso[8], A Pocar[26], A Razeto[5], A Re[1], A Romani[24], R Roncin[5,14], N Rossi[5], S Schönert[2], D Semenov[11], H Simgen[22], M Skorokhvatov[12,20], O Smirnov[13], A Sotnikov[13], S Sukhotin[12], Y Suvorov[27,12], R Tartaglia[5], G Testera[8], J Thurn[19], M Toropova[12], E Unzhakov[11], A Vishneva[13], R B Vogelaar[6], F von Feilitzsch[2], H Wang[27], S Weinz[28], J Winter[28], M Wojcik[17], M Wurm[28], Z Yokley[6], O Zaimidoroga[13], S Zavatarelli[8], K Zuber[19] and G Zuzel[17] (Borexino collaboration)

[1] Dipartimento di Fisica, Università degli Studi e INFN, 20133 Milano, Italy
[2] Physik-Department and Excellence Cluster Universe, Technische Universität München, 85748 Garching, Germany
[3] Chemical Engineering Department, Princeton University, Princeton, NJ 08544, USA
[4] Institut für Experimentalphysik, Universität, 22761 Hamburg, Germany
[5] INFN Laboratori Nazionali del Gran Sasso, 67010 Assergi (AQ), Italy
[6] Physics Department, Virginia Polytechnic Institute and State University, Blacksburg, VA 24061, USA
[7] Physics Department, Princeton University, Princeton, NJ 08544, USA
[8] Dipartimento di Fisica, Università degli Studi e INFN, 16146 Genova, Italy
[9] Lomonosov Moscow State University Skobeltsyn Institute of Nuclear Physics, 119234 Moscow, Russia
[10] Gran Sasso Science Institute (INFN), 67100 L'Aquila, Italy
[11] St. Petersburg Nuclear Physics Institute NRC Kurchatov Institute, 188350 Gatchina, Russia
[12] NRC Kurchatov Institute, 123182 Moscow, Russia
[13] Joint Institute for Nuclear Research, 141980 Dubna, Russia
[14] AstroParticule et Cosmologie, Université Paris Diderot, CNRS/IN2P3, CEA/IRFU, Observatoire de Paris, Sorbonne Paris Cité, 75205 Paris Cedex 13, France
[15] Department of Physics, University of Houston, Houston, TX 77204, USA
[16] Institute for Theoretical and Experimental Physics, 117218 Moscow, Russia

---

[29] Presenter. To whom any correspondence should be addressed.

[17] M. Smoluchowski Institute of Physics, Jagiellonian University, 30059 Krakow, Poland
[18] Kiev Institute for Nuclear Research, 06380 Kiev, Ukraine
[19] Department of Physics, Technische Universität Dresden, 01062 Dresden, Germany
[20] National Research Nuclear University MEPhI (Moscow Engineering Physics Institute), Kashirskoe highway 31, Moscow, 115409, Russia
[21] Kepler Center for Astro and Particle Physics, Universität Tübingen, 72076 Tübingen, Germany
[22] Max-Planck-Institut für Kernphysik, 69117 Heidelberg, Germany
[23] Dipartimento di Fisica e Scienze della Terra Università degli Studi di Ferrara e INFN, Via Saragat 1-44122, Ferrara, Italy
[24] Dipartimento di Chimica, Biologia e Biotecnologie, Università degli Studi e INFN, 06123 Perugia, Italy
[25] Physics Department, Queen's University, Kingston ON K7L 3N6, Canada
[26] Amherst Center for Fundamental Interactions and Physics Department, University of Massachusetts, Amherst, MA 01003, USA
[27] Physics and Astronomy Department, University of California Los Angeles (UCLA), Los Angeles, California 90095, USA
[28] Institute of Physics and Excellence Cluster PRISMA, Johannes Gutenberg Universität Mainz, 55099 Mainz, Germany

E-mail: gioacchino.ranucci@mi.infn.it

Abstract. The Borexino experiment is running at the Laboratori del Gran Sasso in Italy since 2007. Its technical distinctive feature is the unprecedented ultralow background of the inner scintillating core, which is the basis of the outstanding achievements accumulated by the experiment. In this talk, after recalling the main features of the detector, the impressive solar data gathered so far by the experiment will be summarized, with special emphasis to the most recent and prominent result concerning the detection of the fundamental pp solar neutrino flux, which is the direct probe of the engine mechanism powering our star. Such a milestone measurement puts Borexino in the unique situation of being the only experiment able to do solar neutrino spectroscopy over the entire solar spectrum; the counterpart of this peculiar status in the oscillation interpretation of the data is the capability of Borexino alone to perform the full validation across the solar energy range of the MSW-LMA paradigm. The talk will be concluded highlighting the perspectives for the final stage of the solar program of the experiment, centered on the goal to fully complete the solar spectroscopy with the missing piece of the CNO neutrinos. If successful, such a measurement would represent the final crowning of the long quest of Borexino to unravel all the properties of the neutrinos from the Sun.

1. Introduction

Borexino at Gran Sasso is the last player which entered the solar neutrino arena, where thanks to its unprecedented low background it provided breakthrough results in the low energy sub-MeV regime. Having already measured three components of the solar neutrino spectrum over the past years, i.e. ^{7}Be, ^{8}B and pep (providing jointly with this component also a stringent upper limit on the CNO contribution), Borexino has recently crowned its remarkable series of results with the detection of the fundamental pp neutrino flux, coming from the reaction which provides most of the Sun's energy. Therefore Borexino is the first experiment able to perform an almost complete spectroscopy of the whole solar neutrino flux, allowing a thorough data-model comparison [1]. The last step of this investigation will be the attempt to measure the tiny CNO flux, which Borexino will undertake over the next years.

Moreover, Borexino has also detected an unquestionable geoneutrino signal (i.e. anti-neutrinos from the radioactivity inside the Earth), contributing to pave the way to a complete new method to investigate the interior of our planet.

In the following the experimental characteristics which made possible these outstanding accomplishments will be briefly reviewed, together with the features of the results achieved so far.

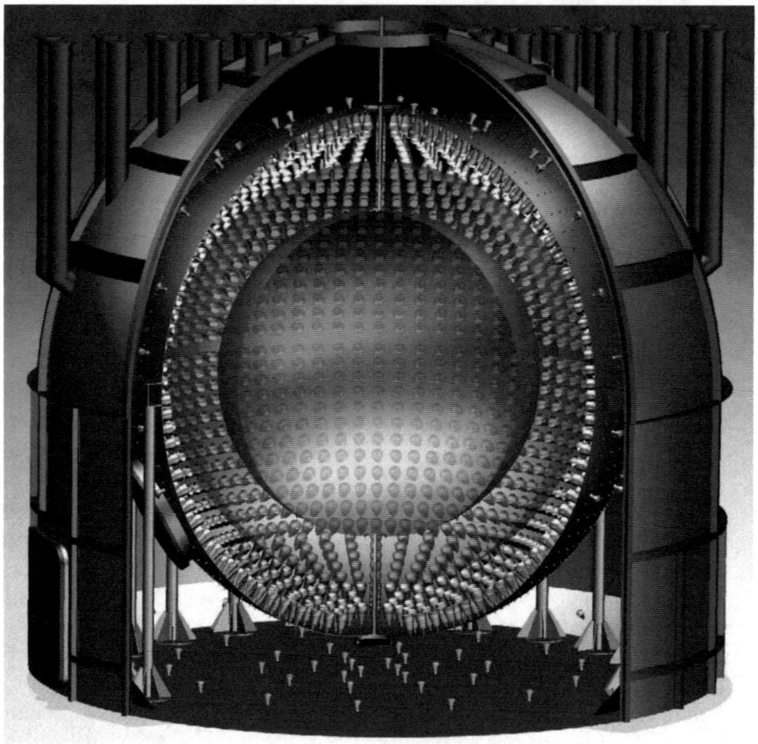

Figure 1. Sketch of the Borexino experiment, highlighting its major components arranged according to a graded shielding design.

2. Characteristics of the detector

Borexino [2] is a scintillator detector which employs as active detection medium a mixture of pseudocumene (PC, 1,2,4- trimethylbenzene) and PPO (2,5-diphenyloxazole, a fluorescent dye) at a concentration of 1.5 g/l. Because of its intrinsic high luminosity (50 times more than in the Cerenkov technique) the liquid scintillation technology is extremely suitable for massive calorimetric low energy spectroscopy. However, the lack of directionality of the method makes it impossible to distinguish neutrino scattered electrons from electrons due to natural radioactivity, as is done in the Cerenkov detector exploiting the association of the neutrino signals to the Sun direction. Therefore the crucial requirement of the Borexino technology is an extremely low radioactive contamination of the detection medium, at fantastic unprecedented levels, definitely below the acceptable limits in the water of SNO and Super-Kamiokande.

To reach ultralow operating background conditions in the detector, the design of Borexino (see figure 1) is based on the principle of graded shielding, with the inner scintillating core at the centre of a set of concentric shells of increasing radiopurity. The scintillator mass (278-ton) is contained in a 125 μm thick nylon Inner Vessel (IV) with a radius of 4.25 m. Within the IV a fiducial mass is software defined through the estimated events position, obtained from the PMTs timing data via a time-of-flight algorithm.

A second nylon outer vessel (OV) with radius 5.50 m surrounds the IV, acting as a barrier against radon and other background contaminations originating from outside. The region between the inner and outer vessels contains a passive shield composed of pseudocumene and 5.0 g/l (later reduced to 3.0 g/l) of DMP (dimethylphthalate), a material that quenches the residual scintillation of PC so that spectroscopic signals arise dominantly from the interior of the IV.

A 6.85 m radius stainless steel sphere (SSS) encloses the central part of the detector and serves also as a support structure for the PMTs. The region between the OV and the SSS is filled with the same inert buffer fluid (PC plus DMP) which is layered between the inner and outer vessels.

Finally, the entire detector is contained in a tank (radius 9 m, height 16.9 m) filled of ultra-pure water. The total liquid passive shielding of the central volume from external radiation (such as that originating from the rock) is thus 5.5 m of water equivalent (m.w.e). The scintillator material in the IV was less dense than the buffer fluid by about 0.1% with the original DMP concentration of 5 g/l; this resulted in a slight upward buoyancy force on the IV, implying the need of thin low-background ropes made of ultra-high density polyethylene to hold the nylon vessels in place.

This modest buoyancy was further reduced of more than a factor 10 by removing via distillation a fraction of the total DMP content in the buffer: the process ended with a final DMP concentration of 3 g/l, still perfectly adequate to suppress the buffer scintillation, while at the same time appropriate to substantially prevent the outward scintillator flow from a small leaking point which developed on the vessel surface about 10 months after the start-up of the data taking.

The scintillation light is viewed by 2212 8" PMTs (ETL 9351) uniformly distributed on the inner surface of the SSS. All but 371 photomultipliers are equipped with aluminium light concentrators designed to increase the collection efficiency of the light from the scintillator, and concurrently minimizing the detection of photons not coming from the active scintillating volume. Residual background scintillation and Cerenkov light that escape quenching in the buffer are thus reduced. The PMTs without concentrators can be used to study this background, as well as help identify muons that cross the buffer and not the inner vessel.

All the materials of the internal components of the detector (stainless steel, phototubes, cables, light concentrators, nylon) were specially selected for extremely low radioactivity. Furthermore, only qualified ultraclean processes were employed for their realization, followed by careful surface cleaning methods.

The final assembly of the elements in the SSS was carried out in clean room conditions: the entire interior of the sphere was converted into a clean room of class 1000, while in front of the main entrance of the sphere itself an on purpose clean room of class 100-1000 was used for all the final cleaning procedures of the equipment. Phototubes and ancillary mounting parts, light concentrators, calibration optical fibres, inner and outer vessels together with the anchoring systems, all passed and were treated through the access clean room, as well as the elaborated scaffolding system which was designed specifically to make it possible to the operators the safe and clean access to the whole sphere surface.

Clearly, for the success of the experiment key elements were also the many liquid purification and handling systems, which were designed and installed to ensure the proper manipulation of the fluids at the exceptional level demanded by Borexino. Without entering in the details of the complex layout and functionality of these plants, some salient points deserve to be mentioned.

The PC specially produced for Borexino, according to a stringent quality control plan developed jointly with the company, was shipped to the Gran Sasso location through special transport tanks cleaned, treated and prepared by us; the first operation there was the transfer of the PC via a dedicated unloading station to four big reservoir tanks, forming the so called Storage Area, were the PC was stored prior to be inserted in the detector. Taken from the Storage Area, the PC was first purified via distillation, then either mixed with PPO for insertion in the Inner Vessel, or mixed with DMP for the insertion in the buffer region. Involved in this set of operations were the main Borexino plants: the Distillation Skids, the Filling Stations, the elaborate Interconnection system which allowed flexible transfers of the liquid throughout the various plants. Furthermore, the PPO was pre-mixed with a limited quantity of PC in a dedicated PPO System, originating a concentrated PPO solution which was then mixed in line with the PC at the level of the Filling Stations.

Other important ancillary plants, fundamental for the success of the operations, are the nitrogen systems, which delivered regular nitrogen, or on site purified nitrogen, or specially produced nitrogen with exceptionally low content of ^{39}Ar and ^{85}Kr, to be used for the crucial manipulations of the liquid in the inner vessel.

Finally, an ultrapure water system was used to produce the water for the cleaning operations, for the fill of the External Tank, and for the preliminary water fill of the Inner Detector.

Obviously, besides the design for the proper operating conditions, the cleanliness of the manufacturing processes used for the assembly of all these plants was of paramount importance for the success of Borexino, as well as the thorough and accurate cleaning procedures that they underwent upon their installation on site.

The stability of the detector and the perfect understanding of its response gained with a detailed MC contrasted with the results from a thorough calibration campaign have been the last ingredients of its success.

When data taking started in May 2007, it appeared immediately that the daunting task of the ultralow radioactivity was successfully achieved, representing *per se* a major technological breakthrough, opening a new era in the field of ultrapure detectors for rare events search.

The exceptional purity obtained implies that, once selected by software analysis the design fiducial volume of 100 tons and upon removal of the muon and muon-induced signals, the recorded experimental spectrum is so clean to show spectacularly the striking feature of the ^{7}Be scattering edge, i.e. the unambiguous signature of the occurrence of solar neutrino detection.

In the next paragraphs the major results of Borexino are succinctly recapitulated.

3. Solar results and physics implications

3.1. The ^{7}Be flux

The latest ^{7}Be result has been published in [3]. Taking into accounts the systematic errors, stemming essentially from the uncertainty in the energy scale and in the fiducial volume selection, the ^{7}Be evaluation is 46 ± 1.5stat(+1.5-1.6)sys counts/day/100 tons: hence, summing quadratically the two errors, a remarkable 5% global precision has been achieved in this critical measurement.

By assuming the MSW-LMA solar neutrino oscillations, the Borexino result can be used to infer the ^{7}Be solar neutrino flux. Using the oscillation parameters from [4], the detected ^{7}Be count rate corresponds to a total flux of $(4.84 \pm 0.24) \times 10^{9}\ cm^{-2}s^{-1}$, very well in agreement with the prediction of the Standard Solar Model [1]. For comparison, the measured count rate in case of absence of oscillations would have been 74 ± 5.2 counts/day/100 tons.

Finally, the resulting electrons survival probability at the ^{7}Be energy is Pee=0.51± 0.07.

3.2. ^{8}B

The distinctive feature of the ^{8}B neutrino flux measurement performed by Borexino [5] is the very low 3 MeV threshold attained, decisively lower than the previous measurements from the Cerenkov experiments.

The measurement is very difficult, since the total background, both of radioactive and cosmogenic origin, in the raw data is overwhelming if compared to the expected signal. The specific background suppression strategy adopted in this case is based on two ingredients: on one hand a careful MC evaluation of the main radioactive contaminants of relevance for this measure, i.e. ^{214}Bi from Radon and the external ^{208}Tl from the nylon wall of the Inner Vessel, and on the other the “in-situ” identification and suppression of the muon and associated cosmogenic signals.

The observed ^{8}B rate in the detector is 0.217±0.038(stat)±0.008(sys) cpd/100ton, corresponding to an equivalent flux Φ(8B)=(2.4±0.4±0.1)×10^{6} $cm^{-2}s^{-1}$; if as for the ^{7}Be case, we take into account the oscillation probability, then the ratio with the flux foreseen by the SSM is 0.88±0.19.

3.3. Pep and CNO

By far the most important background in studying pep and CNO solar neutrino fluxes is the 11C decay, a radionuclide continuously produced in the scintillator by the cosmic muons surviving through the rock overburden and interacting in the liquid scintillator. The beta plus decay of 11C originates a continuous spectrum which sits exactly in the middle of the energy region between 1 and 2 MeV, which is just the window for the pep and CNO investigation.

Actually, to a less extent also the external background induced by the gammas from the photomultipliers is an obstacle, especially above 1.7 MeV.

In [6] a threefold coincidence strategy encompassing the parent muon, the neutron(s) emitted in the spallation of the muon on a ^{12}C nucleus, and the final ^{11}C signal has been devised and described in detail. Such a strategy applied to the Borexino data led to a pep rate of 3.13±0.23 (stat.)±0.23 (syst.) counts per day/100ton [7], from which the corresponding flux can be calculated, assuming the current MSW-LMA parameters, as Φ(pep)= (1.6±0.3) 10^{8} cm^{-2} s^{-1}, in agreement with the SSM: indeed the ratio of this result to the SSM predicted value is f_{pep}=1.1±0.2. The resulting electrons survival probability at the pep energy is Pee=0.51± 0.07; finally, it should be underlined that the significance of the pep detection is at the 97% C.L..

The same analysis, keeping the pep flux fixed at the SSM value, originates a tight upper limit on the CNO flux, i.e. Φ(CNO) $\leq$ 7.4 10^{8} cm^{-2} s^{-1} , corresponding to a ratio with the SSM prediction less than 1.4.

3.4. Pp neutrinos

The measure of the pp flux has been released by the Collaboration in [8], a fundamental additional milestone toward the full solar neutrino spectroscopy by a single solar neutrino experiment. The data used for this measurement were accumulated after an extensive purification campaign that was performed in 2010 and 2011 and reduced, in particular, the content of two isotopes: ^{85}Kr, now close to 0, and ^{210}Bi, which are important backgrounds in the low-energy region.

The pp neutrino rate has been extracted by fitting the measured energy spectrum of the selected events in the 165–590 keV energy window to the expected spectra of the signal and background components. Given the low energy of the pp induced signals in the liquid scintillator, the most important background for this analysis is the ^{14}C content intrinsic to the liquid scintillator itself. Even if the isotopic abundance of ^{14}C in the scintillator is in absolute value very low, at the level of 10^{-18}, this species originates a large count rate of low energy events, which therefore can also produce a pile effect. The associated pile-up spectrum had to be determined with great accuracy, since it extends over the same energy region of the expected pp signals. A data driven methodology was developed to extract with high precision from the data themselves both the ^{14}C and the pile-up spectra.

The main components of the fit used to infer the pp rate are the solar neutrino signal, i.e. the pp component and the low-energy parts of the ^{7}Be, pep and CNO components); the mentioned dominant ^{14}C background and the associated pile-up; and other identified radioactive backgrounds: ^{85}Kr, ^{210}Bi, ^{210}Po and ^{214}Pb. The free fit parameters are the rates of the pp solar neutrinos and of the ^{85}Kr, ^{210}Bi and ^{210}Po backgrounds. The ^{7}Be neutrino rate is constrained at the measured value, within the error, and pep and CNO neutrino contributions are fixed at the levels of the SSM, taking into account the values of the neutrino oscillation parameters.

The solar pp neutrino interaction rate measured by Borexino is 144±13 (stat.) ±10 (syst.) counts per day/100ton. The stability and robustness of the measured pp neutrino interaction rate was verified by performing fits with a wide range of different initial conditions. Once statistical and systematic errors are added in quadrature and the latest values of the neutrino oscillation parameters are taken into account, the corresponding solar pp neutrino flux is $(6.6\pm0.7)\times10^{10}cm^{-2}\ s^{-1}$, and the associated survival probability is 0.64 ± 0.12.

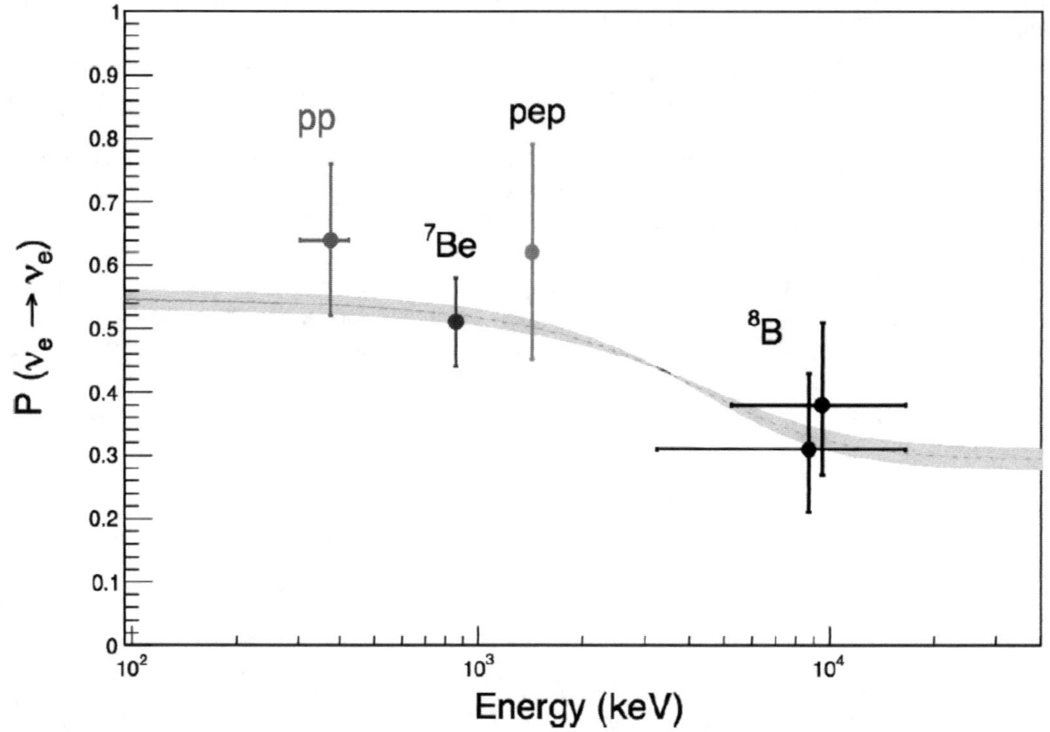

Figure 2. Validation over the full solar neutrino energy range of the MSW-LMA solution provided by Borexino with only its own data.

3.5. MSW-LMA global picture

In figure 2 the MSW predicted Pee (electron neutrino survival probability) is shown (the violet band corresponds to the $\pm1\sigma$ prediction of the MSW-LMA solution), together with several experimental points, all coming from Borexino: i.e. black the ^{8}B from the analyses with two different thresholds, light blue the pep, blue the ^{7}Be, red the very recent pp datum. This figure demonstrates the unique achievement of Borexino, which through an almost complete solar spectroscopy has been in condition, alone, to confirms spectacularly the MSW-LMA solar neutrino oscillation scenario, while providing the first direct measurements of the survival probability in the low energy sub-MeV Vacuum MSW regime.

This striking confirmation is also complemented by the measure of the day-night asymmetry of the ^{7}Be flux [9], which is found equal to Adn = 0.001 ± 0.012 (stat) ± 0.007 (syst), fully consistent with zero and hence with the model prediction. It is worth to mention that, by including this measure in the global fit of all solar neutrino experiments, the otherwise surviving LOW region is completely wiped out, even without including the KamLAND data.

4. Terrestrial neutrinos

As an ultra-pure liquid scintillator, Borexino is a perfect tool to detect anti-neutrinos (geoneutrinos) coming from Earth. The updated geoneutrino signal from Borexino has been reported in [10], for a data taking period of 1353 days. With a fiducial exposure of $(3.69 \pm 0.16)\times10^{31}$ proton × year, after all selection cuts and background subtraction, (14.3 ± 4.4) geo-neutrino events were detected, assuming a fixed chondritic Th/U ratio of 3.9. This result corresponds to a geoneutrino signal of (38.8 ± 12.0)

TNU (Terrestrial Neutrino Unit: = 1 event / year/ 10^{32} protons), disfavouring also the no-geoneutrino hypothesis with a p-value of 6×10^{-6}.

The only sizable background for this measure is represented by the anti-neutrinos from reactors, which we evaluated equal to 31.2(+7-6.1), well in agreement with the expectation (33.3±2.4), thus further confirming the validity of our entire measurement procedure.

5. Conclusion

The most recent solar and terrestrial neutrino results stemmed from Borexino have further reinforced the ultra-low background achievements of this experiment, an exceptional breakthrough in the field of techniques for rare processes search. The ^{7}Be, ^{8}B, pep and the very recently measured pp components have all been detected (together with a tight upper limit on CNO), leading to the validation of the MSW-LMA oscillation paradigm in the entire energy regime of the solar neutrinos, strengthened also by the determination of the absence of day-night asymmetry in the ^{7}Be flux. Very remarkably, Borexino performed this validation with its own data, without the need to resort to the results of other solar neutrino experiments.

Finally, the highly significant measurement of the terrestrial neutrinos not only complements the physics potentiality of the detector, but points towards a future new direction of research in the studies of the interior of the Earth.

Acknowledgments

The author wishes to thank the organizers for the invitation to such an interesting and stimulating conference. Russian colleagues from MEPhI acknowledge partial support from MEPhI Academic Excellence Project (contract No. 02.a03.21.0005, 27.08.2013).

References

[1] Serenelli A M, Haxton W C and Peña-Garay C 2011 *Ap. J.* **743** 20
[2] Alimonti G *et al.* (Borexino Collaboration) 2009 *Nucl. Instrum. Meth.* A **600** 568
[3] Bellini G *et al.* (Borexino Collaboration) 2011 *Phys. Rev. Lett.* **107** 141302
[4] Nakamura K *et al.* (Particle Data Group) 2010 Review of Particle Physics *J. Phys.* G **37** 075021
[5] Bellini G *et al.* (Borexino Collaboration) 2010 *Phys. Rev.* D **82** 033006
[6] Back H *et al.* (Borexino Collaboration) 2006 *Phys. Rev.* C **74** 045805
[7] Bellini G *et al.* (Borexino Collaboration) 2012 *Phys. Rev. Lett.* **108** 051302
[8] Bellini G *et al.* (Borexino Collaboration) 2014 *Nature* **512** 383-6
[9] Bellini G *et al.* (Borexino Collaboration) 2011 *Phys. Lett.* B **707** 22-6
[10] Bellini G *et al.* (Borexino Collaboration) 2013 *Phys. Lett.* B **722** 295

Journal of Physics: Conference Series **675** (2016) 012037 doi:10.1088/1742-6596/675/1/012037

Data acquisition system based on fast waveform digitizers for large neutrino detectors

G Lukyanchenko[1,*] and E Litvinovich[1,2]

[1] National Research Center Kurchatov Institute, Akademika Kurchatova pl. 1, Moscow, 123182, Russia

[2] National Research Nuclear University MEPhI (Moscow Engineering Physics Institute), Kashirskoe highway 31, Moscow, 115409, Russia

E-mail: *egorxe@yandex.ru

Abstract. For large volume neutrino and antineutrino detectors it is crucial to have an efficient data acquisition system capable of digitizing data from thousands of detection channels. Here we present a flexible DAQ system architecture consisting of a large number of fast waveform digitizers and configurable FPGA-based trigger logic. The current implementation of the system is functioning in the Borexino neutrino detector providing zero dead time spectroscopy data in the energy range from 1 up to 100 MeV. Acquisition complex in combination with our custom analysis software is successfully being used for registration of geoneutrinos, as well as search for neutrino signal from GRBs, solar netrino spectroscopy and other applications.

1. Introduction

In this work we present a DAQ complex based on flash ADCs (also called fast waveform digitizers, FWFDs) used to digitize signal waveforms. The complex also includes programmable discriminator and programmable logic unit for trigger processing. Such system could be used for data acquisition as a part of different particle physics detectors. System architecture allows to simultaneously process data from more than 100 individual channels and enables a possibility to use a complex data acquisition control logic. Currently the system described is implemented in Borexino detector and provides neutrino and antineutrino spectroscopy in energy range from 1 up to 100 MeV.

Borexino is an ultra low background neutrino scintillation detector located in underground Laboratori Nazionali del Gran Sasso (Italy) on the depth of 3800 mwe. Detector target is a nylon sphere 4.25 m in radius containing 278 tons of pseudocumene-based liquid organic scintillator. Around the target lies concentric spherical 6.25 m buffer volume filled with dimethylphthalate doped pseudocumene buffer liquid. Scintillation light in the target and the buffer is registered by 2212 low background PMTs. Spheres are contained inside 2100 m^3 water tank (called outer detector, OD) used to shield the target from external radiation and provide a muon veto by means of additional 208 PMTs [1].

The primary Borexino task is a low energy neutrino spectroscopy [2] in sub-MeV range, which led to main data acquisition system based on spectroscopic ADCs been optimized for low energy range and unable to measure events energy above 15 MeV with high accuracy. But to allow study of rare events above 15 MeV of astrophysical neutrino origin, such as supernova neutrinos, DSNB neutrinos, GRB-correlated neutrino flux and others, an additional DAQ system to enable spectroscopy in the higer energy range was required. To address this requirement a FWFD-based DAQ system described in this work was introduced into Borexino alongside the main one. Here we will base our description mainly on Borexino implementation. Nevertheless, it should be noted that DAQ system architecture was developed to be flexible enough not to be limited to Borexino FWFD subsystem, but could be used for different detectors with large number of acquisition channels. Currently the next implementation of this trigger and DAQ architecture based on newer hardware is being adopted in iDream detector for remote neutrino reactor monitoring[3].

2. Borexino FWFD-based DAQ system structure

The main requirements for Borexino high energy DAQ were to provide zero dead time spectroscopy of events with energies up to 50-100 MeV with maximum possible independence of main DAQ. Borexino FWFD-based DAQ system is installed into 4 VME crates with unified address space and consists of (figure 1):

- Analog circuits for PMT signal summing
- 34 VME ADCs with three 400 MHz 8-bit channels each (CAEN V896)
- FPGA-based trigger module with VME-interface
- Clock and trigger distributors
- 16 channel constant fraction VME programmable discriminator (CAEN V812)
- VME single-board control computer

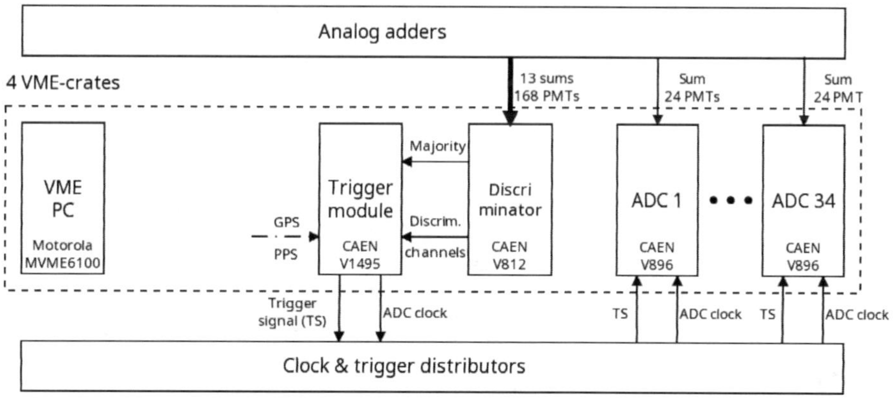

Figure 1. General Borexino FWFD-based DAQ system scheme.

The signal from PMTs comes to the FWFD system via Borexino front end electronic boards common to all Borexino DAQ systems. Front end boards (FEB) provide PMT signal amplification and first stage signal sum of up to 12 PMTs per board. FWFD-system second stage analog summing boards provide separate sums of 2 FEBs, of 14 FEBs and of all detector PMTs. Twofold sums and full detector sum are digitized by individual ADC channels and sums of 14 FEBs are used as discriminator inputs [1].

Usage of fast ADCs to digitize signals from scintillation neutrino detector provide several benefits in comparison to classical spectroscopy ADCs. It allows to eliminate dead time and analyze waveform shape and pulse edges. Flash ADCs used in the FWFD system continuously digitize the input signal to an internal memory buffer and immediately switch to next buffer when the trigger signal arrives. This allows zero-dead time PMT signal acquisition crucial for rare event studies. All 34 VME ADC modules are synchronized by the clock and trigger signals distribution modules which are used to ensure digitization clock phase alignment and simultaneous trigger signal edge. Each ADC module consists of 3 independent 400 MHz 8-bit ADC channels, giving a total of 102 digitization channels. 91 partial detector sum, 2 full detector sums with different amplification and 1 outer detector sum is being digitized, several remaining channels are a hot spare.

As a main trigger source 13 sums of 14 FEBs are connected to inputs of 16 channel constant fraction discriminator with VME interface. Discriminator uses a majority logics, it provides a majority output if several input channels are simultaneously above threshold. Per channel thresholds and majority level are set from DAQ control software running on the control computer with remote access [4].

3. Versatile trigger logic

Central part of the described DAQ system is a versatile trigger module. Trigger module is used to produce a trigger signal on useful detector events and suppress trigger when digitization is not desirable for some reason. Also trigger module allows to store an information about produced trigger and provides an additional functionality such as different event rates counting and estimation of dead time due to external inhibits. To allow maximum performance and flexibility the trigger module is based on FPGA (field programmable gate array) reconfigurable logic chip. Trigger functionality is provided by programming this FPGA chip. FPGA and trigger configuration should be accessible from the control computer, for Borexino implementation the VME-bus was used for these means. Internal trigger layout is present in the figure 2. Trigger module consists of following main blocks all realized inside the FPGA firmware:

- Trigger production unit.
- Trigger info storage unit (FIFO).
- Rate counters and dead time timers.
- GPS time unit.

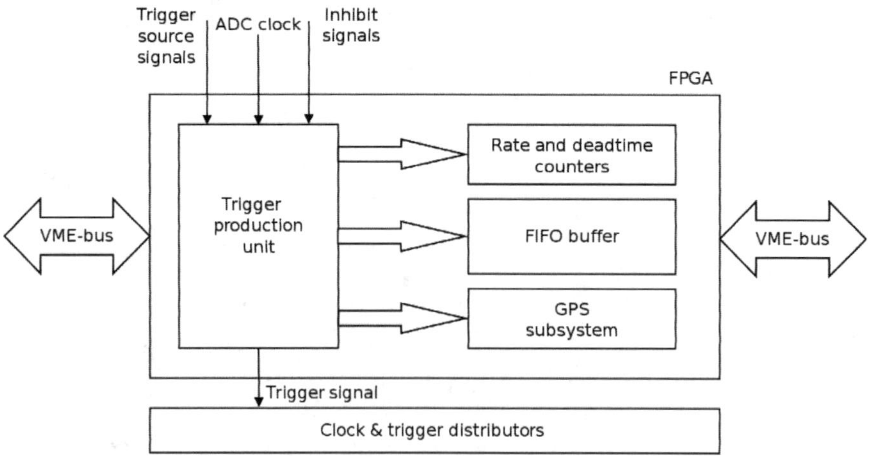

Figure 2. Trigger unit block diagram.

To achieve maximum versatility, the FPGA trigger firmware was programmed in the way to allow adjustment of all submodules parameters via VME-bus registers without firmware reload. Also FPGA firmware itself could be updated from control computer via VME.

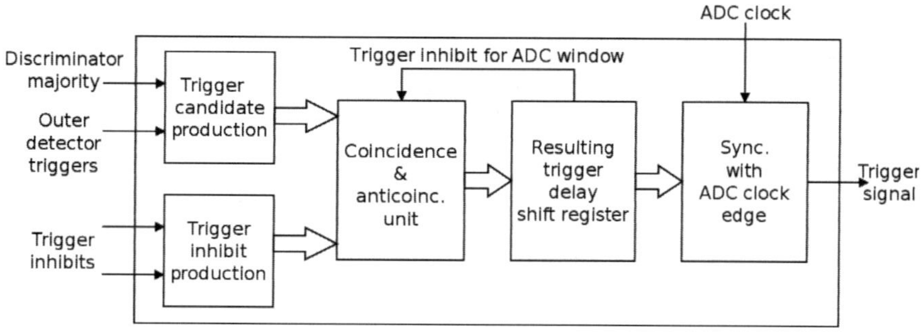

Figure 3. Trigger production logic.

The main integral part of the trigger system is the trigger production unit (figure 3). It produces triggers based on up to 8 trigger inputs in anticoincidence with up to 8 inhibits. Coincidence and anticoincidence time windows and delays for each input along with the additional trigger candidate and inhibit parameters are adjustable on run time. By changing coincidence parameters and signal delays to match processes inside detector it is possible to tune the system for a wide variety of physical tasks. In Borexino trigger sources are a signal from majority discriminator and two different OD triggers. The inhibits are several calibration triggers and noise suppression system output, which are used to reduce undesirable event rate. Also trigger system denies triggers for a configurable time, which is set to be a bit shorter than ADC time window cause it's possible to have several physical events in one ADC window or in nearby windows (in last case windows are "glued" together by analysis software). After trigger passes through coincidence logic, it is delayed on shift register to allow ADCs to fill the time window after the trigger. Several triggers could be processed simultaneously. Information about each produced trigger such as trigger source, GPS time and other flags is stored in buffer (FIFO) size of which is set to match the size of ADC event buffer. FIFO buffer is meant to be read by DAQ software on the control computer to get info about occurred events.

Ease of FPGA reconfiguration also allows to realise additional functionality required for particular setup. In Borexino FWFD system trigger unit is used to produce GPS time tags from precise PPS (pulse per second) signal and to save data from main DAQ trigger system to perform event correlation between two systems later in software. Also rate of all produced trigger types, trigger sources and trigger inhibits is counted and is made accessible via VME registers to allow online monitoring at DAQ time.

4. Conclusion

We've presented a DAQ system architecture built on fast waveform digitizers and FPGA-based trigger. This system is capable of processing data from particle physics detectors with large number of channels and currently provides data for rare neutrino event analysis in Borexino detector and is being integrated into new iDream detector.

Acknowledgments

This work was supported by RFBR grants 14-22-03013, 14-22-03031 and 14-02-31381 and partially supported by MEPhI Academic Excellence Project (contract No. 02.a03.21.0005, 27.08.2013).

References

[1] Alimonti G *et al.* 2009 *Nucl. Instr. and Meth.* A **600** 568-93

[2] Arpesella C *et al.* 2008 *Phys. Lett.* B **658** 101-8
[3] Gromov M *et al.* 2015 *Moscow University Physics Bulletin* **70** 190-4
[4] Litvinovich E *et al.* 2013 (in Russian) *Preprint* IAE-6756/2 (original Russian title: Sistema sbora dannih na osnove bistrih ocifrovshikov formi impulsa)

Upgrade of the ALICE Inner Tracking System

Svetlana Kushpil[1,2] for the ALICE Collaboration
[1] Nuclear Physics Institute of the CAS, Řež, Czech Republic
E-mail: skushpil@ujf.cas.cz

Abstract. ALICE detector was constructed to study the properties of hot and dense hadronic matter formed in relativistic nuclear collisions. During the second long LHC shutdown in 2019-2020, the collaboration plans to upgrade the current vertex detector, the Inner Tracking System (ITS), in order to increase the reconstruction accuracy of secondary vertices and to lower the threshold of particle transverse momentum measurement. The upgrade strategy of ITS is based on the application of new Monolithic Active Pixel Sensors (MAPS) designed in 0.18 μm CMOS technology. The 50 μm thick chip consists of a single silicon die incorporating a 0.18 μm high-resistivity silicon epitaxial layer (sensor active volume) and matrix of charge collection diodes (pixels) with readout electronics. Radiation hardness of the upgraded ITS is one of the crucial moments in the overall performance of the system. A wide set of MAPS structures with different read-out circuits was produced and is being studied by the ALICE collaboration to optimize the pixel sensor functionality. An overview of the ALICE ITS upgrade and the expected performance improvement will be presented together with selected results from a campaign that includes several irradiation and beam tests.

1. General ALICE upgrade

ALICE [1] is an experiment designed to study properties of the Quark-Gluon Plasma using nucleus-nucleus, proton-nucleus and proton-proton collisions at the CERN Large Hadron Collider. ALICE has confirmed the existence of the quark-gluon plasma and extended the precision and kinematic reach of many significant probes measured previously at RHIC and at the SPS. After the second long shutdown of the LHC in 2019–2020, ALICE will focus on the measurement of heavy-flavor hadrons, quarkonia and low-mass dileptons at low transverse momenta [2]. This requires high-precision measurements at low transverse momentum p_T resulting in the need of recording large samples of minimum-bias events. In order to achieve this, ALICE plans to read out all Pb–Pb collisions delivered by the LHC, which corresponds to a maximum collision rate of 50 kHz. The upgraded detector will record an integrated luminosity of 10 nb^{-1} in Pb–Pb collisions which means a factor 100 more compared to the data that will be recorded up to 2018. The ALICE upgrade strategy includes several projects [2]. Two new detectors will be installed: a high resolution, low material budget Inner Tracking System and a Muon Forward Telescope. They will be surrounding a new beampipe with a smaller radius, which will allow the first layer of the ITS to be moved closer to the interaction point resulting in better track impact parameter determination. The readout chambers of the Time Projection Chamber will be replaced by Gas Electron Multiplier detectors and its readout electronics will be upgraded together with the readout electronics of the Transition Radiation Detector, the

[2] Present address: Nuclear Physics Institute of the CAS, Řež, 25068 Czech Republic

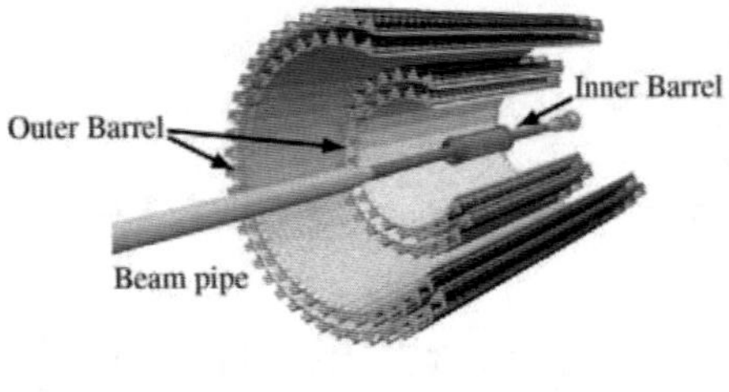

Figure 1. The upgraded ITS.

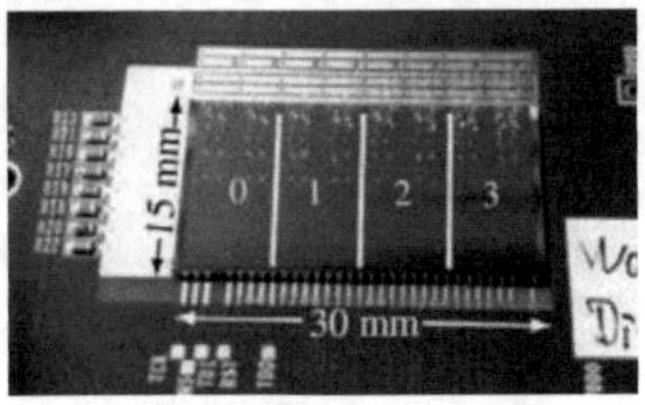

Figure 2. The pALPIDE version 1 chip with the four sub-matrices indicated.

Time Of Flight detector and the Muon Spectrometer. The forward trigger detectors, the online system and the offline reconstruction and analysis framework will also be upgraded.

2. The new Inner Tracking System

The goals of the ITS upgrade [3] are to improve the track impact parameter resolution (by at least a factor of 5 in z direction and by a factor of 3 in $r\phi$ direction), the tracking efficiency and the transverse momentum resolution at very low p_T. In order to achieve these improvements the first layer of silicon detectors will be placed closer to the interaction point (figure 1). The reduction of the beam pipe diameter in the center of ALICE will allow the first ITS layer to moved as close 22 mm to the interaction poitn instead of the present 39 mm. The spatial resolution and the tracking performances will gain from a reduction of the material budget, in particular in the inner layers. The use of MAPS, the optimization of the front-end electronics, and the consequent reduction of the power and signal cables will reduce the total material budget X/X_0 per layer from the present 1.14% to as low as 0.3% for the three innermost layers and to $\sim$ 0.8% for the outer layers. Both track impact parameter and momentum resolutions will improved by increasing the detector granularity. The current design assumes that all layers are segmented in pixels with dimensions of 28 μm × 28 μm. Finally, the tracking efficiency and p_T resolution will also be improved by increasing the number of layers (from present 6 to 7).The seven layers of the new ITS will cover an area of around 10 m^2 and will consist of around 1.25 $\times 10^{10}$ pixels. The layers will be grouped according to the following: the three innermost layers make up the Inner Barrel and the four outermost layers the Outer Barrel (figure 1). All the layers will be segmented into elements called staves which extend the full length of the detector in the direction of the beam. A stave is the smallest operable unit of the detector and has a slightly different design in the case of the Inner and Outer Barrel. In addition it will be possible to remove the detector for maintenance and replace faulty modules during yearly shutdowns. The new ITS will not measure the ionization in the silicon layers. For what concerns the environment in which the new ITS will operate, recent simulations have shown that the overall dose expected for the innermost layer can reach up to 2.7 Mrad (TID) and 1.7 $\times 10^{13}$ 1 MeV n_{eq}/cm^2 (NIEL), including a safety factor of 10, for the full integrated luminosity. The present ITS runs up to a maximum of 1 kHz. The upgraded ITS aims to read out data up to a rate of 50 kHz in Pb–Pb collisions and 400 kHz in pp collisions, with a dead time of about 10%.

3. Pixel chips

The active sensors and the readout electronics have very demanding requirements in terms of granularity, material thickness, readout speed, power consumption and radiation hardness. An intrinsic spatial resolution of 5 μm for the innermost layer requires a pixel pitch of $\sim$ 30 μm. A significant reduction of the material budget, in particular for the innermost layers, to reduce

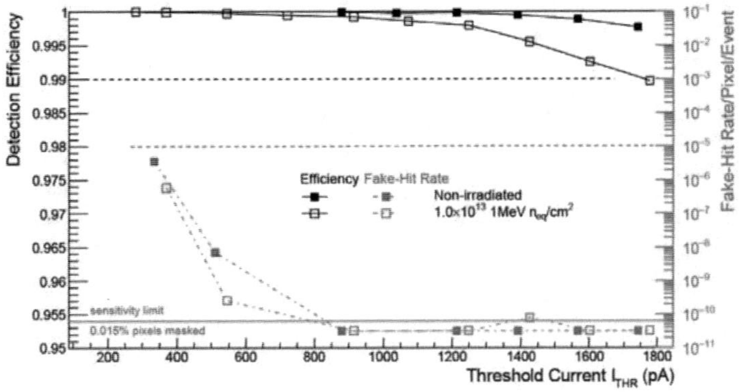

Figure 3. Detection efficiency and fake hit rate before and after irradiation level of 1.0 $\times 10^{13}$ 1 MeV n_{eq}/cm^2. Horizontal lines (0.99 for detection efficiency and 10^{-5} hits/pixel/event for fake hit rate) represent the project goals. Lines, connecting points, are a guide to the eye.

the multiple Coulomb scattering, leads to a silicon thickness of 50 μm at maximum. According to the thermal studies that were carried out on prototype mechanics the power density on the sensor should not exceed 300 mW/cm^2. In order to cope with the assumed interaction rates and dead time in Pb–Pb and pp collisions, the maximum acceptable sensor integration time is $\sim$ 30 μs to limit the event pile-up. Finally, the pixel detectors have to be tolerant against the radiation levels expected in the ITS environment. Based also on the STAR Pixel experience [4], the monolithic CMOS MAPS technology has been chosen for the pixel detectors. The 0.18 μm CMOS technology by TowerJazz has been selected for the implementation of pixel chip for all layers of the new ITS. Due to the small feature size, it is expected to be more robust to the total ionizing dose than other technologies employed so far. The same feature size and the number of metals available are adequate to implement high density, low power digital circuits, minimizing thus the dead area and making the power extraction more efficient. This technology allows the production of chips on wafers with a resistivity of $1 \div 5$ kΩ · cm. In this way a substantial part of the epitaxial layer is already depleted at the reverse bias voltage applied to the collection diodes in the CMOS sensors. This will increase the signal-to-noise ratio and improve the resistance to non-ionizing radiation.

4. Performance of the pALPIDEfs

There are two parallel chip developments ongoing for the ITS upgrade, the ALPIDE and the MISTRAL-O. Here we focus on the ALPIDE architecture. It has 512 × 1024 digital pixels with the size of 28 μm × 28 μm. It is read out using an asynchronous priority encoder network, thanks to which only hit pixels are read out resulting in very fast acquisition of the data from the full matrix. The chip is organized in four sub-matrices consisting of different types of pixels (figure 2). Several prototypes of the ALPIDE family have been designed by the ITS group. These prototypes have been characterized in several test beam campaigns at different test beam facilities (PS, SPS, DESY, BTF Frascati and PAL Pohang). The beam energy at these facilities range is from 60 MeV to 120 GeV and the particles used are electrons, positrons and pions. The following results were obtained at the PS using a 6 GeV negative pion beam. A telescope consisting of seven pALPIDE chips where the outer six chips are the tracking planes and the middle chip is being tested. Here we present the selected results of pALPIDE-2 with 25 μm

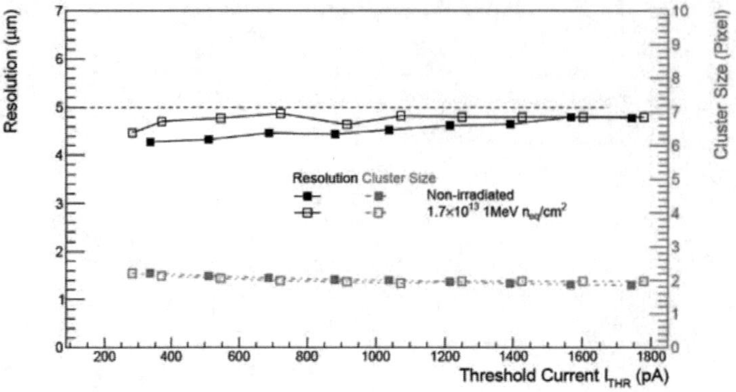

Figure 4. Position resolution and cluster size before and after irradiation level of 1.7 $\times 10^{13}$ 1 MeV n_{eq}/cm^2. Lines, connecting points, are a guide to the eye.

epitaxial layer and -6 V back bias voltage, before and after irradiation. For data analysis EUTelescope software package is used [5]. Figure 3 represents the particle detection efficiency, that is better than 99% for different threshold currents. At the same time fake hit rate is less than 10^{-5} hits/event/pixel. ITS project goals are thus fulfilled at a wide operating range of the threshold current. The good performance is also observed for chips that were irradiated with 10^{13} 1 MeV n_{eq}/cm^2. In figure 4 the resolution stays below 5 µm, which fulfills the project goal, and cluster size is only slightly affected by irradiation to expected NIEL of 1.7 $\times 10^{13}$ 1 MeV n_{eq}/cm^2.

5. Conclusion

The Inner Tracking System of ALICE will be upgraded during the second long LHC shutdown to fulfill the ALICE physics requirements. Its key features are low material budget, an increased number of layers, a greater radial extension with a first layer closer to the interaction point, and a smaller pixel size. The pixel sensors will employ the CMOS MAPS produced using the 0.18 µm TowerJazz technology. First full scale prototypes show a good performance and large operational margin.

Acknowledgments

The work has been supported by the grant LG 13031 of the Ministry of Education of the Czech Republic.

References

[1] Aamodt K *et al.* (ALICE Collaboration) 2008 *J. Inst.* **3** S08002
[2] ALICE Collaboration 2012 *CERN-LHCC-2012-012* LHCC-I-022
[3] ALICE Collaboration 2014 *J. Phys.* G**41** 087002 *CERN-LHCC-2013-024 ALICE-TDR-017*
[4] Schambach J *et al.* 2015 *Phys. Procedia* **66** 514-519
[5] EUTELESCOPE (*https://eutelescope.web.cern.ch*)

Journal of Physics: Conference Series **675** (2016) 012039 doi:10.1088/1742-6596/675/1/012039

Radiation hardness of semiconductor avalanche detectors for calorimeters in future HEP experiments

V Kushpil[1,4,*], V Mikhaylov[1,2], A Kugler[1], S Kushpil[1,4], V P Ladygin[3], O Svoboda[1] and P Tlustý[1]

[1] Nuclear Physics Institute, Academy of Sciences of the Czech Republic, 25068, Řež, Czech Republic

[2] Czech Technical University in Prague, Faculty of Nuclear Sciences and Physical Engineering, 16636 Prague, Czech Republic

[3] Joint Institute for Nuclear Research, 141980 Dubna, Russian Federation

[4] National Research Nuclear University MEPhI (Moscow Engineering Physics Institute), Kashirskoe highway 31, Moscow, 115409, Russia

E-mail: [*]kushpil@ujf.cas.cz

Abstract. During the last years, semiconductor avalanche detectors are being widely used as the replacement of classical PMTs in calorimeters for many HEP experiments. In this report, basic selection criteria for replacement of PMTs by solid state devices and specific problems in the investigation of detectors radiation hardness are discussed. The design and performance of the hadron calorimeters developed for the future high energy nuclear physics experiments at FAIR, NICA, and CERN are discussed. The Projectile Spectator Detector (PSD) for the CBM experiment at the future FAIR facility, the Forward Calorimeter for the NA61 experiment at CERN and the Multi Purpose Detector at the future NICA facility are reviewed. Moreover, new methods of data analysis and results interpretation for radiation experiments are described. Specific problems of development of detectors control systems and possibilities of reliability improvement of multi-channel detectors systems are shortly overviewed. All experimental material is based on the investigation of SiPM and MPPC at the neutron source in NPI Rez.

1. Introduction

The use of modern semiconductor devices as the replacement of traditionally used photomultipliers has three main reasons – cost, insensitivity to magnetic fields and small size. Also it is important to note that the power supply voltage for semiconductor analogues of photomultipliers is 30-100 V in contrary to voltage level of kilovolts required for photomultipliers. Moreover, one can simply place a pre-amplifier near to the photodetector, which together with its small size allows the construction of calorimeters with a high degree of granularity. It gives an opportunity to extend the tracking system up to the sophisticated detector module itself and improve tracking and energy resolution.

On the other hand, one cannot forget about the shortcomings inherent in modern solid-state semiconductor photodetectors with internal gain. Firstly, this kind of apparatus has sensitivity to the detector's external temperature changes. It leads to the need of the additional compensation of detector's operating point voltage to stabilize the detector gain and to minimize its own noise caused by the dark current.

The second drawback is connected to the operating principle of all modern Solid-State Photomultipliers (SSPM). Despite the fact that the market is saturated with a variety of devices, all of them are built on a similar principle. The phenomenon of microplasma breakdown in the heterogeneity known since the 60 s of the last century. The basis of any SSPM is a microscopic cell with mesh size of 3-50 μm representing the PN junction with artificially created inhomogeneity in the centre of the junction, which is the hub of the high electric field in the avalanche region.

Parameters of the material around the heterogeneity are chosen so, that at the operating voltage the probability of the beginning of microplasma breakdown caused by carriers thermogenerated in region of high-field is very small. In other words, there is a certain threshold number for nonequilibrium charge carriers in the depletion region, starting from which the microplasma breakdown occurs with a probability close to 1. The resulting microplasma must have the property of self-extinguishing, i. e. once created it changes the shape and magnitude of the electrostatic field so that avalanche multiplication stops. Various methods can be applied for this purpose, such as connection of a high resistance sequentially to the PN junction, which acts as a voltage divider decreasing the junction voltage with the increase of its current. Method that is more complex is to select the material properties of the cell so, that an avalanche does not develop any further due to the depletion of ionisable impurities in the high-field region.

Based on the description of physical processes, one can see that all SSPM have essentially the same principle of operation – they cannot amplify photons falling on the photodetector, but can amplify photo generated carriers inside silicon. Moreover, the range of SSPM linearity is sufficiently narrow. On the other hand, it is known that the microplasma breakdown areas are the source of wide range spectrum of photons [13], which in turn triggers the neighbouring cells called the phenomenon of after-response or crosstalk.

Still the most significant drawback of SSPM is a quite low radiation hardness compared with photomultipliers, which leads to the need for further study of these devices prior to their use in experiments. However, in cases where the detector loads are low and there are no strict requirements for the material budget the use of these detectors is justified. Commonly, detectors of this kind are zero-angle calorimeter and similar, i.e. detectors placed outside the main detector system and applied for the measurement of the energy distribution of the very forward going projectile nucleons and nuclei fragments. Several examples of such detectors already existing and being constructed for future experiments are further overviewed.

2. Calorimeters NA49, NA61/CERN, FAIR and NIKA

The NA49 is one of the first experiments, where a calorimeter designed with SSPM was applied [1]. Many ideas of construction such a detectors were piloted there further widely used in the construction of new ones. Originally, the NA49 had a Ring Calorimeter and a Veto Calorimeter, designed to measure the energy flow due to the particles produced in nucleus-nucleus collisions. Both calorimeters were of the sampling type composed of sandwiched layers of lead or iron absorber and scintillator plates. The summed light signal from the scintillator plates is proportional to the energy of the particles absorbed in the calorimeter. The device had a cylindrical structure, coaxial with the incident beam, and its sensitive area is circular with an outer diameter of 3.0 m and a 56 cm central hole. It is segmented into 240 cells, 24 sectors in azimuth and 10 rings radially.

During the further upgrades leading to the new being operated nowadays NA61 experiment, the two original forward calorimeters were replaced by the single Projectile Spectator Detector [2]. The PSD calorimeter consists of 44 modules which cover a transverse area of 120×120 cm^2. The central part of the PSD consists of 16 small modules with transverse dimension of 10×10 cm^2, the outer part consists of 28 large 20×20 cm^2 modules. Each module consists of 60 pairs of alternating lead plates and scintillator tiles with 16mm and 4mm thickness, respectively. The stack of plates is tied together with 0.5 mm thick steel tape and placed in a box made of 0.5 mm thick steel. Steel tape and box are spot-welded together providing appropriate mechanical rigidity. The full length of the modules corresponds to 5.7 nuclear interaction lengths. The scintillation light is read out via wavelength shifting fibers by SSPM Avalanche Photo-Diodes (APD). This layout is enough to provide fine granularity corresponding to the beam intensity up to 2×10^5 ions/sec and beam energy up to 150 AGeV.

The Projectile Spectator Detector (PSD) at the Compressed Baryonic Matter experiment of FAIR being built nowadays is a very similar machine with a few minor changes [3]. The calorimeter comprises 44 individual modules, each consisting of 60 lead/scintillator layers with a surface of 20×20 cm^2. Combined with the small beam hole of 6cm diameter it compensates for the very high beam intensity up to 10^9 ions/sec and energy up to 35 AGeV. The Zero Degree Calorimeter at MPD detector within currently being developed NICA collider is quite similar as well [4]. The prior difference is the number of the modules – 84 pieces and its sizes of $5 \times 5 \times 120$ cm^3, as well as the increased beam hole in center of 10×10 cm^2 size. This change provides suitable occupancy of the detected particles and therefore transverse segmentation to fit the experimental conditions of beam intensity up to 1×10^9 ions/sec and energy up to 63 AGeV. The individual geometry of the modules except its transverse sizes for both described detectors is the same as for the NA61 PSD detector.

All afore described detectors are intended to measure the non-interacting nucleons and fragments emitted at very low polar angles in forward direction in nucleus-nucleus collisions. It is used to determine the collision centrality and the orientation of an event plane. Moreover, the precise event-by-event measurement of the energy carried by projectile spectators enables the extraction of the number of interacting nucleons from the projectile with the precision of one nucleon. The high energy resolution of the PSD is important for the study of fluctuations in nucleus-nucleus collisions which are expected to be sensitive to properties of the phase transition between the quark-gluon plasma and hadron-resonance matter. Namely, the PSD provides the precise control over fluctuations caused by the variation of the number of interacting nucleons and thus excludes the "trivial" fluctuations caused by variation of the collision geometry.

For selection SSPM we can use next criteria. Area S1 of SSPM must be selected for maximal load. It is possible in experiment and for linearity of response what we need. For pixel area A and fluence of photons in the bunch Nph from scintillator we must be select S1 from relation $S1/A > K{\cdot}Nph$ for number of events K. The sensitivity of SSPM must be higher compared with scintillator output. For prototype of CBM module 1MIP produces about 20 photons. From this point of view we can estimate what maximal number of event we can register for SSPM with active area 3 mm^2 for size of the cell 10 μm we can obtain $1.5{\cdot}10^4$ and for other size of 50 μm – 600 events. Next important parameter is dead-time Td. The dead-time defines the maximal frequencies of events $Fmax<1/Td$. We can not increase area of SSPM in arbitrary way because of noise and increased after-pulsing. All criteria must be applied to select SSPM with respect to irradiation environment.

3. Investigation of radiation hardness of SiPM and MPPC

The radiation loads for SSPM were quite small in the first experiments. It seems that the passage of charged particles and gamma radiation will not lead to any substantial problems. The simulation by FLUKA for two months of run shown that the neutron absorbed dose reaches the 10^{13} n/cm^2 [5]. The magnitude of absorbed dose is of the order comparable with the concentrations of impurities in the semiconductor detectors. Therefore, one must consider at least an increase of the detector intrinsic noise. For the optimal detector operation, it is important to know how the noise spectrum changes, since then one can theoretically correct the signal bandwidth of the amplifier and cut off the spectrum parts corresponding to noise. Those considerations are leading to the understanding that the method of processing SSPM signal should be different from the conventional processing techniques of other detectors.

Two additional SSPM investigation procedures were developed within our group, namely measurement and analysis of the Capacitance Frequency Characteristics (CF) and study of transient process of switching the PN junction from an open to a closed state in the detector. Results of CF measurement are processed with help of a simplified model, where the ratio of the average lifetime of minority carriers to their average concentration is the main parameter characterizing the change in the junction after irradiation. This parameter provides a qualitative picture of the processes dynamics of the traps generation in silicon. The second technique based on the switching charge measurement gives an information about the lifetime of minority carriers in the base of PN junction.

Investigated APDs were irradiated at the Cyclotron facility of NPI Řež using quasi-monoenergetic 35 MeV secondary neutron beam with doses indicated at corresponding measurement figures [6]. Doses here are measured by the special PIN diode calibrated for a 1 MeV neutrons equivalent dose; the temperature during the irradiation and measurements was 22 ± 0.5 °C [7].

The I-V characteristics of all Ketek PM3375 [8], Zecotek MAPD-3N [9] and Hamamatsu S12572-010P [10] APDs show the increase of the dark current I_{dark} in ~ 10^3 times right after irradiation by dose of 10^{13} n/cm^2 (figure 1) [8]. Due to self-annealing I_{dark} of Zecotek APD increased a bit, but I_{dark} of Ketek APD decreased a bit on the contrary. Hamamatsu APD was initially irradiated with a much smaller dose of 10^{10} n/cm^2 leading to the increase of I_{dark} in ~ 10 times near the operating voltage, which is still a significant change (figure 2).

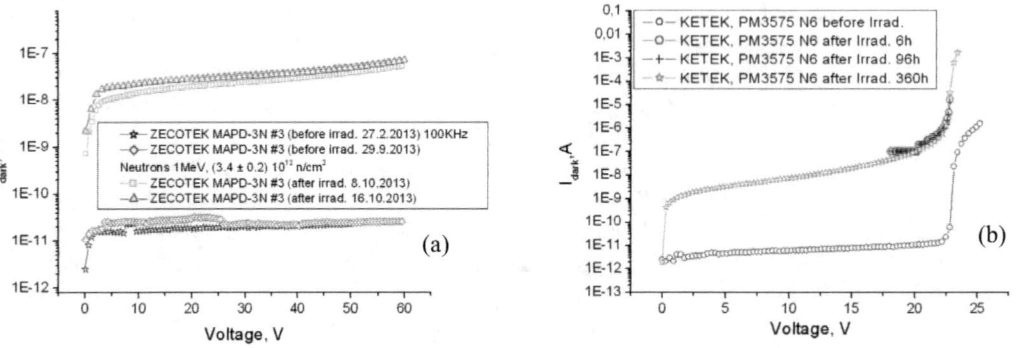

Figure 1. I-V characteristics of APDs before and after irradiation: a) Zecotek; b) Ketek.

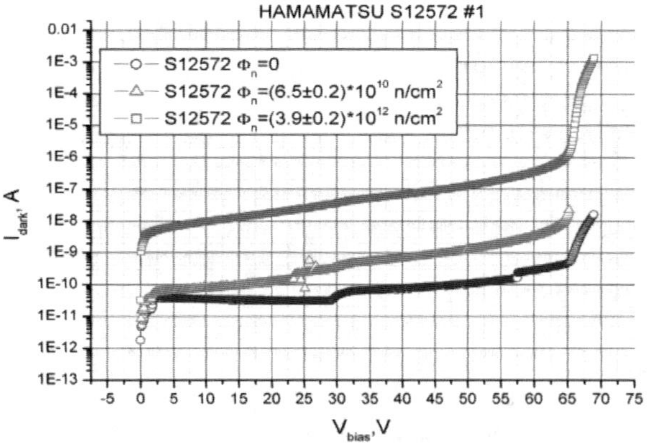

Figure 2. I-V characteristics of Hamamatsu APD before and after irradiation.

New method of C-F data analysis is based on the dynamical process describing variation of nonequilibrium carriers due to recharge of traps' levels in silicon volume depicted in [11]. For the measured capacitance *C,* variation of nonequilibrium carriers' concentration Δn with lifetime τ is described by equation:

$$\frac{\partial \Delta n}{\partial t} = \frac{1}{e} \cdot \frac{\partial J_n}{\partial x} + \frac{\Delta n}{\tau}, \qquad (1)$$

where e is elementary charge and J_n is the current through p-n junction.
Based on this, by applying simple substitutions one can achieve:

$$\frac{1}{C}(f) = \frac{1}{e} \cdot \frac{\langle \tau \rangle}{\langle N_t \rangle} \cdot \Delta\varphi \cdot f, \qquad (2)$$

where N_t *is* traps levels concentration, $\Delta\varphi$ - voltage modulation applied to p-n junction at frequency f.

Achieved equation (2) is describing the dependence of traps' levels in semiconductor volume on the capacitance of p-n junction and can be extremely useful for the C-F characteristics analysis [12].
C-F characteristics of both Zecotek and Ketek APDs are showing the capacitance increase for high frequencies and the capacitance decrease for low frequencies after irradiation (figure 3a,b). Considering the *1/C ~ f* model, this effect can be related respectively to the increase of short-living traps amount and the decrease of long-living traps amount. The delimiting value of traps lifetime to be

considered as short- or long-living is 0.5 μs for Zecotek and 2.5 μs for Ketek. Situation for Hamamatsu APD is different, probably due to its very complicated internal structure. Capacitance and therefore traps amount increased significant for whole frequency range right after the first irradiation with dose of 10^{10} n/cm^2, while no big impact was observed after increase of the dose by roughly two orders of magnitude (figure 4a). A characteristic point can be found corresponding to traps lifetime of 1.8 μs separating the long-living traps having the stable average N_t with many peaks from the short-living traps having the N_t increasing with frequency in a stable fashion (figure 4b).

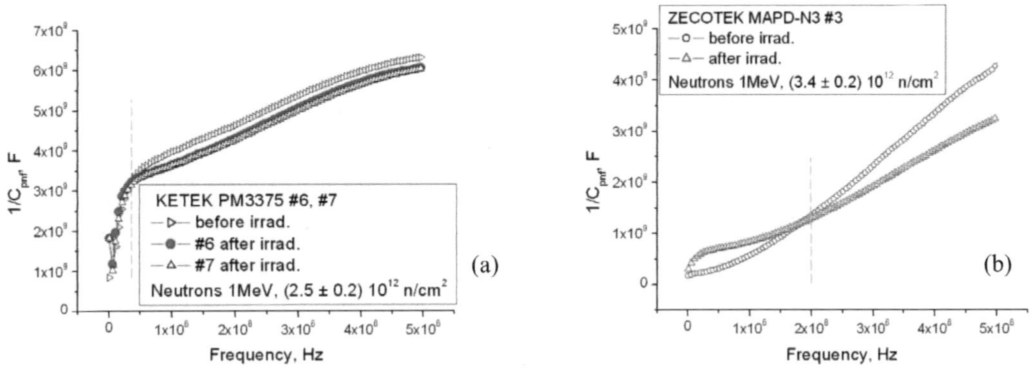

Figure 3. C-F characteristics of APDs before and after irradiation: a) Ketek; b) Zecotek.

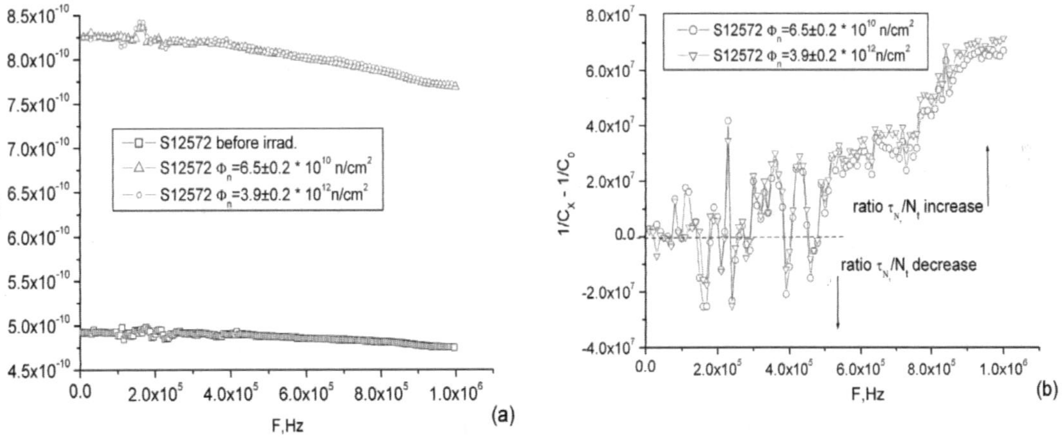

Figure 4. Full C-F (a) and relative C-F (b) of Hamamatsu APD before and after irradiation.

4. Conclusion

The conducted investigation shows that the replacement of PMTs by SSPM can be very prominent solution. Still is not clear whether it is possible to apply SSPM as core of tracking detectors near the beam axis in a high granularity systems probably requiring the core detector structure changes. It is necessary to apply more recourses to provide a stable detector operation. For modern detectors operating in high intensity neutron fields it is important to conduct accurate radiation hardness investigations. The main purpose of those studies should be minimization of the detector noise after irradiation caused by partial deterioration of its parameters due to change of detectors internal structure. Study of the CF characteristics shows that the SSPM noise characteristics depend heavily on the production technology. Controlling the bandwidth of readout electronics is one of the ways to improve signal to noise ratio of the detector. Comparing the Ketek, Zecotek and Hamamatsu SSPMs one can conclude that each of them has specific properties, which allows to optimize its operation by combination of various control methods: control of detector's conditions, bandwidth control and signal filtering. Direct measurement of the SSPM noise spectrum is currently under development and will allow us to specify the further investigation approach to improve detectors parameters during irradiation.

Acknowledgements
The authors thank the NPI cyclotron and neutron generators staff for excellent beam conditions and service, Z. Sadygov and collaborants from Joint Institute for Nuclear Research (Dubna, Russia) for Zecotek APDs, and F. Guber and collaborants from Institute for Nuclear Research (Moscow, Russia) for help with the cosmics setup. This work was supported by the European Community FP7 – Capacities, contract HadronPhysics3 No.283286, LG12007 of the Ministry of Education of the Czech Republic and M100481202 of the Academy of Sciences of the Czech Republic grants, by project MSMT LG14004 of Cooperation Program between JINR and NPI in 2013-2014 and RFBR under grant No. 13-02-00101a. This work was supported by MEPhI Academic Excellence Project (contract №02.a03.21.0005, 27.08.2013).

References

[1] Large Acceptance Hadron Detector for an Investigation of Pb-induced Reactions at the CERN SPS (*http://na49info.web.cern.ch/*)
[2] NA61/SHINE experiment at the CERN SPS (*http://shine.web.cern.ch/*)
[3] The CBM experiment at FAIR facility (*http://www.fair-center.eu/public/experiment-program/nuclear-matter-physics/cbm.html*)
[4] Kekelidze V D *et al.* (NICA Collaboration) Design and Construction of Nuclotron-based Ion Collider fAcility (NICA) *Conceptual Design Report* (*http://nica.jinr.ru*)
[5] Böhlen T T *et al.* 2014 The FLUKA Code: Developments and Challenges for High Energy and Medical Applications *Nuclear Data Sheets* **120** 211–4
[6] Stefanik M *et al.* 2014 Neutron Spectrum Determination of the p (35 MeV) - Be Source Reaction by the Dosimetry Foils Method *Nuclear Data Sheets* **119** 422
[7] Kushpil V, Kushpil S and Huna Z 2012 A simple device for the measurement of kerma based on commercial PIN photo diodes *EPJ Web of Conferences* **24** 07008
[8] Ketek PM3350 Datasheet (*http://www.ketek.net/products/sipm/pm3350/*)
[9] Zecotek WHITE PAPER MAPD Photo-Detectors (*http://www.zecotek.com/media/MAPD-WhitePaper-March-2011.pdf*)
[10] Hamamatsu S12572-010P Datasheet (*http://www.hamamatsu.com/ jp/en/S12572-010P.html*)
[11] Schibli E and Milnes A G 1968 Effects of deep impurities on n+p junction reverse-biased small-signal capacitance *Solid-State Electronics* **11** 323–34
[12] Kushpil V *et al.* 2015 Radiation hardness investigation of avalanche photodiodes for the Projectile Spectator Detector readout at the Compressed Baryonic Matter experiment *Nucl. Instrum. Meth.* A **787** 117–20
[13] Mirzoyan R, Kosyra R and Moser H-G 2009 Light emission in Si avalanches *Nucl. Instrum. Meth.* A **610**(1) 98–100

Journal of Physics: Conference Series **675** (2016) 012040 doi:10.1088/1742-6596/675/1/012040

CNO and *pep* solar neutrino measurements and perspectives in Borexino

S Davini[1,29], M Agostini[2], S Appel[2], G Bellini[3], J Benziger[4], D Bick[5], G Bonfini[6], D Bravo[7], B Caccianiga[3], F Calaprice[8], A Caminata[9], P Cavalcante[6], A Chepurnov[10], D D'Angelo[3], A Derbin[11], L Di Noto[9], I Drachnev[1], A Etenko[12], K Fomenko[13], D Franco[14], F Gabriele[6], C Galbiati[8], C Ghiano[9], M Giammarchi[3], M Goeger-Neff[2], A Goretti[8], M Gromov[10], C Hagner[5], E Hungerford[15], Aldo Ianni[6], Andrea Ianni[8], K Jedrzejczak[17], M Kaiser[5], V Kobychev[18], D Korablev[13], G Korga[6], D Kryn[14], M Laubenstein[6], B Lehnert[19], E Litvinovich[12,20], F Lombardi[6], P Lombardi[3], L Ludhova[3], G Lukyanchenko[12,20], I Machulin[12,20], S Manecki[7], W Maneschg[22], S Marcocci[1], E Meroni[3], M Meyer[5], L Miramonti[3], M Misiaszek[17,6], M Montuschi[23], P Mosteiro[8], V Muratova[11], B Neumair[2], L Oberauer[2], M Obolensky[14], F Ortica[24], M Pallavicini[9], L Papp[2], L Perasso[9], A Pocar[26], G Ranucci[3], A Razeto[6], A Re[3], A Romani[24], R Roncin[6,14], N Rossi[6], S Schönert[2], D Semenov[11], H Simgen[22], M Skorokhvatov[12,20], O Smirnov[13]A Sotnikov[13], S Sukhotin[12], Y Suvorov[27,12], R Tartaglia[6], G Testera[9], J Thurn[19], M Toropova[12], E Unzhakov[11], A Vishneva[13], R B Vogelaar[7], F von Feilitzsch [2], H Wang[27], S Weinz[28], J Winter[28], M Wojcik[17], M Wurm[28], Z Yokley[7], O Zaimidoroga[13], S Zavatarelli[9], K Zuber[19]and G Zuzel[17](Borexino Collaboration)

[1] Gran Sasso Science Institute (INFN), 67100 L'Aquila, Italy
[2] Physik-Department and Excellence Cluster Universe, Technische Universität München, 85748 Garching, Germany
[3] Dipartimento di Fisica, Università degli Studi e INFN, 20133 Milano, Italy
[4] Chemical Engineering Department, Princeton University, Princeton, NJ 08544, USA
[5] Institut für Experimentalphysik, Universität, 22761 Hamburg, Germany
[6] INFN Laboratori Nazionali del Gran Sasso, 67010 Assergi (AQ), Italy
[7] Physics Department, Virginia Polytechnic Institute and State University, Blacksburg, VA 24061, USA
[8] Physics Department, Princeton University, Princeton, NJ 08544, USA
[9] Dipartimento di Fisica, Università degli Studi e INFN, Genova 16146, Italy
[10] Lomonosov Moscow State University Skobeltsyn Institute of Nuclear Physics, 119234 Moscow, Russia
[11] St. Petersburg Nuclear Physics Institute NRC Kurchatov Institute, 188350 Gatchina, Russia
[12] NRC Kurchatov Institute, 123182 Moscow, Russia
[13] Joint Institute for Nuclear Research, 141980 Dubna, Russia
[14] AstroParticule et Cosmologie, Université Paris Diderot, CNRS/IN2P3, CEA/IRFU, Observatoire de Paris, Sorbonne Paris Cité, 75205 Paris Cedex 13, France

[29] Presenter. To whom any correspondence should be addressed.

[15] Department of Physics, University of Houston, Houston, TX 77204, USA
[16] Institute for Theoretical and Experimental Physics, 117218 Moscow, Russia
[17] M. Smoluchowski Institute of Physics, Jagiellonian University, 30059 Krakow, Poland
[18] Kiev Institute for Nuclear Research, 06380 Kiev, Ukraine
[19] Department of Physics, Technische Universität Dresden, 01062 Dresden, Germany
[20] National Research Nuclear University MEPhI (Moscow Engineering Physics Institute), Kashirskoe highway 31, Moscow 115409, Russia
[21] Kepler Center for Astro and Particle Physics, Universität Tübingen, 72076 Tübingen, Germany
[22] Max-Planck-Institut für Kernphysik, 69117 Heidelberg, Germany
[23] Dipartimento di Fisica e Scienze della Terra Università degli Studi di Ferrara e INFN, Via Saragat 1-44122, Ferrara, Italy
[24] Dipartimento di Chimica, Università e INFN, 06123 Perugia, Italy
[25] Physics Department, Queen's University, Kingston ON K7L 3N6, Canada
[26] Amherst Center for Fundamental Interactions and Physics Department, University of Massachusetts, Amherst, MA 01003, USA
[27] Physics and Astronomy Department, University of California Los Angeles (UCLA), Los Angeles, California 90095, USA
[28] Institute of Physics and Excellence Cluster PRISMA, Johannes Gutenberg-Universität Mainz, 55099 Mainz, Germany

E-mail: stefano.davini@gssi.infn.it

Abstract. The detection of neutrinos emitted in the CNO reactions in the Sun is one of the ambitious goals of Borexino Phase-II. A measurement of CNO neutrinos would be a milestone in astrophysics, and would allow to solve serious issues in current solar models. A precise measurement of the rate of neutrinos from the *pep* reaction would allow to investigate neutrino oscillations in the MSW transition region. The *pep* and CNO solar neutrino physics, the measurement in Borexino Phase-I and the perspectives for the new phase are reviewed in this proceeding.

1. Introduction

Borexino is a real-time solar neutrino detector that is designed to detect low energy solar neutrinos [1, 2]. The motivating goal of low energy solar neutrino detection experiments is to directly probe the nuclear reaction processes in the Sun, and explore neutrino oscillations over a broader range of energies than has been done to date.

Mono-energetic (1.44 MeV) *pep* solar neutrinos, produced in the *pp* fusion chain, are an ideal probe to test the transition from vacuum-dominated oscillation to matter-enhanced oscillations predicted by the MSW-LMA model of the neutrino oscillations: the flux of *pep* predicted by from the Standard Solar Model has a small uncertainty (1.2%) due to the solar luminosity constraint. The detection of neutrinos resulting from the CNO cycle would have huge implications in astrophysics: it would be the first direct evidence of the nuclear processes that are believed to fuel massive stars ($>1.5\ M_\odot$). Furthermore, a measurement of the CNO neutrino flux may resolve the solar metallicity problem [3]. The total CNO flux is strongly dependent on the inputs to the solar modelling, being 40% higher in the High Metallicity (GS98) than in the Low Metallicity (AGSS09) solar model [3].

2. The Borexino detector

The features of the Borexino detector are described in detail in [1]. One of the unique features of the Borexino detector is the very low radioactive background. The active detector is 278 tons of two-component liquid scintillator composed of pseudocumene (PC) and 2,5-diphenyloxazole (PPO), a wavelength shifter. The scintillator is contained in a thin nylon vessel, shielded by two PC buffers separated by a second nylon vessel. The scintillator and buffers are contained within a 13.7 m stainless steel sphere that is housed in a 16.9 m domed water tank for additional shielding and muon veto [4].

Neutrinos are detected by their elastic scattering on electrons in the liquid scintillator. The scintillation light is detected with an array of 2200 photomultiplier tubes mounted on the inside surface of the stainless steel sphere. The number of photomultipliers hit is a measure of the energy imparted to the electron, but has no sensitivity to the direction of the neutrino.

Borexino is in data taking since May 2007. Borexino Phase-I covers the period from May 2007 to May 2010. After the purification of the scintillator performed between May 2010 and August 2011, in November 2011 the Phase-II of Borexino started. Borexino Phase-II is expected to last until 2016, before the begining of SOX [5].

Borexino Phase-I solar neutrino results, described in detail in [2], include a high-precision measurement of ^{7}Be neutrinos [6], the measurement of the absence of day-night asymmetry of ^{7}Be neutrinos [7], a measurement of ^{8}B solar neutrinos with a threshold recoil electron energy of 3 MeV [8], and the first time measurement of *pep* solar neutrinos and the strongest constraint up to date on CNO solar neutrinos [9]. Other results include the study of solar and other unknown anti-neutrino fluxes [10], observation of geo-neutrinos [11, 12], measurement of neutrino velocity [13], searches for solar axions [14], and experimental limits on the Pauli-forbidden transitions in ^{12}C nuclei [15].

The direct real-time measurement of the solar neutrinos from the fundamental *pp* reaction [16] is the greatest achievement so far of Borexino Phase-II.

3. First evidence of *pep* solar neutrinos and limits on CNO solar neutrino flux

The Borexino collaboration reported in 2012 the first time measurement of the solar *pep* neutrino rate and the strongest limits on the CNO solar neutrino flux to date [9]. This measurement has been made possible by the combination of the low levels of intrinsic background in Borexino-I, and the implementation of novel background discrimination techniques.

The detection of *pep* and CNO solar neutrinos is challenging: their expected interaction rates are a few counts per day in a 100 ton target, and the main backgrounds, the cosmogenic β^+-emitter ^{11}C and radiogenic ^{210}Bi, are one order of magnitude more intense [9].

^{11}C is produced in the scintillator by cosmic muon interactions with ^{12}C nuclei. In 95% of the cases at least one free neutron is created in the ^{11}C spallation process and then captured in the scintillator [4, 17, 18]. The ^{11}C background can be reduced by applying a space and time veto after coincidences between signals from the muons and the cosmogenic neutrons, discarding exposure that is more likely to contain ^{11}C due to the correlation between the parent muon, the neutron and the subsequent ^{11}C decay (the Three-Fold-Coincidence, TFC). The rejection criteria were chosen to obtain the best compromise between ^{11}C rejection and preservation of exposure [2, 9].

The residual ^{11}C surviving the TFC cut is still a significant background. The pulse shape differences between e^- and e^+ were used to discriminate ^{11}C β^+ decays from neutrino induced e^- recoils and β^- decays. A slight difference in the time distribution of the scintillation signal arises from the finite lifetime of ortho-positronium as well as from the presence of annihilation γ-rays. [19]. A pulse shape parameter was constructed using a boosted-decision-tree algorithm [2, 9].

The analysis is based on a binned likelihood multivariate fit performed on the energy, pulse shape, and spatial distributions of selected scintillation events. The non-uniform radial distribution of the external γ-ray background was included in the multivariate fit and strongly constrained its contribution [2, 9].

The best estimate for the interaction rate of *pep* solar neutrinos in Borexino is $(3.1 \pm 0.6\,(\text{stat}) \pm 0.3\,(\text{syst}))$ counts/(day$\cdot$100ton). If this reduction in the apparent flux is due

Isotope	Specs for LS	BX Phase-I	BX Phase-II
^{238}U	$\leq 10^{-16}$ g/g	$(5.3 \pm 0.5)\times10^{-18}$ g/g	$\leq 8 \times10^{-20}$ g/g
^{232}Th	$\leq 10^{-16}$ g/g	$(3.8 \pm 0.8)\times10^{-18}$ g/g	$\leq 9 \times10^{-19}$ g/g
^{14}C/^{12}C	$\leq 10^{-18}$	$(2.69 \pm 0.06)\times10^{-18}$	unchanged
^{40}K	$\leq 10^{-18}$ g/g	$\leq 0.4 \times10^{-18}$ g/g	unchanged
^{85}Kr	≤ 1 cpd/100t	(30 ± 5) cpd/100t	≤ 7 cpd/100t
^{39}Ar	≤ 1 cpd/100t	$\ll {}^{85}$Kr	$\ll {}^{85}$Kr
^{210}Po	not specified	~ 700 cpd/100t (decaying)	~ 80 cpd/100t (decaying)
^{210}Bi	not specified	$\sim 20 - 70$ cpd/100 t	(25 ± 2) cpd/100t

Table 1. Residual radioactive contamination of the Borexino liquid scintillator before and after the purification performed in 2010-2011. ^{210}Po rate is a factor 100 less than at the begin of data taking. ^{210}Bi is a factor 2 less than in Phase-I.

to ν_e oscillation to ν_μ or ν_τ, we find $P_{ee} = 0.62 \pm 0.17$ at 1.44 MeV. Assuming MSW-LMA solar neutrino oscillations, the Borexino results can be used to measure the *pep* solar neutrino flux, corresponding to $\Phi_{pep} = (1.6 \pm 0.3) \times 10^8$ cm^{-2} s^{-1}, in agreement with the Solar Standard Model [2, 9].

Due to the similarity between the electron-recoil spectrum from CNO neutrinos and the spectral shape of ^{210}Bi decay, whose rate is ~ 10 times greater, Borexino Phase-I only provided an upper limit on the CNO neutrino interaction rate. Assuming MSW-LMA solar neutrino oscillations, the 95% C.L. limit on the solar CNO neutrino flux is 7.7 $\times 10^8$ cm^{-2} s^{-1}. The limit on CNO solar neutrino flux is 1.5 times higher than the flux predicted by the High Metallicity Solar Standard Model [2, 9].

4. Prospects in Borexino Phase-II

A new measurement of *pep* and CNO solar neutrinos is foreseen in Borexino Phase-II. The radioactive background levels in Borexino Phase-II are lower, thanks to a set of scintillator purifications performed after Phase-I. The residual background contaminations are summarised in table 1. The contamination of ^{210}Bi, which was the dominant radioactive background, is reduced of a factor ~ 3. New methods for pulse shape discrimination between e^- and e^+, and for constraining the residual rate of ^{210}Bi, are also under study.

5. Conclusions and outlook

Borexino has achieved the necessary sensitivity to provide, for the first time, evidence of the rare signal from *pep* solar neutrinos and to place the strongest constraint on the CNO solar neutrino flux to date. This result raises the prospect for higher precision measurements by Borexino Phase II, since the next dominant background, ^{210}Bi, has been reduced by scintillator purification.

Acknowledgments

Russian colleagues from MEPhI acknowledge partial support from MEPhI Academic Excellence Project (contract No. 02.a03.21.0005, 27.08.2013).

References

[1] Alimonti G *et al. (Borexino Collaboration)* 2009 *Nucl. Instrum. Meth.* A **600** 568
[2] Bellini G *et al. (Borexino Collaboration)* 2014 *Phys. Rev.* D **89** 112007
[3] Basu S *ASP Conference Series* 2009 **416** 193
[4] Bellini G *et al. (Borexino Collaboration)* 2011 *JINST* **6** P05005

Journal of Physics: Conference Series **675** (2016) 012040 doi:10.1088/1742-6596/675/1/012040

[5] Bellini G *et al. (Borexino Collaboration).* 2013 *JHEP* **1308** 038
[6] Bellini G *et al. (Borexino Collaboration)* 2011 *Phys. Rev. Lett.* **107** 141302
[7] Bellini G *et al. (Borexino Collaboration)* 2012 *Phys. Lett.* B **707** 22
[8] Bellini G *et al. (Borexino Collaboration)* 2010 *Phys. Rev.* D **82** 033006
[9] Bellini G *et al. (Borexino Collaboration)* 2012 *Phys. Rev. Lett* **108** 051302
[10] Bellini G *et al. (Borexino Collaboration)* 2011 *Phys. Lett.* B **696** 191
[11] Bellini G *et al. (Borexino Collaboration)* 2010 *Phys. Lett.* B **687** 299
[12] Agostini M *et al. (Borexino Collaboration)* 2015 *Phys. Rev.* D **92** 031101
[13] Alvarez Sanchez P *et al. (Borexino Collaboration)* 2012 *Phys. Lett.* B **716** 401-5
[14] Bellini G *et al. (Borexino Collaboration)* 2012 *Phys. Rev.* D **85** 092003
[15] Bellini G *et al. (Borexino Collaboration)* 2010 *Phys. Rev.* C **81** 0343317
[16] Bellini G *et al. (Borexino Collaboration)* 2014 *Nature* **512** 383
[17] Bellini G *et al. (Borexino Collaboration)* 2012 *JCAP* **05** 015
[18] Bellini G *et al. (Borexino Collaboration)* 2013 *JCAP* **1308** 049
[19] Davini S 2014 *Eur. Phys.* J. Plus **128** 89

Hierarchi problem

Sergey G Rubin
National Research Nuclear University MEPhI (Moscow Engineering Physics Institute), Kashirskoe highway 31, Moscow, 115409, Russia

E-mail: sergeirubin@list.ru

Abstract. The way to solve the hierarchy problem based on multidimensional gravity is discussed. Various metrics of deformed extra space are produced at the Planck scale. It is shown that the Higgs vacuum value depends on a metric of extra space and hence their different numerical values are realized in various universes. An interval of the Higgs vacuum values is proved include zero value. Our universe belongs to a set of universes the vacuum values of which are close to zero

1. Introduction

Main problem discussed in this paper concerns the smallness of electroweak scale or more definitely the vacuum average of the Higgs field comparing to the Planck scale.

In terms of the multidimensional gravity vacuum state v of the Higgs field depends on a metric G of an extra space, $v = v(G)$. As was shown in [1], [2], [3] the metric continuously varies in wide range. Each specific metric is realized in a specific universe. Our Universe is associated to those extra metric which gives extremely small ratio v/M_{Pl}. Prompt realization of this idea is described below. This approach assumes existence of large number of various universes, see also [4].

2. Multidimensional Lagrangian

Let us consider a space $M = M_4 \times V_n$ with metric

$$ds^2 = G_{AB}dZ^A dZ^B = g_{\mu\nu}(x)dx^\mu dx^\mu + G_{ab}(y)dy^a dy^b \tag{1}$$

Here M, M' are the manifolds with metrics $g_{mn}(x)$ and $G_{ab}(x, y)$ respectively. x and y are the coordinates of the subspaces M and M'. We will refer to 4-dim space M and n-dim compact space M' as a main space and an extra space respectively. Here the metric has the signature (+ - - - ...), the Greek indices $\mu, \nu = 0, 1, 2, 3$ refer to 4-dimensional coordinates). Latin indices run over $a, b, ... = 4, 5....$ Consider the Higgs field action acting in 6-dim manifold

$$S = \frac{m_D^{D-2}}{2} \int d^D Z \sqrt{|G|} \left[f(R) + L_H \right]; \tag{2}$$

$$f(R) = u_1(R - R_0)^2 + u_2 \tag{3}$$

L_H is a Lagrangian of the proto-Higgs field

$$L_H = \{\partial_A H^+ G^{AB} \partial_B H - U(H^+ H)\} \tag{4}$$
$$U = b_2 \left(H^+ H\right) + b_4 \left(H^+ H\right)^2 .$$

Equations of motion are

$$\nabla^2_{4+n} H = b_2 H + 2b_4 \left(H^+ H\right) H; \quad h.c. \tag{5}$$

Classical equations for the metric of extra space have the following form

$$R_{ab} f' - \frac{1}{2} f(R) G_{ab} \nabla^2 - \nabla_a \nabla_b f_R + G_{ab} \nabla^2 f' = \frac{1}{m_D^{D-2}} T_{ab}, \tag{6}$$

$+$ *additional conditions.*

Here the first term proportional to R_4 is omitted due to inequality $R_4 \ll R_n$ assumed throughout the paper. The choice of metric $G_{\theta\theta} = -r(\theta)^2$; $G_{\phi\phi} - r(\theta)^2 \sin^2(\theta)$ leads to the Ricci scalar expressed in terms of the radius $r(\theta)$.

As a result explicit form of equation (6) to be solved numerically is

$$\partial^2_\theta R + \cot\theta \partial_\theta R = -\frac{1}{2} r(\theta)^2 \left(R_0^2 - R^2\right) + u_2/u_1. \tag{7}$$

As the additional conditions let us fix the metric at the point $\theta = \pi$: $r(\pi) = r_\pi; r'(\pi) = 0; R(\pi) = R_\pi; R'(\pi) = 0$. From here on we will use the units $m_D = 1$. Evidently, there is a set of solutions to system (6) depending on additional conditions. Maximally symmetrical extra spaces which are used in great majority of literature represent a small subset of this set.

Let us seek for the solution in the form

$$H(Z) = H(x, y) = \chi(x) W(y) \tag{8}$$

where the function $\chi(x)$ relates to the Higgs field. The proto-Higgs Lagrangian density acquires the form

$$\begin{aligned} L_\chi = \{&W(y)^2 \partial_\mu \chi(x)^+ g^{\mu\nu} \partial_\nu \chi(x) + \chi(x)^+ \chi(x) G^{yy} (\partial_y W)^2 \\ &- b_2 W(y)^2 \chi(x)^+ \chi(x) - b_4 W(y)^4 [\chi(x)^+ \chi(x)]^2 \end{aligned} \tag{9}$$

after substitution (8) into (4).

Equation (5) is transformed into equation

$$\begin{aligned} &\nabla^2_n W(y) + b_2 W(y) = -2b_4 \chi^+(x) \chi(x) W(y)^3 \simeq \\ &-2b_4 \left\langle \chi^+(x) \chi(x) \right\rangle W(y)^3. \end{aligned} \tag{10}$$

with additional conditions

$$W(\pi) = W_\pi; \quad W'(\pi) = 0. \tag{11}$$

Dimensionless average $\langle \chi^2 \rangle$ is supposed to b of order unity. Integrating out extra-coordinates y in action (2) with Lagrangian (9) we obtain the known Higgs action

$$\begin{aligned} S_H = \int d^4x \sqrt{g(x)} \{&\partial_\mu h(x)^+ g^{\mu\nu} \partial_\nu h(x) + \\ \lambda_{eff} v_{eff}^2 h(x)^+ h(x) &- \lambda_{eff} [h(x)^+ h(x)]^2 \} \end{aligned} \tag{12}$$

where

$$\lambda_{eff} \equiv \frac{4B_4}{K^2}, \quad v_{eff}^2 \equiv -\frac{B_2}{K \lambda_{eff}}. \tag{13}$$

and

$$
\begin{aligned}
K &= \frac{m_D^{D-2}}{2} \int d^n y \sqrt{|G(y|)} W(y)^2 \\
B_2 &= \frac{m_D^{D-2}}{2} \int d^n y \sqrt{|G(y)|} \left[b_2 W(y)^2 - G^{yy} (\partial_y W)^2 \right] \\
B_4 &= \frac{m_D^{D-2}}{2} b_4 \int d^n y \sqrt{|G(y)|} W(y)^4 .
\end{aligned}
\tag{14}
$$

The Higgs field h is connected to the proto-Higgs field χ as $h(x) \equiv \sqrt{K}\chi$ After the Higgs discovery its self coupling is known and is expressed in terms of the Higgs particle mass $\lambda_{eff} = m_{Higgs}^2/(2v_{Higgs}) = 125^2/(2\cdot 246^2) = 0.13$. This value determines parameter b_4 according to expressions (13), (14). The main aim of the paper is to find conditions for inequality

$$
v_{eff}^2 \equiv \frac{1}{\lambda_{eff}} |B_2|/K \ll m_D^2 \tag{15}
$$

see (13), representing the Hierarchy problem. As was discussed above it is enough to prove that values v_{eff} is settled in an extremely tight interval including zero value. The left hand side of inequality (15) depends on random additional conditions caused by space-time foam.

Dependence of the Higgs vacuum average v on the boundary value W_π was obtained numerically. The curve intersects zero at point $W_\pi = 0.3$. It means that universes with the vacuum average be arbitrary close to zero do exist. We are in one of such universe.

Acknowledgments
This work was performed within the framework of the Center FRPP supported by MEPhI Academic Excellence Project (contract No. 02.a03.21.0005, 27.08.2013) and was supported by the Ministry of Education and Science of the Russian Federation, Project No. 3.472.2014/K.

References

[1] Rubin S 2015 *EPJC* **75** 333 (*arXive:*gr-gc/1503.05011)
[2] Gani V, Dmitriev A and Rubin S 2015 *Int. J. Mod. Phys.* D **24** 1545001 (*arXive:*gr-gc/1411.4828)
[3] Rubin S 2015 *Gravitation & Cosmology* **21** 143 (*arXive:*gr-gc/1403.2062)
[4] Freivogel B and Susskind L 2004 *Phys.Rev.* D **70** 126007 (*arXive:*hep-th/0408133)

The λ energy dependence deduced from Bose – Einstein correlations of $\pi\pi$ pairs produced in pp collisions

G Alexander[1] **and V Okorokov**[2]

[1]School of Physics and Astronomy, Raymond and Beverly Sackler Faculty of Exact Sciences, Tel-Aviv University, 69978 Tel-Aviv, Israel
[2]National Research Nuclear University MEPhI (Moscow Engineering Physics Institute), Kashirskoe highway 31, 115409 Moscow, Russia

E-mail: gideona@post.tau.ac.il, VAOkorokov@mephi.ru

Abstract. The energy dependence of the chaoticity parameter λ, derived from Bose–Einstein correlations (BEC) of pion-pairs produced in pp collisions, is investigated. Considered is the one dimension (1D) of the BEC analysis in terms of a Gaussian and / or exponential distributions. The experimental data are examined in terms of the relation between the pion sources and the BEC dimension R which in turn are deduced from the charged outgoing particle multiplicity. This approach follows the λ general energy behavior as obtained from the 1D BEC analyzes of the pp collision data. Prediction for the λ dependence on energy is obtained over a multi-TeV energy range based on a model of independent pion sources. The decrease of the λ value with energy expected within the framework of this approach is supported by the experimental findings.

1. Introduction

With the recent operation of the Large Hadron Collider (LHC) at CERN the opportunity to study the one dimension (1D) BEC of identical bosons at very high hadron-hadron collision energies has been utilized by the ALICE, ATLAS and CMS collaborations. In parallel the energy dependence of the BEC dimension R has recently been investigated and found in proton-proton (pp) collisions to increase with $\log(\sqrt{s_{pp}})$ [1, 2, 3]. Here we report on a study of the chaoticity parameter λ energy behavior which determines the strength of the BEC measured effect. To this end we utilize the relation between the BEC dimension R and the number of the groups of pion sources which we stipulate to be proportional to the average charged particle multiplicity produced in the pp reactions. One of the frequently used parametrization for the 1D BEC analysis of hadrons which emerge from a sphere volume, is that of Goldhaber [4], namely

$$C_{Gauss}(Q) = 1 + \lambda_{Gauss} e^{-Q^2 R^2_{Gauss}} , \quad (1)$$

which assumes a static source in the plane-wave approach of a spherical volume with a radial Gaussian distribution of the emitter. The second often used parametrization, which assumes a radial Lorentzian distribution of the source, is given by

$$C_{Expo}(Q) = 1 + \lambda_{Expo} e^{-Q R_{Expo}} , \quad (2)$$

where in both representations $Q^2 = -(p_1 - p_2)^2$ is the difference squared of the 4-momentum vectors of the two correlated identical bosons. The λ factor can vary between 0 and 1. In the 1D analysis the relation between R_{Gauss} and R_{Expo} dimensions can be evaluated from the requirement that the first Q moment in a given BEC distribution will be equal when treated by a fit to a Gaussian distribution or an Exponential one, namely

$$\frac{1}{S_2}\int_{Q_1}^{\infty} Qe^{-Q^2R^2_{Gauss}}dQ = \frac{1}{S_1}\int_{Q_1}^{\infty} Qe^{-QR_{Expo}}dQ, \text{ where } S_{j=1,2} \equiv \int_{Q_1}^{\infty} e^{-(QR)^j}dQ. \quad (3)$$

This relation remains essentially the same as long as the upper integration is higher than 2 GeV. In particular for the value of $Q_1 = 0$ GeV one obtains the known relation

$$R_{Gauss} = R_{Expo}/\sqrt{\pi}. \quad (4)$$

A connection between λ_{Gauss} and λ_{Expo} can be estimated from the relations

$$\int_{Q_1}^{Q_2} \lambda_{Expo}e^{-QR_{Expo}}dQ \simeq \int_{Q_1}^{Q_2} \lambda_{Expo}e^{-QR_{Gauss}\sqrt{\pi}}dQ \simeq \int_{Q_1}^{Q_2} \lambda_{Gauss}e^{-Q^2R^2_{Gauss}}dQ\ . \quad (5)$$

which are valid as long as the upper integration level is above 2 GeV. Where in the range between Q_1 and Q_2 the Gaussian and Exponential BEC parametrization fit equally well the measured Q distribution. In setting $Q_1 = 0$ and $Q_2 = \infty$ GeV one finds that

$$\lambda_{Gauss} \;=\; 2\lambda_{Expo}/\pi\ . \quad (6)$$

It has generally been found that at low Q values, e.g. $Q_{low} \leq 0.1$ GeV, the Gaussian parametrization fails to yield a satisfactory fit to the measured BEC distribution and the Exponential one is describing it much better [13]. Therefore it is not surprising that the measured ratios $\lambda_{Expo}/\lambda_{Gauss}$ deduced from table 1 are significantly different than the one given by equation (6) of $\pi/2$ and are lying in the neighborhood of 2. A relation between the pairs $(R_{Gauss}, \lambda_{Gauss})$ and $(R_{Expo}, \lambda_{Expo})$ obtained in 1D BEC fits to the Q distribution are analyzed in details in [16] where the obtained $\lambda_{Expo}/\lambda_{Gauss}$ ratios are very similar to the measured ones listed in table 1.

2. The BEC parameters dependence on energy

It has been shown that in the 1D BEC there exists a relation between $R_{n=1}$ and R_n where n is the number of the independent similar emitting pion group sources (see e.g. [17]), namely

$$R_n = n(\lambda_n\,/\,\lambda_{n=1})R_{n=1}\ . \quad (7)$$

which relates the one group sources dimension R_1 at low energies with its value R_n for n identical group sources. The $\lambda_{n=1}$ and λ_n parameters are the chaoticity values respectively for $n = 1$ and n group sources. Inasmuch that one aims for the energy behaviour of λ rather than on its absolute value sufficient is to determine the energy dependence of the ratio $\lambda = \lambda_n/\lambda_{n=1}$. Recently the BEC one dimension R dependence on energy was observed in pp collisions to increase from low to high energies and was fitted [1] to yield (s is in TeV^2 units and $s_0 = 1$ TeV^2)

$$R(s) \;=\; \{(1.64 \pm 0.11) + (0.07 \pm 0.01)\ln\varepsilon\}\,R_{n=1}, \quad \varepsilon \equiv s/s_0. \quad (8)$$

Thus the behavior of $\lambda(s)$ in terms of relations (7) and (8) requires a solution of the equation

$$\{(1.64 \pm 0.11) + (0.07 \pm 0.01)\ln\varepsilon\}\,R_{n=1} \;=\; n(s)\lambda R_{n=1}\ , \quad (9)$$

Table 1. A compilation of λ values obtained from the 1D BEC of pion-pairs produced in pp collisions. The statistical and systematic errors are added in quadrature in the table.

1D BEC analysis in pp collisions		Chaoticity parameter		References
Experiment/Collaboration	$\sqrt{s_{pp}}$ [GeV]	λ_{Gauss}	λ_{Expo}	
BNL-E766	7.21	0.466 ± 0.015	—	[5]
NA23	26.0	0.32 ± 0.08	—	[6]
ABCDHW	31.0	0.35 ± 0.04	—	[7]
ABCDHW	44.0	0.42 ± 0.04	—	[7]
AFS	58.1	0.34 ± 0.04	—	[8]
ABCDHW	62.0	0.42 ± 0.08	—	[7]
AFS	63.0	0.40 ± 0.03	0.77 ± 0.07	[9]
SFM	63.0	0.42 ± 0.04	—	[10]
STAR	200	0.353 ± 0.003	—	[11]
ALICE	900	0.31 ± 0.03	0.55 ± 0.05	[12]
ATLAS	900	0.34 ± 0.03	0.74 ± 0.11	[13]
CMS	900	0.315 ± 0.014	0.63 ± 0.03	[14, 15]
CMS	2360	0.32 ± 0.01	0.66 ± 0.09	[14]
ATLAS	7000	0.25 ± 0.02	0.53 ± 0.05	[13]
CMS	7000	—	0.62 ± 0.04	[15]

where $n(s)$ is equal to the number of group of identical sources as a function of energy. As it has already been seen that R increases with the average charged multiplicity, $\langle N_{ch}\rangle$, of the colliding hadrons and since we aim only to estimate the λ dependence on energy it is sufficient to require that the number of sources is proportional to the average charged multiplicity.

A compilation of λ_{Gauss} and λ_{Expo} obtained from the 1D BEC of pion-pairs produced in pp collisions are given in table 1 and shown in figure 1 as a function of energy for the λ_{Gauss}. The data are seen to scatter somewhat in the energy region of 20 to 60 GeV which may well be attributed to problems concerning the reference samples and different analysis criteria. Despite of this discrepancy, a general decrease with energy of the λ_{Gauss} values is apparent.

To evaluate the λ_{Gauss} dependence on energy we follow the formalism outlined in [18, 19] where the hadron-hadron collisions are contributed by a "hard" and a "soft" component. In this approach the charged multiplicity of pp collisions per rapidity η is equal to $dN_{ch}/d\eta = N_{pp}$. Since however it has been found that $dN_{ch}/d\eta$ is independent of η it follows that N_{pp} is proportional to the total charge multiplicity of the pp collision. Inserting the N_{pp} reported expression

$$N_{pp}(s_{pp}) = 3.43\,\varepsilon^{0.11}\ . \tag{10}$$

and $n(s_{pp}) = N_{pp}(s_{pp})$ into equation (9) one obtains to a good approximation the λ behaviour (not absolute values) as a function of energy

$$\lambda_{Gauss}(s_{pp})\ \simeq\ \{0.288 + 0.013\ln\varepsilon\}\varepsilon^{-0.11}\ , \tag{11}$$

which is plotted in figure 1 normalized to the measured value at 200 GeV. As seen equation (11) describes the experimental energy dependence of λ_{Gauss} quite reasonable and predicts a decrease of the chaoticity parameter with the increase of the collision energy. Note that at $\sqrt{s_{pp}} = 14$ TeV, the highest planned LHC energy, the 1D λ_{Gauss} should reach the 0.20 ± 0.04 value. Due to the slow decrease of the chaoticity parameter at high energies the expected λ value and its errors at the current LHC energy of 13 TeV should be similar to that predicted at 14 TeV.

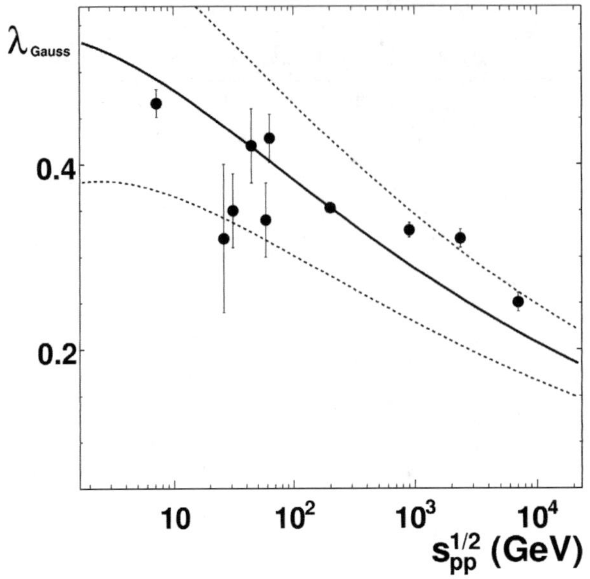

Figure 1. The 1D BEC $\lambda_{Gauss}(\sqrt{s_{pp}})$ as deduced from the measured $\pi\pi$ produced in pp collisions. The λ_{Gauss} values at $\sqrt{s_{pp}}$ close to the 63 GeV and at $\sqrt{s_{pp}} = 900$, 7000 GeV are averages over experiments at $\sqrt{s_{pp}} = 62-63$, 900 and 7000 GeV respectively. The expected $\lambda_{Gauss}(\sqrt{s_{pp}})$ derived in this work is shown by the continuous line, normalized to the measured λ_{Gauss} at 200 GeV where the dotted lines represent its ±1 s.d. limits.

3. Summary

The 1D BEC measured values of λ_{Gauss} are seen to decrease with energy. The approach of relating the R and multiplicity increase with pp collisions energy with the number of sources results in a decrease of the 1D λ_{Gauss} similar to the one obtained experimentally. Prediction for the chaoticity dependence on energy is obtained over a multi-TeV energy range.

Acknowledgments

Thanks are due to E. Levin, C. Pajares and E.K.G Sarkisyan for many helpful comments. The work of one of us (V.O.) was performed within the framework of the Center of Nanostructured Electronics supported by NRNU MEPhI Academic Excellence Project (contract No 02.a03.21.0005, 27.08.2013).

References

[1] Alexander G 2012 *J. Physics* G **39** 085007
[2] Alexander G and Ben Mordechai I 2013 *J. Physics* G **40** 125101
[3] Okorokov V A 2015 *Adv. High Energy Phys.* **2015** 790646
[4] Goldhaber G *et al.* 1959 *Phys. Rev. Lett.* **3** 181 300
[5] Uribe J *et al.* 1994 *Phys. Rev.* D **49** 4373
[6] Bailly J L *et al.* 1989 *Z. Phys.* C **43** 341
[7] Breakstone A *et al.* 1987 *Z. Phys.* C **33** 333
[8] Åkesson T *et al.* 1987 *Phys. Lett.* B **155** 128
[9] Åkesson T *et al.* 1987 *Z. Phys.* C **36** 517
[10] Breakstone A *et al.* 1985 *Phys. Lett.* B **162** 400
[11] Aggarwal M M *et al.* 2011 *Phys. Rev.* C **83** 064905

[12] Aamodt K *et al.* 2010 *Phys. Rev.* D **82** 052001
[13] Sýkora I (on behalf of the ATLAS Collaboration) 2015 *Talk* on the DIS.
[14] Khachatryan V *et al.* 2010 *Phys. Rev. Lett.* **105** 032001
[15] Khachatryan V *et al.* 2011 *J. High Energy Phys.* **0511** 029
[16] Okorokov V A *to be published*
[17] Alexander G 2003 *Rep. Prog. Phys.* **66** 481
[18] Kharzeev D and Nardi M 2001 *Phys. Lett.* B **507** 121
[19] Back B B *et al.* 2002 *Phys. Rev.* C **65** 061901

Search for isobar-analog states of superheavy hydrogen isotopes $^{5\text{-}7}$He

B A Chernyshev, Yu B Gurov, L Yu Korotkova, D S Kuznetsov, S V Lapushkin, M V Tel'kushev and T D Schurenkova
National Research Nuclear University MEPhI (Moscow Engineering Physics Institute), Kashirskoe highway 31, Moscow, 115409, Russia

E-mail: chernyshev@mephi.ru

Abstract. Search for isobar-analog states (IAS) of superheavy hydrogen isotopes $^{5\text{-}7}$H was performed among the high-excited states of helium isotopes $^{5\text{-}7}$He. The excited spectra were measured in stopped pion absorption by light nuclei. The experiment was performed at low energy pion channel of LANL with two-arm multilayer semiconductor spectrometer. Excited states of $^{5\text{-}7}$He were observed in three-body reaction channels on ^{10,11}B nuclei. Several excited levels were observed for the first time. ^{6}He excited state with E_x = 27.0(8) MeV observed in ^{10}B(π^-,pt)X channel is an IAS candidate for ^{6}H with E_r ~ 5.5 MeV. ^{7}He excited state with E_x = 24.8(4) MeV observed in ^{10}B(π^-,pd)X, ^{11}B(π^-,pt)X and ^{11}B(π^-,dd)X channels is an IAS candidate for ^{7}H with E_r ~ 3 MeV.

Introduction

Study of light neutron-rich nuclei is one of the main current subjects of nuclear physics and is a field of great experimental activity [1]. A relatively small number of nucleons makes possible to find a correct microscopic description of their properties and, therefore, to verify existing nuclear models as well as nucleon–nucleon potentials.

Superheavy hydrogen isotopes ^{n}H (n $\geq$ 4) are the most interesting ones due to extremely large neutron–to–proton ratio and a proton 1*s*–shell being open, unlike that of other elements. The experimental search for these exotic nuclei and investigation of their level structure has been continued for a long time. Nevertheless, the available experimental information is still contradictory and is extremely limited, except for ^{4}H [1]. Superheavy hydrogen isotopes are nucleon-unstable ones. ^{5,6}H isotopes were observed in stopped pion absorption by light nuclei [2]. Other results on ^{5}H were obtained in measurements on radioactive ion beams (references to these works are reviewed in [1]). The resonance energy of ^{5}H ground state measured in these experiments falls in the range 1.7 MeV [3] to 5.4 MeV [2]. At first time, ^{6}H was observed in the experiments on heavy ion beams [4, 5]. The resonance energy ($\approx$ 2.7 MeV) obtained in these measurements differs significantly from value (6.6(7) MeV) obtained in the stopped pion absorption [2]. The observation of ^{7}H nucleus is still under examinations since the cross sections of its production are very low [1, 2].

In such situation investigations of new reaction channels of the superheavy hydrogen isotope formation are of great interest. For this purpose we attempted to find isobar analog states (IAS) of superheavy hydrogen isotopes $^{5\text{-}7}$H among the high-excited states of the helium isotopes $^{5\text{-}7}$He. Search for the helium states was performed in the following channels of the stopped pion absorption

^{10}B(π^-,dt)X and ^{11}B(π^-,tt)X – for ^{5}He, ^{11}B(π^-,dt)X, ^{10}B(π^-,pt)X, and ^{10}B(π^-,dd)X – for ^{6}He, ^{10}B(π^-,pd)X, ^{11}B(π^-,pt)X and ^{11}B(π^-,dd)X – for ^{7}He.

Experiment details

The experiment was performed on the meson factory LAMPF using the two arm semiconductor spectrometer [6]. The π^- beam with the energy of 30 MeV passed through the beryllium moderator and stopped in a thin target ($\approx$ 24 mg·cm^{-2}). In one experimental run the measurements were carried out on the targets ^{9}Be, ^{10,11}B and ^{12,14}C.

Charged particles emitted after pion absorption were detected by two semiconductor telescopes located at an angle of 180° with respect to each other. The energy resolution (FWHM) for single-charged particles (p, d, t) was better than 0.5MeV [6]. The error of absolute energy calibration did not exceed 100 keV [2].

A search for the neutron-rich states was carried out in the peaks of the missing mass spectrum (*MM*). In the measurements of any pairs of singly charged particles the *MM* resolution was $\approx$ 1MeV [2]. The error of the *MM* absolute calibration (δMM) did not exceed 100 keV [2].

The spectrometer and experimental technique are described in more detail in [2, 6].

Results and discussion

Search for ^{5}He states was carried out in the reactions: ^{10}B(π^-,dt)X and ^{11}B(π^-,tt)X. The *MM* spectra measured in these reactions are shown in figures 1 and 2. The sum of α-particle and neutron masses is taken to be a reference point. To resolve ^{5}H states we used the method of least squares in describing the experimental spectra by the sum of *n*–particle distributions over phase space ($n \geq 4$) and Breit–Wigner distributions [7].

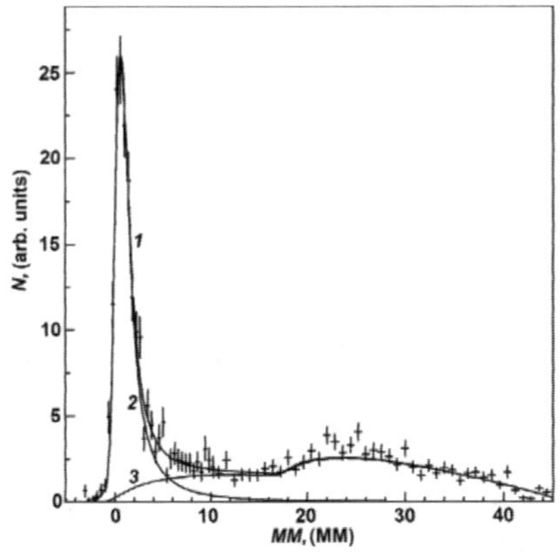

Figure 1. *MM* spectrum for the reaction ^{10}B(π^-,dt)X. Dots with error bars are the experimental data. The solid lines are: 1 – the fit, 2 – Breit–Wigner distribution for ground state, 3 – sum of phase-space distributions.

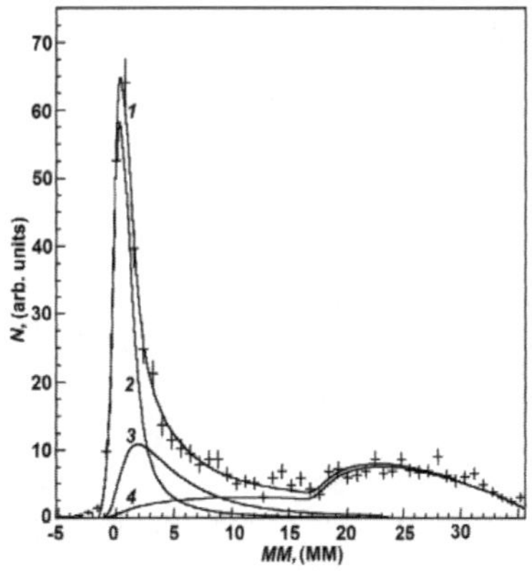

Figure 2. *MM* spectrum for the reaction ^{11}B(π^-,tt)X. Dots with error bars are the experimental data. The solid lines are: 1 – the fit, 2 (3) – Breit–Wigner distribution for ground (first excited) state, 4 – sum of phase-space distributions.

In both spectra peaks due to three-body channels with ^{5}He in the final states are observed. The peak in figure 1 is due to the ^{5}He ground state. The peak in figure 2 is due to the superposition of the ground and first excited states. Values of resonance parameters of these states (E_r – resonance energy, Γ – level width): (0.80(3) MeV, 0.7(1) MeV) and (2.1(1) MeV, 5.6(3) MeV) are in a coincident with data of the compilation [8].

We did not obtained statistically significant evidence on high excited states of ^{5}He. In the figure 1 some excess of the experimental events exists in the range from 22 MeV to 28 MeV. Several states with isospin T = 1/2 lie in this region [8]. Thus we did not obtain any indication on the existence of IAS of ^{5}H with T = 3/2.

Search for ^{6}He excited states was carried out in the following reactions: ^{11}B(π^-,dt)X, ^{10}B(π^-,pt)X, ^{10}B(π^-,dd)X. The *MM* spectra for these reactions are shown in figure 3. The mass of ^{6}He ground state is taken to be a reference point. Peaks formed due to the appearance of the ground state and the excited states in the three-body reaction channels are well seen. Average values of parameters of ^{6}He excited states, obtained in this paper are presented in the table 1. High excited states with E_x > 20 MeV are observed in ^{10}B(π^-,pt)X channel only. These states may have a spin equals to 1 or 2. In last case these states are isobar-analog states of ^{6}H with resonance energy $E_r \approx 0$ MeV and 5.5 MeV. The first value is inconsistent with results of the works on search for ^{6}H [2, 4, 5]. The second value is close to resonance energy of ^{6}H ground state (6.6(7) MeV) obtained in measurements of stopped pion absorption [2].

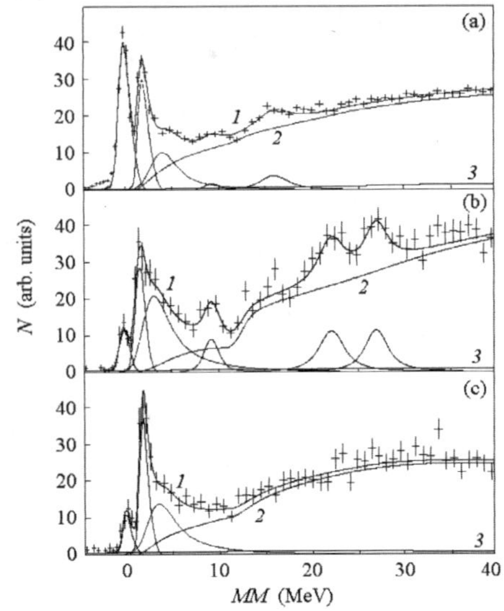

Figure 3. *MM* spectrum for the reaction a) ^{11}B(π^-,dt)X; b) ^{10}B(π^-,pt)X; c) ^{10}B(π^-,dd)X. Dots with error bars are the experimental data. The solid lines are the Breit–Wigner distributions, 1 – fit, 2 – sum of phase-space distributions and 3 – background coming from random coincidences.

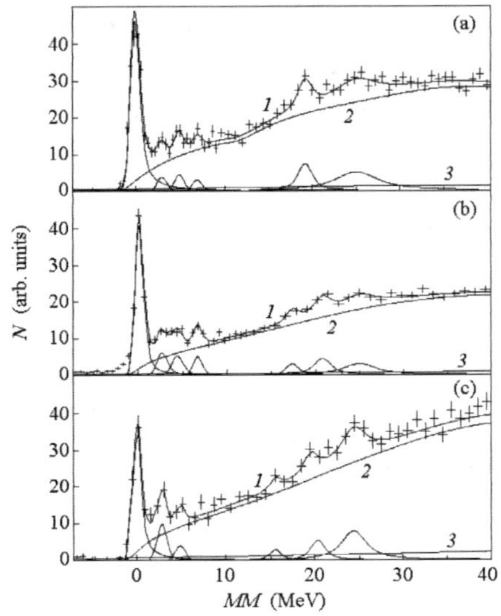

Figure 4. *MM* spectrum for the reaction a) ^{11}B(π^-,pt)X; b) ^{11}B(π^-,dd)X; c) ^{10}B(π^-,dd)X. Dots with error bars are the experimental data. The solid lines are the Breit–Wigner distributions, 1 – fit, 2 – sum of phase-space distributions and 3 – background coming from random coincidences.

Search for ^{7}He excited states was carried out in the following reactions: ^{11}B(π^-,pt)X, ^{11}B(π^-,dd)X, ^{10}B(π^-,pd)X. The *MM* spectra for these reactions are shown in figure 4. The sum of ^{6}He and neutron masses is taken to be a reference point. Peaks formed due to the appearance of the ground state and the excited states in the three-body reaction channels are well seen. Average values of parameters of ^{6}He excited states, obtained in this paper, are presented in the table 1. High excited state with E_x = 24.8 MeV is observed in the all studied channels. This state is the candidate on isobar-analog of ^{7}He with resonance energy $E_r \approx 3$ MeV. From the experimental data on spectroscopy of helium, lithium, and beryllium isotopes it follows that the number of neutrons N = 6 becomes magic rather than N = 8 [1]. Based on the latter, it is reasonable to expect that among superheavy hydrogen isotopes, ^{7}H should be the most bounded one. However, until now any reliable evidence for the existence of this isotope in a stable state and in a resonance state has not been received. Thus, our result for IAS is important for future experimental search for the ^{7}H.

Table 1. Parameters of excited states of ^{6}He and ^{7}He measured in this work.

^{6}He			^{7}He		
E_r, MeV	Γ MeV	Reaction	E_r, MeV	Γ MeV	Reaction
1.80(3)	0.11(2)	1,2,3	3.1(1)	≤0.5	4,5,6
3.5(2)	3.1(4)	1,2,3	4.9(2)	≤0.5	4,5,6
9.3(2)	1.0(4)	1,2	6.7(2)	≤0.5	4,5
15.(9)	3.2(7)	1	16.9(5)	1.0(3)	5,6
22.1(1.0)	2.7(1.4)	2	19.8(3)	1.5(3)	4,5,6
27.0(8)	2.5(1.1)	2	24.8(4)	4.6(7)	4,5,6

Reactions in columns 3 and 6: 1 – ^{11}B(π^-,dt)X; 2 – ^{10}B(π^-,pt)X; 3– ^{10}B(π^-,dd)X; 4 – ^{11}B(π^-,pt)X; 5 – ^{11}B(π^-,dd)X; 6 – ^{10}B(π^-,dd)X.

Conclusion

Excited states of $^{5\text{-}7}$He were observed in three-body channels of stopped pion absorption by boron isotopes ^{10,11}B. We found the two candidate on IAS of superheavy hydrogen isotopes. ^{6}He excited state with E_x = 27.0(8) MeV observed in ^{10}B(π^-,pt)X channel is an IAS candidate for ^{6}H with E_r ~ 5.5 MeV. ^{7}He excited state with E_x = 24.8(4) MeV observed in ^{10}B(π^-,pd)X, ^{11}B(π^-,pt)X and ^{11}B(π^-,dd)X channels is an IAS candidate for ^{7}H with E_r ~ 3 MeV.

Acknowledgments

This work was performed within the framework of the Center of Fundamental Research and Particle Physics supported by MEPhI Academic Excellence Project (contract № 02.a03.21.0005, 27.08.2013).

References

[1] Tanihata I, Savajols H and Kanungo R 2013 *Prog. Part. Nucl. Phys.* **68** 215
[2] Gurov Yu B *et al.* 2009 *Phys. Part. Nucl.* **40** 558
[3] Korsheninnikov A A *et al.* 2001 *Phys. Rev. Lett.* **87** 092501
[4] Aleksandrov D V *et al.* 1984 *Sov. J. Nucl. Phys.* **62** 18
[5] Belozerov A V *et al.* 1986 *Nucl. Phys.* A **460** 352
[6] Gornov M G *et al.* 2000 *Nucl. Instrum. Meth.* A **446** 461
[7] Lane A M and Thomas R G 1958 *Rev. Mod. Phys.* **30** 257
[8] Tilley D *et al.* 2002 *Nucl. Phys.* A **708** 3

^{8}Be and ^{9}B nuclei in dissociation of relativistic ^{10}B and ^{11}C nuclei

D A Artemenkov[1,2], V Bradnova[1], E Firu[3], N K Kornegrutsa[1], M Haiduc[3], K Z Mamatkulov[1], R R Kattabekov[1], A Neagu[3], P A Rukoyatkin[1], V V Rusakova[1], R Stanoeva[4], A A Zaitsev[1,5], P I Zarubin[1,5,6] and I G Zarubina[1,5]

[1] Joint Institute for Nuclear Research, Dubna, Russia
[2] National Research Nuclear University MEPhI (Moscow Engineering Physics Institute), Kashirskoe highway 31, Moscow, 115409, Russia
[3] Institute of Space Science, Magurele, Romania
[4] South-Western University, Blagoevgrad, Bulgaria
[5] P. N. Lebedev Physical Institute of the Russian Academy of Sciences, Moscow, Russia

E-mail: zarubin@lhe.jinr.ru

Abstract. Progress in the study of nuclear clustering in the relativistic ^{10}B and ^{11}C nuclei dissociation in nuclear track emulsion is presented. The contribution of the unbound ^{8}Be and ^{9}B nuclei to their structure is determined on the basis of measurements of the emission angles of relativistic He and H fragments.

1. Introduction

Nuclear track emulsion (NTE) is exposed at the JINR Nuclotron to relativistic nuclei ^{7,9}Be, ^{8,10}B, ^{10,11}C and ^{12}N for the experimental study of the evolution of the cluster structure of light nuclei [reviewed in ref. [1], [2]]. Virtual cluster configurations in an incident nucleus are completely captured in coherent dissociation in which the target nucleus is not destroyed visibly ("white"stars, an example in figure 1). Therefore, the contributions of cluster states to the structure of the nucleus in question can be estimated by the probability of appearance of respective fragment ensembles. The nuclei ^{7}Be, ^{8}Be in the ground (^{8}Be$_{g.s.}$) and first excited states (^{8}Be$_{2+}$), as well as the ^{9}B may serve as a basis for the neighboring nuclei contributing to them with particular probabilities. Verification of this concept can be carried out on the sequence of the nuclei ^{9}Be, ^{10}B, ^{10,11}C and ^{12}N.

Reconstruction of the decays of relativistic ^{8}Be and ^{9}B nuclei is possible by the energy variable $Q = M^* - M$, where $M^{*2} = (\Sigma P_j)^2 = \Sigma(P_i \cdot P_k)$, M the total mass of fragments, and their $P_{i,k}$ 4-momenta defined under the assumption of conservation of an initial momentum per nucleon by fragments. For the "white"stars of ^{9}Be and ^{10}C nuclei the assumption that He fragments correspond to ^{4}He, and H ones in ^{10}C – ^{1}He is justified. Then ^{8}Be and ^{9}B identification is reduced to measurements the opening angles of between the directions of fragment emission.

Distributions over the opening angle Θ_{2He} for pairs of He fragments of "white"stars ^{9}Be→2He and ^{10}C → 2He + 2H (82% of the ^{10}C statistics) produced at energy of 1.2 A GeV are presented

[6] To whom any correspondence should be addressed.

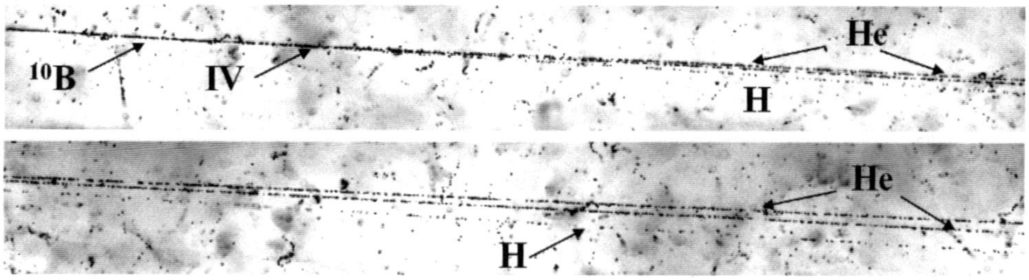

Figure 1. Macro photo of the coherent dissociation event ("white"star) of 1 A GeV ^{10}B nucleus into He nucleus pair and single H nucleus; IV - approximate position of the dissociation vertex ($\Theta_{2\alpha}$ = 5.3 mrad, $Q_{2\alpha}$ = 87 keV, $Q_{2\alpha p}$ = 352 keV)

in figure 2. In both cases the values of Θ_{2He} of 75-80% of the pairs are distributed about equally in the intervals of $0 < \Theta_{n(arrow)} < 10.5$ mrad and $15.0 < \Theta_{w(ide)} < 45.0$ mrad. The remaining pairs are attributed to a "medium"$10.5 < \Theta_m < 15.0$ and "widest"of $15.0 < \Theta_{vw} < 45.0$ intervals. The Q distribution directly connected with Θ_{2He} point out that "narrow"pairs of Θ_n are produced via ^{8}Be$_{g.s.}$ while pairs Θ_w via ^{8}Be$_{2+}$. There is a peak in the interval Θ_m reflecting the level 5/2- (2.43 MeV) of ^{9}Be. Fractions of events in the intervals Θ_n and Θ_w are equal to 0.56 ± 0.04 and 0.44 ± 0.04 for ^{9}Be, while for ^{10}C 0.49 ± 0.06 and 0.51 ± 0.06, i. e. they practically coincide. They indicate to a simultaneous presence of virtual ^{8}Be$_{g.s.}$ and ^{8}Be$_{2+}$ states in the ground states of the ^{9}Be and ^{10}C nuclei. Earlier, basing on the $Q_{2\alpha}$ energy distribution of the triples $2\alpha + p$ from the "white"stars $^{10}\mathrm{C} \to 2\alpha + 2p$ it was concluded that in the structure of the ^{10}C nucleus the core ^{9}B is manifested with a probability of around (30 ± 4)%, and the ^{8}Be$_{g.s.}$ decays are arise only through the ^{9}B decays. These conclusions allow one to interpret a significant fraction of "white"stars produced by ^{11}C and ^{10}B nuclei only on the basis of angular measurements.

2. Dissociation of relativistic ^{11}C nuclei

It is already established that 144 "white"stars produced by the ^{11}C in NTE are distributed over the charge channels in the following way: 2He + 2H (50%), 3He (17%), ^{7}Be + He (13%), He + 4H (11%), B+H (5%), Li + He + H (3%), 6H (2%). The distributions of He fragments over the opening angle Θ_{2He} (Figure 3) show that ^{8}Be$_{g.s.}$ decays are presented in 21% 2He + 2H and 19% in the 3He events. These distributions allow one to assume a strong contribution of ^{8}Be$_{2+}$ decays but it is a subject of future consideration.

The ^{9}B nucleus can exist in ^{11}C as an independent virtual component or as a component of a virtual basis ^{10}B. Decays ^{9}B$_{g.s.}$ in "white"stars $^{10}\mathrm{C} \to$ 2He + 2H are identified in accordance with a limitation on the opening angle between directions of ^{8}Be$_{g.s.}$ and each H fragments $\Theta(^8\mathrm{Be}_{g.s.} + \mathrm{H}) < 40$ mrad (figure 4) [3]. Application of such a condition the "white"stars $^{11}\mathrm{C} \to$ 2He + 2H allows one to identify 20 ^{9}B$_{g.s.}$ decays (figure 4) constituting 30% of events in this charge channel or 18% of the ^{11}C "white"stars.

3. Dissociation of relativistic ^{10}B nuclei

An analysis of the NTE exposure to 1 A GeV ^{10}B nuclei has pointed out that triples 2He + H (about 65%) dominate among "white"stars. However, the nature of this effect has not been studied, being in the "shadow"of studies with radioactive nuclei. In connection with the discussed analysis the balance should be established of probabilities of ^{10}B coherent dissociation via decays ^{8}Be$_{g.s.}$, ^{8}Be$_{2+}$ and ^{9}B. Recently, 250 2He + H "white"stars are selected in an

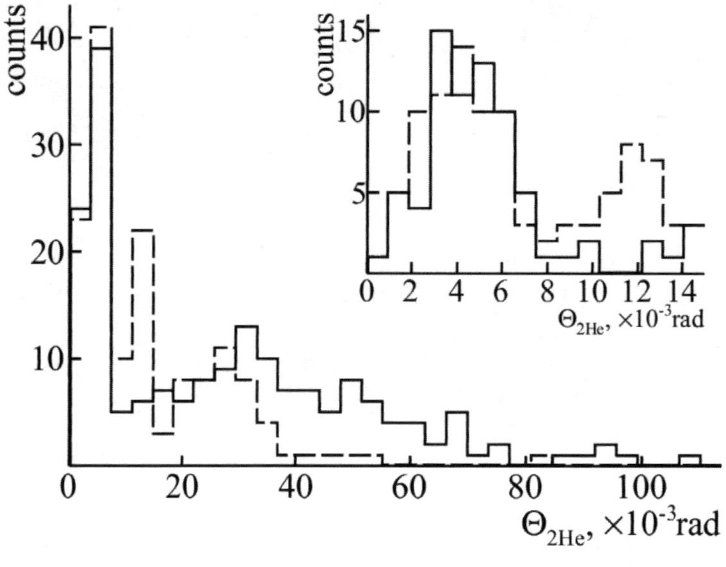

Figure 2. Distributions over the opening angle Θ_{2He} of α-particle pairs in "white"stars $^{10}C \to$ 2He + 2H (solid histogram) and $^{9}Be \to$ 2He (dashed histogram) at 1.2 A GeV; insertion part of the distribution in an interval Θ_n

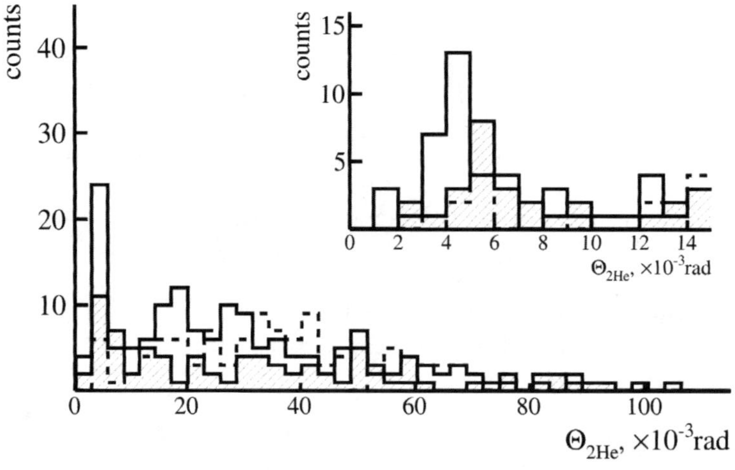

Figure 3. Distribution over the opening angle Θ_{2He} of α-particle pairs in "white"stars $^{11}C \to$ 2He + 2H (solid histogram), $^{11}C \to$ 3He (dashed histogram) at 1.2 A GeV and $^{10}B \to$ 2He + H (hatched histogram) at 1 A GeV; insertion part of the distribution in interval Θ_n.

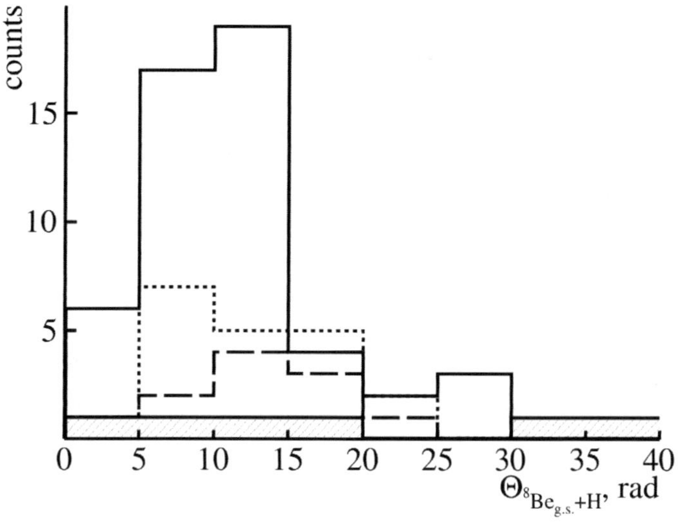

Figure 4. Distributions over the opening angle $\Theta(^8\mathrm{Be}_{g.s.}$ + H) in "white"stars ^{10}C → $^8\mathrm{Be}_{g.s.}$ + 2H (solid histogram), ^{11}C → $^8\mathrm{Be}_{g.s.}$ + 2H (dashed histogram), ^{10}B → $^8\mathrm{Be}_{g.s.}$ + H (hatched histogram) all found stars ^{11}C → $^8\mathrm{Be}_{g.s.}$ + 2H (dotted histogram)

accelerated search. Measurements of the first 84 stars pointed to four decays $^8\mathrm{Be}_{g.s.}$ (figure 3), six of which originated from $^9\mathrm{B}_{g.s.}$ decays (figure 4). Perhaps, decays $^9\mathrm{B}_{g.s.}$ will be the main source decays $^8\mathrm{Be}_{g.s.}$ in ^{10}B "white"stars as in the ^{10,11}C cases. Then the cluster configuration involving the deuteron $^8\mathrm{Be}_{2+}$ + d can be a source of $^8\mathrm{Be}_{2+}$ decays. Measurements of "white"stars ^{10}B, including identification of He and H isotopes by a multiple scattering method, are in progress now. The distributions of the energy of α-particle pairs, $Q_{2\alpha}$ and triples $2\alpha + p$, $Q_{2\alpha p}$ from the ^{10}B →2 $\alpha + p$ events for ongoing experiment are shown in figures 5 and 6 respectively.

Consideration of the nucleosynthesis toward ^{10,11}B, ^{11,10}C and ^{12}N in the "hot break-out"^{7}Be(^{3}He,γ)^{10}C(e$^+$,ν)^{10}B assists to recognize the relationship of their structures. The increase of α-clustering in ^{10}C provides a "window"for the synthesis via the intermediate states ^{9}B + p, $^8\mathrm{Be}_{2+}$ + 2p and ^{6}Be + α. These clusters are preserved in subsequent reactions ^{10}C(e$^+$,ν)^{10}B(p,γ)^{11}C(e$^+$,ν)^{11}B. "Window"of the reaction ^{7}Be(^{4}He,γ)^{11}C allows only association of the ^{7}Be and ^{4}He clusters, also contributing to the ^{11}C and ^{10}B structure. Thus a hidden variety of virtual configurations in the nuclei ^{10,11}C and ^{10,11}B arises. In turn, these nuclei provide a basis for capture reactions of protons or He isotopes (or in neutron exchange) for synthesis of the subsequent nuclei which leads to translation of the preceding structures.

Within the framework of the relativistic approach the following picture of nuclear clustering emerges. As the fundamental elements of its structure atomic nuclei contain a virtual association of nucleons and clusters. Their simplest observable manifestations the lightest nuclei ^{4,3}He and ^{3,2}H having no excited states. Superposition of lightest clusters and nucleons form ^{7}Be, ^{8}Be in the ground and first excited states and ^{9}B ones which, in turn, serve as composing clusters. A balance of possible superpositions in states with an appropriate spin and parity determine binding and ground state parameters of corresponding nuclei. Further joining of nucleons and the lightest nuclei leads to shell type structure. Interlacing cluster and shell degrees of freedom does a group of nuclei in the beginning of the table of isotopes of a "laboratory"of nuclear

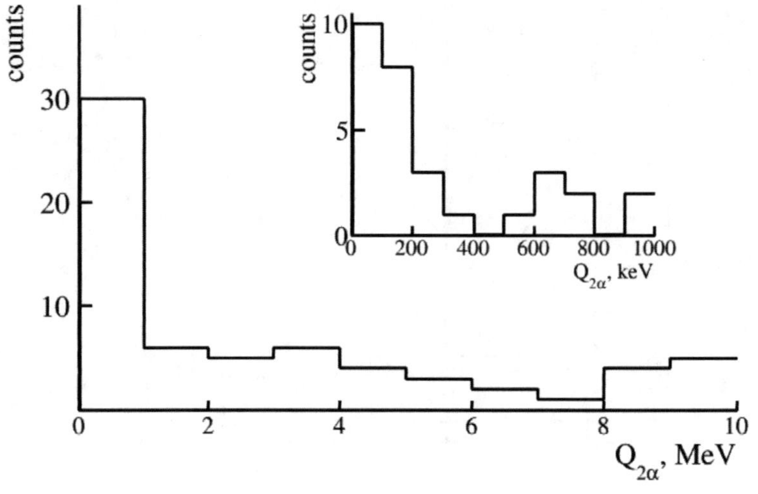

Figure 5. Distribution of the energy $Q_{2\alpha}$ of α-pairs from the $^{10}B \rightarrow 2\ \alpha + p$ events

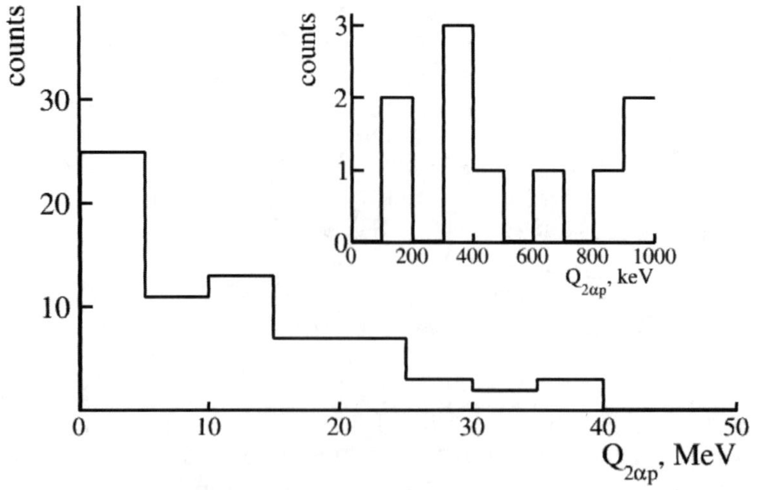

Figure 6. Distribution of the total energy $Q_{2\alpha p}$ of the triples $2\alpha + p$ from the $^{10}B \rightarrow 2\ \alpha + p$ events

quantum mechanics and nuclear astrophysics.

Acknowledgments
The authors are grateful to A. I. Malakhov (JINR), N. G. Polukhina and S. P. Kharlamov (LPI) for their support and critical discussion of the results. This work was supported by the grant from the Russian Foundation for Basic Research 12-02-00067 and 16-02-00062 and grant of plenipotentiary representatives of the government of Bulgaria, Egypt, Romania and the Czech Republic at JINR.

References

[1] Zarubin P I 2013 *Clusters in Nuclei, Vol 3* (Lecture Notes in Physics Vol 875) ed Beck Christian (Cham: Springer Int. Publ.) p 51 (*arXiv*:1309.4881)
[2] Artemenkov D A *et al.* 2015 *Phys. At. Nucl.* **78** 794 (*arXiv*:11411.5806)
[3] Toshito T *et al.* 2008 *Phys. Rev.* C **78** 067602

Search for exotic cluster configurations in ^{14}C nucleus

L Yu Korotkova, B A Chernyshev, Yu B Gurov, V S Karpuhin, S V Lapushkin, R V Pritula and T D Schurenkova
National Research Nuclear University MEPhI (Moscow Engineering Physics Institute), Kashirskoe highway 31, Moscow, 115409, Russia

E-mail: korotkovalara@gmail.com

Abstract. The analysis of 2-dimentional Dalitz' diagram, measured in ^{14}C(π^-, pd)X reaction, allowed to distinguish the pion absorption by ^{3}p intranuclear cluster and to obtain an indication on the existence of ^{3}p + ^{11}Li configuration in ^{14}C nucleus. Highly excited states of ^{12,13}Be isotopes were found with the energy of $E_x \approx 30$ MeV for the first time. It was shown that these states decay as follows ^{12}Be* $\rightarrow$ p + ^{11}Li and ^{13}Be* $\rightarrow$ d + ^{11}Li.

1. Introduction

Carbon isotope ^{14}C is of interest for the study of light neutron-rich nuclei. The Antisymmetrized Molecular Dynamics (AMD) model predicts ^{14}C$_{g.s.}$ to be a system of "3α + 2n" [1]. This structure manifests itself as a triangle of alpha particles and neutrons between [2]. Information about the structure of ^{14}C is also important in studies for the heavier carbon isotopes; ^{14}C is "core" for them [3].

Stopped pion absorption is a multinucleon process, so this reaction is useful in order to search clusters in nuclei. In this paper, we look for indications of the existence of "^{3}p + ^{11}Li" configuration in the ground state of ^{14}C$_{g.s.}$ For this purpose, we performed the analysis of the results of correlation measurements of ^{14}C(π^-, pd)X reaction. The search method is based on the allocation of the events of quasi-free pion absorption by intranuclear cluster ^{3}p (π^- + ^{3}p $\rightarrow$ p + d), where the residual nucleus ^{11}Li is a "spectator". Earlier, this method allowed to obtain indications on the existence of "^{3}He + ^{6}He" configuration in ^{9}Be [4] and "^{5}Li + ^{6}He" - in ^{11}B [5].

Analysis of the inclusive measurements of ^{14}C(π^-, p)X and ^{14}C(π^-, d)X allows to search for highly excited states of the beryllium isotopes ^{12,13}Be. Information about these states is of interest in studies of cluster resonances and its decays, as well as of the deformation of nuclear systems. Research of cascade channels of pion absorption π^- + ^{14}C $\rightarrow$ d + ^{12}Be* $\rightarrow$ d + p + ^{11}Li and π^- + ^{14}C $\rightarrow$ p + ^{13}Be* $\rightarrow$ p + d + ^{11}Li can provide data on excited levels and rare cluster decays of ^{12,13}Be. Earlier, we used this method to found lithium decay modes ^{8}Li $\rightarrow$ t + ^{5}He [6] and ^{9}Li $\rightarrow$ t + ^{6}He [7].

2. Search for «^{3}p + ^{11}Li» configuration in ^{14}C nucleus

Search for "^{3}p + ^{11}Li" configuration was carried out in the correlation measurements of ^{14}C(π^-, pd)X reaction. Results were obtained with a beam of low energy π^--mesons (LEP) using two-arm semiconductor spectrometer [8]. The contribution of the impurity ^{12}C in ^{14}C target was subtracted, using the procedure described in [9]. This method is based on measurements of ^{12}C(π^-, pd)X obtained in the same experimental run. The data obtained are presented in figure 1 as 2-dimensional distribution

of the kinetic energies of pd-pairs (Dalitz' diagram). On the kinematic boundary of the plot one can see the following areas with the formation of three particles in the final state:
1 – 15 MeV $\leq E_d \leq$ 50 MeV, 30 MeV $\leq E_p \leq$ 60 MeV - quasi-free pion absorption by intranuclear cluster ^{3}p (π^- + ^{3}p $\rightarrow$ p + d); where the momentums $p_p \approx p_d$, and the residual nucleus ^{11}Li does not take part in the reaction;
2 – $E_d \approx$ 65 MeV, $E_p \leq$ 15 MeV corresponding to 2-body reaction channel π^- + ^{14}C $\rightarrow$ d + ^{12}Be* with the following decay ^{12}Be$^*\rightarrow$ p + ^{11}Li;
3 – $E_p \approx$ 70 MeV, $E_d \leq$ 10 MeV corresponding to 2-body reaction channel π^- + ^{14}C $\rightarrow$ p + ^{13}Be* with the following decay ^{13}Be$^*\rightarrow$ d + ^{11}Li.

In order to distinguish the reaction channels, responsible for the increase in population density in the region 1 on Dalitz' diagram, we considered the missing-mass spectrum (*MM*) shown in figure 2. The reference point here is the mass of ^{11}Li$_{g.s.}$. Peak in the area of $MM \approx$ 0 MeV is due to 3-body channel of ^{14}C(π^-, pd)^{11}Li reaction with the formation of the ground and the first excited state of ^{11}Li with E_x = 0.9 ± 0.1 MeV. Analysis shows that the events lying in the range of - 1.0 MeV < MM < 2.0 MeV populate mainly the region 1 in figure 1. This indicates that the dominant contribution in forming low-lying states of ^{11}Li is quasi-free absorption channels.

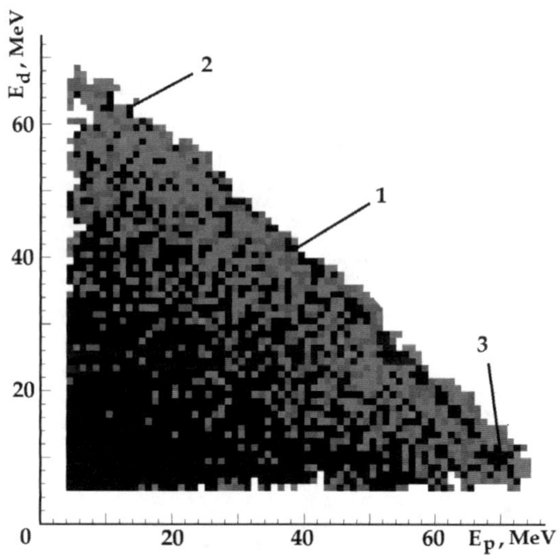

Figure 1. Experimental 2-dimensional distribution of events for ^{14}C(π^-, pd)X reaction. The darker points on the chart correspond to more events.

Additional evidence of "^{3}p + ^{11}Li" cluster configuration in ^{14}C can be obtained from the momentum distribution of the residual nucleus ^{11}Li, shown in figure 3. Satisfactory description of the peak observed in the region of small values of the momentum was achieved in the approximation of a harmonic oscillator:

$$F(p) \sim (p/p_F)^2 \exp(-(p/p_F)^2). \quad (1)$$

The best agreement with the experimental data is achieved with $p_F \approx$ 150 ± 10 MeV / c. This value is typical for intranuclear Fermi motion of clusters, which proves a significant contribution of quasi-free pion absorption by ^{3}p cluster: π^- + ^{3}p $\rightarrow$ p + d.

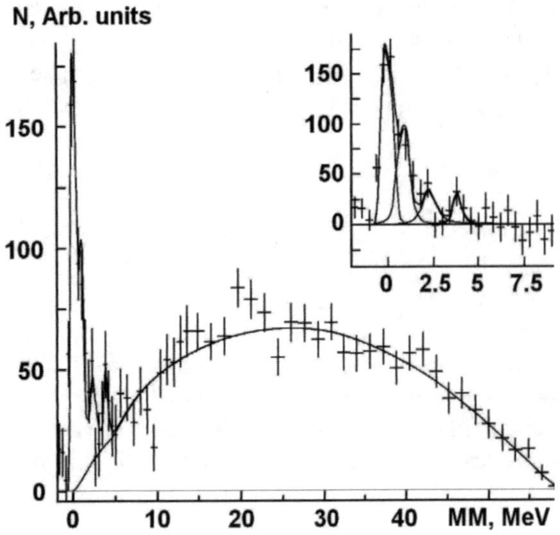

Figure 2. Missing mass spectrum for $^{14}C(\pi^-, pd)X$ reaction. Dots with error bars are the experimental data, the curve is the total fit.

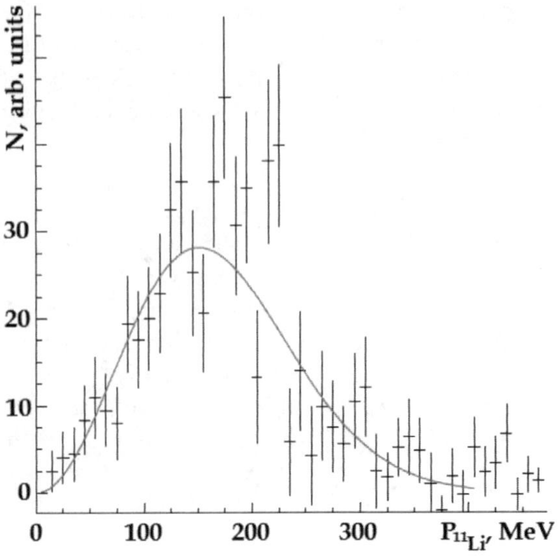

Figure 3. The momentum distribution of the residual nucleus ^{11}Li from $^{14}C(\pi^-, pd)X$ reaction for – 1.0 MeV < MM < 2.0 MeV events. Dots with error bars are the experimental data, the curve is the description of the spectrum using (1).

Thus, the analysis of the spectra shown in figure 2-3, suggests the presence of the configuration "$^3p + {}^{11}Li$" in ^{14}C nucleus. The properties of the intranuclear ^{11}Li system may be different from those of the isotope ^{11}Li in a free state.

3. Search for rare cluster decays of $^{12,13}Be$

In order to find highly excited states of $^{12,13}Be$ beryllium isotopes we analyzed *MM* spectra of $^{14}C(\pi^-, p)X$ and $^{14}C(\pi^-, d)X$ reactions, presented in figure 4. The spectra were obtained for the events from the areas 2 and 3 on the Dalitz' diagram (figure 1).

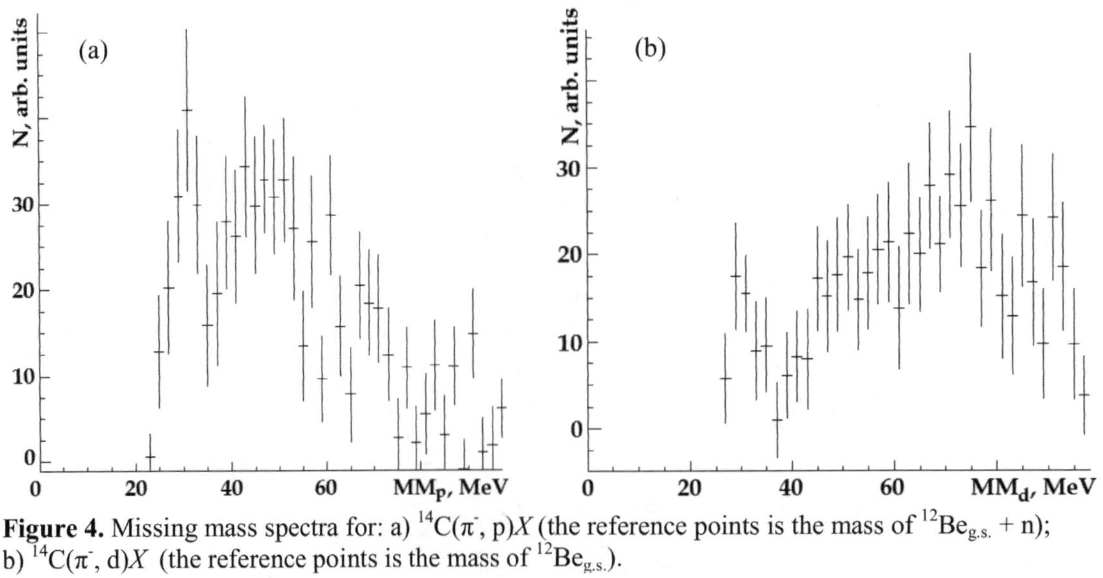

Figure 4. Missing mass spectra for: a) $^{14}C(\pi^-, p)X$ (the reference points is the mass of $^{12}Be_{g.s.} + n$); b) $^{14}C(\pi^-, d)X$ (the reference points is the mass of $^{12}Be_{g.s.}$).

The peaks in figure 4 are due to the 2-body reaction channels where the excited states of beryllium decay as $^{12(13)}$Be → p(d) + ^{11}Li (threshold is 23.20 (20.97) MeV). One can clearly see in both spectra the broad structure in the region of MM ≈ 30 MeV. Due to the lack of statistics, this structure cannot be interpreted as a manifestation of one or more excited levels.

Highly excited states of ^{12}Be, decay into p + ^{11}Li, were also observed in [10]. The authors describe the experimental spectrum with two excited levels with E_x = 25.0 MeV and 28.0 MeV [10]. The average value of the peak position (E_x ≈ 30 MeV), observed on figure 4b, is slightly higher than in [10], however, we do not exclude the presence of E_x = 28.0 MeV state in our spectrum.

The existence of highly excited ^{13}Be levels is first observed in our measurements. Note that the decay ^{13}Be → d + ^{11}Li is also observed for the first time.

4. Conclusion

Search for rare cluster configuration in ^{14}C nucleus was carried out in the correlation measurements of the stopped pion absorption reaction ^{14}C(π^-, pd)X. A study of 2-dimensional energy distribution for the particles (p, d) made possible to distinguish 2-body (π^-+ ^{14}C → p (d) + $^{13(12)}$Be*) and 3- body (π^-+ ^{14}C → p + d + ^{11}Li) reaction channels. Analysis of the 3-body channel, allowed to identify the pion absorption by a system of 3 protons (π^- + ^{3}p → p + d). This result points out on the presence of the configuration "^{3}p + ^{11}Li" in ^{14}C nucleus. The properties of the intranuclear ^{11}Li system may be different from those of the isotope ^{11}Li in a free state. We also investigated the cascade reaction channels π^-+ ^{14}C → d (p) + ^{12}Be → d (p) + p(d) + ^{11}Li. The indications on highly excited states of ^{12,13}Be with the energies E_x ≈ 30 MeV were obtained for the first time.

Acknowledgments

This work was performed within the framework of the Center of Fundamental Research and Particle Physics supported by MEPhI Academic Excellence Project (contract № 02.a03.21.0005, 27.08.2013).

References

[1] Suhara T and Kanada-En'yo Y 2011 *Journ. of Phys.: Conf. Ser.* **321** 012047
[2] von Oertzen W *et al.* 2004 *Eur. Phys. J.* A **21** 193
[3] Tanihata I *et al.* 2013 *Progr. in Part. and Nucl. Phys.* **68** 215
[4] Gurov Yu B *et al.* 2006 *JETP Lett.* **84** 1
[5] Gurov Yu B *et al.* 2014 *Bull. Rus. Acad. Sci. Phys.* **78** 355
[6] Gurov Yu B *et al.* 2015 *Bull. Rus. Acad. Sci. Phys.* **77** 366
[7] Gurov Yu B *et al.* 2015 *Bull. Rus. Acad. Sci. Phys.* **79** 474
[8] Gornov M G *et al.* 2000 *Nucl. Instrum. Meth.* A **446** 461
[9] Chernyshev B A *et al.* 2013 *Eur. Phys. J.* A **49** 68
[10] Charity R J *et al.* 2007 *Phys. Rev.* C **76** 064313

Journal of Physics: Conference Series **675** (2016) 022005 doi:10.1088/1742-6596/675/2/022005

Charged charmoniumlike states at Belle

K Chilikin

Nuclear Research Nuclear University MEPhI(Moscow Engineering Physics Institute), Kashirskoe highway 31, Moscow, 115409, Russia.

E-mail: chilikin@itep.ru

Abstract. Results on charged charmoniumlike states observed in B-meson decays at Belle are reviewed. The review also includes the results on these states from other experiments (BABAR and LHCb).

1. Introduction

Recently, a number of new states containing a $c\bar{c}$ quark pair have been observed. Among these states are charged charmoniumlike state candidates with a minimal quark content that is necessarily exotic: $|c\bar{c}u\bar{d}\rangle$. The results on charged charmoniumlike states observed in B-meson decays are reviewed below.

2. Charged charmoniumlike states

2.1. $Z_c(4430)^+$

The first observed charged charmoniumlike state was the $Z_c(4430)^+$ [1]. It was observed in 2007 by the Belle Collaboration in the decay mode $\bar{B} \to \psi(2S)\pi^+K$, where $\bar{B}$ is $\bar{B}^0$ or B^+ and K is K^- or K^0_S. A one-dimensional analysis was performed; the distribution of the invariant mass of the combination $(\psi(2S)\pi^+)$ was studied outside of the regions of the $K^*(892)$ and $K_2^*(1430)$. The distribution of the $\psi(2S)\pi^+$ invariant mass is shown in figure 1. A narrow peak with the parameters $M = 4433 \pm 4 \pm 2$ MeV$/c^2$, $\Gamma = 45^{+18}_{-13}{}^{+30}_{-13}$ MeV and the significance of 6.5σ was observed.

After that the BABAR Collaboration searched for the $Z_c(4430)^+$ in the decays $\bar{B} \to \psi(2S)\pi^+K$ and $\bar{B} \to J/\psi\pi^+K$ [2] using a different method. The data were divided into intervals in $K\pi^+$ invariant mass. The Legendre moments depending on the K^* helicity angle were determined for each interval, and the expected distribution of $\psi(2S)\pi^+$ invariant mas was reconstructed. The distribution of $M_{\psi(2S)\pi^+}$ observed in data was fitted to a sum of the expected distribution and the $Z_c(4430)^+$ signal. No significant $Z_c(4430)^+$ signal was observed.

The methods of the first Belle analysis and the BABAR analysis did not take into account the interference of the amplitudes of the decays via the $Z_c(4430)^+$ and K^*. To confirm the existence of the $Z_c(4430)^+$ more reliably, the Belle Collaboration performed a Dalitz analysis of the decays $\bar{B} \to \psi(2S)\pi^+K$ using the same data [3]. The observation of the $Z_c(4430)^+$ in Belle data was confirmed with 6.4σ significance. The $Z_c(4430)^+$ parameters have changed: the new values of the mass and width were $M = 4443^{+15}_{-12}{}^{+19}_{-13}$ MeV$/c^2$ and $\Gamma = 107^{+86}_{-43}{}^{+74}_{-56}$ MeV, respectively. Projections of the fit results onto the $M^2_{\psi(2S)\pi^+}$ axis with and without the $Z_c(4430)^+$ are shown in figure 2.

However, a Dalitz analysis is not sensitive to the $Z_c(4430)^+$ quantum numbers because of information loss due to the integration over the angular variables. A full amplitude analysis of $\bar{B}^0 \to \psi(2S)K^-\pi^+$ decays, which has a better sensitivity to the $Z_c(4430)^+$ quantum numbers, has been performed by the Belle Collaboration in [4]. The default model included the contributions of the K^* resonances and the $Z_c(4430)^+$ with quantum numbers 0^-, 1^-, 1^+, 2^- or 2^+. The hypothesis $J^P = 1^+$ was preferred and the hypotheses 0^-, 1^-, 2^- and 2^+ were excluded at the level 3.4σ, 3.7σ, 4.7σ and 5.1σ, respectively. Comparison of the fit results without the $Z_c(4430)^+$ and with the $Z_c(4430)^+$ ($J^P = 1^+$) is shown in figure 3. The resulting mass and width of the $Z_c(4430)^+$ were $M = 4485^{+22+28}_{-22-11}$ MeV/c^2 and $\Gamma = 200^{+41+26}_{-46-35}$ MeV, respectively.

This result was later confirmed by the LHCb Collaboration in [5]. In addition, the LHCb collaboration measured the $Z_c(4430)^+$ amplitude in six bins; the obtained Argand plot confirmed that the phase of the $Z_c(4430)^+$ amplitude rotates as expected for a resonance. The projection of the fit result onto $M^2_{\psi(2S)\pi^+}$ and the Argand plot are shown in figure 4b.

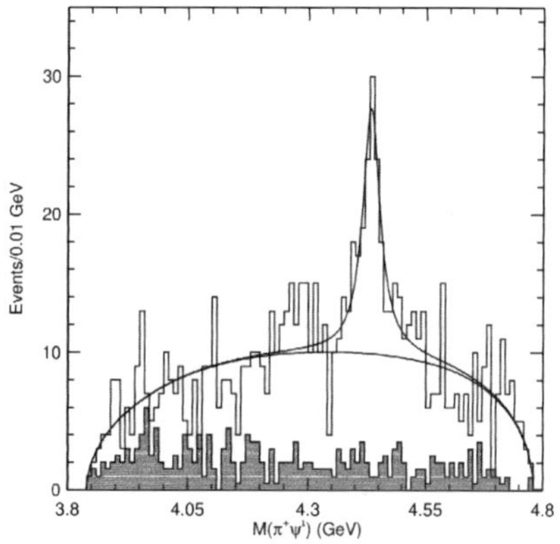

Figure 1. Results of the fit to the invariant mass of the system $\psi(2S)\pi^+$ (from [1] by the Belle Collaboration).

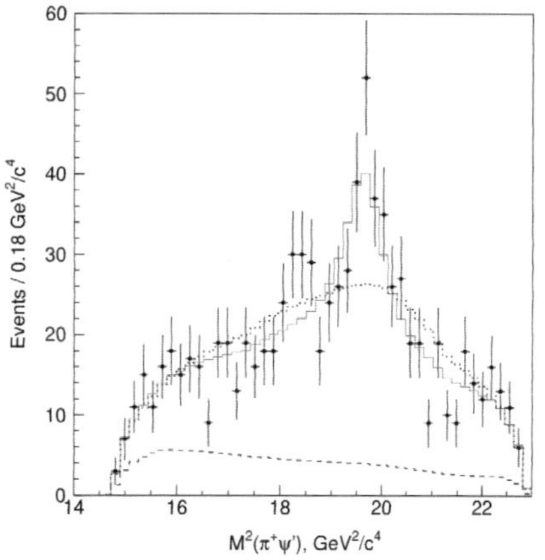

Figure 2. Projection of the Dalitz plot fit results without (dashed line) and with (solid line) the $Z_c(4430)^+$ onto the $M^2_{\psi(2S)\pi^+}$ axis (from [3] by the Belle Collaboration).

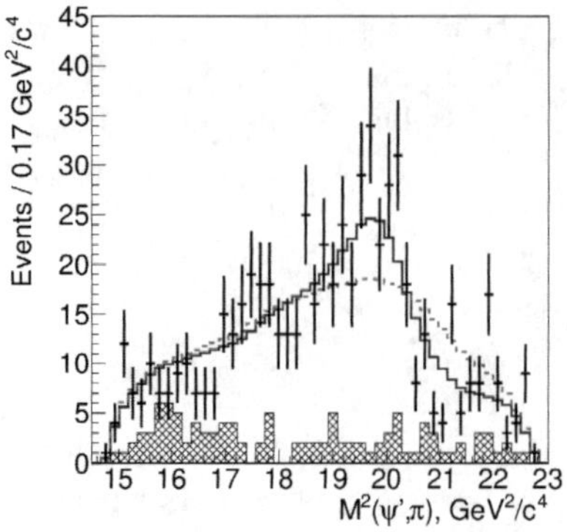

Figure 3. Projection of the four-dimensional fit results without (dashed line) and with (solid line) the $Z_c(4430)^+$ onto $M^2_{\psi(2S)\pi^+}$ (from [4] by the Belle Collaboration).

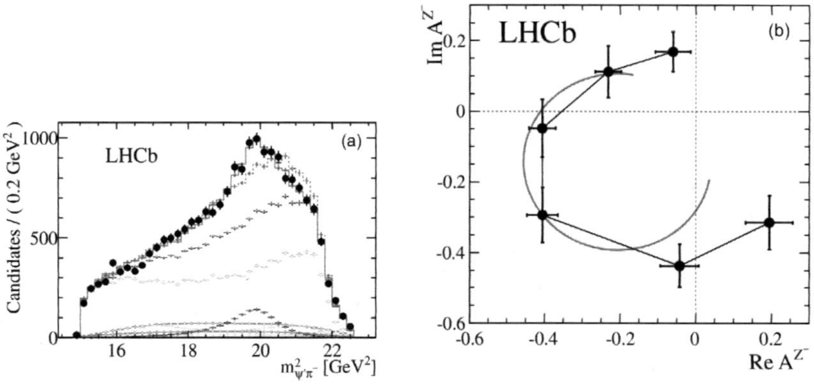

Figure 4. (a) Projection of the four-dimensional fit results with with the $Z_c(4430)^+$ onto $M^2_{\psi(2S)\pi^+}$ with the contributions of individual resonances and the $Z_c(4430)^+$. (b) Argand plot by the LHCb Collaboration [5].

2.2. $Z_c(4050)^+$ *and* $Z_c(4250)^+$

Two charged charmoniumlike states, the $Z_c(4050)^+$ and $Z_c(4250)^+$, were observed by the Belle Collaboration in the decays $\bar{B}^0 \to \chi_{c1} K^- \pi^+$ in [6]. A Dalitz analysis similar to [3] has been performed. Two new states with masses and widths $M = 4051 \pm 14^{+20}_{-41}$ MeV/c^2, $\Gamma = 82^{+21}_{-17}{}^{+47}_{-22}$ MeV and $M = 4248^{+44}_{-29}{}^{+180}_{-35}$ MeV/c^2, $\Gamma = 177^{+54}_{-39}{}^{+316}_{-61}$ MeV, respectively, were observed. The significance of the $Z_c(4250)^+$ was 6.2σ and the significance of the $Z_c(4050)^+$ was 5.0σ. The projections of the fit results with and without new resonances onto $M_{\chi_{c1}\pi^+}$ with the condition 1.0 GeV$^2/c^4 < M^2_{K^-\pi^+} < 1.75$ GeV$^2/c^4$ are shown in figure 5.

The BABAR Collaboration searched for the $Z_c(4050)^+$ and $Z_c(4250)^+$ in [7]. A method similar to [2] was used. No significant signal was observed.

2.3. $Z_c(4200)^+$

The Belle Collaboration performed an amplitude analysis of the decays $\bar{B}^0 \to J/\psi K^- \pi^+$ in [8]. A new charmoniumlike state $Z_c(4200)^+$ with quantum numbers $J^P = 1^+$ was observed.

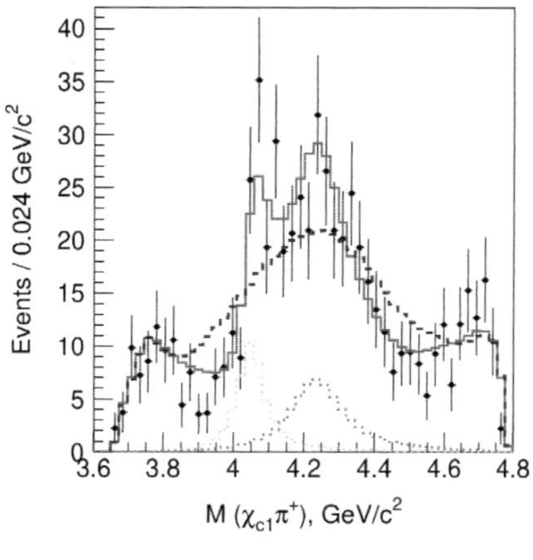

Figure 5. Projection of the Dalitz plot fit results without (dashed line) and with (solid line) the $Z_c(4050)^+$ and $Z_c(4250)^+$ onto the $M^2_{\chi_{c1}\pi^+}$ axis and the $Z_c(4050)^+$ and $Z_c(4250)^+$ contributions (from [6] by the Belle Collaboration).

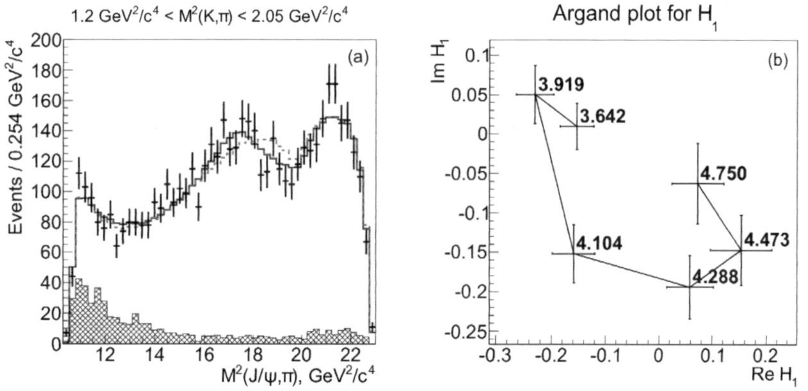

Figure 6. (a) Projection of the four-dimensional fit results without (dashed line) and with (solid line) the $Z_c(4200)^+$ onto $M^2_{J/\psi\pi^+}$ and the $Z_c(4200)^+$. (b) Argand plot for the helicity amplitude H_1 by the Belle Collaboration [8].

Its mass and width were measured to be $M = 4196^{+31+17}_{-29-13}$ MeV/c^2 and $\Gamma = 370^{+70+70}_{-70-132}$ MeV, respectively. The significance of the $Z_c(4200)^+$ was 7.9σ for the default model and $6.2\,\sigma$ including the systematic error. Comparison of the fit results for the models without the $Z_c(4200)^+$ and with the $Z_c(4200)^+$ ($J^P = 1^+$) is shown in figure 6. Other $Z_c(4200)^+$ quantum number hypotheses: 0^-, 1^-, 2^- and 2^+ were excluded at the levels of 6.1σ, 7.4σ, 4.4σ and 7.0σ, respectively. Also, evidence for the decay $Z_c(4430)^+ \to J/\psi\pi^+$ was found.

Similarly to LHCb analysis in [5], the helicity amplitudes of the $Z_c(4200)^+$ were measured in six bins. The absolute value and phase of the helicity amplitude H_1 change in accordance with the expectations for a resonance, while it is not possible to draw any conclusions for the helicity amplitude H_0 due to large relative errors. The Argand plot for the helicity amplitude H_1 is shown in figure 6b.

3. Conclusions

Though the first charged charmoniumlike state, the $Z_c(4430)^+$, is now confirmed, the available experimental information for these states is incomplete. Further measurements of the properties of charged charmoniumlike states and searches for their decays into different final states are required for their theoretical interpretation.

Acknowledgements

This work is supported by Russian Science Foundation grant (project №15-12-30014).

References

[1] Choi S K *et al.* (Belle Collaboration) 2008 *Phys. Rev. Lett.* **100** 142001
[2] Aubert B *et al.* (BABAR Collaboration) 2009 *Phys. Rev.* D **79** 112001
[3] Mizuk R *et al.* (Belle Collaboration) 2009 *Phys. Rev.* D **80** 031104
[4] Chilikin K *et al.* (Belle Collaboration) 2013 *Phys. Rev.* D **88** 074026
[5] Aaij R *et al.* (LHCb Collaboration) 2014 *Phys. Rev. Lett.* **112** 222002
[6] Mizuk R *et al.* [Belle Collaboration] 2008 *Phys. Rev.* D **78** 072004
[7] Lees J P *et al.* (BABAR Collaboration) 2012 *Phys. Rev.* D **85** 052003
[8] Chilikin K *et al.* (Belle Collaboration) 2014 *Phys. Rev.* D **90** 112009

Bottomonium(-like) states at e^+e^- colliders

R V Mizuk

National Research Nuclear University MEPhI (Moscow Engineering Physics Institute), Kashirskoe highway 31, Moscow, 115409, Russia

E-mail: mizuk@itep.ru

Abstract. We review recent results on bottomonium(-like) states from e^+e^- colliders. They include energy scan of the e^+e^- annihilation cross sections into $\Upsilon(nS)\pi^+\pi^-$ and $h_b(nP)\pi^+\pi^-$ final states, studies of transitions from $\Upsilon(4S)$ resonance with emission of η meson and update of $\Upsilon(5S) \to B^{(*)}\bar{B}^*\pi$ analysis from Belle.

1. Introduction

Studies of bottomonium were always considered to be an important test of our understanding of hadron spectroscopy [1]. Indeed, bottomonium represents one of the simplest hadronic systems that can be considered as approximately non-relativistic due to high mass of b quarks. In addition, there are many narrow and well separated states, thus the system provides a rich phenomenological material. Recently, several low-lying spin-singlet states were observed by the Belle experiment [2, 3].

In this review we will concentrate on the properties of highly excited states with masses above the $B\bar{B}$ threshold. During last decade it became clear that such states posess unexpected for quarkonium properties. We will consider energy scans in the $\Upsilon(10860)$ and $\Upsilon(11020)$ regions, η transitions between bottomonium states and properties of charged exotic bottomonium-like states $Z_b(10610)$ and $Z_b(10650)$.

2. Energy scan in the $\Upsilon(10860)$ and $\Upsilon(11020)$ region

In the data collected at the $\Upsilon(10860)$ resonance, the Belle collaboration observed numerous hadronic processes that involve production of bottomonia or charged bottomonium-like states. These are e^+e^- annihilations into $\Upsilon(nS)\pi^+\pi^-$ ($n = 1, 2, 3$) and $\Upsilon(1S)K^+K^-$ [4], $h_b(nP)\pi^+\pi^-$ ($n = 1, 2$) [2], $\chi_{bJ}(1P)\omega$ [5], $\Upsilon(nS)\eta$ ($n = 1, 2$) and $\Upsilon(1D)\eta$, $Z_b(10610, 10650)^+\pi^-$ [6, 7, 8]. The energy dependence of the cross section was measured only for the $\Upsilon(nS)\pi^+\pi^-$ channels [9, 10], with an observation of clear $\Upsilon(10860)$ peak. For other channels the question of whether the final states are produced from resonances, the continuum, or both, remains open.

An energy scan of the $e^+e^- \to \Upsilon(nS)\pi^+\pi^-$ cross sections revealed that they peak also at the $\Upsilon(11020)$ resonance [10]. Thus one can expect that other processes that were observed at $\Upsilon(10860)$ are also enhanced at the $\Upsilon(11020)$.

The rates of all the above processes are anomalously large, which suggests exotic multiquark admixtures in the $\Upsilon(10860)$ and $\Upsilon(11020)$ wave functions [1]. Further studies, in particular the comparison of the $\Upsilon(10860)$ and $\Upsilon(11020)$ properties, may be helpful for a better understanding of highly excited quarkonium(-like) states. Since the $\Upsilon(10860)$ [$\Upsilon(11020)$] is expected to contain a $b\bar{b}$ pair in the $5S$ [$6S$] spin-triplet state, in this paper it is referred to for brevity as the $\Upsilon(5S)$ [$\Upsilon(6S)$].

The Belle experiment reported preliminary results on the energy dependence of the $e^+e^- \to h_b(nP)\pi^+\pi^-$ ($n = 1, 2$) cross sections [11]. The $h_b(nP)\pi^+\pi^-$ final states were reconstructed inclusively using missing mass of the $\pi^+\pi^-$ pairs. The requirement that the reaction proceeds via intermediate Z_b states, $e^+e^- \to Z_b\pi \to h_b(nP)\pi^+\pi^-$, was also applied: $10.59\,\mathrm{GeV}/c^2 < M_{\mathrm{miss}}(\pi^+) < 10.67\,\mathrm{GeV}/c^2$. The measured Born cross sections are shown in figure 1. The

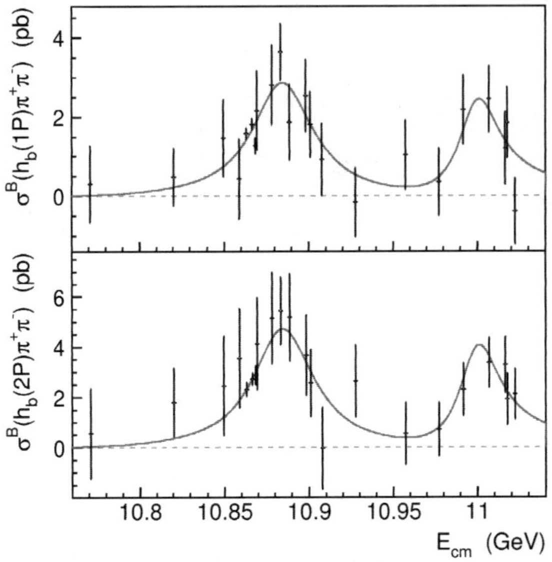

Figure 1. (colored online) Cross sections for the $e^+e^- \to h_b(1P)\pi^+\pi^-$ (top) and $e^+e^- \to h_b(2P)\pi^+\pi^-$ (bottom) processes as a function of center-of-mass energy [11]. Points with error bars are the data, red solid curves are the fit results.

energy dependence of the $h_b(1P)\pi^+\pi^-$ and $h_b(2P)\pi^+\pi^-$ cross sections looks very similar. It shows a two-peak structure without any significant non-resonant continuum contribution. A simultaneous fit to the energy dependence of the $e^+e^- \to h_b(nP)\pi^+\pi^-$ ($n = 1, 2$) cross sections is performed. A fit function is a coherent sum of two Breit-Wigner amplitudes and (optionally) a constant with an energy continuum contribution. The significance of the non-resonant continuum contribution is found to be $1.5\,\sigma$ only. Thus the default fit function does not include the continuum contribution, and it is considered only for estimation of a systematic uncertainty on the resonance parameters. The fit results for the default model are given in table 1. Resonance parameters are consistent with those of the $\Upsilon(5S)$ and $\Upsilon(6S)$ measured in the $\Upsilon(nS)\pi^+\pi^-$ final state, so the observed structures are identified as $\Upsilon(5S)$ and $\Upsilon(6S)$.

The shapes of the $e^+e^- \to h_b(nP)\pi^+\pi^-$ and $e^+e^- \to \Upsilon(nS)\pi^+\pi^-$ [10] cross sections look very similar. The most prominent difference is a smaller relative yield of the $\Upsilon(6S)$ compared to the $\Upsilon(5S)$ in the $e^+e^- \to \Upsilon(nS)\pi^+\pi^-$ cross section. Since the $h_b(nP)\pi^+\pi^-$ channel is produced only via intermediate Z_b states, while in the $\Upsilon(nS)\pi^+\pi^-$ channel there are both Z_b and non-resonant contributions, this difference can be explained if the non-resonant contribution in the $\Upsilon(nS)\pi^+\pi^-$ channel is suppressed at $\Upsilon(6S)$.

In the charmonium region the difference between the shapes of the $e^+e^- \to h_c\pi^+\pi^-$ and $e^+e^- \to J/\psi\pi^+\pi^-$ cross sections is much more pronounced, with possible new peaks and a non-resonant contribution to the $h_c\pi^+\pi^-$ process [12]. There is at this time no clear explanation of

Table 1. The masses and widths of the $\Upsilon(5S)$ and $\Upsilon(6S)$ resonances measured from energy scans of $e^+e^- \to \Upsilon(nS)\pi^+\pi^-$ $(n = 1, 2, 3)$ [10] and $e^+e^- \to h_b(nP)\pi^+\pi^-$ $(n = 1, 2)$ [11].

Parameter	$h_b(nP)\pi^+\pi^-$	$\Upsilon(nS)\pi^+\pi^-$
$M_{\Upsilon(5S)}$, MeV/c^2	$10884.7^{+3.2+8.6}_{-2.9-0.6}$	$10891.1 \pm 3.2^{+0.5}_{-1.5}$
$\Gamma_{\Upsilon(5S)}$, MeV	$44.2^{+11.9+2.2}_{-7.8\ -15.8}$	$53.7^{+7.1+0.9}_{-5.6-5.4}$
$M_{\Upsilon(6S)}$, MeV/c^2	$10998.6 \pm 6.1^{+16.1}_{-1.1}$	$10987.5^{+6.4+9.0}_{-2.5-2.1}$
$\Gamma_{\Upsilon(6S)}$, MeV	29^{+20+2}_{-12-7}	$61^{+9\ +2}_{-19-20}$

this difference between charmonium and bottomonium.

3. Observation of the $\Upsilon(4S) \to h_b(1P)\eta$ transitions

Hadronic transitions between the lowest mass quarkonium levels can be described using the QCD multipole expansion (ME) [13]. In this approach, the heavy quarks emit two gluons that subsequently transform into light hadrons. The $\pi\pi$ and η transitions between the vector states proceed via emission of $E1E1$ and $E1M2$ gluons, respectively. Therefore, η transitions are highly suppressed as they require a spin flip of the heavy quark [14]. Indeed, the ratio of branching fractions

$$\mathcal{R}^{\eta S}_{\pi\pi S}(n, m) = \frac{\mathcal{B}[\Upsilon(nS) \to \eta\Upsilon(mS)]}{\mathcal{B}[\Upsilon(nS) \to \pi^+\pi^-\Upsilon(mS)]}$$

is measured to be small for low-lying states: $\mathcal{R}^{\eta S}_{\pi\pi S}(2, 1) = (1.64 \pm 0.23) \times 10^{-3}$ [15, 16, 17] and $\mathcal{R}^{\eta S}_{\pi\pi S}(3, 1) < 2.3 \times 10^{-3}$ [16].

Above the $B\bar{B}$ threshold, BaBar observed the transition $\Upsilon(4S) \to \eta\Upsilon(1S)$ with the unexpectedly large branching fraction of $(1.96 \pm 0.28) \times 10^{-4}$, corresponding to $\mathcal{R}^{\eta S}_{\pi\pi S}(4, 1) = 2.41 \pm 0.42$ [18]. This apparent violation of the heavy quark spin-symmetry was explained by the contribution of B meson loops or, equivalently, by the presence of a four-quark $B\bar{B}$ component inside the $\Upsilon(4S)$ wave function [19, 20]. At the $\Upsilon(5S)$ energy, the anomaly is even more striking. The spin-flip processes $\Upsilon(5S) \to \pi\pi h_b(1P, 2P)$ are found not to be suppressed with respect to the spin-symmetry preserving reactions $\Upsilon(5S) \to \pi\pi\Upsilon(1S, 2S)$ [2], and all the $\pi\pi$ transitions show the presence of new resonant structures [6] that cannot be explained as conventional bottomonium states. Further insight into the mechanism of the hadronic transitions above the threshold can be gained by searching for the $E1M1$ transition $\Upsilon(4S) \to \eta h_b(1P)$, which is predicted to have a branching fraction of the order of 10^{-3} [23].

Following the approach used for the observation of the $h_b(1P, 2P)$ production at the $\Upsilon(5S)$ energy [2], Belle investigated the missing mass spectrum of η mesons in the $\Upsilon(4S)$ data sample [24]. Figure 2 shows the $M_{\rm miss}(\eta)$ distribution, where the transition $\Upsilon(4S) \to \eta h_b(1P)$ is observed with a statistical significance of 11σ. From the position of the peak, the $h_b(1P)$ mass is measured to be $M_{h_b(1P)} = (9899.3 \pm 0.4 \pm 1.0)$ MeV/c^2, that corresponds to the $1P$ hyperfine splitting $\Delta M_{\rm HF}(1P) = (+0.6 \pm 0.4 \pm 1.0)$ MeV/c^2. The value is compatible with zero, which agrees with the previous Belle measurement [3] and can be interpreted as evidence of the absence of sizable long range spin-spin interactions. The branching fraction of the transition is measured to be $\mathcal{B}[\Upsilon(4S) \to \eta h_b(1P)] = (2.18 \pm 0.11 \pm 0.18) \times 10^{-3}$, in agreement with the available theoretical prediction [23]. This process is found to be the strongest known transition from the $\Upsilon(4S)$ meson to lower bottomonium states.

Exploiting the radiative transition $h_b(1P) \to \gamma\eta_b(1S)$, Belle performed a new measurement of the $\eta_b(1S)$ mass $M_{\eta_b(1S)} = (9400.7 \pm 1.7 \pm 1.6)$ MeV/c^2. This result is in agreement with the value obtained by Belle with the $\Upsilon(5S) \to \pi^+\pi^- h_b(1P) \to \pi^+\pi^-\gamma\eta_b(1S)$ process [3] but exhibits

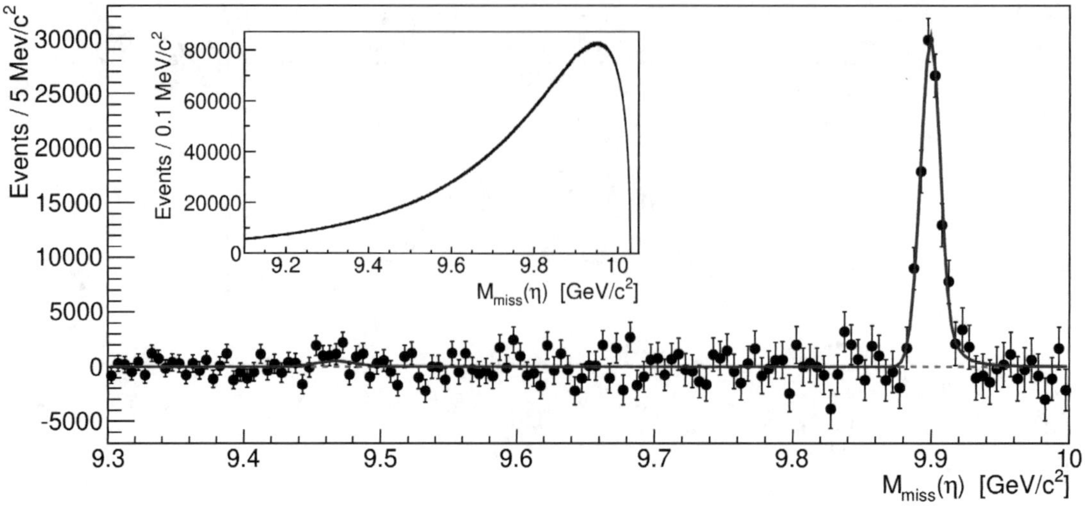

Figure 2. $M_{\rm miss}(\eta)$ distribution after the background subtraction. The solid blue curve shows the fit with the signal PDFs, while the dashed red curve represents the background only hypothesis. The inset shows the $M_{\rm miss}(\eta)$ distribution before the background subtraction.

a discrepancy with the $M1$-based measurements by BaBar and CLEO [25, 26, 27]. From the theoretical point of view, the Belle measurements are in agreement with the predictions of many potential models and lattice calculations [28], including the recent lattice result in Ref. [29].

4. Observation of $Z_b(10610) \to B\bar{B}^*$ and $Z_b(10650) \to B^*\bar{B}^*$

In the $\Upsilon(5S) \to \Upsilon(nS)\pi^+\pi^-$ $(n = 1, 2, 3)$ and $\Upsilon(5S) \to h_b(nP)\pi^+\pi^-$ $(n = 1, 2)$ transitions Belle observed intermediate charged resonances $Z_b(10610)$ and $Z_b(10650)$ that decay to bottomonium and charged pion [6]. The minimal quark content of these states is obviously exotic, it is a four-quark combination $|b\bar{b}u\bar{d}\rangle$. However, the configuration of these quarks was unclear. The two proposed possibilities are $B^{(*)}\bar{B}^*$ molecules or compact diquark-antidiquark objects [1]. The fact that the masses of $Z_b(10610)$ and $Z_b(10650)$ are very close to the $B\bar{B}^*$ and $B^*\bar{B}^*$ thresholds, respectively, supports the molecular hypothesis. Also, measured spin-parity of $J^P = 1^+$ [8] agrees with this hypothesis, as well as other properties of these states [30]. However, the "smoking gun" test is provided by the Z_b decays to the open flavor channels: they should be suppressed in case of the diquark-antidiquark structure and enhanced for the molecules [1].

Belle performed reconstruction of the $\Upsilon(5S) \to B\bar{B}^*\pi$ and $\Upsilon(5S) \to B^*\bar{B}^*\pi$ decays using full reconstruction of one B meson and pion and examining the missing mass spectrum of $B\pi$ combinations. The spectrum of $M_{\rm miss}(\pi)$ for $\Upsilon(5S) \to B\bar{B}^*\pi$ decays shows clear $Z_b(10610)$ signal without significant contribution of the $Z_b(10650)$, while in the case of $\Upsilon(5S) \to B^*\bar{B}^*\pi$ one finds the $Z_b(10650)$ signal. The fact that the $Z_b(10650) \to B\bar{B}^*$ decay is suppressed compared to the $Z_b(10650) \to B^*\bar{B}^*\pi$ decay despite a much larger phase space can not be explained in the diquark-antidiquark pictures, while it qualitatively agrees with the molecular hypothesis. The branching fraction of the open flavor channels is found to be at the level of 80%, thus these channels dominate, which disfavors the tetraquark hypothesis too. Preliminary results of the presented study are available in Ref. [31]. Belle will publish final results soon.

5. Conclusions

Recently it was established that the anomalous production of $e^+e^- \to \Upsilon(nS)\pi^+\pi^-$ and $e^+e^- \to h_b(nP)\pi^+\pi^-$ reactions is related to the $\Upsilon(5S)$ and $\Upsilon(6S)$ resonances, with little or no continuum contribution. The anomalous transitions from $\Upsilon(6S)$ were observed for the first time. Resonant structure studies of the $\Upsilon(6S) \to h_b(nP)\pi^+\pi^-$ decays indicates that they proceed via intermediate $Z_b(10610)$ and/or $Z_b(10650)$ states. One more channel where statistics is probably sufficient to measure energy dependence of the cross section at Belle is $\Upsilon(2S)\eta$. This result is to come. More precise energy scan in the $\Upsilon(5S)$ and $\Upsilon(6S)$ energy regions and above are expected at the new experiment Belle-II, which is under construction.

Belle observed $\Upsilon(4S) \to h_b(1P)\eta$ decays with very high branching fraction and performed a new measurement of the $h_b(1P)$ and $\eta_b(1S)$ parameters, that agree with previous Belle measurements. These results are important for understanding of hadronic transitions from Υ states above open flavor threshold. Also, since at Belle-II a 50 times larger sample of $\Upsilon(4S)$ decays will be collected, it opens new perspectives for studies of spin-singlet states, in particular, the search for exclusive $\eta_b(1S)$ decays.

Belle also updated the study of $Z_b(10610)$ and $Z_b(10650)$ decays to open flavor channels. Dominance of the $Z_b(10610) \to B\bar{B}^*$ and $Z_b(10650) \to B^*\bar{B}^*$ is confirmed, while the $Z_b(10650) \to B\bar{B}^*$ decay is found to be suppressed despite relatively large phase space. These results are in agreement with the hypothesis, that Z_b states have molecular structure, and strongly disfavor the diquark-antidiquark hypothesis. Further improvement in the lineshape measurement is crucial to determine couplings in molecular models [32].

Acknowledgments

The author acknowledge support from Russian Science Foundation under Grant No. 15-12-30014.

References

[1] Brambilla N *et al.* 2014 *Eur. Phys. J.* C **74** 2981
[2] Adachi I *et al.* (Belle Collaboration) 2012 *Phys. Rev. Lett.* **108** 032001
[3] Mizuk R *et al.* (Belle Collaboration) 2012 *Phys. Rev. Lett.* **109** 232002
[4] Chen K F *et al.* (Belle Collaboration) 2008 *Phys. Rev. Lett.* **100** 112001
[5] He X H *et al.* (Belle Collaboration) 2014 *Phys. Rev. Lett.* **113** (14) 142001
[6] Bondar A *et al.* (Belle Collaboration) 2012 *Phys. Rev. Lett.* **108** 122001
[7] Krokovny P *et al.* (Belle Collaboration) 2013 *Phys. Rev.* D **88** 052016
[8] Garmash A *et al.* (Belle Collaboration) 2015 *Phys. Rev.* D **91** 072003
[9] Chen K F *et al.* (Belle Collaboration) 2010 *Phys. Rev.* D **82** 091106
[10] Santel D *et al.* (Belle Collaboration) 2015 *arXiv*:hep-ex/1501.01137
[11] Abdesselam A *et al.* (Belle Collaboration) 2015 *arXiv*:hep-ex/1508.06562
[12] Ablikim M *et al.* (BESIII Collaboration) 2013 *Phys. Rev. Lett.* **111** (24) 242001
Yuan C Z 2014 *Chin. Phys.* C **38** 043001
[13] Gottfried K 1978 *Phys. Rev. Lett.* **40** 598
[14] Voloshin M B 2008 *Prog. Part. Nucl. Phys.* **61** 455
[15] He Q *et al.* (CLEO Collaboration) 2008 *Phys. Rev. Lett.* **101** 192001
[16] Lees J P *et al.* (BaBar Collaboration) 2011 *Phys. Rev.* D **84** 092003
[17] Tamponi U *et al.* (Belle Collaboration) 2013 *Phys. Rev.* D **87** 011104
[18] Aubert B *et al.* (BaBar Collaboration) 2008 *Phys. Rev.* D **78** 112002
[19] Meng C and Chao K T 2008 *Phys. Rev.* D **78** 074001
[20] Voloshin M B 2011 *Mod. Phis. Lett.* A **26** 778
[21] Bondar A *et al.* (Belle Collaboration) 2012 *Phys. Rev. Lett.* **108** 122001
[22] Krokovny P *et al.* (Belle Collaboration) 2013 *Phys. Rev.* D **88** 052016
[23] Guo F K *et al.* 2010 *Phys. Rev. Lett.* **105** 162001
[24] Tamponi U *et al.* (Belle Collaboration) 2015 *Phys. Rev. Lett.* **115** (14) 142001
[25] Aubert B *et al.* (BaBar Collaboration) 2008 *Phys. Rev. Lett.* **101** 071801
[26] Aubert B *et al.* (BaBar Collaboration) 2009 *Phys. Rev. Lett.* **103** 161801

[27] Bonvicini G *et al.* (CLEO Collaboration) 2010 *Phys. Rev.* D **81** 031104
[28] Burns T J 2013 *Phys. Rev.* D **87** 034022
[29] Dowdall R J *et al.* (HPQCD Collaboration) 2014 *Phys. Rev.* D **89** 031502(R)
[30] Bondar A E *et al.* 2011 *Phys. Rev.* D **84** 054010
[31] Adachi I *et al.* (Belle Collaboration) 2009 *arXiv*:hep-ex/1209.6450
[32] Hanhart C *et al.* 2015 *arXiv*:hep-ph/1507.00382

Journal of Physics: Conference Series **675** (2016) 022007 doi:10.1088/1742-6596/675/2/022007

Baryonic B Decays

R Chistov

National Research Nuclear University MEPhI (Moscow Engineering Physics Institute), Kashirskoe hughway 31, Moscow, 115409, Russia

E-mail: chistov@itep.ru

Abstract. In this talk the decays of B-mesons into baryons are discussed. Large mass of B-meson makes possible the decays of the type $B \to baryon\ \overline{baryon}\ (+mesons)$. Experimental observations and measurements of these decays at B-factories Belle and BaBar have stimulate the development of theoretical models in this field. We briefly review the experimental results together with the current theoretical models which describe baryonic B decays.

1. Introduction

Due to large B meson mass, its decays to baryons are possible. The only example of baryonic decay of charm meson is $D_s^+ \to p\bar{n}$. The thing is that baryons should be produced in pairs in the Standard Model and the sum of 2 protons or proton and neutron masses (thresholds) is larger than the D^0 or D^+ masses, respectively.

First observations and studies of baryonic B decays were done by ARGUS [1] and CLEO [2]. During last decade, B-factories Belle and BaBar dominate in this field. Belle and BaBar observed many new baryonic B decay modes [3] and connected with them new phenomena. In this talk we summarize and discuss important features of baryonic B decays, mostly with one or two charm baryons in the final state. For the reference, the discussed baryonic B decay modes are quoted in table 1.

2. New phenomena in baryonic B decays

Below we provide the list of newly found phenomena.

- Hierarchy of branching fractions for multi-body modes: $\mathcal{B}(\bar{B}^0 \to \Lambda_c^+\bar{p}\pi^+\pi^-) >> \mathcal{B}(B^- \to \Lambda_c^+\bar{p}\pi^-) >> \mathcal{B}(\bar{B}^0 \to \Lambda_c^+\bar{p})$ and $\mathcal{B}(\bar{B}^0 \to D^{*+}p\bar{p}\pi^-) >> \mathcal{B}(\bar{B}^0 \to D^{*0}p\bar{p})$. It means that 3-body decays have substantially larger branching fractions than their 2-body counterparts and 4-body decays are usually more frequent than the 3-body modes.
- It was observed that in many 3- and 4-body decay modes the invariant mass of baryon-antibaryon pair is peaking near threshold. This feature is called threshold enhancement.
- Belle observed [4, 5] hierarchy of branching fractions for two-body modes: the decays to two charm baryons have larger branching fraction than the decays to one charm baryon plus light baryon and the latter have larger branching fraction than the decays into two light baryons: $\mathcal{B}(B \to Baryon_{c1}\overline{Baryon_{c2}}) \sim 10^{-3} >> \mathcal{B}(B \to Baryon_c\overline{Baryon}) \sim 10^{-5} >> \mathcal{B}(B \to Baryon\overline{Baryon}) < 10^{-6}$ (see table 1).

Table 1: B decay modes considered in this talk and their world average branching fractions.

Decay	$\mathcal{B}$
$B^- \to \Xi_c^0 \Lambda_c^-$	$(2.16 \pm 0.54) \times 10^{-3}$
$B^0 \to p\bar{\Lambda}_c^-$	$(1.9 \pm 0.2) \times 10^{-5}$
$B^0 \to p\bar{p}$	$< 1.1 \times 10^{-7}$
$B^- \to \Lambda_c^+ \bar{p}\pi^-$	$(2.92 \pm 0.14) \times 10^{-4}$
$\bar{B}^0 \to \Lambda_c^+ \bar{p}\pi^+\pi^-$	$(12.35 \pm 0.72) \times 10^{-4}$

3. Phenomenological explanation of new phenomena

All these phenomena can be understood in the framework of the following dynamical picture [6, 7].

To produce baryon-antibaryon pair in two body B decay, one energetic $q\bar{q}$ pair must be emitted at high invariant mass by a hard gluon (high q^2). So, the 2-body decay amplitude is suppressed by a factor of order α_s/q^2. In the 3-body baryonic $B \to Baryon\overline{Baryon} + meson$ decay, it is possible that the baryon-antibaryon pair is emitted collinearly against the meson. The $q\bar{q}$ pair emitted from a gluon is moving in nearly the same direction, i.e. this gluon is soft and is not suppressed by $1/q^2$. Therefore, the branching fraction of $B \to Baryon\overline{Baryon} + meson$ decay is larger than $B \to Baryon\overline{Baryon}$ decay. It also explains that the baryon-antibaryon mass is peaking near threshold in multi-body modes. Indeed, the baryon-antibaryon pair against the fast meson tends to populate low invariant masses, close to the corresponding baryon-antibaryon threshold.

If we return to the 2-body baryonic B decay modes, we can see the same dynamics which can be understood as follows. In an energetic charm baryon, the momentum is carried by the constituent charm quark. Thus, in $b \to c\bar{c}s$ decays, the energetic c quark will fragment into $Baryon_{c1}$ and $\bar{c}$ into $\overline{Baryon_{c2}}$. Consequently, no hard gluon is needed to produce the energetic $Baryon_{c1}\overline{Baryon_{c2}}$ pair in B decay. But in $B \to Baryon_c\overline{Baryon}$ decay two hard gluons are needed to produce an energetic $\overline{Baryon}$: one hard gluon to kick the spectator quark of the B meson to make it energetic and the other to produce the $q\bar{q}$ pair. Therefore, the decay $B \to Baryon_c\overline{Baryon}$ is suppressed with respect to $B \to Baryon_{c1}\overline{Baryon_{c2}}$ by a factor $\approx \alpha_s^4\ 10^{-2}$. These qualitative statements have been supported by calculations of the decay rates for $\bar{B} \to \Xi_c\bar{\Lambda}_c$ [8] and $\bar{B}^0 \to \Lambda_c^+\bar{p}$ [9]. From the phenomenological side of view, this dynamical suppression can be translated in terms of the thresholds: the invariant mass of two charm baryons in $B \to Baryon_{c1}\overline{Baryon_{c2}}$ is a B mass, but their threshold, i.e. the sum of their masses, is much closer to B meson mass than in B decay to charm baryon and light (anti-)baryon. It explains the $\mathcal{B}(B \to Baryon_{c1}\overline{Baryon_{c2}}) >> \mathcal{B}(B \to Baryon_c\overline{Baryon})$ inequality. The same consideration could be used to explain the $\mathcal{B}(B \to Baryon_{c1}\overline{Baryon}) >> \mathcal{B}(B \to Baryon\overline{Baryon})$ inequality.

The common feature is that the baryon form factor $1/(M_{12}^2 - (m_1 + m_2)^2)$ works in each of three mentioned above phenomena. Here M_{12} is the invariant mass of baryon-antibaryon pair and $(m_1 + m_2)$ is its threshold [10, 11].

4. Summary

In conclusion, the baryonic beauty meson decays into (charm) baryons observed and studied at B-factories Belle and BaBar have enriched our understanding of realization of QCD at middle energies (~ 5 GeV $\sim M_B$). Hopefully, this field of flavour physics will develop further with new experimental data at LHCb and Belle II.

Journal of Physics: Conference Series **675** (2016) 022007 doi:10.1088/1742-6596/675/2/022007

Acknowledgments
This work was performed within the framework of the Center of Nanostructured Electronics supported by MEPhI Academic Excellence Project (contract No. 02.a03.21.0005, 27.08.2013) and is supported by the Russian Ministry of Education and Science contracts 14.A12.31.0006.

References

[1] Albrecht H *et al.* (ARGUS Collaboration) 1989 *Z. Phys.* C**42** 519
[2] Crawford G D *et al.* (CLEO Collaboration) 1992 *Phys. Rev.* D**45** 752-70
[3] Olive K A *et al.* (Particle Data Group) 2014 *Chin. Phys.* C**38** 090001
[4] Gabyshev N *et al.* (Belle Collaboration) 2003 *Phys. Rev. Lett.* **90** 121802
[5] Chistov R *et al.* (Belle Collaboration) 2006 *Phys. Rev.* D**74** 111105(R)
[6] Hou W-S and Soni A 2001 *Phys. Rev. Lett.* **86** 4247-50
[7] Suzuki M 2007 *J. Phys.* G**34** 283-298
[8] Cheng H-Y, Chua C-K and Hsiao Y-K 2009 *Phys. Rev.* D**79** 114004
[9] He X-G *et al.* 2007 *Phys. Rev.* D**75** 034011
[10] Chua C-K and Hou W-S 2003 *Eur. Phys. J.* C**29** 27-35
[11] C-K Chua *et al.* 2002 *Phys. Rev.* D**66** 054004

Journal of Physics: Conference Series **675** (2016) 022008 doi:10.1088/1742-6596/675/2/022008

Charm Baryon Spectroscopy

R Chistov

National Research Nuclear University MEPhI (Moscow Engineering Physics Institute), Kashirskoe hughway 31, Moscow, 115409, Russia

E-mail: `chistov@itep.ru`

Abstract. B-factories Belle and BaBar during its operation made not only measurements connected with B-meson decays but also numerous observation and measurements in charm physics. In particular, their results on charm baryon decays and spectroscopy have enlarged and enriched the current picture of heavy flavour hadrons. In this talk we overview current status of charm baryons and their excited states.

1. Introduction

In this talk we summarize present status of excited charm baryons, decaying strongly or electromagnetically. We quote their masses (or mass difference between excited baryon and the corresponding ground state), natural widths, decay modes and assigned quantum numbers.

2. Excited Λ_c^+ baryons

Table 1 shows summary on the excited Λ_c^+'s. First two states, $\Lambda_c(2595)^+$ and $\Lambda_c(2625)^+$ were observed by ARGUS and CLEO, confirmed in other experiments and now are well established. Their masses occured to be close to the predicted two lowest mass orbitally excited Λ_c^+'s with total angular momentum of light quarks L=1. Therefore, the overall quantum numbers for these two excited charm baryons are $J^P = (\frac{1}{2})^-$ and $J^P = (\frac{3}{2})^-$. Recently, CDF [1] performed world best mass measurements: $M(\Lambda_c(2595)^+) = 2592.25 \pm 0.24 \pm 0.14$ MeV/c^2, $M(\Lambda_c(2625)^+) = 2628.11 \pm 0.13 \pm 0.14$ MeV/c^2.

Next two states, $\Lambda_c(2765)^+$ and $\Lambda_c(2880)^+$, were observed by CLEO [2] in $\Lambda_c^+\pi^+\pi^-$ final state. Later, BaBar [3] observed that the latter state has also the D^0p decay mode. It is the first example where excited charm baryon decays into charm meson and light baryon. (Usually, excited charm baryons decay onto charm baryon and light mesons.) In that analysis BaBar observed for the first time else one state, $\Lambda_c(2940)^+$, decaying into D^0p. By looking at D^+p final state, they found no signals which results in the conclusion that the $\Lambda_c(2880)^+$ and $\Lambda_c(2940)^+$ are really Λ_c^+ excited states, not Σ_c. Belle experiment found that $\Lambda_c(2880)^+$ decays also through the $\Sigma_c(2445)^{++/0(*)}\pi^{-/+}$ mode [4]. Belle reported the result of an angular analysis that favors the 5/2 for the $\Lambda_c(2880)^+$ spin hypothesis. The measured ratio of branching fractions $\mathcal{B}(\Lambda_c(2880)^+ \to \Sigma_c(2520)\pi^\pm)/\mathcal{B}(\Lambda_c(2880)^+ \to \Sigma_c(2455)\pi^\pm) = (0.225 \pm 0.062 \pm 0.025)$ together with theoretical predictions based on HQS [5, 6] favour even parity.

The open questions in the present excited Λ_c^+ family are the experimental determination of quantum numbers for almost all states and the nature of $\Lambda_c(2765)^+$ state: whether it is excited Σ_c^+ or Λ_c^+.

Table 1: Summary of excited Λ_c^+ baryons family.

Charmed Baryon Excited State	Mode	Mass or ΔM, MeV/c^2	Natural Width, MeV/c^2	J^P
$\Lambda_c(2595)^+$	$\Lambda_c^+\pi^+\pi^-$, $\Sigma_c\pi$	2595.4 ± 0.6	$3.6^{+2.0}_{-1.3}$	$1/2^-$
$\Lambda_c(2625)^+$	$\Lambda_c^+\pi^+\pi^-$, $\Sigma_c\pi$	2628.1 ± 0.6	< 1.9	$3/2^-$
$\Lambda_c(2765)^+$	$\Lambda_c^+\pi^+\pi^-$, $\Sigma_c\pi$	2766.6 ± 2.4	50	??
$\Lambda_c(2880)^+$	$\Lambda_c^+\pi^+\pi^-$, $\Sigma_c\pi$, $\Sigma_c(2520)\pi$, D^0p	2881.53 ± 0.35	5.8 ± 1.1	$5/2^+$ (experimental evidence)
$\Lambda_c(2940)^+$	D^0p, $\Sigma_c\pi$	$2939.3^{+1.4}_{-1.5}$	17^{+8}_{-6}	??

Table 2: Summary of excited $\Sigma_c^{++,+,0}$ baryons family.

Charmed Baryon Excited State	Mode	Mass or ΔM, MeV/c^2	Natural Width, MeV/c^2	J^P
$\Sigma_c(2520)^{++}$	$\Lambda_c^+\pi^+$	$231.99 \pm 0.10 \pm 0.02$	$14.77 \pm 0.25^{+0.18}_{-0.30}$	$3/2^+$
$\Sigma_c(2520)^{+}$	$\Lambda_c^+\pi^+$	231.0 ± 2.3	< 17 @ 90% CL	$3/2^+$
$\Sigma_c(2520)^{0}$	$\Lambda_c^+\pi^+$	$231.98 \pm 0.11 \pm 0.04$	$15.41 \pm 0.41^{+0.20}_{-0.32}$	$3/2^+$
$\Sigma_c(2800)^{++}$	$\Lambda_c^+\pi^+$	$514.5^{+3.4+2.8}_{-3.1-4.9}$	75^{+18+12}_{-13-11}	
$\Sigma_c(2800)^{+}$	$\Lambda_c^+\pi^0$	$505.4^{+5.8+12.4}_{-4.6-2.0}$	62^{+37+52}_{-23-38}	
$\Sigma_c(2800)^{0}$	$\Lambda_c^+\pi^-$	$515.4^{+3.2+2.1}_{-3.1-6.0}$	61^{+18+22}_{-13-13}	$3/2^-(?)$
	$\Lambda_c^+\pi^-$	$560 \pm 8 \pm 10$	86^{+33}_{-22}	

3. Excited Σ_c baryons

Table 2 summarizes the excited $\Sigma_c^{++,+,0}$ baryons. Triplet of $\Sigma_c(2520)^{++,+,0}$ baryons, decaying into $\Lambda_c^+\pi$ is well established. Recently Belle [7] precisely measured the mass differences (between $\Sigma_c(2520)$ and Λ_c^+) and widths of the double charged and neutral members of this triplet to be $\Delta M(\Sigma_c(2520)^{++}) = (231.99 \pm 0.10 \pm 0.02)$ MeV/c^2, $\Gamma(\Sigma_c(2520)^{++}) = (14.77 \pm 0.25^{+0.18}_{-0.30})$ MeV/c^2 and $\Delta M(\Sigma_c(2520)^0) = (231.98 \pm 0.11 \pm 0.04)$ MeV/c^2, $\Gamma(\Sigma_c(2520)^0) = (15.41 \pm 0.41^{+0.20}_{-0.32})$ MeV/c^2, respectively.

The triplet of excited Σ_c baryons, $\Sigma_c(2800)$ states, was observed by Belle [8]. Based on measured mass and theoretical predictions [9, 10] these states are identified as members of the predicted Σ_{c2} $3/2^-$ triplet. From the study of resonant substructure in $B^- \to \Lambda_c^+\bar{p}\pi^-$ decays, BaBar found significant signal in $\Lambda_c^+\pi^-$ with the mean value higher by about 3 σ from obtained by Belle (see table 2). The widths from two measurements, Belle and BaBar, are consistent.

4. Excited Ξ_c baryons

Due to the measurements and observations at B-factories Belle and BaBar, the picture of excited $\Xi_c^{+,0}$ and Ω_c^0 baryons dramatically enriched. Table 3 summarizes the excited $\Xi_c^{+,0}$ and Ω_c^0 baryons. Several states with masses above 2900 MeV/c^2 and decaying into $\Lambda_c^+K^-$, $\Lambda_c^+K^{-/0}\pi^{+/-}$ and $\Sigma_c(2455/2520)K$ have been observed. Some of these states are seen by both Belle [11, 12] and BaBar [13] and are believed to be well established. These are the $\Xi_c(2980)^+$ and $\Xi_c(3080)^{+,0}$ states which provide the first example when the excited charm strange baryon is decaying into charm baryon and strange meson. All previous excited charm strange baryons were observed to decay into charm strange baryon and pion(s).

Next new state, $\Xi_c(2930)^0$, decaying into $\Lambda_c^+K^-$ is seen by BaBar [14] and still needs to be confirmed. The $\Xi_c(3055)^+$ found in $\Sigma_c(2455)^{++}\pi^-$ final state by BaBar [13], has been recently confirmed by Belle [12]. In Table 3 the width and mass measurements for the $\Xi_c(3055)^+$ are not averaged but quoted separately for BaBar and Belle. But $\Xi_c(3123)^+$ claimed by BaBar [13] in $\Sigma_c(2520)^{++}\pi^-$ final state hasn't been recently confirmed by Belle [12] with twice larger statistics. One should also note that recently Belle [12] has performed most precise measurement of $\Xi_c(2645)^+$ width (see Table 3).

5. Excited Ω_c^0 baryons

The excited Ω_c^0 double charm baryon are seen by both BaBar [15] and Belle [16], the $\delta M = M(\Omega_c^{*0}) - M(\Omega_c^0)$ are in good agreement agreement in both experiments (see table 3) and consistent with most theoretical predictions [17, 18, 19, 20].

Table 3: Summary of excited $\Xi_c^{+,0}$ and Ω_c^0 baryons families.

Charmed Baryon Excited State	Mode	Mass or ΔM, MeV/c^2	Natural Width, MeV/c^2	J^P
$\Xi_c'^+$	$\Xi_c^+\gamma$	2575.6 ± 3.1		$1/2^+$
$\Xi_c'^0$	$\Xi_c^0\gamma$	2577.9 ± 2.9		$1/2^+$
$\Xi_c(2645)^+$	$\Xi_c^0\pi^+$	$2645.9^{+0.6}_{-0.5}$	$2.6 \pm 0.2 \pm 0.4$	$3/2^+$
$\Xi_c(2645)^0$	$\Xi_c^+\pi^-$	2645.9 ± 0.5	< 5.5	$3/2^+$
$\Xi_c(2790)^+$	$\Xi_c'^0\pi^+$	2789.1 ± 3.2	< 15	$1/2^-$
$\Xi_c(2790)^0$	$\Xi_c'^+\pi^-$	2791.8 ± 3.3	< 12	$1/2^-$
$\Xi_c(2815)^+$	$\Xi_c^+\pi^+\pi^-$, $\Xi_c(2645)^0\pi^+$	2816.6 ± 0.9	< 3.5	$3/2^-$
$\Xi_c(2815)^0$	$\Xi_c^0\pi^+\pi^-$, $\Xi_c(2645)^+\pi^-$	2819.6 ± 1.2	< 6.5	$3/2^-$
$\Xi_c(2930)^0$	$\Lambda_c^+K^-$	2931.6 ± 6	36 ± 13	??
$\Xi_c(2980)^+$	$\Lambda_c^+K^-\pi^+$, $\Sigma_c^{++}K^-$, $\Xi_c(2645)^0\pi^+$	2971.4 ± 3.3	26 ± 7	??
$\Xi_c(2980)^0$	$\Xi_c(2645)^+\pi^-$	2968.0 ± 2.6	20 ± 7	??
$\Xi_c(3055)^+$	$\Sigma_c^{++}K^-$	3054.2 ± 1.3 (BaBar) $3058.1 \pm 1.0 \pm 2.1$ (Belle)	17 ± 13 (BaBar) $9.7 \pm 3.4 \pm 3.3$ (Belle)	??
$\Xi_c(3080)^+$	$\Lambda_c^+K^-\pi^+$, $\Sigma_c^{++}K^-$, $\Sigma_c(2520)^{++}K^-$	3077.0 ± 0.4	5.8 ± 1.0	??
$\Xi_c(3080)^0$	$\Lambda_c^+K_S^0\pi^-$, $\Sigma_c^0K_S^0$, $\Sigma_c(2520)^0K_S^0$	3079.9 ± 1.4	5.6 ± 2.2	??
$\Omega_c(2770)^0$	$\Omega_c^0\gamma$	2765.9 ± 2.0	$70.7^{+0.8}_{-0.9}$	$3/2^+$

6. Summary

In conclusion, interesting feature recently discovered by BaBar and Belle is that now we know that transitions between families are possible (between Ξ_c and Λ_c^+ families of excited baryons). Also, highly excited Λ_c^+ baryons are found to have a possible decay mode into charm meson and proton.

Acknowledgments

This work was performed within the framework of the Center of Nanostructured Electronics supported by MEPhI Academic Excellence Project (contract No. 02.a03.21.0005, 27.08.2013) and is supported by the Russian Ministry of Education and Science contracts 14.A12.31.0006.

References

[1] Aaltonen T *et al.* (CDF collaboration) 2011 *Phys. Rev.* D**84** 012003
[2] Artuso M *et al.* (CLEO collaboration) 2001 *Phys. Rev. Lett.* **86** 4479-82
[3] Aubert B *et al.* (BABAR collaboration) 2007 *Phys. Rev. Lett.* **98** 012001
[4] Mizuk R *et al.* (Belle collaboration) 2007 *Phys. Rev. Lett.* **98** 262001
[5] Isgur N and Wise M B 1991 *Phys. Rev. Lett.* **66** 1130-3
[6] Cheng H-Y and Chua C-K 2007 *Phys. Rev.* D**75** 014006
[7] S.-H. Lee *et al.* (Belle collaboration) 2014 *Phys. Rev.* D**89** 091102(R)
[8] R. Mizuk *et al.* (Belle collaboration) 2005 *Phys. Rev. Lett.* **94** 122002
[9] Copley L, Isgur N, and Karl G 1979 *Phys. Rev.* D**20** 768

[10] Pirjol D and Yan T-M 1997 *Phys. Rev.* D**56** 5483-5510
[11] Chistov R *et al.* (Belle collaboration) 2006 *Phys. Rev. Lett.* **97** 162001
[12] Kato Y *et al.* (Belle collaboration) 2014 *Phys. Rev.* D**89** 052003
[13] Aubert B *et al.* (BABAR collaboration) 2008 *Phys. Rev.* D**77** 012002
[14] Aubert B *et al.* (BABAR collaboration) 2008 *Phys. Rev.* D**77** 031101
[15] Aubert B *et al.* (BABAR collaboration) 2006 *Phys. Rev. Lett.* **97** 232001
[16] Solovieva E *et al.* (Belle collaboration) 2009 *Phys. Lett.* B**672** 1-5
[17] Rosner J L 1995 *Phys. Rev.* D**52** 6461-5
[18] Glozman L Y and Riska D 1996 *Nucl. Phys.* A**603** 326-44
[19] Jenkins E E 1996 *Phys. Rev.* D**54** 4515-31
[20] Burakovsky L, Goldman J T and Horwitz L 1997 *Phys. Rev.* D**56** 7124-32

Journal of Physics: Conference Series **675** (2016) 022009 doi:10.1088/1742-6596/675/2/022009

Flavor physics at Super B-factories era

P Pakhlov and T Uglov

National Research Nuclear University MEPhI (Moscow Engineering Physics Institute), Kashirskoe highway 31, Moscow, 115409, Russia

E-mail: pakhlov@itep.ru

Abstract. We review numerous results from the B-Factories, obtained last decade. They provide currently strong constraints on New Physics extensions beyond the Standard Model. We discuss the physics program at Super B-factory, a next generation asymmetric collider with the luminosity almost two orders of magnitude higher than those achieved at the existing colliders, and its capability in cooperation with LHC of new insights into New Physics phenomena.

1. Introduction

Up to now the Standard Model (SM) of the particle physics remains one of the best experimentally verified theories. For almost 50 years since its establishment the SM managed to overcome all experimental tests and precisely describes all processes in a wide energy range up to the scale probed at energy frontier experiments at LHC. Moreover, the SM predicted the existence of new processes not only in particle physics but also in cosmology and astrophysics. The observation by Atlas and CMS of the Higgs boson [1], the last fundamental particle of the SM, that escaped the detection for decades, marked a triumph of the Standard Model. Furthermore, by now LHC confirmed measured properties of Higgs boson are consistent with the SM expectations.

Despite the great success in describing the matter and forces of nature, the SM persists to remain a not complete theory. First of all, it is more a theoretical framework built from experimental observations rather than a fundamental theory based on the first principles. Furthermore, it fails to resolve intrinsic problems, such as instability of the fundamental weak scale against radiative corrections, and remains unanswered many fundamental questions, such as origin of gauge groups, fermion masses and mixing hierarchy etc. These suggest that the SM is only an effective theory, which does not remain valid up to an arbitrarily high energy scale. On the contrary, it is widely believed that the New Physics (NP) beyond the SM can be observed at precision or energy frontier experiments in the near future.

The quark sector of the SM is especially rich in puzzles and the largest contributor in terms of number of free parameters. The measuring of quark mixing parameters provides a major test of the Cabibbo-Kobayashi-Maskawa (CKM) description of flavor changing currents and CP violation [2]. Although CKM mixing does provide a source for CP violation – one of the Sakharov's conditions for the Universe evolution [3], the magnitude of the matter-antimatter asymmetry cannot be explained by the CKM mechanism only. This may indicate that some hidden mechanism resulting at larger CP violation exists at higher energies. Flavor Physics is a promising tool for NP searches through quantum loop effects. Rare decays, neutral meson-antimeson mixing and CP violation are potentially subjected to NP virtual corrections.

2. Future Flavor experiments

The major information about CKM matrix is obtained in B meson study, thus providing the stringent test of the CKM mechanism. Carrying out these studies was the main motivation for construction of two B-Factory experiments, Belle and BaBar. Since 1999 both experiments have performed many precise and *independent* measurements of the CKM parameters. The underlying idea was to check the overall consistency of the CKM framework; any discrepancy between measurements could be interpreted as potential NP effects. While with the current precision of data no significant deviations were found, it was demonstrated that Flavor Physics is indeed sensitive to TeV scale effects.

Future experiments are scheduled to address open questions and perform more precise tests:

- At LHC the main contributor into flavor studies is LHCb experiment with luminosity $2\times10^{32}\,\mathrm{cm}^{-2}\mathrm{s}^{-1}$ corresponding the $B\overline{B}$ production rate 10^{12} per year.
- Two projects of asymmetric energy e^+e^- collider to operate at the $\Upsilon(4S)$ resonance, Super KEKB (Japan) and SuperB (Italy), were elaborated, but only the first one survived.
- BEPCII charm factory and discussed Super-c-tau factory will perform searches for NP in charm sector and provide complimentary information required for B physics.

Together with possible direct observation of NP at LHC, these experiments can help to uncover its flavor structure. Furthermore, even if nothing will be found at LHC, the Flavor experiments still have a chance to discover NP, thus perform an important role in the particle physics program.

2.1. Advantages of Super B-factory

Although LHCb experiment successfully started data taking and produced by now many excellent results in B physics, there are strong arguments to push for the next generation of unprecedented high luminosity e^+e^- machine, Super KEKB. The number of produced and triggered $B\overline{B}$ pairs at LHCb exceeds those expected at Super KEKB, however e^+e^- machine provides very clean environment and minimal trigger bias, which is essential for many important observables. Belle II experiment at Super KEKB has advantages in a study of the decays that involve reconstruction of photons and K_L, which are the key modes for many measurements. Even more important are neutrino modes, such as $B\to\tau\nu$, $B\to D^{(*)}\tau\nu$, $B\to K^{(*)}\nu\bar{\nu}$. They probe charged Higgs and SUSY and can be studied at e^+e^- machines only. The measurement of $|V_{ub}|$ through the decay $b\to u\ell^+\nu$ is also important task to constrain the CKM mechanism.

The flavor tagging is more efficient at Super B-Factory which compensates lower production rate in indirect CP violation studies. The high reconstruction efficiency and low trigger bias help to reduce substantially systematic uncertainties in many types of measurements, such as Dalitz analysis. At last, e^+e^- machine allows to accumulate huge statistics of $\tau^+\tau^-$ data.

Evidently, Super B-Factory has a complementary physics program, and is required in addition to LHCb to cover fully the flavor sector. We illustrate methods to measure the key observables using Belle results, and extrapolate the sensitivity for Belle II. The target SuperKEKB luminosity is $8\times10^{35}\,\mathrm{cm}^{-2}\mathrm{s}^{-1}$, corresponding to $50\,\mathrm{ab}^{-1}$ data sample accumulated within 5 years of operation.

3. Unitarity Triangle measurements at Super B-factory

The unitary relations of the CKM matrix can be represented as Unitarity Triangles (UT) in the complex plane. One of these triangles, that visualizes the relation between the first and third columns of the CKM matrix, $V_{td}V_{tb}^* + V_{cd}V_{cb}^* + V_{ud}V_{ub}^* = 0$, is the most important for CP violation studies in B_d decays. Many important results described below can be depicted as constraints in this UT. The main questions addressed to these measurements: are the measured angles consistent with sides? are the angle from loop and tree decays consistent? By now the consistency is good, but the present accuracy can not excludes the NP effects at 1% level.

3.1. β from $B^0 \to (c\bar{c})K^0$ decays

The most precise determination of the angle β is provided by the measurement of the mixing-induced CP violation in $B^0 \to (c\bar{c})K^0$ decays:

$$A_{CP}(\Delta t) = \frac{N(\overline{B}^0 \to (c\bar{c})K^0) - N(B^0 \to (c\bar{c})K^0)}{N(\overline{B}^0 \to (c\bar{c})K^0) + N(B^0 \to (c\bar{c})K^0)} = \sin 2\beta \sin(\Delta m_d \Delta t) + A\cos(\Delta m_d \Delta t). \quad (1)$$

These modes are dominated by the $b \to c\bar{c}s$ tree diagram. The penguin contribution has the same weak phase within few per cent accuracy, which makes direct CP violation vanishing to a very good approximation. Besides theoretical clarity, these channels also offer experimental advantages because of the large branching fractions and the presence of narrow resonances in the final state, which provides a powerful suppression of combinatorial background.

The most precise determination of mixing-induced CP violation in $B^0 \to (c\bar{c})K^0$ decay was provided by Belle [4]. Figure 1 shows time-dependent asymmetries for both CP-odd and CP-even final states at Belle. The sign of the asymmetry for opposite CP eigenvalues is flipped, as expected. The measured parameters, $\sin 2\beta = 0.667 \pm 0.023 \pm 0.012$ and $A = 0.006 \pm 0.016 \pm 0.012$, are consistent with the SM expectations, in particular with zero direct CP violation. It is

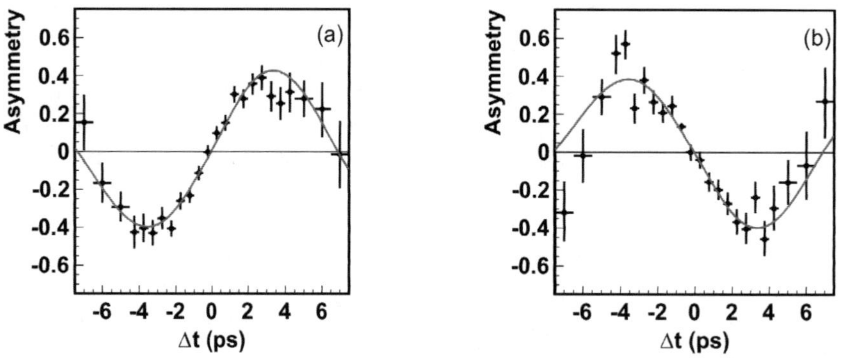

Figure 1. The asymmetry for all CP-odd modes (a) and the CP-even mode (b) at Belle.

expected that similar reconstruction and flavor tagging efficiencies, and similar vertex resolution will be achieved by Belle II. As the systematic errors are well under control using the control samples, they are expected to scale with statistics. Belle II can achieve accuracy of 0.3° [5] in measurement of β, which will provide a solid reference point to search for evidence of NP.

3.2. Measurement of the angle α

The angle α can be determined from a time-dependent CP asymmetry in charmless $b \to u\bar{u}d$ decays. The weak decay phase of $b \to u$ transition (figure 2a) is related to V_{ub}, thus CP asymmetry in a pure $b \to u$ modes is equal to $\sin 2\alpha$. However, a penguin diagram (figure 2b) contributes with a different phase. This causes a deviation of the magnitude of indirect CP asymmetry from $\sin 2\alpha$ and a non-zero direct CP asymmetry. Following the idea of M. Gronau and D. London [6] the angle α is extracted using the isospin relation among branching fractions and CP asymmetries of $B^0 \to \pi^+\pi^-$, $B^0 \to \pi^0\pi^0$, and $B^+ \to \pi^+\pi^0$ decays. The method allows to constrain the contribution from the penguin amplitude generally with an eight-fold ambiguity.

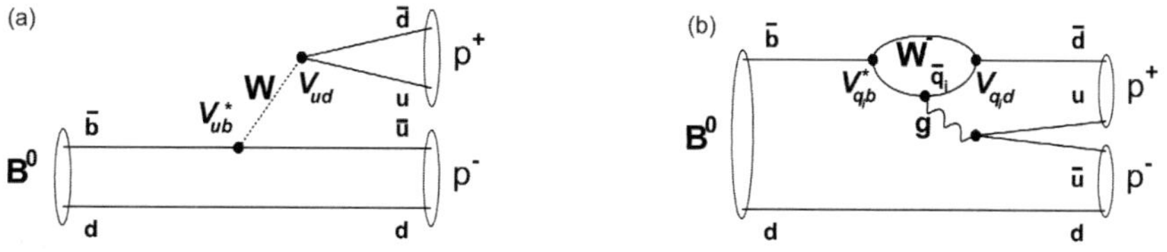

Figure 2. Feynman diagrams for $B^0 \to \pi^+\pi^-$ decays.

The decay $B \to \pi\pi$ has the simplest two-body topology, however, the large observed direct CP violation, and large branching fraction for $B^0 \to \pi^0\pi^0$ [7] suggest that penguin contribution to this final state is large, thus complicating the extraction of α. Using the full data set Belle measured the CP asymmetry in $B^0 \to \pi^+\pi^-$ quite precisely [8]: $A = 0.33 \pm 0.06 \pm 0.03$ and $S = -0.64 \pm 0.08 \pm 0.03$ (the world's most precise measurement!), but, because of trigonometrical ambiguities in extraction of α this mode allows only to exclude some intervals (figure 3, red line).

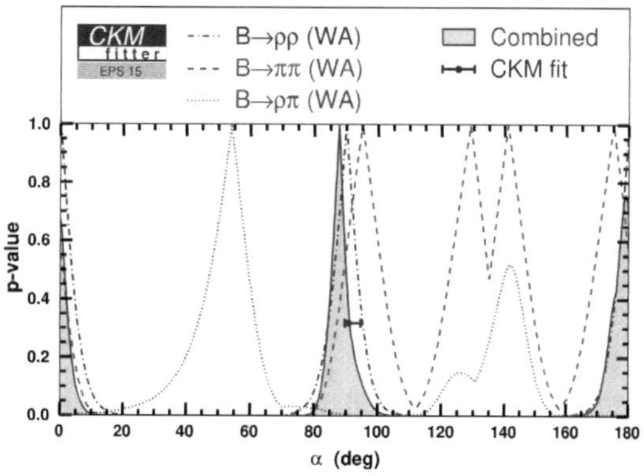

Figure 3. Difference $1 - CL$ for a range of α from $\pi\pi$, $\rho\rho$ and $\rho\pi$ analyses as averaged by CKMfitter Group [10].

We are lucky to get much higher sensitivity in α from the $B \to \rho\rho$ decays, where the penguin contribution turns out to be small. While this final state is more difficult for experimental study because of two wide vector mesons in the final state, it provides in the moment the best constrain on α (figure 3, blue line). Determination of α is also possible using the decays $B \to \pi\rho$, in spite of the final state is not CP eigenstate. The measurement of four isospin amplitudes is required which leads to 12 unknowns in the isospin pentagon. The problem is simplified and (what is more important) the ambiguity introduced by geometry of isospin triangles is removed with the time-dependent Dalitz analysis of the $B^0 \to (\pi\rho)^0$ decays. Belle performed such study using part of the available data and obtained the constraint $68^\circ < \alpha < 95^\circ$ at 68% CL [9].

The current world average of α including measurements of these three modes from Belle and BaBar, $(87.6^{+3.5}_{-3.3})^\circ$ [10], is quite precise (figure 3). The α measurements necessarily involve neutral modes, thus LHCb can hardly improve the accuracy solely. On the contrary, Belle II can reduce significantly the α errors as they are predominantly reducible. The expected accuracy is better than 2° using the $\pi\pi$ and $\rho\rho$ modes with $50\,\mathrm{ab}^{-1}$ data (though with two-fold ambiguity) [5]. Even better precision ($\sim 1^\circ$) with a single solution can be achieved with $\pi\rho$ mode.

3.3. Measurement of the angle γ

The angle γ relies on the measurement of direct CP violation in $B^+ \to D^0 K^+$ decays caused by interference between the two amplitudes with different CKM phases (figure 4). The interference

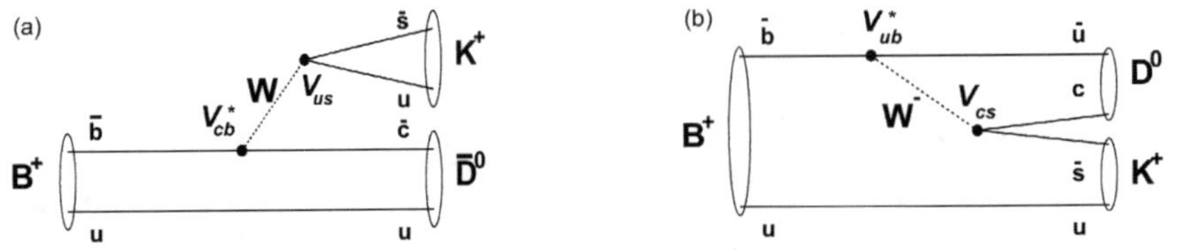

Figure 4. Feynman diagrams for $B^+ \to D^0 K^+$ decays.

appear only if both D^0 and $\overline{D}^0$ mesons decay to a common final state. The method is theoretically clean due to absence of loop contributions. However, the color suppressed amplitude (figure 4b) is a factor of ~ 10 smaller, hence resulting in a small CP asymmetry. There are three methods to measure γ: GLW method [11] uses D^0 decays in CP-eigenstates like K^+K^-; ADS method [12] is based on doubly-Cabibbo suppressed decays like $D^0 \to K^+\pi^-$; GGSZ method [13] is based on a Dalitz analysis of three body D^0 decays such as $D^0 \to K_S\pi^+\pi^-$.

The GGSZ method provides the highest statistical power. Here the amplitude for $B^\pm \to D^0(K_S\pi^+\pi^-)K^+$ decay as a function of Dalitz plot variables $m^2_\pm = m^2(K_S\pi^\pm)$ is given by

$$f_{B^\pm} = f_D(m^2_\pm, m^2_\mp) + r_B e^{\pm\gamma + i\delta} f_D(m^2_\mp, m^2_\pm), \qquad (2)$$

where $f_D(m^2_+, m^2_-)$ is the amplitude of the $D^0 \to K_S\pi^+\pi^-$ decay, r_B is two amplitudes ratio, and δ is a strong phase difference. Once f_D is fixed, a simultaneous fit to $B^\pm$ data allows to extract γ, r_B and δ separately. The f_D can be determined from a large sample of flavor-tagged $D^{*+} \to D^0\pi^+$ decays produced in e^+e^- annihilation. However, with this approach the description of f_D is based on a model, that includes interfering resonances in $K_S\pi^+$, $K_S\pi^-$ and $\pi^+\pi^-$ systems. Using GGSZ method with a model-dependent f_D description Belle's measurement [14] yielded $\gamma = (78.4^{+10.8}_{-11.6} \pm 3.6 \pm 8.9)^\circ$, where the last error comes from the model uncertainty.

While the statistical errors of GGSZ method will be reduced with increased data sample at Super KEKB, the accuracy will be still limited by the model uncertainty. The new approach was tested by Belle [15] to fight this seeming irreducible limitation of the method. Instead of using parametrized f_D function, Belle substituted the Dalitz plot distribution taken directly from the data obtained by CLEO [16] from the decays of quantum-correlated $D^0\overline{D}^0$ pairs produced in the $\psi(3770)$. Belle obtained $\gamma = (77.3^{+15.1}_{-14.9} \pm 4.1 \pm 4.3)^\circ$, where the last error is due to limited precision of CLEO sample. The large model uncertainty is replaced by purely statistical error of CLEO data, which in future can be reduced with the BESIII or Super-c-tau factory data.

The model-independent approach offers an perspective course for studies at Belle II and LHCb. Here the final state is charged and LHCb has an advantage of larger statistics, while Belle II will contribute to the improved systematics. The expected accuracy at Belle II is 2° [5].

3.4. UT summary

Belle and BaBar performed a plenty of analysis to constrain the UT sides as well. The most important Belle II contribution is to constrain V_{ub} from the $b \to u\ell^+\nu$ studies. To improve the accuracy here the contribution from theory, in particular from the lattice QCD, is also critical.

The present status of the UT studies is illustrated with figure 5 produced by the CKMfitter group [10]. Each colored band corresponds to a different kind of process, and looking closely we can see quite nice agreement between independent constrains. Thanks to Belle, BaBar and

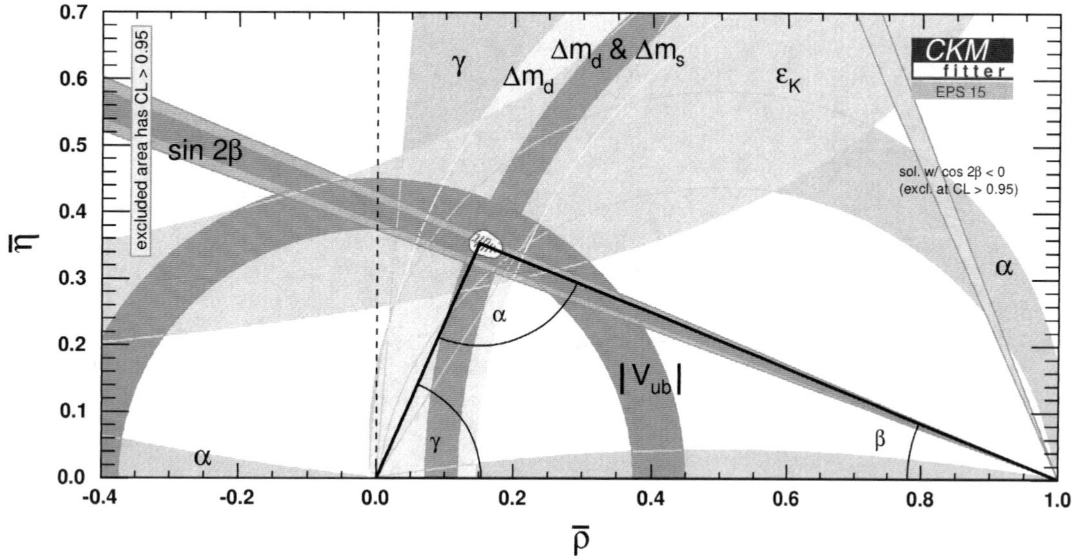

Figure 5. Constraints on the UT as compiled by CKMfitter group [10].

LHCb the square of allowed area for the UT upper apex position is reduced by two order of magnitudes compared to the pre B-factory era. Almost the same factor of improvement in precision is expected after joint efforts of LHCb and Belle II, which hopefully reveals the inconsistency and indicates on NP effects.

4. *CP* violation in penguin dominated modes

It is widely believed that B penguin decays can serve as one of the most sensitive probe for NP due to possible non-SM contribution (*e.g.* from SUSY) in the loop diagram. In particular, manifestations of NP contribution in the penguin modes can be revealed as deviations of CP violation parameters from the SM expectations. In $b \to s\bar{s}s$ hadronic decays the SM weak phase is the same as in the $B^0 \to (c\bar{c})K^0$ transition. Therefore, the main task is to check whether the penguin CP violation parameter $\beta^{\rm eff}$ is equal to β, and the direct CP violation is absent ($A=0$). The SM corrections to these relations are expected to be very small, $\sim$1%.

Earlier Belle CP measurements in $B^0 \to \phi K_S$ [17] showed an exciting $3.5\,\sigma$ deviation of $\beta^{\rm eff}$ from β. The later update [18] showed already quite good but disappointing agreement. Now all penguin modes agree well with the SM expectations, and the precision of $\sin 2\beta^{\rm eff}$ is still statistically limited, typically $0.1 - 0.2$ for different mode. The most precise measurement is obtained with $B^0 \to \eta' K_S$ decay. Obtaining of 1% level sensitivity which provides already a real probe for the NP requires studies at Super KEKB and LHCb. Again, because of a photon in the final state, Belle II has an advantage over LHCb to achieve better sensitivity in $B^0 \to \eta' K_S$.

5. Rare B decays

Precise measurements of rare decays, *i.e.* processes suppressed in the SM, are sensitive to NP at scales exceeding those achievable at LHC. This is demonstrated by the following studies involving both loop and tree decay diagrams.

5.1. Radiative penguin decays

The dominant SM contribution to $b \to s\gamma$ decays is from a loop involving the t-quark and W^+. The measurement of inclusive $\mathcal{B}(b \to s\gamma)$ gives access to the value of V_{ts}, but, what is more important, provides a tool to search for and constrain physics beyond the SM. Indeed, the SM particles in the loop may be replaced by hypothetical particles such as charged Higgs or SUSY particles resulting in observable deviation of the decay rate.

The inclusive $b \to s\gamma$ rate suffers from the minimal theoretical uncertainty. The SM $\mathcal{B}(b \to s\gamma)$ is calculated including NNLO corrections with a $\sim 7\%$ precision. Ideally, the inclusive photon spectrum should be measured over the entire energy range, but practically, its lower part is hardly accessible due to insurmountably large background. The world average for $E_\gamma > 1.6$ GeV, $\mathcal{B}(b \to s\gamma) = (3.55 \pm 0.24 \pm 0.09) \times 10^{-4}$ [20], is consistent with the theoretical prediction, and have been used to constrain NP scenarios, *e.g.*, the charged Higgs is bounded from this measurement to be heavier than 295 GeV. At Belle II one could expect a measurement of $\mathcal{B}(b \to s\gamma)$ with a relative accuracy of 6% [5], which matches the anticipated precision of the theoretical predictions. Belle II can also measure $\mathcal{B}(b \to d\gamma)$ with $\sim 25\%$ accuracy that can serve to check consistency of the $|V_{td}/V_{ts}|$ value obtained in this measurement and in the ratio of B_s and B_d mixing strengths.

5.2. $B \to \tau\nu$

The pure leptonic $B \to \tau\nu$ decay is sensitive to the NP including the SM extensions with charged Higgs, that could significantly suppress or enhance the branching ratios for these decays. Experimentally it is a real challenge to identify modes with τ lepton due to multiple neutrinos in the final state. At the (Super) B-factories, the exclusive production of a B meson pair with no extra particles allows to tag the signal decay by reconstruction of all particles originating from the accompanying B meson. Using the hermecity of the detector it is possible to identify the signal as absence of the energy deposited in the detector not associated with the tag and signal particles (see figure 6). This mode is hardly accessed at LHCb.

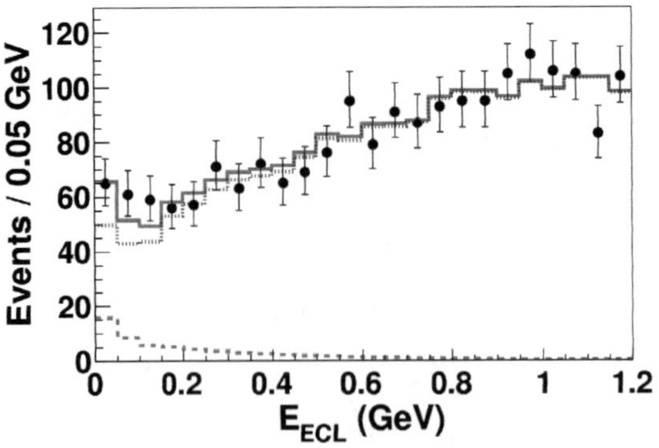

Figure 6. Distributions of deposited energy in the electromagnetic calorimeter at Belle for $B \to \tau\nu$ candidates [21]. The signal is seen as an excess of events with low energy deposition.

The first evidence for $B \to \tau\nu$ was reported by Belle [21] in 2006 and now Belle can see this decay with a $4.6\,\sigma$ significance [22]. The measured by Belle branching ratio with hadronic and semileptonic tag, $\mathcal{B}(B \to \tau\nu) = (0.91 \pm 0.19 \pm 0.11) \times 10^{-4}$ [22], sets constraints on the parameters of various models involving charged Higgs bosons. For large $\tan\beta$ the excluded region are more stringent than those obtained from the direct searched at LHC. At Belle II accumulation of luminosity helps to reduce both statistical and systematic errors by a factor of ~ 7 and bound further the constrain on the charged Higgs mass provided that the SM calculations will be also improved with the help of lattice QCD.

5.3. Semileptonic decays with τ

The SM predicts $B\to D^{(*)}\tau\nu$ branching fractions to be $\gtrsim 1\%$, *i.e.* strictly speaking these modes are not rare. The results are usually presented in terms of $R(D^{(*)}) \equiv \mathcal{B}(B\to D^{(*)}\tau\nu)/\mathcal{B}(B\to D^{(*)}\ell\nu)$, which is independent of the V_{cb} and form factor parameterization and can be compared with the SM expectations.

The decays $B\to D^{(*)}\tau\nu$ are observed now at Belle and BaBar with high significance, and their $R(D^{(*)})$ values are consistent [23]. Recently LHCb also reported the measurement of $R(D^{*+})$. The world averaged results, $R(D^*)=0.322\pm0.018\pm0.012$ and $R(D)=0.391\pm0.041\pm0.028$, show a $3.9\,\sigma$ disagreement with the SM [20]. It is worth to mention that they also disfavor the type II two-Higgs doublet model. With larger statistics of Super KEKB besides the improved accuracy of $R(D^{(*)})$ ($\sim 2\%$), the q^2 and the angular distributions of the τ and $D^{(*)}$ decays could also provide useful information for testing the SM and constraining NP models.

6. Conclusion

The Belle II experiment has an important mission to search for NP in the flavor sector exploiting a huge jump in luminosity and a plenty of independent measurements. If the NP will be observed at LHC before the start of Belle II, the flavor sector of NP still need to be constrained, which is only possible with both LHCb and Super B-factory complimentary programs. The Belle II experiment has a broad physics program uncovered by LHCb, as many important measurements can not be made at hadronic machines.

The SuperKEKB commissioning will start in 2016 while the construction of the Belle II detector is ongoing and expected to start data taking 2017. The aim of the Belle II project is to accumulate $50\,\mathrm{ab}^{-1}$, corresponding to about 55 billion $B\overline{B}$ pairs by the year 2023. The projected sensitivities with this data are below 1° for β, and $\sim 1^\circ$ for α and γ. The accuracy in CP violation studies, branching fractions and kinematics characteristics in rare B decays will be also improved by an order of magnitude. The examples described above are only a small part of possible measurements to be performed with the Belle II experiment. A more detailed overview can be found in Ref. [5].

Acknowledgments

The work is supported by the Russian Ministry of Education and Science under the contract 14.A12.31.0006.

References

[1] Aad G *et al.* (ATLAS Collaboration) 2012 *Phys. Lett.* B **716** 1
Chatrchyan S *et al.* (CMS Collaboration) 2012 *Phys. Lett.* B **716** 30
[2] Cabibbo N 1963 *Phys. Rev. Lett.* **10** 531
Kobayashi M and Maskawa T 1973 *Prog. Theor. Phys.* **49** 652
[3] Sakharov A D 1967 *Pisma Zh. Eksp. Teor. Fiz.* **5** 32
[4] Adachi I *et al.* (Belle Collaboration) 2012 *Phys. Rev. Lett.* **108** 171802
[5] Aushev T *et al.* 2010 *arXiv*:1002.5012
[6] Gronau M and London D 1990 *Phys. Rev. Lett.* **65** 3381
[7] Abe K *et al.* (Belle Collaboration) 2005 *Phys. Rev. Lett.* **94** 181803
[8] Dalseno J *et al.* (Belle Collaboration) 2013 *Phys. Rev.* D **88** 092003
[9] Kusaka A *et al.* (Belle Collaboration) 2008 *Phys. Rev.* D **77** 072001
[10] Charles J *et al.* (CKMfitter Group) 2005 *Eur. Phys. J.* C **41** 1 (*http://ckmfitter.in2p3.fr/*)
[11] Gronau M and London D 1991 *Phys. Lett.* B **253** 483
Gronau M and Wyler D 1991 *Phys. Lett.* B **265** 172
[12] Atwood D, Dunietz I and Soni A 1997 *Phys. Rev. Lett.* **78** 3257

Atwood D, Dunietz I and Soni A 2001 *Phys. Rev.* D **63** 036005
[13] Giri A *et al.* 2003 *Phys. Rev.* D **68** 054018
Bondar A 2002 *Proc. of BINP Analysis Meeting on Dalitz Analysis*
[14] Poluektov A *et al.* (Belle Collaboration) 2010 *Phys. Rev.* D **81** 112002
[15] Aihara H *et al.* (Belle Collaboration) 2012 *Phys. Rev.* D **85** 112014
[16] Briere R A *et al.* (CLEO Collaboration) 2009 *Phys. Rev.* D **80** 032002
[17] Abe K *et al.* (Belle Collaboration) 2003 *Phys. Rev. Lett.* **91** 261602
[18] Chen K F *et al.* [Belle Collaboration] 2007 *Phys. Rev. Lett.* **98** 031802
[19] Limosani A *et al.* (Belle Collaboration) 2009 *Phys. Rev. Lett.* **103** 241801
[20] Amhis Y *et al.* (Heavy Flavor Averaging Group) *arXiv*:1207.1158 (*http://www.slac.stanford.edu/xorg/hfag*)
[21] Ikado K *et al.* (Belle Collaboration) 2006 *Phys. Rev. Lett.* **97** 251802
[22] Kronenbitter B *et al.* (Belle Collaboration) 2015 *Phys. Rev.* D **92** 051102
[23] Lees J P *et al.* (BaBar Collaboration) 2012 *Phys. Rev. Lett.* **109** 101802
Huschle M *et al.* (Belle Collaboration) 2015 *arXiv*:1507.03233
Aaij R*et al.* (LHCb Collaboration) 2015 *arXiv*:1506.08614

The charged Z_c states from rescattering in conventional B decays

P Pakhlov and T Uglov

National Research Nuclear University MEPhI (Moscow Engineering Physics Institute), Kashirskoe highway 31, Moscow, 115409, Russia

E-mail: uglov@itep.ru

Abstract. In this paper we discuss a possible interpretation of the charged charmonium-like states observed in B decays as a rescattering effect. This approach allows to avoid introduction of exotic hadrons into the theory, while explains the unusual peaking structures 4observed recently in three body B decays. In particular, the peaks in the $\psi(2S)\pi^+$ mass spectrum are ascribed to the rescattering process with kinematics governed by the presence of conventional radial excitations, the $D_s^{(*)\prime}$ resonances, in the hidden intermediate state. Our predictions are compared to the LHCb results.

The Belle discovery in 2003 [1] of an unexpected narrow structure in $J/\psi\pi\pi$ invariant mass established a new era of seemingly well understood quarkonium physics. This resonance, called the $X(3872)$, is the first charmonium-like state, that was found not to fit into the conventional quarkonium model. Since the observation of the $X(3872)$ the list of exotic charmonium is updated with few dozens of XYZ states [2]. None of the various either exotic (tetraquarks, molecular states, charmonium hybrids and hadrocharmonium) or conventional (potential models revision *etc.*) explanations of these states can adequately describe the full range of the observed phenomena.

First charged charmonium-like state, the $Z^+(4430)$, was observed by Belle [3] in 2007 as a resonance structure in the $\psi(2S)\pi^+$ spectrum from $B \to \psi(2S)\pi^+K^-$ decay. While BaBar has not confirmed this observation [5], it also could not rule out the $Z^+(4430)$ existence. Recent LHCb analysis puts an end to the controversy by observation of the $Z^+(4430)$ with extremely high significance and study of the resonant-like properties of the peak [6], including the complex phase motion. Being interpreted as a real resonance containing a $c\bar{c}$ pair, this state necessarily consists of at least an extra pair of light quark-antiquark due to its non-zero charge. The most popular interpretations of the $Z^+(4430)$ are tetraquark [7], hadrocharmonium [8], and $D\bar{D}^*$ molecules [9], and non-resonant interpretations such as the "cusp effect" [10], rescattering via the chain $\bar{B} \to D^{*-}D_1(2420)K \to \psi(2S)\pi^+K$ [11], and the initial single pion emission mechanism [12].

Recently we have suggested another interpretation of the $Z^+(4430)$ [13, 14] as an effect of rescattering from the open charm channel into a final state with a charmonium and a pion. The peaking structures in $\psi(2S)\pi^+$ spectrum are explained by the presence of the conventional resonance ($D_s^{\prime -}$ meson) in the hidden intermediate state (conventional B decay into a pair of anti-charm and charm-strange mesons). The Feynman diagrams illustrating this process are shown in figure 1. The main feature of the proposed model is a non-vanishing rescattering

$(\bar{D}^{*0}D^{*+}) \leftrightarrow (\psi(2S)\pi^+)$ amplitude over a wide range of $(\bar{D}^{*0}D^{*+})$ masses. It turns out, that the suggested approach is able to explain the Breit-Wigner-like shape of the peak in $\psi(2S)\pi^+$ spectrum, giving a similar complex phase motion. We note that the quantum numbers of $Z^+(4430)$, $J^P = 1^+$, were predicted within this approach before they were measured [13] and then have been confirmed by Belle [15, 16] and LHCb [6]. Other structures in $\psi(2S)\pi^+$ spectrum, in particular a broad bump near $M \sim 4200\,\text{MeV}$, discovered by the LHCb collaboration and interpreted as another Z^+ resonance, were also predicted.

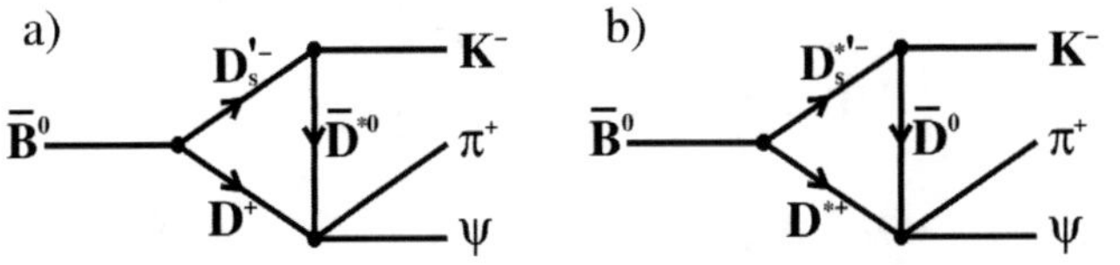

Figure 1. Triangle Feynman diagrams for $\bar{B} \to \psi(2S)\pi^+K^-$ decay with $D_s^{\prime -}$ (a) and $D_s^{*\prime -}$ (b) mesons in the hidden intermediate state.

Within the proposed approach B meson decays via the tree diagram to the D_s^{*-} radial excitations ($D_s^{(*)\prime -}$) and $D^{(*)+}$ meson. These Cabibbo-favored decays are expected to have a large branching fractions, moreover one of them was observed by Belle [17] with $\mathcal{B}(B^- \to D_{s1}^*(2700)^- D^0) \times \mathcal{B}(D_{s1}^*(2700)^- \to \bar{D}^0 K) \sim 10^{-3}$. Although other modes have not been observed yet, the large inclusive $B \to D^{(*)}\bar{D}^{(*)}K$ fractions [18] suggest the branching fraction for $B \to D_s^{(*)\prime -}D^{(*)+}$ decays to be large as well. In the three body final state $D^{(*)}\bar{D}^{(*)}K$ from these modes, two D mesons are produced with a low relative momenta ($v/c \approx 0.3 - 0.5$). Therefore one can expect the non vanishing rescattering of two charmed mesons into charmonium and a light meson. The diagrams, corresponding to the two decay chains, that only could contribute to the $\psi(2S)\pi^+K$ final state due to parity conservation,

$$\bar{B} \to D_s^{\prime -}D^+, \text{ followed by } D_s^{\prime -} \to \bar{D}^{*0}K^-, \tag{1}$$

and

$$\bar{B} \to D_s^{*\prime -}D^{*+}, \text{ followed by } D_s^{*\prime -} \to \bar{D}^0 K^-, \tag{2}$$

are shown in figure 1a,b. Amplitudes, corresponding to these triangle diagrams, are calculated within on-shell approximation taking into account the $D_s^{(*)\prime -}$ Breit-Wigner amplitude. We assume the s-wave rescattering dominates. The full decay amplitude has the following form in the helicity formalism:

$$\mathcal{A}_{M_{Z^+(4430)} \equiv M_{(\bar{D}D)^{*+}}} = \sum_{\lambda} \int \mathcal{A}_{BW}(M_{D_s^{(*)\prime -}}) D^J_{0,\lambda}(\theta_{\text{dec}})\; D^1_{\lambda,0}(\theta_{\text{rot}})\; D^1_{0,0}(\theta_{\text{form}})\; dM_{D_s^{(*)\prime -}}, \tag{3}$$

where J is the $D_s^{(*)\prime -}$ spin; θ_{dec} is the decay angle of the $D_s^{(*)\prime -}$ (the angle between the $\bar{B}$ and $\bar{D}^{(*)0}$ in the $D_s^{(*)\prime -}$ rest frame); θ_{rot} is the rotation angle of the $\bar{D}^{*0}$ spin from the $D_s^{(*)\prime -}$ frame for the reaction (1) or the $\bar{B}$ frame for the reaction (2) to the $Z^+(4430)$ frame; θ_{form} is the formation angle of $Z^+(4430)$, *i.e.* the angle between the $\bar{B}$ and D^* in the $Z^+(4430)$ rest frame. The first Wigner D-function is responsible for the proper angular distribution of the $D_s^{(*)\prime -}$ decay (in the case considered in (2), only the zero helicity projection works). The second function, $D^1_{0,\lambda}(\theta_{\text{rot}})$, describes the D^* spin rotation from the frame where it is produced to the frame where it is absorbed. Finally, the $D^1_{0,0}(\theta_{\text{form}})$ corresponds to the proper formation of the spin-1 $Z^+(4430)$ pseudostate from the vector (D^*) and the pseudoscalar (D). Two variables, $M_{D_s^{(*)\prime -}}$ and θ_{dec}, fully describe the three-body kinematics, thus $M_{(\bar{D}D)^{*+}}$, $\theta_{\rm rot}$ and $\theta_{\rm form}$ are functions of these two variables.

We have checked the plausibility of our hypothesis using the LHCb data. The distribution of $M^2(\psi(2S)\pi^+)$ for the $1.0 < M^2(K^-\pi^+) < 1.8\,{\rm GeV}^2$ interval borrowed from [6] is shown in figure 2a. In this spectrum the contributions from $K^*(890)$, $K_2^*(1430)$ and non-resonant three-body decays as determined by the LHCb fit have been subtracted incoherently. This remaining spectrum we attribute to the rescattering contribution and have performed a toy fit with expected rescattering shape to the residual spectrum. Ignoring of the phase-dependent interference effects is not quite correct procedure, but we can not perform a better way as we have only access to the published LHCb one-dimensional plots. We thus use it for illustration of plausibility of our approach only. In our fit the amplitudes of the reactions (1) and (2) (the later consists of two contributions with $\lambda = 0$ and 1) are free parameters. These amplitudes were calculated numerically from the equation 3. The fit results are shown in figure 2a: the solid line represents the fitting function; the dashed, dotted and dashed-dotted curves show the contributions from the process (1) and (2) with λ=1 and λ=0, respectively.

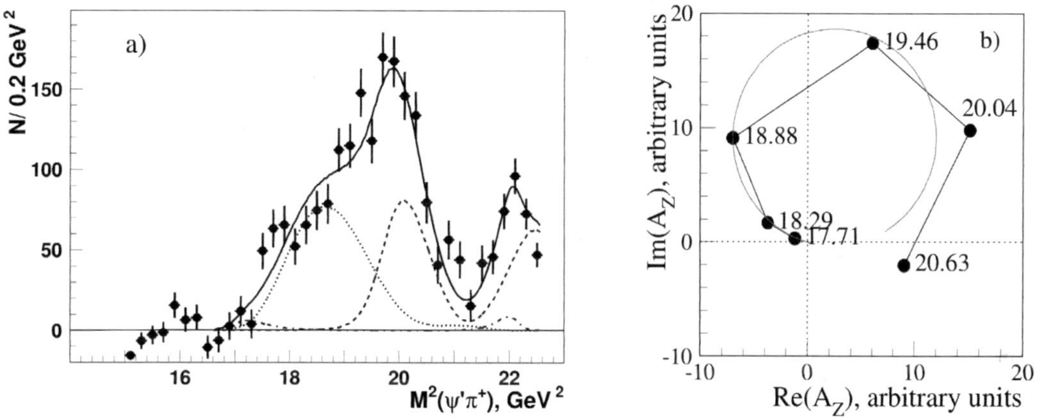

Figure 2. a) The distribution of $M^2(\psi(2S)\pi^+)$ after incoherent subtraction of contributions from $K^*(890)$, $K_2^*(1430)$ and non-resonant three-body decays extracted from the LHCb data [6]. The solid line represents fit to the data points. The dashed, dotted and dashed-dotted curves represent contributions from the process (1) and (2) with $\lambda = 1$ and $\lambda = 0$, respectively. b) Argand diagram for the toy fit results near the $Z^+(4430)$ region.

The phase of the Breit-Wigner-like structure near $M^2(\psi(2S)\pi^+) \sim M^2_{Z^+(4430)}$, obtained from our fit, is shown as Argand diagram in figure 2b in the same $M^2(\psi(2S)\pi^+)$ bins as in LHCb paper [6]. Initial phase is set to an arbitrary value, while in the LHCb case it is fixed from the 4-dimensional fit. The Breit-Wigner-like variation around the peak comes from the $D_s^{(*)\prime -}$ decay's Breit-Wigner amplitude. However, the direction of the phase motion is flipped with respect to those of usual Breit-Wigner: it tends to rotate clockwise in the Argand diagram. Since there is an ambiguity ($\mathcal{A} \leftrightarrow \bar{\mathcal{A}}$) in the extraction of the $Z^+(4430)$ amplitude from the measured $|\mathcal{A}(Z^+) + \mathcal{A}({\rm non-}Z^+)|^2$, our result does not contradict to the LHCb data.

Using the toy fit described above, one can extract the parameters of the $D_s^{\prime -}$, which turn out to be well statistically constrained: $M = (2614 \pm 4)\,{\rm MeV}$, $\Gamma = (92 \pm 10)\,{\rm MeV}$. One of the main sources of the systematic uncertainties in this study comes from the effect of interference with the $K^{*(*)}$ background. This source was estimated by ascribing different phases to the amplitudes of $K^*(890)$, $K_2^*(1430)$ and s-wave three-body phase space and performing another fit with varying $D_s'^-$ mass and width. Variations of the best fit $D_s'^-$ parameters depending on the $K^{*(*)}$ phases are estimated to be ±10 MeV for the $D_s'^-$ mass and $^{+20}_{-13}$ MeV for its width. Thus, parameters of $D_s'^-$ meson are: $M = (2614 \pm 4^{+20}_{-13})$ MeV, $\Gamma = (92 \pm 10 \pm 10)$ MeV.

The discussed approach could be ultimately tested by performing 4-dimensional fit by either Belle, BaBar or LHCb for $B \to \psi(2S)\pi K^-$ decays using amplitudes (3) instead of resonance-like Z^+'s. While this analysis could be complicated, there is another simple way to test our hypothesis: the Z^+-like structures should appear in the distributions of $M_{(D^*_\perp \bar{D})^+} \times \cos^2(\theta_{\rm form})$ in either $B \to \bar{D}^{*0} D^+ K^-$ or $B \to \bar{D}^0 D^{*+} K^-$ decays, or in both. The $M_{(D^*_\perp \bar{D})^+} \times \cos^2(\theta_{\rm form})$ is the $(D^*_\perp \bar{D})^+$ combination mass spectrum corrected in each bin for the fraction of the D^* transverse component in the $(D\bar{D})^{*+}$ rest frame, and also the 1^+ formation factor $D^2(\theta_{\rm form}) = \cos^2(\theta_{\rm form})$.

If our hypothesis is correct, once the rescattering amplitude is known from B decays, it should be universal in other processes, where a $D\bar{D}$ pair produced. Among such process are the charmed meson production in the e^+e^- annihilation, $\gamma\gamma$ fusion *etc.* In particular, the process $e^+e^- \to D\bar{D}^*\pi$ can result via rescattering into $e^+e^- \to J/\psi(\psi(2S))\pi\pi$ final state, which is observed and remains puzzling for almost a decade [2]. Unfortunately, the former process has not been studied in details experimentally: the cross section of $e^+e^- \to D\bar{D}^*\pi$ has been measured only by Belle [19] with not sufficient precision. Moreover, to check that rescattering hypothesis works to explain the $e^+e^- \to J/\psi(\psi(2S))\pi\pi$ cross section, one needs the measured helicity decomposition of the $D\bar{D}^*\pi$ final state.

On the other hand, the recent observation of charmonium-like pentaquark by LHCb [20] suggests the similar explanation of the peaking structure observed in the $J/\psi p$ invariant mass distribution. Indeed, if the rescattering $\Lambda_c D \to J/\psi p$ is not small the underlying process that leads to the peaking structure in $J/\psi p$ mass could be the conventional decay $\Lambda_b \to \Lambda_c D_s^{*\prime} \to \Lambda_c DK$.

In summary, we suggest a model, which is able to describe the observed features in the $\psi(2S)\pi^+$ spectrum from the $B \to \psi(2S)\pi^+ K^-$ without introducing of exotic multiquark states. A toy fit to the published LHCb data demonstrates the plausibility of our approach. A critical test for this hypothesis is suggested.

Acknowledgments

This work was performed within the framework of the Center of Nanostructured Electronics supported by MEPhI Academic Excellence Project (contract No. 02.a03.21.0005, 27.08.2013) and was supported in part by RFBR under the contract 15-32-20810\15.

References

[1] Choi S-K *et al.* (Belle Collaboration) 2003 *Phys. Rev. Lett.* **91** 262001
[2] Pakhlova G V, Pakhlov P N and Eidelman S I 2010 *Phys. Usp.* **53** 219
[3] Choi S-K *et al.* (Belle Collaboration) 2008 *Phys. Rev. Lett.* **100** 142001
[4] Mizuk R *et al.* (Belle Collaboration) 2009 *Phys. Rev.* D **80** 031104
[5] Aubert B *et al.* (BaBar Collaboration) 2009 *Phys. Rev.* D **79** 112001
[6] Aaij R *et al.* (LHCb Collaboration) 2014 *Phys. Rev. Lett.* **112** 222002
[7] Liu X-H, Zhao Q and Close F E 2008 *Phys. Rev.* D **77** 094005
[8] Dubynskiy S and Voloshin M B 2008 *Phys. Lett.* B **666** 344
[9] Liu X *et al.* 2008 *Phys. Rev.* D **77** 034003
[10] Bugg D V 2008 *J. Phys.* G **35** 075005
[11] Rosner J L 2007 *Phys. Rev.* D **76** 114002
Meng G Z *et al.* 2009 (CLQCD Collaboration) *Phys. Rev.* D **80** 034503
[12] Chen D-Y and Liu X 2011 *Phys. Rev.* D **84** 094003
[13] Pakhlov P 2011 *Phys. Lett.* B **702** 139
[14] Pakhlov P and Uglov T 2015 *Phys. Lett.* B **748** 183
[15] Chilikin K *et al.* (Belle Collaboration) 2013 *Phys. Rev.* D **88** 074026
[16] Chilikin K *et al.* (Belle Collaboration) 2014 *Phys. Rev.* D **90** 112009

[17] Brodzicka J *et al.* (Belle Collaboration) 2008 *Phys. Rev. Lett.* **100** 092001
[18] Olive K A *et al.* (Particle Data Group) 2014 *Chin. Phys.* C **38** 090001
[19] Pakhlova G *et al.* (Belle Collaboration) 2009 *Phys. Rev.* D **80** 091101
[20] Aaij R *et al.* (LHCb Collaboration) 2015 *Phys. Rev. Lett.* **105** 115

Journal of Physics: Conference Series **675** (2016) 022011 doi:10.1088/1742-6596/675/2/022011

Recent PHENIX results on hard probes and direct photon production

V Riabov[1,2] for the PHENIX collaboration
[1] PNPI of NRC "Kurchatov Institute", Orlova rostcha-1, Gatchina, 188300, Russia
[2] National Research Nuclear University MEPhI (Moscow Engineering Physics Institute), Kashirskoe highway 31, Moscow, 115409, Russia

E-mail: riabovvg@mail.pnpi.spb.ru

Abstract. A hot and dense matter called strongly interacting quark-gluon plasma (sQGP) is created in heavy ion collisions at RHIC energies. Detailed study of the properties of this new state of matter is a driving force of recent research at RHIC. In these proceedings we present most recent PHENIX results for system size and energy dependence of hadron and jet production at high transverse momentum in heavy ion collisions at RHIC. We also report latest results for direct photon production including soft direct photon yields and anisotropic flow.

1. Introduction

The main purpose of recent researches at RHIC is a detailed study of the properties of strongly interacting quark-gluon plasma (sQGP). Such a new state of matter is believed to be produced in central heavy ion collisions [1]. Properties of sQGP are studied by measuring yields and angular distributions of various particles produced in the final state. The main difficulty in this research is that measured particles are produced at different stages of the interaction and their properties relevant to sQGP stage can be smeared by later hadronic interactions. Reliable information about properties of sQGP could be gained by measuring as many signatures of sQGP as possible in dependence on size of the interacting system and collision energy.

In this proceeding we present the most recent PHENIX results for two very informative observables which are hadrons and jets at high transverse momentum (p_T) as well as soft direct photons. Hard scattered partons loose energy via rescattering and radiation traversing the medium produced in heavy ion collisions resulting in jet quenching. Thus measurement of high p_T hadron and jet production is a way to probe opacity and density of the produced medium. The electromagnetic probes such as direct photons are not affected by strong nuclear forces. Thus they are extremely valuable in constraining the time evolution of the medium.

2. Hadron production at high and intermediate transverse momentum

Production of identified hadrons and jets has been measured in p+p and heavy ion collisions. Potential collective effects in nuclear collisions are numerically studies with nuclear modification factor R_{AA}, which is a ratio of the yields measured for the same observable in A+A and p+p collisions scaled by the corresponding number of binary inelastic nucleon-nucleon collisions (N_{coll}), which is estimated using Glauber model [2].

Figure 1 shows nuclear modification factors (R_{dAu}) measured for reconstructed jets in d+Au collisions at $\sqrt{s_{NN}}$ = 200 GeV [3]. Plot on the figure 1a corresponds to minimum bias collisions while plot on the figure 1b shows centrality dependence of R_{dAu}. In minimum bias collisions R_{dAu} is consistent with unity within uncertainties in the entire measurement range thus providing strong constraints on the initial state effects over a wide kinematic range. Results in minimum bias collisions are consistent with calculations that account for modification of parton distribution functions in nuclei with EPS09 parameterization [4]. Models that introduce energy loss agree with data better when assume smaller momentum transfers [5]. Figure 1b shows strong centrality dependence of R_{dAu} for inclusive jets. Observed centrality dependence presents a challenge for theoretical models, one of possible explanations is proton colour fluctuations [6]. Presented measurements in d+Au and future measurements in other collision systems like p+Au, ^{3}He+Au will help to test various theoretical models for centrality dependence of jet modification in small collision systems.

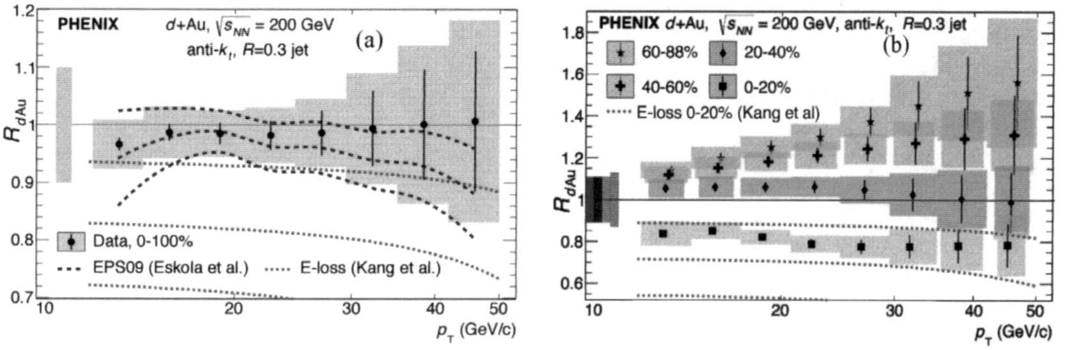

Figure 1. Nuclear modification factors measured for jets in d+Au collisions at $\sqrt{s_{NN}}$ = 200 GeV. Figure 1a corresponds to minimum bias collisions; results are presented in comparison with theoretical model calculations. Figure 1b shows centrality dependence of the nuclear modification factor.

Figure 2 shows preliminary PHENIX results for nuclear modification factors (R_{AA}) measured for reconstructed jets in Cu+Au collisions at $\sqrt{s_{NN}}$ = 200 GeV. In most peripheral collisions R_{AA} is somewhat larger than unity but is consistent with it at high transverse momentum. The R_{AA} deceases with increasing centrality and in most central collisions production of jets is suppressed by about a factor of two. Suppression does not show any significant dependence on p_T at all centralities.

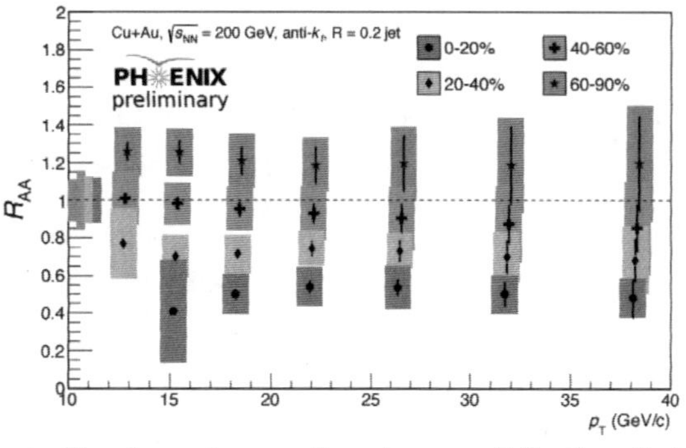

Figure 2. Centrality dependence of nuclear modification factor measured for reconstructed jets in Cu+Au collisions at $\sqrt{s_{NN}}$ = 200 GeV.

Another important result is collision energy dependence of jet quenching. Nuclear modification factors were measured for π^0's in Cu+Cu and Au+Au collisions at $\sqrt{s_{NN}}$ = 22-200 GeV [7, 8]. Additional information comes from charged hadron measurements at the LHC in Pb+Pb collisions at $\sqrt{s_{NN}}$ = 2.76 TeV [9]. These measurements show suppression of hadron production at collision energy

$\sqrt{s_{NN}} > 39$ GeV. Measurements at $\sqrt{s_{NN}} = 22$ GeV show no suppression consistent with previous measurements at the SPS where suppression was not observed also. Measurements also showed very similar suppression at high p_T of neutral pions in Au+Au collisions at $\sqrt{s_{NN}}$ = 39-200 GeV; similar suppression of neutral pions in Au+Au collisions at $\sqrt{s_{NN}}$ = 200 GeV and charged hadrons in Pb+Pb collisions at $\sqrt{s_{NN}}$ = 2.76 TeV. Similarity of the nuclear modification factors measured in heavy ion collisions at different energies means that this parameter is biased by the shapes of particle production spectra. Instead one can look at fractional momentum loss [1,10]. Figure 3 shows dependence of fractional momentum loss on charged particle multiplicity as a measure of energy density. Regardless of similarity of nuclear modification factors the fractional momentum losses are quite different. Momentum losses increase by ~ 50% from central Au+Au collisions at $\sqrt{s_{NN}}$ = 200 GeV to most central Pb+Pb collisions at $\sqrt{s_{NN}}$ = 2.76 TeV and by about a factor of six from 62 GeV to 2.76 TeV. Momentum losses monotonously increase with multiplicity and all measured points merge to one universal curve at high values of particle multiplicity independent of collision energy. One can also see that the same energy points corresponding to different nuclei follow the same scaling.

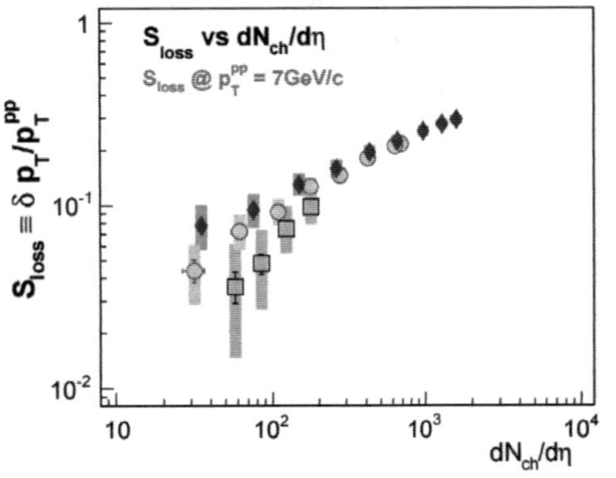

Figure 3. Fractional momentum loss in different collision systems as a function of charged particle multiplicity.

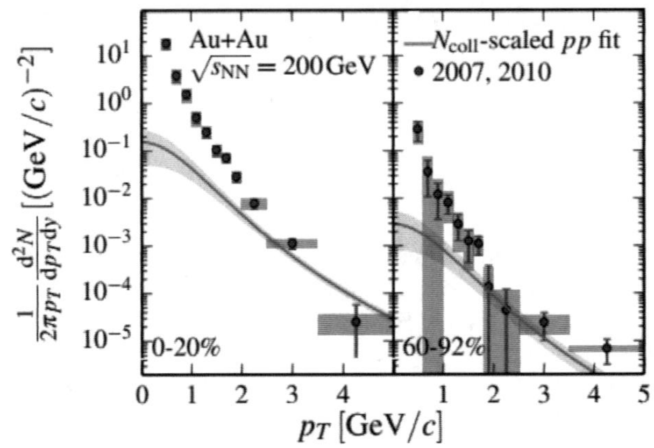

Figure 4. Direct photon p_T spectra in Au+Au collisions at $\sqrt{s_{NN}}$ = 200 GeV. Green curves correspond to N_{coll} scaled p+p.

3. Direct photons

Recently PHENIX measured centrality dependence of direct photon production in Au+Au collisions at $\sqrt{s_{NN}}$ = 200 GeV [11]. In figure 4 results are compared to yields measured in p+p collisions at the same energy of $\sqrt{s}$ = 200 GeV and scaled by N_{coll}. Direct photon yields in Au+Au collisions significantly exceed scaled p+p yields at all centralities. Fits of excess yields to exponential function return slope parameters equal to ~ 240 MeV with no centrality dependence within uncertainties. At the same time photon yields depend on centrality and scale with number of participating nucleons (N_{part}) as $N_{part}{}^{\alpha}$, where $\alpha = 1.48 \pm 0.08$ (stat.) $\pm$ 0.04 (syst.).

Important result is a measurement of direct photon anisotropy at low transverse momentum [12]. Figure 5 show dependence of elliptic (v_2) and triangular (v_3) flow on transverse momentum measured in different centrality bins. Flow was measured using two different methods; results shown with black and green markers agree well in the overlap region. Values of v_2 and v_3 are very similar to that previously measured for light hadrons. Similar to hadrons v_3 shows very weak dependence on collision centrality.

4. Conclusion

New measurements of jets and soft direct photons confirm previous findings and present a challenge for theoretical models. Theoretical models need to assume very early emission of direct photons from

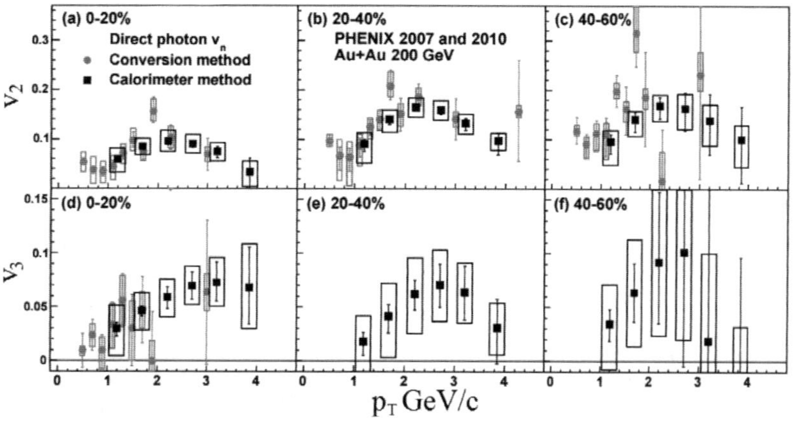

Figure 5. Elliptic and triangular flow for soft direct photons in Au+Au collisions at $\sqrt{s_{NN}}$ = 200 GeV. Black and green markers to correspond to different measurement techniques.

the interacting system describe large excessive yields of direct photons. Large values of flow coefficients point to late emission of direct photons when temperature is lower but collective flow has time to develop. These two observations are difficult to reconcile within the same models and there is still no satisfactory description yet available.

Acknowledgments
This work was partially supported by MEPhI Academic Excellence Project (contract No. 02.a03.21.0005, 27.08.2013).

References

[1] Adcox K *et al.* 2005 *Nucl.Phys.*A **757** 184-283
[2] Adcox K *et al.* 2001 *Phys. Rev. Lett.* **86** 3500
[3] Adare A *et al.* 2015 *arXiv*:1509.04657
[4] Eskola K J et al. 2009 *J. High Energy Phys.* **04** 065
[5] Kang Z B *et al.* 2012 *Phys. Lett.* B **718** 482-487
[6] Bzdak A *et al.* 2015 *arXiv*:1408.3156
[7] Adare A *et al.* 2008 *Phys. Rev. Lett.* **101** 162301
[8] Adare A *et al.* 2012 *Phys. Rev. Lett.* **109** 152301
[9] Aamodt K *et al.* 2011 *Phys.Lett.* B **696** 30-39
[10] Adare A *et al.* 2015 *arXiv*:1509.06735
[11] Adare A *et al.* 2015 *Phys. Rev.* C **91** 064904
[12] Adare A *et al.* 2015 *arXiv*:1509.07758

Computer analysis of nuclear track emulsion exposed to thermal neutrons and Cf source

K Z Mamatkulov[1], I Ambřozová[2], D A Artemenkov[1,3], V Bradnova[1], D V Kamanin[1], R R Kattabekov[1], L Majling[2], A Marey[4], O Ploc[2], V V Rusakova[1], R Stanoeva[5], K Turek[2], A A Zaitsev[1,6], P I Zarubin[1,6] and I G Zarubina[1,6]

[1] Joint Institute for Nuclear Research, Joliot-Curie 6, Dubna, Moscow region,141980, Russia
[2] Nuclear Physics Institute, of the ASCR, v.v.i., Řež 130, 250 68 Řež, Czech Republic
[3] National Research Nuclear University MEPhI (Moscow Engineering Physics Institute), Kashirskoe highway 31, Moscow, 115409, Russia
[4] Minoufia University, Tala't street, EL azhar territory, Menouf city, Menofia, Egypt
[5] South-West University, Ivan Michailov st. 66, 2700 Blagoevgrad, Bulgaria
[6] P. N. Lebedev Physical Institute of the Russian Academy of Sciences, Leninskij Prospekt 53, Moscow, 119991, Russia

E-mail: zarubin@lhe.jinr.ru

Abstract. Application of the nuclear track emulsion technique (NTE) in radioactivity and nuclear fission studies is discussed. It is suggested to use a HSP-1000 automated microscope for searching for a collinear cluster tri-partition of heavy nuclei implanted in NTE. Calibrations of α-particles and ion ranges in a novel NTE are carried out. Surface exposures of NTE samples to a Cf-252 source started. Planar events containing fragments and long-range α-particles as well as fragment triples only are studied. Splittings induced by thermal neutrons are studied in boron-enriched emulsion. Use of the image recognition program "ImageJ" for obtaining characteristics of individual events and for events from the large scan area is presented.

1. Introduction

Possessing an excellent sensitivity and spatial resolution nuclear track emulsions preserve their position as a universal and relatively cheap detector for surveying and searching investigations into nuclear and particle physics [1-3]. Unique opportunities of NTE deserve further use in fundamental and applied research in state-of-art accelerators and reactors, as well as with sources of radioactivity, including natural ones. Application of NTE is especially justified in those pioneering experiments in which nuclear particle tracks cannot be reconstructed with the help of electronic detectors.

The NTE technique is based on intelligence, vision and performance of researchers using traditional microscopes. Despite widespread interest, its labor consumption causes limited samplings of hundreds of measured tracks which present as a rule only tiny fractions of the available statistics. Implementation of computerized and fully automated microscopes in the NTE analysis allows one to bridge this gap. These are complicated and expensive devices of collective or even remote usage allow one to describe unprecedented statistics of short nuclear tracks.

To make such a development purposeful it is necessary to focus on such a topical issue of nuclear physics the solution of which can be reduced to simple tasks of recognition and measurement of tracks in NTE to be solved with the aid of already developed programs. One of the suggested problems is a search for the possibility of a collinear cluster tri-partition [4]. The existence of this phenomenon could be established in observations of such a type of ternary fission of heavy nuclei in which a lightest fragment is emitted in the direction of one of the heavy fragments.

Despite distinct observability of fission fragments they can not be fully identified in NTE. However, NTE is valuable due to combination of the best angular resolution and maximum sensitivity. Besides, it is possible to measure the lengths and thicknesses of tracks, and, thus, to classify the fragments. As an initial stage, to provide statistics of ternary fissions it is suggested to analyze a sufficient NTE area exposed to ^{252}Cf source with an appropriate density of tracks of α-particles and spontaneous fission fragments. Such an approach will be developed by a NTE with an admixture of the ^{252}Cf isotope [5,6]. Another option is exposure of NTE manufactured with a ^{235}U isotope addition by thermal neutrons.

A large-scale NTE scanning is suggested to be performed on the microscope HSP-1000 [7] of the Department of radiation dosimetry (DRD) of Nuclear Physics Institute of the ASCR, v. v. i.. The use of the NTE resolution will be full if the microscope will be adapted to operate with lenses of the highest magnification. Development of algorithms for automatic search and analysis of short tracks of heavy ions in NTE will be required. On the experimental side, ion ranges in NTE must be calibrated in the α-decay and fission energy scale [8]. Progress of the preparatory phase of the proposed study is summarized below.

2. Exposure to Cf source

Surface exposures of NTE samples in DRD were performed by a manually moving ^{252}Cf source. Most likely, the ^{252}Cf isotope decays by emission of α-particles of energy of 5-6 MeV, the tracks of which mainly populate an exposed sample. This isotope also undergoes a spontaneous fission to a pair or even triple of fragments with probabilities of 3%, and about 0.1%, respectively.

In the surface exposure should not be observed more than two ternary fission fragments as the third one is emitted in the contacting source side. The sign of a ^{252}Cf exposure consists in presence of α-particle tracks from ternary fission events whose ranges significantly exceed the decay α-particle ranges. This channel dominates in the ^{252}Cf ternary fission having a 90% probability.

Search for heavy ion tracks on surface of the NTE samples exposed to the ^{252}Cf source is carried out on the KSM microscope with a 15$\times$ objective. Usually MBI-9 microscopes are used for this stage. Using of KSM to search for very rare fission events has eased immediate transitions to their precise measurements with a 90$\times$ objective. Planar triples are found consisting of a pair of fragments and a long-range α-particle as well as of fragments only. Their examples are given in figure 1. It is worth to emphasize a remarkable fact of the observation of triples in NTE but not only pairs of fragments. For such a full observation of triples their vertices should be dipped to a depth not less than a typical track thickness. Figure 2a shows the distribution of vertices of Cf fissions into three fragments over NTE layer depth which has an average value of 2.7 $\pm$ 0.3 μm (RMS equal to 2.2 μm). Perhaps this effect is due to the binding of Cf atoms of AgBr micro crystals and their drift. Apparently, the source surface protection with initial thickness of the 50 μg/cm^2 gold deposition (according to the source passport) does not prevent such a penetration.

In 45 events of a true ternary fission, i. e., not containing α-particles the ranges of all fragments (figure 1, mid and right photo) are measured. Effective criteria for a fission into three heavy fragments is their amount range (figure 2b), which has an average value of 15.5 $\pm$ 0.9 μm when RMS 6.2 μm. In addition, the opening angles between the fragments are measured

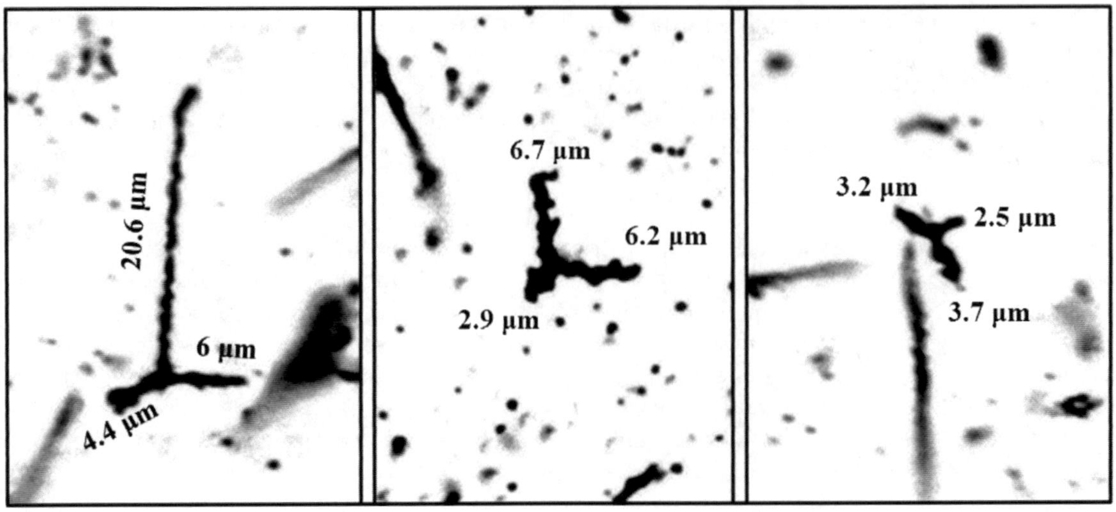

Figure 1. Examples of observed events of ternary fission; track lengths are specified. Left photo: long-range α-particle (long arrow), fragment (middle arrow). Mid and right photo: three fragment tracks.

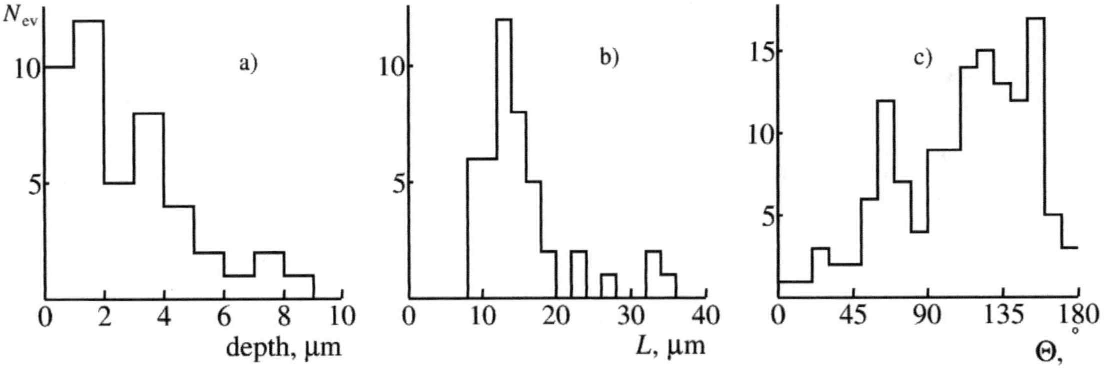

Figure 2. Distributions for the ^{252}Cf fissions into three fragments over depth in NTE layer (a), amount ranges of three fragments (b) and opening angles between fragments (c).

in these events (figure 2c). Their distribution is characterized by a mean value 111 $\pm$ 3° when RMS 39°.

3. Experience of automatic measurements

The initial experience of a computer analysis of tracks in NTE is obtained using the ImageJ program [9], available online. Boron-enriched NTE samples was exposed with thermal neutrons. Steps and results of image recognition of such event analysis are presented in figure 3 (a-d).

By computer analysis (ImageJ recognition) from scanned area of 2.64 mm^2 were found about 245 events of type of n_{th} + ^{10}B $\rightarrow$ ^{7}Li + ^{4}He (figure 3d). A mean Li range (at RMS) by distribution is equal to 1.8 $\pm$ 0.01 (0.15) μm, and the ^{4}He one is 4.3 $\pm$ 0.3 (0.5) μm. Which corresponds with results of 30 manual measured events – $< L(\text{Li}) >$ = 1.9 $\pm$ 0.1 (0.5) μm, $< L(\text{He}) >$ = 4.1 $\pm$ 0.1 (0.6) μm.

Thus, the present study focused on the NTE return in practice of nuclear experiment will serve

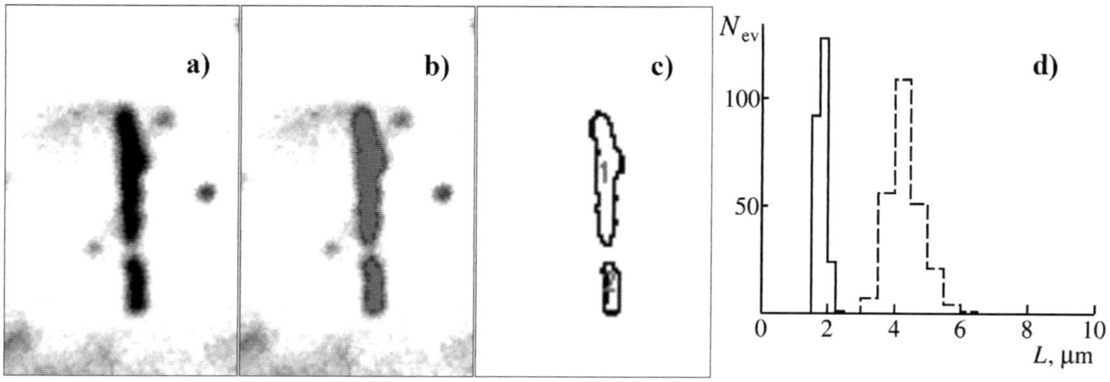

Figure 3. Example of disintegration of boron nucleus by thermal neutron to the Li and He n_{th} (a) and steps of image recognition via the ImageJ program (a-c). Distribution of mean range of Li (solid line) and He (dotted line) (d).

as a prototype of solution of an impressive variety of problems. Macro photos of the discussed exposures and videos based on them are available on the BECQUEREL project website [10].

Acknowledgements

The authors express gratitude to Profs. T. Hussein and M. Ghoneim (Cairo University), H. El-Samman of the Minoufia University, Yu. V. Pyatkov (Moscow Engineering Physics Institute), A. I. Malakhov (JINR) and N. G. Polukhina (Lebedev Physics Institute FIAN) for discussions and support. This work was supported by grants of plenipotentiary representatives of governments of Bulgaria, Czech Republic, Egypt and Romania at JINR.

References

[1] Powell C F, Fowler P H and Perkins D H 1959 *The Study of Elementary Particles by the Photographic Method* (London - New York: Pergamon Press)
[2] Barkas W H 1963 *Nuclear Research Emulsions* (New York - London: Academic Press)
[3] Goldschmidt-Cremont Y 1953 *Annu. Rev. Nucl. Sci.* 141
[4] Kamanin D V and Pyatkov Y V 2014 Clusterization in ternary fission *Clusters in Nuclei Lecture Notes in Physics* **875** 183
[5] Titterton E W and Brinkley T A 1960 Rare Modes in the Spontaneous Fission of Californium-252 *Nature* **187** 228
[6] Muga M L, Bowman H R and Thompson S G 1961 Tripartition in the Spontaneous-Fission Decay of ^{252}Cf *Phys. Rev.* **121** 270
[7] *http://www.odz.ujf.cas.cz/home/resources/microscope-hsp-1000*
[8] Mamatkulov K Z *et al.* 2015 *Proc. Conf. of Fundamental Research and Particle Physics 18-20 February 2015 Moscow arXiv:*1508.02707
[9] Image processing and analyses in Java *http://rsb.info.nih.gov/ij/*
[10] The BECQUEREL Project *http://becquerel.jinr.ru/*

Recent results of the femtoscopic measurements from RHIC and LHC

G A Nigmatkulov

National Research Nuclear University MEPhI (Moscow Engineering Physics Institute), Kashirskoe highway 31, Moscow, 115409, Russia

E-mail: ganigmatkulov@mephi.ru

Abstract. The two-particle correlations at low relative momenta (correlation femtoscopy) are sensitive to quantum statistics and allow to measure the space-time picture of the system evolution created in heavy-ion (HI) and particle collisions. The spatio-temporal parameters extracted from HI collisions describe the system at the last moment of the collision evolution — kinetic freeze-out and provide the essential information about the formation of the quark-gluon plasma. The measurements at many facilities showed the dependencies of the spacial scales from the event multiplicity and particle transverse mass,m_T. In this paper we show recent results obtained at RHIC and LHC energies and compare them to the theoretical expectations.

1. Introduction

The main goal of the ultrarelativistic heavy-ion experiments is to create and study the strongly interacting matter at high temperature or density with color degrees of freedom that was predicted by the quantum chromodynamics (QCD) and called quark-gluon plasma (QGP) [1–3]. The experiments at Relativistic Heavy Ion Collider (RHIC) showed that the matter formed in such collisions is a strongly coupled fluid of deconfined quarks and gluons (sQGP) [4–7] at high temperature and low chemical potential, μ_B. The medium in the collision zone is expected to achieve local equilibrium and exhibit approximately hydrodynamic flow (transverse and longitudinal expansion) [8,9]. The hydrodynamic approach successfully describes the momentum distributions of hadrons produced in heavy-ion collisions (HIC) [10, 11], however it gives a poor description of the spatial distributions of decoupling hadrons [12]. The study of the spatial extent may provide model constrains on the nuclear equation of state and the initial temperature [13].

The information about the space-time structure and the dynamical properties of the emitting source can be extracted using momentum correlations of particles at small relative velocities [14]. The increase of the source radii for more central collisions is attributed to the increasing volume of the emission region. In addition to the geometrical effects, the space-momentum correlations may be affected by the collective flow [15]. For instance, the systematic decrease of the source radii with mean transverse pair momentum, k_T, was attributed to the longitudinal and transverse flow [16]. The important check of the collective behavior is that all particle species created in the collision should follow the transverse mass scaling — the decrease of the femtoscopic radii with increasing transverse mass $m_T = \sqrt{k_T^2 + m^2}$. In this paper we present recent femtoscopic measurements of charged pions and kaons produced in HIC as a function of the m_T performed at Large Hadron Collider (LHC) and top RHIC energies.

2. Correlation femtoscopy

The space-time structure and dynamical properties of the emitting source can be extracted using momentum correlations of particles at small relative momenta via measuring two-particle correlation function. Experimentally, the two-particle correlation function of identical particles with momenta $\mathbf{p_1}$ and $\mathbf{p_2}$ is defined as $C(\mathbf{q}) = A(\mathbf{q})/B(\mathbf{q})$, where $A(\mathbf{q})$ is the measured distribution of the momentum difference, $\mathbf{q} = \mathbf{p_1} - \mathbf{p_2}$, of particles from the same event. The quantity $B(\mathbf{q})$ is obtained by mixing particles from different events.

In order to extract the size of the emitting source one may fit the correlation function using the Bowler-Sinyukov formula:

$$C(q) = N\left(1 - \lambda + \lambda K(q)(1 + e^{-R_{inv}^2 q^2})\right), \tag{1}$$

where N is the normalization factor, λ is the correlation strength, and R_{inv} is the one-dimensional size of the source. The $q = \sqrt{(\mathbf{p_1} - \mathbf{p_2})^2 - (E_1 - E_2)^2}$ is the relative four-momentum of the particles from the pair. The quantity $K(q)$ is the squared Coulomb wave function integrated over the source of the given radius. In order to study the length scales in the beam and transverse directions one may decompose the relative momentum $\mathbf{q}$ according to the Bertsch-Pratt (or "out-side-long") convention [17,18]. The momentum difference is calculated in the longitudinally co-moving system (LCMS), where the longitudinal pair momentum vanishes. The components of the relative momentum, q_l, q_o and q_s, correspond to the directions along the beam, transverse pair momentum, and the perpendicular to those two, respectively. In the three-dimensional case the Eq. 2 will be given by:

$$C(q) = N\left(1 - \lambda + \lambda K(q)(1 + e^{-R_o^2 q_o^2 - R_s^2 q_s^2 - R_l^2 q_l^2})\right), \tag{2}$$

where R_o, R_s and R_l are the spacial scales of the source. The ratio R_o/R_s is sensitive to the emission duration and the longitudinal component, R_l, is sensitive to the system evolution time.

3. Femtoscopic measurements at RHIC and LHC

The femtoscopic measurements of $\pi^{\pm}\pi^{\pm}$, $K^{\pm}K^{\pm}$, $K_S^0 K_S^0$, pp and $\bar{p}\bar{p}$ correlations from Pb-Pb collisions at $\sqrt{s_{NN}}$ =2.76 TeV were performed by ALICE Collaboration [19]. The femtoscopic radii were extracted from one-dimensional correlation functions in terms of invariant relative momentum. Figure 1 shows the extracted source radii for different particle species as a function of m_T. It is seen that in the overlapping transverse pair momentum regions, the femtoscopic radii for pions, kaons and protons are consistent with each other within uncertainties. The line represents the hydrokinetic model prediction (HKM) [20]. The HKM values are specifically from $K^{\pm}K^{\pm}$. The HKM incorporates realistic initial conditions, hydrodynamic evolution, microscopic transport and resonance decays [21, 22]. The measured charged kaon R_{inv} show very good agreement with the theoretical predictions.

Recently, the PHENIX Collaboration performed the three-dimensional analysis of the femtoscopic radii of charged pions and kaons produced in Au+Au collisions at $\sqrt{s_{NN}}$ =200 GeV [23]. Figure 2 shows the ratio of the transverse components of the extracted source radii, R_o/R_s, of the $\pi^{\pm}\pi^{\pm}$ and $K^{\pm}K^{\pm}$ as a function of m_T. This ratio is sensitive to the emission duration of particles. The values for kaons are larger than those for pions for all m_T ranges and collision centralities that indicates the longer emission duration time for kaons than for pions. In addition, the difference for R_o and R_s between charged pions and kaons at the same m_T is larger in more central collisions. It is shown that the m_T scaling breaks for those radii, but k_T scaling works well for all radii.

The HKM model reproduces the data for kaons well, but not for pions [25]. In addition to the HKM, the data for pions in most central collisions were compared with (3+1)-D viscous hydrodynamic model calculations [24]. This model employs a Glauber initial condition and η/s =0.08. The model prediction follows the general trends in the data.

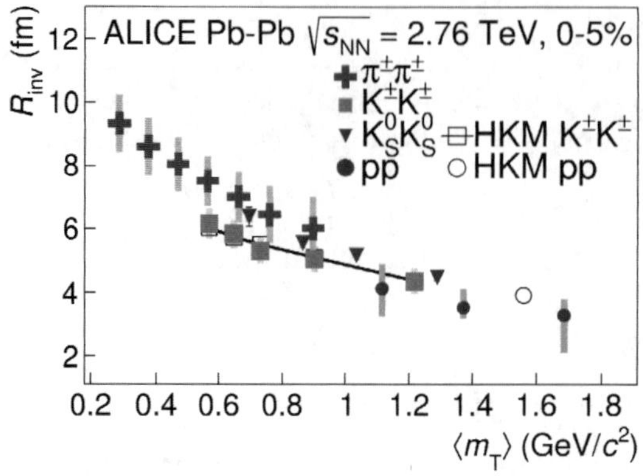

Figure 1. Comparison of the measured source radii with the HKM model measured by ALICE Collaboration for Pb-Pb collisions (0–5% centrality). Statistical (thin lines) and systematic (boxes) uncertainties are shown [19].

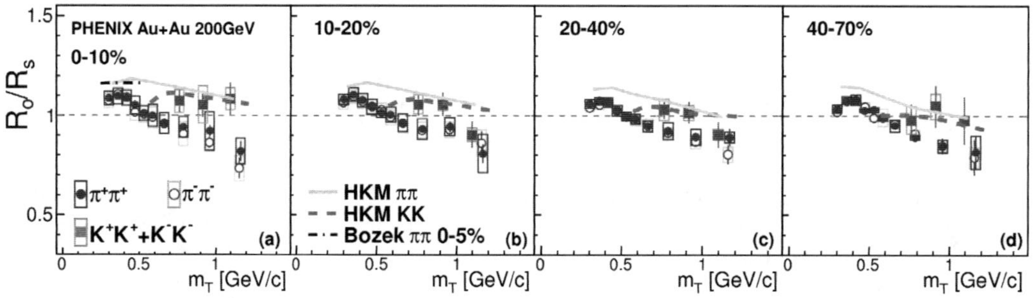

Figure 2. The ratio of R_o and R_s of the charged pions and kaons as a function of m_T measured by PHENIX Collaboration in Au+Au collisions at top RHIC energy [23]. Solid and dashed lines correspond to the predictions from the hydrokinetic model for pions and kaons, respectively. The dash-dot lines correspond to the predictions from the viscous-hydrodynamic model (Bozek) [24].

4. Conclusion

The recent results of the two-particle femtoscopy of charged hadrons obtained by the ALICE and PHENIX Collaborations are presented. The one-dimensional source radii of charged pions, kaons and (anti)protons measured at the LHC energies exhibit the transverse mass scaling within uncertainties and consistent with the hydrodynamic model predictions. On the other hand, the measurements at top RHIC energy show the breaking of the m_T scaling for pions and kaons for R_o/R_s. It is observed that the ratio R_o/R_s is larger for kaons than that for pions. This indicates the different emission duration of π and K. The hydrokinetic model fails to describe the data for pions in more central collisions. The presented results of recent femtoscopic measurements at top RHIC and LHC energies and their comparisons with the theoretical expectations show the non-trivial behavior of the created medium.

Acknowledgements

This work was partially supported by MEPhI Academic Excellence Project (contract №02.a03.21.0005, 27.08.2013).

References

[1] Collins J S and Perry M J 1975 *Phys. Rev. Lett.* **34** 1353
[2] Shuryak E V 1980 *Phys. Rep.* **61** 71
[3] McLerran L D and Svetitsky B 1981 *Phys. Lett.* B **98** 195
[4] Arsene I *et al.* 2005 *Nucl. Phys.* A **757** 1
[5] Back B B *et al.* 2005 *Nucl. Phys.* A **757** 28
[6] Adams J *et al.* 2005 *Nucl. Phys.* A **757** 102
[7] Adcox K *et al.* 2005 *Nucl. Phys.* A **757** 184
[8] Heinz U W and Kolb P F 2002 *Nucl. Phys.* A **702** 269
[9] Shuryak E V 2009 *Prog. Part. Nucl. Phys.* **62** 48
[10] Ollitrault J Y 2008 *Eru. J. Phys.* **29** 275
[11] Kestin G and Heinz U W 2009 *Eur. Phys. J.* C **61** 545
[12] Pratt S 2009 *Nucl. Phys.* A **830** 51c
[13] Lisa M A *et al.* 2005 *Annu. Rev. Nucl. Part. Sci.* **55** 357
[14] Lednický R 2004 *Phys. Atom. Nucl.* **67** 72
[15] Retiere F and Lisa M A 2004 *Phys. Rev.* C **70** 044907
[16] Mount E *et al.* 2011 *Phys. Rev.* C **84** 014908
[17] Bertsch G, Gong M and Tohyama M 1988 *Phys. Rev.* C **37** 1896
[18] Pratt S 1986 *Phys. Rev.* D **33** 1314
[19] Adam J *et al.* 2015 (*Preprint arXiv*:1506.07884)
[20] Shapoval V M *et al.* 2014 *Nucl. Phys.* A **929** 1
[21] Sinyukov Y M, Akkelin S V and Hama Y 2002 *Phys. Rev. Lett.* **89** 052301
[22] Akkelin S V *et al.* 2008 *Phys. Rev.* C **78** 034906
[23] Adare A *et al.* 2015 (*Preprint arXiv*:1504.05168)
[24] Božek P 2014 *Phys. Rev.* C **89** 044904
[25] Karpenko I A and Sinyukov Y M 2010 *Phys. Rev.* C **81** 054903

Intercomparison of flow measurements at RHIC experiments

S S Vdovkina

National Research Nuclear University MEPhI (Moscow Engineering Physics Institute), Kashirskoe highway 31, Moscow, 115409, Russia

E-mail: svvdovkina@gmail.com

Abstract. The measurements of collective flow effects in particle production have provided invaluable insights on the transport properties of the strongly interacting matter produced in relativistic heavy-ion collisions at RHIC. The detailed comparison of flow measurements from PHENIX and STAR experiments at RHIC have been presented and discussed. For elliptic flow v_2 of charged hadrons from Au+Au collisions at 200 GeV the two data sets overlap excellently for centralities > 20%, they increasingly diverge at small centralities, with a 30% difference between STAR an PHENIX in the 0-5% centrality bin. For v_3 values the agreement is much worse and coming from the difference in STAR measurements. More investigations are needed to understand the reason for such differences.

1. Introduction

Ultra-relativistic heavy ion collisions offer the opportunity to study the strongly interacting matter under extreme conditions. In nuclear collisions at the Relativistic Heavy Ion Collider (RHIC) with energies up to $\sqrt{S_{NN}} = 200$ GeV and at the Large Hadron Collider (LHC) at energies of $\sqrt{S_{NN}} = 2.76$ TeV, nuclear matter is heated to temperatures up to 200-500 MeV. At these high energies quantum chromodynamics predict a cross-over transition to a new state of matter, so called quark-gluon plasma (QGP) [1]. The most important observables supporting the discovery of the QGP are the high values of elliptic flow and the suppression of high p_T hadrons that can be explained within assuming a strongly coupled quark-gluon liquid to be formed [2].

The collective expansion, also called flow, originates from the initial pressure gradients in the created hot and dense matter. These pressure gradients transform the initial spatial deformations and inhomogeneities of the created matter into momentum anisotropies of the final state particle production, which are experimentally characterized by so-called flow harmonics v_n [3]. Anisotropic flow can probe the properties of the system created in heavy-ion interactions, such as the equation of state (EoS) and transport coefficients like the shear viscosity to the entropy density ratio (η/s).

Every physical quantity which can be experimentally measured is important for a verification of a theory and therefore should be obtained properly. A good measurement should be reproducible; in particular, it should be done in such a way that one can easily compare results from different experiments, using different detectors.

2. Flow analysis methods

There are several methods to measure the magnitude of the flow signal v_n. The various methods have different sensitivities to non-flow effects and v_n fluctuations. Non-flow refers to correlations between particles which are independent of the initial geometry of the collision system, such as resonance decays, HBT correlations, final state interactions and jets.

- *Event Plane method*
  The Event Plane (EP) method [5] uses the anisotropic flow itself to determine the event plane. Within this method, one first identifies the event plane angle (Ψ_n) in each event using a specific detector at forward rapidity and then calculates the correlation of particles near mid rapidity as $v_n = \langle cos[n(\phi - \Psi_n)]\rangle / Res\{\Psi_n\}$, where $Res\{\Psi_n\}$ is the event plane resolution factor. Note that detectors evaluating Ψ_n and detectors measuring the azimuthal angle of the particle ϕ should have the rapidity gap to suppress the auto- and non-flow correlations. Two-particle cumulant method based on the similar idea.
- *The 4-particle cumulant method*
  The advantage of the cumulant method is that the multi-particle cumulant removes the contribution of non-flow correlations from lower-order correlations [6]. The flow contribution can be obtained by subtracting the 2-particle correlation from the 4-particle correlation.
- *Lee-Yang Zero (LYZ) method*
  As opposed to the four-particle cumulant method which is sensitive to the correlations of four particles, LYZ method [7] is sensitive to the correlations of all the particles. Thus it is supposed to remove non-flow correlations to all orders.

Event-by-event fluctuations of v_n affect each method of measurement differently. The event plane and two-particle cumulant methods have a positive contribution while the multi-particle methods (4-particle cumulant and LYZ) have a negative contribution. As a result of fluctuations, an event-plane measurement of v_n yields an ambiguous measure lying somewhere between the event-averaged mean value $\langle v_n \rangle$ and the root-mean-square value $\langle v_n^2 \rangle^{1/2}$. Where exactly depends on the event plane resolution $Res\{\Psi_n\}$, which strongly depends on the experimental setup.

3. Comparison of PHENIX and STAR flow results

An important step toward reducing the systematic uncertainty associated with flow measurements at RHIC, is a detailed comparison of differential flow results $v_n(p_T$, centrality) obtained by STAR and PHENIX collaborations.

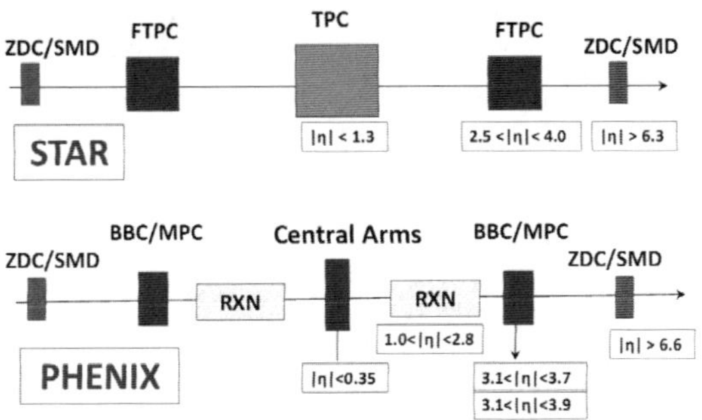

Figure 1. The pseudo-rapidity η acceptance of the PHENIX and STAR detectors used for event-plane reconstruction, along with the mid-rapidity detectors used for charged hadron measurements.

The flow measurements in the PHENIX collaboration were performed using the EP method, where one correlate the azimuthal angles ϕ of the charged tracks in the PHENIX central arms ($|\eta| \leq 0.35$) with the azimuth of the estimated n-order event plane Ψ_n, determined via hits in the two beam-beam counters (BBCs), reaction plane hodoscopes (RXNs) and zero-degree calorimeters (ZDCs). The respective η coverage for these event-plane detector pairs are: $1.0 < |\eta_{\mathrm{RXN}}| < 2.8$, $3.1 < |\eta_{\mathrm{BBC}}| < 3.9$ and $|\eta_{\mathrm{ZDC}}| > 6.5$, see figure 1. The event centrality is determined by using the total charge deposited in the BBCs.

The STAR collaboration uses both: multi-particle methods (4-particle cumulant and LYZ method) and EP method. In EP method, one correlate the azimuthal angles ϕ of the charged tracks reconstructed in the STAR Time Projection Chamber (TPC) ($|\eta| < 1.0$) with with the azimuth of the estimated n-order event plane Ψ_n, determined via tracks in TPC ($|\eta_{TPC}| < 1.0$) and FTPC ($2.5 < |\eta_{FTPC}| < 4.0$), hits in ZDCs ($|\eta_{ZDC}| > 6.5$), see figure 1. The centrality determination is based on the number of charged tracks in the TPC with track quality cuts. To suppress non-flow effects for EP measurements using TPC one apply $\Delta\eta$ gap between tracks used for EP construction and tracks used for the measurements of v_n.

In order to make proper comparison for EP method we should use the event plane detectors with the same η coverage and similar event plane resolution. This is the case for STAR/PHENIX ZDC/SMD and for STAR FTPC/PHENIX BBC. Unfortunately, we have only one colliding system and one beam energy for the proper comparison of the published elliptic flow (v_2) results between STAR and PHENIX: the v_2 values of charged hadrons from Cu+Cu collisions at $\sqrt{S_{NN}} = 200$ GeV obtained using the EP from STAR FTPC [9] and PHENIX BBC [10]. Figure 2 compares the centrality dependence of the ratio of v_2 obtained using EP from STAR FTPC and PHENIX BBC for four bins in p_T. The difference between two data sets are less than 5-10% within a typical systematic uncertainty of the measurements of 5%. So even with the similar $\Delta\eta$-gap and event plane resolution there are some differences between two experiments.

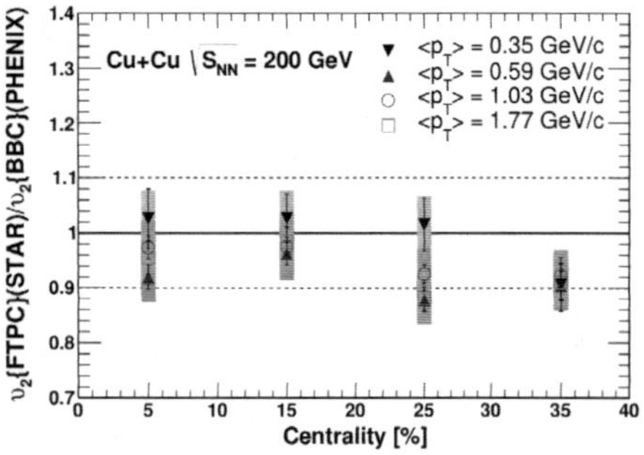

Figure 2. Centrality dependence of the ratio of STAR $v_2(p_T)$ values [9] obtained using EP from STAR FTPC to the v_2 values [10] obtained using EP from PHENIX BBC. The data are for charged hadrons from Cu+Cu collisions at $\sqrt{S_{NN}} = 200$ GeV. The 5% systematic errors are shown.

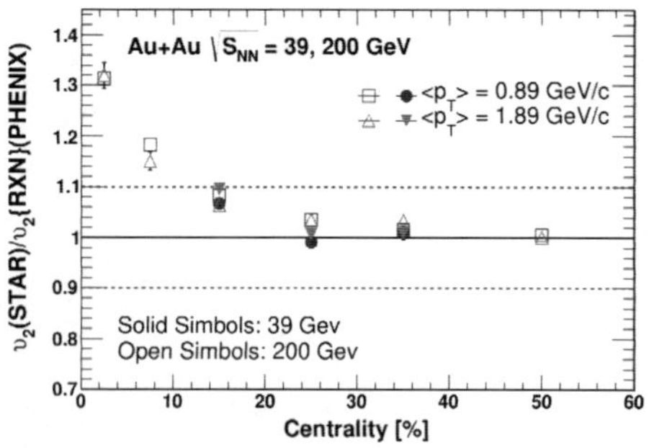

Figure 3. Centrality dependence of the ratio of STAR $v_2(p_T)$ values obtained using EP from STAR TPC to the v_2 values obtained using EP from PHENIX RXN. The data are for charged hadrons from Au+Au collisions at $\sqrt{S_{NN}} = 39$ and 200 GeV.

However, it is hard to make proper comparison for the v_2 measurements of charged hadrons emerged from Au+Au collisions at $\sqrt{S_{NN}} = 200$ GeV. This is the reference point for the Beam Energy Scan program at RHIC and for comparison with LHC results. Here we can make the comparison between v_2 results obtained using the PHENIX EP RXN ($1.0 < |\eta_{RXN}| < 2.8$) and STAR TPC EP with $\Delta\eta > 0.5$. The open symbols from figure 3 shows the centrality dependence of the ratio of STAR $v_2(p_T)$ values to the v_2 values obtained from PHENIX for charged hadrons at $\sqrt{S_{NN}} = 200$ GeV. While the two data sets overlap excellently for centralities > 20%, they increasingly diverge at small centralities, with a 30% difference between STAR an PHENIX in the 0-5% centrality bin. The PHENIX EP measurements are less affected by non-flow effects due to larger $\Delta\eta$ gap. However, it is hard to explain the difference by non-flow effects only as we see the same difference between STAR and PHENIX results for Au+Au collisions at $\sqrt{S_{NN}}$ = 39 GeV. The excess of the STAR over the PHENIX data is almost uniform in p_T and could be explained by a small shift in the centrality definitions between the experiments.

Figure 4 compares v_3 values obtained with STAR TPC and FTPC event plane detectors to PHENIX RXN EP v_3 measurements. One can see a huge difference (40-50%) between TPC and FTPC data and < 10% difference between STAR TPC to PHENIX RXN results on the other hand. To understand this enormous difference between STAR results more comparisons should be made, e.g. for the elliptic flow between FTPC and TPC EP.

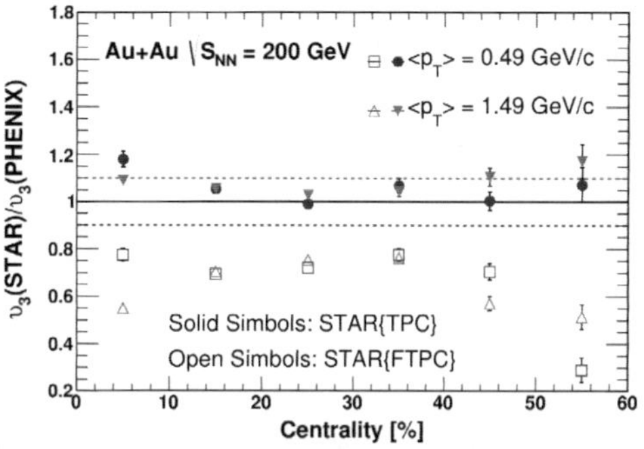

Figure 4. The centrality dependence of the ratio of the v_3 values of STAR{TPC}(solid) and STAR{FTPC}(open) to the PHENIX $v_3(p_T)$ points fit for the two p_T bins at $\sqrt{S_{NN}}$ = 200 GeV Au+Au collisions. The points are extracted from [11].

4. Conclusion

The comparison of flow measurements from PHENIX and STAR experiments was presented and discussed. The divergences between experimental values are present for every kind of juxtaposition. Precise comparisons require careful attention to the details of each of various measurements, and this will be important in the future when trying to extract more precise quantitative information from experimental data.

Acknowledgments

This work was performed within the framework of the Center of Nanostructured Electronics supported by MEPhI Academic Excellence Project (contract № 02.a03.21.0005, 27.08.2013).

References

[1] Aoki Y *et al.* 2006 *Nature* **443** 675.
[2] Adams J *et al.* 2005 *Nucl. Phys.* A **757** 102.
[3] Voloshin S and Zhang Y 1996 *Phys.* C **70** 665.
[4] Bhalerao R S and Ollitrault J Y 2006*Phys. Lett.* B **641** 260.
[5] Poskanzer A M and Voloshin S A 1998 *Phys. Rev.* C **58** 1671.
[6] Borghini N, Dinh P M and Ollitrault J Y 2001 *Phys. Rev.* C **63** 054906.
[7] Bhalerao R S, Borghini N and Ollitrault J Y 2004 *Phys. Lett.* B **580** 157.
[8] Abelev B I *et al.* 2008 *Phys. Rev.* C **77** 054901.
[9] Abelev B I *et al.* 2010 *Phys. Rev.* C **81** 44902.
[10] Adare A *et al.* 2007 *Phys.Rev.Lett.* **98** 162301.
[11] Adamczyk L *et al.* 2013 *Phys. Rev.* C **88** 14904.

Scaling properties of collective effects at RHIC

A S Zaytsev
National Research Nuclear University MEPhI (Moscow Engineering Physics Institute), Kashirskoe highway 31, Moscow, 115409, Russia
E-mail: zaitsev136@gmail.com

Abstract. Azimuthal anisotropy is one of the key observables to study the properties of matter created in high energy heavy-ion collisions at RHIC and the LHC. The collective behaviour is quantified in terms of anisotropy coefficients v_n measured with respect to their corresponding event planes. Predictions from the viscous hydrodynamics for the scaling of the anisotropic flow coefficients v_n with eccentricity, system size and transverse energy are tested using the recent data from PHENIX Collaboration.

1. Introduction

Full characterization of the transport properties of the strongly interacting matter (the quark-gluon plasma – QGP) produced in heavy ion collisions, is a central goal of the experimental heavy ion programs at both the RHIC and the LHC. One of the key observables sensitive to these properties is collective flow – the anisotropy of the particle emission in the plane transverse to the beam direction [1]. This anisotropy can be quantified as a function of particle transverse momentum p_T and collision centrality (cent) or the number of participant nucleons N_{part}, by the Fourier coefficients v_n:

$$v_n = \langle \cos n(\phi - \Psi_n) \rangle, \tag{1}$$

where ϕ is the azimuthal angle of an emitted particle, and ψ_n are the azimuths of the estimated participant event planes. The brackets denote averaging over particles and events [2, 3].

Examination of the different scaling properties of the collective flow coefficients can be useful for testing viscous hydrodynamic model compatibility for the QGP matter. It also provides important constraints for distinguishing between different models of initial conditions, as well as a route to constrain the temperature dependence of the ratio of shear viscosity η to entropy density s: η/s[4].

2. Acoustic scaling

Collective flow measurements provide a great method to evaluate one of the key transport properties of the QGP – ratio of shear viscosity and entropy density η/s. There is a qualitative difference between the radial flow and higher angular harmonics. While the former monotonously grows with time, driven by the outward pressure gradient with a fixed sign, the latter are basically sounds, or damped oscillators. One effects of viscosity on sounds is damping of their amplitude. Several works based on the assumption of the acoustic nature of the anisotropic flow suggest that viscous corrections to v_n / ε_n grow exponentially as n^2 and $1/\bar{R}$ [5, 6, 8]:

$$\frac{v_n}{\varepsilon_n} \propto \exp\left[-Cn^2 \frac{\eta}{s} \frac{1}{T\bar{R}} \right], \tag{2}$$

where C is some constant, T is the temperature, $\bar{R}$ is the transverse size of the collision zone and ε_n is the n-th order eccentricity moment of the fireball initial state. For a given n, this equation indicates a characteristic linear dependence of $\ln(v_n/\varepsilon_n)$ on $1/\bar{R}$, with a slope proportional to η/s. This scaling pattern is confirmed by state-of-the-art viscous hydrodynamic calculations [7] shown in figure 1. Figure 1b indicates a clear sensitivity of the linear fit slopes to η/s. Figure 1c shows dependence of $\ln(v_n/\varepsilon_n)$ on n^2 which is also linear [8].

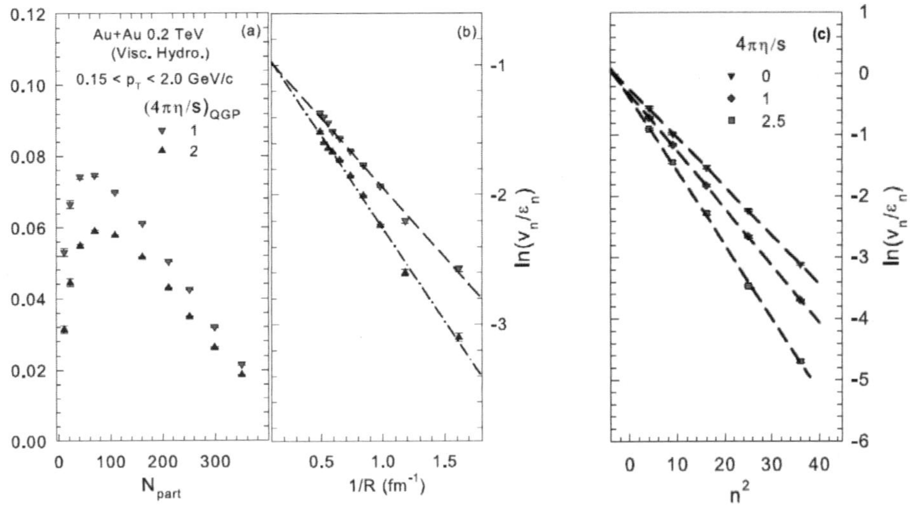

Figure 1. (a) v_2 vs. $N_{\rm part}$ from viscous hydrodynamic calculations [7] for two values of specific shear viscosity as indicated. The results are for $0.15 < P_T < 2.0$ GeV/c for Au+Au collisions at $\sqrt{s_{NN}}$ = 200 GeV. (b) $\ln(v_2/\varepsilon_2)$ vs. $1/\bar{R}$ for the v_2 values shown in (a). (c) $\ln(v_n/\varepsilon_n)$ vs. n^2 for different values of specific shear viscosity. The dashed and dot-dashed curves are linear fits.

The data used for testing the acoustic scaling are taken from measurements by the PHENIX collaboration for Cu+Cu and Au+Au collisions at $\sqrt{s_{NN}}$ = 62.4 and 200 GeV. To obtain v_2 values the event-plane method was used. Monte Carlo Glauber (MC-Glauber) simulations were used to compute participant eccentricity $\varepsilon_2(\text{centrality})$ and $\bar{R}(\text{centrality})$ from the two-dimensional profile of the density of sources in the transverse plane [9]; $1/\bar{R} = \sqrt{1/\sigma_x^2 + 1/\sigma_y^2}$, where σ_x and σ_y are the respective root-mean-square width of the density distribution.

Figures 2a,b show v_2 and v_2/ε_2 (respectively) vs. number of the collision participants obtained from the PHENIX collaboration data for p_T= 1.0 – 2.0 GeV/c[10]. For ideal fluids one should observe the eccentricity scaling of v_2: at a given energy v_2/ε_2 should be independent on the transverse size of the system $\bar{R}$. If the equilibration is incomplete the eccentricity scaling is broken and v_2/ε_2 also depends on the Knudsen number $K = \lambda/\bar{R}$ (λ is the mean free path) and should follow the equation (2). Figure 2c shows the dependence of $\ln(v_n/\varepsilon_n)$ on $1/\bar{R}$ for these data, which is well described by linear fits (solid lines). The slopes are found to be the same within errors for Au+Au collisions at $\sqrt{s_{NN}}$ = 62.4 and 200 GeV. However, they are different for Au+Au and Cu+Cu at $\sqrt{s_{NN}}$ = 200 GeV, which may indicate the larger viscous damping for smaller colliding systems. The slope is larger for Cu+Cu at 62.4 GeV than for 200 GeV. The later observation may suggest that the beam energy scan at RHIC performed for several colliding systems will provide more information which will help to constrain the temperature dependence of η/s.

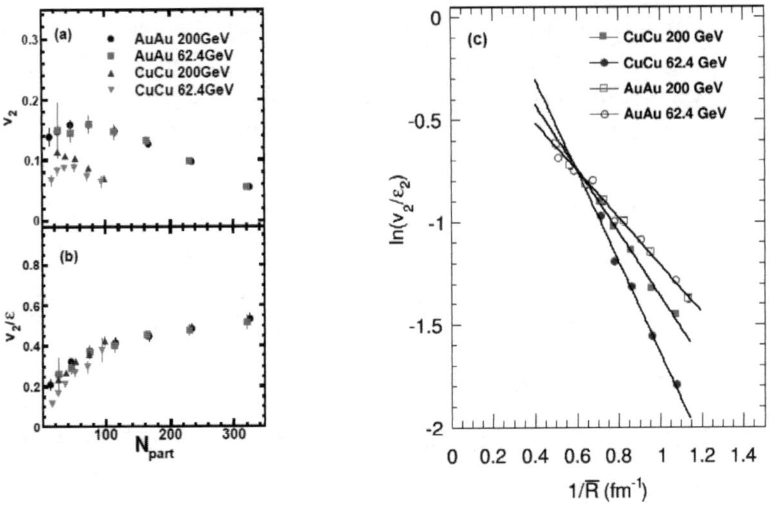

Figure 2. (a), (b) v_2 and v_2/ε_2 vs. number of the collision participants from the PHENIX data. The results are for 1.0<P_T< 2.0 GeV/c for Au+Au and Cu+Cu collisions at $\sqrt{s_{NN}}$ = 62.4 and 200 GeV.(c) ln(v_2/ε_2) vs. $1/\bar{R}$ for the values shown in (b). The solid lines are linear fits.

3. Constituent quark scaling

The collective flow coefficients measured for different identified hadrons also shows specific scale properties. Figure 3 shows a comparison of $v_2(p_T)$, $v_3(p_T)$ and $v_4(p_T)$ for $\pi^{\pm}$, $K^{\pm}$ and $p+\bar{p}$ emerged from 0%–50% central Au+Au collisions $\sqrt{s_{NN}}$ = 200 GeV. The data were obtained by the PHENIX collaboration via the event-plane (solid points) and two-particle correlation (open points) methods. It is noticeable that in the low-p_T region the anisotropy appears largest for the lightest hadrons whereas in the intermediate-p_T region ($3 < p_T < 4$ GeV/c) this mass dependence partly reverses, such that the anisotropy is greater for the (anti-)baryons ($n_q = 3$) than for the mesons ($n_q = 2$) at the same p_T [11].

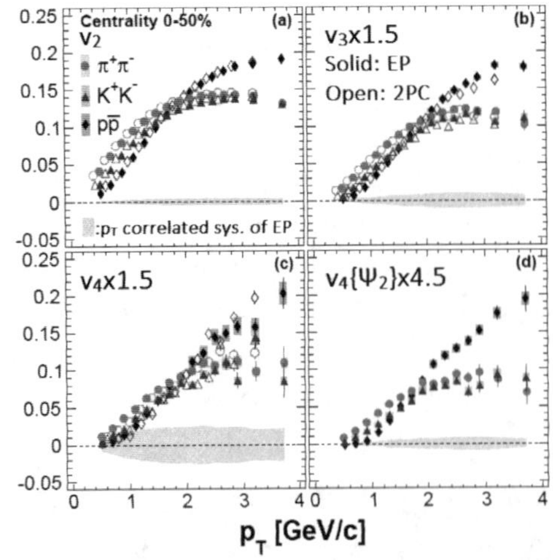

Figure 3. Fourier coefficients (v_2, v_3, v_4 and v_4 with respect to the second harmonic event plane in plots (a) – (d) respectively) for charge combined $\pi^{\pm}$, $K^{\pm}$ and $p+\bar{p}$ at midrapidity for 0%–50% central Au+Au collisions at $\sqrt{s_{NN}}$ = 200 GeV. The green bands indicate the p_T-correlated systematic uncertainties of the $\pi^{\pm}$ results from the event-plane method. The shaded boxes around the data pointsare p_T-uncorrelated systematic uncertainties [11].

The baryon-meson splitting in the intermediate-p_T region can be taken as an indication that the number of constituent valence quarks n_q is an important determinant of the final-state hadron flow in this range. “Constituent quark number scaling” has been suggested to test this assumption, i.e. the scaling of v_2/n_q for identified hadrons to transverse kinetic energy per constituent quark number KE_T/n_q

[12, 13]. For higher order harmonics following dependence was observed: $v_n(p_T) \propto (v_2)^{n/2}$ [4, 14], which may stem from the hydrodynamic evolution of the medium [2, 15], the quark coalescence process [16] and the acoustic nature of anisotropic flow [4]. These observations lead to universal (for each harmonic order *n*) quark number scaling – $v_n/(n_q)^{n/2}$ vs. KE_T/n_q. The adherence of the data to this empirical scaling is shown in figure 4 [11]. $v_4\{\Psi_2\}$ denotes that the fourth harmonic coefficient was evaluated with respect to the second harmonic event plane. The scaled values lie on a single curve for each harmonics for all the hadron species within a ±15% range.

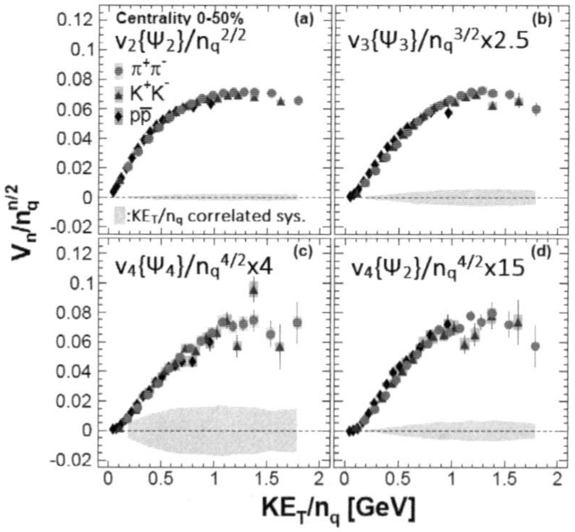

Figure 4. Quark-number scaling for 0%–50% central Au+Au collisions at $\sqrt{s_{NN}}$ = 200 GeV. n_q is the constituent valence quark number of each hadron. Differences between (a) – (d) plots are the same as in figure 3 [11].

4. Conclusion

The test for several scaling relations is presented for anisotropic flow at RHIC using the recent data from the PHENIX Collaboration. The presence of scaling variables provides a very transparent indications of the dynamics underlying a given physical process.

Acknowledgments

This work was performed within the framework of the Center of Nanostructured Electronics supported by MEPhI Academic Excellence Project (contract No 02.a03.21.0005, 27.08.2013).

References

[1] Lacey R A 2002 *Nucl. Phys.* A **698** 559
[2] Ollitrault J-Y 1992 *Phys. Rev.* D **46** 229
[3] Adare A *et al.* 2010 *Phys. Rev. Lett.* **105** 062301
[4] Lacey R A *et al.* 2011 *arXiv*:1105.3782
[5] Staig P and Shuryak E 2011 *Phys. Rev.* C **84** 034908
[6] Shuryak E 2014 *arXiv*:1412.8393
[7] Song H *et al.* 2011 *Phys.Rev.* C **83** 054910
[8] Lacey R A *et al.* 2014 *Phys. Rev. Lett.* **112** 082302
[9] Lacey R A *et al.* 2010 *arXiv*:1009.5230
[10] Adare A *et al.* 2014 *arXiv*:1412.1043
[11] Adare A *et al.* 2014 *arXiv*:1412.1038
[12] Adare A *et al.* 2007 *Phys. Rev. Lett.* **98** 162301
[13] Adare A *et al.* 2012 *Phys. Rev.* C **85** 064914
[14] Aad G *et al.* 2012 *Phys. Rev.* C **86** 014907
[15] Gardim F G *et al.* 2012 *Phys. Rev. Lett.* **109** 202302
[16] Kolb P F *et al.* 2004 *Phys. Rev.* C **69** 051901

A practical parametrisation of line shapes of near-threshold resonances

C Hanhart[1], Yu S Kalashnikova[2], P Matuschek[1], R V Mizuk[2,3], A V Nefediev[2,3] and Q Wang[1]

[1]Forschungszentrum Jülich, Institute for Advanced Simulation, Institut für Kernphysik (Theorie) and Jülich Center for Hadron Physics, D-52425 Jülich, Germany
[2]Institute for Theoretical and Experimental Physics, 117218, B.Cheremushkinskaya 25, Moscow, Russia
[3]National Research Nuclear University MEPhI (Moscow Engineering Physics Institute), Kashirskoe highway 31, Moscow, 115409, Russia

E-mail: `nefediev@itep.ru, q.wang@fz-juelich.de`

Abstract. A practical parametrisation for the line shapes of near threshold resonance(s) is derived in the framework of a coupled-channel model which includes an arbitrary number of elastic and inelastic channels as well as a bare pole term. The parameters have a direct relation to phenomenology and can be employed to study the nature of the near-threshold states. The resulting analytical parametrisation is therefore ideally suited to investigate the full information content provided by the measurements and to establish a link between the experimental data and their theoretical interpretation.

1. Introduction

Since 2003 when Belle Collaboration announced the first observation of the meson-like state $X(3872)$ in the spectrum of charmonium [1], many other charmonium- and bottomonium-like states were observed which do not fit into the quark-model picture. At present there is no consensus in the literature regarding their nature.

Some of the new states reside in the vicinity of strong S-wave thresholds and they are seen in both open-flavour (elastic) and hidden-flavour (inelastic) final states. Since vast and detailed information is becoming available from the existing experiments and even more precise data are expected from future high statistics and high precision ones [2, 3, 4, 5], adequate theoretical parametrisations for the data analysis are urgently called for. It is important that such parametrisations are consistent with the requirements of unitarity and analyticity and allow one to exploit the full information content provided by the measurements. This is possible only via a simultaneous analysis of the data in all available channels.

2. Coupled-channel problem and a practical parametrisation for the line shapes

We present a coupled-channel approach based on the Lippmann-Schwinger equation (LSE) for the t-matrix, t. The potential is taken in a form consistent with phenomenology [6],

$$\hat{V} = \begin{array}{c} \quad \text{Pole} \qquad \beta=\overline{1,N_e} \qquad i=\overline{1,N_{in}} \\ \begin{pmatrix} 0 & f_\beta & \lambda_i (k_i^{in})^{l_i} \\ f_\alpha & v_{\alpha\beta} & g_{i\alpha}(k_i^{in})^{l_i} \\ \lambda_j (k_j^{in})^{l_j} & g_{j\beta}(k_j^{in})^{l_j} & 0 \end{pmatrix} \end{array} \begin{array}{l} \text{Pole} \\ \alpha=\overline{1,N_e} \\ j=\overline{1,N_{in}} \end{array} . \tag{1}$$

It contains various types of interaction between the bare pole (labeled as "0", for example, its position is M_0), the set of N_e elastic open-flavour channels $(Q\bar{q})(q\bar{Q})$ (here Q and q denote a heavy and a light quark, respectively) labeled by Greek letters, and a set of N_{in} inelastic hidden-flavour channels $(Q\bar{Q})(q\bar{q})$ referred to by Latin letters. The direct interactions in the inelastic channels are neglected, since they are expected to be weak.

The elastic potential $v_{\alpha\beta}$ is approximated by a constant matrix, and the elastic-inelastic transition form factors are taken in a separable form. Here k_i^{in} is the relative momentum in the i-th inelastic final state (the corresponding reduced mass and the threshold are denoted as μ_i^{in} and $m_{th_i^{in}}$, respectively). The omission of rescatterings within the inelastic channels allows us to disentangle the latter from the elastic channels and from the pole term. Thus we define

$$V_{00} = -\sum_i \lambda_i^2 J_i, \quad V_{\alpha 0} = f_\alpha - \sum_i g_{i\alpha} J_i \lambda_i, \quad V_{\alpha\beta} = v_{\alpha\beta} - \sum_i g_{i\alpha} J_i g_{i\beta},$$

where the real part of the inelastic loop integral

$$J_i = \int |\boldsymbol{q}|^{2l_i} S_i(\boldsymbol{q}) d^3q \to \frac{i(2\pi)^2}{\sqrt{s}} m_{th_i^{in}} \mu_i^{in} (k_i^{in})^{2l_i+1} \tag{2}$$

is omitted since it only renormalises parameters of the interaction. For convenience we define the effective potentials

$$V_{\alpha\beta}^{\text{eff}} = v_{\alpha\beta} - G_{\alpha\beta} - V_{\alpha 0} G_0 V_{0\beta}, \quad V_{\alpha 0}^{\text{eff}} = V_{\alpha 0}(1 + G_0 V_{00}), \; G_{\alpha\beta} = \sum_i g_{i\alpha} J_i g_{i\beta},$$

where $G_0 = 1/(M_0 - M + V_{00} - i0)$ and arrive at a pair of decoupled LSE

$$\begin{aligned} t_{\alpha\beta} &= V_{\alpha\beta}^{\text{eff}} - \sum_\gamma V_{\alpha\gamma}^{\text{eff}} J_\gamma t_{\gamma\beta}, \; t_{\alpha 0} = V_{\alpha 0}^{\text{eff}} - \sum_\beta V_{\alpha\beta}^{\text{eff}} J_\beta t_{\beta 0}, \\ J_\alpha &= \int S_\alpha(\boldsymbol{p}) d^3p = (2\pi)^2 \mu_\alpha (\kappa_\alpha + i k_\alpha) \equiv R_\alpha + i I_\alpha, \end{aligned} \tag{3}$$

with μ_α and k_α being the reduced mass and the relative momentum in the α's elastic channel, respectively, $k_\alpha = \sqrt{2\mu_\alpha(M - m_{th_\alpha}) + i\epsilon}$, where m_{th_α} is the position of the α-th elastic threshold. With the help of the formalism of [7, 8] the solution of equation (3) can be written in the form

$$t_{\alpha\beta} = t_{\alpha\beta}^w + \frac{\phi_\alpha \bar{\phi}_\beta}{M - M_0 + \mathcal{G}_0}, \quad t_{\alpha 0} = \frac{M - M_0}{M - M_0 + \mathcal{G}_0} \phi_\alpha,$$

where

$$t^w = t^v + \psi[\mathcal{G} - G^{-1}]^{-1} \bar{\psi}$$

$$\psi_{\alpha\beta} = \delta_{\alpha\beta}\chi_\alpha - t_{\alpha\beta}^v J_\beta, \quad \bar{\psi}_{\alpha\beta} = \delta_{\alpha\beta}\chi_\alpha - J_\alpha t_{\alpha\beta}^v,$$

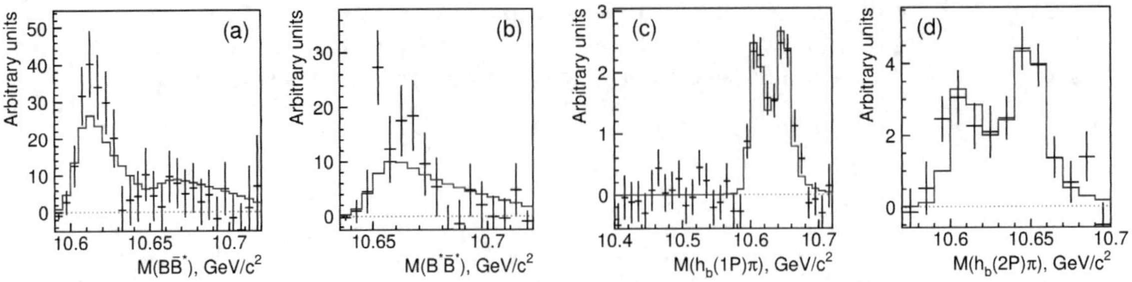

Figure 1. Fitted line shapes of the $Z_b(10610)$ and $Z_b(10650)$ in the $B^{(*)}\bar{B}^*$ and $\pi h_b(mP)$ $(m = 1, 2)$ channels.

$$\mathcal{G}_{\alpha\beta} = J_\alpha \psi_{\alpha\beta} = \bar{\psi}_{\alpha\beta} J_\beta = \delta_{\alpha\beta} J_\alpha - J_\alpha t^v_{\alpha\beta} J_\beta,$$

$$\phi_\alpha = V_{\alpha 0} - \sum_\beta t^w_{\alpha\beta} J_\beta V_{\beta 0}, \quad \bar{\phi}_\alpha = V_{0\alpha} - \sum_\beta V_{0\beta} J_\beta t^w_{\beta\alpha},$$

$$\mathcal{G}_0 = \sum_i \lambda_i^2 J_i + \sum_\alpha V_{0\alpha} J_\alpha \phi_\alpha = \sum_i \lambda_i^2 J_i + \sum_\alpha \bar{\phi}_\alpha J_\alpha V_{\alpha 0}.$$

Thus the entire problem is reduced to inverting a matrix as small as $N_e \times N_e$ — all other components of the t-matrix are algebraically related to $t_{\alpha\beta}$.

Since our knowledge of most resonance properties comes from production experiments, we build the production amplitude in the channel x as

$$\mathcal{M}_x = -\sum_\beta$$ 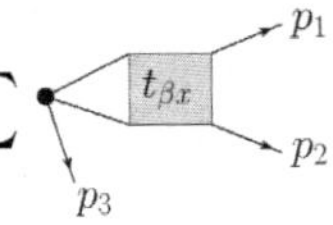 $$= -\sum_\beta \mathcal{F}_\beta J_\beta t_{\beta x}, \tag{4}$$

where it was assumed that the production proceeds through the N_e point-like elastic sources $\mathcal{F}_\alpha$ and the Born term was neglected that is justified by the existence of near-threshold poles in the t-matrix. Neglecting the FSI with particle 3 and partially integrating over the phase space one can arrive at the formulae

$$\frac{d\mathrm{Br}^e_\alpha}{dM} = \Lambda \Big| \sum_\beta \xi_\beta t_{\beta\alpha} \Big|^2 p_3 k_\alpha, \quad \frac{d\mathrm{Br}^{in}_i}{dM} = \Lambda \Big| \sum_\alpha \xi_\alpha t_{\alpha i} \Big|^2 p_3 (k_i^{in})^{2l_i+1} \tag{5}$$

for the elastic and inelastic differential rates, where the allowed parameter range for M is given by $M_{\min} = m_1 + m_2$ and $M_{\max} = M_{\mathrm{tot}} - m_3$, p_3 is the spectator momentum, and new parameters $\Lambda = \mathcal{F}_1^2$ and $\xi_\alpha = \mathcal{F}_\alpha / \mathcal{F}_1$ were introduced. Since for all elastic channels the range of forces is described by the same physics, it is natural to use $R_\alpha = (2\pi)^2 \mu_\alpha \kappa$.

3. Line shapes for the $Z_b(10610)$ and $Z_b(10650)$ states

As an application for our approach we consider the isovector 1^+ states $Z_b(10610)$ and $Z_b(10650)$ residing near the $B\bar{B}^*$ and $B^*\bar{B}^*$ thresholds, respectively, that are produced in $\Upsilon(5S)$ decays $\Upsilon(5S) \to \pi Z_b$ and are seen in seven decay channels: $Z_b \to B\bar{B}^*$, $B^*\bar{B}^*$, $\pi\Upsilon(nS)$ $(n = 1, 2, 3)$, and $\pi h_b(mP)$ $(m = 1, 2)$ [9, 10]. The quantum numbers of the final quarkonia fix the angular momenta of the inelastic final states to $l = 0$ for all $\pi\Upsilon(nS)$ final states and to $l = 1$ for the $\pi h_b(mP)$ final states. As an experimental input we use the background-subtracted and efficiency-corrected distributions in M for the $B^{(*)}\bar{B}^*$ and $\pi h_b(nP)$ channels [11, 9] with floating normalisation in each channel, and the ratios of total branching fractions: $\mathrm{Br}^e_{B\bar{B}^*}/\mathrm{Br}^e_{B^*\bar{B}^*}$, $\mathrm{Br}^{in}_i/\mathrm{Br}^e_{B^*\bar{B}^*}$ [11, 9, 12, 13], where index i runs over all five inelastic channels $\pi\Upsilon(nS)$ and $\pi h_b(mP)$. We do not use the information on the $Z_b^{(\prime)}$ line shapes in the $\pi\Upsilon(nS)$ channels, since in the one-dimensional fit it is not possible to take into account correctly the interference with the nonresonant continuum, which is significant in the $\Upsilon(5S) \to \pi^+\pi^-\Upsilon(nS)$ transitions. To take into account the experimental resolution we convolve all the distributions with a Gaussian function with $\sigma = 6$ MeV.

The fact that the b-quark mass $m_b \gg \Lambda_{\rm QCD}$ allows us to use heavy quark spin symmetry (HQSS) to reduce the number of parameters. In particular [14, 15],

$$\xi = \frac{g_{[\pi\Upsilon(5S)][B^*\bar{B}^*]}}{g_{[\pi\Upsilon(5S)][B\bar{B}^*]}} = -1, \quad \frac{g_{[\pi\Upsilon(nS)][B^*\bar{B}^*]}}{g_{[\pi\Upsilon(nS)][B\bar{B}^*]}} = -1, \quad \frac{g_{[\pi h_b(mP)][B^*\bar{B}^*]}}{g_{[\pi h_b(mP)][B\bar{B}^*]}} = 1. \tag{6}$$

In the same limit the direct interaction in the $B^{(*)}\bar{B}^*$ system can be parametrised in terms of two parameters γ_s and γ_t [6]. The bare pole is not included. Results of the simultaneous fit are shown in figure 1. Since the suggested parametrisation follows from the Lippmann-Schwinger approach for the coupled-channel problem then all unitarity and analyticity constraints for the t-matrix are fulfilled automatically. The inelastic channels enter additively that makes it particularly simple to extend the basis of the model and to include extra inelasticities. Finally, the developed parametrisation with the strict HQSS constraints imposed provides a good (CL=47%) description of the experimental data for the bottomonium-like states $Z_b(10610)$ and $Z_b(10650)$.

4. Conclusions

We presented an approach which allowed us to build a practical parametrisation for the line shapes of near-threshold states consistent with all requirements of unitarity and analyticity. While additional effects such as finite widths of the constituents and the final state interaction with the spectator may also play a role, they can be included on top of the interactions considered in this work so that the presented parametrisation is expected to be realistic that is exemplified by the simultaneous fit to the data on the bottomonium-like states $Z_b(10610)$ and $Z_b(10650)$.

Acknowledgments

R Mizuk and A Nefediev are supported by the Russian Science Foundation (Grant No. 15-12-30014).

References

[1] Choi S K *et al.* (Belle Collaboration) 2003 *Phys. Rev. Lett.* **91** 262001
[2] Abe T *et al.* 2010 (Belle-II Collaboration) *arXiv*:1011.0352v1
[3] Drutskoy A G *et al. Eur. Phys. J.* A **49** 7
[4] Asner D M *et al.* 2009 *Int. J. Mod. Phys.* A **24** S1
[5] Lutz M F M *et al.* (PANDA Collaboration) 2009 *arXiv*:0903.3905v1
[6] Hanhart C *et al.* 2015 *arXiv*:1507.00382v1
[7] Baru V *et al.* 2010 *Eur. Phys. J.* A **44** 93
[8] Hanhart C, Kalashnikova Yu S and Nefediev A V 2011 *Eur. Phys. J.* A **47** 101
[9] Bondar A *et al.* (Belle Collaboration) 2012 *Phys. Rev. Lett.* **108** 122001
[10] Adachi I *et al.* (Belle Collaboration) 1012 *arXiv*:1209.6450v2
[11] Bondar A 2015 *Proc. XXVII Int. Symp. on Lepton Photon Interactions at High Energies (Ljubljana)* in press
[12] Adachi I *et al.* (Belle Collaboration) 2012 *Phys. Rev. Lett.* **108** 032001
[13] Garmash A *et al.* (Belle Collaboration) 2015 *Phys. Rev.* D **91** 7 072003
[14] Bondar A E *et al.* 2011 *Phys. Rev.* D **84** 054010
[15] Voloshin M B 2011 *Phys. Rev.* D **84** 031502

Journal of Physics: Conference Series **675** (2016) 022017 doi:10.1088/1742-6596/675/2/022017

Chiral extrapolation of the $X(3872)$ binding energy

V Baru[1,2], E Epelbaum[1], A A Filin[1], J Gegelia[3,4] and A V Nefediev[2,5]

[1]Institut für Theoretische Physik II, Ruhr-Universität Bochum, D-44780 Bochum, Germany
[2]Institute for Theoretical and Experimental Physics, 117218, B.Cheremushkinskaya 25, Moscow, Russia
[3]Forschungszentrum Jülich, Institute for Advanced Simulation, Institut für Kernphysik (Theorie) and Jülich Center for Hadron Physics, D-52425 Jülich, Germany
[4]Tbilisi State University, 0186 Tbilisi, Georgia
[5]National Research Nuclear University MEPhI (Moscow Engineering Physics Institute), Kashirskoe highway 31, Moscow, 115409, Russia

E-mail: nefediev@itep.ru, arseniy.filin@ruhr-uni-bochum.de

Abstract. The role of pion dynamics in the $X(3872)$ charmonium-like state is studied in the framework of a renormalisable effective quantum field theory approach and they are found to play a substantial role in the formation of the X. Chiral extrapolation from the physical point to unphysically large pion masses is performed and the results are confronted with the lattice predictions. The proposed approach overrides the gap between the lattice calculations and the physical limit in m_π.

1. Introduction

In spite of its relatively long history, the meson-like state $X(3872)$ in the spectrum of charmonium [1] remains not fully understood theoretically. According to the Particle Data Group, this state resides very close to the neutral $D\bar{D}^*$ threshold [2],

$$E_B = M_{D^0} + M_{\bar{D}^{*0}} - M_X = (0.12 \pm 0.26)\ \mathrm{MeV}, \tag{1}$$

so it is natural to assume its wave function to have a large molecular admixture. The 1^{++} quantum numbers of the X determined recently by the LHCb Collaboration [3, 4] are consistent with its interpretation as an S-wave $D^0\bar{D}^{*0}$ bound state, however the binding mechanims in the X still remain an open question. In view of the fact that the 3-body ($D^0\bar{D}^0\pi^0$) threshold resides below the 2-body ($D^0\bar{D}^{*0}$) threshold and the splitting between these thresholds is quite small, approximately 7 MeV, the role played by pions in the X should be important and the 3-body dynamics is expected to strongly affect the properties of the X. The small binding energy (1) allows for an effective field theory (EFT) formulation of the problem — see, for example, [5] for the pionless EFT based on pure contact $D\bar{D}^*$ interactions or [6, 7] for the so-called X-EFT approach with pions treated perturbatively. In [8] the properties of the $X(3872)$ molecular state were studied in a heavy-meson EFT framework with nonperturbative pions including all relevant 3-body scales, and the dependence of the X binding energy on the pion mass was studied in [9]. Another important issue related to the 3-body forces in the X is the problem of the nonperturbative renormalisation of the 3-body Faddeev-type equations describing the

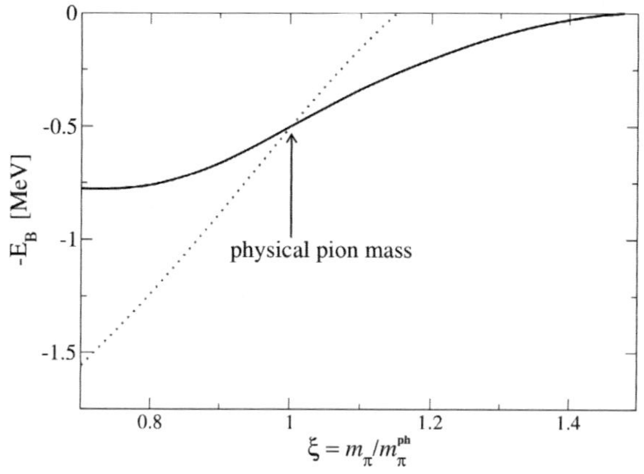

Figure 1. Pion mass dependence of the $X(3872)$ binding energy at LO. Solid line: the full calculation; dashed line: the static approximation.

interaction between heavy mesons in the X. The renormalisable approach proposed in [10] for the deuteron allows one to get rid of the finite cut-off artifacts in the leading order (LO) amplitude and to make a prediction for the m_π-dependence of the binding energy at this order, which then acquires NLO corrections. This approach can be applied to the $X(3872)$ that allows one to extrapolate the X binding energy to $m_\pi > m_\pi^{\rm ph}$ and therefore to establish a link between the theory and the lattice calculations traditionally performed for unphysically large pion masses [11, 12, 13].

2. Coupled-channel three-body equations

The key idea of the approach proposed in [10], that is the relativised formulation of the coupled channel equations, was generalised recently in [14] for the $D\bar{D}^*$ scattering, so the corresponding t-matrix elements $a_{00}^{nn'}(\boldsymbol{p},\boldsymbol{p}')$ and $a_{c0}^{nn'}(\boldsymbol{p},\boldsymbol{p}')$ in the C-even channel have the form

$$\left\{\begin{array}{l} a_{00}^{nn'}(\boldsymbol{p},\boldsymbol{p}') = \lambda_0 V_{00}^{nn'}(\boldsymbol{p},\boldsymbol{p}') - \sum\limits_{i=0,c} \lambda_i \int d^3k V_{0i}^{nm}(\boldsymbol{p},\boldsymbol{k}) \dfrac{1}{\Delta_i(\boldsymbol{k})} a_{i0}^{mn'}(\boldsymbol{k},\boldsymbol{p}'), \\ a_{c0}^{nn'}(\boldsymbol{p},\boldsymbol{p}') = \lambda_c V_{c0}^{nn'}(\boldsymbol{p},\boldsymbol{p}') - \sum\limits_{i=0,c} \lambda_i \int d^3k V_{ci}^{nm}(\boldsymbol{p},\boldsymbol{k}) \dfrac{1}{\Delta_i(\boldsymbol{k})} a_{i0}^{mn'}(\boldsymbol{k},\boldsymbol{p}'), \end{array}\right. \quad (2)$$

where $\lambda_0 = \langle 0|\vec{\tau}_1\cdot\vec{\tau}_2|\bar{0}\rangle = \langle c|\vec{\tau}_1\cdot\vec{\tau}_2|\bar{c}\rangle = 1$ and $\lambda_c = \langle 0|\vec{\tau}_1\cdot\vec{\tau}_2|\bar{c}\rangle = \langle c|\vec{\tau}_1\cdot\vec{\tau}_2|\bar{0}\rangle = 2$ are the isospin factors for the π^0- and $\pi^\pm$-exchange, respectively. The OPE potential $V_{ij}^{nn'}(\boldsymbol{p},\boldsymbol{p}')$ can be extracted from the lowest-order $D^*D\pi$ interaction Lagrangian taken in the form [7]

$$\mathcal{L} = \frac{g_c}{2f_\pi}\left(\boldsymbol{D}^{*\dagger}\cdot\boldsymbol{\nabla}\pi^a\tau^a D + D^\dagger\tau^a\boldsymbol{\nabla}\pi^a\cdot\boldsymbol{D}^*\right), \quad (3)$$

where the coupling constant g_c is related to the $D^{*0}\to D^0\pi^0$ decay width as

$$\Gamma(D^{*0}\to D^0\pi^0) = \frac{g_c^2 m_0 q^3}{24\pi f_\pi^2 m_{*0}}, \quad q = \lambda^{1/2}(m_{*0}^2, m_0^2, m_{\pi^0}^2)/(2m_*). \quad (4)$$

Here and in what follows, m_*, m, and m_π denote the masses of the D^* meson, D meson, and pion, respectively. Charged and neutral states are distinguished by an additional index, for example m_{*c} versus m_{*0}. Because of the P-wave nature of the $D^*\to D\pi$ decay the OPE

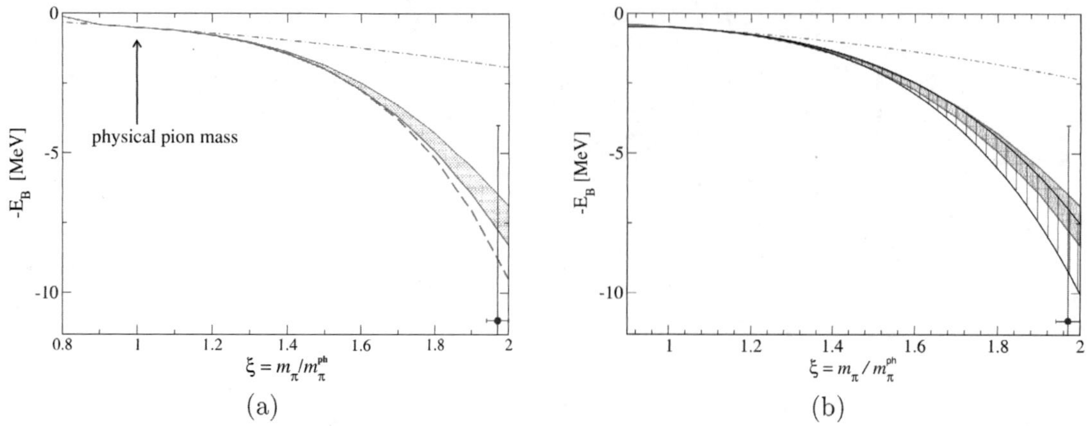

Figure 2. Pion mass dependence of the $X(3872)$ binding energy at NLO. The result obtained for the m_π-dependent contact interaction only is shown with the dashed-dotted line.

contains a short-range part and is therefore well-defined only in connection with the short-range contact term C_0 [9],

$$V_{ij}^{nn'}(\boldsymbol{p},\boldsymbol{p}') = g_c^2(\boldsymbol{p}+\boldsymbol{p}')^n(\boldsymbol{p}+\boldsymbol{p}')^{n'}F_{ij}(\boldsymbol{p},\boldsymbol{p}') + C_0\,\delta^{nn'}, \tag{5}$$

where indices n and n' are contracted with the corresponding indices of the D^* polarisation vectors and $F_{ij}(\boldsymbol{p},\boldsymbol{p}')$ contains the relativised $D\bar{D}\pi$ and $D^*\bar{D}^*\pi$ propagators [14]. The energy E is counted from the neutral two-body threshold,

$$M = m_{*0} + m_0 + E. \tag{6}$$

3. Behaviour of the X binding energy for unphysical pion masses

At LO the contact interation is adjusted to reproduce the binding energy at the physical pion mass (for definiteness we set $E_B(m_\pi^{\rm ph}) = 0.5$ MeV), and the m_π-dependence of the E_B stems only from the pionic effects in the OPE potential and from those in the renormalised selfenergy loops. As a result, at LO of our EFT the dependence $E_B(m_\pi)$ can be predicted in a parameter-free way — see Fig. 1. So at LO the binding energy decreases with the pion mass growth and the 3-body effects are important for the dynamics of the system.

In NLO one includes corrections quadratic in m_π, also for the contact term. Notice that a stronger bound X for unphysically large m_π's is only possible if the short-range part of the interaction changes the slope of the binding energy at the physical point. Thus lattice predictions of a stronger bound X for $m_\pi > m_\pi^{\rm ph}$ indicate the importance of the short-range dynamics in the X. Since the nature of the short-range interaction is obscure, its dependence on m_π can only be guessed with the help of the ideas of naturalness. In particular, one can resort to the perturbative treatment of the m_π-dependent contact operator at NLO [14] (dashed line in figure 2(a)), use the heavy-meson formulation with the finite cut-off $\Lambda \in [500\text{ MeV}, 700\text{ MeV}]$ [9] (black hatched band in figure 2(b)) or employ the nonperturbative resonance saturation approach [14] (red dotted band at both plots in figure 2). The dashed-dotted line in both plots in figure 2 corresponds to the calculation without pions. From figure 2 one can see a good agreement between different approaches to the naturalness. The predicted value of the binding energy at $m_\pi = 266$ MeV appears to be consistent with the lattice result from reference [13] shown as the blue dot with an error bar. As in LO, one can observe a strong influnce of the pion dynamics on the system.

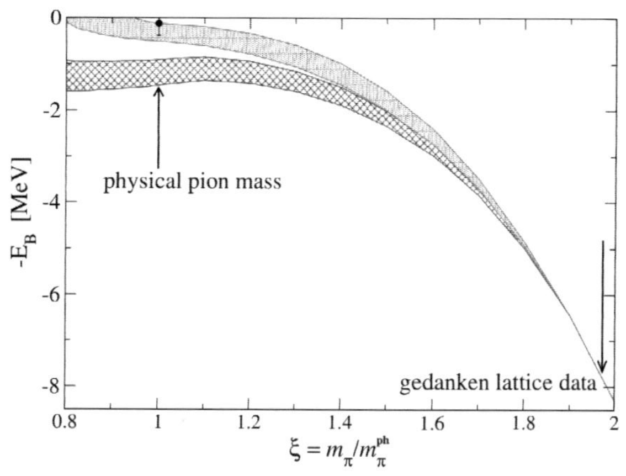

Figure 3. Pion mass dependence of the $X(3872)$ binding energy. The red dotted band is for the full calculation with dynamical pions at NLO while the blue crossed band is for the static OPE.

Alternatively one can start from the lattice results obtained for unphysically large pion masses and use them as an input to adjust the short-rang part of the potential. Then the binding energy can be extrapolated to the physical point and confronted with the experimental data. This approach is illistrated in figure 3 with the help of gedanken lattice data. The role of the 3-body dynamics is also clearly seen in this figure.

4. Concluions

In conclusion we emphasise that our approach opens the way to override the gap between the unphysically large pion masses used on the lattices and the physical limit for the $X(3872)$ charmonium-like state. It can also be adapted to other near-threshold states, especially to those in which the three-body dynamics plays an important role.

Acknowledgments

This work is partially supported by the DFG (Grant No. CRC 110) and by MEPhI Academic Excellence Project (contract No. 02.a03.21.0005, 27.08.2013). Work of V Baru is supported by the DFG (Grant No. GZ: BA 5443/1-1 AOBJ: 616443). Work of A Nefediev is supported by the Russian Science Foundation (Grant No. 15-12-30014).

References

[1] Choi S K *et al.* (Belle Collaboration) 2003 *Phys. Rev. Lett.* **91** 262001
[2] Beringer J *et al.* (Particle Data Group) 2012, *Phys. Rev.* D **86** 010001
[3] Aaij R *et al.* (LHCb Collaboration) 2013, *Phys. Rev. Lett.* **110** 222001
[4] Aaij R *et al.* (LHCb Collaboration) 2015, *Phys. Rev.* D **92** 1 011102
[5] AlFiky M T, Gabbiani F and Petrov A A 2006 *Phys. Lett.* B **640** 238
[6] Voloshin M B 2004 *Phys. Lett.* B **579** 316
[7] Fleming S *et al.* 2007 *Phys. Rev.* D **76** 034006
[8] Baru V *et al.* 2011 *Phys. Rev.* D **84** 074029
[9] Baru V *et al.* 2013 *Phys. Lett.* B **726** 537
[10] Epelbaum E and Gegelia J 2012 *Phys. Lett.* B **716** 338
[11] Prelovsek S and Leskovec L 2013 *Phys. Rev. Lett.* **111** 192001
[12] Lee S H *et al.* (Fermilab Lattice and MILC Collaborations) 2014 *arXiv*:1411.1389v1
[13] Padmanath M, Lang C B and Prelovsek S 2015 *Phys. Rev.* D **92** 3 034501
[14] Baru V *et al.* 2015 *arXiv*:1509.01789v1

Yields of p, d, t formed in stopped pion absorption by intranuclear clusters

B A Chernyshev, Yu B Gurov, L Yu Korotkova, S V Lapushkin, T I Leonova, R V Pritula and T D Schurenkova
National Research Nuclear University MEPhI (Moscow Engineering Physics Institute), Kashirskoe highway 31, Moscow, 115409, Russia

E-mail: rvpritula@gmail.com

Abstract. Results of the study of spectra and yields of p, d, t formed in stopped pion absorption by nuclei are presented. Data on charged particle formation have been obtained for 17 isotopes in wide atomic mass range ($6 < A < 209$). Empirical model was proposed to describe A-dependences of primary p, d, t yields.

1. Introduction

Study of charged particle formation in the reaction of stopped pion absorption by nuclei allows to gain insight into mechanisms of particle production in nuclear reactions at the intermediate energies (up to 1 GeV) as well as understanding of cluster structures of nuclei. Because of energy and momentum conservation, pion absorption is multinucleon process. The particles formed in such processes are referred to as "primary" ones and their spectra and yields may be connected to properties of intranuclear structures. Therefore A-dependences of yields of the primary particles are of great interest.

Energy spectra of particles produced in this reaction were measured in many experimental works [1-9]. Nevertheless, these results cannot be used to obtain systematic set of data because of significant differences in the absolute normalization, especially in the high-energy region ($E > 70$ MeV), which is known to be populated mostly by primary particles.

In this work the yields of charged particles p, d, t are analyzed using the unique set of data. These data were obtained in the PNPI RAS experiment. Phenomenological model [10] was applied to estimate the contributions of primary particles.

2. Experiment details

The experiment was carried out on the pion channel of the PNPI RAS synchrocyclotron using the charged particle semiconductor spectrometer [11].

The pion beam with the momentum of 100 MeV/c passed through the carbon moderator and two monitor surface-barrier Si(Au) detectors and then stopped in the 440 μm thick target installed on 45° angle with respect to the beam axis. Charged particles produced by absorption in the target were detected by two multilayer telescopes. The energy resolution was 0.6 MeV. Application of the "active target" method [10] permitted high precision of the spectra normalization – 7% throughout the experiment. The energy spectra of p (d, t) were measured in the range from 5 MeV (10 MeV) to the kinematic thresholds of reactions. The measurements were carried out on large set of target ^{6,7}Li, ^{9}Be,

^{10,11}B, ^{12}C, ^{28}Si, ^{40}Ca, ^{59}Co, ^{93}Nb, ^{114,117,120,124}Sn, ^{169}Tm, ^{181}Ta, ^{209}Bi.

3. Results and analysis

The measured spectra showed similar smooth declining form [10]. The exceptions were triton spectra measured on light nuclei, where there is a distinct peak possibly connected to pion absorption by alpha-clusters. All spectra were successfully described using our phenomenological model [10]. It then allowed to estimate contributions of different processes in the total particle formation and thus to calculate corresponding yields for each of studied nuclei. These results were used to obtain A-dependences of yields of primary particles, which are presented in the figure 1 for p, d and t.

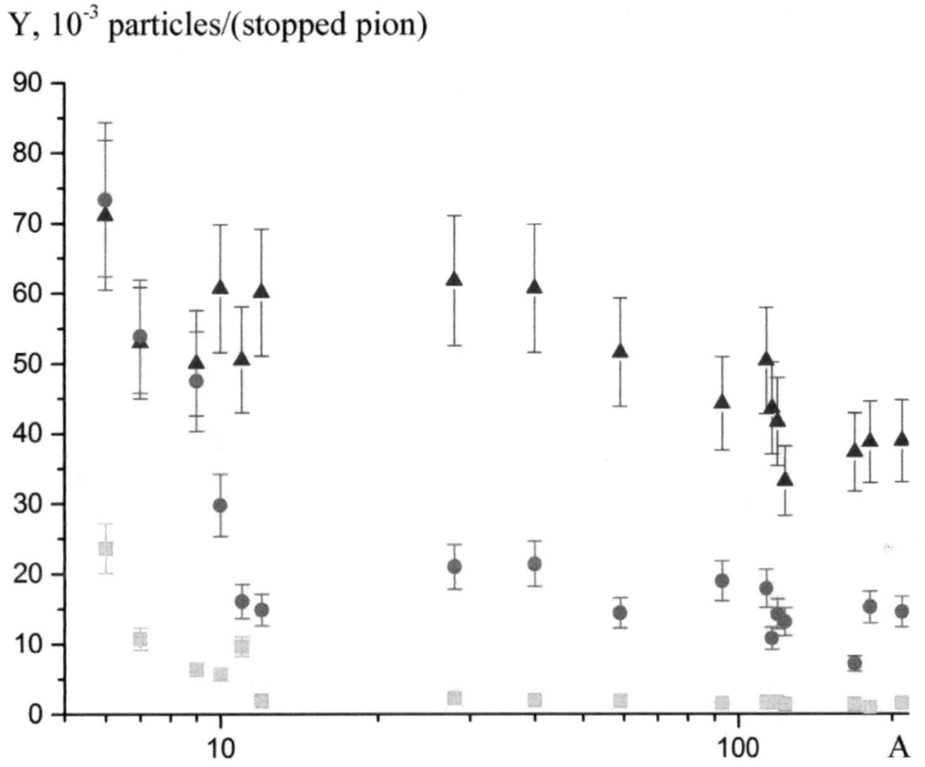

Figure 1. Yields of primary p (▲), d (●), t (■).

The yields are used to develop a set of expressions in order to describe the yields of the primary particles. The empirical relations can then be used as a generalization of our results on the entire mass number region.

We assume that the value of the yield Y^{p}_{prime} is proportional to the probability of the pion absorption on pp-pair [12 – 14]. Thus, assuming that absorption by heavier clusters is insignificant, A-dependence can be described as follows:

$$Y^{p}_{prime} = C_p \cdot \frac{\exp(-\beta_p \cdot A^{1/6})}{1 + 2N \cdot R'/(Z-1)}, \quad (1)$$

where C_p – free parameter.

R' is the ratio of the elementary widths of the pion absorption on pn and pp pairs. We supposed it to be independent of parameters of the target nucleus in the range 6 < A < 209.

The factor $\exp(-\beta_p A^{1/6})$ is the probability for the primary particle to leave nucleus without interactions, where β_p is a parameter dependent only on the type of primary particle.

In figure 2 the results of calculations using relation (1) are compared to primary proton yields with $\beta_p = 0.5$ и $C_p = 0.14$. The values of these parameters are chosen to achieve the best agreement with the data. The empirical expression (1) reproduces well the A-dependence of the primary proton yields throughout all the studied range of mass numbers with the error ~ 15%. The value of the ratio of the elementary widths is R' = 3.3±0.5. This result is in a good agreement with work [15] where authors analyzed the inclusive and correlation measurements of spectra of protons produced in pion absorption on ^{9}Be, ^{12}C, ^{28}Si, Cu and Ge.

Such result provides an estimation of important physical parameter R' and allows to determine the yields of primary protons in the reaction of pion absorption on any nucleus.

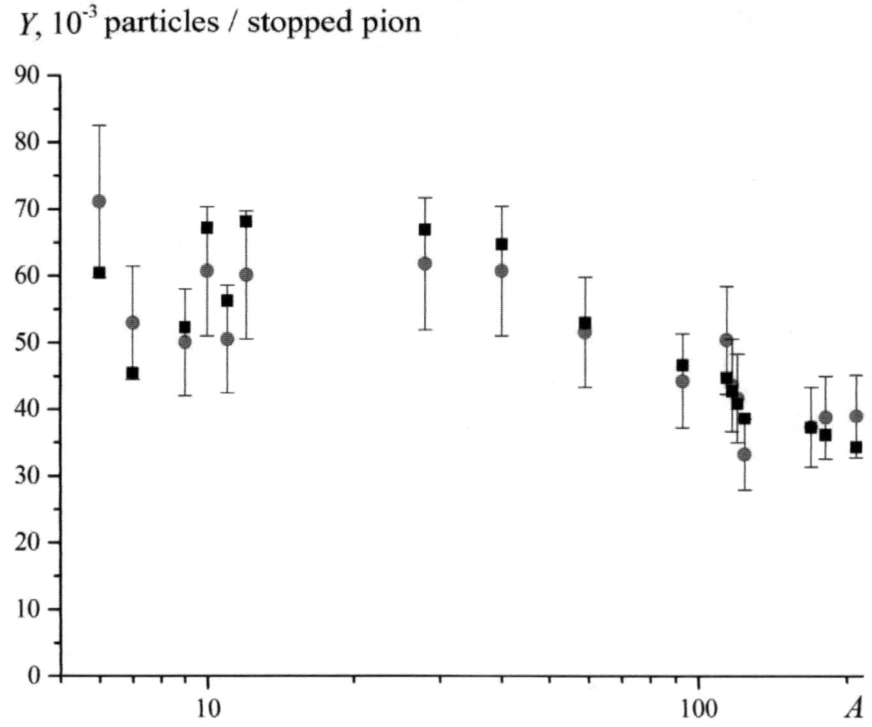

Figure 2. Proton yields: • — primary part, ■ — calculations via (1).

Similar method is used to derive expressions for d and t A-dependences of yields of d and t. Probability of the primary d or t leaving the nucleus without further interactions is taken in the similar exponential form with β parameter chosen separately for each particle type. The estimation of the probability of pion absorption on corresponding clusters (^{3,4}He) is much more complicated than in the case of protons, so simple empirical expressions were chosen which provided the best agreement with the data. Expressions for d and t yields are:

$$Y_{pr}^{d} = C_d \frac{Z(Z-1)N}{A-3} \exp(-\beta_d A^{1/6}), \tag{2}$$

$$Y_{pr}^{t} = C_t \frac{Z(Z-1)N(N-1)}{(A-4)^2} \exp(-\beta_t A^{1/6}). \qquad (3)$$

These relations reproduce primary d and t yields with ~15% precision, but only for medium and heavy nuclei from the mass number range of 28 < A < 209. For the light nuclei satisfactory description was not obtained, possibly because of their complex cluster structure.

4. Conclusions

A-dependences of the yields of the primary p, d and t were described with an accuracy of ~ 15% over the wide range of mass numbers: 6 < A < 209 for protons, 28 < A < 209 for deuterons and tritons.

Acknowledgments

This work was performed within the framework of the Center of FRPP supported by MEPhI Academic Excellence Project (contract № 02.a03.21.0005, 27.08.2015).

References

[1] Budyashov Yu G *et al.* 1972 *JETP* **62** 21
[2] Schleputz F W *et al.* 1979 *Phys. Rev.* C **19** 135
[3] Mechtersheimer G *et al.* 1979 *Nucl. Phys.* A **324** 379
[4] Randoll H *et al.* 1982 *Nucl. Phys.* A **381** 317
[5] Sennhauser U *et al.* 1982 *Nucl. Phys.* A **382** 429
[6] Pruys K S *et al.* 1981 *Nucl. Phys.* A **352** 388
[7] Isaak H P *et al.* 1983 *Nucl. Phys.* A **392** 385
[8] Cernigoi C *et al.* 1983 *Nucl. Phys.* A **411** 382
[9] Cernigoi C *et al.* 1986 *Nucl. Phys.* A **456** 599
[10] Korotkova L Yu *et al.* 2012 *Bull. Russ. Acad. Sci. Phys.* **76** 460
[11] Gornov M G *et al.* 1984 *Nucl. Instrum. Meth.* A **225** 42
[12] Butsev V S *et al.* 1980 (in Russian) *Phys. Elem. Part. Atom. Nucl.* **11** 900 (original Russian title: *Fizika elemtarnyh chastits i atomnogo yadra)*
[13] Weyer H G 1990 *Phys. Rep.* **195** 295
[14] Lee T-S 2002 *Ann. Rev. Nucl. and Part. Sci.* **22** 23
[15] Gornov M G *et al.* 1988 (in Russian) *Nucl. Phys.* **47** 959 (original Russian title: *Yadernaya Fizika*)

Physics at ILC

A G Drutskoy
National Research Nuclear University MEPhI (Moscow Engineering Physics Institute), Kashirskoe highway 31, Moscow, 115409, Russia

E-mail: Drutskoy@itep.ru

Abstract. The physics case for the International Linear Collider (ILC) project is discussed. The project is proposed to be built in Japan and a final decision has to be taken in 2016 on a government level. The initial data taking with e^+e^- beams is planned at the center-of-mass energy of 500 GeV with following runs at 250 GeV and 350 GeV. Potential physics studies at ILC are discussed with a special attention to the measurements, which are expected to have a better sensitivity at ILC in respect to the LHC experiments.

1. Introduction

The linear e^+e^- collider at a few hundred GeV center-of-mass (CM) energy is a next mega-size project, which has to be built in high energy physics after the LHC [1]. Realization of such project, so-called International Linear Collider (ILC), was recently proposed in Japan. The basic physics goal, which has to be reached at ILC, is a detailed study of Higgs boson behaviours. The Higgs boson was discovered at Large Hadron Collider (LHC) a few years ago and by now several decay channels of this particle have been observed. However, the number of the studied channels and their statistical significance is limited at LHC and will be limited even after modernization. In contrast, the linear collider will allow to reconstruct several tens of thousands of the Higgs bosons. It will provide an opportunity to study already observed channels with a high statistical sensitivity and to discover new rare decays of the Higgs boson. The linear collider will allow to measure precisely different parameters of the Higgs boson, such as its mass, width and *CP* quantum number admixture [2].

Besides the detailed studies of the Higgs boson, the linear collider is well suited for precise studies of the top quark, *Z* and *W* bosons and searches for physics Beyond the Standard Model (BSM) [3]. In particular, a high rate is expected for the $e^+e^- \to t\bar{t}$ process, providing a huge sample of top quarks, taking into account the process production cross section of about 1 pb at the 350 GeV CM energy. ILC has several advantages over LHC, such as cleanliness (low backgrounds), a high production rate for heavy particles, simplicity of experimental event topology and a high accuracy of underlying processes theoretical calculations.

2. Proposed linear collider in Japan

In 2013 the candidate site for the linear collider construction was proposed in Kitakami, Japan. The general layout for ILC is shown in figure 1. The basic scenario supposes the data taking start at the 500 GeV CM energy with 31 km long collider. The later upgrade assumes increasing the size to 50 km and the energy to 1 TeV. The expected beam size at the interaction point is 6 nm x 500 nm x 300 mm.

The ILC luminosity should reach ~2 x 10^{34} $cm^{-2}s^{-1}$ with a possible increase by a factor two; the beam polarization is expected to be > 80% for e^- and 30-40% for e^+ beams.

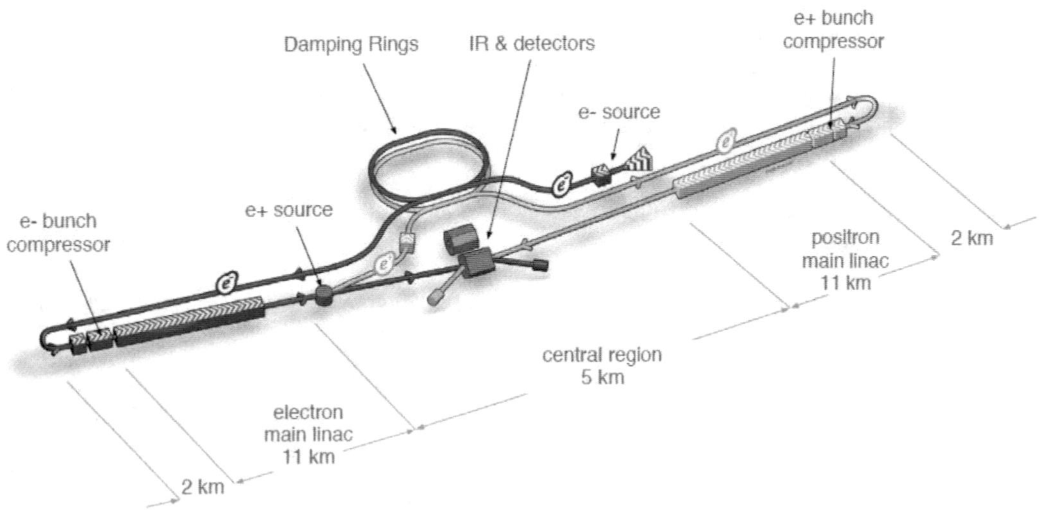

Figure 1. Basic layout of the International Linear Collider.

The current scenario assumes to take about 500 fb^{-1} integrated luminosity at the beginning of the data taking, with following collection of 200 fb^{-1} at 350 GeV and of 500 fb^{-1} at 250 GeV. The physics at 500 GeV CM energy is more comprehensive, comparing with 250 GeV and 350 GeV, due to three-body channels opening, such as $e^+e^- \rightarrow t\bar{t}H$. To increase the cross section for the later channel the CM energy can be initially raised to 550 GeV.

3. ILC detectors: ILD and SiD

The two detectors (figure 2) are proposed for the ILC project: International Large Detector (ILD) and Silicon Detector (SiD).

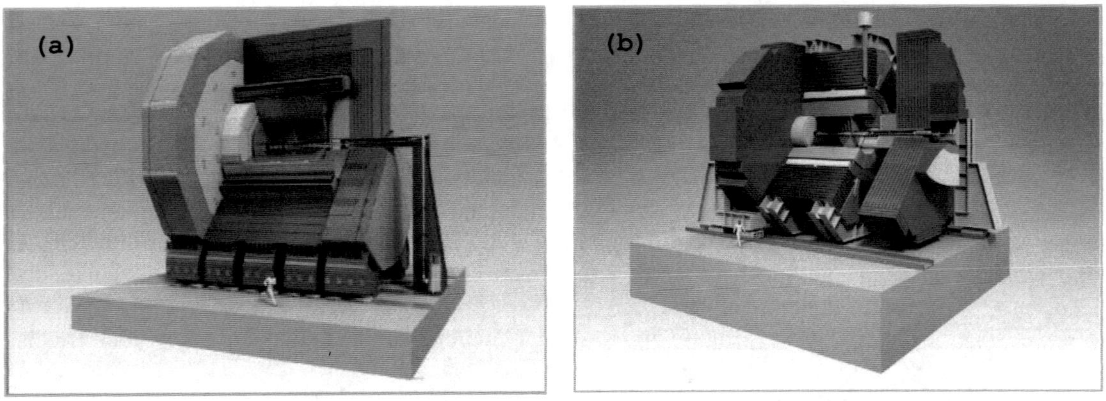

Figure 2. General 3D view of the detectors: a) ILD; b) SiD.

ILD detector combines excellent tracking and finely-grained calorimetry systems. This gives ILD the ability to reconstruct the energy of individual particles, known as the Particle Flow approach. The precision that can be achieved by ILD is ideal for studies in particle physics which call for accurate measurements of particles and their properties.

SiD is a multi-purpose detector, optimized for the broad range of physics opportunities at ILC. It is a compact, cost-constrained detector made possible with a 5 Tesla magnetic field and silicon tracking. The SiD concept incorporates Si/W electromagnetic calorimetry and all-Si tracking in a detector design which attempts to optimize physics performance, constrain costs, and be robust against physics and machine backgrounds. The highly granular calorimeter is optimized for particle flow analysis.

4. Physics case for the ILC experiments

4.1 Higgs boson measurements

The mail goal of the ILC experiments is measurements of the Higgs couplings. It can be done using three major Higgs boson production processes: $e^+e^- \to ZH$ ("higgsstrahlung"), $e^+e^- \to \nu_e \bar{\nu}_e H$ ("W fusion"), and $e^+e^- \to e^+e^- H$ ("Z fusion"). These processes have large production cross sections at the region of the 230-500 GeV CM energy (figure 3). For each of these processes we can identify all of the major Higgs decays modes, such as $H \to b\bar{b}$, WW^*, $c\bar{c}$, $\tau\tau$ and $\gamma\gamma$, with high efficiency.

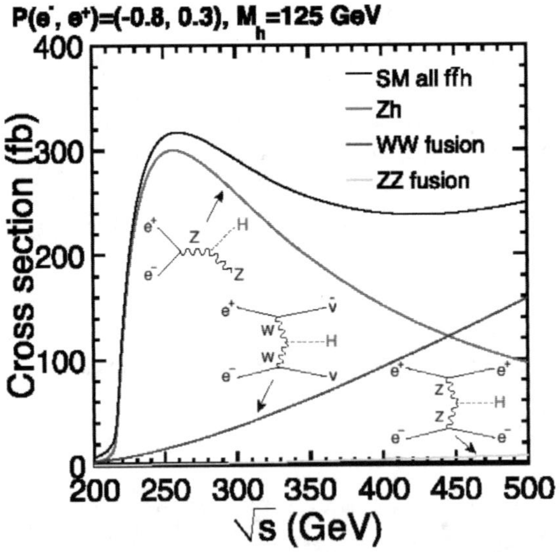

Figure 3. Cross sections for the three major Higgs production processes as a function of CM energy.

The higgsstralung process $e^+e^- \to ZH$ has a special advantage. By reconstructing the Z boson at two lepton mode its recoil mass can be calculated. The recoil mass to Z can be used to observe a clear Higgs signal without looking at the Higgs decay at all. This has three important consequences. First, it gives an opportunity to determine the total width of the Higgs boson and to obtain the absolute normalization of the Higgs couplings. Second, it allows to obtain the rate of the Higgs decays to invisible or exotic states. Third, it provides the way for a very precise determination of the mass of the Higgs boson.

Using the absolute normalization of the Higgs couplings the partial couplings to different final states can be measured in a model-independent way with a high accuracy. The accuracy of the most couplings measurements can reach the precision of 1% or better in the course of the ILC program. Running ILC at 550 GeV rather that 500 GeV would improve the precision from 9% to

3% using the channel $e^+e^- \to t\bar{t}H$. Such measurements are very important and can be used for indirect searches for a New Physics. Different BSM models will result in different biases from the couplings, expected in the Standard Model (figure 4).

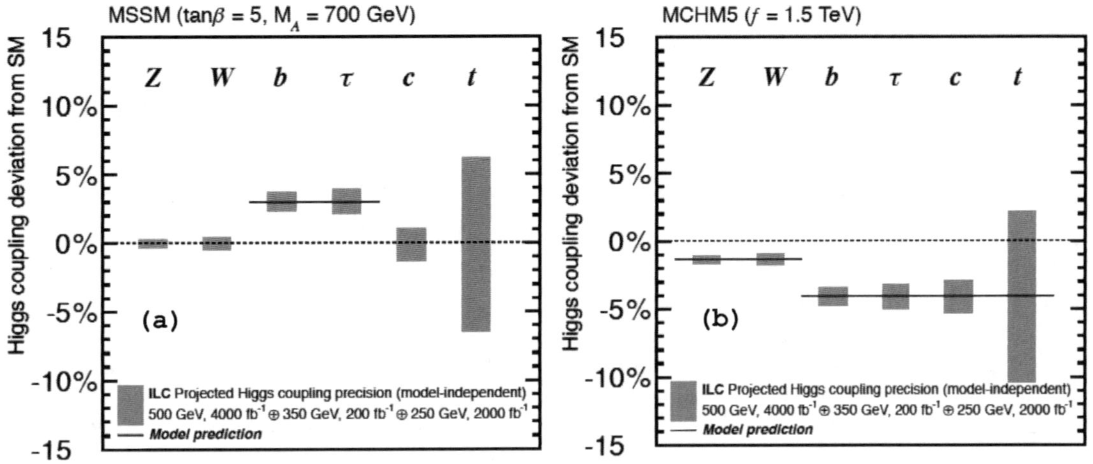

Figure 4. Two examples of BSM models and their predicted effects on the pattern of Higgs boson couplings. a) a supersymmetric model; b) a model with Higgs boson compositeness. The error bars indicate the 1σ uncertainties expected from the model-independent fit to the full ILC data set.

4.2 Top quark measurements.

Comprehensive studies of the top quark can be performed with the linear collider. This includes precise measurements of the top quark mass, width, CP-violating effects [4] and couplings. Top production cross section close to threshold of two top masses has to be measured to obtain a precise value of mass. Also the $e^+e^- \to t\bar{t}$ process is sensitive to the intermediate Z' boson contribution, additionally to the regular Z boson. It is also important to measure the forward-backward asymmetry in top production processes to resolve the tension between experiment and theory, observed for that parameter at Tevatron.

4.3 Other measurements

An important part of the ILC program is a search for new particles, which are not presented in the Standard Model. The QCD processes can be studied and compared with theoretical predictions with a high accuracy. Potentially *CP*-violation effects can emerge in various processes, indirectly indicating to BSM contributions. The linear collider has a strong physics program, which is complementary to the LHC physics program in many aspects.

References

[1] Baer H *et al.* 2013 *arXiv*:1306.6352
[2] Moortgat-Pick G *et al.* 2015 *arXiv*:1504.01726
[3] Fujii K *at al.* 2015 *arXiv*:1506.05992
[4] Atwood D *et al.* 2001 *Phys. Rept.* **347** 1

Recent Standard Model measurements at CMS

M Dordevic on behalf of the CMS Collaboration
CERN, Route de Meyrin 385, CH-1211 Geneva 23, Switzerland
E-mail: milos.dordevic@cern.ch

Abstract. Recent results of Standard Model physics using 7 and 8 TeV data recorded by the CMS detector are reviewed. This overview includes studies of vector boson production, results on V+jets production with light and heavy flavours, multiboson measurements and anomalous couplings searches and also the latest results on jet production and properties. The most recent 13 TeV results are presented as well. The outlined results are compared to the prediction of the Standard Model.

1. Introduction

The most recent results of the Standard Model (SM) measurements are presented in this overview. The full statistics from the LHC Run 1 was used, corresponding to an integrated luminosity of around 5 fb^{-1} and 20 fb^{-1} at the energy of 7 and 8 TeV, respectively. As of 2015, the LHC Run 2 has started with proton-proton collisions at the world-record energies of 13 TeV, with the initial goal to deliver around 100 fb^{-1} by the end of 2018. One of the first results from the LHC Run 2 is the measurement of the charged hadron pseudorapidity distribution and it is presented at the end of this overview.

2. Vector boson studies

2.1. Angular coefficients of Z bosons at 8 TeV

The first measurement of the five most significant angular coefficients of Z boson produced in proton-proton collisions and decaying to muon pairs was performed [1] . These coefficients govern the kinematics of Z boson decay and follow from the vector and axial vector (V-A) structure of boson-fermion couplings. The parameters A_0, A_1 and A_2 are related to the polarisation of the Z boson, while A_4 and A_5 are sensitive to the V-A couplings of muons. The contribution of the qg Compton process is significant in proton-proton collisions. The angular coefficients were measured in eight q_T and two $|y|$ bins. The measurement of the $A_0 - A_2$ for rapidity of $|y| < 1$ is shown in figure 1 and compared with MadGraph (LO) [2], POWHEG (NLO) [3] and FEWZ (NNLO) [4, 5, 6] predictions. The violation of the Lam-Tung relation [7] ($A_0 = A_2$) is observed, where $A_0 > A_2$, especially at the high q_T. This is anticipated by QCD calculations beyond the leading order.

2.2. Final state radiation in Z decay at 7 TeV

The study of final state radiation photons in Z boson decaying to muon pairs was performed at the 7 TeV [8]. The differential cross section was measured as a function of the photon transverse momentum and its separation from the nearest muon. This result will support future measurements of the W mass, the study of $Z\gamma$ production and searches for new physics with photons in the final state. The differential cross section $d\sigma/dE_T$ was measured for photons close to a muon ($0.05 < \Delta R_{\mu\gamma} < 0.5$) and far from the muon ($0.5 < \Delta R_{\mu\gamma} < 3$). The differential cross sections $d\sigma/dE_T$ and $d\sigma/d\Delta R_{\mu\gamma}$ were measured for the events with large (> 50 GeV) and small (< 10 GeV) transverse momentum of the Z boson. The prediction from POWHEG interfaced to PYTHIA [9, 10] reproduces the data well, with the discrepancies below 5% for the events with $5 < E_T < 50$ GeV and $0.05 < \Delta R_{\mu\gamma} < 2$, that is within the uncertainties. The measured differential cross section $d\sigma/dE_T$, for the case of $q_T > 50$ GeV, and its comparison with the simulation is shown in figure 2.

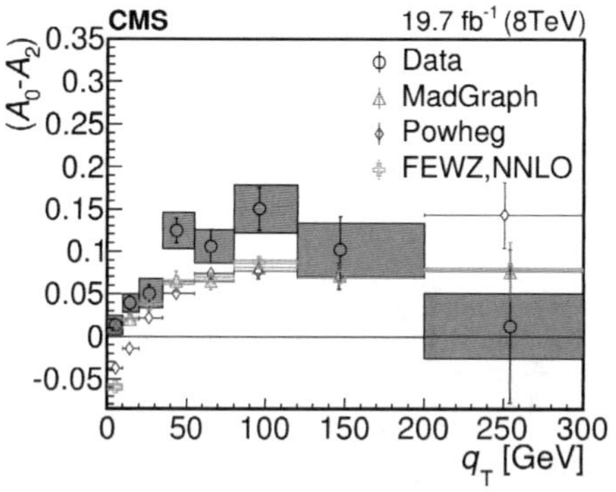

Figure 1. The difference between the angular coefficients A_0 and A_2 for $|y < 1|$, as a function of q_T, and its comparison to the prediction of MadGraph, POWHEG and FEWZ event generators. The vertical bars represent the statistical uncertainties and the boxes the systematic uncertainties of the measurement.

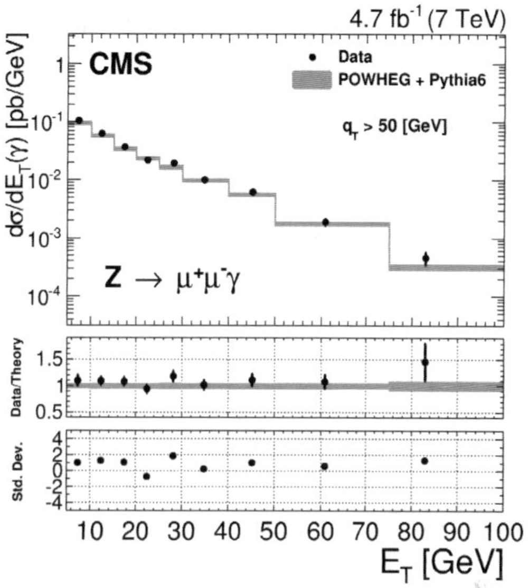

Figure 2. The measured $Z(\mu\mu)\gamma$ differential cross section as a function of the photon transverse momentum, for $q_T > 50$ GeV. The shaded bands represent the POWHEG+PYTHIA calculation including the theoretical uncertainties and at bottom is the ratio of data to the MC expectation.

3. Vector boson + jets studies

3.1. Z/γ^*+jets to γ+jets cross section ratio at 8 TeV

The ratio of the differential cross sections for the processes Z/γ^*+jets and $\gamma+jets$ was measured [11]. For high transverse momentum of the vector boson and at the LO, the effects due to the mass of Z boson are small and this ratio is expected to become constant. This is important for the studies of final states with hard jets and large missing transverse energy, such is the search for supersymmetry using the M_{T2} variable [12], performed by the CMS Collaboration. In this study, the $Z \to \nu\nu + jets$ background was modelled by measuring the $\gamma + jets$ and scaling it by the simulated ratio of the $Z \to \nu\nu + jets$ to $\gamma + jets$. The measurement of this ratio can help to reduce the uncertainties related to background estimation in these searches. The comparison was made with the prediction of MadGraph and BlackHat [13] generators and this is shown in figure 3. It was observed that both generators are overestimating the data by 20% at the high transverse momentum.

3.2. Electroweak W production cross section at 8 TeV

The electroweak production cross section of W boson in association with two jets was performed at the 8 TeV [14]. The study of this process could be used as a potential probe for the anomalous triple gauge couplings (aTGC) and it presents an important background to Higgs boson searches in the vector boson fusion (VBF) topology. The events with isolated electron or muon, two jets well separated in rapidity and significant missing transverse energy are selected. To distinguish between the signal and background events a Boosted Decision Tree (BDT) is used and an unbinned maximum likelihood fit to the dijet invariant mass distribution, shown in figure 4, is performed in order to extract the number of signal events from data. A good agreement between the measured cross section and the prediction of MadGraph interfaced to PYTHIA6 was observed.

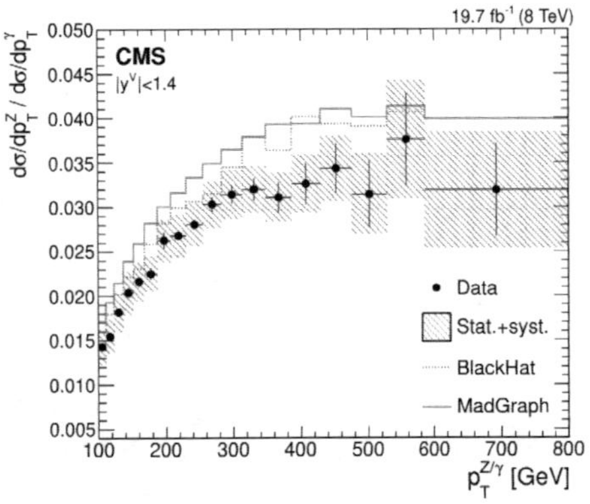

Figure 3. The ratio of the differential cross sections for the $Z/\gamma^* + jets$ and $\gamma + jets$, shown for the central bosons ($|y^V| < 1.4$) and comparison to BlackHat and MadGraph generators. The hatched (gray) band represent the total uncertainty of the measurement.

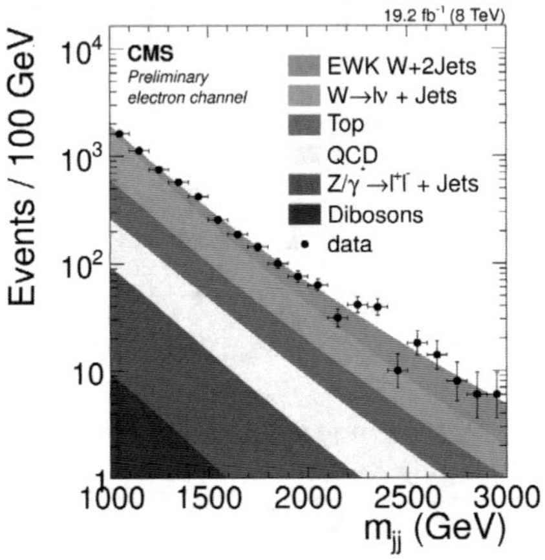

Figure 4. Distribution of the invariant mass of the dijet pair associated to the W boson, shown in the electron channel. Overall, the applied approach produces a high quality model of the data.

4. Multiboson studies

4.1. $Z\gamma \to \nu\nu\gamma$ production at 8 TeV

The $Z(\nu\nu)\gamma$ production was studied at the 8 TeV [15]. This process has six times larger branching ratio (BR) than for a decay to a particular lepton pair and also the acceptance for neutrino is higher. The cross section is measured in the kinematic region with high missing transverse energy, high photon transverse momentum and pseudorapidity of photon restricted to the barrel region of the CMS detector, where purity is the highest. The measured cross section was found to be in a good agreement with the NNLO calculation by Grazzini et. al. No evidence was found for the aTGC couplings and 95% CL limits were set on the h_3^V and h_4^V parameters of $ZZ\gamma$ and $Z\gamma\gamma$ couplings. The expected and observed confidence level contours in the (h_3, h_4) parameter space are shown in figure 5. These are the most stringent limits to date on neutral aTGCs.

Journal of Physics: Conference Series **675** (2016) 022020 doi:10.1088/1742-6596/675/2/022020

4.2. W^+W^- *production at 8 TeV*

The production cross section for the WW process is the highest among the diboson pairs. This process was studied at the 8 TeV by selecting two electrons or muons with the high transverse momentum, large missing transverse energy and zero or one jet with high transverse momentum [16]. The measured cross section was found to be in a good agreement with the NNLO prediction [17]. The differential cross section was measured as a function of various kinematic variables and overall good agreement was found in comparison to the predictions of MadGraph, MC@NLO [18] and POWHEG generators. Search for the aTGC was performed in the framework of the effective field theory [19]. No evidence of the anomalous WWZ and WWγ vertices was observed. The aTGC limits were extracted using the reconstructed dilepton invariant mass distribution, shown in figure 6. An improved measurement of the coupling constant c_{WWW}/Λ^2 was obtained.

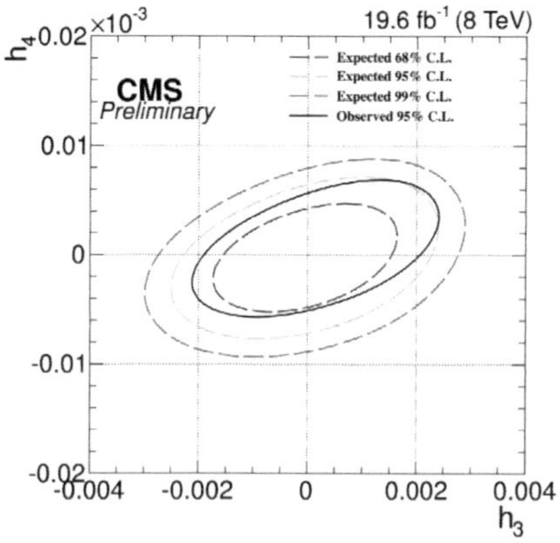

Figure 5. The observed and expected contours in (h_3, h_4) parameter space showing two-dimensional 95% CL limits on $ZZ\gamma$ couplings.

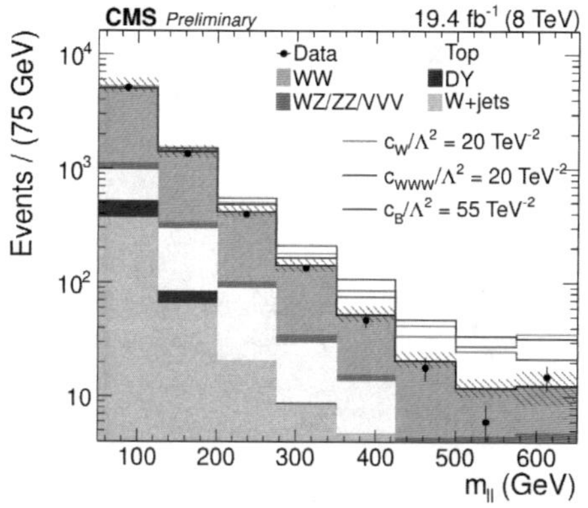

Figure 6. The invariant mass of the dilepton pair, shown for data, simulated SM and aTGC predictions. The last bin includes all the events with $m_{ll} > 575$ GeV.

5. Jet physics studies

5.1. Dijet azimuthal decorrelation at 8 TeV

The dijet azimuthal decorrelations between the two jets with the largest transverse momenta are measured [20]. At the LO in pQCD the two final-state partons balanced in transverse momentum are produced in back-to-back topology and non-perturbative effects of the multiparton interaction perturb this correlation only mildly. The production of a third jet leads to a decorrelation in the azimuthal angle. The measurement of the dijet azimuthal decorrelation enables to study the multijet production process while considering only the two jets with the highest transverse momentum. Figure 7 shows the normalised dijet cross section measured as a function of the dijet azimuthal angular separation in seven regions of the leading jet transverse momentum and compared to pQCD calculations for 3-jet production with up to four outgoing partons ($\propto \alpha_S^4$), that provide LO and NLO predictions for ($\pi/2 < \Delta\phi_{Dijet} < 2\pi/3$) and ($2\pi/3 < \Delta\phi_{Dijet} < \pi$), respectively. From the LO dijet MC event generators PYTHIA6, HERWIG++ [21] and POWHEG, PYTHIA8 shows the best agreement, while a very good description was observed using the multijet improved MadGraph generator interfaced to PYTHIA6 for showering, hadronisation and multiparton interactions, as shown in figure 8.

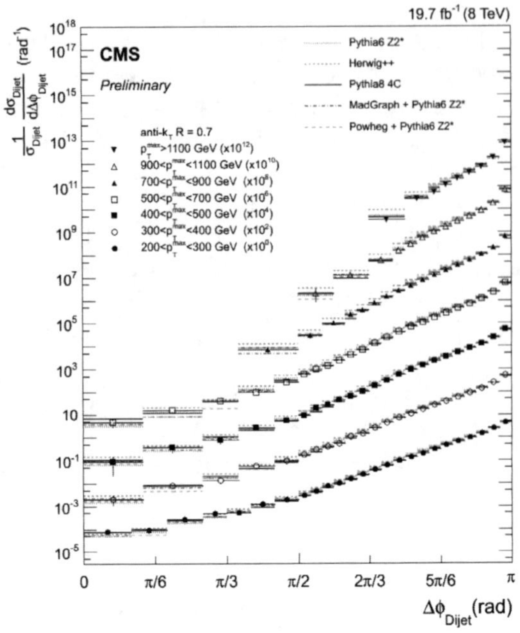

Figure 7. The normalised dijet cross section differential in $\Delta\phi_{Dijet}$ for seven p_T^{max} regions and compared to different event generators.

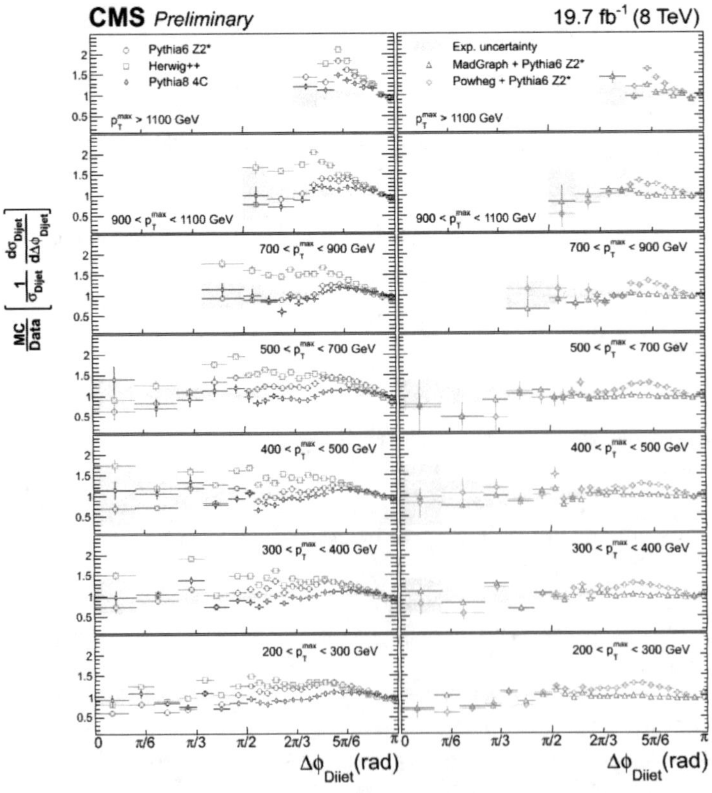

Figure 8. Ratios of the MC predictions to the normalised dijet cross section differential in $\Delta\phi_{Dijet}$, presented for all the p_T^{max} regions.

6. Forward and small-x QCD studies

6.1. Pseudorapidity distribution of charged hadrons at 13 TeV

The pseudorapidity distribution of charged hadrons was measured at the energy of 13 TeV, when the CMS operated at zero magnetic field [22]. The yields of charged hadrons represent a first step towards exploring the physics of a new energy frontier. The special low-intensity beam configuration was used, with 0.2-5% proton-proton interaction probability per bunch crossing. Products of secondary interactions are excluded and contributions from prompt leptons are removed. Two reconstruction techniques were used, based on the hits in the CMS pixel detector. The tracklet method uses hit pairs and has the background estimated from data, while the track method is based on identifying pixel hit triplets that fall on a straight line. The two methods agree within 2% in the central and with less than 3% in the forward pseudorapidity region. The charge hadron density was measured in the most central region of $|\eta| < 0.5$. A good agreement with EPOS generator using LHC tune [23] was obtained, while the PYTHIA8 with CUETP8S1 [24] underestimates the data in the forward pseudorapidity region, as it is presented in figure 9. The comparison of the center-of-mass dependence of $dN_{ch}/d\eta$ at 13 TeV was made with the previous measurements at lower energies (ISR [25, 26], UA5 [27, 28], PHOBOS [29] and ALICE [30, 31]). It was observed that the PYTHIA8 and EPOS LHC globally reproduce the energy dependence of hadron production in inelastic pp collisions, as shown in Figure 10.

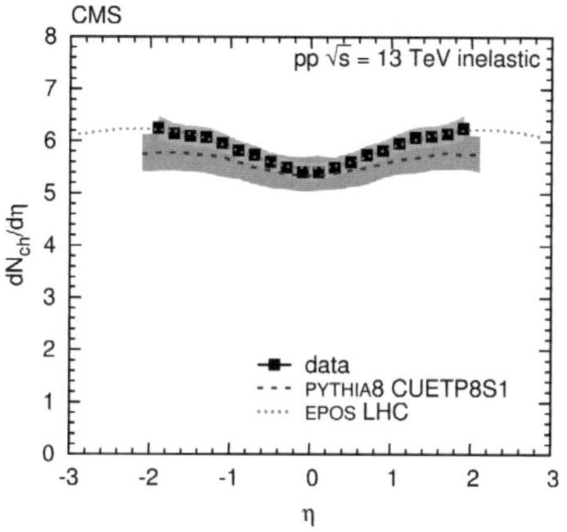

Figure 9. Distribution of the pseudorapidity density of charged hadrons for $|\eta| < 2$ and comparison with the PYTHIA8 and EPOS LHC event generators. The blue band is the envelope of CUETP8S1 tune parametric uncertainties.

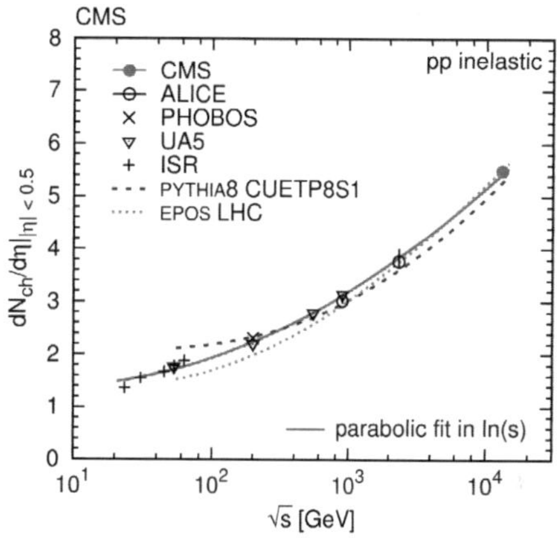

Figure 10. Center-of-mass energy dependence of $dN_{ch}/d\eta|_{|\eta|<0.5}$ observed at the 13 TeV, compared to PYTHIA8 and EPOS LHC predictions, and also lower energy ISR, UA5, PHOBOS and ALICE data.

7. Conclusion

Recent SM results from CMS using LHC Run 1 data provided extensive tests of the electroweak and strong interactions. Vector bosons were studies with an unprecedented precision and pQCD was tested in the regions never been probed before. Measurements of vector bosons associated with jets are improved and new PDF constraints can be derived. Multiboson studies are performed in the various final states and some of the world leading aTGC and aQCG limits were extracted. Jet production was studied in the extended phase space. The pseudorapidity distribution of charged hadrons at 13 TeV is presented as well.

References

[1] CMS Collaboration 2015 *CMS PAS SMP-14-015*
[2] Alwall J *et al.* 2011 *JHEP* **1106**

[3] Alioli S *et al.* 2008 *JHEP* **07** 060
[4] Melnikov K and Petriello F 2006 *Phys. Rev.* D **74** 114017
[5] Gavin R *et al.* 2011 *Comput. Phys. Commun.* **182** 2388-2403
[6] Li Y and Petriello F 2012 *Phys. Rev.* D **86** 094034
[7] Lam C S and Tung W K 1979 *Phys. Lett.* B **80** 228
[8] CMS Collaboration 2015 *Phys. Rev.* D **91** 092012
[9] Sjstrand T, Mrenna S and Skands P 2006 *JHEP* **05** 026
[10] Sjstrand T, Mrenna S and Skands P 2008 *Comput. Phys. Comm.* **178** 852
[11] CMS Collaboration 2015 *CMS PAS SMP-14-005*
[12] CMS Collaboration 2012 *JHEP* **10** 018
[13] Berger C F *et al.* 2009 *Phys. Rev.* D **80** 074036
[14] CMS Collaboration 2015 *CMS PAS SMP-13-012*
[15] CMS Collaboration 2015 *CMS PAS SMP-14-019*
[16] CMS Collaboration 2015 *CMS PAS SMP-14-016*
[17] Gehrmann T *et al.* 2013 *Phys. Rev. Lett.* **113** 212001
[18] Frixione S and Webber B R 2002 *JHEP* **0206** 029
[19] Degrande C 2013 *Annals Phys.* **335** 21
[20] CMS Collaboration 2015 *CMS PAS SMP-14-015*
[21] Corcella G *et al.* 2001 *JHEP* **0101** 010
[22] CMS Collaboration 2015 *arXiv*:1507.05915
[23] Werner K, Liu F M, and Pierog T 2006 *Phys. Rev.* C **74** 044902
[24] CMS Collaboration 2015 *CMS-PAS-GEN-14-001*
[25] Rossi A M *et al.* 1975 *Nucl. Phys.* B **84** 269
[26] Aachen-CERN-Heidelberg-Munich Collaboration 1977 *Nucl. Phys.* B **129** 365
[27] UA5 Collaboration 1986 *Z. Phys.* C **33**
[28] UA5 Collaboration 1987 *Phys. Rept.* **154** 247
[29] PHOBOS Collaboration 2004 *Phys. Rev. Lett* **93** 082301
[30] ALICE Collaboration 2010 *Eur. Phys. J.* C **68** 89
[31] ALICE Collaboration 2010 *Eur. Phys. J.* C **68** 345

Journal of Physics: Conference Series **675** (2016) 022021 doi:10.1088/1742-6596/675/2/022021

Charge-dependent azimuthal correlations of secondary particles in heavy ion collisions

V Okorokov and P Parfenov

National Research Nuclear University MEPhI (Moscow Engineering Physics Institute), Kashirskoe highway 31, 115409 Moscow, Russia

E-mail: VAOkorokov@mephi.ru, P.Parfenov@gsi.de

Abstract. The P/CP symmetry breaking in quantum chromodynamics (QCD) could be realized via transitions between local fluctuations of gauge fields. Azimuthal correlations which characterize the asymmetry of the emitted charged particles with respect to the reaction plane in non-central nucleus-nucleus collisions are the promising tools for experimental study of local P/CP violation in the strong interactions. The preliminary estimations of correlators within the model of chiral magnetic effect are presented for types of nuclei and collision energies corresponded to RHIC and the LHC beams for two various nuclear densities, namely, for approach of the hard sphere and for the two-component Fermi model. Besides of the correlator estimations for the symmetric collisions, the preliminary results for magnetic field in asymmetric $Cu + Au$ collisions are also shown.

1. Introduction

The Lagrangian of QCD contains natural terms that can break the P/CP symmetry [1]. There is a fundamental interrelation between geometry and fundamental properties of QCD Lagrangian. The vacuum of QCD is a very complicated matter with rich structure, which can corresponds to the fractal-like geometry [2]. The non-trivial topology of QCD vacuum opens the possibility for existence of metastable domains which possess of various properties with respect to the discrete P/CP symmetries. In the presence of background Abelian electromagnetic field decays of such domains or classical transitions (sphalerons) between them in deconfinement phase of color charges with restored chiral symmetry can lead to the separation of secondary charged particles toward the magnetic field. This phenomenon called also chiral magnetic effect (CME) is the experimental manifestation of the local topologically induced P/CP parity violation in the strong interactions [3].

2. Magnetic field in heavy-ion collisions

The magnetic field in nucleus-nucleus collisions can be obtained, for instance, using the Lienard–Wiechert potentials. This approach is described in detail elsewhere [3]. In the paper the strength of the magnetic field (eB) in collisions of various ion beams are compared for two parameterizations of nuclear density, namely, the sphere with sharp boundary, i.e. rigid sphere [3] and the following equation within the framework of two-component Fermi model

$$\rho_i^{\mathrm{F}}(\mathbf{x}'_\perp) = N_i^{\mathrm{F}}\varepsilon[1+\exp(-\varepsilon/a)]^{-1}\theta_i(\varepsilon^2), \quad \varepsilon \equiv \sqrt{R_i^2 - [\mathbf{x}'_\perp \pm \mathbf{b}/2 \mp (R_2^2 - R_1^2)\mathbf{e}_\perp/2b]^2}, \qquad (1)$$

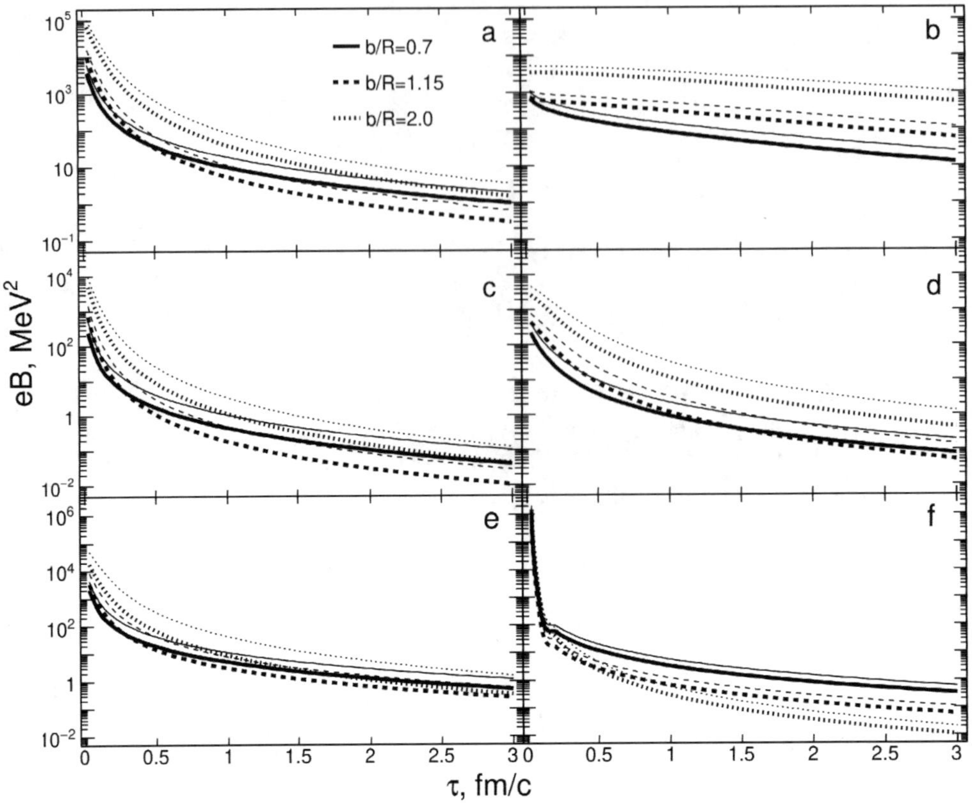

Figure 1. Dependence of the strength of magnetic field on proper time in nucleus-nucleus collisions at the origin ($x'_\perp = 0$) and central pseudorapidity ($\eta = 0$): $Au + Au$ at $\sqrt{s_{NN}} = 200$ (a) and 7.7 GeV (b); $Cu + Cu$ at $\sqrt{s_{NN}} = 200$(c) and 62.4 GeV (d); $Cu + Au$ at $\sqrt{s_{NN}} = 200$ GeV (e), $Pb + Pb$ at $\sqrt{s_{NN}} = 2.76$ TeV (f). The solid lines show $eB(\tau)$ for central collisions, dashed lines – for mid-central, dotted lines – for peripheral. Thin lines correspond to the limit case of the rigid sphere, thick lines – to the Fermi parametrization.

$$1/N_i^{\rm F} = 2\pi a \left\{R_i^2 \ln(1+\kappa_i) + 2a\left[R_i {\rm Li}_2\left(-e^{\kappa_i}\right) - a{\rm Li}_3\left(-e^{\kappa_i}\right)\right] - 1.5a^2\zeta(3)\right\}.$$

Here b is the impact parameter, R_i – the radius of the nucleus moved along of the z-axis in the positive ($i = 1$) and negative ($i = 2$) directions, a is the cutoff factor, $\kappa_i \equiv R_i/a$, ${\rm Li}_j(x)$ and $\zeta(x)$ are the polylogarithm of j order and Riemann zeta function respectively, the nuclear densities are normalized such that $\int d\mathbf{x}'_\perp \rho_i^{\rm F}(\mathbf{x}'_\perp) = 1$ as well as in [3]. It would be stressed that the equation (2) is valid for asymmetric, in general case, nucleus-nucleus collisions. The strength of Abelian magnetic field is studied for two parameterizations of nuclear density for wide set of nucleus-nucleus collisions at RHIC and the LHC energies ($\sqrt{s_{NN}}$), namely, for $Cu + Cu$ at $\sqrt{s_{NN}} = 22.6, 62.4$ and 200 GeV, $Cu + Au$ at $\sqrt{s_{NN}} = 200$ GeV, $Au + Au$ at $\sqrt{s_{NN}} = 7.7 - 200$ GeV, $Pb + Pb$ at $\sqrt{s_{NN}} = 2.76$ TeV and $U + U$ at $\sqrt{s_{NN}} = 193$ GeV. The preliminary eB dependence on proper time (τ) obtained for several collision types and $\sqrt{s_{NN}}$ is shown in figure 1 for central ($b/R = 0.7$), mid-central ($b/R = 1.15$) and peripheral ($b/R = 2.0$) interactions. At large τ as expected in accordance with [4] the eB increases significantly with the decrease of the collision energy in both the $Au+Au$ (figures 1a, b) and the $Cu+Cu$ (figures 1c, d) with opposite behaviour of peak eB values at very small τ. The eB shows the much more sharper decrease for $Pb+Pb$ collisions at the LHC energy for small τ (figure 1f) than that for highest RHIC energy $\sqrt{s_{NN}} = 200$ GeV. As seen the transition from the simplest case of the rigid sphere to the more realistic Fermi distribution for nuclear density leads to some decrease of the strength of magnetic field especially for RHIC energies.

3. Correlators of the charged particles

In the framework of CME model the theoretical correlator is given by [4, 5]

$$\langle \mathbf{K}^{\mathrm{T}}_{\alpha\beta} \rangle = \frac{\pi^2}{16} \frac{\langle \Delta_\alpha \Delta_\beta \rangle}{\langle N_\alpha N_\beta \rangle}, \quad \langle \Delta_\alpha \Delta_\beta \rangle = \kappa \alpha_S \left(\frac{\pi R}{2} \sum_f q_f^2 \right)^2 \varepsilon_{\alpha\beta} \int_{S_\perp} d^2 x_\perp \xi_{\alpha\beta}(x_\perp) \int_{\tau_i}^{\tau_f} d\tau \tau \left[eB(\eta, \tau) \right]^2 . \quad (2)$$

Here $\varepsilon_{\pm\pm} = -0.5\varepsilon_{\pm\mp} = 0.5$, $\xi_{\pm\pm}(x_\perp) \equiv \sum_{i=+,-} \xi_i^2(x_\perp, \lambda)$, $\xi_{\pm\mp}(x_\perp, \lambda) \equiv \prod_{i=+,-} \xi_i(x_\perp, \lambda)$, α_S – renormalized strong coupling constant, N_α – multiplicity of particles with electric charge sign α in event, q_f – electric charge (in units of e) of quark with flavor f, Δ_α is the difference between total charges with fixed sign α on each side of the reaction plane, numerical coefficient $\kappa \sim 1$, $x_\perp$ – position of nucleon from one of the colliding nuclei in the plane perpendicular to the beam axis, $S_\perp$ – area of overlap region of colliding nuclei in the transverse plane, functions $\xi_\pm(x_\perp, \lambda)$ take into account the suppression effect via screening length λ [3]. The correlators $\langle \Delta_\alpha \Delta_\beta \rangle$ are calculated for set of collision types and energies indicated above. The variation of the type of nucleus and $\sqrt{s_{NN}}$ allow the investigation of the energy and atomic number dependencies of $\langle \Delta_\alpha \Delta_\beta \rangle$ as well as the influence of both the parametrization of nuclear density and the λ on correlator values. The preliminary results are shown in figures 2 and 3.

The larger λ leads to noticeable increase of same-charge correlator $\langle \Delta_\pm^2 \rangle$ in $Cu+Cu$ collisions at some fixed $\sqrt{s_{NN}}$ (figure 2) as well as for heavy ion collisions ($Au+Au$). On the other hand

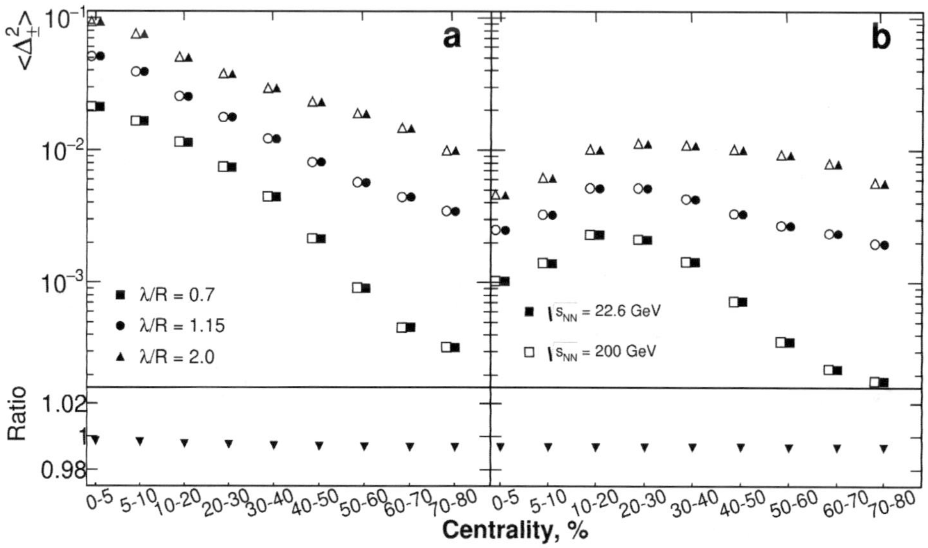

Figure 2. Centrality dependence of correlators for same-charged particle pairs in $Cu+Cu$ collisions for rigid sphere (a) and Fermi parametrization (b) for nuclear density at different λ/R and initial energies. The low panels show the ratio of correlator values at $\sqrt{s_{NN}} = 22.6$ GeV to the $\langle \Delta_\alpha \Delta_\beta \rangle$ at $\sqrt{s_{NN}} = 200$ GeV. Due to the structure of correlators this ratio is independent on λ.

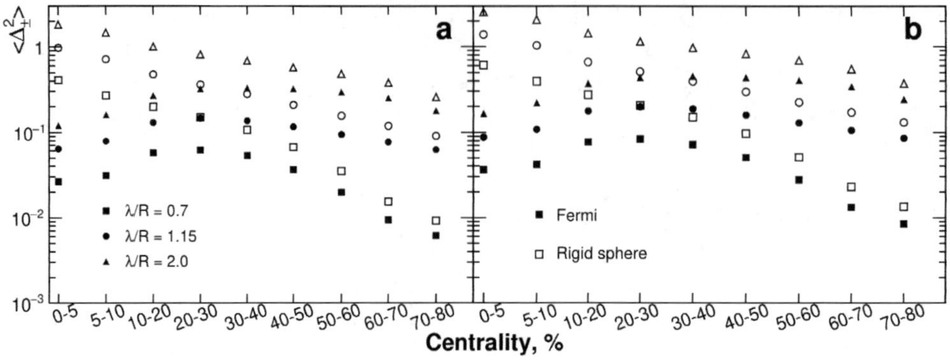

Figure 3. Centrality dependence of the correlator $\langle\Delta^2_{\pm}\rangle$ for $Pb+Pb$ collisions at $\sqrt{s_{NN}} = 2.76$ TeV (a) and for $U+U$ at $\sqrt{s_{NN}} = 193$ GeV (b) for various screening lengths and parameterizations for nuclear density.

as seen in figure 2 the values of $\langle\Delta^2_{\pm}\rangle$ weakly depend on collision energy for wide range of $\sqrt{s_{NN}} = 22.6 - 200$ GeV. Furthermore the dependence of $\langle\Delta^2_{\pm}\rangle$ on $\sqrt{s_{NN}}$ in $Cu+Cu$ collisions is similar to that for interactions of heavy ions ($Au+Au$). It should be noted that the growth of multiplicity of secondary particles with increase of $\sqrt{s_{NN}}$ leads to the some decrease of the magnitudes of $\langle\mathbf{K}^{\mathrm{T}}_{\alpha\beta}\rangle$ as well as of corresponding experimental observables for higher energy nucleus-nucleus collisions. Therefore the energy dependence of $\langle\Delta^2_{\pm}\rangle$ obtained here agrees with expectations from [3, 4, 5]. In general the correlator values for same-charged particle pairs are smaller for Fermi parametrization than that for approach of rigid sphere at corresponding centralities. But the type of parametrization for nuclear density influences on the $\langle\Delta^2_{\pm}\rangle$ mostly for central collisions and this influence decreases for more peripheral collisions (figures 2, 3). The same-charge correlator values change weakly with increasing of atomic number in $Pb+Pb$ and $U+U$ interactions but the $\langle\Delta^2_{\pm}\rangle$ shows the noticeable increase at transition from moderate ($Cu+Cu$) to heavy ($Au+Au$, $Pb+Pb$, $U+U$) nucleus-nucleus collisions.

4. Conclusion

The strength of magnetic field in $A+A$ collisions as well as multiplicity-independent correlators within the framework of the CME are calculated for two parameterizations of nuclear density, namely, for rigid sphere and for two-component Fermi model. The calculations are made for wide set of beam types and collision energies. The more realistic Fermi distribution leads to some decrease of the strength of magnetic field especially for RHIC energies. For the first time the estimations of the magnetic field are calculated in asymmetric $Cu+Au$ collisions. The same-charge correlator magnitudes depend on collision energy weakly for $Cu+Cu$ and $Au+Au$, but this correlator increases noticeably with growth of atomic number from moderate to heavy nuclei. Comparison of correlator values with experimental observables is in the progress.

Acknowledgments

The work of V.O. was performed within the framework of the Center of Nanostructured Electronics supported by NRNU MEPhI Academic Excellence Project (contract No 02.a03.21.0005, 27.08.2013).

References

[1] Olive K A *et al.* 2014 *Chin. Phys.* C **38** 090001

[2] Okorokov V A and Sandrakova E V 2013 *Univ. J. Phys. Appl.* **1** 196
[3] Kharzeev D E, McLerran L D and Warringa H J 2008 *Nucl. Phys.* A **803** 228
[4] Okorokov V A 2009 *arXiv*: 0908.2522 [nucl-th]
Okorokov V A 2013 (In Russian) *Nucl. Phys. and Engin.* **4** 805 (original title of the Russian journal: *Yadernaya Fizika i Enginiring*)
[5] Okorokov V A 2013 *Int. J. Mod. Phys.* E **22** 1350041

Study of nuclear multifragmentation induced by ultrarelativistic μ-mesons in nuclear track emulsion

D A Artemenkov[1,2], V Bradnova[1], E Firu[3], N K Kornegrutsa[1], M Haiduc[3], K Z Mamatkulov[1], R R Kattabekov[1], A Neagu[3], P A Rukoyatkin[1], V V Rusakova[1], R Stanoeva[4], A A Zaitsev[1,5], P I Zarubin[1,5] and I G Zarubina[1,5]

[1] Joint Institute for Nuclear Research, Dubna, Russia
[2] National Research Nuclear University MEPhI (Moscow Engineering Physics Institute), Kashirskoe highway 31, Moscow, 115409, Russia
[3] Institute of Space Science, Magurele, Romania
[4] South-Western University, Blagoevgrad, Bulgaria
[5] P. N. Lebedev Physical Institute of the Russian Academy of Sciences, Moscow, Russia

E-mail: zarubin@lhe.jinr.ru

Abstract. Exposures of test samples of nuclear track emulsion were analyzed. The formation of high-multiplicity nuclear stars was observed upon irradiating nuclear track emulsions with ultrarelativistic muons. Kinematical features studied in this exposure of nuclear track emulsions for events of the muon-induced splitting of carbon nuclei to three α-particles are indicative of the nuclear-diffraction interaction mechanism.

Possessing an excellent sensitivity and spatial resolution, nuclear track emulsions preserve their position as a universal and relatively cheap detector for surveying and searching investigations into nuclear and particle physics. The use of this classic procedure in beams from modern accelerators and reactors proved to be quite successful. In a number of important problems, the completeness of observations that is ensured by nuclear track emulsions remains inaccessible for electronic detection methods.

Test track-emulsion samples manufactured by the MICRON production unit of the Slavich Company JSC [1] are being presently irradiated within the Becquerel project [2]. The samples in question are created by casting emulsion layers 50 to 200 μm thick onto glass substrates. The basic properties of this nuclear track emulsion are close to those of BR-2 nuclear track emulsions, which are sensitive to relativistic particles. The production of BR-2 nuclear track emulsions had been performed for more than four decades and was completed about ten years ago. The product emulsion [1] has already been used in the range-based spectrometry of α-particles [3, 4].

Test irradiations were aimed primarily at a general quality control and a control of the emulsion sensitivity to relativistic particles, as well at a comparison of ranges of slow nuclei that have strongly ionizing low energies with the values calculated on the basis of the SRIM simulation code [5]. The present article combines the results obtained by analyzing recent exposures of nuclear track emulsions to ultrarelativistic muons. So wide a variety of experimental implementations, including those in [3, 4], became possible owing to the use of the new nuclear track emulsion, whose properties permitted applying the same strategy to coordinate measurements for tracks of length between several microns and several tens of microns. The

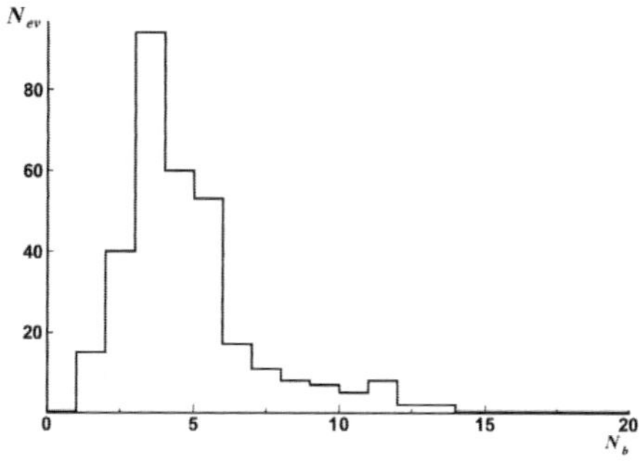

Figure 1. Distribution of nuclear stars with respect to the number of strongly ionizing b particles, N_b, in a track-emulsion sample irradiated with muons.

video data on the interactions studied in the track emulsion are available on the website quoted in [2].

The deep-inelastic scattering of ultrarelativistic muons is a commonly recognized means for studying the parton structure of nucleons and nuclei. The irradiation of track emulsions with these particles makes it possible to study concurrently the multifragmentation of nuclei under the effect of a purely electromagnetic probe. Multiphoton exchanges or transitions of virtual photons to vector mesons may serve as a fragmentation mechanism. At CERN, a track-emulsion sample was exposed to 160-GeV muons. Earlier, comparable analysis was performed for NTE longitudinally exposed to 150 GeV [6]. The objective of the exposure described here was to study experimental loads in the vicinity of the beam axis and to assess preliminarily the character of muon interactions.

The samples under study were placed in front of the target of the COMPASS experiment at a distance of about 25 cm from the beam axis (halo), where the intensity amounted to about 10^6 particles per centimetre squared per cycle. The track-emulsion samples 9×12 cm in area and about 100 μm in thickness were oriented both along and across the beam. A nine-hour irradiation in the case of the transverse orientation proved to be the most favourable for our analysis. So long-term an irradiation was possible owing to the smallness of the cross section for muon interaction and to a small effect of beam ionization in relation to the longitudinal arrangements of layers. The duration of the irradiation run was constrained for fear of overloads with tracks from interactions in the glass substrate. In principle, this duration could be increased by two orders of magnitude without causing complications for the ensuing analysis.

Scanning led to finding, in irradiated track emulsions, about 300 stars containing not less than three target fragments. The topology of stars was determined by the number of strongly ionizing b particles. Figure 1 shows the distribution of N_b. Although the solid angle within which tracks could be observed was limited, the formation of high-multiplicity stars involving almost one-half of the charge of heavy nuclei in the composition of the emulsion could be proven.

Seventy-two stars containing only triples of b particles stopped in the track emulsion were associated with the breakup process $^{12}C \to 3\alpha$. Alpha-particle ranges and spatial emission angles were determined on the basis of coordinate measurements for tracks. The mean α-particle range was 23.1 ± 0.6 μm (RMS $=8.4$ μm). The α-particle energy was estimated on the basis of the SRIM model. Its mean value proved to be 5.3 ± 0.1 MeV (RMS $=1.3$ MeV).

The fact that the interpretation of this group of events is unambiguous makes it possible to assess the character of their production on the basis of the total transverse momentum P_T of α-particle triples. The distribution of P_T (see figure 2) is characterized by the mean

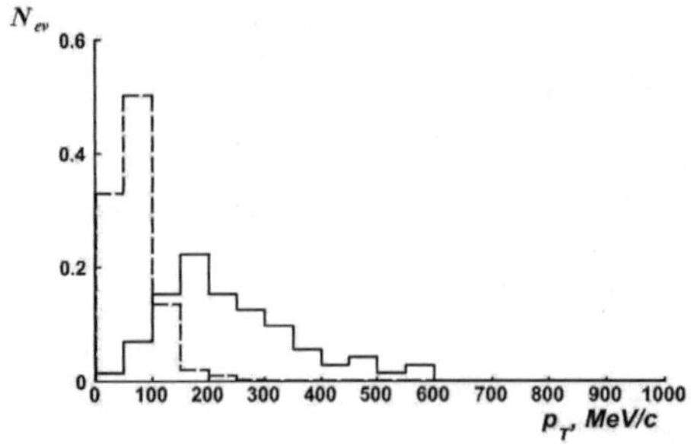

Figure 2. Total-transverse-momentum (P_T) distribution of (solid-line histogram) 72 triples of α-particles in a track emulsion irradiated with muons and (dashed-line histogram) 400 triples of α-particles from the reaction n(14.1 MeV) $+^{12}$C $\rightarrow 3\alpha + n$ [3] (normalization to the number of events).

value of 241 $\pm$ 28 MeV/c (RMS = 123 MeV/c). It is described by the Rayleigh distribution at the parameter value of 190 $\pm$ 13 MeV/c. These parameter values are typical of the nuclear diffraction interaction. In the case of purely electromagnetic exchange, the P_T distribution would be concentrated in the range of $P_T <$ 100 MeV/c. It is useful to compare this distribution of P_T with the substantially narrower distribution of P_T for the reaction n(14.1 MeV)$+^{12}$C $\rightarrow 3\alpha + n$ (see figure 1 in[3]). The latter is characterized by the mean value of 69 $\pm$ 4 MeV/c (RMS = 38 MeV/c) and by the Rayleigh distribution parameter value of 55 $\pm$ 28 MeV/c.

The distribution of the total energy of α-particle triads, $Q_{3\alpha}$, in figure 3 is substantially broader than that in the case of the reaction n(14.1 MeV) $+^{12}$C $\rightarrow 3\alpha$ [7], the latter revealing distinctly the clustering features of the ^{12}C nucleus. In the case being considered, the distribution of $Q_{3\alpha}$ is concentrated above the α-cluster levels of excitation of the ^{12}C nucleus. The distribution of the energy of α-particle pairs, $Q_{2\alpha}$, in figure 4 does not reveal any similarity in the spectra for the exposures being considered either. In the exposure to muons, there is virtually no signal from the decays of the ground state ^{8}Be$_{g.s.}$ in the range of $Q_{2\alpha} <$ 200 keV (see inset in figure 4), which manifest themselves as narrow α-particle pairs [7]. The distribution of $Q_{2\alpha}$ does not exhibit a peak from the decays of the first excited state ^{8}Be^{2+} at 3 MeV. Moreover, this distribution of $Q_{2\alpha}$ proves to be substantially broader than that in the case of the reaction n(14.1 MeV) $+^{12}$C $\rightarrow 3\alpha$ [7].

By and large, the $P_T, Q_{3\alpha}$ and $Q_{2\alpha}$ distributions for the irradiation of nuclear track emulsions with muons are indicative of a hard character of the process without manifestations of the well-known structural features of the ^{12}C nucleus, including the formation of α-particle triples in the continuum region. We emphasize that the contribution of ^{12}C breakup with a threshold of 7.36 MeV should have inevitably manifested itself in the channel being discussed ($N_b = 3$). However, the circumstance that it is the nuclear diffraction mechanism rather than the soft electromagnetic mechanism that manifests itself for this channel, which possesses the minimum threshold, seems unexpected and deserves a theoretical analysis. The corroboration of this conclusion is of importance for interpreting not only multifragmentation under the effect of ultrarelativistic muons. It may also serve as a basis for interpreting the multifragmentation of relativistic nuclei in peripheral interactions not leading to the formation of target fragments (white stars).

These observations, which are of a preliminary character, indicate that a full-scale investigation of a complete muon-induced disintegration of nuclei on the basis of multilayered assemblies from thick layers of substrate-free nuclear track emulsion is highly promising. In order

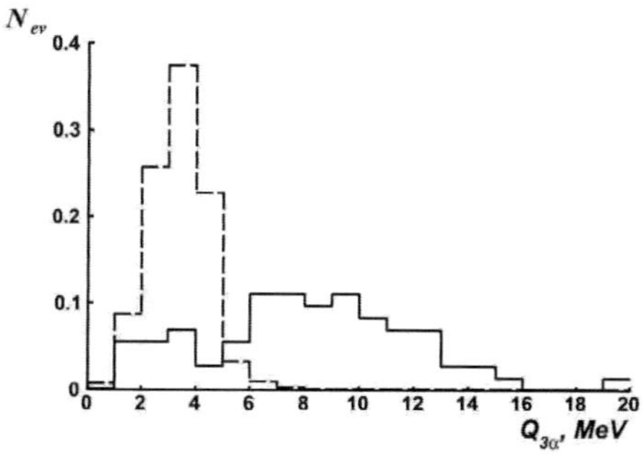

Figure 3. Distribution of the total energy of α-particle triples, $Q_{3\alpha}$, in a track emulsion irradiated with (solid-line histogram) muons and (dashed-line histogram) neutrons in the reaction $n(14.1\ \text{MeV}) + ^{12}\text{C} \rightarrow 3\alpha + n$ [3] (normalization to the number of events).

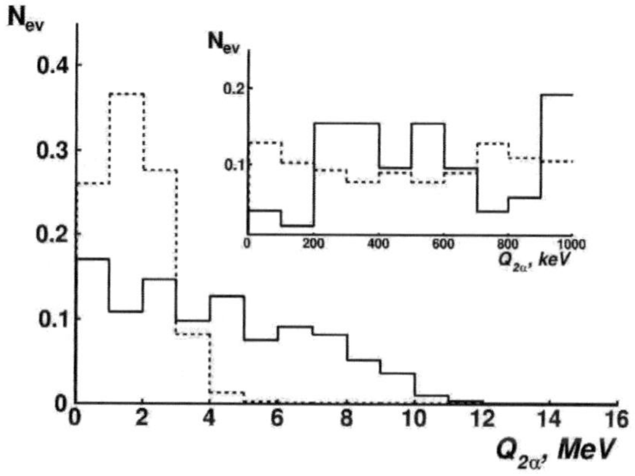

Figure 4. Distribution of the energy of α-particle pairs, $Q_{2\alpha}$, in a track emulsion irradiated with (solid-line histogram) muons and (dashed-line histogram) neutrons in the reaction $n(14.1\ \text{MeV}) + ^{12}\text{C} \rightarrow 3\alpha + n$ [3] (normalization to the respective number of pairs).

to interpret reliably data obtained upon exposing track-emulsion layers to muons, it is necessary to test the hadron-background level at the places where the emulsion layers were irradiated.

References

[1] Slavich Company JSC, *www.slavich.ru, www.newslavich.com*
[2] The BECQUEREL Project *http://becquerel.jinr.ru/*
[3] R R Kattabekov *et al.* 2013 *Phys. At. Nucl.* **76** 1219 (*arXiv*:1310.2080)
[4] N K Kornegrutsa *et al.* 2013 *Phys. At. Nucl.* **76** 84
[5] J F Ziegler, J P Biersack and M D Ziegler 2008 *SRIM - The Stopping and Range of Ions in Matter* (Chester: SRIM Co.)
[6] P L Jain, K Sengupta and G Singh 1988 *Nuclear Physics* B **301** 517
[7] R R Kattabekov *et al.* 2013 *Phys. At. Nucl.* **76** 88

Limits on Higgs boson couplings in Effective field theory

N Belyaev[1] and T Reid[2]
[1]National Research Nuclear University MEPhI (Moscow Engineering Physics Institute), Kashirskoe highway 31, Moscow, 115409, Russia
[2]Manhattan College, 4513 Manhattan College Parkway, Riverdale, NY 10471, USA

E-mail: Nikita.Belyaev@cern.ch

Abstract. We review the Effective Field Theory (EFT) to make projections on physics beyond the Standard Model in the Higgs sector. We provide relations between the non-Standard Model couplings of the Strongly-Interacting Light Higgs (SILH) effective Lagrangian implemented in the eHDecay package and the corresponding terms of the spin-0 Higgs Characterisation model's effective Lagrangian used with the aMC@NLO Monte Carlo generator. Constraints on BSM couplings are determined on the basis of existing experimental limits on Higgs boson width and branching ratios.

1. Introduction

In the summer of 2012, the CMS and ATLAS Collaborations at the LHC reported the discovery of a new resonance in searches for the Standard Model Higgs boson. To study this discovery in greater detail, both collaborations conducted analyses of the full LHC Run-I data [1-2]. It was shown that the new particle has a mass around 125 GeV and that the dominant production mechanism is gluon fusion. The decay of the new resonance into pairs of two gauge bosons (γγ, ZZ, and WW) was observed. These production and decay modes identified the discovered particle as a neutral boson. The subsequent measurements of its couplings to fermions and bosons have shown the compatibility of these results with the expectations of the Standard Model related to the Higgs boson [3-4].

In the Standard Model, the electroweak symmetry breaking through the Higgs mechanism requires the presence of a neutral Higgs boson with spin 0 and positive CP-parity. Many theories beyond the Standard Model require an extended Higgs sector which also contains CP-odd physical eigenstates of Higgs boson as far as CP-even. In this case, mixing between CP-even and CP-odd eigenstates would be permissible and the Higgs boson observed experimentally may thus have mixed CP-parity. This is a very important possibility because the CP-violation effects in the Standard Model are too small to explain the baryon asymmetry of the Universe.

Detailed studies of Higgs boson candidate spin and parity performed by ATLAS and CMS collaborations have shown that the dominant hypothesis of spin and parity is $J^{CP} = 0^{++}$ [5-6]. These studies were performed with 25 fb^{-1} datasets obtained by each major experiment of the LHC and allow us to obtain the upper limit on a possible CP-odd contribution.

In this paper, we review the framework of the Effective field theory implemented in the eHDecay package in order to obtain limits on Higgs boson couplings by using the experimental limits on Higgs boson partial and total decay widths.

2. Effective field theory basis

The Effective field theory (EFT) is an approximate model for a physical phenomenon designed without the intention to describe the causal mechanism of the phenomenon. It includes the appropriate degrees of freedom to describe physical phenomena occurring at a chosen length scale or energy scale, while ignoring substructure and degrees of freedom at shorter distances. For the purpose of making projections about the characteristics of non-SM Higgs bosons, EFTs are very useful, as they can provide a framework for simulation that isn't strictly bound to a physical reality.

In this paper, we analyzed the Higgs effective Lagrangian implemented in a widely used Higgs Characterization model [7] for the aMC@NLO Monte Carlo generator. Such Lagrangians can be written for all reasonable spins (0, 1 and 2) of the Higgs boson. We review spin 0 effective Lagrangian describing the interaction of the Higgs boson with the gauge bosons of the Standard Model. The structure of this Lagrangian is described at (1), (2), and (3).

$$
\begin{aligned}
L_0^V = \Big\{ & c_\alpha \kappa_{SM}\left[\frac{1}{2} g_{HZZ} Z_\mu Z^\mu + g_{HWW} W_\mu^+ W^{-\mu}\right] \\
& -\frac{1}{4}\left[c_\alpha \kappa_{H\gamma\gamma} g_{H\gamma\gamma} A_{\mu\nu} A^{\mu\nu} + s_\alpha \kappa_{A\gamma\gamma} g_{A\gamma\gamma} A_{\mu\nu} \tilde{A}^{\mu\nu}\right] \\
& -\frac{1}{2}\left[c_\alpha \kappa_{HZ\gamma} g_{HZ\gamma} Z_{\mu\nu} A^{\mu\nu} + s_\alpha \kappa_{AZ\gamma} g_{AZ\gamma} Z_{\mu\nu} \tilde{A}^{\mu\nu}\right] \\
& -\frac{1}{4}\left[c_\alpha \kappa_{Hgg} g_{Hgg} G_{\mu\nu}^a G^{a,\mu\nu} + s_\alpha \kappa_{A\gamma gg} g_{Agg} G_{\mu\nu}^a \tilde{G}^{a,\mu\nu}\right] \\
& -\frac{1}{4}\frac{1}{\Lambda}\left[c_\alpha \kappa_{HZZ} Z_{\mu\nu} Z^{\mu\nu} + s_\alpha \kappa_{AZZ} Z_{\mu\nu} \tilde{Z}^{\mu\nu}\right] \\
& -\frac{1}{2}\frac{1}{\Lambda}\left[c_\alpha \kappa_{HWW} W_{\mu\nu}^+ W^{-\mu\nu} + s_\alpha \kappa_{AWW} W_{\mu\nu}^+ \tilde{W}^{-\mu\nu}\right] \\
& -\frac{1}{\Lambda} c_\alpha\Big[\kappa_{H\partial\gamma} Z_\nu \partial_\mu A^{\mu\nu} + \kappa_{H\partial Z} Z_\nu \partial_\mu Z^{\mu\nu} \\
& + \left(\kappa_{H\partial W} W_\nu^+ \partial_\mu W^{-\mu\nu} + h.c.\right)\Big]\Big\} X,
\end{aligned} \tag{1}
$$

where Λ is the cutoff scale and $c_\alpha (\equiv \cos\alpha)$ is cosine of the mixing angle between 0^+ and 0^-. The (reduced) field strength tensors are defined as follows:

$$
\begin{aligned}
& V_{\mu\nu} = \partial_\mu V_\nu - \partial_\nu V_\mu \quad \left(V = A, Z, W^\pm\right), \\
& G_{\mu\nu}^a = \partial_\mu G_\nu^a - \partial_\nu G_\mu^a + g_s f^{abc} G_\mu^b G_\nu^c,
\end{aligned} \tag{2}
$$

and the dual tensor is:

$$
\tilde{V}_{\mu\nu} = \frac{1}{2}\varepsilon_{\mu\nu\rho\sigma} V^{\rho\sigma}. \tag{3}
$$

Lagrangian (1) contains CP-even and CP-odd Higgs boson eigenstates which could be mixed by setting the α parameter to, alternately, zero and one. Case $c_\alpha = 1$ corresponds to a pure Standard Model CP-even Higgs boson state while $c_\alpha = 0$ corresponds to a pure BSM CP-odd Higgs boson state.

3. Limits on the Higgs couplings

In order to obtain limits on the Higgs boson couplings the eHDecay package was used [8]. This tool allows us to calculate the Higgs boson decay width and branching ratios with the given Lagrangian parameters. For example, figure 1 and figure 2 show how the Higgs boson partial decay width and branching ratio depend on a $C_{\gamma\gamma}$ coupling (assuming $C_{Z\gamma} = C_{ZZ} = C_{WW} = C_{gg} = 1$).

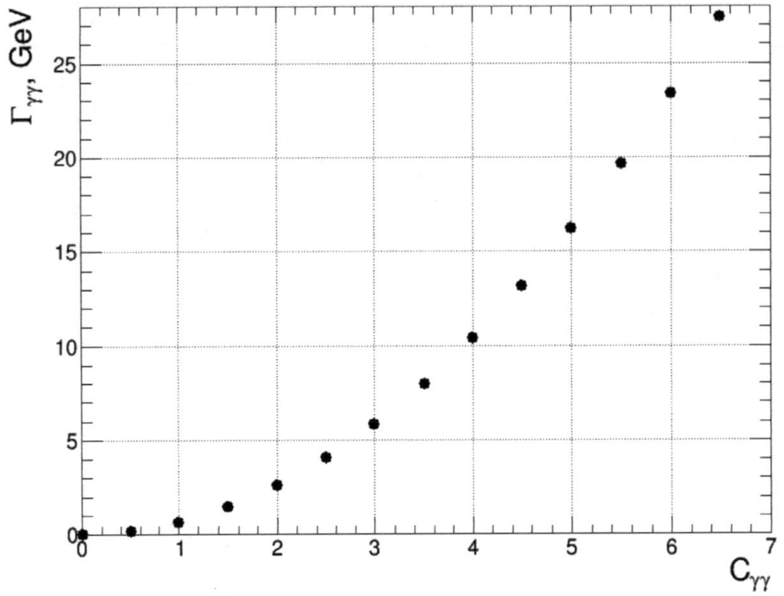

Figure 1. Higgs boson decay widths calculated via eHDecay as function of $C_{\gamma\gamma}$ coupling.

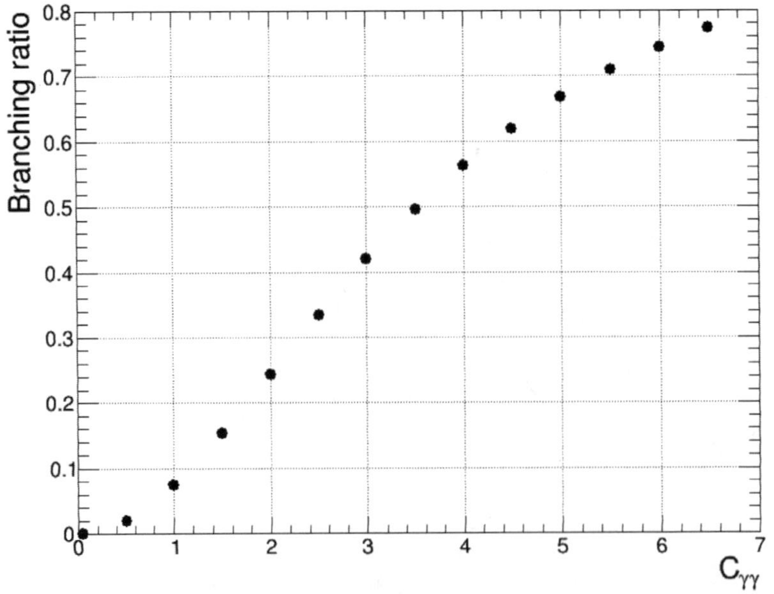

Figure 2. Higgs boson branching ratios calculated via eHDecay as function of $C_{\gamma\gamma}$ coupling.

During this study the following couplings were reviewed: $C_{\gamma\gamma}$, $C_{Z\gamma}$, C_{ZZ}, C_{WW}, C_{gg}, C_{ZdZ} and C_{WdW}. These couplings can be transformed to couplings used in Lagrangian (1) by the mapping formulas indicated at table 1 and table 2, where α_{EM} and α_S are the electromagnetic and strong interaction constants respectively, v is the Higgs vacuum expectation value, θ_W is the Weinberg angle and $c = \sqrt{\frac{\alpha_{EM} G_F m_Z^2}{8\sqrt{2}\pi}}$.

Table 1. Mapping formulas for C_Z, C_W, C_{ZZ}, C_{WW} and $C_{\gamma\gamma}$ couplings.

C-coupling	C_Z	C_W	C_{ZZ}	C_{WW}	$C_{\gamma\gamma}$
k-coupling	$c_\alpha k_{SM}$	$c_\alpha k_{SM}$	$-\frac{v}{2\Lambda} c_\alpha k_{HZZ}$	$-\frac{v}{2\Lambda} c_\alpha k_{HWW}$	$-\frac{47\alpha_{EM}}{36\pi} c_\alpha k_{H\gamma\gamma}$

Table 2. Mapping formulas for $C_{gg}, C_{ZdZ}, C_{WdW}, C_{Zd\gamma}$ and $C_{Z\gamma}$ couplings.

C-coupling	C_{gg}	C_{ZdZ}	C_{WdW}	$C_{Zd\gamma}$	$C_{Z\gamma}$
k-coupling	$\frac{\alpha_S c_\alpha k_{Hgg}}{6\pi}$	$-\frac{v}{\Lambda} c_\alpha k_{H\partial Z}$	$-\frac{v}{\Lambda} c_\alpha k_{H\partial W}$	$-\frac{v}{\Lambda} c_\alpha k_{H\partial\gamma}$	$-c_\alpha k_{HZ\gamma} \frac{c(94cos^2\theta_W - 13)}{18\pi}$

Currently, LHC data exhibits an upper limit on the total Higgs boson decay width of 22 MeV [9]. Upcoming limits on partial decay widths are much more loose ($\Gamma_{ZZ} < 46\ MeV$ and $\Gamma_{WW} < 52\ MeV$). Obtained limits on coupling regions corresponding to experimental restrictions are indicated in table 3 (for C_{ZZ}, C_{WW}, C_{ZdZ} and C_{WdW}, the condition of positivity of the branching ratio was also involved).

Table 3. Coupling regions corresponding to the tightest experimental limits on a Higgs boson decay width.

Coupling	C_{ZZ}	C_{WW}	C_{gg}	$C_{\gamma\gamma}$	$C_{Z\gamma}$	C_{ZdZ}	C_{WdW}
Maximum	1.97	1.32	1.6593	5.827	12.687	1.01	0.90
Minimum	$-3.85 \cdot 10^5$	$-3.27 \cdot 10^4$	-1.6715	-5.820	-12.673	$-1.99 \cdot 10^5$	$-2.20 \cdot 10^4$

4. Conclusion

In this paper, the Effective field theory approach was used in order to calculate the limits for Higgs boson couplings from the experimental restrictions on the Higgs boson decay width. Mapping formulas which establish a connection between eHDecay and Higgs Characterization EFT bases were also obtained and can be used to recalculate obtained limits in terms of k-couplings of a widely used Higgs Characterisation model.

Acknowledgements

For his guidance and insight, we would like to thank Dr. R. Konoplich. The work of N. Belyaev was performed within the framework of the Center for Fundamental Research and Particle Physics supported by MEPhI Academic Excellence Project (contract № 02.a03.21.0005, 27.08.2013). The work of T. Reid is partially supported by the US National Science Foundation under Grant No. PHY-1402964.

References

[1] ATLAS Collaborations 2012 *Phys. Lett.* B **716** 1-30
[2] CMS Collaborations 2012 *Phys. Lett.* B **716** 30-61
[3] ATLAS Collaborations 2013 *Phys. Lett.* B **726** 88-119
[4] CMS Collaborations 2015 *Phys. Rev.* D **92** 012004
[5] CMS Collaboration 2013 *Phys. Rev. Lett.* **110** 081803
[6] ATLAS Collaborations 2013 *Phys. Lett.* B **726** 120-144
[7] Artoisenet P *et al.* 2013 *JHEP* **11** 043
[8] Contino R *et al.* 2014 *Comput. Phys. Commun.* **185** 3412-23
[9] CMS Collaboration 2014 *Phys. Lett.* B **736** 64-85

International Conference on Particle Physics and Astrophysics (ICPPA-2015) IOP Publishing
Journal of Physics: Conference Series **675** (2016) 022024 doi:10.1088/1742-6596/675/2/022024

Monitoring of physics performance of ILC Software based on Higgs Recoil Mass

E Volkova[1,*] and G Voutsinas[2]

[1] National Research Nuclear University MEPhI (Moscow Engineering Physics Institute), Kashirskoe highway 31, Moscow, 115409, Russia
[2] DESY, 22607, Hamburg, Germany

E-mail: *barkentin@mail.ru, georgios.voutsinas@desy.de

Abstract. We discuss a part of ILC software development, that allow us to make automated testing of ILC results. The testing code consists of automated everyday Higgs recoil mass analysis and compares Higgs recoil mass with one of the previous day result. This code uses the result of generation, Mokka simulation and Marlin reconstruction of ILC events.

1. Introduction

The ILD detector is one of two detectors of the International Linear Collider [1], that will collide electrons and positrons at energies of initially 500 GeV, upgradeable to 1 TeV. The ILC software will consists of ≈60 packages, developed by people in different institutes. A key task in software development is the automated testing on different levels of the process: at compile and build time, at runtime and on the results level. The first testing tasks are more technical, while the latter one relies on physics and statistics understanding.

The physics performance will be monitored via the simulation study of the Higgs boson production for processes in which the Higgs is produced together with a well measurable di-muon system using the current proposal of the ILD detector. This part of ILC software can be included in the highly builds, comparing everyday the obtained value of the Higgs recoil mass with the one of the previous day. The higgs recoil mass reconstruction was conducted based on Marlin and Mokka modeling of ILD.

More information about ILC is in [2].

2. Higgs recoil mass

The relevant process for the present study is the recoil reaction $e^+e^- \to HZ \to Hf\bar{f}$ (where f=leptons and quarks), also called Higgs-strahlung. Higgs recoil mass was reconstructed in the $HZ \to \mu^+\mu^- X$ process, because the track quality for muons is better and the Z-boson mass peak is more clear than for electrons [3].

All data analysed in this discussion have been produced by the GEANT4 generation (E_{cms} = 250 GeV), Mokka detector simulation and Marlin reconstruction. The generated Higgs boson mass is $120GeV/c^2$. 500 fb^{-1} have been reconstructed with the use of Marlin software and two different reconstruction algorithms: DBD tracking and cellular automaton [4]. The final program will compare resalts of two parts with 5000 events every day. On development stage the programm reconstructed 2000 events and all interim results were showed for 2000 events.

The first step of analysis was muon tracks selection and recovery of Z-boson from two muons, and the result is shown in figure 1.

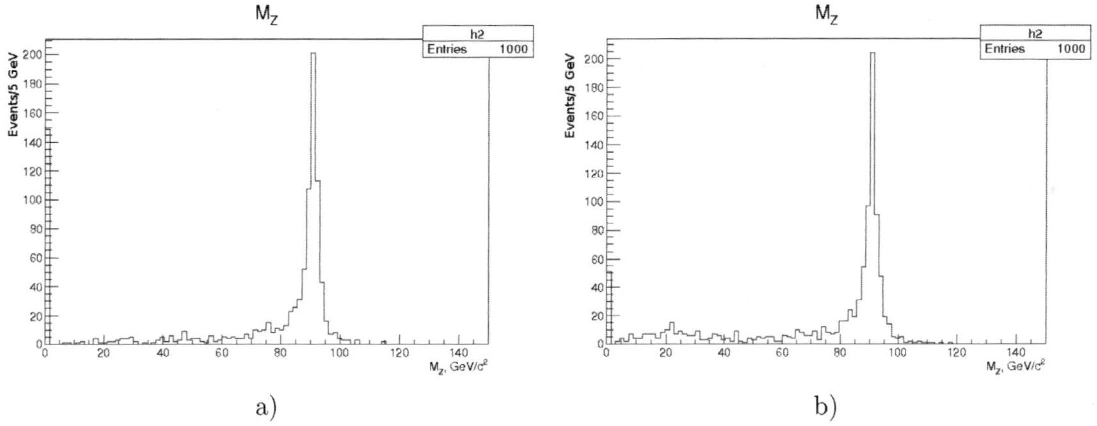

Figure 1. Z-boson mass of two data parts: a) for first part of date; b) for secondpart of date.

The Higgs recoil mass was calculated via the relative kinematic equation:

$$M_{recoil}^2 = s + M_Z^2 - 2E_Z\sqrt{s}, \tag{1}$$

Then the selection of one muon pair from each event was included for peak improvement. The idea of selection is to choose a muon pair that has the most close mass to the well known Z-boson mass [5]. The results for Higgs recoil mass with muons selection are shown in figure 2.

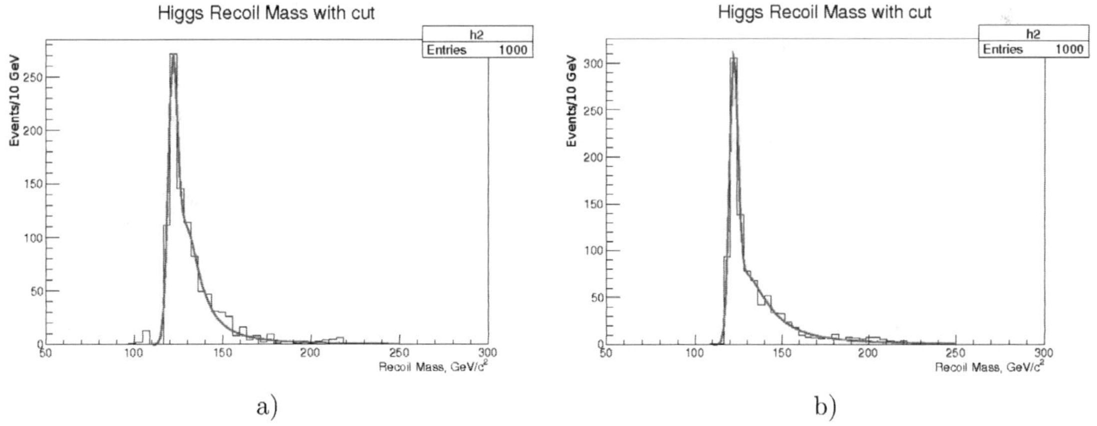

Figure 2. Higgs recoil mass with Z-boson select: a) for first part of date; b) for secondpart of date.

The energy of the incoming beams is smeared with an energy spread of 0.3%, that is a one of the sources of tail. The another one is initial state radiation and beamstrahlung [6].

Fit function for resulting spectra consists of Gaussian for the Peak with Gaussian and Exponent functions for tail.

Then for peak improvement the $70\text{GeV}/c^2 < M_z < 110\text{GeV}/c^2$ cut and rebin was made, and the final result for two parts of 5000 events is represented in figure 3.

This work was conducted in the framework of the DESY summer student program.

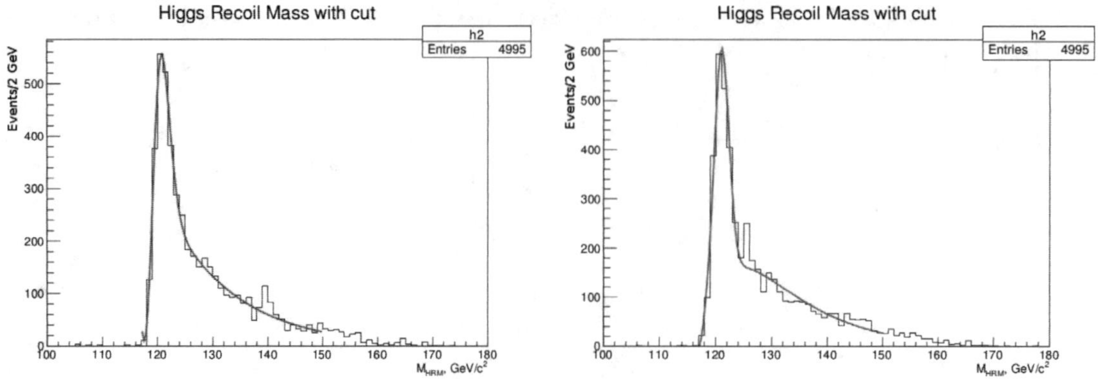

a) b)

Figure 3. Final Higgs recoil mass: a) for first part of date M_{reco} =120.76±0.06; b) for secondpart of date 120.76±0.06.

3. Conclusion

The testing code of ILC software was created and will be use in automated testing of ILC results via higgs recoil mass analyse of a lot of pairs independent parts, that will consist of 5000 events. The results of the program are shown on table 1, total error $= 0.5\text{GeV}/c^2$, the difference of values $= 0.3\text{GeV}/c^2$.

Table 1. Higgs recoil mass for two different parts with 1000 events.

Mean	Err
121.9 GeV/c^2	±0.5GeV/c^2
122.17 GeV/c^2	±0.19GeV/c^2

The result for parts with 5000 events is shown on table 2, total error $= 0.08\text{GeV}/c^2$.

Table 2. Higgs recoil mass for two different parts with 5000 events.

Mean	Err
120.76 GeV/c^2	±0.06GeV/c^2
120.76 GeV/c^2	±0.06GeV/c^2

The next step is to improve the fitting of the distribution. After that, the presented in the discussion part of ILC software can be included in the ILC software diagnostic tools.

References

[1] T Behnke *et al.* 2013 The International Linear Collider Technical Design ReportVolume 4: Detectors *arXiv*:1306.6329

[2] K Fujii, *et al.* 2015 Physics Case for the International Linear Collider *arXiv*:hep-ex 1506.05992
[3] H Li, R Pschl and F. Richard 2009 HZ Recoil Mass and Cross Section Analysis at ILD LC-PHSM-2009-006 *arXiv*: 1202.1439
[4] F Gaede *et al.* 2014 Track reconstruction at the ILC: the ILD tracking software, in proceedings of 20th International Conference on Computing in High Energy and Nuclear Physics (CHEP2013) *Journal of Physics* **513** (2014) 022011
[5] J Beringer *et al.* 2012 (Particle Data Group), Review of Particle Physics *Phys. Rev.* D **86** 010001
[6] T Barklow and P Chen 1992 Beamstrahlung Spectra in Next Generation Linear Colliders SLAC-PUB-5718 *Rev. DAPNIA/SPP 92-02*

The radiative decay of a neutrino with a magnetic moment in a strong magnetic field with taking account of the positronium influence on the photon dispersion

A F Mosichkin

Division of Theoretical Physics, Department of Physics, Yaroslavl State P G Demidov University, Sovietskaya 14, 150000 Yaroslavl, Russia

E-mail: anatoly_mosichkin@mail.ru

Abstract. We study the process of the radiative decay of the neutrino with a magnetic moment in a strong magnetic field, with taking account of the influence of the positronium on the photon dispersion. The positronium contribution into the photon polarization operator leads to an essential modification of the photon dispersion law, and of the neutrino radiative decay amplitude. It has been shown that the probability of the neutrino radiative decay essentially increases under an influence of the positronium on the photon dispersion.

1. Introduction

Neutrino processes such as radiative decay of the neutrino $\nu \to \nu\gamma$ also called Cherenkov neutrino process, are of great interest for a long time. The process $\nu \to \nu\gamma$ in a magnetic field has been investigated in the case of a relatively weak field [1], of a strong field [2] and for the field of arbitrary intensity [3]. In all these papers, the situation was analysed of low-energy neutrinos compared to the mass of the electron. The case of more significant neutrino energies should be more interesting in the light of possible astrophysical applications. In this case, the radiative corrections become substantially enhanced, leading to a renormalization of the photon wave function and a significant deviation from the vacuum dispersion law. These factors were taken into account for the first time in [4]. In the most of the previous papers, the analyses of the neutrino radiative decay were performed with taking account only of the contribution of free electrons and positrons in the loop correction to the photon polarization operator. The influence of a bounded electron-positron pair on the photon polarization operator was taken into account in [5] for the case of the standard electromagnetic interaction of a neutrino, arising at the loop level and thus being strongly suppressed. In this paper, we study the process of the radiative decay $\nu \to \nu\gamma$ of the neutrino with a magnetic moment, with taking account of the positronium influence on the photon polarization operator.

2. The amplitude of the process $\nu \to \nu\gamma$

The interaction Lagrangian corresponding to the process $\nu \to \nu\gamma$ for a neutrino with a magnetic moment has the form:

$$L = -\frac{i\mu_\nu}{2}\Big(\bar{\nu}(x)\sigma_{\alpha\beta}\nu(x)\Big)F^{\alpha\beta}, \tag{1}$$

where $\sigma_{\alpha\beta} = \frac{1}{2}(\gamma_\alpha\gamma_\beta - \gamma_\beta\gamma_\alpha)$, $F^{\alpha\beta} = \partial^\alpha A^\beta - \partial^\beta A^\alpha$, μ_ν is the magnetic moment of a neutrino. The 4-momentum conservation law for this process takes the form: $p = p' + q$. The neutrino field operator $\bar{\nu}(x)$ and the photon field operator A^α are defined as an expansion over the de Broglie plane waves. The amplitude of the radiative decay of a neutrino with a magnetic moment has the form:

$$M = -\frac{i\mu_\nu}{2}\Big(\bar{\nu}(p')\sigma_{\alpha\beta}\nu(p)\Big)\Big(q^\alpha\varepsilon_\lambda^{*\beta}(q) - q^\beta\varepsilon_\lambda^{*\alpha}(q)\Big). \tag{2}$$

The photon polarization operator in a magnetic field has four eigenvectors, two of which correspond to the physical photon polarization vectors [6]. It is known that in the limit of a strong magnetic field, only photos of the so-called second mode can be born in a radiative decay of a neutrino. The polarization vector of the second mode photon has the form:

$$\varepsilon_\alpha^{(2)} = \sqrt{Z}\frac{(q\tilde{\varphi})_\alpha}{\sqrt{q_\parallel^2}}. \tag{3}$$

Here, the factor $Z = \left(1 - \partial\Pi/\partial q_\parallel^2\right)^{-1}$ is caused by the effect of the renormalization of the photon wave function, where Π is the eigenvalue of the photon polarization operator of the second mode, $\varphi^{\alpha\beta} = F^{\alpha\beta}/B$ is the dimensionless tensor of the external electromagnetic field, and $\tilde{\varphi}^{\alpha\beta} = (1/2)\varepsilon^{\alpha\beta\mu\nu}\varphi_{\mu\nu}$ is the dual tensor. It is convenient to split the 4-momentum of the photon into two components. In the frame where the z axis is directed along the external magnetic field, these are the vector in the Minkowski subspace (0; 3): $q_\parallel^\alpha = (\omega; 0; 0; k_3)$ and the vector in a plane perpendicular to the external magnetic field (1; 2): $q_\perp^\alpha = (0; k_1; k_2; 0)$.

3. The probability of the process $\nu \to \nu\gamma$

The probability of the neutrino radiative decay is determined by the integral over the phase space:

$$W = \int \frac{(2\pi)^4|M|^2\delta^4(p - p' - q)}{8E} \cdot \frac{d^3k}{(2\pi)^3\omega} \cdot \frac{d^3p'}{(2\pi)^3E'} \tag{4}$$

After simple transformations, the expression for the amplitude squared of this process can be reduced to the form:

$$|M|^2 = -4\mu_\nu^2 Z\frac{q^2}{q_\parallel^2}(p\varphi p')^2. \tag{5}$$

After substitution the amplitude square (5) into (4) and integration, the decay probability can be reduced to a single integral:

$$W = \frac{\mu_\nu^2 E^3}{\pi^2 m_e^2}\int\limits_0^1 \frac{x^2(-\Pi)}{1 - \partial\Pi/\partial q_\parallel^2}\left(\sqrt{1 - x^2} - x\arccos x\right)dx, \tag{6}$$

where $x = |q_\perp|/(2E)$. The eigenvalue of the photon polarization operator and its derivative depend on $q_\perp^2$ and $q_\parallel^2$ that are linked together by the dispersion law:

$$q_\parallel^2 - q_\perp^2 = \Pi(q_\perp^2, q_\parallel^2). \tag{7}$$

Thus the eigenvalue of the photon polarization operator and its derivative can be expressed through the value $q_\perp^2$.

4. The photon polarization operator with taking account of a positronium

It is necessary to calculate the contribution of the free and bounded electron-positron pair into the photon polarization operator, in order to calculate the probability of the radiative decay of a neutrino with a magnetic moment, with taking account of the positronium influence on the photon dispersion. It is convenient to use the relation between the photon decay probability into some final state of X, and the imaginary part of the polarization operator [7]:

$$Im(\Pi) = -\omega W_{\gamma \to X}. \tag{8}$$

Under the state of X, we mean here the free and bounded electron-positron pair. The photon polarization operator with taking account of the positronium can be written as:

$$\Pi = -2\alpha(eB)e^{-\frac{q_\perp^2}{2eB}}\Big[\frac{1}{\pi}H(z) + \frac{\lambda z}{1-\lambda^2 - z}\Big], \tag{9}$$

where $H(z) = \int\limits_0^1 dx/[1 - z(1-x^2) - i0] - 1$, $z = q_\|^2/(4m_e^2)$, $\lambda = \alpha/(2\nu)$, α is a fine structure constant, the parametr ν determines the energy levels of the positronium in a strong magnetic field: $\varepsilon = \alpha^2 m_e/(4\nu^2)$. The expression for the probability of the radiative decay of the neutrino with a magnetic moment can be written as:

$$\frac{W}{W_0} = \xi^{\frac{3}{2}} \int\limits_0^1 \frac{dx x^2 e^{-2\xi x^2}\left(\frac{1}{\pi}H(z) + \frac{\lambda z}{1-\lambda^2-z}\right)\left(\sqrt{1-x^2} - x \arccos x\right)}{1 + \frac{\alpha\beta}{2}e^{-2\xi x^2}\frac{d}{dz}\left(\frac{1}{\pi}H(z) + \frac{\lambda z}{1-\lambda^2-z}\right)}. \tag{10}$$

The variables x and z are linked by the dispersion equation:

$$z - \xi\beta x^2 = -\frac{\alpha\beta}{2}e^{-2\xi x^2}\Big(\frac{1}{\pi}H(z) + \frac{\lambda z}{1-\lambda^2-z}\Big), \tag{11}$$

where $\xi = E^2/(eB)$, $\beta = B/B_e$, $B_e = 4.41 \times 10^{13} G$, $W_0 = 2\alpha\beta^{\frac{5}{2}}\mu_\nu^2 m_e^3/\pi^2$.

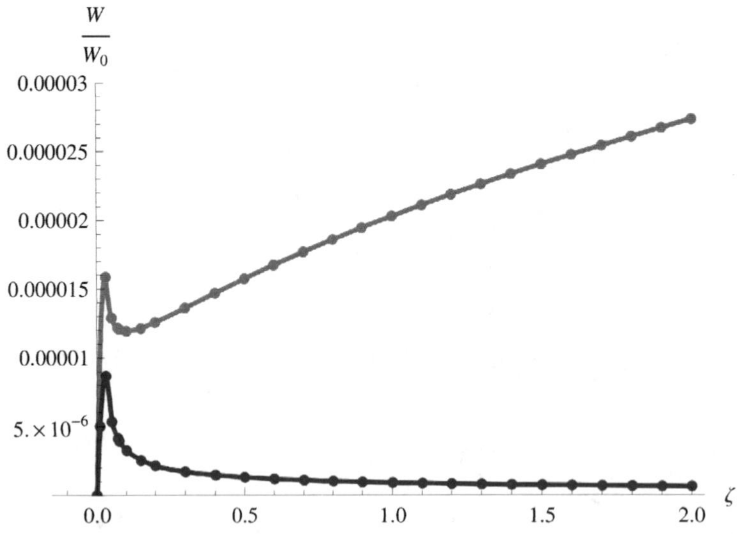

Figure 1. The probability of the radiative decay of a neutrino.

In figure 1, the plot is shown of the probability of the radiative decay of a neutrino with a magnetic moment as a function of the field parameter. The lower curve describes the decay probability without taking account of the positronium influence on the photon dispersion, while in the upper curve the positronium influence is taken into account.

5. Conclusions

In this paper, we have studied the process of the radiative decay of a neutrino with a magnetic moment in a strong magnetic field with taking account of the positronium influence on the photon dispersion. The result of a numerical calculation of the decay probability is presented in Fig. 1 which shows that taking account of the positronium influence on the photon dispersion leads to a significant change in the behaviour of the dependence $W(\xi)$ and its essential increase.

Acknowledgments

The study was supported by the Russian Foundation for Basic Research (Project No. 14-02-00233-a).

References

[1] Galtsov D V and Nikitina N S 1972 *Sov. Phys. JETP* **35** 1047 (Original Russian text: 1971 *ZhETF* **62** 2008)
[2] Skobelev V V 1976 *Sov. Phys. JETP* **44** 660 (Original Russian text: *ZhETF* **71** 1263)
[3] Ioannisian A N and Raffelt G G 1997 *Phys. Rev.* D **55** 7038
[4] Gvozdev A A, Mikheev N V and Vasilevskaya L A 1977 *Phys. Lett.* B **410** 211
[5] Anikin R A and Mikheev N V 2013 *Phys. At. Nucl.* **76** 1541 (Original Russian text: *Yad. Fiz.* **76** 1610)
[6] Batalin I A and Shabad A E 1971 *JETP* **33** 483 (Original Russian text: *ZhETF* **60** 894)
[7] Weldon H A 1983 *Phys. Rev.* D **28** 2007

International Conference on Particle Physics and Astrophysics (ICPPA-2015) IOP Publishing
Journal of Physics: Conference Series **675** (2016) 032001 doi:10.1088/1742-6596/675/3/032001

H, He, Li and Be Isotopes in the PAMELA-Experiment

W Menn[19], O Adriani[5,6], G C Barbarino[7,8], G A Bazilevskaya[9], R Bellotti[10,11], M Boezio[2], E A Bogomolov[12], M Bongi[4,6], V Bonvicini[2], S Bottai[6], A Bruno[10,11], F Cafagna[11], D Campana[8], P Carlson[13], M Casolino[3,15], G Castellini[16], C De Donato[3], C De Santis[14], N De Simone[3], V di Felice [3,4] V Formato[1,2], A M Galper[17], A V Karelin[17], S V Koldashov[17], S Koldobskiy[17], S Y Krutkov[12], A N Kvashnin[9], A Leonov[17], V Malakhov[17], L Marcelli[14], M Martucci[14,18], A G Mayorov[17], M Merge'[3,14], V V Mikhailov[17], E Mocchiutti[2], A Monaco[10,11], N Mori[6], R Munini[1,2], G Osteria[8], F Palma[3,14], B Panico[8], P Papini[6], M Pearce[13], P Picozza[3,14], M Ricci[18], S B Ricciarini[16], R Sarkar[2], V Scotti[7,8], M Simon[19], R Sparvoli[3,14], P Spillantini[5,6], Y I Stozhkov[9], A Vacchi[2], E Vannuccini[6], G Vasilyev[12], S A Voronov[17], Y T Yurkin[17], G Zampa[2], N Zampa[2], M S Potgieter [21] and E E Vos [21]

[1] University of Trieste, Department of Physics, I-34147 Trieste, Italy
[2] INFN, Sezione di Trieste I-34149 Trieste, Italy
[3] INFN, Sezione di Rome "Tor Vergata", I-00133 Rome, Italy
[4] Agenzia Spaziale Italiana (ASI) Science Data Center, I-00044 Frascati, Italy
[5] University of Florence, Department of Physics, I-50019 Sesto Fiorentino, Florence, Italy
[6] INFN, Sezione di Florence, I-50019 Sesto Fiorentino, Florence, Italy
[7] University of Naples "Federico II", Department of Physics, I-80126 Naples, Italy
[8] INFN, Sezione di Naples, I-80126 Naples, Italy
[9] Lebedev Physical Institute, RU-119991, Moscow, Russia
[1]0 University of Bari, Department of Physics, I-70126 Bari, Italy
[11] INFN, Sezione di Bari, I-70126 Bari, Italy
[12] Ioffe Physical Technical Institute, RU-194021 St. Petersburg, Russia
[13] KTH, Department of Physics, and the Oskar Klein Centre for Cosmoparticle Physics, AlbaNova University Centre, SE-10691 Stockholm, Sweden
[14] University of Rome "Tor Vergata", Department of Physics, I-00133 Rome, Italy
[15] RIKEN, Advanced Science Institute, Wako-shi, Saitama, Japan
[16] IFAC, I-50019 Sesto Fiorentino, Florence, Italy
[17] National Research Nuclear University MEPhI (Moscow Engineering Physics Institute), Kashirskoe highway 31, Moscow, 115409, Russia
[18] INFN, Laboratori Nazionali di Frascati, Via Enrico Fermi 40, I-00044 Frascati, Italy
[19] Universität Siegen, Department of Physics, D-57068 Siegen, Germany
[20] INFN, Sezione di Perugia, I-06123 Perugia, Italy

E-mail: menn@pamela.physik.uni-siegen.de

Abstract. On the 15th of June 2006, the PAMELA satellite-borne experiment was launched from the Baikonur cosmodrome and it has been collecting data since July 2006. The apparatus comprises a time-of-flight system, a silicon-microstrip magnetic spectrometer, a silicon-tungsten electromagnetic calorimeter, an anti-coincidence system, a shower tail counter scintillator and a neutron detector. The scientific objectives addressed by the mission are the measurement of the antiprotons and positrons spectra in cosmic rays, the hunt for antinuclei as well as the determination of light nuclei fluxes from hydrogen to oxygen in a wide energy range and with very high statistics. In this paper the identification capability for light nuclei isotopes using two different detector systems (Time-of-Flight and multiple dE/dx measurements in the calorimeter) and preliminary results of the isotopic ratios will be presented.

1. Introduction

Measurements of the isotopic composition of elements of the cosmic radiation provide significant constraints on cosmic ray source composition and cosmic ray transport and acceleration in the galaxy. The isotopes of Li, Be, and B in cosmic rays are pure spallation products of primary cosmic rays, mainly C, N, and O, when they interact with interstellar matter during their propagation in the Galaxy. The rare isotopes ^{2}H and ^{3}He in cosmic rays are believed to originate mainly from the interaction of high energy protons and helium with the galactic interstellar medium. The energy spectrum of these components carries fundamental information regarding the propagation of cosmic rays in the galaxy which are competitive with those obtained from other secondary to primary measurements such as B/C. The PAMELA experiment has been observing galactic cosmic rays since July 2006 at an altitude ranging from $\sim$ 350 km to $\sim$ 600 km on-board of the Russian Resurs-DK1 satellite which executes a quasi-polar orbit (70° inclination). The results presented here are based on the data set collected by PAMELA between from July 2006 and December 2007 for the hydrogen and helium isotopes, and between July 2006 and September 2014 for the lithium and beryllium isotopes.

2. The PAMELA instrument

The PAMELA apparatus is composed of several sub-detectors: ToF system, anti-coincidence system (CARD, CAS, CAT), magnetic spectrometer with microstrip silicon tracking system, W/Si electromagnetic imaging calorimeter, shower-tail-catcher scintillator (S4) and neutron detector. A detailed description of the PAMELA instrument and an overview of the mission can be found in [19]. The core of the instrument is a magnetic spectrometer, made of a permanent magnet (0.43 T) and a silicon tracking system (resolution in the bending side 4 μm) for a maximum detectable rigidity of 1 TV. The momentum resolution is better than 4% between 2 GV and 20 GV. The Time of Flight (ToF) system is divided in 6 layers, arranged in three planes, each plane composed of two layers. The first plane (S1) is placed on top of the instrument, the second and third plane are placed above (S2) and below (S3) the spectrometer. The W/Si sampling imaging calorimeter comprises 44 single-sided silicon strip detector planes interleaved with 22 plates of tungsten absorber. Each tungsten layer has a thickness of 0.74 radiation lengths (2.6 mm) and it is sandwiched between two printed circuit boards, which house the silicon detectors as well as the frontend and digitizing electronics. Each silicon plane consists of 3x3, 380 μm thick, 8x8 cm^2 detectors, segmented into 32 strips with a pitch of 2.4 mm. The orientation of the strips for two consecutive silicon planes is shifted by 90 degrees, thus providing 2-dimensional spatial information. The total depth of the calorimeter is 16.3 radiation lengths and 0.6 nuclear interaction lengths.

3. Data analysis

3.1. Event selection

Requirements were set on the event quality to select positively charged particles with a precise measurement of the absolute value of the particle rigidity and velocity. The particle charge for hydrogen and helium particles was identified using the ionization measurements provided by the

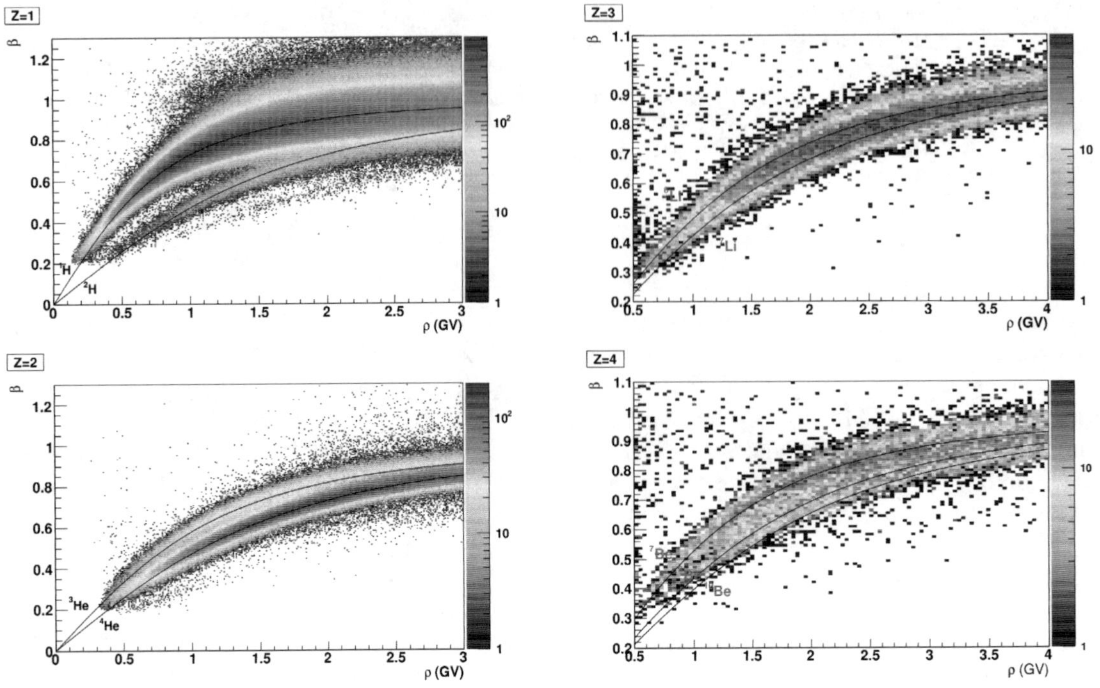

Figure 1. β vs. rigidity for $Z = 1$ (top left), $Z = 2$ (bottom left), $Z = 3$ (top right) and $Z = 4$ (bottom right) particles.

silicon sensors of the magnetic spectrometer. The requirements are identical to the selection in [1] and we refer to this paper for more details. Lithium and beryllium events have been selected by means of ionization energy losses in the ToF system. Charge consistency has been required between S12 and ⟨S2⟩ and ⟨S3⟩ (the arithmetic mean of the ionizations for the two layers constituting S2 and S3, respectively).

3.2. Isotope separation in the PAMELA instrument

In each sample an isotopic separation at fixed rigidity is possible since the mass of each particle follows the relation $m = RZe/\gamma\beta c$, where R is the magnetic rigidity, $Z \times e$ is the particle charge, and γ is the Lorentz factor. The particle velocity β can either be provided directly from the timing measurement of the ToF system, or indirectly from the energy loss in the calorimeter, which follows β via the Bethe-Bloch formula $dE/dx \propto \frac{Z^2}{\beta^2}$ (neglecting logarithmic terms). In figure 1 we show β as derived by the ToF system vs. the particle rigidity for the four different charge samples. The black lines in the figure represent the expectations for each isotope.

The isotopic analysis of nuclei with the calorimeter is restricted to events which do not interact inside the calorimeter, selecting events by applying cuts on the ratio between the energy deposited in the strip closest to the track and the neighboring strip on each side and the total energy detected. In a single silicon layer, the energy loss distributions shows a Landau tail which degrades the resolution of the dE/dx measurement. Using a truncation method, the 50% of samples with larger pulse amplitudes are excluded before taking the mean of the dE/dx measurements, thus reducing the effect of the Landau tail. The mean dE/dx for each event vs. the rigidity measured with the magnetic spectrometer for the four different charge samples is shown in figure 2.

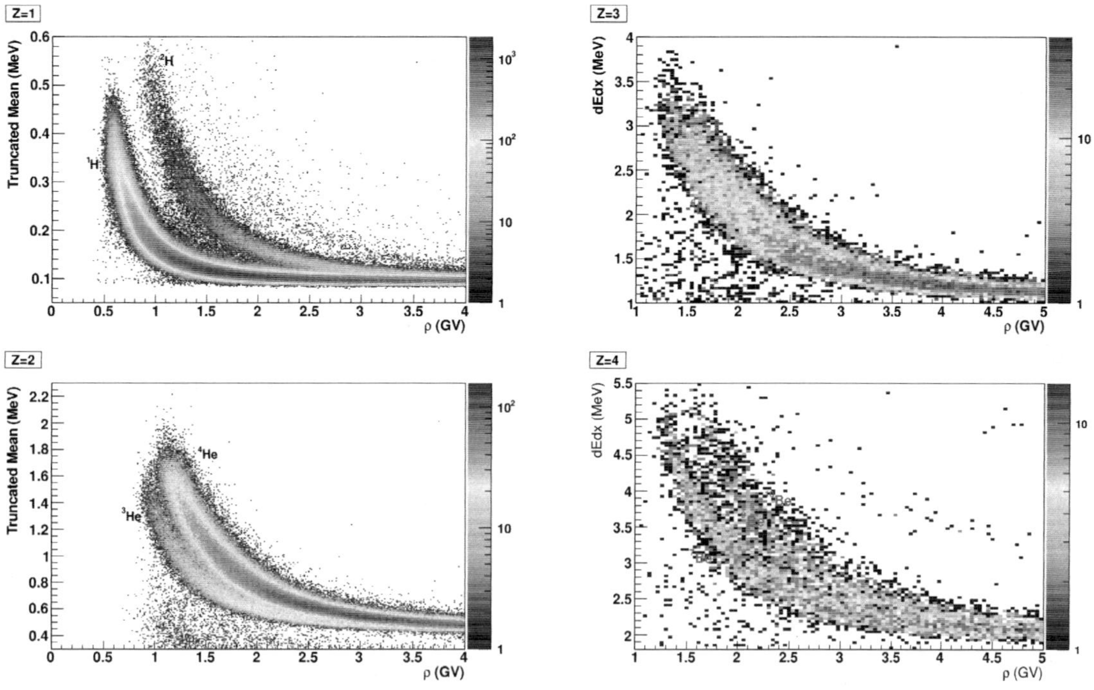

Figure 2. Mass separation for for $Z = 1$ (top left), $Z = 2$ (bottom left), $Z = 3$ (top right) and $Z = 4$ (bottom right) particles using the "truncated mean"-method.

3.3. Raw isotope numbers

For $Z = 1$ and $Z = 2$ particles the counts in each rigidity range were derived in a similar manner as in [1] by fitting gaussians to the $1/\beta$ distributions. $1/\beta$ distributions were chosen since the shape of a $1/\beta$ distribution is gaussian. The dE/dx distributions of the calorimeter have a non-gaussian shape, therefore one has to model the expected distributions of the observable quantities and then perform likelihood fits. The expected dE/dx distributions for each isotope are created using the full Monte Carlo simulation of the PAMELA apparatus based on the GEANT4 code, which has been already described in [1]. For more details see [15], [16]. For the ToF analysis of $Z = 3$ and $Z = 4$ particles we chose a slightly different approach and created $1/\beta$ distributions using the simulation, and then performed the Likelihood analysis. As an example we show in figure 3 the distributions for ToF and calorimeter for $Z = 4$ in the 1.9 - 2.1 GV rigidity range.

3.4. Flux Determination

To derive each isotope flux the number of selected events derived in the previous section had to be corrected for the selections efficiencies, particle losses, contamination and energy losses. For $Z = 1$ and $Z = 2$ particles most of the corrections could be directly taken from [1]. The analysis of $Z = 3$ and $Z = 4$ particles is in an earlier state, so only isotopic ratios and no fluxes are presented. The efficiency for the calorimeter is derived using simulated data for higher energies and flight data for lower energies, using the redundant detectors to select the isotopes.

4. Results and discussion

In figure 5 and 6 we show the hydrogen and helium isotope fluxes (top) and the ratios of the fluxes (bottom) measured with the ToF or the calorimeter, compared to previous measurements.

Journal of Physics: Conference Series **675** (2016) 032001 doi:10.1088/1742-6596/675/3/032001

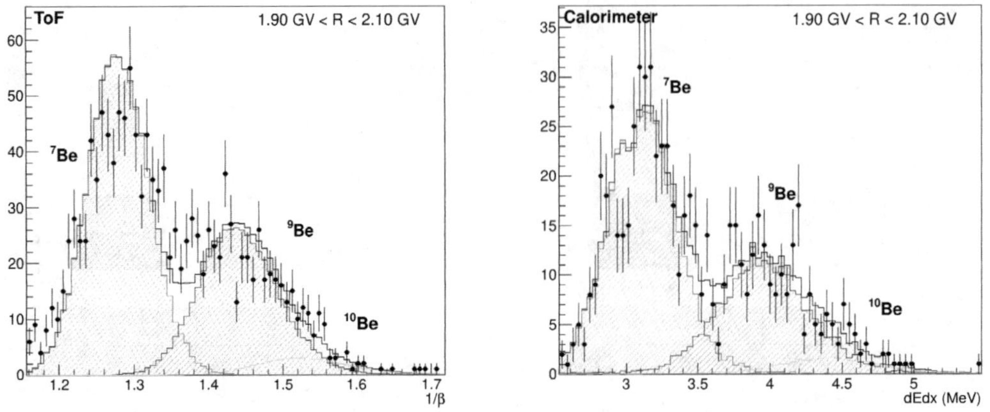

Figure 3. $1/\beta$ distributions of the ToF (left) and truncated mean dE/dx of the calorimeter (right) for beryllium in the 1.9 - 2.1 GV range.

In this context it is important to know that all the former measurements shown in figures 5 and 6, except AMS-01, are from balloon-borne experiments and thus effected by the non-negligible background of atmospheric secondary particle production.

In figure 4 we show some preliminary results for the ratios ^{7}Li/ ^{6}Li and ^{7}Be/(^{9}Be+^{10}Be) derived with the ToF (blue circles) or the calorimeter (red circles) together with other measurements.

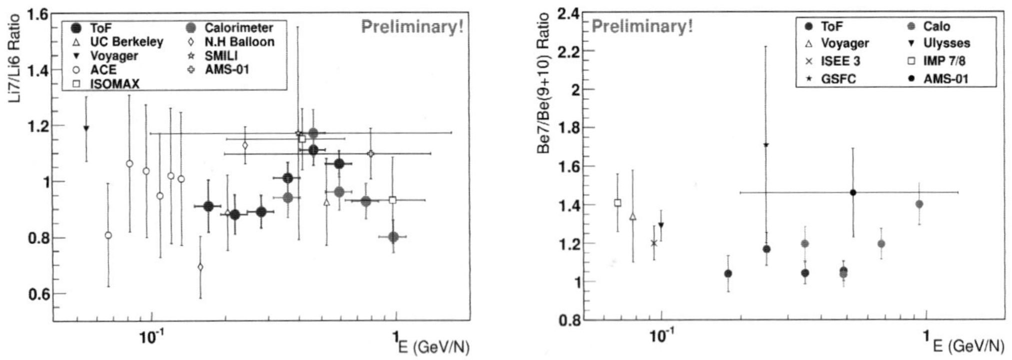

Figure 4. Results for the ratios ^{7}Li/ ^{6}Li and ^{7}Be/(^{9}Be+^{10}Be) derived with the ToF (blue circles) or the calorimeter (red circles). Error bars show statistical uncertainty only. Previous experiments: AMS-01 [4], CRIS/ACE [10], ISOMAX [13], SMILI [5], Voyager [24], UC Berkeley [7], N. H. Balloon [22], Ulysses [8], ISEE-3 [26], IMP 7/8 [11], GSFC [12].

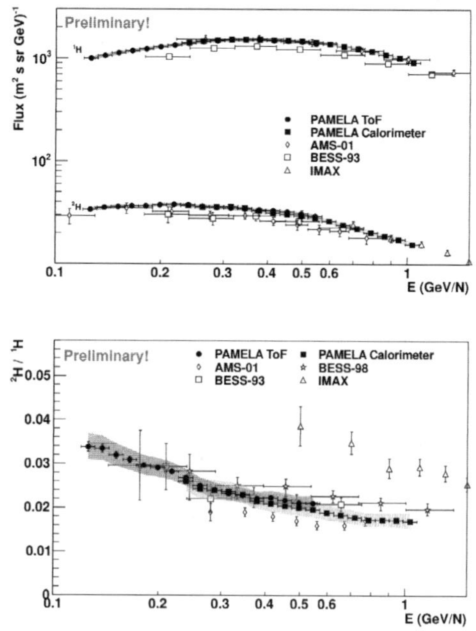

Figure 5. ^{1}H and ^{2}H absolute fluxes (top) and their ratio (bottom) derived with the ToF (circles) or the calorimeter (squares). Previous experiments: AMS-01 [3, 4, 14], BESS-93 [21], BESS-98 [18], IMAX [9]. Error bars show the statistical uncertainty while shaded areas show the systematic uncertainty.

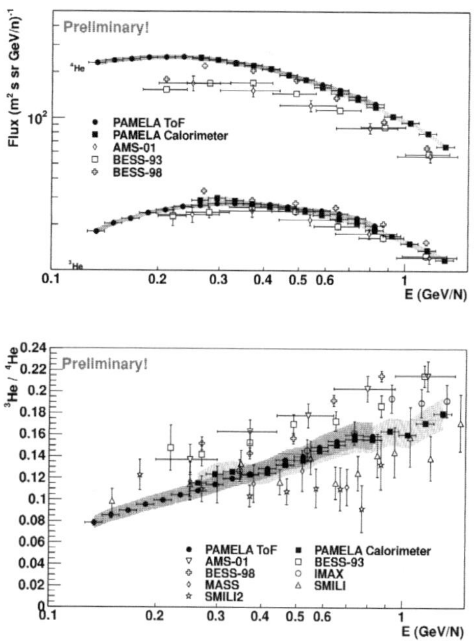

Figure 6. ^{4}He and ^{3}He absolute fluxes (top) and their ratio (bottom) derived with the ToF (circles) or the calorimeter (squares). Previous experiments: AMS [4], BESS-93 [21], BESS-98 [17], IMAX [20], SMILI-1 [6], SMILI-2[25], MASS [23].

References

[1] Adriani O *et al* 2013 *Astrophys. J.* **770** 2
[2] Adriani O *et al* 2014 *Astrophys. J.* **791** 30
[3] Aguilar M, Alcaraz J *et al* 2002 *Phys. Rep.* **366** 331
[4] Aguilar M *et al* 2011 *Astrophys. J.* **736** 105
[5] Ahlen S P *et al* 2000 *Astrophys. J.* **534** 757
[6] Beatty J J *et al* 1993 *Astrophys. J.* **413** 268
[7] Buffington A, Orth C D and Mast T S 1978 *Astrophys. J.* **226** 355
[8] Connell J J 1998 *Astrophys. J.* **501** L59
[9] de Nolfo G A *et al* 2000 *AIP Conf. Proc.* **528** 425
[10] de Nolfo G A *et al* 2006 *Adv. Space Res.* **38** 155864
[11] Garcia-Munoz M, Mason G M and Simpson J A 1977 *Astrophys. J* **217** 859
[12] Hagen F A, Fisher A J, and Ormes J F 1977 *Astrophys. J.* **212** 262
[13] Hams T *et al* 2004 *Astrophys. J.* **611** 892-905
[14] Lamanna G *et al* 2001 *Proc. 27th Int. Cosmic Ray Conf. (Hamburg)* 1617
[15] Menn W *et al* 2013a *J. Phys.: Conf. Ser.* **409** 012030
[16] Menn W *et al* 2013 *Proc. 33nd Int. Cosmic Ray Conf. (Rio de Janeiro)* 0233
[17] Myers Z D *et al* 2001 *Proc. 27th Int. Cosmic Ray Conf. (Hamburg)* 1805
[18] Myers Z D *et al* 2005 *Adv. Space Res.* **35** 151-5
[19] Picozza P *et al* 2007 *Astropart. Phys.* **27** 296
[20] Reimer O *et al* 1998 *Astrophys. J.* **496** 490
[21] Wang J Z, Seo E S *et al* 2002 *Astrophys. J.* **564** 244
[22] Webber W R and Kish J 1979 *Proc. 16th Int. Cosmic Ray Conf. (Kyoto)* **1** 389

[23] Webber W R *et al* 1991 *Astrophys. J.* **380** 230
[24] Webber W R, Lukasiak A and McDonald F B 2002 *Astrophys. J.* **568** 210-5
[25] Wefel J P *et al* 1995 *Proc. 24th Int. Cosmic Ray Conf. (Rome)* **2** 630
[26] Wiedenbeck M E and Greiner D E 1980 *Astrophys. J.* **239** L139

Study of spatial and energy characteristics of relativistic electron bursts in magnetosphere with robust methods

T R Zharaspayev*, **S Yu Aleksandrin and S V Koldashov**

National Research Nuclear University MEPhI (Moscow Engineering Physics Institute), Kashirskoe highway 31, Moscow, 115409, Russia

E-mail: **masscry@gmail.com*

Abstract. Electron bursts are well-known phenomena of fast increase in particle fluxes in near-Earth space. Powerful local geophysical events like earthquakes or thunderstorms can induce precipitation of electrons with defined energy spectrum from the radiation belt, which would be registered as fast increase in particle count rate on board the low orbit satellite. Using particle burst energy spectrum evolution in time one can detect the area of particles precipitation. Background particles are registered by instruments too and can't be separated from burst particles. High level of background particles can have large impact on detection of the area of particles precipitation. A robust regression method to solve problem of background particles is introduced and compared with standard method of linear regression. Results of comparison between various data analysis methods in application to study of spatial and energy characteristics of relativistic electron bursts in the Earth magnetosphere are presented in this work. Robust method proved to be optimal for data analysis of energy spectrum evolution in time for search of zones of local radiation belt disturbances.

1. Introduction

The shape of Earth's magnetic field forms a trap for charged particles. It is called the Earth radiation belt. Some geophysical phenomena (earthquakes, thunderstorms) are known to be the cause of particle precipitations below the Earth radiation belt to heights of low orbital satellites. Precipitated particles take part in three types of motion: rotation around geomagnetic field line, bouncing between mirror points and longitudinal drift around the Earth. Satellite experiment below radiation belt with particle detection instrument crossing disturbed L-shell would detect precipitated particles as short time increase in count rate of particles. This well-known phenomenon is called particle burst and has been studied for long time in many satellite experiments [1, 2, 3, and references therein].

ARINA experiment is conducted on board the low-orbit russian sattelite Resurs-DK1 with altitude 350-600 km and inclination 70° since 2006 and VSPLESK experiment were carried out from 2008 to 2013 on board the International Space Station on altitudes 350-400 km with inclination 51°[3]. ARINA and VSPLEKS instruments based on multilayer scintillation telescope-spectrometer detector with 10 layers of plastic scintillator plates, instruments can detect 3-30 MeV electrons and protons in energy range 30-100 MeV.

If one assumes that particle precipitation occurs instantly in single point-like region(locally), one can expect that precipitated particles would drifts along disturbed L-shell as a cloud of particles and the shape of this cloud will be mainly defined by energy spectrum of precipitated particles. Study of energy spectrum and time characteristics of particles of cloud in principle will determine parameters of precipitation area.

This article discusses features of various data analysis methods in application to study of electron bursts in the Earth magnetosphere using numerical models of local particles precipitation and background particle fluxes.

2. Particle longitudinal drift process simulation

As it was mentioned above, longitudinal drift velocity of particles depends on their energy. The time(T_{dr}) for particle to perform a complete rotation around the Earth defined by formula (1) (based on article [4]):

$$T_{dr} = K_1 \frac{1+\varepsilon}{\varepsilon(1+\varepsilon)} \frac{K_2}{L} \tag{1}$$

where ε is the ratio of the kinetic energy of the particle to its rest energy, L - drift shell coordinate.

$$K_1 = \frac{4}{3}\pi \frac{Z|e|}{m_0} \frac{M}{c^3} \tag{2}$$

M - geomagnetic field momentum, m_0 - electron mass. The K_2 depends on latitude of particle's mirror point and varies from 1.0 to 1.5, where $K_2 = 1.0$ on geomagnetic equator.

$$t_{dr}(\Delta\lambda, \varepsilon) = T_{dr}(\varepsilon) \frac{\Delta\lambda}{360} \tag{3}$$

where $\Delta\lambda$ - longitudinal distance between points of precipitation and registration and $\varepsilon = E/E_0$, where E is particle energy.

Using this model, one can assume that registered particles of cloud would group along the curve $t_{dr}(\Delta\lambda, \varepsilon)$ (3), where arrival time to detector from precipitation region depends on value of $\Delta\lambda$ and ε.

In this work, using Monte-Carlo method two sets of electrons were generated and followed all the way to their registration. The first set are electrons precipitated locally and instantly from radiation belt (see figure 1), and the second one consists of background albedo electrons (poisson process, random in time) (see figure 2). It is natural to search time of registration on particle energy dependence in form of plot where on x-axis particle energy is plotted, and on y-axis its registration time. Main objective of this research is to find burst electrons from merged sets of background and burst electrons, and determine $\Delta\lambda$ parameter.

3. Data analysis methods

In this work two regression analysis methods were used: Linear regression method (Ordinary Least Squares) and Robust regression method. General formula for this type of analysis is shown on (4)[5, 6]:

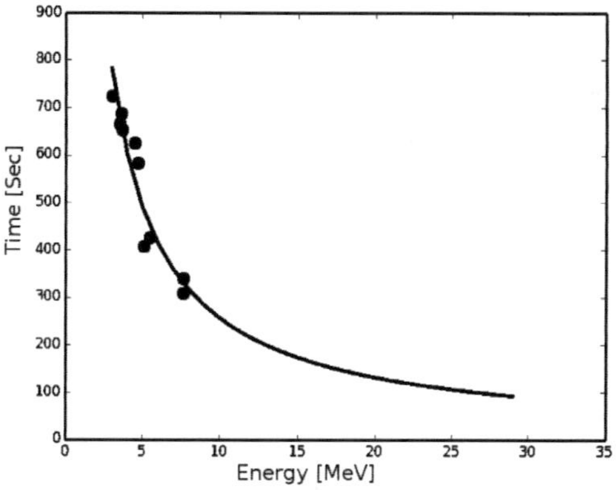

Figure 1. Registered burst electrons [as ●] and $t_{dr}(\Delta\lambda, \varepsilon)$ curve [as —] (Monte-Carlo simulation results)

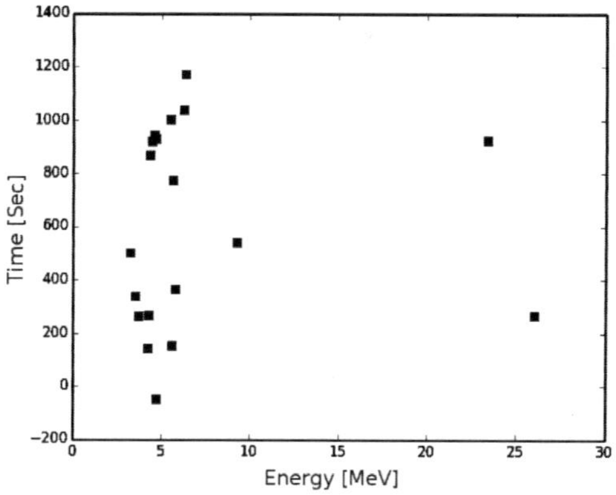

Figure 2. Registered background albedo electrons (as ■) (Monte-Carlo simulation results)

$$R_{method} = \sum_{i=1}^{n} w_i \cdot \frac{(t_i - t_{dr}(E_i, \Delta\lambda))^2}{\sigma_i^2} \tag{4}$$

In order to reconstruct curve and find longitudinal distance between zones of precipitation and registration, a global minimum of (4) have to be found, where t_i - time of particle registration, E_i - particle energy, σ_i - standard deviation of measured particle energy.

In case of ordinary least squares method parameter w_i is equal for all registered particles. In case of robust regression, using some assumptions or additional information about dataset, one need to create some weight function to provide w_i.

Using this weight function, one assumes that some of particles are part of burst and others are part of background.

4. Minimum linear distance

As it can be seen from figure 2, albedo electrons lays chaotically in relative to the curve, withal burst electrons lay close to the curve. Using this knowledge we assume that the particle lying far from the curve is has less probability to be part of the burst.

$$d_i = \sqrt{\frac{(E_i - E_{min})^2}{A * E_{lim}} + \frac{(t_i - t_{dr}(E_{min}, \Delta\lambda))^2}{B * t_{lim}}} \tag{5}$$

$$w_i = \frac{1}{1 + e^{-(\beta_0 + \beta_1 d_i)}} \tag{6}$$

In this work we used minimum linear distance from curve to each point (5) to measure burst electron likeness using logistic function (6) with defined parameters β_0 and β_1, where value $w_i = 1$, means that particle i - is burst electron, and for value $w_i = 0$ particle i - is background albedo electron. In this work first approximation curve is calculated using linear regression method, then we calculate w_i of particles using their minimum linear distance to first approximation curve. Using robust regression of parameter, one can minimize impact of background albedo electrons on $\Delta\lambda$ calculation. As one can see from figure 3, linear method failed to converge to original curve, in the same time, robust curve fits almost perfect to it. The results of that exact simulation round are the following: $\Delta\lambda_{robust} = 175 \pm 11$ with real $\Delta\lambda_{sim} = 180$.

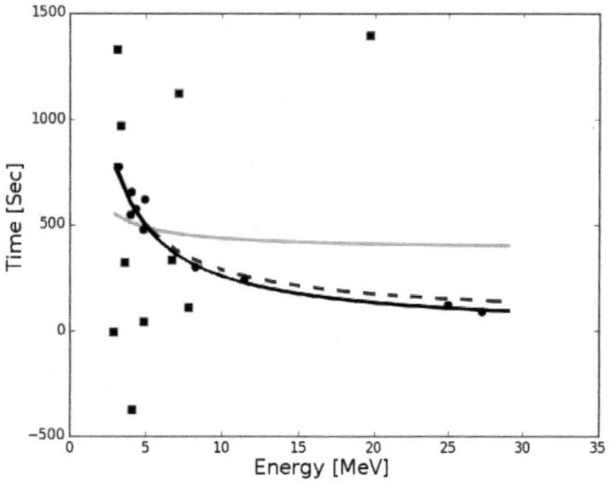

Figure 3. Results of t_{dr} curve reconstruction. Burst particles are shown as ● and background particles as ■. First (linear)[—] and second (robust) [— —] curve approximations, with original (see figure 1) [—] curve

5. Linear and robust regression method comparison

Regression methods were compared using a Monte-Carlo simulation of particle precipitation, followed by their longitudinal drift and registration by instrument on board satellite. Some number of background particles were injected and registered with burst particles. Instruments uncertanties of energy detection also took into account.

Methods were compared by varying parameters such as count of precipitated particles and background particles, their energy spectrum, level of instrument uncertainies.

As can be seen from figure 4, linear regression very fast deviates from original curve, nevertheless robust regression fits to original curve within the errors until the level of background particles less than 80% of precipitated (burst) particles.

Robust regression method show no deviation when background to burst ration is lower than 50%, but linear regression show deviation greater than 10% even in conditions where background to burst ratio is 20%.

Both methods have tendency to higher deviation, when background to burst ratio increases, but when ratio reaches 170% linear method show same deviation, because the method converges to straight line.

On figure 5 fit parameter deviation from original parameter dependency on total count of registered particles is shown. As one can see, errors become smaller when number of particles become bigger, but because with growing number of burst particles, number of background particles grows too. The whole image become less clearer and minimal linear distance heuristic performs poorer, than on small particles count.

Fit deviation dependency of robust method have a minimum at 20 particles, because in situation with low particle count, each background particle misinterpreted as burst particle have higher impact on fit, as particle count rises, effect from misinterpreted particles drops and the method shows the best results, after that as number of particles rises, cumulative effect of misinterpreted particles grows shifting fit results from original.

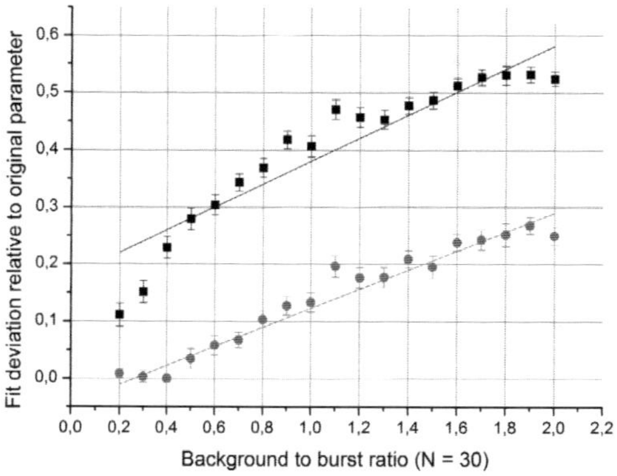

Figure 4. $\Delta\lambda$ parameter of reconstructed t_{dr} curve deviation from original $\Delta\lambda$ parameter dependency on precipitated particles(burst) on background particles ratio. [Linear ■, Robust ●]

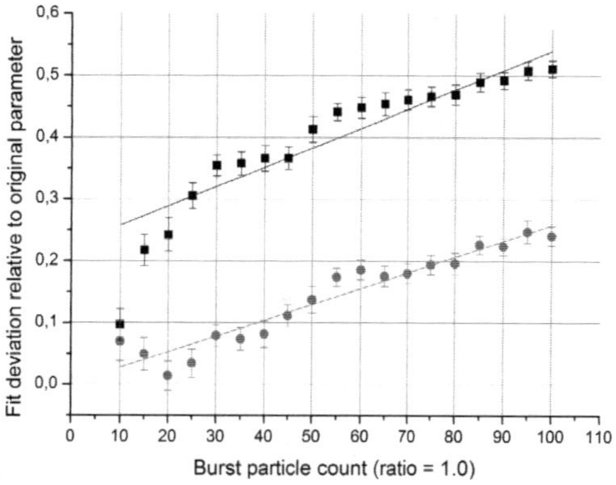

Figure 5. $\Delta\lambda$ parameter of reconstructed t_{dr} curve deviation from original $\Delta\lambda$ parameter dependency on total count of registered particles. [Linear ■, Robust ●]

6. Conclusion

Using Monte-Carlo simulation the comparison between linear and robust regression methods have been carried out. Using robust regression method we managed to improve quality of fit and minimize the impact of background particles on analysis.

Robust regression method based on minimal linear distance heuristic with current algorithm settings in average make correct results (deviation less than 10%) in application to longitudinal drift distance determination before following conditions are met: background to burst ration is less than 80%, total number of registered burst and background particles less than 100.

Acknowledgments

This work was performed within the framework of the Center FRPP supported by MEPhI Academic Excellence Project (contract 02.a03.21.0005, 27.08.2013).

References

[1] Aleksandrin S Y *et al.* 2003 *RAS Izvestiya, physical series* **67** 79–83
[2] Aleksandrin S Y *et al.* 2003 *Annales Geophysicae* **21** 597–602
[3] Aleksandrin S Y *et al.* 2010 Study of local radiation belt perturbations in the satellite experiments ARINA and VSPLESK *Proceedings of the 31st National Conference on Cosmic Rays* (Moscow: Moscow State University)
[4] Elliot H 1963 *Reports on Progress in Physics* **26** 145–80
[5] Hastie T, Tibshirani R and Friedman J 2001 *The Elements of Statistical Learning* (Springer New York Inc.)
[6] Ososkov G A, Polyanskij A and Puzynin I V 2002 (in Russian) *Particles and nucleus physics* **33** 676–740 (original Russian title: *Fizika ehlementarnyh chastic i atomnogo yadra*)

Journal of Physics: Conference Series **675** (2016) 032003 doi:10.1088/1742-6596/675/3/032003

Trapped positrons observed by PAMELA experiment

V V Mikhailov[1],O Adriani[2], G Barbarino[3], G A Bazilevskaya[4], R Bellotti[5], M Boezio[6], E A Bogomolov[7], M Bongi[2], V Bonvicini[6], S Bottai[2], A Bruno[5], F S Cafagna[5], D Campana[3], R Carbone[3], P Carlson[8], M Casolino[9], G Castellini[10], L Consiglio[3], C De Santis[9], N De Simone[9], V. Di Felice[9], A M Galper[1], A V Karelin[1], S V Koldashov[1], S Koldobsky[1], S Yu Krutkov[7], A N Kvashnin[4], A A Leonov[1], V V Malakhov[1], L Marcelli[9],M Martucci[9,12], A G Mayorov[1], W Menn[11],M Merge[9,13] E Mocchiutti[5], A Monaco[5], N. Mori[10],R Munini[6], G Osteria[3], P Papini[2], F Palma[9,13], B Panico[3], M Pearce[8], P Picozza[9,13], M Ricci[12], S B Ricciarini[2], R Sarkar[6], V Scotti[3] M Simon[11], R Sparvoli[9,13], P Spillantini[2], Yu I Stozhkov[4], A Vacchi[6], E Vannuccini[2], G I Vasiliev[7], S A Voronov[1], Yu T Yurkin[1], G Zampa[5], N Zampa[6]

[1] National Research Nuclear University MEPhI (Moscow Engineering Physics Institute), Kashirskoe highway 31, Moscow, 115409, Russia
[2]INFN,Sezione di Florence and Physics Department of University of Florence, Italy
[3]INFN, Sezione di Naples and Physics Department of University of Naples Federico II
[4]Lebedev Physical Institute, Russia
[5]INFN, Sezione di Bari Physics and Department of University of Bari, Italy
[6]INFN,Sezione di Trieste, Italy
[7]Ioffe Physical Technical Institute, Russia
[8]KTH, Department of Physics, and The Oskar Klein Centre for Cosmoparticle Physics, Sweden
[9]INFN, Sezione di Rome "Tor Vergata", Italy
[10]INFN, IFAC, Italy
[11]INFN, Physics Department of Universitaet Siegen, Germany
[12]INFN, Laboratori Nazionali di Frascati, Italy
[13] University of Rome Tor Vergata, Department of Physics, Italy
[14]Agenzia Spaziale Italiana (ASI) Science Data Center, I-00044 Frascati, Italy

E-mail: vvmikhajlov@mephi.ru

Abstract. Measurements of electron and positron spatial distributions in energy range from 80 MeV to several GeV below the geomagnetic cutoff rigidity were carried out using the PAMELA magnetic spectrometer. The instrument is installed on board the Resurs-DK satellite which was launched June 15th 2006 on an elliptical orbit with the inclination 70 degrees and the altitude 350-600 km. The procedure of trajectories calculations in the geomagnetic filed gives a way to separate stably trapped and short lived albedo components produced in interactions of cosmic ray protons with the residual atmosphere. The work presents spatial distributions of trapped, quasitrapped and short-lived albedo electrons and positrons in the near Earth space. Electron to positron ratio points out on different production mechanism of trapped and quasitrapped particles.

1. Introduction

High energy galactic cosmic rays interact in the Earth atmosphere resulting producing secondary particles. The flux of secondary particles is composed mainly by electrons, protons, neutrons and gamma-rays. An appreciable fraction of charged secondaries can travel backward in space along the Earth's magnetic field lines and can reach the satellite altitudes forming a high energy radiation halo in the Earth vicinity. Flux of secondary electrons was first calculated in paper [1]. Namely in this work it was considered process of decay of charged pions produced in cosmic ray proton interactions. Pions decay throw $\pi^{\pm} \rightarrow \mu^{\pm} \rightarrow e^{\pm}$ chain to electrons and positrons. Because production rate is proportional to cosmic ray intensity *Icr*, residual atmosphere density $\rho(h)$ and particle's time of live T is $\propto 1/\rho(h)$ then resulting intensity of secondary particles $J(h) \propto Icr \times \rho \times T$ will be constant with altitude in wide altitude range forming a halo around the Earth. Nuclear interactions of trapped protons of energies >300 MeV with upper atmosphere constituents could be considered for the production of positrons and over electrons in the innermost magnetosphere. Calculations performed in paper [2] predict the existence of a positron belt in a narrow region around L-shell = 1.2. The flux ratio of positrons to electrons e+/e- is estimated to be ~ 4 for energies of 40-1500 MeV. This mechanism exhibits sharply decreasing spectrum. At 200-300 MeV energies, positron fluxes from the trapped proton source are still comparable with the cosmic ray born positrons, but at greater energies production by trapped protons is negligible.

The magnetic spectrometer PAMELA was launched onboard the Resurs-DK1 satellite on the 15th of June 2006 and it is continuously taking data till present time. The satellite has a quasi-polar (70^0 inclination) elliptical orbit at an altitudes between 350 and 600 km. Preliminary results of PAMELA observation of secondary electron and positron fluxes near the Earth, which were made in first year of the flight , were reported in paper [3]. At present time about 50TB data were downlinked for analysis during 8 year of the work. Additional accumulated statistics allows to study secondary electrons and positrons generation and propagation in magnetosphere with more details than in previous experiments. Moreover, the backtracking procedure which determines the movement of particles before their detection was applied for data analysis. This procedure offers new opportunities to compare data with models and allows to determine individual particles origin [4, 5]. This work presents PAMELA spectrometer measurements of spatial distributions of secondaries electrons and positrons made using data obtained from 2006 to 2014.

2. PAMELA spectrometer

The instrument consists of a Time-of-Flight system (TOF), an anticoincidence system, a magnetic spectrometer, an electromagnetic calorimeter, a shower tail catching scintillator and a neutron detector . The TOF system gives the main trigger for particle acquisition, measures the absolute value of the particle charge and its flight time while crossing the apparatus (the accuracy is better than 350 psec). A rigidity is determined by the magnetic spectrometer, composed by a permanent magnet with a magnetic field intensity 0.4 T and a set of six micro-strip silicon planes. The electron and positron identification is provided by the calorimeter, a series of mini-strip silicon layers interleaved by tungsten planes (16.3 radiation and 0.6 nuclear interaction lengths deep). Particles not cleanly entering the PAMELA acceptance are rejected by the anticoincidence system. Using of the TOF system, the magnetic spectrometer and additional analysis of the calorimeter information allows extracting leptons and measuring their energy (from 50 MeV to several hundred GeV) effectively. The acceptance is about 21.6 cm^2sr [6].

The main axis of PAMELA points to a local zenith. Orbit characteristics allow measuring particles with pitch angles (the angle between the particle velocity and the magnetic field vector) of about 80-90^0 in the equatorial region. The angle between the main axis of PAMELA and magnetic field decreases with latitude, so for middle and high latitudes it is possible observe

Journal of Physics: Conference Series **675** (2016) 032003 doi:10.1088/1742-6596/675/3/032003

smaller values of pitch angles.

3. Data analysis

For each registered event the following parameters were measured or calculated: the number of tracks and energy losses in the magnetic spectrometer planes; the rigidity and the track length (by fitting the track in the magnetic field); the time of flight. A particle velocity was calculated using the time of flight and the track length. Moreover a set of variables dealing with point of interaction, transversal and longitudinal profiles was calculated using the calorimeter [6].

Electrons and positrons were identified using information about dE/dx energy losses in the spectrometer planes to determine charge Z=1 , shower properties in the electromagnetic calorimeter, particle velocity and a rigidity. The misidentification of protons and pions is the largest source of background. Particle identification based on the calorimeter data can be tuned to rejection power 10^4 - 10^5 for protons and pions, while selecting > 80-90% of the electrons or positrons over all energy range.

Total accumulated statistic for electrons and positron is about 4×10^6 in whole energy range.

Gathering power of the instrument was estimated with Monte-Carlo simulation with official PAMELA Collaboration software [6]. An efficiency of the instrument may change with time and must be taken into account carefully in a data proceeding. It was verified from experimental data itself by using different combination of information from imaging calorimeter, magnetic spectrometer and time of flight system.

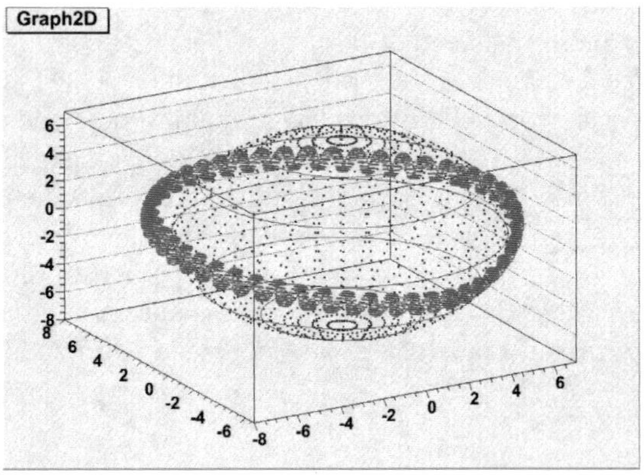

Figure 1. An example of a particle trajectory with energy 1.06 GeV trapped in equatorial region.

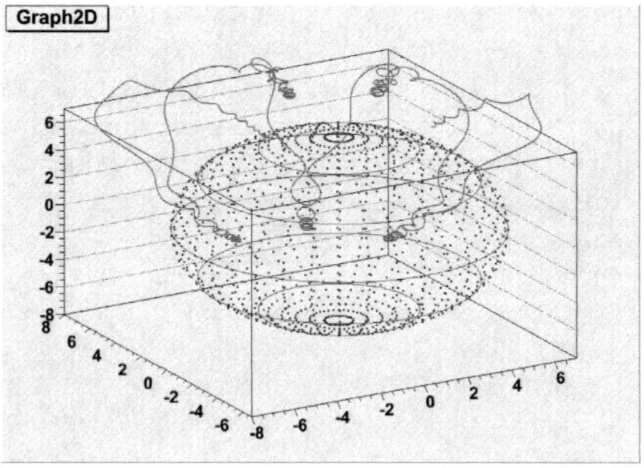

Figure 2. An example of a quasitrapped particle trajectory with energy 1.06 GeV.

For electrons and positrons differential spectra, a longitudinal analysis was performed for set of latitudes intervals from equator to pole region [3, 7].These results show features of spectra for different geomagnetic regions due to their origin. Primary particles are observed mainly in polar region and above cut-off rigidity near equator. Secondary "short lived" component prevails near equator below geomagnetic cut-off at low rigidities and trapped particles can be measured in South Atlantic Anomaly with pitch-angles about 90 degree.

Using a geographical coordinates and an orientation of PAMELA as a function of time, the McIlvain geomagnetic coordinates L-shell and B were calculated for every event. For the calculation, the IGRF model (http://nssdcftp.gsfc.nasa.gov/models/geomagnetic/igrf) of the Earth magnetic field was used. Then backtracking procedure was applied for every identified events to obtain trajectory of events up to 35 second before detection and it was stopped if particles touched the Earth atmosphere on 40 km altitude or escaped magnetosphere. Boundary

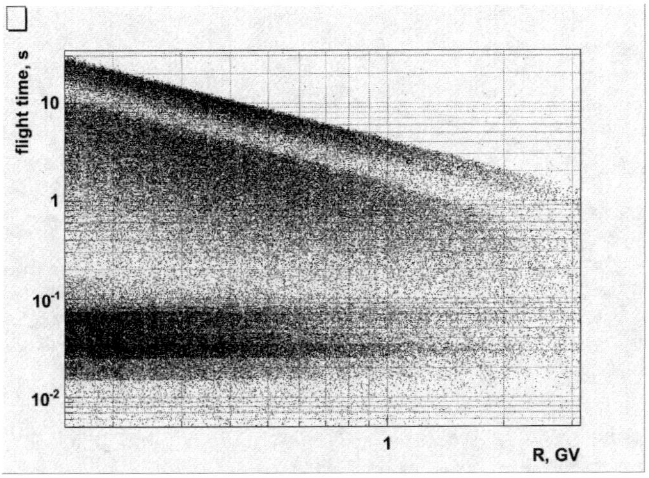

Figure 3. Flight time vs partice's rigidity.

escaping was chosen to be 20000 km. Trajectory helps to determine particle origin for every individual event. See in figure 1 an example of trajectory of trapped electron with energy about 1 GeV. Primary electrons and positrons were excluded from analysis together with particles in transition region near geomagnetic cut-off, applying condition $R<10/L^3$ [GV].

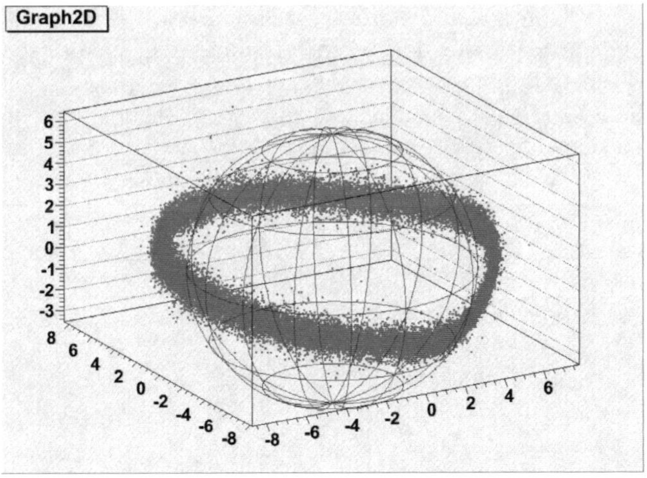

Figure 4. Particles positions 10 second before detection for trapped positrons and electrons.

Calculations of particle trajectories in geomagnetic field provides additional information about spatial distribution of secondary fluxes. This procedure gives possibility to explore fluxes in enlarged regions outside the satellite orbits. Tracking of positrons and electrons with rigidity below cut-off clearly separate measured data set in three different categories: 1) particles with very short time of live (less then 1 s) which absorbed in opposite hemisphere after first bounce; 2) particles with time of live about 1 s, points of origin for this sample lies far from the satellite position on the other longitudes. This is so-called quasi-trapped particles; 3) long lived particles with time more then tens seconds. In this case impossible to determine the point of particle origin because they makes many revolution around the Earth and may stay trapped for hours, days and ,perhaps, years.

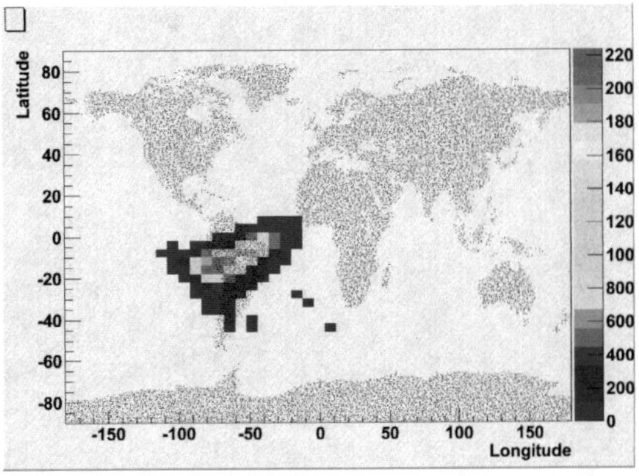

Figure 5. Regions where trapped positrons were observed by PAMELA.

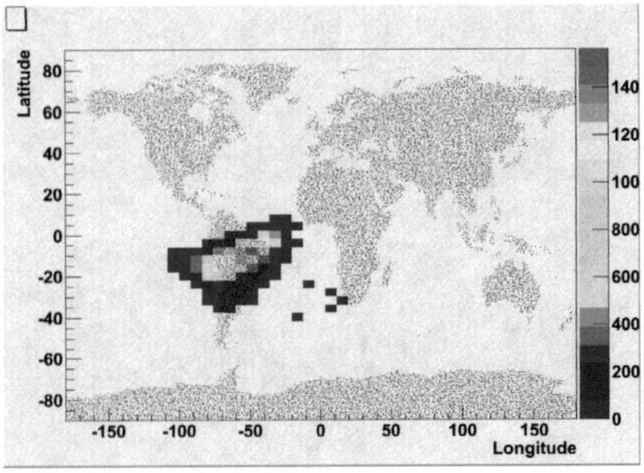

Figure 6. Regions where trapped electrons were observed by PAMELA.

4. Results

The track knowledge give possibility to separate particles of different origin measured in the same region. In figure 2 time of flight in magnetosphere versus of rigidity in magnetosphere is shown for equatorial region (L <2). Typically albedo particles reach the orbit of the satellite for time about 0.1 second. For relativistic electrons and positrons this flight time do not dependent from particle's rigidity. But if electron or positron has appropriate pitch-angle to be mirrored by magnetic field, flight time is increasing dramatically due to drift around the Earth (see figure 1b). Drift speed is increasing with particle increasing of rigidity and so flight time of quasitrapped particles is decreasing with rigidity as it is seen from figure. Such behavior of secondary albedo and quasitrapped electrons and positrons correspond to that was observed in AMS-01 experiment [5] where they were called as short-lived and long-lived second leptons.

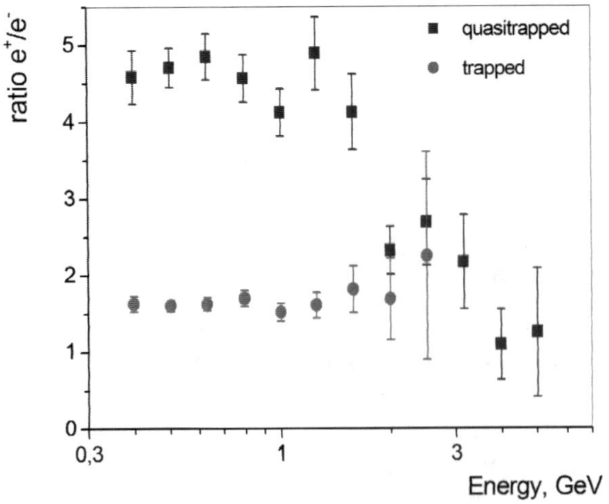

Figure 7. e^+/e^- ratio vs energy for quasitrapped and trapped components

There is also a significant part of track with very large livetime which can not be determined by tracing procedure. Figure 3 shows calculated spatial position of such trapped electrons and positrons they had 10 second before detection. In figure 4 points of measuring of trapped electrons and positrons are shown. Maximum of count is observed at L-shell about 1.18 and it is increasing with local magnetic field B decreasing below 0.215 G. Inside this region positron to electron ratio is maximal near West part of the South Atlantic Anomaly. In agreement with our previous analysis [7] positron fraction of very long lived trapped particles is much smaller the simple reentrant albedo one (see figure 7) .

5. Summary

Secondary electron and positron fluxes have complex spatial structure caused by geomagnetic field, production cross-section and atmospheric absorption of produced secondaries. Using backtracing procedure over $\sim 10^6$ events three components of secondary positrons were clearly separated: albedo, quasitrapped and stably trapped. Measured positron to electron ratio in the PAMELA experiment points out on different production mechanism of trapped and quasitrapped particles .

Acknowledgments

We acknowledge support from, Russian Federal Space Agency (Roscosmos), the Italian Space Agency (ASI), Deutsches Zentrum fur Luft- und Raumfahrt (DLR), the Swedish National Space Board, the Swedish Research Council. Russian colleagues also acknowledge partial support from Russian Scientific Foundation (grant 14-12-00373).

References

[1] N L Grigorov Paper of AS USSR 1977 **24** 810
[2] A A Gusev U B Jayanthet I M Martin et al. Journal of Geophys. Res. 2001 **106** A26111
[3] O Adriani et al. [PAMELA collaboration] Journal of Geophys. Res. 2009 A12218 doi:10.1029/2009JA014660
[4] V Plyaskin Astropart. Phys. 2008 **30** 18
[5] J Alcaraz et al. Phys. Lett. B,2000 **484** 10
[6] P Picozza A M Galper G Castellini et al. [PAMELA collaboration] Astropart. Phys. 2007 **27** 296-315
[7] O Adriani et al.[PAMELA collaboration] Phys.Rep. 2014 **544** 4 323-370 doi: 10.1016/j.physrep.2014.06.003

Journal of Physics: Conference Series **675** (2016) 032004 doi:10.1088/1742-6596/675/3/032004

Evaluation of the Antiproton Flux from the Antineutrino Electron Scattering

V V Alekseev[1], K M Belotsky[2], Yu V Bogomolov[1], R I Budaev[2], O A Dunaeva[1], A A Kirillov[1,2], A V Kuznetsov[1,3,*], M N Laletin[2], A D Lukyanov[1], V V Malakhov[2], A G Mayorov[1,2], M A Mayorova[2], A F Mosichkin[1], A A Okrugin[1], S A Rodenko[2] and A M Shitova[1]

[1] Department of Physics, Yaroslavl State P.G. Demidov University, 150000 Sovietskaya st., 14, Yaroslavl, Russia

[2] National Research Nuclear University MEPhI (Moscow Engineering Physics Institute), Kashirskoe highway 31, Moscow, 115409, Russia

E-mail: *avkuzn@uniyar.ac.ru

Abstract. Recent experiments in high energy cosmic ray physics, PAMELA and AMS-02, excite a new interest to the mechanisms of generation of galactic antiparticles. In spite of the fact that global picture coincides with the predictions of the standard model, there are some black spots stimulating scientists to involve into research a particularly new physics like dark matter. In the present work, we make an attempt to estimate the impact of standard neutrino processes into the total flux of secondary antiprotons detected by contemporary experiments.

1. Introduction

From the time of the first detection of antiprotons in cosmic rays in 1979 [1, 2], the experiments were sufficiently enlarged and improved. Contemporary detectors based on the satellites could help to open the window to a new physics. Despite the fact that the entire data from modern accelerator experiments coincides with the predictions of the standard model, a lot of open questions exist on the border of cosmology and particle physics, mainly on a possible particle origin of dark matter.

Recently, there was a message from AMS 02 collaboration [3] where new results on the $\bar{p}/p$ ratio at high rigidity was presented. The antiproton spectrum appeared to be surprisingly hard. It should be emphasized that the information was preliminary and the corresponding article is not yet proclaimed but nevertheless there appeared an avalanche of articles dedicated to probable explanations of the announced discrepancy. Most of them relate on the physics of dark matter and probable candidates of it (see, for example, [4] for the recent status on dark matter detection). There are also some articles [5, 6, 7] where authors claim the possibility of intra-galactic sources of antimatter like supernova explosions.

It is well known that neutrino physics plays the crucial role in the evolution of supernovae. The aim of this paper is to estimate a possible contribution of neutrino processes to the production rate of secondary antiprotons that could be detected by existing experiments. For

[3] To whom any correspondence should be addressed.

the first step, we restrict ourselves with the neutrino-electron process $\bar{\nu}_e + e \to \bar{p} + n$. The reason of choosing such a process is an attractive opportunity to perform the analysis by making calculations which are almost completely analytical. Other possibilities for antiproton production via electroweak processes in the Galaxy are also discussed.

2. A general approach to the calculation of the secondary antiproton flux

Let us evaluate the contribution of the neutrino-electron process $\bar{\nu}_e + e \to \bar{p} + n$ to the total antiproton flux measured in an experiment. For simplicity, we neglect the effects of antiproton propagation through the Galaxy and assume that all antiprotons, produced in the selected volume, reach the detector. The total flux of any type of cosmic ray particles is defined by the formula

$$\Phi = \frac{\mathrm{d}N}{\mathrm{d}E\,\mathrm{d}t\,\mathrm{d}s\,\mathrm{d}\Omega}, \tag{1}$$

where $\mathrm{d}N$ is the number of particles detected per the energy interval $\mathrm{d}E$, per the time interval $\mathrm{d}t$, per the detector area interval $\mathrm{d}S$, and arrived from the solid angle interval $\mathrm{d}\Omega$. On the other hand, a number of particles produced in the reaction $1 + 2 \to 3 + X$ in the volume $\mathrm{d}V$, per the time interval $\mathrm{d}t$, is expressed via the total cross-section σ of the reaction: $\mathrm{d}N_3 = \sigma v\, n_1 n_2\, \mathrm{d}V\, \mathrm{d}t$, where v is the so-called relative velocity, and $n_{1,2}$ are the particle densities. If one needs to have a distribution of the produced particles over the energies and the solid angles, the differential cross-section $\mathrm{d}\sigma/(\mathrm{d}E\,\mathrm{d}\Omega)$ should be taken instead of the total cross-section. The particle densities $n_{1,2}$ should be expressed via the particle fluxes. Finally, the total antiproton flux caused by the reaction $\bar{\nu}_e + e^- \to \bar{p} + n$ takes the form:

$$\Phi_{\bar{p}} = \int \mathrm{d}r\, \mathrm{d}E_{\bar{\nu}}\, \mathrm{d}\Omega_{\bar{\nu}}\, \frac{1}{v_{\bar{\nu}}}\, \Phi_{\bar{\nu}}(E_{\bar{\nu}})\, \mathrm{d}E_e\, \mathrm{d}\Omega_e\, \frac{1}{v_e}\, \Phi_e(E_e)\, \frac{(P_{\bar{\nu}} P_e)}{E_{\bar{\nu}}\, E_e}\, \frac{\mathrm{d}\sigma(\bar{\nu}_e + e^- \to \bar{p} + n)}{\mathrm{d}E_{\bar{p}}\, \mathrm{d}\Omega_{\bar{p}}}\, c\,, \tag{2}$$

where c is the speed of light and the integration over r involves the detectable part of the Galaxy. Note that the discussed approach can be used for other processes or astrophysical sources of elementary particles.

3. Antineutrino electron process

The antineutrino-electron process $\bar{\nu}_e + e^- \to \bar{p} + n$ resulting in antiproton and neutron, is a crossed process to the standard neutron beta-decay. To calculate the differential cross-section, we start with the matrix element in the following form:

$$\mathcal{M} = \frac{G_{\mathrm{F}} V_{ud}}{\sqrt{2}} \left[\bar{u}_n \Big(g_V \gamma_\mu + g_A \gamma_\mu \gamma_5\Big) u_{\bar{p}}\right] \times \left[\bar{v}_{\bar{\nu}} \gamma^\mu (1 - \gamma_5) v_e\right], \tag{3}$$

where g_V and g_A are the vector and axial-vector constants, $|V_{ud}| \simeq 0.974$ and G_{F} is the Fermi constant. After some standard evaluations we get an averaged squared amplitude in the form:

$$\overline{|\mathcal{M}|^2} = 32\, G_{\mathrm{F}}^2 |V_{ud}|^2 \left[(1+\alpha)^2 (P_n P_\nu)(P_p P_e) + (1-\alpha)^2 (P_p P_\nu)(P_e P_n) + (1-\alpha^2) m_n m_p (P_\nu P_e)\right], \tag{4}$$

where $\alpha = |g_A/g_V| \simeq 1.27$.

The differential cross-section is expressed via the averaged squared amplitude as follows:

$$\mathrm{d}\sigma(\bar{\nu}_e + e^- \to \bar{p} + n) = \frac{\overline{|\mathcal{M}|^2}}{32\pi^2 s} \delta(E_\nu + E_e - E_{\bar{p}} - E_n) \frac{\mathrm{d}^3 P_{\bar{p}}}{E_n E_{\bar{p}}}. \tag{5}$$

For our present purposes, we do not integrate the obtained expression because it enters the equation (2).

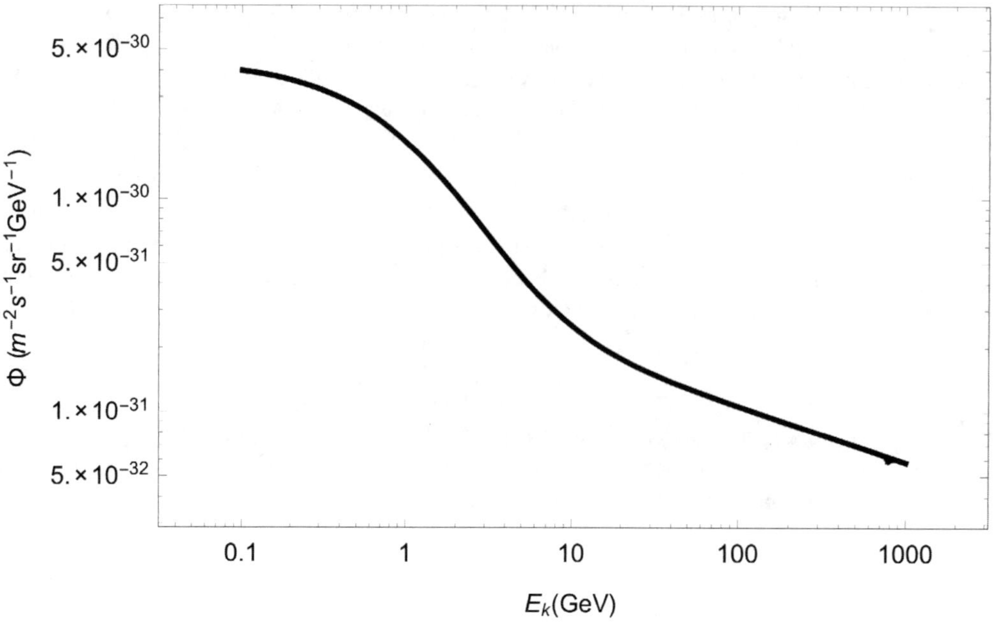

Figure 1. Antiproton flux from $\bar{\nu_e} + e \to \bar{p} + n$ as a function of antiproton kinetic energy.

4. Antiproton flux estimation

Estimation of the total antiproton flux from the neutrino process is based on the equation (2), where we use the cross-section of the process given in the previous section and the following values of the electron antineutrino flux:

$$\Phi_{\bar{\nu}} \simeq 6.85 \times 10^{-7} E_{\nu}^{-2.46}\,\mathrm{cm}^{-2}\,\mathrm{s}^{-1}\,\mathrm{sr}^{-1}\,\mathrm{GeV}^{-1}, \tag{6}$$

that is one sixth of the diffuse astrophysical neutrino spectrum derived in [8] (see also [9] for previous estimations). As for initial electrons, our analysis shows that a consideration of slow-moving or being at rest electrons is inappropriate in this case because of a large threshold energy for neutrinos which result in high antiproton energies much large than the interval we are interested in. Thus, we use the electron flux $\Phi_e \simeq 3 \times 10^{-2} E_e^{-3.2}\,\mathrm{cm}^{-2}\,\mathrm{s}^{-1}\,\mathrm{sr}^{-1}\,\mathrm{GeV}^{-1}$ which is a rough approximation of the linear part of the cosmic electron flux presented in [10, 11, 12]. Despite the fact that accurate expressions for neutrino and electron fluxes depend on the the actual range of measured energies, these values afford to perform a zero-order estimation and make a conclusion on the effect size.

In order to estimate the antiproton flux one has to insert Eq. (5) into Eq. (2). Further integration could be significantly simplified by introducing new variables: i) the vector of the initial momentum: $\mathbf{Q} = \mathbf{p}_{\bar{\nu}} + \mathbf{p}_e = \mathbf{p}_{\bar{p}} + \mathbf{p}_n$; ii) the polar angle θ_1 between the vectors $\mathbf{p}_{\bar{p}}$ and $\mathbf{Q}$; iii) the polar angle θ_2 between the vectors $\mathbf{Q}$ and $\mathbf{p}_e$; iv) the azimuthal angle between the planes formed by the vectors $\mathbf{p}_{\bar{p}}$, $\mathbf{Q}$ and $\mathbf{Q}$, $\mathbf{p}_e$. The integration over the azimuthal angle can be easily performed analytically.

The result of integration is presented in the figure as the dependence of the final antiproton flux on the antiproton kinetic energy $E_k = E_p - m_p$.

The numerical values of the obtained antiproton flux for all adequate range of antiproton energies are rather small if compared with the measured antiproton flux $\sim 10^{-5}\,\mathrm{m}^{-2}\,\mathrm{s}^{-1}\,\mathrm{sr}^{-1}\,\mathrm{GeV}^{-1}$ at energies $\sim 10^2\,\mathrm{GeV}$ [13]. We can conclude that the process in question will not give any significant contribution to the total antiproton flux. Nevertheless the result is valuable as the first attempt of the calculation of this kind. One should pay attention on the approximate character of the total estimation, but more strictly recorded estimates including for example details of propagation should not drastically change the total conclusion of the paper.

There are several more common processes involving neutrino and antiproton, mainly inclusive processes $\bar{\nu_e} + p \to \bar{p} + X$. One can expect significantly greater flux of antiprotons from that processes because the flux of cosmic protons essentially exceeds the cosmic electron flux, but the small values of constants of interaction should nullify this benefit.

5. Conclusion

We have estimated the flux of antiprotons from the neutrino process that could in principle be detected by contemporary experiments.

One of the first attempts to connect the problem of antiprotons distribution through the Galaxy with neutrinos was made in Ref. [14]. The author roughly estimated the neutrino flux from antiprotons and concluded that the obtained value has been too small to be detected. It was written in Ref. [14] that the calculations would free others from the need to remake the calculations because the effect was negligible. We should clarify that the usefulness of this research is, in our opinion, first of all, that the analysis of the specific process of antiproton production was conducted almost entirely by analytical calculations, so one can use this technique for the analysis of other mechanisms of antiproton production with other fluxes of initial particles and with the other dynamics of the interaction, which is defined by the cross-section of a particular process, such as the annihilation of dark matter particles.

Acknowledgments

The research is financed by the grant of the Russian Science Foundation (Project No. 15-12-10039). The work of MEPhI group (K.M.B., R.I.B., M.N.L., V.V.M., M.A.M., S.A.R.) on the parametrization of the cosmic electron spectrum and on a detailed comparison of the flux of antiprotons from the neutrino process with the flux of antiprotons measured by the PAMELA detector was performed within the framework of the Center of FRPP supported by MEPhI Academic Excellence Project (contract No. 02.a03.21.0005, 27.08.2013)

References

[1] Bogomolov E A *et al.* 1979 *16th International Cosmic Ray Conference* vol 1, ed Saburo Miyake (Tokyo: Institute for Cosmic Ray Research, University of Tokyo) p 330
[2] Golden R L *et al.* 1979 *Phys. Rev. Lett.* **43** 1196
[3] Kounine A 2015 Talk at the AMS Days at CERN (*http://indico.cern.ch/event/381134/*)
[4] Bi X-J, Yin P-F and Yuan Q 2014 *arXiv*:hep-ph/1409.4590
[5] Ahlers M, Mertsch P and Sarkar S 2009 *Phys. Rev.* D **80** 123017
[6] Savchenko V, Kachelriess M and Semikoz D V 2015 *Astrophys. J. Lett.* **809** L23
[7] Kohri K *et al.* 2015 *arXiv*:astro-ph.HE/1505.01236
[8] Aartsen M G *et al.* (IceCube Collaboration) 2015 *Phys. Rev.* D **91** 022001
[9] Halzen F 2013 *arXiv*:hep-ph/1311.6350
[10] Adriani O *et al.* 2011 *Phys. Rev. Lett.* **106** 201101
[11] Mikhailov V V *et al.* 2011 *Bull. RAS: Phys.* **75** 316 (Original Russian text: *Izv. RAN. Ser. Fiz.* **75** 345)
[12] Sparvoli R 2013 Recent results from the space experiment PAMELA *XIII Lomonosov Conference, Moscow*
[13] Adriani O *et al.* 2012 *JETP Lett.* **96** 621 (Original Russian text: *Pis'ma v ZhETF* **96** 693)
[14] Ramadurai S 1986 *Astrophysical Letters* **25** 35

Journal of Physics: Conference Series **675** (2016) 032005 doi:10.1088/1742-6596/675/3/032005

The measurement of the dipole anisotropy of protons and helium cosmic rays with the PAMELA experiment

A V Karelin[12], O Adriani[1,2], G C Barbarino[3,4], G A Bazilevskaya[5], R Bellotti[6,7], M Boezio[8], E A Bogomolov[9], M Bongi[1,2], V Bonvicini[8], S Bottai[2], A Bruno[6,7], F Cafagna[7], D Campana[4], P Carlson[13], M Casolino[11], G Castellini[14], C De Donato[10], C De Santis[11], N De Simone[10], V Di Felice[10], V Formato[8,†], A M Galper[12], S A Koldobskiy[12], S V Koldashov[12], S Y Krutkov[9], A N Kvashnin[5], AA Leonov[12], V Malakhov[12], L Marcelli[11], M Martucci[11,15], A G Mayorov[12], W Menn[16], M Mergè[10,11], V V Mikhailov[12], E Mocchiutti[8], A Monaco[6,7], N Mori[2], R Munini[8,17], G Osteria[4], F Palma[10,11], B Panico[4], P Papini[2], M Pearce[13], P Picozza[10,11], M Ricci[15], S B Ricciarini[14], R Sarkar[18,††], V Scotti[3,4], M Simon[16], R Sparvoli[10,11], P Spillantini[1,2], Y I Stozhkov[5], A Vacchi[8,19], E Vannuccini[2], GI Vasilyev[9], S A Voronov[12], Y T Yurkin[12], G Zampa[8], N Zampa[8]

[1]University of Florence, Department of Physics, I-50019 Sesto Fiorentino, Florence, Italy
[2]INFN, Sezione di Florence, I-50019 Sesto Fiorentino, Florence, Italy
[3]University of Naples "Federico II", Department of Physics, I-80126 Naples, Italy
[4]INFN, Sezione di Naples, I-80126 Naples, Italy
[5]Lebedev Physical Institute, RU-119991, Moscow, Russia
[6]University of Bari, Department of Physics, I-70126 Bari, Italy
[7]INFN, Sezione di Bari, I-70126 Bari, Italy
[8]INFN, Sezione di Trieste, I-34149 Trieste, Italy
[9]Ioffe Physical Technical Institute, RU-194021 St. Petersburg, Russia
[10]INFN, Sezione di Rome "Tor Vergata", I-00133 Rome, Italy
[11]University of Rome "Tor Vergata", Department of Physics, I-00133 Rome, Italy
[12] National Research Nuclear University MEPhI (Moscow Engineering Physics Institute), Kashirskoe highway 31, Moscow, 115409, Russia
[13]KTH, Department of Physics, and the Oskar Klein Centre for Cosmoparticle Physics, AlbaNova University Centre, SE-10691 Stockholm, Sweden
[14]IFAC, I-50019 Sesto Fiorentino, Florence, Italy
[15]INFN, Laboratori Nazionali di Frascati, Via Enrico Fermi 40, I-00044 Frascati, Italy
[16]Universität Siegen, Department of Physics, D-57068 Siegen, Germany
[17]University of Trieste, Department of Physics, I-34147 Trieste, Italy
[18]Indian Centre for Space Physics, 43 Chalantika, Garia Station Road, Kolkata 700084, West Bengal, India
[19]University of Udine, Department of Mathematics and Informatics, I-33100 Udine, Italy
[20]Centre for Space Research, North-West University, 2520 Potchefstroom, South Africa
[†]Now at INFN, Sezione di Perugia, I-06123 Perugia, Italy
[††]Previously at INFN, Sezione di Trieste, I-34149 Trieste, Italy

E-mail: karelin@hotbox.ru

Abstract. In the framework of the PAMELA experiment the features of the large-scale anisotropy have been measured within the energy range 1-20 TeV/n. The measurements were carried out with the use of the calorimeter on the base of the hypothesis about the existence of a dipole anisotropy. The amplitude and phase of the dipole were obtained. The results are in agreement with the ground-based observations.

1. Introduction

All the results of the ground-based experiments [1-7] agree in the fact that there is a large-scale anisotropy and that at the first order it is a dipole, the amplitude and phase of which depend on energy. These parameters are different in the various measurements in the energy range 1-20 TeV. However, their values have a crucial role for the processes that might be involved for an anisotropy explanation. For example, in the work [8] the anisotropy is explained by a single nearby source. The same source is responsible for the positron and antiproton spectra as they are seen in the PAMELA [9,10] and AMS [11,12] experiments.

Beside to ground based measurements the satellite ones are able to measure the anisotropy in the both hemispheres and do not undergo by any atmospheric influences. But unlike the ground-based measurements, where it is possible to collect a large number of events ($\sim$ tens of millions) to observe the effect of anisotropy (at an intensity level of 10^{-3}) in the angular distribution of the primary cosmic rays (mostly protons), the satellite PAMELA experiment provides with much lower statistics due to the small size of its aperture.

The aim of this study was to measure the phase and amplitude of the dipole anysotropy of cosmic rays on the basis of information collected in the PAMELA experiment within the energy range 1–20 TeV/n during Jun 2006 – Nov 2014.

2. The PAMELA experiment

The PAMELA magnetic spectrometer was launched in the summer of 2006 and has been operating since that time [13]. The main scientific goals of the experiment are the study of particle and antiparticle fluxes in a wide energy range. PAMELA apparatus consist of several various detectors positioned around a magnetic spectrometer. While the tracker is able to measure the deflection of particles in the magnetic field up to energies of about 1 TeV, another PAMELA subdetector, a calorimeter, could be used to extend the measured energy range. Furthermore the calorimeter allows to measure arrival particle direction over a wide range of angles. The calorimeter consists of 44 silicon planes, with 96 strips in each one; interleaved with 22 tungsten layers. In neighboring silicon planes, strips are orthogonal providing topological and longitudinal information of the shower development.

3. The data analysis

To measure the particle direction the shower axis inside the calorimeter was reconstructed. The iterative procedure allowed to restore the axis along the primary particle track throughout the 44 planes [14]. This procedure is a fit of the centres of gravity of energy released in each plane of each view. The events with reconstructed shower axis were further selected by the cuts based on the total energy deposition in the calorimeter. A cut threshold was set at the level of 180000 mips corresponding to particles within the energy range 1–20 TeV/n.

The anisotropy of cosmic rays is obtained by comparing two distributions: the reference and the experimental ones. The reference distribution of events corresponds to case when the cosmic ray direction distribution is perfectly isotropic taking into account a real exposure and a registration efficiency of the apparatus as well as its mass distribution throughout a structure. The reference distribution was simulated by a shuffling method [15] and contains of

100 times more events than the experimental one. These distributions were constructed in the equatorial coordinate system. In our case, we consider a single coordinate only: a right ascension, which varies from 0 to 360 degrees. All events distributed along a declination coordinate were summarized.

Since the existence of the dipole anisotropy based on the ground-based observations is not questioned, so far the task, as mentioned above, was to obtain its characteristics: the amplitude and phase. We have approximately 600,000 coordinate values of right ascensions, reflecting the direction of incidence of the particles in the experimental data.

The basis of the analysis is the hypothesis that there is the dipole in the experimental distribution. For this reason the size of an integration bin equal to 180 degrees was chosen. In our approach this bin reproduces an integration radius in HEALPIx method for the anisotropy study in 3D map [16]. All events within this bin were integrated. The bin centre was shifting through the entire range from 0 to 360 degrees with 5 degree step and the integration procedure was repeated at each step. Then we took an average value for each 6 bins (we had 72 values in total) and got 12 values which are shown in figure 1. The obtained dipole is demonstrated in terms of relative intensities: *Ir* is the experimental one and *Is* is the reference one. The number of events in the reference distribution was normalized to the experimental one.

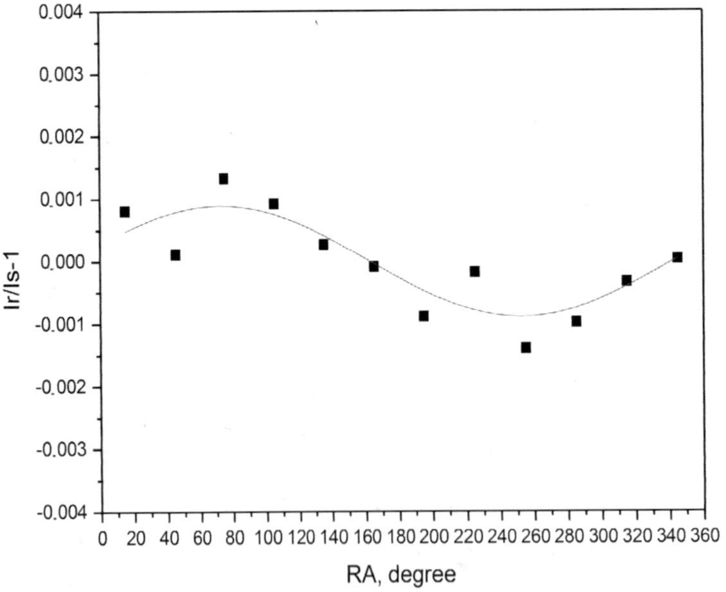

Figure 1. The *Ir*/*Is*-1 depending on RA. *Ir* is the real intensity, *Is* is the simulated isotropic intensity. The red line is a fit by the sine wave.

4. The results

The results were obtained by an approximation of the values in figure 1 by a sine function with the Levenberg-Marquardt algorithm [17]. The amplitude A and phase ϕ (a position of the maximum) were obtained from this approximation: A=0.0013±0.0003, ϕ =70±10 degree. For the value of amplitude the factor k=2/π (the difference between amplitude of the real dipole and the one obtained by the approximation) was taking into account. Uncertainties are the results of this approximation. It should be noticed that the obtained results are in good agreement with Ice-Cube experimental data [6].

Acknowledgments
We acknowledge support from The Italian Space Agency (ASI), Deutsches Zentrum fur Luftund Raumfahrt (DLR), The Swedish National Space Board, The Swedish Research Council, The Russian Space Agency (Roscosmos). This work has been done with support of the Russian Science Foundation grant N 14-12-00373 and the RF President grant MK-4599.2014.2.

References

[1] Guillian G, Hosaka J and Ishihara K 2005 *29th ICRC, Pune, India* **6** 85
[2] Amenomori M, Ayabe S and Cui S W 2005 *Astrophys. J.*. **626** 29
[3] Abdo A A *et al.* 2009 *Astrophys. J.* **698** 2121
[4] Abbasi R *et al.* 2011 *Astrophys. J.* **740** 16
[5] BenZvi S 2014 *ASTRA Proc.* **1** 33
[6] Santander M 2013 *30rd ICRC, Rio de Janeiro, Brazil*
[7] Lidvansky A S *et al.* 2007 *30th ICRC, Merida, Mexico*
[8] *arXiv*:1505.02720 [astro-ph.HE]
[9] Adriani O *et al.* 2012 *JETP Letters* **96** 10 621
[10] Adriani O *et al.* 2013 *Phys. Rev. Lett.* **111** 081102
[11] Aguilar M *et al.* 2013 *Phys. Rev. Lett.* **110** 141102
[12] AMS-02 Collaboration, talks given at the AMS Days at CERN, April 15-17, 2015
[13] Adriani O *et al.* 2014 *Physics Reports* **544** 4 323
[14] Borisov S V *et al.* 2013 *Physics Reports* **56** 1 1
[15] Karelin A V *et al.* 2013 *J. Phys.: Conf. Ser.* **409** 012029
[16] URL: *http://healpix.sourceforge.net/documentation.php*
[17] Levenberg K 1944 *Quarterly of Applied Mathematics* **2** 164

The May 17, 2012 solar event: back-tracing analysis and flux reconstruction with PAMELA

A Bruno[1,25], O Adriani[2,3], G C Barbarino[4,5], G A Bazilevskaya[6], R Bellotti[1,7], M Boezio[8], E A. Bogomolov[9], M Bongi[2,3], V Bonvicini[8], S Bottai[3], U Bravar[10], F Cafagna[7], D Campana[5], R Carbone[8], P Carlson[11], M Casolino[12,13], G Castellini[14], E C Christian[15], C De Donato[12,17], G A de Nolfo[15], C De Santis[12,17], N De Simone[12], V Di Felice[12,18], V Formato[8,19], A M Galper[16], A V Karelin[16], S V Koldashov[16], S Koldobskiy[16], S Y Krutkov[9], A N Kvashnin[6], M Lee[10], A Leonov[16], V Malakhov[16], L Marcelli[12,17], M Martucci[17,20], A G Mayorov[16], W Menn[21], M Mergè[12,17], V V Mikhailov[16], E Mocchiutti[8], A Monaco[1,7], N Mori[2,3], R Munini[8,19], G Osteria[5], F Palma[12,17], B Panico[5], P Papini[3], M Pearce[11], P Picozza[12,17], M Ricci[20], S B Ricciarini[3,14], J M Ryan[10], R Sarkar[22,23], V Scotti[4,5], M Simon[21], R Sparvoli[12,17], P Spillantini[2,3], S Stochaj[24], Y I Stozhkov[6], A Vacchi[8], E Vannuccini[3], G I Vasilyev[9], S A Voronov[16], Y T Yurkin[16], G Zampa[8], N Zampa[8] and V G Zverev[16]

[1] Department of Physics, University of Bari "Aldo Moro", I-70126 Bari, Italy
[2] Department of Physics and Astronomy, University of Florence, I-50019 Sesto Fiorentino, Florence, Italy
[3] INFN, Sezione di Florence, I-50019 Sesto Fiorentino, Florence, Italy
[4] Department of Physics, University of Naples "Federico II", I-80126 Naples, Italy
[5] INFN, Sezione di Naples, I-80126 Naples, Italy
[6] P.N. Lebedev Physical Institute of the Russian Academy of Sciences, RU-119991 Moscow, Russia
[7] INFN, Sezione di Bari, I-70126 Bari, Italy
[8] INFN, Sezione di Trieste, I-34149 Trieste, Italy
[9] Ioffe Physical Technical Institute, RU-194021 St. Petersburg, Russia
[10] Space Science Center, University of New Hampshire, Durham, NH, USA
[11] KTH, Department of Physics, and the Oskar Klein Centre for Cosmoparticle Physics, AlbaNova University Centre, SE-10691 Stockholm, Sweden
[12] INFN, Sezione di Rome "Tor Vergata", I-00133 Rome, Italy
[13] RIKEN, Advanced Science Institute, Wako-shi, Saitama, Japan
[14] IFAC, I-50019 Sesto Fiorentino, Florence, Italy
[15] Heliophysics Division, NASA Goddard Space Flight Ctr, Greenbelt, MD, USA
[16] National Research Nuclear University MEPhI (Moscow Engineering Physics Institute), Kashirskoye highway 31, RU-115409 Moscow, Russia
[17] Department of Physics, University of Rome "Tor Vergata", I-00133 Rome, Italy
[18] Agenzia Spaziale Italiana (ASI) Science Data Center, I-00133 Rome, Italy
[19] Department of Physics, University of Trieste, I-34147 Trieste, Italy
[20] INFN, Laboratori Nazionali di Frascati, I-00044 Frascati, Italy

[25] To whom any correspondence should be addressed.

[21] Department of Physics, Universität Siegen, D-57068 Siegen, Germany
[22] Indian Centre for Space Physics, 43 Chalantika, Kolkata 700084, West Bengal, India
[23] Previously at INFN, Sezione di Trieste, I-34149 Trieste, Italy
[24] Electrical and Computer Engineering, New Mexico State University, Las Cruces, NM, USA

E-mail: alessandro.bruno@ba.infn.it

Abstract. The PAMELA space experiment is providing first direct observations of Solar Energetic Particles (SEPs) with energies from about 80 MeV to several GeV in near-Earth orbit, bridging the low energy measurements by other spacecrafts and the Ground Level Enhancement (GLE) data by the worldwide network of neutron monitors. Its unique observational capabilities include the possibility of measuring the flux angular distribution and thus investigating possible anisotropies associated to SEP events. The analysis is supported by an accurate back-tracing simulation based on a realistic description of the Earth's magnetosphere, which is exploited to estimate the SEP energy spectra as a function of the asymptotic direction of arrival with respect to the Interplanetary Magnetic Field (IMF). In this work we report the results for the May 17, 2012 event.

1. Introduction

The PAMELA satellite experiment, operating since 2006 June, is providing precise and detailed observations of the cosmic-ray radiation environment in low Earth orbit [1]. In particular, the instrument offers the rare opportunity to directly measure the SEP energy spectra and angular distribution from 80 MeV to several GeV during solar cycles 23 and 24. Recently, PAMELA reported its measurement of the May 17, 2012 event, associated to to first GLE of the new solar cycle [2]. This paper presents the results of the SEP analysis with the emphasis on the methods developed to estimate the fluxes as a function of the asymptotic arrival direction.

2. The PAMELA Experiment

PAMELA (a Payload for Antimatter Matter Exploration and Light-nuclei Astrophysics) is a space-borne experiment designed for a precise measurement of the charged cosmic radiation in the kinetic energy range from some tens of MeV up to several hundreds of GeV [3]. The Resurs-DK1 satellite, which hosts the apparatus, was launched into a semi-polar (70 deg inclination) and elliptical (350÷610 km altitude) orbit on June 15, 2006; in 2010 it was changed to an approximately circular orbit at an altitude of about 580 km. The spacecraft is 3-axis stabilized; its orientation is calculated by an onboard processor with an accuracy better than 1 deg. Particle directions are measured with a high angular resolution ($<$ 2 deg). Details about apparatus performance, proton selection, detector efficiencies and measurement uncertainties can be found elsewhere (e.g. [6]).

3. Geomagnetic Field Models

The analysis of SEP events reported in this work is based on the IGRF-11 [7] and the TS07D [8, 9] models for the description of the internal and external geomagnetic field sources, respectively. The TS07D is a high resolution dynamical model of the storm-time geomagnetic field in the inner magnetosphere, based on recent satellite measurements. Consistent with the data-set coverage, the model is valid up to about 30 Earth's radii (Re). Solar wind and IMF parameters are obtained from the high resolution (5-min) Omniweb database [10].

4. Trajectory Analysis

Using spacecraft ephemeris data (position, orientation, time), and the particle rigidity (R = momentum/charge) and direction provided by the PAMELA tracking system, trajectories of all detected protons were reconstructed in the Earth's magnetosphere by means of an accurate tracing program [4] based on the aforementioned geomagnetic models. The asymptotic arrival directions were evaluated with respect to the IMF direction, with polar angles α and β denoting the pitch-angle and the gyro-phase angle, respectively.

5. Flux Evaluation

Fluxes were obtained by averaging data over single polar passes. In order to account for possible anisotropies, SEP intensities were evaluated as a function of particle rigidity R and asymptotic pitch-angle α:

$$\Phi(R,\alpha) = \frac{N(R,\alpha)}{2\pi \int\limits_{\Delta R} dR \int\limits_{\Delta\alpha} d\alpha \int\limits_{T} dt H(R,\alpha,t)}, \tag{1}$$

where N is the number of proton counts corrected by selection efficiencies, T is the livetime and H is the apparatus *effective area* (cm^2):

$$H(R,\alpha) = \frac{sin\alpha}{2\pi} \int_0^{2\pi} d\beta \left[A(R,\theta,\phi) \cdot cos\theta\right], \tag{2}$$

where A is the apparatus response function, and θ and ϕ are the polar angles denoting the particle direction in the PAMELA frame [5]. The used approach is analogous to the one developed for the measurement of geomagnetically trapped protons [11], but in this case the transformation between local (θ,ϕ) and magnetic (α,β) angles can not be obtained by simple trigonometric operations since it depends on particle propagation in the geomagnetic field; thus, back-tracing methods were exploited. To assure a high resolution, ~2800 trajectories (uniformly distributed inside PAMELA field of view - FoV) were reconstructed in the magnetosphere for 1-sec time steps along the satellite orbit and 22 rigidity values between 0.39÷4.09 GV. At a later stage, results were extended over the full FoV through a bilinear interpolation. Since the PAMELA semi-aperture is ~20 deg, the observable pitch-angle range is relatively small (a few deg) except in regions close to the geomagnetic cutoff, where trajectories become chaotic and corresponding asymptotic directions rapidly change with particle rigidity and looking direction; this ends up increasing measurement uncertainties. Consequently, these zones were excluded from the analysis.

6. Results

Figure 1 reports PAMELA's vertical asymptotic directions of view during the first polar pass that registered the event (01:58 – 2:20 UT), for different values of particle rigidity (color code). Left panels show the reconstructed directions in terms of GEO (top) and GSE (bottom) coordinates, for different rigidity range (color code). The spacecraft position is indicated by the grey curve. The contour curves represent values of constant pitch angle with respect to the IMF direction, denoted with crosses. It can be noted that IMF direction is almost perpendicular to the sunward direction. As PAMELA is moving (eastward) and changing its orientation along the orbit, observed asymptotic directions rapidly vary performing a (clockwise) loop over the region above Brazil. Right panels in the same figure display the distributions of asymptotic directions as a function of UT, and particle rigidity (top) and pitch-angle (bottom); the color codes refer to the corresponding pitch-angle and rigidity values, respectively. Solid curves in the top-right panel denote the estimated Störmer vertical cutoff for the PAMELA epoch (~ $14.3/L^2$ GV).

Preliminary results of PAMELA's differential proton fluxes averaged over the first polar pass are shown in figure 2. The top panel reports the rigidity spectra for different pitch-angle bins (color code); for comparison, the dashed line in the top panel represents the galactic proton background (included in the results) measured by PAMELA during two days prior to the

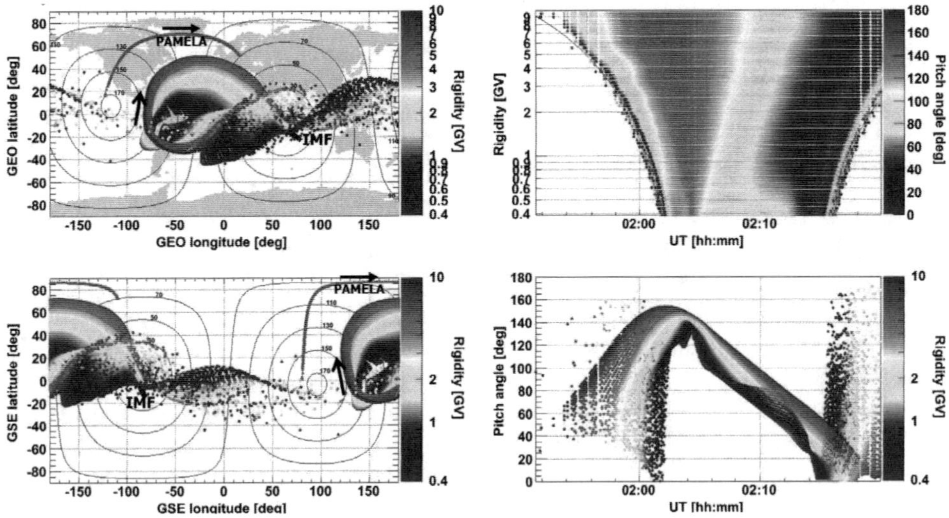

Figure 1. PAMELA's vertical asymptotic directions of view during the first polar pass that registered the event. See the text for details.

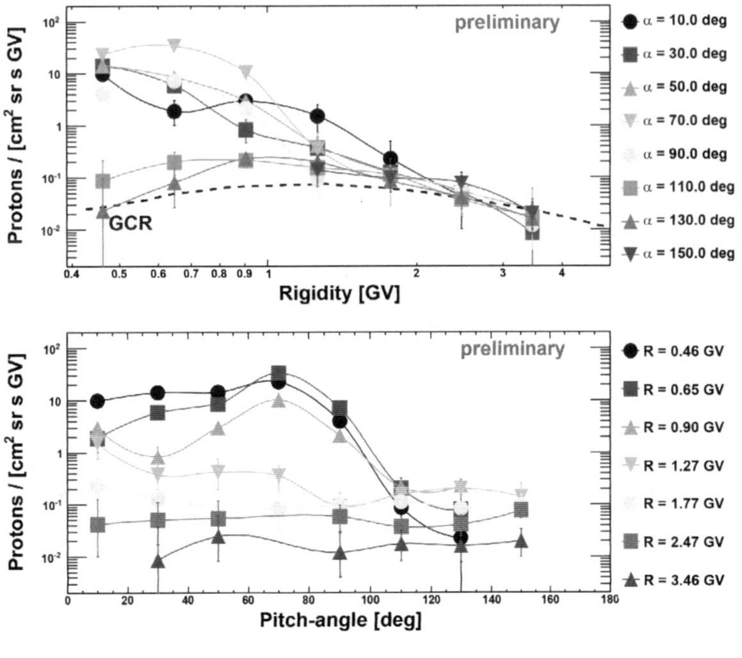

Figure 2. Top: proton rigidity spectra measured by PAMELA during the first polar pass, for different pitch-angle bins (color code). Bottom: corresponding pitch-angle profiles, for different rigidity bins (color code).

SEP arrival, which was estimated to be isotropic. The bottom panel displays corresponding pitch-angle profiles, for different rigidity bins (color code). The vertical error bars include only statistical uncertainties.

Two populations with very different pitch angle distributions can be noted: a low-energy component ($\lesssim$ 1 GV) confined to pitch angles <90 deg and exhibiting significant scattering or redistribution; and a high-energy component (1÷2 GV) that is beamed with pitch angles <30 deg and relatively unaffected by dispersive transport effects, consistent with neutron monitor observations. The presence of these simultaneous populations can be explained by postulating a local scattering/redistribution in the Earth's magnetosheath, as discussed in [2].

7. Conclusions

The PAMELA space experiment is providing detailed measurements of the SEP fluxes with energies from about 80 MeV to several GeV in near-Earth orbit, bridging the low energy data by other spacecrafts and the GLE data by the worldwide network of neutron monitors. Its unique observational capabilities include the possibility of measuring the SEP angular distribution. The analysis is supported by an accurate back-tracing simulation based on a realistic modeling of the Earth's magnetosphere. The results of the calculation for the May 17, 2012 event are discussed in this paper. The trajectory analysis enables the investigation of flux anisotropies, providing fundamental information for the characterization of SEPs. It will prove to be a vital ingredient for the interpretation of the solar events observed by PAMELA during solar cycles 23 and 24.

References

[1] Adriani O *et al.* 2014 Physics Reports **544**(4) 323–70 (*DOI:10.1016/j.physrep.2014.06.003*)
[2] Adriani O *et al.* 2015 PAMELA's Measurements of Magnetospheric Effects on High Energy Solar Particles *ApJ* **801**(1) L3 (*DOI:10.1088/2041-8205/801/1/L3*)
[3] Picozza P *et al.* 2007 *Astropart. Phys.* **27** 296–315 (*DOI:10.1016/j.astropartphys.2006.12.002*)
[4] Bruno A *et al.* 2015 Solar energetic particle events: trajectory analysis and flux reconstruction with PAMELA *PoS(ICRC2015)085*
[5] Bruno A *et al.* 2014 Back-tracing and flux reconstruction for solar events with PAMELA *arXiv*:1412.1765
[6] Adriani O *et al.* 2013 Time dependence of the proton flux measured by PAMELA during the 2006 July – 2009 December solar minimum *ApJ* 765(2) 91 (*DOI:10.1088/0004-637X/765/2/91*)
[7] Finlay C C *et al.* 2010 International Geomagnetic Reference Field: the eleventh generation *Geophysical Journal International* **183** 1216–30
[8] Tsyganenko N A and Sitnov M I 2007 Magnetospheric configurations from a high-resolution data-based magnetic field model *J. Geophys. Res.* **112** A06225
[9] Sitnov M I *et al.* 2008 Dynamical data-based modeling of the storm-time geomagnetic field with enhanced spatial resolution *J. Geophys. Res.* **113** A07218
[10] King J H and Papitashvili N E 2004 Solar wind spatial scales in and comparisons of hourly Wind and ACE plasma and magnetic field data *J. Geophys. Res.* **110**(A2) A02209 (*http://omniweb.gsfc.nasa.gov/*)
[11] Adriani O *et al.* 2015 Trapped proton fluxes at low Earth orbits measured by the PAMELA experiment *ApJ* **799**(1) L4 (*DOI:10.1088/2041-8205/799/1/L4*)

International Conference on Particle Physics and Astrophysics (ICPPA-2015) IOP Publishing
Journal of Physics: Conference Series **675** (2016) 032007 doi:10.1088/1742-6596/675/3/032007

Features of re-entrant albedo deuteron trajectories in near Earth orbit with PAMELA experiment

S A Koldobskiy[12,20], O Adriani[1,2], G C Barbarino[3,4], G A Bazilevskaya[5], R Bellotti[6,7], M Boezio[8], E A Bogomolov[9], M Bongi[1,2], V Bonvicini[8], S Bottai[2], A Bruno[6,7], F Cafagna[7], D Campana[4], P Carlson[13], M Casolino[11], G Castellini[14], C De Donato[10], C De Santis[11], N De Simone[10], V Di Felice[10], V Formato[8,†], A M Galper[12], A V Karelin[12], S V Koldashov[12], S Y Krutkov[9], A N Kvashnin[5], AA Leonov[12], V Malakhov[12], L Marcelli[11], M Martucci[11,15], A G Mayorov[12], W Menn[16], M Mergè[10,11], V V Mikhailov[12], E Mocchiutti[8], A Monaco[6,7], N Mori[2], R Munini[8,17], G Osteria[4], F Palma[10,11], B Panico[4], P Papini[2], M Pearce[13], P Picozza[10,11], M Ricci[15], S B Ricciarini[14], R Sarkar[18,††], V Scotti[3,4], M Simon[16], R Sparvoli[10,11], P Spillantini[1,2], Y I Stozhkov[5], A Vacchi[8,19], E Vannuccini[2], GI Vasilyev[9], S A Voronov[12], Y T Yurkin[12], G Zampa[8] and N Zampa[8]

[1]University of Florence, Department of Physics, I-50019 Sesto Fiorentino, Florence, Italy
[2]INFN, Sezione di Florence, I-50019 Sesto Fiorentino, Florence, Italy
[3]University of Naples "Federico II", Department of Physics, I-80126 Naples, Italy
[4]INFN, Sezione di Naples, I-80126 Naples, Italy
[5]Lebedev Physical Institute, RU-119991, Moscow, Russia
[6]University of Bari, Department of Physics, I-70126 Bari, Italy
[7]INFN, Sezione di Bari, I-70126 Bari, Italy
[8]INFN, Sezione di Trieste, I-34149 Trieste, Italy
[9]Ioffe Physical Technical Institute, RU-194021 St. Petersburg, Russia
[10]INFN, Sezione di Rome "Tor Vergata", I-00133 Rome, Italy
[11]University of Rome "Tor Vergata", Department of Physics, I-00133 Rome, Italy
[12] National Research Nuclear University MEPhI (Moscow Engineering Physics Institute), Kashirskoe highway 31, Moscow, 115409, Russia
[13]KTH, Department of Physics, and the Oskar Klein Centre for Cosmoparticle Physics, AlbaNova University Centre, SE-10691 Stockholm, Sweden
[14]IFAC, I-50019 Sesto Fiorentino, Florence, Italy
[15]INFN, Laboratori Nazionali di Frascati, Via Enrico Fermi 40, I-00044 Frascati, Italy
[16]Universität Siegen, Department of Physics, D-57068 Siegen, Germany
[17]University of Trieste, Department of Physics, I-34147 Trieste, Italy
[18]Indian Centre for Space Physics, 43 Chalantika, Garia Station Road, Kolkata 700084, West Bengal, India
[19]University of Udine, Department of Mathematics and Informatics, I-33100 Udine, Italy
[†]Now at INFN, Sezione di Perugia, I-06123 Perugia, Italy
[††]Previously at INFN, Sezione di Trieste, I-34149 Trieste, Italy

E-mail: `sakoldobskiy@mephi.ru`

[20] To whom any correspondence should be addressed.

Abstract. The results of trajectory reconstruction for re-entrant albedo deuteron fluxes obtained in the PAMELA experiment are presented in this work. PAMELA is an international experiment aimed on measurements of cosmic ray particle fluxes in wide energy range. In particular, analysis of PAMELA data gives possibility to identify deuterons. Classification of re-entrant albedo deuterons with energies from 70 to 400 MeV/nucleon depending on theirs reconstructed lifetimes and generation zones is presented here at first time.

1. Introduction

The particles registered in near Earth orbit can be attributed to galactic cosmic ray (GCR) particles generated in astrophysical sources, albedo particles originated in interactions of high energy GCR with the nuclei of the residual atmosphere and radiation belt particles trapped by Earth's magnetic field. For the first time the registration of deuteron albedo radiation was mentioned in [1, 2]. The results of the unique attempt to measure the spectrum of deuterons under radiation belt in the AMS-01 experiment were presented in [3]. The last and the most accurate measurements were implemented in PAMELA experiment [4]. PAMELA instrument was installed on board Resurs DK-1 satellite, which was launched on June 15, 2006. Detailed description of the PAMELA experiment can be found in [5, 6, 7, 8]. Here we present the original and unique study of albedo deuteron generation zones.

2. Particle identification

For the identification of detected particle nature, as well as the recovery of its energy the multivariate correlation analysis of the signals from different detectors was used [4, 8, 9]. This identification method was updated because trajectory reconstruction in Earth's magnetic field requires reliable determination of each deuteron event. If the type of particle will be determined wrong, the trajectory will be reconstructed wrong as well. So only deuterons with kinetic energy from 70 to 400 MeV/nucleon were selected from full data sample. Nevertheless it is impossible to reject secondary deuterons born in the spectrometer material by nuclear interactions with incoming CR protons or helium so from 3% to 10% of selected events are protons.

Measured magnetic rigidities were corrected for the energy losses in ToF and MC, evaluated with Monte-Carlo GEANT4 [10] simulations of the particle interaction with instrument material.

3. Deuteron trajectory reconstruction routine

Trajectory reconstruction was performed solving equations of particle motion in Earth's magnetic field with numerical integration methods [11], using the geomagnetic field model International Geomagnetic Reference Field (IGRF) [12]. Maximum value of drift time was chosen to be 10 seconds.

Deuteron trajectories were reconstructed back from the registration point until they intersected the boundaries of "dense" atmosphere, which was assumed at an altitude of 30 km. These particles were classified as re-entrant albedo deuterons.

4. Trajectory reconstruction results

Data sample from July 2006 to January 2008 was used and more than 20.000 re-entrant deuteron events with energies from 100 to 400 MeV/nucleon were selected.

In the figure 1 the calculated lifetimes for albedo deuterons are presented. It's clear that about 90% of them have lifetimes less than 2 seconds before their registration, it's "splash" albedo particles. Let's name that these particles are short-lived. In contrary, long-lived deuterons with lifetimes more than 2 seconds constitute about 10% of albedo deuteron population.

Journal of Physics: Conference Series **675** (2016) 032007 doi:10.1088/1742-6596/675/3/032007

In the figure 2 the coordinates of re-entrant albedo deuteron generation zones depending from their reconstructed lifetime and measured energies are given. Only number of registered particles is shown, because it's not possible to calculate spectra due to low statistics for different pitch-angles energies. It is shown that re-entrant albedo deuterons are localized in different geomagnetic zones depending from their measured and reconstructed attributes. Short-lived particles are generated uniformly almost in broad longitude and latitude intervals excluding zone of South Atlantic Magnetic Anomaly (SAA) and polar regions.

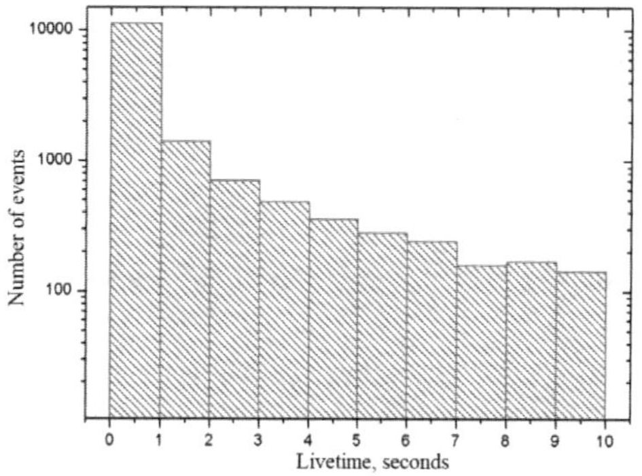

Figure 1. Lifetime distribution for re-entrant albedo deuterons.

But if we consider populations of particles with bigger lifetimes then we will find that these particles were not born uniformly. Instead there are two zones of their generation: Zone I in the East of SAA and Zone II near the polar regions.

Another interesting fact is that these particles belong to different sub-populations of re-entrant albedo deuterons, just like for protons registered in PAMELA experiment [13].

Long-lived deuterons born in Zone I are an "quasitrapped" particles, these particles have features of their motion in Earth's magnetosphere very close to radiation belt particles. It is clear from figure 2 that there are an longitudinal and energetic dependences for quasitrapped deuterons. If we take bigger lifetimes and bigger energies these particles will be born in zones closer to SAA. Long-lived deuterons born in Zone II are an "pseudotrapped" particles, they are born in zone of geomagnetic penumbra and their magnetic rigidities are close to cutoff ones [14]. These particles have complicated trajectories in Earth's magnetic field.

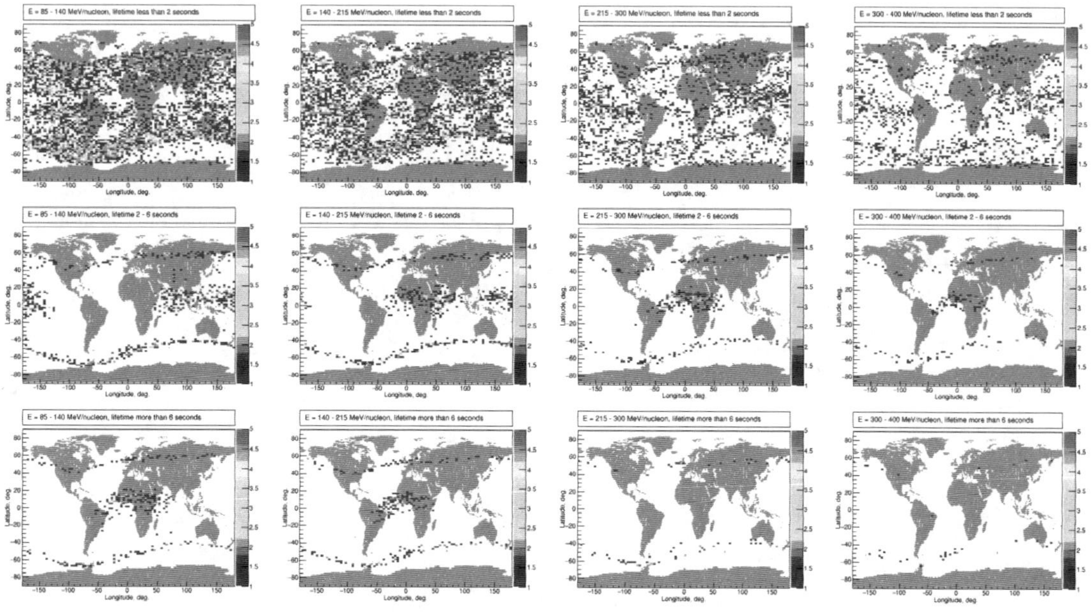

Figure 2. Coordinates of re-entrant albedo deuteron generation zones depending on their reconstructed lifetime and measured energies.

5. Conclusion

As a result of this work the trajectory reconstruction for re-entrant albedo deuterons with kinetic energies from 70 to 400 MeV/nucleon was performed. Three populations of albedo particles with completely different character of motion in Earth's magnetic field were completely distinguished.

Acknowledgments

This work performed within the framework of the Center FRPP supported by MEPhI Academic Excellence Project (contract 02.03.21.0005, 27.08.2013) with financial support of the Russian Science Foundation (grant 14-12-00373).

References

[1] Looper M D, Blake J B and Mewaldt R A 1998 *Adv. Space Res.* **21**(12) 1679
[2] Looper M D *et al.* 1996 *Rad. Meas.* **26** 967
[3] Lamanna G *et al.* 2001 *Proc. of the 27th Intern. Cosmic Ray Conf.* 1614
[4] Koldobskiy S A *et al.* 2015 *Bull. Russ. Acad. Sci., Phys.* **79**(3) 294
[5] 5 Picozza P *et al.* 2007 *Astropart. Phys.* **27** 296
[6] 6 Ricciarini S *et al.* 2008 *Nucl. Instrum. Methods Phys. Res. A* **582** 892
[7] 7 Papini P *et al.* 2008 *Nucl. Instrum. Methods Phys. Res. A* **588** 259
[8] 8 Voronov S A, Danilchenko I A and Koldobskiy S A 2011 *Instr. And Exp. Tech.* **54** 752
[9] Adriani O *et al.* 2013 *Astrophys. J.* **770**(1) 2
[10] Agostinelli S *et al.* 2003 *Nucl. Instrum. Meth. A* **506** 250
[11] Smart D F and Shea M A 2005 *Adv. Space Res.* **36** 2012
[12] Finlay C C *et al.* 2010 *Geophys. J. I.* **183** 1216
[13] Adriani O, Barbarino G C, Bazilevskaya G A *et al.* 2015 *J. Geophys. Res.-Space* **120**(5) 3728
[14] Hofmann D J and Sauer H H 1968 *Space Sci. Rev.* **5-6**(8) 750

Journal of Physics: Conference Series **675** (2016) 032008 doi:10.1088/1742-6596/675/3/032008

Observation of global electromagnetic resonances by low-orbiting satellites

V V Surkov
National Research Nuclear University MEPhI (Moscow Engineering Physics Institute), Kashirskoe highway 31, Moscow, 115409, Russia

E-mail: surkovvadim@yandex.ru

Abstract. Penetration of Schumann resonances energy from the Earth-ionosphere resonance cavity into the circumterrestrial space is examined. This study focuses on estimates of Alfvén wave amplitude and spectra in the frequency range of 7–50 Hz which can be observed by low-orbiting satellites. Differences in Schumann resonances observation conditions between the nighttime and sunlit sides of the ionosphere are analyzed. Particular emphasis has been placed on the ionospheric Alfvén resonator (IAR) excited by both global thunderstorm activity and individual lightning discharges. IAR spectra in the frequency range of 0.5–10 Hz are calculated for ionospheric altitudes. The calculated spectral amplitudes of IAR and Schumann resonances are compatible with C/NOFS satellite observations. To explain a shift of IAR resonant frequencies observed during C/NOFS satellite passage through terminator region, the IAR model is developed in which an interference of Alfvén waves reflected from the ionospheric E-layer and the IAR upper boundary is taken into account.

1. Introduction

The electromagnetic energy stemming from the global lightning activity of the planet is believed to be the major source for excitation of Schumann resonances (SR) in the Earth-ionosphere resonance cavity. The SR signature can be detected in the frequency range $7-50$ Hz from any place of the Earth away from thunderstorm and man-made electromagnetic sources [1]. The resonant energy transfer from the world-wide electric storms and magnetospheric-ionospheric plasma convective flow can be also accumulated in ionospheric Alfvén resonator (IAR). The lower boundary of IAR coincides with the high-conducting ionospheric E-layer while the IAR upper boundary is due to Alfvén wave reflection from the strong gradient of background plasma density above the maximum of the F-layer [2]. Typically the ground-based observations of the IAR signatures exhibit the spectral resonance structure (SRS) in the frequency range $0.5-5$ Hz.

Recent ELF electric field measurements onboard low-orbiting C/NOFS satellite revealed a distinct picture of the first five SR [3]. Moreover several spectrograms of the electric field meridional and zonal components with "fingerprint" pattern manifest itself similar to typical IAR SRS, although these spectral peaks lie in the frequency band ~1–15 Hz [4]. These abnormal spectra were detected by the C/NOFS satellite in the near-equatorial region near the terminator. A possibility for the SR energy penetration from the Earth-ionosphere resonant cavity into the upper ionosphere has been recently examined by Surkov et al. [5]. The feasibility of the IAR signature detection by low-orbiting satellites with magnetic or electric sensors has been discussed by Plyasov et al. [6].

2. Estimates of Schumann resonance amplitudes in the ionosphere

In order to compare the magnitudes of SR at the ionospheric altitudes and on the Earth surface, we use the simplified plane stratified model which consist of the infinitely conductive ground $(z<-h)$, neutral atmosphere $(-h<z<-h_0)$, conducting atmosphere $(-h_0<z<0)$, conducting gyrotropic E-layer of the ionosphere $(0<z<l)$ and the F-layer $(l<z)$. In this model the F-layer is the semi-space filled with cold collisionless plasma. The origin of the reference frame is placed at the interface between an ionospheric E-layer and the atmosphere, z-axis is vertically upward, x axis is directed from West to East, and y-axis is directed from South to North. At first the individual cloud to ground (CG) lightning discharge flowing vertically upward is considered as a primary source for the electromagnetic perturbations.

To treat the electric and magnetic fields in the E-layer, Maxwell's equations are needed, which, in their full form, are given by

$$\nabla\times\mathbf{B}=\mu_0\left(\sigma_{\parallel}\mathbf{E}_{\parallel}+\sigma_P\frac{\mathbf{B}_0\times(\mathbf{E}_{\perp}\times\mathbf{B}_0)}{B_0^2}+\sigma_H\frac{\mathbf{B}_0\times\mathbf{E}_{\perp}}{B_0}\right) \tag{1}$$

$$\nabla\times\mathbf{E}=i\omega\mathbf{B} \tag{2}$$

Here $\mathbf{B}_0$ is the unperturbed geomagnetic field; $E_{\parallel}$ $\mathbf{E}_{\parallel}$ and $\mathbf{E}_{\perp}$ denote components of electric field parallel and transverse to $\mathbf{B}_0$, respectively; $\sigma_{\parallel}$, σ_P, and σ_H are parallel, Pedersen, and Hall conductivities of the ionospheric plasma, respectively. All electromagnetic fields is assumed to vary with time as $\exp(-i\omega t)$, where ω denotes a frequency.

In the F-layer, instead of equation (1) we have to use the following:

$$\nabla\times\mathbf{B}=-\frac{i\omega}{c^2}\left(\varepsilon_{\parallel}\mathbf{E}_{\parallel}+\varepsilon_{\perp}\frac{\mathbf{B}_0\times(\mathbf{E}_{\perp}\times\mathbf{B}_0)}{B_0^2}\right) \tag{3}$$

where $\varepsilon_{\parallel}$ and $\varepsilon_{\perp}=(c/V_A)^2$ are parallel and transverse components of the ionospheric plasma permittivity tensor, respectively; V_A is Alfvén wave velocity and c denotes a speed of light in vacuum.

In what follows we study axially symmetric problem assuming that the geomagnetic field $\mathbf{B}_0$ is vertical. Due to the axial symmetry of this model it is suitable to search for the solution of equations (1)–(3) with the proper boundary conditions in the form of Bessel transform with respect to polar radius for all field components. It should be noted that on the ground the azimuthal component b_φ excited by a lightning discharge is usually much greater than the radial component b_r. Analysis of this problem have shown that the ratio of azimuthal magnetic perturbations in ionosphere and on the ground can be roughly estimated through the height-integrated Pedersen and Hall conductivities, Σ_P and Σ_H, as follows [5]:

$$n_b=\left|\frac{b_\varphi(0)}{b_\varphi(-h)}\right|=\left|\frac{i+(1+\alpha_P)k_A h}{1+\alpha_P-ik_A h\left\{(1+\alpha_P)^2+\alpha_H^2\right\}}\right| \tag{4}$$

where $\alpha_P = (\Sigma_P + \Sigma_a)/\Sigma_w$, $\alpha_H = \Sigma_H/\Sigma_w$, $\Sigma_w = (\mu_0 V_A)^{-1}$ is the Alfvén wave conductance, Σ_a is the height-integrated atmospheric conductivity and $k_A = \omega/V_A$ denotes the Alfvén wave number. Substituting the nighttime atmospheric and ionospheric parameters: $\Sigma_a = 0.15$ S, $\Sigma_P = 0.4$ S, $\Sigma_H = 0.6$ S, and $h = 50$ km into equation (4), gives $n_b \approx 0.5$. For daytime conditions ($\Sigma_a = 1.5$ S, $\Sigma_P = 5$ S, and $\Sigma_H = 7.5$ S) this estimate is on one order of magnitude smaller than the above one.

Analogous ratio of the radial electric field in the ionosphere and on the ground is given by

$$\left|\frac{e_r(0)}{e_z(-h)}\right| \approx \frac{V_A(1+\alpha_P)}{c\left[(1+\alpha_P)^2 + \alpha_H^2\right]} \tag{5}$$

Whence it follows that this ratio is about 2×10^{-3} for the night-time ionosphere. Although this estimate decreases by one order of magnitude for the daytime conditions. The C/NOFS satellite measurements have shown that the electric field amplitude of the first SR in the ionosphere is ~ 0.3 μV/(m Hz$^{1/2}$), that is $3-4$ orders lower than the amplitude of the SR vertical electric component $e_z(-h)$ on the ground. Thus, our estimates of SR amplitudes in the ionosphere are in a good agreement with the C/NOFS satellite observations.

3. Analysis of IAR spectra observed by low-orbiting satellite

A simplified plane-stratified model of IAR has been used to study electromagnetic perturbation inside the IAR. This model is different from that used above by incorporating upper IAR boundary at height about 10^3 km. To model the increase in Alfvén wave velocity with height, we assume that the Alfvén velocity has a jump across the upper IAR boundary. The Hall conductivity of the ionospheric E-layer leads to the coupling between shear Alfvén and fast magnetosonic modes inside the IAR. The major source for IAR excitation is supposed to be either a separate intense lightning stroke or stochastic global thunderstorm activity.

In the framework of the above model the ratio between the magnetic disturbance inside IAR and on the ground can be roughly estimated as $B_{\mathrm{IAR}}/B_{\mathrm{g}} = \omega h V_{AM}/V_{AI}^2$ [6], where V_{AI} and V_{AM} are Alfvén wave velocities in the IAR and the magnetosphere, respectively. According to this estimate the peak disturbance inside the IAR can be an order of magnitude larger than that in the atmosphere. In the frequency range of interest, i.e. $0.5-5$ Hz, the spectral amplitudes of electric power spectra are estimated as $W_E \sim 0.1-0.2$ μV/m·Hz$^{1/2}$. This value is thus consistent with the observations during nighttime conditions since the electric field power spectra detected by C/NOFS satellite at the altitude range of $400-850$ km were $\sim 0.05-0.1$ μV/m·Hz$^{1/2}$ [3].

However the dynamic spectra recently observed by the C/NOFS satellite can hardly be interpreted on the basis of standard IAR theory because these spectra were shifted to higher frequencies (up to 15 Hz) as compared with typical SRS on the ground. To study this effect in more detail, we develop a simple one-dimensional model of the field-aligned Alfvén wave propagation inside the IAR. Analysis of this problem has shown that interference between the primary Alfvén wave and the waves reflected from IAR boundaries can greatly affect the power spectra of electromagnetic perturbations detected by satellite. Let H be the length of field line piece bounded by the resonator boundaries and z be the satellite coordinate measured along the field line. The interference effect results in "modulation" of power spectra in such a way that the spectra have a quasi-oscillatory shape with "periods" $f_1 = V_{AI}/(2z)$ or $f_2 = V_{AI}/\left[2(H-z)\right]$ depending on the satellite coordinate z and the direction of the primary Alfvén wave propagation. The implication here is that f_1 and f_2 are equal to the reciprocal propagation delay time between the primary and reflected Alfvén waves. Both these frequencies are greater than the fundamental IAR frequency $f_0 = V_{AI}/(2H) \approx 0.5-1$ Hz.

Model simulations have demonstrated that the interference effect can mask the spectral peaks due to excitation of the IAR eigenoscillations. Moreover, the changes in coordinate z due to the satellite movement along orbit can result in temporal variations of the power spectra obtained on the basis of Fast Fourier Transform (FFT) technique. To explain the increase of the spectral peak frequencies detected by C/NOFS satellite when the satellite descended from 650 to 450 km altitude [3], we suppose that the distance (measured along the field line) between the satellite and the ionospheric E-layer; that is the IAR lower boundary, decreased with time thereby producing both the increase in f_1 and the shift of spectral peaks to the higher frequencies.

4. Conclusion

The above theoretical analysis is indicative of the feasibility of detection of both Schumann resonances and IAR signatures in space by modern magnetometers and electric field sensors onboard low-orbiting satellites. However it seems likely that the observed "fingerprint" spectrograms resembling SRS are not typical for IAR eigenmodes because these spectrograms are subjected to the interference effects between primary and reflected Alfvén waves propagating in IAR. The observation onboard low-orbiting satellite can thus provide us with additional information about ULF/ELF electromagnetic fields and noises at the ionospheric altitude.

Acknowledgements

This study was supported by the grant number 13-05-12091 from RFBR and by MEPhI Academic Excellence Project (contract No. 02.a03.21.0005, 27.08.2013).

References

[1] Nickolaenko A P and Hayakawa M 2002 *Resonances in the Earth-Ionosphere Cavity* (Netherlands: Kluwer Acad. Dordrecht)

[2] Surkov V and Hayakawa M 2014 *Ultra and Extremely Low Frequency Electromagnetic Fields* (Springer Geophysics Series XVI Springer) 486 p

[3] Simões F A, Pfaff R F, and Freudenreich H T 2011 Satellite observations of Schumann resonances in the Earth's ionosphere *Geophys. Res. Lett.* **38**(22) L22101 *(doi:10.1029/2011GL049668)*

[4] Simões F *et al.* 2012 Detection of ionospheric Alfvén resonator signatures in the equatorial ionosphere *J. Geophys. Res.* **117** A11305 *(doi:10.1029/2012JA017709)*

[5] Surkov V V *et al.* 2013 Penetration of Schumann resonances into the upper ionosphere *J. Atmos. Solar-Terr. Phys.* **97** 65–74 *(doi: 10.1016/j.jastp.2013.02.015)*

[6] Plyasov A A, Surkov V V, Pilipenko V A, Fedorov E N, and Ignatov V N 2012 Spatial structure of the electromagnetic field inside the ionospheric Alfvén resonator excited by atmospheric lightning activity *J. Geophys. Res.* **117** A09306 *(doi:10.1029/2012JA017577)*

The GAMMA-400 gamma-ray telescope for precision gamma-ray emission investigations

N P Topchiev[1,*], A M Galper[1,2], V Bonvicini[3], O Adriani[4], R L Aptekar[5], I V Arkhangelskaja[2], A I Arkhangelskiy[2], A V Bakaldin[2], L Bergstrom[6], E Berti[4], G Bigongiari[7], S G Bobkov[8], M Boezio[3], E A Bogomolov[5], L Bonechi[7], M Bongi[4], S Bottai[4], G Castellini[9], P W Cattaneo[10], P Cumani[3,16], O D Dalkarov[1], G L Dedenko[2], C De Donato[11], V A Dogiel[1], N Finetti[4], D Gascon[12], M S Gorbunov[8], Yu V Gusakov[1], B I Hnatyk[13], V V Kadilin[2], V A Kaplin[2], A A Kaplun[2], M D Kheymits[2], V E Korepanov[14], J Larsson[15], A A Leonov[1,2], V A Loginov[2], F Longo[3], P Maestro[7], P S Marrocchesi[7], M Martinez[16], A L Men'shenin[17], V V Mikhailov[2], E Mocchiutti[3], A A Moiseev[18], N Mori[4], I V Moskalenko[19], P Yu Naumov[2], P Papini[4], J M Paredes[12], M Pearce[15], P Picozza[11], A Rappoldi[10], S Ricciarini[9], M F Runtso[2], F Ryde[15], O V Serdin[8], R Sparvoli[11], P Spillantini[4], Yu I Stozhkov[1], S I Suchkov[1], A A Taraskin[2], M Tavani[20], A Tiberio[4], E M Tyurin[2], M V Ulanov[5], A Vacchi[3], E Vannuccini[4], G I Vasilyev[5], J E Ward[16], Yu T Yurkin[2], N Zampa[3], V N Zirakashvili[21] and V G Zverev[2]

[1] Lebedev Physical Institute, RU-119991 Moscow, Russia
[2] National Research Nuclear University MEPhI (Moscow Engineering Physics Institute), Kashirskoe highway 31, Moscow, 115409, Russia
[3] Istituto Nazionale di Fisica Nucleare, Sezione di Trieste, I-34149 Trieste, Italy
[4] Istituto Nazionale di Fisica Nucleare, Sezione di Florence, I-50019 Sesto Fiorentino, Florence, Italy
[5] Ioffe Physical Technical Institute, RU-194021 St. Petersburg, Russia
[6] The Oskar Klein Centre, Department of Physics, Stockholm University, AlbaNova University Center, SE-106 91 Stockholm, Sweden
[7] Department of Physical Sciences, Earth and Environment, University of Siena and Istituto Nazionale di Fisica Nucleare, Sezione di Pisa, Italy
[8] Scientific Research Institute for System Analysis, RU-117218 Moscow, Russia
[9] Istituto di Fisica Applicata Nello Carrara, I-50019 Sesto Fiorentino, Florence, Italy
[10] Istituto Nazionale di Fisica Nucleare, Sezione di Pavia, I-27100 Pavia, Italy
[11] Istituto Nazionale di Fisica Nucleare, Sezione di Rome "Tor Vergata", I-00133 Rome, Italy
[12] Departament d'Astronomia i Meteorologia, Institut de Ciències del Cosmos, Universitat de Barcelona, Spain
[13] Taras Shevchenko National University, Kyiv, 01601 Ukraine
[14] Lviv Center of Institute of Space Research, Lviv, 79060 Ukraine
[15] KTH Royal Institute of Technology, Department of Physics and the Oskar Klein Centre, AlbaNova University Center, SE-10691 Stockholm, Sweden
[16] Institut de Física d'Altes Energies, Bellaterra, Spain
[17] Research Institute for Electromechanics, RU-143502 Istra, Moscow region, Russia

[18] NASA Goddard Space Flight Center and CRESST/University of Maryland, Greenbelt, Maryland 20771, USA
[19] Hansen Experimental Physics Laboratory and Kavli Institute for Particle Astrophysics and Cosmology, Stanford University, Stanford, CA 94305, USA
[20] Istituto Nazionale di Astrofisica IASF and Physics Department of University of Rome Tor Vergata, I-00133 Roma, Italy
[21] Pushkov Institute of Terrestrial Magnetism, Ionosphere, and Radiowave Propagation, Troitsk, Moscow region, Russia

E-mail: *tnp51@yandex.ru

Abstract. The GAMMA-400 gamma-ray telescope with excellent angular and energy resolutions is designed to search for signatures of dark matter in the fluxes of gamma-ray emission and electrons + positrons. Precision investigations of gamma-ray emission from Galactic Center, Crab, Vela, Cygnus, Geminga, and other regions will be performed, as well as diffuse gamma-ray emission, along with measurements of high-energy electron + positron and nuclei fluxes. Furthermore, it will study gamma-ray bursts and gamma-ray emission from the Sun during periods of solar activity. The GAMMA-400 energy range is expected to be from ~20 MeV up to TeV energies for gamma rays, up to 10 TeV for electrons + positrons, and up to 10^{15} eV for cosmic-ray nuclei. For 100-GeV gamma rays, the GAMMA-400 angular resolution is ~0.01° and energy resolution is ~1%; the proton rejection factor is ~5×10^5. GAMMA-400 will be installed onboard the Russian space observatory.

1. Introduction

The main goal for the GAMMA-400 mission is to perform a sensitive search for signatures of dark matter particles in high-energy gamma-ray emission. This task was set for the GAMMA-400 project by Nobel Laureate Academician V.L. Ginzburg in the end of 1980's [1, 2] and in his list of very important issues in modern cosmology at the beginning of XXI Century noted the issue of dark matter and its detection [3]. Within the framework of this project, which has now since expanded internationally, the design and construction of a future, complex, precision gamma-ray telescope is being carried out [4-7].

It should be noted, up to now, neither in indirect measurements of EGRET [8], currently operating Fermi-LAT [9], ground-based gamma-ray facilities [10] nor in direct measurements of various ground-based experiments including experiments at LHC, any direct evidence of hypothetical dark matter particles were not found.

GAMMA-400 is designed to search for signatures of dark matter in the fluxes of gamma-ray emission and electrons + positrons. Precision investigations of gamma-ray emission from Galactic Center, Crab, Vela, Cygnus, Geminga, and other regions will be performed, as well as diffuse gamma-ray emission, along with measurements of high-energy electron + positron and nuclei fluxes. Furthermore, it will study gamma-ray bursts and gamma-ray emission from the Sun during periods of solar activity.

2. The GAMMA-400 gamma-ray telescope

The GAMMA-400 physical scheme is shown in figure 1. GAMMA-400 consists of plastic scintillation anticoincidence top and lateral detectors (ACtop and AClat), converter-tracker (C), plastic scintillation detectors (S1 and S2) for time-of-flight system (ToF), two-part calorimeter (CC1 and CC2), lateral detectors (LD), plastic scintillation detectors (S3 and S4), and neutron detector (ND).

The anticoincidence detectors surrounding the converter-tracker are used to distinguish gamma quanta from significantly larger number of charged particles (e.g., in the region of 10-100 GeV, the flux ratios for gamma rays to electrons and protons are ~ $1:10^2:10^5$ [11]). The gammas recognizing provides by the counting and triggers signals formation system due special algorithms of backsplash rejection [12].

All scintillation detectors consist from two independent 1-cm layers. The time-of-flight system, where detectors S1 and S2 are separated by approximately 500 mm, determines the top-down direction of arriving particles. The additional scintillation detectors S3 and S4 improve hadron and electromagnetic shower separation.

The converter-tracker consists of 13 layers of double (x, y) silicon strip coordinate detectors (pitch of 0.08 mm). The first three and final two layers have no tungsten while the middle eight layers are interleaved with tungsten conversion foils. Using the first three layers without tungsten allows us to measure gamma rays down to approximately 20 MeV. The total converter-tracker thickness is about 1 X_0 (where X_0 is the radiation length). The converter-tracker information is used to precisely determine the conversion point and the direction of each incident particle.

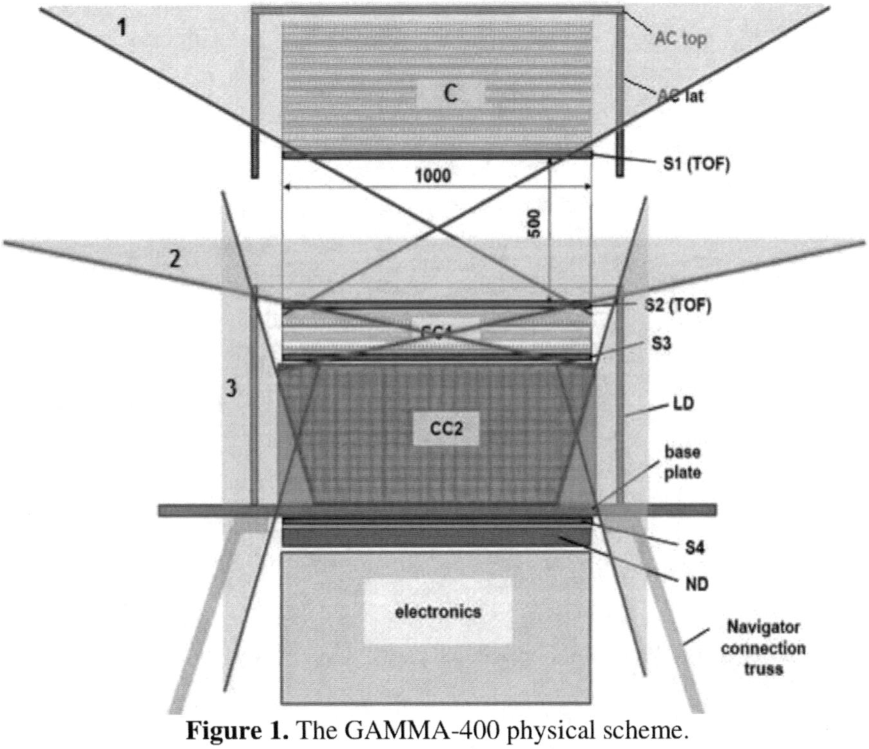

Figure 1. The GAMMA-400 physical scheme.

The two-part calorimeter measures particle energy. The imaging calorimeter CC1 consists of 2 layers of double (x, y) silicon strip coordinate detectors (pitch of 0.08 mm) interleaved with planes from CsI(Tl) crystals, and the electromagnetic calorimeter CC2 consists of CsI(Tl) cubic crystals with dimensions of 36 mm × 36 mm × 36 mm. The thickness of CC1 and CC2 is 2 X_0 and 23 X_0, respectively. The total calorimeter thickness is 25 X_0 or 1.2 λ_0 (where λ_0 is nuclear interaction length) when detecting vertical incident particles and 54 X_0 or 2.5 λ_0 when detecting laterally incident particles. Using a deep calorimeter allows us to extend the energy range up to several TeV for gamma rays, 10 TeV for electrons, and to reach an energy resolution of approximately 1% above 100 GeV.

Neutron detector is used to separate hadron and electromagnetic showers.

The GAMMA-400 initial steps to modeling of principal possibilities to detect particles from vertical and lateral directions are shown in figure 1 for several fields of view: (1) for gamma rays and electrons + positrons in the main aperture, (2) for gamma-rays and electrons + positrons in the additional one; and for gamma-rays, electrons + positrons, and nuclei in the lateral aperture (3). The results shows the best angular resolution of ~0.01° within ±30° for vertical particles (E_γ = 100 GeV) in the main aperture and ~0.2° for gammas with E_γ = 100 GeV in the additional one [13] and the possibility to extent energy range up to several TeV for gamma rays, 10 TeV for electrons in the lateral aperture [12].

3. The GAMMA-400 performance and comparison with Fermi-LAT and ground-based facilities

The GAMMA-400 performance has been presented earlier in [4-7]. At present, using additional first three double layers (without tungsten) allows us to achieve the best angular resolution in the energy range from ~20 MeV up to TeV in spite of some improved the Fermi-LAT performance with Pass 8 [14].

Table 1. Comparison of the Fermi-LAT and GAMMA-400 parameters.

	Fermi-LAT	**GAMMA-400**
Orbit	Circular, 565 km	Highly elliptical, 500-300000 km (without the Earth's occultation)
Operation mode	Sky-survey (3 hours)	Point observation (up to 100 days)
Source exposition	1/7	1
Energy range	20 MeV - 300 GeV (γ, e)	~20 MeV - 1 TeV (γ) 1 GeV - 10 TeV (e)
Effective area (E_γ > 1 GeV)	~6500 cm^2 (total) ~4000 cm^2 (front)	~4000 cm^2
Coordinate detectors - readout	Si strips (pitch 0.23 mm) digital	Si strips (pitch 0.08 mm) analog
Angular resolution	~4° (E_γ = 100 MeV) ~0.2° (E_γ = 10 GeV) ~0.1° (E_γ > 100 GeV)	~2° (E_γ = 100 MeV) ~0.1° (E_γ = 10 GeV) ~0.01° (E_γ > 100 GeV)
Calorimeter - thickness	CsI(Tl) ~8.5 X_0	CsI(Tl) + Si ~25 X_0
Energy resolution	~10% (E_γ = 10 GeV) ~10% (E_γ > 100 GeV)	~3% (E_γ = 10 GeV) ~1% (E_γ > 100 GeV)
Proton rejection factor	~10^3	~5x10^5
Mass, kg	2800	4100
Telemetry downlink volume	15 Gbytes/day	100 Gbytes/day

Table 1 presents a comparison of the Fermi-LAT and GAMMA-400 parameters. GAMMA-400 has numerous advantages:

- highly elliptical orbit (without the Earth's occultation and away from the radiation belts) allows us to observe with the full aperture of ±60° different gamma-ray sources continuously over a long period of time with the exposition greater by a factor of 7 than for Fermi-LAT operating in the sky-survey mode;
- thanks to a smaller pitch (by a factor of 3) and analog readout in the coordinate silicon strip detectors, GAMMA-400 has an excellent angular resolution above ~20 MeV;
- due to the deep (~25 X_0) calorimeter, GAMMA-400 has an excellent energy resolution and can more reliably to detect gamma rays up to several TeV and electrons + positrons up to 10 GeV for vertically incident events. When measuring lateral events (electrons + positrons and nuclei) it can detect diffuse gamma-ray emission and cosmic rays using only the CC2 calorimeter;
- owing to the better gamma-ray separation from cosmic rays (in contrast to Fermi-LAT, the presence of a special trigger with event timing, time-of-flight system, two-layer scintillation detectors), GAMMA-400 is significantly well equipped to separate gamma rays from the background of cosmic rays and backscattering events.

GAMMA-400 will have also the better angular and energy resolutions in the energy region 10-1000 GeV in comparison with current and future space- and ground-based instruments: H.E.S.S. [15], MAGIC [16], VERITAS [17], HAWC [18], CTA [19], DAMPE [20], and CALET [21] (figure 2, table 2) and it allows us to fill the gap between the space- and ground-based instruments.

Table 2. Comparison of parameters for the current, future space- and ground-based instruments.

	SPACE-BASED GAMMA-RAY INSTRUMENTS				**GROUND-BASED GAMMA-RAY INSTRUMENTS**			
	Fermi-LAT	**DAMPE**	**CALET**	**GAMMA-400**	**H.E.S.S.**	**MAGIC**	**VERITAS**	**CTA**
Particles	γ, e	e, nuclei, γ	e, nuclei, γ	γ, e, nuclei	γ	γ	γ	γ
Operation period	2008-	2015	2015	~2023	2012-	2009-	2007-	~2020
Energy range, GeV	0.02-300	5-10000	10-10000	**0.02-10000**	> 30	> 50	> 100	> 20
Angular resolution ($E_\gamma > 100$ GeV)	0.1°	0.1°	0.1°	**~0.01°**	0.07°	0.07° (E_γ = 300 GeV)	0.1°	0.1° (E_γ = 100 GeV) 0.03° (E_γ = 10 TeV)
Energy resolution ($E_\gamma > 100$ GeV)	10%	1.5%	2%	**~1%**	15%	20% (E_γ = 100 GeV) 15% (E_γ = 1 TeV)	15%	20% (E_γ = 100 GeV) 5% (E_γ = 10 TeV)

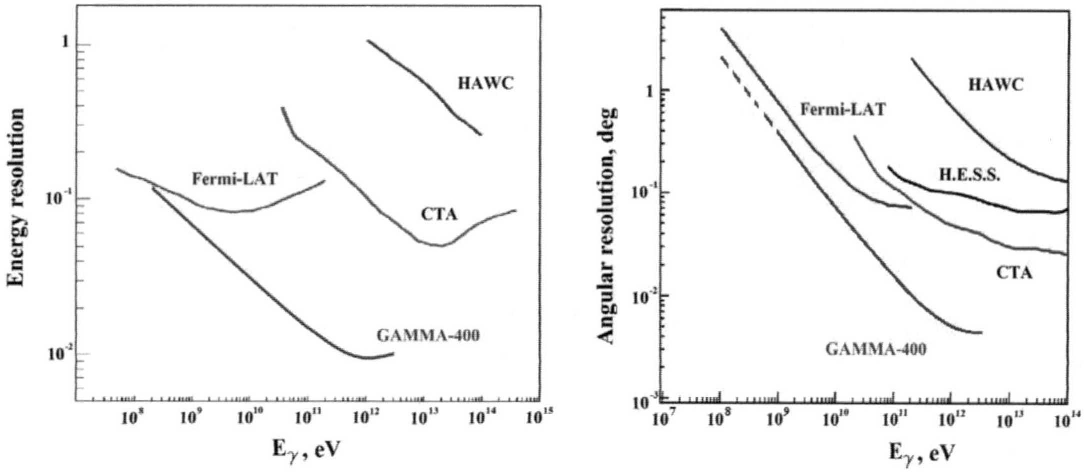

Figure 2. Comparison of energy and angular resolutions for GAMMA-400, Fermi-LAT, H.E.S.S., HAWC, and CTA (figure 2 from [24] was used).

GAMMA-400 will study continuously over a long period of time different regions of Galatic plane, for example, Galactic Center, Crab, Vela, Cygnus, Geminga with the full aperture of ±60°. In particular, using the gamma-ray fluxes obtained by Fermi-LAT [22, 23], we can expect that GAMMA-400 when observing the Galactic Center with aperture of ±45°during 1 year will detect: 57400 photons for E_γ > 10 GeV; 5240 photons for E_γ > 50 GeV; 1280 photons for E_γ > 100 GeV; 535 photons for E_γ > 200 GeV.

4. GAMMA-400 gamma-ray observatory

In addition to the gamma-ray telescope, the GAMMA-400 scientific complex includes the KONUS-FG gamma-ray burst monitor and two star sensors for determining the gamma-ray telescope axis with an accuracy of ~5″, along with two magnetometers for measuring the magnetic field. The GAMMA-400 gamma-ray observatory will be installed onboard of the Navigator space platform, which is designed and manufactured by the Lavochkin Association.

Using the Navigator space platform gives the GAMMA-400 experiment a highly unique opportunity for the near future gamma- and cosmic-ray science, since it allows us to install a very large scientific payload (a mass of 4100 kg, power consumption of 2000 W, and a telemetry downlink of 100 GB/day, with lifetime more than 7 years), which will provide GAMMA-400 with the means to significantly contribute as the next generation instrument for gamma-ray astronomy and cosmic-ray physics.

The GAMMA-400 experiment will be initially launched into a highly elliptical orbit (with an apogee of 300,000 km and a perigee of 500 km, with an inclination of 51.4°), with 7 days orbital period. Under the action of gravitational disturbances of the Sun, Moon, and the Earth after ~6 months the orbit will transform to about an approximately circular one with a radius of ~200,000 km and will not suffer from the Earth's occultation and shielding by the radiation belts. A great advantage of such an orbit is the fact that the full sky coverage will always be available for gamma-ray astronomy, since the Earth will not cover a significant fraction of the sky, as is usually the case for low Earth orbit. Therefore, the GAMMA-400 source pointing strategy will hence be properly defined to maximize the physics outcome of the experiment. The launch of the GAMMA-400 space observatory is planned for the middle of the 2020s.

References

[1] Dogiel V *et al.* 1988 *Space Sci. Rev.* **49** 215
[2] Ginzburg V *et al.* 1995 *Adv. Space Res.* **15** 93
[3] Ginzburg V 2004 *Phys. Usp.* **47** 1155
[4] Galper A *et al.* 2013 *Adv. Space Res.* **51** 297
[5] Galper A *et al.* 2013 *AIP Conf. Proc.* **1516** 288
[6] Adriani O *et al.* 2014 *arXiv:*1412.4239
[7] Topchiev N *et al.* 2015 *Bull. RAS. Physics* **79** 417
[8] Wim de Boer 2005 *CERN COURIER* 29491
[9] Ackermann M *et al.* 2015 *arXiv:*1506.00013
[10] Doro M 2014 *arXiv:*1404.5017
[11] Moiseev A A *et al.* 2007 *Astroparticle Phys.* **27** 339
[12] Arkhangelskaja I V *et al* 2015 *Physics Procedia* **74** 212
[13] Leonov A A *et al* 2015 *Physics Procedia* **74** 183
[14] URL: *http://www.slac.stanford.edu/exp/glast/groups/canda/lat_Performance.htm*
[15] Balzer A *et al.* 2013 *arXiv:*1311.3486
[16] Mazin D *et al.* 2014 *arXiv:*1410.5073
[17] Ong R E *et al.* 2014 *Adv. Space Res.* **53** 1483
[18] Westerhoff S *et al.* 2014 *Adv. Space Res.* **53** 1492
[19] CTA Consortium 2011 *Experimental Astronomy* **32** 93
[20] Dong Yi-Fan *et al.* 2015 *arXiv:*1503.00415
[21] Mori N 2013 *Nucl. Phys.* B *(Proc. Suppl.)* **239-240** 199
[22] Ackermann M *et al.* 2013 *arXiv:*1306.6772
[23] URL: *http://fermi.gsfc.nasa.gov/ssc/data/access/*
[24] Funk S *et al.* 2013 *Astroparticle Phys.* **43** 348

Perspectives of the GAMMA-400 space observatory for high-energy gamma rays and cosmic rays measurements

N P Topchiev[2,*], A M Galper[1,2], V Bonvicini[7], O Adriani[8], R L Aptekar[3], I V Arkhangelskaja[1], A I Arkhangelskiy[1], A V Bakaldin[1], L Bergstrom[9], E Berti[8], G Bigongiari[10], S G Bobkov[6], M Boezio[7], E A Bogomolov[3], S Bonechi[10], M Bongi[8], S Bottai[8], G Castellini[11], P W Cattaneo[12], P Cumani[7], O D Dalkarov[2], G L Dedenko[1], C De Donato[13], V A Dogiel[2], N Finetti[8], M S Gorbunov[6], Yu V Gusakov[2], B I Hnatyk[14], V V Kadilin[1], V A Kaplin[1], A A Kaplun[1], M D Kheymits[1], V E Korepanov[15], J Larsson[16], A A Leonov[1,2], V A Loginov[1], F Longo[7], P Maestro[10], P S Marrocchesi[10], A L Men'shenin[4], V V Mikhailov[1], E Mocchiutti[7], A A Moiseev[17], N Mori[8], I V Moskalenko[18], P Yu Naumov[1], P Papini[8], M Pearce[16], P Picozza[13], A Rappoldi[12], S Ricciarini[11], M F Runtso[1], F Ryde[16], O V Serdin[6], R Sparvoli[13], P Spillantini[8], Yu I Stozhkov[2], S I Suchkov[2], A A Taraskin[1], M Tavani[19], A Tiberio[8], E M Tyurin[1], M V Ulanov[3], A Vacchi[7], E Vannuccini[8], G I Vasilyev[3], Yu T Yurkin[1], N Zampa[7], V N Zirakashvili[5] and V G Zverev[1]

[1] National Research Nuclear University MEPhI (Moscow Engineering Physics Institute), Kashirskoe highway 31, Moscow, 115409, Russia
[2] Lebedev Physical Institute, Russian Academy of Sciences, Moscow, Russia
[3] Ioffe Institute, Russian Academy of Sciences, St. Petersburg, Russia
[4] Research Institute for Electromechanics, Istra, Moscow region Istra, Russia
[5] Pushkov Institute of Terrestrial Magnetism, Ionosphere, and Radiowave Propagation, Troitsk, Moscow region, Russia
[6] Scientific Research Institute for System Analysis, Russian Academy of Sciences, Moscow, Russia
[7] Istituto Nazionale di Fisica Nucleare, Sezione di Trieste and Physics Department of University of Triest, Italy
[8] Istituto Nazionale di Fisica Nucleare, Sezione di Firenze and Physics Department of University of Florence, Firenze, Italy
[9] Stockholm University, Department of Physics; and the Oskar Klein Centre, AlbaNova University Center, Stockholm, Sweden
[10] Department of Physical Sciences, Earth and Environment, University of Siena and Istituto Nazionale di Fisica Nucleare, Siena, Italy
[11] Istituto di Fisica Applicata Nello Carrara - CNR and Istituto Nazionale di Fisica Nucleare, Sezione di Firenze, Firenze, Italy
[12] Istituto Nazionale di Fisica Nucleare, Sezione di Pavia, Pavia, Italy
[13] Istituto Nazionale di Fisica Nucleare, Sezione di Roma 2 and Physics Department of University of Rome Tor Vergata, Rome, Italy
[14] Taras Shevchenko National University of Kyiv, Kyiv, Ukraine
[15] Lviv Center of Institute of Space Research, Lviv, Ukraine
[16] KTH Royal Institute of Technology, Department of Physics, and the Oskar Klein Centre, AlbaNova University Center, Stockholm, Sweden

Journal of Physics: Conference Series **675** (2016) 032010 doi:10.1088/1742-6596/675/3/032010

[17] CRESST/GSFC and University of Maryland, College Park, Maryland, USA
[18] Hansen Experimental Physics Laboratory and Kavli Institute for Particle Astrophysics and Cosmology, Stanford University, Stanford, USA
[19] Istituto Nazionale di Astrofisica IASF and Physics Department of University of Rome Tor Vergata, Bologna, Italy

*leon@ibrae.ac.ru

Abstract. The GAMMA-400 gamma-ray telescope is intended to measure the fluxes of gamma-rays and cosmic-ray electrons and positrons in the energy range from 100 MeV to several TeV. Such measurements concern the following scientific tasks: investigation of point sources of gamma-rays, studies of the energy spectra of Galactic and extragalactic diffuse emission, studies of gamma-ray bursts and gamma-ray emission from the Sun, as well as high precision measurements of spectra of high-energy electrons and positrons. Also the GAMMA-400 instrument provides the possibility for protons and nuclei measurements up to knee. But the main goal for the GAMMA-400 mission is to perform a sensitive search for signatures of dark matter particles in high-energy gamma-ray emission. To fulfill these measurements the GAMMA-400 gamma-ray telescope possesses unique physical characteristics in comparison with previous and present experiments. The major advantage of the GAMMA-400 instrument is excellent angular and energy resolution for gamma-rays above 10 GeV. The GAMMA-400 experiment will be installed onboard of the Navigator space platform, manufactured by the NPO Lavochkin Association. The expected orbit will be a highly elliptical orbit (with apogee 300.000 km and perigee 500 km) with 7 days orbital period. An important profit of such an orbit is the fact that the full sky coverage will always be available for gamma ray astronomy.

1. Introduction

The GAMMA-400 observatory will allow to address a broad range of science topics, like search for signatures of dark matter, studies of Galactic and extragalactic gamma-ray sources, Galactic and extragalactic diffuse emission, gamma-ray bursts and charged cosmic rays acceleration and diffusion mechanism up to the knee. GAMMA-400 will be able to study gamma-rays, from 100 MeV up to several TeV, electrons up to 10 TeV, protons and nuclei up to the knee (10^{15}–10^{16} eV). It will search for possible dark matter signal thanks to an unprecedented energy resolution that will permit to detect features associated with dark matter annihilation or decay in the spectra of Galactic Center region [1].

The GAMMA-400 will be installed onboard the Russian Space Observatory.

2. The GAMMA-400 gamma-ray telescope

The GAMMA-400 physical scheme is shown in figure 1. Starting from the top, the telescope is composed by the following layers:

- an anticoincidence system (AC) is composed by plastic scintillators, located both on top and on the lateral side of the apparatus. AC is essentially used to veto charged particles.
- a converter-tracker system (C) consists of 13 layers of double (x, y) silicon strip coordinate detectors (pitch 0.08 mm). The first three and final two layers have no tungsten while the middle eight layers are interleaved with tungsten conversion foils. The thickness of tungsten is 0.1 X_0. The total converter-tracker thickness is about 1 X_0. Using the first three layers without tungsten allows to measure gamma rays from approximately 20 MeV. The system is used to convert the photons and precisely reconstruct the photon direction by the detection in the silicon layers the conversion e+/e^- pair.
- a Time of Flight system (ToF) is formed by plastic scintillators S1 and S2, separated by approximately 500 mm. This system is used both to generate the trigger for the apparatus and to reject albedo particles by measuring their velocity.

- an imaging calorimeter CC1. CC1 is composed by 2 layers of CsI(Tl) crystals, each 1 X_0 deep, interleaved with 2 silicon microstrip layers of the same type used for the converter-tracker. The thickness of CC1 is $2X_0$. The imaging calorimeter is used to significantly improve the photon angular resolution by precisely measuring the conversion e^+/e^- pair after a large 50 cm lever arm below the converter-tracker.
- the scintillation detectors S3 and S4 improve hadrons/electron separation.
- deep, isotropic and homogeneous electromagnetic calorimeter (CC2) is composed by 28×28×12 small cubic CsI(Tl) crystals (with 3.6 cm side), developed on the basis of the CaloCube project R&D [4]. The overall dimensions of the calorimeter are the following: 1 m × 1 m × 0.48 m, corresponding to $54.6\ X_0 \times 54.6\ X_0 \times 23.4\ X_0$ and $2.5\ \lambda_I \times 2.5\ \lambda_I \times 1.1\ \lambda_I$. Using a deep calorimeter allows us to extend the energy range up to several TeV for gamma rays and to reach an energy resolution of approximately 1% above 100 GeV. The overall mass (~2000 kg), coupled to the large dimensions, will allow detection of high energy hadrons, up to 10^{15} eV, with an effective geometric factor of the order of 4 $m^2 \times sr$, giving the possibility to directly probe on orbit the knee region. This is accomplished by detecting particles not only on the top surface, but also on the lateral side, hence significantly increasing the overall geometrical factor.
- a neutron detector (ND), located below the CC2, is used to improve the proton/electron separation.

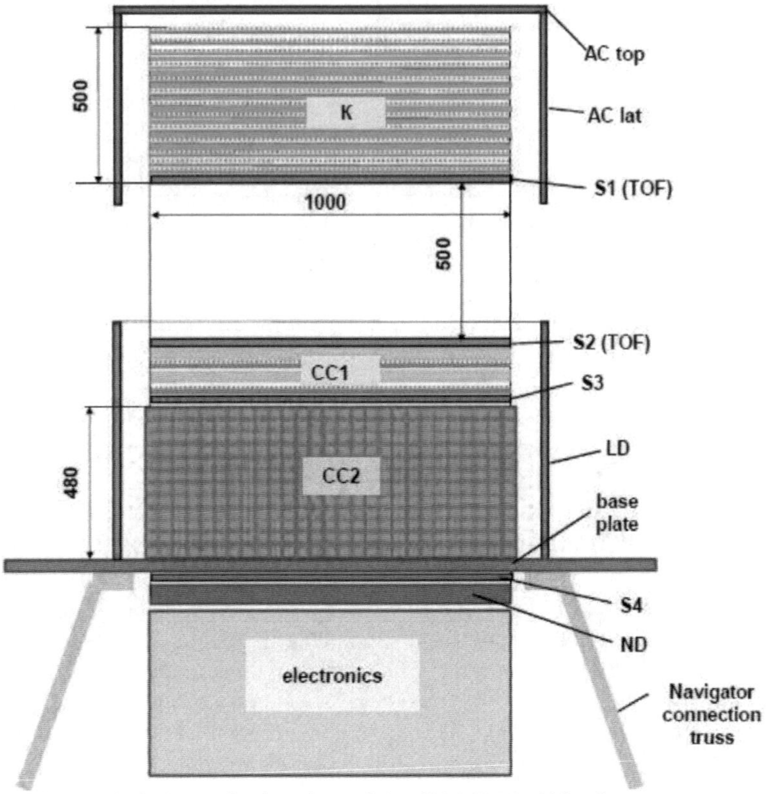

Figure 1. Schematic drawing of the GAMMA-400 telescope.

3. The GAMMA-400 gamma-ray telescope performance

To calculate GAMMA-400 instrument performance the following main trigger for gamma registration was used: the signal from energy release in S1 has to be before than the signal in S2 and the signal in

AC system has to be less than the value of threshold selected in such way to suppress the back splash influence. For what concern the instrument's performances in the gamma detection. Figures 2 and 3 show respectively the effective area, the angular and the energy resolution of the GAMMA-400 telescope.

In figure 2a the dependence of effective area from initial energy of vertical gamma rays and in figure 2b the dependence of effective area from the incidence angle of the initial 100-GeV gamma are presented. These curves were obtained taking into account the main trigger signal of GAMMA-400: S1 & S2 & (no AC). Time and segmentation methods are used to reduce the influence of backscattering particles created when incident γ-rays interact with the calorimeter's matter and produce low energy photons, electrons… moving in the opposite direction [5]. The effective area of GAMMA-400 is approximately twice lower than total effective area of Fermi-LAT [6,7]. However, for a full comparison of the performances it is necessary to take into account the optimized choice of the pointing strategy for important sources or for the whole sky survey, considering that the full sky is available to GAMMA-400, due to the peculiar orbit, very far away from any possible earth occultation. Moreover, the high energy part of the photon spectrum can be investigated with the calorimeter CC2 only, with an angular resolution better than two degrees. This way the acceptance can be significantly increased, reaching a level of ~4 m^2.

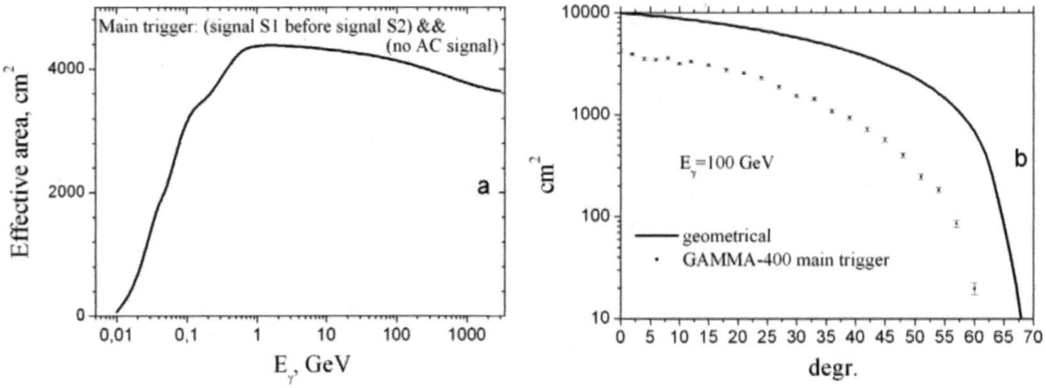

Figure 2. Effective area of GAMMA-400 in the photon detection. a) The dependence of effective area from initial energy of gamma with vertical incidence. b) The dependence of effective area from the incidence angle of the initial 100 GeV gamma.

The angular resolution of GAMMA-400 is comparable to the Fermi-LAT in the 100 MeV - few GeV range, while it is significantly better above few GeV (figure 3a). This is accomplished by a proper combination of the excellent spatial resolution and the long lever arm available in between the converter-tracker and the CC1 calorimeter. The low energy part (less than 500 MeV) of GAMMA-400 angular resolution curve was obtained for gamma events converted in the upper thin layers of converter-tracker system. Taking into account the power-law behaviour of gamma spectrum, the reducing of effective area is not so important for such energies.

The performance of the instrument for what concern the energy resolution is demonstrated in figure 3b. It is clearly better than Fermi-LAT one above few hundreds MeV, reaching a resolution better than 2% for the multi GeV range. This is possible thanks to the excellent performances of the homogeneous and deep calorimeter, able to fully absorb electromagnetic shower generated by multi-GeV photons.

The characteristics of GAMMA-400 instrument will allow exploring high energy gamma from any space region in more details in comparison with FERMI-LAT. To demonstrate the advantage of GAMMA-400 capabilities the center of Galaxy region ($-1<l<0.76$; $0.285<b<0.3$) was selected concerning possible location of Dark Matter particles [8, 9]. In this region there are four point sources identified in third source FERMI-LAT catalog [10]. Using information about the energy spectrum and total intensity for these sources in the energy range 10-100 GeV, and diffuse flux from background models [11] it is possible to simulate the energy distribution for gamma events identified in the vicinity of Galactic center during FERMI-LAT mission. This distribution is shown in figure 4 (green color). The value of energy bin for this distribution was selected equal 10 GeV according to the value 0.1 of FERMI-LAT relative energy resolution about 100 GeV. If there is some energetic gamma line with energy 85 GeV, providing 25 gamma events for FERMI-LAT effective area during total time of the mission, then energy distribution will change (blue color in figure 4a). The magnified view of sum energy distribution is shown in figure 4b. It is evidently that it is impossible to identify some additional source with enough statistic significance having such energy resolution. But if the energy distribution would be build with energy bin 2 GeV according to energy resolution of GAMMA-400, than additional source could be easy identified. In figure 5a the energy distributions for the ordinary gamma (red color) and for the gamma from line (blue color) with the value of bin 2 GeV are shown. In figure 5b the magnified view of sum energy distribution is shown. In four bins with additional gamma from line 85 GeV there are 25 events from the additional source and 7 events from the ordinary gamma. So, the level of significance 4 sigma is provided and the additional flux of gamma ~ 10^{-10} cm^{-2} sec^{-1} can be identified.

FERMI (http://www.slac.stanford.edu/exp/glast/groups/canda/lat_Performance.htm)

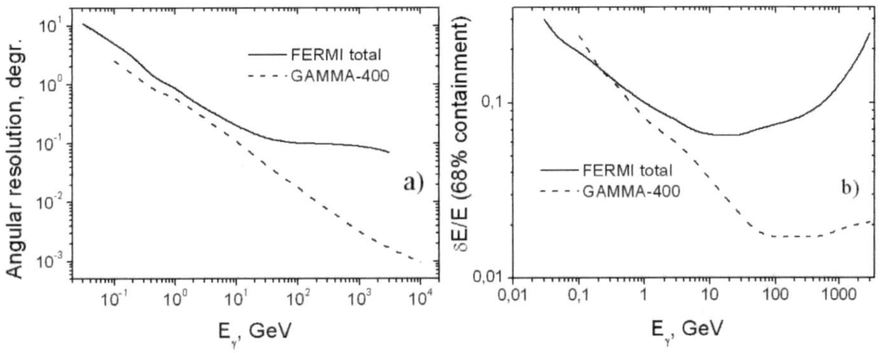

Figure 3. The angular and energy resolutions of GAMMA-400 in the photon detection. a) The dependence of angular resolution from initial energy of gamma. b) The dependence of energy resolution from initial energy of gamma. The characteristics of FERMI-LAT are also presented.

The GAMMA-400 observatory is also an optimal tool for the very high energy hadron detection. For this item two critical parameters exist in the detection system: the acceptance, that should be as large as possible, to have the possibility to measure a significant amount of the 10^{15} eV particles, and the hadronic energy resolution. The large area homogeneous and isotropic calorimeter, accepting particles from all sides, gives us for the first time the possibility to directly observe in space such high energy hadrons. Figure 6 shows the expected number of high energy protons and helium nuclei for a 10 years exposure of the GAMMA-400 experiment, assuming the Polygonato model [12], with an expected p/e rejection factor better than 10^5 [3], and with a realistic reconstruction efficiency of the order of 40%. It is expected to collect a statistic larger than 100 events with energy ~10^{15} eV both for protons and helium, allowing a detailed probe of the interesting knee region, that has never been directly investigated with an in orbit detector up to now.

Journal of Physics: Conference Series **675** (2016) 032010 doi:10.1088/1742-6596/675/3/032010

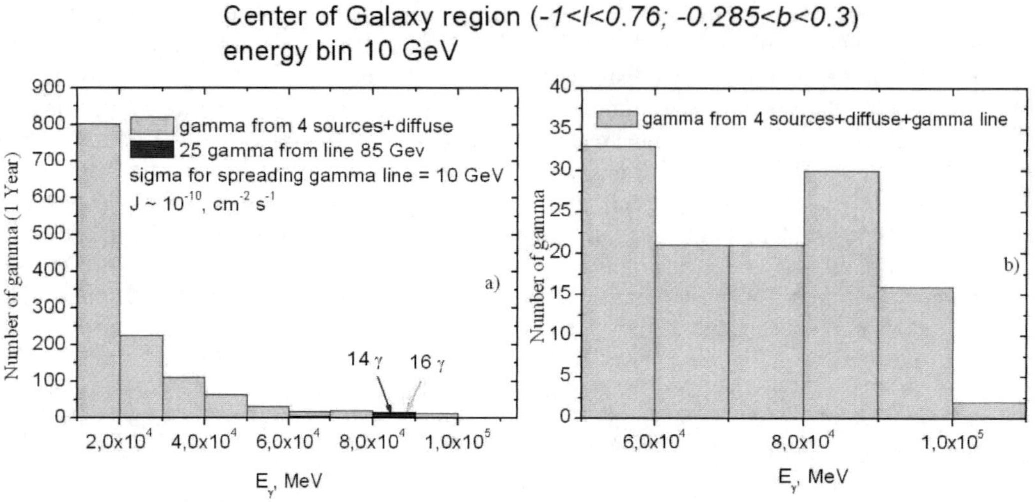

Figure 4. a) The energy distribution with the value of bin equal 10 GeV for the experimental data obtained during FERMI-LAT mission in the vicinity of Galactic center for energy range 10-100 GeV and selected for maximum zenith angle less, than 90^0 (green color) and additional contribution from line 85 GeV (blue color) b) The magnified view of the sum energy distribution.

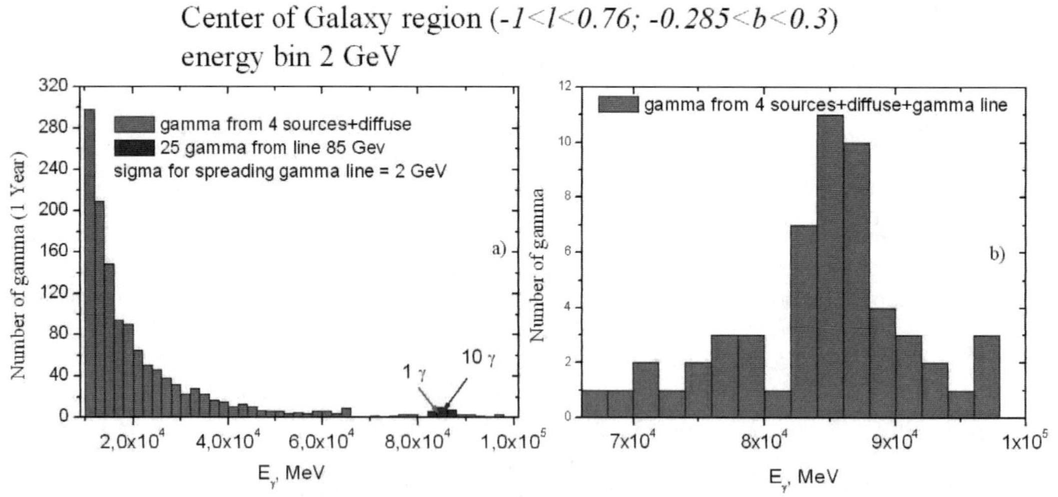

Figure 5. a) The energy distribution with the value of bin equal 2 GeV for the experimental data as in figure 4a (red color) and additional contribution from line 85 GeV (blue color). b) The magnified view of the sum energy distribution.

Protons and Helium (Polygonato model)											
Effective GF (m^2 sr)	σ(E)/E	E>0.1 PeV		E>0.5 PeV		E>1 PeV		E>2 PeV		E>4 PeV	
		p	He	p	He	p	He	p	He	p	He
~4	35%	7.8×10^3	7.4×10^3	4.6×10^2	5.1×10^2	1.2×10^2	1.5×10^2	28	43	5	10

Figure 6. Expected number of proton and helium events in 10 years data taking, according to the Polygonato model.

4. Conclusion

The GAMMA-400 instrument has been designed for the optimal detection of gamma rays in a broad energy range (from 100 MeV up to several TeV), with excellent angular and energy resolutions. The observatory will also allow precise and high statistic studies of the electron component in the cosmic rays up to the multi TeV region, as well as protons and nuclei spectra up to the knee region.

References

[1] Galper A M *et al.* 2013 *Bull. RAS. Physics* **77**(11) 1339
[2] Galper A M *et al.* 2013 *Adv. Space Res.* **51** 297
[3] Leonov A A *et al.* 2015 *Adv. Space Res.* **56** 1538
[4] Adriani O *et al.* 2013 *Proceedings of the International Conference on Calorimetry for the High Energy Frontier (CHEF)*
[5] Ginzburg V L *et al.* 2009 *Bull. RAS. Physics* **73**(5) 664
[6] Atwood W B *et al.* 2009 *Astropart J.* **697** 1071
[7] Abdo A A *et al.* 2009 *Astropart Phys.* **32** 193
[8] Hooper D & Goodenough L 2011 *Physics Letters* B **697** 412
[9] Kaplinghat M *et al.* 20115 *Phys. Rev. Lett.* **114** 21
[10] *FERMI-LAT catalog (http://fermi.gsfc.nasa.gov/ssc/data/access/lat/4yr_catalog/)*
[11] URL: *http://fermi.gsfc.nasa.gov/ssc/data/access/lat/BackgroundModels.html*
[12] Horandel J 2004 *Astropart Phys.* **21** 241

Method to select gamma rays with energy above 50 GeV against a charge-particle background in the GAMMA-400 space telescope

M D Kheymits[1,3], A A Leonov[1], A M Galper[1,2], V Kadilin[1], I V Arkhangelskaya[1], A I Arkhangelskiy[1], Yu V Gusakov[2], V G Zverev[1], V Kaplin[1], P Yu Naumov[1], M F Runtso[1], S I Suchkov[2], N P Topchiev[2], Yu T Yurkin[1], A Bakaldin[1] and O Dalkarov[2]

[1] National Research Nuclear University MEPhI, Kashirskoe Shosse 31, Moscow, 115409, Russia

[2] P. N. Lebedev Physical Institute of the Russian Academy of Sciences, Leninskij Prospekt 53, Moscow, 119991, Russia

E-mail: mcsaksik@gmail.com

Abstract. Studying high-energy (>50 GeV) cosmic gamma radiation raises a problem of selection of neutral gamma rays from a background of charged particles. The problem is embarrassed by the *backsplash* effect. The backsplash consists, in the main, of low-energy (1 MeV) secondary photons moving backwards and is produced by any high-energy gamma quantum. A charged-particle rejection method using the anticoïncidence and time-of-flight systems is proposed. Charged-particle events are distinguished from those being triggered by high-energy gamma-rays producing backsplash. The method is based on the time separation of signals. It allows us to keep the gamma-ray detection efficiency high up to high energies.

1. Introduction

Modern status of fundamental research in cosmology, astronomy, high-energy physics, and cosmic-ray acceleration and propagation raises series of problems unsolvable without involving results of extraterrestrial high-energy γ-ray astronomy. The problem of dark matter origin is a representative example.

Typical gamma-ray telescope (such as GAMMA-1 [1], AGILE [2], EGRET [3], Fermi/LAT [4]) for tens of MeV – hundreds of GeV range is a pair-production telescope with three main components: a tracker to determine the direction of each incident particle, a calorimeter to determine its energy, and an anticoïncidence detector surrounding the tracker to distinguish gamma-ray events from the much more numerous charged-particle events. Tracker has high-Z foils (typically tungsten) for pair conversion interleaved with position-sensitive detectors (silicon strip detector planes) for the direction determination. Tracker provides highly efficient trigger for low energy

[3] To whom any correspondence should be addressed.

gammas, but raise problem of charged-particle rejection. To retain the rejection at high energies one can use trigger generated by much faster scintillation detectors.

A high-energy gamma-ray particle produces an electromagnetic shower in the telescope. Its major part develops inside the calorimeter. The more energetic the γ-ray particle being detected is, the more particles are produced in the shower, including the ones moving upwards, anti-parallel to the incident direction, and being referred to as the *backsplash*. It consists mainly of γ-ray photons with energy $\sim 1\,\mathrm{MeV}$. This soft-spectrum flux 'soils' upper detectors and imitate a detection of a charged particle. Backsplash may dramatically reduce efficiency above tens of GeV. Triggering of AC caused by backsplash prevents detection of high-energy γ-radiation and makes detector efficiency vanish at high energy. To solve the problem, a new method of charged-particle rejection by measuring AC triggering time is being proposed.

In this work, a scheme of gamma-ray telescope's time-of-flight (ToF) and anticoïncidence (AC) systems is proposed. It provides measurement of time of particle's flight via AC and ToF detectors. The proposed procedure of event time selection keeps high-energy gamma-ray detection efficiency high.

This method is used for the GAMMA-400 instrument [5–7], nevertheless it still can be applied to most γ-ray telescopes with time-sensitive AC and ToF systems.

2. Anticoïncidence and time-of-flight systems

GAMMA-400 anticoïncidence (AC) system covers converter on top and on each lateral side. Its time-of-flight (ToF) system consists of two detectors — S1 and S2 — with the same design. Each detector of AC and ToF is composed of two layers, each consisting of a number of scintillation paddles. Paddles of the two layers are shifted across, so that gaps between those of one layer overlap gaps in another. This overlapping provides utterly high detector efficiency. Each one is made of polystyrene-based scintillating material. Scintillation light is detected with photosensors placed on opposite edges.

Photosensors' outputs are connected to a *time-coördinate compensation circuit* (TCC), which allows time of particle's interaction with scintillator material to be determined via compensation of dependency of flash detection time on place of its origin [8].

In this paper, the upper horizontal plane of AC is considered.

3. Time-selection method

Consider the following event. A gamma-ray photon goes down with no interaction in AC and produces an e^-e^+-pair in C. The pair continues its motion down. Being charged particles,

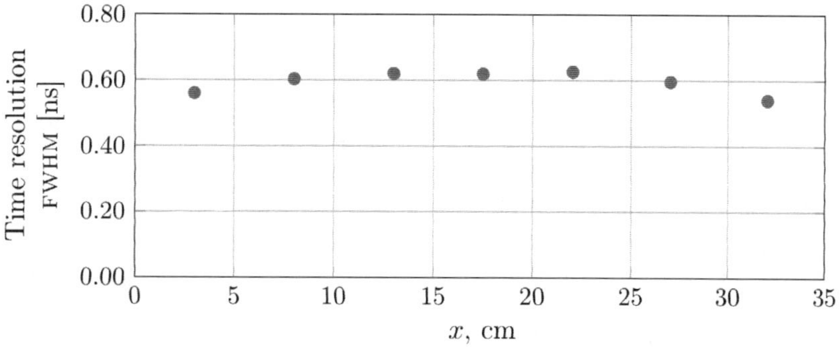

Figure 1: Time resolution vs. source position

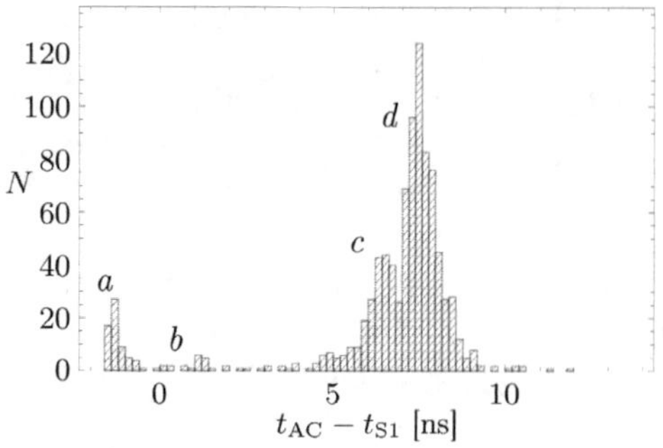

Figure 2: Distribution of the difference between AC and S1 triggerings when detecting gamma rays; a: conversion in AC, b: backsplash from C, c: from CC1, d: from CC2,

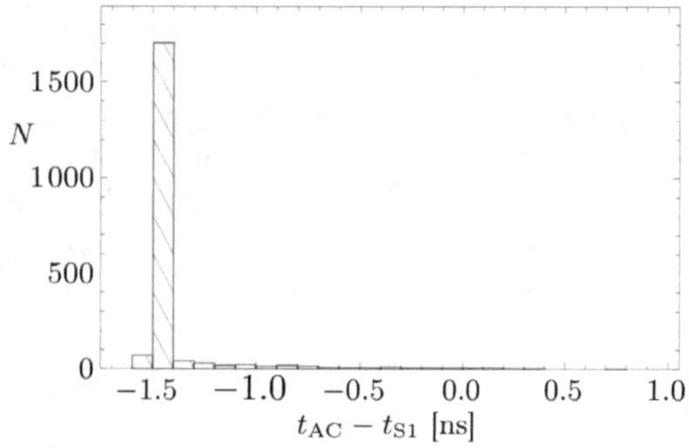

Figure 3: Distribution of the difference between AC and S1 when detecting protons (note expanded horizontal scale)

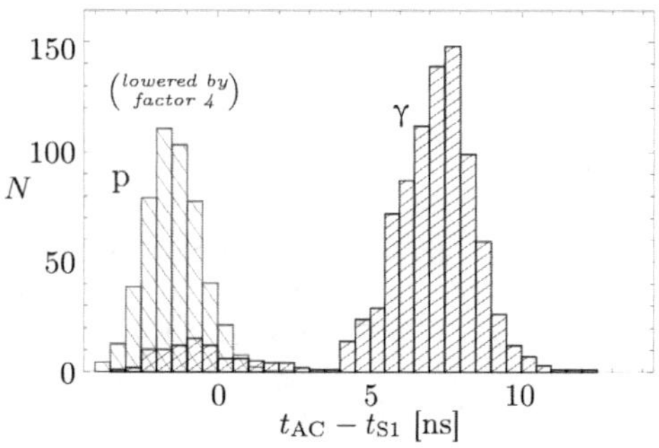

Figure 4: Normally blurred distribution of the difference between AC and S1 triggerings when detecting gamma rays and protons. Standard deviation of the blur is a pythagorean sum of AC and S1 time resolutions. For convenience of comparison, heights of columns on proton histogram are lowered fourfold.

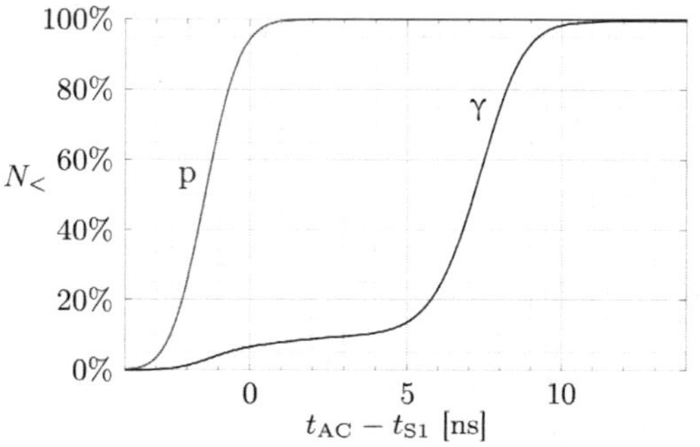

Figure 5: Cumulative distribution functions of S1-to-AC lag. Value of $N_<$ at abscissa t refers to fraction of events where the lag is less than t. Both gamma rays and protons are plotted.

e^- and e^+ (or their products) give a signal in S1, then in S2, come in the calorimeter, and develop a shower. The latter produces backsplash, whose particles trigger a signal in AC.

Since both AC–S1 and S1–CC1 distances are about half-meter long, time to move particles from S1 to CC1, to form a shower, and to bring a backsplash back to AC amounts to several nanoseconds, whereas in case of detection of a primary charged particle by the instrument (in the same direction), AC is triggered as soon as this particle enters it. Thus, time lags till AC is triggered, from the moment a primary charged particle enters the telescope and from the moment a primary gamma-ray particle does, differ. After the charged particle gets into the instrument, signal in AC appears with a small delay only determined by light propagation in the scintillator, whereas an interval until signal in AC appears after the gamma enters is composed of the time of backslash particles to be produced and to move up to AC and equals to several nanoseconds.

Thus, a signal in AC appears several nanoseconds later that in the case of flight of a charged particle in the same direction.

Detectors of AC and ToF systems are proved to have time resolution $\delta t \approx 0.6$ ns (see fig. 1).

The 100 GeV gamma-ray detection by the instrument has been simulated using Monte-Carlo method. Detector triggering time is defined as the earliest triggering time of all the constituent paddles, which in turn are defined by TCC mathematical model.

Events are selected by the set of detector signals indicating the presence of conversion in C. The distribution of difference between AC triggering time and S1 triggering time is plotted in fig. 2. Four peaks can be seen in the histogram: *a*) the left one (-1.5 to 0 ns) corresponds to gamma's interaction inside AC; *b*) the next low peak (0 to 3 ns) corresponds to the backsplash from the converter; *c*) (4 to 7 ns) and *d*) (7 to 10 ns) are the backsplash from the two parts of the calorimeter — CC1 and CC2, respectively.

The leftmost peak is the only one in the case of a charged particle since it always triggers AC first. It is shown by proton-event simulation in fig. 3. Tails appear due to the case of several interactions in AC (early backsplash or δ-electron). In these instances, TCC technique undervalues time: $\hat{t}_0 < t_0$.

The histograms mentioned above leave out of account inherent time resolution of AC and S1 detectors. If we blur the lag normally (Gaussianly) by $\delta t \times \sqrt{2} \approx 0.85$ ns we get histograms shown in fig. 4. It is the case to be expected in real instrument. The events of the two types are seen to be separated fairly enough even though being blurred.

Cumulative plots (see fig. 5) show that excluding AC from anticoïncidence scheme $t_{\text{th}} \approx 3$ ns past S1 triggering leads to sure identification and classification of γ-ray events even in case of backsplash. Calculation yields that t_{th} being at least 2.9 ns leads to protons' admixture to gammas selected not more than 10^{-5}, while gamma-ray loss being 10%.

4. Conclusion

The proposed procedure of using time-sensitive detectors to distinguish events of charged-particle detection and γ-ray particles that produced backsplash triggered AC makes possible to effectively reject charged particles — primarily, protons — without significant loss of γ-ray detection efficiency. In the GAMMA-400 γ-ray telescope, this efficiency is at least 90% for high-energy gammas. Implementation of the technique using scintillation detectors, with SiPMs, is shown to be possible.

References

[1] Akimov V *et al.* 1988 *Space Science Reviews* **49** 111–124

[2] Tavani M *et al.* 2008 (*Preprint* 0807.4254v1)

[3] Hartman R C 1988 *Space Science Reviews* **49** 69–84

[4] Atwood W B *et al.* 2009 *Astrophys. J.* **697** 1071–1102

[5] Ginzburg V L *et al.* 2009 *Bulletin of the Russian Academy of Sciences: Physics* **73** 664–666

[6] Galper A M *et al.* 2013 *Advances in Space Research* **51** 297–300

[7] Topchiev N P, Galper A M *et al.* 2015 *Bulletin of the Russian Academy of Sciences. Physics* **79** 417–420

[8] Rothermel H 1976 *Nucl. Instr. and Meth.* **137** 219–223

[9] Thompson D J *et al.* 1993 *Astrophys. J. Suppl.* **86** 629

[10] Kaplin V A, Runtso M F, Topchiev N P and Fradkin M I 2008 *Proceedings of the MEPhI scientific session, 2008* **9** 150–152

Method of incident low-energy gamma-ray direction reconstruction in the GAMMA-400 gamma-ray space telescope

M D Kheymits[1,3] A A Leonov[1] V G Zverev[1] A M Galper[1,2] I V Arkhangelskaya[1] A I Arkhangelskiy[1] S I Suchkov[2] N P Topchiev[2] Yu T Yurkin[1] A V Bakaldin[1] and O D Dalkarov[2]

[1] National Research Nuclear University MEPhI, Kashirskoe Shosse 31, Moscow, 115409, Russia

[2] P. N. Lebedev Physical Institute of the Russian Academy of Sciences, Leninskij Prospekt 53, Moscow, 119991, Russia

E-mail: mcsaksik@gmail.com

Abstract. The GAMMA-400 gamma-ray space-based telescope has as its main goals to measure cosmic γ-ray fluxes and the electron-positron cosmic-ray component produced, theoretically, in dark-matter-particles decay or annihilation processes, to search for discrete γ-ray sources and study them in detail, to examine the energy spectra of diffuse γ-rays — both galactic and extragalactic — and to study gamma-ray bursts (GRBs) and γ-rays from the active Sun. Scientific goals of GAMMA-400 telescope require fine angular resolution. The telescope is of a pair-production type. In the converter-tracker, the incident gamma-ray photon converts into electron-positron pair in the tungsten layer and then the tracks are detected by silicon-strip position-sensitive detectors. Multiple scattering processes become a significant obstacle in the incident-gamma direction reconstruction for energies below several gigaelectronvolts. The method of utilising this process to improve the resolution is proposed in the presented work.

1. Introduction

The GAMMA-400 scientific complex [1, 2] is designed to study cosmic γ-ray emissions in the energy range of 0.1 GeV to 3 TeV and acquire data that will allow us to determine the nature of dark matter in the Universe to develop a theory of the origin of high-energy cosmic rays. The GAMMA-400 instrument is designed to resolve energy spectra peculiarities, which are expected in gamma-ray emission from areas where hypothetical dark-matter particles decay or annihilate producing gamma rays. Temporal study of high-energy gamma-ray emission from discrete sources may shed light upon the nature of particle acceleration in these sources.

The design of the GAMMA-400 telescope is described in [2].

A method of gamma-ray direction reconstruction used in GAMMA-400 for higher energies [7] encounters a problem of low efficiency below 300 MeV since low-energy gammas suffer multiple scattering, which is not considered. This paper describes approach to reconstruct low-

[3] To whom any correspondence should be addressed.

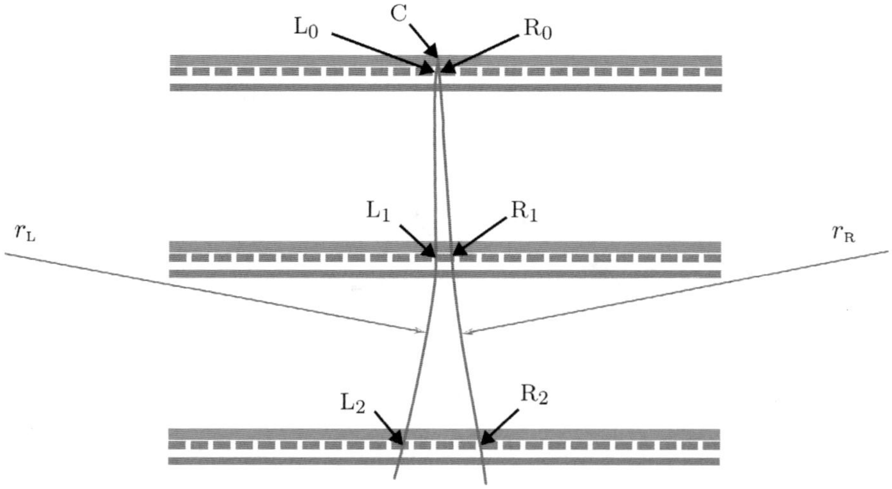

Figure 1. Schematic illustration of using the method involved. Here shown are the points in the converter-tracker used in the reconstrucion algorithm. C is a point in a tungsten layer where a conversion occured. L_i and R_i are strips where an electron or a positron hits a converter's silicon layer. Three consecutive layers directly below the converting tungsten layer are considered.

energy gamma-ray direction using only converter planes and utilising multiple-scattering process.

2. Reconstruction method

Events were selected with signals in all layers of time-of-flight system, with particles arriving from upper hemisphere in the instrument aperture.

When the energy of the incident gamma ray particle is $\lesssim 10\,\text{GeV}$, an electron-positron pair produced in the primary conversion can be readily tracked individually.

Due to multiple scattering, electrons and positrons are deflected from initial direction and their trajactories can be fitted by a circular arc. As Monte-Calrlo simulation shows, mean curvature of an arc decreases due to multiple scattering with rise of kinetic energy because of more sizeable scattering of low-energy particles.

Beforehand, the relation between mean (actually, statistical mode by histogram) of radius of curvature and electron (positron) energy is calculated using Monte-Carlo simulation. The relation appears to be $r/\text{m} = 5.5\,E/\text{GeV}$.

It is clear that tracks of pair particles of higher energy (smaller curvature) lie closer to initial gamma-ray track. The idea is to find a somewhat 'weighted bisector' of pair tracks in each orientation with weights being estimated by a track curvature. We use here the fact that mean curvature due to multiple scattering decreases with rise of kinetic energy.

The method is the following.

- Find tungsten layer W in which a primary conversion occurs (no tracker signal above and >3 consecutive directly below in each X/Y orientation).
- Define the silicon layer (two strip planes) immediately below W to be number $i = 0$; others are assigned numbers 1 and 2 (see fig. 1).

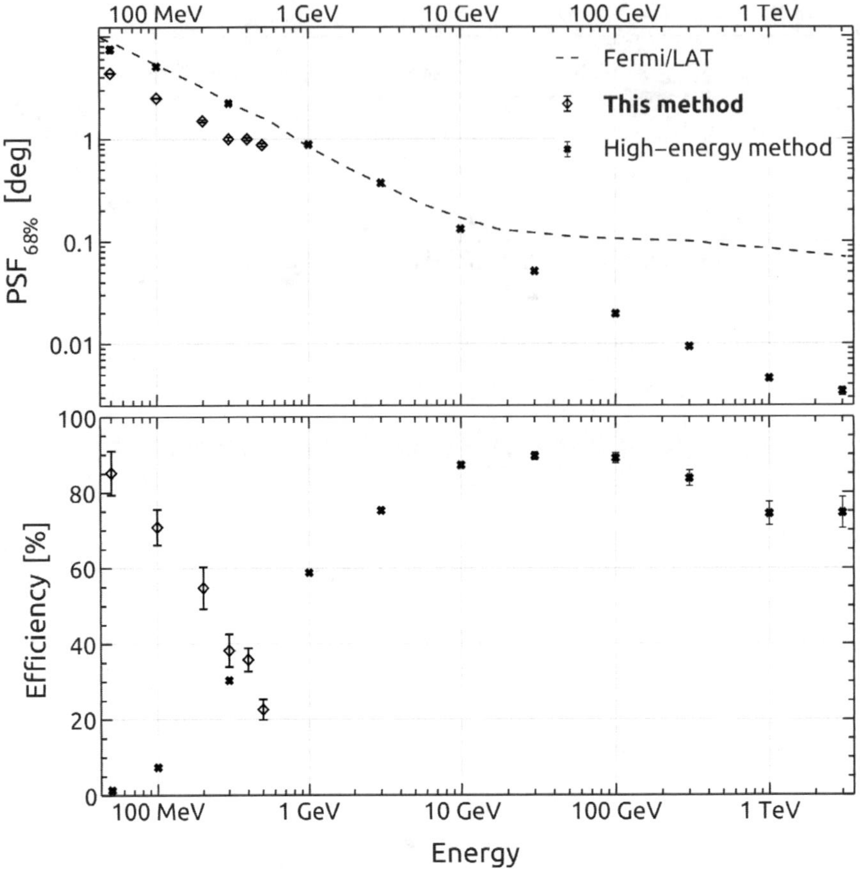

Figure 2. Angular resolution (PSF 68% containment) and reconstruction efficiency versus initial-gamma energy, as obtained by the method involved (diamonds ⋄). Also shown are: the result by the method used with higher energies [7] (crosses ✖) and the Fermi-LAT data [8] (dashed line - - -).

- Obtain coördinate X_{C} of the conversion point C. To do this, first approximations of the two directions are computed via weighted linear fit. We fit points L_0, L_1, L_2 (last weight being 10 times smaller), and the fitted line intersects W layer at point C_{L} (C_{R} got similarly).
- Radii of curvature r_{L} and r_{R} (see fig. 1) are defined to be the radii of circles passing the three points, $(\mathrm{L}_0, \mathrm{L}_1, \mathrm{L}_2)$ and $(\mathrm{R}_0, \mathrm{R}_1, \mathrm{R}_2)$ respectively. They serve as a rough estimation of electron (positron) energy using pre-calculated relation between r and E. Thus, we have now estimations E_{L} and E_{R}. Now, point C is defined by

$$X_{\mathrm{C}} = \frac{E_{\mathrm{L}} X_{\mathrm{C_L}} + E_{\mathrm{R}} X_{\mathrm{C_R}}}{E_{\mathrm{L}} + E_{\mathrm{R}}}$$

- Now that we have got the 4th point on both tracks, the linear fit described above is re-done with new point triples: $(\mathrm{C}, \mathrm{L}_0, \mathrm{L}_1)$ and $(\mathrm{C}, \mathrm{R}_0, \mathrm{R}_1)$. Just as before, the lower point's weights are reduced tenfold. Two straight lines are obtained with angles $\alpha_{\mathrm{L}}^{\mathrm{X}}$ and $\alpha_{\mathrm{R}}^{\mathrm{X}}$ from the vertical. The angle

$$\alpha_{\mathrm{X}} = \frac{E_{\mathrm{L}} \alpha_{\mathrm{L}}^{\mathrm{X}} + E_{\mathrm{L}} \alpha_{\mathrm{R}}^{\mathrm{X}}}{E_{\mathrm{L}} + E_{\mathrm{R}}}$$

together with similarly evaluated α_Y is considered as a direction of the incident gamma $(\tan(\alpha_X), \tan(\alpha_Y), 1)$.

In fig. 2, angular resolution (PSF 68% containment) is shown. It is sligthly better then that for the standard method [7] used for high energies and has relatively high efficiency at low energies ($\leq$ 300 MeV). The Fermi-LAT PSF size [8] is shown for a comparison.

3. Conclusion

The mehod proposed leads to fine angular resolution with having reconstruction efficiency relatively high at energies below 300 MeV.

References

[1] Galper A M, Adriani O, Aptekar R L *et al.* 2013 *Adv. Space Res.* **51** 297

[2] Topchiev N P, Galper A M *et al.* 2015 *Bulletin of the Russian Academy of Sciences. Physics* **79** 417–420

[3] Atwood W B, Abdo A A, Ackerann M *et al.* 2009 *Astrophys. J.* **697** 1071

[4] Abdo A A, Ackerann M, Ajello M *et al.* 2009 *Astropart. Phys.* **32** 193

[5] Ackermann M, Ajello M, Allafort A *et al.* 2013 **209** 34

[6] Tavani M, Barbiellini G, Argan A *et al.* 2009 *Astron. Astrophys.* **502** 995

[7] Galper A M *et al.* 2011 *B. Lebedev Phys. Inst.+* **38** 191

[8] Wood M, Caputo R, Rando R, Charles E, Digel S and Baldini L 2015 Fermi LAT Performance URL `http://www.slac.stanford.edu/exp/glast/groups/canda/lat_Performance.htm`

[9] Galper A M, Adriani O, Aptekar R L *et al.* 2012 *AIP Conf. Proc.* **1516** 288

[10] Galper A M, Adriani O, Aptekar R L *et al.* 2013 *Bull. Russ. Acad. Sci. Phys.* **77** 1339

Journal of Physics: Conference Series **675** (2016) 032013 doi:10.1088/1742-6596/675/3/032013

The structure of control and data transfer management system for the GAMMA-400 scientific complex

A I Arkhangelskiy[1], S G Bobkov[2], O V Serdin[2], M S Gorbunov[2] and N P Topchiev[3]
[1] National Research Nuclear University MEPhI (Moscow Engineering Physics Institute), Kashirskoe highway 31, Moscow, 115409, Russia
[2] Scientific Research Institute of System Analysis of the Russian Academy of Sciences, Nahimovskij Prospect 36/1, Moscow, 117218, Russia
[3] P.N. Lebedev Physical institute of the Russian Academy of Sciences, Leninskij Prospekt 53, Moscow, 119991, Russia

E-mail: AIArkhangelkiy@mephi.ru

Abstract. A description of the control and data transfer management system for scientific instrumentation involved in the GAMMA-400 space project is given. The technical capabilities of all specialized equipment to provide the functioning of the scientific instrumentation and satellite support systems are unified in a single structure. Control of the scientific instruments is maintained using one-time pulse radio commands, as well as program commands in the form of 16-bit code words, which are transmitted via onboard control system and scientific data acquisition system. Up to 100 GByte of data per day can be transferred to the ground segment of the project. The correctness of the proposed and implemented structure, engineering solutions and electronic elemental base selection has been verified by the experimental working-off of the prototype of the GAMMA-400 scientific complex in laboratory conditions.

1. Introduction

The GAMMA-400 space project [1-3] is intended for precision investigation of the cosmic gamma-emission in the energy band from ~20 MeV up to several TeV, e^-/e^+ fluxes from ~1 GeV up to ~10 TeV, cosmic-ray nuclei fluxes with energies up to ~10^{15} eV and gamma-ray bursts in the energy band of 10 keV–15 MeV. The GAMMA-400 space observatory will be launched on the NAVIGATOR service platform [4] designed by LAVOCHKIN ASSOCIATION at the beginning of the next decade on the elliptical orbit with following initial parameters: an apogee ~300000, a perigee ~500 km, a rotation period ~7 days and inclination of 51°. The expected observatory operational lifetime is more then 7 years. The planned scientific complex main technical parameters are: weight ~4100 kg, power consumption ~2000 W, total scientific downlink transmission up to 100 GByte/day.

2. The information structure of GAMMA-400 scientific instrumentation

The GAMMA-400 scientific instrumentation can be subdivided into two main parts: GAMMA-400 scientific complex and NAVIGATOR service platform systems directly providing the operation of scientific equipment. Figure 1 illustrates the structure of the integrate control and data transfer management system presented a distributed system for acquisition and pre-processing data from scientific measuring systems (SMS) and providing the control and fine tuning of their functioning.

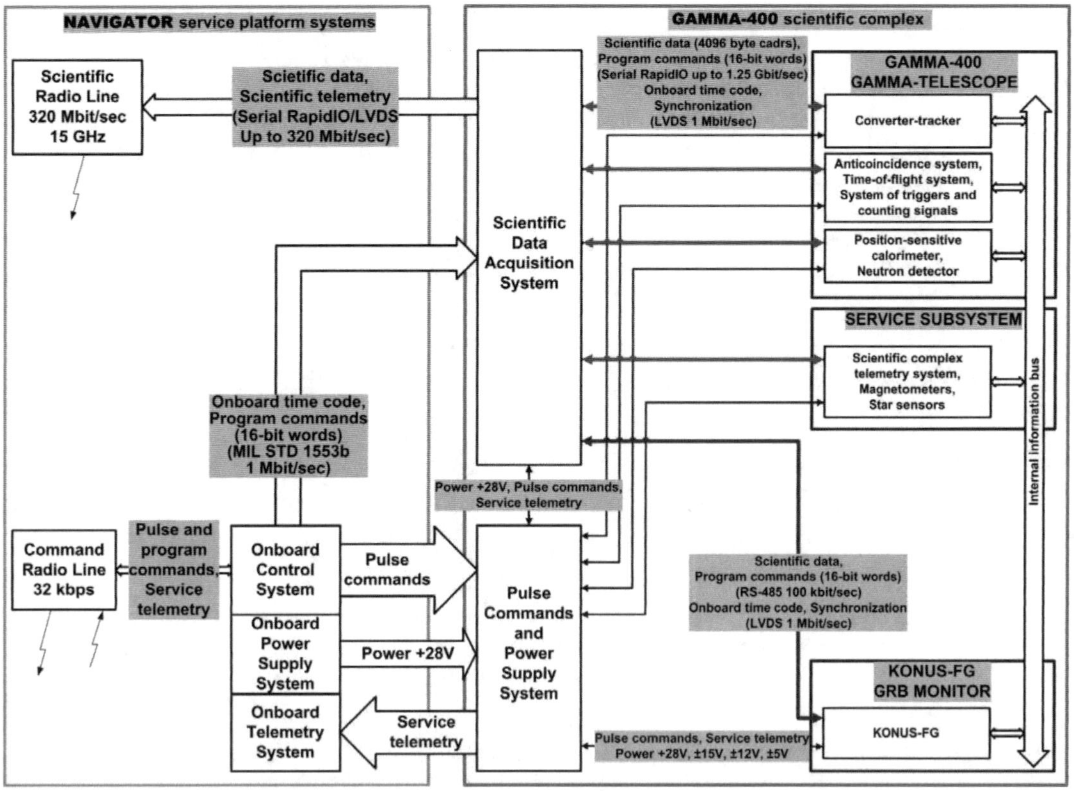

Figure 1. Functional diagram of GAMMA-400 control and data transfer management system.

The GAMMA-400 scientific complex consists of the following subsystems [5-7]:

- gamma-telescope GAMMA-400;
- gamma-ray burst monitor KONUS-FG;
- service subsystem, including two magnetometers for measuring the magnetic field along the satellite trajectory, two star sensors for determining the GAMMA-400 axes with accuracy of approximately 5" and scientific complex telemetry system for cyclic registration up to 65535 scientific complex housekeeping parameters;
- scientific data acquisition system (SDAS) [8] for acquisition and pre-processing data from SMS, storage it in non-volatile mass memory (1 TByte), scientific data and telemetry transfer into high-speed scientific radio line (up to 320 Mbit/sec) for their transmission to the ground segment of the project and control information reception from spacecraft onboard control system via multiplexed channel MIL-STD-1553B, its decoding and transfer into SMS;
- pulse commands and power supply system, providing secondary power supply for scientific complex (+28V, ±15V, ±12V, ±5V), one-time pulse radio commands sharing and their transmitting to SMS and transit of the SMS telemetry parameters to the onboard telemetry system.

The next systems of NAVIGATOR service platform are used for direct providing the functioning of scientific measuring systems:

- command radio line for receiving one-time pulse radio commands and 16-bit program control commands from control ground station and transmit them through onboard control system (OCS) to SMS. Precise timing of scientific data is maintained in SDAS by using onboard time

Journal of Physics: Conference Series **675** (2016) 032013 doi:10.1088/1742-6596/675/3/032013

code from OCS, forming a 32-bit serial onboard time code transmitted to each SMS via LVDS interface along with reference synchronization signals 1 Hz, 1kHz and 1 MHz;

- scientific radio line for scientific data transmitting to the data-acquisition ground stations at a frequency of 15 GHz. The nominal data transfer rates are 160/118.8 or 320/237.6 Mbit/sec with/without coding respectively. Data encryption is used to reduce the quantity of transmission failures (level of 10^{-9} of malfunction probability in the downlink data transmission) by means of the cascade coding technique involving the Reed-Solomon and convolution codes;
- onboard telemetry system for direct (bypassing of SCTS and SDAS) monitoring of analog and digital parameters, contacts and resistance thermo sensors (up to 100 of the most important parameters for operational diagnostics of the scientific complex health);
- onboard power supply system, providing overall power supply for scientific complex (+28V).

For the control and fine tuning of scientific measuring systems functioning one can use:

- up to 100 one-time pulse radio commands in the form of either voltage pulse with duration ~0.1÷0.3 sec with amplitude equal to satellite power or "dry contacts" – switch on/switch off;
- up to 65535 programming control commands (16-bits control words) for each SMS.

The main part of the scientific data is transferred in the form of 4096 bytes length digital arrays from scientific measuring systems to the SDAS, which maintains data reception from four (three for gamma-telescope subsystems and one for service subsystem) sources of digital arrays via high-speed Serial RapidIO (1.25 Gbit/sec) channels. For data transfer from low-information (~100 kbit/sec) KONUS-FG the serial interface RS-485 is used. Data are loaded upon a request for service from SMS.

Two main operating regimes can be specified for GAMMA-400 scientific complex:

- event registration regime, corresponding to the registration of events according to the master trigger signals generated by system of triggers and counting signals formation. In this regime the data acquisition from all detecting system of the gamma-telescope is performed;
- patrol regime, realized all time independently of event registration. In this regime the periodic data acquisition from detecting system of gamma-telescope is performed. The period and set of interrogated detecting systems can be changed by the control commands.

In table 1 the information data (unpacked) requirements for GAMMA-400 scientific measurement systems for main operating regimes are presented.

Table 1. Estimated dataflow from scientific complex in the main operation regimes.

Scientific measurement system	Event regime (bytes/event)	Patrol regime (bytes/day)
Gamma-telescope GAMMA-400	325 k	800 M
GRB monitor KONUS-FG	-	200 M
Service subsystem	-	70 M
TOTAL	**325 k**	**1070 M**

Figure 2 and 3 represents the functional and timing diagram of the Serial RapidIO interface between SDAS and SMS of scientific complex. Each SMS includes uniform interface module, based on high-reliability, fault tolerance system-on-chip 1907VM038 with fast internal input-output channels [9]. Interface module includes additional DDR2 RAM (512 MB) for buffering scientific data arrays, received from SMS via 33 MHz 8-bit synchronous parallel bus and NAND FLASH memory for executable program storage.

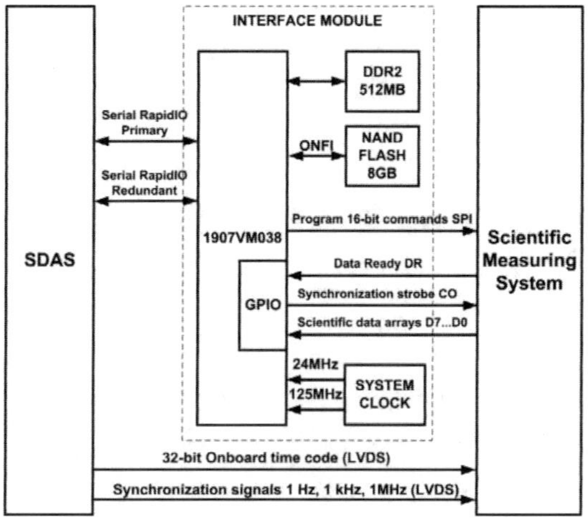

Figure 2. Functional diagram of the Serial RapidIO interface between SDAS and scientific measuring systems of scientific complex.

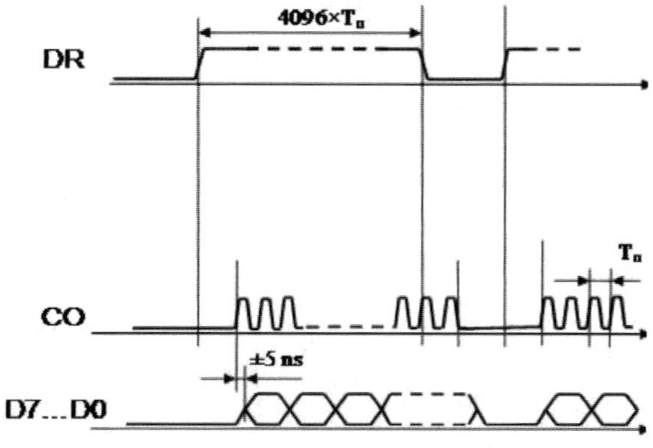

Figure 3. Timing diagram of the scientific data interchange between SDAS and SMS of scientific complex. **DR** – "data ready" request from SMS, **CO**-33 MHz synchronization strobe, **D7...D0** - 8-bit parallel scientific data (all signals are LVTTL).

3. Conclusion

During the GAMMA-400 project development, the prototype of scientific complex was created [10]. Experimental working-off of the main construction units in laboratory conditions has justified the engineering solutions and electronic elemental base selection for design of scientific complex equipment. The structure of the onboard control complex and the data transmission system from scientific instruments for the GAMMA-400 apparatus satisfies all scientific complex requirements. The use of the distributed system involving the scientific complex and service systems of the NAVIGATOR service module provides possibility for adaptive and operational management of the parameters of scientific instruments, as well as acquisition and transmission to the ground segment of the project complete streams of data and operational integrity of the scientific complex.

Acknowledgments
This work performed within the framework of the Center FRPP supported by MEPhI Academic Excellence Project (contract No. 02.a03.21.0005, 27.08.2013). Also, the work was partly supported by Russian Foundation for Basic Research (RFBR), Grants No. 14-29-09268 and No. 14-29-09207.

References

[1] Topchiev N P, Galper A M, Bonvicini V *et al* 2015 *Bull. Russ. Acad. Sci. Phys.* **79(3)** 417
[2] Galper A M, Topchiev N P, Bonvicini V *et al* 2015 *Physics Procedia* **74** 177
[3] Galper A M, Adriani O, Arkhangelskaja I V *et al* 2013 *Advances in Space Research* **51** 297
[4] Syrov A S, Smirnov V V, Sokolov V N *et al* 2015 *Cosmonautics and Rocket Engineering* **3** 58
[5] Galper A M, Adriani O, Aptekar R L *et al* 2013 *Bull. Russ. Acad. Sci. Phys.* **77(11)** 1339
[6] Galper A.M., Adriani O, Aptekar R L *et al* 2013 *AIP Conf. Proc.* **1516** 288
[7] Ginzburg V L, Kaplin V A, Runtso M F *et al* 2009 *Bull. Russ. Acad. Sci. Phys.* **73(5)** 664
[8] Bobkov S G, Serdin O V, Gorbunov M S *et al* 2015 The scientific data acquisition system of the GAMMA-400 space project *J. Phys.:Conf. Series This volume*
[9] Bobkov S G, Serdin O V, Arkhangelskiy A I *et al* 2015 *Physics Procedia* **74** 224
[10] Arkhangelskiy A I, Arkhangelskaja I V, Kheymits M D *et al* 2015 *Physics Procedia* **74** 191

The scientific data acquisition system of the GAMMA-400 space project

S G Bobkov[1], O V Serdin[1], M S Gorbunov[1], A I Arkhangelskiy[2] and N P Topchiev[3]
[1] Scientific Research Institute of System Analysis of the Russian Academy of Sciences, Nahimovskij Prospect 36/1, Moscow, 117218, Russia
[2] National Research Nuclear University MEPhI (Moscow Engineering Physics Institute), Kashirskoe highway 31, Moscow, 115409, Russia
[3] P.N. Lebedev Physical institute of the Russian Academy of Sciences, Leninskij Prospekt 53, Moscow, 119991, Russia

E-mail: AIArkhangelkiy@mephi.ru

Abstract. The description of scientific data acquisition system (SDAS) designed by SRISA for the GAMMA-400 space project is presented. We consider the problem of different level electronics unification: the set of reliable fault-tolerant integrated circuits fabricated on Silicon-on-Insulator 0.25 mkm CMOS technology and the high-speed interfaces and reliable modules used in the space instruments. The characteristics of reliable fault-tolerant very large scale integration (VLSI) technology designed by SRISA for the developing of computation systems for space applications are considered. The scalable net structure of SDAS based on Serial RapidIO interface including real-time operating system BAGET is described too.

1. Introduction

The GAMMA-400 space project [1-5] is intended for investigation of cosmic γ-emission in the energy band from ~20 MeV up to several TeV, electron/positron fluxes from ~1 GeV up to ~10 TeV and cosmic-ray nuclei fluxes with energies up to $\sim 10^{15}$ eV by means of GAMMA-400 gamma-telescope representing the core of the scientific complex. The investigation of gamma-ray bursts in the energy band of 10 keV–15 MeV are possible too by means of KONUS-FG apparatus included in the complex.

The scientific data acquisition system (SDAS) represents the information kernel of scientific complex and must be a high-reliable element. The SDAS is a distributed system for acquisition, pre-processing and registration data from scientific measuring systems (SMS) of scientific complex. Data encryption is used to reduce the quantity of transmission failures. In order to increase the reliability, SDAS is made using a scheme with two hot- and cold-reserved subsystems. All SMS-SDAS data exchange signals are double redundant. Each redundant line assigns with its own allocated data transceiver. Additional reliability level of SDAS is achieved by minimization of high integrity chips amount.

2. The scientific data acquisition system of the GAMMA-400 project

The main functions of the scientific data acquisition system (SDAS) are following:

- the data acquisition from SMS of scientific complex (up to 100 GByte per day);
- preliminary data processing of scientific information and storage it in nonvolatile mass memory (1 TByte total);

- scientific information transfer into high-speed (320 Mbit/sec) scientific radio line for its transmission to the data-acquisition ground stations;
- control information reception from spacecraft onboard control system via MIL-STD-1553B interface, its decoding and transfer into SMS;
- receiving signals from onboard time and frequency standard system and onboard control system and generating high-stable reference synchronization signals (1 Hz, 1 kHz and 1 MHz) and 32-bit onboard time code for precise timing of scientific data and transmitting them to each SMS via LVDS interface.

The functional diagram of SDAS is shown in figure 1. The SDAS consists of three types of modules: BTSSNI-201 (the central block with processing unit), BTSSNI-301 (data storage unit) and the control module BTSSNI-001. The main purpose of BTSSNI-001 is the conversion of the primary supply channel obtained from the power management system and the control of the supply channels for the modules. The primary and redundant modules with the control module BTSSNI-001 are set to the unifying board that is placed into the package. Processor elements are joined to the net using high-speed and high-reliable 1x 4x 1.25 Gbit/sec Serial RapidIO interface.

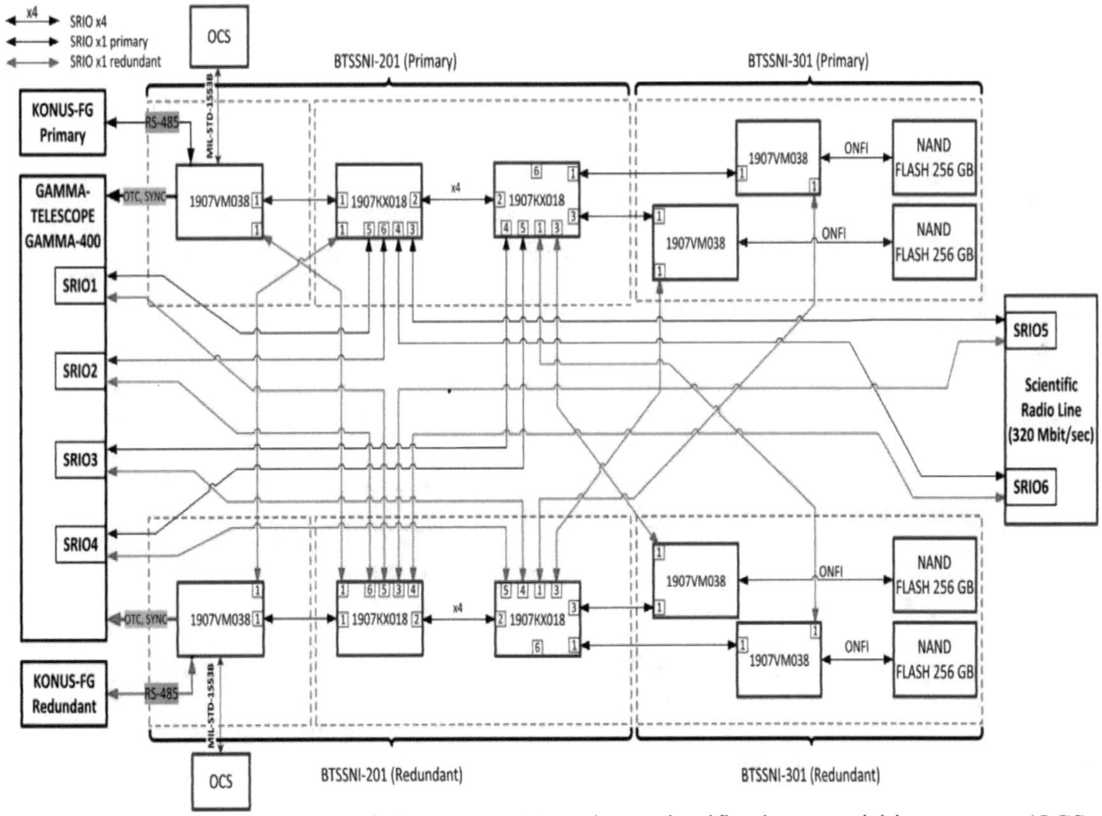

Figure 1. Functional diagram of GAMMA-400 project scientific data acquisition system (OCS - onboard control system, OTC – 32-bit onboard time code, SYNC – high-stable synchronization signals (1 Hz, 1 kHz, 1 MHz) , SRIO - Serial RapidIO interface, digits in squares are the identifiers of SRIO ports).

Data acquisition from SMS is divided into two data flows. The data acquisition from the gamma-telescope of scientific complex and scientific telemetry system is provided using four high-speed data acquisition Serial RapidIO channels with total throughput of ~70 MByte/sec. The scientific data is transferred to the processing block, which consists of 1907VM038 device. This microprocessor provides both processed data storage and the fetch of data from the storage using six-port Serial RapidIO switch 1907KX018 to transfer it to the high-speed scientific radio line. The data acquisition from low-informative ~100 kbit/sec KONUS-FG device is realized via RS-485 interface. The data storage unit is designed using the system-in-package NAND flash memory. For the reliable transferring during the whole scientific complex lifetime (7-10 years), we use the redundant 1 TByte data storage.

The connection of SDAS with the onboard control system is provided using the redundant multiplexed channel MIL-STD-1553B and one-time pulse commands. The diagnostics of different SDAS nodes and control of temperature of "heating" VLSI are provided by onboard telemetry system.

Table 1 presents the distinctive features of radiation-tolerant VLSI designed by SRISA implemented in the SDAS.

Table 1. The distinctive features of VLSI implemented in SDAS of GAMMA-400 project [6].

Parameter	1907VM038	1907KX018
Clock rate	100 MHz	100 MHz
Serial RapidIO performance	1 channel 1x 4x 1.25 Gbit/sec	6 channels 1x 4x 1.25 Gbit/sec
SpaceWire performance	4 channels 400 Mbit/sec	-
SEL	none	none
SEE LET_{th} (without error correction)	~6 MeV·cm^2/mg	~6 MeV·cm^2/mg
σ_{sat}	$< 3\cdot10^{-8}$cm^2/bit	$< 3\cdot10^{-8}$cm^2/bit
TID_{Si}	$\geq$ 50 krad	$\geq$ 50 krad
Operating temperature	-60 ÷ +125°C	-60 ÷ +125°C
Power consumption	$\leq$ 8 W	$\leq$ 6 W
Package	DBGA 675	DBGA 399

System-on-chip 1907VM038 based on processor core KOMDIV-32 [7-10] is used for various onboard systems. It consists of 32-bit central processing unit (CPU), 128-bit arithmetic co-processor CP2, system controller with DDRII, SPI, 1.25 Gbit/sec Serial RapidIO, SpaceWire (4 channels), I^2C, GPIO, UART. Clock frequency is 100 MHz, the throughput of RAM is 512 MByte/sec.

The Serial RapidIO switch VLSI 1907KX018 is used for the connection of various switches and systems-on-chip. It consists of six Serial RapidIO ports; the transferring environment is configured independently: LP-Serial 4X or 1X. Maximum transferring speed is 1.25 Gbit/sec (per channel). It has the commutation table for each port, performance control system and built-in error-correcting block. The switch can directly connect up to 256 devices in the system, which is usually enough for the onboard systems. Individual commutation tables allow to flexibly configuring the transferring of data packages. The performance control system is used for the defining of characteristics of data flow in channel, overload detection, the localization of locking. The defining of the data flow is provided at the stage of net topology setting simulation and experimentally. The possible reasons of overloads can be the degradation of channels due to faults. Early detection of faults decreases the time needed to enable the reserved path. It is very useful that the switch can define the locked channels for the reserved path choosing. Maximum power consumption when all six ports in 4X mode are used is less than 6 W. In 1X mode or using the individual switching-off of an individual port's transmitter, the power consumption is reduced. The supply voltages are 2.5 V and 3.3 V. The high fault-tolerance is achieved using the error-correcting codes for the data package memory, the redundant commutation tables, the layout and circuit techniques at the standard cells level. The switch has the extended ability of the performance diagnostics. The state of diagnostics registers is available through I^2C interface, Also, the separate fault signals can be sent to the discrete signal port for the further analysis by the repairing system.

As operating system, real-time OS (RTOS) Baget 3.0 [11, 12] was chosen. This RTOS is developed based on the following general approaches:

- use of standards (ARINC 653 and POSIX 1003.1 for programming interface; C standard for C language and libraries);
- portability;
- advanced facilities for tracing, logging, diagnostics. and error handling (health monitor):
- flexible scheduling;
- object-oriented approach:
- scalability (configuration tools);
- instrumental software for developing and debugging user cross-applications;
- large number of environmental packages for creating graphics applications, databases, and mapping systems.

For the best portability, OS is divided on three main parts: main part independent on hardware, written on C language (the biggest); the second part, dependent only on central processor type, written on C or Assembler language (much smaller); modules support part containing modules drivers.

During software design, cross-development technology is used whereby, using the host computer with a general-purpose OS, the source codes and OS libraries are stored as well as the compilation and build of the boot image that is executed on the target computer under the control of RTOS are performed. For debugging in terms of the source code, a remote debugger is used.

3. Conclusion

The unification of space-grade chips is achieved at the core level, interfaces and the data transferring systems. The system is made hierarchical to reduce the power consumption and to increase the reliability. The upper level is switchable Serial RapidIO environment with 1.25 GByte/sec per channel. The lower level is low-speed multiplexed MIL-STD-1553B and RS-485 channels. The computation system hardware level is designed based on VPX standard.

During the current stage of GAMMA-project development, the prototype of SDAS was created. Experimental working-off of the main construction units in laboratory conditions has justified the engineering solutions and electronic elemental base selection for its design. The overall performance (>320 Mbit/sec) and power consumption (<80 W) of the SDAS satisfies all scientific complex requirements.

Acknowledgments

This work performed within the framework of the Center FRPP supported by MEPhI Academic Excellence Project (contract No. 02.a03.21.0005, 27.08.2013). Also, the work was partly supported by Russian Foundation for Basic Research (RFBR), Grants No. 14-29-09268 and No. 14-29-09207.

References

[1] Topchiev N P, Galper A M, Bonvicini V *et al* 2015 *Bull. Russ. Acad. Sci. Phys.* **79(3)** 417
[2] Galper A M, Adriani O, Arkhangelskaja I V *et al* 2013 *Advances in Space Research* **51** 297
[3] Galper A M, Adriani O, Aptekar R L *et al* 2013 *Bull. Russ. Acad. Sci. Phys.* **77(11)** 1339
[4] Galper A M , Adriani O, Aptekar R L *et al* 2013 *AIP Conf. Proc.* **1516** 288
[5] Ginzburg V L, Kaplin V A, Runtso M F *et al* 2009 *Bull. Russ. Acad. Sci. Phys.* **73(5)** 664
[6] Bobkov S G, Serdin O V, Arkhangelskiy A I *et al* 2015 *Physics Procedia* **74** 224
[7] Bobkov S G 2012 *Information Technologies* **12** 2 (In Russian)
[8] Sokolov I A, Stepchenkov Yu A, Bobkov S G *et al* 2014 *Informatics and its Applications* **8(1)** 45 (In Russian)
[9] Bobkov S G 2014 *Vestnik Rossiiskoi Akademii Nauk* **84(11)** 1010 (In Russian)
[10] Bobkov S G, Eremin A V, Kataev A F *et al* 2013 *Software & Systems* **4** 49 (In Russian)
[11] Godunov A N 2010 *Software & Systems* **4** 16 (In Russian)
[12] Godunov A N and Soldatov V A 2014 *Programming and Computer Software* **40(5)** 259

International Conference on Particle Physics and Astrophysics (ICPPA-2015) IOP Publishing
Journal of Physics: Conference Series **675** (2016) 032015 doi:10.1088/1742-6596/675/3/032015

Gamma-quanta onboard identification in the GAMMA-400 experiment using the counting and triggers signals formation system.

I V Arkhangelskaja[1], A I Arkhangelskiy[1], E N Chasovikov[1], A M Galper[1,2], M D Kheymits[1], A E Murchenko[1], Y T Yurkin[1]
[1]National Research Nuclear University MEPhI (Moscow Engineering Physics Institute), Kashirskoe highway 31, Moscow, 115409, Russia
[2]Lebedev Physical Institute, Russian Academy of Sciences, Leninskiy prospekt 53, Moscow, 119991, Russia

E-mail: irene.belousova@usa.net

Abstract. GAMMA-400 (Gamma Astronomical Multifunctional Modular Apparatus) will be the new generation satellite gamma-observatory. Gamma-telescope GAMMA-400 consists of anticoincidence system (top and lateral sections - ACtop and AClat), the converter-tracker (C), time-of-flight system (2 sections S1 and S2), position-sensitive calorimeter CC1 makes of 2 strips layers and 2 layers of CsI(Tl) detectors, electromagnetic calorimeter CC2 composed of CsI(Tl) crystals, neutron detector ND, scintillation detectors of the calorimeter (S3 and S4) and lateral detectors of the calorimeter (LD). All detector systems ACtop, AClat, S1-S4, LD consist of two BC-408 based sensitive layers of 1 cm thickness each. Three apertures provide events registration both from upper and lateral directions. The main aperture provides the best angular (all strip layers information analysis) and energy (energy deposition in the all detectors studying) resolution. Gamma-telescope GAMMA-400 is optimized for the gamma-quanta and charged particles with energy 100 GeV detection with the best parameters in the main aperture. Triggers in the main aperture will be formed using information about particle direction provided by time of flight system and presence of charged particle or backsplash signal formed according to analysis of energy deposition in combination of both layers anticoincidence systems ACtop and AClat individual detectors. For double-layer ACtop taking into account both amplitude and temporal trigger marker onboard analysis only 2.8 % photons will be wrongly recognized as electrons or protons for 100 GeV particles. The part of charged particles mistakenly identified as gammas is $\sim 10^{-5}$ using described algorithms. For E~3 GeV less than 3% photons will be wrongly recognized as charged particles and fraction of wrongly identified charged particles will be also $\sim 10^{-5}$. In the additional aperture the particles identification is provided by analysis of signals corresponding to energy deposition in the individual detectors S2, S3 and fast signals from CC1 individual detectors discriminators. Low energy (0.2 - 10 MeV) photons in the lateral aperture recognizing by using simple anticoincidence signals from the individual detectors of LD. Gamma-quanta of higher energies are identified using energy deposition in the individual detectors of S3, S4, LD and fast signals from CC2 individual detectors discriminators. The results of anticoincidence system individual detectors thresholds modelling are discussed.

1. GAMMA-400 short description

GAMMA-400 (Gamma Astronomical Multifunctional Modular Apparatus) will be the new generation satellite γ-observatory. Its main scientific goals will be [1, 2]: discrete astrophysical sources observations, indirect dark matter origin study by the gamma-ray astronomy methods, diffuse background γ-emission investigations, high energy GRB emission research, research of high energy light nuclei fluxes, the study of high energy e−e+ fluxes. Also it is possible to study solar flares using GAMMA-400 lateral aperture. GAMMA-400 physical scheme is shown at figure 1.

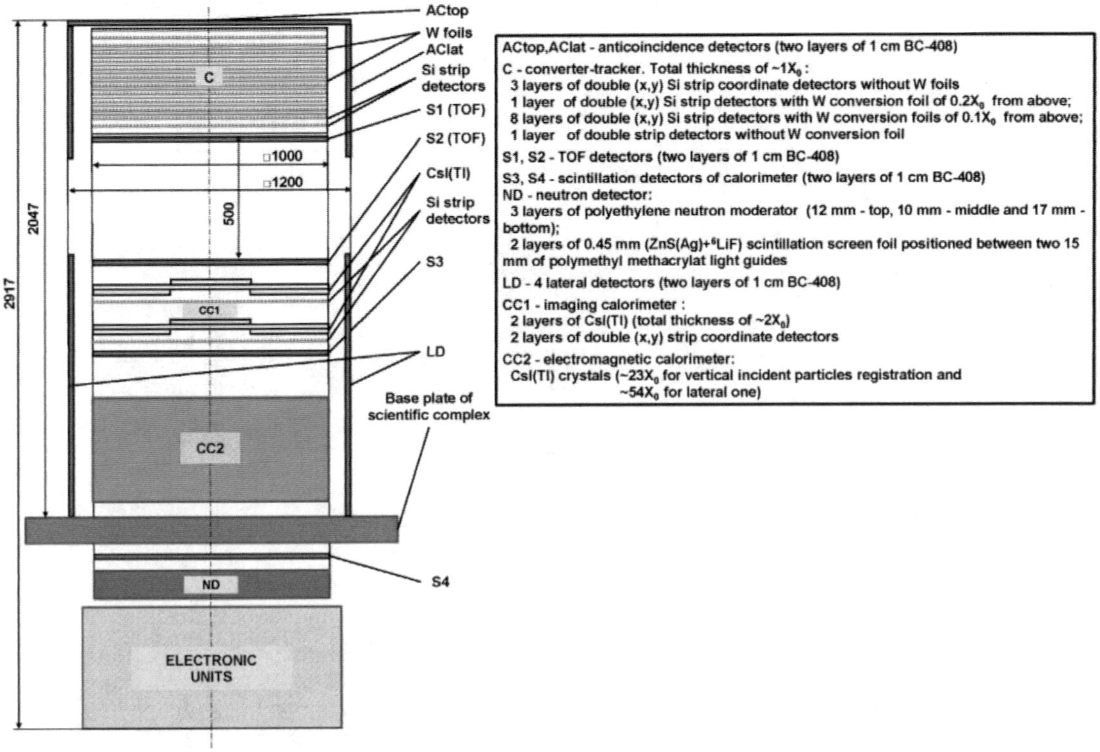

Figure 1. The physical scheme of the GAMMA-400.

Gamma-telescope GAMMA-400 consists of the converter-tracker (C), time-of-flight system (2 sections S1 and S2 composed of BC-408 detectors), anticoincidence system (BC-408 based top and lateral sections - ACtop and AClat), position-sensitive calorimeter CC1 makes of 2 strips layers (pitch of 0.08 mm) and 2 layers of CsI(Tl) detectors, electromagnetic calorimeter CC2 composed of CsI(Tl) crystals, BC-408 based scintillation detectors of the calorimeter (S3 and S4), BC-408 based lateral detectors of the calorimeter (LD), neutron detector ND [3, 4]. All BC-408 based detector systems consist of two sensitive layers of 1 cm thickness each. The converter-tracker is composed of 13 layers of double (x, y) silicon strip coordinate detectors (pitch of 0.08 mm) [1]. The first three and final one layers are without tungsten while the middle nine layers are preceded with tungsten conversion foil each: the first one with thickness 0.2 X_0 and the next eighth with 0.1 X_0 one (where X_0 is radiation length). The thickness of CC1 and CC2 is 2 X_0 (0.1 λ_0) and 23 X_0 (1.1 λ_0) respectively (where λ_0 is nuclear interaction length). The total calorimeter thickness is 25 X_0 or 1.2 λ_0 for vertical incident particles registration and 54 X_0 or 2.5 λ_0 for laterally incident ones. Together with the γ-telescope GAMMA-400, the space observatory will include non shown at figure 1 devices: two star sensors for determining the GAMMA-400 axes with accuracy of approximately 5", two magnetometers, and the KONUS-FG gamma-ray burst monitor (4 position and 2 spectrometric detectors) [4].

2. Gamma-quanta onboard identification in different apertures

Three apertures provide events registration both from upper and lateral directions. Photons distinguish provides by the system of counting and triggers signals formation [5]. The main aperture supply the best angular (all strip layers information analysis) and energy (energy deposition in the all detectors studying) resolution. Gamma-telescope GAMMA-400 is optimized for the gamma-quanta and charged particles with energy ~1.0×10^2 GeV detection with the best parameters in the main aperture [4]: the angular resolution ~0.01°, the energy resolution ~1%, and the proton rejection factor ~5×10^5. In the main aperture triggers will be formed using information about particle direction provided by TOF system due special TOF signals matrix [5] and presence of charged particle or backsplash. Figure 2 shows trigger creation scheme and detectors response analysis matrix for triggers and trigger markers signals formation for the main aperture. In this aperture gamma-telescope operated in event by event registration mode.

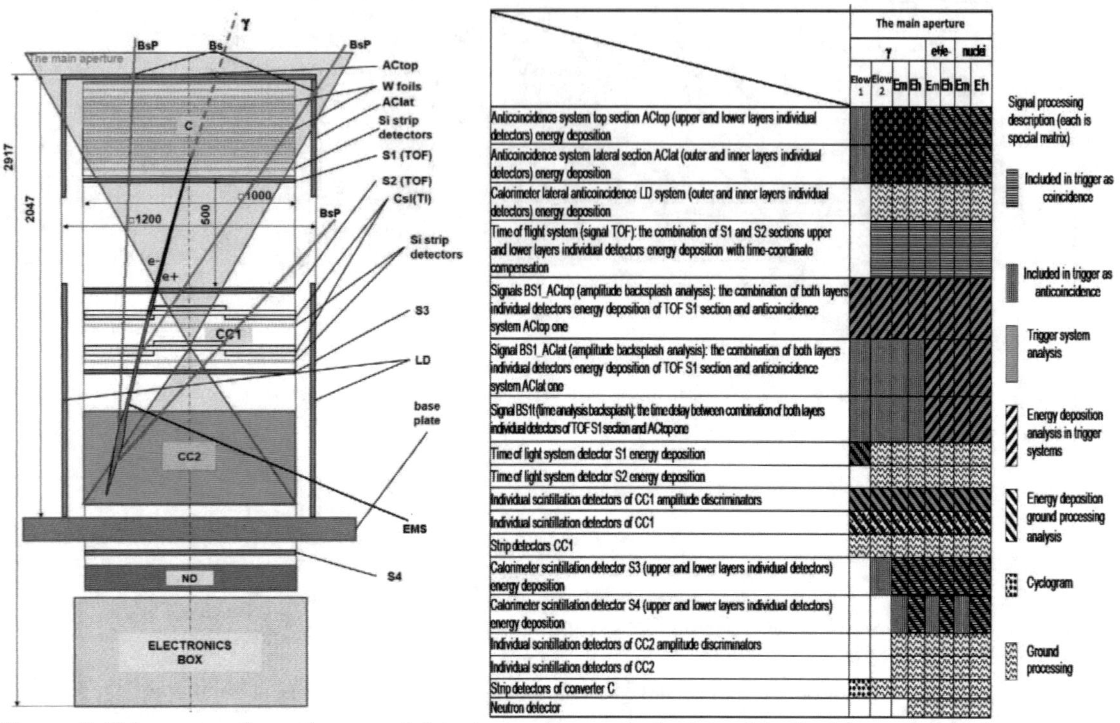

Figure 2. Trigger creation scheme and detectors response analysis matrix for triggers and trigger markers signals formation for the main aperture. Backsplash particles marked by red lines, primary gamma-quantum and its produced shower marked by magenta color.

GAMMA-400 is able to registered gamma-rays and electrons (positrons) in the energy range from 0.1–3.0×10^3 GeV in the main aperture. Also this aperture allows investigating high energy light nuclei fluxes characteristics. The GAMMA-400 effective area in the main aperture is ~4000 cm^2 at Eγ > 1 GeV. Both additional and lateral apertures energy resolution is ~2% for electrons, positrons, light nuclei and gamma-quanta in energy range E>10 GeV.

Most part of backsplash produced in calorimeters and conversion foils. Results of backsplash from calorimeters modelling for the main aperture are presented at figure 3. Also at some situations backsplash formation region is located in ACtop lower layer just below of upper layer individual detector on which particle is impinged. The energy deposition in ACtop upper layer from such kind of events is less than 1 MeV. Moreover, part of backsplash particles absorbed in conversion foils, all detectors' supports and TOF detectors. Also the backsplash electrons could be absorbed or produced photons in the first conversion foil and only sufficiently lower energy gamma component propagate through ACtop lower layer. Additionally, sufficient amount of backsplash particles tracks are outside ACtop detector system due to long distance between ACtop and calorimeters. Usually backsplash particles delayed at individual time Δt_BS correspondingly twice distance between ACtop and place of backsplash production. For GAMMA-400 this time is approximately 3.5 ns for most particles.

Results of gamma-quanta direct interaction modeling for energies 100 GeV and 3 GeV are shown at figure 4. Data examination has shown only 2.8 % photons will be wrongly recognized as electrons or protons for double-layer ACtop taking into account both amplitude and temporal trigger marker onboard analysis for 100 GeV particles. The part of charged particles mistakenly identified as gammas is $\sim 10^{-5}$ using described algorithms. For E~3 GeV less than 3% photons will be wrongly recognized as charged particles and fraction of wrongly identified charged particles will be also $\sim 10^{-5}$.

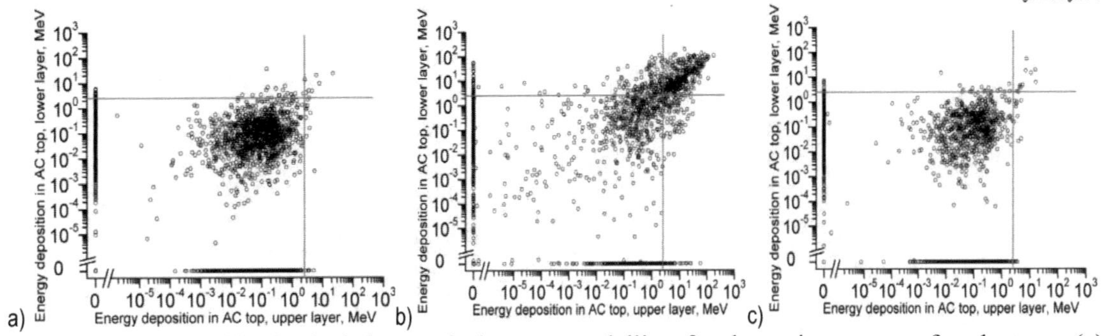

Figure 3. Results of backsplash from calorimeters modelling for the main aperture: for electrons (a), protons (b) and gammas (c) with energy 3 GeV.

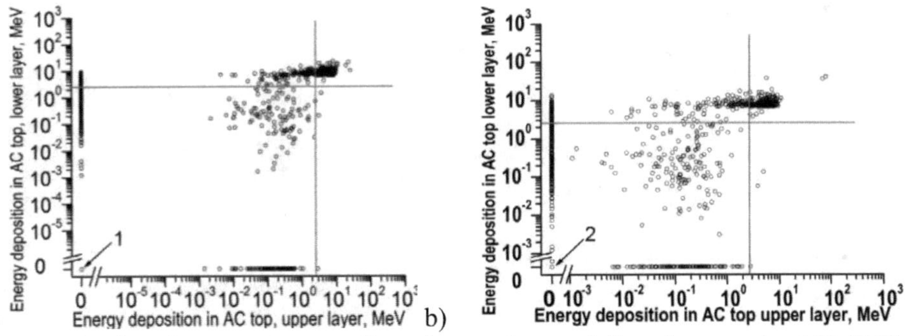

Figure 4. Results of gamma-quanta direct interaction modeling for energies 100 GeV (a) and 3 GeV (b). Lines indicates thresholds for electrons and protons with the same energies than gammas. Areas marked (1) and (2) contains events with energy deposition less than 1 keV in both ACtop layers constituted ~60% and ~70% correspondingly.

In the additional aperture gamma-telescope operated in event by event registration mode too (see panels (a) at the figures 5 and 6). Particles registration starts with signal of CC1 fast discriminators in anticoincidence with TOF pulse. Up-down particle direction indicates due fast signals from detectors CC1 and S3, anticoincidence in this aperture provides by S2, LD and S4. The angular resolution is provided by strip layers in the CC1 and for gamma-quanta is from $\sim 5^{\circ}$ to $\sim 4^{\circ}$ in energy range 1.0 MeV - 0.1 GeV and from $\sim 4^{\circ}$ to $\sim 0.7^{\circ}$ for 0.1 - 1.0 GeV one. For particles with E >1.0 GeV the angular resolution is $\sim 0.7^{\circ}$. The additional aperture energy resolution provides due to energy deposition analysis in S2, S3, S4, LD and calorimeters (CC1 and CC2).

Gamma-quanta, electrons/positrons and light nuclei with energy E>10 GeV also registered in the lateral aperture. This aperture allows detecting of low-energy gammas in the range 0.2 - 10 MeV and photons with energy of 10 MeV – 10 GeV. The energy resolutions in these cases are 8% - 2% and 2% correspondingly according to GAMMA-400 "Technical Project" stage results.

In the lateral aperture gamma-telescope operated in event by event registration mode only for particles with E> 100 MeV.

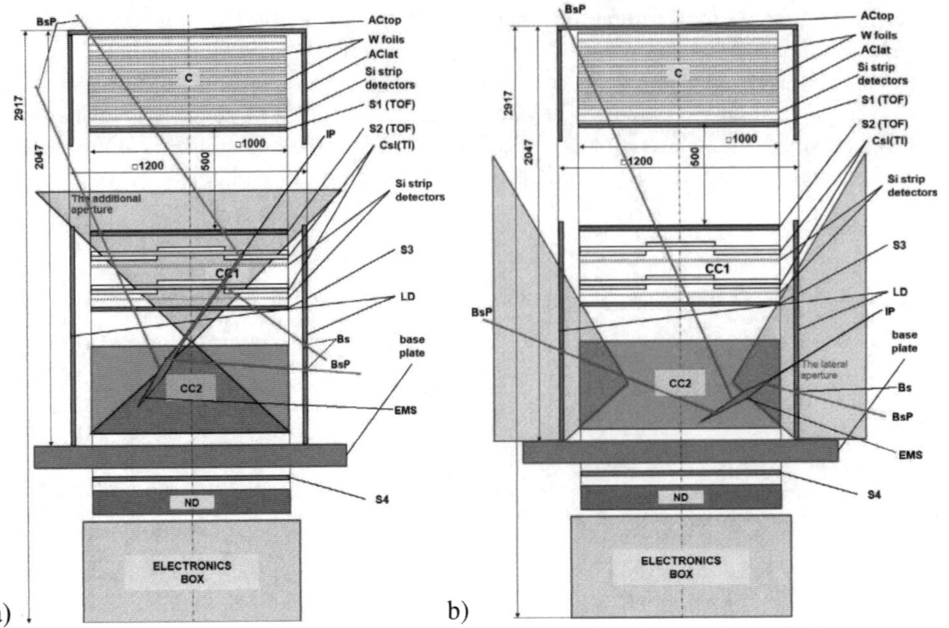

a) b)

Figure 5. Trigger creation scheme for the additional (a) and lateral (b) apertures. Backsplash particles marked by red lines, primary gamma-quantum and its produced shower marked by magenta color.

a)

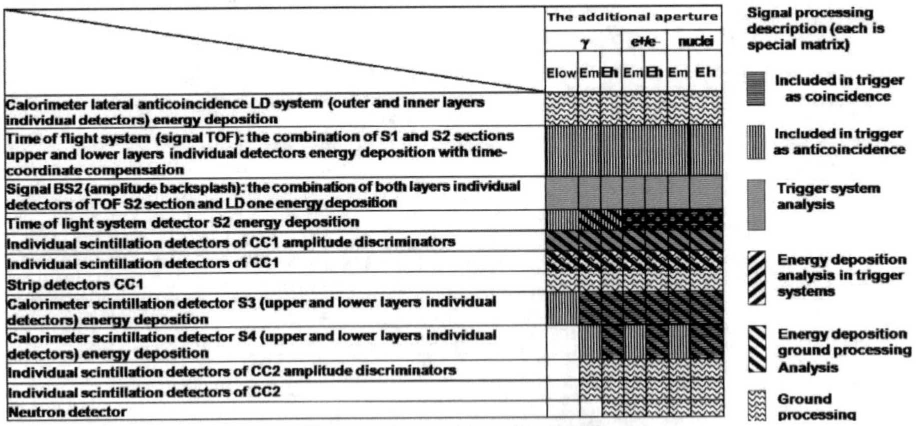

b)

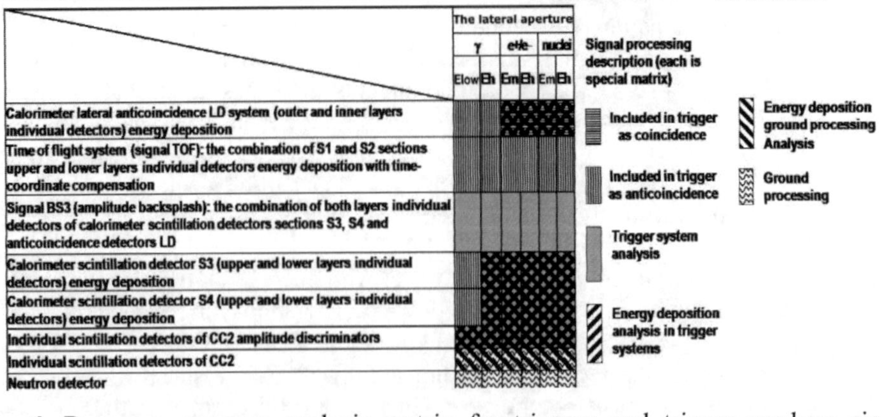

Figure 6. Detectors response analysis matrix for triggers and trigger markers signals formation formation matrixes for the additional (a) and lateral (b) apertures.

Event detection in its aperture begins with signal of CC2 individual detectors fast discriminators in anticoincidence with TOF pulse and CC1 individual detectors fast discriminators. Higher energies γ-quanta recognised using energy deposition in the individual detectors of LD, S3, S4 and fast signals from CC2 individual detectors discriminators (see panels (b) at the figures 5 and 6).

Low energy photons recognizing by using simple anticoincidence signals from the individual detectors of LD and registered in spectral mode with acquisition timescales depends on detectors count rate. Angular resolution will ~5° for only for gamma-quanta with energies 0.2 - 10 MeV in the lateral aperture obtained due individual detectors of CC2 count rate analysis only for non-stationary events (GRB, solar flares and so on). The applied method looks like BATSE (Burst And Transient Source Experiment) detector onboard Compton Gamma Rays Observatory algorithm for transient sources (see [6] and references therein) differ from occultation analysis technique using in this experiment too.

3. Conclusion

Three apertures provide events registration both from upper and lateral directions. The main aperture provides the best angular (all strip layers information analysis) and energy (energy deposition in the all detectors studying) resolution. Gamma-telescope GAMMA-400 is optimized for the gamma-quanta and charged particles with energy 100 GeV detection with the best parameters in the main aperture.

Triggers in the main aperture will be formed using information about particle direction provided by time of flight system and presence of charged particles or backsplash signal formed according to analysis of energy deposition in combination of both layers anticoincidence systems ACtop and AClat individual detectors due the system of counting and triggers signals formation [5].

For double-layer ACtop less than 3% photons will be wrongly recognized as electrons or protons for ~3-100 GeV particles. The part of charged particles mistakenly identified as gammas will be $\sim 10^{-5}$ using described algorithms.

In the additional aperture the particles identification is provided by analysis of signals corresponding to energy deposition in the individual detectors S2, S3 and fast signals from CC1 individual detectors discriminators. Low energy (0.2 - 10 MeV) photons in the lateral aperture recognizing by using simple anticoincidence signals from the individual detectors of LD. Gamma-quanta of higher energies are identified using energy deposition in the individual detectors of S3, S4, LD and fast signals from CC2 individual detectors discriminators.

Acknowledgments

This work performed within the framework of the Center FRPP supported by MEPhI Academic Excellence Project (contract No. 02.a03.21.0005, 27.08.2013).

References

[1] Topchiev N P *et al.* 2015 *Bull. Russ. Acad. Sci. Phys.* **79(3)** 417
[2] Galper A M *et al.* 2015 *Physics Procedia* **74** 177
[3] Galper A M *et al.* 2013 *Bull. Russ. Acad. Sci. Phys.* **77(11)** 1339
[4] Galper A M *et al.* 2013 *Advances in Space Research.* **51** 297
[5] Arkhangelskaja I V *et al* 2015 *Physics Procedia* **74** 212
[6] Hakkila J 1990 *BATSE software for the analysis of the gamma ray burst spatial distribution.* Alabama Univ. Research Reports: 1990 NASA/ASEE Summer Faculty Fellowship Program 5p. SEE N91-18967 10

Modeling of low-energy charged particles passage through GAMMA-400 gamma-telescope thermal insulation and two-layer plastic scintillation detectors used as anticoincidence shield

E N Chasovikov, I V Arkhangelskaja, A I Arkhangelskiy, M D Kheymits, Yu T Yurkin
National Research Nuclear University MEPhI (Moscow Engineering Physics Institute), Kashirskoe highway 31, Moscow 115409, Russia

E-mail: e.chasovikov@gmail.com

Abstract. The results of low-energy charged particles passage through GAMMA-400 gamma-telescope thermal insulation and two-layer plastic scintillation detectors used as anticoincidence shield are presented. An existing GEANT4 GAMMA-400 model is used. Effects of thermal insulation on charged particle passage are investigated. These results will be used to testing the effect of low-energy charged particles flux on GAMMA-400 gamma-quanta registration capabilities. Sufficiently large energy deposition in two-layer plastic anti-coincidence scintillation detectors might interfere with high-energy particle registration and identification. However, GAMMA-400 detection capabilities are not affected by this, as the energy deposition in the lower layer of S3 is less than 1.5 MeV in all simulated cases. This value is less than threshold for high energy particles identification start (2.5 MeV). It makes impossible to incorrectly identify a low-energy charged particle energy deposition as backsplash from a high-energy gamma-quantum.

1. Introduction

GAMMA-400 is a space gamma-ray telescope project currently being developed by the international collaboration of the same name. It's main scientific objectives include study of high energy cosmic rays, galactic and extragalactic gamma-ray sources, gamma-ray bursts and dark matter [1-4]. This gamma-ray telescope will be covered by thermal insulation to provide thermal balance.

In order to determine quantitative properties of interactions of elementary particles with detector material and structural materials of the spacecraft a mathematical model is required.

GAMMA-400 orbit is located both within and outside magnetosphere. During its lifetime, GAMMA-400 will be exposed to low-energy charged particle flux [5], which may detrimentally affect it's anticoincidence system.

The main objective of this paper is to model the interaction between low-energy charged particles and thermal insulation and energy deposition in gamma-ray telescope detection systems.

2. Implementation

An existing Geant4 environment computer model of GAMMA-400 with all engineering elements included was used [6].

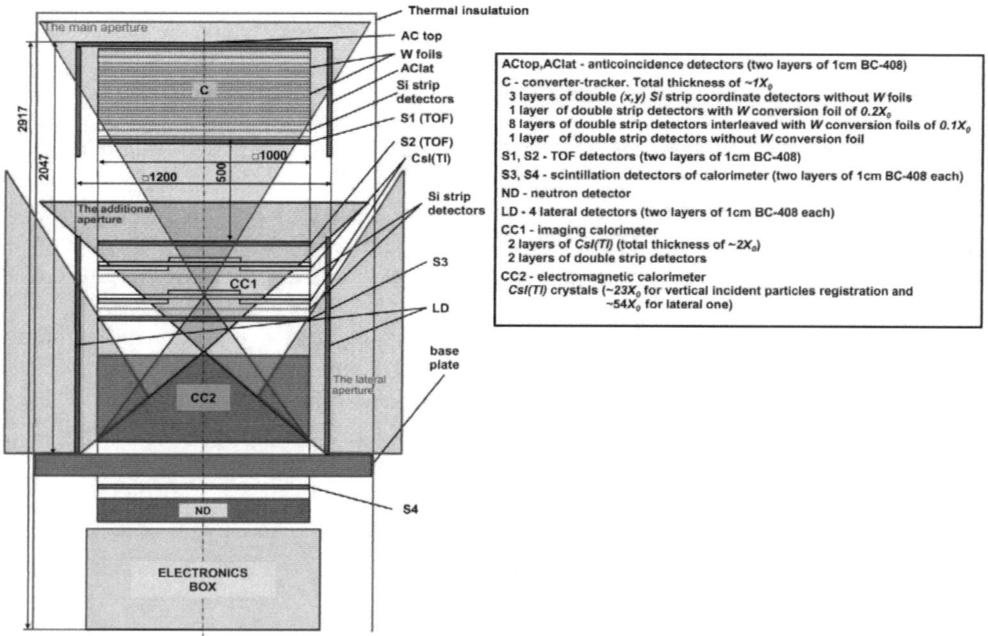

Figure 1. Principal scheme of GAMMA-400 with thermal insulation included

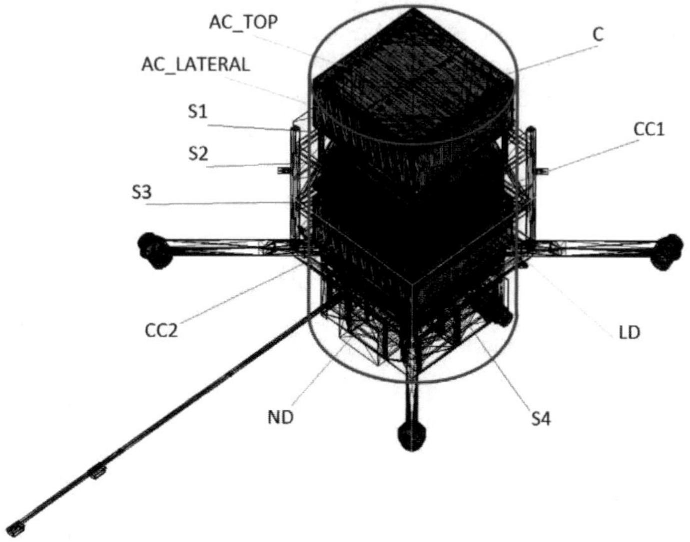

Figure 2. Layout of gamma-telescope GAMMA-400 in Geant4 environment

The principal scheme of GAMMA-400 is shown figure 1. This gamma-telescope consists of the following systems:

- Anticoincidence system top section ACtop
- Anticoincidence system lateral section AClat
- Calorimeter lateral anticoincidence LD System
- Time of flight system detector S1
- Time of flight system detector S2

- Position-sensitive calorimeter CC1
- Electromagnetic calorimeter CC2
- Scintillation detectors of calorimeter S3 and S4
- Neutron detector

All plastic detection systems (AC top, AC lat, LD, S1, S2, S3, S4) are composed of two layers of individual detectors [7].

In the main aperture triggers will be formed using information about particle direction provided by TOF system due special TOF signals matrix analyzed in the system of counting and triggers signals formation [8] for high-energy particles and presence of charged particle or backsplash due analysis of energy deposition in ACtop, S1, S2, S3.

The thermal insulation is modeled as a cylinder consisting of 1 layer of polyamide-6 ([-NH-(CH2)5-CO-]n), density 1.1g/cm^3, thickness 0.7mm and 2 aluminum foil layers, thickness 0.01mm each. Gamma-telescope GAMMA-400 layout in Geant4 environment is shown in figure 2 [6].

Passage of the following particle sets was simulated for the main aperture of gamma-telescope:

1. 10000 of 1 MeV electrons
2. 10000 of 3 MeV electrons
3. 10000 of 10 MeV electrons
4. 10000 of 30 MeV protons
5. 10000 of 100 MeV protons.

Sets 1 and 2 are selected to observe the difference between behavior of electrons interacting with matter only by Compton scattering and electrons being able to create photons due to bremsstrahlung. Set 3 represents the upper limit of trapped electron energy [9]. Set 4 is used to simulate low-energy protons expected to be absorbed by AC top upper layer [10]. Set 5 represents a proton capable of forming time-of-flight system signal [8].

Table 1. Amount of particles with non-zero energy deposition in plastic detectors.

Detector layer	e- 1MeV	e- 3MeV	e- 10MeV	Proton 30 MeV	Proton 100 MeV
AC top upper level	161	9834	9992	9936	9996
AC top lower layer	75	401	9640	38	9936
S1 upper layer	0	17	122	2	256
S1 lower layer	1	16	123	0	225
S2 upper layer	0	4	77	5	100
S2 lower layer	0	11	75	3	85
S3 upper layer	0	4	16	0	7
S3 lower layer	0	1	12	0	3

Table 2. Maximal energy deposition in plastic detectors, MeV.

Detector layer	e- 1MeV	e- 3MeV	e- 10MeV	Proton 30 MeV	Proton 100 MeV
AC top upper level	0.199	2.668	9.343	20.983	96.570
AC top lower layer	0.261	1.659	6.069	5.823	79.943
S1 upper layer	0	0.766	3.371	1.364	17.906
S1 lower layer	0.450[a]	0.875	3.948	0	21.944
S2 upper layer	0	0.322	2.831	0.467	13.779
S2 lower layer	0	0.320	1.926	0.975	2.468
S3 upper layer	0	0.255	0.904	0	1.225
S3 lower layer	0	0.041[a]	0.705	0	0.021

[a] These values are based only on single incidence, and should be interpreted as mean values of Landau distribution

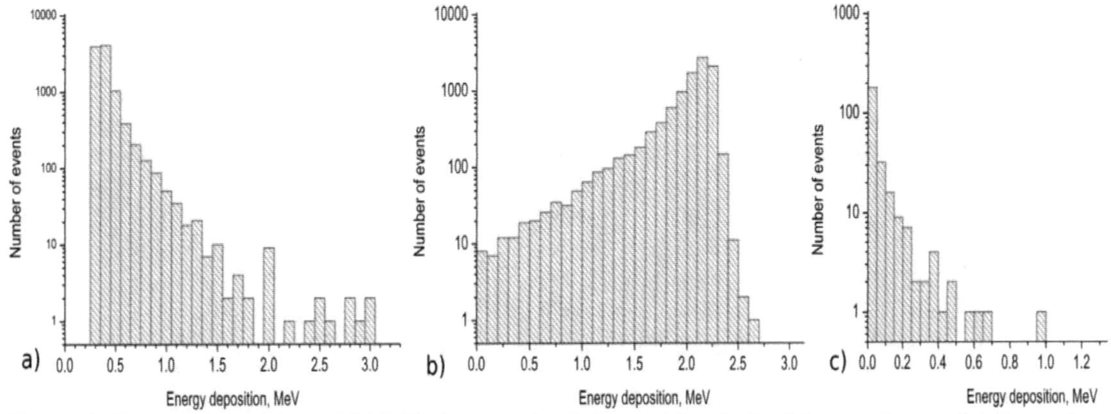

Figure 3. Energy deposition of 3 MeV electrons in a) thermal insulation b) upper layer of AC top c) lower layer of AC top.

As example, the energy deposition of 3 MeV electrons is represented in figure 3. The histograms in figure 3 take into account only particles interacting with a given layer.

The precise amount of particles with non-zero energy deposition is listed in table 1. Table 2 contains the maximal values of energy deposition in plastic detectors. As shown in table 1, the majority of electrons with energy of up to 3 MeV and protons with energy of up to 30 MeV didn't reach the lower layer of AC top, and 97% of all simulated particles were absorbed before entering S1. According to the data in table 2, the energy deposition in the lower layer of S3 is less than 1.5 MeV in all simulated cases, which makes it impossible to incorrectly identify a low-energy charged particle energy deposition as backsplash from a high-energy γ-quantum [8].

Based on data from GEOTAIL [11, 12] and Ulysses experiments [13, 14], known charged particle energy spectra [15, 16] and area of GAMMA-400 individual plastic detectors, the expected particles fluxes will be ~1.3×10^3 s^{-1} (for electrons with E>1 MeV) and ~10 s^{-1} (for protons with energy at least 30 MeV). This is not enough to cause significant dead time in GAMMA-400 detector systems because of energy deposition in S3 is less than threshold for high energy particles identification start (2.5 MeV [8]) – see tables 1, 2.

3. Conclusion

Passage of low-energy charged particles through GAMMA-400 gamma-telescope thermal insulation and two-layer plastic scintillation detectors used as anticoincidence shield was modeled.

The majority of electrons with energy of up to 3 MeV and protons with energy of up to 30MeV were absorbed before entering the lower layer of AC top.

The energy deposition in the lower layer of S3 is less than 1.5 MeV in all simulated cases. This value is less than threshold for high energy particles identification start (2.5 MeV [8]). It makes impossible to incorrectly identify a low-energy charged particle energy deposition as backsplash from a high-energy gamma-quantum.

Acknowledgments

This work performed within the framework of the Center FRPP supported by MEPhI Academic Excellence Project (contract No. 02.a03.21.0005, 27.08.2013).

References

[1] Topchiev N P *et al.* 2015 *Bull. Russ. Acad. Sci. Phys.* **79(3)** 417
[2] Topchiev N P *et al.* 2015 arXiv:1507.06246

[3] Galper A M *et al.* 2013 *Bull. Russ. Acad. Sci. Phys.* **77(11)** 1339
[4] Galper A M *et al.* 2013 *Advances in Space Research.* **51** 297
[5] Khyzhniak E V *et al.* 2015 *Phys. Proc.* **74** 238
[6] Chasovikov E N *et al.* 2015 *Phys. Proc.* **74** 206
[7] Runtso M F *et al.* 2015 *Phys. Proc.* **74** 220
[8] Arkhangelskaja I V *et al.* 2015 *Phys. Proc.* **74** 212
[9] Baker D N *et al.* 1986 *AGARD, The Aerospace Environment at High Altitudes and its Implications for Spacecraft Charging and Communications* **4-1** 406
[10] Sternheimer R M *et al* 1982 *Phys. Rev. B* **26(11)** 6067
[11] URL: http://pwg.gsfc.nasa.gov/geotail.shtml
[12] URL: http://cdaweb.gsfc.nasa.gov/istp_public/.
[13] URL: http://solarsystem.nasa.gov/missions/profile.cfm?MCode=Ulysses
[14] URL: http://ulysses.jpl.nasa.gov/science/data.html
[15] Hayakawa S 1969, *Cosmic Ray Physics* (New York , Wiley-Interscience)
[16] Olive K A *et al* 2014 Chin. Phys. C **38** 090001

Investigation of the dynamic range of calorimeter scintillation detector for space gamma-ray telescope

M F Runtso, P Yu Naumov, P P Naumov and A A Solodovnikov

National Research Nuclear University MEPhI (Moscow Engineering Physics Institute), Kashirskoe highway 31, Moscow, 115409, Russia

E-mail: MFRuntso@mephi.ru

Abstract. An arrangement of the GAMMA-400 space gamma-ray telescope that currently is under the ground testing, suggests implementation of fast two-layer calorimeter scintillation detector system S3 with large dynamic range for electromagnetic showers detection in the main operation mode of the device. The S3 constructive features are demonstrated. The experimental method and basic diagram of the ground prototype dynamic range investigation are described.

1. Introduction

The international project GAMMA-400 includes the gamma-ray telescope for the energy range from 10^2 MeV up to 3×10^3 GeV onboard of the space satellite. This device will investigate the point gamma-ray sources and the peculiarities in the energy spectrum of diffuse gamma emission including one might be produced by the annihilation of so-called "dark matter" particles [1].

The gamma-ray telescope [1] layout is shown in the figure 1.

It has the sensitive area of about 1 m^2 and consists of fast plastic scintillation anticoincidence detector AC to forbid the charged particle detection due special algorithms used in the system of counting and triggers signals formation [2], multilayer (tungsten + silicon strip detectors) converter C for the conversion of gammas to the electron-positron pair and for their trajectories visualization. The time-of-flight system (S1 and S2 fast plastic scintillator counters with 50 cm flight base) generates the signal for up-to-down moving charged particles. Two sections of heavy CsI(Tl) calorimeter (CC1 and CC2) are interleaved by two-layer fast scintillation detector S3 indicating the stage of shower development (the main mode for the S3 detector implementation in the telescope system). The system of the particle's leakage detector S4 and the neutron detector ND for clear electron-proton discrimination are located below the calorimeter.

The electronic box is located under the last detector to exclude any matter on the particle trajectory.

All of gamma-telescope components will be mounted on the base plate and installed at the space platform named NAVIGATOR.

We are considering here the prototype version for gamma-telescope described in [3].

During the converting of high-energy gamma-quanta in the converter C and interaction process in calorimeter, electromagnetic cascade develops, so the number of charged particles in the scintillation detector system S3 considerably increases. This requires a large dynamic range of the S3 detector system.

So, first of all, for the S3 dynamic range increasing the detector system consists of two layers of plastic scintillator Bircon-408, has total sensitive area of 1 m^2 and layer thickness of 10 mm. For better light collection the S3 layers are subdivided into modules of 100 mm wide. Light is detected by basic group of 6 silicon photomultipliers (SIPM) [4] located on the both ends of the scintillator module. SIPM SensL MicroFB-60035-SMT was selected. This type of fast SIPM device has the package dimensions 7.0 x 7.0 mm, signal amplification factor is about 3×10^6, and signal rise time is about 1 ns. SIPM can operate at low voltage (less than 30 V) and can be exploited in extended temperature range satisfied to hard implementation conditions.

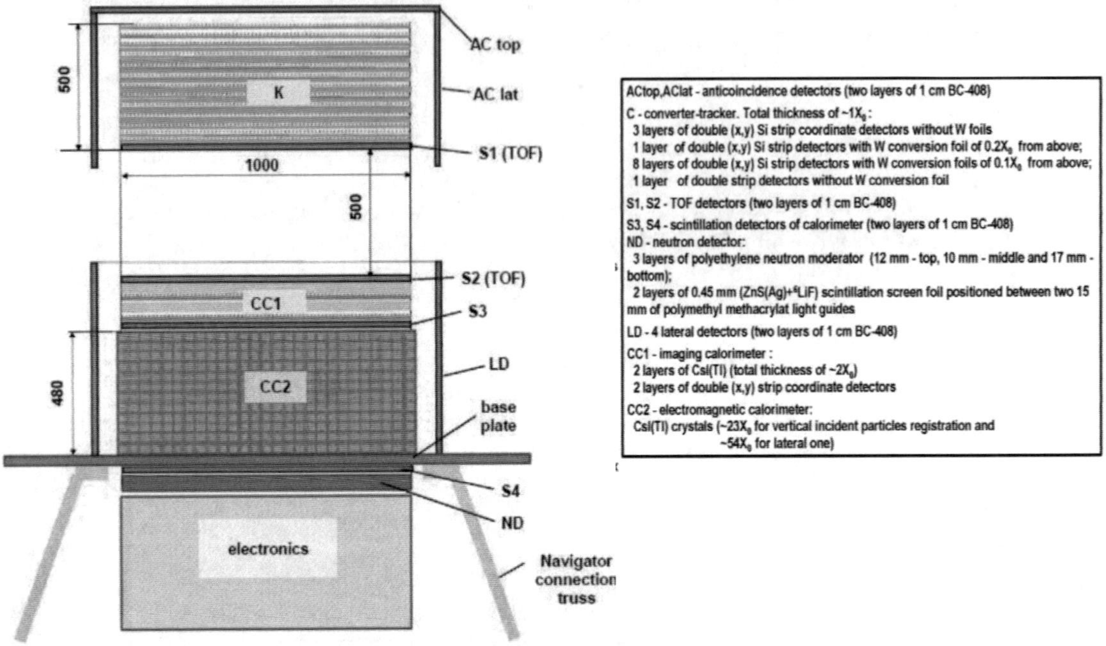

Figure 1. **Layout of the gamma-ray telescope GAMMA-400.**

The modern SIPMs are good photosensors for compact detector devices in space applications. But it has a limited signal dynamic range connected with so-called saturation effect and, so, the range investigation for real experimental conditions must be carried out.

2. Measurement of the S3 dynamic range

Preliminary, Monte-Carlo simulation was done using a program we developed especially to compute the physical characteristics of the GAMMA-400 telescope by simulating the passage of particles through device matter on the base of Geant4 simulation toolkit [5]. The simulation shows that for gammas with the energy 1000 GeV the maximum energy deposition in the S3 detector corresponds to about 1000 minimum ionizing particles (MIP). It should be noted what the average energy deposition in whole thickness of S3 detector system from the single charged relativistic particle is about 4 MeV.

We especially have chosen the SIPM with maximum number of cells to increase the dynamic range. Next, we suggest to use an additional registration channel to expand the dynamic range. This channel will collect signals from the group of two SiPMs that are located on the each end of scintillator. For this group we use optical filters for the attenuation of the light signal.

In the figure 2 the simplified diagram of the S3 detector dynamic range measurements is shown.

This experimental method has the two purposes. First of all we obtain the mean value of energy deposition from one particle in one layer of detector (1 MIP) to use it for the energy scale calibration.

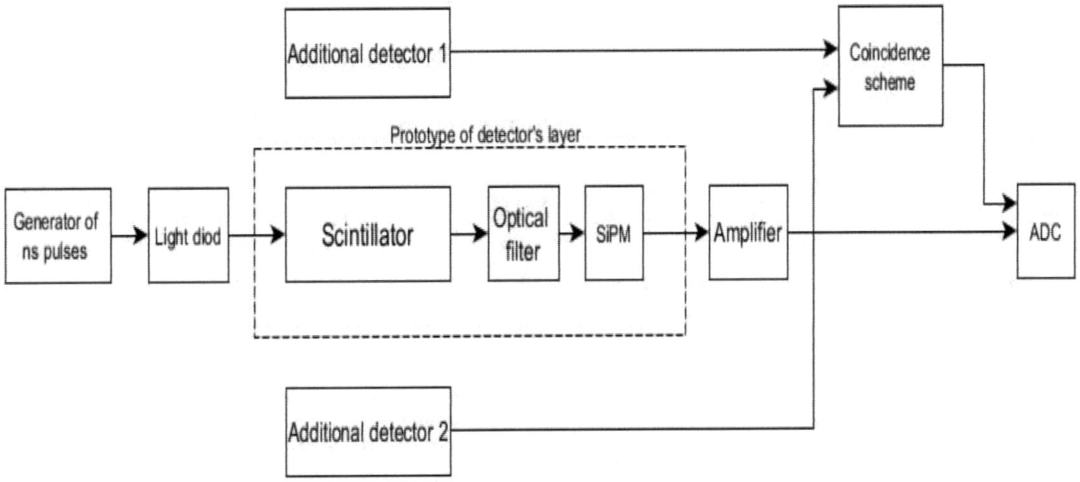

Figure 2. Diagram of the measurement setup.

The second purpose is the measurement of SiPM amplitude dependence from the light splash intensity (see below).

The set of optical filters were used for attenuation of light splash intensity. Thus, we can simulate the particle registration process with higher energy (for example, above 1000 GeV).

In the figure 3 we illustrate the results of the dynamic range measurements for the S3 one layer detector system.

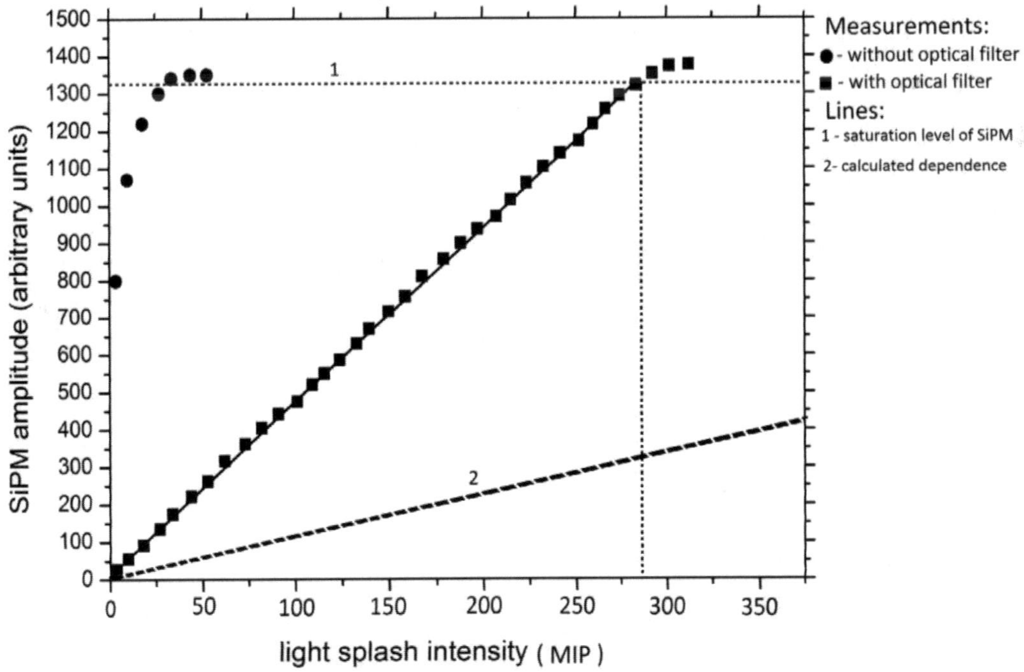

Figure 3. Results of the S3 one layer detector dynamic range measurements.

The brightening of photodetectors was implemented with ultraviolet light-emitting diode triggered by nanosecond pulse signal generator. Dotted plot shows the dependence of signal amplitude from

SiPM without optical filter, rectangles show the same dependence with optical filter. Initial coefficient of attenuation equals 100. But we can do it more if needed (see below).

Measurements without optical filter allow us to achieve saturation level of SiPM amplitude. It is about 1200 arbitrary units (in the units of 15 bit ADC channels). Strait horizontal line 1 corresponds to the SiPM saturation level and it is the upper limit for dynamic range at the current experimental conditions.

The ground S3 detector calibration in MIP units was carried out with the use of atmospheric cosmic rays fluxes (muon particles mainly). The calibration scale of the pulse generator was obtained with the help of photomultiplier Planacon XP85012 with high quality of amplification linearity and large signal dynamic range. So, we obtained the linear dependence on scintillator illumination with optical filter (see squared line in the figure 3) in MIP units.

We achieved the dynamic range of the S3 one layer detector equal to 280 MIP (see vertical line in the figure 3). From the figure 3 it can be seen also, if we shall increase the attenuation of filter to about 4 times, the S3 one layer dynamic range can expand up to 1000 MIP. This is the value that we can calculate for the 1000 GeV gamma quanta detection (see line 2 in the figure 3) and we shall have some reserve before SIPM saturation. So, we can increase the optical filter attenuation to the value corresponding to gamma-quanta energy more than 1000 GeV.

3. Conclusion

The preliminary measurements of the S3 one layer prototype dynamic range are presented. The resulting range for the charged particle shower detection from 1 to 1000 MIP might be obtained that corresponds to the design requirements.

Acknowledgements

This work was performed within the framework of the Fundamental Research and Particle Physics Center supported by MEPhI Academic Excellence Project (contract N 02.a03.21.0005, 27.08.2013).

References

[1] Adriani O *et al* 2013 Status of the GAMMA-400 project *Advances in Space Research* **51 (2)** 297

[2] Arkhangelskaja I V *et al* 2015 The system of counting and triggers signals formation for gamma-telescope GAMMA 400 *Physics Procedia* **74** 212

[3] Arkhangelskiy A I *et al* 2015 The prototype of GAMMA-400 apparatus *Physics Procedia* **74** 191

[4] Buzhan P *et al* 2003 *Nucl. Instr. and Meth. A* **504** 48

[5] Agostinelli S *et al* 2003 Geant4 -simulation toolkit *Nucl. Instr. and Meth. A* **506 (3)** 250

On the possibility to use semiconductive hybrid pixel detectors for study of radiation belt of the Earth.

A. Guskov[1,*], **G. Shelkov**[1], **P. Smolyanskiy**[1], **A. Zhemchugov**[1]

[1] Joint Institute for Nuclear Research, Dubna

E-mail: *avg@jinr.ru

Abstract. The scientific apparatus GAMMA-400 designed for study of electromagnetic and hadron components of cosmic rays will be launched to an elliptic orbit with the apogee of about 300 000 km and the perigee of about 500 km. Such a configuration of the orbit allows it to cross periodically the radiation belt and the outer part of magnetosphere. We discuss the possibility to use hybrid pixel detecters based on the Timepix chip and semiconductive sensors on board the GAMMA-400 apparatus. Due to high granularity of the sensor (pixel size is 55 mum) and possibility to measure independently an energy deposition in each pixel, such compact and lightweight detector could be a unique instrument for study of spatial, energy and time structure of electron and proton components of the radiation belt.

1. Timepix detectors

Pixel hybrid detectors consisted of the Timepix chip [1], developed at CERN, coupled with a semiconductive sensor are used as a particle detector in a variety of disciplines from the study of cosmic rays to biomedical imaging. Good spatial (pixel size is 55 μm) and energy (energy deposition can be measured individually in each pixel) resolution, low level of noise, compact size and radiation hardness are doubtless advantages of this type of detectors. Thin plates (0.3-1.0 mm) of silicon (Si) or gallium arsenide compensated with chromium (GaAs) with 256×256 square pixels are used as a sensor. GaAs sensors are developed and produced in the Tomsk State University (Tomsk, Russia)[4, 5, 6, 7]. The Joint Institute for Nuclear Research (Dubna, Russia) has wide experience in the production and operation of such kind of the detectors since 2008 [2, 3].

Layout of the Timepix detector is presented in Fig. 1. Sensitivity of the detector starts from about 6 keV for γ-quants, from 30 keV for electrons and from 500 keV for protons. Interaction of an incoming particle with the material of the sensor produces a cluster of energy deposition. Analysis of geometrical shape of such cluster and energy distribution over pixels provides possibility to classify clusters as produced by soft photons, low-energetic electrons, non-relativistic protons, neutrons, heavy ions and minimum ionizing particles (ultrarelativistic charged particles). Fig 2. shows response of the Timepix detector with 0.3 mm silicon sensor placed in the vicinity of the interaction point of 2 GeV deutron beam with lead target. Long straight horizontal lines correspond to the beam halo deutrons.Typical responses of the Timepix detector, equipped with 0.3 mm silicon sensor, irradiated with 350 keV electrons and with 2 MeV protons, coming to the surface at 90° angle, are presented on Fig. 3 (left) and (right) correspondently. Since for lowenergetic particles full energy deposition occurs inside the sensor, the energy of incoming electrons and protons can be measured at least up to 0.3 MeV and 6 MeV in 0.3 mm silicon and up to 1.4 MeV and 17 MeV in 1 mm gallium arsenide sensors correspondently.

Basic properties of the Timepix detectors are presented in Tab. 1.

Figure 1. Layout of the Timepix detector

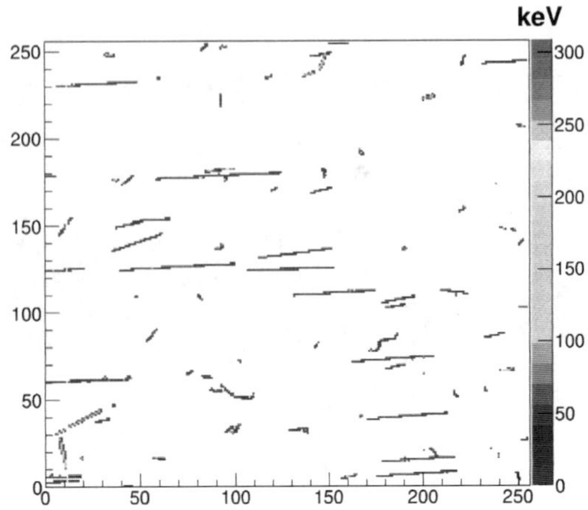

Figure 2. Response of the Timepix detector equipped with 0.3 mm silicon sensor placed in the vicinity of the interaction point of 2 GeV deutron beam with lead target.

Table 1. Basic parameters of the Timepix detector.

Parameter	Value
Mass of the detector and infrastructure (without cabling), g	200
Dimensions $L \times W \times H$, cm	$15 \times 5 \times 3$
Power consumption, W	2.5
Radiation hardness of Timepix chip, MGy	4.6 [8]
Radiation hardness of Si sensor, MGy	0.1-0.5
Radiation hardness of GaAs sensor, MGy	1.5 [9]
Electric field strength, V/μm	0.5-1.0
Minimal frame length, ms	0.01
Amount of data, kB per frame with occupancy 10%	20
Maximal frame rate, kHz	100
Sensitivity range, keV	>6(γ),>30(e),>500(p)
Working area, cm^2	2

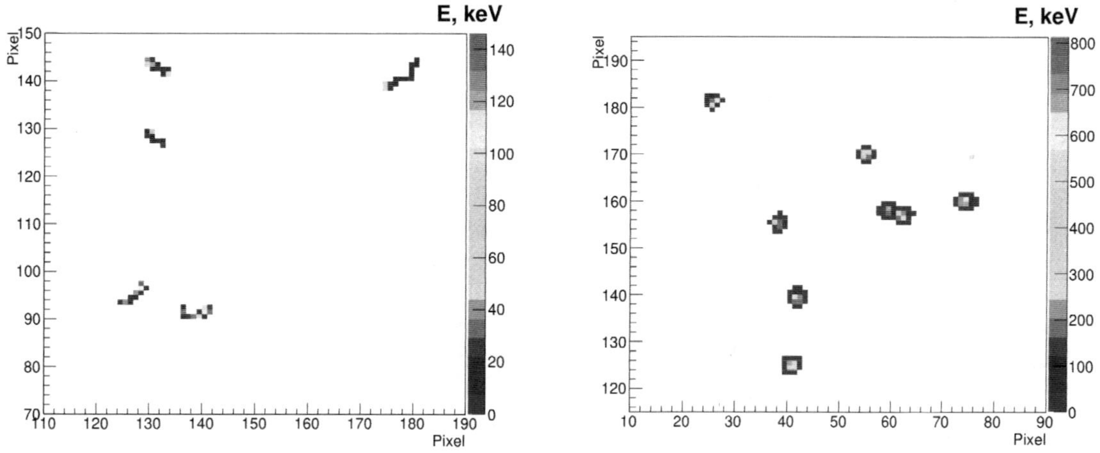

Figure 3. Typical responses of the Timepix detector equipped with 0.3 mm silicon sensor and irradiated with 350 keV electrons (left) and with 2 MeV protons (right). Angle of incidence is 90°. Only a part of the frame is shown.

2. Timepix detectors at GAMMA-400

The magnetically confined radiation zones around the Earth populated with electrons and protons and named radiation belts are still the objects of intense study. On the one hand the detailed information about structure, intensity, composition, energy spectrum of the particle flows, their variation in time and correlation with the solar activity provides possibility to better estimate the radiation hazard for spacecraft. On the other hand this information is important for the tests of the models describing capture, acceleration, transport, precipitation and loss of radiation belt particles (see, for instance, [10, 11]. Possible correlation between high-energy particle flux in the radiation belts and seismic activity of the Earth is the special point of attention in resent years since such a correlation can help to predict earthquakes [12, 13].

Our proposal is to place one or a few Timepix detectors onboard the GAMMA-400 apparatus [14, 15] to monitor continuously the flux of low energy charged particles along the orbit of the apparatus. There is already a positive experience to use the Timepix detectors in space. It is used on board the International Space Station to accurately monitor radiation doses from various sources [16]. NASA has used the Timepix detector during the first test flight of the "Orion module" - part of the next generation of manned US spacecraft [17]. It was also installed on board the ESA Proba V satellite launched in 2013, where it first operated out of the shielding in a vacuum without any problems with thermoregulation [18].

To minimize the perturbation of the radiation environment by the body of the apparatus the sensors could be attached to the external constructions together with the magnetometers or the star sensors out of the thermoshielding. To keep possibility to detect protons with energy above 1 MeV and electrons with energy above 50 keV the shielding of the working area should be on the level of about 2 mg/cm^2 of a light material like mylar or aluminum foil. Readout and control of the Timepix detector can be performed via the dedicated FPGA-based service board. The service board can be connected to the SDAS by means of a high-speed interconnect like RapidIO. Tunable HV power supply controlled by the service board is necessary to bias the sensor of the Timepix detector up to -500 V.

The initial elliptic orbit of the apparatus will have the following parameters: the apogee of about 300 000 km, the perigee of about 500 km, the inclination 51.8^o and period of about 7 days. Since the perigee lies low enough, the apparatus will periodically cross both external and internal radiation belts and be out of the magnetosphere. The Timepix detector is able to monitor effectively the flux of protons and electrons separately basing on the cluster shape and energy distribution analysis if the sensor occupancy is below $\sim 10\%$. That corresponds to the flux Φ_0 of about 3×10^7 $c^{-1}cm^{-2}$ and the minimal frame length of 10 μs. The maximal general flux of protons with energy above 1 MeV and electrons in with energy above 50 keV through a solid angle of 2π in the radiation belts was estimated according to the AE-8/AP-8 model [19, 20]. It is about 2×10^8 $c^{-1}cm^{-2}$ for maximal and about 7×10^7 $c^{-1}cm^{-2}$ for minimal solar activity. So at the part of the orbit where particle flux is lower than Φ_0 the Timepix detector could operate as a precise monitor of the electron and proton flux with variable time window to keep reasonable occupancy at the level of a few per cent, while in the region of higher flux it could act as a rough indicator of the general flux using the minimal time window. The measurement of the energy spectrum could also be available for low energy component dominating in the outer part of the radiation belt.

In parallel with monitoring of the radiation belt the Timepix detector could be a unique instrument for investigation of charged particles flux in the outer region the Earth's magnetosphere. Such data can be combined with an information about the solar activity and an information from the magnetometer planned to be installed at GAMMA-400. Study of energetic component of the solar wind out of the magnetosphere and its interaction with the magnetic field of our planet could be another promising task for the Timepix detector. This program will be especially important when the orbit will become more circular with a radius of about 100 000 km.

Acknowledgments

This work is supported by the Ministry of Education and Science of Russian Federation under the contract No. 14.618.21.0001.

References

[1] Llopart X *et al.* 2007 *NIM A* **581** 485
[2] Tlustos L, Shelkov G and Tolbanov O 2011 *NIM A* **633** 103
[3] Butler A *et al. Physics of Particles and Nuclei Letters* **12** 59
[4] Ayzenshtat G *et al.* 2002 *NIM A* **487** 96
[5] Tyazhev A *et al.* 2003 *NIM A* **509** 34
[6] Ayzenshtat G *et al.* 2003 *NIM A* **509** 268
[7] Ayzenshtat G *et al.* 2004 *NIM A* **531** 121
[8] Plackett R *et al.* 2009 *Proc. TWEPP-09* p157
[9] Afanaciev K*et al.* 2012 JINST **7** 11022.
[10] Summers D *et al. Dynamics of the Earth's Radiation Belts and Inner Magnetosphere* (Washington: American Geophysical Union)
[11] Lemaire J *et al.* 2013 *Geophysical Monograph Series* **97**
[12] Gal'per A *et al.* 1995 *Advances in Space Research* **15** 131
[13] Hayakawa M (2015) *Earthquake Prediction with Radio Techniques* (Singapore: John Wiley & Sons)
[14] Topchiev N *et al.* 2015 *arXiv*:1507.06246
[15] Topchiev N *et al.* 2015 *Bulletin of the Russian Academy of Sciences. Physics* **79** 3 417
[16] Stofflea N *et al.* 2015 *NIM A* **782** 143
[17] Bahadori A *et al.* 2015 NASA/TP-2015-218575
[18] Francois M *et al.* 2014 *International Journal of Remote Sensing* **35** 7 2548
[19] Sawyer D and Vette J 1976 *AP-8 trapped proton environment for solar maximum and solar minimum* NASA TM-X-72605
[20] Vette J 1991 *The AE-8 trapped electron model environment* NSSDC/WDC-A-R&S 91-24

Neutrino processes $\nu\bar{\nu} \to e^- e^+$ and $\nu \to \nu e^- e^+$ in a strong magnetic field

A V Kuznetsov[1], **D A Rumyantsev**[1,*] **and V N Savin**[2,**]

[1] Division of Theoretical Physics, Department of Physics, Yaroslavl State P G Demidov University, Sovietskaya 14, 150000 Yaroslavl, Russia

[2] Division of Physics, Yaroslavl Higher Military School of Air Defence, Moskovskiy Prosp. 28, 150001 Yaroslavl, Russia

E-mail: avkuzn@uniyar.ac.ru, *rda@uniyar.ac.ru, **vs_post07@mail.ru

Abstract. The processes of neutrino production of electron–positron pairs, $\nu\bar{\nu} \to e^- e^+$ and $\nu \to \nu e^- e^+$, in a magnetic field of arbitrary strength, where electrons and positrons can be created in the states corresponding to excited Landau levels, are analysed. The results can be applied for calculating the efficiency of the electron–positron plasma production by neutrinos in the conditions of the Kerr black hole accretion disc considered by experts as the most possible source of a short cosmological gamma burst.

1. Introduction

An intense electromagnetic field makes possible the processes which are forbidden in a vacuum such as the neutrino production of an electron–positron pair $\nu \to \nu e^- e^+$. The list of papers devoted to an analysis of this process and the collection of the results obtained could be found e.g. in [1]. In most cases, calculations of this kind were made either in the crossed field approximation, or in the limit of a superstrong field much greater than the critical value of $B_e = m_e^2/e \simeq 4.41 \times 10^{13}$ G (we use natural units $c = \hbar = k_B = 1$), when the electrons and positrons are born in states corresponding to the ground Landau level. However, there exist physical situations of the so-called moderately strong magnetic field, $p_\perp^2 \geq eB \gg m_e^2$, when electrons and positrons mainly occupy the ground Landau level, however, a noticeable fraction may be produced at the next levels.

The indicated hierarchy of physical parameters corresponds to the conditions of the Kerr black hole accretion disk, regarded by experts as the most likely source of a short cosmological gamma-ray burst. The disc is a source of copious neutrinos and anti-neutrinos, which partially annihilate above the disc and turn into $e^\mp$ pairs, $\nu\bar{\nu} \to e^- e^+$. This process was proposed and investigated in many details (for the list of references see e.g. [2, 3]) as a possible mechanism for creating relativistic, $e^\mp$-dominated jets that could power observed gamma-ray bursts. In [2], in addition to $\nu\bar{\nu}$ annihilation, the contribution of the magnetic field-induced process $\nu \to \nu e^- e^+$ to the neutrino energy deposition rate around the black hole was also included for the first time. The authors [2] concluded in part, that the process $\nu \to \nu e^- e^+$ could dominate over the basic process $\nu\bar{\nu} \to e^- e^+$. They used the result for the energy deposition rate in the process $\nu \to \nu e^- e^+$ obtained in [4, 5] in the crossed field limit, while in those physical conditions (B to 180 B_e, E_ν to 25 MeV) the approximation of a crossed field is poorly applicable (as well as the approximation of a superstrong field when e^-e^+ are created in the ground Landau level). The next Landau levels can be also excited, as we have shown in our paper [3]. Furthermore, the authors [2] considered the process $\nu\bar{\nu} \to e^-e^+$ without taking account of the magnetic field influence.

Thus, the aim of this paper is the study of the processes $\nu\bar{\nu} \to e^-e^+$ and $\nu \to \nu e^-e^+$ in the physical conditions of the moderately strong magnetic field, where the electrons and positrons would be born in the states corresponding to the excited Landau levels. Possible astrophysical applications are discussed.

2. Neutrino process $\nu \to \nu e^-e^+$ in a strong magnetic field

The total probability of the process $\nu \to \nu e^-_{(n)}e^+_{(\ell)}$, when the electron and the positron are created in the nth and ℓth Landau levels, is, in a general case, the sum of the probabilities of the four polarization channels:

$$W_{n\ell} = W_{n\ell}^{--} + W_{n\ell}^{-+} + W_{n\ell}^{+-} + W_{n\ell}^{++} \,. \tag{1}$$

For each of the channels, the differential probability over the final neutrino momentum per unit time, after integration over the momenta of the electron and positron, is reduced to one nontrivial integral:

$$\mathrm{d}W_{n\ell}^{ss'} = \frac{\beta\, \mathrm{d}^3P'}{(2\pi)^4 16EE'} \int \frac{\mathrm{d}p_z}{\varepsilon_n\, \varepsilon'_\ell}\, \delta(\varepsilon_n + \varepsilon'_\ell - q_0)\, |\mathcal{M}_{n\ell}^{ss'}|^2 \,, \tag{2}$$

where $\varepsilon_n = \sqrt{M_n^2 + p_z^2}$, $M_n = \sqrt{m_e^2 + 2\beta n}$, $\beta = eB$. The energy of the initial neutrino should exceed a certain threshold value. In the reference frame, where the momentum of the initial neutrino directed at an angle θ to the magnetic field, the threshold energy is given by:

$$E\,\sin\theta \geq M_n + M_\ell \,. \tag{3}$$

Some details of calculations can be found in our paper [3].

The probability of the $\nu \to \nu e^-e^+$ process defines its partial contribution into the neutrino opacity of the medium. The estimation of the neutrino mean free path with respect to this process gives the result which is too large [1] compared with the typical size of any compact astrophysical object, where a strong magnetic field could exist. However, a mean free path does not exhaust the neutrino physics in a medium. In astrophysical applications, we could consider the values that probably are more essential, namely, the mean values of the neutrino energy and momentum losses, caused by the influence of an external magnetic field. These values can be described by the four-vector of losses Q^α,

$$Q^\alpha = E \int q^\alpha\, \mathrm{d}W = -E\,(\mathcal{I}, \mathbf{F}) \,. \tag{4}$$

where q is the difference of the momenta of the initial and final neutrinos, $q = P - P'$, $\mathrm{d}W$ is the total differential probability of the process. The zeroth component of Q^α is connected with the mean energy lost by a neutrino per unit time due to the process considered, $\mathcal{I} = \mathrm{d}E/\mathrm{d}t$. The space components of the four-vector (4) are similarly connected with the mean neutrino momentum loss per unit time, $\mathbf{F} = \mathrm{d}\mathbf{P}/\mathrm{d}t$. It should be noted that the four-vector of losses Q^α can be used for evaluating the integral effect of neutrinos on plasma in the conditions of not very dense plasma, where an one-interaction approximation of a neutrino with plasma is valid.

In [2], the formula for the energy deposition rate was taken, which was calculated in the crossed field limit [4, 5]. However, in the region of the physical parameters used in [2] (B to 180 B_e, E_ν to 25 MeV), the approximation of a crossed field is poorly applicable, as well as the approximation of a superstrong field when e^-e^+ are created in the ground Landau level. The contribution of the next Landau levels which can be also excited, should be taken into account. In [3], the results are presented of our calculation of the mean neutrino energy losses caused by the process $\nu \to \nu e^- e^+$ in a moderately strong magnetic field, i.e. in the conditions of the Kerr black hole accretion disk. It was shown that the crossed field limit gives the overstated result which is in orders of magnitude greater than the sum of the contributions of lower excited Landau levels. On the other hand, the results with $e^- e^+$ created at the ground Landau level give the main contribution to the energy deposition rate, and almost exhaust it at $B = 180 B_e$.

This would mean that the conclusion [2] that the contribution of the process $\nu \to \nu e^- e^+$ to the efficiency of the electron-positron plasma production by neutrino exceeds the contribution of the annihilation channel $\nu\bar{\nu} \to e^- e^+$, and that the first process dominates the energy deposition rate, does not have a sufficient basis. A new analysis of the efficiency of energy deposition by neutrinos through both processes, $\nu\bar{\nu} \to e^- e^+$ and $\nu \to \nu e^- e^+$, in a hyper-accretion disc around a black hole should be performed, with taking account of our results [3] for the process $\nu \to \nu e^- e^+$.

3. The strong magnetic field influence on the process $\nu\bar{\nu} \to e^- e^+$

The local energy-momentum deposition rate due to the process $\nu\bar{\nu} \to e^- e^+$ is defined by the equation [6]:

$$Q^\alpha_{\nu\bar{\nu}} = \int \frac{\mathrm{d}^3 p}{(2\pi)^3} \, f_\nu(p) \int \frac{\mathrm{d}^3 p'}{(2\pi)^3} \, f_{\bar{\nu}}(p') \, (p^\alpha + p'^\alpha) \, \frac{(pp')}{EE'} \, \sigma(\nu\bar{\nu} \to e^- e^+) \,, \tag{5}$$

where σ is the cross-section of the process, p and p' are the four-momenta of the neutrino and antineutrino, $f_\nu(p)$ and $f_{\bar{\nu}}(p')$ are the local distribution functions depending on the distribution functions at the surface of the black hole accretion disc, and on the details of propagation.

In a strong magnetic field, the cross-section takes the form:

$$\sigma(\nu\bar{\nu} \to e^- e^+) = \sum_{f=e,\mu,\tau} \, \sum_{n,\ell} \sigma(\nu_f \bar{\nu}_f \to e^-_{(n)} e^+_{(\ell)}) \,, \tag{6}$$

where the upper limit of summation over n, ℓ is defined by the condition $(M_n + M_\ell)^2 \le (p+p')^2_\parallel$, ($q^2_\parallel = q_0^2 - q_z^2$, if z is along $\mathbf{B}$).

Unlike the cross-section in vacuum where it depends on the Mandelstam parameter S only, the cross-section in a magnetic field depends on the set of kinematic variables, e.g.: energies E, E', two polar angles and one azimuth angle. In figure 1, we take for the sake of illustration the case $E = E'$, and take certain angles. The dependence is presented of σ (solid line) and σ_{vac} (dashed line) on E. The cross-section has a peculiar "sawtooth" profile due to the square-root singularities [7], which is similar to the profile of the process $\gamma^* \to e^- e^+$ width in a strong field [8, 9]. After averaging over small intervals $E \pm \Delta E$, the dependence becomes smoother. It can be seen that in calculations of the energy-momentum deposition rate by integration over the neutrino and antineutrino momenta, the field influence appears to be inessential.

4. Conclusions

- The processes $\nu \to \nu e^- e^+$ and $\nu\bar{\nu} \to e^- e^+$ are investigated in the magnetic field of an arbitrary strength, when $e^- e^+$ can be produced in the excited Landau levels.
- The neutrino energy losses due to the process $\nu \to \nu e^- e^+$ are calculated. The results should be used for calculations of the efficiency of the $e^- e^+$ plasma production by neutrinos in the conditions of the Kerr black hole accretion disk. In these conditions, the crossed field limit gives the overstated result which is in orders of magnitude greater than the sum over the lower Landau levels.

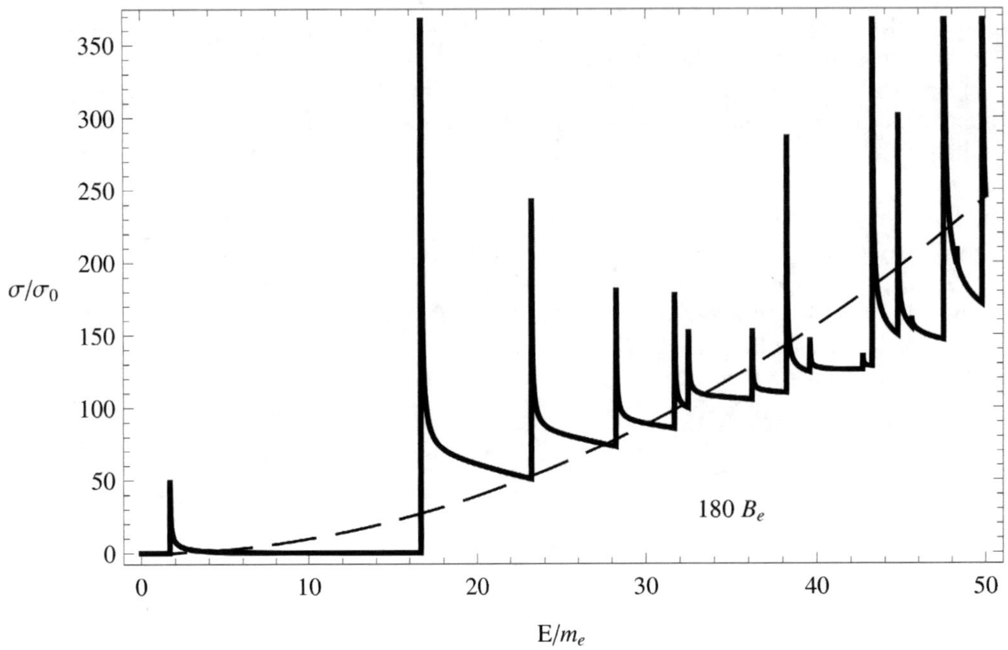

Figure 1. Cross-section of the process $\nu\bar{\nu} \to e^- e^+$ in the case $E = E'$ and for fixed angles: the energy dependence in the field (solid line) and in vacuum (dashed line); $\sigma_0 = 4\, G_{\rm F}^2\, m_e^2/\pi$ is the so-called typical weak cross-section.

- The cross-section of the process $\nu\bar{\nu} \to e^- e^+$ in a strong field, has a peculiar "sawtooth" profile, which is close to the vacuum cross-section after averaging. In calculations of the energy-momentum deposition rate by integration over the neutrino and antineutrino momenta, the field influence appears to be inessential.

Acknowledgments

The study was performed with the support by the Project No. 92 within the base part of the State Assignment for the Yaroslavl University Scientific Research, and was supported in part by the Russian Foundation for Basic Research (Project No. 14-02-00233-a).

References

[1] Kuznetsov A and Mikheev N 2013 *Electroweak Processes in External Active Media* (Berlin, Heidelberg: Springer-Verlag)
[2] Zalamea I and Beloborodov A M 2011 *Mon. Not. R. Astron. Soc.* **410** 2302
[3] Kuznetsov A V, Rumyantsev D A and Savin V N 2014 *Int. J. Mod. Phys.* A **29** 1450136
[4] Kuznetsov A V and Mikheev N V 1997 *Phys. Lett.* B **394** 123
[5] Kuznetsov A V and Mikheev N V 1997 *Phys. At. Nucl.* **60** 1865 (Original Russian text: *Yad. Fiz.* **60** 2038)
[6] Birkl R, Aloy M A, Janka H-Th and Müller E 2007 *Astron. Astrophys.* **463** 51
[7] Klepikov N P 1954 (In Russian) *Zh. Eksp. Teor. Fiz.* **26** 19
[8] Daugherty J K and Harding A K 1983 *Astrophys. J.* **273** 761
[9] Baier V N and Katkov V M 2007 *Phys. Rev.* D **75** 073009

Vertical convection in turbulent accretion disk and light curves of X-ray Nova A0620-00

Konstantin Malanchev[1,2]

[1] M.V. Lomonosov Moscow State University, Faculty of Physics, Leninskie gory, GSP-1, Moscow, 119991, Russia

[2] M.V. Lomonosov Moscow State University, Sternberg Astronomical Institute, Universitetsky pr., 13, Moscow, 119234, Russia

E-mail: `malanchev@physics.msu.ru`

Abstract. We describe phenomenon of X-ray Novae in a model of non-stationary accretion α-disk with account for irradiation and vertical convection in outer disk region. We extended the commonly used disk vertical structure model by adding viscous turbulent energy generation in mixing length theory. This model was used to simulate both optical and X-ray light curves of the 1975 outburst of X-ray Nova A0620-00.

1. Introduction

X-ray novae are close binary systems with relativistic star (a black hole or a neutron star) and a low-mass Roche-lobe-filling star (see reviews [1], [2]). We will consider systems that contains a black hole and a low-mass main-sequence star and show the so-called fast rise and exponential decay (FRED) X-ray light curves. Light curves of such systems can be described by non-stationary evolution of a standard accretion disk ([3], [4]). Analytical solutions of this problem are presented in [5], [6] and [7]. These solutions cannot take into account some of the features of accretion disk structure.

First, outer parts of the disk could be cooler than 10^4 K and have a zone of partially ionized hydrogen where opacity law differs from that in a hotter region of the disk. Second, vertical convection appears in the zone of partially ionized hydrogen ([8], [9]). Third, the outer part of the disk with temperature 10^4 K could be irradiated by X-rays from the inner part of the disk that has a temperature of $10^5 \div 10^6$ K.

Here we describe a numerical model that includes these features and apply it to both X-ray and optical light curves of 1975 outburst of X-ray nova A0620–00. In this work we discuss only the behaviour of the descending branches of the light curves.

2. Viscous evolution of accretion disk

Let us write the continuity equation for a geometry thin axially symmetric accretion disk:

$$\frac{\partial \Sigma_0}{\partial t} = -\frac{1}{r}\frac{\partial}{\partial r}(\Sigma_0 v_r r), \qquad (1)$$

where $\Sigma_0 = \int_{-\infty}^{\infty} \rho dz$ is the surface density of the disk at radius r, z is the axial coordinate, v_r is the radial velocity of the accreting matter.

We assume that the angular momentum is taken away only from outer radius of the disk, in other words, there is no wind, and tidal torque from the secondary star is efficient only for the outer edge of the disk [10]. Then equation of the angular momentum transfer reads

$$\Sigma_0 v_r \frac{\partial(\omega r^2)}{\partial r} = -\frac{1}{2\pi}\frac{1}{r}\frac{\partial F}{\partial r}, \tag{2}$$

where $\omega = \sqrt{GM_x/r^3}$ is the angular velocity of accreting matter, G is the gravitation constant, M_x is the mass of the black hole, $F = 2\pi r^2 \times \int_{-\infty}^{\infty} \alpha P \mathrm{d}z$ is the viscous torque in standard accretion disk [4], α is the turbulent viscous parameter, P is the pressure.

Combination of equations (1) and (2) yields the diffusion equation:

$$\frac{\partial \Sigma_0}{\partial t} = \frac{1}{4\pi}\frac{(GM_x)^2}{h^3}\frac{\partial^2 F}{\partial h^2}, \tag{3}$$

where $h = \sqrt{\omega r^2}$ is the specific Keplerian angular momentum.

The last equation is a second order differential equation, so we need two boundary conditions to solve it. The first boundary condition is for the inner edge of the disk where matter falls down on the black hole almost without producing viscous torques:

$$F_{\text{in}} = 0. \tag{4}$$

The second boundary condition is due to the assumption that during an outburst the mass transfer rate from the secondary star to the disk is much smaller than accretion rate on the black hole:

$$\left.\frac{\partial F}{\partial h}\right|_{\text{out}} = 0. \tag{5}$$

Equation (3) with boundary conditions (4, 5) composes a system of equations for two unknown functions: $F(t,h)$ and $\Sigma_0(t,h)$. In the next section we will find relations between these functions by solving the vertical structure equations.

3. Vertical structure equations

Equations of the vertical structure of the disk are similar to equations of stellar radial structure. Let us write and discuss them one by one.

3.1. Hydrostatic equilibrium

We simulate the disk on timescales much larger the free fall time and therefore we can assume the hydrostatic equilibrium:

$$\frac{\mathrm{d}P}{\mathrm{d}z} = \rho g_z, \tag{6}$$

where ρ is the density, $g_z \simeq \omega^2 z$ is the free fall acceleration.

3.2. Surface density

The second equation comes from surface density definition:

$$\frac{\mathrm{d}\Sigma}{\mathrm{d}z} = \rho. \tag{7}$$

3.3. Energy generation

There are two sources of energy in the outer region of the disk: viscous heating αP and thermalization of X-ray photons coming from the inner region of the disk ϵ_{X}:

$$\frac{\mathrm{d}Q}{\mathrm{d}z} = \alpha P + \epsilon_{\mathrm{X}}. \tag{8}$$

We used the same approach as in paper [11] to calculate ϵ_{X}.

3.4. Energy transfer

When the surface temperature of the disk is higher than 10^4 K all the energy is transfered by radiation. But if there is a vertical convection in the cold disk region then it could transfer part of energy in the direction from the central plane to the disk surface.

$$\frac{\mathrm{d}T}{\mathrm{d}z} = \frac{g_z \rho T}{P} \nabla, \tag{9}$$

where $\nabla \equiv \mathrm{d}\log T/\mathrm{d}\log P$ is the actual logarithmic gradient. In this work we describe the vertical convection by mixing length model in accretion disks from paper [12].

3.5. Equation of state

We use equation of state of perfect gas to link thermodynamical quantities:

$$P = \frac{\rho \mathcal{R} T}{\mu(\rho, T)}, \tag{10}$$

where $\mathcal{R}$ is the gas constant, $\mu(\rho, T)$ is the molecular weight of the gas. We assume that accretion matter is a mixture of neutral hydrogen, ionized hydrogen and neutral impurity, so molecular weight μ depends on thermodynamical state of the gas. In addition we assume that this mixture has solar composition [13].

4. Simulation of the light curves

Solution of the vertical structure equations (6 — 10) provides us with relation between the surface density Σ_0 and the viscous torque F. One can use this relation to solve equation (3) and find temporal evolution of the disk and, as a result, the light curves of the outburst.

We assuming the initial condition in the form:

$$F(h) = \frac{2}{\pi}(h_{\mathrm{out}} - h_{\mathrm{in}})\dot{M}_0 \times \sin\left(\frac{\pi}{2}\frac{h - h_{\mathrm{in}}}{h_{\mathrm{out}} - h_{\mathrm{in}}}\right), \tag{11}$$

where $\dot{M}_0$ is the initial accretion rate on the black hole, h_{in} and h_{out} are specific angular momenta at the inner and outer edges of the disk, respectively. Here the initial accretion rate $\dot{M}_0$ is the accretion rate at the moment of the peak of X-ray luminosity.

For our simulation of light curves of A0620–00 we used parameters of the binary system from [14] and [15]: mass of the secondary star is 0.4 solar masses, mass of the black hole M_{X} is 6.6 solar masses, orbital inclination is 51°, orbital period is 0.323 days, the Kerr parameter of the black hole is 0.2.

Results of our simulations are shown in Fig. 1 and Fig. 2. It is seen that the X-ray light curve (Fig. 2) has the secondary peak. We explains it with an additional mass transfer from the secondary star to the disk 43 days after the X-ray luminosity maximum. We need to add 30% of the disk mass to it to explain the observed feature.

The late part of the model optical light curve (Fig. 1) lies lower than observations. It could be explained by adding some light from the secondary star. The star could be irradiated by the X-ray photons from the accretion disk and heated. The dashed line on Fig. 1 shows the light curve for the disk with the star that has a constant luminosity.

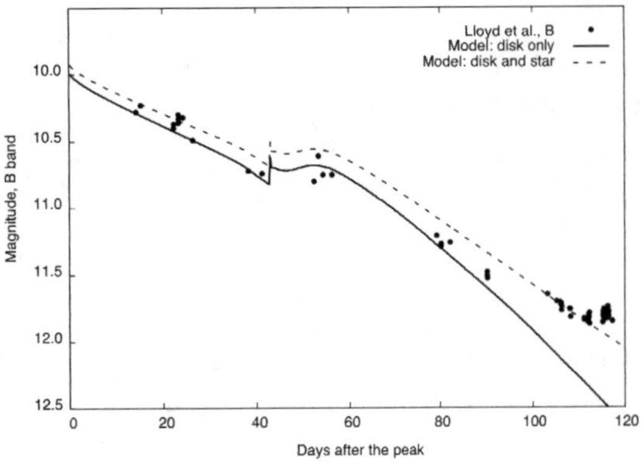

Figure 1. Optical light curve of X-ray nova A0620–00. Dots indicates observations [16]. The solid line shows the model with flux from the disk, the dashed line shows the model with optical star with magnitude $m = 13^{\mathrm{m}}$.

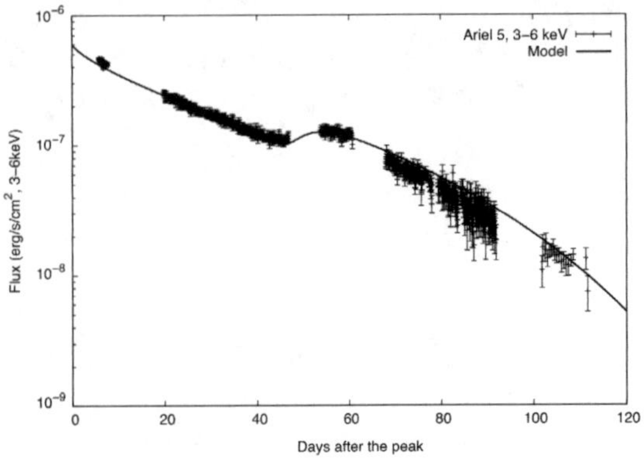

Figure 2. Soft X-ray ($3 \div 6$ keV) light curve of X-ray nova A0620–00. Dots indicates observations [17], the solid line indicates model.

5. Results

We showed that the standard model of disk accretion can explain behaviour of light curves of X-ray nova both in the optical and X-rays. In the context of this theory we found that during 1975 outburst of X-ray nova A0620–00 the α parameter of the disk was approximately 0.5. We also found that the distance to the source in our model should be somewhat smaller than in [14] and is approximately 0.85 kpc.

Acknowledgments

We thank N.I. Shakura, G.V. Lipunova, A.V. Mescheryakov, and K.A. Postnov for fruitful discussions. The author acknowledges support from Russian Foundation for Basic Research grant 14-02-91345. The work was made on equipment funded by M.V. Lomonosov Moscow State University Program of Development.

References

[1] Cherepashchuk A M 2000 Space Sci. Rev. **93** 473–580
[2] Postnov K A and Yungelson L R 2014 *Living Reviews in Relativity* **17** 3
[3] Shakura N I 1972 AZh **49** 921
[4] Shakura N I and Sunyaev R A 1973 A&A **24** 337–355
[5] Lyubarskij Y E and Shakura N I 1987 *Soviet Astronomy Letters* **13** 386
[6] Lipunova G V and Shakura N I 2000 A&A **356** 363–372
[7] Lipunova G V 2015 ApJ **804** 87
[8] Hoshi R 1979 *Progress of Theoretical Physics* **61** 1307–1319
[9] Meyer F and Meyer-Hofmeister E 1981 A&A **104** L10
[10] Ichikawa S and Osaki Y 1994 PASJ **46** 621–628
[11] Mescheryakov A V, Shakura N I and Suleimanov V F 2011 *Astronomy Letters* **37** 311–331
[12] Malacnhev K L and Shakura N I 2015 *Astronomy Letters* **41** 797–808
[13] Asplund M, Grevesse N, Sauval A J and Scott P 2009 ARA&A **47** 481–522
[14] Cantrell A G, Bailyn C D, Orosz J A, McClintock J E, Remillard R A, Froning C S, Neilsen J, Gelino D M and Gou L 2010 ApJ **710** 1127–1141
[15] Gou L, McClintock J E, Steiner J F, Narayan R, Cantrell A G, Bailyn C D and Orosz J A 2010 ApJ **718** L122–L126
[16] Lloyd C, Noble R and Penston M V 1977 MNRAS **179** 675–681
[17] Kaluzienski L J, Holt S S, Boldt E A and Serlemitsos P J 1977 ApJ **212** 203–210

Continuum correlations in accreting X-ray pulsars

M I Gornostaev[1,2], K A Postnov[1,2] and D Klochkov[3]

[1] Faculty of Physics, M. V. Lomonosov Moscow State University, Leninskie Gory, Moscow 119991, Russia

[2] Sternberg Astronomical Institute, Moscow M. V. Lomonosov State University, Universitetskij pr., 13, Moscow 119992, Russia

[3] Institute of Astronomy and Astrophysics, Karl-Eberhard University, Tübingen, Sand 1, D-72076, Germany

E-mail: mgornost@gmail.com; kpostnov@gmail.com

Abstract. Two-dimensional structure of accretion columns in the radiation-diffusion limit is calculated for two possible geometries (filled and hollow cylinder) for mass accretion rates $\dot{M}$ ranging from 10^{17} to 1.2×10^{18} g s^{-1}. The observed spectral hardening in the transient X-ray pulsars with increasing $\dot{M}$ can be reproduced by a Compton-saturated sidewall emission from optically thick magnetized accretion columns with taking into account the emission reflected from the neutron star atmosphere. At $\dot{M}$ above some critical value $\dot{M}_{cr} \sim (6-8) \times 10^{17}$ g s^{-1}, the height of the column becomes such that the contribution of the reflected component to the total emission starts decreasing, which leads to the saturation and even slight decrease of the spectral hardness. Hollow-cylinder columns have a smaller height than the filled-cylinder ones, and the contribution of the reflected component in the total emission does not virtually change with $\dot{M}$ (and hence the hardness of the continuum monotonically increases) up to higher mass accretion rates than $\dot{M}_{cr}$ for the filled columns.

1. Observations

After the discovery of X-ray pulsars in 1971 [1] it was realized that in bright pulsars the radiation plays a crucial role in braking of the accreting matter onto the surface of a neutron star with strong magnetic field. It is now well recognized that once the radiation density above the polar cap starts playing the role in the accreting matter dynamics, an optically thick accretion column above the polar cap is formed [2] (see [3] for the latest investigation). The characteristic height of the column increases with accretion rate, most of the emission escapes through sidewalls, and it can be expected that above some X-ray luminosity the dependence of the observed properties of the continuum emission on the X-ray luminosity can become different from that in the low-luminosity regime.

Indeed, a similar bimodality is observed in the dependence of the X-ray continuum hardness on luminosity. This dependence can be studied using, for example, data from all-sky monitors such as RXTE/ASM and MAXI, i.e. without dedicated spectroscopic observations which are necessary to measure CRSFs. Following this approach, we measured the hardness ratio as a function of luminosity in the accreting pulsars GX 304−1, 4U 0115+63, V 0332+53, EXO 2030+375, A 0535+26 and MXB 0656−072 using the data from different energy bands of RXTE/ASM. The result is shown in Fig. 1. To convert the ASM count rates into X-ray luminosities, we used published distances to the sources (GX 304−1: ∼2 kpc, [4]; 4U 0115+63:

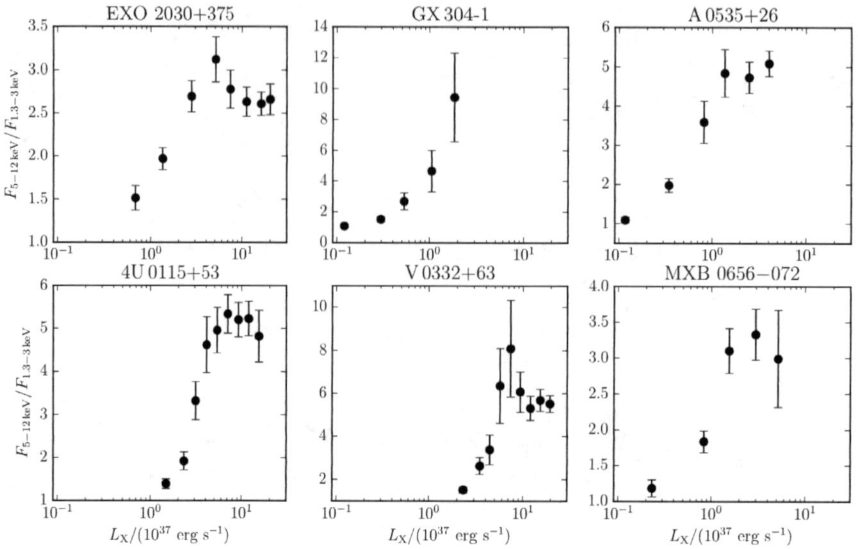

Figure 1. The ratio of the fluxes in 5–12 keV and 1.33–3 keV ranges (referred to as "hardness ratio" in text) measured with RXTE/ASM for six transient accreting pulsars as a function of the total ASM flux in the 1.33–12 keV range.

∼7 kpc, [5], V 0332+53: ∼7 kpc, [6]; EXO 2030+375: ∼7 kpc, [7]; A 0535+26: ∼ 2 kpc, [8]; MXB 0656-072: ∼ 3.9 kpc, [9]) and their broadband X-ray spectra from the archival pointed observations with INTEGRAL and RXTE available to us. One can see that at lower fluxes the hardness ratio increases with flux. At a certain flux, however, a flattening of the hardness ratio is observed in 4U 0115+63, V 0332+53, EXO 2030+375, A 0535+26 and MXB 0656−072. This 'turnover' occurs at the flux roughly corresponding luminosities $(3-7)\times 10^{37}$ erg s^{-1}. In GX 304−1, which does not show such a turnover, this 'critical' luminosity has simply not been reached during the outbursts registered by RXTE/ASM.

A similar dependence of spectral properties on the X-ray flux at luminosities below $\sim 10^{37}$ erg s^{-1} was also reported from spectroscopic observations of the transient Be-X-ray pulsar GRO J1008-57 [10].

At the lowest energies covered by RXTE/ASM, the spectrum of accreting pulsars is significantly affected by photo-electric absorption caused mostly by matter in the vicinity of the source. Such an absorption naturally influences the derived hardness ratios and cannot be accounted for without a high-resolution spectral analysis using pointed observations. Neutral absorbing matter surrounding the accretor might correlate with the mass transfer rate in the binary and might therefore be proportional to the X-ray luminosity of the pulsar. Such an effect might at least contribute to the observed positive correlation between the hardness ratio and flux. However, we do not expect any break in the slope of the correlation between the absorption column density and flux at $L_X \sim 10^{37}$ erg s^{-1} (although we cannot exclude this possibility). We therefore identify the observed breaks with those reported by [11] at similar X-ray luminosities based on the data from RXTE/PCA observations taken at higher energies which should be unaffected by the photo-electric absorption.

2. Numerical simulations of accretion column in radiation-diffusion approximation

2.1. Boundary conditions

We introduce cylindrical coordinates r, φ, z centred at the columns axis and $z = 0$ at the neutron star surface. The initial velocity of the falling matter at high altitude above the surface of the neutron star is $v_0 = 10^{10}$ cm s^{-1}. At the cylinder base ($z = 0$), the velocity is $v = 0$. As a boundary condition at the column side surface we used the relation between the radial energy flux to the energy density U in the form $F_r(r_0, z) = 2cU(r_0, z)/3$, which roughly corresponds to the conditions expected in the scattering atmospheres in the Eddington approximation.

2.2. Basic equations

The cylindrical symmetry of the problem makes it essentially two-dimensional. The steady-state momentum equation (ignoring gravity, which is very good approximation as discussed e. g. in [2]) for accretion braking by the radiation with energy density U reads

$$(\boldsymbol{S} \cdot \nabla)\, \boldsymbol{v} = -\frac{1}{3}\nabla U, \tag{1}$$

where $\boldsymbol{S} = \rho \boldsymbol{v} = \text{const}$ is the mass continuity equation. The integration of these equation yields

$$U = 3S(v_0 - v). \tag{2}$$

Following [12], the energy equation can be written as

$$\nabla \cdot \boldsymbol{F} = -\boldsymbol{S} \cdot \nabla \left(\frac{v^2}{2}\right), \tag{3}$$

where we has neglected the flow of internal energy of the falling matter.

The radiative transfer equation in the diffusion approximation is

$$\boldsymbol{F} = -\frac{c}{3\varkappa\rho}\nabla U + \frac{4}{3}U\boldsymbol{v}, \tag{4}$$

where the inflow of gas is considered to be stationary.

From equation (4) in cylindrical coordinates we find the components of the radiation energy flux:

$$F_r = \frac{-\frac{c}{3\varkappa_\perp\rho}\frac{\partial U}{\partial r}}{1 + \frac{1}{3\varkappa_\perp\rho}\frac{1}{U}\left|\frac{\partial U}{\partial r}\right|}, \tag{5}$$

$$F_z = \frac{-\frac{c}{3\varkappa_\parallel\rho}\frac{\partial U}{\partial z} - \frac{4}{3}Uv}{1 + \frac{1}{3\varkappa_\parallel\rho}\frac{1}{U}\left|\frac{\partial U}{\partial z}\right|}. \tag{6}$$

where the coefficients $\left(1 + \frac{1}{3\varkappa_\perp\rho}\frac{1}{U}\left|\frac{\partial U}{\partial r}\right|\right)^{-1}$ and $\left(1 + \frac{1}{3\varkappa_\parallel\rho}\frac{1}{U}\left|\frac{\partial U}{\partial z}\right|\right)^{-1}$ are introduced to use the modified diffusion approximation in the optically thin regions where the change of the radiation energy density on the photon mean free path becomes significant (see [13]). Equations (1) – (6) describe the accretion shock mediated by radiation in the diffusion limit and provide the structure of the zone where most of the accretion energy responsible for the formation of the emergent spectrum is released. The 'sinking zone' downstream the radiation shock, where the matter settles down subsonically and advects thermal energy towards the NS surface, should be treated separately (see [2, 13]). Different physical effects can appear in this zone, e.g. 'photon bubble oscillations' discussed in [14, 15].

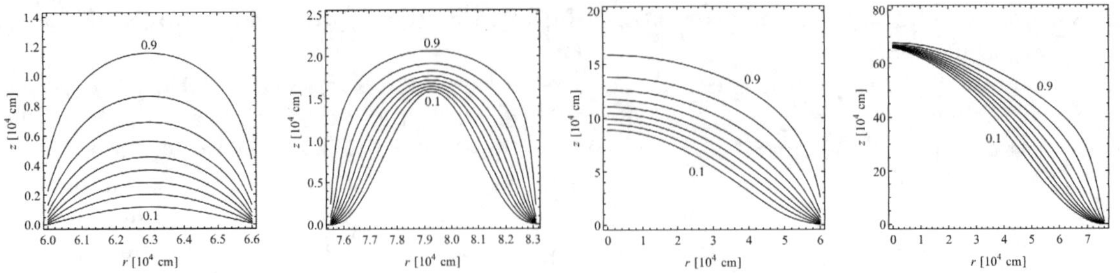

Figure 2. Contours of constant Q (0.9 to 0.1, from top to bottom). The hollow cylinder geometry with $b = 0.1$ calculated for $\dot{M}_{17} = 1$ (a) and $\dot{M}_{17} = 5$ (b). The filled cylinder geometry for $\dot{M}_{17} = 1$ (c) and $\dot{M}_{17} = 5$ (d).

2.3. Equations in dimensionless form

The convenient variables are dimensionless velocity of matter squared $Q = v^2/v_0^2$, the dimensionless column radius $\tilde{r} = \varkappa_T S_{17} r/c$ and height $\tilde{z} = \varkappa_T S_{17} z/c$, where $S_{17} = S(\dot{M}_{17} = 1) = 8.79 \times 10^6$ erg s^{-1} cm^{-2} (in the case of filled column) and $\dot{M}_{17} = \dot{M}/(10^{17}\mathrm{g\,s^{-1}})$ is dimensionless mass accretion rate onto one pole. For the assumed value of $r_0(\dot{M}_{17} = 1)$ (here and below r_0 is the current column radius) the dimensionless factors almost exactly equal to 10^{-4} cm^{-1}. For the scattering cross-sections we shall assume $\varkappa_\perp = \varkappa_T$ and $\varkappa_T/\varkappa_\parallel = 10$.

With this notation and dimensionless variables, the system of equations (2), (3), (5), (6) is equivalent to one non-linear partial differential equation (cf. [12])

$$\tilde{K}_\perp \left[\frac{\partial^2 Q}{\partial \tilde{r}^2} + \frac{1}{\tilde{r}}\frac{\partial Q}{\partial \tilde{r}}\right] + \tilde{K}_\parallel \left[\frac{\varkappa_T}{\varkappa_\parallel}\frac{\partial^2 Q}{\partial \tilde{z}^2} - 8\frac{\partial}{\partial \tilde{z}}\left(\sqrt{Q} - Q\right)\right] - \frac{\partial Q}{\partial \tilde{z}} = 0, \qquad (7)$$

where the coefficients $\tilde{K}_\perp = \left(1 + \frac{v_0}{6c(1-\sqrt{Q})}\left|\frac{\partial Q}{\partial \tilde{r}}\right|\right)^{-1}$ and $\tilde{K}_\parallel = \left(1 + \frac{\varkappa_T}{\varkappa_\parallel}\frac{v_0}{6c(1-\sqrt{Q})}\left|\frac{\partial Q}{\partial \tilde{z}}\right|\right)^{-1}$ are calculated at each consecutive iteration using values of Q obtained in the previous iteration.

2.4. Numerical procedure

Equation (7) was solved numerically using the finite difference approximation method. A time-relaxation scheme was used. In this approach, the equation in the form $L[Q] = 0$, where L is the corresponding differential operator, has been replaced by a parabolic equation of the form $\frac{\partial Q}{\partial t} - L[Q] = 0$. The steady-state solution of the last equation coincides with the solution of the original elliptic equation. An explicit finite-difference scheme was used to solve the parabolic equation.

2.5. Results of numerical calculations

Numerical simulations were carried out for accretion rates onto one pole in the range 10^{17} – 1.2×10^{18} g s^{-1} for the hollow cylinder with wall thickness br_0 (b=0.1) and the filled cylinder geometry. The column structure at different accretion rates is shown in Fig. 2. Note that at a given mass accretion rates and the NS magnetic field, the height of the braking zone in the hollow cylinder case is smaller than that in the case of the filled cylinder, as expected from the dependence of the radiation diffusion time t_d on the wall thickness, $t_d \propto b$.

3. Hardness ratio variations

3.1. Saturated Compton spectrum of sidewall emission

The LTE treatment of the spectrum of the sidewall emission from accretion columns is clearly a strong oversimplification. To take into account the scattering on electrons in the optically thin boundary of the column, we can use the model calculations of the radiation transfer problem in a semi-infinite plane-parallel atmosphere with strong magnetic field. In this case the emerging spectrum will be formed in the saturated Compton regime, with the mode 1 photons (extraordinary, i.e. polarized perpendicular to the $\boldsymbol{k}-\boldsymbol{B}$-plane, where $\boldsymbol{k}$ is the photon wavevector and $\boldsymbol{B}$ is the magnetic field vector) predominantly escaping through sidewalls [16]. The intensity of mode 2 (ordinary, i.e. polarized in the $\boldsymbol{k}-\boldsymbol{B}$ plane) photons is comparable to that of mode 1 photons only at large angles to the normal, $\mu' \sim 0$, and thus insignificantly contribute to the total flux from the unit surface area due to the geometrical factor.

In this regime, the specific intensity of extraordinary photons at energies far below the cyclotron resonance can be written to an accuracy of a few per cents as [16]

$$I'_\nu = I_\nu(1+2\mu') = \frac{2\sqrt{3}}{5}(1+2\mu')\frac{h\nu\nu_g^2}{c^2\tau_0}e^{-\frac{h\nu}{kT}}\,, \tag{8}$$

where $\mu' = \cos\theta'$ is the angle between the normal to the atmosphere and the escaping radiation direction in the plasma reference frame, τ_0 is the characteristic optical depth of the problem, $\nu_g = \frac{eB}{2\pi m_e c}$ is the electron gyrofrequency in the magnetic field. By expressing the characteristic optical depth τ_0 through the emergent radiation flux Φ using Eqs. (34-37) from [16], we find

$$I_\nu = \frac{3}{10\pi}\left(\frac{h\nu}{kT}\right)^2\frac{\Phi}{\nu}e^{-\frac{h\nu}{kT}}\,. \tag{9}$$

Therefore, we are in the position to calculate the emerging spectrum from the column in this approximation using the solutions for the column structure obtained above. To do this, it is sufficient to substitute the total radial energy flux at each height of the column $\Phi \to F_r(z)/2$ ($F_r(z)$ (6) is known from our numerical calculations), and temperature $T \to T(z)$ estimated deep inside the column (at the optical depth ~ 1), so that $T(z) = (U(z)/a_r)^{1/4}$). Note here that at large optical depth the number density of ordinary and extraordinary photons are equal, with the extraordinary photons becoming overwhelmingly dominant in the optically thin outer layers. Therefore, only half of the radial energy flux, $F_r(z)/2$, should be taken into account in the spectral calculations.

3.2. Account for the reflected component

Now we should take into account the fact that the electrons in the optically thin part of the column are moving with high velocity $v_0 \sim 1/3c$, and thus the emission beam of the column (8) should be Doppler-boosted towards the neutron star surface and reflected from the neutron star atmosphere, as discussed in [17] and [18]. The reflection coefficient in the case of single Compton electron scattering in strong magnetic field is determined by the ratio $\lambda(\nu) = \kappa_{sc}/(\kappa_{sc}+\kappa_{abs})$, where κ_{sc} and κ_{abs} is the scattering and absorption coefficients for photons in the strong magnetic field, respectively. Note that the calculated X-ray albedo turns out to be very similar for both extraordinary and ordinary photons and is virtually insensitive to the photon incident angle to the magnetic field. Below we shall use the mode-averaged value of X-ray albedo for the angle $\pi/4$. It is seen that soft photons are mostly absorbed, while hard photons are reflected by the neutron star atmosphere. Therefore, the reflected spectrum is harder than the incident one.

Let α be the angle between the photon wavevector and the direction to the NS center. Due to the boosting, part of radiation from the column will be intercepted by the neutron star surface and reflected. The critical angle $\alpha_{\max}$ within which the radiation from the height z above the surface will be intercepted by the neutron star (in the Schwarzschild metric) is [18]

$$\sin\alpha_{\max} = \frac{R_{NS}}{R_{NS}+z}\sqrt{\frac{1-r_S/(R_{NS}+z)}{1-r_S/R_{NS}}}, \tag{10}$$

where $r_S = 2GM/c^2$ is the Schwarzschild radius. Suppose from now that α is the angle between photon wavevector and the plasma bulk velocity vector. Then for a point on the accretion column sidewall of radius r_0 on the hight z the critical angle reads

$$\alpha^*(z) = \arcsin\left(\frac{R_{NS}}{\sqrt{(R_{NS}+z)^2+r_0^2}}\sqrt{\frac{1-r_S/\sqrt{(R_{NS}+z)^2+r_0^2}}{1-r_S/R_{NS}}}\right) - \arctan\frac{r_0}{R_{NS}+z}. \tag{11}$$

Let Σ be the surface of the column, $\Sigma'(\varphi, z)$ be a point on the surface and $d\Sigma = r_0 d\varphi dz$ be the elementary area. The flux from the surface element at frequency ν (in conventional notations) reads $f_\nu(\Sigma') = dE(\Sigma')/(dt d\nu d\Sigma)$ and can be separated into three parts in accordance to the angle relative to the observer. Photons escaping with angles $\alpha > \alpha^*$ to the column are directly seen as the proper column radiation f_ν^{col}. The second part includes the radiation f_ν^{ref} intercepted by the neutron star and reflected from the neutron star surface with X-ray albedo $\lambda(\nu)$. The third part f_ν^{abs} is the radiation absorbed and re-radiated by the neutron star atmosphere. Then we can write:

$$\begin{aligned} f_\nu(\Sigma') &= \int_{-1}^{1} \frac{df_\nu(\Sigma')}{d\cos\alpha} d\cos\alpha = \int_{-1}^{\cos\alpha^*} \frac{df_\nu(\Sigma')}{d\cos\alpha} d\cos\alpha + \int_{\cos\alpha^*}^{1} \frac{df_\nu(\Sigma')}{d\cos\alpha} d\cos\alpha \\ &= \int_{-1}^{\cos\alpha^*} \frac{df_\nu(\Sigma')}{d\cos\alpha} d\cos\alpha + \lambda(\nu) \int_{\cos\alpha^*}^{1} \frac{df_\nu(\Sigma')}{d\cos\alpha} d\cos\alpha + (1-\lambda(\nu)) \int_{\cos\alpha^*}^{1} \frac{df_\nu(\Sigma')}{d\cos\alpha} d\cos\alpha \\ &= f_\nu^{col}(\Sigma') + f_\nu^{ref}(\Sigma') + f_\nu^{abs}(\Sigma'). \end{aligned} \tag{12}$$

The integrands [18] in the cylindrical coordinates with account for the axial symmetry ($I_\nu(\Sigma') = I_\nu(z)$, giving by (8)) have the form:

$$\frac{df_\nu(\Sigma')}{d\cos\alpha} = I_\nu(z)\frac{2D^3}{\gamma}\sin\alpha\left(1+\frac{\pi D}{2}\sin\alpha\right), \tag{13}$$

where the Doppler factor is $D = 1/(\gamma(1-\beta\cos\alpha))$, $\gamma = 1/\sqrt{1-\beta^2}$ is the plasma Lorentz factor and $\beta = v/c$. Here we have taken into account that the specific flux (per unit frequency interval) is D times as small as the integral flux. Then for the total radiation from the column at frequency ν we obtain:

$$\begin{aligned} L_\nu &= \iint_\Sigma f_\nu(\Sigma') d\Sigma = 2\pi r_0 \int_0^{z_{max}} f_\nu(z) dz \\ &= 2\pi r_0 \int_0^{z_{max}} \left(f_\nu^{col}(z) + f_\nu^{ref}(z) + f_\nu^{abs}(z)\right) dz \\ &= L_\nu^{col} + L_\nu^{ref} + L_\nu^{abs}, \end{aligned} \tag{14}$$

where z_{max} is the upper limit of the computational area. The observed X-ray flux from one column is $F_\nu = (L_\nu^{col} + L_\nu^{ref})/(4\pi d^2)$, where d is the distance to the source. For the illustrative

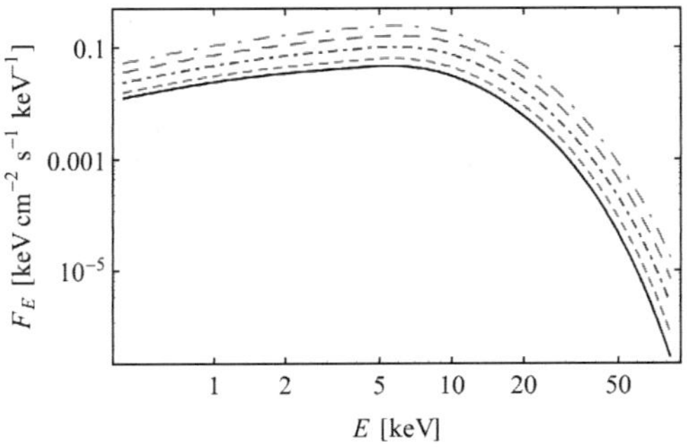

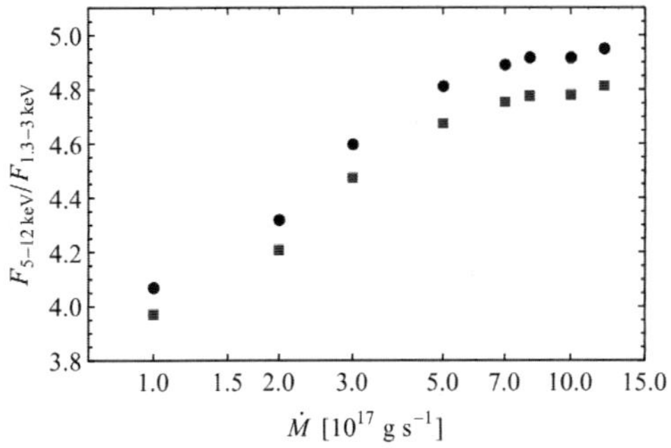

Figure 3. Left: the total spectrum of direct sidewall and reflected from the neutron star atmosphere from the optically thick filled accretion column for mass accretion rates $\dot{M}_{17} = 2$, 3, 5, 8 and 12 (from bottom to up, respectively), $R_{NS} = 13$ km. Right: the hardness ratio HR of the total spectrum. Shown are calculations for the neutron star radius $R_{NS} = 10$ km (squares) and $R_{NS} = 13$ km (circles). The single-scattering albedo $\lambda(\nu)$ from NS atmosphere calculated for electron number density $n_e = 5 \times 10^{25}$ cm^{-3}, electron temperature $T = 3$ keV, magnetic field $B = 3 \times 10^{12}$ G, photon incident angle to the magnetic field $\pi/4$.

purposes, we assume $d = 5$ kpc. Taking into account of the radiation absorbed by the neutron star atmosphere L_ν^{abs} will add more soft photons to the total spectrum.

Since we are observing both the direct and reflected radiation from the column, the total change in the hardness of the spectrum will depend on the fraction of the reflected radiation in the total flux. This, in turn, depends on the height of the column. We have calculated the total flux as the sum of the direct and reflected component as a function of $\dot{M}$ with taking into account the photon ray propagation in the Schwarzschild metric of the neutron star with a fiducial mass $M = 1.5M_\odot$ and radius $R = 10$ and 13 km. The magnetic field of the neutron star is set to 3×10^{12} G, so that the CRSF energy is about 35 keV. The result is presented in Fig. 3. It is seen that the hardness ratio HR $= \int_{5\,\mathrm{keV}}^{12\,\mathrm{keV}} F_\nu d\nu \Big/ \int_{1.3\,\mathrm{keV}}^{3\,\mathrm{keV}} F_\nu d\nu$ first increases with the mass accretion rate (X-ray luminosity), but starting from $\dot{M}_{cr} \sim (6-8) \times 10^{17}$ g s^{-1} it gets saturated and even decreases. This is in agreement with observations shown in Fig. 1.

For the hollow cylinder accretion columns which have smaller height (see Fig. 2), the fraction of the reflected component in the total emission does not virtually change with mass accretion rate within the calculated range, therefore the hardness ratio of the total spectrum monotonically increases with $\dot{M}$ in this range. Note also that in this case the spectrum is harder than in the case of the filled column because the reflected (harder) emission dominates in the total spectrum.

4. Summary and conclusions

Therefore, our model calculations lead to the following conclusions.

1. The spectral hardening in X-ray pulsars with positive CRSF energy dependence on X-ray flux can be explained by increasing of the Comptonization parameter y in the slab atmosphere of the accretion mound. It is in this regime that positive correlation of the cyclotron line is observed, for example, in Her X-1 [19].

2. At high accretion rates, the radiation-supported optically thick accretion column grows above the polar cap, the sidewall emission from the column is formed by extraordinary photons in the saturated Compton regime. The spectrum of this emission gets harder with increasing mass accretion rate.

3. With further increasing mass accretion rate the height of the column increases, such that the fraction of radiation reflected from the neutron star atmosphere starts decreasing. As the reflected radiation is harder than the incident one, the spectrum of the total emission (direct plus reflected) practically stops hardening (of course, the points in the Fig. 3 have some uncertainty).

4. In the frame of this model, the saturation of the spectral hardness in the case of a hollow cylinder geometry of the accretion column can be achieved at much higher accretion rates (roughly, scaled with the relative thickness of the column wall, b), because the characteristic height of the column in this case is correspondingly smaller than that of the filled column.

5. Variations in the pulse profile form and polarization properties during the transition from the Coulomb-braking low-luminosity regime to the radiation-braking high-luminosity state should occur with changing mass accretion rate (X-ray luminosity) in accreting X-ray pulsars. The profile change has already been reported during the low-states in Vela X-1 [20].

References

[1] Giacconi R, Gursky H, Kellogg E, Schreier E and Tananbaum H 1971 *ApJL* **167** L67
[2] Basko M M and Sunyaev R A 1976 *MNRAS* **175** 395–417
[3] Mushtukov A A, Suleimanov V F, Tsygankov S S and Poutanen J 2015 *MNRAS* **447** 1847–1856
[4] Parkes G E, Murdin P G and Mason K O 1980 *MNRAS* **190** 537–542
[5] Negueruela I and Okazaki A T 2001 *A&A* **369** 108–116 (*Preprint* astro-ph/0011407)
[6] Negueruela I, Roche P, Fabregat J and Coe M J 1999 *MNRAS* **307** 695–702 (*Preprint* astro-ph/9903228)
[7] Wilson C A, Finger M H, Coe M J, Laycock S and Fabregat J 2002 *ApJ* **570** 287–302 (*Preprint* astro-ph/0201227)
[8] Steele I A, Negueruela I, Coe M J and Roche P 1998 *MNRAS* **297** L5 (*Preprint* astro-ph/9803113)
[9] McBride V A, Wilms J, Coe M J, Kreykenbohm I, Rothschild R E, Coburn W, Galache J L, Kretschmar P, Edge W R T and Staubert R 2006 *A&A* **451** 267–272 (*Preprint* astro-ph/0602113)
[10] Kühnel M, Müller S, Kreykenbohm I, Fürst F, Pottschmidt K, Rothschild R E, Caballero I, Grinberg V, Schönherr G, Shrader C, Klochkov D, Staubert R, Ferrigno C, Torrejón J M, Martínez-Núñez S and Wilms J 2014 Luminosity dependent accretion state change in GRO J1008-57 *European Physical Journal Web of Conferences* (*European Physical Journal Web of Conferences* vol 64) p 6003
[11] Reig P and Nespoli E 2013 *A&A* **551** A1 (*Preprint* 1212.5944)
[12] Davidson K 1973 *Nature Physical Science* **246** 1–4
[13] Wang Y M and Frank J 1981 *A&A* **93** 255–268
[14] Arons J 1992 *ApJ* **388** 561–578
[15] Klein R I, Arons J, Jernigan G and Hsu J J L 1996 *ApJL* **457** L85
[16] Lyubarskii Y É 1986 *Astrophysics* **25** 577–586
[17] Lyubarskii Y E and Syunyaev R A 1988 *Soviet Astronomy Letters* **14** 390
[18] Poutanen J, Mushtukov A A, Suleimanov V F, Tsygankov S S, Nagirner D I, Doroshenko V and Lutovinov A A 2013 *ApJ* **777** 115 (*Preprint* 1304.2633)
[19] Staubert R, Shakura N I, Postnov K, Wilms J, Rothschild R E, Coburn W, Rodina L and Klochkov D 2007 *A&A* **465** L25–L28 (*Preprint* astro-ph/0702490)
[20] Doroshenko V, Santangelo A and Suleimanov V 2011 *A&A* **529** A52 (*Preprint* 1102.5254)

Duration distributions for gamma-ray bursts registered in various experiments since VENERA11/KONUS up to Fermi/GBM.

I V Arkhangelskaja
National Research Nuclear University MEPhI (Moscow Engineering Physics Institute), Kashirskoe highway 31, Moscow, 115409, Russia

E-mail: irene.belousova@usa.net

Abstract. Gamma-ray bursts duration distributions properties for events registered by experiments CGRO/BATSE, VENERA11/KONUS, VENERA12/KONUS, Swift/BAT, GRANAT/PHEBUS, Suzaku/WAM, RHESSI and Fermi/GBM are considered. GRBs observed since 1967 and now several thousands of events were listed in more than 30 catalogues. Gamma-ray bursts duration distribution was the first analysed using data of BATSE instrument onboard the CGRO. The GRBs duration distribution analysis had shown the existence of two bursts classes: long and short separated by t_{90} = 2 s. But results of similar distributions for bursts observed by other detectors have shown shifting of boundary between short and long events from value of 2 s. For example, Swift/BAT GRBs subset analysis gives the value of ~1 s for this separator point. Moreover, t_{90} has dependence from instrument registered this burst – it is function of detector sensitivity threshold and operation energy band. For instance, the duration of GRB060418 burst t_{90} is ~52 s according to Swift/BAT data and only 36 s according to RHESSI data. Therefore, the type of GGB (whether it short or long) should be defined only taking into account distinctive features of instrument detected this event. Also attributes of third intermediate GRBs subgroup appearance in events subsets for various detectors are discussed. Firstly this subgroup was found some years ago in BATSE GRB duration and duration-hardness distributions.

1. Introduction

Gamma-ray bursts (GRBs) were first registered by Vela series satellites launched in 1969 and 1970 [1]. The first GRB with duration ~1s was observed on July 2, 1967 by detectors onboard Vela–4A in energy range 0.1 - 1 MeV [2]. GRBs characteristics vary in very large intervals. For example, bursts duration lies in the interval 10^{-2} –10^{3} s, registered near the Earth fluence varies in the range 10^{-8} –10^{-3} erg/(cm^2×s). Several thousand GRBs were detected up to now (see, for example, [3]) by more than 30 instruments onboard various satellites in both hear-Earth and interplanetary space.

2. GRBs properties based on various satellite data The examples of GRBs subsets observed in several experiments are presented in table 1.

The first detailed GRB catalogue (1B) was obtained as a result of BATSE experiment onboard the Compton Gamma Ray Observatory (CGRO) [4]. Gamma-ray bursts duration distribution was the first analysed using its data.

Table 1. GRBs detected by various satellites.

Satellite	Operation period [a]		Detector name	Energy band	Time resolution	Number of GRBs detected [b]
	start	end				
Vela satellite series [1, 2, 5]	October 1963	March 1972	CsI scintillator counter	0.2 – 1.0 MeV (Vela 5a, 5b) 0.3 – 1.5 MeV (Vela 6a, 6b)	0.05s	73
VENERA 11-12	September 1978	April 1980	KONUS [3, 6]	50.0-200 keV	15.625 ms	141
			Signe 2M (SNEG-2MZ) [3, 7]	80-800 keV	1/512 s	~50
Pioneer Venus Orbiter [3, 8]	December 1978	August 1992	Gamma Ray Burst Detector (OGBD)	0.2 to 2.0 MeV	64 ms	~270
Granat	December 1989	December 1996	PHEBUS [9]	10^2 keV – 1.6 MeV	~10 ms	206
			GIGMA [10]	35 – 1300 keV		~90
Ulysses [11]	October 1990	June 2009	Solar X-Rays and Cosmic Gamma-Ray Bursts (HUS/GRB)	15.0 - 150 keV (scintillation counters) 5.0 - 15 keV (solid state detectors)	From 0.25 to 2s depending on telemetry rate	1889
CGRO	April 1991	June 2000.	BATSE (LAD and SD) [12]	20 keV-2.0 MeV	2 μs	~2700
			OSSE [13]	50 keV-10 MeV	4 ms	Several GRBs
			COMPTEL [14]	0.7–30 MeV	0.1s	~30
			EGRET (TASCS) [15]	1.0 - 2×10^2 MeV	1s	~50
Wind [16]	November 1994	operated	KONUS	50.0-200 keV	64 ms	2556
HETE-2 [17]	October 2000	March 2006	Fregate	7 - 4×10^2 keV	20 ms	~300
			WXM	2.0-30 keV	80 ms	~80
			SXC	1.5-12 keV	200 ms	
INTERGRAL [18]	October 2002	operated	Spectrometer SPI	20 keV - 15 MeV	single photon counting	~300
			Imager IBIS	15 keV - 10 MeV	-	Several GRBs
			JEM-X	3- 35 keV	-	
Suzaku [19]		operated	HXD-WAM	12-300 keV	4 ms	~750
RHESSI [20]	February 2002	February 2010	RHESSI	3 keV–20 MeV	2 ms	522 [20, 21]
Swift	November 2004	operated	BAT [22, 23]	15.0 - 150 keV	200 μs	1128
Agile [24]	April,	operated	GRID	30 MeV – 30 GeV	10 ms	Several

	2007		MCAL	18 – 60 keV		tens GRBs
Fermi [25]	July, 2008	operated	GBM [26]	~8 keV ÷ ~ 30 MeV	2 μs	1710
			LAT [27]	~ 20 MeV to >300 GeV	26.5 μs	104

[a] For RHESSI interval of processed data is listed.

[b] This table lists total amount of observed GRBs while figures 3 – 5 shows only bursts with defined duration t_{90}.

2.1. The dependence of burst duration from detector distinctive features

According to [4], bursts duration was described by the time intervals t_{50} and t_{90} in which the integrated counts from the burst increase from 25% to 75% and from 5% to 95% correspondingly. The GRBs duration distribution analysis had shown the existence of two bursts classes: long and short separated by $t_{90} = 2$ s.

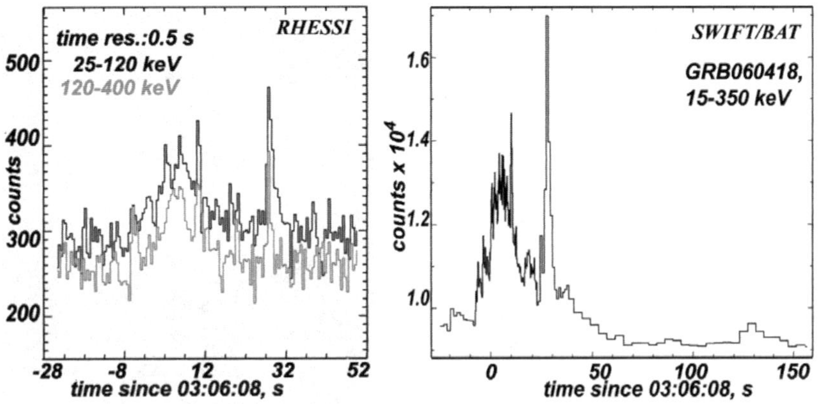

Figure 1. Temporal profiles for GRB060418 on RHESSI [20] and SWIFT [22] data. The duration of this burst $t_{90\ GRB060418\ SWIFT/BAT} \approx 52$ s and $t_{90\ GRB060418\ RHESSI} \approx 36$ s.

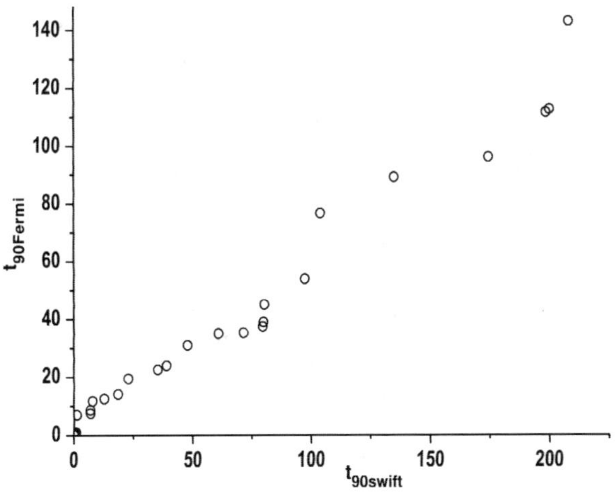

Figure 2. GRBs duration on SWIFT/BAT and Fermi/GBM data for subset of bursts simultaneously registered by these instruments.

Unfortunately t_{90} depends on instrument registered this burst. The detector sensitivity threshold and operation energy band influences to its value. Figure 1 presents the temporal profiles for GRB060418 by RHESSI and SWIFT data as illustration of this fact. The duration of this burst was $t_{90\_GRB060418\_SWIFT/BAT} \approx 52$ s according to SWIFT data [22] and $t_{90\_GRB060418\_RHESSI} \approx 36$ s according to

Journal of Physics: Conference Series **675** (2016) 032022 doi:10.1088/1742-6596/675/3/032022

RHESSI data [20]. This difference was caused by GRB spectral behavior and differences of sensitivity threshold and operation energy band between RHESSI and SWIFT.

The other example of burst duration ambiguity is shown at the figure 2. The subset of 28 GRBs observed simultaneously by SWIFT/BAT and Fermi/GBM was analysed: for 14 of them $t_{90\_SWIFT/BAT} > t_{90\_Fermi/GBM}$, for 9 bursts the opposite situation occurred and only for 5 GRB these values are comparable [22, 26]. So, we must take into account these differences in our investigation of various GRBs distributions in duration.

2.2. Short and long GRBs on data of various satellites

The GRBs duration distribution analysis on BATSE data since 1B catalogue [4] up to finally updated 5B [12] has shown the existence of 2 bursts classes: long and short separated by t_{90} = 2 s – see figure 3a.

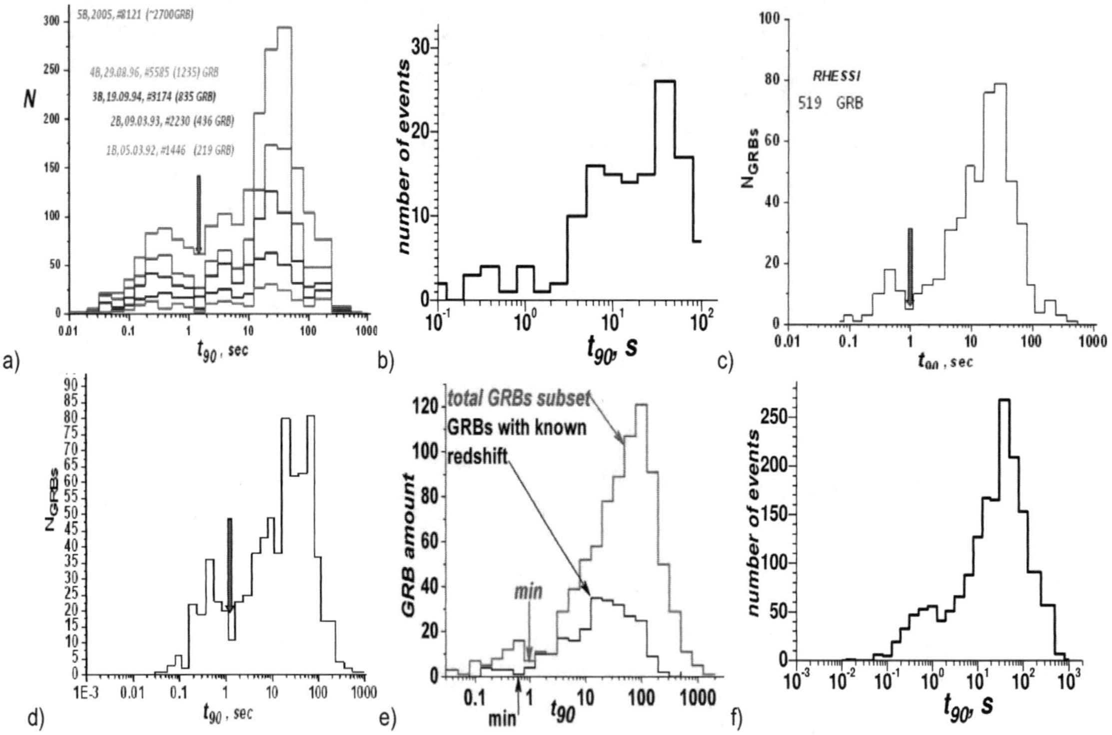

Figure 3. Duration distributions for bursts from subsets registered by BATSE (a), WIND/VENERA_11-12 (b), RHESSI (c), Suzaku/HXD-WAM (d), Swift/BAT (e), Fermi/GBM (f).

Taking into account obtained in subsection 2.1 ambiguity of GRBs duration we suppose for bursts observed in other experiments this boundary shifting. Figure 3 also shows duration distributions for bursts from subsets registered by WIND/VENERA_11-12 (b), RHESSI (c), Suzaku/HXD-WAM (d), Swift/BAT (e), Fermi/GBM (f). The analysis of several subset duration distributions presented at figure 3 allows concluding the difference of boundary between short and long events from value of 2 s obtained from CGRO/BATSE bursts catalogue. Durations of 1 s more likely correspond to this separator point for Swift/BAT and RHESSI GRBs subsets distributions.

2.3. The dependence of burst duration from redshift

Figure 4 presents duration distributions for Swift GRBs with known redshift (274 bursts from 1128) without and with events t_{90} correction to z. The analysis of figure 3 shows that value of 1 s more likely correspond to this separator point for Swift/BAT GRBs subset duration distributions without correction on redshift.

However real cosmological sources time properties should investigated only taking into account its redshift. The figure 4 indicates value ~ 0.6 s more likely as separator point between short and long GRBs after its duration correction on redshift for Swift/BAT catalogue events.

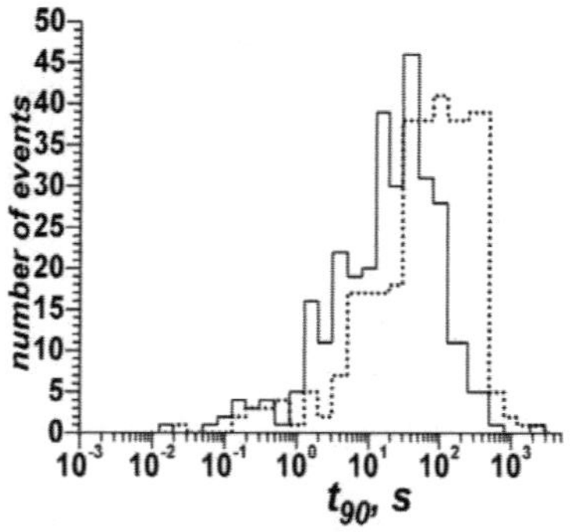

Figure 4. Swift GRBs (274 bursts with known redshift) subset duration distributions without (dotted histogram) and with (solid one) correction on redshift.

Thus, the data analysis allows concluding sufficiently changing of duration distribution and shifting the separator point between short and long GRBs again after bursts duration correction to redshift.

2.4. Intermediate GRB subgroup

The subgroup of intermediate GRBs was first found in 1999 [28] during 4B current BATSE catalogue (lately published as 5B one [12]) analysis. Later this type GRBs are widely discussed - see, for example, [29].

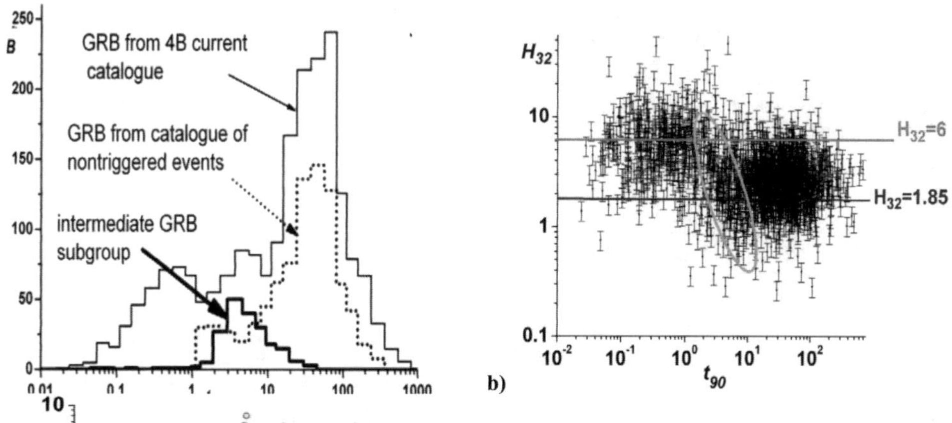

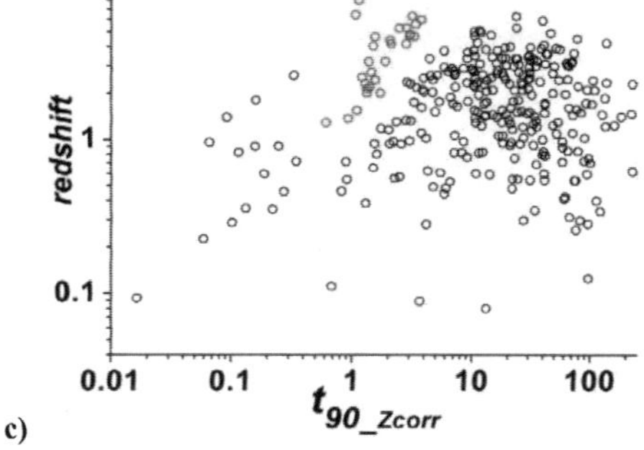

Figure 5. Intermediate GRBs subgroup appearance in events subsets for various detectors: a) BATSE current catalogue duration distribution, b) BATSE current catalogue distribution on hardness and duration, c) Swift/BAT distribution on redshift and duration (this subgroup marked by red colour).

Intermediate GRBs subgroup appearance in events subsets for various detectors is presented at figure 5. This subgroup is absent in nontriggered event catalogue [30, 31] (see dotted histogram at panel a) but clearly seen in BATSE current catalogue distribution on hardness and duration (marked by orange ellipse), and Swift/BAT distribution on redshift and duration (this subgroup indicated by red colour).

3. Conclusion

GRBs observed since 1967 and now several thousands of events were listed in more than 30 catalogues. Gamma-ray bursts duration distribution was first analyzed using data of BATSE instrument onboard the CGRO. The GRBs duration distribution analysis has shown the existence of two bursts classes: long (t_{90} more than 2 s) and short (t_{90} less than 2 s). But results of similar distributions analysis for bursts observed by other detectors have shown shifting of boundary between short and long events from value of 2 s. For example, Swift/BAT and RHESSI GRBs subsets analysis gives the value of ~1 s for this separation point. Moreover, t_{90} depends on instrument registered this burst. The detector sensitivity threshold and operation energy band influences to its value. Therefore, the type of GRB (whether it short or long) should be defined only taking into account distinctive features of instrument detected this event and GRB redshift. After taking into account characteristics GRB duration correction on redshift third intermediate GRBs subgroup appears also in Swift/BAT burst subset too.

Acknowledgments
This work performed within the framework of the Center FRPP supported by MEPhI Academic Excellence Project (contract No. 02.a03.21.0005, 27.08.2013).

References

[1] Klebesadel P et al 1973 *Astrophys. J. Lett.* **182** 85
[2] Strong I B et al 1974 *Astrophys. J. Lett.* **188** 1
[3] *http://nssdc.gsfc.nasa.gov/nmc/*
[4] Fishman G J et al 1994 *Astrophys. J. Suppl. Serie* **92** 229
[5] *http://heasarc.gsfc.nasa.gov/docs/heasarc/missions/vela6.html*
[6] Mazets E P et al 1979 *Soviet Astronomy Letters* **5 (Mar.-Apr)** 87
[7] Diachkov A V et al 1982 *Advances in Space Research* **3#4** 211
[8] Colin L et al 1977 *Space Sci. Rev.* **20** 451
[9] Tkachenko A Yu et al 2002 *Astronomy Letters* **28#6** 353
[10] Burenin R A et al 2000 *Astronomy Letters* **26#7** 413
[11] Hurley K et al 1992 *Astron. Astrophys. Suppl. Series* **92#2**: 401
[12] Paciesas W S et al 1999 *Astrophys. J. Suppl. Series* **122**: 465
[13] Share G H et al 1999 *BAAS* **31** 711
[14] *http://wwwgro.unh.edu/bursts/cgrbdata.html*
[15] Dingus B L et al 1994 *AIP Conf. Proc.* **307** 22
[16] Cline T L et al 2003 *AIP Conf. Proc* **662** 143
[17] *http://space.mit.edu/HETE/Bursts/*
[18] Lonjou V et al 2005 *Nuclear Instruments and Methods in Physics Research* **A 554** (1-3) 320
[19] *http://www.astro.isas.jaxa.jp/suzaku/HXD-WAM/WAM-GRB/*
[20] *http://vizier.cfa.harvard.edu/viz-bin/VizieR?-source=J/A+A/498/399*
[21] *http://www.rhessi-grb.info/*
[22] Barthelmy S D et al 2005 *Space Sci. Rev.* 1**20** 143
[23] *http://swift.gsfc.nasa.gov/archive/grb_table/fullview/*
[24] Tavani M et al 2009 *Astron. Astrophys.* **502** 995
[25] Atwood W B et al 2009 *Astrophys. J.* **697** 1071
[26] *http://heasarc.gsfc.nasa.gov/W3Browse/fermi/fermigbrst.html*
[27] *http://fermi.gsfc.nasa.gov/ssc/observations/types/grbs/lat_grbs/*
[28] Belousova I V et al 1999 *Astronomy Reports* **43 #11** 734
[29] Horváth I et al. 2006 *Astron. Astrophys.* **447#1** 23
[30] Kommers J M et al *Preprint* astro-ph/0012519v1 28Dec 2000
[31] Stern B E et al 2000 *Astrophys.J.Lett.* **538** L21

Latitudinal profiles of solar protons in the Earth's magnetosphere

L Lazutin
Space Science Division, Scobeltsyn Institute for Nuclear Physics, Moscow State University, Leninskie gory, 1, Moscow, 119991, Russia

E-mail: lll@srd.sinp.msu.ru

Abstract. Dynamics of the latitudinal profiles penetrating into magnetosphere solar protons is studied using particle spectrometers data on board of the low latitude satellite CORONAS-F with orbit inclination ~83°. Formations of several different types of the profiles during magnetic storms are considered.

1. Introduction

Low energy (<100 MeV) solar protons penetrate into the Earth's magnetosphere through the magnetotail with the deepness depending of the particle energy and the magnetic disturbance. Investigations of the penetration boundary both experimental and theoretic have long history (see for example [1-5]).

In the present work dynamics of the latitudinal profiles, dependence on particle energy and magnetic activity were studied instead of single point penetration boundary. Several particle spectrometers were installed onboard the low latitude satellite CORONAS-F [6] with orbit inclination ~83°. Data obtained by these detectors used in the presented article.

2.Types of latitudinal profiles

2. 1. Definitions.

Example of the latitudinal profiles of three energy ranges: 1-5, 14-26 and 50-90 MeV with typical energy shift are presented on Fig. 1a. McIlwain L-index for a low magnetic activity is used here as a equivalent of corrected geomagnetic latitudes. Point of transition from polar cap to lower proton flux is shown as "A", 'PB' presents a commonly used position of penetration boundary, and "F" indicates the latitude where solar proton counts meets with counter's background galactic cosmic rate.

2.2. Night sector.

In the magnetotail and outer night-time quasitrapping region protons have isotropic distribution, and precipitation flux registered by satellite is equal to the interplanetary one. Larmor radius of the proton ρ there is larger than magnetic field line curvature radius Rc. When the ratio $\varepsilon = \rho / Rc$, named parameter of adiabasity, reaches critical level, (between 0.1 and 0.6, [7,8]) some particles became trapped and precipitating flux began to decrease. Thus nighttime profiles formation depends on a local magnetosphere structure. Nigh-time field line stretching during magnetic storm shifts profiles earthward. Proton penetration mode is stochastic – some protons precipitates earlier, some later. Because of that a relation to the results of trajectory calculation must be regarded by care.

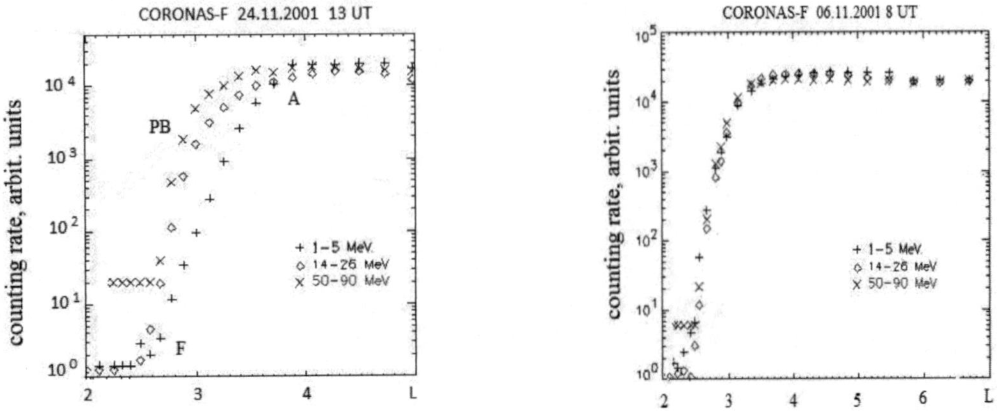

Figure 1. Examples of the latitudinal profiles measured by CORONAS-F with energy dispersion (left) and coincidence of all profiles at the end of main magnetic storm phase (right). Errors are inside experimental points.

At the end of the main phase of a strong magnetic storms stretching became so intense that particles of all energies from 1 to 50 MeV freely penetrate down to the dipole field lines where all became stopped independent of energy. Figure 1b shows an example of such profiles.

2.3. Day time sector

Figures 2a and b present examples of comparison of day and night profiles. There is theoretical possibility for penetration on the day sector through casp region, but strong magnetic field gradient at the transition region does not allow for low energy protons to penetrate deep to close field lines. More reliably for them to arrive at the day side by the magnetic drift.

During magnetic storms our L-coordinate does not reflect drift trajectory. It must be calculated based on conservation of the second adiabatic invariants. In a paper [8] such trajectories were found for a model of disturbed magnetosphere and it was shown that protons from some night high latitude region of isotropy will arrive to the dayside well trapped, with decreased level of the precipitating flux. As a result, day profiles are more prolonged and precipitation boundaries are located at the higher latitudes.

Increased proton energy leads to the penetration closer to the Earth. Difference between magnitude and magnetic field line configuration became smaller, and smaller became difference between day and night profiles.

3. Magnetic storms

During the main phase of the magnetic storms all profiles, day and night, became shifted to the lower latitudes with a steeper intensity decrease. Latitudinal shift is very individual at different magnetic storms. We will point out only at one peculiarity at the end of the main phase of a strong magnetic storm. As one can see on figure 1b, all night time profiles coincide for the whole registered energy range from 1 to 50-100 MeV. Proposed in [9] explanation based on a distortion of regular magnetic field decrease with latitudes. Particles of all energies freely penetrate outer tail-like field lines, lingering together at the steep dipole magnetic field lines.

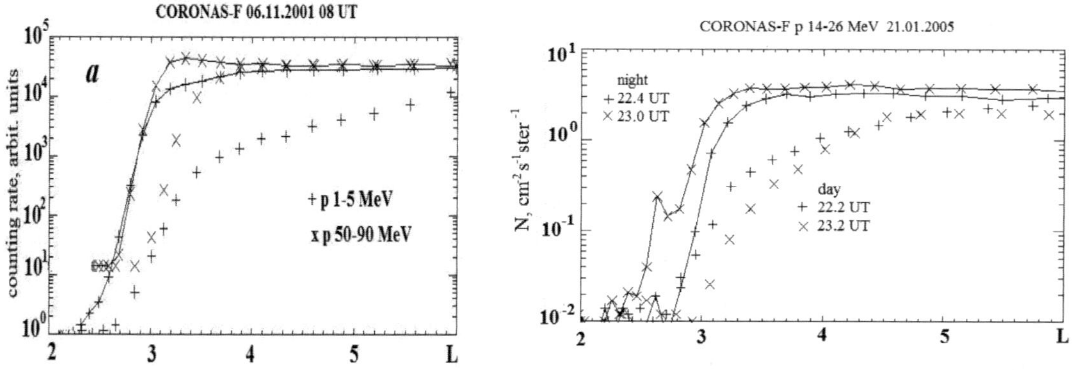

Figure 2. Two examples of day (separated symbols) and night (connected by lines for better visualization without any suppositions about linear approximation applicability) profiles. Errors are inside experimental points.

4. Solar wind pressure

Fast increase of a solar wind pressure with arrival of enhanced flux may cause latitudinal shift of a proton profiles before the beginning of the storm main phase. Such effect of magnetosphere distortion was found by Meng [10] by active aurora dynamics and later such shift was mentioned in October 2003 storm case study [11].

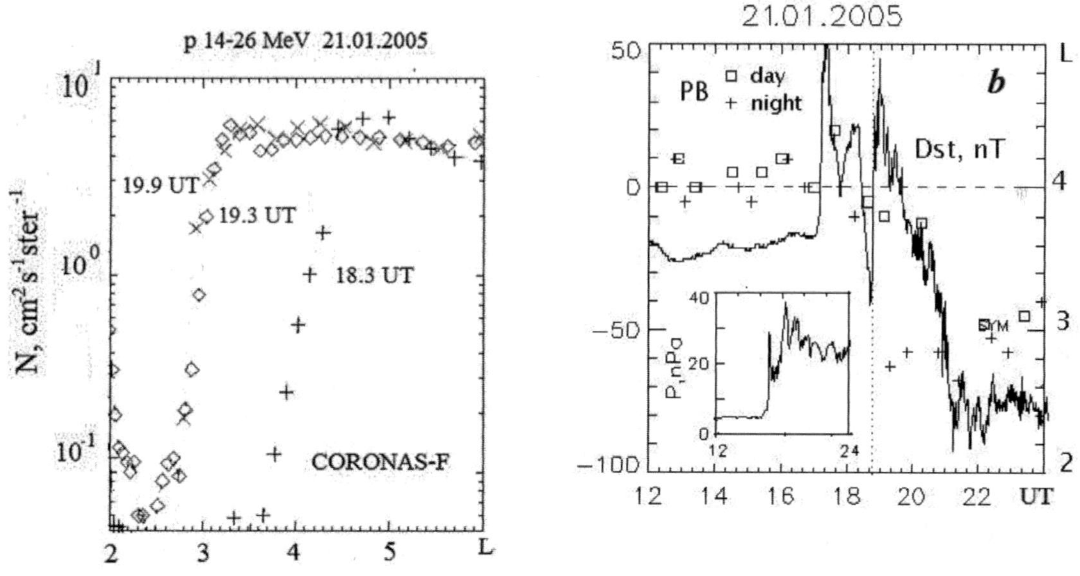

Figure 3. Shift of the profiles caused by solar wind pressure increase (a) and magnetic storm development (b), with solar wind pressure plot in the inserted box. Errors are inside experimental points.

Figure 3a shows an example of such shift, while figure 3b described Dst development, proton penetration boundary position and solar wind pressure in inserted box. Such shift may be explained by the cross-tail current which closed pressure induced magnetopause currents and created tailward extends magnetic field lines. Importance of the solar wind pressure for the night penetration boundary position was found previously by [12].

5. Conclusion

- Solar protons penetrate to the night side with isotropic pitch-angle distribution until particle larmor radius is larger than magnetic field line curvature. Closer to the Earth precipitating flux started to decrease because of partial trapping and only small fraction of particles reach background boundary.
- To the day side protons comes by the magnetic drift with a trajectory, which conserved second adiabatic invariant. As a result of magnetic field asymmetry, protons from isotropic night side position may arrive to the dayside trapping region with decreased precipitation flux. As a result, dayside profile are more prolonged and their mechanism of formation is different because depends of the whole magnetic field configuration.
- With the increase of the proton energy and toward the end of storm main phase protons penetrate closer to the Earth, profile became steeper and difference between day and night profiles became smaller. At the end of the main phase of a strong magnetic storms profiles coincides for all energy range because of the specific magnetic field configuration.
- Essential earthward shit of the nighttime proton profiles may be observed before the beginning of the main phase caused by the rise of the solar wind pressure.

Acknowledgements
I am very grateful to all colleagues from Space Science Division who created CORONAS-F satellite and spectrometers installed onboard it, received and prepared data and participated in scientific analysis for a help and cooperation.

References

[1] Smart D *et al* 2000 *Space Sci. Rev.* **93** 281
[2] Ivanova T *et al* 1985 *Geomagnetism and Aeronomy* **25** 1 7
[3] Kuznetsov S *et al* 2007 *Solar System Research* **41** 4 348
[4] Morrill G and Quenby J 1971 *Planet. Space Sci.* **19** 1541
[5] Dmitriev A *et al* 2010 *J. Geophys. Res.* **115** doi:10.1029/2010JA015380.
[6] Kuznetsov S *et al* 2002 *Adv. Sp. Res.* **30** 1857
[7] Young S *et al* 2008 *J. Geophys. Res.* **113** doi:10.1029/2006JA012133
[8] Kuznetzov S and Ribakov A 2003 *Geomagnetism and Aeronomy* **43** 3 315
[9] Lazutin L *et al* 2011 *Geomagnetism and Aeronomy* **51** 2 198
[10] Meng C. 984 *J. Geophys. Res.* **89** 227
[11] Panasyuk M *et al* 2004 *Cosmic research* **42** 5 489
[12] Nesse Tyssøy H and Stadsnes J 2015 *J. Geophys. Res. Space Physics* **120** 553

Journal of Physics: Conference Series **675** (2016) 032024 doi:10.1088/1742-6596/675/3/032024

Jovian electrons as an instrument of investigation of the interplanetary medium structure

E Daibog[1], K Kecskemety[2], L Lazutin[1], Yu Logachev[1]
1. Space Science Division, Scobeltsyn Institute for Nuclear Physics, Moscow State University, Russian Federation
2. Wigner Research Centre for Physics, Budapest, Hungary

E-mail: daibog@srd.sinp.msu.ru

Abstract. Electrons accelerated in the Jupiter magnetosphere are usually registered by near-earth spacecraft under optimal magnetic connection between the Earth and Jupiter, taking place once in 13 months (Earth- Jupiter synodic period). During the period of minimal solar activity between 23 and 24 cycles in 2007-2009 electrons of MeV energies were observed practically at each of 14 solar rotations (more than a year), which requires extremely long quasistationary state of inner heliosphere with constant Earth- Jupiter connection. To explain this situation the model with long living magnetic trap, co-rotating with the Sun, was suggested. Passing by the Jupiter this trap captures electrons, which then are registered by subsequent passing of the trap by the Earth.

1. Introduction

The first detection of Jovian electrons (J-electrons) took place in 1973 by Pioneer-10 near Jupiter [1]. It was noted in [2] and confirmed by further investigations aboard IMP-8 [3,4] that under quiet Sun Jupiter is the main origin of MeV electrons near the Earth. Jovian electrons also were registered in 1973-74 at the Mars-7 [5] and Prognoz-3 [6]. It was discovered that Jupiter is the source of lower energy electrons (above 40 keV) as well. Most natural to assume that the Jovian electrons reach the Earth along interplanetary magnetic field lines mainly in the period of direct connection between Earth and Jupiter [2, 7]. In the period when there was no direct Earth-Jupiter connection (the latter occurs every thirteen months – synodic Earth-Jupiter period), the appearance of Jovian electrons near Earth and their 27-day variations over the following 5–6 months were observed.

The interest to Jovian electrons resumed after the extremely quiet minimum of solar activity between 23 and 24 cycles in 2007-2008 when they were observed near the Earth for 14 successive solar rotations (see figure 1). It became clear, that the existing model of propagation of Jovian electrons to the Earth is unable to explain many peculiarities of their fluxes that initiated us to suggest another propagation model.

2. Observations

During very long quiet period in 2007-2009 a series of Jovian electron peaks in 14 successive rotations were observed. The optimum magnetic connection between Earth and Jupiter at Vsw = 450 km/s took place in the middle of this period (February 15 2008). Before and after this time the source of electrons is also Jupiter but with less efficiency depending on the conditions of perpendicular particle propagation near Jupiter.

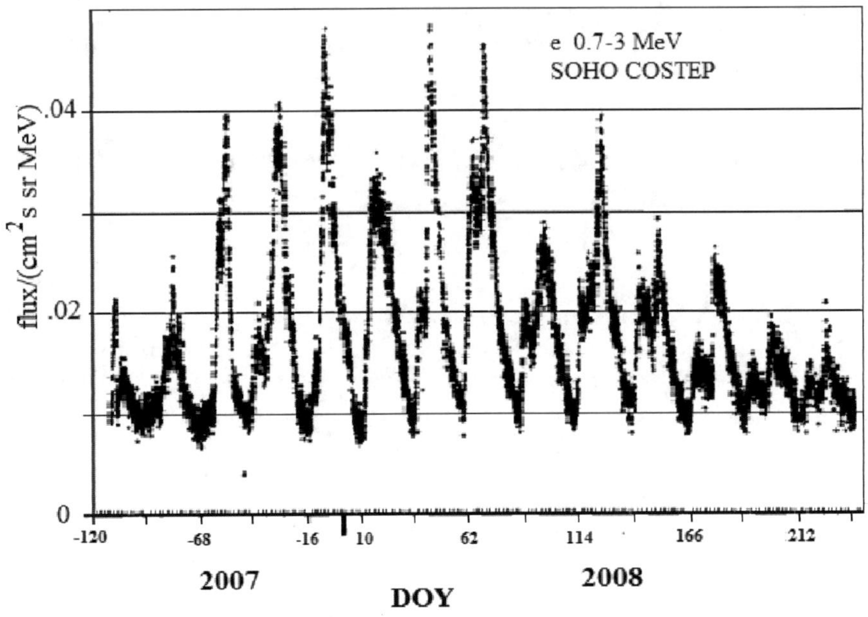

Figure 1. Fluxes of Jovian electrons 0.7-3 MeV (SOHO, COSTEP) during 14 consecutive solar rotations in 2007-2008, a long period with steady-state interplanetary medium, observed in flux variations of Jovian electrons.

3. The model of a trap

The longer term occurrence of electrons near Earth can be explained by the possible formation of magnetic traps in interplanetary space that are filled with electrons when passing near Jupiter and then identified when passing near Earth (see figure 2). According to the proposed model, Jovian electrons can be observed in any period irrespective of the phase of the Earth-Jupiter connection, while the 27-day variation of these electron fluxes is explained by the rotation of traps along with the Sun.

During solar activity minimum of 23-24 cycles, between 2007 and 2009, the Sun was characterized by an extremely quiet weather that affected the state of interplanetary space, inside the inner heliosphere (<5 AU). Solar activity was very low and practically only a low background of ions was observed whereas Jovian electrons were well defined.

Magnetic traps filled with the Jovian electron fluxes observed during fourteen solar rotations in 2007–2008 should have existed for the entire indicated period. The creation of magnetic traps depends on the solar wind speed. Time profile of the solar wind speed remained almost unchanged over the course of the solar rotations, and the magnetic traps that formed were stable as well [8].

The suggested trap model rotating with the Sun is schematically shown in figure 2 where the magnetic field lines for slow (300 km/s) and fast (600 km/s) solar wind are represented; those are fitted in a way to intersect beyond the Jupiter orbit.

To observe trapped Jovian electrons in the vicinity of the Earth such a trap must satisfy a few essential conditions. These are large extent and long time stability. Except that beyond the Jupiter orbit it must have a magnetic mirror to prevent free escape of electrons, which ensure long existence of Jovian electrons in the trap

The 27-day periodicity of nearly constant structures of solar wind suggests the existence of a long-lasting trap filled by electrons during its passage by Jupiter at every solar rotation and subsequent registration at the Earth.

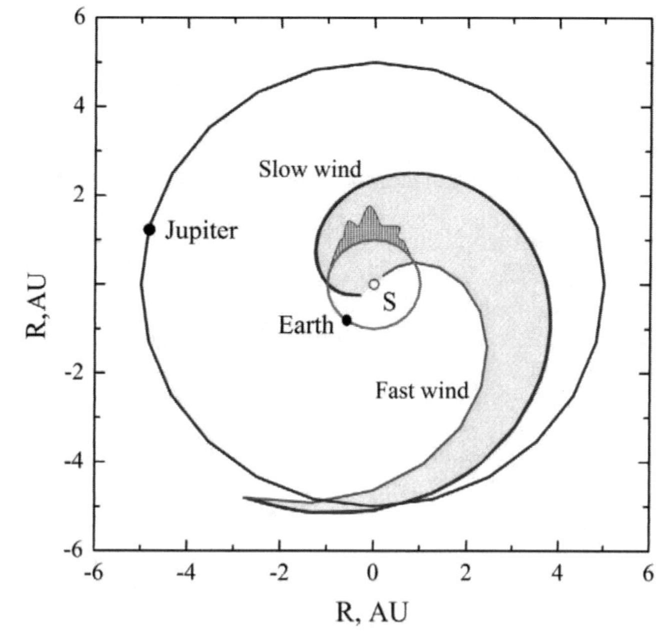

Figure 2. Schematic magnetic trap, filled up with electrons under every passage by the Jupiter. The possible electron time profile at the Earth orbit is shown.

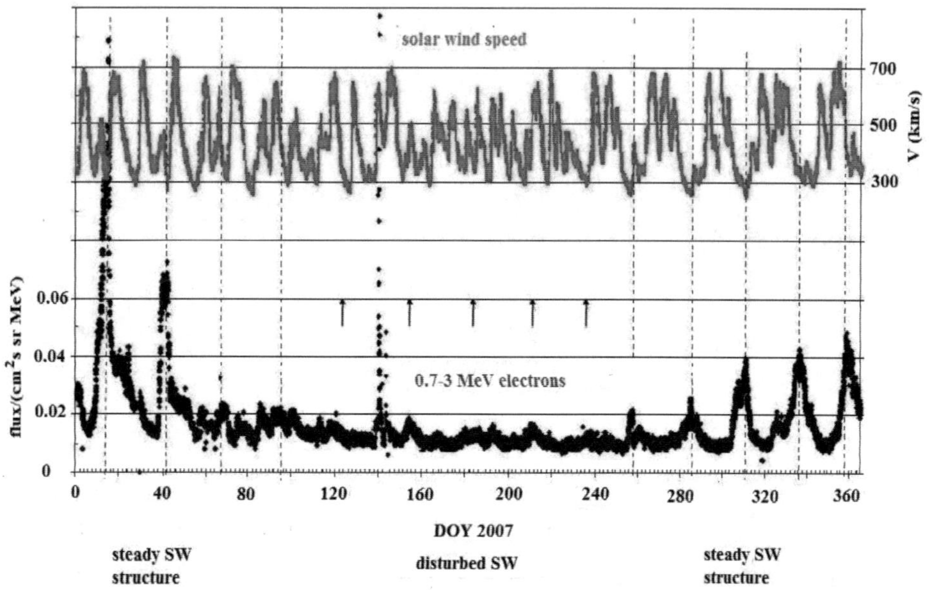

Figure 3. Jovian electrons and solar wind speed during 2007. Vertical dashed lines and arrows mark illusive electron maxima, separated by 26-27 days

As mentioned above, the structure of solar wind through the whole period of 14 MeV electron risings in 2007-2008 was very stable. When the structure of the solar wind is disturbed, the trap becomes unstable or is not formed at all, and J-electrons near the Earth are not observed (see figure 3). Here before DOY 80 and after DOY 260 of 2007 a steady structure of solar wind speed was observed which caused the formation of the trap and registration of Jovian electrons near the Earth. In the middle of 2007 the SW speed profile is not regular and Jovian electrons are absent even under nearly optimum magnetic connection. Thus one can conclude that in two first rotations the flux tubes were filled by electrons, while the next 9 were empty followed by another 14 increases of J-electrons, shown in figure 1.

What was the difference between 2007 and 2008? Optimal Earth-Jupiter magnetic connection under SW speed 450 km/s was correspondingly Jan 15 2007 and Feb 15 2008. After optimal connection in 2007 J-electrons were observed only during one solar rotation while in 2008 6-7 rotations. The geometry is the same but SW speeds dramatically differ, and situation with electrons is also radically different.

4. Summary

- The hypothesis of the model of long-living magnetic traps, co-rotating with the Sun, proposed earlier in [8], is confirmed. This model explains 27-day variations of low energy electrons, continuously accelerated in the Jovian magnetosphere. Passing by the Jupiter this trap captures Jovian electrons, which then are registered during subsequent passing of the trap by the Earth.
- Stability of the solar wind structures is a factor of main importance. When the structure of the solar wind is disturbed, the trap becomes unstable or is not formed at all, and J-electrons near the Earth are not observed.

Acknowledgements
The work was supported by the Russian Foundation for Basic Research, project no. 14-02-00196. The authors thank G.M. Surova for the help in preparing the manuscript.

References

[1] Chennet D L., Conlon T F, Simpson J A 1974 *J. Geophys. Res.* **79** 3551.
[2] Teegarden B J et al 1974 *J. Geophys. Res.* **79** 3615.
[3] Krimigis S M , Sarris E T , Armstrong T P 1975 *Proc. of 14th ICRC*, **2** 752.
[3a] Krimigis S M , Sarris E T , Armstrong T P 1975 *Geophys. Res. Lett.* **2** 561.
[4] Mewaldt R A , Stone T C , Vogt R E 1976 *J. Geophys. Res.* **81** 2397.
[5] Alekseev N V et al. 1982 *Izv. AN USSR, ser. fiz.* **46** 1695.
[6] Alekseev N V et al. 1983 *Izv. AN USSR, ser. fiz.* **47** 1810
[7] Trainor J H et al. 1974 J. *Geophys. Res.* **79** 129.
[8] Daibog E, Kecskemety K, Logachev Yu 2013 *J. of Phys. Conf. Series* **409** 012162

Journal of Physics: Conference Series **675** (2016) 032025 doi:10.1088/1742-6596/675/3/032025

The background model in the energy range from 0.1 MeV up to several MeV for low altitude and high inclination satellites.

I V Arkhangelskaja, A I Arkhangelskiy
National Research Nuclear University MEPhI (Moscow Engineering Physics Institute), Kashirskoe highway 31, Moscow, 115409, Russia

E-mail: irene.belousova@usa.net

Abstract. The gamma-ray background physical origin for low altitude orbits defined by: diffuse cosmic gamma-emission, atmospheric gamma-rays, gamma-emission formed in interactions of charged particles (both prompt and activation) and transient events such as electrons precipitations and solar flares. The background conditions in the energy range from 0.1 MeV up to several MeV for low altitude orbits differ due to frequency of Earth Radiation Belts – ERBs (included South Atlantic Anomaly - SAA) passes and cosmic rays rigidity. The detectors and satellite constructive elements are activated by trapped in ERBs and moving along magnetic lines charged particles. In this case we propose simplified polynomial model separately for polar and equatorial orbits parts: background count rate temporal profile approximation by 4-5 order polynomials in equatorial regions, and linear approximations, parabolas or constants in polar caps. The polynomials' coefficients supposed to be similar for identical spectral channels for each analyzed equatorial part taken into account normalization coefficients defined due to Kp-indexes study within period corresponding to calibration coefficients being approximately constants. The described model was successfully applied for the solar flares hard X-ray and gamma-ray emission characteristic studies by AVS-F apparatus data onboard CORONAS-F satellite.

1. Introduction

The gamma-ray background physical origin for X-ray and gamma-emission detectors installed onboard low altitude satellites defined by: diffuse cosmic gamma-emission, atmospheric gamma-rays, gamma-emission formed in interactions of charged particles (both prompt or activation) and Earth albedo neutrons [1, 2]. Prompt interactions caused due cosmic rays, transient events (magnetospheric electrons precipitations and charged particles originated in solar flares) and particles trapped in Earth Radiation Belts (ERBs) including South Atlantic Anomaly (SAA) region. Activation component occurs during ERBs and SAA passing and its influence lasts correspondingly half-life of nuclei formed in trapped particles interactions with detectors and construction elements. Respectively different prevalent background components, low altitude satellite orbit divided into areas of 4 types: polar cups and equatorial, ERBs, SAA regions. Events recognising and analysis is possible only in equatorial regions and polar cups.

The background conditions in the energy range from 0.1 MeV up to several MeV for low altitude orbits differ due frequency of ERBs and SAA regions passes and cosmic rays rigidity. Usually for low inclination orbit satellites a combination of Legendre polynomials used for background approximations – see, for example, [2, 3]. Its coefficients defined by cutoff rigidity for orbit inclination. Effects of detectors material and construction elements activation in ERB and SAA regions or moving along magnetic lines charged particles taken into account due activation analysis because of satellite pass through this areas once per several days. In such cases most part of orbit might be classified as equatorial without loss of generality in spite of background count rate periodical structure – see, for example, figure 1.

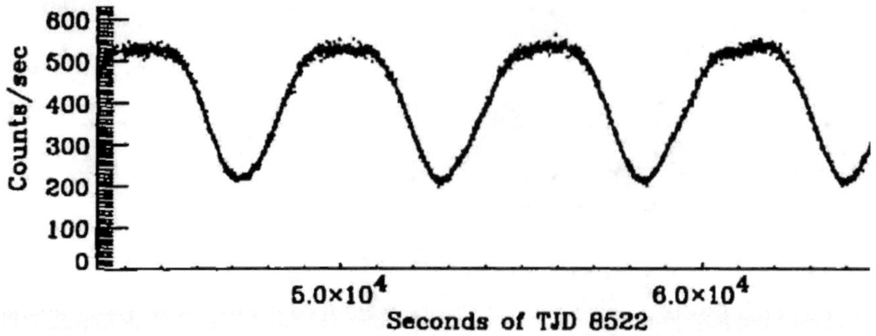

Figure 1. Modelled using Legendre polynomials (solid curve) and measured (points) background count rate for BATSE experiment onboard CGRO satellite in the energy band 30-40 keV. Adopted from [3].

High inclination satellites enter SAA once per some days at least but it pass through ERB several times per day. Therefore such situation requires background model corrections due additional periodic components for each activated nuclei. It makes background model very complicated.

We suppose another solution. Taking into account the absence of event recognising possibility during SAA and ERBs passing we propose simplified polynomial model separately for polar and equatorial orbits parts.

2. The AVS-F apparatus short description

The AVS-F apparatus (Amplitude-Time Spectrometry of the Sun) (see, in particular, [4, 5] and references therein) was installed onboard the specialized automatic station CORONAS-F [6] operated since July, 2001 up to December, 2005. The apparatus was intended for the solar flares hard X-ray and gamma-ray emission characteristic studies and for gamma-ray bursts detection.

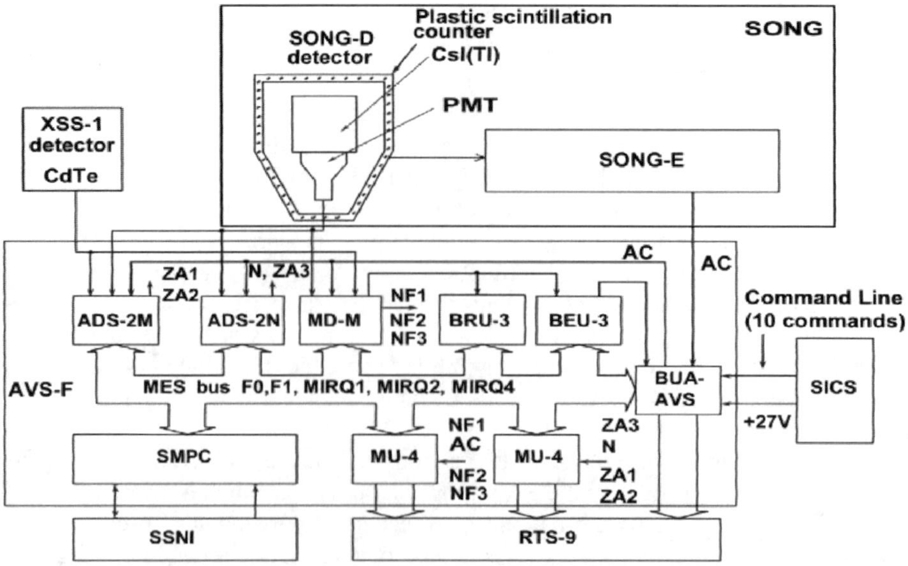

Figure 2. Functional diagram of the AVS-F experiment.

AVS-F instrument was the system of electronics for onboard data acquisition from the SONG-D (Solar Neutrons and Gamma-quanta) [7] scintillation detector developed by SINP MSU, and X-ray semiconductor spectrometer XSS-1 [8] constructed by MEPhI and IKI RAS in cooperation. AVS-F apparatus was developed basing on the block system of electronics and included electronics crate with power supplies and a set of functional blocks. AVS-F experiment functional diagram is presented in figure 2. SONG-D was the detector with 200 mm diameter and 100 mm height based on CsI(Tl) for hard X-ray and γ-ray analysis, fully surrounded by plastic anticoincidence shield. XSS-1 was CdTe detector for X-ray analysis with size 4.9 mm × 4.9 mm. According to the preflying calibration data energy regions of the AVS-F apparatus were 3-30 keV in X-ray band (detector XSS-1), 0.1-10 MeV in low energy range and 2-80 MeV in high energy one (detector SONG-D). The onboard calibration was made every month due to 11 background lines: positron line of 0.511 MeV and characteristic for this apparatus background lines formed during the interaction of the protons and neutrons with the material of the detector and structural materials of the satellite. During the whole period of apparatus operation its energy ranges were varieties: the amplification coefficient was changed approximately on ~1.8 % per month and the detector threshold was increased at ~1 % per month [9, 10]. The energy resolution of the system was ~13.0% (~86 keV) for γ-quanta with energy 0.662 MeV from ^{137}Cs [10].

3. Background modeling for AVS-F apparatus

The orbit inclination of CORONAS-F satellite was ~82.5° and it passed through the ERBs 4 times per each 90 min and SAA once per several days during at least 4 subsequently orbits. For such orbit the γ-events observations are possible only in equatorial orbit parts or polar caps and this orbits regions passing duration is not enough for de-excitation of activation lines. In this case we propose simplified polynomial model separately for polar and equatorial orbits parts: background count rate temporal profile approximation by 4-5 order polynomials in equatorial regions, and linear approximations, parabolas or constants in polar caps.

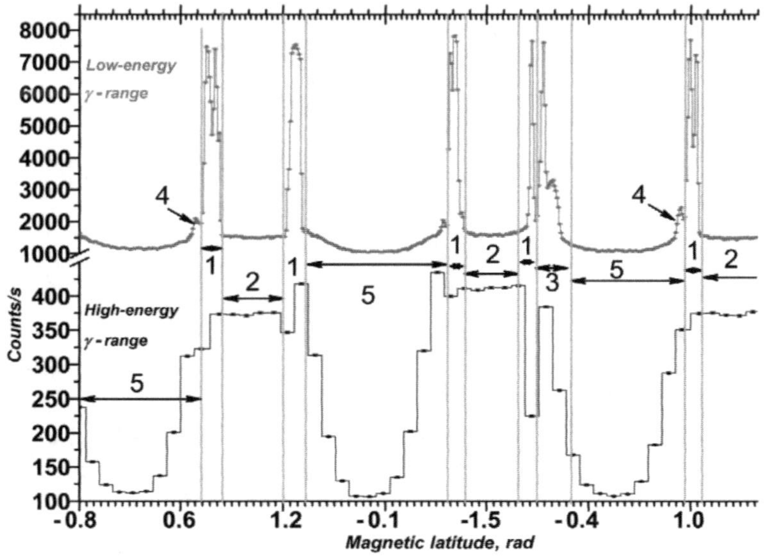

Figure 3. Typical example of AVS-F counts rate latitudinal profile in the low-energy and high-energy gamma-bands. Adopted from [5].

Typical dependence of the summarized AVS-F counts rate on geomagnetic latitude in the low- and high-energy gamma-bands is presented in figure 3. There are well-defined ERBs, polar caps, SAA and equatorial regions (1, 2, 3 and 5 correspondingly) in the high-energy and low-energy bands. Also the areas of precipitated from the external Earth radiation belt electrons bremsstrahlung registration [11] in the low-energy gamma-band were identified (4). Averaged count rates in each spectral channel (for the low- energy and high-energy ranges) were approximated by fourth- and five-degree polynomials (the example see at figure 4) at the equatorial orbit regions and by parabolic curves and linear functions at the polar ones ([5] and references therein). Procedure of event temporal profiles behaviour investigation by the AVS-F apparatus data includes background subtraction in each spectral channel both for SONG-D (82 channels of the low-energy range and 64 channels of the high-energy range) and XSS-1 (32 channels of the X-ray range) detectors following mentioned above method.

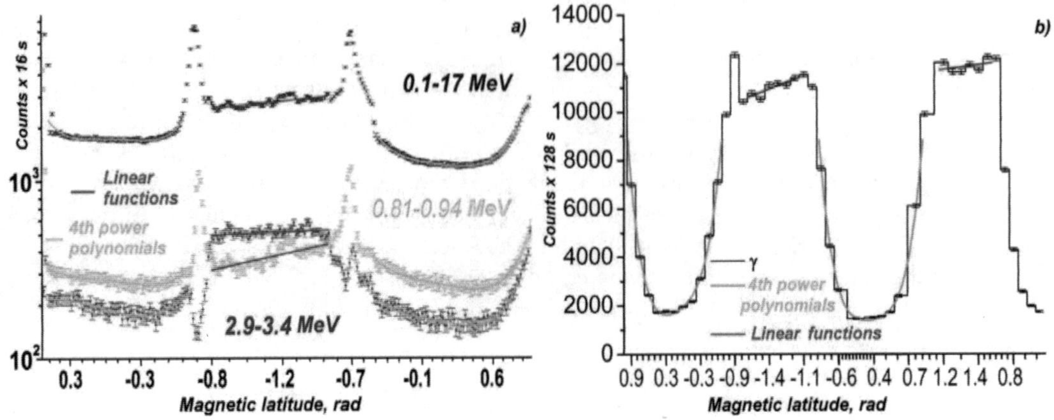

Figure 4. Typical latitudinal profiles of the AVS-F counts rate and their approximation in the low-energy (a) and high-energy (b) gamma-bands. Profiles on the equatorial parts of the CORONAS-F orbit are approximated by fourth-degree polynomials (orange curves), on the polar parts by linear functions (red lines).

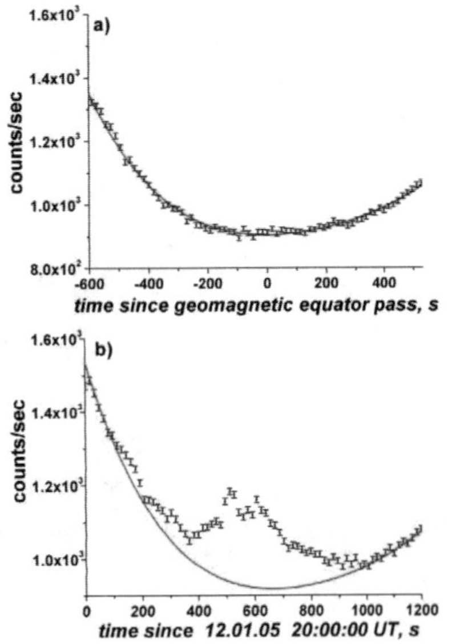

Figure 5. Averaged through 4 orbits AVS-F background count rate in the energy band 0.1-20 MeV with its 4-degree polynomial approximation (a) and obtained polynomial in combination with data for January 12, 2005 solar flare (b).

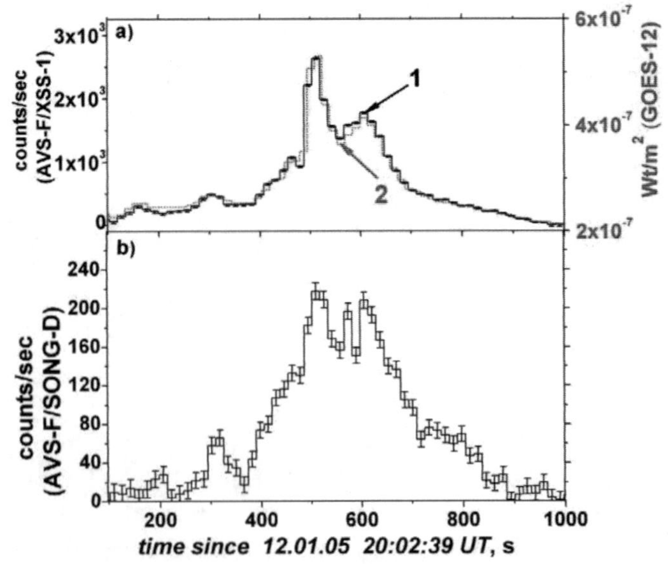

Figure 6. AVS-F data with background subtraction for January 12, 2005 solar flare from XSS-1 with combination of GOES-12 temporal profile (a) and from SONG-D (b). Primary GOES-12 data with 3s time resolution was averaged to 15 s for visualization.

Figure 4 presents the mean typical latitudinal profiles of the AVS-F counts rate and their approximation in the low-energy (a) and high-energy (b) gamma-bands for October 2003. Profiles on the equatorial parts of the CORONAS-F orbit are approximated by the fourth-degree polynomials, ones on the polar cups fitted by the linear functions. The polynomials' coefficients supposed similar for identical spectral channels for each analyzed equatorial part taken into account normalization coefficients defined due to Kp-indexes study within period corresponding calibration coefficients being approximately constants. Also taking into account the minimum count rate estimation in the geomagnetic equatorial region within geomagnetic latitude $\pm 5^{\circ}$ at the different orbits parts in supposition of its linear dependence on Kp - index averaged on 5 bins in time interval from –24 to –12 hours before the time of geomagnetic equator passing. The examples of minimum count rate estimations for background orbit parts during the end of October 2003 are listed in table 1. The estimated and measured values confirms in the interval $\pm 1\sigma$.

Table 1. Background orbit parts characteristics for the end of October 2003 (adopted from [12].

Date	Start	End	Averaged counts rate[a]	Kp series[b]	<Kp>	Estimated min count rate
23	21:52:44	22:22:51	810±7	2 2 2 2 2	2.0	820±20
24	19:36:05	20:06:12	931±8	2 2 2 2 4	2.4	910±20
24	21:09:00	21:39:08	1043±8	2 2 2 4 5	3.0	1030±20
25	20:25:08	20:55:20	979±8	3 3 2 2 4	2.8	990±20
27	20:30:22	21:00:31	1017±8	3 3 4 3 2	3.0	1030±20
30	19:53:01	20:19:21	1681±10	8 7 6 5 5	6.2	1690±20

[a] Averaged counts rate on equator in the region of geomagnetic latitude $\pm 5^{\circ}$ in the energy band 0.1–17 MeV.
[b] Kp series from -24 to -12 hours before the time of geomagnetic equator passing.

The other example of background approximation presented at figure 5 for time period January 2005. Figure 6 shows such approximation type using for background subtraction during January 12, 2005 solar flare. This faint solar event with GOES class B4.6 lasts from 20:08 UT till 20:13 UT with maximum at 20:11 UT in energy range 1-8 A on GOES-12 data [13]. The polynomial from figure 5 used for background subtraction on SONG-D data and the result have shown at panel (b) at figure 6. Analogues polynomial used for XSS-1 data processing – see panel (a) at figure 6. For this flare the count rate temporal profile in the X-ray band on AVS-F data with background subtraction coincided at the 99% significance level with one on GOES-12 data. This fact also confirms applicability of presented model.

4. Conclusion

The gamma-ray background physical origin for low altitude orbits defined by: diffuse cosmic gamma-emission, atmospheric gamma-rays, gamma-emission formed in interactions of charged particles (both prompt and activation) and transient events such as electrons precipitations and solar flares. The background conditions in the energy range from 0.1 MeV up to several MeV for low altitude orbits differ due frequency of ERBs (included SAA) passes and cosmic rays rigidity. The detectors and satellite constructive elements are activated by trapped in ERBs and moving along magnetic lines charged particles.

In this case we propose simplified polynomial model separately for polar and equatorial orbits parts: background count rate temporal profile approximation by 4-5 order polynomials in equatorial regions, and linear approximations, parabolas or constants in the polar caps. The polynomials' coefficients supposed similar for identical spectral channels for each analyzed equatorial part taken into account normalization coefficients defined due to Kp-indexes study within period corresponding to calibration coefficients being approximately constants. Also taking into account the minimum count rate estimation in the geomagnetic equatorial region at the different orbits parts in supposition of its linear dependence on Kp - index averaged on 5 bins in time interval from −24 to −12 hours before the time of geomagnetic equator passing. The described model was successfully applied for the solar flares hard X-ray and gamma-ray emission characteristic studies by AVS-F apparatus data onboard CORONAS-F satellite. Presented model can be applied to detectors onboard other high inclination low altitude satellites within period corresponding to devices calibration coefficients being approximately constants.

Acknowledgments
This work performed within the framework of the Center FRPP supported by MEPhI Academic Excellence Project (contract No. 02.a03.21.0005, 27.08.2013).

References
[1] Gehrels P M 1992 *Nucl. Instr. And Meth.* **A313** 513
[2] Dean A J et al 1991 *Space. Sci. Rev.* **57** 109
[3] Rubin B C et al 1996 *Astron. Astrophys. Suppl.* **120** 687
[4] Arkhangelskiy A I et al 2007 *Cosmic Research* **45 # 3** 278.
[5] Kotov Yu D et al 2014 The Study of the Cosmic Gamma-Emission Nonstationary Fluxes Characteristics by the AVS-F Apparatus Data *The Coronas-F Space Mission Key Results for Solar Terrestrial Physics* ed V D Kuznetsov (Berlin Heidelberg:Springer-Verlag) chapter 9 pp 175-256
[6] Kuznetsov V D 2005 *Solar System Research* **39 # 6** 433.
[7] Kuznetsov S N et al 2002 *Adv. Space Res.* **30 #7** 1857
[8] Pankov V M et al 2006 *Solar System Research* **40 # 4** 314
[9] Arkhangelskaja I V et al 2006 *Solar System Research* **40 # 4** 133
[10] Arkhangelskaja I V et al 2003 (in Russian) *Space science and technology* **9 # 2** 20 (original Russian title: *Kosmicheskaya nauka i technika*, printed in Ukraine)
[11] Arkhangelskaja I V et al 2008 *Solar System Research* **42 # 6** 536
[12] Arkhangelskaja I V et al 2015 Physics Procedia **74** 281
[13] *http://sec.noaa.gov/ftpdir/indices/2005_events/20050112events.txt*

International Conference on Particle Physics and Astrophysics (ICPPA-2015) IOP Publishing
Journal of Physics: Conference Series **675** (2016) 032026 doi:10.1088/1742-6596/675/3/032026

Solar magnetograms editing using discrete Morse theory

N G Makarenko[1], **D O Park**[2], **V V Alexeev**[3]

[1]Chief scientist, Pulkovo Observatory, Pulkovskoye shosse 65, Saint-Petersburg, Russia
[2]Researcher, Pulkovo Observatory
[3]Researcher, Yaroslavl State University, Sovetskaya ul. 14, Yaroslavl, Russia

E-mail: ng-makar@mail.ru

Abstract. Solar magnetic field structure in active regions is complicated for modeling. We construct by magnetogram data a special structure – discrete Morse-Smale complex. We perform simplification of complex using persistence pair elimination. Finally, we obtain a simplified complex that stores main topological features of the magnetogram data and neighbourhood relation between them.

1. Introduction

Matching of the observed magnetic fields in Active Regions (AR) of the Sun with models is a complicated issue. Actually, magnetograms produced by space observatories are too complex [4] to compare with simple models (like a magnetic charge). Thus, simplification of the magnetograms is necessary. However, the greater part of AR formed by the background noise with small intensity values. Influence of the sunspot on the topology (for example, the Euler characteristic) of the entire AR is low [7]. Moreover, AR includes parts of both polarities with distinct field extrema values. Critical points of the scalar field are characterized by the Morse index — the number of negative eigenvalues of Hessian. Thus, in the two-dimensional case, minima have index zero, maxima — index two, and saddle points — index one. We introduce a discrete structure (cell complex), matching minima with vertices, saddle points with edges, and maxima with faces of this structure. After that, we construct a discrete gradient field which consists of ascending and descending integral lines. If intersections of the integral lines are transversal, we obtain cells of the so-called Morse-Smale complex. Every cell contains exactly one minimum, one maximum, and two saddle points. Euler characteristic (χ) defined as an alternating sum of critical point numbers of the different indices. Hence simplification methods must reduce critical point number in a way that do not change χ. Editing, or simplifying of field consists of sequentially eliminations of the persistent pairs "saddle - minimum" or "saddle - maximum". The global field topology during simplification remains unchanged.

2. Related Work

Discrete Morse theory is a useful tool of the topological data analysis. There are many applications of this in astrophysics [11], geomorphology, particle physics [3] etc.

Discrete Morse theory was stated by Forman [2]. Lewiner [6] was the first who presented technique for computation of a discrete gradient field that agrees with the flow behavior of a scalar function. Sousbie [11], Robins et al. [10] presented faster algorithms using various techniques. Gyulassy et al. [3] proposed a probabilistic approach. Edelsbrunner et al. [1] introduced a topological persistence and persistent pairs cancellation concepts.

In this parer, we show the result of a Morse-Smale complex construction and simplification for a bipolar field (obtained by processing of SDO/HMI data).

3. Preliminaries

Morse theory concerned on the behavior of the smooth scalar functions defined over generic manifolds. At the first, we present some basic definitions, and then extend theory for discrete domains.

3.1. Morse Theory

Consider a smooth function $f : \mathbb{M} \to \mathbb{R}$ defined over two-dimensional manifold $\mathbb{M}$.

Denote as ∇f the *gradient* of f. A point $p \in \mathbb{M}$ is *critical* if $\nabla f(p) = 0$, and it is *degenerate* if a determinant of Hessian matrix $\mathcal{H}_f(p) = d^2f/dxdy(p)$ takes zero value.

The function f is a *Morse function* if all its critical points are non-degenerate. In addition, require that all values of f in critical points are different. For such a function, *Morse Lemma* states [8] that there exists a chart (x, y) in a neighbourhood U of critical point p such that $x(p) = 0$, $y(p) = 0$ and $f(x, y) = f(p) \pm x^2 \pm y^2$. The number of minus signs in this equation gives an *index (order)* of critical point. One can think about index as the number of independent downward directions. In 2D case, critical points of index 0 are called *minima*, index 1 – *saddles*, index 2 – *maxima*.

Integral line is a curve which is tangent to gradient field in every point. Precisely, it is a curve $l(t) \subset \mathbb{R}^2$ such that $dl(t)/dt = \nabla f$. *Origin* of integral line is $\lim_{t\to-\infty} l(t)$ and *destination* is $\lim_{t\to+\infty} l(t)$. Actually, each integral line has origin and destination in critical points.

Integral lines of f have the following properties:

(i) every point $p \in \mathbb{M}$ belongs to exactly one integral line,

(ii) every two integral lines are fully distinct (but may have the same origin or destination).

From (i) and (ii) follows that the domain of f can be decomposed into regions containing integral lines with common origin or destination. *Ascending (descending) manifold* is a set of points whose integral lines have common origin (destination). Decomposition of $\mathbb{M}$ into ascending (descending) manifolds is called a Morse complex. A critical point with index i has an i-dimensional descending manifold and $2 - i$-dimensional ascending manifold. 1-manifolds of a Morse-Smale complex are called *arcs*.

Morse function f is called *Morse-Smale* if intersection of its ascending and descending manifolds are transversal. We obtain a *Morse-Smale complex* by intersection of ascending and descending manifolds.

3.2. Discrete Morse Theory

Supposing the magnetic field intensity were a Morse function, which critical points correspond to the topological features of the field and arcs define neighbourhood relation between critical points.

Regrettably, a magnetogram data is given over discrete domain and do not comply with criteria of Morse function. Thus, we need to give a generalization of Morse theory for functions defined over discrete space.

For details of such a generalization, namely *discrete Morse theory*, we refer to Forman [2].

How to define correctly a discrete Morse function by given magnetogram values and how to construct a gradient field are highly intricate issues. We use approach stated by Sousbie

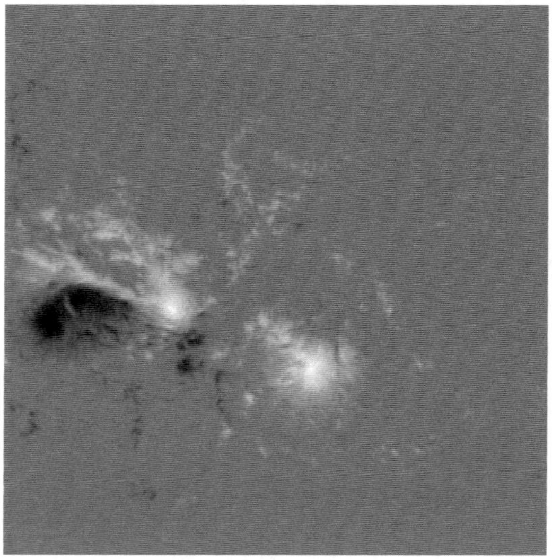

Figure 1. Initial 400 × 400 SDO/HMI magnetogram.

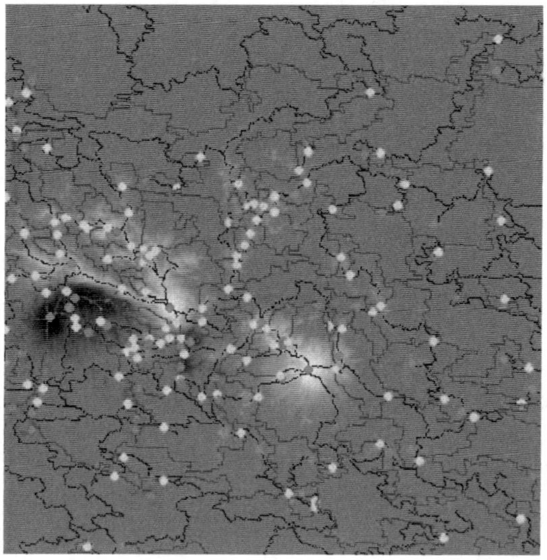

Figure 2. Intermediate result of simplification. Red are maxima, blue are minima, green are saddles. There are 76 maxima, 44 minima, and 117 saddle points (the Euler characteristic $\chi = 3$).

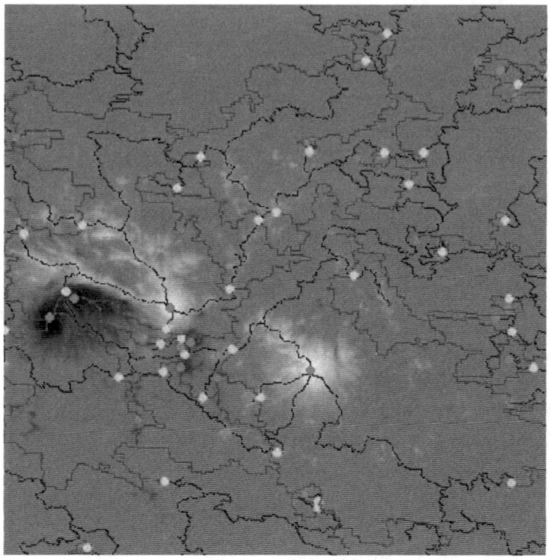

Figure 3. Intermediate result of simplification. There are 29 maxima, 14 minima, and 37 saddle points ($\chi = 6$).

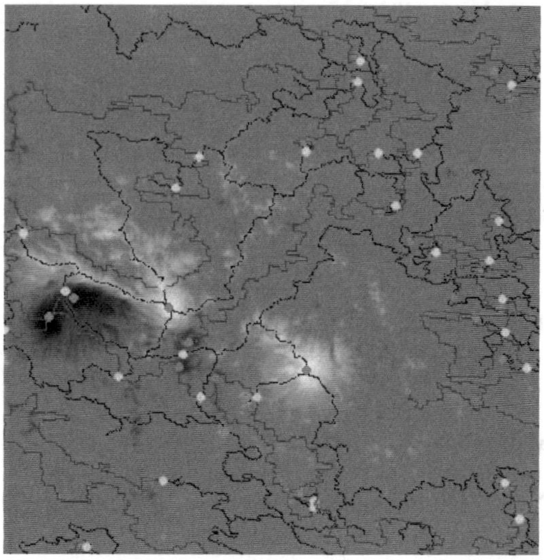

Figure 4. DMC obtained by persistence simplification. Remaining critical points are correspondent to the key topological features. There are 23 maxima, 11 minima, and 29 saddle points ($\chi = 5$).

in [11] implemented for 2D grid (initially, the proposed algorithms were designed for simplicial complex). Some details of implementation we get from Robins et al [10].

3.3. Persistence Simplification
Only a small part of identified topological features (correspondent to critical points of the computed discrete Morse complex) are seemed significant, most of them appear to be noise. the way to eliminate minor topological features is *the persistent pair identification and cancellation.*

For each arc (σ_0, σ_1) of the discrete Morse complex (DMC) we associate a number $|f(\sigma_0) - f(\sigma_1)|$, which is called *persistence* of arc.

Each step of simplification consists of cancellation of pair "maximum – saddle" or "minimum – saddle" connected by arc with the smallest persistence value. After cancellation performed, we recover arcs of DMC incoming the removed vertices.

For more detailed discussion about topological persistence and its applications to discrete Morse theory, we refer to Edelsbrunner et al [1].

4. Results
We have applied a simplification algorithm on the SDO/HMI magnetogram data. We use 2-periodical boundary conditions obtained by reflection and gluing initial rectangle borders in torus $\mathbb{T}^2$. Process of the Morse-Smale simplification presented on Figures 1 – 4. Initial MS-complex contains 5503 maxima, 9330 minima, and 14799 saddle points. MS-complex after editing (Fig. 4) contains 23 maxima, 11 minima, and 29 saddle points.

Our implementation of this algorithm is in C++. Pictures are generated using OpenCV library [9].

5. Future work
Firstly, we are intended to obtain a sequence of edited fields for Active Regions corresponding to flares. After that, we want to look at the dependence of general predictors of field complexity [5] on the simplification level. An other problem is comparison of MS-complexes constructed by the bipolar field with a scalar energy field $B^2/2$.

Acknowledgments
The reported study was supported by RFBR, research project No. 15-32-51276

References
[1] Edelsbrunner H *et al* 2002 Topological persistence and simplification *Discrete Comp. Geometry* **28** 511–33
[2] Forman R 1998 Morse theory for cell complexes *Adv. in Mathematics* **134** 90–145
[3] Gyulassy A *et al* 2015 *Topological and Statistical Methods for Complex Data* 205–18
[4] Knyazeva I S, Makarenko N G 2014 *Astronomy Reports* **58** 145–155
[5] Kniazeva I S, Makarenko N G, Urtiev F A 2015 Geomagnetism and Aeronomy. **55** 1134–40.
[6] Lewiner T 2002 Constructing discrete Morse functions. *Master's thesis, Department of Mathematics, PUC-Rio*
[7] Makarenko N G, Knyazeva I S, Karimova L M 2010 *Astronomy Reports* **54**, 747–755
[8] Matsumoto Y 2002 An introduction to Morse theory *American Mathematical Soc.*
[9] OpenCV library *opencv.org*
[10] Robins V *et al* 2011 *IEEE Transactions on Pattern Analysis and Machine Intelligence* **33** 1646–58
[11] Sousbie T 2010 *arXiv:1009.4015v1*

Journal of Physics: Conference Series **675** (2016) 032027 doi:10.1088/1742-6596/675/3/032027

Generalized Laplacian for magnetograms of solar active region as possible predictor of strong flare.

D.M. Volobuev
Researcher, *Pulkovo Observatory*,
Pulkovskoye shosse 65, Saint-Petersburg, Russia

E-mail: dmitry.volobuev@mail.ru

N.G. Makarenko[1]
Chief scientist, *Pulkovo Observatory*,

E-mail: ng-makar@mail.ru

I.S. Knyazeva
Researcher, *Pulkovo Observatory*,

E-mail: iknyazeva@gmail.com

Abstract. Search for predictors of strong flare produced in solar active region (AR) is important application of solar physics. Here we consider the sequence of magnetogram (LOS SDO/HMI instrument) for AR 2034, 2035 and 2036 (April 2014). All three AR were observed on the Sun at about the same time, characterized by low probability of flare events according to official forecasts of NOAA, but 2036 still produced X1-flare near the center of solar disc (April 18). We propose that Generalized Laplacian is a descriptor which could help predict this and similar events. The Laplacian is associated with the flow of Ricci curvature and with topological invariants of the observed field - Betti numbers for compact manifolds. Using discrete version of Morse theory, we consider each pixel of energy flux (B^2) image as a simplex and calculate its combinatorial Bochner Laplacian. It was found that maximum of Laplacian is located near AR polarity inversion line. Evolution of total spatial variation of the Laplacian has a number of maxima in time for each of examined AR. However, the maxima in AR 2035 and AR 2034 have relatively low amplitude, while the highest maximum prefaced X1 flare in AR 2036 by about 29 hours.

1. Introduction

Current forecasts of AR flaring activity are based on tens of predictors (see e.g. [1]). Their choice is motivated by the rich observational phenomenology and heuristic considerations about the difficulties in changing the geometry and/or topology of the AR, anticipating the flare. Recent comparative analysis has shown that efficiency of different solar flare predictors is not greater than efficiency of

[1] To whom any correspondence should be addressed.

Journal of Physics: Conference Series **675** (2016) 032027 doi:10.1088/1742-6596/675/3/032027

alternative model based on the morphology of the 60 class in McIntosh AR classifier [2]. Possible reason for lacking of these predictions is that different descriptors are linear correlated and consequently they lack to complete the mathematical morphology.We consider here generalized (Bochner) laplacian as new descriptor which could be used in prediction models. This differential geometry originated descriptor is metric invariant and has direct link to global topology (Betti numbers) [5]. It can be calculated from the image if the pixel is considered as a simplex (figure 1) after calculation of generalized curvature and rough laplacians [3,4]

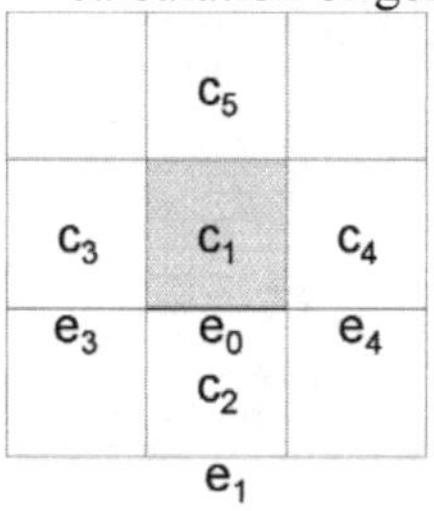

Figure 1. The pixel of image considered as a simplex with edges e and cells c. Subscripts corresponds to neighbouring pixels numbering used in equation (1).

Rough Laplacian $\Box_2(c_1,c_2)=\dfrac{\omega(e_0)}{\sqrt{\omega(c_1)\omega(c_2)}}$ and 2-curvature $F_2=\sum_{e<c_1}\left(\dfrac{\omega(e)}{\omega(c_1)}-\dfrac{\omega(e)}{\sqrt{\omega(e_0)\omega(e_1)}}\right)$ can be defined for 2-cell (figure 1). Here combinatorial weights $\omega(c)$ of cell is the amplitude of magnetic energy (ME) at given pixel and $\omega(e)$ is average ME of pixels adjacent along the edge e. Boundary pixels are cropped. Combinatorial Bochner 2-Laplacian is defined as the difference

$$B_{2c}=\Box_2-F_2 \qquad (1)$$

2. Data

SDO/HMI line-of-sight (LOS) magnetograms **http://hmi.stanford.edu/** (figure 2) were used here for ARs 2034, 2035 and 2036. These ARs were thoroughly studied [2] for their nearly equal complexity but different flaring activity which is challenging for forecast methods. The evolution was considered 5 days before and 2 days after X1 flare event at hourly time intervals.

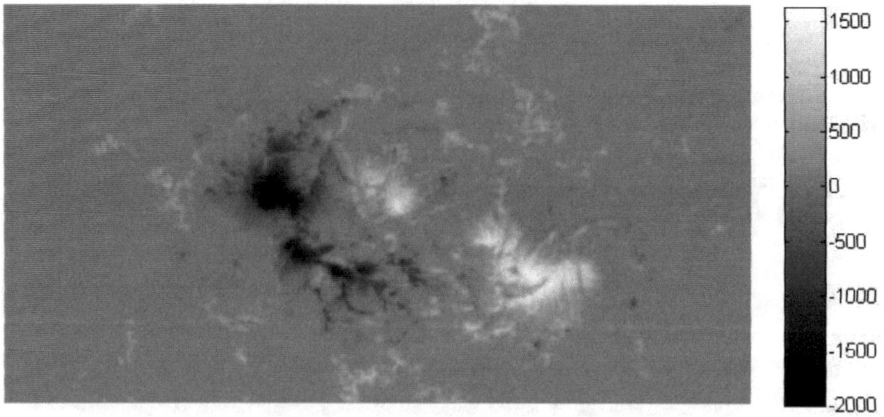

Figure 2. Example of AR 2036 magnetogram, 4 days before X1-flare. Scaled in Gauss.

3. Method

Each magnetogram B (figure 2) in hourly sequence was squared to get unsigned ME i.e. B^2. ME was normalized by 2000x2000 Gauss^2 to scale it approximately into [0 1]. Bochner laplacian (1) was calculated for each ME- images. Example of this Bochner laplacian scaled logarithmically for ME of magnetogram in figure 2 is shown in figure 3. Total spatial variation of Bochner laplacian

$$Var\left\{B_{2c}(B^2)\right\} = \max\left\{B_{2c}(B^2)\right\} - \min\left\{B_{2c}(B^2)\right\} \tag{2}$$

was calculated as a measure of topological complexity for magnetogram.

4. Results and Discussion

Big Bochner laplacian of ME (figure 3) underlines unsigned magnetic flux whereas a small one selects magnetic polarity inversion line. These both features are important for AR description [1] but difficult to eliminate for complex AR with diffusive polarity inversion line like this one. Total variation of the field is often considered as a measure of complex geometry object. It shows different evolution for AR 2034 (maximum C-class flaring), AR 2035 (M1 - flare) and AR 2036 (X1 - flare). Biggest total variation of Bochner laplacian is found for AR 2036 (figure 4) with local big maximum in time at 23 hours before X1 flare. Some smaller local maxima were found for AR 2035 and much smaller maxima for AR 2034, so these three ARs are correctly sorted by their flaring activity. AR 2035 flare behind the limb was even stronger than AR 2036 flare but it was 7 days later and its evolution is difficult or impossible to observe near and behind the limb.

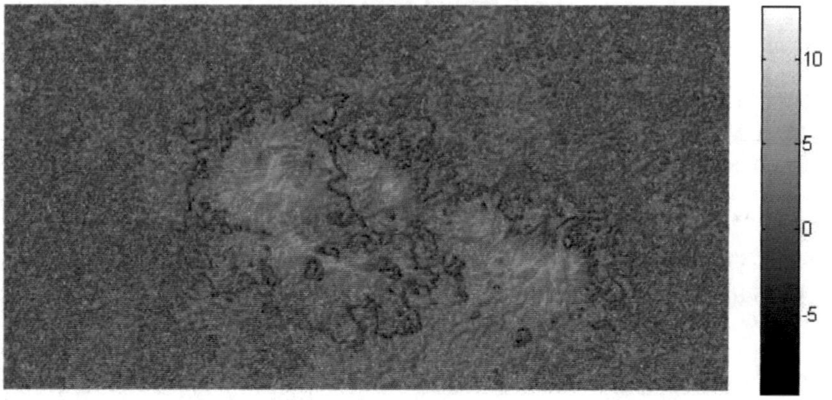

Figure 3. Bochner laplacian of magnetic energy for AR 2036, 4 days before X1-flare, logarithmic scale.

Proposed total variation of Bochner laplacian of magnetic energy can likely be considered as a candidate to be a precursor of strong flare. However this result is very preliminary and should further be checked on extended statistics of flaring and non-flaring ARs.

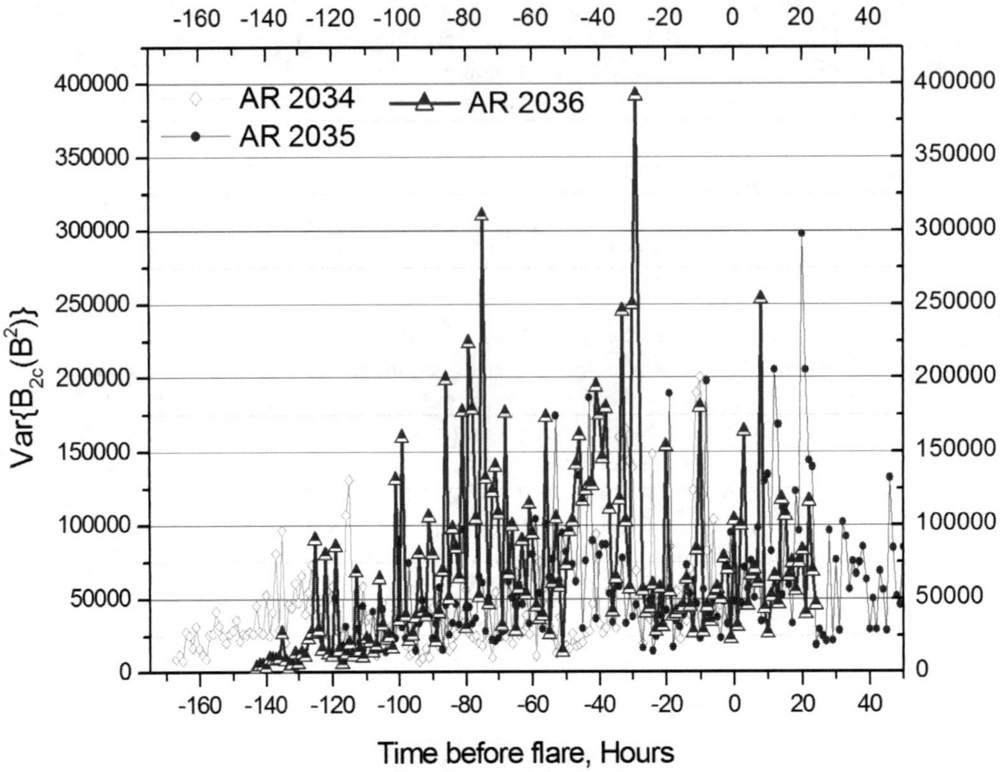

Figure 4. Evolution of total variation of Bochner laplacian for ARs 2034, 2035 and 2036. Zero epoch is April 18, 2014, 13:00. Flaring activity data was taken from http://solardemon.oma.be/science/flares.php

Acknowledgments
This work was supported by Russian Fund for Basic Research, grant 15-01-09156 and by Ministry of Education and Science of the Republic of Kazakhstan, grant 2308/GF3.

References

[1] Barnes G, Leka K D 2008, *Astrophys. J.*, **688**, L107
[2] Knyazeva I S, Makarenko N G, Urt'ev F A 2015, *Geomagnetism and Aeronomy*, **55,** 1134
[3] Saucan E, Wolansky G, Appleboim, E, Zeevi Y Y 2009, In: *Image and Signal Processing. CISP'09. 2nd International Congress on*, 1
[4] Forman, R 2003, *Discrete and Computational Geometry*, **29**, 323.
[5] Bochner S, 1948, *Annals of Mathematics*, *Second Series*, **49**, 379.

The new approach for dynamical regimes detection in geomagnetic time series

I S Knyazeva[1,2], N G Makarenko[1], Y A Kuperin[2] and L A Dmitrieva[2]

[1] Pulkovo observatory,Pulkovskoe sh. 65,Saint-Petersburg, Russia
[2] St.Petersburg State University, 7/9 Universitetskaya nab, St.Petersburg , Russia

E-mail: i.knyazeva@spbu.ru

Abstract. We suggest the new approach that could be used in the task of geomagnetic indexes forecasting. This approach based on the topological structure of the time series. The hourly data of different geomagnetic indexes were analyzed. We split the whole series to the calm parts, and geomagnetic storm preceding parts and find that the main measure of the topological structure of the series which called topological persistence differ significantly between these two groups.

1. Introduction

The task of magnetic storm forecasting remains an important problem of the space weather investigation. Approaches to storm forecasting depends on forecasting horizon. It could be divided by long-term (more than 7 days), medium-term (several hours to several days) and short term (several hours)[1]. The first two approaches based on forecasting of solar activity which lead to magnetic storm and the last one, short term forecast based on the information from spacecraft in the Sun-Earth libration point. Short term forecast is rather exact, but the alert time (less than 1 hour) is too short for practical use and the level of forecasting rate falls down rapidly with the growth of forecasting horizon.

It is widely accepted that the magnetic storm preceded by geoeffective pattern in solar wind speed and proton density and interplanetary magnetic field (IMF) parameters. It is known that for severe storm the IMF directed southward (B_z is negative) and the pressure is raised (solar wind speed and proton density raised). But this rule is not strictly [1]. G.Mansilla [2] and other authors shows absence of high linear correlation between the peak values of solar wind parameters with peak Dst index. This fact points to more complex geoeffective structure in solar wind parameters. The complex topological structure of time series could be described in terms of topological complexity. The topological data analysis is quite a new area in area such as data mining and computer vision. But its popularity growth rapidly. In this work we show one possible application for detection different dynamical regimes in geomagnetic time series.

2. Pre storm dynamics of solar wind and IMF

Building of geomagnetic storm prognosis on the base of geoeffective structures recognition is a fruitfull method (from Khabarova [1]). But the main problem in complexity of these structures. There are no strict conditions which always fulfilled before the storm started. It is known that for the most severe storm precede a strong increase of solar wind speed accompanied long-lasting negative B_z level, but for moderate storm these are not true. By the way, as shown in papers [1] and [3] prediction of the moderate storm as an important task as strong storm forecasting. Moreover, phase shift (time) between geoeffective changes in solar wind parameters and interplanetary B_z changes from 0.5 to 5 hours, see [4]. Also, these structures differ during the solar cycle, because of difference physics properties of solar-interplanetary-magnetosphere coupling [5]. That's why it is hard to give a formal description of geoeffective conditions which must be performed before the storm, and consequently to make a good prediction system.

Below we provide a short intuitive introduction to an approach which allows to describe variations of some parameter during a specific time in several key features. This approach was borrowed from computational topology and know calls Topological Data Analysis (TDA).

2.1. Brief introducation to topological data analysis (TDA)

We would like to introduce the approach from computational topology which could describe the inner structure of time series. The full information about computational topology could be found in Edelsbrunner and Harer book [6]. Very clear introduction also could be found in [7]. To illustrate the main ideas, consider the graph of a continuous function $f(x) \in R$ and its sublevel set $A_t = f^{-1}(-\infty, t)$, i.e. points at which the function is below a specified level, see Figure 1. To study the topology of a function, one can measure how the topology of its sublevel sets evolves when the threshold is continuously and monotonically changing from lower levels to high. It is clear that for $t_1 > t_2, A_{t_2} \subset A_{t_1}$. Such a sequence of nested intervals is called filtration, precisely lower-star filtration. Sublevels consist of points and intervals in the horizontal axis. They are called the connected components, which splits the graph of its domain. Let consider the level 1.13 at Figure 1 and interval $[a, b]$ consist from points $f(x)$ which is below this level. Its boundaries $\partial[a, b] = b - a$ can be combined with each other by sliding them along the interval. Formally, this means that they are topologically equivalent: $a\ b : b = a + \partial[a, b]$. The number of independent components for each level calls Betti number β_0. From figure 1 it could be seen that minimums of the function creates a connected component, and an adjacent maximum reduces the number of components. The level at which new component created calls birth of component, and level at which component was destroyed calls death of component. This fact is depicted in so called persistence diagram (see Figure 1, right. Here, on abscissa is the birth of all components, and on the ordinate - the level of death. Thus, at the minimum point $(3.30, 0.76)$ new component appears, which collapses at adjacent maximum $(1.40, 2.26)$. Corresponding to this event point $(0.76, 2.26)$ is shown on the persistence diagram. In practice it is clear that small fluctuations in the graph correspond to points near the diagonal. Diagonal corresponds to a zero life expectancy. Long-lived or persistent components lies far from the diagonal. Stability Theorem states that for two functions that are close in terms of supremum metric corresponding persistence diagrams are also close.

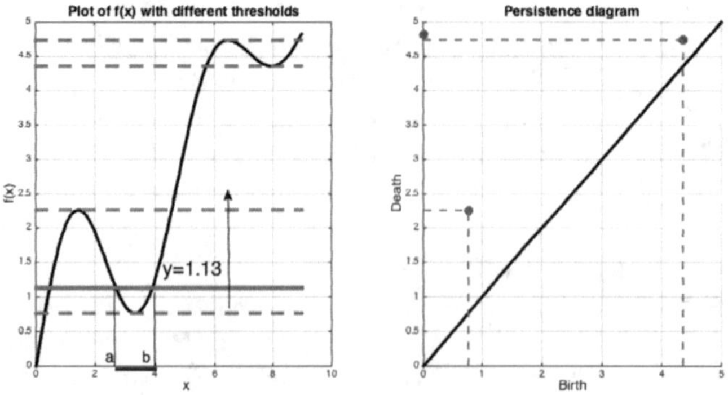

Figure 1. Example of lower-level set filtration of continuous function and corresponding persistent diagram

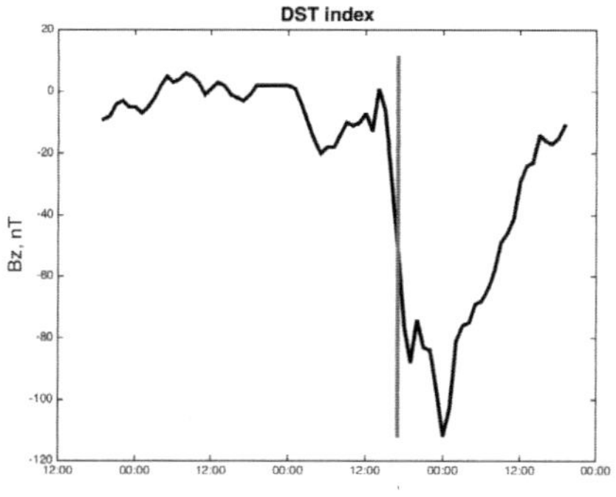

Figure 2. Dynamic of hourly Dst index. Vertical line indicates the beginning of the storm. We consider here that the point is a start of storm if the Dst index drops below -80 nT and lasts under more than 3 hours.

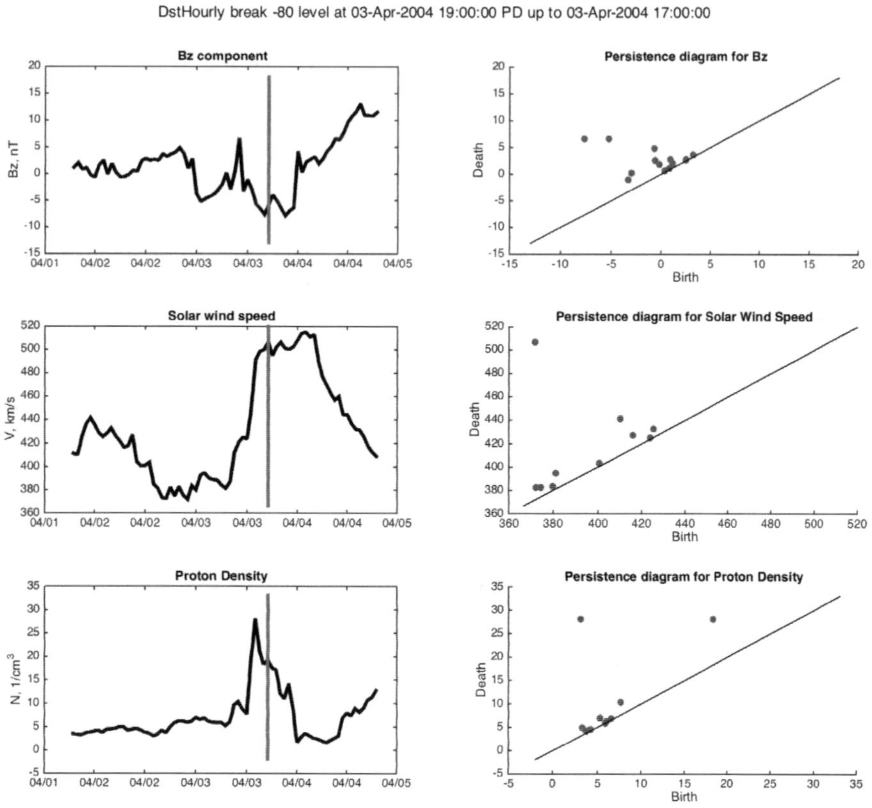

Figure 3. IMF B_z component, Solar wind speed and Solar proton density dynamic and pre-storm Persistence Diagram (PD). Vertical line indicates the start of the storm.

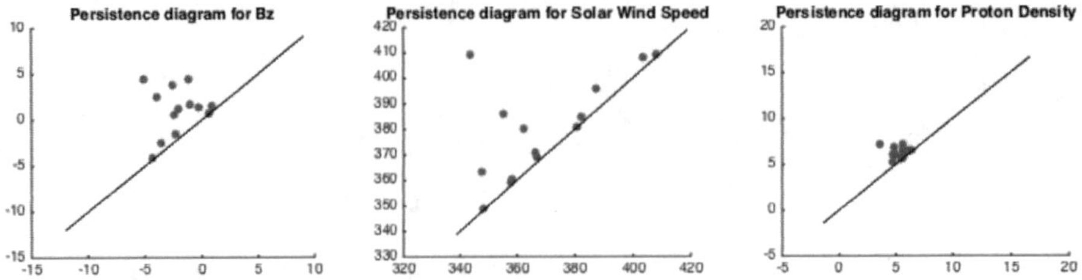

Figure 4. Persistence diargam of Bz, Solar wind speed and proton density for non storm case.

2.2. TDA of solar wind and IMF B_z parameters

For solar wind speed, solar wind proton density and IMF B_z component hourly data were obtained from the website http://www.srl.caltech.edu/ACE/ASC/level2/index.html. Dst index data were downloaded from http://spidr.ngdc.noaa.gov/spidr/. At this work the period from 2001 to 2007 years was analysed. This period corresponds to solar maxima and declining phase of 23 Solar cycle. The solar wind and IMF data were processed as follows.

- The 48 hours period was taken up to 2 hours before the storm start for all three parameters. We consider here that the point is a start of strong storm if the Dst index drops below -80 nT and lasts below more than 3 hours and moderate storm if the Dst index drops below -50 nT and not exceed -100 nT during the whole storm phase.
- Persistence diagrams of solar wind speed, solar wind density and B_z component for this period were computed
- Separately were randomly chosen 100 fragments lasting 48 hours for each three parameters without storm during time fragments and after 24 hours from the end of the fragment.

At Figure 2 is shown an example of behaviour of the Dst index during the storm on 03 April, 2004. It could consider as a strong storm. The vertical line indicates the start of the storm. At Figure 3 corresponding series for solar wind speed, density and B_z component and persistence diagram is shown. We can see that for all three series, there are points at the diagrams which lie far from the diagonal, the location of these point describe the structure of the underlying series. For example, for B_z component appearence of negative movement could be seen as a separate two point in the left side of the diagram. The sharp increase in solar wind and proton density also could be seen as a separate long-living components. We find that for non-storm cases the picture of persistence diagram differs, typical view is shown at Figure 4.

3. Discussion and conclusion

In this work we represented new possible approach for pre-storm structure analysis. The persistence diagrams for pre-storm parts of time series of solar wind speed, proton density and B_z component were estimated. It turned up that persistence diagrams for pre-storm and calm periods differs significantly. That's mean that features from persistence diagram could be used in prediction systems, for example, such as described in the paper [8]

Acknowledgments

This work was partly supported by RFBR grant # 15-01-09156

References

[1] Khabarova O V 2007 *Sun and Geosphere* **2(1)** 32-37
[2] Mansila G A 2008 it Physica Scripta **78**(4) 045902
[3] Romanova N V, Pilipenko V A, Yagova N V and Belov A V 2005 *Cosmic Res* **43**(3) 186-193

[4] Kane R P and Echer E 2007*J. Atmos. Sol. Terr. Phys* **69** 1009-1020
[5] Gonzalez W D, Tsurutani B T and Clua de Gonzalez A L 1999 *Space. Sci. Rev.* **88** 529-62
[6] Edelsbrunner H and Harer J 2009 *Computational Topology, An Introduction* (AMS) p 241
[7] Ghrist R *Elementary Applied Topology* (ed. 1.0, Createspace) p 269
[8] Shugai J S, Dolenko S A, Persiantsev I G and Orlov Y V 2006 *Pat. Recogn. and Im. Anal.***16**(1) 79-81

Journal of Physics: Conference Series **675** (2016) 032029 doi:10.1088/1742-6596/675/3/032029

The long-term oscillations in sunspots and related inter-sunspot sources in microwave emission

I A Bakunina[1,*], **V E Abramov-Maximov**[2] **and V V Smirnova**[2]

[1] National Research University Higher School of Economics, 25/12 B. Pecherskaja ul., Nizhny Novgorod 603155, Russia

[2] Central Astronomical Observatory at Pulkovo, Russian Academy of Sciences, Pulkovskoe chaussee 65/1, St Petersburg 196140, Russia

E-mail: *rinbak@mail.ru

Abstract. This work presents the microwave long-term oscillations with periods of a few tens of minutes obtained from Nobeyama radioheliograph (NoRH) at frequency 17 GHz. In two active regions the fluctuations of radio emission of different types of intersunspot sources (ISS) (compact and extended) were compared with the fluctuations in magnetic fields of sunspots. Common periods in variations of microwave emission of different type of sources and magnetic field of sunspots were discovered. The delay of 17 minutes was revealed for oscillations of the extended ISS with respect to variations of magnetic field of its tail sunspot. The model of the sunspot magnetic structure based on the concept of three magnetic fluxes for explanation of this fact is discussed.

1. Introduction

An analysis of long quasi-periodic oscillations (LQPO) in the microwave emission generated at different sunspots, detected by different techniques with the use of different, spatially separated instruments, Siberian Solar Radio Telescope (SSRT) and Nobeyama Radioheliograph (NoRH) performed in [1] revealed significant LQPO with periods 30–170 min in both SSRT (5.7 GHz) and NoRH (17 GHz) data for several sunspots during different days of observations with non stable periods at both frequencies. The detection of the same or similar periods in the time series of different radio-parameters (intensity, polarization and degree of circular polarization) recorded at different frequencies and so at different heights in the sunspot's magnetosphere, suggest a global nature of long-term oscillations. In particular, the oscillations may be present in variations of sunspot's magnetic field. In this case the slow variations of the magnetic field can be explained by global eigen oscillations of sunspots [2].

In [3] a significant correlation between fluctuations in radio sources above sunspots and variations of strengths of their magnetic fields according NoRH and SDO/HMI are shown and common periods were discovered as: 30–40 min, 60–70 min, 100–110 min and 150–200 min. The typical shifts between the time series of the maximum magnetic field strength in spots and the overlying radio source, were found of about 50 min [4].

The attempt to study surroundings of spots on the subject of LQPO was made in [5, 6] where data on the variations of millimeter (37 GHz) radio emission of solar active regions were analyzed in comparison with data on the variations of the magnetic field strength of sunspots and there was shown the presence of a slow quasi-periodic oscillations of flux density radio emission of near-sunspot and intersunspot radio sources and oscillations of the magnetic field strength of photospheric sunspots with equal periods 200–400 min. It was also noted that temporal variations in the series of radio emission fluxes are with respect to the corresponding variations of the magnetic field strength of the closest spot, and the delay time consists 15–35 min. The observed delay times in this paper are not associated with the times of Alfven wave propagation or fast magnetoacoustic wave propagation from the photosphere to the radio source (if they were related, the delay times would be no larger than 1–2 min), but variations of the millimeter radiation source with respect to variations of the magnetic field and time delay between them are associated with the eigen oscillations of the lower-lying sunspot and their physical interpretation based on a novel concept of the magnetic structure of sunspots (the model of three magnetic fluxes) which explains the formation of intersunspots sources [7].

In [7] ISS were divided into two groups: radio sources that can exist during one day or even more over the neutral line of the photospheric magnetic field in the intersunspot area (extended ISS) and ISS above the neutral line which passes through the umbra of a complex sunspot (compact ISS). Although the latter sources are in fact above-sunspots sources, it was used a single name for all sources above the neutral line, namely intersunspot (microwave) sources, or ISS [7].

In the present study data on the fluctuations of centimeter (17 GHz) radio emission of solar active regions are analyzed in comparison with data on the variations of the magnetic field strength of photospheric sunspots. The radio observations were carried out with the Nobeyama Radioheliograph (NoRH, Japan) [8]. Data on the sunspot magnetic fields were provided with the Michaelson Doppler Imager (MDI) [9] instrument onboard the Solar and Heliospheric Observatory (SOHO) [10]. We focus on comparison the fluctuations of radio emission of different types of sources: compact and extended intersunspot sources (ISS) with respect to the sunspot's magnetic field oscillations.

2. Data analysis and results

2.1. Oscillations in the compact ISS

Consider the compact ISS above complex sunspot. As a rule, the emergence of such ISS is accompanied by a series of powerful flares. Investigating the fluctuations in such ISS is extremely difficult because of the ambiguous interpretation - there is a lot of radio bursts in ISS of such type, it often "noises" before a powerfull flare, it is difficult to distinguish the quiet period of its existence.

Compact ISS in AR 09415 10.04.2015 was observed at frequency 17 GHz in the head part of the active region above the leading sunspot of dual magnetic polarity (see figure 1, left panel).

Figure 2 (right panel) demonstrates significant magnetic field oscillations with periods of 15 and 128–200 minutes.

Figure 2 for the right part of the north polarity of the complex sunspot doesn't show significant common periods for the variations of ISS's radio emission and magnetic field, but figure 3 shows the behavior of the magnetic peak strength of the south polarity with a period of about 3 hours and a similar oscillations of the degree of circular polarization of radio emission for the part of ISS which is located above the magnetic field of south polarity of the head sunspot. The delay about 60 min is observed for the radio-emission with respect to oscillations of magnetic field.

2.2. Oscillations in the extended ISS

Consider the example of an extended ISS in AR 09455, 12.05.2001 (Figure 1, right panel).

Figure 4 demonstrates significant fluctuations of magnetic field with periods of 16, 64 and ~200 min.

Figure 5 shows time profiles both the maximum of intensity of magnetic field and maximum of intensity of ISS and cross-correlation function for them. One can see good cross-correlation

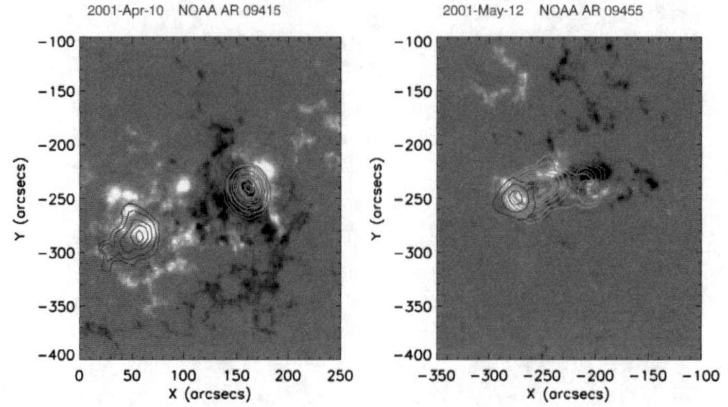

Figure 1. Left: NoRH,17 GHz, partial image, AR 09415 10.04.2001. Right: Extended ISS in AR 09455 12.05.2001. ISS observed at frequency 17 GHz above neutral line of magnetic field of the whole AR. NoRH,17 GHz, partial image. Grey map (background) shows SOHO/MDI magnitograms (black - S-, white - N-polarity of the magnetic field). Countours of red color- negative circular polarization, blue - positive (TV, brightness temperature in a circular polarization); green countours - TI (brightness temperature in intensity).

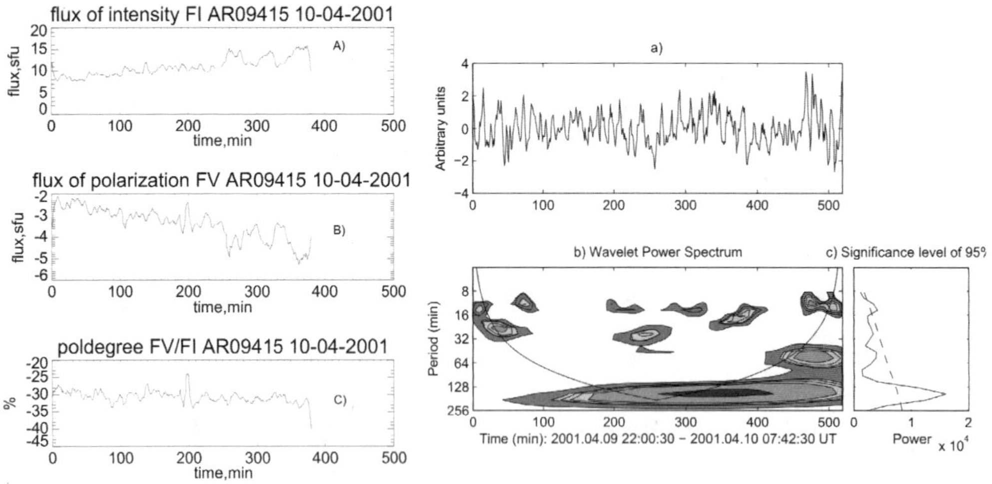

Figure 2. Left: Time profile a) of the intensity flux FI of radio emission, b) of the circular polarization flux FV, c) the degree of circular polarization FV/FI above the head of AR 09415 (the right part of the head spot), April 10, 2001, with 10 minute cadence. Start time is 00:00:03 UT, End time is 06:20:03 UT. Right: Time profile and wavelet spectrum of the peak strength of the magnetic field of north polarity (in Gs) at the head of AR 09415 (the right part of the head spot), April 10, 2001, according to the data of SOI MDI/SOHO with 1 minute cadence. Start time is 22:00:30 UT, end time is 07:42:30 UT.

with coefficient ~0.6 with the time-delay of 17 min. It is in a good accordance with the results described in [5, 6].

Journal of Physics: Conference Series **675** (2016) 032029 doi:10.1088/1742-6596/675/3/032029

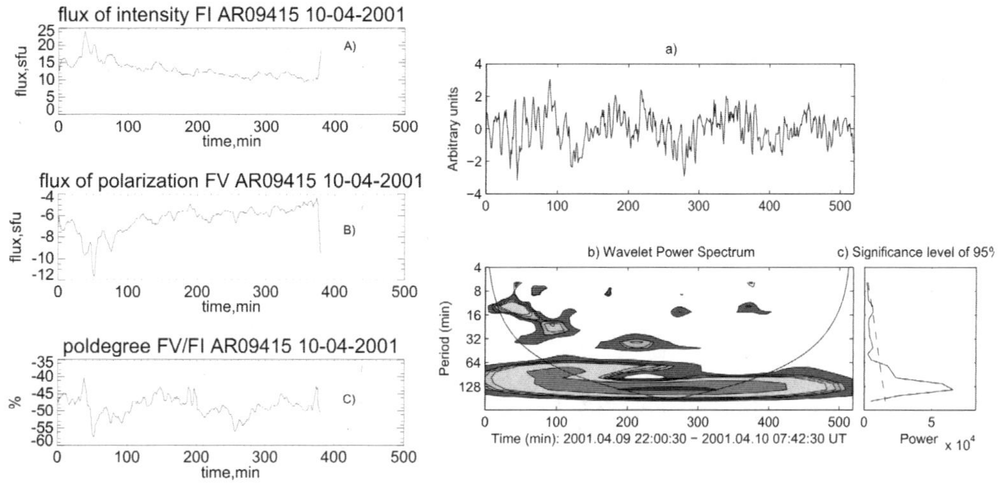

Figure 3. Left: Time profile a) of the intensity flux FI of radio emission, b) of the circular polarization flux FV, c) the degree of circular polarization FV/FI above the head of AR 09415 (the left part of the head spot), April 10, 2001, with 10 minute cadence. Start time is 00:00:03 UT, End time is 06:20:03 UT. Right: Time profile and wavelet spectrum of the peak strength of the magnetic field of south polarity (in Gs) at the head of AR 09415 (the left part of the head spot), April 10, 2001, according to the data of SOI MDI/SOHO with 1 minute cadence.Start time is 22:00:30 UT, end time is 07:42:30 UT.

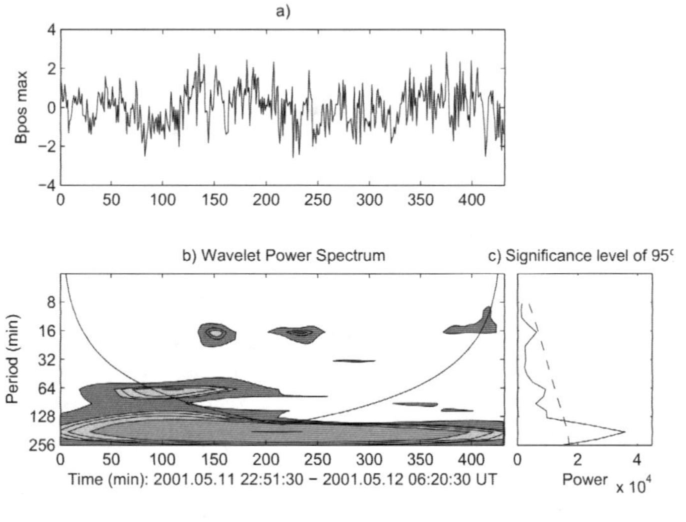

Figure 4. Time profile and wavelet spectrum of the peak strength of the magnetic field of north polarity (in Gs) for the following sunspot of AR 09455, 12.05.2001, according to the data of SOI MDI/SOHO with 1 minute cadence.

3. Discussion and conclusions

The model of ISS suggested in [11] and based on the model of the shallow spot [2] suggests that the nearspot and interspot radio sources are originally connected to a spot by the magnetic field

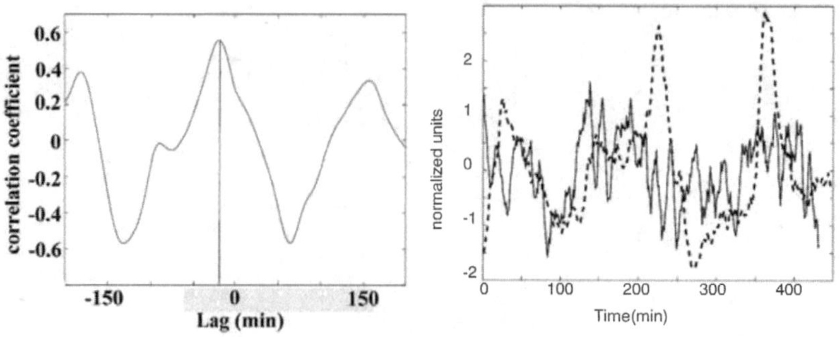

Figure 5. Left: cross-correlation function for the maximum of intensity of magnetic field of positive polarity of the following sunspot and maximum of intensity of ISS 12.05.2001. Right: Time profiles both the maximum of intensity of magnetic field of positive polarity of the following sunspot and maximum of intensity of ISS 12.05.2001.

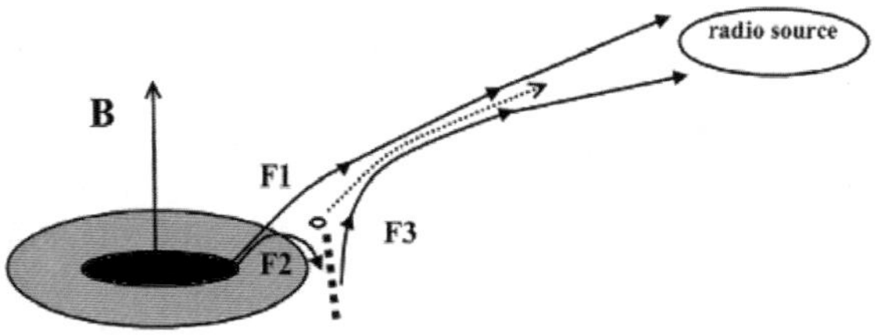

Figure 6. Magnetic structure of sunspot and formation of intersunspot microwave radiosources [11].

lines coming from the spot umbra and are formed due to thermalization of accelerated particles produced in small scale current layers at the periphery of the spot. Quasiperiodic variations of the magnetic field of the spot as a slowly oscillating massive coherent object [2] are transferred from the spot to the radio source along the magnetic field lines with the speed of sound and modulate the radiation of the source.

As can be seen from the figure 6, the magnetic pipe-loop in the top of which ISS is situated, formed from about half of the lines of force of the flux F1, originating in the sunspot's umbra and lines of force of the flux F2, reconnected with the flux F3. Thus, ISS initially closely linked to the spot, and quasi-periodic variations in the magnetic field of the spot, as an integral object, appear as quasi-periodic boundary conditions changings in the basis of the magnetic loop of ISS.

These changes are transmitted along the magnetic field lines in two ways: by means of fast magnetoacoustic waves with a speed that in the chromosphere is very high, about 1 $Mm/second$, and slow magnetoacoustic waves traveling along the magnetic field with the speed of sound which is about 10 km/s, i.e. 100 times less than the velocity of Fast Magnetic Sound Waves (FMSW). When there is the gyro-cyclotron or gyrosynchrotron emission of radio waves, which depends on the magnetic field, the source will feel changings of the magnetic field in the spot very quickly, in a time of the order

$$\frac{L}{V(FMSW)} = \frac{2 \times 10^9}{2 \times 10^8} = 10s$$

For changes of thermal radio emission of an extended ISS the temperature, density characteristics of ISS have to be changed. And this is possible only when the Slow Magnetic Sound Waves comes up from the outer boundary of the spot, and along the magnetic loop a new hydrostatic distribution will be installed. This time is 100 times longer, i.e. 15–17 minutes. These delays are found in millimeter radio emission for ISS in [5, 6].

We considered microwave radiation at a frequency of 17 GHz for two active regions of 23 cycle of solar activity (compact (AR 09415) and extended (AR 09455) ISS) and the magnetic field of related sunspots on the magnetogramms SOHO/MDI. ISS at 17 GHz considered in the current study demonstrate commonly identified periods with the magnetic field in 16, 64 and 200 minutes. This is in agreement with the works [6] and [3]. We found the time delays for compact ISS for radio emission with respect to magnetic field of about 60 min and it is approximately the same as for sunspot's microwave sources [4]. We found that the time delay for extended ISS is 17 min and it is agreed with time delays found in [5, 6] for millimeter ISS (18–35 min)

Acknowledgments
This work was supported by the Russian Foundation for Basic Research, projects no. 13-02-00714 and 14-06-00253. We thank P.H.Scherrer and the SOHO team for providing the SOHO/MDI data. We thank NoRH team for providing the NoRH data. Wavelet software was provided by C.Torrence and G.Compo, and is available at URL: http://paos.colorado.edu/research/wavelets/

References
[1] Bakunina I A *et al.* 2013 *Publ. Astron. Soc. Japan* **65** S13
[2] Solov'ev A A and Kirichek E A 2014 *Astrophys. Space Sci.* **352** 23
[3] Abramov-Maximov V E *et al.* 2013 *Publ. Astron. Soc. Japan* **65** S12
[4] Nagovitsyn Yu A, Nagovitsyna E Yu and Abramov-Maximov V E 2013 *Astronomy Reports* **57**(8) 636 (original Russian text: *Astronomicheskii Zhurnal* 2013 **90**(8) 692)
[5] Smirnova V V *et al.* 2013 *Astron. Astrophys.* **552** 23
[6] Smirnova V V *et al.* 2015 *Geomagnetism and Aeronomy* **55**(7), in press
[7] Bakunina I A *et al.* 2015 *Sol. Phys.* **290** 37
[8] Nakajima H *et al.* 1994 *Proc.IEEE* **82** 705
[9] Scherrer P H *et al.* 1995 *Solar Phys.* **162** 129
[10] Domingo V, Fleck B and Poland A I 1995 *Solar Phys.* **162** 1
[11] Solov'ev A A 2015 *Geomagnetism and Aeronomy* **55**(7), in press

Analysis of lower atmosphere pressure field response for short-time cosmic ray variations by Multifield Comparison Measure method

I V Artamonova[1,2], D M Volobuev[3], N G Makarenko[3]

[1] Voeikov Main Geophysical Observatory, St.-Petersburg, Russia

[2] St.-Petersburg State University, St.-Petersburg, Russia

[3] Central Astronomical Observatory of the Russian Academy of Sciences at Pulkovo, St.-Petersburg, Russia

E-mail: artirin@yandex.ru

Abstract. Pressure variation in lower atmosphere which take place after intensive solar proton events and Forbush-decreases of galactic cosmic rays (GCRs) are analyzed for the period 1980-2006. There were plotted groups of charts (multifields) for 48 solar proton events with energies of particles $Ep > 90$ MeV and for 48 Forbush-decreases of GCRs with amplitudes $dN/N > 2.5\%$. These multifields revealed a growth of matrix norm over North Atlantic region and North of European part of Russia during days following the bursts of solar protons and Forbush-decreases of GCRs, respectively. These results confirm hypothesis about relation of regional cyclogenesis processes with short-term variations of solar and galactic cosmic rays.

1. Introduction

The investigation of solar activity influence on the variations of lower atmosphere parameters now is actual task of Solar-Earth physics. Variations of solar and galactic cosmic rays now are considered as most probable agents of solar activity, which may influence the Earth's lower atmosphere. Cosmic rays with energies from ~100 MeV to several GeV can penetrate into the Earth's atmosphere up to troposphere or lower stratosphere level and also their intensity is strongly modulated by solar activity (for example, [1]).

Veretenenko and Thejll [2-3], Artamonova and Veretenenko [4-5] showed that short-time variations of cosmic rays during solar proton events (SPEs) and Forbush-decreases of galactic cosmic rays (FD of GCRs) may influence the evolution of extratropical baric systems: cyclones and anticyclones, respectively. The physical mechanism of these processes is unclear and now is under consideration. The most probable hypothesis by author's opinion is one supposed by B. Tinsley (for example, [6]).

Conclusions of these works were based on analysis of atmospheric pressure variation during short-time cosmic ray disturbances using superposition epoch method. It is known that this mathematical method is correct for analysis of a one-dimensional values. In a multidimensional case it requires to use the Pareto sets [7]. As for a two-dimensional atmospheric pressure field, anomalies of one sign can

have various localization and when averaging (superposition of fields) the "indistinct" picture will be received. So, identification of maxima can be made only within some interval of an optimality.
In this work we analyze atmospheric pressure variations observed during solar proton events and Forbush-decreases of GCRs using Multifield Comparison Measure method (MCM [8-10]). MCM method allows to receive more objective estimation of connection between several two-dimensional scalar fields (multi-fields) in comparison with superposition epoch method.

2. Multifield Comparison Measure method

The Multifield Comparison Measure method based on analysis of matrix norm calculated for partial gradients of all analyzed fields in every point of manifold.

Let introduce the definition of matrix norm according to [8]. Let A be a $n\times m$ matrix of real numbers. The norm of the matrix A is defined as:

$$\|A\| = \max_{\|x\|=1,\ x\in\mathbb{R}^n} \|Ax\| \quad (1)$$

where $\|x\|$ represents the Euclidean norm of vector $x \in \mathbb{R}^2$.

Let multi-field $F = \{f_1, f_2, f_3, \dots, f_m\}$ be a set of m smooth scalar functions defined on a compact two-dimensional manifold $\mathbb{R}^2$. The derivative at point $p(x,y) \in \mathbb{R}^2$ is written as a matrix of partial derivatives:

$$dF(p) = \begin{bmatrix} \frac{df_1}{dx}(p) & \frac{df_1}{dy}(p) \\ \vdots & \vdots \\ \frac{df_m}{dx}(p) & \frac{df_m}{dy}(p) \end{bmatrix} \quad (2)$$

According to [8-10] *multifield comparison measure* η_p^F at point p defined as norm of the matrix $dF(p)$, i.e. $\eta_p^F = \|dF(p)\|$. From the definitions of the multifield comparison measure and the matrix norm, we have:

$$\eta_p^F = \left(\max_{x\in\mathbb{R}^n,\|x\|=1} x^T\left(dF(p)\right)^T\left(dF(p)\right)x\right)^{\frac{1}{2}} \quad (3)$$

where upper index T means transposing.

Let apply the spectral theorem from linear algebra and rewrite matrix product $(dF(p))T(dF(p))$ as $U^T\Lambda U$, where U is an orthogonal matrix and Λ is a diagonal matrix consisting of the eigenvalues of $(dF(p))T(dF(p))$ as entries in its diagonal:

$$\eta_p^F = \left(\max_{x\in\mathbb{R}^n,\|x\|=1} x^T U^T \Lambda U x\right)^{\frac{1}{2}} \quad (4)$$

Since U is orthogonal matrix, i.e. $U^T U = I$, and x is any unit vector, we can rewrite the expression (4) as:

$$\eta_p^F = \left(\max_{x\in\mathbb{R}^n,\|x\|=1} x^T \Lambda\, x\right)^{\frac{1}{2}} = \max\ \{\sqrt{\lambda}\}, \quad (5)$$

where λ is a diagonal element of matrix Λ, i.e. eigenvalues of $(dF(p))T(dF(p))$.

So, to compare the several two-dimensional scalar fields by MCM method [8-10] the next algorithm should be used. Firstly, to compose the multi-field matrix $F = \{f_1, f_2, f_3, \dots, f_m\}$ with m analyzed scalar fields as arguments, secondly, to calculate the matrix of partial derivatives $dF(p)$ in point $p(x, y)$ for every scalar field f_i and thirdly, to find the maximum norm of obtained matrix. The matrix norm calculated by this algorithm satisfies three important properties: symmetry, coordinate system independence and stability. The proof of these properties is presented in [8]. The results of calculations curried out by MCM method in the MatLab are presented in the next section.

3. Experimental data analysis

In this work we analyzed the atmospheric pressure variations in the Northern hemisphere observed during Forbush-decreases of GCRs and intensive solar proton events. The NCEP/NCAR reanalysis data for the isobaric level 1000 hPa were used. For this investigation 48 Forbush-decreases with amplitude $\delta N/N > 2.5$ % for the cold half of the year (October-March) 1980-2006 were chosen based on the Apatity neutron monitor data(geomagnetic latitude 63^{0}N). The list of 48 solar proton events (with energies of particles $Ep > 90$ MэB) for the October-March 1980-1995 was taken from the work [3].

Both for Forbush-decreases, and for solar proton events the maximum of partial derivative matrix norms for two sets of 48 atmospheric pressure fields were calculated on the grid 2.5°x2.5°. Dates of FDs and SPEs beginnings were considered as zero days. Results of calculations are presented on figure 1 and figure 2: the maximum of partial derivative matrix norms for undisturbed level – on the 9th day before FD and SPE beginnings (a), the maximum of partial derivative matrix norms for days of maximum pressure field deviations, observed in connection with FDs and SPEs – 1st and 4th days after corresponding event onsets (b), maps of difference of matrix norms for disturbed and undisturbed conditions (c).

(a) (b) (c)

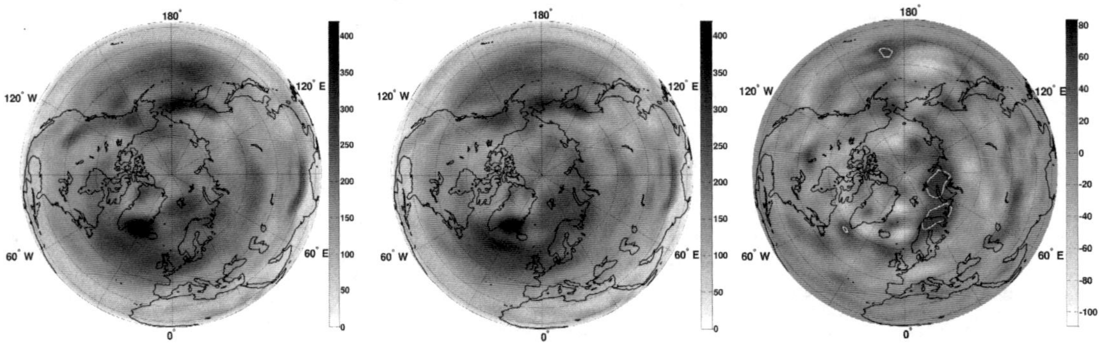

Fig. 1 Maximum of matrix norm of atmospheric pressure field on the level 1000 hPa in the course of 48 Forbush-decreases of galactic cosmic rays: (a) – undisturbed level, (b) – 4th day after FD beginning, (c) – difference between (b) and (a), white lines shows areas where matrix norm deviations from undisturbed level more than 3σ.

(a) (b) (c)

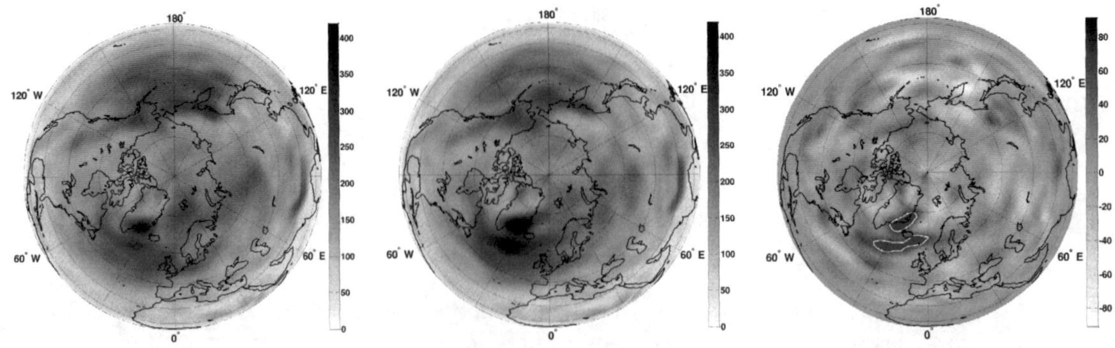

Fig. 2 Maximum of matrix norm of atmospheric pressure field on the level 1000 hPa after 48 solar proton events: (a) - undisturbed level, (b) - 1st day after SPE beginning, (c) - difference between (b) and (a), white lines shows areas, where matrix norm deviations from undisturbed level more than 3σ.

As it is seen from figures 1 (c) и 2 (c), in case of Forbush-decreases partial derivative matrix norm revealed growth over North Atlantic, Scandinavia, North of European part of Russia, North-West Siberia and partly over Arctic Ocean. In case of solar proton events area partial derivative matrix norm growth located only over North Atlantic region. These results are agree with conclusions made in previous works: the intensification of North Atlantic cyclones after solar proton events [2-3] and the blocking anticyclones formation over Scandinavia, Northern Europe and North of European part of Russia in the course of Forbush-decreases [4-5].

So, the results of atmospheric pressure field variation analysis by Multifield Comparison Measure method confirms earlier calculations made using superposition epoch method. Similarity of results obtained by both methods can be explained by the features of the lower atmosphere thermobaric field structure in the Northern hemisphere caused by orography (existence of the centre of atmospheric action in North Atlantic - the Greenland minimum) and the intensification of natural processes of cyclogenesis in the North Atlantic region in the course of short-time variations of cosmic ray intensity.

4. Conclusions

The relations between short-time variations of solar and galactic cosmic rays and pressure in the lower atmosphere on the isobaric level 1000 hPa using Multifield Comparison Measure method were analyzed. It was shown that both Forbush-decreases of GCRs and solar proton events are accompanied with significant growth of partial derivative matrix norms of atmospheric pressure field over the North Atlantic, Northern Europe and North of European part of Russia, this can be explained by intensification of the regional cyclogenesis processes during short-time variations of cosmic rays. The obtained results confirm the conclusions of previous works [2-5] and show that short-time variation of cosmic rays may influence the evolution of extratropical baric systems.

References

[1] Bazilevskaya G A 2005 *Adv. Space Res.* **35** 458
[2] Veretenenko S V, Thejll P 2004 *J. Atmos. Sol.–Terr. Phys.* **66** 393
[3] Veretenenko S V, Thejll P 2005 *Adv. Space Res.* **35** 470
[4] Artamonova I V, Veretenenko S V 2011 *J. Atmos. Sol.–Terr. Phys.* **73** 366
[5] Artamonova I V, Veretenenko S V 2014 *Adv. Space Res.* **454** 291
[6] Tinsley B A 2000 *Space Sci. Rev.* **94** 231
[7] Huettenberger L, Heine C, et al. 2013 *Computer Graphics Forum* **32** 341
[8] Nagaraj S, Natarajan V, Nanjundiah R S A 2011 *Computer Graphics Forum* **30** 1101
[9] Hansen et al (eds) 2014 *Scientific visualization* (London: Springer-Verlag) p 139
[10] Edelsbrunner H, Harer J, Natarajan V, Pascucci V 2004 *Proc. IEEE Conf. Visualization* p 275

Pyroelectric neutron generator for calibration of neutrino and dark matter detectors

A S Chepurnov[1], V Y Ionidi[1], O O Ivashchuk[2], A S Kubankin[2,3], A N Oleinik[2], A V Shchagin[2,4]
[1] Skobeltsyn Institute of Nuclear Physics, Lomonosov Moscow State University, Moscow, Russia
[2] Laboratory of Radiation Physics, Belgorod State University, Belgorod, Russia
[3]Lebedev Physical Institute, Moscow, Russia
[4] Kharkov Institute of Physics and Technology, Kharkov, Ukraine

E-mail: aschepurnov@yandex.ru

Abstract. Pyroelectric crystals, such as $LiNbO_3$ or $LiTaO_3$ being under influence of a temperature gradient can produce an electric field up to 10^5 kV/cm. It was experimentally confirmed that a crystal installed in a chamber with a residual gas pressure of about 1 mTorr could be used to generate X-Ray radiation with an energy up to 100 keV The same setup could be used to generate s 2.45 MeV neutrons if the target is deuterated and residual gas is D2. Due to such properties as On/Off mode of operation and the absence of radioactive materials, pyroelectric neutron generators seem to be a promising tool for calibration of neutrino and dark matter and other low background detectors. We propose the application of the controlled pyroelectric neutron generator for calibration of such detectors.

The pyroelectric effect consists on the generation of electric charges on the surface of the pyroelectric crystal, perpendicular to the polar z-axis, during a change of the crystal temperature. Such property is found both in crystalline and in ceramic materials. The charge polarity depends on the temperature gradient (cooling or heating) and the crystal z-axis orientation. The generated charge is compensated by free charge from air in a very short time at atmospheric pressure. However, the charge is not compensated in low pressure conditions then high electric potential up to 100 kV can be generated on the crystal surface. This is the cause of the polarization of molecules from the residual gas, as well as of the emission and acceleration of electrons from the pyroelectric crystal surface.

The possibility of X-rays generation using the Pyroelectric effect for $LiNbO_3$ and $LiTaO_3$ pyroelectric crystals was confirmed experimentally in [1] for the first time. Two different possibilities can be realized for X-rays production when a grounded target is located in front of the charged surface of the crystal in vacuum. If the pyroelectric crystal surface has a negative polarity as it is presented in figure1a, then the electrons from polarized molecules and the crystal surface are accelerated to the grounded target producing bremsstrahlung and characteristic radiation from the target. When the crystal has a positive charge polarity as it is shown in figure1b, the electrons from polarized molecules are accelerated towards the crystal surface producing bremsstrahlung and characteristic radiation from the crystal surface. It should be noted that the ions of the residual gas are also accelerated in the opposite direction relative to the electrons propagation, but the ions X-ray yield is substantially smaller in comparison with the yield produces by the electrons. Several geometries can be realized to develop pyroelectric X-ray sources [2-4].

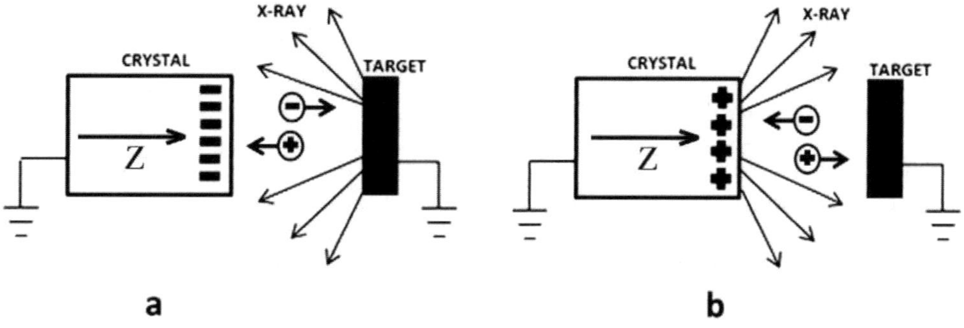

Figure 1. The two cycles of X-ray generation by pyroelectric crystal: a) crystal negative polarity; b) crystal positive polarity.

The considered pyroelectric effect can be used for neutron generation also. If deuterium gas is used instead of air as the residual gas and the target contains deuterium, then 2.45 MeV neutrons can be generated during a nuclear D-D reaction [5]. The neutrons yield depends on the energy and current of accelerated deuterium ions, which interact with the deuterated target. The tungsten tip installed on the crystal surface is usually applied to increase the probability of the deuterium ionization. Figure 2 schematically illustrates the mechanisms of neutron generation using the pyroelectric effect. Neutron generation using the pyroelectric effect was widely investigated in recent years [6-7].

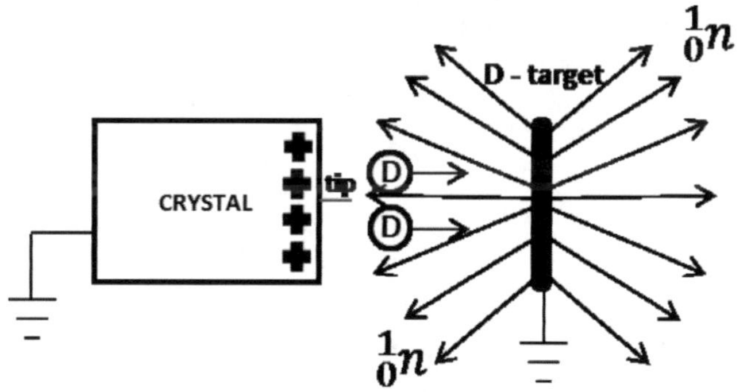

Figure 2. The neutrons generation using the pyroelectric effect.

Neutron calibration is one of the critical procedure performed time to time for most of neutrino and dark matter search experiments [8]. One of the main reason is that neutrons are main backgrounds for such experiments which fake expected physical events. The sources for calibration should have an ultra-low background, a compact size and the possibility to control the intensity. Radioactive materials and high-voltage power supplies should be avoided in such sources.

Our proposal consists in the application of neutron sources based on the pyroelectric effect described above for calibration of low-background detectors. Such sources will have a typical size of several cubic centimeters, they do not contain any radioactive substances and could be manufactured low background and they don't need external high voltage power supply. It is also important that the output neutron intensity can be controlled.

The pyroelectric neutron source can be installed stationary into a neutrino or dark matter detector. When the source is tuned off, it does not produce any radiation and does not hamper operation of the detector. The source can be tuned on by connecting of a low voltage power supply that should provide variation of the temperature of the pyroelectric crystal. The tuned on source produces 2.45 MeV neutrons. The scattering of the neutrons on atomic nuclei of the detector produces recoil nuclei that can be used for calibration of the detector. Besides of neutrons, the source can produce X-rays that can hamper the calibration. The X-rays should be absorbed by walls of the chamber of the source.

The neutrons source intensity is of several thousand neutrons per thermal cycle. These values are enough for low background detector calibration. At the moment, the work on improving the pyroelectric source characteristics is being performed.

Acknowledgments
A.V. Shchagin is thankful to F. Goennenwein for preliminary discussions about applicability of the pyroelectric neutron source for calibration of neutrino and dark matter detectors. Also, authors are thankful to S.I. Bashko and his team from ISSP RAS for production of tungsten tips. This study was supported financially by the Russian Foundation for Basic Research, projects 14-22-0301 ofi_m and the Ministry of Education and Science of the Russian Federation, project 3.2009.2014/K.

References

[1] Brownridge J D 1992 Pyroelectric x-ray generator *Nature* **358** 277–8
[2] Nagaychenko V I *et al.* 2007 (In Russian) Journal of Surface Investigation. X-ray, Synchrotron and Neutron Techniques **3** 81 (Original Russian title: Poverkhnost'. Rentgenovskie, Sinkhrotronnye i Neitronnye Issledovaniya)
[3] Brownridge J D and Shafroth S M 2007 *New Research on Lasers and Elecrooptics,* ed. by William T Arkin (New York: Nova Science Publishers) p 37
[4] URL: www.amptek.com/coolx.html
[5] Naranjo B, Gimzewski J K and Putterman S 2005 *Nature* **434** 1115
[6] Geuther J, Danon Y and Saglime F 2006 *Phys. Rev. Lett.* **96** 054803
[7] Gillich D *et al.* 2009 *Nucl. Instrum. Meth.* A **602** 306
[8] Bondar A E *et al.* 2013 (in Russian) Proposal for neutron scattering systems for calibration of dark matter search and low-energy neutrino detectors *Vestnik of NSU: Physics Series* **8** 27

Cosmic ray physics with TeV muons in large volume detectors

Patrick Berghaus for the IceCube Collaboration[1]

National Research Nuclear University MEPhI (Moscow Engineering Physics Institute), Kashirskoye highway 31, Moscow, 115409, Russia

E-mail: berghaus@icecube.wisc.edu

Abstract. Large volume detectors such as IceCube, designed for the detection of astrophysical neutrinos register cosmic ray-induced atmospheric muon bundles at a rate of several thousand events per second. Due to the large amount of surrounding material, the effective energy threshold for muons reaching the detector typically lies at approximately one TeV. Through careful evaluation of event profiles it is possible to address cosmic ray and particle physics issues in an unprecedented energy region. Results from the analysis of one year of IceCube data are presented and their implications discussed.

1. Introduction

For underground detectors placed within mountains or below earth or sea level, the amount of material between the surface and the instrumented volume is of the order of several thousand tons per square meter. The only charged particles produced in cosmic ray air showers remaining at detector depth are muons, and even for these the energy threshold lies in the TeV region.

Large-volume underground detectors have for a long time been used for cosmic ray physics. The most common and simplest measurement is the zenith angle distribution of atmospheric muons. Although the exact functional form depends on the geometry of the surrounding material, the distance that particles have to traverse before reaching the detector usually increases towards the horizon as a function of the zenith angle θ_{zen}. The muon energy spectrum can be derived by relating threshold energy to the varying distance between surface and detector. Measurements during the early 1990s, summarized in [1], often showed significant excesses over model predictions, a situation that was rectified in 1998 by the LVD collaboration [2], who demonstrated that the muon flux was in fact consistent with air shower simulations based on a primary nucleon flux following a simple power law of the form $E^{-2.78\pm0.05}$.

The IceCube array, designed primarily for the detection of astrophysical neutrinos, is the first particle detector with an instrumented volume of the order of a cubic kilometer [3]. The consequent increases in effective area and observable track length represent a significant qualitative advance for the study of cosmic ray physics. In recognition of this fact, the underground detector was complemented with the surface array *IceTop*, whose primary stated purpose is to measure the electromagnetic component of air showers in coincidence with high-energy muon bundles traversing the main detector volume [4].

[1] http://icecube.wisc.edu

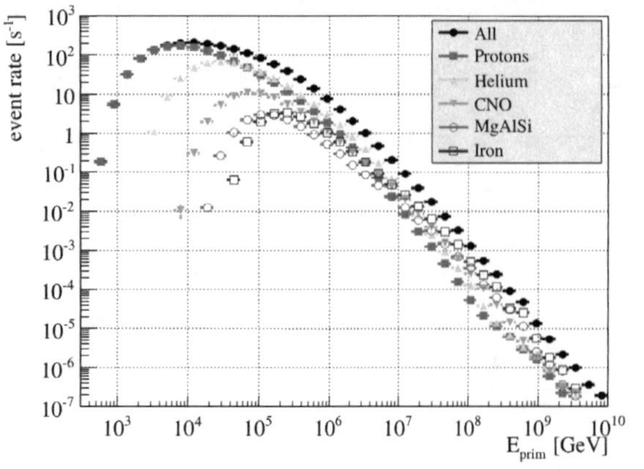

Figure 1. Atmospheric muon event yield in IceCube in dependence of primary type simulated with CORSIKA [5] and weighted according to the Hillas 3a model [6].

Upcoming projects such as the pair of deep-sea neutrino detectors collectively referred to as KM3Net [7] currently have no provision for a surface array, and cosmic ray studies will have to rely exclusively on the large underground volume. One analysis opportunity, pioneered by the IceCube collaboration, is the measurement of cosmic ray anisotropies using the enormous statistics provided by atmospheric muon events registered at a rate of approximately 3,000 s^{-1} [8]. Another, less directly connected to cosmic ray physics but with important implications for the study of nuclear interactions, is the measurement of showers containing individual muons with high transversal momentum [9].

That this is by far not the limit of possiblities can be seen from figure 1. The energy spectrum of cosmic ray primary particles responsible for air shower events in IceCube covers over six orders of magnitude, starting at about 10 TeV and reaching up to more than 10 EeV. Within this region lie several regions sparsely covered by existing detectors, and features of major interest, such as the knee and the ankle, which are not yet fully understood. The following is a brief description of techniques for the exploitation of atmospheric muon data in cosmic ray studies, taking advantage of sophisticated analysis methods developed for use in the latest generation of high-energy neutrino detectors.

2. IceCube Results

2.1. Event Types

Atmospheric muon events at high energies can be roughly divided into two basic types, illustrated in figure 2. In one case, the shower contains at least one unusually energetic muon, produced by a hadron decay at the early stage of shower development. In the other, a very energetic primary produces a bundle of very high multiplicity but with a rather modest median energy of a few hundred GeV at detector depth. As the integrated energy deposition in either case can be very similar, and the coarse instrumentation of the detector does not permit the resolution of individual muon tracks, any distinction between the two types must be based on the differential energy loss profile.

In practice, the simplest classification method relies on the identification of very strong individual energy losses, the majority of which are caused by bremsstrahlung [11]. While the selection efficiency for high-energy muon events is limited by the comparatively low probability of such an interaction taking place within the detector volume, the method is quite robust and has the added advantage that the reconstructed energy of the particle shower can be used as an estimator for the muon energy.

A clean sample of high-multiplicity bundles can be obtained by inverting the selection criteria, keeping only tracks without exceptional energy losses. This does not strictly exclude the presence of high-energy muons, but assures that their contribution to the overall energy deposition is not unusually high. The two categories are illustrated in figure 3 in terms of the leading muon energy fraction. A full description of selection and analysis procedures can be found in [10].

2.2. Muon Bundles

The multiplicity of muon bundles N_μ can be approximated by the simple proportionality $N_\mu \propto E_{\text{prim}}^{\alpha} \cdot A^{1-\alpha}$, where E_{prim} is the energy of the primary nucleus and A its atomic mass. From an experiental perspective, the number of muons is proportional to the energy deposition in the detector, provided an event sample is selected that excludes exceptional catastrophic energy losses.

While energy and mass of incident cosmic rays cannot be measured simultaneously based solely on the muon bundle multiplicity in the deep detector, it is possible to derive a qualitative description of the behavior of the average mass in dependence of primary energy by adding external information from air shower arrays as illustrated in figure 4. The experimental measurement is in this case presented under two assumptions about the primary composition,

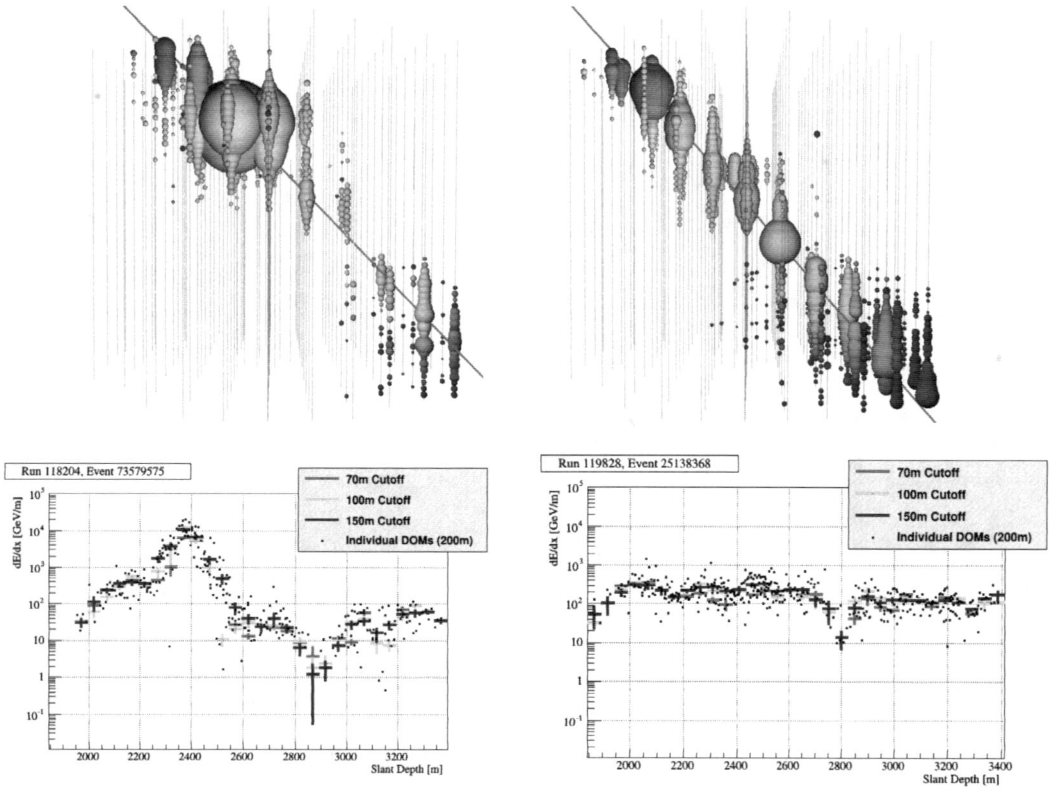

Figure 2. Example for atmospheric muon events in IceCube experimental data. Top: Event display. The size of the spheres is proportional to the number of photo-electron registered by an optical module, their color indicates the arrival time of the first photon (red: early, blue: late). Bottom: Reconstructed differential energy deposition along track. The x-axis indicates the distance to the surface, measured along the track. The dip near x=2800 m is an artifact caused by the presence of a horizontal ice layer with very high dust concentration. Left: High energy muon, identified by the large stochastic loss near x=2400 m. Right: High-multiplicity bundle with smooth energy deposition.

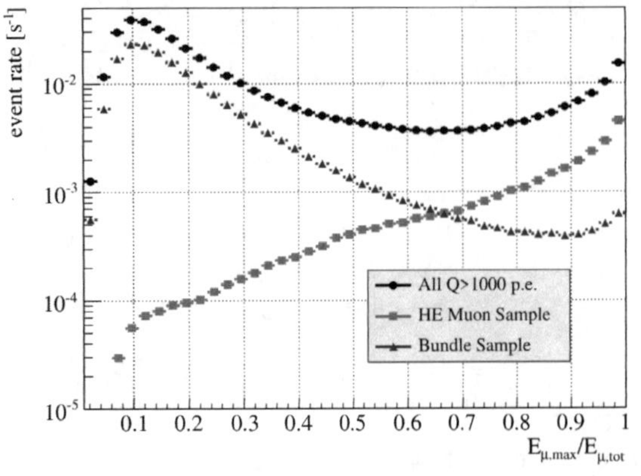

Figure 3. Event samples used for atmospheric muon physics analyses in IceCube [10]. The parameter on the x-axis correponds to the true fraction of total bundle energy carried by the leading muon for simulated data with more than 1,000 registered photo-electrons in the detector.

reflecting the extreme cases of pure proton and iron fluxes. The position of the all-particle measurement relative to the two hypotheses corresponds approximately, but not rigorously, to the average logarithmic mass of cosmic rays at a given energy.

In the interesting and poorly explored region beyond 10^{17} eV, the measurement indicates a uniformly heavy primary composition, consistent with data from KASCADE-Grande [12] and preliminary findings by IceCube using events registered in coincidence between surface and deep detector arrays [13].

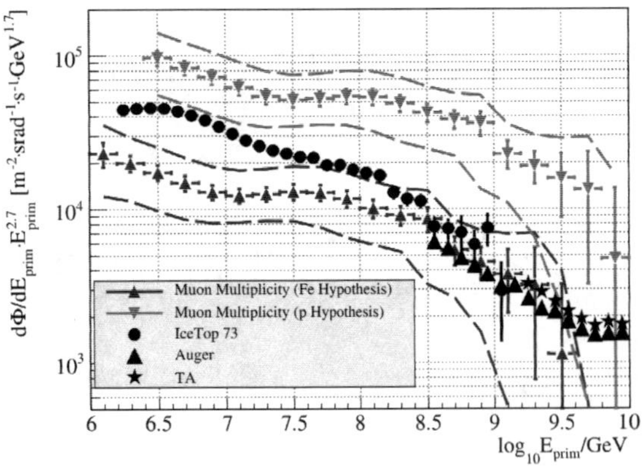

Figure 4. Interpretation of muon multiplicity spectrum by relation to all-particle flux measurements from IceCube [14] and other detectors [15, 16].

2.3. High-Energy Muons

The analysis of high-energy muons provides an opportunity for the study of cosmic rays as well as particle physics, particularly the forward production of heavy quarks in nuclear collision. Recent theoretical calculations for the atmospheric lepton flux component from charm decays contain large uncertainties [17, 18] and in the case of muons the situation is further complicated by rare electromagnetic decays of light vector mesons [19], not present in neutrino fluxes.

The first limit for the contribution of prompt decays of short-lived hadrons to the atmospheric lepton flux was set by the LVD collaboration [20]. An analysis using data taken with the BUST detector at Baksan showed an indication for a strong excess at high energies whose underlying cause is not fully understood [21], and which has so far not been corroborated independently. A measurement of the prompt flux based on atmospheric neutrinos is difficult due the presence of a poorly constrained background from astrophysical sources [22].

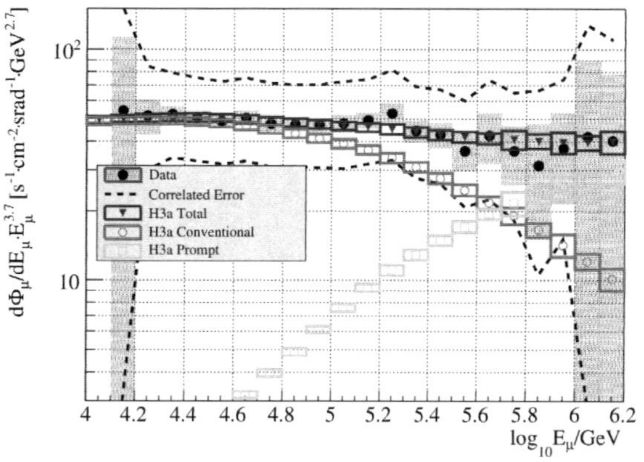

Figure 5. Experimentally measured all-sky spectrum of high-energy muons using one year of IceCube data compared to prediction based on H3a flux model [6]. Between 15 TeV and 1.5 PeV the energy spectrum approximately follows a power law with index $\gamma_\mu = -3.78 \pm 0.02(stat.) \pm 0.03(syst.)$. The absence of a spectral steepening is likely due to the emergence of a flux component from prompt hadron decays, here illustrated by the green curve.

The muon energy spectrum derived from IceCube data is shown in figure 5. Since the ratio between lepton and parent nucleon energy is generally about one order of magnitude, the measurement covers the region of the knee, where most current primary flux models predict a sharp cutoff in the all-nucleon spectrum. The fact that the measured spectrum is consistent with an unbroken power law can be interpreted as an indication for the presence of a flux component from prompt meson decays whose spectrum is not attenuated by re-interactions with atmospheric nuclei and therefore harder than that originating in decays of pions and kaons.

It would in principle be possible to separately constrain cosmic ray nucleon flux and magnitude of the prompt component by adding information about the angular distribution of high energy muon events. However, careful examination of experimental data reveals the presence of an unexplained systematic effect, most likely related to the exclusively downward orientation of optical modules in IceCube and the very short scattering length for photons in the refrozen ice of the drill holes [10].

3. Outlook

The current results demonstrate that atmospheric muon measurements in large volume detectors are a valuable tool for cosmic ray physics. Using the constantly accruing amount of experimental data in IceCube, it will soon be possible to repeat the muon bundle multiplicity measurement with statistics increased by one order of magnitude. Physics issues that can be addressed are the nature of the recently identified structure of the cosmic ray flux beyond the knee [14], the composition of cosmic rays around the ankle with potential implications for the interpretation of the cutoff at EHE energies [23], and investigation of the apparent excess in the number of muons for air showers at energies around 10^{19} eV as reported by the Pierre Auger collaboration [24].

Water-based detectors currently under construction do not have the disadvantages inherent to the IceCube array and its potential future extensions. Specifically, optical modules contain multiple photomultiplier tubes and are designed for uniform directional acceptance. The very long scattering length of photons in water as compared to the complex and varying optical properties of the Antarctic ice will reduce systematic uncertainties and make track reconstruction simpler and more reliable. An accurate simultaneous measurement of both atmospheric prompt leptons and cosmic ray nucleon flux around the knee should be within reach of one or both of the Mediterranean water detectors. Its validity can be further supported by demonstrating consistency with a simple angular distribution measurement, and by comparing results from the two individual arrays against each other.

References

[1] Sinegovsky S I *et al.* 2010 *Int. J. Mod. Phys.* A **25** 3733 (*arXiv*:0906.3791 [astro-ph.HE])
[2] Aglietta M *et al.* (LVD Collaboration) 1998 *Phys. Rev.* D **58** 092005 (*arXiv*:hep-ex/9806001)
[3] Karle A (IceCube Collaboration) *arXiv*:1401.4496 [astro-ph.HE]
[4] Abbasi R *et al.* (IceCube Collaboration) 2013 *Nucl. Instrum. Meth.* A **700** 188 (*arXiv*:1207.6326 [astro-ph.IM])
[5] URL: *https://web.ikp.kit.edu/corsika/*
[6] Gaisser T K, Stanev T and Tilav S 2013 *Front. Phys. China* **8** 748 (*arXiv*:1303.3565 [astro-ph.HE])
[7] Margiotta A (KM3NeT Collaboration) 2014 *JINST* **9** C04020 (*arXiv*:1408.1132 [astro-ph.IM])
[8] Abbasi R *et al.* (IceCube Collaboration) 2012 *Astrophys. J.* **746** 33 (*arXiv*:1109.1017 [hep-ex])
[9] Abbasi R *et al.* (IceCube Collaboration) 2013 *Phys. Rev.* D **87** 012005 (*arXiv*:1208.2979 [astro-ph.HE])
[10] Aartsen M G *et al.* (IceCube Collaboration) *arXiv*:1506.07981 [astro-ph.HE]
[11] Koehne J H *et al.* 2013 *Comput. Phys. Commun.* **184** 2070
[12] Apel W D *et al.* 2013 *Astropart. Phys.* **47** 54 (*arXiv*:1306.6283 [astro-ph.HE])
[13] The IceCube Collaboration, T. Feusels, ICRC 2015, PoS 334
[14] Aartsen M G *et al.* (IceCube Collaboration) 2013 *Phys. Rev.* D **88**(4) 042004 (*arXiv*:1307.3795 [astro-ph.HE])
[15] Aab A *et al.* (Pierre Auger Collaboration) *arXiv*:1307.5059 [astro-ph.HE]
[16] Abu-Zayyad T *et al.* (Telescope Array and Pierre Auger Collaborations) *arXiv*:1310.0647 [astro-ph.HE]
[17] Garzelli M V, Moch S and Sigl G *arXiv*:1507.01570 [hep-ph]
[18] Bhattacharya A *et al.* 2015 *JHEP* **1506** 110 (*arXiv*:1502.01076 [hep-ph])
[19] Fedynitch A *et al.* *arXiv*:1503.00544 [hep-ph]
[20] Aglietta M *et al.* (LVD Collaboration) 1999 *Phys. Rev.* D **60** 112001 *arXiv*:hep-ex/9906021)
[21] Bogdanov A G *et al.* 2012 *Astropart. Phys.* **36** 224 (*arXiv*:0911.1692 [astro-ph.HE])
[22] Aartsen M G *et al.* (IceCube Collaboration) 2015 *Astrophys. J.* **809**(1) 98 *arXiv*:1507.03991 [astro-ph.HE])
[23] Watson A A, 2014 *Rept. Prog. Phys.* **77** 036901 (*arXiv*:1310.0325 [astro-ph.HE])
[24] Aab A *et al.* (Pierre Auger Collaboration) 2015 *Phys. Rev.* D **91**(3) 032003
2015 *Phys. Rev.* D **91**(5) 059901 (*arXiv*:1408.1421 [astro-ph.HE])

Search of predictors of geoeffective heliospheric events by means of muon hodoscope URAGAN

V V Shutenko, I I Astapov, N S Barbashina, A N Dmitrieva, R P Kokoulin, K G Kompaniets, A A Petrukhin and I I Yashin
National Research Nuclear University MEPhI
(Moscow Engineering Physics Institute), 115409, Moscow, Russia

E-mail: vvshutenko@mephi.ru

Abstract. The flux of charged primary cosmic rays (PCR) is modulated by changes of interplanetary magnetic field which depend on various heliospheric processes. After interaction of PCR with nuclei of atoms of the Earth's atmosphere, these modulations are transferred to a flux of secondary muons. Muon hodoscope URAGAN allows to trace changes not only of intensity of the muon flux, but also its angular distribution in a wide range of zenith angles (0-80°). Some results of searching of predictors of geoeffective heliospheric events by means of muon hodoscope URAGAN are presented.

1. Introduction

The flux of charged primary cosmic rays (PCR) is modulated by changes of interplanetary magnetic field which depend on various heliospheric processes. The Earth magnetosphere is substantially subject to influence of dynamic processes in the interplanetary environment: strong disturbances of the interplanetary magnetic field result in reorganization of the magnetic field inside magnetosphere, to development of geomagnetic storms and substorms. Disturbances of the interplanetary magnetic field are propagated from the Sun to the Earth with a speed of the solar wind 300-1500 km•s^{-1}, but the velocity of propagation of PCR is close to the light velocity. The PCR flux modulated by the disturbance reaches the near-Earth space considerably earlier than this disturbance. After interaction of PCR with nuclei of atoms of the Earth's atmosphere, these modulations are transferred to the flux of secondary muons. Muon hodoscope URAGAN (MH URAGAN) [1] allows to trace changes not only of the intensity of the muon flux, but also its angular distribution in a wide range of zenith angles (0-80°). The results of a search of predictors of geoeffective heliospheric processes with the help of the analysis of variations of the muon flux are considered.

2. Initial data

As the initial data for searching the predictors, the URAGAN data and OMNI database were used.

At preparation of the data of MH URAGAN in real time, which are presented at the site of the Unique Scientific Facility 'Experimental complex NEVOD' [2], time series of hourly average values and their statistical errors for various characteristics of a registered muon flux are formed:

- *Isum* – count rate with barometric correction, s^{-1};
- A_Z – vertical projection of the vector of the local anisotropy with barometric correction;
- A_S – horizontal projection of the vector of the local anisotropy to the north-south axis;

- A_E – horizontal projection of the vector of the local anisotropy to the west-east axis.

The secondary characteristics are formed concerning the average values obtained during 24 hours:

- r_h – horizontal projection of the vector of the relative anisotropy;
- r_S – horizontal projection of the vector of the relative anisotropy to the north-south axis;
- r_E – horizontal projection of the vector of the relative anisotropy to the west-east axis.

Among the characteristics of the muon flux, the most convenient are those which have no long-term trends. But variations of characteristics because of the daily rotation of the Earth, in our case cannot be neglected. Therefore the following characteristics were examined: r_h, r_S, r_E.

From the time series of hour values r_h. r_S, r_E, the following secondary quantities (here symbol X designates r_h or r_S, or r_E) were calculated:

1) δX_1 – the ratio of the difference of the current and the preceding hour values of the characteristic X to the statistical error of this difference;
2) δX_{24} – the ratio of the difference of the current and 24 hours prior to it hour values of the characteristic X to the statistical error of this difference;
3) σX_{24} – root-mean-square deviation of the hour values of the characteristic X over the period 24 hours (including the current hour).

As indicators of the condition of the heliosphere and magnetosphere of the Earth, the following parameters of the interplanetary magnetic field (IMF) and geomagnetic activity from database OMNI were used:

- Vsw – hourly average value of the speed of the solar wind, km•s^{-1};
- $BzGSE$ – hourly average value of z-projection of IMF in GSE coordinate system, nT;
- Dst – hourly average equatorial index of the geomagnetic activity, nT;
- Kp – the planetary geomagnetic activity index.

For the analysis, the URAGAN data and OMNI data from February, 2007 till December, 2013 were used. The following criteria of the perturbed state in OMNI data were used: $Vsw \geq 500$ km•s^{-1}, $BzGSE \leq -7$ nT, $Dst \leq -30$ nT, $Kp \geq 4$.

3. Searching predictors

Joint processing of time series of the URAGAN and OMNI data in a moving time window of 5 day was carried out. First for consideration the intervals in which in the first 4 days there were no disturbances, and on the last day $Kp \geq 4$ have been taken. As there were found only 65 such intervals, for the increase in statistics the intervals in which in preceding four days there could be disturbances on Vsw, $BzGSE$, Dst have been added, too. Thus, for the visual analysis, 130 5-day intervals in which the last day value $Kp \geq 4$ was observed, have been selected.

In the selected intervals the behaviour of horizontal projections of the vector of the relative anisotropy (r_h, r_S, r_E) and their derivative quantities were analyzed (δr_{h1}, δr_{S1}, δr_{E1}, δr_{h24}, δr_{S24}, δr_{E24}, σr_{h24}, σr_{S24}, σr_{E24}). During the analysis, those projections for which in the first two days of the examined interval their values were appreciably smaller, than in the subsequent two, or qualitative by different, were marked. In figure 1, examples of the images of time series prepared for the visual analysis are shown.

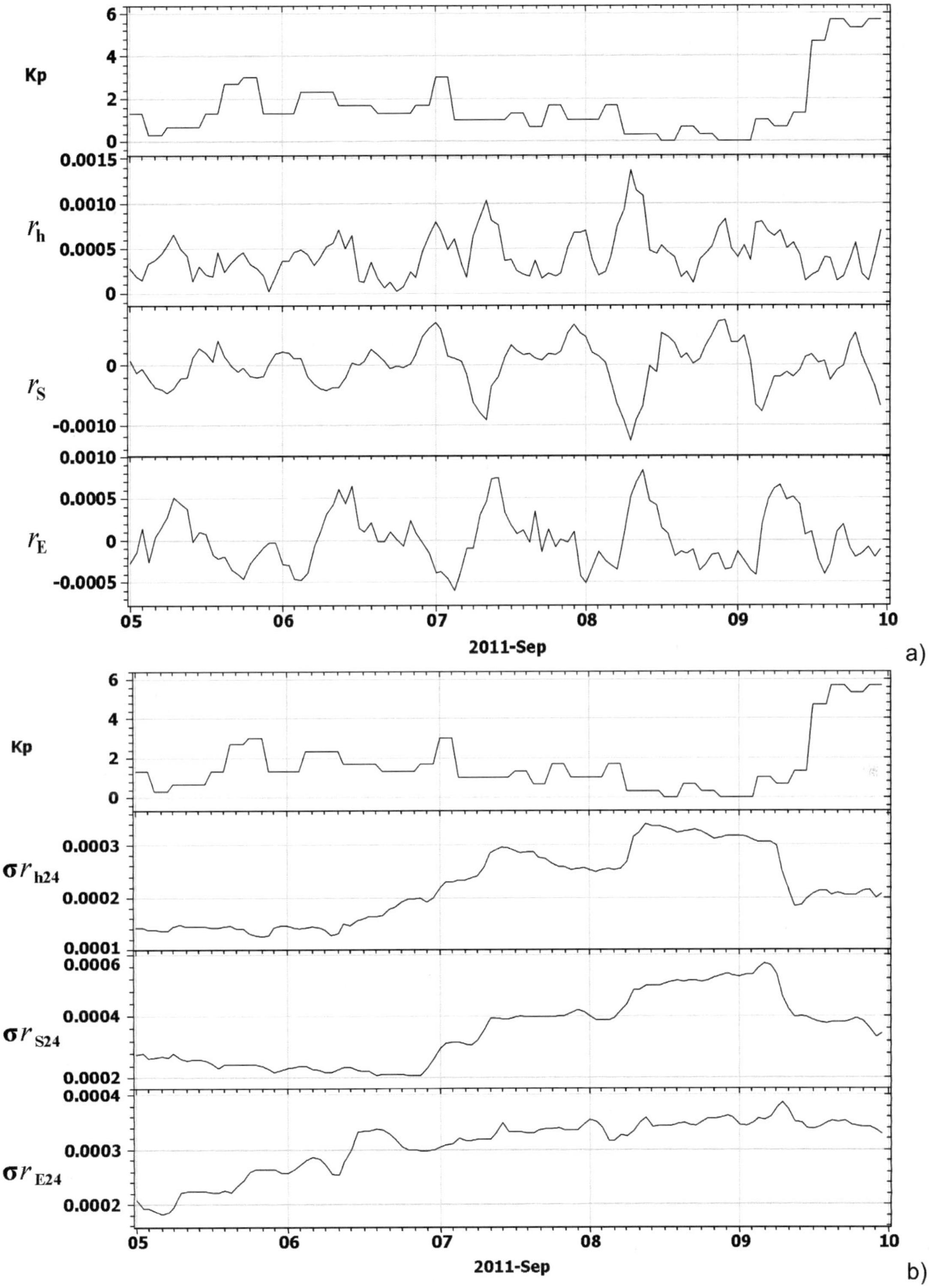

Figure 1. Examples of images of time series for the interval from 05-09-2011 till 09-11-2011: a) – Kp and (r_h, r_S, r_E); b) – Kp and (σr_{h24}, σr_{S24}, σr_{E24}).

For each event, according to the information from weekly bulletins NOAA Space Weather Highlights (ftp://ftp.swpc.noaa.gov/pub/warehouse/) and http://www.spaceweather.com/, probable sources of disturbances have been determined (CH – Coronal Hole, CIR – Co-rotating Interacting Region, SSB – Solar Sector Boundary, SSBC – Solar Sector Boundary Crossing, HSS – High-Speed Stream, CME – Coronal Mass Ejection, Bz – Z-projection IMF). As the visual analysis of the time series δr_{h1}, δr_{S1}, δr_{E1}, δr_{h24}, δr_{S24}, δr_{E24} has not given any result, the results of the visual analysis only for (r_h, r_S, r_E), and (σr_{h24}, σr_{S24}, σr_{E24}) have been formulated.

For the selected 65 intervals, in 15 cases the source of the disturbance was CME, and for selected 130 intervals, in 37 cases the source of the disturbance was CME. In total, before the geomagnetic storm with $Kp \geq 4$, the qualitative changes in time series of the URAGAN data were observed:

1) with all sources of disturbance:
   - (r_h, r_S, r_E) – in 28 cases from 130 (21%), and in 13 cases from 65 (20%);
   - (σr_{h24}, σr_{S24}, σr_{E24}) – in 64 cases from 130 (49%), and in 31 cases from 65 (47%).
2) with CME as the source of the disturbance:
   - (r_h, r_S, r_E) – in 11 cases from 37 (29%), and in 7 cases from 15 (46%);
   - (σr_{h24}, σr_{S24}, σr_{E24}) – in 22 cases from 37 (59%), and in 10 cases from 15 (66%).

4. Conclusion

As a result of the performed search, as predictors geoeffective heliospheric processes the root-mean-square deviations (σr_{h24}, σr_{S24}, σr_{E24}) of hour values r_h, r_S, r_E for the period 24 hours, including the current hour, can be used. Each of these predictors can be formulated, for example, as follows: increase of the value for a threshold size (for example, 10^{-4}) within two days. That is, in a moving time window with duration of 48 hours a difference between the value at the end of the window (current value) and the minimal value (in a whole window) is determined. If this difference is more or equal to threshold value, this can mean the presence of the predictor of geoeffective heliospheric processes.

Acknowledgments

This work was performed at the Unique Scientific Facility 'Experimental complex NEVOD' within the framework of the Centre of Fundamental Research and Particle Physics supported by the MEPhI Academic Excellence Project (contract 02.a03.21.0005 of 27.08.2013).

References

[1] Barbashina N S et al 2008 *Instrum. Exp. Tech.* **51 2** 180
[2] Yashin I I et al 2015 *J. Phys.: Conf. Ser.* **632** 012086

Atmospheric effects in the intensity of muon bundles and geometrical mechanism of their formation

R P Kokoulin, A G Bogdanov, A N Dmitrieva, E V Romanenkova, V V Shutenko, E A Yurina

National Research Nuclear University MEPhI (Moscow Engineering Physics Institute), Kashirskoe shosse 31, Moscow, 115409, Russia

E-mail: rpkokoulin@mephi.ru

Abstract. Temporal changes in the intensity of muon bundles produced as a result of interactions of primary cosmic ray particles with energies of the order of 10^{15} eV and detected at the ground level are analyzed. Seasonal variations, barometric and temperature effects, and correlations with the altitudes of various levels of residual pressure are considered. It is shown that muon bundle intensity variations are well explained in frame of a simple mechanism related with changes of the muon lateral distribution function at the observation level caused by geometrical changes of the effective altitude of the formation of the bundles.

1. Introduction

The rate of the events generated as a result of interactions of cosmic rays in the atmosphere and detected on the Earth's surface is subject to variations caused by the atmospheric conditions [1]. The correct understanding of the influence of atmospheric effects is important for comparison of data of the experiments conducted in different geographical points and conditions, and for introduction of necessary corrections. It is important to note however that the values of the meteorological effects as well as the physical processes responsible for their formation are different for different cosmic ray components (muon, hadronic, electron-photon) and for events of different classes.

In present paper, variations in the intensity of muon bundles detected at the ground level with the coordinate-tracking detector DECOR [2] are considered. The bundle is the event with a simultaneous passage of several genetically related muons through the setup. The main sources of muon bundles are decays of pions and kaons generated in nuclear cascades initiated in the atmosphere by high energy primary cosmic ray particles. Already in the first experiments on muon bundle detection with DECOR conducted in 2004 – 2007, significant changes of the rate of registered events were noticed [3]. For the first time, experimental estimates of the barometric and temperature coefficients for muon bundle intensity were obtained, which appeared unexpectedly high. As a possible interpretation of the results, a new mechanism related with a geometrical transformation of the lateral distribution function of EAS muons for a varying temperature profile of the atmosphere was suggested.

Unfortunately, preceding series of measurements with DECOR were relatively short and did not exceed 6 – 7 month; in particular, no data were taken in summer periods (from July to September). Starting from 2012, nearly continuous measurements are conducted which create necessary pre-requisites for the analysis of variations on a qualitatively new level. Below, the results of the analysis of the data on the intensity of muon bundles detected during three years of observations are presented.

2. Experimental data and seasonal variations

The coordinate-tracking detector DECOR [2] is a part of the NEVOD experimental complex. It is deployed in the galleries of the NEVOD building in MEPhI (Moscow, 55.7° N, 37.7° E, ~ 165 m a.s.l.) and consists of 8 assemblies (supermodules, SMs), each with area of 8.4 m^2. The supermodule includes 8 vertical planes of plastic streamer tube chambers equipped with two-coordinate external strip readout system. Spatial and angular accuracies of muon track reconstruction in the SM are better than 1 cm and 1°, respectively.

Selection of muon bundles in DECOR is based on the assumption that the tracks of muons generated in the atmosphere (at large distances from the setup) are nearly parallel. At a trigger level, the events with at least three (of eight) hit SMs are recorded. At the off-line data processing, the geometry reconstruction of the tracks in the SMs is performed, and candidate events with at least 3 quasi-parallel tracks (coinciding in direction within a 5°-cone) detected in 3 different SMs of DECOR are selected. Distribution of the events in zenith angle is rather wide; about 90% of the detected bundles lie in the range from 21° to 65° with a median value near 42°. According to earlier estimates [4], these bundles are mainly formed as a result of interactions of primary cosmic ray particles with typical energies of $10^{15} – 10^{16}$ eV.

Experimental data are accumulated as a sequence of separate sets (runs) with duration of up to ~ 40 h. Short runs (less than 10 h, comprising about 3% of the total statistics) have been excluded from the present analysis. As a result, we left with 590 runs corresponding to the total live time of 17,760 h, containing about 4.4 million muon bundles registered over the period from May 03, 2012 to April 08, 2015. For every run, the number of the selected muon bundles, live registration time, start/stop moments, average atmospheric pressure at the setup location and other parameters have been determined. The mean rate of muon bundle events during observations equals to 248.1 per hour. The measured time dependence of the rate of muon bundle detection over the period of observations is shown in figure 1. Every point represents one run data; the error bars indicate statistical uncertainties.

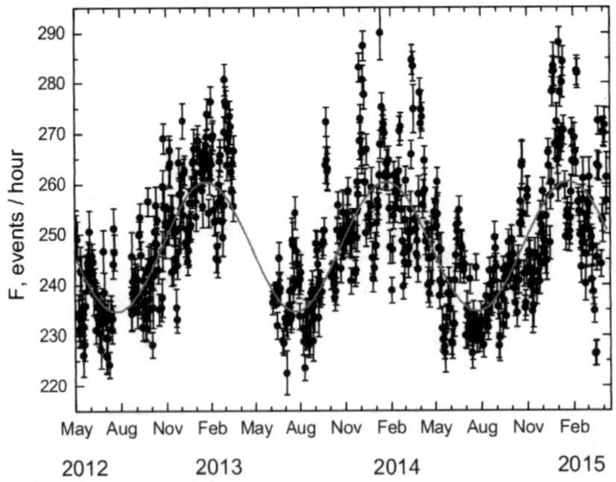

Figure 1. Temporal variations of the muon bundle detection rate. The smooth curve represents the first annual harmonic.

As follows from the figure, clear seasonal variations repeated every year are observed in the event rate. The smooth curve in the figure represents the first annual harmonic obtained by means of a weighted least square fitting of the data. On the average, the intensity of muon bundles detected in winter and in summer differs by more than 10%; the maximum intensity is typically observed in the second half of January, and the minimal one in July.

3. Barometric and temperature effects

On the background of smooth seasonal variations in the event intensity, short-term deviations from an overall harmonic dependence with duration of the order of few days are observed. It is natural to assume that these deviations are related with the changes of the atmospheric conditions. In order to estimate the influence of barometric and temperature effects, we fitted the data with a following phenomenological formula:

$$F = F_0(1+\beta_P\Delta P)(1+\beta_T\Delta T);\ \Delta P = P - \langle P\rangle;\ \Delta T = T - \langle T\rangle. \tag{1}$$

Here $\langle P\rangle$ and $\langle T\rangle$ are average values of atmospheric pressure and temperature for the observation period. As a temperature characteristic of the atmosphere, we use here the mass average temperature of the air calculated from the retrospective data of the GDAS (Global Data Assimilation System) model [5]. Mean values of the ground pressure and the mass average temperature were equal to 746.7 mm Hg and 247.0 K, respectively. Barometric and temperature coefficients β_P and β_T were found iteratively. Iteration procedure rapidly converges, and the estimates of the parameters for the second and third iterations differ by less than their statistical errors. Correlations of muon bundle detection rate with atmospheric pressure and mass average temperature air are shown in figures 2 and 3.

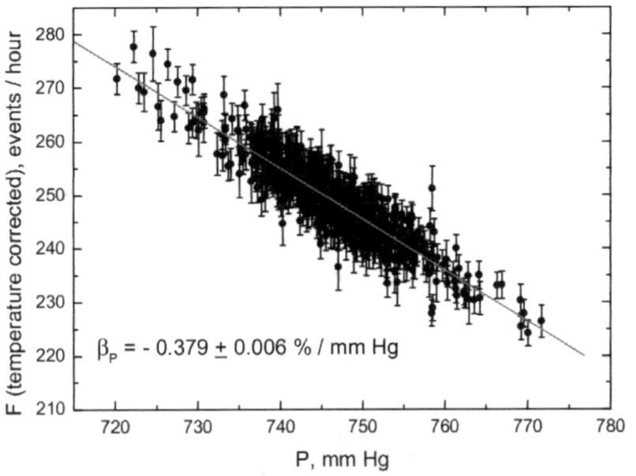

Figure 2. Correlations of the event rate (corrected for the temperature effect) with barometric pressure at the observation point.

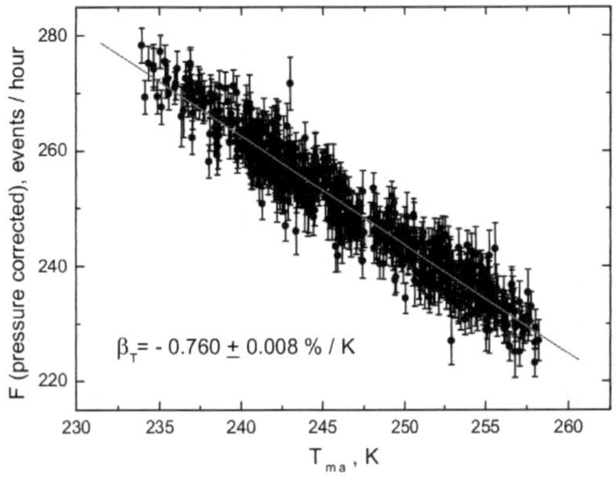

Figure 3. Correlations of the event rate (corrected for the barometric effect) with the mass average air temperature T_{ma}.

4. Correlations with the altitude of the fixed pressure levels

Among other parameters, information on the altitude of the isobar surface for a set of residual pressure levels P_L is available in the GDAS data [3]. We have analysed correlations of the muon bundle detection rate with the altitude of various pressure levels and fitted the observed dependence as:

$$F = F_0(1+\beta_H\Delta H);\ \Delta H = H - \langle H\rangle. \tag{2}$$

Here $\langle H\rangle$ is the average altitude of a certain pressure level, and β_H is the corresponding coefficient of a linear regression. The results for several values of the residual pressure are summarized in table 1. As seen from the table, the most close correlations (R_H = - 0.971, which evidences for an almost functional dependence) are obtained for the residual pressure of 500 mbar. Comparison of the data with a linear fit for this pressure level is given in figure 4.

Table 1. Correlations of the muon bundle intensity with the altitude for several fixed pressure levels. Indicated errors of regression coefficient are statistical.

P_L, mbar	$\langle H\rangle$, km a.s.l.	R_H	β_H, %/ 10 m
100	16.07	- 0.850	- 0.1324 ± 0.0014
300	9.08	- 0.956	- 0.2016 ± 0.0019
500	5.53	- 0.971	- 0.3067 ± 0.0028
700	2.98	- 0.947	- 0.4477 ± 0.0042
900	0.99	- 0.746	- 0.4940 ± 0.0059

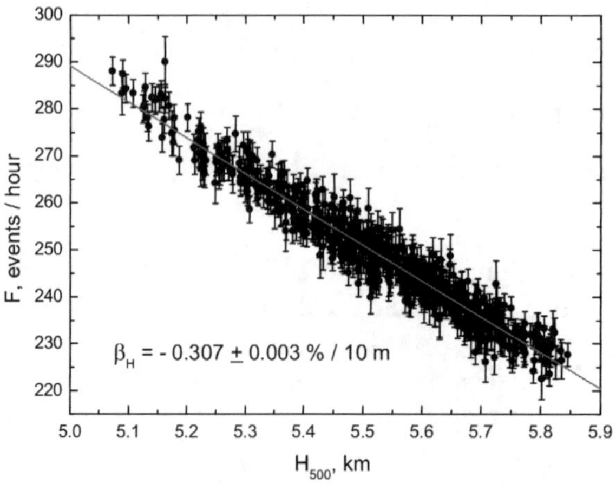

Figure 4. Correlations of the event rate with the altitude of the 500 mbar residual pressure level. Line: linear fit defined by equation (2).

5. Geometrical mechanism of muon bundle intensity variations

Taking into account that the energies of muons in the bundles are of the order of tens GeV, the observed values of barometric and temperature coefficients cannot be explained by a usual mechanism of absorption and decay of low energy particles, typical for single muons detected at the ground level [6]. An alternative approach to interpretation of the variations in the muon bundle intensity was suggested in [3]. Detection of the bundles with a detector, which is small compared to the lateral size of EAS in muon component (of the order of few hundred meters at the ground), corresponds to selection of events according to the local muon density *D*. Under some simplifying assumptions, the integral local muon density spectrum may be estimated as [4]:

$$F(\geq D) = N_0 D^{-\beta} \int [\rho(E_0, \mathbf{r})]^{\beta} dS \ , \qquad (3)$$

where $\rho(E_0, \mathbf{r})$ is the lateral distribution function (LDF) of muons in a plane orthogonal to the shower axis, E_0 is some effective primary energy, **r** is the point in the EAS cross section (corresponding to the detector location), and β is the spectrum slope.

Changes in the effective altitude of muon bundle formation *H* relative to the mean value $<H>$ lead to a scaling geometrical transformation of the muon LDF, and the intensity (3) will follow the law $F \sim (<H>/H)^{\alpha}$, where $\alpha = 2(\beta - 1)$. An illustration of the influence of the temperature of the atmosphere on the muon LDF is given by figure 5.

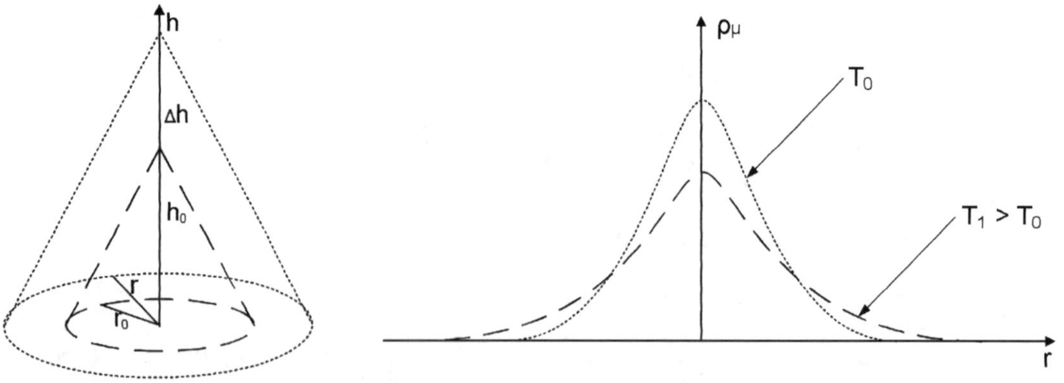

Figure 5: Illustration of the increase of the radial spread of muons (left) and of changes of the lateral distribution function of EAS muons (right) at the heating of the atmosphere [3].

For a fixed pressure, the geometrical altitude is proportional to the temperature, therefore we may expect the temperature dependence as $F \sim (<T>/T)^{\alpha}$. On the other hand, for a constant temperature, the altitude of a fixed pressure level P_L varies with the pressure at the ground P as $H \sim \ln(P/P_L)$, and assuming that the bundles are formed at altitudes corresponding to a residual pressure P_L, we obtain the expected dependence of the intensity on the ground pressure as $F \sim [\ln(<P>/P_L)/\ln(P/P_L)]^{\alpha}$. Taking into account the experimental estimate of the local muon density spectrum slope β = 1.96±0.02 derived from the muon multiplicity distribution [4], and assuming that most of the muon bundles are formed near the pressure level P_L = 500 mbar (see section 4), one can easily find the values of the variation parameters that follow from the considered geometrical mechanism (see the last column of table 2). A good agreement with the results of the experimental data analysis is obvious.

Table 2. Comparison of experimental values of the parameters describing muon bundle intensity variations with predictions of the geometrical model.

Parameter	Measured value	Geometrical model
β_P, % / mm Hg	- 0.382 ± 0.006	- 0.374 ± 0.008
β_T, % / K	- 0.760 ± 0.008	- 0.777 ± 0.016
β_{H500}, % / 10 m	- 0.307 ± 0.003	- 0.359 ± 0.007

6. Conclusion

A detailed study of the influence of atmospheric effects on the intensity of muon bundles detected at the ground level has been performed. A simple geometrical model of variation mechanism related with changes of the effective altitude of the formation of muon bundles in combination with the phenomenology of local muon density spectra explains well both qualitatively and quantitatively the observed values of barometric and temperature coefficients and correlations with the altitudes of fixed residual pressure levels.

Acknowledgments
This work was performed at the Unique Scientific Facility 'Experimental complex NEVOD' within the framework of the Center Fundamental Research and Particle Physics supported by MEPhI Academic Excellence Project (contract 02.a03.21.0005 of 27.08.2013).

References

[1] Dorman L I 1972 Meteorological effects in cosmic rays (Moscow: Nauka)
[2] Barbashina N S *et al* 2000 *Instrum. Experim. Techniques* **43** 743
[3] Tolkacheva N V *et al* 2011 *Bull. Russ. Acad. Sci. Physics* **75** 377
[4] Bogdanov A G *et al* 2010 *Phys. Atom. Nucl.* **73** 1852
[5] NOAA Air Resources Laboratory (ARL) 2015 http:/ready.arl.noaa.gov/gdas1.php
[6] Dmitrieva A N *et al* 2011 *Astropart. Phys.* **34** 401

Journal of Physics: Conference Series **675** (2016) 032035 doi:10.1088/1742-6596/675/3/032035

Investigation of the energy characteristics of EAS muon component with the NEVOD-DECOR setup

A G Bogdanov[1], N S Barbashina[1], L I Dushkin[1], V V Kindin[1], R P Kokoulin[1], K G Kompaniets[1], G Mannocchi[2], A A Petrukhin[1], E V Romanenkova[1], O Saavedra[3], G Trinchero[2], V A Khomyakov[1], S S Khokhlov[1], D V Chernov[1], V V Shutenko[1], E A Yurina[1] and I I Yashin[1]

[1] National Research Nuclear University MEPhI (Moscow Engineering Physics Institute), Kashirskoe highway 31, Moscow, 115409, Russia
[2] Istituto di Fisica dello Spazio Interplanetario – INAF, 10133 Torino, Italy
[3] Dipartimento di Fisica dell' Universita di Torino e INFN, 10125 Torino, Italy

E-mail: agbogdanov@mephi.ru

Abstract. Investigations of the energy characteristics of muon component with the increase of the primary cosmic rays energy can be a key to solving 'muon puzzle' – the problem of excess of EAS muons (observed in several experiments at high – ALEPH, DELPHI – and ultrahigh energies – DECOR, Pierre Auger Observatory) in comparison with the expected flux. The measurements results of the energy deposit of inclined muon bundles in water depending on the zenith angle and the local density of muons are presented. As a measure of the energy deposit, the total number of photoelectrons registered by PMTs of the Cherenkov water calorimeter NEVOD was used. The local density of muons, which gives an estimate of the energy of primary particles was obtained from the data of coordinate-tracking detector DECOR. The experimental data are compared with the results of calculations based on simulations of the muon component of EAS by means of the CORSIKA code.

1. Introduction

The studies of the muon component of extensive air showers (EAS) are of particular interest because its characteristics are closely associated with both the energy spectrum, mass composition of primary cosmic rays (on the one hand) and the properties of nucleus-nucleus interactions (on the other hand). If the increase of the EAS muons number with the rise of the primary particles energy is usually explained in terms of the heavier mass composition of primary cosmic rays, the appearance of very high energy muons should testify in favor of the inclusion of new physical processes (new state of matter). One of the possible approaches to search very high energy muons (and to validate the existing hadron interaction models) is the measurement of the muon component energy deposit in the detector material depending on the primary particle energy [1], since total muon energy loss almost linearly depends on muon energy: $dE_\mu/dX \sim a+bE_\mu$.

Such studies are currently being conducted in the NEVOD-DECOR experiment on registration of inclined muon bundles. Cherenkov water detector (CWD) NEVOD measures the energy deposit of muons, and coordinate-tracking detector DECOR measures the number of muons and their direction. It is important to note that the earlier developed new method of EAS investigations by local muon density spectra (LMDS) [2] allows to estimate the energy of the primary particles according to the DECOR data. The point is that, though the contribution the events with a fixed local density is given by the showers with different primary energies detected at different (random) distances from the shower axis, due to a fast decrease in the cosmic ray flux with the increase in energy the effective interval of primary particle energies is relatively narrow.

2. Experimental complex (setup)

Experimental complex NEVOD-DECOR includes Cherenkov water detector NEVOD [3,4] with volume 2000 m^3 and coordinate-tracking detector DECOR [5] with total area 70 m^2.

The detecting system of the NEVOD detector is formed by a spatial lattice, in the knots of which 91 quasi-spherical optical modules (QSMs) registering the Cherenkov radiation from any direction with practically the same efficiency are located. The distances between the modules are 2.5 m along the detector and 2 m across it and over the depth. Each QSM consists of 6 low-noise FEU-200 photomultipliers with flat 15 cm diameter photocathodes directed along rectangular coordinate axes. The use of 2-dynode signal readout from PMT provides a wide dynamic range of the measurements ($1 - 10^5$ photoelectrons) and allows to carry out calorimetric studies, in particular, to investigate the energy characteristics of EAS muon component.

The DECOR setup represents a modular multilayer system of plastic streamer tube chambers – 8 vertical assemblies (supermodules, SMs) arranged in the galleries of the building from three sides of the water calorimeter NEVOD. Each SM has an effective area 8.4 m^2 and consists of 8 planes of streamer tube chambers with resistive cathode coating, the planes are equipped with a two-coordinate external strip readout system. Spatial and angular accuracy of the muon track location in the supermodule is better than 1 cm and 1°, respectively.

Selection of events by the triggering system is based on coincidences between the signals from DECOR supermodules. An example of a muon bundle event registered in the NEVOD-DECOR setup and the main parts of the experimental complex are shown in figure 1.

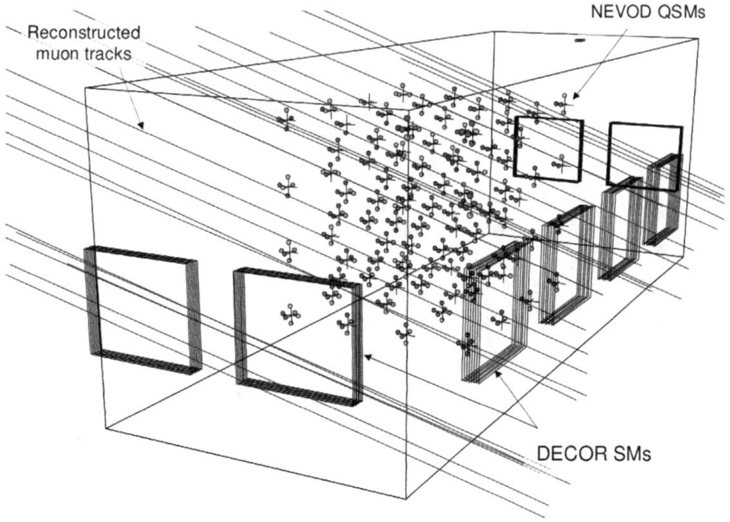

Figure 1. An example of the muon bundle event in the NEVOD-DECOR complex.

3. Experimental data

In the present work, data of two long series of measurements: from May 2012 to March 2013 and from July 2013 to April 2015 with inclined muon bundles detected by the experimental complex NEVOD-DECOR have been analyzed. Total live observation time is equal to 17439 h. In these data, 29335 events with muon multiplicity $m \geq 5$ and zenith angles $\theta \geq 55^{\circ}$ were found. Additionally, from the first part of the experimental material (for 3253 h) muon bundles arriving at lower zenith angles were selected ($40^{\circ} \leq \theta < 55^{\circ}$, 15084 events). Multi-muon events were selected in two 60°-wide sectors of azimuth angle, where most of DECOR SMs (six of eight) were screened with the CWD volume. Average threshold muon energy for such selection criteria is close to 2 GeV.

The local density of muons D in the events was estimated from the DECOR detector data. Taking into account the spectrum slope and Poisson fluctuations of the number of muons that hit the setup, it was calculated as: $D = (m - \beta)/S$, where m is muon multiplicity, S is the total area of six DECOR SMs for a given direction, and $\beta \approx 2.1$ is the integral slope of the LMDS in the considered muon density and zenith angle region [2].

As a measure of the muon bundle energy deposit in the CWD, the sum of the all PMTs signals (ΣN_{pe}, in photoelectrons) of the NEVOD detector was used. It was assumed that the number of Cherenkov photons, which are generated as a result of the muons passage through water calorimeter, is approximately proportional to the total muon energy loss in the detector material. In the first approximation, the total energy deposit in the CWD is nearly proportional to the local muon density, therefore in the further analysis we use the specific energy deposit $\Sigma N_{pe}/D$ (i.e., the CWD response normalized to the muon density in the event).

4. Results of data analysis

The measured dependence of the average specific energy deposit on the zenith angle is presented by the points in figure 2. At moderate zenith angles (40 – 55°), rapid decrease of the measured energy deposit can be explained as an atmospheric suppression of the residual contribution of electromagnetic and hadron EAS components. At large zenith angles ($\theta > 60^{\circ}$) practically only the muon component is detected on the ground level and the average specific energy deposit increases with the zenith angle, thus reflecting the increase of the mean muon energy in the bundles: from 150 GeV at $\theta = 60^{\circ}$ to 500 GeV at $\theta = 85^{\circ}$. The solid curves in the figure represent the results of calculations of the expected dependence of the specific energy deposit for muon bundles, obtained by simulating the EAS muon component by means of the CORSIKA code (v.7.40) [6] for primary protons and iron nuclei; the combination of SIBYLL-2.1 and FLUKA2011 hadron interaction models was used. On the basis of the modeled showers, two-dimensional lateral distribution functions (LDF) of muons were constructed, and the mean energies of muons in the bundles, taking into account the features of the LMDS technique [2], were found. The specific energy loss was calculated for the mean energy of muons in the bundles as the sum of ionization and radiation loss (bremsstrahlung, pair production and photonuclear interactions) by means of interpolation of tabulated data [7]. Finally, the calculated dependence was normalized to the experimental data on the number of photoelectrons at the zenith angle of 60°. As seen from the figure, the experimental dependence is in satisfactory agreement with the expectation.

In figure 3, the experimental values of the average specific CWD response for muon bundles detected at zenith angles $\theta \geq 55^{\circ}$ (where the residual contribution of EAS components is small) are presented as a function of the muon density. In fact (for a fixed range of zenith angles), this is a measurement of the dependence of the energy deposit on the energy of primary particles. Arrows in the bottom part of the figure indicate typical (mean logarithmic) energies of primary particles which give the contribution to the formation of the events with corresponding values of muon density. The curves in the figure represent the expected dependence of the energy deposit on the local muon density, calculated on the basis of simulations with CORSIKA for 60° zenith angle. The same calibration coefficient as in figure 2 was applied in order to normalize calculated results to the data. Results of simulations demonstrate a trend to a decrease of the mean muon energy in the bundles with the increase of the primary particle energy. On the contrary, the data exhibit a hint to some increase of the energy deposit at densities of muons greater than 1 particle/m^2 (effective primary energies of 10^{17} eV and above). The reasons for these deviations may be both of physical and methodical origin, therefore, a detailed analysis of experimental procedure and calculation results as well as the increase in the statistics will be continued.

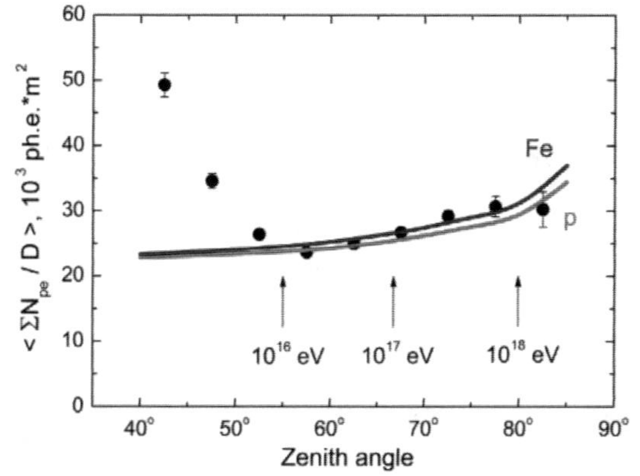

Figure 2. Dependence of the average specific energy deposit of muon bundles in the Cherenkov water detector on zenith angle. Points are measurement results, curves represent the expected dependencies obtained from CORSIKA-based simulations.

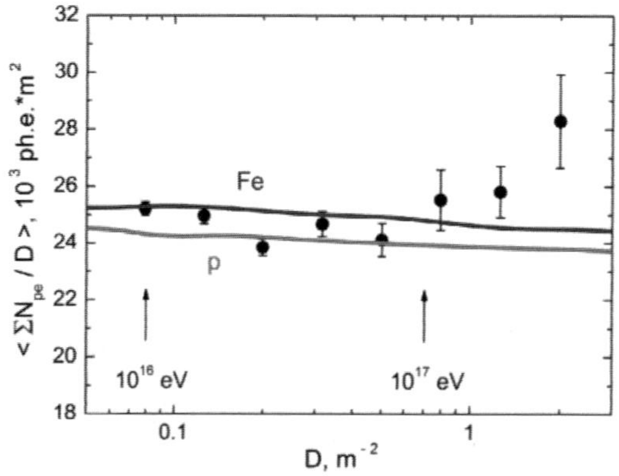

Figure 3. Dependence of the average specific energy deposit on the local muon density. Arrows indicate typical energies of primary particles. Points are measurement results, curves represent the expected dependencies obtained from CORSIKA-based simulations.

5. Conclusion

The experiment on the investigation of the energy characteristics of inclined muon bundles formed as a result of interactions of primary cosmic ray particles with energies $10^{16} - 10^{18}$ eV is being conducted at the Experimental Complex NEVOD. The results of the measurements of zenith-angular dependence of the average specific energy deposit in the Cherenkov water detector are in a reasonable agreement with CORSIKA-based simulations of the EAS muon component and confirm the increase of the mean energy of muons in the bundles at large zenith angles. An indication for an increase of the average specific energy deposit compared to the expectation at primary energies above 10^{17} eV has been found. However, further increase of experimental statistics and a careful analysis of possible systematic effects are necessary.

Acknowledgments

This work was performed at the Unique Scientific Facility ‘Experimental complex NEVOD’ within the framework of the Centre of Fundamental Research and Particle Physics supported by the MEPhI Academic Excellence Project (contract 02.a03.21.0005 of 27.08.2013).

References

[1] Petrukhin A A 2014 *Nucl. Instrum. Meth.* A **742** 228
[2] Bogdanov A G *et al* 2010 *Phys. Atom. Nucl.* **73** 1852
[3] Aynutdinov V M *et al* 1998 *Astrophys. Space Sci.* **258** 105
[4] Khokhlov S S *et al* 2011 *Astrophys. Space Sci. Trans.* **7** 271
[5] Barbashina N S *et al* 2000 *Instrum. Experim. Techniques* **43** 743
[6] Heck D *et al* 1998 *Forschungszentrum Karlsruhe Report FZKA 6019*
[7] Groom D E, Mokhov N V and Striganov S I 2001 *Atom. Data and Nucl. Data Tables* **78** 183

Journal of Physics: Conference Series **675** (2016) 032036 doi:10.1088/1742-6596/675/3/032036

Spatial distribution of Cherenkov light from cascade showers in water

V A Khomyakov, A G Bogdanov, V V Kindin, R P Kokoulin, A A Petrukhin, S S Khokhlov, V V Shutenko and I I Yashin
National Research Nuclear University MEPhI (Moscow Engineering Physics Institute), Kashirskoe Shosse 31, Moscow, 115409, Russia

E-mail: VAKhomyakov@mephi.ru

Abstract. The analysis of spatial distribution of Cherenkov light generated by cascade showers in Cherenkov water detector (CWD) NEVOD was performed. Showers generated by nearly horizontal muons were selected. Muons tracks were reconstructed with high precision using the coordinate tracking detector DECOR. For the first time, the dependence of Cherenkov light intensity on the depth of a shower at different distances from its axis was measured.

1. Introduction

Cherenkov water detectors are the most widely developing instruments for research of the characteristics of cosmic rays, for studies of muons and neutrinos with extremely high energies. The energy of these particles is estimated through cascade showers. The topical experimental task is to investigate the distribution of the Cherenkov light from cascade showers in water. Its decision will allow to refine approaches to selection of cascades and reconstruction of their parameters, and possibly also to improve the models of high energy cascade development which are implemented in scientific software packages. So far this problem has not been studied experimentally. The operating large-scale Cherenkov detectors such as Baikal, ANTARES, IceCube, have the distances between the measuring modules that exceed several times the longitudinal size of the shower in water. It limits the possibility to obtain the detailed picture of the Cherenkov light generated by the cascade.

In the present work the light field from the cascade showers was studied at the Unique Scientific Facility 'Experimental complex NEVOD' [1] located in National Research Nuclear University MEPhI. The Cherenkov detector, which is a part of the complex NEVOD, has the dense spatial lattice of detecting modules. This enables to obtain the distribution of Cherenkov light with the step of detalization about 0.5 m, which is determined by the size of the detecting modules.

The water reservoir of the CWD NEVOD has the size $9\times9\times26$ m^3. The detecting system is a spatial lattice with quasi-spherical modules (QSMs) at its nodes that register the Cherenkov radiation from relativistic charged particles (the total number of QSMs in the lattice is 91). Each QSM consists of six FEU-200 photomultipliers with flat photocathodes oriented along the orthogonal coordinate system axes.

The small step of the spatial lattice (2.5 m along the detector, 2 m across it, and 2 m in depth) and the wide dynamic range of the registered signals (1–10^5 photoelectrons for each photomultiplier [2]) allow cascade showers to be investigated by measuring the full cascade curve in individual events.

2. Selection of cascade showers

The events with cascades generated in the water reservoir of the CWD NEVOD by the nearly horizontal muons were studied. Tracks of muons were reconstructed using the coordinate-tracking detector DECOR [3]. Each track was registered by the pair of supermodules located at the opposite short sides of the reservoir. Such a way the registered muons have a narrow range of zenith (85–90 degrees) and azimuthal (±15 degrees from the longitudinal CWD axis) angles. The mean energy of such muons is about 100 GeV.

The problem of reconstructing the cascade curve for showers with a known axis position was solved in [4]. The reconstruction of experimental cascade curves is based on recalculating the responses from photomultipliers that detect the Cherenkov radiation from the shower particles (the direction of which is assumed to coincide with the track of the parent particle and with the axis of the shower) into the number of these particles. It is also assumed that the Cherenkov photons are emitted at the same angle to the axis.

The part of the axis of the shower that is inside the detecting system is divided into bins. The number of charged particles falling into each bin which can be ‘seen’ by PMTs of the detector is measured. The length of the bin is one radiation length that equals to 36.1 g/cm^2 for water. The energy of the shower is reconstructed by fitting the experimental dependence of number of radiating cascade particles on the depth with a function that is one-dimensional cascade curve approximation [5].

The events with showers with the reconstructed energies in the range from 100 GeV to 500 GeV were selected from the experimental series lasting 12 thousand hours of live time. The final sample size is 635 events.

3. Distribution of light from cascade showers in water

The dependence of QSM response on its distance from the axis of the shower and on the depth of the shower (measured along its axis) for the selected events is examined. The reconstructed point of the maximum of the shower t_{max} (the age of shower $s = 1$) is taken as the origin for the depth axis.

The square root of the sum of squared amplitudes of triggered PMTs is taken as the response of the module; as shown in [6], this value does not practically depend on the incident light direction for the quasi-spherical module with 6 PMTs. The responses of the modules in all events are normalized to reconstructed energy values:

$$B = \frac{\varepsilon_0}{\varepsilon}\sqrt{\sum A_i^2}\ ,$$

where ε_0 is the normalizing energy of 200 GeV (close to the average energy of showers in the sample);
ε is the reconstructed energy of the shower in event;
A_i is the amplitude of the i-th PMT in a QSM (in photoelectrons, ph.e.).

The obtained average spatial distribution of Cherenkov light is shown in figure 1.

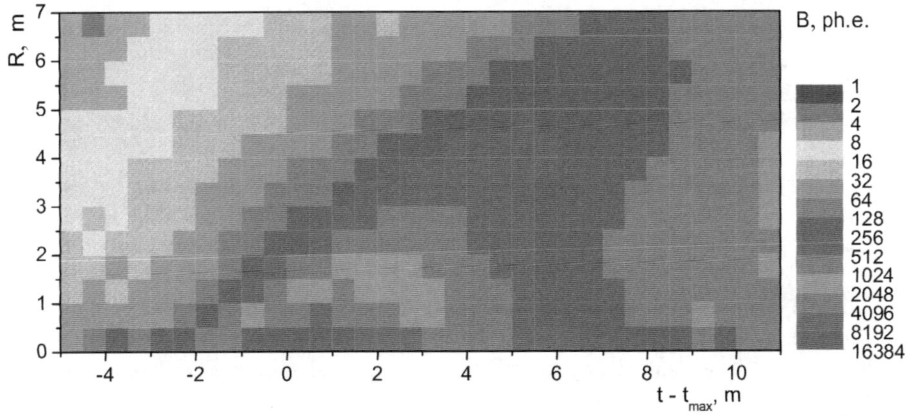

Figure 1. The dependence of Cherenkov light intensity on the depth of the shower (t-t_{max}) at different distances (R) from its axis.

The graph shows that the propagation of the light from the cascade has a good directivity; the light cone corresponding to the angle of Cherenkov radiation in water ($\theta_C \approx 41^\circ$) is clearly visible. However, it is also evident that the light spreads not strictly under the Cherenkov angle to the axis of shower, but almost in all directions. It is mainly caused by the deviation of the cascade particles from the direction of the shower axis due to their multiple scattering. It is also necessary to take into account the light attenuation in water for a more detailed analysis.

Since the parameters of the light attenuation are the same for cascades and for single muons, it is reasonable to consider the ratio of light intensity for cascades (B) to the intensity measured for events with single muons (B_μ). The spatial distribution of this value is presented in figure 2.

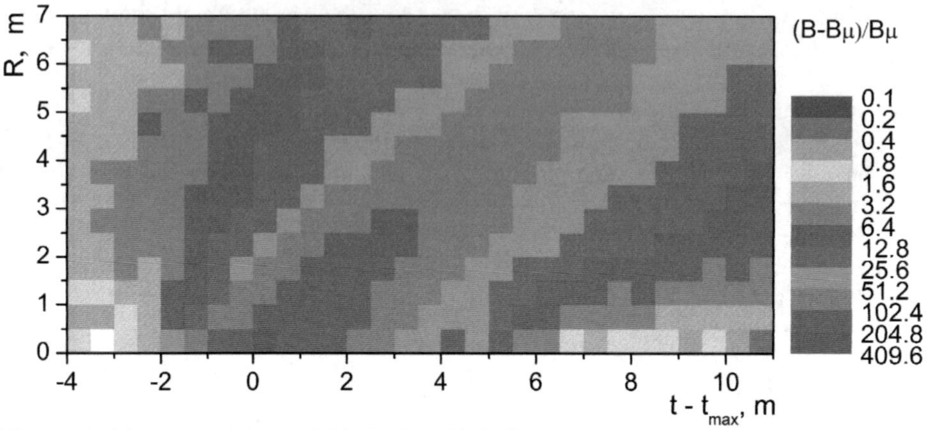

Figure 2. The dependence of Cherenkov light intensity from the showers normalized to the intensity from single muons.

For a further analysis of the influence of cascade particle scattering on the distribution of Cherenkov light, the dependence of $(B\text{-}B_\mu)/B_\mu$ value on the depth of shower for different distances was considered. Figure 3a shows longitudinal profiles of $(B\text{-}B_\mu)/B_\mu$ value for three intervals of distances from the axis of shower. Shifts and broadening of the profiles with the increasing distance are seen.

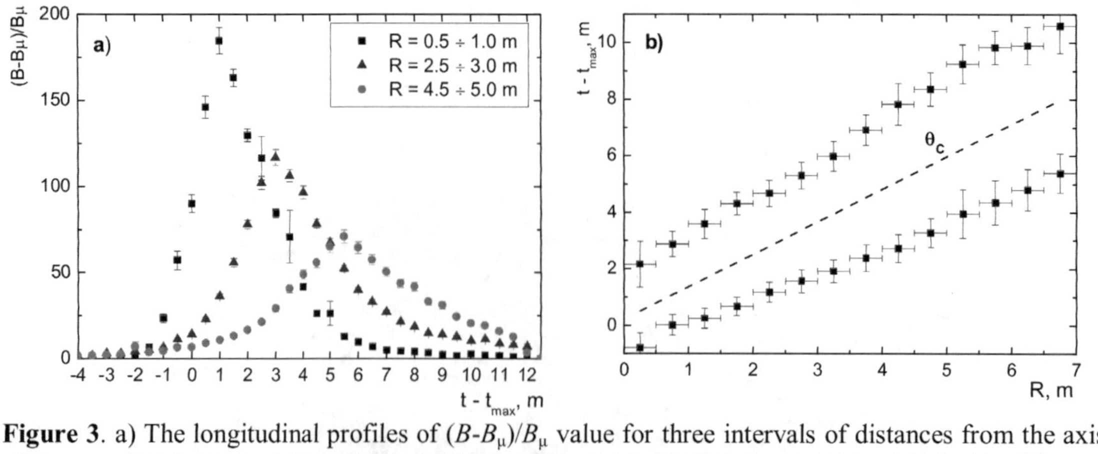

Figure 3. a) The longitudinal profiles of $(B\text{-}B_\mu)/B_\mu$ value for three intervals of distances from the axis of shower. b) The boundaries of longitudinal profiles at half of their maximum height for different distances from the axis. The dashed line shows the direction of Cherenkov radiation from particles moving strictly along the axis of the shower.

In details, these features are illustrated in plot 3b, where the points limiting the width of the profile at half maximum are shown as a function of the distance from the cascade axis. The plot shows that the width of the profile at half maximum at the distance of 7 m is almost doubled.

4. Conclusion

For the first time, the spatial distribution of Cherenkov light from cascade showers generated by muons in water has been experimentally measured at the CWD NEVOD with a dense lattice of quasi-spherical measuring modules. It has been found that the volume of highest light intensity has a compact shape with a radius of 1.5–2 m, which can be used for selection of cascades in the events with a large energy deposit in the Cherenkov detector.

Analysis of the results shows a good directivity of the light from the cascade at the angle close to the angle of Cherenkov radiation in water (approximately 41°). This confirms the possibility of using the one-dimensional cascade model for the reconstruction of cascade curve by the response of the detector NEVOD at small distances from the axis (comparable to the length of the cascade).

On the other hand, the broadening of the Cherenkov cone related with the multiple scattering of cascade particles at larger distances from the shower axis is observed. These results can be used for verification and improvement of models of development of electromagnetic cascades in water and also for development of criteria of selection of cascade showers in Cherenkov water detectors and for reconstruction of their parameters.

Acknowledgements

This work was performed at the Unique Scientific Facility 'Experimental complex NEVOD' within the framework of the Center Fundamental Research and Particle Physics supported by MEPhI Academic Excellence Project (contract 02.a03.21.0005 of 27.08.2013). We also acknowledge the support of the Russian Foundation for Basic Research (project no. 15-02-07763).

References

[1] Aynutdinov V M *et al* 1997 *Astrophys. Space Sci.* **258** 105
[2] Khokhlov S S *et al* 2011 *Astrophys. Space Sci. Trans.* **7** 271
[3] Barbashina N S *et al* 2000 *Instrum. Experim. Techniques* **43** 743
[4] Khokhlov S S *et al* 2015 *J. Phys.: Conf. Ser.* **632** 012038
[5] Hayakawa S 1969 *Cosmic Ray Physics* (NY : J.Wiley & Sons)
[6] Khomyakov V A *et al* 2015 *Physics Procedia* **74** 442

Journal of Physics: Conference Series **675** (2016) 032037 doi:10.1088/1742-6596/675/3/032037

The Taiga project

I I Yashin[9], I I Astapov[9], N S Barbashina[9], A G Bogdanov[9], V Boreyko[10], N M Budnev[2], M Büker[6], M Brückner[11], A Chiavassa[4], O B Chvalaev[2], A V Gafarov[2], N Gorbunov[10], V Grebenyuk[10], O A Gress[2], A Grinyuk[10], O G Grishin[1], A N Dyachok[2], S N Epimakhov[6], T V Eremin[1], D Horns[6], A L Ivanova[2], N N Kalmykov[1], N I Karpov[1], Y A Kazarina[1], V V Kindin[9], N V Kirichkov[1], S N Kiryuhin[1], R P Kokouli[9], K G Kompaniets[9], E N Konstantinov[2], A V Korobchenko[2], E E Korosteleva[1], V A Kozhin[1], M Kunnas[6], L A Kuzmichev[1], V V Lenok[2], B K Lubsandorzhiev[3], N B Lubsandorzhiev[1], R R Mirgazov[2], R Mirzoyan[5,2], R D Monkhoev[2], R Nachtigall[6], A L Pakhorukov[2], M I Panasyuk[1], L V Pankov[2], A A Perevalov[2], A A Petrukhin[9], V A Platonov[2], V A Poleschuk[2], M Popescu[12], E G Popova[1], A Porelli[8], V V Prosin[1], V S Ptuskin[7], G I Rubtsov[3], M Rueger[8], V G Rybov[2], V S Samoliga[2], P S Satunin[7], A Saunkin[2], V Yu Savinov[2], Yu A Semeney[2], B A Shaibonov (junior)[3], A A Silaev[1], A A Silaev (junior)[1], A V Skurikhin[1], M Slunecka[8], C Spiering[8], L G Sveshnikova[1], V A Tabolenko[2], A Tkachenko[10], L G Tkachev[10], M Tluczykont[6], A D Veslopopov[2], E P Veslopopova[2], D M Voronov[2], R Wischnewski[8], K O Yurin[9], A V Zagorodnikov[2], V N Zirakashvili[7] and V L Zurbanov[2]

[1]Skobeltsyn Institute of Nuclear Physics MSU, Moscow, Russia
[2]Institute of Applied Physics ISU, Irkutsk, Russia
[3]Institute for Nuclear Research of RAS, Moscow, Russia
[4]Dipartimento di Fisica dell' Universita di Torino and INFN, Torino, Italy
[5]Max-Planck-Institute for Physics, Munich, Germany
[6]Institut fur Experimentalphysik, University of Hamburg, Germany
[7]IZMIRAN, Troitsk, Moscow, Russia
[8]DESY, Zeuthen, Germany
[9]National Research Nuclear University MEPhI (Moscow Engineering Physics Institute), Kashirskoe highway 31, Moscow, 115409, Russia
[10]Joint Institute for Nuclear Research, Dubna, Russia
[11]Institute for Computer Science, Humboldt-University Berlin, Germany
[12]ISS, Bucharest, Romania

E-mail: iiyashin@mephi.ru

Abstract. The TAIGA project is aimed at solving the fundamental problems of gamma-ray astronomy and physics of ultrahigh energy cosmic rays with the help of the complex of detectors, located in the Tunka valley (Siberia, Russia). TAIGA includes a wide-angle large area Tunka-HiSCORE array, designed to detect gamma-rays of ultrahigh energies in the range 20 – 1000 TeV and charged cosmic rays with energies of 100 TeV – 100 PeV, large area muon detector to improve the rejection of background EAS protons and nuclei and a network of imaging atmospheric Cherenkov telescopes for gamma radiation detection. We discuss the goals and objectives of the complex features of each detector and the results obtained in the first stage of the HiSCORE installation.

1. Introduction

Over the last decade, the ground-based high energy gamma-ray astronomy became the most dynamically developing field in very high energy (VHE) astrophysics. Detection of gamma rays as messengers from powerful processes in Galactic and extragalactic sources has some advantages compared to detection of charged cosmic rays since they keep their emission direction from sources, and unlike neutrinos, they are easy to detect. Most of the knowledge about sources of high energy gamma rays has been obtained from arrays of Imaging Atmospheric Cherenkov Telescopes (IACT), the most famous HEGRA [1], H.E.S.S. [2], MAGIC [3] and VERITAS [4]. The method proposed by A M Hillas in 1985 [5] allows distinguishing, with high efficiency, Extensive Air Showers (EAS) produced by gamma rays from those initiated by charged cosmic rays. The sensitivity level of next generation of IACTs is optimized for the energy range from 100 GeV up to tens TeV [6]. Gamma radiation with energies above 100 GeV was detected from more than 150 sources of different types, but gamma rays with energies higher than 10 TeV are detected only from about 10 sources. The detection of sources of hard gamma rays with a spectrum up to several hundreds TeV would be direct evidence of acceleration of cosmic rays in these sources up to the knee in the all particle cosmic ray spectrum. Such galactic PeV accelerators, or Pevatrons [7], have not been detected up to now. An important argument in favor of the search of the gamma-ray with energies above 100 TeV beacame the detection of neutrino with PeV energies by IceCube [8]. However, to access the energy range from 10 TeV up to several 100 TeV, a detection area of the order of 10 km^2 or more is required. Alternatively to imaging gamma telescopes, the shower-front sampling technique allows to install large effective areas and also naturally provides large viewing angles of the instrument. Its operating principle is based on the sampling of the density and timing (arrival-time and spread) of the air shower-front with distributed arrays of detector stations [9]. The TAIGA (Tunka Advanced Instrument for Gamma-ray and cosmic ray Astrophysics) is a new hybrid detector system for ground-based gamma-ray astronomy for energies from a few TeV to several PeV, and for cosmic-ray studies from 100 TeV to several hundreds of PeV. TAIGA will also be located in the Tunka valley, where since 2009 the Tunka-133 Cherenkov EAS detector is in operation. The TAIGA observatory includes HiSCORE – an array of wide-angle integrating air Cherenkov stations, furthermore an array of IACTs, an array of particle detectors, both on the surface and underground, and the Tunka-133 EAS detector. Tunka-133 detects the Cherenkov radiation emitted in the atmosphere by charged EAS particles and currently consists of 175 optical detectors located on an area of 3 km^2 [10]. The individual stations consist of single large PMTs (diameter 20 cm or 37 cm).

2. Tunka-HiSCORE

Tunka-HiSCORE represents an array of wide field of view (FOV) (~0.6 sr) integrating air Cherenkov detector stations, placed at distances of 150 – 200 m from each other, covering an area of initially ~1 km^2, and ~100 km^2 at a later phase of the experiment. Each station consists of 4 neighbored PMTs of size either 20 cm or 25 cm in diameter. The PMTs have Winston-cone shape light-guides, which increase their light collection area by factor of four (see figure 1). DAQ and time synchronization system consists of 8-channel optical station board (OSB) for digitization of signals from anodes and dynodes of 4 PMTs of the optical station and synchronization boards (SB) placed in the DAQ center. All boards are designed on the basis of DRS-4 chip and FPGA Xilinx Spartan-6. OSBs and SB are connected via single-mode optical fibers. To synchronize all stations of the array to sub-ns precision a hybrid approach combining a custom-made synchronization technique (100 MHz clocks distributed over separate fibers from the array center) and the new White Rabbit Ethernet-based are used [11].

The accuracy of time synchronization is <1 ns. The dead time of OBS is smaller than 0.5 ms. Methods for gamma-ray and cosmic-ray data reconstruction for HiSCORE are similar to ones used for Tunka-133. The PMT outputs are summed up. The detector stations measure the light amplitudes and their arrival time differences over a distance of few hundred meters. This approach allows one to reconstruct with a high precision the arrival direction and the axis position of the EAS as well as the energy E_0 and shower maximum height X_{max}. The expected accuracy of the core location measurement is 5 – 6 m, that of the arrival direction about 0.1 degree, for primary energy, ~ 10% and for X_{max} ~ 15 – 20% [12]. The data can be used to reconstruct in detail the cosmic ray spectrum and its composition and will also improve gamma/hadron separation.

Figure 1. The Tunka-HiSCORE optical station. Additional window: the layout of one PMT segment.

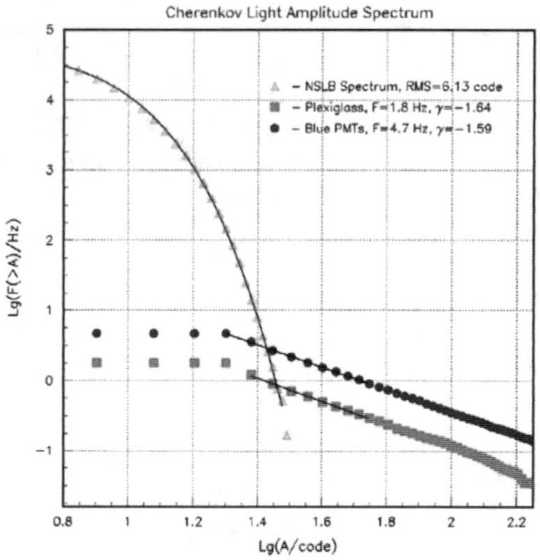

Figure 2. Integral amplitude spectra of the night sky and Cherenkov light flashes from EAS. Triangles: NSLB spectrum. Squares: spectrum from the OS covered with plexiglass. Circles: spectrum from the OS without plexiglass. Errors are within the data points [13].

The integral amplitude spectrum of Cherenkov light flashes from air showers obtained during first tests of HiSCORE optical station (OS) is shown in figure 2 [13]. The spectrum of the night sky light background (NSLB) fluctuation from summator is also given. The value of the sky background fluctuations in RMS units is 6.13, which corresponds to a voltage of 9.2 mV.

From 2013 till the end 2014, Tunka-HiSCORE operated as a 9-station prototype array. The array was arranged on a regular grid of 3×3 stations, with a side length of 300 m [12]. During first experimental series the test with a bright wide-angle LED light source ~200 m outside the array has been performed to check single station performances and full array event reconstruction quality [11]. For the calibration experiment, 45°-inclined mirrors were installed on top of the stations. The event reconstruction precision estimate is obtained by fitting the relative trigger times of the 9 stations: RMS < 0.5 ns. The EAS reconstruction with the HiSCORE-9 prototype with model fit gave RMS ~ 0.6 ns.

The configuration of the Tunka-HiSCORE stations as of 2014 is shown in figure 3 [12]. It comprises 28 detectors at 100 m spacing forming a super-cell structure. The total area of the setup is 0.25 km^2. To increase the observation time for the Crab Nebula, all optical boxes are tilted towards South by 25 degrees. During the winter season 2014-2015, the main task was to adjust the detector and ensure reliable operation.

3. Tunka-GRANDE

The timing array of Tunka-HiSCORE alone has a poor gamma-hadron separation at low energies (10–100 TeV), with a quality factor of the order of one below 100 TeV and only reaching 2 above few hundreds TeV. To improve the selection efficiency, an underground muon detector is foreseen to determine the type of a primary particle.

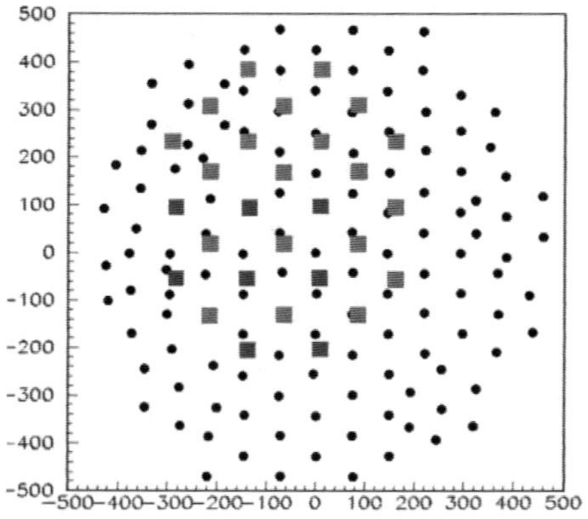

Figure 3. The layout of the Tunka-HiSCORE as of 2014-2015. Black circles: Tunka-133 EAS array; black squares: HiSCORE as a 9-station prototype array; red squares: engineering array (2014 – 2015).

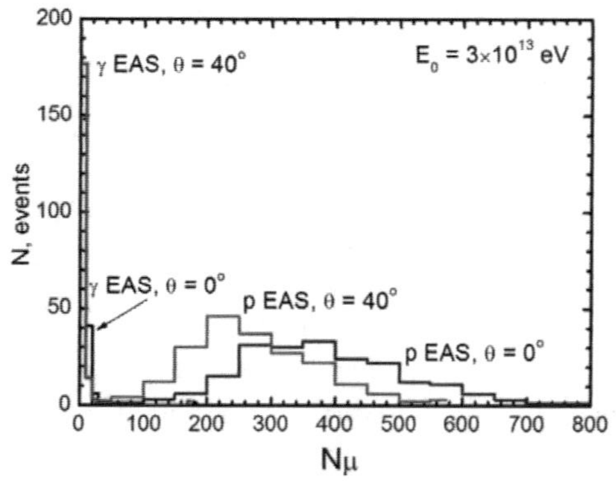

Figure 4. The distribution of EAS with primary energy $E_0 = 3\times10^{13}$ eV produced by protons and γ-quanta (2) over the number of muons for zenith angles θ = 0° and θ = 40° at the altitude of observation.

To formulate the main requirements for the muon detectors of the Tunka-HiSCORE facility, preliminary calculations were performed using the CORSIKA program, version 6.990 (FLUKA 2011.2+SIBYLL 2.1 models of interaction; EGS4 for electromagnetic interactions). The standard U.S. atmospheric model was used, and the Earth's magnetic field was disregarded [14]. The threshold energies of the secondary particles were 100 MeV for hadrons and muons, 2 MeV for electrons, and 1 MeV for γ-quanta. The altitude of observation was 150 m above sea level. EAS from primary protons and γ-quanta of cosmic rays with fixed energies $E_0 = 10^{13}$, 3×10^{13}, and 10^{14} eV were simulated for zenith angles θ = 0°, 20° and 40°. For each set of parameters, 200 showers were simulated. The simulated distributions over the number of muons in showers from primary protons and γ-quanta with $E_0 = 3\times10^{13}$ eV (zenith angles, 0° and 40°) are shown in figure 4.

As seen from the figure, EAS from protons and γ-quanta are separated reliably. The number of muons at this energy is hundreds of particles on average, so the area of the muon detector must be at least 0.2 – 0.3% that of the entire facility in order to organize multiple coincidences [14]. Therefore, the overall area of shielded muon detectors should be 2000 – 3000 m^2 of the total area of the Tunka-HiSCORE EAS detector (1 km^2). To suppress the background from γ-rays in ten times relative to the signal from muons, the muon detectors must be buried under a layer of soil 5 – 10 rad. units (1 – 2 m of bulk soil).

For the first stage of the muon array, scintillation detectors formerly operated as part of the EAS-TOP and the KASCADE-Grande arrays are used [12]. This so-called Tunka-GRANDE detector consists of 19 scintillation stations, each of them with a surface and an underground part. The stations are located at distances about 20 m from the centers of the Tunka-133 clusters. Each station includes 12 scintillation counters (80×80×4 cm^3 each) with a total area of 8 m^2 size and is equipped with underground muon detector with 8 identical counters in each with total area 5 m^2. The electronics of the scintillation systems are similar to those of the Tunka-133 array and are packed into containers along with ground-based particle counters.

4. The Tunka-IACT

The key advantage of the TAIGA gamma-ray observatory is the hybrid detection of EAS Cherenkov radiation by the wide-angle array Tunka-HiSCORE and the narrow-angle imaging detectors Tunka-IACT [15]. At present, the first prototype of the telescope is being deployed in the Tunka valley. Mirrors of the telescope will have an area of about 10 m^2, and a focal length of 4.75 m. The camera will consist of 547 photomultipliers with the total field of view (FOV) 9.72°, and the FOV of one pixel is 0.36°. Preliminary results of simulations show that common operation of the HiSCORE array and a net of IACTs can be a very effective, inexpensive and quick way to expand gamma-ray astronomy in the unexplored ultrahigh energy region, as well as to increase the sensitivity of HiSCORE at lower energies and to improve the reliability of separating gamma-initiated EAS against the background of charged cosmic rays. The timing detectors allow increasing the distance between the telescopes since they deliver part of the information which otherwise could be only obtained from IACT stereo operation. The basic idea is that the two setups will complement each other by combining time measurements of showers with shower imaging [15]. It gives the possibility to increase the spacing of our telescopes to about 600 m (see figure 5). The showers are seen by only one telescope at a time, but the reconstruction of these EAS from the HiSCORE timing array provides an information about the core position and incident angle of the shower front; this enables an increase of telescope spacing without loss of gamma-hadron separation power and therefore makes it possible to cover large areas with fewer telescopes [16].

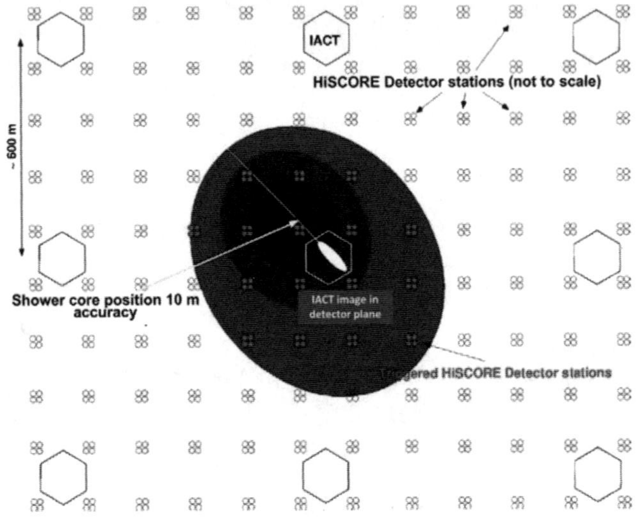

Figure 5. Principle for the combination of IACT and shower front sampling array observation [16]

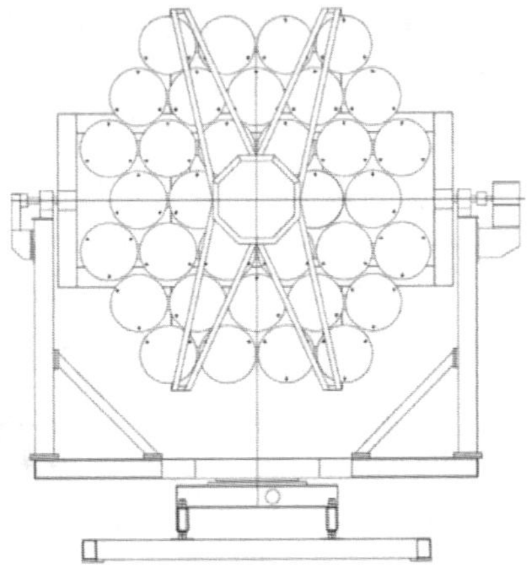

Figure 6. Mechanical design of the first TAIGA imaging telescope.

Each telescope will be composed of a mosaic, 34-segment mirror in Davis-Cotton design, with a 60 cm diameter of individual mirrors. The full diameter of the mirror is 4.3 m, its focal length 4.75 m (see figure 6). At present, the camera with its 547 hexagonal-shaped pixels on 14 hexagonal rings around a central pixel (figure 7a) is being assembled.

As a photodetector, we use the XP1911 PMT with a window of 15 mm diameter. Each PMT is equipped with a Winston cone with an entrance size of 30 mm and an exit spot of 15 mm diameter. The sensitive area of the camera has a diameter of about 81 cm. The entire array of pixels is divided into clusters of 28 PMTs in each. Each cluster includes an electronic board MAROC, the basic element of which is a 64-channel chip ASIC MAROC3 [17] (see figure 7b).

In the figure 7b, the layout of a cluster block with 28 photomultipliers is shown. Such a block is the basic camera unit. Structurally, the electronics of the base block is divided into several boards: four divider boards of the 7 PMT groups (7HEX), four PMT power supply plates, four daughter plates with the DAC for the high-voltage source control, and the ADC for the measurement of the PMT current and signal processing carried out by a PMT DAQ board based on MAROC3.

5. Conclusions

The TAIGA Collaboration made a significant progress towards creating a new observatory for gamma-ray astronomy of ultrahigh energies in Tunka valley. The observatory is targeted to the search of gamma-ray sources with PeV energies and the study of the flux of UHE primary cosmic rays on the basis of a new approach. With the hybrid detector concept, combining the air Cherenkov timing-array technique with the air Cherenkov imaging technique, TAIGA will deploy a unique detector, optimizing the sensitivity to gamma-rays above 10 TeV.

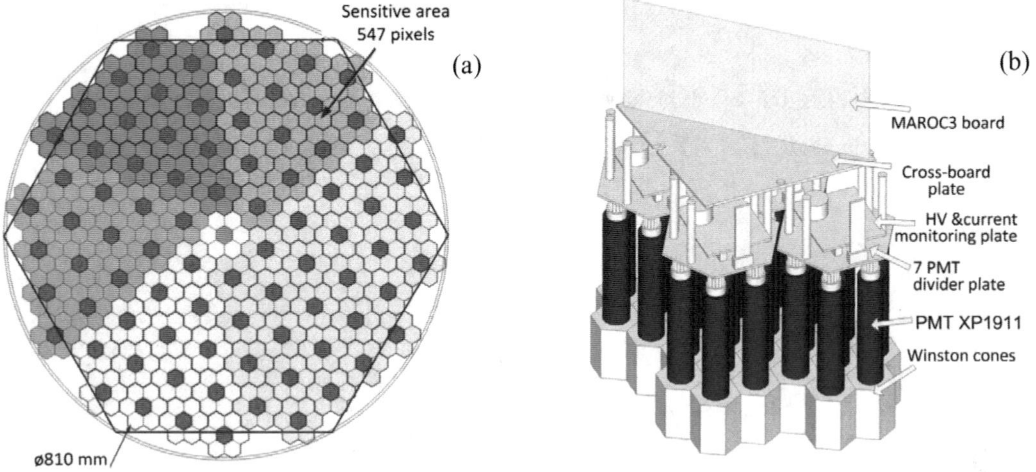

Figure 7. a) The matrix of pixels of the Tunka-IACT camera. b) The basic PMT block of TAIGA-IACT camera.

Acknowledgments
The paper was performed at the Centre of Fundamental Research and Particle Physics supported by the MEPhI Academic Excellence Project (contract 02.a03.21.0005 of 27.08.2013). We also acknowledge the support of the Russian Federation Ministry of Education and Science (agreements N 14.B25.31.0010, task No 3.889.2014/K), the Russian Foundation for Basic research (13-02-00214, 13-02-12095, 15-02-10005, 15-02-05769), the Helmholtz association (grant HRJRG-303), and the Deutsche Forschungsgemeinschaft (grant TL 51-3).

References
[1] Aharonian F *et al.* 2006 *Astronomy & Astrophysics* **454** 775
[2] Aharonian F *et al.* 2011 *Astropart. Phys.* **34** 738
[3] Albert J *et al.* 2006 *Astrophys. J.* **639** 761–65
[4] Acciari V *et al.* 2011 *ApJ Letters* **730** L20
[5] Hillas M 1985 *Proc. 19th Cosmic Ray Conf.(La Jolla, USA)* vol. 3 p 445
[6] Actis M *et al.* 2011 *Experimental Astronomy* 121
[7] Gabici S 2007 *ApJ Letters* **665** L131
[8] Aartsen M G *et al.* (IceCube Collaboration) 2013 *Phys. Rev. Lett.* **111** 021103
[9] Tluczykont M 2011 *Advances in Space Research* **48** 1935
[10] Berezhnev S F 2012 *Inst. Meth. in Phys. Res.* A **692** 98
[11] Porelli A 2015 *J. Phys.: Conf. Ser.* **632** 012041
[12] Budnev N 2015 *J. Phys.: Conf. Ser.* **632** 012034
[13] Gress O 2013 *Nucl. Inst. Meth. in Phys. Res.* A **732** 290
[14] Yashin I I 2015 *Bull. Russ. Acad. Sci. Phys.* 7**9** 392
[15] Tluczykont M 2015 *J. Phys.: Conf. Ser.* **632** 012042
[16] Kunnas M 2015 *J. Phys.: Conf. Ser.* **632** 012040
[17] The MAROC page of the *Weeroc SAS* company site: http://www.weeroc.com/products/maroc

Characteristics of the Forbush decrease of 22 June 2015 measured by means of the muon hodoscope URAGAN

N S Barbashina, N V Ampilogov, I I Astapov, V V Borog, A N Dmitrieva, A A Petrukhin, O A Sitko, V V Shutenko and E I Yakovleva
National Research Nuclear University MEPhI (Moscow Engineering Physics Institute), Kashirskoe highway 31, Moscow, 115409, Russia

E-mail: NSBarbashina@mephi.ru

Abstract. Results of the studies of cosmic ray muon flux variations during the powerful Forbush effect registered by the muon hodoscope URAGAN on June 22, 2015 are presented. From the muon flux angular distribution, the dependence of the intensity decrease amplitude on the primary particle energy in the region above 10 GeV has been obtained. The changes of this dependence at different phases of the Forbush effect development have been studied. Based on the analysis of spatial and angular variations of the muon flux, the values of the local anisotropy vector parameters, as well as unique muon snapshots (muonographies) have been obtained. The characteristics of the heliospheric and magnetospheric disturbances during the considered event have been analyzed.

1. Introduction

Muon hodoscope URAGAN [1] has been operated as a part of the Unique Scientific Facility "Experimental complex NEVOD" [2] since 2006. One of the tasks being solved using the hodoscope is a muon diagnostics of the near-Earth space. One of the brightest examples of the solar activity influence on the cosmic rays are drastic decreases of the secondary particle flux detected at the Earth's surface by various detectors – the so-called Forbush decreases (FD). To study the FD detected in the hodoscopic mode, a special technique [3, 4] has been developed. This technique provides a wide range of characteristics of the cosmic ray muon flux variations during the FD: integral, energy, spatial-angular and temporal. As a result of the analysis of these characteristics, we obtain the FD amplitudes, the power indices of the energy dependence of the cosmic ray intensity decrease amplitude in the energy range of 10 – 30 GeV, their temporal dependences at different phases of the FD development, the muon snapshots (muonographies), the horizontal projections of the muon flux relative anisotropy vector during the FD, and the correlations between its projections to the North-South and East-West directions. Such studies are carried out for all FD registered by the URAGAN.

In this work, the above-mentioned characteristics of one of the most powerful FD (June 22, 2015) registered by the URAGAN are presented.

2. Information about the event of June 22, 2015

The decrease in the URAGAN counting rate on June 22, 2015 is the second in the FD amplitude registered over the period from 2006 to 2015. Figure 1 shows the cosmic ray flux variations measured by the muon hodoscope URAGAN and Moscow neutron monitor (MOSC) in the period from June 15 to July 1, 2015.

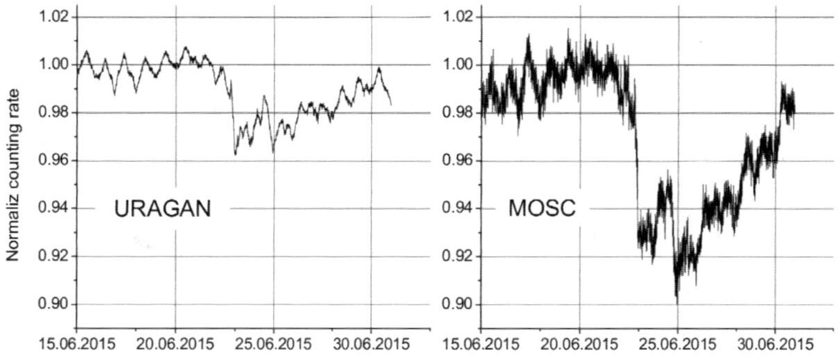

Figure 1. Variations of the cosmic ray flux according to the URAGAN (moun) and MOSC (neutron) data during the period from June 15 to July 1, 2015.

It is clearly seen that the decrease of the counting rate in the muon flux is substantially less than in the neutron component. It is due to the fact that muons are sensitive to higher primary energies than neutrons.

Analysis of the Sun has shown that the flare activity preceding the FD was high. Most likely, the variations in the URAGAN counting rate were caused by the flares of M-class on 21.06.2015 at 01:02 (M2.0) and at 02:06 (M2.6) and by the subsequent CMEs (halo II type) at 02:36 with velocity of up to 1560 km/s and the CME at 02:48 with velocity of up to 1950 km/s [5].

The near-Earth space was also in disturbed conditions during the analyzed period. Figure 2 shows the variations of the interplanetary magnetic field (B_t, B_z), the solar wind velocity (V_{SW}) and the muon counting rate.

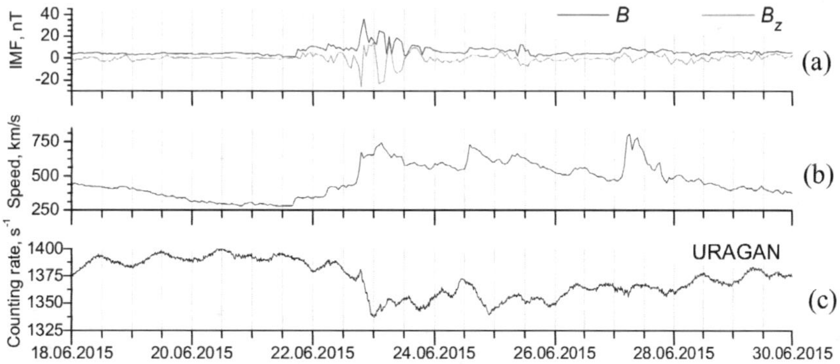

Figure 2. a) Induction of the interplanetary magnetic field B and Z component; b) solar wind velocity; c) URAGAN counting rate.

According to ACE [6] it can be seen that a strong disturbance of the interplanetary magnetic field begins simultaneously with the increasing of the solar wind velocity at 16.00 on June 22. The IMF parameters B_t and B_z change to 30 and -30 nT, correspondingly. The solar wind velocity changes from 400 to 700 km/s. Geomagnetic disturbances were registered on the Earth: as SSC on June 21, 2015 at 17:00, by *Dst*-index – on June 23 at 05:00, by *Kp*-index – on June 22 at 18:00.

3. Results of the URAGAN data analysis during the FD of June 22, 2015

In the muon flux, as also in the neutron flux according to the MOSC data, the decrease beginning was registered by the muon hodoscope on 22.06.2015 at 19:10. The amplitude of the URAGAN integral counting rate decrease (determined according to the technique described in [3]) amounted to 2.41% ± 0.16%. The amplitude of the MOSC counting rate decrease was 5.60% ± 0.20%. The ratio of the URAGAN and MOSC decrease amplitudes $A_{FD}^{URG} \approx 0.43 A_{FD}^{MOSC}$.

Study of the event energy characteristics [4, 7] is based on the analysis of dependences of the FD amplitudes (A_{FD}) in five zenith-angular intervals (0°–17°, 17°–26°, 26°–34°, 34°–44° and over 44°) on the mean logarithmic energy (E_{ln}) of primary protons which contribute differently to the muon hodoscope counting rate variations (14.7 GeV, 15.5 GeV, 17.5 GeV, 19.7 GeV and 25.3 GeV, correspondingly). Such dependences are approximated by function E^{α}. Registration of the muon flux in hodoscopic mode enables to estimate the amplitude spectrum index α with a small time increment at different phases of the event development: decrease, minimum and recovery.

Figure 3a shows the dependence of A_{FD} on E_{ln} and the estimated value of α. And figure 3b shows the temporal variations of the amplitude spectrum index.

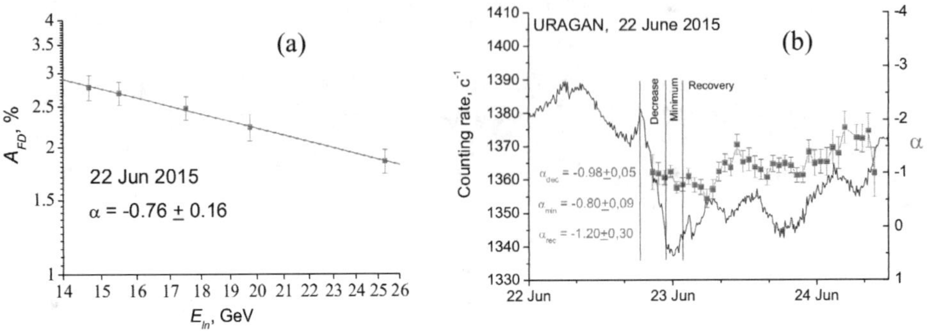

Figure 3. FD on June 22, 2015: a) the dependence of A_{FD} on E_{ln}; b) temporal variations of the amplitude spectrum index.

The earlier analysis of all FDs registered by the URAGAN has shown that, on average, the power index (α) of the energy dependence is close to –1. As can be seen from the figure 3a, the amplitude spectrum for the FD of June 22, 2015 is 'harder' ($\alpha = -0.76 \pm 0.16$) and the amplitudes for zenith angles above 44 degrees (for the mean logarithmic energy of primary particles exceeding 25 GeV) are about 2%. Temporal variations of the index α (figure 3b) for the investigated event show that at the phase of the decrease its average value is -0.98 ± 0.05, at the minimum $\alpha = -0.80 \pm 0.09$ and during the recovery period $\alpha = -1.20 \pm 0.30$. From the obtained data it can be seen that at the counting rate minimum the amplitude spectrum is harder than at the phase of the decrease, but during the recovery phase a softer spectrum is observed.

As a result of the analysis of spatial and angular characteristics of the muon flux registered in the hodoscopic mode, the muon snapshots (muonographies) of the FD of June 22 have been obtained. These muonographies illustrate the two-dimensional dynamics of muon flux variations during the analyzed FD. Figure 4 shows such muonographies for the phase of the decrease.

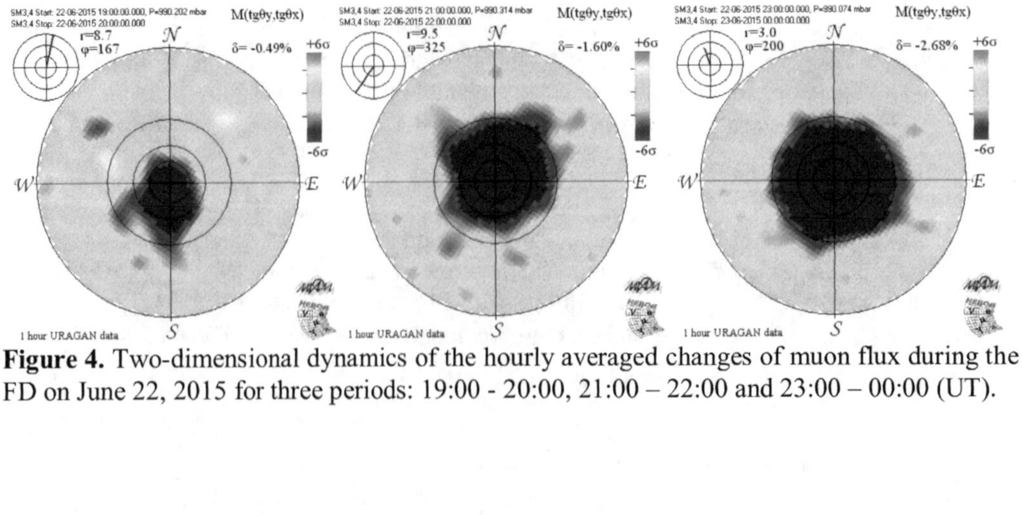

Figure 4. Two-dimensional dynamics of the hourly averaged changes of muon flux during the FD on June 22, 2015 for three periods: 19:00 - 20:00, 21:00 – 22:00 and 23:00 – 00:00 (UT).

For the quantitative description of the muonographies shots, correlations between the projections of the muon flux relative anisotropy vector on the North-South and East-West directions have been plotted [8] (figure 5).

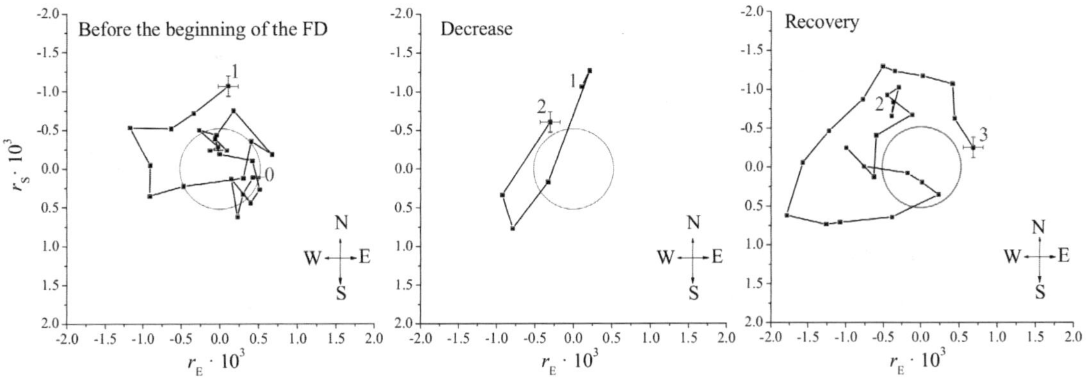

Figure 5. Correlations between r_S and r_E for the FD on June 22, 2015 from 19:00 to 23:00 UT at three phases of the event development (typical errors of the r_S and r_E are shown in points 1, 2 and 3).

The circles in the figures 5 correspond to 4σ variations of the r_S and r_E equal to 0.52×10^{-3} in a quiet period. An exceedance of the circle limits indicates an appearance of a strong anisotropy. We see that this event had a strong anisotropy at all phases. At the first phase, a large anisotropy is observed from the north-west direction, in the second phase there is a moving from north to south and vice versa, and at the phase of the recovery again in the north-west direction.

4. Conclusion

The FD of 22 June 2015 was investigated by means of the muon hodoscope technique, and the following results were obtained: amplitude of this FD was about 2.4%; dependence of the amplitude on the energy of primary cosmic rays gives a 'hard' spectrum ($\alpha = -0.76$), and the FD amplitude for the mean logarithmic energy of primary particles more than 25 GeV is about 2%; this FD had a strong anisotropy at all phases: before FD and at the recovery a large anisotropy was observed from the north-west direction, in the phase of the decrease it was moving from north to south and vice versa.

Acknowledgments

This work was performed at the Unique Scientific Facility 'Experimental complex NEVOD' within the framework of the Center Fundamental Research and Particle Physics supported by MEPhI Academic Excellence Project (contract 02.a03.21.0005 of 27.08.2013).

References

[1] Barbashina N S *et al* 2008 *Instrum. Experim. Techn.* **51** 180
[2] Saavedra O 2013 *J. Phys.: Conf. Ser.* **409** 012009
[3] Barbashina N S *et al* 2009 *Bull. Rus. Acad. Sci.: Physics* **73** 343
[4] Barbashina N S *et al* 2011 *Bull. Rus. Acad. Sci.: Physics* **75** 814
[5] CACTus catalogue: http://sidc.be/cactus
[6] OMNI Database: ftp://nssdcftp.gsfc.nasa.gov/spacecraft_data/omni
[7] Barbashina N S *et al* 2013 *J. Phys.: Conf. Ser.* **409** 012189
[8] Barbashina N S *et al* 2015 *J. Phys.: Conf. Ser.* **632** 012049

Journal of Physics: Conference Series **675** (2016) 032039 doi:10.1088/1742-6596/675/3/032039

The registration system of the coordinate-tracking setup on the drift chambers

E A Zadeba[1], A A Borisov[1, 2], R M Fakhrutdinov[1, 2], R P Kokoulin[1], K G Kompaniets[1], A S Kozhin[1, 2], A S Ovechkin[1], A A Petrukhin[1], V V Shutenko[1] and I I Yashin[1]

[1] National Research Nuclear University MEPhI (Moscow Engineering Physics Institute), Kashirskoe highway 31, Moscow 115409, Russia

[2] RF SSC Institute of High Energy Physics, Protvino 14228, Russian Federation

E-mail: EAZadeba@mephi.ru

Abstract. The large-scale coordinate-tracking detector for registration of near-horizontal muon flux generated by ultra-high energy cosmic rays is being developed in MEPhI. Detector is based on the multiwire drift chambers from the neutrino experiment at the IHEP U-70 accelerator. Their key advantages are a large effective area (1.85 m^2), good coordinate and angular resolution with a small number of measuring channels. Detector will be operated as a part of the experimental complex NEVOD, in particular, its registration system allows joint operation with Cherenkov water detector (CWD) and coordinate detector DECOR. Coordinate tracking unit on the drift chambers (CTUDC) is mounted on the opposite sides of CWD. It consists of two coordinate planes containing 8 drift chambers and represents a prototype of a full-size setup. Registration system of the CTUDC is based on the E-MISS electronics developed in IHEP, its principle of operation is presented.

1. Introduction

The aim of the large-scale coordinate-tracking detector is solving the problem of the excess of muon bundles that increases with the energy of the primary cosmic rays [1, 2], what can be caused by both cosmo- or nuclear-physical reasons [3]. The only characteristics that responds differently to changes in the composition of cosmic rays and the inclusion of new physical processes, but has not been investigated to the present time, is the energy of the muon component of extensive air showers [3]. Such studies are performed at the experimental complex NEVOD-DECOR [4]; however, the coordinate detector DECOR does not cover the entire aperture of the water Cherenkov detector and does not exclude the possibility of passing of several muons between the individual supermodules of the detector. Besides, the size of its cells limits the possibility of separating two or more particles at small distances (less than 3 cm). The new coordinate-tracking detector [5] based on drift chambers will increase the coverage of the side aperture of the Cherenkov water detector (CWD) NEVOD and significantly improve the resolution of close tracks.

The detector is based on the multiwire drift chambers from the neutrino experiment at the IHEP U70 accelerator [6]. Overall sizes of the chambers are 4000×508×112 mm^3. Sensitive area is 1.85 m^2 that is 91% of the chamber area. The chamber represents an aluminum box with ends limited by plexiglas plugs, to which wires, gas inlets and high voltage connectors are mounted. The drift chamber is filled with a gas mixture of 94% Ar and 6% CO_2. A uniform electric field is formed by field-forming wires with a pitch of 5 mm, the potential on them varies linearly from 0 to 12 kV. The drift chamber has four signal wires alternately shifted by ± 0.75 mm in the drift direction relative to the centre of the chamber; such configuration allows to reconstruct the projection of the track on the plane perpendicular to the wires. Thus, for the reconstruction of tracks in space at least two non-parallel drift chambers are required. Spatial accuracy of the chamber is 1 mm, the angular resolution is about 0.03 rad, two close tracks can be separated at a distance of 3 mm.

The signals from the wires are processed by an on-board amplifier-shaper that forms 75 ns pulses (depending on the signal length) in the LVDS levels.

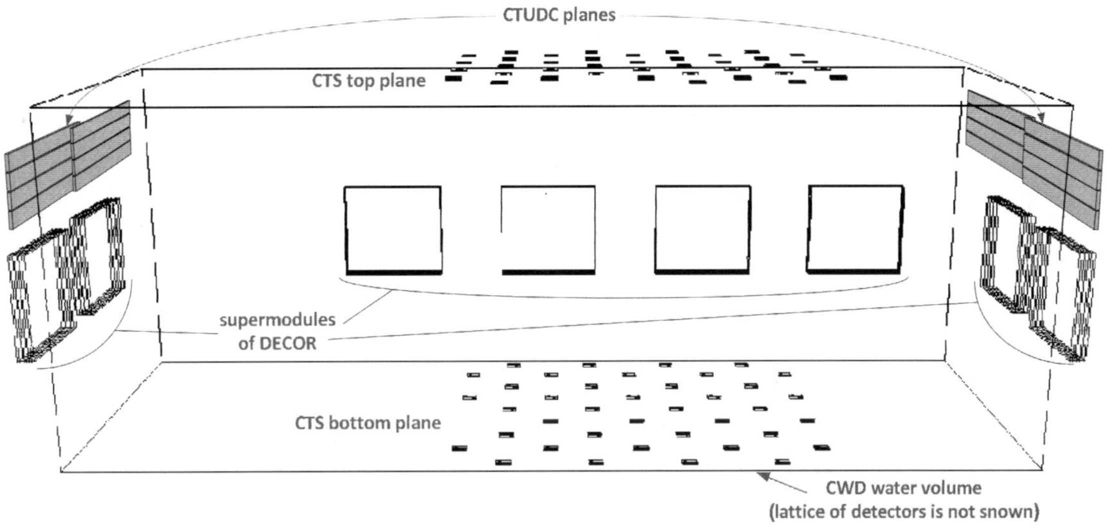

Figure 1. Arrangement of the CTUDC planes relative to the DECOR and CWD.

2. CTUDC

Coordinate-tracking unit based on drift chambers (CTUDC) consists of two vertical coordinate planes installed at the different sides of the CWD in the short galleries of the third floor of the NEVOD building, one floor above the DECOR supermodules (figure 1). Such location allows the registration of near horizontal tracks as by CTUDC (triggered by CWD) so by joint operation with DECOR, that will significantly increase the range of muon track zenith angles from 85–95° to 80–100°.

Each plane consists of 8 drift chambers installed in two rows, overlapped by 30 cm to exclude dead zones in the chamber ends; it causes a 4° angle between the planes and CWD wall. The effective area of the plane is 14.8 m^2, area of two DECOR supermodules located a floor below is 17.5 m^2, so the total area of coordinate detectors for registration of near-horizontal particles coming along the CWD almost doubles. One of the main goals of this setup is to examine conditions of the joint operation with CWD and DECOR.

3. Registration system

CTUDC is designed for joint operation with the triggering system of the experimental complex NEVOD (NEVOD TS) that binds registration systems of CWD, DECOR, calibration telescope system (CTS) and array of neutron detectors PRISMA. Triggering system has a rather fast data handling: the period between the passage of a particle through the working volume of CWD and the trigger formation is about 500 ns. On the other hand, the maximum drift time of the electrons in the drift chamber is 6 μs, so registration system and DAQ of CTUDC cannot be directly integrated into the NEVOD TS and should be implemented separately with possibility of off-line interconnection between the NEVOD and CTUDC data.

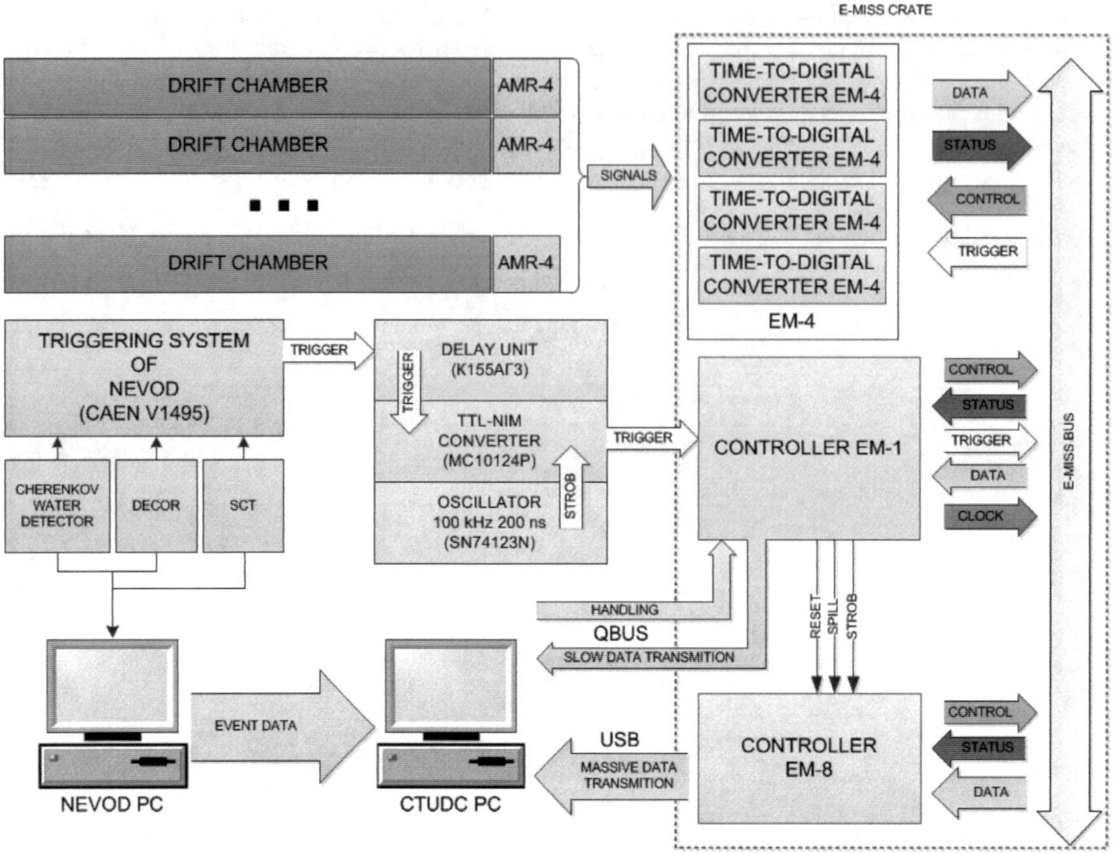

Figure 2. Block-diagram of the CTUDC registration system.

Registration system of CTUDC is based on the E-MISS electronics (a block diagram is shown in figure 2) developed for accelerator experiments in IHEP. It includes a 128-channel time-to-digital converter EM-4, main crate controller EM-1 and standalone controller EM-8.

Primary processing of signals from sense wires of drift chamber is carried out by the 4-channel shaper-amplifier AMR-4 (mounted on the end face of the DC) with a single adjustable threshold for all channels. It forms 75 ns LVDS pulses that pass through several commutation blocks to a TDC EM-4. There are 64 channels in 16 chambers, it lefts a half of TDC channels unused so after the expansion of the drift chamber setup to the TREK [6] CTUDC channels can be included into its full registration system.

The EM-4 represents four 32-channel independent TDCs based on a FPGA ACEX with 32×1024 words memory in each. TDC has 8 μs match window with a 10-bit width time measurement, that corresponds to 7.8125 ns LSB. Such discretization is acceptable taking into account that spatial accuracy of the drift chamber is about ~1 mm, that corresponds to 24 ns drift time.

Trigger signal of the NEVOD TS acts as the time stamp for TDC, it is produced with implementation of any trigger condition of CWD and DECOR (configured separately for each detector). All delays in the triggering system are finely tuned, so a time between the passage of the first particle in the event and the all-system trigger does not depend on the type of completed trigger condition and is approximately 225 ns, it takes 100 ns for a cable delays and a hardware jitters and 125 ns for the triggering system. All signals from the drift chambers should be received by the TDC before the arrival of the NEVOD trigger, so the delay of the trigger is necessary. For this purpose, a special block for VME crate was developed, it combines a galvanic isolation by rapid optocoupler, TTL to

NIM converter (MC10124P) and 8 μs delay (SN74123N). Reception of the trigger signal is carried out by EM-1 controller that distributes it via MISS bus to EM-4 TDCs.

After the generation of the trigger signal in the NEVOD TS, the system transmits all information about inner triggers to the NEVOD central computer that starts (via Ethernet) to gather all amplitude information from the CWD and a configuration of triggered DECOR streamer tubes. Simultaneously the NEVOD central computer sends a network packet to CTUDC that contains main information about the event: the number and time of the event, types of CWD and DECOR inner trigger signals. The network packet is started with a 4-byte sequence at which the CTUDC central computer determines the type of the received packet, decodes it and starts to collect information from the E-MISS system.

Readout from the E-MISS system can be performed via Q-BUS from EM-1 or via USB from EM-8 that significantly differ. The EM-1 is designed for a full bus control; it is the only module that can be handled from the computer. It has three trigger inputs and three controllable NIM outputs, that are used for EM-8 control. The memory of module is two 1K 16-bit blocks (one for address and one for data) that is insufficient to read full memory of the EM-4 (4×32×1024 bits) so readout of a TDCs is performed for each event. This is acceptable for low rate measurements, but it is impossible to measure DC noise in such mode.

The EM-4 does not have continuous mode of operation for noise measurements, so an oscillator based on two monostable multivibrators is used for production of a trigger with 100 kHz rate. Each trigger opens 8 μs match window every 10 μs. EM-1 controller is not able to handle such high rates unlike standalone controller EM-8 that has memory for 16M of 32-bit words and can accumulate data. On the other hand, EM-8 cannot be controlled directly via USB and is not able to perform readout and data transmission independently, so three NIM outputs of EM-1 are used to control this module. Another feature of joint operation of two controllers is the possibility to reduce dead time (during transmission of data to computer) due to their alternate readout from the TDC. Full-sized setup TREK will consist of 264 drift chambers and it will produce a high load for registration system (including 10 EM-4 modules), especially for noise measurements, so an upgrade for the EM-8 is in progress.

Central computer of CTUDC gathers information from the E-MISS, and after the primary data processing the software forms the event. Later these events will be off-line joined with the events saved at the NEVOD central computer by means of time stamps and event numbers.

4. Conclusion
The unique coordinate-tracking detector CTUDC based on drift chambers from IHEP neutrino experiment is developed in MEPhI. Detector has a complex registration system that allows its joint operation with other systems of the experimental complex NEVOD. Registration system is based on the E-MISS electronics from IHEP that fulfills the requirements of the experiment. Currently the detector is on the calibration stage. It has been already cross-calibrated with the coordinate detector DECOR and operates jointly with the Cherenkov water detector.

Acknowledgments
This work was performed at the Unique Scientific Facility "Experimental complex NEVOD" within the framework of the Center Fundamental Research and Particle Physics supported by MEPhI Academic Excellence Project (contract 02.a03.21.0005 of 27.08.2013). The work is also supported by the grant of the Russian Foundation for Basic Research (project 13-02-12207-ofi-m-2013).

This work is supported by the grant of the Russian Foundation for Basic Research (project 13-02-12207-ofi-m-2013) and was performed at the Unique Scientific Facility "Experimental complex NEVOD" within the framework of the Centre of Fundamental Research and Particle Physics supported by the MEPhI Academic Excellence Program (contract № 02.a03.21.0005).

References
[1] Barbashina N S *et al.*, 2007 *Nucl. Phys. B (Proc. Suppl.)* **165** 317

[2] Bogdanov A G *et al.* 2010 *Physics of Atomic Nuclei* **73** 852
[3] Petrukhin A A 2011 *Nucl. Phys. B (Proc. Suppl.)* **212-213** 235
[4] Kovylyaeva E A *et al.* 2014 *Bull. Lebedev Phys. Inst.* **41** 218
[5] Zadeba E A *et al.* 2014 *JINST* **9** C08018
[6] Bozhko N I *et al.* 1986 *Nucl. Instrum. Meth.* A **243** 388

Analysis of powerful heliospheric non-geoeffective event of the 28 April, 2015 in muon flux

I I Astapov[1], N S Barbashina[1], I S Veselovsky[1,2,3], N V Osetrova[1], A A Petrukhin[1] and V V Shutenko[1]

[1] National Research Nuclear University MEPhI (Moscow Engineering Physics Institute), Kashirskoe highway 31, Moscow, 115409, Russia

[2] Skobeltsyn Institute of Nuclear Physics, Lomonosov Moscow State University (MSU), 119991, Moscow, Russia

[3] Space Research Institute (IKI), Russian Academy of Sciences, 117997, Moscow, Russia

E-mail: IIAstapov@mephi.ru

Abstract. The coronal mass ejection (CME) that occurred on April 28, 2015 is analyzed. The passage of the ejection did not cause geoeffective disturbances in the near-Earth space. At the same time, the CME had a significant impact on the flux of cosmic rays registered on the Earth's surface by the muon hodoscope URAGAN.

1. Introduction

Cosmic rays represent a powerful instrument for the study of a variety of dynamic processes in the interplanetary space, including coronal mass ejections. The disturbances caused by the solar activity have a direct impact on the fluxes of primary and, therefore, secondary cosmic rays. In recent years, a new direction of investigation of the processes in the interplanetary space – muon diagnostics [1] – becomes more and more popular. Muon diagnostics is based on the analysis of the muon flux registered by the wide-aperture coordinate detectors – muon hodoscopes – which allow real-time reconstruction of tracks of all muons registered from the celestial hemisphere [2, 3].

The largest ejection of the 24th solar cycle occurred on July 23, 2012. This CME had a high velocity that reached ~ 3000 km/s. At the same time, its front was directed opposite from the Earth and was registered only by the STEREO-A satellite [4]. Despite the non-geoeffective character of the CME in July 2012, it had a strong effect on the cosmic ray flux detected on the Earth's surface by the muon hodoscope URAGAN [5].

The paper presents an analysis of another powerful coronal mass ejection that occurred at the end of April 2015 at the phase of the solar activity decrease. The maximum rate of emission reached ~ 2000 km/s. At the same time, the geomagnetic conditions in the near-Earth space are characterized as quiet. So this event, as the CME of July 23, 2012, was of non-geoeffective character.

In this paper, the results of the analysis of data of the muon hodoscope URAGAN – the part of the Unique Scientific Facility 'Experimental complex NEVOD' (MEPhI, Moscow) – which allows to explore not only the integral counting rate of the flux of registered muons, but also its local anisotropy, are provided.

2. Muon hodoscope URAGAN and experimental data

Muon hodoscope URAGAN [2] (55.7°N, 37.7°E, 173 m above sea level) is the coordinate detector that allows to investigate the variations of the muon flux angular distribution on the Earth's surface. Muons retain the direction of the primary particles motion, which allows to study primary cosmic rays in the interplanetary space. URAGAN consists of four independent supermodules (SM). Each SM is assembled of eight layers of gas-discharge chambers (streamer tubes) equipped with two-coordinate system of external readout strips which provides a high spatial and angular accuracy of muon track detection (correspondingly, 1 cm and 1°) in a wide range of zenith (0°-80°) and azimuthal (0°-360°) angles in the real time mode. Data are accumulated by minute intervals and contain matrices of two-dimensional angular distribution of the muon flux.

To study the URAGAN integral counting rate, 10-min data summed over all modules and corrected for the barometric and temperature effects are used [6]. For the study of two-dimensional variations of muon flux registered by the muon hodoscope URAGAN, a local anisotropy vector $\vec{A}$ [7] which is the sum of the unit vectors of particle tracks normalized by the total number of tracks is used. Local anisotropy vector $\vec{A}$ indicates the average arrival direction of muons which is close to the vertical. To study its deviations from the mean value $<\vec{A}>$, the relative anisotropy vector which represents the difference between the current vector and the average anisotropy vector calculated over a long period of time is used: $\vec{r} = \vec{A} - <\vec{A}>$. For the analysis of muon flux variations, the horizontal projection of the relative anisotropy vector r_h which characterizes the 'side impact' on the muon flux angular distribution is used: $r_h = \sqrt{r_x^2 + r_y^2}$.

Its value depends on the state of the heliosphere and can be 10^{-4}–10^{-3}. To estimate the statistical significance of the observed deviations, the experimental data of 2009 (minimum of solar activity) were used. For 2009 the mean value is 2.45×10^{-4} and rms-deviation is 1.28×10^{-4}. Taking into account that distribution of r_h is close to the Rayleigh's one, for values more than $r_h/\sigma_{rh} > 5$ the probability of random deviation is less than 1%.

3. Coronal mass ejections on 28 April, 2012

The considered coronal mass ejection of the halo-III class occurred on April 28, 2015 at 17:48. The CME average speed was ~ 1700 km/s [8]. Figure 1 shows the snapshot of the CME according to the data of the LASCO C2 coronagraph which is installed on the SOHO spacecraft [9]. It is worth noting that the satellites STEREO-A and -B have come together at one point on the back side of the Sun and information from them is absent in the database CACTus SECCHI-A and -B for the 2015 [10].

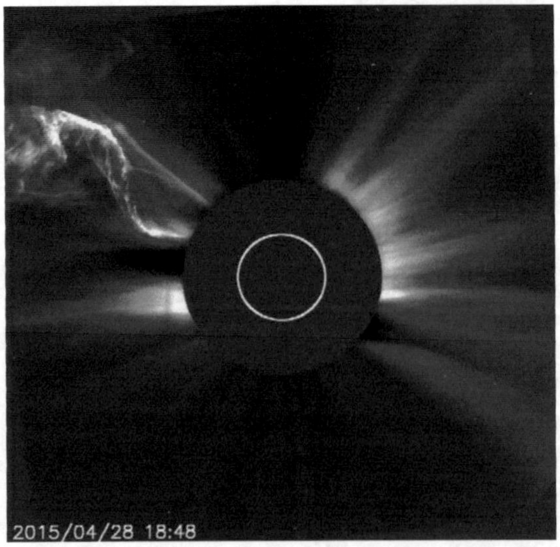

Figure 1. Snapshots of the coronal mass ejection on April 28, 2015 according to the data of the LASCO C2 coronagraph.

Figure 2 shows data on the state of the interplanetary magnetic field, solar wind and the Earth's magnetosphere (Kp- and Dst- indices) according to the OMNI database [11] for the period from April 20 to May 10, 2015. The vertical solid line indicates the beginning of the CME, the dashed lines show the times of passage by the CME of the distances of 1 and 2 AU. From the graphs it is seen that in this period the disturbances in the interplanetary space and in the Earth's magnetosphere were not observed.

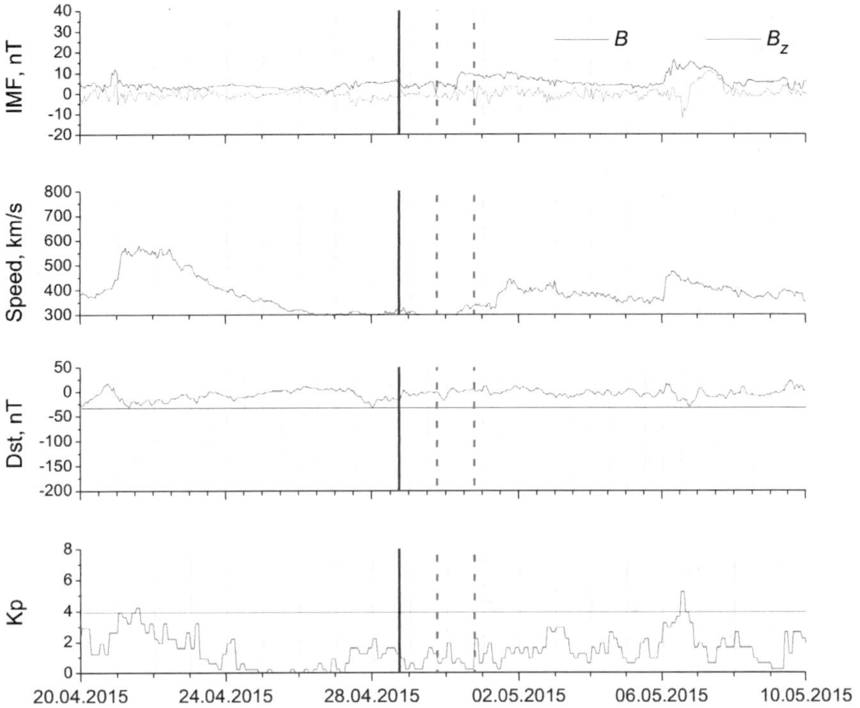

Figure 2. From the top downwards: induction of the interplanetary magnetic field *B* and *Z* component, solar wind velocity, *Dst* and *Kp* indices of geomagnetic disturbance.

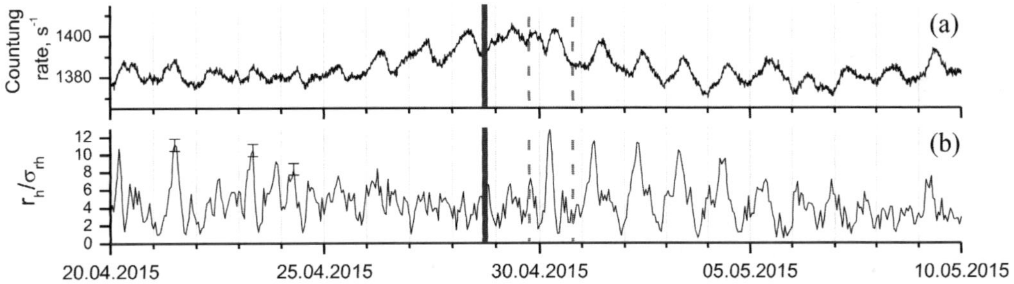

Figure 3. a) Counting rate of the muon hodoscope URAGAN; b) horizontal projection of the relative local anisotropy vector of muon flux (typical errors for several peak values of the parameter r_h/σ_{rh} are shown).

Figure 3 shows the muon hodoscope URAGAN data in the same period (from April 20 to May 10, 2015). The graph on the top presents the integral counting rate and the graph on the bottom shows the projection of the relative muon flux local anisotropy vector. The vertical lines mark the periods of time similar to figure 2.

From the upper graph in figure 3 it can be concluded that during the passage of the ejection through the heliosphere in the period from April 28 to May 2, 2015 no abrupt changes (for example, Forbush decreases) in the counting rate behavior were observed. However, it can be seen that since April 26 the increase of the muon hodoscope URAGAN integral counting rate is seen during the quiet state of the interplanetary magnetic field and from April 30 to May 5 its decrease is observed.

Analysis of the local anisotropy parameter provides additional information about the variations of the cosmic ray flux. In the period from April 25 to April 28 the maximal (peak) values of r_h/σ_{rh} were ~ 6.4. On April 30 the drastic increase (in 2 times) of the r_h/σ_{rh} parameter is took place. The peak value of r_h/σ_{rh} was 12.8. After that, during the following four days a gradual decrease was observed. Table 1 shows the peak values of the projections of the relative local anisotropy vector r_h/σ_{rh}. It is seen that there was their gradual decrease by ~ 10% approximately every 25 hours. Thus, at each revolution of the Earth around its axis the CME impact on the muon flux registered by the hodoscope is detected.

Table 1. Peak values of r_h/σ_{rh} parameter.

Time of the peak r_h/σ_{rh}	Maximum value r_h/σ_{rh}
30.04.2015 06:00	12.8
01.05.2015 07:00	11.4
02.05.2015 08:00	11.2
03.05.2015 08:00	10.2
04.05.2015 10:00	9.5
05.05.2015 09:00	7.1

4. Conclusion

Registration of muons in the hodoscopic mode enables to investigate with a single setup not only the intensity of the cosmic ray flux, but their local anisotropy sensitive to the changes in the heliosphere. Such non-geoeffective ejections have little effect on the muon flux counting rate, in contrast to the local anisotropy parameters, that provide more detailed information about variations of the cosmic ray muon flux sensitive to the disturbances in the heliosphere caused by the passage of the CMEs.

Acknowledgments

This work was performed at the Unique Scientific Facility “Experimental complex NEVOD” within the framework of the Center of Fundamental Research and Particle Physics supported by the MEPhI Academic Excellence Project (contract 02.a03.21.0005 of 27.08.2013).

References

[1] Yashin I I *et al* 2013 *J. Phys.: Conf. Ser.* **409** 012192
[2] Barbashina N S *et al* 2008 *Instrum. Experim. Techn.* **51** 180
[3] Yashin I I *et al* 2015 *Adv. Space Res.* doi:10.1016/j.asr.2015.06.003
[4] Russell C T *et al* 2013 *Astrophys. J.* **770** 38
[5] Astapov I I *et al* 2015 *Adv. Space Res.* doi:10.1016/j.asr.2015.03.002
[6] Dmitrieva A N *et al* 2013 *J. Phys. Conf. Ser.* **409** 012130
[7] Astapov I I *et al* 2015 *Adv. Space Res.* doi:10.1016/j.asr.2015.05.039
[8] The CACTus LASCO CME list: *http://sidc.oma.be/cactus/catalog.php*
[9] SOHO data: *http://sohowww.nascom.nasa.gov/*
[10] The CACTus COR2 CME list: *http://secchi.nrl.navy.mil/cactus/*
[11] OMNI Database, 2015: *http://omniweb.gsfc.nasa.gov/*

Detector station and registering system of the NEVOD-EAS array cluster

I A Shulzhenko[1], N V Ampilogov[1], I I Astapov[1], A Chiavassa[2,3], S S Khokhlov[1], K G Kompaniets[1], V Yu Kutovoy[1], O I Likiy[1] and I I Yashin[1]

[1] National Research Nuclear University MEPhI (Moscow Engineering Physics Institute), Kashirskoe highway 31, Moscow, 115409, Russia

[2] Dipartimento di Fisica dell' Università degli Studi di Torino, 10125 Torino, Italy

[3] Sezione di Torino dell' Istituto Nazionale di Fisica Nucleare – INFN, 10125 Torino, Italy

[4] Istituto di Fisica dello Spazio Interplanetario – INAF, 10133 Torino, Italy

E-mail: IAShulzhenko@mephi.ru

Abstract. The design features of the detector stations of the cluster type shower array NEVOD-EAS which is now under construction on the basis of the Unique Scientific Facility 'Experimental complex NEVOD', as well as the operation principle of the cluster registering system are discussed.

1. Introduction

At the present time, the cluster-type detector NEVOD-EAS [1–4] for the registration of extensive air showers (EAS) in the energy range of 10^{15}–10^{17} eV corresponding to the 'knee' region of the primary cosmic ray energy spectrum is under construction on the basis of the Unique Scientific Facility 'Experimental complex NEVOD' (MEPhI, Moscow).

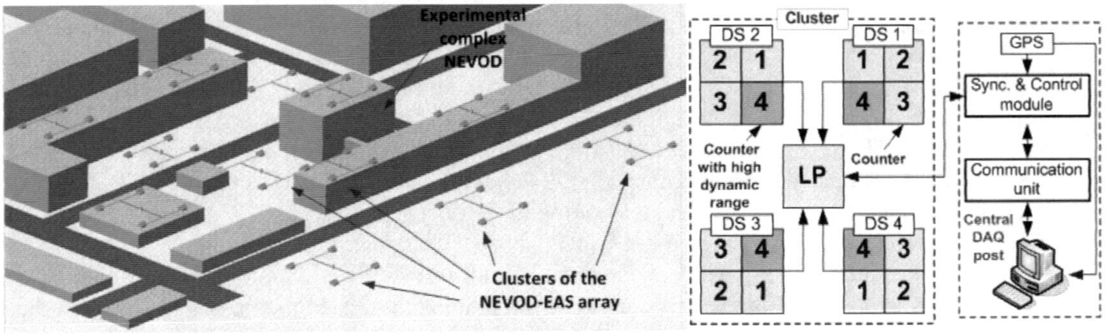

a) b)

Figure 1. The layout (a) and the structure (b) of the NEVOD-EAS array.

The new shower array will enable to define the size, the axis position and the arrival direction of the EAS, in order to significantly increase the accuracy of energy estimation of the primary particles responsible for generation of muon bundles registered by the NEVOD [5,6] and DECOR [7] detectors.

The NEVOD-EAS detector with a total area of ~ 2×10^4 m^2 will include 12 clusters of 16 scintillation counters in each for the detection of the EAS electron-photon component. Counters of the cluster are grouped into four detector stations (DS) arranged in one plane in the vertices of a rectangle with the characteristic side lengths of about ~ 20 m. Information from the DS of the cluster is fed to the registering system located inside the Local Post (LP). Registering system performs digitization of the DS analog signals using the FADC, selection of events according to the specified triggering conditions and transfers data to the Central DAQ Post. Distances between the clusters are ~ 50 m. The structure and the layout of the NEVOD-EAS array are shown in figure 1.

2. Detector station of the NEVOD-EAS array

Detector station consists of four scintillation counters of the extensive air shower electron-photon component (figure 2a).

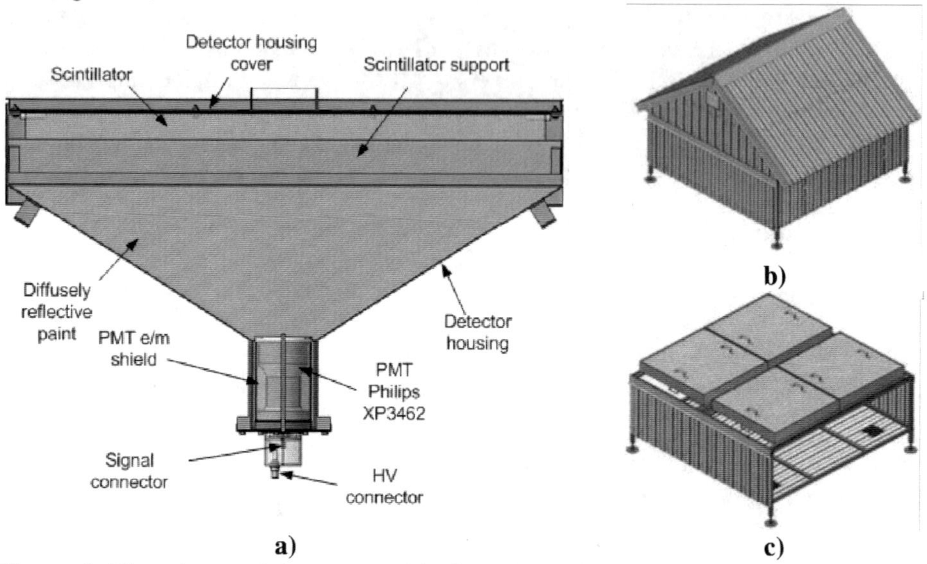

Figure 2. The scheme of the counter (a), the DS housing (b) and the layout of counters inside the DS housing (c).

Scintillation counters were previously used in the KASCADE-Grande [8] and EAS-Top [9] experiments. Main elements of the counter construction are: NE102A scintillator with the dimensions of $800\times800\times40$ mm^3 and Philips XP3462 PMT installed inside the pyramidal stainless steel housing. The PMT is placed under the scintillator at a distance of 30 cm. The inner surface of the counter pyramid is painted with a special diffuse-reflective coating to improve the collection of light from the flashes in the scintillator caused by the passage of particles through its sensitive volume. The construction of the NEVOD-EAS counters allows installation of one additional Philips XP3462 PMT. For the protection against the environmental influences, the DS counters are enclosed inside the special external housings (figure 2b). The area of the detector station is about 2.5 m^2.

Three of the four DS counters are equipped with one ('standard') photomultiplier Philips XP3462 which is used for the time and density measurements during the EAS registration. The gain of the standard PMTs is chosen so that the response of scintillation counters to the passage of near-vertical single muon was practically the same (~ 13 pC).

The fourth DS counter is equipped with two (a 'standard' and an 'additional') photomultipliers. The additional PMT ensures the wide range of linearity of measuring signals at the high particle densities (~ 10000 particles/m^2) and has a gain 90 times smaller than the standard one. The additional PMT response to the passage of the near-vertical muon is ~ 0.15 pC.

The techniques and results of studies of the PMT and scintillator characteristics, as well as the techniques of the NEVOD-EAS counter assembly and adjustment are described in [3] and [4].

The photomultipliers are powered with special block based on the TRACO PHV 12-2.0K2500N DC/DC-converter. This block is powered by the source of the constant voltage +12 V which is transformed into a constant high voltage in the range from -100 to -2000 V. The PMT power supply output voltage is adjusted using the multi-turn resistor. The operating voltage of the detector station is a maximal supply voltage of five DS PMTs. The supply voltage providing the required gain of the remaining DS PMTs is adjusted using the ballast resistors soldered into the PMT dividers.

Since the scintillator light yield depends on the ambient temperature, it is necessary to consider the temperature coefficient for the processing of counter amplitude information. To measure the DS internal temperature during the NEVOD-EAS operation monitoring, the K-type thermocouple Fluke 80PK-11 which is designed for measuring temperatures in the range of -30°C to 105°C is used.

The signals from the DS counters are transferred to the cluster LP using the RG-174 cables. The PMT high voltage power supply is powered via the twisted pair cable. The receiving of the DS thermocouple information is also performed with the twisted pair cable.

3. NEVOD-EAS array cluster DAQ system

The main elements of the cluster registering system (figure 3) are: block of primary data processing, 4-channel summator-multiplexor (SM), PET-7019 input-output module which controls the SM and receives information from the DS thermocouples.

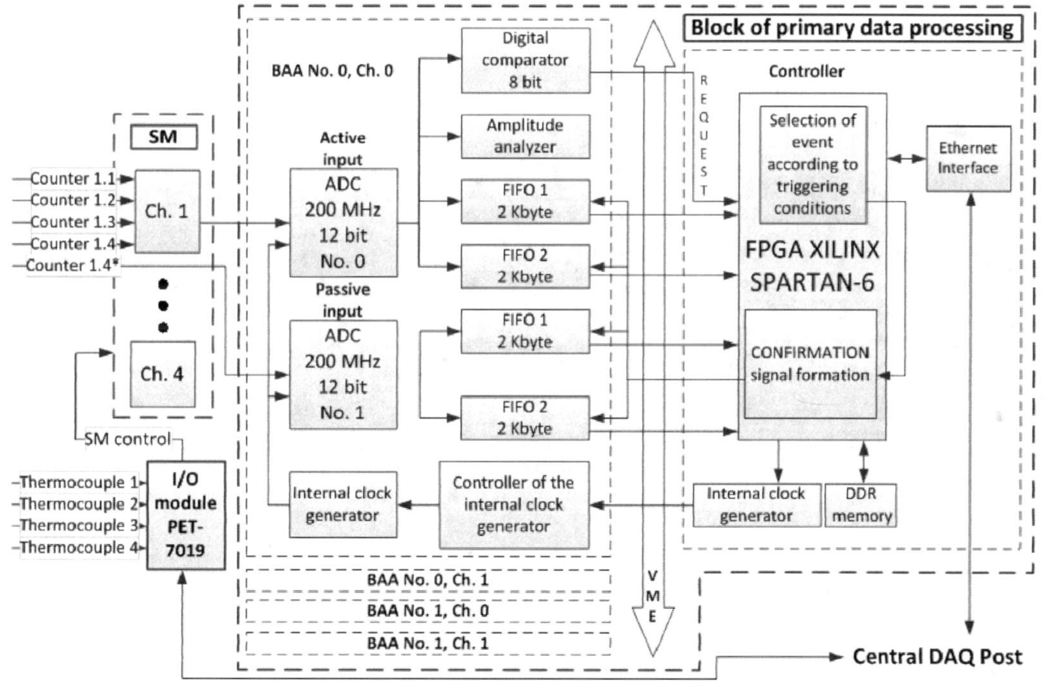

Figure 3. The scheme of the NEVOD-EAS cluster registering system.

The block of primary data processing ensures digitizing of analog signals from the DS of the cluster, selection of events according to the specified triggering conditions (coincidence of up to 4 detector stations within a time gate with a duration from 10 ns to 2 ms) and data transmission to the Central DAQ Post. It consists of two 2-channel boards of amplitude analysis (BAA) and a controller which are designed with the FPGA Xilinx Spartan-6. The communication between the controller and BAA is performed using the VME-bus. Each BAA channel has an active and a passive inputs. At each input, the signals are digitized by the 12-bit FADC in the amplitude range from -2.5 V to 2.5 V with a sampling frequency of 0.2 GHz. The registration threshold on the BAA channel active input is set using the 8-bit DAC. Each event is timestamped with an accuracy of up to 10 ns.

Registering system of the NEVOD-EAS cluster can operate in two modes: exposition and monitoring. Switching between two modes is performed using the 4-channels SM. Each SM channel has 4 inputs and serves one detector station. In the exposition mode, the analog signals from 4 standard PMTs of the DS are summed on the corresponding SM channels. The summed signals from the SM outputs are fed to the active inputs of the corresponding BAA channels. The additional PMT signals are fed directly to the corresponding BAA passive inputs. In the monitoring mode, only the defined input of each SM channel is turned on. That enables to register amplitude and charge spectra from the selected DS counters and, therefore, to control their parameters.

The control of the SM is performed using 4 discrete outputs of the PET-7019 I/O-module. Also this module has several analog inputs which are connected to the thermocouples installed inside the DS and LP housings. For the communication with the I/O-module the Modbus TCP protocol is used.

Control command receiving and data transmission are performed by the NEVOD-EAS cluster registering system via the fiber-optic communication line.

4. Conclusion

The NEVOD-EAS shower array designed on the basis of the scintillation detector stations with the dynamic range of up to ~ 10^4 particles/m^2 combined by the intracluster registering system will allow determination of the size, position of the axis and the arrival direction of EAS with the energies in the range of 10^{15}-10^{17} eV. New data obtained in the conjunctive operation of the NEVOD-EAS array with other installations of the Unique Scientific Facility 'Experimental complex NEVOD' will enable to reduce the uncertainty of the PCR particle energy estimations with the NEVOD and DECOR detectors. The NEVOD-EAS array is now under mounting on the roofs of the MEPhI campus laboratory buildings (see figure 4).

Figure 4. The NEVOD-EAS array clusters on the roof of the MEPhI laboratory building.

Acknowledgements

This work was performed at the Unique Scientific Facility 'Experimental complex NEVOD' under the Agreement on Russian–Italian cooperation within the framework of the Centre of Fundamental Research and Particle Physics supported by MEPhI Academic Excellence Project (contract 02.a03.21.0005 of 27.08.2013).

References

[1] Shulzhenko I A *et al* 2013 *J. Phys.: Conf. Ser.* **409** 012098
[2] Shulzhenko I A *et al* 2013 *Bull. Russ. Acad. Sci. Phys.* **77** 641
[3] Shulzhenko I A *et al* 2015 *Bull. Russ. Acad. Sci. Phys.* **79** 389
[4] Yashin I I *et al* 2015 *J. Phys.: Conf. Ser.* **632** 012029
[5] Aynutdinov V M *et al* 1997 *Astrophysics and Space Science* **258** 105
[6] Khokhlov S S *et al* 2011 *Astrophys. Space Sci. Trans.* **7** 271
[7] Barbashina N S *et al* 2000 *Instrum. Exp. Tech.* **43** 743
[8] Aglietta M *et al* 1990 *Nucl. Phys.* B (Proc. Suppl.) **16** 493
[9] Chiavassa A *et al* 2003 *Proc. 28th ICRC (Tsukuba)* vol 2 p 992

First results of the cosmic ray muon variation study by means of the scintillation muon hodoscope

N V Ampilogov, I I Astapov, N S Barbashina, V V Borog, A N Dmitrieva, A A Kovylyaeva, K G Kompaniets, A A Petrukhin, V V Shutenko, I I Yashin

National Research Nuclear University MEPhI
(Moscow Engineering Physics Institute), 115409, Moscow, Russia

E-mail: nvampilogov@mephi.ru

Abstract. The new scintillation muon hodoscope to study cosmic ray muon flux variations was created in MEPhI. The basic characteristics of the hodoscope (sensitivity area, precision of the muon track reconstruction, 'live' time etc.) are comparable with other hodoscopes (TEMP and URAGAN) of MEPhI. Modular design is a distinctive feature of the detector, supplying relativity easy transportability, and low maintenance requirements give a possibility of a long-term autonomic operation. First results of the cosmic ray muon variation study by means of the scintillation muon hodoscope are presented and discussed.

1. Introduction

Passing through the heliosphere, primary cosmic ray flux is modulated by processes related with solar activity, that leads to the muon flux variations on the Earth surface. Also, the intensity of the muon flux at the surface is sensitive to different powerful thermodynamic atmospheric processes. Thus, measuring of muon flux intensity and temporal-angular distributions gives us a possibility to investigate different phenomena in heliosphere, magnetosphere and atmosphere. The method which provides this way of remote monitoring is named a muon diagnostics [1].

For implementation of the method of muon diagnostics, the coordinate-tracking detectors of muons – hodoscopes – are necessary. Taking into account the experience of the muon hodoscope design and operation: TEMP (1995) [2] and URAGAN (2005) [3], a new scintillation muon hodoscope (ScMH) was created (MEPhI, 2010 – 2014) [4]. ScMH is capable to register small variations of the muon flux passing through the detector from all directions of the upper hemisphere. The basic characteristics of the ScMH (sensitivity area, precision of the muon track reconstruction, 'live' time etc.) are comparable with TEMP and URAGAN. But the ScMH has some competitive difference: modular design, supplying relativity easy transportability, and low maintenance requirements which give a possibility of long-term autonomic work.

2. Scintillation muon hodoscope

2.1. The setup

A new scintillation muon hodoscope is a coordinate-tracking muon detector which consists of identical detectors – supermodules (SM). Each SM has a sensitive area about 11.6 m^2, a good spatial (about 2.5 cm) and angular (better than 2°) accuracy, a wide zenith angle aperture of muon registration 0° – 75°. A long narrow strip (with dimensions 10.6 × 26.3 × 3460 mm^3) of plastic scintillator with wavelength shifting (WLS) fiber optics is a basic detection element of ScMH [5]. SM consists of four X – Y coordinate planes (CP) mounted in a common assembly. General view of the ScMH SM is presented in figure 1.

Each CP consists of two coordinate layers with orthogonally oriented strips, which provides measurements of X and Y coordinates of muon track passing point. The coordinate layer consists of two adjacent modules – basic non-dismountable modules (BM) of the detector, consisting of 64 scintillation strips in a common housing. BM has a single photodetector to receive light pulses from all its strips. One end of each WLS fiber is positioned opposite the corresponding cell of 64-channel multi-anode photomultiplier tube (MAPMT).

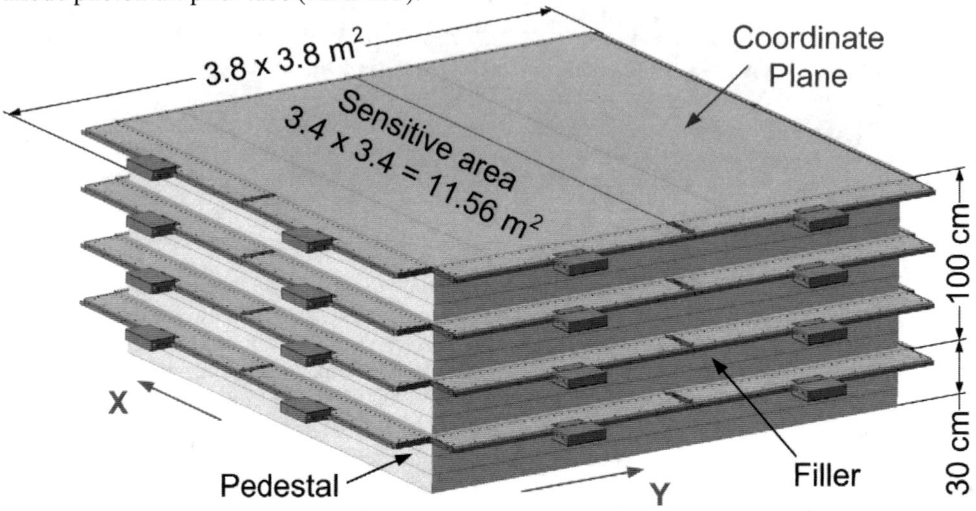

Figure 1. A general view of the wide-aperture scintillation muon hodoscope.

Data acquisition (DAQ) system of ScMH supermodule consists of the following sub-systems: front-end electronic boards for MAPMT signal procession (based on MAROC2 chip); a central DAQ controller for processing of data from all BMs and trigger formation (based on Cyclone III FPGA Development Kit); a PC for data collection, processing and final data storage.

2.2. Data of scintillation muon hodoscope

Experimental data of ScMH represent a sequence of one-minute frames (the live time is about 54 s), which includes: matrices of angular distribution of muons, monitoring data (counting rates of basic modules), number of reconstructed tracks and number of tracks passing through all coordinate planes of the hodoscope, statistics of triggered strips ('hits'), live time of the frame, atmospheric pressure and temperature in the facility hall averaged during a minute, etc.

The time rows of the muon tracks reconstruction efficiency of several coordinate layers of ScMH SM02, with temperature in facility hall and atmospheric pressure are presented in figure 2. The muon track reconstruction efficiency is estimated for each coordinate layer thanks to DAQ system which has a soft condition of master trigger – 5 triggered coordinate layers from 8. The efficiency is calculated as a ratio of muon counting rate with a corresponding coordinate layer to all muon counting rate of SM. This value allows estimating of operation efficiency of each coordinate layer and corresponding basic modules relative to other coordinate layers and quality as a whole. Figure 2 shows that the SM02 of the ScMH operated stably during August 2015, wherein pressure changed from 989 mbar to 1008 mbar, temperature in the facility hall was 23.5°C with ±1.5°C scatter.

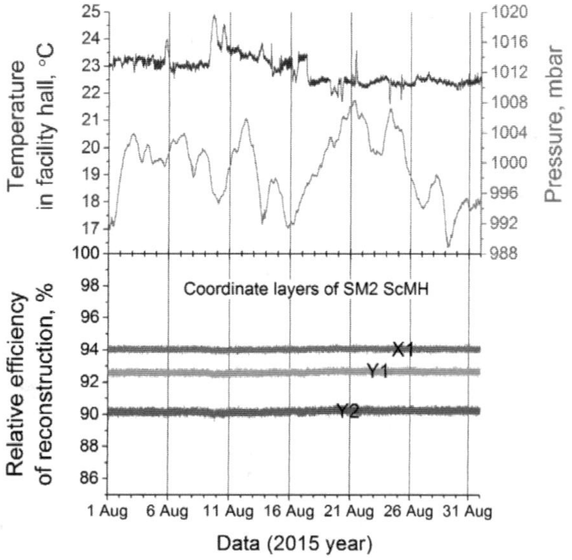

Figure 2. The time rows the muon tracks reconstruction efficiency of several coordinate layers of ScMH SM02 (down), with temperature in the facility hall and atmospheric pressure (up).

3. Detection of various phenomena with scintillation muon hodoscope

3.1. The Forbush decrease in data of scintillation muon hodoscope

The counting rate of the muon hodoscope is sensitive to different processes – in planetary scale, of terrestrial and extraterrestrial origin. The hourly average counting rates of the Moscow Neutron Monitor (min^{-1}), scintillation muon hodoscope (s^{-1}) and muon hodoscope URAGAN (s^{-1}) [6, 7] during second part of June 2015 are presented in figure 4. The responses of these detectors during Forbush decrease of 22 June 2015 are clearly visible in the figure. The counting rates of scintillation muon hodoscope and muon hodoscope URAGAN decreased by about 3%. The counting rate of the Moscow Neutron Monitor decreased by about 6.5%.

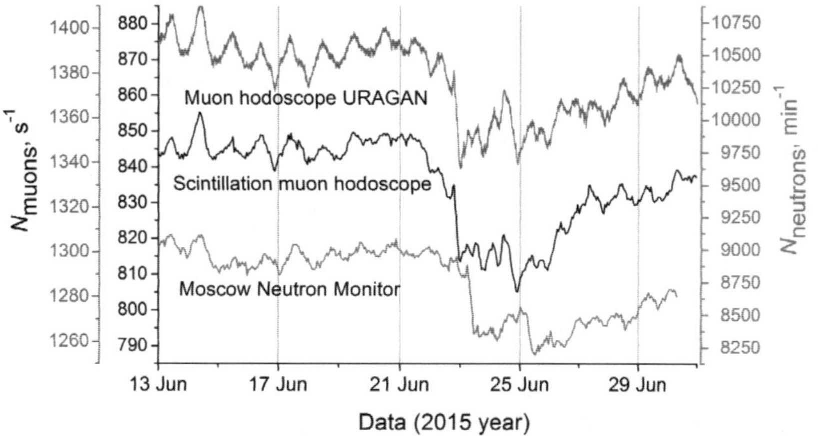

Figure 4. The hourly average counting rates of muon hodoscope URAGAN, scintillation muon hodoscope and Moscow Neutron Monitor during Forbush decrease of 22 June 2015.

3.2. The thunderstorm in data of scintillation muon hodoscope

Active turbulent atmospheric processes, for example thunderstorms, cause sharp changes of anisotropy of the muon flux. To visualize the dynamics of these changes, the 'muon snapshots' (muonographies) are used [8]. To improve the contrast of the images, the original matrix is normalized to the averaged during preceding 24 hours. It gives the matrix in standard deviations of detected number of events for every cell. To suppress the high frequency deviation, the filtration procedure is applied for the matrices. Examples of 5-min muon images of ScMH and URAGAN hodoscopes, obtained during thunderstorm of 14 Jun 2014 are shown in figure 5. As it follows from the figure, these detectors exhibited a same decrease level of muon flux from same direction – about 0.47% of scintillation muon hodoscope and about 0.43% of muon hodoscope URAGAN.

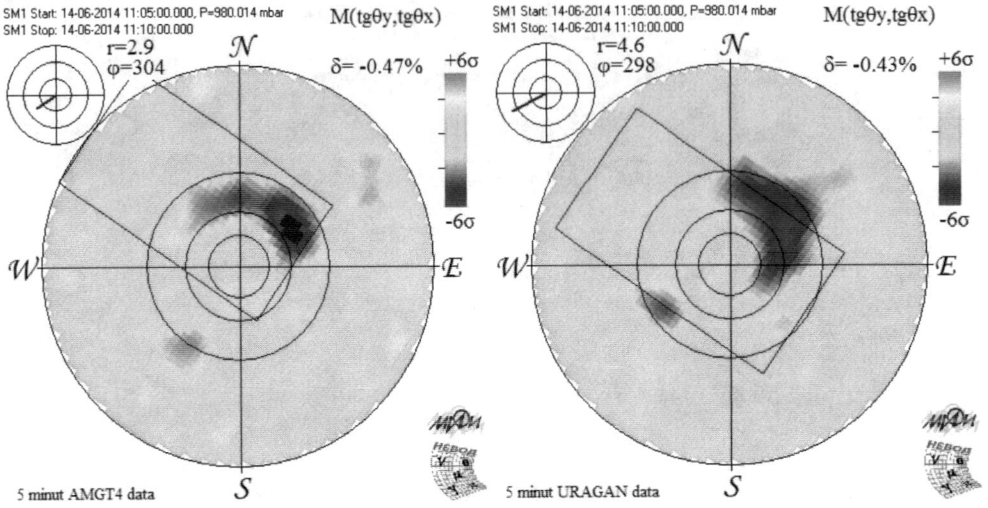

Figure 5. The muon images of scintillation muon hodoscope (left) and muon hodoscope URAGAN (right) during thunderstorm of 14 Jun 2014.

4. Conclusions

The results of the analysis of the scintillation muon hodoscope operation show that the constructive and technological solutions used during the detector design allowed to create the detecting system of scintillation muon hodoscope which ensures required parameters and stability of operation during long time. The created scintillation muon hodoscope is sensitive to events of changing of primary cosmic ray flux and to local events caused by atmospheric processes.

Acknowledgments

This work was performed at the Unique Scientific Facility 'Experimental complex NEVOD' within the framework of the Centre of Fundamental Research and Particle Physics supported by the MEPhI Academic Excellence Project (contract № 02.a03.21.0005 of 27.08.2013).

5. References

[1] Barbashina N S *et al* 2007 *Bull. Russ. Acad. Sci. Phys.* **71** 1041
[2] Borog V V *et al* 1995 *Proc. 24th Int. Cosmic Ray Conf. (Rome)* vol 4 p 1291
[3] Barbashina N S *et al* 2008 *Instrum. Experim. Techn.* **51** 180
[4] Ampilogov N V *et al* 2011 *Astrophys. Space Sci. Trans.* **7** 435
[5] Astapov I I *et al* 2011 *Proc. 32nd Int. Cosmic Ray Conf. (Beijing)* vol 11 p 447
[6] Shutenko V V *et al* 2009 *Bull. Russ. Acad. Sci. Phys.* **73** 347
[7] Barbashina N S *et al* 2013 *J. Phys.: Conf. Ser.* **409** art. 012189
[8] Mikhaylenko A S *et al* 2011 *Bull. Russ. Acad. Sci. Phys.* **75** 827

Project of the URAN array for registration of atmospheric neutrons

D M Gromushkin[1], N S Barbashina[1], F A Bogdanov[1], R P Kokoulin[1], V V Ovchinnikov[1], A A Petrukhin[1], Yu V Stenkin[1,2], S S Khokhlov[1], I A Shulzhenko[1] and I I Yashin[1]

[1] National Research Nuclear University MEPhI (Moscow Engineering Physics Institute), Kashirskoe highway 31, Moscow, 115409, Russia
[2] Institute for Nuclear Research of RAS, 117312, Moscow, Russia

E-mail: DMGromushkin@mephi.ru

Abstract. The project of a new setup is directed at the registration of atmospheric neutrons (URAN) generated by hadronic component of extensive air showers (EAS). The setup includes 72 en-detector which simultaneously register two major EAS components: electromagnetic by the group passage of charged particles and hadron component by the thermal neutrons. The neutrons and charged particles are detected using a specialized scintillation composition made of granulated alloy of crystals based on the ZnS(Ag) powder with an admixture of B_2O_3.

1. Introduction

The main purpose of the proposed project is the creation of the setup for the detection of the extensive air shower (EAS) neutron component in the range of PeV-energies of cosmic rays (CR), where the 'knee' (the steepening) of the size spectrum of EAS registered on the Earth's surface is observed. Neutrons are the only EAS component unexplored by now. At the same time, the neutron component which is formed by the interaction of the shower hadron core with the atom nuclei of the atmosphere carries important information about the EAS development [1]. Creating of a detector for its investigation will enable to obtain novel data for understanding the processes of the EAS development and for correct interpretation of the EAS study results from the viewpoint of estimation of the primary cosmic ray energy spectrum and composition. An important feature of the setup should be the possibility of simultaneous registration of the EAS electron-photon and hadron components using the same detector. This would remove many of systematic uncertainties in comparing the results of measurements of these components [1].

2. The URAN detector

The URAN setup includes 72 en-detectors for simultaneous registration of electron-photon (e) and the neutron (n) EAS components. The en-detectors are combined into independent cluster structures of 12 detectors. Clusters are located on two roofs of the laboratory buildings (3 clusters on each roof). The layout of the URAN setup is shown in figure 1. The characteristic distance between the detectors is 4–5 meters. The total area of the setup is ~ 10^3 m^2.

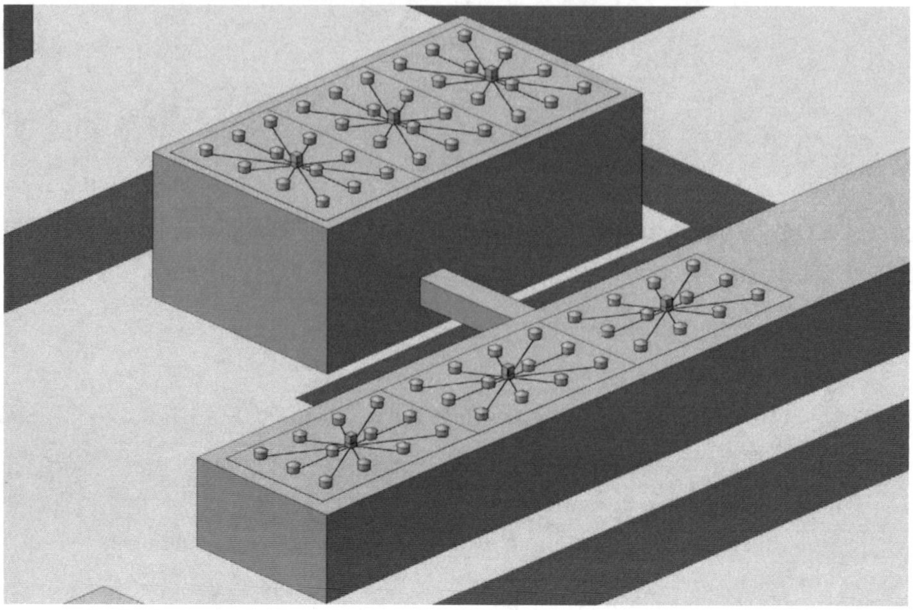

Figure 1. The layout of the URAN detector counters.

To ensure the operation of the detectors and primary cluster data processing, the Local Posts (LP) are used. The structure of the URAN cluster is shown in figure 2. The cluster LP contains: the 12-channel high-voltage power supply which allows remote variation of the supply voltage of any detector; the low-voltage (±12 V) power supply which ensures operation of the preamplifiers and integrator-amplifiers; two 12-channel boards of amplitude analysis; the crate for the boards of amplitude analysis; the mediaconverter and the LP thermostabilization system. One LP can support operation of two clusters.

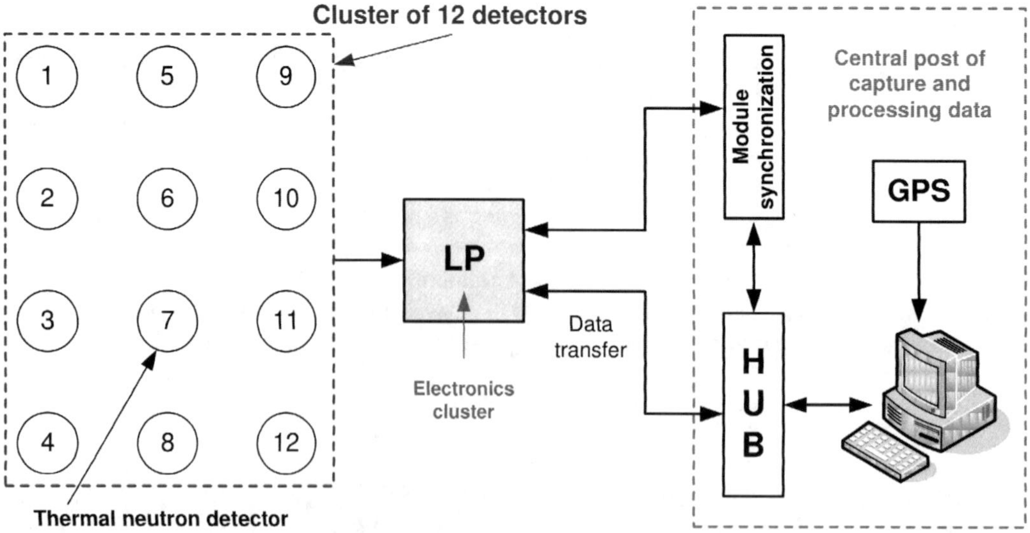

Figure 2. The cluster structure of the URAN detector.

Transferring of analog signals from the detector to the LP is performed via coaxial cables housed inside the metal hoses intended to protect them against environmental influences and mechanical damages. Apart from the coaxial cables, the hoses contain the cables for the PMT high-voltage and amplifier low-voltage power supply. Data transmission from the LP to the Central DAQ Post located in one of the laboratory buildings is performed via fiber-optic communication lines.

The Central Post of control and data acquisition ensures operation of the URAN setup in the exposition and monitoring modes, controls all Local Posts of the setup and stores experimental data. Main elements of the Central DAQ Post are: the central computer, the network equipment, the mediaconverters (Ethernet→Op. Link) and the module of external synchronization. Time synchronization of clusters with an accuracy of 10 ns is performed using global positioning systems (GPS/GLONASS).

At the central computer, information about registered events is processed by a special software and is stored in files that contain event timestamps and parameters of the signals from the detectors of triggered clusters. The software also allows automatic control of all elements of the cluster Local Post of primary data acquisition.

3. En-detector

For the registration of the EAS neutron component, a new type of detector well-proven in the PRISMA-32 setup (MEPhI, Moscow) [2] is used. The scheme of the detector with the outer enclosure is shown in figure 3. The detector housing represents a light-tight black plastic tank with a volume of 200 liters (570×740 mm). To improve the light collection, the diffusely reflecting cone is used. The $ZnS(Ag)+B_2O_3$ scintillator is located on the base of the cone. The FEU-200 photomultiplier is placed at the top of the cone. The PMT with a high voltage divider mounted on the tank cover using a special holder forms the detecting block. The effective area of the scintillator is ~ 0.36 m^2.

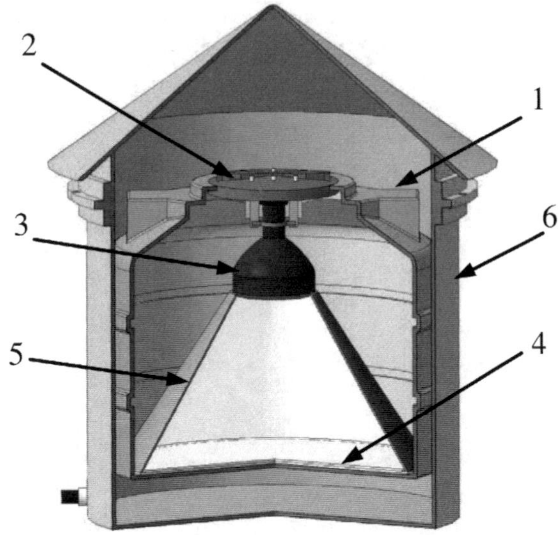

Figure 3. En-detector design; 1 - light protecting housing, 2 – lid, 3 – FEU-200 PMT, 4 – scintillator layer, 5 - light reflecting cone, 6 – outer metal enclosure.

As the detectors will be used in outdoor conditions, the external galvanized steel cylindrical enclosure (diameter of 800 mm, height of 1100 mm and wall thickness of 0.9 mm) with a conical roof has been developed for their protection.

4. Registering electronics of the URAN detector

The cluster registering electronics will ensure the operation of detectors and signal digitizing and will be located in the Local Posts. Electronics of one cluster includes two 12-channel boards of amplitude analysis developed according to Euromechanics 6U standard. One board is designed for digitizing analog signals from the 12th dynode output (with an integration time of 1 μs) and allows selection of events according to various triggering conditions (triggering of 1 or several counters within a specified time gate), registration of delayed neutrons during 20 ms and data transmission from a cluster to a Central DAQ Post. The second board is intended for digitizing analog signal from the 7th dynode output. Using of two dynodes provides wide dynamic range of measured energy releases produced mainly by the EAS electron component.

For the 12th dynode measurements, the board of amplitude analysis with an integrated controller, sampling frequency of 200 MHz and 5 μs time gate is used. The sampling frequency can be switched to 1 MHz. The time gate can be increased up to 20 ms in steps of 1 μs. For the 7th dynode, the board of amplitude analysis with an integrated controller and sampling frequency of 1 MHz is used.

5. Conclusion

The URAN setup will allow to obtain novel data about extensive air showers and their hadron component. The spectrum of EAS in the number of thermal neutrons, the function of the lateral distribution of hadrons in the EAS and other parameters of hadron component will be obtained. Additionally, the URAN setup will become an integral part of existing and being created detectors of the Unique Scientific Facility 'Experimental complex NEVOD' which already includes the detectors NEVOD, DECOR, CTS, as well as the NEVOD-EAS and TREK detectors that are now under development and construction.

Acknowledgments

This work was performed at the Unique Scientific Facility 'Experimental complex NEVOD' within the framework of the Center of Fundamental Research and Particle Physics supported by the MEPhI Academic Excellence Project (contract 02.a03.21.0005 of 27.08.2013).

References

[1] Stenkin Yu V 2009 *Nucl. Phys.* B *(Proc. Suppl.)* **196** 293
[2] Gromushkin D M *et al* 2015 *Physics of Atomic Nuclei* **78** 349
[3] Gromushkin D *et al* 2014 *JINST* **9** C08028

Journal of Physics: Conference Series **675** (2016) 042001 doi:10.1088/1742-6596/675/4/042001

Estimation of reliability of linear point structures revealed in two-dimensional distributions of experimental data

O V Falomkina[1,3], Yu V Pyatkov[2,3], Yu P Pyt'ev[1] and D V Kamanin[3]

[1] Faculty of Physics, M. V. Lomonosov Moscow State University, Leninskie Gory, Moscow, 119991, Russia

[2] National Research Nuclear University MEPhI (Moscow Engineering Physics Institute), Kashirskoe highway 31, Moscow, 115409, Russia

[3] Joint Institute for Nuclear Research, Dubna, Moscow region, 141980, Russia

E-mail: olesya.falomkina@gmail.com, yvp_nov@mail.ru

Abstract. In the experiments at the FOBOS spectrometer [1] dedicated to study the spontaneous fission of the 248Cm and 252Cf nuclei in the mass correlation distribution of fission fragments new unusual structures bounded by magic clusters were observed for the first time. The structures were interpreted as a manifestation of a new exotic decay called collinear cluster tri-partition (CCT). These pioneer results were confirmed and detailed later in the series of experiments at different time-of-flight spectrometers [2]. Interpretation of the results obtained needs estimation of the statistical reliability of the structures mentioned above. The report presents the results of the solution to the problem of statistical reliability estimation on the basis of morphological image analysis [3].

1. Introduction

In this work based on the results of RFBR grants 14-02-93960, 14-07-00409-a we continue to study a previously unknown type of nuclear transformations found by project authors – collinear cluster tri-partition (CCT). In series of experiments at different time-of-flight spectrometers we have observed multiple manifestations of a new type of multibody decay of low excited heavy nuclei called by us collinear cluster tri-partition (CCT) [1], [2]. The results were obtained predominantly in the frame of the "missing mass" approach. It means that only two from at least three decay partners were actually detected whereas a total mass of these fragments being less the mass of mother system serves a signature of a multibody decay. Thus analyzing of the mass-mass distribution of the fission fragments detected in coincidence let reveal us the decay mode absolutely unknown in the past. Unfortunately there is an essential random background linked with the scattered fragments from the conventional binary fission in the mass region of possible manifestation of the CCT. We have find that due to the physics behind (clustering) the CCT events form some regular predominantly linear or/and almost rectangular structures in the mass-mass plots (see the example of "rectangular" structures at figure 1). The main goal of this work is to estimate the statistical reliability of these structures in the mass correlation distribution of fission fragments or, in other words, to estimate probability of "random" realization of the

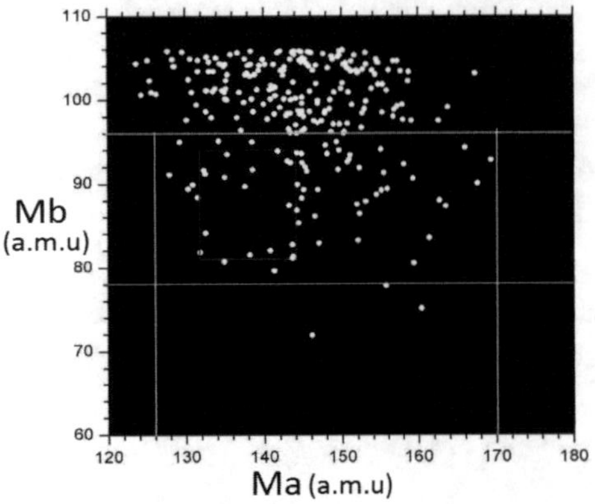

Figure 1. Example of a "rectangular" structure for estimating of its reliability. The chosen "field of view" is shown by a gray line: Ma varies from 125 a.m.u. to 169 a.m.u, Mb varies from 79 a.m.u. to 96 a.m.u.

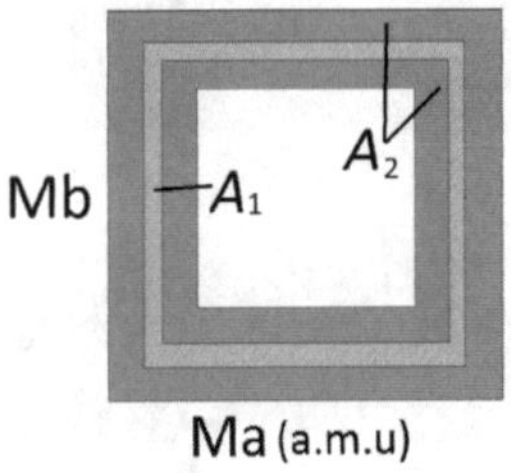

Figure 2. Example of domains A_1, $A_2 \subset \Omega$ of constant brightness in a model image of a "rectangular" structure.

corresponding structures. Previously we have developed an approach [4] based on morphological image analysis [3] which will be used to solve this task.

2. Method of morphological image analysis

Let us briefly consider some notions of the method of morphological image analysis [4]. Let $\tilde{f}$ be an experimentally obtained signal (mass-mass distribution), which can be represented in the form

$$\tilde{f} = f + \nu, \tag{1}$$

where f is a signal possibly containing several structures of our interest, and ν is additive noise. An image of a signal $f(\cdot)$ is understood as a numerical square-integrable function defined on a subset X of the Euclidean plane $\mathcal{R}^2$. The domain X is termed the field of view, and the value $f(x)$ of the function $f(\cdot)$ at the point $x \in X$ is termed the brightness of the point x of the field of view X. In the case under consideration, $X = \{x_1, \ldots, x_n\}$ and, correspondingly, the images $\tilde{f}(\cdot)$, $f(\cdot)$, and $\nu(\cdot)$ from (1) are defined at the same points and are elements of the Euclidean plane $\mathcal{R}^n$. As to the error $\nu \in \mathcal{R}^n$, we will assume that it is a random image with zero mathematical expectation $\mathbf{E}\nu = 0$ and the correlation operator $\sigma^2 I$, where $I \in \{\mathcal{R}^n \to \mathcal{R}^n\}$ is the unit operator and σ^2 is unknown.

Let us denote the "rectangular" structure image by $\omega(\cdot)$ and assume it to be defined on a variable-size mobile subset Ω of the field of vision X. The form of the image $\omega(\cdot)$ will be understood as the set of images

$$V_\omega = \{\omega(\cdot),\, \omega(x) = c_1\chi_{A_1}(x) + c_2\chi_{A_2}(x),\, c_1 \geq c_2,\, c_1, c_2 \in \mathcal{R}^1,\, x \in \Omega\}, \tag{2}$$

where $\chi_{A_i}(x) = \begin{cases} 1, & x \in A_i, \\ 0, & x \notin A_i, \end{cases}$ $x = 1, 2$. V_ω is a convex closed cone in $\mathcal{R}^2$ and in $\mathcal{R}^n$. In this definition A_1 and A_2 are different subsets of constant brightness in Ω. According to this definition, the form of the image of an object contains all images of this object differing in the brightnesses on subdomains of constant brightness in Ω.

Figure 2 shows domains A_1 and $A_2 \subset \Omega$ with a constant brightness in the image of a "rectangular" structure. In this figure, the field of view is divided into subdomains A_1 and A_2.

The "rectangular" structure directly corresponds to the domain A_1, and the points surrounding it correspond to the domain A_2. The form (in the common sense of this word) and size of the domains A_1 and A_2 are specified (postulated) a priori by the researcher. The proposed method enables one to verify the correctness of this postulate. In the given case, the form of the linear structure was determined on the basis of figure 1. The brightnesses over the domains A_1 and A_2 are assumed to be constant. The fact that the brightness at the image point belonging to the "rectangular" structures must be higher than that at the points surrounding it is reflected in the condition $c_1 \geq c_2$ in expression (2).

The projection (see below) of some image $g(\cdot)$ defined on Ω onto a form V_ω is understood as the image $(P_{V_\omega}g)(\cdot)$, which uniquely exists, because V_ω is a convex closed cone (see [3]): $(P_{V_\omega}g)(x) = \hat{c}_1\chi_{A_1}(x) + \hat{c}_2\chi_{A_2}(x)$, $x \in \Omega$, where $\hat{c}_1$ and $\hat{c}_2$ are solutions to the following minimization problem:

$$\int\limits_\Omega (g(x) - \hat{c}_1\chi_{A_1}(x) - \hat{c}_2\chi_{A_2}(x))^2\,dx = \min_{c_1,c_2\in\mathcal{R}^1,\,c_1\geq c_2} \int\limits_\Omega (g(x) - c_1\chi_{A_1}(x) - c_2\chi_{A_2}(x))^2\,dx.$$

Below, we will omit, for brevity, the sign $(\cdot)$ in the notation of images.

Let us consider the problem of separating the "rectangular" structure in the framework of the above-formulated model of signal detection as a problem of testing the statistical hypothesis H that the image f contains a fragment f_ω that can be represented in the form

$$H:\ \exists f_\omega = g+\nu,\ \exists t\in T,\ g\in t(V_\omega),\ \nu\in\mathcal{N}(0,\sigma^2 I),\ ||\nu||^2 \ll ||g||^2, \tag{3}$$

where the form of g coincides with (2) within a shift and scale transformation, $t \in T$ is a shift and scale transformation, T is the class of such transformations, $||z||^2 = \int_\Omega z(x)\,dx$, and z is some image. The alternative K is that such fragments are absent. For solving the problem of testing this hypothesis, the following functional [3] is used:

$$j(z) = \frac{||(I-P_{V_\omega})z||^2}{||(P_{V_\omega}-P_{V_U})z||^2}. \tag{4}$$

In (4) z is some image and $P_{V_U}z$ is a projection of image z onto the form U of the uniform field of view: $U = \{u(\cdot),\ u(x) = \text{const}\chi_\omega(x),\ x\in\Omega\}$. Functional (4) possesses the following properties [4]. If there is a fragment f_ω satisfying condition (3) but not representable in the form of the the "noised uniform field of view"

$$f_\omega = g+\nu,\ \exists t\in T,\ g=t(U), \tag{5}$$

then the value of functional (4) is small. If there is a fragment f_ω satisfying condition (5), then the functional $j(f_\omega)$ has the order $O(1)$. If, at last, fragments f_ω satisfying condition (4) or (5) are absent, then the functional $j(f_\omega)$ again has the order $O(1)$.

The decision rule has the following form: a hypothesis H is accepted if, by a shift and scale transformation, a fragment f_ω can be found such that $j(f_\omega) \leq A$, where A is an empirically determined constant, and it is rejected if such a fragment is absent. The value of functional (4) characterizes the "distance"' between an image z and an image of the form (2). Functional (4) is invariant with respect to brightness and contrast transformations. In order to determine the constant A, the value of the constant was found (based on real data) at which the image of a determined "rectangular" structure satisfied the researcher. In our case, that was A = 1.35. Then, the reliability of the obtained value was tested by means of a model experiment. We used 100000 model images of an additive noise with the parameters corresponding to the real experiment. Then an empirical distribution of the values of functional (4) at the specified

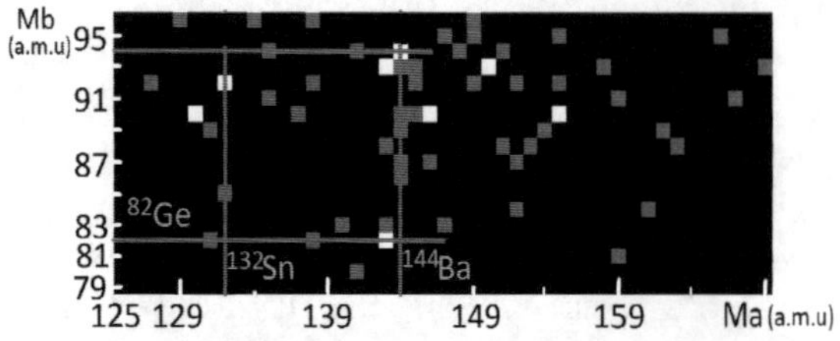

Figure 3. Image for "rectangular" searching taken from figure 1. Gray lines were added manually to show "magic" numbers.

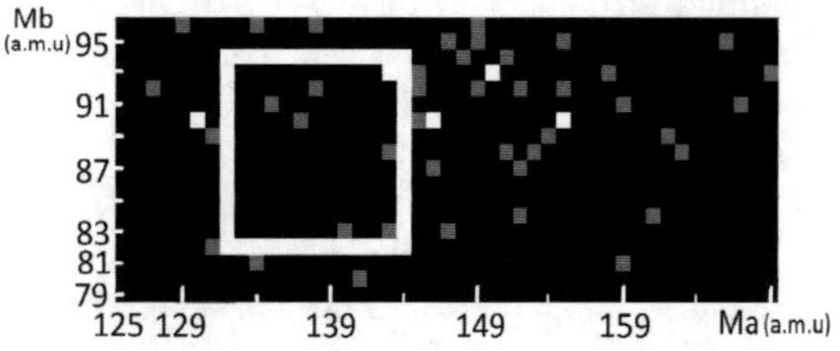

Figure 4. The "rectangular" structure found by morphological method.

noise level was constructed. On the basis of this distribution, the probability $P(j \leq A) \approx 0.02$ was found. This probability is the probability of false acceptance of a hypothesis against the nearest "homogeneous field of view" alternative. Properties of the functional (4) imply that this probability estimates an upper bound for the probability of false acceptance of the hypothesis against the alternative that such a fragment is absent. This criterion is analogous to the principle of the locally homogeneous strongest criterion [5]. According to the statistical analysis made above, the probability, that the "rectangular" structure found on real data (see figures 3, 4), is generated by the noise is small and approximately equals to 2%.

3. Conclusion

In order to obtain the quantitative estimates of the "rectangular" structures statistical reliability an approach based on morphological image analysis has been developed. In the framework of this approach, the probability of a random realization (due to the presence of noise) of a structure or its scaled copy is estimated. Only the presence of such an estimate justifies the physical interpretation of the revealed structures, removing the question of whether these structures are statistical artifacts or not.

Acknowledgments

This work was supported by RFBR grants 14-02-93960, 14-07-00409-a and 14-07-00441-a. This work was performed in part within the framework of the Center Fundamental Research and Particle Physics supported by MEPhI Academic Excellence Project (contract No. 02.a03.21.0005, 27.08.2013).

References

[1] Ortlepp H *et al.* 1998 *Nucl. Instrum. Meth.* A **403** 65
[2] Kamanin D Pyatkov Yu 2013 *Clusterization in Ternary Fission* (Lecture Notes in Physics vol 3) ed C Beck **875** pp 183–246
[3] Pyt'ev Yu 1993 Morphological Image Analysis *Pattern Recognition and Image Analysis* **3** 1 pp. 19–28
[4] Pyatkov Yu *et al.* 2011 *Pattern Recognition and Image Analysis* **21** 82
[5] *Handbook of Statistics* vols 1–18, ed C Rao (Amsterdam: North Holland/American Elsevier)

X-ray fluorescence analysis of low concentrations metals in geological samples and technological products

I A Lagoida and A V Trushin*
National Research Nuclear University MEPhI (Moscow Engineering Physics Institute), Kashirskoe highway 31, Moscow, 115409, Russia

E-mail: *info@techade.ru

Abstract. For the past several years many nuclear physics methods of quantitative elemental analysis have been designed. Many of these methods have applied in different devices which have become useful and effective instrument in many industrial laboratories. Methods of a matter structure analysis are based on the intensity detection of the X-ray radiation from the nuclei of elements which are excited by external X-ray source. The production of characteristic X-rays involves transitions of the orbital electrons of atoms in the target material between allowed orbits, or energy states, associated with ionization of the inner atomic shells. One of these methods is X-ray fluorescence analysis, which is widespread in metallurgical and processing industries and is used to identify and measure the concentration of the elements in ores and minerals on a conveyor belt. Samples of copper ore with known concentrations of elements, were taken from the Ural deposit. To excite the characteristic X-rays radionuclide sources ^{109}Cd, with half-life 461.4 days were used. After finding the calibration coefficients, control measurements of samples and averaging of overall samples were made. The measurement error did not exceed 3%.

1. Introduction

Quantitative X-ray fluorescence (XRF) analysis involves the conversion of measured fluorescent intensities to the concentration of the analytes [1]. XRF raw data is the number of counts of element-specific fluorescent X-ray energies received in an XRF instrument detector. This raw data is visualized in a spectrum graph, where the x-axis represents element-specific fluorescent energies, and the y-axis represents counts or pulses. In a first approximation, the concentration for analyte in the samples is equal:

$$C_i = KI_i,$$

where C_i is the concentration of the analyte, I_i –intensity which have been measured for each analyte element, K – calibration coefficient.

In this work the results of Wavelength Dispersive X-ray Fluorescence (WDXRF) were used for further calibration. WDXRF analysis is more accurate then XRF, but there is no possibility to use the WDXRF analysis in the manufacturing conditions for rapid analysis on the conveyor. Because this technique implies the homogeneity of the samples which is achieved with careful sample preparation. Also, methodology for conducting this analysis requires a great deal of precision in the construction of the experimental geometry. XRF analysis can give experimental information faster and doesn't require careful preparation of samples. Therefore, it is preferable on the mining factories.

The average value, which is given at the end of the paper, was considered for the following reason – enrichment processes includes adding reactants in quantity grams per ton, depending on the concentration of copper in the large volume of ore. That's why, adding of the reactants is required availability of the correct average value of the copper concentration in the active volume.

2. Implementation

Copper samples taken from the Ural deposits were chosen to perform the experiment. These samples were marked from 1 to 17 milled to a powder and sent to the Moscow Institute of Experimental Mineralogy for defining the concentrations of elements in this samples use WDXRF analysis. The results of this analysis are shown in table 1, measurement error did not exceed 5%.

Table 1. The results of WDXRF analysis.

sample	mass fraction of copper, %	sample	mass fraction of copper, %	sample	mass fraction of copper, %
1	0.66	8	0.84	15	0.88
2	0.8	9	0.6	16	1.35
3	0.71	10	1.1	17	1.46
4	1.05	11	1.5		
5	0.52	12	1.6		
6	0.47	13	3.13		
7	0.7	14	2.15		

For sample excitation four sources ^{109}Cd, with an activity equal 40 mCi were used. Radionuclide ^{109}Cd distributed within the ceramic matrix which is placed into the insert of tungsten alloy. Typical dimensions of the source are D=8 mm, H=5 mm, d=5 mm, h=1.2 mm. Structure of the ^{109}Cd source is shown on figure 1.

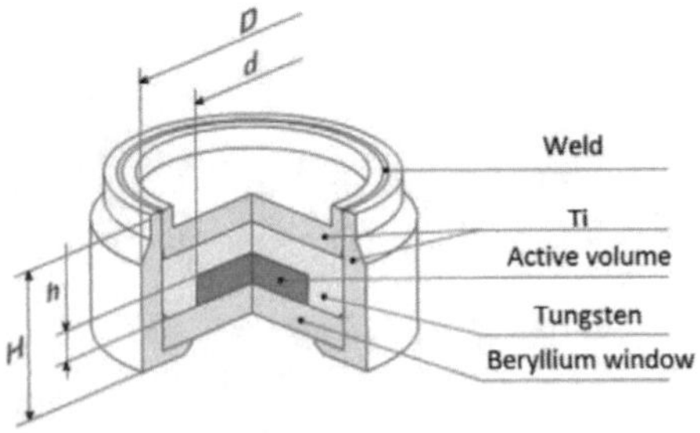

Figure 1. Structure of ^{109}Cd source.

The half-life of sources is 461.4 ±1.2 days. Photon energy range is:

Ag $XK_{\alpha 2}$ 22keV
Ag $XK_{\alpha 1}$ 22.2keV
Ag $XK_{\beta 2}$ 25keV
Ag $XK_{\beta 1}$ 25keV

Amptek Si-Pin detector with resolution 210 eV for Cu Kα was used for detection of characteristics X-rays. The count rate of characteristic lines for all the samples were measured for 3 minutes. Error of measuring of the spectral line intensityfor Cu Kα did not exceed 4%. Some resulting spectrum is shown below (see figure 2). Values of measured intensities for Cu Kα line for all 17 samples are shown in table 2. The intensities are normalized by a factor 42.7.

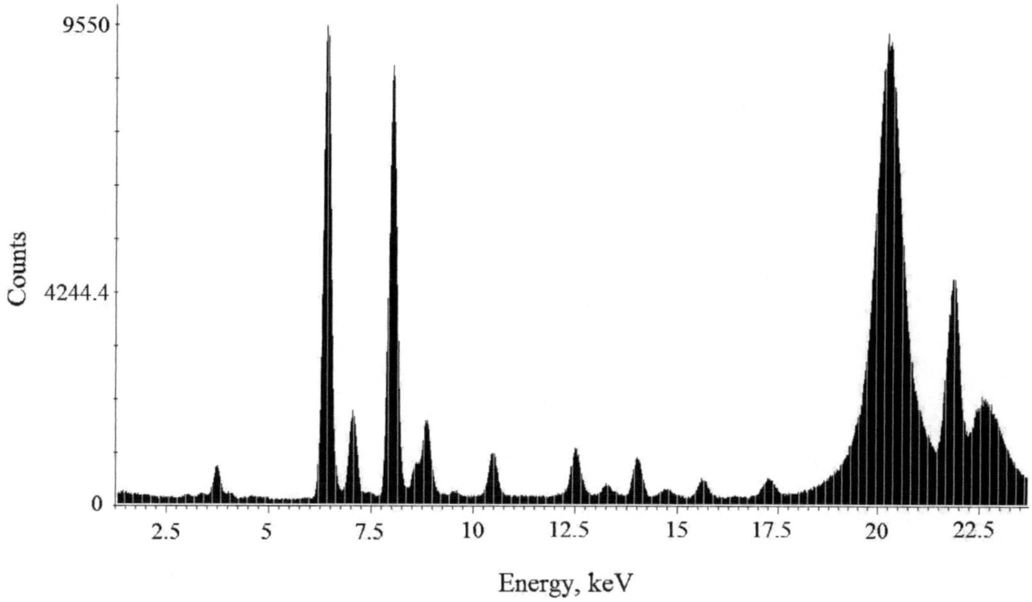

Figure 2. Spectrum of the sample 13 ($C_{cu}(\%) = 3.13$).

Based on the data from tables 1 and 2 the calibration curve has been constructed (see figure 3).

Table 2. Measured Cu Kα intensities (Error did not exceed 4%).

sample	I_i (counts per sec)	sample	I_i (counts per sec)	sample	I_i (counts per sec)
1	2.7	8	3.2	15	3.5
2	3.5	9	2.4	16	4.7
3	3.0	10	3.9	17	5.7
4	4.2	11	6.2		
5	2.2	12	6.6		
6	1.9	13	12.5		
7	2.8	14	8.8		

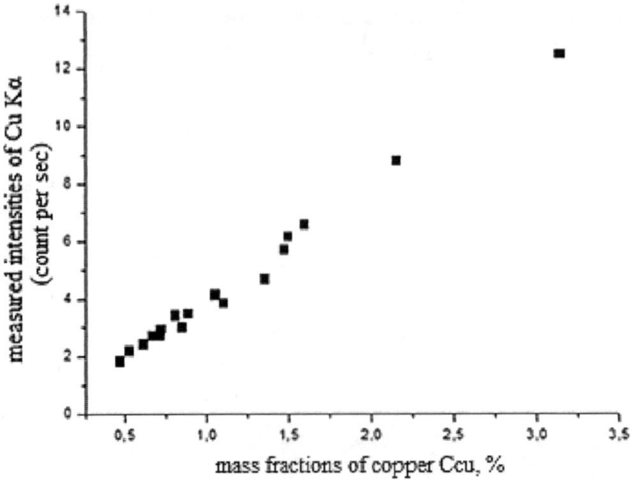

Figure 3. Calibration curve. The dependence of measured copper intensities on the mass fractions of copper in the samples (see table 1 and 2). Experimental errors are within the data points.

Dependence is described by the equation (1), approximation error is equal 3%.

$$I_{cu} = 4.08C_{cu} - 0.07. \quad (1)$$

Hereafter control measurements of samples intensities have been made. After that, the values of the mass fractions of copper for all samples were calculated using equation (1), error didn't exceed 11%. The average values for the mass fraction of copper have been calculated using following formula:

$$C_{mean} = \sum_{N=1}^{17} \frac{C_N}{N}.$$

Average value of the mass fractions of copper for WDXRF analysis is equal: $C_{meanWDXRF}(\%) = 1.15 \pm 0.07$. Whereas, average value of the mass fractions of copper for XRF analysis is equal: $C_{meanXRF}(\%) = 1.13 \pm 0.03$. Errors here were defined not as error of an average for several measurements but as error of indirect measurements to demonstrate applicability of the method described in article.

3. Conclusion

The dependence of the measured copper intensities on the copper mass fractions was obtained and it will be used in manufacturing processes for determining the concentration of copper in the ore coming from the mine to the conveyor belt.

Acknowledgments

This work was partially supported by MEPhI Academic Excellence Project (contract No. 02.a03.21.0005, 27.08.2013).

References

[1] Rene E Van Grieken and Andrzej A Markowicz 2001 *Handbook of X-Ray Spectrometry Second Edition* (New York: Marcel Dekker, Inc)

[2] B Beckhoff *et al.* 2006 *Handbook of Practical X-Ray Fluorescence Analysis* (Berlin: Springer)

Journal of Physics: Conference Series **675** (2016) 042003 doi:10.1088/1742-6596/675/4/042003

Development of wide-ranged diamond-based detector unit for gamma radiation measurement

M A Baranova[1,2], A V Boyko[1], S B Chebyshev[1], I I Cherkashin[1], V P Kireev[1] and V I Petrov[1]

[1]JSC SNIIP, Raspletina St., 5, Moscow, 123060, Russia

[2]National Research Nuclear University MEPhI (Moscow Engineering Physics Institute), Kashirskoe highway 31, Moscow, 115409, Russia

E-mail: mariya_baranova@list.ru, aboyko@sniip.ru, sbchebyshev@aem-group.ru, iicherkashin@sniip.ru, vkireev@sniip.ru, vpetrov@sniip.ru

Abstract. In the article the description of wide-ranged diamond-based detector unit is given. Characteristics of the diamond detector were studied in current and in impulse mode. As well it was studied how detector`s sensitivity depends on power doze within the limits from 10^{-3} to 0,4Gy/h (impulse mode) and from $10^{-1}$to $2\cdot10^{4}$Gy/h (current mode). On the basis of the obtained data it is possible to estimate about the possibility of using the detector to prevent emergency accident on a nuclear power plant and for everyday control at a nuclear power plant.

1. Introduction

The most important task of radiation control at a nuclear power plant is the earliest detection of radiation accidents, monitoring of their progression and isolation in process and finally analyzing the information about the event and accident consequence assessment[1].

In compliance with IAEA Safety Standards NPP must be provided with Irradiance measuring systems which remain their functional abilities under condition of severe beyond-the-design basis accident and enabling to get information necessary for the drafting of recommendations for civil protection in emergency situations and recovery from an accident.

Also usage of emergency wide-range detector unit can solve the following problems:

- verifiable detection of radiation accidents;
- data acquisition of reactor core condition;
- loss of coolant data acquisition;
- control and prediction of radiation situation in containment; Etc.

Usually equipment is located within the reactor containment. It shall be located to provide maximum coverage of the areas being monitored and to minimize shielding effects from other equipment or structures[2].

For solving those problems we need specially sophisticated detecting units with various abilities of detecting which can be used at the nuclear power plant for everyday control as well as to detect significances during an accident at a nuclear power plant.

Traditionally detectors are made of materials which are proved to be vulnerable and not able to resist high levels of radiation and high temperatures. Detectors based on diamond crystal, which have a high radiation resource and keep working at critical temperatures are considered to be quite perspective[3].

2. Diamond detector

During an accidental situation some special conditions can block the work of electronic schemes. To solve that problem the wide-range emergency detecting blocks consist of two parts: the first part is the detecting node which is disposed in an emergency zone and the second part is the electronic equipment for measuring and analysing the information which is disposed out of the accidental zone (figure 1). Taking into account the length of the cable between detector and remote electronics – it should be minimized.

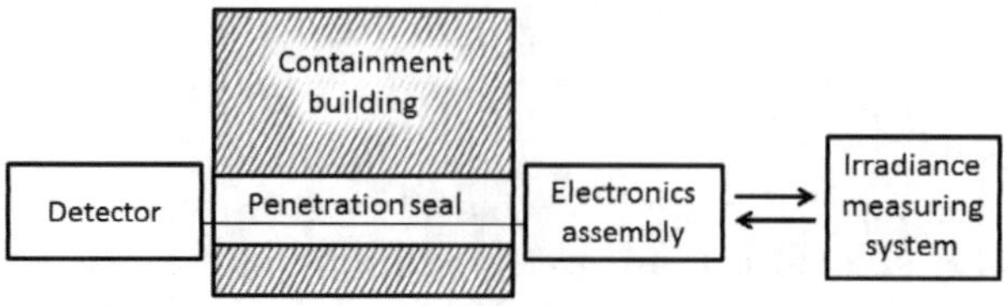

Figure 1. Configuration of detecting block towards the protecting shell.

In the layout of the detecting device (figure 2) from the power supply (PS) to the diamond detector (DD) in an outer block of detecting power 50V was provided (if we reduce bias voltage the detector won`t collect an average charge from the inside levels of the crystal, as soon as if we enlarge the power – the speed of the counting does not grow significantly, but serious electronic breakdown appear. That is why power 50V is the most obtainable). Depending on the operating mode signal from the detector goes to the ADC or preamp. Current, appearing under the influence of ionizing radiation in the detector through the work of digital converter (ADC) was converted into the chain of impulses coming to the microcontroller (M). In impulse mode signal from detector arrives to preamplifier (A). The chain of impulses displayed as the speed of counting impulses per second, which is given as an average number of impulses per 1, 10, 100, 1000 seconds was taken out to the node of indication (D). As the measured value was taken as an average speed of counting in a direct proportion to an average current, appearing in the detector.

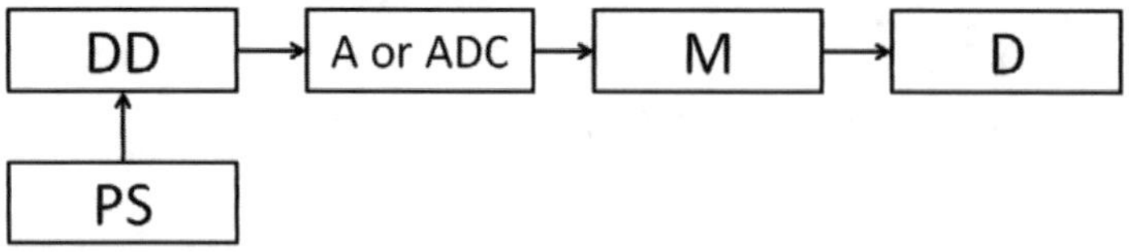

DD – diamond detector
PS – power supply
ADC – digital converter
A –preamplifier
M – microcontroller
D – indicator

Figure 2. Block-scheme of the detector unit.

Diamond has several outstanding properties which make it very attractive for radiation detector applications. Compared to silicon which is most preferred semiconductor material for wide range of radiation detection applications, it has higher bandgap, higher resistivity and e–h mobilities, and better resistance to corrosion. Because of its valuable properties, it is considered to be a promising material for applications involved in high radiation, high temperature, harsh environments, and for fast timing applications [4].

The operating principle of the diamond detector issimilar to the operating principle of the solid-state ionization camera. Passing through the diamond crystal, charged particles cause ionization inside of it. As the result of ionization free charge carriers (conduction electrons and holes) are produced and they move under the influence of the electric field towards the definite electrodes. The same as a usual ionization camera operate, the diamond detector has got two signal operating modes from the detector: impulse mode and current mode.

3. Experiment and results

Current mode scientific tests of the diamond detector proved, that lineal pattern of the measuring characteristics is kept within the dose power limits from 10^{-1} to $2 \cdot 10^{4}$Gy/h (figure 3). And because of linearity is keeping we can make a conclusion that it is possible to register radiation about 10^{5}Gy/h. So the rough measurement range is from 10^{-1} to 10^{5}Gy/h.

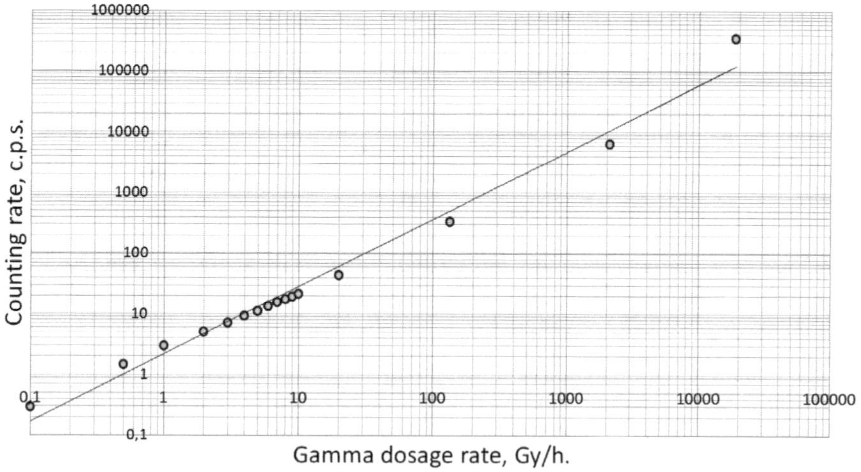

Figure 3. Count rate on gamma dose rate dependence diagram (current mode).

But is this dose rate does not detect an accident at an early stage. There was checked the linearity of counting speed of the diamond-based detector unit in impulse mode within the limits of powers from 10^{-3} Gy/h to 0,4 Gy/h (figure 4). From the results received it is possible to conclude, that within the limits of the given powers the linearity does not change. Increasing the exposure time can expand the range of measurements till 10^{-4} Gy/h.

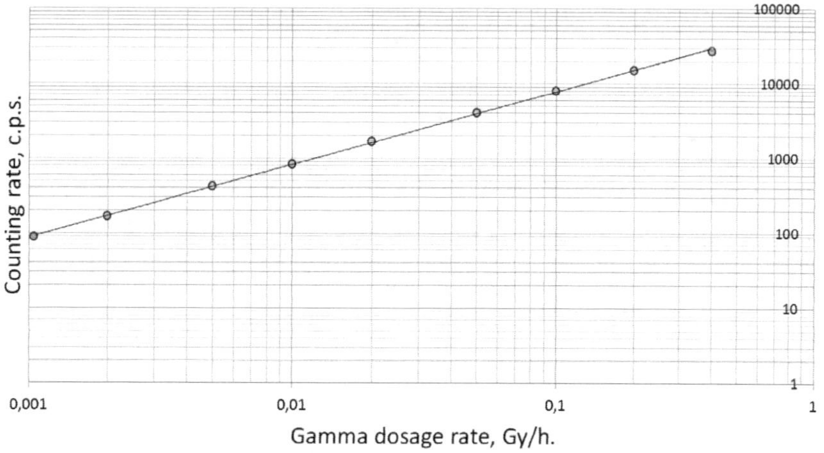

Figure 4. Count rate on gamma dose rate dependence diagram (impulse mode).

4. Conclusion

During the research work the characteristics of the diamond detector were studied in current and in impulse mode. As well it was studied how detector`s sensitivity depends on power doze within the limits from 10^{-3} to 0,4 Gy/h (impulse mode) and from 10^{-1} to $2 \cdot 10^{4}$Gy/h (current mode). The data of the research allow us to make the following conclusion: dependence of the average speed of counting from the power of photon radiation has a lineal character within the limits 10^{-4} to 10^{5} Gy/h. Non-lineal character of measuring characteristics is kept within the limits of 10%.

The results received prove that it is principally possible to create a wide-ranged unit on the basis of diamond detectors for detecting the power and dose of photon radiation absorbed.

References

[1] Knizhnik A S, Fedorchenko S N and Chebyshov S B 2015 (in Russian) *ANRI Journal* **2**(81) 61-6 (Original Russian title: *ANRI*)
[2] IEC 60951-3 1989 part 3
[3] AmosovV N *et al.* 2012 (in Russian) *Applied Physics* **3** 79-85 (Original Russian title: *Prikladnaya fizika*)
[4] Amit Kumar *et al.* 2015 *Nucl. Instrum. Meth.* A **785** 55–60.

Boronated Scintillator Detector for Use in Space with Ionization Calorimeters

G I Britvich[1], S K Chernichenko[1], M A Demichev[2], I I Gnezdilov[3], V I Mukhin[3*] and A V Soukhih[1]
[1] Institute for High Energy Physics (IHEP), ploschad Nauki 1, Protvino, Moscow region, 142281, Russia
[2]Joint Institute for Nuclear Research (JINR), Joliot-Curie 6, Dubna, Moscow region, 141980, Russia
[3] National Research Nuclear University MEPhI (Moscow Engineering Physics Institute), Kashirskoe highway 31, Moscow, 115409, Russia

E-mail: y.gnezdilov@gmail.com, mdemichev@jinr.ru, *g.abc256@gmail.com

Abstract. Boronated Scintillator Detector (BSD) for use in space with ionization calorimeters was suggested. BSD improved e/h showers separation, which are initiated in the ionization calorimeter in interaction it with high energy particles. Improve the rejection is based on the hadron-induced showers tend to be accompanied by significantly more neutron activity than electromagnetic showers. The detector is composed of natural boron-loaded (5%) castable plastic scintillation plates. To collect light using wavelength-shifting (WLS) fibers. The experiment showed that the photoelectron yield is ~ 40 ph.el./MeV with using PMT EMI 9954KB. Simulation on GEANT4 was obtained neutron detection efficiency. The simulation was conducted in the assumption that neutrons have the spectrum ^{252}Cf and fall plane-parallel on the entry surface of the detector.

1. Introduction

In many physics experiments where ionization calorimeters are employed, the requirement of an accurate energy measurement is accompanied by the requirement of very high hadron-electron rejection power. Normally the latter requirement is achieved by designing a high-granularity calorimeter with sufficient depth so that the showers can fully develop. Particle showers initiated by nuclei in the calorimeter have a profile different from an electron-induced electromagnetic cascade, and the hadron rejection power deriving from this difference can be significantly enhanced by making use of the thermal neutron activity at late times relative to the start of the shower. A feature of the use the BSD in space with ionization calorimeters is the low contribution of gamma background, which is a sum of 60% the diffusion space of gamma radiation [1] plus 40% gamma albedo from the atmosphere [2] and short time of detection neutron. It is a few hundred microseconds. Indeed hadron-induced showers tend to be accompanied by significantly more neutron activity than electromagnetic showers. In order to precisely characterize the neutron component associated with the different types of showers, a detailed Monte Carlo simulation has been performed using the FLUKA simulation package. The CALET BGO homogeneous calorimeter, consisting of a BGO tower (60×60×30 cm^3) with a vertical depth of approximately 30 cm has been simulated as a "test detector". For each simulated event the produced neutrons have been propagated till they reached the outside of the calorimeter. Neutron yield outside the CALET calorimeter shown on figure 1 for 1 TeV interacting protons and 400 GeV electrons (1 TeV proton showers give the same calorimetric signal as 400 GeV electrons in BGO on average).

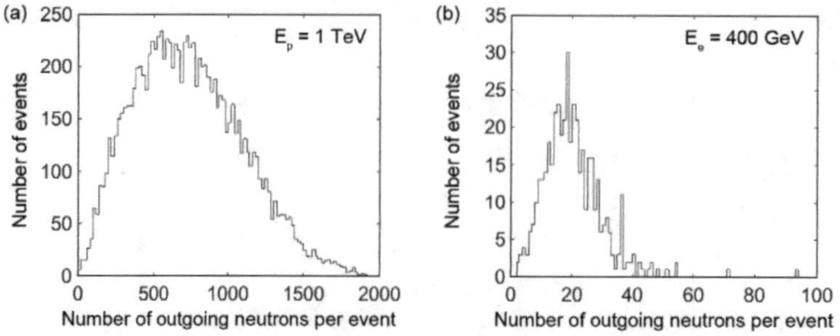

Figure 1. Neutron yield outside the CALET calorimeter for (a) 1 TeV interacting protons and (b) 400 GeV electrons [3]

The rejection factor achievable for hadronic showers can be as high as ~ 10^3 considering the neutron counting. The bulk of neutrons originate from the excitation and de-excitation of nuclei and exhibit a maximum in a several MeV energy region. Many neutrons undergo moderation before escaping and their energy is consequently degraded down to the few eV energy region. Some neutrons can also be produced promptly in the hadronic interactions along the shower core, with an energy reaching that of the primary proton. The highest energy neutrons (E > 10 MeV) arrive simultaneously with the charged component of the shower, while the low energy and more abundant component arrives into the neutron detector with a delay ranging from 10 to 1000 ns and, thus, can be easily identified. The figure for electromagnetic showers is similar but with a much reduced contribution for the prompt neutron emission [3]. Similar results were obtained for the hadronic and electromagnetic interaction with the CsI(Tl) homogeneous calorimeter (100×100×43 cm^3) which is mounted on the GAMMA-400 gamma-ray telescope [4].

2. Overview of the known the boronated scintillator detectors for use in space with ionization calorimeters

2.1. ISS-CREAM BSD

The first Boronated Scintillator Detector (BSD) is a detector sub-system for the future Cosmic Ray Energetics and Mass for the International Space Station (ISS-CREAM) mission [5]. It aims to complement the instrument's tungsten calorimeter in identifying cosmic-ray electrons above 100 GeV. Such electrons are of significant scientific interest, but their identification is complicated by the overwhelmingly more abundant hadronic cosmic rays, hence making significant hadronic rejection power of paramount importance. Particle showers initiated by nuclei in the ISS-CREAM calorimeter have a profile different from an electron-induced electromagnetic cascade, and the hadron rejection power deriving from this difference can be significantly enhanced by making use of the thermal neutron activity at late (>400 ns) times relative to the start of the shower. Indeed hadron-induced showers tend to be accompanied by significantly more neutron activity than electromagnetic showers. The BSD endeavors to measure this late thermal neutron shower activity by detecting the boron capture of these thermal neutrons in a boron-loaded plastic scintillator, located underneath the calorimeter. BSD represents a plastic scintillation block Eljen Technologies EJ-254 with natural boron-loaded weight percentage of 5% which is made in the form of right prism with square basis the size of 60×60×3.8 cm^3. Light collection of scintillations is realized by 12 items Hamamatsu R1924A (ø22 mm) photomultiplier tubes, that are situated in groups of 6 items on opposite lateral faces of the scintillation block.

Results test of the BSD in the CERN H2 beam line shown in figure 2, where shows the resulting energy deposit distributions for electrons and pions in both the ISS-CREAM BSD and the calorimeter. The BSD's ability to distinguish electrons from hadrons is immediately clear. Even for a broad range of calorimeter signal values, the BSD's electron distribution tends to cluster at lower values of BSD signal than the pion distribution. As will be seen, these differences sharpen even more when the calorimeter's signal range is limited to emulate an energy range selection of the sort that will be carried out with flight data.

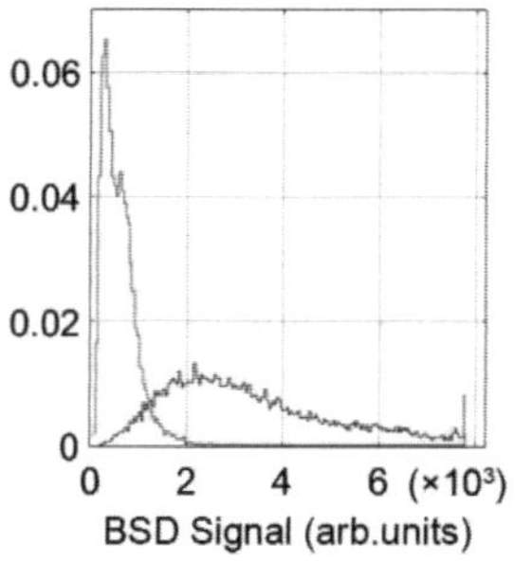

Figure 2. BSD signals for 75 GeV, 100 GeV, 125 GeV, 150 GeV, and 175 GeV electron runs (red) and 250 GeV, 300 GeV, and 350 GeV pion runs (blue). Histograms are normalized by total number of counts [5].

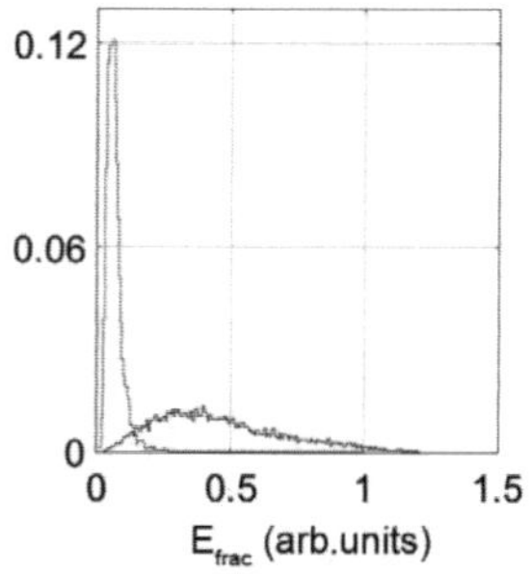

Figure 3. E_{frac} distributions for 100 GeV – 125 GeV electrons (red) and 250 GeV – 350 GeV pions (blue). Histograms are normalized by total number of counts [5].

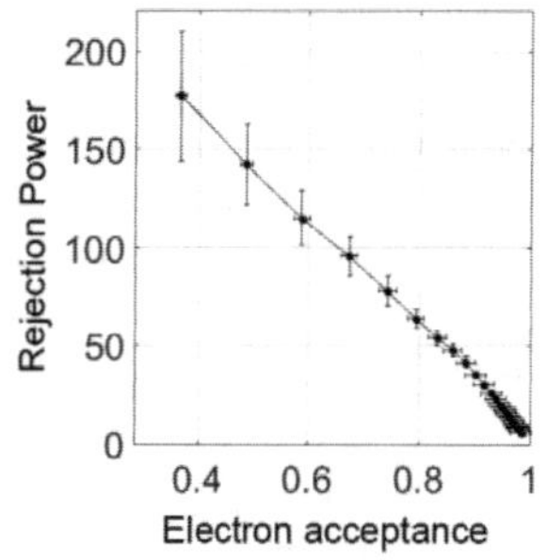

Figure 4. Rejection power versus electron acceptance for 100 GeV – 125 GeV electrons and 250 GeV– 350 GeV pions [5].

For the purpose of deriving rejection power, the parameter E_{frac} is defined as

$$E_{frac} = \textit{(BSD Delayed Neutron Signal) / (Calorimeter Energy Deposit)} \quad (1)$$

A plot of E_{frac} in a specific range of calorimeter energy deposit is shown in figure 3, with electrons shown in red and pions shown in blue. Clearly, the majority of pions fall well outside of the range of E_{frac} values for electrons, most of which lie below E_{frac} = 0:1. Thus, in choosing candidate electron populations for spectral studies, a higher efficiency by reducing the maximum allowable value of E_{frac}. Below a given value of E_{frac}, hadronic rejection power for the BSD was defined as

$$\textit{Rej. Power} = \textit{(Fraction of Electrons Accepted) / (Fraction of Hadrons Not Rejected)} \quad (2)$$

Figure 4 shows a plot of rejection power versus electron acceptance derived by gradually raising the maximum allowable value of E_{frac} in figure 3. Error bars are statistical only. Here, the rejection power using only the BSD and the total energy deposit in the calorimeter is near 140 at 50% electron acceptance, which was the BSD target design performance specification. Rejection power from other handles on electron/hadron discrimination (e.g. shower shape) will combine with the BSD's neutron detection to create ISS-CREAM's overall planned rejection power of at least 10^4.

2.2. DAMPE BSD

The second BSD is a detector sub-system for the future DArk Matter Particle Explore (DAMPE) mission [6]. DAMPE is a powerful space telescope for high energy gamma-ray, electron and cosmic rays detection. The ionization DAMPE calorimeter consist of about 31 radiation lengths thickness, made up of 14 layers of BGO bars in a hodoscopic arrangement. The neutron detectors is added to the bottom of the calorimeter. Finally, in order to detect delayed neutron resulting from hadron shower and to improve the electron/proton separation power a neutron detector is placed just below the calorimeter. BSD consists of 4 plastic scintillation block Saint-Gobain Corporation BC-454 with natural boron-loaded weight percentage of 5%. The dimension of each scintillator is 30×30×1 cm^3, total size of the BSD is 60×60×1 cm^2. Total rejection power at last 10^5.

3. The design of the BSD using boron-loaded plastic scintillation plates and wavelength-shifting (WLS) fibers for collect light

3.1. Design

The third BSD is a detector sub-system for the future space projects related to the registration of high-energy particles and gamma rays, for example, GAMMA-400 [4], NEYTRONIY-100 [7]. These projects demand neutron detectors with larger area of sensitive surface not less than 1 m^2. As shown in figure 5, the proposed BSD with the size of 100×100×10 cm^3 contains 800 scintillation plates with the size of 25×5×0.5 cm^3, which contain natural boron with the weight concentration up to 3% (SC-331 plastic scintillator). Scintillation plates were made by casting under the pressure according to the technology developed in Institute of High Energy Physics. Each plate has 168 holes under WLS fibers with the diameter of 1 mm. WLS fibers inserted into holes of the scintillator provides light collection for one photodetector. PMT or SiPM may be used as a photodetector. Using WLS fibers allows making neutron detectors which have great light-collecting power. In order to increase neutron detection efficiency the detector additionally contains 2 layers of hydrogen-containing moderator with the total thickness of 5 cm between which scintillation plates were placed.

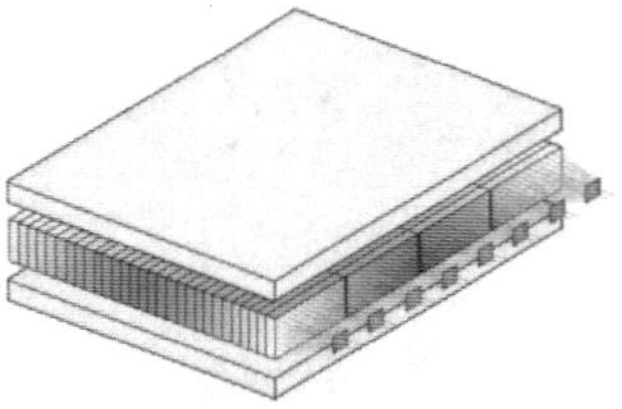

Figure 5. BSD with WLS fibers, SiPMs and moderator (yellow).

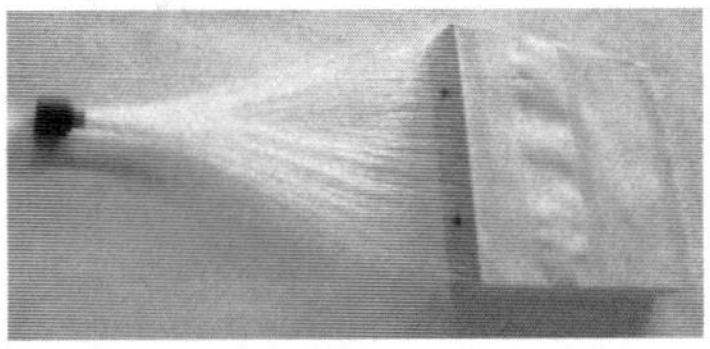

Figure 6. Experimental module detector with WLS fibers (75x25x5 cm^3).

3.2. Experiment

To measure quantum yield experimentally, an experimental module which is shown in figure 6 with the size of 75×25×5 cm^3 was made when using PMT. The experimental module of the detector contained scintillation plates without boron wrapped into light-reflecting material Tyvek. Light collection was realized by WLS fibers Kuraray Y-11(200) with the emission peak of 476 nm. EMI 9954KB was used as PMT with the spectral sensitivity prolonged into the green area and with the quantum detection efficiency of 16% on the wavelength of 526 nm. According to the method described in the work [8], the amount of photoelectrons was measured from the energy of gamma-quanta absorbed in the detector. From the results of the measurements shown in figure 7, it follows that quantum yield is equal to 55.17 ph.el./MeV with the error of 4%. Since the light output of scintillation plates with boron is 25% lower, quantum yield of the BSD is ~ 41 ph.el./MeV. When using SiPM SensL ArrayC-60035-4P (12×12 mm^2) as a photodetector quantum yield of the BSD is ~ 2 times higher. Because the light output of SC-331 plastic scintillator from ^{10}B (n,α) ^6Li reaction products was equivalent to the light output from a β-particle with an energy of about 110–130 keV, the quantum yield is ~ 10 ph.el. for one absorbed neutron. The obtained amount of photoelectrons is enough for signal extraction against the noise background.

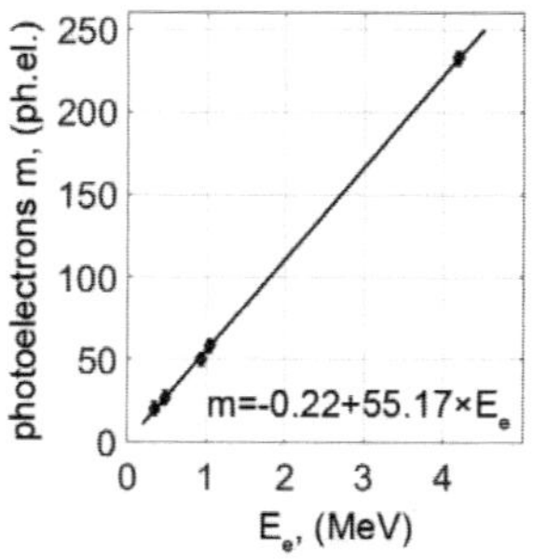

Figure 7. The dependence of the number of photoelectrons (ph.el.) from the energy gamma rays that absorbed in the scintillation detector.

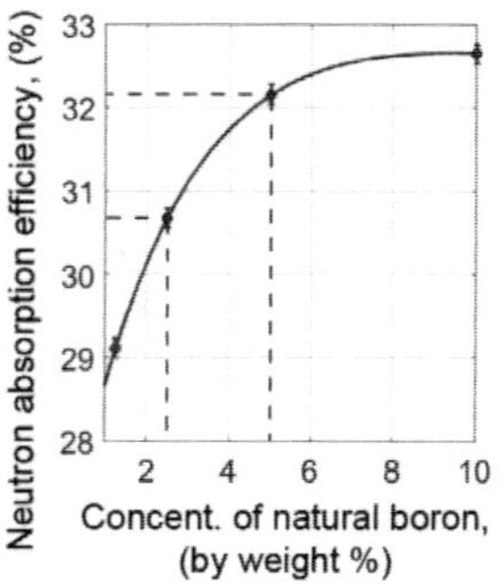

Figure 8. The dependence of neutron absorption efficiency from the concentrations of natural boron in BSD.

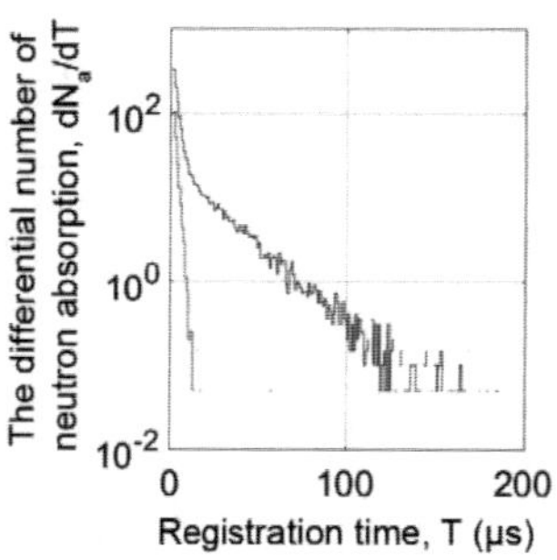

Figure 9. The differential distribution of neutron absorbing time (after 400 ns) with an integration time of 1 μs for 5000 neutrons that crossed ISS-CREAM BSD (red) and BSD (blue).

3.3. Simulation

Neutron detection efficiency was obtained by the simulation on GEANT4. The simulation was conducted in the assumption that neutrons have the spectrum ^{252}Cf and fall plane-parallel on the entry surface of the detector. The energy spectrum of the neutron fission of ^{252}Cf is a good approximation to the spectrum of evaporated neutrons produced in the interaction of high-energy e and h with a calorimeter. Table 1 shows the estimated efficiency of neutron detection of ISS-CREAM BSD and the BSD without or with hydrogen-containing moderator.

Table 1. Calculated neutron absorption efficiency

	Neutron absorption efficiency	
	w/o moderator	with moderator
ISS-CREAM BSD	4.2%	24.7%
BSD	4.3%	29.8%

Including the moderator into the detector design increases neutron detection efficiency with ^{252}Cf spectrum by 6–7 times. The graph in figure 8 shows that the increasing of natural boron (from 2.5% to 5.0%) does not much affect neutron absorption efficiency. The decay time of Y-11(200) under laser excitation is 8.5 ns. The decay time of SC-331 plastic scintillator is 2.2 ns. Thus, WLS fibers delay rise time of rise-up portion of the scintillation impulse and shifts maximum highlighting time in comparison with the scintillator without boron. Since the proposed BSD is composed of 4 scintillation blocks with the size of 100×25×5 cm^3 and contains 4 separate detection channels, its operating speed does not exceed 3 ns. Figure 9 shows the differential distribution of neutron absorption time (after 400 ns) with the integration time of 1 μs for 5000 neutrons that crossed BSD. Thus, BSD allows detecting neutron fluxes with the fluence of ~ 5 n/cm^2 without miscounts.

4. Conclusions

The described BSD has neutron detection efficiency by 7 times more than ISS-CREAM which increases rejection power. Using thin scintillation plates and wavelength-shifting fibers for light collection allows making BSD with an area of sensitive surface with the size of a few square meters and low power consumption. Thus, BSD is a high efficiency detector for use in space with ionization calorimeters.

Acknowledgements
The authors acknowledge Irina Popova for the creating conditions for the article writing and Azat Fattakhov in its assistance.

References

[1] Ajello M *et al.* 2008 *The Astrophysical Journal* **689** 666–77
[2] Petry D 2005 *High Energy Gamma-Ray Astronomy* 709–714 (*arXiv*:astro-ph/0410487v1)
[3] Adriani O *et al.* 2011 *Nuclear Instruments and Methods in Physics Research* **A628** 332–8
[4] Leonov A A *et al.* 2014 *Proceedings of Science* (*arXiv*:1412.1486v2)
[5] Anderson T *et al.* 2013 *Proceedings of the 33rd Int. Cosmic Ray Conf.* (Rio de Janeiro)
[6] Chang Jin 2014 *Chin. J. Space Sci.* **34**(5) 550–7
[7] Kalmykov N N *et al.* 2013 (in Russian) *Nuclear Physics* **76** 82–9
[8] Britvich G I *et al.* 2013 *Preprint* IHEP 2013-23

Silicon detectors with boron converters of different geometrical modifications for fast neutrons registration

E V Ryabeva[1] , V V Kadilin[1], G L Dedenko[1] and Tant Zin[2]
[1] National Research Nuclear University MEPhI (Moscow Engineering Physics Institute), Kashirskoe highway 31, Moscow, 115409
[2] Defence Service Science and Technology Research Center (DSSTRC), Pyin Oo Lwin, Myanmar

E-mail:EVRyabeva@mephi.ru

Abstract. In this article the development of the model for fast neuron detector is considered. The construction of detector is based on alternating layers of silicon as sensing material of charged particles and boron (enriched with ^{10}B) as neutron converter located in moderator Simulation in GEANT4 showed that detector with one layer of silicon and one layer of boron has very low detection efficiency for fast neutron (approximately 4%).. The study of the possibility to increase detector efficiency due to geometry optimization of it is presented It is shown that efficiency depends on the number of layers and their location in the materials of detectors and moderator. It was found that the optimal geometry is that where the layers of boron-silicon are angularly related to each other. It is shown that the detection efficiency for a model with set of boron-silicon layers angularly related to each other is up may reach 6%.

1. Introduction

In the article the model of fast neutron detector with small size and low power consumption is considered. The detector consists of neutron converter – layers of boron enriched in the isotope ^{10}B and layers of pure silicon as the detector of the charged particles. The article offers the optimum size of the layers of boron and silicon, and the calculated detection efficiency of the detector for optimized model. The detection efficiency is ≈4% for the model of the detector with a thickness of 1 micron layer of boron and silicon layer 5.5 mm with parallel layers The detection efficiency for a model where the set of boron-silicon layers are angularly related to each other is up to 6%.

2. Neutron flux to be registered by detector

The proposed detector is focused primarily on the fast neutrons detection. Modern powerful neutron sources are based on the nuclear reactions of the charged particles beam (including high-energy) with the target material [1]. The problem of registration of secondary neutrons produced by high-energy electrons and hadrons arises during the astrophysical observations in outer space in the high-energy gamma-range [2]. The mechanism of the neutron origin in the sources mentioned above is similar and consequently an energy range of the resulting neutron fluxes are similar. The average energy of evaporated neutrons is ~ 2 MeV, they have an isotropic angular distribution. The energy range of an evaporated neutrons is close to the range of fission neutrons with peak located at the energies from 0.5 MeV to 1 MeV.

In this paper the software package GEANT4 [3] is used to simulate the interaction of neutrons with a neutron detector and determine the detection efficiency using different designs of the detectors.

3. The principle of the neutron detection

One of the most popular reaction to convert slow neutrons into directly detected particles is the ^{10}B(n,α) [4]. The reaction may be written:

$$^{10}_{5}B + ^{1}_{0}n \rightarrow \begin{cases} ^{7}_{3}Li + ^{4}_{2}\alpha & 2.792\text{MeV (ground state)} \\ ^{7}_{3}Li^{*} + ^{4}_{2}\alpha & 2.792\text{MeV (excited state)} \end{cases}$$

Silicon detector is proposed to be used as monitor of charged particles that are born in this reaction. Given that 94% of all reactions is on the way in which the lithium nucleus born in the excited state and the reaction products are emitted in the exactly opposite directions it is possible to calculate E_{Li} = 0.84 MeV and $E\alpha$ = 1.47 MeV, the energy of reaction (2310 keV) is shared between the alpha particle and a lithium nucleus. That is why the discrimination threshold in the model was set on energy 50 keV.

4. Overview of the similar design

The options for a detector based on layers of boron and silicon semiconductor were discussed in the earlier studies [5–7]. It was showed that the simplest version of detector geometry, where the layers of the semiconductor and converter are irradiated normally by the incident neutrons [5] has a low detection efficiency for thermal neutrons (about 4%) at the discrimination threshold 300 keV. As practice showed, the measured values of detection efficiency are usually lower than calculated, since it requires a higher level of discrimination that would exclude gamma rays. The modifications of the geometric arrangement of the layers were proposed to increase the efficiency. For example, in [7] the detector surface is fabricated with grooves into which the converter material is deposited. It is shown that such a structure can greatly improve the detection efficiency. Study presented in this article also aimed at searching for the simple optimum geometry of the detector layers of boron and silicon which can improve the efficiency of neutron detection without increasing of a cost of the detector.

5. Calculation of the geometry for the detector simplest version

Simplest versions of neutron detector have a vertical and horizontal parallel layers of boron and silicon placed in polyethylene as moderator for fast neutrons (figure 1 and figure 2) and have a combined thickness of 5 cm and dimensions 80x80 cm. The thickness of the polyethylene layer is 2.5 cm in construction with horizontal layers (Boron + Silicon) and 2 cm in construction with vertical arrangement of layers (Boron + Silicon). The material of the converter – Boron – is enriched with the isotope ^{10}B to 90%.

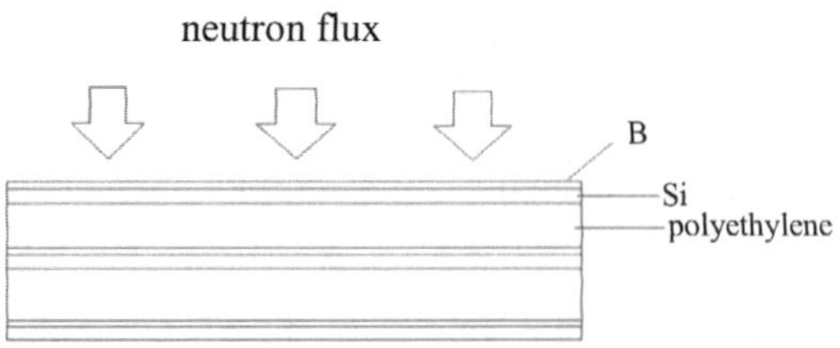

Figure 1. Design of detector with horizontal arrangements of layers (B+Si).

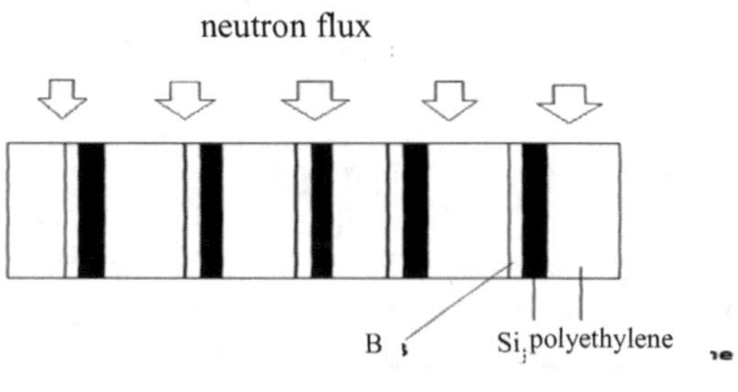

Figure 2. Design of detector with vertical arrangements of layers (B+Si).

6. The optimum thickness of layers of silicon and boron

First and foremost to optimize the detector efficiency and geometry it is necessary to determine the optimum thickness for the silicon and boron layer. To do this, the mean free path of the particles in the materials of the proposed structure has been calculated and compared with the literature [4]. The converter layer must be kept thin because of the limited range of the reactions product in solid. For alpha particles it is equal to 3.6 μm in boron, 1.6 μm in lithium and 5.5 μm in silicon. Particles range in solid boron for alpha particles – 3.6 μm and for lithium – 1.6 μm. For silicon layer particles range for alpha particles is 5.5 μm. It is proposed to choose the thickness of the boron from 1 μm to 2.5 μm in order to alpha particles and nuclei of lithium will be able to leave a layer of boron. For silicon layer is advisable to choose a thickness 5.5 μm, so that all the alpha-particles are completely absorbed in it. The values of the neutron detection efficiency with different thicknesses of boron converter (1 μm and 2.5 μm) were calculated. In these calculations, the thickness of the silicon layer was 5.5 μm. The number of layers (Boron+ Silicon): 3 layers in model with horizontal layers (Boron+ Silicon) and 80 layers in the model with a vertical arrangement of layers. The detection efficiency value was found in the range 3.80 – 4.56% for all designs. The efficiency is slightly better for the design with boron layer thickness 1 μm than with the thickness 2.5 μm in both construction of the neutron detector. The design with vertical layers get a slightly better detection efficiency.

7. The detector surrounded by layers of boron and silicon

Further calculations were performed for the embodiment of the neutron detector, surrounded by layers of boron and silicon on all sides as it is showed in figure 3. The thickness of the silicon layer is fixed at 5.5 μm. Simulation gave the values of the neutron detection efficiency in range (4.1–4.2)±0.1%.

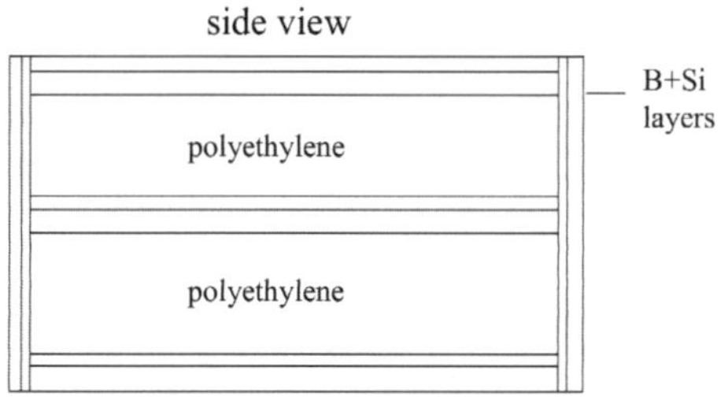

Figure 3. The design of the neutron detector, surrounded by layers of boron and silicon on all sides.

The additional layers (or grater thickness) do not provide a significant increase in the efficiency of the neutron detection. However, we must consider the fact that the increase in the number of layers or increase the thickness of converters and recorders leads to the growth in the size and price of the neutron detector.

8. The detector with layers of silicon and boron angularly related to each other

The construction of the neutron detector with layers of silicon and boron angularly related to each other as it is shown in figures 4 and 5. The layers of the converter and sensor laid at the angle of 45° to each other. The energy discrimination level was equal 50 keV. The simulated efficiency of the detectors was found to be not more than 6.0±0.5%.

Journal of Physics: Conference Series **675** (2016) 042005 doi:10.1088/1742-6596/675/4/042005

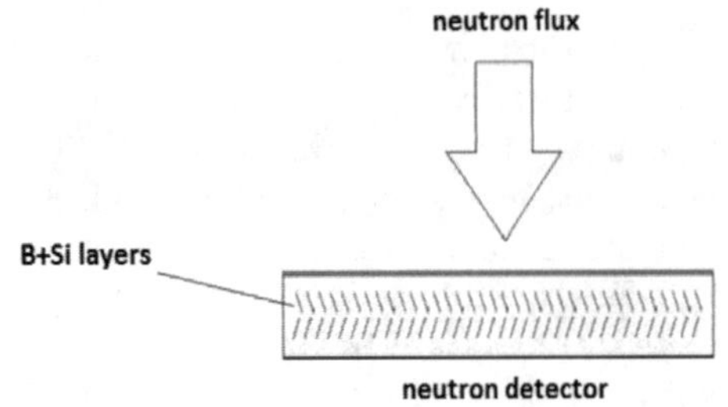

Figure 4. Design of neutron detector. The layers of silicon and boron angularly relating to each other are placed in polyethylene as moderator.

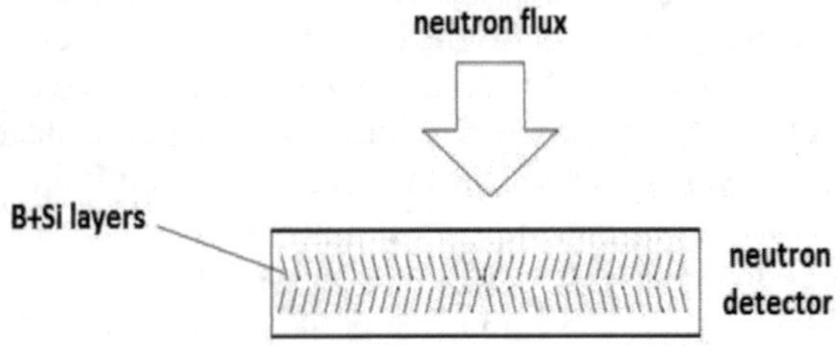

Figure 5. Design of neutron detector. The layers of silicon and boron angularly relating to each other are placed in polyethylene as moderator. The angles are different.

The dependence of the efficiency on the energy discrimination level and the angle between the layers were simulated. These dependences are shown in figures 6 and 7. The simulation of efficiency for various models and various discrimination threshold showed that the improvement of the registration efficiency is not significant when the discrimination threshold and the angle between the layers are changed.

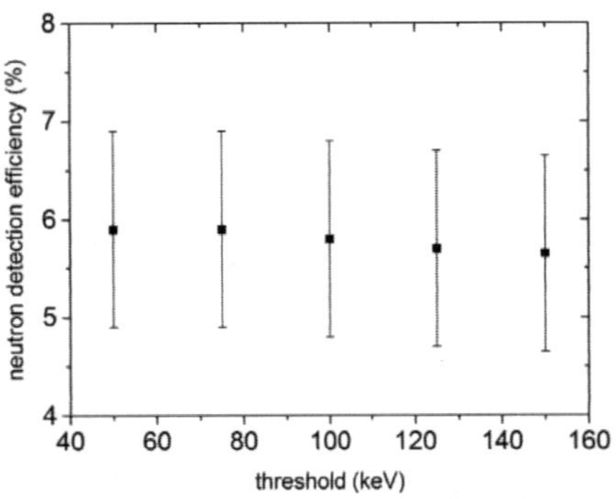

Figure 6. The dependence of the detection efficiency on discrimination threshold energy release.

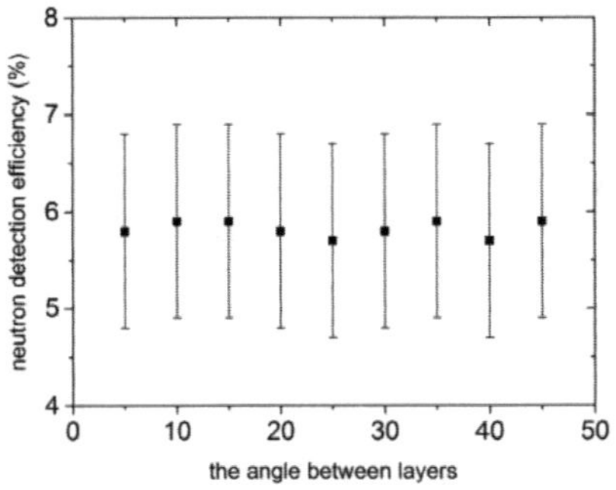

Figure 7. The dependence of the detection efficiency on the angle between layers B and Si.

9. Conclusion

According to the results of the calculations for neutron detector it is advisable to limit the thickness of the boron layer to 1 μm and the thickness of the silicon to 5.5 μm. Registration part of the detector has a limited size, because it is a semiconductor detector. Due to the lack of sophisticated electronics and systems for collecting the optical signal power consumption of the detector will be minimal. The main drawback of such a detector is the low detection efficiency. In the design of the neutron detector with a horizontal and vertical layout of the boron and silicon layers detection efficiency is about 4%, which confirmed the previously published data [4].

The detection efficiency of neutron detector can be increased by changing the geometry of the detector. The efficiency of the detector, wherein the layers are angularly related to each other, reaches ≈6%.

Acknowledgments
Russian colleagues also acknowledge partial support from MEPhI Academic Excellence Project (contract № 02.a03.21.0005, 27.08.2013).

References
[1] Report 2005 Development opportunities for small and medium scale accelerator driven neutron sources *Report of a technical meeting held in Vien*na, 18-21, IAEA-TECDOC-1439
[2] Topchiev N P *el al.* 2015 The experiment "Gamma-400": Status and Prospects *Bulletin of the Russian Academy of Sciences: Physics, ISSN: 1062-8738* **79**(3) 417-20
[3] URL: *http://geant4.web.cern.ch/geant4/UserDocumentation/UsersGuides/ForApplicationDeveloper*
[4] Knoll G F 2010 *Radiation Detection and Measurements* (John Wiley & Sons, 4-th Edition) p 521
[5] McGregor D S *et al.* 2003. Design Considerations for Thin Film Coated Semiconductor Thermal Neutron Detectors, Part I: Basics Regarding Alpha Particle Emitting Neutron Reactive Films *Nucl. Instrum. Meth.* A **500** 272-308
[6] Wielunski M *et al.* 2004 Spectral identification of thin-film-coated and solid-form semiconductor neutron detectors *Nucl. Instrum. Meth.* A **517** 180-7
[7] Jahan Q *et al.* 2007 Neutron Dosimeters Employing High-Efficiency Perforated Semiconductor Detectors *Nucl. Instrum. Meth.* B **263** 183-5

Application peculiarities of magnetic materials for protection from magnetic fields

P Wai[1,*], V Dmitrenko[1], S Grabchikov[2], K Vlasik[1], A Novikov[1], D Petrenko[1], V Trukhanov[2], S Ulin[1], Z Uteshev[1], V Chernysheva[1] and A Shustov[1]

[1] National Research Nuclear University MEPhI (Moscow Engineering Physics Institute), Kashirskoe highway 31, Moscow, 115409, Russia

[2] Scientific-Practical Materials Research Centre, National Academy of Sciences of Belarus, Minsk, Belarus

E-mail: *phyowainyunt@gmail.com

Abstract. In different materials for magnetic shields, the maximum permeability is achieved for different values of the magnetic field. This determines the choice of material. So for protection from magnetic fields strength of 10 – 150 A/m it is advisable to apply the amorphous ribbon 84KXCP. For stronger fields (more than 400 A/m) it is recommended to use MFS based on $Ni_{20}Fe_{80}$. Use of these materials allows creating an effective shield working in a wide range of magnetic field strengths.

1. Introduction

In modern physics experiments and installations detectors (detection units) scientific equipment is used that is susceptible to constant magnetic fields. For example, accelerators (ATLAS CERN) widely use photomultiplier tubes (PMTs) that are very sensitive to magnetic fields, which are adjacent to the superconducting magnet generating a magnetic field with a strength of several Tesla (10^4 Gauss). The same situation is observed in the equipment installed on the spacecraft. For example, the "Pamela" experiment, which is a magnetic spectrometer, uses a magnetic system based on the strengths of the constant magnets and the PMT. In the equipment used in nuclear medicine the situation is the same.

The most difficult case of protection from exposure to external fields is the shielding from static magnetic fields [1, 2]. Usually, the magnetostatic shielding is based on the principle of shunting the magnetic field by ferromagnetic material [3, 4], the main essence of which is enclosing the force lines inside the material with a low resistance to magnetic flux.

One of the main and most effective ways to protect against EMR is shielding. Therefore, the work on creation of highly effective shield against alternating electromagnetic radiation, as well as against the constant magnetic fields on the basis of new materials and technologies is very important.

To provide a high shielding effectiveness in a wide range of frequencies it is desirable to apply the multilayer metallic shields, consisting of layers with high permeability and layers with high electrical conductivity. The shielding effectiveness in such materials increases due to the fact that there is a multiple reflections of the EMI between the layers in multilayer shields consisting of metals with different characteristic impedance. The reflection effect is observed at interfaces with different wave resistances.

2. Method and experiment

Magnetic shield is formed by electrodeposition method of soft magnetic alloys $Fe_{20}Ni_{80}$ [5]. Quantitative evaluation of the shielding effectiveness E conducted by ratio of the voltage measurements or magnetic field induction in the protected area of space in the absence of the shield H (or B) and in the presence of his H_1 (or B_1) [6]:

$$E=B/B_1=H/H_1$$

The figure 1 shows the dependences of the shielding effectiveness of magnetostatic field for the shields based in electrodeposited alloy $Ni_{80}Fe_{20}$ and amorphous ribbon 84KXCP. For magnetic fields from 15 to 135 A/m the effectiveness of the shield of ribbon 84KXCP is higher than the one of the shield of the alloy $Ni_{80}Fe_{20}$. In the interval from 135 to 2700 A/m shields based on alloy $Ni_{80}Fe_{20}$ are more preferred, although in the sense of μ_a values they are inferior to shields based on amorphous ribbon 84KXCP. From this experiment it follows that the shielding effectivenesses for different materials are substantially different. For the amorphous alloys 84KXCP with high μ_a values and low values of coercivity $H_c = 0.7$ A/m the position of maximum of the magnetic permeability curve ($\mu = f$ (H)) should be in the range of magnetic fields lower than 1A/m. Lower values of $H\mu_a$ of shields based on amorphous ribbon 84KXCP compared to shields based on alloy $Ni_{80}Fe_{20}$ (figure 1) determine their high effectiveness in magnetic field range from 15 to 135 A/m (figure 1). For values of H $\geq$100 A/m all 84KXCP material goes into magnetic saturation and the efficiency of the shield decreases sharply. Based on the above results, we can conclude the following: for optimum protection from static magnetic fields it is necessary to have clearly defined parameters of these fields; the selection of the shielding materials should be taken into account as long as the main magnetic characteristics - initial and maximum magnetic permeability, saturation magnetization, residual magnetization and the maximum field strength of the magnetic permeability. Parameter $H\mu_a$ is a very important characteristic to evaluate and predict the effectiveness of magnetic shields.

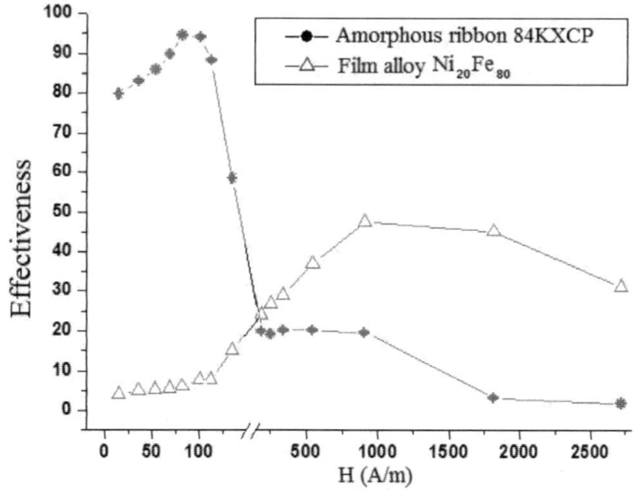

Figure 1. Dependence of the effectiveness of shields. Experimental errors located within the points. (Lines connecting points are guided to the eyes only).

It should also be noted that the maximum efficiency of the final cylindrical shields is achieved for external magnetic fields greater than the value of H corresponding to the maximum μ. For example, for the amorphous ribbon 84KXCP and alloy $Ni_{80}Fe_{20}$ the values H at which μ has a maximum, are equal to 1 A/m and 80 A/m, respectively. Maximum shielding effectiveness of the amorphous ribbon 84KXCP is achieved for the magnetic field of 80 A/m and for the alloy $Ni_{80}Fe_{20}$ – 900 A/m. This is due to the different topologies of the magnetic flux in the sample during the measurement of parameters μ and E. In the first case we have a uniform, closed magnetic flux in a ring-shaped sample. In the second – two magnetic fluxes directed toward each other in a sample of cylindrical shape. Some inconsistencies in the results of measurement of the alloys $Ni_{80}Fe_{20}$ shielding effectiveness are caused by significant differences in the size and shape of the testing shields.

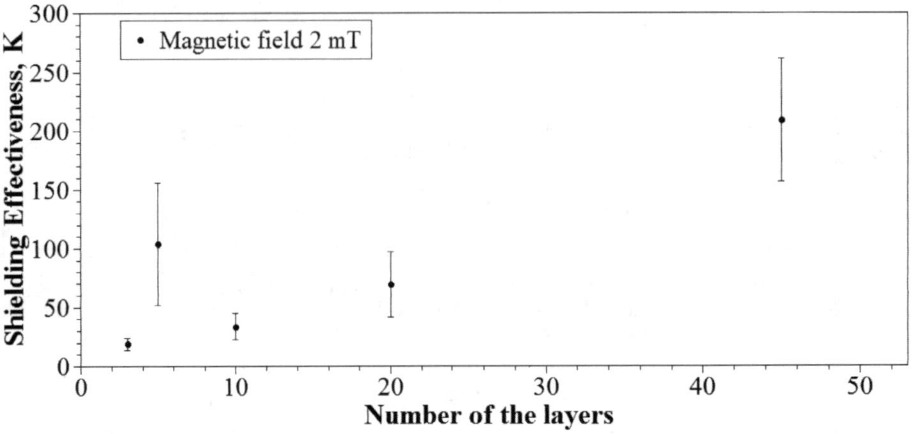

Figure 2. Dependence of the shielding effectiveness against a constant magnetic field induction on the number of layers (at a constant total thickness of the soft magnetic material).

The shielding effectiveness increases with the number of layers in the shielding screen (at a constant total thickness of the soft magnetic material). The figure 2 shows the shielding effectiveness versus number of layers dependence for the magnetic field of 2 mT. The result obtained for the shield with five layers, possibly due to an error in the deposition of the magnetic layers of this shield.

3. Conclusion

Thus, this paper is devoted to the comparison of the magnetic shields based on the amorphous ribbon 84KXCP and alloy $Ni_{80}Fe_{20}$. Experimental results on the shielding effectiveness made of the $Ni_{80}Fe_{20}$ alloy and the industrial material 84KXCP show that for magnetic fields in the range from 100 to 2700 A/m shields based on alloy $Ni_{80}Fe_{20}$ are preferred over shields based on amorphous ribbon 84KXC.

Acknowledgments

Authors wish to acknowledge the Center of Fundamental Researches and Particle Physics. This work was partially supported by MEPhI Academic Excellence Project (contract No. 02.a03.21.0005, 27.08.2013).

References

[1] Fufaeva L and Timofeev A 1976 (in Russian) Shielding effectiveness of constant magnetic field by multilayer shields *Proc. of MAI* **364** 58-63 (Original Russian title: Effectivnost ekranirovaniya postoyannyh magnitnyh polei mnogoslonymi plonochnymi ekranami *Sbornik trudov MAI* **364** 58-63)

[2] Glonrgin U 1967 (in Russian) *Elements of the theory and calculation of magnetostatic fields of ferromagnetic bodies* (Leningrad: Shipbuilding) p 180 (Original Russian title: *Elementy teorii i rascheta magnitostaticheskih polei ferromagnitnyh tel.* L.: Sudostroenie)

[3] Volin M 1981 (in Russian) *Parasitic processes in electronic equipment* (Moscow: Radio and Communications) p 296 (Original Russian title: *Parazitnye processy v radiojelektronnoi apparature.* M.: Radio i svjaz')

[4] Chernyshenko A 1983 (in Russian) *Construction of microwave devices and screens* (Moscow: Radio and Communications) p 400 (Original Russian title: *Konstrukcii SVCh ustrojstv i ekranov*. M.: Radio i svjaz')

[5] Grabchikov C, Sosnovskar L and Sharpa T 2009 (in Russian) Multilayer electromagnetic shield *RB patent no* 11843 *from 2009.01.28* (Original Russian title: Mnogoslojny elektromagnitny ekran *Patent RB №11843 ot 2009.01.28*)

[6] Shapiro D 1975 (in Russian) *Basic theory of electromagnetic shielding.* (Leningrad: Publisher Energy) p 112 (Original Russian title: *Osnovy teorii elektromagnitnogo ekranirovanija.* L.: Izdatel'stvo energija)

Measurement of Radon concentration by Xenon gamma-ray spectrometer for seismic monitoring of the Earth

A Novikov*, S Ulin, V Dmitrenko, K Vlasik, O Bychkova, D Petrenko, Z Uteshev and A Shustov
National Research Nuclear University MEPhI (Moscow Engineering Physics Institute), Kashirskoe highway 31, Moscow, 115409, Russia

E-mail: *nas133@gmail.com

Abstract. A method for earthquake precursors search based on variations of ^{222}Rn concentration determined via intensity measurement of ^{222}Rn daughter nuclei gamma ray emission lines by means of xenon gamma-ray spectrometer is discussed. The equipment description as well as the first experimental data are presented.

1. Introduction

Prediction of earthquakes today is a very important topic of interest of the modern science. One of the most promising methods for earthquake prediction today is based on measurement of concentration of radon-222 (^{222}Rn), emitted from the rock. Preliminary research shows that a certain increase in ^{222}Rn concentration can act as a precursor of a possible earthquake (several days in advance) [1]. One of the most perspective methods for the task is the registration of gamma-ray emission lines of ^{222}Rn daughter nuclei, in particular ^{214}Bi and ^{214}Pb.

Notably, there have already been attempts to employ scintillation detectors for this purpose [2]. However, this kind of equipment has a number of disadvantages. First of all, insufficient energy resolution that, for example, does not allow to resolve the 609 keV energy line of ^{214}Bi and the 583 keV line of ^{208}Tl. Secondly, this type of detector is sensitive to temperature variations, which greatly limits the options for installation locations.

Therefore, for reliable isolation of gamma-ray emission lines of ^{222}Rn daughter nuclei a gamma-ray spectrometer with good energy resolution capable of stable operation in various environments is required. These requirements can be satisfied by detectors based on high-pressure xenon [3].

2. Equipment description

In this study a gamma-spectrometer based on a two-liter cylindrical xenon gamma-ray detector (XGD) was used. Its main physical and technical characteristics are given in table 1.

In March 2015 this equipment was installed in a seismic research laboratory of the North-Osetian branch of the Russian Academy of Sciences Geophysical Survey (Vladikavkaz). This room did not have a ventilation system and the temperature and the humidity were kept at a constant level, which eliminated influence of external factors on the radon concentration. The continuous seismic monitoring by means of the equipment described above at the seismic station in Vladikavkaz was organized in such a way that all of the information was in the on-line mode transmitted to the Radiation laboratory at NRNU MEPhI, where it was processed, and where the operation of the measurement equipment was controlled.

Table 1. XGD main physical and technical characteristics.

Characteristics		
Mass, kg	6.0	
Working medium mass, g	580	
Working volume, cm^3	2000	
Gas density, g/cm^3	0.29	
Dimensions, cm	(Diameter) (Length)	14.0 38.5
Power consumption, W	< 10	
Registered γ-rays energy range, MeV	0.5 – 3	
Energy resolution at 662 keV, %	2.1±0.5	
Operating temperature range, °C	0 ÷ 100	

3. Experimental results

The measurement of Rn activity was carried out from 19.03.2015 to 13.05.2015 and spectra were registered each hour. In figure 1 a total energy spectrum for 24 hours is presented. In this spectrum gamma-ray lines of ^{40}K, daughter nuclei of radon ^{222}Rn (^{214}Bi and ^{214}Pb), and also ^{232}Th (one of the products of decay of which is ^{220}Rn) are present. For seismic monitoring it is necessary to distinguish between gamma-ray lines of daughter nuclei of radon and thoron, in particular 583 keV and 609 keV, which is enabled by the high energy resolution of XGD.

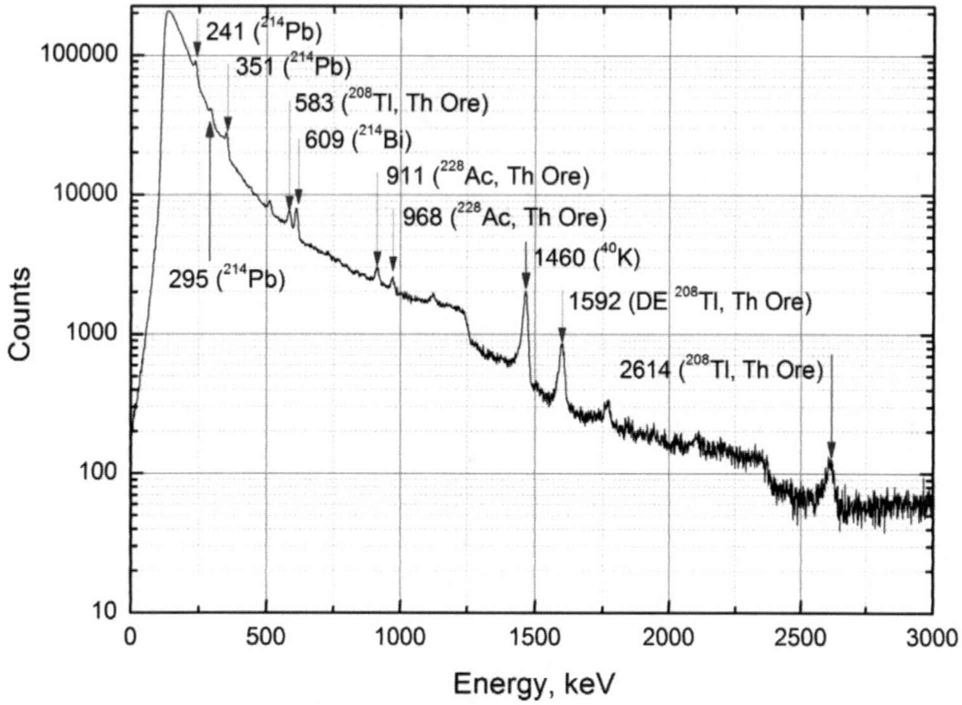

Figure 1. Total energy spectrum for 24 hours.

As an example, the time variation of the intensity of the 609 keV (^{214}Bi) gamma-ray line is shown in figure 2.

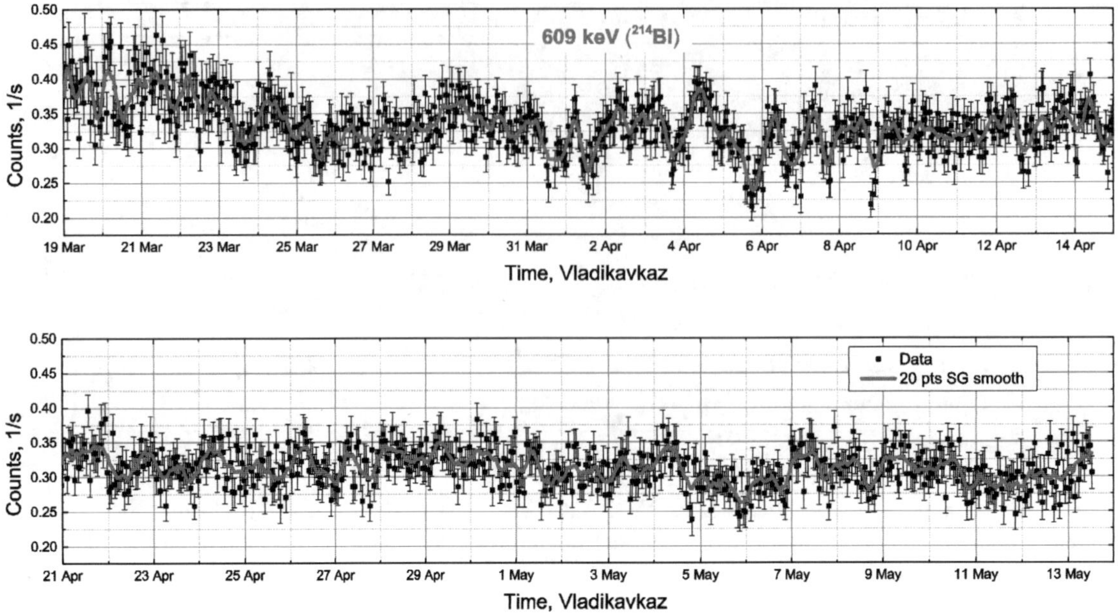

Figure 2. Time variation of the intensity of the 609 keV (^{214}Bi) gamma-ray line.

In the course of the experiment daily variations of gamma-ray lines intensities for daughter nuclei of ^{222}Rn were observed. To account for them time dependences of the intensities during night-time were plotted. To do that the collected energy spectra were averaged over the period of 10 hours (22:00 – 07:00). The obtained plots could be used for qualitative analysis of intensity variations for the gamma-ray lines in question and therefore for seismic monitoring.

In accordance with the data presented at the web-site of The European-Mediterranean Seismological Centre an analysis of earthquakes with regard to their epicenter and magnitude was carried out. Two areas were chosen, 200 km and 700 km away from the measurement equipment. The first area was used to account for earthquakes with magnitude $\geq$ 3.5, and the second one for more powerful earthquakes with magnitude $\geq$ 4.

During two months of measurements in the studied region of seismic activity six earthquakes marked in figure 3 were registered. The "star" symbol on the map corresponds to the location of the equipment. Table 2 contains information on each of the events.

Table 2. Information on the earthquakes that took place in the chosen region during measurement period.

№	Magnitude	Date	Time, UTC	Coordinates		Distance, km	Depth, km
				N	E		
1	ML 3.6	21.03.15	17:15:30	42.53	43.23	131	2
2	Mw 5.0	22.03.15	22:45:24	40.32	52.14	690	30
3	ML 4.2	30.03.15	9:54:01	40.90	48.55	399	6
4	mb 4.1	06.04.15	1:26:32	40.22	45.08	316	2
5	mb 3.5	14.04.15	0:34:08	43.66	45.66	105	5
6	ML 4.0	28.04.15	14:01:02	38.91	43.52	468	15

Figure 3. Locations of the earthquakes and measurement equipment.

Figure 4 presents the time dependence of the night-time intensity of the 609 keV gamma-ray line. The time of the beginning of each earthquake is marked with an arrow. According to the plot, before each earthquake a growth of intensity of a ^{222}Rn daughter nuclei line is observed, however to confirm these results more long-term measurements in areas with higher seismic activity are necessary.

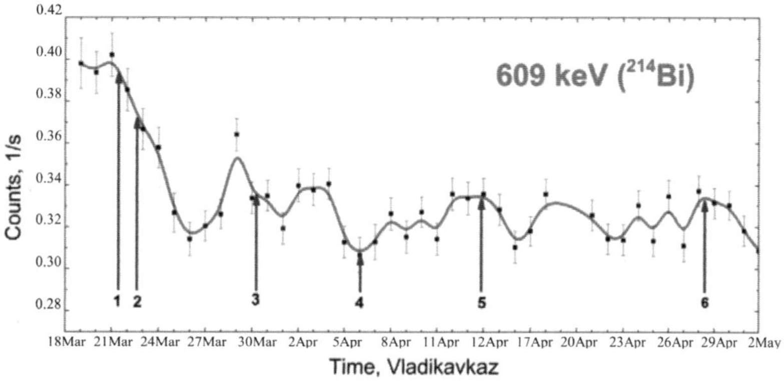

Figure 4. Time variation of the night-time intensity of the 609 keV (^{214}Bi) gamma-ray line.

4. Conclusion

A method for search of possible earthquake precursors based of measurement of ^{222}Rn concentration by means of intensity measurements of gamma-ray lines of its daughter nuclei using high-pressure xenon gamma-ray spectrometer is shown. First experiments employing such method that have been carried out show the possibility of its utilization.

For more reliable distinction of the possible precursors detailed statistics and placement of equipment in locations with high variation of ^{222}Rn concentration and high seismic activity are needed. For more detailed information on the expected time and location of the earthquake a large number of gamma-spectrometric devices in various locations connected into a wide network is necessary.

Acknowledgments

This work was partially supported by MEPhI Academic Excellence Project (contract № 02.a03.21.0005, 27.08.2013).

References

[1] Outkin V I *et al.* 2013 Radon monitoring as a possible indicator of tectonic events *Earth Syst. Dynam. Discuss.* **4** pp 93–107

[2] Giuliani G, Attanasio A and Fioravanti G 2013 Gamma detectors for continuous monitoring of radon *J. Int. Environmental Application & Science* **8**(4) pp 541–50

[3] Novikov A S *et al.* 2014 Xenon detector with high energy resolution for gamma-ray line emission registration *Proc. SPIE* **9213** 921318

Performances investigation and material selection of PMT magnetic shields for the space experiments with GRIS and PING-M instruments

R M Faradzhaev, Yu A Trofimov, E E Lupar and V N Yurov
National Research Nuclear University MEPhI (Moscow Engineering Physics Institute), Kashirskoe highway 31, Moscow, 115409, Russia

E-mail: faradzhaev_rodio@mail.ru

Abstract. PMT performances significantly change under the influence of magnetic field. Even the relatively weak geomagnetic field, which typically value is about 0.5 gauss, has an appreciable effect. Gain variations of PMT with 76 mm photocathode diameter may reach 10-20 % depending on spatial orientation. Therefore, it is necessary to apply magnetic shields for PMT response stability enhancement. The performances investigation of magnetic shields made of steel, permalloy and amorphous metallic alloy ribbon for PMT with 51 and 76 mm photocathode diameters was carried out. Based on obtained data the choice of magnetic shield was made.

1. Introduction

Photomultiplier tubes (PMTs) are very sensitive to the external magnetic field and their characteristics significantly change under the influence of magnetic fields. Even the relatively weak geomagnetic field, which typically value is about 0.5 gauss, has an appreciable effect [1]. Magnetic shields of different materials such as permalloy or amorphous metallic alloy (AMA) ribbon can be used to reduce the influence of magnetic field on PMT gain. PMT shields also have additional functions such as protection from ambient light and eliminating the influence of an external electric field [2].

In this study PMT gain variations under the influence of geomagnetic field with different configurations of magnetic shields have been investigated. The main purpose of investigation is selection the material and configuration of the PMT magnetic shields for GRIS [3] and PING-M [4] instruments that currently under development in the Astrophysics Institute of NRNU MEPhI.

A new type of magnetic shield made of AMA ribbons KNSR with 40 μm thickness made by Central Design Bureau of special radio materials [5] is presented.

It is well known that metals and alloys have a crystal lattice and strictly ordered arrangement of atoms. The amorphous metallic alloys feature is the absence of strict periodicity in the amorphous state. It is considered that the strict periodicity for unlimited long distance in the positional relationship of atoms absence in amorphous metallic alloys leads to isotropy of the magnetic properties [6]. However, amorphous alloys have anisotropic magnetic properties in reality, and exposed to thermal processing to decrease it. An important advantage of amorphous alloys is the opportunity to mix various chemical elements with over a wide range concentration changing, which improved physical, chemical and mechanical properties [7]. Therefore various designs of the amorphous alloys have a number of especial properties: high strength and hardness, high permeability, low coefficient of temperature dependence, etc. In this way amorphous metals are increasingly used in different fields due to its magnetically soft properties.

2. Experimental setup

In this study scintillation detectors (figure 1) based on Hamamatsu R6233-100 and R6231-100 PMTs with diameter of photocathode 76 mm and 51 mm respectively, and the cylindrical $LaBr_3$(Ce) experimental sample with dimension ∅25x25 mm, grown at the ISSP RAS [8] were used. The investigated shields wholly cowered PMTs bodies.

PMT output is directly connected to the pulse signal processor "Green Star" SBS-77 that had served as an amplifier–shaper and the amplitude-to-digital converter [9]. Spectra were accumulated and analyzed in the ESBS-7x program. Measurements were carried out with radioactive source ^{137}Cs.

Independent measurements of magnetic field suppression by the magnetic shields have been carried out using setup based on three-axis digital magnetometer Honeywell HMR2300 (figure 2). This magnetometer consists of three magneto-resistive sensors which oriented in orthogonal directions to measure the X, Y and Z vector components of a magnetic field. HMR2300 can measure magnetic field in the range of ±2 gauss and has accuracy 0.01 – 0.52% of full scale in the range of ±1 gauss [10].

Figure 1. The $LaBr_3$ (Ce) crystal with R6231-100 and R6233-100 PMTs detector assembly.

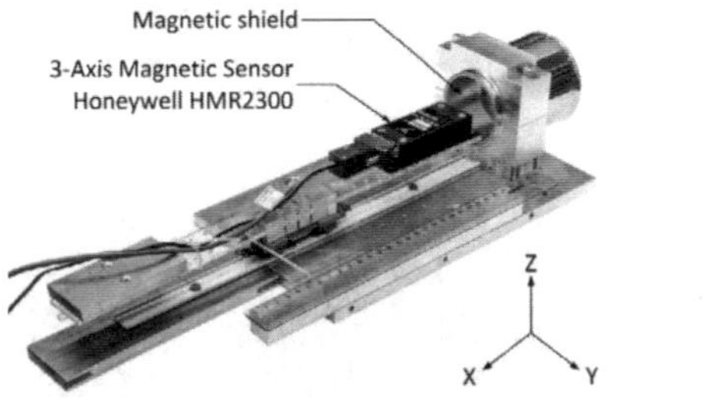

Figure 2. HMR2300 magnetometer based setup.

3. Types of magnetic shields

The magnetic shields of different shapes and designs for R6231-100 and R6233-100 PMTs have been investigated. Following samples of shields were studied for R6233-100 PMT: aluminum shield (A), steel shield (B), shield based on multilayer film nanostructures (C), the AMA ribbons shield (D) and Hamamatsu E989-15 shield (E). The Aluminum shield (F), two AMA ribbons shield (G and H) and Hamamatsu E989-05 shield (I) for R6231-100 PMT were investigated as well.

The A and F shields does not protected from the magnetic field practically. These only have mechanical and light protection purposes.

The B shield made of steel grate St3 with low carbon concentration and its composition similar to the technically pure iron. Permeability of St3 is relatively low. The shield has 2 mm thickness and 314 g weight.

The C shield consist of 10 layers of magnetic material (permalloy FeNi) with thickness 50 μm alternating with 9 layers of 5 μm copper (Cu) electrolytically deposited on an aluminum base. The weight of the shield with the aluminum base is 182 g.

The D, G, and H shields made of AMA ribbons which are fixed on an aluminum base. In order to provide electrical contact of the ribbons their edges were glued with conductive adhesive. The D shield has the 112 g weight (about 25 g weight of AMA) and made of two AMA ribbons layers. The second layer glued on the first one in orthogonal direction. The G shield made of a single layer of AMA ribbons, which is parallel to the axis of PMT and the H shield consists of one layer of the ribbons perpendicularly oriented to the axis of the PMT. The weight for both G and H shields is 75 g, and the weight of AMA ribbon is about 10 g.

The E and I standard Hamamatsu shields made of 0.8 mm permalloy, which have the weight 200 g and 135 g respectively.

4. Methods and results of the measurements

4.1. Measurements with three-axis digital magnetometer

The investigations of the geomagnetic field suppression by the magnetic shields were carried out with three-axis digital magnetometer HMR2300, which was fixed on a trolley moving on guide rails (figure 2). Experimental setupis constructed of nonmagnetic materials – aluminum and plastic. The distance between magnetic sensor and shields fixed with pointer attached to the trolley.

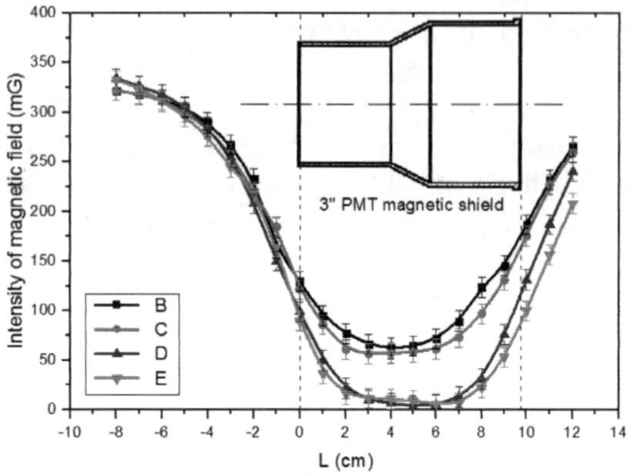

Figure 3. The suppression of geomagnetic field by B, C, D and E shields of R6233-100 PMT (lines connecting points are guide to the eyes only).

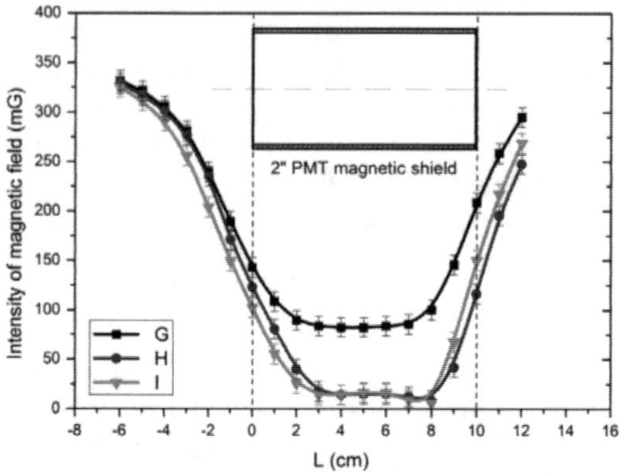

Figure 4. The suppression of geomagnetic field by G, H and I shields of R6231-100 PMT (lines connecting points are guide to the eyes only).

The value of the geomagnetic field was 330-350 mG at the point of measurements. The figures 3 and 4 show that the suppression of geomagnetic field by B and C shields of R6233-100 PMT is about 3 times like G shield made for R6231-100 PMT. The geomagnetic field suppression of the rest shields is not less than 100 times, which is the lower limit of digital magnetometer HMR2300. Measurements analysis of three components of the field showed that G shield (with parallel AMA ribbons) has two order of magnitude worse suppression of the magnetic field transverse component (directed perpendicular to the axis of the shield) than H shield with perpendicular AMA ribbons. Suppression of the other components for these shields is comparable. The sensitivity of magnetometer is not enough to determine truly suppression value of the magnetic field for D, E, H and I shields. Additional information can be obtained by direct measurements of magnetic field suppression degree of the shields in various designs using the detector assemblies.

4.2. Investigations of PMTs gain variations for different configurations of magnetic shields under influence of the geomagnetic field

The measurements of ^{137}Cs source peak position (662 keV) in spectrum at different orientations of detector axes (total 24 orientations for one measurement series) were carried out to investigate the PMT gain variations under the influence of the geomagnetic field. The Z axis is directed along the PMT axis. The X axis directed perpendicular to the Z axis in the plane of the photocathode. The combination of 6 main orientations were used in measurements, there are: up, down, north, south, west, east. One of the Z axis directions was selected and the X axis rotated around the Z with an interval of 90 degrees. Then direction of Z axis was changed and with X axis performed the same actions. For example, figure 5 shows the up-north direction (when the Z axis is directed up and the X-axis to the north). The relative deviations of ^{137}Cs peak from the mean value were calculated for each of the orientation position in the geomagnetic field:

$$Si = \frac{Aev - Ai}{Aev},$$

where A_{ev}– mean peak position, i – number of measurements in the series.

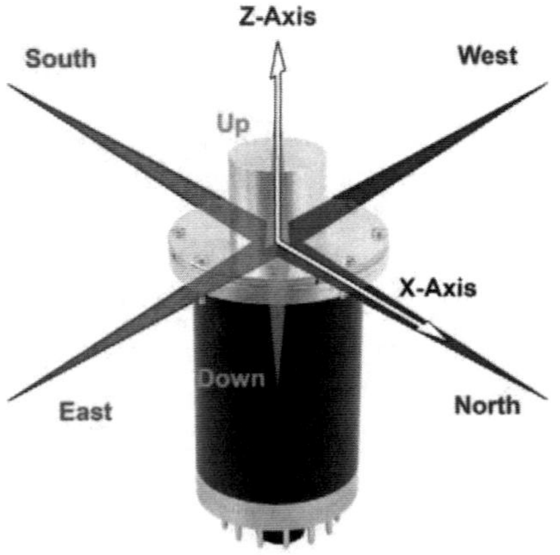

Figure 5. Detector orientation: up – north (Z axis is directed to the up and X-axis to the north).

Table 1. Measurement results.

PMT	Shields	δ_g ,%
R6233-100	A (Aluminum)	16±4
	B (Steel)	3.7±0.8
	C (Multilayer film)	5.0±1.3
	D (2 AMA ribbon)	0.7±0.2
	E (Ham. permalloy)	0.5±0.1
R6231-100	F (Aluminum)	10±4
	G(parallel AMA)	2.0±0.8
	H(perpendicular AMA)	0.25±0.05
	I (Ham. permalloy)	0.20±0.01

According to the relative deviation the mean gain variation (δ_g) for each shield calculated:

$$\delta_g = 2\sqrt{\frac{1}{n}\Sigma S_k^2},$$

where n=8; S_k– relative deviation of orientations, which doesn't includeany combinations with up and down directions, since deviations in these orientations are relatively small. The measurement results are shown in table 1.

Data obtained by magnetometer and results represented in table 1 show that the shield G with the parallel arrangement of AMA ribbons has not less than an order of magnitude worse magnetic field suppression than shields with perpendicular arrangement of the ribbon such as D and H, which ones show suppression comparable to permalloy Hamamatsu shields.

5. Conclusion

The shield selection implemented as a result of balanced compromise between the suppression of the external field and the mass-dimensional characteristics of the shields. The geomagnetic field suppression of steel and multilayer film nanostructures shields is not enough. The shields with AMA ribbons (D, H) and permalloy Hamamatsu shields (E, I) demonstrated comparable performances. The amorphous metallic alloy ribbon shields can be made of any size and shape (which is convenient in terms of the detector design). As opposed to a permalloy shield, the light weight ribbons can be fixed on mechanical construction of a detector setup.

Acknowledgements

We would like to express gratitude to A V Kochemasov, I V Rubtsov, Yu E Artemov for their participation in the development of the experimental setup. This work was partially supported by MEPhI Academic Excellence Project (contract № 02.a03.21.0005, 27.08.2013).

References

[1] De Vore P *et al.* 2014 Light-weight flexible magnetic shields for large-aperture photomultiplier tubes *Nucl. Instrum. Meth.* A **737** 222

[2] Hamamatsu Photonics K.K. Editorial Committee 2006 *Photomultiplier tubes: Basics and Applications, 3rd Edition* (Iwata: Hamamatsu Photonics K. K. Electron Tube Division)

[3] Kotov Yu D *et al.* 2015 Solar gamma-ray spectrometer GRIS onboard the International Space Station *J. Advances in Space Research* **56** 1797

[4] Kuznetsov V D 2011 *Interhelioprobe Project. Workshop Proceedings Tarusa* (Moscow) (Original Russian title: *Proekt Intergeliozond. Trudi rabochego soveshaniya Tarusa* (Moskva: Rotaprint IKI RAN 192))

[5] Central Design Bureau of special radio materials: *http://ckbrm.ru*

[6] Kikalo B 2006 *Atomic structure of amorphous alloys and its evolution* (Moscow: Moscow institute of steel and alloys) p 338 (Original Russian title: *Atomnaya struktura amorfnih splavov i ee evolucia* (Moskva: Moskovskiy institut stali i splavov))

[7] Pavlenko T P and Tokar M N 2013 Analysis and study of amorphous alloys properties *J. Electrical engineering and electromechanics* **5** 45

[8] Institute of Solid State Physics RAS: *http://issp.ac.ru*

[9] Belousov P V *et al.* Structure of Spectrometric complex SCS "Green Star": *http://greenstar.ru*

[10] URL: *https://honeywell.com*

Software tool for xenon gamma-ray spectrometer control

I V Chernysheva, A S Novikov, A E Shustov, V V Dmitrenko, Sone Pyae Nyein, D Petrenko, S E Ulin, Z M Uteshev and K F Vlasik
National Research Nuclear University MEPhI (Moscow Engineering Physics Institute), Kashirskoe highway 31, Moscow, 115409, Russia

E-mail: ivchernysheva@gmail.com

Abstract. Software tool «Acquisition and processing of gamma-ray spectra" for xenon gamma-ray spectrometers control was developed. It supports the multi-windows interface. Software tool has the possibilities for acquisition of gamma-ray spectra from xenon gamma-ray detector via USB or RS-485 interfaces, directly or via TCP-IP protocol, energy calibration of gamma-ray spectra, saving gamma-ray spectra on a disk.

1. Introduction

Software tool «Acquisition and processing of gamma-ray spectra" was designed to control the operation of xenon gamma-ray spectrometer [1] by computer, using special unit of digital electronics [2] for signal processing as emulator of the analyzer.

Software tool was developed in the CodeGear ™ RAD Studio 2007 [3] using the programming language DELPHI. Software tool is a set of interactive windows which allow acquisition, saving to disk and reading from disk gamma-ray spectra and spectra of background, visualizing them, carrying out an energy calibration and processing of gamma-ray spectra. This article will review the software tool functions for management of gamma-ray spectrometer.

2. Software tool as emulator of the analyzer

Software tool «Acquisition and processing of gamma-ray spectra" is composed of a main window, window of spectra and auxiliaries windows for determine processing and analyzing options of the spectrum. The software tool supports multiple-window acquisition, view and processing of the spectra.

Software tool has the following functions as emulator of the analyzer:

- selection of an interface for data transfer from xenon gamma-ray spectrometer to a computer – USB or COM;
- setting a threshold for detecting a signal from the xenon gamma-ray spectrometer;
- enabling and disabling charge sensitive amplifier of xenon gamma-ray spectrometer;
- start, stop, pause, resume commands for spectra acquisition;
- autosave spectra at scheduled intervals;
- stop command can be done interactively or automatically at predetermined real-time or live-time;
- automatic acquisition the predetermined number of spectra;
- control of xenon gamma-ray spectrometer through the connection Socket Server-Client; the external program Socket-Client controls the start, stop, time of acquisition and processing the

spectrums analysis by means of the relevant commands transferred to the software tool which runs in the mode of Socket-server;

- display on the screen process of spectrum acquisition and its parameters in real time;
- saving and loading spectra from disk PC.

3. Control of spectrum acquisition

3.1. Spectrum acquisition

To activate a new spectrum it is necessary to call the command "Acquisition the new spectrum" through the "File" main menu or the corresponding toolbar button. It opens a child window, intended to acquisition the gamma spectrum using xenon gamma-ray spectrometer (figure 1).

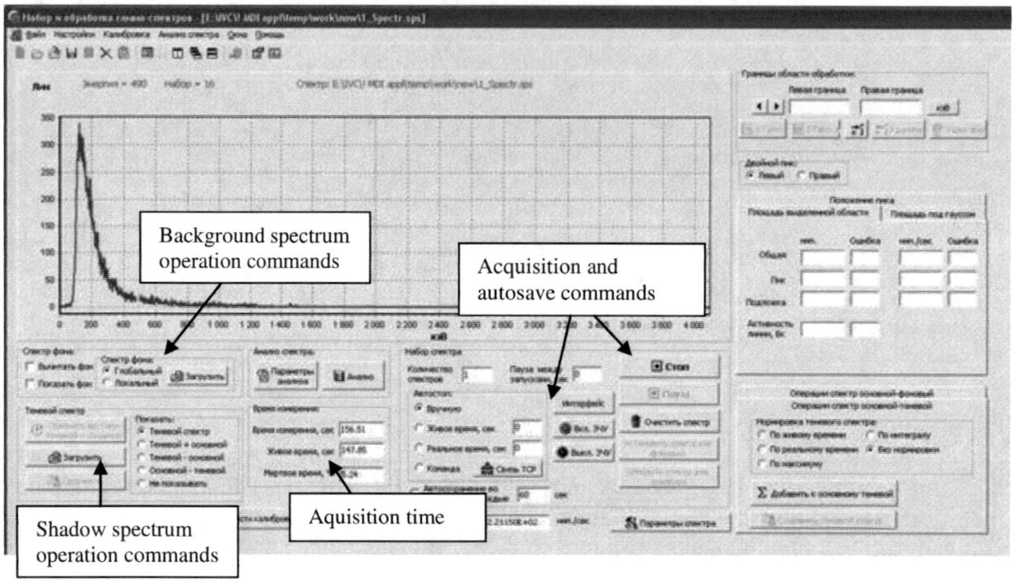

Figure 1. Main window of software.

Before the start of acquisition it is possible to select the data interface using window "Select the port for data transmission» (figure 2), which allows you to choose USB or COM-port interface.

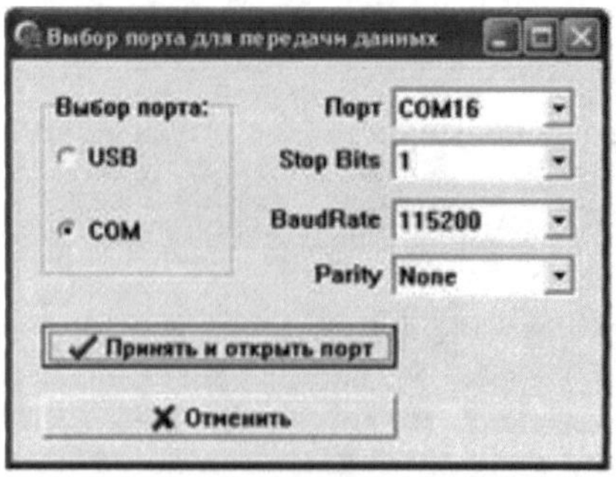

Figure 2. Select the port for data transmission.

If a connection is available, the port is opened for communication with xenon gamma-ray spectrometer. In the case of a successful connection to xenon gamma-ray spectrometer button

"Acquisition" becomes available to press. To manage a spectrum acquisition from xenon gamma-ray spectrometer use the following commands (see main window in figure 1):

- « Acquisition » – start a new spectra acquisition after clear all data from the previous spectra,
- «Stop» – stop of acquisition;
- «Pause" – pause of acquisition;
- «Resume" – to resume (continue) acquisition after pause;
- «Clear spectrum" – reset spectrum and all associated data;
- «Open spectrum for acquisition" – lets to download the available spectrum from a file using the standard dialog box and continue its acquisition;
- «Set the spectrum to background" – a command is only available when the acquisition is stopped and allows to declare the present spectrum as the spectrum of the background and to use it as a background during the future work;
- "Autosave during an acquisition" processes automatic saving spectrum into the file through every user-specified time interval;
- "Autostop" allows choosing the stop options of acquisition: by hand, at the specified of real-time or live-time and from external Socket-client commands. In the latter case the software tool works in a mode Socket-server.

3.2. The Communications Socket Server-Client in the mode of to an external commands

The connection socket server-client must be established for using external commands. To do this, one should click the button "Connection", open window "Client-Server» (figure 3) and select the host address and port through for the connection server-client.

Figure 4 shows the connection installed. Following commands are available for xenon gamma-ray spectrometer:

- kgd01 – start of acquisition,
- kgd02 – stop of acquisition,
- kgd03 – request result of spectra analysis by the total activity exceeding a predetermined threshold value. The answer is sent automatically.
- kgd04:xxx – set time of acquisition, where xxx – time of acquisition in milliseconds.

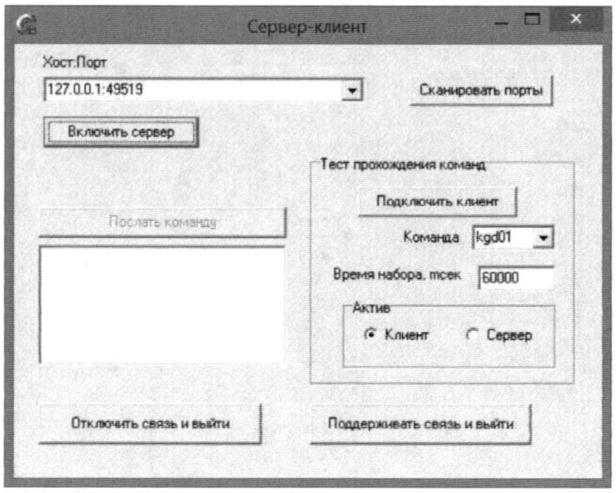

Figure 3. Window of connection Socket Client-Server. Waiting for connection.

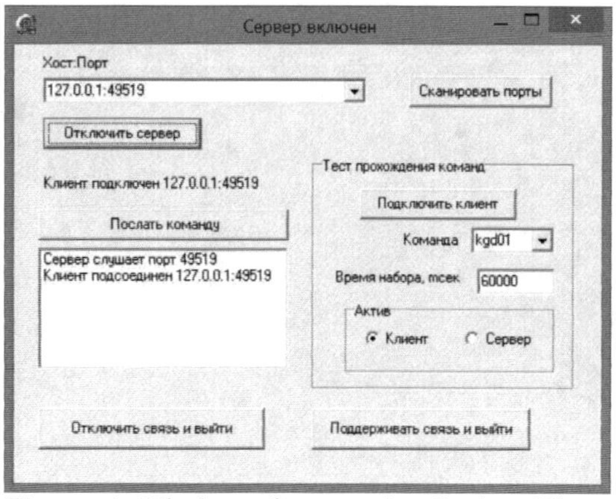

Figure 4. Window of connection Socket Client-Server. The server is turned on; the client is connected to the server.

The software tool also allows determining the name of the spectrum and comments, the data about the type of radioactive source. The user can send commands into block of digital electronics xenon gamma-ray spectrometer: write detection threshold, set rejection and integration time of the signal, enable/disable the charge sensitive amplifier detector.

During the acquisition, it is possible to carry out the energy calibration of the spectra. To do this, it is necessary to enter the value of the corresponding channel energy of the two peaks. Calibration parameters can be as loaded from file as saved to it. After calibration, the spectrum can be redrawing in keV values.

4. Conclusion

Software tool «Acquisition and processing of gamma-ray spectra" was designed to control the operation of xenon gamma-ray spectrometer by computer using a special unit of digital electronics for signal processing as emulator of the analyzer via USB or RS-485 interfaces, directly or via TCP-IP protocol.

Software tool is composed of a main window, window of spectra and auxiliary windows for determining processing and analyzing options of a spectrum and supports multiple-window acquisition, view and processing of the spectra.

Acknowledgments

Authors wish to acknowledge the Center of Fundamental Researches and Particle Physics. This work was partially supported by MEPhI Academic Excellence Project (contract No. 02.a03.21.0005, 27.08.2013).

References

[1] Novikov A *et al.* 2014 Xenon detector with high energy resolution for gamma-ray line emission *Proc. SPIE* **9213** 921318

[2] Novikov A *et al.* 2014 New modification of xenon gamma-ray detector with high-energy resolution *Optical Engineering Journ.* **53** 021108

[3] URL: http://cc.embarcadero.com/item/25015#

Journal of Physics: Conference Series **675** (2016) 042010 doi:10.1088/1742-6596/675/4/042010

Detectors on base of scintillation structures for registration of volumetric activities of gaseous and liquid media gamma radiation

V V Kadilin , V N Yurov, E V Ryabeva, V T Samossadny, E E Lupar , Yu A Trofimov, S V Kolesnikov, S B Chebishev and V O Nebolsin
National Research Nuclear University MEPhI (Moscow Engineering Physics Institute), Kashirskoe highway 31, Moscow, 115409

E-mail:EVRyabeva@mephi.ru

Abstract. The main aim of this research is the development and prototyping of the ionizing radiation detectors for the diagnosis of the physical processes used for monitoring the radiation situation at the thermal or fast neutrons reactors. In this article we present the experimental verification of applicability of the scintillation detectors based on $LaBr_3(Ce)$ and $YAlO_3(Ce)$. The experimental studies of the gamma-ray detection with several designs of the crystal scintillation detectors in gas and liquid are considered. It was shown that the measurement range in the liquid medium at the duration of one measurement of 100 seconds for ^{137}Cs equals from $3.79 \cdot 10^2$ Bq/l to $1.08 \cdot 10^8$ Bq/l for detector prototype based on $YAlO_3(Ce)$.

1. Introduction

The aim of the research is the development and prototyping of ionizing radiation detectors for the diagnosis of the physical processes used for monitoring the radiation situation at the thermal or fast neutrons reactors. It is necessary to control the level of radiation exposure to humans and the environment in real-time measurements and to obtain a reasonable long-term forecasts of radiation exposure [1].

The proposed detector should have the highly sensitive radiometric channels and extended measurement ranges. The developed detectors must be capable of measuring the activity of fission gases ^{85}Kr, ^{85m}Kr, ^{88}Kr, ^{133}Xe, ^{135}Xe in the range from 10^4 Bq/m^3 to $3.7 \cdot 10^{14}$ Bq/m^3 $\pm$ 15% ensuring the identification of radionuclide in controlled gas-sample. It must provide the registration of gamma radiation of liquids to ensure the monitoring of volume activity of ^{137}Cs with limiting measured value $5 \cdot 10^7$ Bq/kg $\pm$ 15% and identification of radionuclide ^{134}Cs, ^{136}Cs, ^{138}Cs, ^{131}I, ^{133}I, ^{24}Na. The energy resolution for isotope ^{137}Cs should be no worse than 4–5%. The upper value of the operating temperature must be at least 80^0C. The dependence of the detection efficiency and energy resolution of the detector on the temperature must be insignificant [1,2].

Crystal detectors made from inorganic scintillators have high efficiency and high performance. They could be used in high temperature range, have good resistance to external mechanical influences and could be sufficiently large. The fist stage of the study was the choice and verification of applicability of the scintillators. After study of information on scintiallators [3] the following materials were chosen: $LaBr_3(Ce)$ and $YAlO_3(Ce)$.

2. Prototypes of the scintillation detectors based on $LaBr_3$(Ce) and $YAlO_3$(Ce)

The detection unit for verification of applicability of the crystals for described tasks was developed.

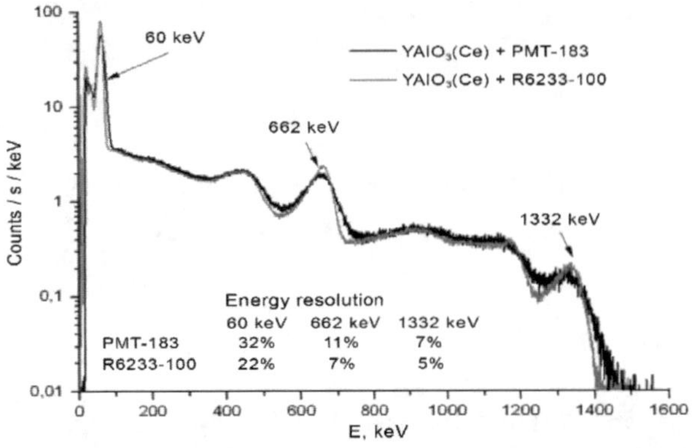

Figure 1. Energy spectrum with $YAlO_3$(Ce) scintillator.

For every scintillator the following experiments were done. The scintilator was irradiated by gamma-source with known spectra. The obtained waveform and the gamma spectra were recorded. The obtained spectra were analyzed and parameters of sensitivity were calculated. For testing the source ^{137}Cs isotope was used with activity ~ $0.6 \cdot 10^5$ Bq in the area of 4π. Figure 1 shows the spectra obtained by the crystal $YAlO_3$(Ce). The best energy resolution of the crystal was 5–7% for energy 1332 keV.Given the fact that this type of crystal has a low emission time – 27 ns, has a weakly pronounced temperature dependence of the light output and is not hygroscopic, it can be considered as a candidate for the implementation of the tasks of identification and activity measurements of radionuclide to a level of $3.7 \cdot 10^{17}$ Bq/m^3.

Energy resolution for crystal $LaBr_3$(Ce) at 662 keV line was equal to 4.2%. The scintillator $LaBr_3$(Ce) has a great light yield and, at the same time, small decay time (about 16 ns). A feature of the scintillator crystals $LaBr_3$(Ce) is relatively high intrinsic activity as a result of the decay of a radioactive isotope ^{138}La. To determine the contribution of the external background detector was surrounded by a layer of lead 10 cm thick. The measurements showed that the intrinsic background of crystal sample is $7.5 \cdot 10^{-3}$ counts/s/keV in the energy range 500–700 keV.

Figure 2. Prototype scintillation detector LABR1.

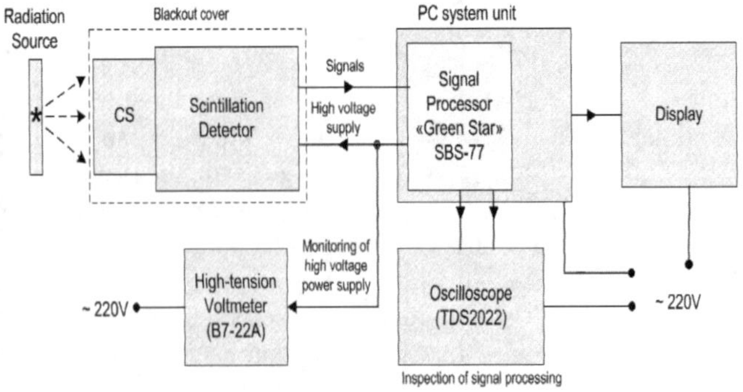

Figure 3. The test stand scheme.

3. Research of crystal scintillation detectors prototypes for the measurement of the activity of the gas environment

The source of ionizing radiation was set at the distance from the scintillation crystal's center. Pulse signals from PMT were processed by special board SBS-77 (firm "Green Star" [4]) installed in the system unit of a personal computer (PC). Form of signal is under the control by oscilloscope TDS-2022. The voltmeter V7-22A controls the PMT voltage. Modes of operation of SBS-77 and accumulated spectra are displayed on a PC monitor. The measurements of the gas activity were carried out. The sealed box with input and output connections for pumping gas through the test system was created for tests. The design of box allows to place inside it any of created detector prototypes. The scheme of the measuring stand is shown in figure 3.

The energy resolution of crystalline scintillation detectors LABR1, LABR2 and YAP1 were measured in the testing center for scientific equipment of Institute of Astrophysics NRNU MEPhI. Results of measurement are shown in figure 4.

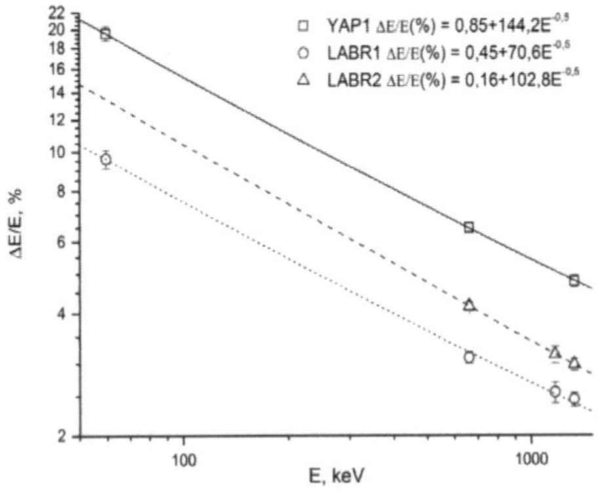

Figure 4. The energy resolution on the energy for the three detector prototypes.

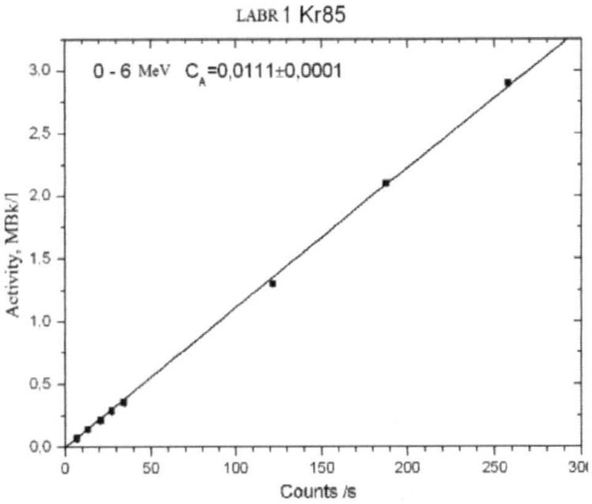

Figure 5. LABR1 counting rate on ^{85}Kr volumetric activity for full spectrum measuring range. Errors are within the data points.

The results confirmed the possibility of any of the prototypes made the sure identification of the radionuclides in the mixture of gaseous fission products ^{85}Kr, ^{85m}Kr, ^{88}Kr, ^{133}Xe, ^{135}Xe.
Crystal scintillation detectors based on $LaBr_3(Ce)$ (LABR1 and LABR2) may provide identification of the nuclides ^{137}Cs, ^{134}Cs, ^{136}Cs, ^{138}Cs, ^{131}I, ^{133}I, ^{24}Na in media. Energy resolution of the crystal scintillation detector based on the $YAlO_3$ (Ce) YAP1 is not sufficient for such measurements.

The experiments showed linear dependence of the counting rate in complete absorption peak on the activity of gaseous ^{85}Kr (see figure 5). It is possible to estimate the minimal ^{85}Kr concentration which can be measured by the proposed detectors.

The *upper limit* is determined by the processing time of interaction which for the currently available processors it equal about 10^{-6} s. On the base of the experimentally obtained data the following values of the activity may be calculated for LABR1 and YAP1:

$$A_{upperLABR1} = 2.9\cdot 10^9\ 10^6/250 = 1.16\cdot 10^{13}\ \mathrm{Bq/m^3},$$

where $2.9\cdot 10^9$ [Bq/m^3]/250[counts/s] is the rate for activity per one detector count (see figure 7), 10^6 – limit for detecting count rate. Using the same approach for YAP1:

$$A_{upperYAP1} = 2{,}86\cdot 10^9\cdot 10^6/160 = 1.8\cdot 10^{13}\ \mathrm{Bq/m^3}.$$

If the measurement's time is 1s, number of events at the peak of total absorption will be about $5\cdot 10^4$, which is enough to ensure accuracy about 1%.

The *lower limit* is determined by measuring the efficiency of the detector, the intensity of the background and the time of measurement. $I_{bacground}$=7.5·10^{-3}counts/s/keV for LABR1. When the measurement time 100s, the limit of the activity to be measured A_{lower}=2.1·10^{7} Bq/m^{3} for the prototype LABR1 and A_{lower}=4.5·10^{7} Bq/m^{3} for the prototype of YAP1.

4. The ability of models of crystal scintillation detector to register gamma radiation of liquids

The supposed liquid medium in most cases is water. The main peculiarity for registration of uniformly distributed radiation in the liquid medium from the radiation of the gas is a relatively high rate of absorption and scattering it in the medium itself.

In the fist approximation, the activity of uniformly distributed source may be presented:

$$A_1 = \frac{A_0}{\mu_{H_2O} l}(1 - e^{-\mu_{H_2O} l})$$

where A_1 – activity emerging from the source [Bq]; A_o – activity in the source [Bq]; μ – linear absorption factor [cm^{-1}]; l – thickness of the radiating source [cm].

The measurement with crystalline scintillation detectors should be conducted in a non-contact method and under moderate temperature conditions. This type of measurements improves the reliability of electronic components, and makes it easy to carry out preventive and calibration work with radiation detectors. The portion of unscattered radiation depending on liquid media and gamma energy may reach the detector in non-contact method. For ^{137}Cs isotope portion of unscattered radiation emitted from measuring volume of radioactive water with thickness 10 cm through 2 mm wall is approximately equal to 0.6.

The spectra of the following isotopes with the following main energy lines were experimentally investigated and analysed: 661.7 keV (^{137}Cs); 604.7 and 795.8 keV (^{134}Cs); 340.5, 818.5 and 1048.07 keV (^{136}Cs); 462.7 and 1048.07 keV(^{138}Cs); 364.48 keV(^{131}I); 529.8 and 875.3 keV(^{133}I);1368 and 2754 keV(^{24}Na). From the analysis of data on energy resolution it can be concluded that all listed lines can be identified with the detectors based on the $LaBr_3$(Ce) LABR1 and LABR2. Prototype based on the $YAlO_3$(Ce), YAP1 better be used for an energy of 662 keV due to its small thickness. Also because of lower energy resolution, it isn't able to identify the radionuclide ^{131}I in the presence of ^{136}Cs and ^{24}Na radionuclides.

For the prototype of the detector LABR1 measuring range in the liquid medium when the duration of one measurement – 100s for ^{137}Cs may be calculated on the base of results for gaseous media and taking into account correlated factors on portion of transmitted radiation. The calculated boundaries for measuring range are the following:

A_{lower} =1.77 10^{2} Bq / l; A_{upper} =6.96 10^{7} Bq / l.

For the prototype of the detector YAP1 measuring range in the liquid medium at the duration of one measurement of 100s for ^{137}Cs are the following:

A_{lower} =3.79·10^{2} Bq/l; A_{upper} =1.08·10^{8}Bq/l.

5. Conclusion

The prototyping and experimental investigation of the detectors based on crystal $YAlO_3$(Ce) and $LaBr_3$(Ce) are conduced. It is shown that when the measurement time is 100s the top level of measured activity of the gaseous ^{85}Kr medium is about 10^{7} Bq/m^{3}. The measuring range for models of detectors based on $YAlO_3$(Ce) and $LaBr_3$(Ce) in a liquid medium (water) for a duration of one measurement 100s for ^{137}Cs is as follows: A_{lower} = 3.79 10^{2} Bq/l and A_{upper} =1.08·10^{8} Bq/l and Alower=1.77·10^{2} Bq/l and Aupper =6.96 10^{7} Bq/l for prototypes correspondently.

Acknowledgments

Authors wish to acknowledge partial support from MEPhI Academic Excellence Project (contract № 02.a03.21.0005, 27.08.2013).

References

[1] Regulations 2003 *Sanitary regulations for design and operation of nuclear power plant* RF, SR NP-03

[2] Chebyshev S B *et al.* 2008 (in Russian) *Nuclear instrument engineering Vol 1 Instruments for ionization radiation measurements* (original Russian title: *Yadernoe Priborostroenie Vol **1** Pribory dlya izmerenia ionizirushich izluchenii)*

[3] Materials of Saint-Gobain Crystals (*www.crystals.saint-gobain.com*)

[4] Belousov P V *et al. Structure of Spectrometric complex SCS "Green Star"* (*www.greenstar.ru*)

Journal of Physics: Conference Series **675** (2016) 042011 doi:10.1088/1742-6596/675/4/042011

Interdisciplinary glossary — particle accelerators and medicine

V V Dmitrieva, V S Dyubkov, V G Nikitaev and S E Ulin
National Research Nuclear University MEPhI (Moscow Engineering Physics Institute), Kashirskoe highway 31, Moscow, 115409, Russia

E-mail: vvdmitriyeva@mephi.ru

Abstract. A general concept of a new interdisciplinary glossary, which includes particle accelerator terminology used in medicine, as well as relevant medical concepts, is presented. Its structure and usage rules are described. An example, illustrating the quickly searching technique of relevant information in this Glossary, is considered. A website address, where one can get an access to the Glossary, is specified. Glossary can be refined and supplemented.

1. Introduction

Number fields of science as well as its applications are founded and evolved swiftly today. It results in new concepts, professional terminology, which can significantly complicate the understanding and information perception by specialists in the interdisciplinary fields. In particular this situation takes place in the case of specialist cooperation from medicine and particle accelerator fields. Particle accelerators in that case are effectively used for disease diagnostics and therapy.

Special problems appear when students and postgraduates study special medicine and particle accelerator courses at Universities, in which rather difficult, and often impossible, to teach courses simultaneously on physics and medicine at the highest level [1].

Interdisciplinary Glossary was made up with the aim of improving the knowledge in the field of particle accelerators and medicine. We hope that Interdisciplinary Glossary will be useful for students and specialists.

2. Glossary structure

High-tech nuclear medicine centers are built up for the provision of high quality medical diagnosis and therapy in Russia today, where experts in various scientific fields and application areas work together with doctors.

Interdisciplinary Glossary which includes a set of hard and electronic copies as well as database of Glossary and contains terminology and explanation in genetics, biology, radio-biology, radiochemistry, radiation safety, radiopharmaceutical, oncology, information technology, particle accelerators, physical methods and means of radiation for medical imaging is required to provide communication between all members on issues concerning common activity in the field of application of methods and tools for nuclear medicine and radiation therapy for cancer diagnostics and treatment.

In particular, particle accelerators are used for radioactive isotope production as well as radio-diagnostics and oncological diseases therapy, sterilization of medical instruments and transplant tissues.

Particle accelerators for medicine are classified into linear and cyclic on the one hand and into electron and proton/ion on the other hand. Generally, electron linear and cyclic accelerators are used for radiotherapy, whereas proton and ion accelerators are mainly used for nuclear medicine purposes [2, 3]. It is well known that radiation therapy uses different types of ionizing radiation (photons, electrons, protons, ions, neutrons, π-mesons). The type and energy of ionizing radiation in complex treatment are determined by a stage and malignant neoplasm prevalence [4].

Interdisciplinary Glossary includes about 1000 terms and explanations, namely:

- 180 on particle accelerators,
- 200 on oncology,
- 200 on computing,

and other terms and explanations concern radiobiology, radiochemistry, radiation safety, radiopharmaceutical etc.

Glossary content has been edited by experts of the corresponding fields of knowledge. Glossary content is sorted by subject and alphabet.

Glossary is based on relational database management system (DBMS) MySQL platform. Operation principle of the DBMS MySQL is similar to any other DBMS operation principle that uses SQL as the command language for creating/deleting databases, tables, replenishing the tables with a data, sampling data [5].

Open source web-application phpMyAdmin was installed for easy data management. phpMyAdmin is a LAMP (**L**inux, **A**pache, **M**ySQL, and **P**HP) application that provides a web-based interface for the administration of MySQL databases. phpMyAdmin has tools for visual table creation, and also allows one to create, modify and delete tables by means of the SQL toolkit.

Flowchart of the developed application is shown in figure 1.

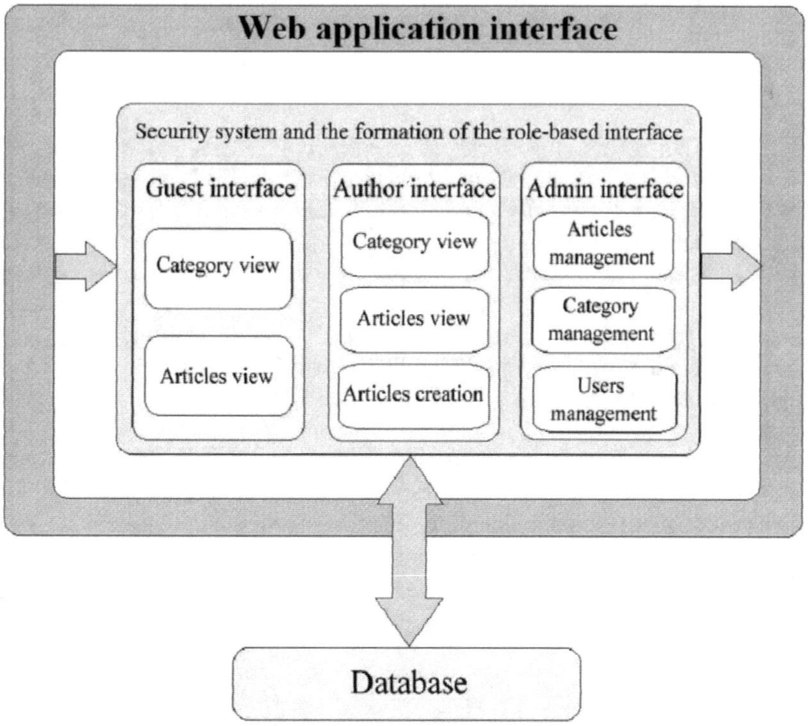

Figure 1. Web-application structure.

An *administrator* interface consists of:

- category management that allows one to create, delete, and modify Glossary categories/subcategories;
- article management that allows one to create, delete, and edit articles included in all Glossary categories/subcategories;
- user management that allows one to assign/remove user rights to certain categories.

A *moderator* interface consists of:

- category management that allows one to create, delete, and modify Glossary categories/subcategories, which is assigned with administrator rights;
- article management that allows one to create, delete and edit articles included in Glossary category/subcategory, to which the moderator has access.

An author interface allows an author to add and edit his articles. A visitor interface allows a guest to view articles of all existing categories.

3. Conclusion

Modern trends in technologies, the availability of information technologies and task complexity force researchers to involve other specialists in different fields of science into investigations [6, 7].

In order to provide interdisciplinary interactions, printed reference books, that contain terms of a special field of knowledge, are used usually. However, it takes a lot of time to find required information in that case.

Multimedia interdisciplinary Glossary on nuclear medicine and radiation therapy which was developed in NRNU MEPhI and based on information technology, allows one to overcome such difficulties. Finally, Glossary allows one to get the modern scientific and technological information effectively.

References

[1] Ulin S E *et al.* 2009 (in Russian) *Physical Methods of Medical Imaging* (original Russian title: *Fizicheskie metody meditsinskoy introskopii*) (Moscow: MEPhI publisher) p 308

[2] Chernyaev A P 2013 (in Russian) *Particle Accelerators in the Modern World* (original Russian title: *Uskoriteli v sovremennom mire*) (Moscow: Moscow State University press) p 368

[3] Karzmark C J, Ninan C S and Tanabe E 1993 *Medical Electron Accelerators* (New York: McGraw-Hill) p 332

[4] Podgoršak E B 2010 *Radiation Physics for Medical Physicists* (Berlin: Springer-Verlag) p 772.

[5] Kolisnichenko D N 2010 (in Russian) *PHP5/6 and MySQL 6. Development of Web applications* (original Russian title: *PHP5/6 i MySQL 6. Razrabotka Web-prilozheniy)* (St. Petersburg: bhv-Petersburg) p 560

[6] Katyshev Yu V, Novikov D L and Polerov E A 1965 English-Russian dictionary on particle accelerators (in Russian), ed. V P Dmitrievskiy (original Russian title: Anglo-russkiy slovar po uskoritelyam zaryazhennyh chastits) (Moscow: Soviet encyclopedia) p 324

[7] Gohberg G B 1977 *Charged particles acceleration. Terminology*, issue 89 (in Russian), (original Russian title: *Uskorenie zaryazhennyh chastits: terminologiya*) (Moscow: Nauka)

Simulations of the cosmic-veto system for the COMET experiment

Oleg Markin* and Evgueny Tarkovsky
on behalf of the COMET collaboration

National Research Nuclear University MEPhI (Moscow Engineering Physics Institute), Kashirskoe highway 31, Moscow, 115409, Russia

E-mail: *rboleg@yandex.ru

Abstract. In the COMET experiment searching for the muon-electron conversion not conserving lepton numbers, a scintillator-strip-based veto system will be used to prohibit the COMET detector from fake signals initiated by cosmic muons. In order to verify the efficiency of the system, we have built its computer model and carried out various simulations. To tune the model, experimentally measured data are utilized. The simulations give the inefficiency of the cosmic-muon registration being below 0.0001, which meets requirements of the experiment. In addition, simulations of neutrons originating from muon captures and traversing a shield beneath the cosmic-veto system have been carried out using the Geant4 toolkit. A Geant4 application has been written with an appropriate detector design and possible spectrum of neutrons' energy. Design of the shield has been optimized to ensure the time loss concerned with fake veto signals caused by the background neutrons is tolerable. Materials of shield's layers are chosen, and optimum thicknesses of the layers are computed.

1. Introduction

The discoveries of recent years suggest the possibility of extending the Standard Model. Many of the extended models predict the charged lepton flavor violation (CLFV). The search for new physics through CLFV complements that being undertaken at LHC. One of the processes appropriate to experiments seeking for CLFV is the neutrinoless muon-electron conversion in the field of nuclei. When a negative muon is stopped by some material, it is trapped by an atom, forming a muonic atom. The fate of the muon is then to either decay in orbit or be captured by a nucleus. In the context of physics beyond the Standard Model, the exotic process of muon capture accompanied by neutrinoless electron emission is also expected, $\mu^- N \rightarrow e^- N$. The final state of the nucleus could be either the ground state or one of the excited states. In general, the transition to the ground state dominates, which is called coherent capture. The rate of the coherent capture is enhanced by a factor approximately equal to the number of nucleons in the nucleus.

The latest search for $\mu^- - e^-$ conversion was performed by the SINDRUM-II collaboration at the PSI. They set the current upper limit on the branching ratio of $\mu^- Au \rightarrow e^- Au$ at $7 \cdot 10^{-13}$. The COMET Phase-I experiment is seeking to measure the coherent neutrinoless muon-electron transition in aluminum nuclei with a single event sensitivity of $3.1 \cdot 10^{-15}$, assuming running period of 110 days. The experiment will utilize a dedicated 8 GeV proton beam extracted from J-PARC's Main Ring via a new beamline. Muons will be produced in decays of pions produced in collisions of the proton beam with a graphite target. The yield of low momentum muons is enhanced using a pion capture solenoid. Muons are momentum- and charge-selected using a muon transport solenoid before being stopped in an aluminum target at the center of a Cylindrical Drift Chamber (CDC) in a 1 T magnetic field.

The COMET Phase-I detector will be augmented with prototypes of the Phase-II straw tracker and the electron calorimeter. As well as providing valuable experience with the detectors, they will be used to characterize the beam and measure backgrounds to the neutrinoless muon conversion signal to ensure that the Phase-II single event sensitivity of $2.6 \cdot 10^{-17}$ can be achieved. Another experiment that also searches for CLFV is the MEG experiment at PSI. This uses a final state different to that of the COMET, $\mu^+ \to e^+ + \gamma$, yet data of the two experiments can be combined to discriminate between different theoretical models.

The event signature of the coherent $\mu^- - e^-$ conversion in a muonic atom is a mono-energetic single electron emitted from the conversion with an energy of $E_e \simeq m_\mu - B_\mu$, where m_μ is the muon mass and B_μ is the binding energy of the muonic atom. From an experimental point of view, the $\mu^- - e^-$ conversion is a very attractive process. Firstly, the electron energy is far above the end-point energy of the free muon decay spectrum. Secondly, since the event signature is a mono-energetic electron, no coincidence measurement is required. The search for this process has the potential to improve sensitivity by using a high muon rate without suffering from accidental background events, which would be serious for other processes, such as $\mu^+ \to e^+\gamma$ and $\mu^+ \to e^+e^+e^-$ decays.

The energy of emitted electrons $E_e \simeq m_\mu$ coincides with the endpoint of muon decay in orbit (DIO), which is the only relevant intrinsic physics background. Since the energy distribution of DIO falls steeply above $m_\mu/2$, this background can be suppressed using a good energy resolution of the COMET detector. There are several other potential sources of background events in the energy region around 100 MeV, involving beam particles or cosmic rays.

Beam-related background events may originate from muons, pions or electrons from the beam. Apart from DIO, muons can produce background events by decay in flight or radiative muon capture. The restriction on the muon momentum to be less than 70 MeV/c allows the background caused by in-flight decay of muons to be avoided. Pions can produce background events by radiative pion capture. Gamma rays from both captures produce electrons mostly through e^+e^- pair production at the aluminum muon-stopping target. However, since the lifetime of muonic atoms is $\sim 1\mu$s, a pulsed beam allows the prompt background events to be removed by performing measurements in a delayed time window.

The cosmic-ray background should be carefully eliminated because of the tiny number of signal events expected in the experiment. Cosmic-ray muons may decay in flight or interact with the materials around the area of the muon-stopping target, producing signal-like electrons in the detector region. For the Phase-I, only the area of the CDC has to be protected against cosmic muons to prevent the muon-induced background. Both passive and active shielding should be used to protect the region around this detector. The passive shielding consists of a thick concrete ceiling as the experiment is underground. Beneath the ceiling, Cosmic Veto Counters (CVC) will be located, whose inefficiency should as small as 0.0001 [1]. In our studies, we develop design and optimize parameters of the system using computer simulations.

2. The cosmic-veto system

Veto on cosmic muon propagation can be effectively organized using scintillator strip modules surrounding the COMET Phase-I detector [1]. Similar strip modules are used in the Belle II detector where the scintillator light is collected through Kuraray WLS fibers to Hamamatsu silicon photodetectors MPPC. To date properties of such a detecting system as well as the radiation hardness of MPPC are studied well enough [2].

The CVC system consists of eight super-modules, two above the CDC, two on either side and one for the front and rear ends. The super-modules will be assembled from modules that are formed from 15 strips made of polystyrene-based scintillator and placed in parallel. Each strip have the same profile and a length of about 220 cm (110 cm for vertical modules). To provide a high registration efficiency, a super-module comprises four layers of modules. We have shown that the desirable value of the inefficiency can be achieved using a coincident signal in any two of the four strip layers, which allows us to reject fake signals caused by the MPPC noise as these rarely coincide within a short resolution time window [3].

3. Efficiency of scintillator counters

First we have carried out computer simulations of inefficiency of the CVC. In these simulations, we only model a horizontal slab of the COMET veto counter, expecting lower inefficiency for vertical slabs due to longer average muon range. The ionization produced by muons in a scintillator strip is proportional to their range in the strip, so the signal of a muon depends on its entrance position and direction as well as on the sizes of inactive zones in the detector. A layer of the CVC consists of scintillator strips of 40 mm width and 7 mm thickness. Surfaces of strips are chemically processed to provide a diffuse reflection of the scintillator light. Processing adds 0.2 mm to the strip width on average. Besides, we assume additional 0.1 mm separation of strips because of their imperfect geometry, so two neighboring strips are separated by a 0.3 mm inactive zone.

Similar to the Belle II KLM detector design, two neighboring modules are separated by an aluminum holder of 5 mm width, which also represent an inactive zone. This wide zone is particularly harmful for muon registration because its width is comparable with the average cosmic muon range in 7 mm thick strips, so that muon signals can vanish. To keep an appreciable muon signal, both inter-strip and inter-module inactive zones of different layers should be shifted w.r.t. each other.

Our simulations are done for both a single layer and a stack of four layers for three configurations described below. In the first one, each subsequent layer is moved w.r.t. the previous one the chosen distance in the same transverse direction. In the second configuration, the second and the fourth layers are moved w.r.t. others the same distance doubled. Finally in the third configuration, the third and the fourth layers are moved w.r.t. others the same distance doubled. Thus, in the latter two configurations an offset of layers equals to an average offset in the the first configuration, see figure 1.

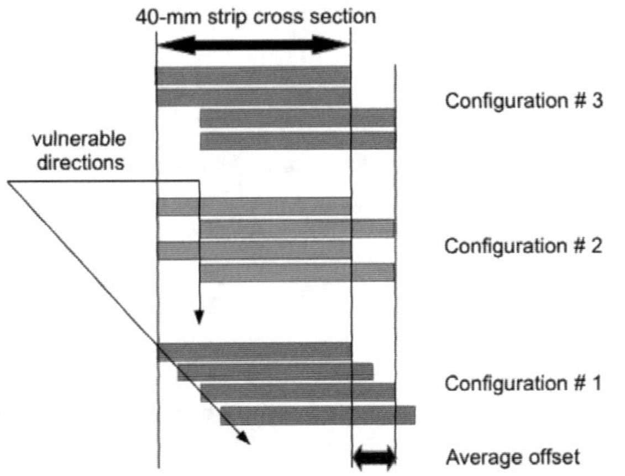

Figure 1. Strip layouts implemented in the simulation.

On its way from a muon entrance point to an MPPC, the collected scintillator light is significantly attenuated. To take into account the attenuation, we use experimentally measured muon signals in a 220 cm long strip. Our simulation show that in a single layer, about one third of unregistered muons goes through inter-module inactive zones. Others are not registered due to occasionally small signals, predominantly at inter-strip inactive zones in the rear part of strips where the attenuation of light is higher. Because of inactive zones in layers, an average range of muons contributing to signal amounts 10 mm contrary to 10.5 mm in a continuous layer of scintillator. This takes place under the condition that only the largest of two range parts contributes to the signal in the case when a muon crosses an inactive zone.

As for the complete stack of four layers, in the second and the third configurations at the offset of 40 mm, positions of inter-strip inactive zones of all layers coincide. At the offset about 43 mm the inter-strip inactive zones of two layers appear below (or above) of inter-module inactive zones of other layers, right on the symmetry line of the zones. Such offsets abruptly increase the inefficiency. Apparently, the first configuration is preferable because (i) positions of its inter-strip inactive zones do not coincide and (ii) they can only be crossed by a muon trajectory at large polar angle where the muon flux is lower.

The simulations have demonstrated that the inefficiency of the first configuration at the readout threshold of eleven photo-detector's pixels is less than 0.0001 when the average offcet of layers exceeds about 20 mm. In addition to the 11-pixel threshold that is a possible level of noise signals' amplitudes after a long exposition of MPPCs to a neutron background of the experiment [2], separate simulations have been done for the readout thresholds equal to 10, 9, 8, and 7 pixels. These have shown the muon registration inefficiency decreases by an order of magnitude when the threshold value is reduced by two pixels.

4. Noise signals

The noise signals of MPPC above the readout threshold can fake muon passing through the counters, which results in wasted time of data taking. This is a result of random coincidences of the noise signals that appear in two strips of two layers within resolution time and could not be distinguished from signals of a real muon in those layers. Veto time windows should be formed around such noise signals, and data from the veto windows should be excluded from the physics analysis, while in fact there was no cosmic muon.

During data taking, the frequency of noise signals grows with irradiation of photo-detectors by background neutrons that damage the p-n conjunction of silicon photo-detectors [2]. We have shown that operating at the threshold above seven pixels keeps the fraction of the lost time at the level of few percents throughout full time of data taking. A disadvantage of working at high thresholds is lower muon registration efficiency. However, our simulations of the efficiency value have demonstrated that it still meets COMET requirements even at the 11-pixel threshold [3].

5. Signals of background neutrons

Neutrons emitted from the muon-stopping target can also cause fake veto signals in the CVC. Besides, they could damage silicon photo-detectors being used in the CVC [3]. On average, neutrons appear in six of ten muon captures by atomic nuclei of a target made of aluminum, which results in $\sim 10^9$ neutrons per second for the COMET Phase-I.

There are two major ways in which neutrons affect the CVC: through gammas from neutron captures by materials of the detector and kicking protons and ions out directly in the CVC. We have studied the latter way in detail. Since there is a threshold for signal amplitudes of the CVC photo-detectors, signals of low-energy neutrons are being rejected. Affordable values of the threshold lie from one tenth to one fifth of the average MIP signal in CVC [3], so the threshold corresponds to 0.2–0.4 MeV of energy deposited in scintillator by ionization. We perform simulations of neutron signals using the Geant4 toolkit, version 10.0, with the FTFP_BERT_HP physics list.

First of all, the simulations have shown that the stopping power of charged projectiles created by fast and medium-energy neutrons in polystyrene is much higher than that for MIP [4]. As a result of the Birks' law, for neutron's energy below 1 MeV, light yield of the projectiles lies below the readout threshold values quoted above. Therefore, such neutrons are not typically seen by the CVC. Furthermore, the sub-millimeter range of the charged projectiles in strips makes coincident signals in two strips extremely unlikely.

Then, we have carried out simulations of a shield between the CDC and CVC. The most convenient shielding material by far is concrete, this is comparatively chip, strong, available in bricks, and allows pouring, which simplifies the construction of complicate profiles. It shields quite effectively against low-energy and fast neutrons, moderating and then capturing them. However, the spectrum of neutrons from muon captures stretches up to tens of MeV. It takes more elastic collisions to moderate energetic neutrons, hence a longer free path, i.e., higher penetration ability, and this trend persists at higher energies. Furthermore, energetic neutrons interact inelastically and cause spallation, giving birth to high-energy projectiles.

For these reasons, a more sophisticated structure should be designed for COMET inner shielding as follows. Iron is not that good in construction of such a heavy self-supporting structure as the COMET inner shield. Iron still makes a much better job in moderating medium-energy neutrons, though at the cost of a sea of fast neutrons. The fast neutrons can be effectively moderated through elastic scattering in polyethylene, which can, however, result in neutron captures on hydrogen accompanied by emission of 2.2 MeV gammas undesirable for the CVC. Gammas emitted in the neutron captures should be shielded by such a high-Z material as lead. Thus, the three layers of different materials represent a better choice for the shield.

In order to optimize the thickness of each layer, we have simulated neutrons impinging on a three-layer slab of 4×4 m^2 area, using an experimentally known energy spectrum [5]. Materials of layers are iron in the first, polyethylene in the second and lead in the third one. Behind the slab, there are a 10 mm thick air gap and then four 7 mm thick scintillator layers interleaved with 3 mm thick polystyrene pads. We vary thickness of the three shield's layers, keeping the total shield's thickness equal to 45 cm and recording spectra of neutrons and gammas in the air gap as well as the response of each of the four scintillator layers.

For instance, for a combination of 25 cm of iron, 10 cm of polyethylene, and 10 cm of lead, the number of coincident signals above 5 pixels is equal to 240, which is $2.4 \cdot 10^{-4}$ of the number of incident neutrons. As the expected rate of muon-capture neutrons at the shield slab is about $1.7 \cdot 10^8$ per second, one gets about $4 \cdot 10^4$ coincident signals per second in the scintillator layers behind the shield. These signals fake cosmic-muon signals. Then, for the whole CVC with four side half-slabs, the rate of the fake signals is about 120 kHz. Thus, if we apply a 50 ns veto window, we have as little as 0.6 % of date-taking time lost due to neutrons from muon captures.

Iron shields against medium- and high-energy neutrons more effectively than concrete, but it is less convenient from the point of view of engineering. A good compromise is the use of reinforced concrete containing scrap iron. Therefore, reinforced concrete as a material of the first shield layer has been implemented in the next simulations, volume fractions of iron and concrete being about 0.33 and 0.66 respectively. Additionally, a 10 cm thick return yoke of solenoid has been placed in front of the inner shield in the simulation. For several combinations of thicknesses of the three shield materials, table 1 shows the number of fake signals as well as neutrons and gammas behind the shield. These simulations have shown that the shield should be comprised of 25 to 30 cm of the iron-concrete mix, 10 cm of polyethylene and 5 cm of lead.

Such a shield reduces the flux of fast and more energetic neutrons at photo-detectors of the CVC by two orders of magnitude, so that no substantial radiation damages of photo-detectors is expected throughout full period of data taking. The optimal configuration of the shield is of an arched shape, which accounts for signal attenuation along strips and better protects photo-detectors that are located at outer ends of strips.

Table 1. Summary of neutron and gamma abundance behind a shield comprised of layers of reinforced concrete, polyethylene and lead.

Reinforced concrete, cm	Polyeth., cm	Lead, cm	# of neutrons behind shield	# of gammas behind shield	# of coincident signals in counters
25	10	5	9600	5800	242
30	5	5	14000	5200	200
30	10	5	6000	4100	142
35	–	5	21400	4000	183
40	–	5	13000	2600	114
50	–	–	6100	11300	381

6. Conclusion

The COMET experiment provides a large window on new physics beyond the Standard Mode. Its first phase allows us to probe the CLFV at a new level. The detector of the experiment have to be surrounded by a cosmic-veto system, which is optimized in our studies. Modeling of the system has demonstrated viability of the necessary high cosmic-muon registration efficiency. Simulations of the efficiency value have shown that it meets COMET requirements even at the 11-pixel threshold. Operating at high thresholds has an advantage of low rate of fake signals caused by both the photo-detector noise and background neutrons, which results in shorter time lost during data taking.

To optimize shielding against the background neutrons, Geant4 simulations of the system have been performed including details of signal coincidence in a scintillator strip veto counter. The simulations have shown that a mix of iron and concrete performs quite well as a base material for the inner shield of the system. In the conjunction with the solenoid yoke, a shield composed of reinforced concrete, polyethylene and lead drastically reduces the neutron flux at the counter strips. The time loss concerned with fake veto signals of neutrons can be reduced to a tolerable value of 1 % of data-taking time by a 40–50 cm thick shield even at the 5-pixel threshold of photo-detectors.

Acknowledgements

This work was performed within the framework of the Center FRPP supported by MEPhI Academic Excellence Project (contract No. 02.a03.21.0005, 27.08.2013).

References

[1] The COMET Collaboration 2009 *Conceptual Design Report for Experimental Search for Lepton Flavor Violating* $\mu - e$ *Conversion at Sensitivity of* 10^{-16} *with a Slow-Extracted Bunched Proton Beam* (COMET) J-PARC P21
[2] Tarkovsky E 2011 Performance of a scintillating strip detector with G-APD readout *Nucl. Instrum. Meth.* A **628** 372–5
[3] Markin O and Tarkovsky E 2014 Simulations of the COMET veto counter *arXiv* :1402.5522
[4] Markin O 2015 Inner Shielding of the COMET Cosmic Veto System *arXiv* :1505.07378
[5] Kozlowski T *et al.* 1985 Energy spectra and asymmetries of neutrons emitted after muon capture *Nucl. Phys.* A **436** 717–32

Journal of Physics: Conference Series **675** (2016) 042013 doi:10.1088/1742-6596/675/4/042013

Research of work stability of diamond detectors used in SCR DDIR

R F Ibragimov[1], E M Tyurin[1], V V Kadilin[1], V A Kolyubin[2], K V Zaharchenko[2] and P G Nedosekin[2]

[1]National Research Nuclear University MEPhI (Moscow Engineering Physics Institute), Kashirskoe highway 31, Moscow, 115409, Russia

[2]Industrial - Technology Center "UralAlmazInvest", Ivano-Franko street, Moscow, 121108, Russia

E-mail: rfibragimov@mephi.ru, vvkadilin@mephi.ru

Abstract. In this work we study influence of various factors on stability of ionizing radiation detectors installed in the cosmic ray spectrometer (SCR) based on diamond detectors of ionization radiation (DDIR). Diamond detectors for SCR are made of single crystals of synthetic diamond type IIa. Diamond detectors were studied successively in three different experiments. Checking detector stability with ambient temperature increased up to 70 degrees Celsius was the first experiment. At next we change the geometry of detector irradiation by rotating nuclear source around it and measuring changes in detector count rate. And last one experiment was about checking the phenomenon of polarization by prolonged detector irradiation by ionizing radiation of various types and energies. The study revealed the presence of the strong influence of the polarization effect on the work of diamond detectors for registration of ionizing particles with short mean free path (in our experiment they were the alfa-particles of ^{238}Pu). In this work correspondence of the experimental results of the "rotation" the source around the detector with the data obtained by simulation in GEANT-4 was shown.

1. Introduction

The detectors of ionizing radiation (IR) based on the diamond sensitive elements are getting more popular in various spheres of human activity in recent years. The idea to use diamond as a sensitive element for the detectors of ionizing radiation appeared in the 1960s, and has been determined by such properties of diamond as high radiation, temperature and chemical resistance [1]. However, difficulty with choosing the crystals with desired properties (mainly due to the different degrees of natural diamond purity) has impeded the intensive development of this direction, and therefore use of diamonds was very costly. Currently advances in growing the artificial diamond single crystals by chemical vapour deposition process (CVD) led to simplification of using diamonds in various applications in science and technology. For example, thanks to its high velocity of charge carriers and radiation resistance, diamond is now widely used as a replacement for silicon sensitive elements in devices monitoring the position of the beam of charged particles in accelerators, preventing the deflection of a beam from a predetermined path [2]. High radiation resistance and its stability at temperatures up to 200 degrees Celsius enabled to use diamond detectors in the vertical neutron chamber in the International Thermonuclear Experimental Reactor (ITER) [3]. This work is devoted to the use of diamond detectors of ionizing radiation in space technology, in particular in the Cosmic Ray Spectrometer developed by the Industrial - Technology Center "UralAlmazInvest". It is expected that the spectrometer will allow to study the cosmic radiation of following types: electrons, protons and heavy ions.

2. Experement setup

In this work we studied the DDIR ELSC-1 based on artificial type IIa diamond for saving stability of it characteristics under various conditions. The following conditions for the detector had been considered: increased ambient temperature; the detector irradiation at different angles of incidence by the IR on the surface of DDIR (that was checked with a simulation of diamond detectors, made by GEANT-4); a long irradiation time of the detector by different types of IR (checking the influence of polarization). Brief description of the ionizing radiation sources used in this study is shown in table 1.

Table 1. Radiation sources.

Name of the nucleus and type of radiation	Energy spectrum
^{238}Pu, alfa-particle	One line with energy near to 5499.9 keV
^{90}Sr-^{90}Y, beta-particle	Continued energy range with maximum value near to 2.28 MeV, and energy of maximum of activity near to 545.9 keV.

3. The heating of the detector and changing the geometry of irradiation

ELSC-1 with a ^{238}Pu was placed in special heater, where temperature was up to 70 degrees Celsius. Then twelve spectra (300 seconds in duration each) were recorded with PC-compatible board, while decreasing detector and source temperature by every 5 degrees Celsius. We studied the changes in the total count rate of the detector and in the count rate at the peak of total absorption. The results of data obtained during several series of such measurements are shown in figure 1. Statistics of pulses set has made main contribution to the measurement error.

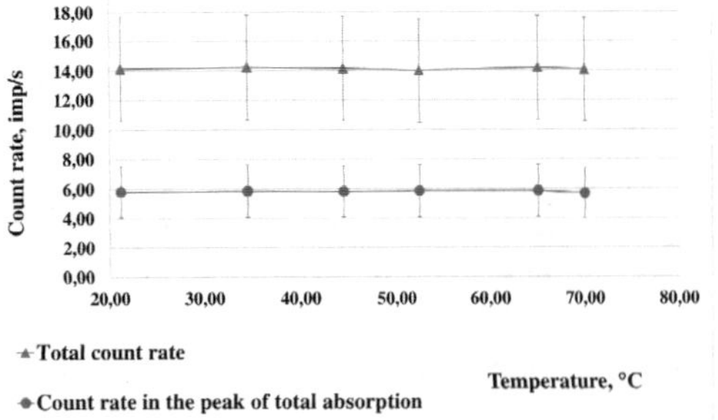

Figure 1. Influence of different temperature on the count rate of DDIR ELSC-1, ^{238}Pu (lines connecting points are guide to the eyes only).

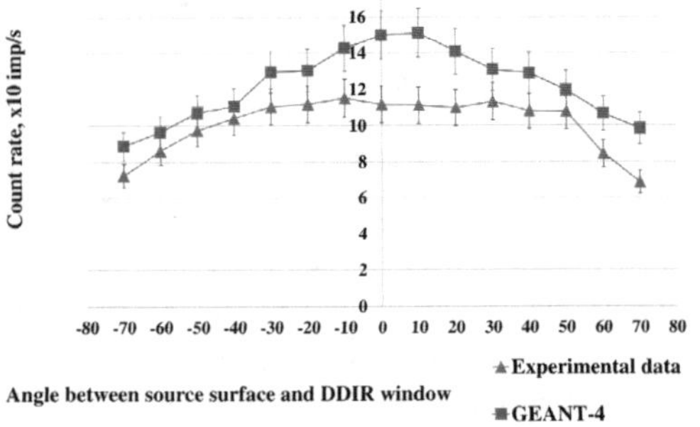

Figure 2. Influence of different experiment geometry on the count rate of DDIR, ^{90}Sr-^{90}Y, (lines connecting points are guide to the eyes only).

As we can see, there are no changes in count rate of ELSC-1 by detecting ionizing radiation of ^{238}Pu, even with ambient temperature near to 70 degrees Celsius.

The study of ELSC-1 detector behavior in a changing geometry of irradiation was the next experiment. The angle between the incident flux of IR and the surface of ELSC-1 varied in increments of 10 degrees, and after each change the spectrum of ionizing source ^{90}Sr-^{90}Y was recorded. Also, a mathematical model was created in the GEANT-4 environment, which imitates a similar experiment. Comparison of the results is shown in figure 2.

As seen from the experimental results, there is a slight decrease in the count rate when the angle of incidence on the detector surface reaches the value of 70 degrees. This means that a decrease in detection efficiency due to reduction of the effective area of the detector under the influence of radiation, prevails over the increase of efficiency, which is caused by increasing of the thickness of the detector. Comparison of experimental results with mathematical models showed similar results, and in this connection we can make one conclusion: a mathematical model of a diamond detector created in GEANT-4 is fully confirmed by the behavior of the detector in the real world.

4. Checking the polarization phenomena

The most significant drawback of diamond detectors today, is their susceptibility to the influence of the polarization effect. This effect is manifested as a decrease in charge collection efficiency of the detector (which reduces the counting rate and the distortion of the spectrum [4]). Checking of the polarization effect in ELSC-1 detector was carried out using a long irradiation of the detector by two types of ionizing radiation. There was continuous recording of twelve spectra, 300 seconds in duration each (total time of measurement is one hour). Comparison of the ELSC-1 readings at the beginning and at the end of the measurement cycle allows us to see whether there is any polarization. The comparison of the results for different radiation sources are shown in figures 3 and 4.

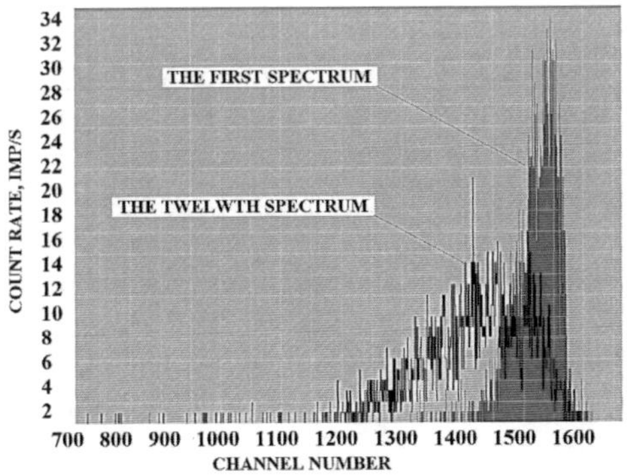

Figure 3. Influence of different temperature on the count rate of DDIR ELSC-1, ^{238}Pu.

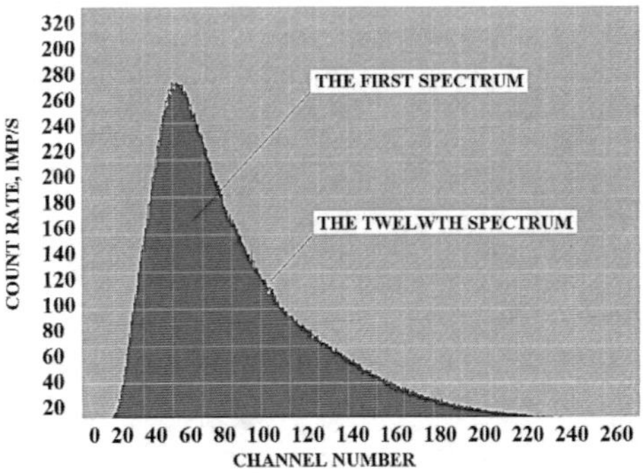

Figure 4. Influence of different experiment geometry on the count rate of DDIR, ^{90}Sr-^{90}Y.

We saw noticeable effect of polarization in case of registration of radiation with short mean free path. We investigated some ways to eliminate polarization (currently known method of turning off the power supply of detector and a method of changing the polarity of the bias voltage [5]). It was found that turning off the power supply of the ELSC-1 for 30 seconds is the better way to eliminate the polarization than reversing polarity of the bias voltage for the same time. In this regard the special scheme for CRS was made that allowed spectrometer to work in the following measurement mode: 60 seconds of measurement, 10 seconds depolarization by turning off the bias voltage. Currently, CRS is at the final stage of manufacturing.

5. Conclusion

It has been found that the DDIR ELSC-1 remains stable when the ambient temperature increases up to 70 degrees Celsius.

We have found a slight change in the counting rate by increasing the angle of incidence of electron radiation on the surface of a diamond detector from 0 to 70 degrees, which confirmed the legitimacy of the use of a mathematical model of the detector set up in GEANT-4.

It is shown that with the help of periodic power outage from the detector for short periods of time (about 30 seconds) the instability of the diamond detector associated with the effect of polarization is almost entirely eliminated.

Acknowledgments

This work was partially supported by MEPhI Academic Excellence Project (contract No. 02.a03.21.0005, 27.08.2013).

References

[1] Grilj V *et al.* 2013 *Nucl. Instrum. Meth.* B **306** 194
[2] Cardarelli R and DiCiaccio A 2013 *Nucl. Instrum. Meth.* A **718** 309
[3] Amosov V *et al.* 2011 (in Russian) *Appl. Phys.* **4** 14-17 (original Russian title: *Prikladnaya fizika*)
[4] Manfredotti C *et al.* 2006 *Diam. Rel. Mater.* **15** 1105
[5] Ibragimov R *et al.* 2014 (In Russian) *Bulletin of the Moscow Engineering Physics Institute* **4** 1-7 (original Russian title: *Vestnik Moscovskogo Ingenerno-Phyzicheskogo Instituta*)

Liquid-xenon detector under the intensive pulse irradiation conditions

M A Kirsanov
National Research Nuclear University MEPhI (Moscow Engineering Physics Institute), Kashirskoe highway 31, Moscow, 115409, Russia

E-mail: makirsanov@mephi.ru

Abstract. The effect of intense pulsed irradiation on the operation of the liquid xenon spectrometer was studied. The ionization chamber filled with liquid xenon was irradiated by bremsstrahlung pulses of the microtron. The pulse repetition rate was 400 Hz. The absorbed dose ranged from 10^{-7} to 0.1 Gy per pulse. Stable operation of the liquid xenon spectrometer in the intervals between the pulses of the accelerator was shown for a long time.

1. Introduction

At the present time the usage of the high-intensity pulsed sources of radiation, such as particle accelerators, synchrotrons, neutron generators and the pulse generators of the x-ray radiation are developing rapidly. The use of radiation detectors in intense radiation fields generated by such sources is limited by their radiation resistance.

The intense radiation can significantly affect the operation of the detectors and lead to a temporary malfunction of the detectors. Furthermore reversible effects after prolonged use under conditions of intense radiation accumulates unremovable defects which lead to deterioration of characteristics of the detectors. The usage of the noble gases as the working detector substance in the presence of a high-intensity radiation has good prospects. The detectors based on noble gases (both in the gaseous and liquid states) have the highest radiation resistance to prolonged intense irradiation [1]. As the noble gases have an atomic structure the radiation defects in traditional understanding do not arise in them and it is possible to use them long time without replacement. However, at high doses of radiation in ionization detectors based on liquid noble gases a special type of disruption may occur that is associated with the formation of a space charge and the flow of ion current.

To obtain experimental information on the effects of high-intensity pulsed radiation on the characteristics of the liquid xenon spectrometer such detector was located in the beam of bremsstrahlung of electron accelerator.

2. Processes in the liquid xenon detector under the intensive pulse irradiation conditions

The study of the liquid xenon detector operation was carried out on the beam of the microtron MT-25 in the Flerov Laboratory of Nuclear Reactions, Joint Institute for Nuclear Research. The parameters of the microtron: the energy of the electrons in the beam of 10-25 MeV, average beam current of up to 15 μA, the pulse repetition rate of 400 Hz, pulse duration of 2.5 μs. At its output, the electron beam bombarded a 2-mm-thick tungsten target with a 3-cm-thick aluminum absorber and generated bremsstrahlung with an appropriate spectrum and angular distribution.

Inside the cryostat there was a multifunction camera that contained researched ionization chamber, the xenon cleaning unit and the control unit of xenon purity. For chamber cooling in cryostat the vapors of the boiling nitrogen were used. The temperature of the chamber could be varied in the range 165-250 K. Most of the measurements performed at T=230 K. Purification of xenon was carried out using the electric spark method [2]. Liquid xenon is purified to a such level that the electron can traverse the distance between electrodes 10 times in average before capture.

In the first series of measurements the study of processes arising in liquid xenon under the influence of intense pulsed radiation was carried out. In this case, the research camera is a flat ionization chamber with two electrodes 1.8 cm in diameter, the area of the electrodes $S = 2.54$ cm^2 and the distance between the electrodes $d = 0.53$ cm. Working chamber volume $V = 1.35$ cm^3 [3]. The absorbed dose ranged from 10^{-7} to 0.1 Gy per pulse.

A space charge is produced in the chamber under high-intensity irradiation. If the formation time of the space charge significantly less than the time of his disappearance, it is possible to use the concept of the critical dose. The critical dose corresponds to a transition from the normal operation of the ionization chamber to the regime of space charge. The critical dose is given by the expressions [4]

$$D_{cr} = \varepsilon\varepsilon_0 wU / (fd^2 e\rho),$$

where e is the elementary electric charge, ε is the permittivity, ε_0 is the electric constant, ρ is the substance density, w is the average energy of production of an electron-ion pair, U is the potential difference between the electrodes, d is the interelectrode distance, and f is the ions yield coefficient from a track. Under our conditions, $D_{cr} = 19$ µGy at $U = 5$ kV.

The behavior of electrons in the chamber depends on the magnitude of the absorbed dose per pulse of radiation. When the dose is less than the critical the electrons leave the volume of the chamber within a few microseconds. When the irradiation dose above critical, the space charge reduces the electric field to zero in large parts of the chamber. In this case, the loss of electrons occurs due to the recombination. This process can last hundreds of microseconds. After the restoration of the electric field the electrons in a few microseconds leaving the chamber. In both cases after a complete collection of the electrons the positive ions remain in the chamber. The positive ions slowly drift to the cathode.

Experimental results showed that under the conditions of this work, the escape of electrons from the chamber occurs for a time shorter than the interval between pulses of radiation. Accumulation of electrons in the chamber does not occur.

Because the ions have a low mobility, they accumulate in the chamber even at low intensity of pulse irradiation. Therefore, the registration of gamma-quanta in the interval between pulses of the accelerator occurs on the falling ion current. In this case, the following factors give the deterioration of the energy resolution of gamma spectrometer: 1) fluctuations in the ion current; 2) a change in the ion current over time between pulses of radiation; 3) the presence of the internal field of the space charge of positive ions. Each of these factors gives the contribution of ω_i to the total relative energy resolution of the detector. The values of these factors are determined by the expressions [4]:

$$\omega_1 = 2.35\frac{\sqrt{D\rho VfT_e / 2t_{int}w}}{Ef / w}; \quad \omega_2 = \frac{2.35}{\sqrt{12}}\frac{D\rho V}{E}\frac{RCT_{dif}}{T_+^2}; \quad \omega_3 = \frac{2.35}{\sqrt{12}}\frac{f_c - f_a}{f_0}, \tag{1}$$

where D is the absorbed dose; V is the chamber volume; T_e is the electron drift time; t_{int} is the interval between pulses; E is the photon energy; R is the load resistance; C is the equivalent capacitance of the detector; T_{dif} is the differentiation time of the amplifier; T_+ is the ion drift time, f_c and f_a is the coefficient of ion yield from the track near the cathode and anode, respectively; f_0 is the ion yield in the absence of irradiation pulses.

The dependence of the contributions to the energy resolution ω_i on the dose per irradiation pulse is shown in figure 1. The calculations were performed for the photon energy $E = 0.662$ MeV.

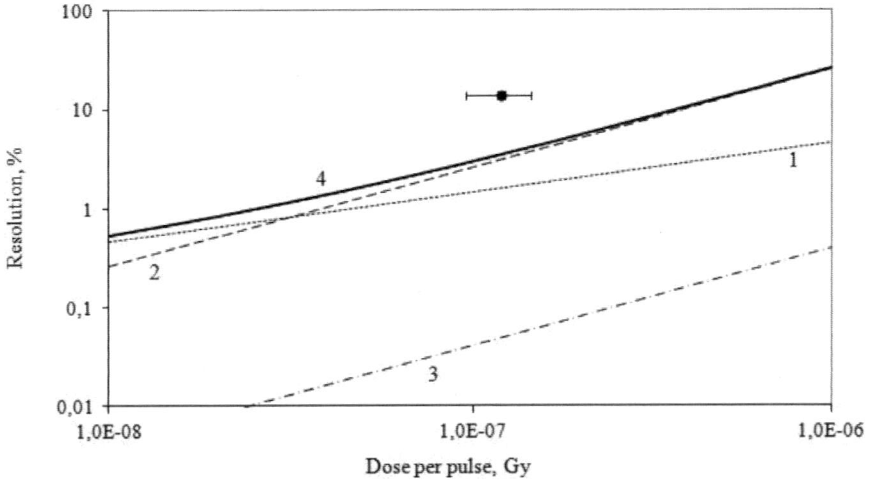

Figure 1. Contributions of different factors to the energy resolution of the our chamber: 1 - ion current fluctuations; 2 - changes in the ion current with time; 3 - electric field of the ions; 4 - total energy resolution.

From the equations (1) it is evident that each of these factors depend on the absorbed dose and the volume of the detector in different ways. In our detector (small volume) factors 1 and 2 have a major role, the impact factor of 3 is very small. In a large detector ($V = 10^3$ cm^3) factor 1 retains a major role, a factor 3 is more important than factor 2. In our detector, the total contribution of the ions to the energy resolution at a dose of 0.1 μGy is 3%, and at a dose of 1 μGy reaches 26%. Thus the accumulation of ions deteriorates the energy resolution of the detector at doses much lower than the critical dose.

3. The operation of the detector in spectrometric mode in the intervals between the pulses of intense radiation

In the second series of measurements we studied the spectrometric properties of the liquid xenon detector during operation in the intervals between the pulses of intense radiation . In this case, research chamber was a cylindrical ionization chamber, the working volume of which was equal to $V = 2.4$ cm^3 [5]. The detector was surrounded by a protective shield consisting of lead and borated polyethylene.

The energy released in the spectrometer during the accelerator pulse was higher than the energy of the detected quanta (~1MeV) by several orders of magnitude. Developed electronic circuit blocking of the preamplifier in the momentum of the accelerator pulse allowed the detector to operate in a spectrometric mode, in the intervals between pulses. Under our conditions, the blocking time was 0.5 ms. In test measurements, the detector registered gamma-quanta from ^{137}Cs source. The corresponding spectra are shown in figure 2. When the accelerator is off the energy resolution of the detector was 9.5%. The registration of gamma-quanta in the intervals between pulses of the accelerator occurs in the presence of ion current. The ion current deteriorates the energy resolution of the detector. When the accelerator is running resolution of the detector was 14%. This experimental point is shown in figure 1. In the cylindrical geometry the resolution for the above factors increases approximately two times. This means that the experimental result does not contradict our calculations.

This detector has been used to study short lived isomer ^{207m}Pb ($T_{1/2} = 0.8$ s), which is under the action of gamma rays formed in the reaction ^{208}Pb(γ,n)^{207m}Pb. The yield of the ^{207m}Pb isomer as a function of the energy of the bremsstrahlung spectrum maximum was measured in [6].

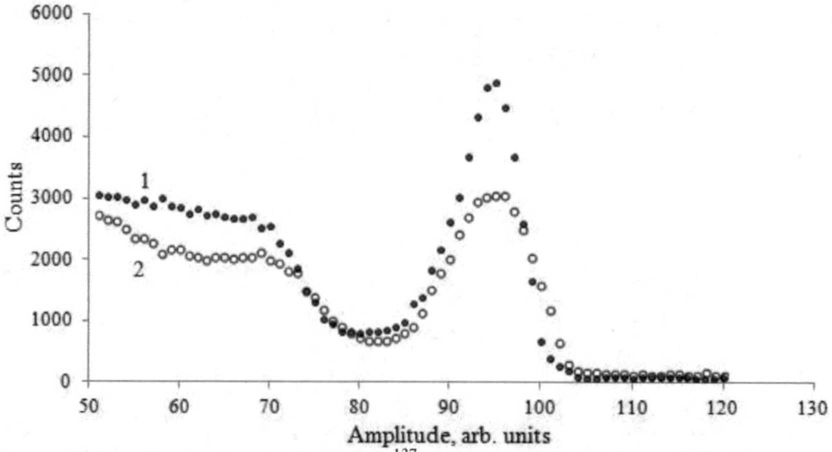

Figure 2. Amplitude spectra from the ^{137}Cs source for two cases: (1) the accelerator is turned off and (2) the accelerator is turned on.

The total absorbed dose in the liquid xenon detector during operation on the accelerator MT-25 for 2 years was 10^7 Gy. There were no change in the spectrometric properties of the detector.

4. Conclusion

The ionization detectors based on noble gases have the high radiation resistance. The effect of intense pulsed irradiation on the operation of the liquid xenon detector was studied. The electrons formed under pulsed irradiation are not accumulated in the detector volume. Due to the low mobility positive ions are accumulated in the detector form a space charge creating an ion current. This leads to a deterioration of the energy resolution of the detector, which depends on the absorbed dose per pulse and the volume of the detector.

It was shown that the liquid xenon detector is able to operate in the spectrometric mode during the intervals between pulses of intense radiation. Cylindrical liquid xenon detector was applied to study short lived isomer ^{207m}Pb.

Acknowledgment

The author is grateful to I. M. Obodovskiy for supporting, Y. P. Gangrskiy and A. G. Belov for the possibility of operating on the microtron MT-25, V. G. Subbotin and S. Iliev for the development of a special preamplifier.

This work was performed within the framework of the Center for FRPP supported by MEPhI Academic Excellence Project (contract № 02.a03.21.0005, 27.08.2013).

References

[1] Kirsanov M A *et al.* 1991 (in Russian) *Atomic Energy* **70** 131 (Original Russian title: *Atomnaya Energiya*)
[2] Pokachalov S G *et al.* 1993 *Nucl. Instrum. Meth.* A **327** 159
[3] Kirsanov M A and Obodovskiy I M 2008 *Instr. Exp. Techn.* **51** 358
[4] Kirsanov M A and Obodovskiy I M 2010 *Instr. Exp. Techn.* **53** 185
[5] Kirsanov M A *et al.* 1991 (in Russian) *Instruments and Experimental Techniques* **1** 75 (Original Russian title: *Pribory i Tekhnika Eksperimenta*)
[6] Kirsanov M A, Obodovski I M and Gangrski Yu P 1993 *Nucl. Instrum. Meth.* A **327** 48

Fast timing and trigger Cherenkov detector for collider experiments

V A Grigoryev[1], V A Kaplin[1,4], T L Karavicheva[2], A S Konevskikh[2], A B Kurepin[2], V A Loginov[1], Yu A Melikyan[1], I V Morozov[2], A I Reshetin[2], D V Serebryakov[2], A I Shabanov[2], M Slupecki[3], W H Trzaska[3] and E M Tykmanov[1]
[1] National Research Nuclear University MEPhI (Moscow Engineering Physics Institute), Kashirskoe highway 31, Moscow, 115409, Russia
[2] Institute for Nuclear Research of the Russian Academy of Sciences, V-312, 60-letiya Oktyabrya prospect 7a, Moscow, 117312, Russia
[3] University of Jyväskylä, P.O. Box 35, FI-40351 Jyväskylä, Finland

E-mail: kaplinv@mail.ru

Abstract. Analysis of fast timing and trigger Cherenkov detector's design for its use in collider experiments is presented. Several specific requirements are taken into account – necessity of the radiator's placement as close to the beam pipe as possible along with the requirement of gapless (solid) radiator's design. Characteristics of the Cherenkov detector's laboratory prototype obtained using a pion beam at the CERN Proton Synchrotron are also presented, showing the possibility of obtaining sufficiently high geometrical efficiency along with good enough time resolution (50 ps sigma).

1. Introduction

Cherenkov detectors are often used in collider experiments as a tool for timing measurements [1, 2]. As a general trend, requirements to such detectors become stricter. In particular, modern trigger Cherenkov detectors should have as high particle detection efficiency as possible, i.e. such devices should have relatively large area and cover the entire measurement region with no blank spots (including the area in close proximity to the beam pipe).

Even in case of using photomultipliers with large-area photocathodes, it is hardly possible to meet the needs mentioned above, because the photocathode's size is smaller than the external dimensions of the photomultiplier's front surface, and a non-detecting edges would always remain, reducing the detector's homogeneity. Moreover, usage of the square-shaped photomultipliers complicates the coverage of the area around the beam pipe and simple changing the radiators forms and sizes leads to drastic decrease in the signal's amplitudes at certain detector's regions.

In order to solve these problems, we propose to place the Cherenkov detector facing backwards to the beam direction. This could help to greatly improve the detector's homogeneity without the loose in its timing parameters, what is demonstrated by the presented below results of the FIT Cherenkov detector's prototype testing (FIT - Fast Interaction Trigger detector - is currently being developed for the upgraded ALICE experiment).

[4] To whom any correspondence should be addressed.

2. The experimental setup

A Cherenkov detector module based on the XP85012 multianode (8x8 anodes array) microchannel plate photomultiplier is going to be used as a basic element of the FIT detector. The first testing procedures carried out under laboratory conditions showed that the photomultiplier is suitable for the use in the FIT detector [3].

One of the distinctive features of this photomultiplier is its relatively large size (53x53 mm^2 sensitive area). This fact must be taken into account while creating a timing detector with ~50 ps time resolution, because there could be problems with the signal delay time variability depending on the point of the particle hitting. In order to study the detector's characteristics in the backward configuration and in case of oversized radiators usage a special research was made at the CERN PS (Proton Synchrotron) relativistic pion beam employing an additional coordinate-sensitive detector, specially developed for this purpose.

To increase the detector's granularity, it was planned to use four independent quartz radiators with a single MCP-PMT. A photo of the tested Cherenkov detector module with XP85012 photomultiplier and four radiators is shown in figure 1. The readout could be done from the common output and from individual "quadrant's" output (one quadrant stands for a square of 16 interconnected anodes).

Figure 1. The Cherenkov detector module.

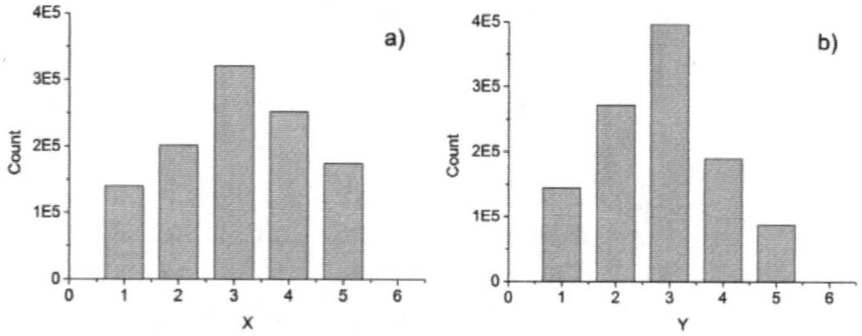

Figure 2. X-dimension (a) and Y-dimension (b) of the beam profile, measured by the specially developed coordinate-sensitive detector.

The measurements were done in a wide beam – the detection of the XY-position of the particle penetration through the tested Cherenkov module was provided by an additional coordinate-sensitive scintillation detector, employing two identical arrays of 5 plastic scintillator strips each with silicon photomultiplier readout. Such detector with the sensitive area equal to the area of the tested detector allows one to detect the position of a particle's penetration with ~1 cm accuracy. In figure 2a,b the beam profile measured with the described XY-detector is shown.

For the data-acquisition system triggering, two auxiliary Cherenkov counters based on FEU-187 PMTs were used. During the off-line data-processing the timing reference for all signals was set to the level of 50% of amplitude (virtual analogue of a constant fraction discriminator). In figure 3a the spectrum of time intervals between the moments of triggering of the two start Cherenkov counters is shown, in figure 3b – a similar spectrum is shown for the time intervals between the tested MCP-PMT signal coming from a common output and the averaged start detectors' moment of triggering.

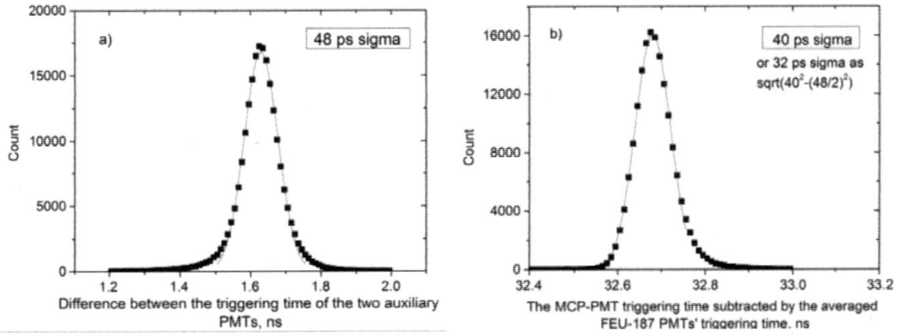

Figure 3. Time spectra, representing the auxiliary Cherenkov counters' combined time resolution (a) and an example time spectrum of the tested MCP-PMT (b).

Time resolution of the interval between the two start Cherenkov detectors is equal to 48 ps or a twice smaller value – 24 ps - in case of the offline averaging of the timing parameters of the both counters' signals. After the subtraction of the averaged start detectors' contribution, the resulting time resolution of the tested MCP-PMT is 32 ps for the typical case, presented in figure 3b.

3. The edge effects influence

In order to study the effects which occur in case of a particle's penetration through the MCP-PMT's window edges, which are not sensitive to light, two sets of radiators were produced. The first set includes four 21.5x21.5x20 mm^3 radiators which could cover the total 53x53 mm^2 MCP-PMT's sensitive area, when combined. The second set of four 54.5x54.5x20 mm^3 radiators could cover the whole 59x59 mm^2 are of the MCP-PMT window.

Figure 4 shows amplitude spectra measured from the individual quadrant's readout channels and the common output channel for the MCP-PMT with the first set of radiators, covering the MCP-PMT's photocathode area only. The left plot (a) is for the front beam conditions, while the right plot (b) is for the backward beam. Both plots show reasonably good amplitude resolution for all curves and no odd peaks. Figure 5 shows similar amplitude spectra for the case of usage of the second set of radiators, covering the whole area of the MCP-PMT window. As one could see from the presented data, the amplitude spectra measured for the case of the front beam (a) are far from being satisfactory because of an odd low-amplitude peak, induced by particles hitting the MCP edge. At the same time, the backward beam spectra (b) are characterized by acceptable amplitude resolution, loosing only one-thirds of typical signal amplitude.

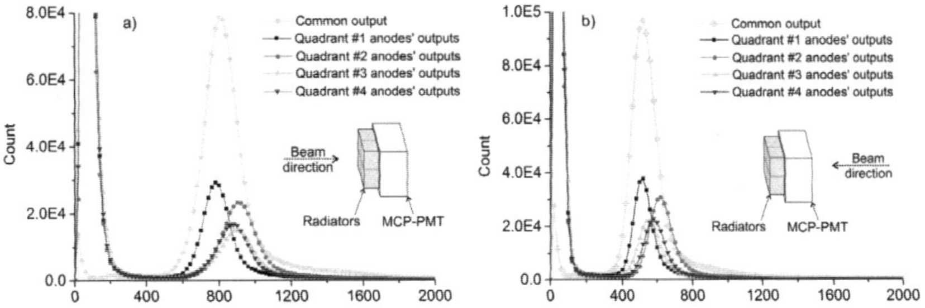

Figure 4. Amplitude spectra of different readout channels of the MCP-PMT with the set of radiators, which covers only the photocathode's area (53x53 mm^2 in total) for the front (a) and backward (b) beam direction.

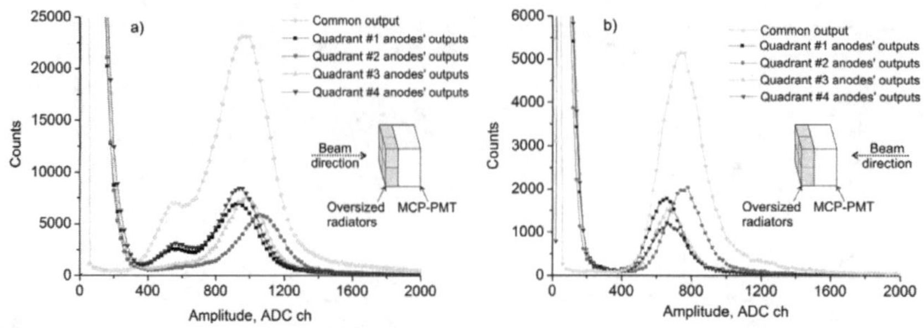

Figure 5. Amplitude spectra of different readout channels of the MCP-PMT with the set of radiators, covering the whole MCP-PMT window's area for the front (a) and backward (b) beam direction.

The timing parameters for all the configurations described are acceptable for the requirements of the experiment: the time resolution for the quadrant's timing readout is ~30 ps, or ~50 ps in the case of the common output readout (because of much larger readout area).

4. Study of a Cherenkov detector parameters with an oversized radiator

In order to study the possibility to use oversized radiators which protrude from the MCP-PMT borders a single 53x53x20 mm^3 radiator was produced and placed on the MCP-PMT window shifted sideways. Figure 6 shows amplitude spectra obtained for a front beam hitting the center of the radiator (a) and its edge (b), which protrudes for 5 mm from the edge of the MCP-PMT's sensitive area. For this configuration, the mean amplitude is decreased twice. In figure 7 a summary of the results of similar measurements is shown for the backward beam.

In the case of the backward beam, amplitude decreases insignificantly even for 15 mm radiator's shift. The time resolution of the device worsens from 36 to 41 ps (sigma) while shifting the radiator from 5 to 15 mm – all of these values are acceptable for the projected purpose.

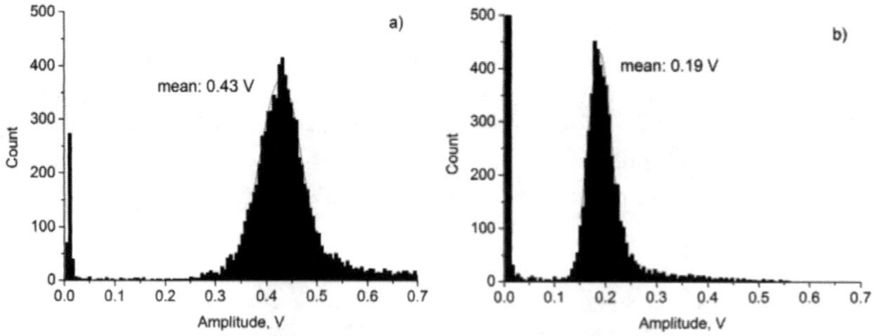

Figure 6. Amplitude spectra, obtained for a front beam hitting the center of the radiator (a) and its edge (b), which protrudes for 5 mm from the border of the MCP-PMT's sensitive area.

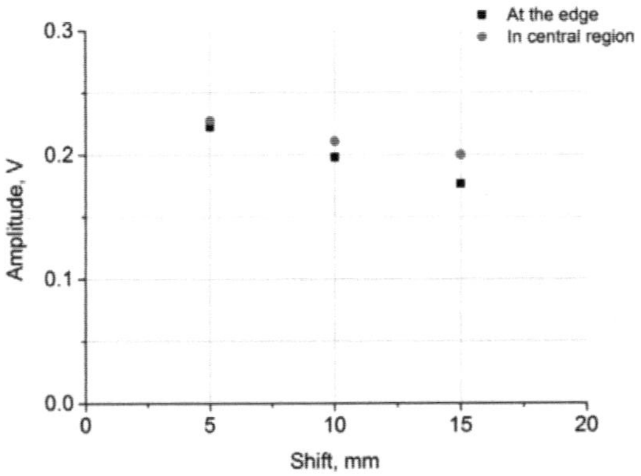

Figure 7. Summary of the results of measurements of average signal's amplitude measured with the common output for the backward beam while the radiator was shifted. Error bars lie within the data points size.

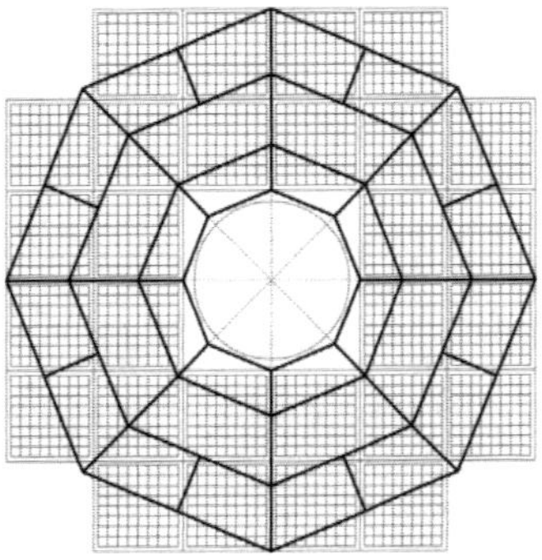

Figure 8. Possible configuration of a Cherenkov detector for a collider experiment. Thick lines – separate radiators' borders, thin lines – separate MCP-PMTs with indicated 8x8 anodes subdivision.

5. Conclusion

The results obtained show that turning the Cherenkov detector backwards to the beam direction might improve the homogeneity of the detector's amplitude parameters without any significant deterioration of its timing properties. Backward Cherenkov detector configuration also allows using oversized radiators, which could cover the unavoidable non-sensitive regions at the photodetectors' edges. These facts could help one to develop a highly effective detector that could cover the whole area required. For example, in the case of the collider experiments the resultant detector configuration could look like presented in figure 8.

Acknowledgments

This work was performed within the framework of the Fundamental Research and Particle Physics Center supported by NRNU MEPhI Academic Excellence Project (contract № 02.a03.21.0005, 27.08.2013) and Russian state grant RFMEFI61014X0003.

References

[1] Ikematsu K *et al.* 1998 *Nucl. Instrum. Methods* A **411** 238-48
[2] Yurevich V I 2013 *Physics of Particles and Nuclei Letters* **10** 258-68
[3] Abelev B *et al.* CERN-LHCC-2013-019 / LHCC-TDR-015

Study of the Planacon XP85012 photomultiplier characteristics for its use in a Cherenkov detector

V A Grigoryev[1], V A Kaplin[1,4], T L Karavicheva[2], A B Kurepin[2], E F Maklyaev[1], Yu A Melikyan[1], D V Serebryakov[2], W H Trzaska[3] and E M Tykmanov[1]
[1] National Research Nuclear University MEPhI (Moscow Engineering Physics Institute), Kashirskoe highway 31, Moscow, 115409, Russia
[2] Institute for Nuclear Research of the Russian Academy of Sciences, V-312, 60-letiya Oktyabrya prospect 7a, Moscow, 117312, Russia
[3] University of Jyväskylä, P.O. Box 35, FI-40351 Jyväskylä, Finland

E-mail: kaplinv@mail.ru

Abstract. Main properties of the multi-anode microchannel plate photomultiplier to be used in a Cherenkov detector are discussed. The laboratory test results obtained using irradiation of the MCP-PMT photocathode by picosecond optical laser pulses with different intensities (from single photon regime to the PMT saturation conditions) are presented.

1. Introduction

In fast scintillation and Cherenkov detectors operating in high magnetic fields fine-mesh vacuum phototubes were used in the latest devices [1] due to their capability to operate in magnetic fields up to about 0.5 T. The future experiments demand higher fields combined with much smaller dimensions (length) of the photosensors. The MCP based PMTs were not used not only because of their high price, but also because of their arbitrary small photocathode area and poor ratio of the photocathode area to the total area of the entrance window.

The new Planacon XP85012 photomultiplier has a large area photocathode (53 x 53 mm^2) having the overall face area of 59 x 59 mm^2 and 28 mm thickness only [2]. The PMT is characterized by a multi-anode structure (8 x 8 anodes array), which provides not only for good coordinate resolution, but also for good time resolution (typically 35 ps) due to small sizes of the anodes. At the same time, in large area Cherenkov detectors there is no reason to take signals from each anode independently because it brings to an enormous number of electronic channels. The anodes should be reasonably united. We made our studies of the PMT timing properties combining up to 16 anodes into a single electronic channel. Besides this, the PMT has a separate positive output uniting all 64 anodes which also can be used.

Our measurements were done using a picosecond laser (405 nm) with the pulses duration of 30 ps, pre-amplifier ORTEC VT120 (x20 gain), LECROY WR 620Zi oscilloscope and standard NIM and CAMAC modules. For precise measurements we used also constant fraction discriminator (Canberra 454) and time-to-amplitude converter (Canberra 2145).

[4] To whom any correspondence should be addressed.

2. Amplitude measurements

Figure 1 shows the PMT signals waveforms detected by different number of electrically interconnected anodes (1, 2 and 4). One can see that the anodes interconnection brings to pulses' broadening. However, the leading edge of the pulses remains practically constant.

Figure 2 shows the dependence of the PMT amplitude versus the intensity of the light pulses. It is linear in the dynamic range more than 1:100 which is enough in most experimental cases. Cherenkov detector module based on the XP85012 multianode microchannel plate photomultiplier is going to be used as a basic element of the FIT detector. The first testing procedures carried out under laboratory conditions showed that the photomultiplier is suitable for the use in the FIT detector [3].

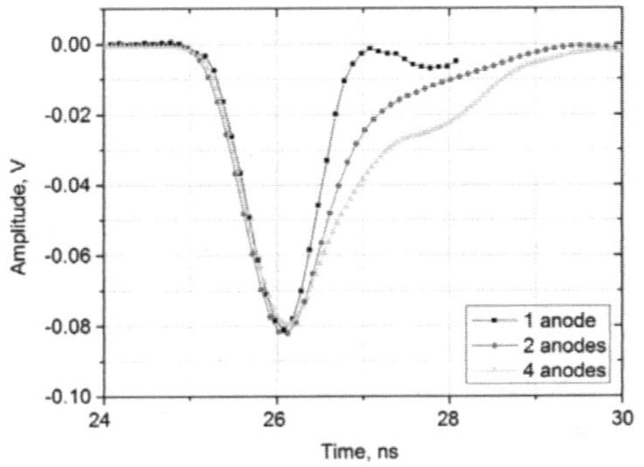

Figure 1. Signal waveforms from different number of the PMT anodes (1, 2 and 4), interconnected electrically.

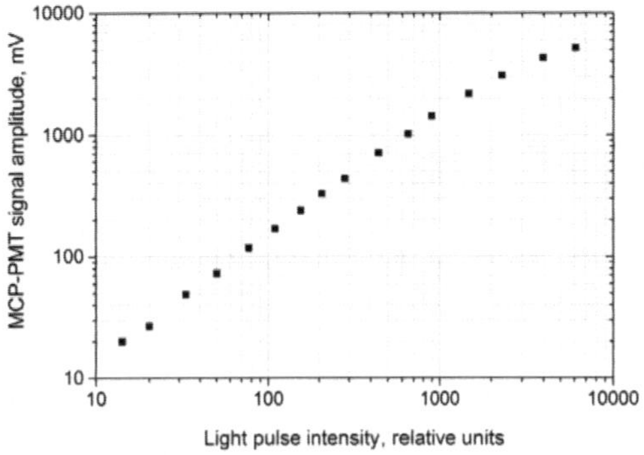

Figure 2. Dependence of the PMT signal amplitude (detected by 16 anodes) on the laser pulses' intensity (U_b = 1300 V, A_{MIP} = 50 mV).

Single photoelectron amplitude spectra are presented in figure 3 being measured with 1, 4 and 16 anodes. The amplitude resolution is much better in the case of 16 anodes which can be explained by a more complete electrons collection from the multichannel plate.

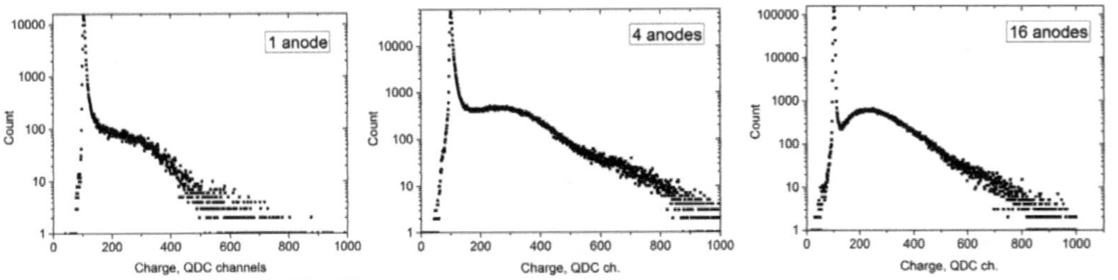

Figure 3. Single photoelectron amplitude spectra measured by 1, 4 and 16 interconnected anodes.

3. Timing measurements

Figure 4 shows the transit time spread for single photoelectron signals. The resulting time resolution values for signals detected by only one anode (50 ps sigma) corresponds to analogous results (35-40 ps) published elsewhere after certain corrections for the laser pulse duration and the timing properties of the analog electronics used. The time resolution obtained for several interconnected anodes is 1.5-2.0 times worse. As the leading edge of the signal from the interconnected anodes is not changed, we suppose that the time resolution worsening arises due to increased capacity of the anode and due to the decrease in the signal-to-noise ratio for this case.

A set of measurements of time resolution at different light pulse intensities up to 100 photoelectrons coming from the PMT's photocathode was done. Time resolution was measured with

Journal of Physics: Conference Series **675** (2016) 042016 doi:10.1088/1742-6596/675/4/042016

LeCroy WR 620 Zi digital oscilloscope for 1, 4 and 16 interconnected anodes with individual anodes' readout at different levels of the discriminator's thresholds. The same procedure was done for 16 interconnected anodes in the case of the common output readout. The summary of the results obtained for optimum levels of the discriminator's thresholds is shown in figure 5. As one can see, in all cases the resolution is better than 25 ps.

The time resolution measured by the NIM and CAMAC electronic circuits versus the amplitude of the light pulse is given in figure 6.

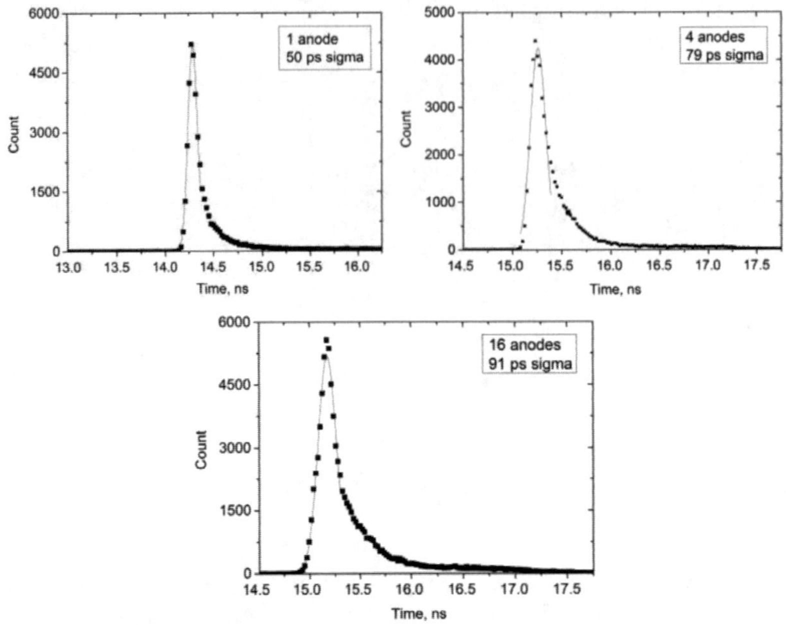

Figure 4. Time distribution spectra for single photon light pulses measured by 1, 4 and 16 interconnected anodes.

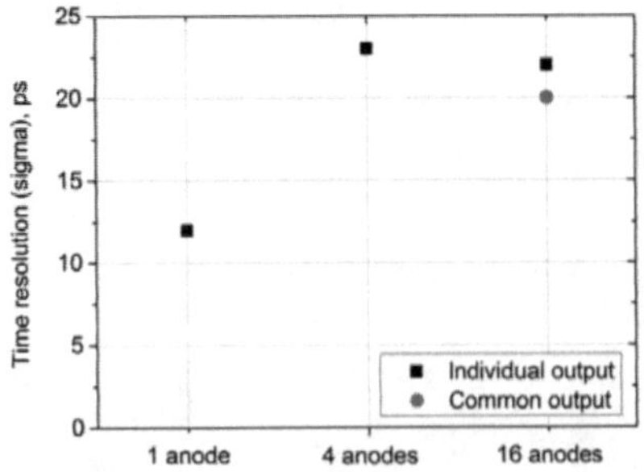

Figure 5. Summary of the time resolution measurements under the detection of 100 photoelectrons light pulses with different number of interconnected anodes for individual and common outputs. The measurements were carried out with LeCroy WR 620Zi digital oscilloscope.

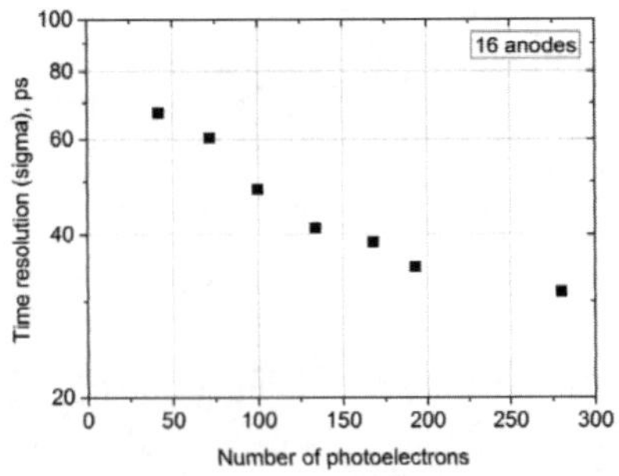

Figure 6. Time resolution as a function of light pulse intensity measured by the analog electronic circuit.

4. Cross-talks in the XP85012 PMT

As it was mentioned above, the PMT of this type offers the possibility to pick up a signal from a common output which unites all anode outputs. Anyhow the presence of this additional output leads to a strong cross-talk arising between individual anode's outputs during registration of a real signal. Figure 7 gives an example of such a cross talk: left picture represents a pulse from a common output (bottom waveform) and from 16 interconnected individual outputs (top waveform) during a real signal detection by these 16 interconnected adjacent channels only. Right picture shows pulses from the same combination of outputs while the whole area of the photocathode is illuminated except of the region with the 16 adjacent anodes under study. This cross talk may be a serious problem when registering simultaneously a number of particles by different groups of interconnected anodes.

The possible way out is to take a signal only from a common output, which is not sensitive to the cross-talks of such type. The other solution under consideration is to make a PMT without the common output by modernizing the PMT's voltage divider and power supply circuits PCB.

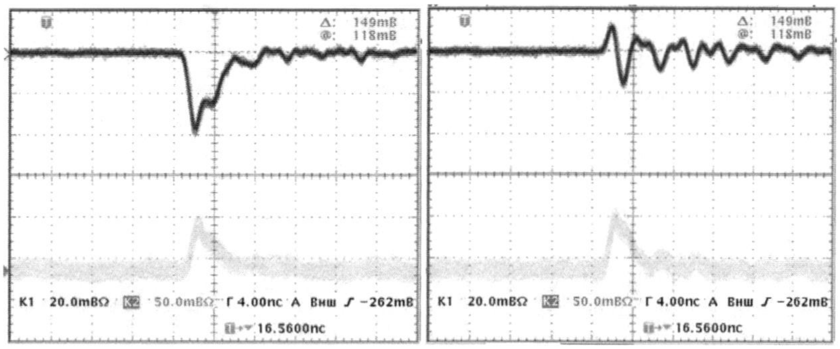

Figure 7. Waveform curves from 16 interconnected individual outputs (top curves) and from the common output (bottom curves). Please, refer to the text above for more detailed description.

5. Conclusion

The Planacon XP85012 MCP-PMT is a good candidate for its application to a Cherenkov detector for collider experiments because of its capability to operate at high magnetic fields, relatively large (53x53 mm^2) photocathode area along with compact total dimensions and reasonable price. Special measurements carried out with analog and digital electronics show the possibility to achieve acceptable amplitude dynamic range (1:100) and time resolution: up to 25 ps in case of direct measurements with digital electronics and up to 50 ps in case of measurements with full analog electronics circuit at 100 ph.e. illumination level.

The cross-talk problem constrains the use of the MCP-PMT in default electronic configuration: even in the case of a few particles simultaneous detection with the device, it would be impossible to use the individual outputs for timing purposes. The possible way out – the common output channel removal from the PMT's PCB – is going to be tested by us in the nearest future.

Acknowledgments

This work was performed within the framework of the Fundamental Research and Particle Physics Center supported by NRNU MEPhI Academic Excellence Project (contract № 02.a03.21.0005, 27.08.2013) and Russian state grant RFMEFI61014X0003.

References

[1] Ikematsu K *et al.* 1998 *Nucl. Instrum. Meth.* A **411** 238-48
[2] Yurevich V I 2013 *Physics of Particles and Nuclei Letters* **10** 258-68
[3] Abelev B *et al.* CERN-LHCC-2013-019 / LHCC-TDR-015

Investigation of possibility of creation of radiation resistance sensors for physical information based on fiber materials

P B Baskov[1], S B Chebyshov[2], V V Kadilin[2], V V Sakharov[1] and I V Mosyagina[1]
[1]Leading Research Institute of Chemical Technology of the State Atomic Energy Corporation "Rosatom", Kashirskoe highway 33, Moscow, 115409, Russia
[2]National Research Nuclear University MEPhI (Moscow Engineering Physics Institute), Kashirskoe highway 31, Moscow, 115409, Russia

E-mail: Baskov58@mail.ru

Abstract. The results of physical and material science and technological development of new materials of radiation photonics - nano- and microstructure of radiation-sensitive and radiation-resistant optical glass and fibers based on quartz are presented in the report. The possibility of their application in neutron diagnostics devices of nuclear power objects are considered. Component and construction options for the radiation-sensitive fiber and glass materials (with isotopes ^{10}B, ^{6}Li, Gd, ions of Nd^{3+}, Ce^{3+} etc.), in which radiation resistance is achieved through the organization of areas of "drain" and annihilation of radiation-induced defects are considered.

The scintillation detection device based on crystalline and organic materials have a high thermal neutron detection efficiency, but their field of application is limited by low temperatures (150-200 ^{0}C) and the ionizing streams (up to 10^7 n/cm^{2}s). Diagnostic panel and portable devices are created on the basis of multi-component silicate fibers (Nucsafe inc.). The further expansion of possibilities of application of fiber optic materials for the radiation instrumentation (including the reactor's instrumentation) can be achieved on the following principles:

- shift from silicate to multi-component optical fibers based on quartz to provide thermal resistance up to 1000 ^{0}C;
- multi-component silica glass with the implementation of the principle of radioluminescent detection of ionizing radiation due to the introduction in the glass matrix neutron-sensitive isotopes (^3Li,^3He,^{10}B, etc.) and the scintillating ions (Ce^{3+}, Nd^{3+}, $Pr^{3+}$and others);
- shift from homogeneous to heterogeneous materials, with a given phase and the structural organization of the luminescent radiation content to increase the radiation resistance of the material.

2D- and 3D-structuring of the functional detection of volume used as a basis for designing high-film-glass-fiber-optic radiation luminescent materials. The technological basis is the low-temperature processes of thermal destruction synthesis of nano- and micro-sized functional coatings, combined with layering processes and processes of chemical vapor synthesis [1,2]. As a result material is amorphous-nanocrystalline (crystallite size of 20-40 nm) and multilayer (interlayer distance ~ 10-30 nm) structure.

Determining for the rate of annihilation processes are defect diffusion coefficient (D), its environmental resistance movement and the distance to the annihilation region (L) (1):

$$1/\tau_{\text{анн}} \approx D/L^2 \quad (1)$$

If the organization of a structure is the complex-phase than the coefficient D is an additive quantity and includes diffusion into the bulk crystalline and amorphous homogeneous environment (D_1 и D_2), motion along the grain boundaries (when nanocrystalline organization) (D_3) and migration of interlayer "surface" (when layering) (D_4), at that $D_1<D_2<<D_3<D_4$. Organization of amorphous and nano-crystalline structure (3D-structuring) of the material ensures the implementation of the abnormally high diffusion coefficients. Multilayer (2D-structuring) allows you to monitor and technologically "organize" the distance to a possible plane annihilation radiation-induced defect (L).

Uranium-oxide coats (radiators) of ionization chambers fission became the basis for the development of principles of 2D- and 3D-structuring. Nanolayers U_3O_8 (depth 20-40 nm) with the enrichment isotope ^{235}U 90% were obtained by low-temperature synthesis of oxide coatings from solutions of uranium carboxylates [3]. Plane organization of (2D-structuring with the inter-planar distance of 20-40 nm) material remained under cyclic repeating the synthesis process to ensure receipt of the radiator depth 2 mkm. Radiographic research has shown amorphous-nanocrystalline (3D) structure of the radiator. On the basis of the technological process we have been made and accepted to complete the uranium-oxide electrodes for the 4th unit of Beloyarsk NPP.

Methodology 2D- and 3D-structuring of the functional volume of material to provide icnreasing of its radiation strength and operational "durability" has been applied to the radiation-fluorescent optical fibers based on quartz. The combining technology with chemical vapor deposition process of structuring at the facility stretching and low-temperature chemical and structural modification of the surface is a technological basis for production of functional fiber optic materials. Radiation-fluorescent fiber-optic materials are the following types of isolated:

- radiation-luminescent quartz layers with a depth of 100 ÷ 300 microns, doped with REE: ^6Li, ^{10}B or Gd;
- ultra-thin film coating converts the neutron Gd_2O_3, U_3O_8, ThO_2, on the quartz layer doped with REE ions.

Figure 1a shows the photograph of the micro-layer on the inner surface of the quartz tube doped ^{10}B and Ce^{3+}. After consolidation in the extractor rod made of silica fibers, which are then, depending on the desired parameters for dimensions and sensitivity of the material in hexagonal structured elements inside the quartz tube and reuse overtighten in multi-core heterogeneous optical fiber (figure 1b). Figure 1b shows a photograph obtained for multi-core heterogeneous optical fibers based on silica glass doped with ^6Li and Ce^{3+}.

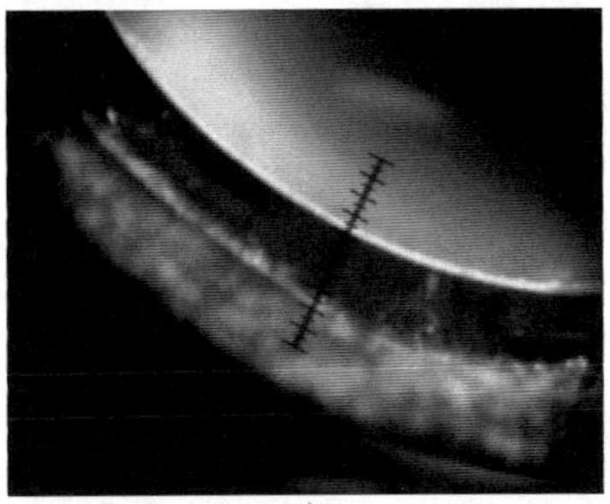
a)

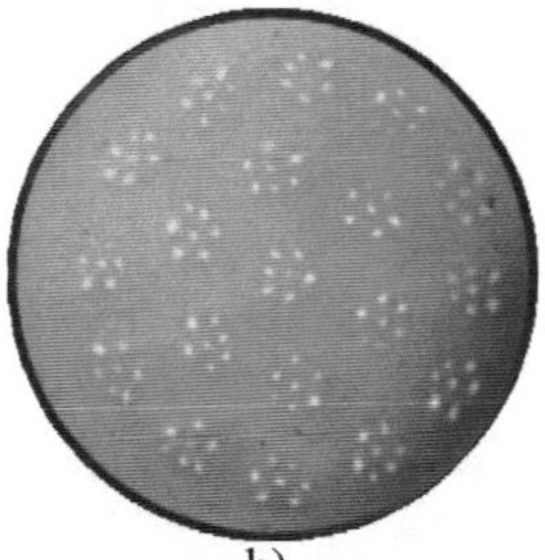
b)

Figure 1. Micro-sized (~ 70 microns)SiO_2- layer, based on silica glass doped ^{10}B and Ce^{3+}(a) and photograph of multi-core heterogeneous optical fibers $Li(6)Ce^{3+}SiO_2$ (b)

A special way to build a hybrid radiation resistant sensor is needed to solving problems of neutron monitoring in areas with ultra-rigid temperature, vibration, and radiation conditions (mode of design and beyond design basis accident in a nuclear installation). At the moment the sensors of reactor physical quantities are based on the principle of removal and transmission of the electrical signal. Efficiency and accuracy of the signal is broken at the time of emergency situations.

The capabilities of radioluminescent and optical converters are connected in the hybrid sensor. For example, in a hybrid chamber the fission informative signal is optical radiation produced by ionization of the chamber working gas under the action of fission fragments of neutron-sensitive cores. The output photon signal made through fiber-optic light guide. Main advantages of it:

- direct conversion of the energy of nuclear reactions in the optical signal;
- informative signal transmission through optical fibers;
- interfacing with fiber optic communication lines (FOCL).

Application of functional materials (glass fibers) on the radiation instrumentation always involves a compromise between their performance and operational characteristics. Extreme conditions require to shift from a homogeneous volume of silicate glass to quartz multi-component 2D- and 3D-nano-structured materials. Appropriate chemical processes 2D- and 3D-structuring based on methods MCVD, low-temperature processes, chemical and structural modification of the surface and high-temperature consolidation.

Principles of 2D- and 3D-structuring of the functional scope used for uranium-oxide radiators of ionization fission chamber, multi-component multi-core quartz radiation-fluorescent optical fibers. A variant of the hybrid sensor neutron fields enabling the transition to an electro-independent autonomous diagnostic systems of physical quantities.

Acknowledgments
The work performed under the state contract as part Federal Target Program "Nuclear Power Technologies of New Generation".

References

[1] Baskov P B *et al.* 2012 (In Russian) *J. Russian Journal of General Chemistry* **56** 36-43 (original Russian title: *Rossiiskii himicheskii jurnal*)
[2] Sakharov V V *et al.* 2012 (In Russian) *J. Proseedings of Universities. Nuclear Power* **4** 130-42 (original Russian title: *Izvestiya visshih uchebnih zavedenii. Yadernaya energetika*)
[3] Sakharov V V *et al.* 2014 (In Russian) *J. Nanoengineering* **2**(32) 3-8 (original Russian title: *Nanoinjeneriya*)

Experimental testing of heavy ions mass search procedure in the measurements with PIN diodes

Yu V Pyatkov[1,2,*], D V Kamanin[2], N A Kondratyev[2], A O Strekalovsky[2], S Ilić[3,2], A A Alexandrov[2], I A Alexandrova[2], N Mkaza[4], E A Kuznetsova[2], V Malaza[4], G V Mishinsky[2], O V Strekalovsky[2] and V E Zhuchko[2]
[1] National Research Nuclear University MEPhI (Moscow Engineering Physics Institute), Kashirskoe highway 31, Moscow, 115409, Russia
[2] Joint Institute for Nuclear Research, Str. Joliot-Curie 6, Dubna, 141980, Russia
[3] University Novi Sad, Faculty of Science, Physics Department, Trg D. Obradovića 4, 21000 Novi Sad, Serbia
[4] University of Stellenbosch, Faculty of Military Science, Military Academy, Private Bag X1, Matieland, 7602 South Africa

E-mail: [*]yvp_nov@mail.ru

Abstract. We discuss the quality of heavy ions (HI) mass reconstruction in the wide range of HI energies and masses using Si PIN diodes for measuring both energy and time-of-flight. The results are based on the experimental data obtained at the IC-100 accelerator in the Flerov Laboratory of the JINR (Dubna, Russia).

1. Introduction

Using of the semiconductor detectors for spectrometry of heavy ions is known to have several methodological obstacles, which are necessary to be overcome. First of them is pulse-height defect (PHD), which is manifested in the form of apparently less energies than particles actually have. Second is plasma delay effect (PD), which prevents precise measurements (TOF technique) due to creation of plasma in detectors, which is obstructing the charge collection. Correct accounting for both effects needs rather complicated procedure of the FF mass reconstruction. The task becomes extremely complicated if we deal with heavy ions in the wide range of energies and masses far from those typical for conventional binary fission. This is a case of our experiments dedicated to studying of the collinear cluster tri-partition (CCT) of heavy nuclei [1, 2].

The problem of adequate reconstruction of the FFs parameters such as velocity, energy and mass measured with the help of semiconductor detectors has a long lived history. The detailed review of physical treating and parametrizationz of PHD and PD for heavy ions can be find in [3]. We have not find in the literature [4–9] suitable algorithm of mass reconstruction to be adequate to our methodically more complicated task namely once again: *wide range of both masses and energies of heavy ions, necessity to take into account not only PHD but PD as well due to short flight path used. It means essential distortion of TOF due to PD.*

2. Experiment

A special experiment aimed at testing of the mass reconstruction procedure was performed at the IC-100 accelerator in Flerov Laboratory of the JINR. The beam of ^{132}Xe ions with the energy of about

1MeV/A bombarded thin metal targets (foils) made of Al, Ti, Cu, Nb, Ag and Au. Both the scattered Xe ions and the ions knocked out from the foils were detected by a time-of-flight spectrometer. The layout of the experiment is presented in figure 1.

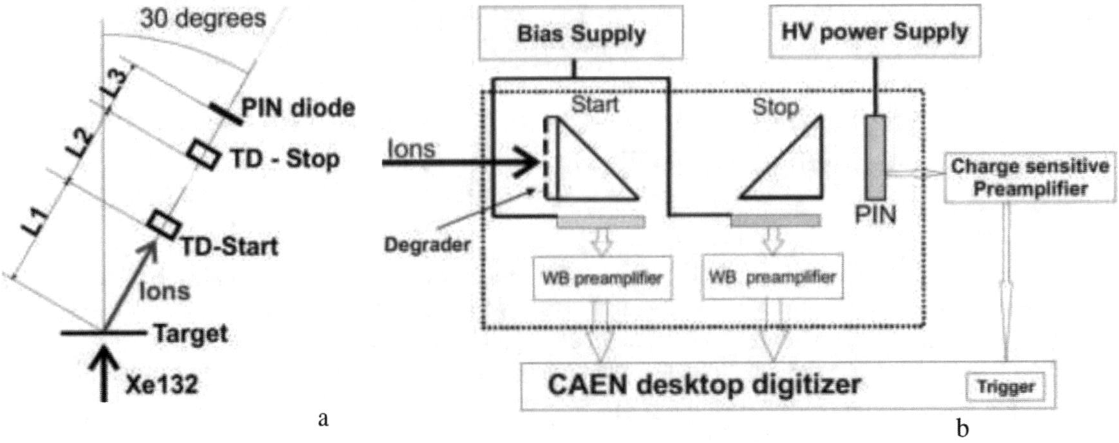

Figure 1. Layout of the experimental setup. Position of the detectors (*a*) and electronics used (*b*). The flight passes do not exceed correspondingly $L1$ = 500 mm, $L2$ = 173 mm, $L3$ = 31 mm.

The time-of-flight spectrometer consists of two microchannel plates based time pick-of detectors TD-Start, TD-Stop and one PIN diode. Evidently, microchannel plates based detectors are free from plasma delay effect that allows using them for measuring of "true" time-of-flight. The PIN diode serves for measuring of both the FF energy (E) and time-of-flight (TOF). The data acquisition system consists of the fast digitizer CAEN DT5742 and a personal computer. In order to measure the knocked out ions of four different energies in the range of (7–115)MeV in the same run a special multi-section degrader was placed before the start detector.

3. Data processing

It is known the energy E of the registered FF to be the sum of the detected energy E_{det} and the pulse-height defect denoted as a function of M and E by $R(M,E)$:

$$E = E_{\text{det}} + R(M,E), \tag{1}$$

where the detected energy of fission fragments is given by:

$$E_{\text{det}}[\text{MeV}] = E[\text{ch}] \cdot dE/dk + E_0$$

dE/dk and E_0 are the linear calibration parameters.

The parameterization for the pulse-height defect in equation (1) was chosen in the version proposed by Mulgin et al. [10] as the following empirical expression:

$$R(M,E) = \frac{\lambda \cdot \text{E}}{1 + \varphi \cdot \dfrac{E}{M^2}} + \alpha \cdot ME + \beta \cdot E\,, \tag{2}$$

where $\{\lambda, \varphi, \alpha, \beta\}$ are the parameters. In addition we know that:

$$E = \frac{M \cdot V^2}{1.9297}, \tag{3}$$

where E is the energy of the FF in MeV, M is the mass of the FF in amu and V is the velocity of the FF in cm/ns. The velocity for this purpose is calculating using the parameters obtained from the linear time calibration.

In order to find the correct values of the parameters $\{\lambda, \varphi, \alpha, \beta\}$ a special iteration procedure has been designed. This procedure consists in obtaining the numerical solution of the following equation:

$$G(\{\lambda, \varphi, \alpha, \beta\}, M, V) = 0$$

Combining equation (1), (2) and (3), we obtain:

$$G = \frac{MV^2}{k} - [E_{\mathrm{det}} + \frac{\lambda \cdot \frac{MV^2}{k}}{1+\varphi \cdot \frac{V^2}{Mk}} + \alpha \cdot \frac{M^2V^2}{k} + \beta \cdot \frac{MV^2}{k}] = 0\,, \tag{4}$$

where $k = 1.9297$.

The correct value of the velocity V could be obtained [6] in the following form:

$$V = \frac{L}{\gamma V M^{2/3} + \mathrm{tof}_{lin}}\,, \tag{5}$$

where tof_{lin} – experimental TOF value obtained using linear calibration, L – the length of the flight-path, γ-additional parameter to be find together with $\{\lambda, \varphi, \alpha, \beta\}$. At each fixed vector of parameters we find numerical solution of the set of expressions (4) and (5) in other words, quasi-mass M found to be the root of the equation (4).

After processing an amount of data, a spectrum of quasi-mass is obtained. The vector of parameters changes by the MINUIT package in order to minimize the following criterion function by changing the parameters $\{\lambda, \varphi, \alpha, \beta\}$:

$$F = [(<ML_T> - <ML>)^2 + (<MH_T> - <MH>)^2] + \mu \sum_{M_{TE}} \frac{(Y(M_{TE}) - Y_T(M_{TE}))^2}{Y(M_{TE})} \tag{6}$$

Where μ is a free parameter that is chosen by the user and it is used as an input parameter to the MUNUIT minimization procedure. This parameter plays a role of specific relative weight of the second term in the criterion function F. The values $<ML>$ and $<MH>$ are average masses of light and heavy fragments calculated from the experimental mass spectrum $Y(M_{TE})$. In the above equation the known values from literature are denote by "T". It is worth noting that the first square bracket term in equation (6) is sensitive to the difference between the centers of the mass peaks for the fission fragments while the second term is responsible for the agreement in shapes between the experimental mass spectrum $Y(M_{TE})$ and the mass spectrum from literature $Y_T(M_{TE})$.

Using of the fast flash-ADC (Amplitude to Digital Convertor) CAEN DT5742 let us to obtain digital images of the signals from all the detectors and to fix the relative position of the signals at the time axis (time diagram). Then in the off-line mode estimations of the ion energy and time-of-flight can be obtained.

Here we present the results when a time reference point on the PIN diode signal was calculated applying new algorithm proposed by us earlier [11]. New off line time reference method consists in approximation of the initial part of a leading edge of the PIN diode signal by a parabolic curve under condition that the parabola top lays on the mean value of the base line. The parameters of the curve are estimated by a minimization of a χ^2 using some points at the leading edge over three sigma (standard deviation of the base line values) level. Experimental testing proves this method to give unbiased time reference corresponding to real beginning of the signal [11]. Due to this fact we can use formally formula (5) putting $\gamma \sim 0$.

4. Results

Masses of the ions detected in the experiment at the IC-100 accelerator were calculated in the frame of the procedure presented above in section 3. Parameters λ and φ in formula (4) were chosen according recommendations of work [10]. The results are presented in figure 2.

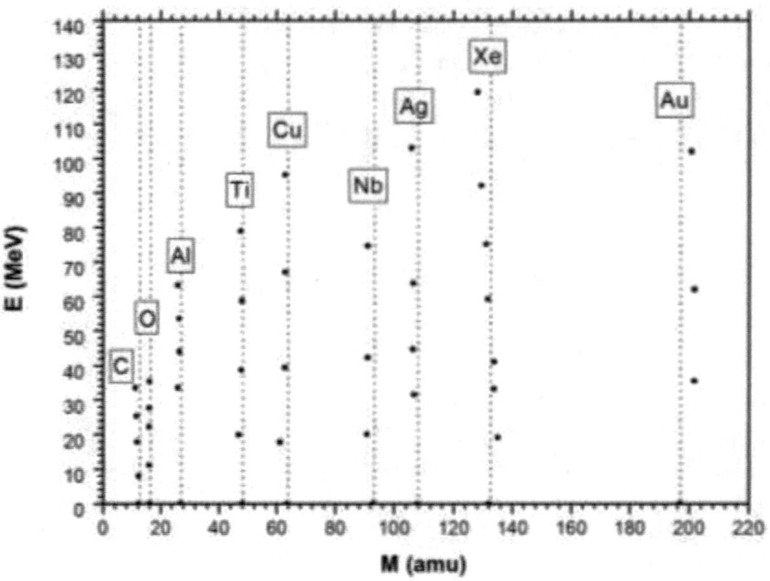

Figure 2. Estimated masses of the detected ions at different energies. Mean mass of natural mix of isotopes for each element involved is marked by a vertical line with a label.

As can be inferred from the figure estimated mass values are in good agreement with the expected ones in a wide range of masses and energies. Systematic deviation of the experimental points from the vertical line for Xe beam ions is presumably due to the influence of the randomly scattered component of the beam. The mass of the heaviest ions of Au is overestimated on approximately four units.

5. Conclusions

Using of both new "parabolic" time pick-of algorithm and mass reconstruction procedure based on the PHD parametrization proposed in [10] let us to reproduce quite satisfactory HI masses in a wide range of masses and energies.

Acknowledgments

This work was supported in part by the RFBR grants 14-02-93960, 16-02-00693. This work was performed in part within the framework of the Center Fundamental Research and Particle Physics supported by MEPhI Academic Excellence Project (contract № 02.a03.21.0005, 27.08.2013).

References

[1] Pyatkov Yu V *et al.* 2010 *Eur. Phys. J.* A **45** 29
[2] Pyatkov Yu V *et al.* 2012 *Eur. Phys. J.* A **48** 94.
[3] Akimov Yu A *et al.* 1967 (In Russian) *Semiconductor detectors of nuclear particles and their application* (original Russian text: *Poluprovodnikovye detektory yadernyh tchastits i ih primenenie.* (Moscow: Atomizdat))
[4] Kim Y S *et al.* 1993 *Nucl. Instrum. Meth.* A **329** 403.
[5] Schmitt H W, Kiker W E and Williams C W 1965 *Phys. Rev.* **137** 13837.
[6] Alexandrov A A *et al.* 1991 *Nucl. Instrum. Meth.* A **303** 323.
[7] Neidel H O and Henschel H 1980 *Nucl. Instrum. Meth.* **178** 137.
[8] Beck C *et al.* 1989 *Phys. Rev.* C **39** 2202.
[9] Kaufman B *et al.* 1974 *Nucl. Instrum. Meth.* **115** 47.
[10] Bohne W *et al.* 1985 *Nucl. Instrum. Meth.* A **240** 145.
[11] Mulgin S I *et al.* 1997 *Nucl. Instrum. Meth.* A **388** 254.

International Conference on Particle Physics and Astrophysics (ICPPA-2015) IOP Publishing
Journal of Physics: Conference Series **675** (2016) 042019 doi:10.1088/1742-6596/675/4/042019

The GMT detector alignment in the STAR experiment

N Ermakov (for the STAR Collaboration)

National Research Nuclear University MEPhI (Moscow Engineering Physics Institute), Kashirskoe highway 31, Moscow, 115409, Russia

E-mail: coffe92@gmail.com

Abstract. The Solenoidal Tracker At RHIC (STAR) uses the Time Projection Chamber (TPC) to perform tracking and particle identification. In order to improve the corrections (such as space charge) and monitor non-static distortions of the TPC, GEM-based chambers (GMT) were installed at eight locations outside the TPC where they will provide optimal sensitivity to the distortions. In order to reach this goal, the ionization clusters were measured by using the ADC signals in each module. The positions of clusters and their deviations from track projections enabled alignment of the GMT modules with respect to TPC to an accuracy $\sim 200\mu m$.

1. Introduction

The physics program of Relativistic Heavy Ion Collider (RHIC) aims to study the nuclear matter at extreme conditions. The Solenoidal Tracker At RHIC (STAR) [1] allows to measure properties of the matter created in the heavy–ion collisions by using various detectors. The main detector of STAR is the Time Projection Chamber (TPC) [2]. It allows to perform particle tracking and identification in a wide momentum range. With increase of beam luminosities, it has become crucial to reduce systematic uncertainties due to the TPC alignment and distortion corrections (for instance, due to the space charge [3]). In order to reach this goal eight GEM Chambers to Monitor the TPC Tracking (GMT) were placed outside TPC at the Time-Of-Flight (TOF) [4] radius. To utilize the GMT module's design resolution of $\sim$150μm in both in-plane axes, we aim to align the modules with respect to the TPC to a similar level. Using the positions and deviations of the ionization clusters that were reconstructed from the ADC signals one can perform the alignment procedure.

2. Results and Discussions

A charged particle that passes the GMT modules will produce primary ionization along its track. The ionization electrons are registered by strips and pads that correspond to the beam (V) and transverse to the beam (U) directions, respectively. Using the information about the sizes of strips, pads and spaces between them allows to reconstruct coordinates of the signal. Analog signal from each strip and pad is converted to digital by Analog–to–Digital Converter (ADC). The ADC spectrum contains a non–zero minimum value that corresponds to zero energy deposited, called the pedestal. In the current analysis it was assumed that the first five events are pedestals. Then, for each subsequent event, the calculated pedestal was subtracted from

ADC signal. The ADC's signal are not centered at zero outside of the cluster locations due to some low-level unresolved issues with the method which have been deemed inconsequential for the purposes of using the devices for calibrations. The TSpectrum package [5] was used to find peak channels in the ADC signals (see red triangles on the figure 1) as a function of channel position. If the peak with the maximum amplitude was 3σ above the background then it was accepted as a cluster. The clusters were fit by the Gaussian distribution. In the case when no clusters were found, the ADC spectrum was added to the pedestal distribution. Figure 1 shows an example of the ADC spectra for two found clusters in both V (a) and U (b) directions within a single module. During the work, it was found that the module 6 does not work. The clusters in

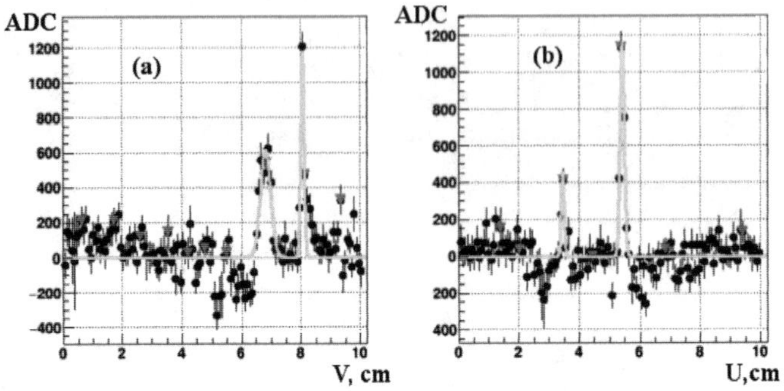

Figure 1. The example of the ionization cluster selection in the GMT module after the pedestal subtraction in the beam (a) and transverse (b) directions. The line represents fit of the clusters (see text). The red triangles correspond to peaks that were found by TSpectrum.

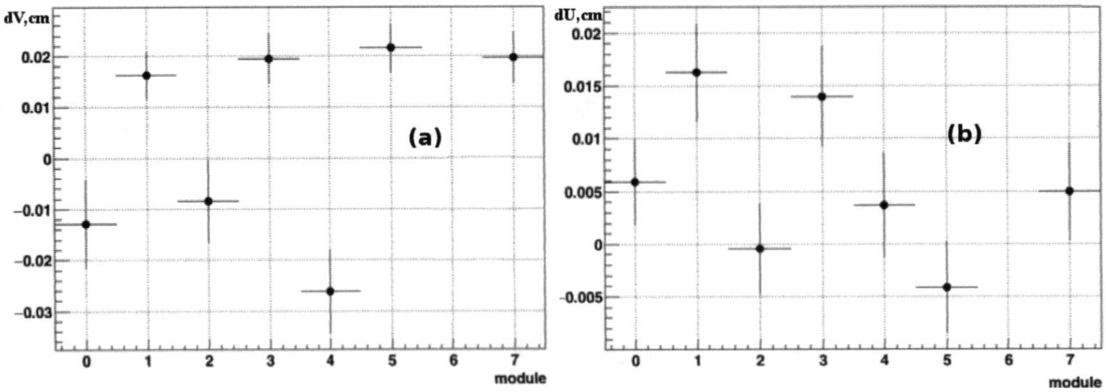

Figure 2. The alignment of the module in the beam (a) and transverse (b) directions with respect to TPC.

both projections were matched by the amplitude of the ADC signals. In order to align the GMT modules with respect to TPC, the positions of clusters were compared with projections of tracks from TPC. Reasonable precision for the alignment may be achieved by using high statistics. The method was tested on tracks from total 0.6 million Au+Au collisions at $\sqrt{s_{NN}} = 14.5$ GeV collected in 2014. The $\sim 5 \times 10^4$ tracks that produced hits in GMT modules were selected and allowed to align the GMT modules with respect to TPC with accuracy better than $\sim 200 \mu$m.

Figure 2 shows the alignment of the GMT modules with respect to TPC in both U (a) and V (b) directions. As a result of different statistics in each module the accuracy of the alignment depends on the module position. In future with higher statistics, it will be possible to improve the alignment further and take into account other degrees of freedom in the alignment, such as rotations of the GMT modules.

3. Conclusion

The GMT modules will allow improved monitoring and correction of TPC distortions. In order to use the GMT modules for this purposes, alignment of the GMT modules is necessary, and was performed by comparing position of ionization clusters with the projections of TPC tracks. The projections of TPC tracks were compared with cluster positions. This allowed to align the GMT modules with respect to TPC with accuracy better than $\sim 200\mu$m.

Acknowledgments

This work was partially supported by MEPhI Academic Excellence Project (contract No. 02.a03.21.0005, 27.08.2013).

References

[1] Ackermann K H *et al.* 2003 *Nucl. Instrum. Meth.* A **499** 624
[2] Anderson M *et al.* 2003 *Nucl. Instrum. Meth.* A **499** 659
[3] Van Buren G *et al.* 2006 *Nucl. Instrum. Meth.* A **566** 22
[4] Llope W J *et al.* 2005 *Nucl. Instrum. Meth.* B **241** 306
[5] URL: `https://root.cern.ch/root/html530/TSpectrum.html`

Journal of Physics: Conference Series **675** (2016) 042020 doi:10.1088/1742-6596/675/4/042020

Highly granular hadron calorimeter: software compensation and shower decomposition

M Chadeeva[1,2], on behalf of the CALICE collaboration

[1]National Research Nuclear University MEPhI (Moscow Engineering Physics Institute), Kashirskoe highway 31, Moscow, 115409, Russia

[2]State Scientific Center of Russian Federation ITEP (Alikhanov Institute for theoretical and experimental physics), B. Cheremushkinskaja ul. 25, Moscow, 117218, Russia

E-mail: marina@itep.ru

Abstract. The highly granular analogue hadron calorimeter was developed and constructed by the CALICE collaboration. The active layers of the calorimeter are assembled from scintillator tiles with individual readout by silicon photomultipliers and are interleaved with absorber plates. The response and resolution of the calorimeter equipped with steel absorber was intensively tested in single particle beams. The application of software compensation techniques developed for the scintillator-steel prototype allows for reduction of the stochastic term of the single particle resolution from $58\%/\sqrt{E/GeV}$ to $45\%/\sqrt{E/GeV}$. The detailed study and decomposition of the longitudinal and radial profiles of hadron-induced showers in the energy range from 10 to 80 GeV are presented and compared to GEANT4 simulations.

1. Introduction

The experimental program proposed for future high-energy lepton colliders such as the ILC or CLIC puts unprecedented requirements on the detector systems [1]. To achieve a jet energy resolution of better than 4% at 100 GeV, the particle flow algorithm (PFA) was developed, which is based on the high granularity of a calorimeter system [2]. The CALICE collaboration has developed, constructed and tested a set of highly granular electromagnetic and hadron calorimeter prototypes. The main goals of this activity were the validation of calibration procedures, study of the calorimeter properties and experimental test of the particle flow concept. This study is focused on the properties of the analogue scintillator-steel hadron calorimeter prototype (Fe-AHCAL) [3]. The experimental response, energy resolution and shower profiles have been analysed and compared to GEANT4 simulations [4]. The high longitudinal and radial granularity of the prototype allows the hadronic shower development to be studied with high spatial resolution using shower decomposition.

2. Experimental setup, data and simulations

The CALICE Fe-AHCAL is a sampling structure of 38 active layers interleaved with 2 cm thick steel absorber plates. Each active layer has an area of 90×90 cm^2 and is assembled from 5 mm thick scintillator tiles of varied transverse sizes, from 3×3 cm^2 in the centre to 12×12 cm^2 in the periphery. The tiles (cells) are individually read out by silicon photomultipliers (SiPM). The longitudinal depth of the Fe-AHCAL is about 5.3 nuclear interaction lengths. The response of each calorimeter cell is calibrated using the visible signal of a minimum-ionising particle (MIP) as described in ref. [3]. To reject noise, only cells with a visible energy above a threshold of 0.5 MIP are used in the analysis.

The experimental data analysed here were collected at CERN SPS in 2007 with beams of charged hadrons with momenta from 10 to 80 GeV. The CALICE setup at CERN is described in detail in ref. [7] and comprised the silicon-tungsten electromagnetic calorimeter (Si-W ECAL) [5], the Fe-AHCAL, and the scintillator-steel tail catcher and muon tracker (TCMT) [6]. The depth of the Si-W ECAL is one nuclear interaction length at normal incidence and it has a very fine transverse segmentation equivalent to 1×1 cm^2 cells. The TCMT is also a sampling calorimeter, it has 16 active layers assembled from scintillator strips with SiPM readout. The total longitudinal depth of the calorimeter setup amounts to $\sim$11 nuclear interaction lengths.

The event selection procedure rejects muon-like and double particle events. The signal from the Čerenkov counter placed upstream of the calorimeters was used for the off-line separation of pions from protons in the positive hadron beams. For the current analysis, the Si-W ECAL was used as a tracker and pre-shower detector to select events with the first inelastic hadron interaction (shower start) at the beginning of the Fe-AHCAL. Such a requirement also helps to minimise the longitudinal leakage. The algorithm of the shower start identification is described in ref. [7]. The simulated single pion samples for the comparison with the experimental results were produced using different physics lists from GEANT4 version 9.6. The simulated samples were digitised taking into account the SiPM response, light crosstalk between neighbouring scintillator tiles in the same layer, and calorimeter noise extracted from data.

3. Energy resolution with software compensation

The reconstructed energy for each event is calculated as the sum of reconstructed energies in ECAL, Fe-AHCAL and TCMT as described in ref. [7]. The energy distributions were fit by Gaussian function in the range ±2 r.m.s. of the sample mean to extract the mean reconstructed energy and resolution for the given energy. The response of the Fe-AHCAL to pions is observed to be linear in the energy range from 10 to 80 GeV. Two techniques of software compensation were developed, which allow the off-line correction of the reconstructed energy on an event-by-event basis [7]. The local software compensation technique uses local energy density information for a cell-by-cell re-weighting of energy deposits. The global software compensation technique exploits the distribution of cell energies to derive one overall weighting factor for the energy in the given event. Neither of the described techniques requires an a priori knowledge of the particle energy. The intrinsic energy resolution of the Fe-AHCAL for hadrons is measured to be $58\%/\sqrt{E/GeV}$ with a constant term of 1.6%. Both software compensation techniques show similar performance resulting in reduction of the stochastic term to $45\%/\sqrt{E/GeV}$. In contrast to data, the simulations predict larger improvement of the resolution with increasing energy.

4. Hadronic shower decomposition

The substructure of hadronic showers in the Fe-AHCAL can be studied using detailed longitudinal and radial shower profiles. The important advantage of the high granularity is the possibility to deconvolve the energy density distribution from the distribution of the shower start position. The longitudinal profiles from shower start represent the average visible energy in units of MIP measured per calorimeter layer, the first bin corresponding to the physical layer where the shower start is identified. An example longitudinal profile extracted from the experimental data for 40 GeV pions is shown in figure 1a where the bin width is $\sim$0.14 in units of the nuclear interaction length $\lambda_{\mathrm{I}}^{\mathrm{eff}}$. The parametrisation of the longitudinal development of hadronic showers with a sum of two gamma distributions was proposed in ref. [8] as a natural extension of the parametrisation of electromagnetic shower profiles. In the frame of this approach, the longitudinal profile can be described with the two-component function

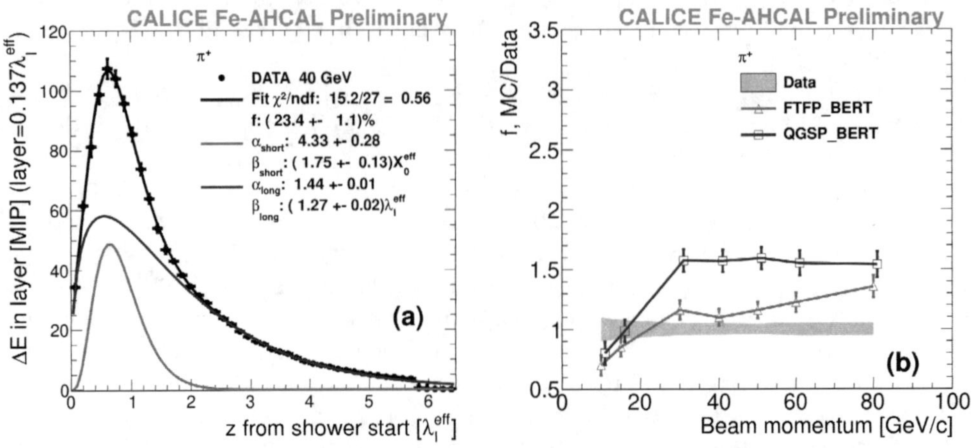

Figure 1. (a) Longitudinal profile of showers induced by positive pions from the test beam at 40 GeV in the Fe-AHCAL (black circles), fit to the profile with equation (1) (black curve) and its components corresponding to the "short" (red) and "long" (blue) contributions. (b) Ratio of the "short" component fraction extracted from simulations using the FTFP_BERT (red) and QGSP_BERT (blue) physics lists to that extracted from data versus beam momentum; the grey band shows the systematic uncertainty for data. Figures from ref. [9].

$$\Delta E(z) = A \cdot \left\{ \frac{f}{\Gamma(\alpha_{\text{short}})} \cdot \left(\frac{z}{\beta_{\text{short}}}\right)^{\alpha_{\text{short}}-1} \cdot \frac{e^{-\frac{z}{\beta_{\text{short}}}}}{\beta_{\text{short}}} + \frac{1-f}{\Gamma(\alpha_{\text{long}})} \cdot \left(\frac{z}{\beta_{\text{long}}}\right)^{\alpha_{\text{long}}-1} \cdot \frac{e^{-\frac{z}{\beta_{\text{long}}}}}{\beta_{\text{long}}} \right\}, \tag{1}$$

where A is the scaling factor, f is the fractional contribution of the "short" component with the shape parameter α_{short} and the slope parameter β_{short}; α_{long} and β_{long} are the shape and the slope parameters of the "long" component.

The fit to the longitudinal profile with equation (1) is performed up to $4.5\lambda_I^{\text{eff}}$. Figure 1a shows the fit result together with the separate contributions of the "short" and "long" components. The slope parameter of the "short" component is ~1.5 in units of radiation lengths X_0^{eff}, which is similar to that of electromagnetic showers. A comparison of shapes of the "short" and "long" components leads us to the conclusion that they can be roughly interpreted as the electromagnetic and hadronic substructures within a hadronic shower. The estimates of the shape and slope parameters for both components agree within uncertainties between data and simulations in the energy range studied. At the same time, the parameter f rises faster in simulations than in data with increasing energy and is significantly overestimated by simulations from 30 GeV and above. The ratio of simulations to data for the parameter f is shown in figure 1b. The overestimation of this observable by simulations can be caused by the overestimation of π^0 production with increasing energy in the GEANT4 hadronic models.

Radial profiles describe the transverse distribution of the energy density within a shower integrated over the longitudinal direction. They are calculated as an average energy density at the distance r from the shower axis within a ring of width Δr. The unprecedented transverse granularity of the Fe-AHCAL allows the bin width for radial profiles $\Delta r = 10$ mm. The ratio of simulations to data of radial profiles is shown in figure 2 for 40 GeV pions and protons. The simulations significantly overestimate the energy density near the shower axis for pions. The

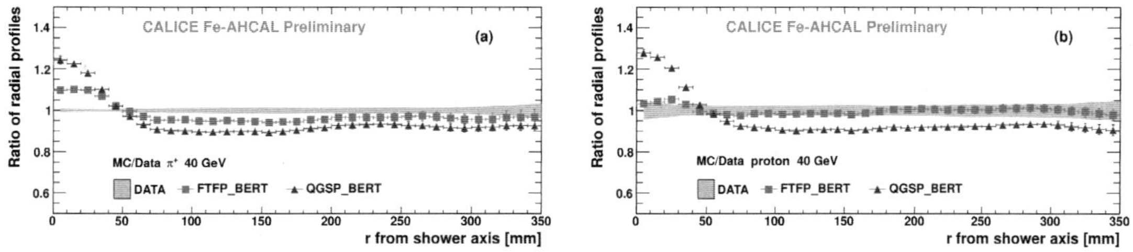

Figure 2. Ratio of radial profiles of showers induced by positive pions (a) and protons (b) at 40 GeV in the Fe-AHCAL extracted from simulations using FTFP_BERT (red) and QGSP_BERT (blue) physics lists to that extracted from test beam data; the grey band shows the systematic uncertainty for data. Figures from ref. [9].

FTFP_BERT physics list gives better predictions for pions and reproduces the behaviour for protons within uncertainties in the energy range studied.

The radial profiles can be parametrised with the sum of two exponential functions, which describe the behaviour in the shower core and far from the shower axis (in the halo region). A comparison of the fits to radial profiles reveals a good agreement of halo slope parameter between data and simulations. The observed underestimation of the core slope parameter is $\sim$5%(10%) by the FTFP_BERT (QGSP_BERT) physics list above 15 GeV.

5. Conclusion

The development of hadronic showers in the highly granular scintillator-steel analogue hadron calorimeter has been studied in the energy range 10—80 GeV. The software compensation techniques developed for the Fe-AHCAL help to reduce the contribution from stochastic fluctuations to the energy resolution by 10—20% for single hadrons. In the frame of the particle flow approach, the software compensation applied to identified clusters in the calorimeter can contribute to the improvement of the jet energy resolution. A comparison of the longitudinal and radial shower development between data and GEANT4 simulations has been performed based on the shower decomposition. From the detailed comparison one can conclude that the observed discrepancy between data and GEANT4 simulations is most likely due to unprecise modelling of the energy dependence of π^0 production.

Acknowledgments

The work is partially sponsored by the Center of Nanostructured Electronics (MEPhI Academic Excellence Project, contract No. 02.a03.21.0005, 27.08.2013), the Russian Ministry of Education and Science contracts 4465.2014.2 and 14.A12.31.0006 and by the RFBR grant 14-02-00873A.

References

[1] Fujii K *et al.* 2015 Physics Case for the International Linear Collider *arXiv*:1506.05992 [hep-ex]
[2] Thomson M A 2009 *Nucl. Instrum. Meth.* A **611** 25 (*arXiv*:0907.3577[hep-ex])
[3] Adloff C *et al.* [CALICE collaboration] 2010 *JINST* **5** P05004 (*arXiv*:1003.2662 [physics.ins-det])
[4] Agostinelli S *et al.* [Geant4 Collaboration] 2003 *Nucl. Instrum. Meth.* A **506** 250
[5] Adloff C *et al.* [CALICE collaboration] 2008 *JINST* **3** P08001 (*arXiv*:0805.4833 [physics.ins-det])
[6] Adloff C *et al.* [CALICE collaboration] 2012 *JINST* **7** P04015 (*arXiv*:1201.1653 [physics.ins-det])
[7] Adloff C *et al.* [CALICE collaboration] 2012 *JINST* **7** P09017 (*arXiv*:1207.4210 [physics.ins-det])
[8] Bock R K, Hansl-Kozanecka T and Shah T P 1981 *Nucl. Instr. and Meth.* **186** 533
[9] The CALICE collaboration 2014 *Preprint* CALICE Analysis Notes CAN-048 (https://twiki.cern.ch/twiki/pub/CALICE/CaliceAnalysisNotes/CAN-048.pdf)

Design and characterization of high-speed CMOS pseudo-LVDS transceivers

S V Kondratenko
National Research Nuclear University MEPhI (Moscow Engineering Physics Institute), Kashirskoe highway 31, Moscow, 115409, Russia

E-mail: utakser@mail.ru

Abstract. High-speed transceiver for on-board systems of data collection and processing need to meet additional requirements, such as low power consumption and increased radiation hardness. It is therefore necessary to compare and search for alternative variants of transceivers on the physical layer, where high transfer speed is not achieved at the cost of a significant increase in power consumption or a limitation of transmission distance by the size of a printed circuit board. For on-board applications, it is also necessary to solve the problem of increasing the radiation hardness without going to expensive types of technology. In this paper, we studied some variants of implementation of pseudo-LVDS transceivers and analyzed their achievable quantitative characteristics. According to the results of calculations and analysis of the literature, specialized transceivers of this type, intended for the manufacture or manufactured according to the bulk CMOS technology processes in the range of 250–80 nm, can provide data speeds up to 6 Gbps at a specific power consumption of less than 4 mW/Gbps.

1. Introduction

On-board systems for data collection and processing obtained in the course of physical experiments can be highly specialized for data collection only and use the land powerful computing resources or based on close cooperation with the productive onboard computer network. Accordingly, these systems can use very different traffic connections and apply a variety of technical implementation – from the low-speed serial interfaces such as CAN to Gigabit Ethernet, and different transmission media – from classic cabling and wireless channels to fiber optics. Large amounts of data have to be processed in the shortest possible time. This leads to the need for the physical layer represented by a new generation of high-speed transceivers with speeds up to several Gbps. Even if this speed is not required directly in a physical experiment, it is useful to have appropriate margin for two reasons. First, this margin may be considered as the foundation for the future. Secondly, it is a mean of ensuring the operation of such transceivers in comfort, far from limiting speeds conditions in view of an adequate margin for the impact of external factors (including radiation) and the natural degradation of the technical characteristics of the onboard equipment operated by a long time. This logic is true, of course, if the speed margin is provided not at the cost of substantial and continuous increase in power consumption, as is typical, for example, for CML interface transceivers. Due to its low power consumption combined with a number of other advantages the transceivers with LVDS interface are currently widely used in the on-board systems [1,2]. LVDS-transceivers` disadvantage is the relatively low speed of transmission (typically – from 0.2 to 0.4 Gbps, the maximum standard – about 2 Gbps). It makes sense to analyze the features and specifications of transceivers with pseudo-LVDS interface that may provide speed higher 6 Gbps (according to the published data [3-6]).

2. Properties of pseudo-LVDS transmitters and receivers

Analysis of the literature [3-6] shows that there is currently no clear definition of pseudo-LVDS interface, although the term is often used by different authors. This situation is obviously connected with various deviations from LVDS-standard [7], which often aim to increase the allowable speed in comparison with the one specified in the standard. The above-mentioned restriction (2 Gbps) is connected, in turn, with a built-in LVDS-standard restriction on slew-rate of the signals from the driver outputs, which limits radiated emissions. Important advantages of LVDS-interface are follows:

1. The insensitivity of the transmission quality to the impedance mismatch between the transmitter (driver) output and the transmission line characteristic impedance. In fact, a pair of LVDS transmitter – LVDS receiver in the classic version implements a transmit mode type of current loop ensuring impedance matching conditions at the receiver input.
2. Small swing of the output signals of the driver (typically ±350 mV), thereby reducing power consumption and achieving high performance due to the reduction of parasitic capacitances recharge problems.
3. The relatively low slew rate of signal change at the driver output (~ 0.3V / 0.3ns = 1 V / ns) and the differential nature of processed and transmitted signals, which leads, as mentioned, to the reduction in the level of noise generated primarily by the driver in general purpose circuits (bus power and ground).
4. The lack of binding of the driver`s output signal levels to the supply voltage and active maintenance of their common (in-phase) level near 1.2 V (or 1.25V at the formation of a reference voltage by division of the supply voltage 2.5V). At the direct communication with the receiver this provides maximum available symmetrical value range ±1V for common-mode noise on the transmission line.

The question is whether we can preserve these benefits during the transition to a non-standard pseudo-LVDS interface with increased speed without violating the basic properties of LVDS and pseudo-LVDS transceivers compatibility on the electric level? Possible and known solution is to construct a pseudo-LVDS driver with impedance matching at the output to the transmission line characteristic impedance due to the built-in resistor 100 Ohms, and increased in two times output current to restore the value of the standard scope of output signals. Figure 1 shows the structure of the described pseudo-LVDS transmitter. For standard LVDS transmitter is no built-in resistor R, and the nominal output current I = 3.5 mA (instead of 2×I). In the transition to such a pseudo-LVDS transmitter listed above LVDS interface advantages (except claim 3 and partly to claim 2) are reserved, and the following additional advantages are provided:

- It may be used as a direct communication with the receiver and communication through coupling capacitors, that is important when there is a relatively large distance between the transmitter and receiver due to a possible difference of ground potential.
- Output impedance nominally equal to the characteristic impedance of the transmission line, thereby reducing reflections and allows operation at higher speeds and, at the same speed, the ability to work on a long line.
- Improved performance due to the reduction of influence of the stray parasitic capacitors at the transmitter output / receiver inputs and increased current in the output stage of the LVDS driver, which recharges these capacitors.

The disadvantage of a such pseudo-LVDS driver is the increased current consumption, which makes up a significant portion of the total current consumption of the transmitter and receiver.

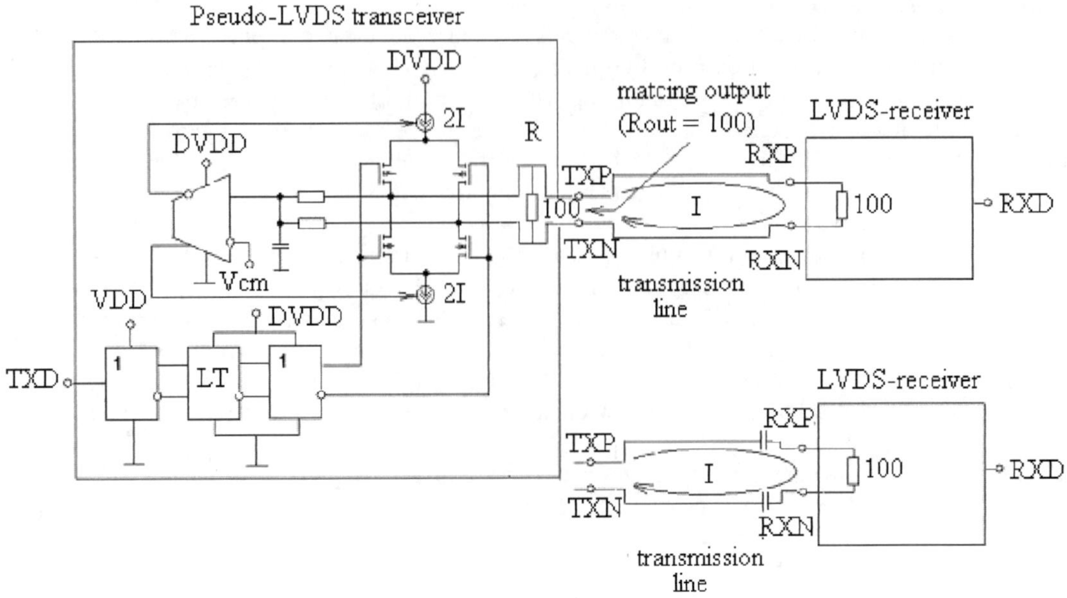

Figure 1. The variant of pseudo-LVDS transceiver (LT – Level Translator).

In comparison with the LVDS-transmitters LVDS-receivers do not have such pronounced features as operate overall function of comparison of differential input signal, coming from the transmission line, to zero regardless of the absolute levels of these signals, as well as receivers that implement other interface types.

3. The characteristics of pseudo-LVDS transceivers

Table 1 shows the characteristics of a number of pseudo-LVDS devices taken from published sources and based on the results obtained by the design within this work.

Table 1. The characteristics of pseudo-LVDS devices.

№	Pseudo-LVDS device	Main deviations from LVDS standard	Speed, bps	Relative power consumption, mW/Gbps	Technology	Ref.
1	Transmitter (driver)	Amplitude of output differential signal is 0.2V, voltage of power source is 1.2V, voltage-mode of transmission	3G	3.8	CMOS 0.18 μm	[3]
2	Transmitter	Amplitude of output differential signal is 0.196 V, voltage of power source is 1.2V	6G	4.2	CMOS DRAM 80 nm	[4]
3	Driver[1]	Pseudo-differential output, voltage of power sources are 1V and 1.4V	105M	N/A	N/A	[5]

№	Pseudo-LVDS device	Main deviations from LVDS standard	Speed, bps	Relative power consumption, mW/Gbps	Technology	Ref.
4	Driver[2]	Pair of CMOS buffers with resistive divider at the outputs	622M	34	CMOS 0.18 μm[3]	[6]
5	Transmitter true-LVDS	-	1.25G	12	CMOS 90 nm [3]	-
6	Transmitter	Matching 100 Ohm resistor at the outputs, increased driver output current	2.5+G	14		
7	Transmitter	Voltage of power source is 1.2V, capacitive decoupling with the receiver	5+G	3.4		

[1] configurable output buffer composed of ADC type of LTC2284
[2] emulation of LVDS driver composed of FPGA Virtex-E series
[3] also 0.25 μm peripherals used

From the analysis of the table 1 it follows that a significant reduction in power consumption can be achieved when using LVDS-drivers based on a reduced compare with peripheral core voltage. Calculations have shown that receivers, designed as a pair to LVDS or pseudo-LVDS transmitters, do not reduce the overall speed and consume significantly less power.

4. Conclusion

LVDS transceivers, commonly used in on-board systems of data collection and processing, may be modified to substantially increase the transmission speed while maintaining their main advantages in comparison with the high-speed transceivers that implement other types of interfaces at the physical layer. Three variants of pseudo-LVDS transmitters carried by bulk CMOS 90nm technology were analyzed. It is shown, that for these transmitters operating in a tandem with designed receivers there is the possibility to reach speed more than 5 Gbit/s at a comparable or lower (relative to published data [3-6]) specific power consumption.

Acknowledgments

Author acknowledge partial support by MEPhI Academic Excellence Project (contract № 02.a03.21.0005, 27.08.2013).

References

[1] Parkes S 2012 *SpaceWire User's Guide* (STAR-Dundee Limited) p 115
[2] Kondratenko S V *et al.* 2010 *Proceedings of the 3rd International SpaceWire Conference (SpaceWire-2010)* p 239
[3] Bae J-H *et al.* 2010 *IEICE Transactions on Electronics* **1** 132
[4] Kim S, Kong B S and Lee C G 2006 *Solid-state circuits conference (ESSCIRC 2006)* **9** 484
[5] 2007 *High Speed ADC Products. High Performance Analog ICs* (Linear Technology Corporation) p 12
[6] 2000 *LVDS Comparision. APEX 20KE vs. Virtex-E Devices* PIB29 (Altera Corporation) p 24
[7] TIA/EIA STANDARD 2001 *Electrical Characteristics of Low Voltage Differential Signaling (LVDS) Interface Circuits* ANSI/TIA/EIA-644-A-2001 (Arlington: Telecommunications Industry Association)

Journal of Physics: Conference Series **675** (2016) 042022 doi:10.1088/1742-6596/675/4/042022

Performance of the pair spectrometer of the GlueX experiment

A Somov[1], S Somov[2] and I Tolstukhin[2,*]
[1] Thomas Jefferson National Accelerator Facility, Jefferson avenue 12000, Newport News, Virginia 23606, USA
[2] National Research Nuclear University MEPhI (Moscow Engineering Physics Institute), Kashirskoe highway 31, Moscow, 115409, Russia

E-mail: somov@jlab.org, s.v.somov@mail.ru, *ivantol@jlab.org

Abstract. The description of the pair spectrometer of the GlueX detector at Jefferson Lab and its performance during the first beam commissioning runs are presented. We measured the amount of light collected from each channel of the pair spectrometer hodoscopes and the time resolution of the pair spectrometer counters.

1. The pair spectrometer of the GlueX experiment

GlueX is the new experiment in the experimental Hall D at Thomas Jefferson Accelerator Facility [1]. The main goal of GlueX is to search for hybrid mesons with exotic quantum numbers using a beam of linearly polarized photons incident on a liquid hydrogen target. Beam photons are produced in a radiator by 12 GeV electrons via the coherent bremsstrahlung process [2]. One of the key components of the Hall D photon beam line is the magnetic pair spectrometer [3], which is used to monitor the beam flux to measure the spectrum of the collimated photon beam. The pair spectrometer reconstructs the energy of a beam photon by measuring the momentum of the $e^{\pm}$ pair produced in a thin converter. The pair spectrometer consists of a thin converter, a dipole magnet with the magnetic field of 1.8 T and two layers of scintillator detectors: 16 coarse counters and the high-granularity hodoscope. Detectors are organized in two arms, each arm covers the $e^{\pm}$-energy range between 3 GeV and 6.25 GeV. The hodoscope is an array of thin scintillator tiles with the thickness of 1 mm and 2 mm. The light from each tile is collected by means of wavelength shifting fibers and is detected by a Hamamatsu silicon photomultiplier (SiPM) with a sensitive area of 3×3 mm^2. SiPM pulses are digitized using a flash ADC. The hodoscope energy resolution varies between 12 MeV and 20 MeV for 6 GeV and 12 GeV reconstructed photons, respectively. The coarse scintillator counters are used to produce a trigger by requiring a coincidence of hits in the two detector arms. They should also provide time measurements with a resolution better than 250 ps, which is required to identify the electron beam bunch where the beam photon is emitted. The counters are instrumented with Hamamatsu R6427-01 PMTs.

---

* To whom any correspondence should be addressed.

2. Beam line performance of the pair spectrometer

The distribution of the maximum flash ADC amplitude measured for each hodoscope scintillator tile is presented in figure 1. The amount of light collected from a tile corresponds to about 60 – 90 SiPM pixels.

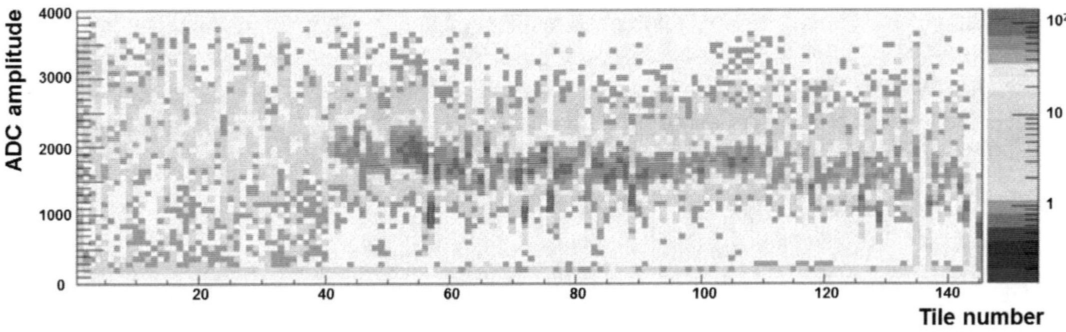

Figure 1. Maximum flash ADC amplitude measured for hodoscope scintillator tiles.

Energy spectrum of beam photons measured with the pair spectrometer during the first Hall-D commissioning run is presented in figure 2. Peaks in the energy spectrum (regions enhanced with linearly polarized photons) correspond to photons produced in coherent bremsstrahlung process from the diamond crystal. During the commissioning phase the Jefferson Lab accelerator provided a 5 GeV electron beam, which energy was about a factor of two smaller than the nominal beam energy of 12 GeV. The time resolution measured with the coarse counters was found to be better than 150 ps. The light yield corresponds to more than 900 photo electrons detected with the PMT.

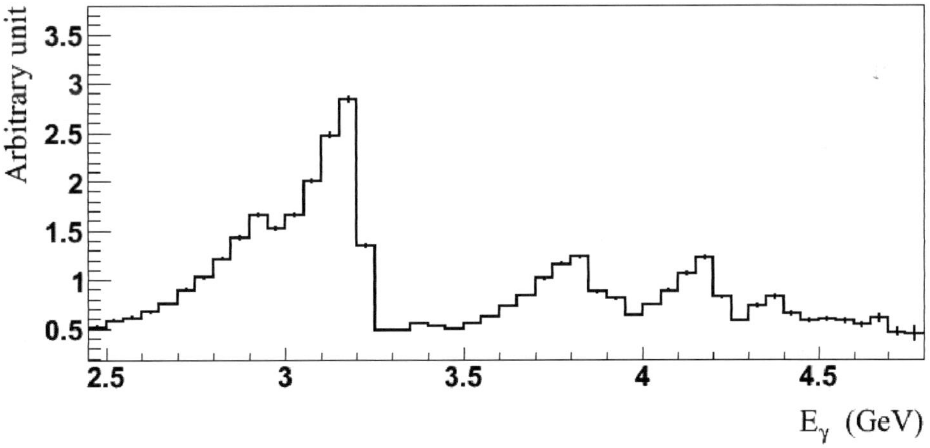

Figure 2. Photon energy spectrum measured with the PS hodoscope during the first commissioning run with a diamond radiator.

3. Conclusion

We have presented the performance of the pair spectrometer for GlueX experiment during the first commissioning beam. The energy spectrum of the photon beam, measured by pair spectrometer, produced in coherent bremsstrahlung process from the diamond crystal. The amount of light collected from each channel of the pair spectrometer hodoscopes corresponds to about 60 – 90 SiPM pixels. The time resolution measured with the pair spectrometer counters was better than 150 ps.

Acknowledgements
This work was supported by the US Department of Energy contract No. DE-AC05-06OR23177, under which Jefferson Science Associates, LLC operates the Thomas Jefferson National Accelerator Facility. Authors would like to thank their colleagues from the GlueX collaboration for their help.

References
[1] The GlueX Collaboration *https://www.gluex.org*
[2] Kaune W *et al.* 1975 Inclusive Cross-Sections for Pion and Proton Production by Photons Using Collimated Coherent Bremsstrahlung *Phys. Rev.* D **11** 478
[3] Barbosa F *et al.* 2015 Pair spectrometer hodoscope for Hall D at Jefferson Lab *Nucl. Instrum. Meth.* A **795** 376–80

Journal of Physics: Conference Series **675** (2016) 042023 doi:10.1088/1742-6596/675/4/042023

Xenon gamma-ray spectrometer for radioactive waste controlling complex

S Ulin, A Novikov, V Dmitrenko, K Vlasik, K Krivova, D Petrenko, Z Uteshev, A Shustov and E Petkovich

National Research Nuclear University MEPhI (Moscow Engineering Physics Institute), Kashirskoe highway 31, Moscow, 115409, Russia

E-mail: seulin@gmail.com

Abstract. Xenon detector based gamma-ray spectrometer for a radioactive waste sorting complex and its characteristics are described. It has been shown that the "thin-wall" modification of the detector allows better registration of low-energy gamma rays (tens of keV). The spectrometer is capable of operation in unfavorable field conditions and can identify radionuclides of interest in less than 1 second.

1. Introduction

At present one can observe an intensive development of the nuclear power industry, the so called "nuclear renaissance", which was significantly held back by after the Chernobyl accident. At the present the new nuclear power units are being launched and new nuclear power plants (NPP) are being built. It becomes clearer each day that further development of humanity is impossible without nuclear energy.

On the other hand, the increasing number of NPP and their units, as well as discontinuation of operation of the old ones, lead to the growth of radioactive waste (RAW) that needs to be sorted according to its composition and concentration of radionuclides to determine their burials' locations.

These days it is not enough to simply determine a level of radioactivity of the RAW. It is necessary to determine with high accuracy the composition and concentration of particular radionuclides in every RAW sample. To solve this problem highly sensitive gamma-spectrometers with high energy resolution have to be utilized for effective sorting of RAW [1,2].

It needs to be noted that usage of modern gamma-spectrometers in radioactive waste controlling complexes will bring about a great economic effect due to separation of RAW with regards to their activity and composition. As it is known, the cost of burial of the RAW with long half-life and great activity is largely different from the cost of low-activity samples burial [3]. Therefore, it is currently quite important to create new gamma-spectrometric devices that would sort RAW in accordance with their activity and composition that have to have high sensitivity, stability towards external vibro-acoustic influence, are capable of operation in field conditions, meaning that they have to possess necessary impermeability and protection.

In this paper a high-pressure xenon gamma-spectrometer created at NRNU MEPhI that will be employed in a RAW controlling complex is described.

2. Equipment description

The xenon gamma-spectrometer (XGS) contains a xenon gamma-ray detector (XGD) made of a cylindrical pulse ionization chamber with Frisch grid. It has an increased useful sensitive volume of 4 liters and the total mass of the working medium is 1.2 kg.

To widen the energy range of the registered gamma-rays, especially at low energies (30–50 keV), body of the detector is made of 0.5 mm wide stainless steel. Taking into consideration the fact that xenon in the detector is under pressure of 40 atm., it is necessary to assure strength of the thin-wall ionization chamber. This goal is achieved by means of usage of composite material Kevlar or glass fiber [4,5]. The latter option was chosen for this work.

Notably, usage of this composite material decreases the mass of the XGD nearly by factor of 3 in comparison with the thick-wall model, the total mass of such device is 5 kg. XGD main physical and technical characteristics are given in table 1, its image is in figure 1.

Table 1. XGD main physical and technical characteristics.

Characteristics		
Mass, kg	5.0	
Working medium mass, kg	1.2	
Working volume, cm^3	4000	
Dimensions, mm	(Diameter)	119
	(Length)	400
Wall thickness, mm	0.5	
Operating temperature range, °C	0 ÷ 100	
Level of acoustic influence, dB	0 ÷ 100	

Besides xenon detector, XGS includes a digital electronics unit (DE) for electric pulse processing and a power supply (PS). DE enables processing of every electric pulse coming from XGD and forms gamma-ray energy spectra [4,6]. The total mass of the equipment is 6 kg.

Figure 1. Xenon gamma-ray detector.

Journal of Physics: Conference Series **675** (2016) 042023 doi:10.1088/1742-6596/675/4/042023

3. Results of XGS testing

Testing of the XGS was carried out using radiation sources from the standard kit. Processing of the electrical pulses and spectra formation took place in the DE and further processing of the spectrometric information was performed by means of special software enabling gamma-ray spectrum identification.

Figure 2 presents a gamma-ray spectrum of ^{241}Am source without background subtraction. A 59.6 keV line is clearly visible in the image. This shows that a thin-wall XGD allows reliable registration of gamma rays with energies of several tens of keV, which is extremely important for identification of fissile materials.

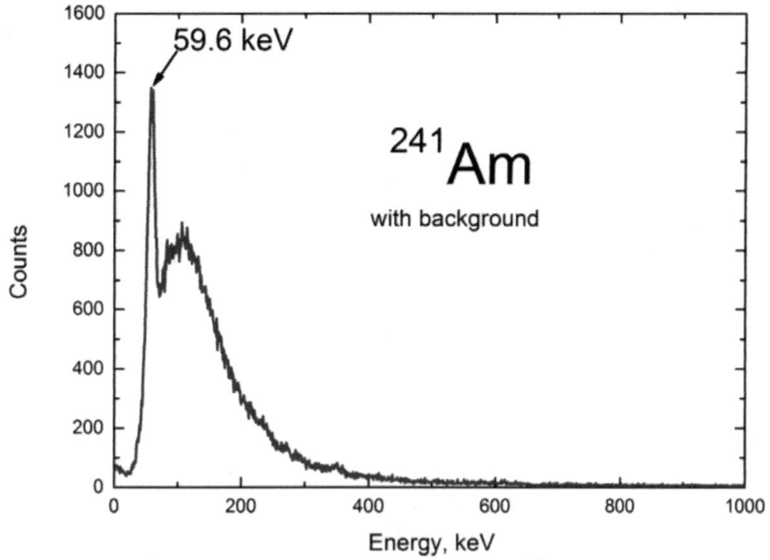

Figure 2. Gamma-ray energy spectrum of ^{241}Am source.

Figure 5 shows energy spectrum of ^{137}Cs gamma-ray source with subtracted background. In this case energy resolution for 662 keV line was (2.3±0.3)%.

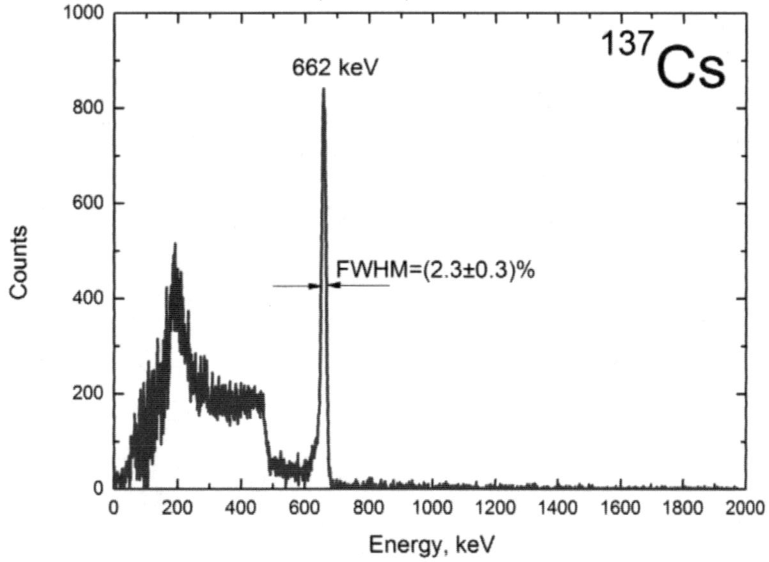

Figure 3. Gamma-ray energy spectrum of ^{137}Cs source.

During trials the possibility to discover and identify various radionuclides with short exposition time was tested. It was shown that ^{137}Cs source with activity of ~ 9 kBq can be discovered and identified by XGS from a distance of 5 cm from the detector in (5–10) seconds. Discovery and identification were carried out using special software implementing the 3σ background excess method as well as the documented requirements for minimum activities of individual nuclides. Processing time for each spectrum was less than 1 second. Here the activities of the studied sources were in the range of (4÷10) kBq.

4. Conclusion

A modified high-pressure xenon gamma-ray detector with digital electronics implementing a new method of electric pulse processing that allows utilization of the XGS in various models of radioactive waste control complexes was created. Software for XGS that carries out processing of gamma-ray spectra and identification of various radionuclides in less than 1 second was developed. The described XGS can be installed as part of the radioactive waste control complex.

Acknowledgments

This work was partially supported by MEPhI Academic Excellence Project (contract № 02.a03.21.0005, 27.08.2013).

References

[1] Sobolev A I *et al.* 2013 Russia: experience of radioactive waste (RAW) management and contaminated site clean-up *Radioactive Waste Management and Contaminated Site Clean-Up: Processes, Technologies and International Experience* (Sawston, Cambridge: Woodhead Publishing Limited) 345–85

[2] Filss P 1995 Relation between the activity of a high-density waste drum and its gamma count rate measured with an unshielded Ge-detector *Appl. Radiat. Isotopes* **46** 805–12

[3] Napoleon A *et al.* 2008 *The Real Costs of Cleaning Up Nuclear Waste: A Full Cost Accounting of Cleanup Options for the West Valley Nuclear Waste Site* (Cambridge, MA: Synapse Energy Economics, Inc.)

[4] Novikov A S *et al.* 2014 New modification of xenon gamma-ray detector with high energy resolution *Opt. Eng.* **53**(2) 021108

[5] Novikov A S *et al.* 2014 Xenon detector with high energy resolution for gamma-ray line emission registration *Proc. SPIE* **9213** 921318

[6] Novikov A S *et al.* 2013 Xenon gamma-ray detector's electrical signals digital processing technique *Proc. of XXIV Int. Symp. on Nuclear Electronics & Computing* 192–197

Journal of Physics: Conference Series **675** (2016) 042024 doi:10.1088/1742-6596/675/4/042024

Thermal neutrons registration by xenon gamma-ray detector

A E Shustov, I V Chernysheva, V V Dmitrenko, A G Dukhvalov, K V Krivova, A S Novikov, D V Petrenko, K F Vlasik, S E Ulin and Z M Uteshev
National Research Nuclear University MEPhI (Moscow Engineering Physics Institute), Kashirskoe highway 31, Moscow, 115409, Russia

E-mail: AEShustov@mephi.ru

Abstract. Experimental results of thermal neutrons detection by high pressure xenon gamma-ray spectrometers are presented. The study was performed with two devices with sensitive volumes of 0.2 and 2 litters filled with compressed mixture of xenon and hydrogen without neutron-capture additives. Spectra from Pu-Be neutron source were acquired using both detectors. Count rates of the most intensive prompt neutron-capture gamma-ray lines of xenon isotopes were calculated in order to estimate thermal neutrons efficiency registration for each spectrometer.

1. Introduction
There is a wide range of neutron detectors based on different materials: scintillators, semiconductors, gas mixtures with high neutron capture cross-section additive (^{3}He, ^{10}BF$_3$, ^{6}Li) and so on. Some of materials are able to simultaneously register neutron and gamma-rays, e.g. ^{6}LiI(Eu). In our previous paper [1] we showed that the high pressure xenon gamma-ray spectrometer (HPXe) can be used for thermal neutron detection. Natural xenon is a mixture of isotopes with neutron capture cross-section acceptable for neutron detection by means of prompt gamma-ray registration. In this research we have improved our analyzing technique for neutron efficiency calculation and test using HPXe with sensitive volumes of 0.2 and 2.0 litters.

2. Xenon gamma-ray spectrometers
The xenon spectrometer is a gamma-ray detector based on cylindrical ionization chamber [2] operating in pulse mode. It is filled with high pressure (40 atm.) xenon and hydrogen mixture and has the Frisch grid. Main parameters of HPXe 2.0L and 0.2L are shown in table 1.

Table 1. Parameters of Xenon gamma spectrometers.

Detector	Energy resolution at 662 keV (%)	Xenon mixture density (g·cm^{-3})	Cathode diameter (cm)	Grid diameter (cm)	Anode diameter (cm)	Gamma-rays energy range (MeV)
HPXe 0.2L	2.4	0.4	3.8	1.2	0.4	0.03–2
HPXe 2.0L	2.0	0.3	11.3	4	2	0.05–3

HPXe 2.0L has high energy resolution 2.0% for 662 keV that is better than commonly used scintillator detectors based on NaI or CsI crystals.

3. Geometry of experiments

Detectors were exposed to thermal neutrons flux. Neutrons emitted by Pu-Be source were moderated using polyethylene container with 11 cm thick wall. The geometry scheme of the experiment is shown in figure 1.

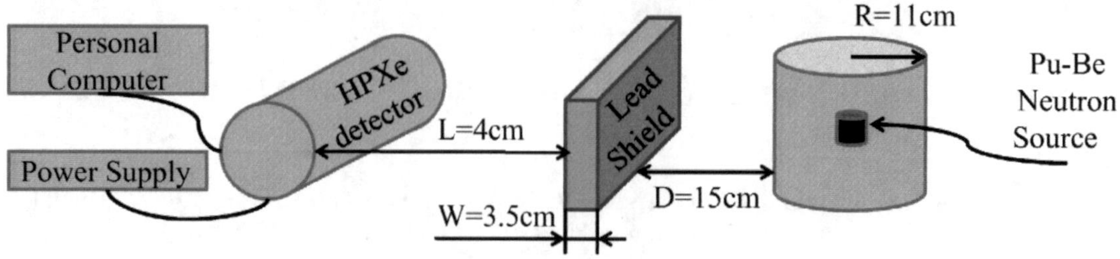

Figure 1. Scheme of experiment.

To suppress gamma-rays emitted during neutrons capture reactions inside polyethylene container, 3.5 cm thick lead was used. Pu-Be source has $4.6 \cdot 10^6$ Bq activity. Detector was placed at 4 cm from Pb shield and 33.5 cm from neutron source. Measurements by ^{6}LiI(Eu) detector placed at the same point gave value of flux 27 neutrons$\cdot$sec$^{-1}\cdot$cm^{-2}.

4. Gamma-ray spectra from Pu-Be source

Before the experiment a gamma-ray background in laboratory without neutron source for both detectors was measured. After that measurements of Pu-Be neutron source surrounded by moderator and Pb shield were done. The results are shown in figures 2 and 3.

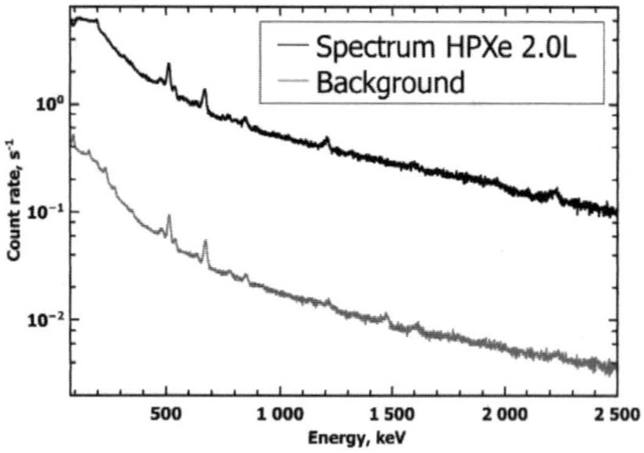

Figure 2. Experimental spectra of neutron source and background measured by HPXe 2.0L detector.

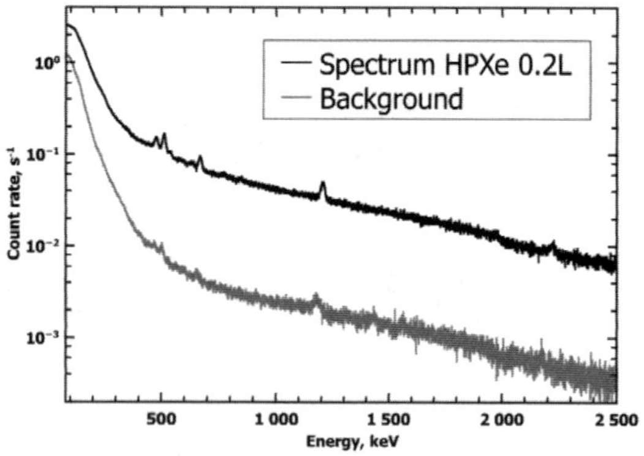

Figure 3. Experimental spectra of neutron source and background measured by HPXe 0.2L detector.

Spectra with subtracted background for each detector at most interesting region are shown in figures 4 and 5.

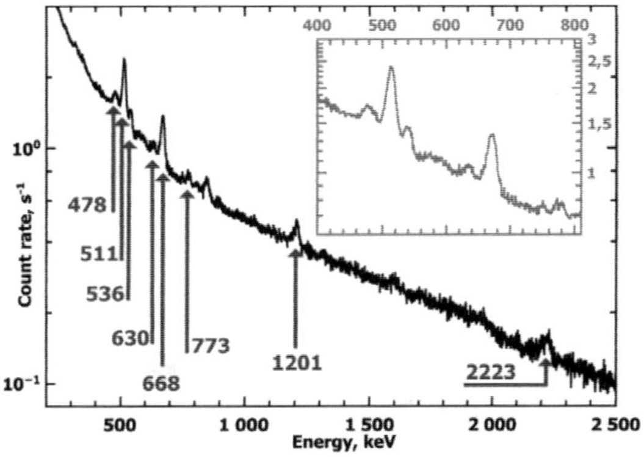

Figure 4. Experimental spectrum of neutron source without background measured by HPXe 2.0L.

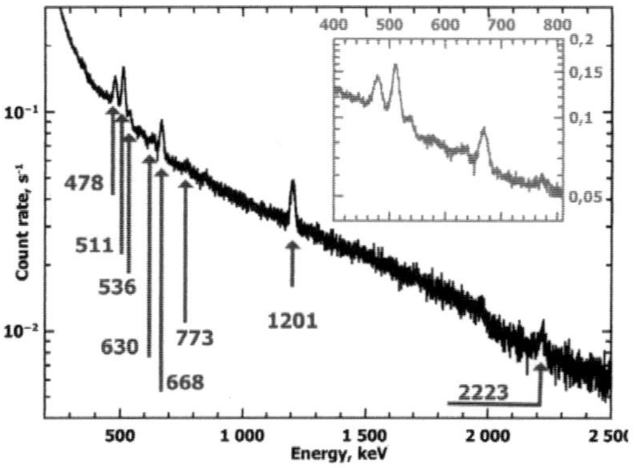

Figure 5. Experimental spectrum of neutron source without background measured by HPXe 0.2L.

Detailed analysis of gamma-ray lines and results are discussed in next section.

5. Thermal neutrons detection by the xenon spectrometer

Natural mixture of xenon consists of nine isotopes. Some of them have high thermal neutron capture cross-section. Abundance and neutron capture cross-section for xenon isotope are presented in table 2.

Table 2. Natural xenon abundance and neutron cross-section for isotopes.

Isotope	^{124}Xe	^{126}Xe	^{128}Xe	^{129}Xe	^{130}Xe	^{131}Xe	^{132}Xe	^{134}Xe	^{136}Xe
Abundance, ε (%)	0.09	0.09	1.92	26.44	4.08	21.18	26.89	10.44	8.87
Thermal neutron capture cross-section, σ_0 (barn)	165	3.8	5.2	21	4.8	85	0.45	0.27	0.26
Effective cross-section, $\sigma_0 \cdot \varepsilon$ (barn)	0.15	0.003	0.1	5.55	1.96	18.0	0.12	0.028	0.023

After capture reaction nuclei usually emit gamma-ray. In our previous research [1] we used the most intensive 668 keV gamma-ray line for neutron detection. This line is a result of neutron radiative capture reaction with ^{131}Xe. Other isotopes also emit gamma-rays due to interaction with neutrons. So these gamma lines seen in spectrum can be used for neutron detection. Multiline detection approach is a more reliable method that allows us to avoid errors and gives more evidence of neutron flux detection. Detailed information about prompt gamma-rays from Xe isotopes is shown in table 3.

Table 3. Energies of neutron capture prompt gamma-rays and partial elemental gamma-ray cross-sections $\sigma_\gamma^Z(E_\gamma)$.

Energy (keV)	483.7	**536.2**	586.2	600.2	**630.3**	**667.8**	**772.7**	1317.9	1985.7	6467.1
Isotope	^{131}Xe	**^{129}Xe**	^{129}Xe	^{131}Xe	**^{131}Xe**	**^{131}Xe**	**^{131}Xe**	^{131}Xe	^{131}Xe	^{131}Xe
$\sigma_\gamma^Z(E_\gamma)$ (barn)	0.55	**1.71**	0.48	0.52	**1.41**	**6.7**	**1.78**	0.89	0.54	1.33
Related intensity (%)	8.21	**25.52**	7.16	7.76	**21.05**	**100**	**26.57**	13.28	8.06	19.85

In this research we used 668, 630, 536 and 773 keV gamma-ray lines for calculation neutron efficiency detections that have high partial elemental gamma-ray cross-sections $\sigma_\gamma^Z(E_\gamma)$ [3]. $\sigma_\gamma^Z(E_\gamma)$ is defined by product $\varepsilon \cdot P(E_\gamma) \cdot \sigma_0$, where $P(E_\gamma)$ is the absolute probability of gamma emission. By area calculation under the peaks we can estimate efficiency using information from ^{6}LiI(Eu) measurements.

^{6}LiI(Eu) detector placed at same point gave the value of flux 27 neutrons·sec^{-1}·cm^{-2}. Final values are shown in tables 4 and 5.

Table 4. Count rate under the peaks and efficiency registration for HPXe 0.2L detector.

Energy (keV)	Count rate (cps)	Efficiency registration (%)
536.2	0.137±0.016	0.027±0.003
630.3	0.091±0.015	0.018±0.003
667.8	0.404±0.014	0.079±0.003
772.7	0.078±0.012	0.015±0.002

Table 5. Count rate under the peaks and efficiency registration for HPXe 2.0L detector.

Energy (keV)	Count rate (cps)	Efficiency registration (%)
536.2	2.7±0.2	0.058±0.004
630.3	2.0±0.3	0.042±0.005
667.8	9.3±0.2	0.199±0.005
772.7	1.41±0.13	0.030±0.003

Total efficiency registration of thermal neutrons by 0.2 litters xenon gamma-ray detector is (0.138±0.006)%. In paper [1] efficiency of 0.2L detector estimated using only 668 keV gamma-ray line was (0.079±0.003)%. Total efficiency registration of thermal neutrons by 2.0L xenon gamma-ray detector is (0.329±0.009)%.

6. Conclusion

The results of thermal neutrons registration by the high pressure xenon detector showed that spectra contain prompt gamma-ray lines from radiative neutron capture reactions of xenon isotopes. Spectrometers can be used as a neutron detector with an efficiency (0.138±0.006)% and (0.329±0.009)% for detectors with volume 0.2 and 2.0 litters, respectively. Improved analyzing technique based on multiline detection allows to increase efficiency of HPXe 0.2L detector from 0.078 to 0.138%.

Acknowledgments

Authors wish to acknowledge the Center of fundamental researches and particle physics. This work was partially supported by MEPhI Academic Excellence Project (contract No. 02.a03.21.0005, 27.08.2013).

References

[1] Dmitrenko D *et al.* 2012 Detection of neutrons and gamma-rays by a xenon pulsed ionization chamber *Instrum. Exp. Tech.* **55** 419

[2] Novikov A *et al.* 2014 Xenon detector with high energy resolution for gamma-ray line emission *Proc. SPIE* **9213** 921318

[3] International Atomic Energy Agency 2007 *Database of Prompt Gamma Rays from Slow Neutron Capture for Elemental Analysis* (Vienna) p 124

Journal of Physics: Conference Series **675** (2016) 042025 doi:10.1088/1742-6596/675/4/042025

Unit for solar hard X-Ray polarimetry PING-P: structure, control, data acquisition

E M Kruglov[1], V P Lazutkov[1], M I Savchenko[1], I I Shishov[1], A S Glyanenko[2, 3] and V N Yurov[2]

[1] Ioffe Institute, Politekhnicheskaya 26, St Petersburg, 194021, Russia
[2] National Research Nuclear University MEPhI (Moscow Engineering Physics Institute), Kashirskoe highway 31, Moscow, 115409, Russia

E-mail: e.kruglov@mail.ioffe.ru, vadim.lazutkov@mail.ioffe.ru, mikhail.savchenko@mail.ioffe.ru, i.shishov@mail.ioffe.ru, ASGlyanenko@mephi.ru, VNYurov@mephi.ru

Abstract. Architecture and operational concept description of the PING-P unit are presented. In-flight calibration and stabilization procedures are listed. Data and technology characteristics are presented. Two key functional modes, the "Patrol" and the "Burst", are described. Spectral data: single detector spectra, double coincidence spectra, calibration spectra, automatic calibration LED amplitude spectra. A burst detection system which is aimed to detect bursts in a pseudo-stationary flux. In-flight control procedure is described. Raw data acquisition, processing and storing techniques implemented in the PING-P unit are considered. Also a prompt data processing procedure and data transmission protocols are reviewed.

1. Introduction

The PING-M experiment is a continuation and development of the experiment to search for polarization of hard x-ray radiation from solar flares "PENGUIN-M" [1,2] onboard "CORONAS-PHOTON" satellite. The PING-M device is a part of the scientific instruments complex in project "Interhelioprobe" [3]. It consists of two modules: spectrometry unit (PING-PIRS) and polarimetry unit (PING-P). The PING-M device is designed to observe the Sun from close distances either in the ecliptic plane or at some heliolatitude, as well as to provide measurements close to the Sun.

2. Structure, purpose and technical characteristics of PING-P

The PING-P unit is intended for:

- measuring degree of linear polarization of hard x-ray radiation of solar flares in 20–150 keV energy range.
- hard x-ray radiation solar flares spectrometry in energy range up to 45 keV.

PING-P unit consists of two parts: the detector and instrument compartments (see figure 1). The unit body has a cylindrical shape with a square base. In the structure detection part of the unit consists of two main elements:

[3] To whom any correspondence should be addressed.

- Active hard x-rays scatterers (DS1...DS3), representing an Assembly of three paraterphenyl (PTF) crystals with a diameter of 40 mm and a height of 40 mm. Each crystal is glued from the bottom to the photomultiplier tubes (PMT). The PMT work provide: voltage divider (VD), preamplifiers (PU) and high-voltage power supply (HVS). These elements are not shown in figure 1. The PMT assemblies are identical for all scintillation detectors.
- Detectors build of the scattered photons (DP1...DP6), covering the lens and consisting of 6 crystals CsI(Tl) with a diameter of 45 mm and thickness of 5 mm in the cases with beryllium entrance windows along with relevant PMT assemblies.

The unit upper part (detector compartment) contains the Assembly of detectors and closed thin (1 mm) metal hood from alloy AMG made by the method of rolling. On assembling, for active scatterer detectors it's possible to install entrance beryllium window, but in order to save weight with thickness of 1 mm foil.

The lower part of the unit (the instrument compartment) includes: two interchangeable the signal processing units (SPU) ("cold reserve"); 2 units of low-voltage converters LVC circuit board ("cold reserve"); 2 units of pulse regulators ("cold reserve"); LED circuit unit ("cold reserve"). At the lower part of the block is placed external connectors to interconnection with unit PING-PIRS.

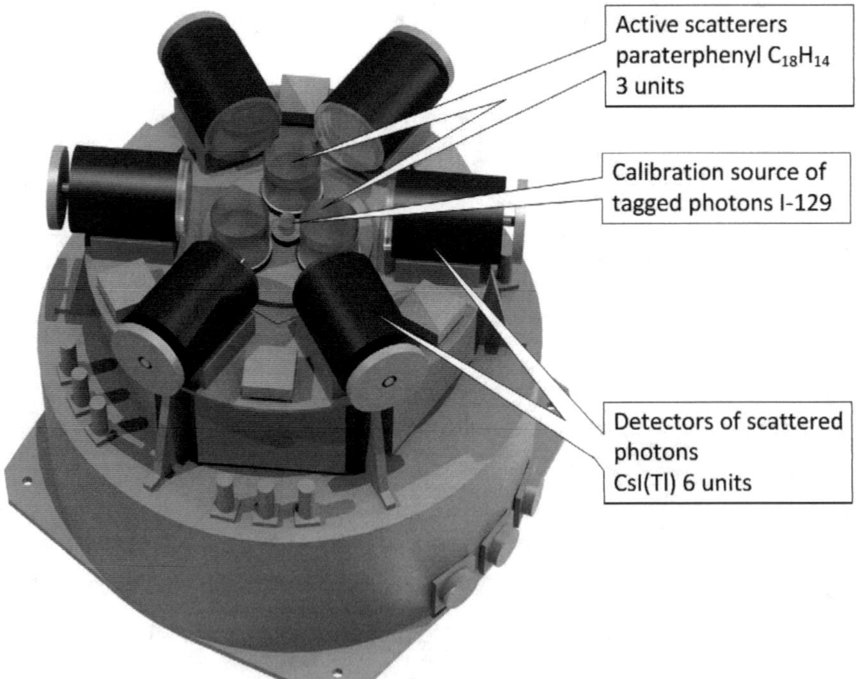

Figure 1. The PING-P polarimeter assembly (bottom - electronics compartment).

The PING-P dimensions, max –320x320x300 mm. The PING-P unit mass is not more than 10 kg.

3. Technical characteristics of PING-P unit

3.1. Main technical data

External power supply is stable DC voltage (12±0.6) V from the PING-PIRS block. The unit power consumption is not more than 10.0 W. Daily volume of scientific information, transmitted to unit PING-PIRS, no more than 30 MB.

3.2. The unit PING-P modes of operation

PING-P unit works in three modes: "Patrol", "Flare" and "Programs".

In "Patrol" mode the acquisition, processing and transmission amplitude spectra and polarization hard x-ray radiation matrices with a frequency of 1 every 60 seconds. In the "Flare" mode – once in 1s.

The PING-P unit switches from "Patrol" mode to "Flare" mode by PING-PIRS command. The switching is based on the signal "FLARE" generated by the PING-P unit and using also PING-PIRS data.

The measurement parameters (criteria of generation of "FLARE" signal, duration of Flare, the frequency of polling) can be changed by Ground command. Mode "Programs" for PING-P is possible only from "Patrol" mode. This operation is performed by Ground command.

3.3. The PING-P unit data transmission

The information exchange between the PING-P and PING-PIRS units is done digitally, using differential synchronous sequential intersection, symmetric in both sides. Galvanic isolation is not provided. Data transmission electrical standard – LVDS with SPI protocol, synchronous serial, arrays with CRC16 checksum. The format of data block in modes "Patrol" and "Flare" is the same. The PING-P unit passes to the PING-PIRS unit scientific, technical, service and control information and status information from stabilization and calibration systems. The composition and volumes of the transmitted information are presented in table 1.

Table 1. Characteristics of data flows.

Data	Volume, bytes	Frequency of transmitting, times per min	
		"Patrol"	"Flare"
Main Physical data (polarization matrix, spectra from detectors, data about fluxes)	6036	1	60
Technical data (temperature, "dead" time, status data from any parts etc.)	82	1	60
Service data (calibration spectra, additional spectra from etc)	4514	1	60
Total	10632		

4. The PING-P unit structure

The PING-P unit electronics can be divided into 2 main functional groups: The analog electronics for PMT operation and digital electronics. Unit PING-P provides the following tasks:

- data acquisition from 10 scintillation detectors block;
- the 2-loop system the stabilization of the energy scale of the instrument on the reference signals of the LEDs and isotope sources (the parameters correction on commands possibility);
- allocation and the analysis of double coincidence of signals from DP (detector of scattered photons) and DS (active scatterer) detectors;
- sorting and primary processing of data;
- storage of 1 frame of data;
- data transfer to block PING-PIRS on PING-PIRS demand;
- receive control commands from PING-PIRS on serial SPI bus according to a communications Protocol;
- the reception of 4 teams switch sets of electronics on a radio channel;

- output signal "FLARE" generation for the specified criteria (with the possibility of changing the criteria for the teams) and its transmission to PING-PIRS unit.

4.1. The PING-P unit analog electronics
Analog electronics of each of the 10 PMT (detectors DS, DP and DK - calibration source detector) include: voltage dividers circuit board (VD) PMT, high voltage supply (HVS) circuit board, preamplifiers (PU) block. All PU, HVS and LEDs are doubled in a circuit array. Switching on a reserve complete set can be made by a command. Each complete set is switched independently.

4.2. The PING-P unit digital electronics
The SPU circuit board concentrates the greater part of electronics of PING-P unit.

The processor performs following main functions: responsible for the initial data acquisition for different periods of reading, generates data for transmitting to the telemetry, parses the incoming command and control signals and produces the necessary control action, logs and stores the necessary parameters in non-volatile ROM, analyzes current information from the detectors to detect outbreaks, and generates a signal FLARE. It consists of two parts with different functions (PLIS1 and PLIS2).

PLIS1 realizes next functions: generates frequency 40 MHz for the ADC, performs the detection of pulses in the data stream of the ADC with a frequency of 40 MHz and measuring their amplitudes, performs self-tuning scintillation detectors on signals LEDs, producing pulse width modulated signals to control the high voltage power source (TRS) and closing thereby the internal LED circuit for each detector, PLIS1 generates clock signals TRS and the excitation signal for the pulse generator of the LED, analyzes the cases of coincidence of pulses from different detectors and generates the necessary signals for reception of amplitude spectra. Data about the amplitude of the pulses, required to be registered, is transferred in PLUS bus connection (from "familiar" naming the PLIS matrix).

PLIS2 realizes next functions: data exchange with the Central processor and the external RAM provides a RAM in the accumulation of the amplitude spectra and the processor access, produces for frequency 40 MHz for processor and 1 Hz, provides delivery of telemetry data in block PING-PIRS by implementing the negotiated exchange of information and receives control signals from block PING-PIRS.

5. Conclusion
It was developed architecture and operational concept PING-P polarimetry unit of PING-M device to international space project “Interhelioprobe”.

Acknowledgements
The authors would like to thank all colleagues from Astrophysics institute of NRNU “MEPhI” and Ioffe Physical-Technial Institute RAS for interest and support in this work.

This work was partially supported by MEPhI Academic Excellence Project (contract № 02.a03.21.0005, 27.08.2013).

References
[1] Kotov Yu D *et al.* 2011 *Bulletin Of The Russian Academy Of Sciences. Physics* **75**(6) 744-7
[2] Kotov Yu D *et al.* 2011 *Solar System Research* **45**(2) 135-45
[3] Kuznetsov V D and Zelenyi L M 2009 *Geomagnetism and Aeronomy* **49**(8) 1137-47

Journal of Physics: Conference Series **675** (2016) 042026 doi:10.1088/1742-6596/675/4/042026

Study of silicon photosensor applicability for scintillator detectors

V M Khilya and S A Voronov
National Research Nuclear University MEPhI (Moscow Engineering Physics Institute), Kashirskoe highway 31, Moscow, 115409, Russia

Abstract. The aim of the present work is the creation a prototype of anticoincidence system AC for gamma-telescope GAMMA-400. The detectors of AC are developed on the basis of plastic scintillator and silicon photomultipliers. This work is focuses on research of applicability of silicon photomultipliers SiPM by company SensL, type 60000 with BC-408 plastics for the prototype of anticoincidence system detector ACtop. In frame of project the assembly for measuring of the SiPM characteristics such as the linearity, boundary of saturation, the time resolution was developed. The final stage of work was the integration of the prototype of anticoincidence detector.

1. Introduction

Gamma-ray astronomy is science domain to study the properties of space objects such as pulsars, supernova remnants, active galactic nuclei, black holes, quasars and other astrophysical subjects by their gamma-emission – see, for example, [1]. Registration of γ-quanta involves the determination of its energy, the time and direction of their arrival. The special instruments are developed for this purpose named gamma-ray telescopes. Gamma-ray telescope is the instrument determining the aspect and time of photon arrival and measures it`s energy. Registration of photons with energy more than 100 MeV based on the generation of electron-positron pairs. This process proceeds by interaction of gamma-quants with the material of the gamma-ray telescope converter – see, for instance, [2]. Up to now the experiments in gamma-ray astronomy were carried out onboard satellites, such as: SAS 2 [3], GAMMA-1 [4], Egret [5], Fermi-LAT [2,6].

2. The GAMMA – 400 experiment

The international project GAMMA-400 intended for registration of γ-emission in the energy range $(0.1–3.0)\times10^3$ GeV [7]. The gamma-telescope GAMMA-400 has a traditional physical layout, it consists of converter, surrounded by anti-coincidence detectors, time of flight system, two calorimeters, and other detectors [8]. The principle of instrument operation is as follows: gamma-quantum is not registered in anticoincidence system, then it is converted into electron-positron pair in the converter-tracker, and then observed by the detectors of telescope. Anticoincidence detectors are used for rejection of the cosmic ray charged particles [9, 10]. The efficiency for charged particles has to be at least 0.99999, because the flux of high energy protons (10^4 $1/m^2s$), several orders of magnitude more than the gamma ray flux (10^{-1} $1/m^2s$) [11]. Proton flux is the main background of charged particles incident on the surface of gamma-ray telescope. Also low energy protons with flux $\sim10^7$ $1/m^2s$ [12, 13] should taken into account because of GAMMA-400 orbit mostly will in the outer Earth magnetosphere – see [14] and references therein in particular.

This work is devoted to the study of applicability of silicon photomultipliers SiPM by company SensL, type 60000 with BC-408 plastics for the prototype of anticoincidence system detector ACtop. The anticoincidence system consists of upper detector ACtop and 4 lateral ones AClat. Each detector is a set of scintillator bands arranged in two layers so as to maximize the rejection of charged particles. Each band is seen by photosensors from the ends of the scintillator. ACtop prototype consists of one scintillators band and two boards with 6 silicon photomultipliers (SiPM) on each. Besides the high protons flux registration efficiency and sensitive area ~10^4см^2 detector should have a high time resolution better than 10^{-8}s. This is necessary for the registration of backsplash particles from the calorimeters. This increases the efficiency of instrument. Time resolution of SiPM has to be less than 500 picoseconds for this experiment. The type 60000 SiPM by SensL company were used instead of usual vacuum photomultipliers due to their low supply voltage, insensitivity to magnetic fields, high time resolution and low sensitivity to temperature and supply voltage variations [15]. SiPM is the photodetector based on matrix of the microphotodiodes, formed on a common substrate, operated in Geiger mode – see, for example, [16].

3. The experimental technique and data analysis

The measurements were carried out by assembly, the scheme is shown in figure 1.

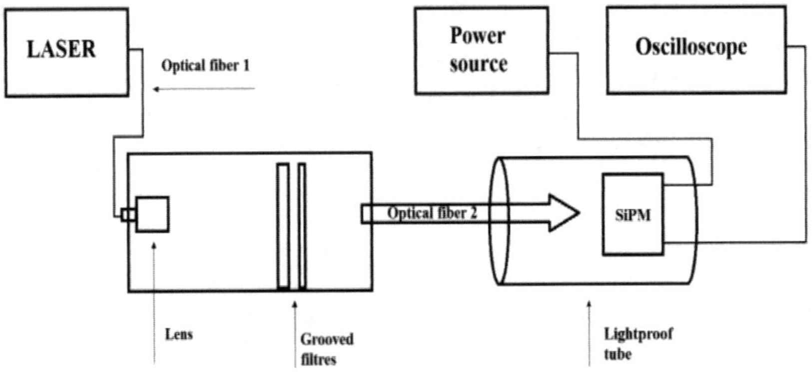

Figure 1. Scheme for measuring amplitude properties.

A beam of light coming from the laser to the input of the lens camera via optical fiber 1. The lens and grooved filters are in camera. Since in the optical fiber light has dispersion, an output beam is defocused. So at the input of the measuring path the lens was installed for light collecting. The focal distance of the lens is 15 cm. This value was chosen thus the light beam is strongly focused on the input of waveguide cable 2. After the lens it is necessary to install the set of grooved filters with different absorption for the light attenuation. Details of the camera are blacked to prevent light reflection. Then, already attenuated beam of light crosses on the fiber cable 2 input, and from the output of the optical fiber 2 reaches on the sensitive surface of SiPM. Laser Picosecond optical pulse generator PLS-405-660 (monochromatic, λ = 405 nm) was selected to generate light flashes to study of the SiPM properties. It emits a narrow beam of light precisely along the optical fiber. SiPM was biased by a low-voltage power supply and the SiPM signal is taken to an oscilloscope type TDS5000B. The intensity of the laser was attenuated using calibrated optical glasses.

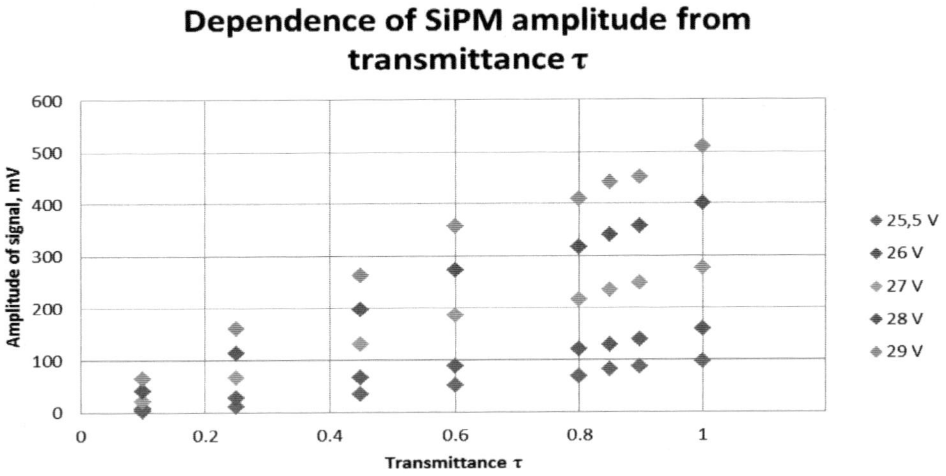

Figure 2. Dependence of SiPM amplitude from transmittance.

The measured signal amplitude dependence on the transmission coefficient of grooved filters is shown in figure 2 for different supply voltages. Same dependences were obtained for the whole SiPM set. This figure demonstrated the linearity in the wide range of incident light intensity. The pulse with minimal amplitude corresponds to the amplitude caused by scintillation of relativistic singly charged particle in scintillator of 1 cm thickness. The amplitude was measured for higher light intensity to determine the saturation area of the sensor. Dependence of amplitude from laser intensity is shown in figure 3. The graph shows the saturation region. For example, the saturation corresponds to the signal with amplitude of 640 mV at a supply voltage 29 V.

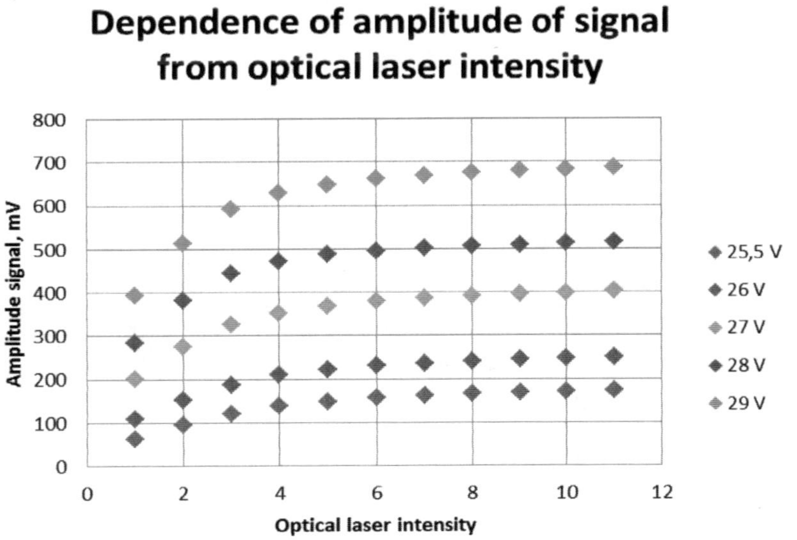

Figure 3. Dependence of SiPM amplitude on laser intensity.

The analysis of time resolution was carried out by means of the assembly shown in figure 4. The light signal is sent simultaneously onto two photosensors. The delay between the signals of pair of SiPM was measured. 2000 events were recorded for each pair of photomultipliers. Figure 5 shows a typical distribution of delays between SiPM couple pulses.

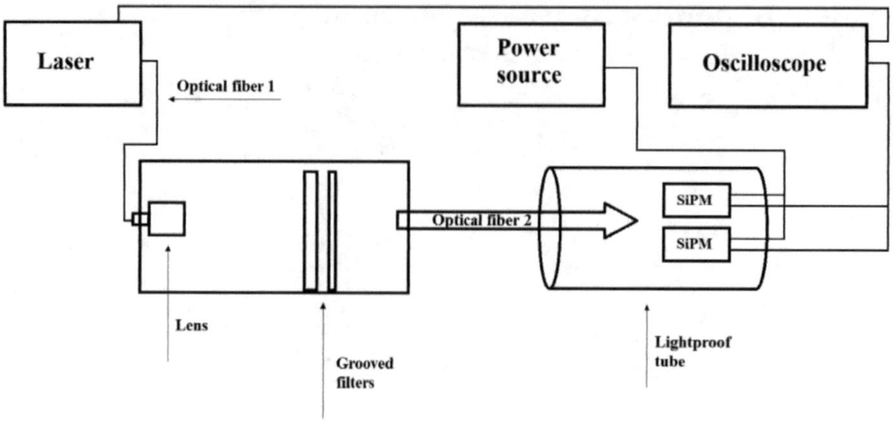

Figure 4. Assembly for measuring time resolution.

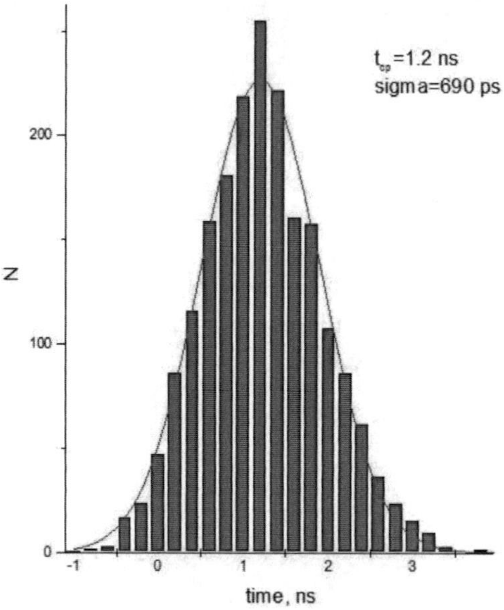

Figure 5. Distribution of delays between the signals from two SiPM

The obtained value of the standard deviation of the distribution was found to be σ = 690 ps. It corresponds to the 490±12 ps time resolution of SiPM.

4. Conclusion

Thus, following conclusions were reported. Firstly, each SiPM of the set is interchangeable, since their amplitude characteristics are similar. Second, the average time resolution of silicon photomultipliers is equal to 490±12 ps. The resulting value satisfies the requirement of at least 3ns needs to backsplash particles rejection in gamma-ray telescope GAMMA-400 [17]. Considering the measurements results, it was concluded that type 60000 SiPM by SensL company can serve for a replacement of the vacuum photomultipliers in anticoincidence detector of gamma-ray telescope.

References

[1] Olive K A *et al* 2014 *Chin. Phys. C* **38** 090001

[2] Swensen E *et al.* 2001 *GLAST Tracker Mechanical Design* PDR Design Review Document. HTN–102070-0002
[3] Derdeyn S M *et al.* 1972 *Nucl. Instr. Meth.* **98** 557
[4] Galper A M *et al.* 1986 *Soviet Astronomy* **30**(5) 508
[5] Thompson D J *et al.* 1993 *Astrophys. J. Suppl.* **86** 629
[6] Ackermann M *et al.* (Fermi/LAT Collaboration) *Astrophys. J. Suppl.* 2012 **203**(4) 2
[7] Topchiev N P *et al.* 2015 The GAMMA-400 Experiment: Status and Prospects *Bulletin of the Russian Academy of Sciences* **79**(3) 417
[8] Galper A M *et al.* 2015 *Physics Procedia* **74** 177
[9] Runtso M F *et al.* 2015 *Physics Procedia* **74** 220
[10] Arkhangelskaja I V *et al.* 2015 *Physics Procedia* **74** 212
[11] Hayakawa S 1969 *Cosmic Ray Physics* (New York , Wiley-Interscience)
[12] URL: *http://ulysses.jpl.nasa.gov/science/data.html*
[13] URL: *http://cdaweb.gsfc.nasa.gov/istp_public/*
[14] Khyzhniak E V *et al.* 2015 *Physics Procedia* **74** 238
[15] SensL Product Selection Guide *http://sensl.com/downloads/ds/PB-Product_Selection_Guide.pdf*
[16] Dolgoshein B *et al.* 2006 *Nucl. Instrum. Meth.* A **563** 368-76
[17] Chasovikov E N *et al* 2015 *Physics Procedia* **74** 206

Selective detector of cosmic particles based on diamond sensitive elements

A A Altukhov[1], K V Zaharchenko[1], V A Kolyubin[1], S A Lvov[1], P G Nedosekin[1], E M Tyurin[2*], R F Ibragimov[2], V V Kadilin[2] and I V Nikolaev[3]
[1]Industrial-Technological Center «UralAlmazInvest» Ivana Franko str. 4, Moscow, 121108, Russia
[2]National Research Nuclear University MEPhI (Moscow Engineering Physics Institute), Kashirskoe highway 31, Moscow, 115409, Russia
[3]«Rocket and space Corporation «Energia» S. P. Korolev», Lenina str., 4A, Korolev, Moscow region, 141070, Russia

E-mail: *tempost@mail.ru

Abstract. The article describes the device for selective registration of electrons, protons and heavy ions fluxes from the solar and galactic cosmic rays in the twelve energy ranges, built on a base of diamond detector. The use of the diamond detectors allowed for the creation a device for registration of cosmic particles fluxes at the external spacecraft surface with the resource not less than 20 years. Selective detector is aimed for continuous monitoring of radiation situation on board the spacecrafts, in order to predict the residual life of their work and prompt measures to actively protect the spacecraft when the flow of cosmic particles is sharply increased.

1. Introduction

Portable devices creation for continuous monitoring of dose and spectral characteristics the cosmic rays on an external spacecrafts surface is an actual task. This device solves the problem of increase the duration of a space systems operation and increase of crews safety on the spaceships. Continuous monitoring the cosmic rays with definition of spectral composition makes possible to obtain information about the change dynamics in the field of cosmic rays on the spacecraft orbit. This allows predicting the development of radiation impact on equipment and crew. Such information makes it possible to apply the necessary anti-radiation efforts: hibernate the insurance electronic equipment and use of additional measures for crew radiation protection.

Monitoring the dose and spectral characteristics of the cosmic rays fields on spacecraft board allows calculating more accurate the life-cycle costs and planned replacement, to ensure the safe operation of the spacecrafts constellation. The use of diamond detectors for such devices creation is caused by the high radiation resistance (up to 500 Mrad) and increase of exploitation resource [1,2].

This paper is focused on the selective-board recorder of cosmic rays developed by Industrial-Technological Center «UralAlmazInvest», with participation of specialists from National Research Nuclear University “MEPhI” and RSC "Energia". The main purpose of this device is a continuous monitoring of the radiation situation on the spacecraft board, including the registration of the absorbed dose from electron and proton component of cosmic rays. The detector also provides measurement of radiation flux density in 12 energy ranges (electrons in four energy ranges; protons in the four energy ranges, heavy charged particles in four ranges of linear energy transfer (LET)).

2. Description of the construction of the selective detector

As mentioned earlier, the basic operating principle of the selective detector for cosmic rays spectra measurement is based on the particles separation in the energy range using a multi-detector registration scheme with selective filters [3].

Figure 1 shows the external view of the selective detector unit, and figure 2 shows the principal composition of the device.

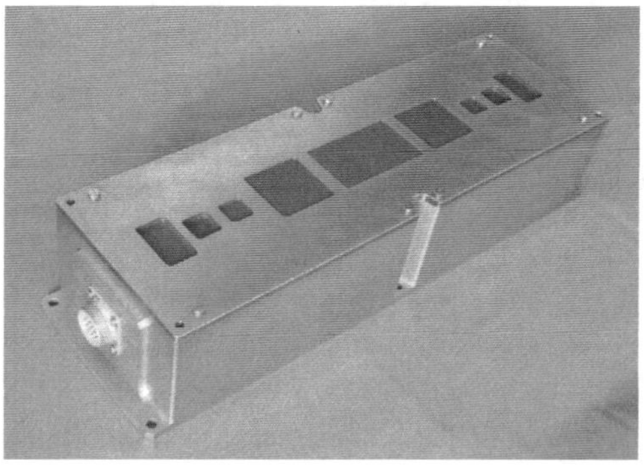

Figure 1. External view of the selective detector unit.

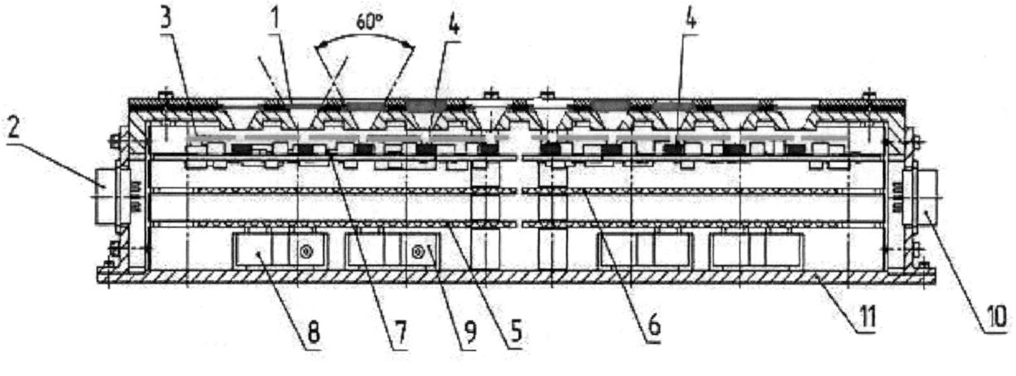

Figure 2. The design of selective detector of the cosmic rays 1 - selective filters; 2 - power connector; 3 - electrostatic screen; 4 - DSE; 5 - power supply board assembly; 6 - host interface board; 7 - board detection unit; 8 - the filter unit; 9 - power supply; 10 - connector interface; 11 - case.

The device case was made from an aluminum alloy, in which two identical spectrometer registration units were installed, one of which is working device, and the second one is used as a backup. The double backup registering node ensures a longer life of the device for up to 20 years. The device case has equivalent thickness value, near to 6 mm of aluminum (thickness including the heat sinks and fixing construction elements). This thickness value provides reliable radiation protection of the device electronic circuit.

Spectrometric unit includes the amplification module board – 7, the digital processing module board – 6, and the power supply module board – 5. Diamond sensitive elements (DSE) were mounted on the amplification module, placed under collimation windows of the spectrometer case. Collimation window are covered with selective filters, providing separation of cosmic rays in energy ranges. The power connector – 2 and the interface connector – 10 are installed at the end faces of the case.

The detectors used are the caseless DSE, shown in figure 3. DSE consists of diamond plate with the deposited aluminum plates mounted on the dielectric base. The electrodes of the diamond plate with gold conductors are connected to the ground terminals. DSE is lacquered for sealing. This design allows placement of DSE close to the signal preamplifier input, ensures minimization of the electronic noise on the gain path, and minimization of the electromagnetic interference effect.

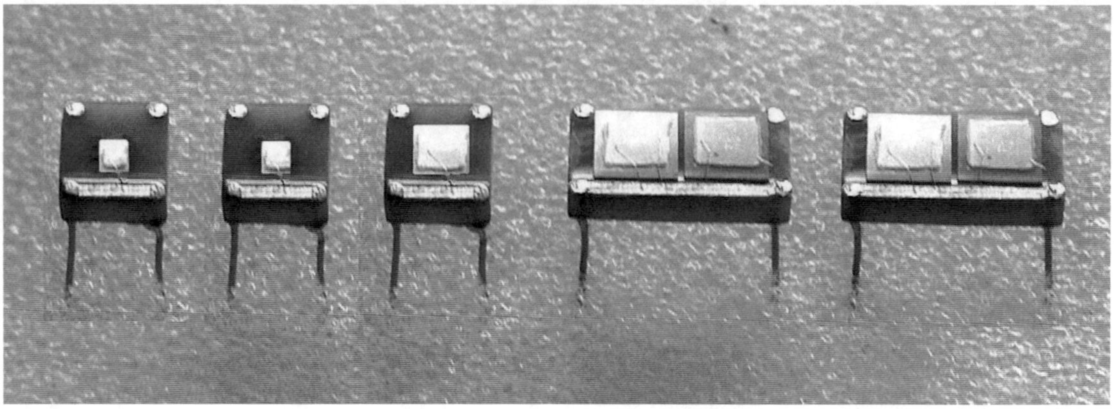

Figure 3. External view of the DSE.

The DSEs have different area of sensitive surface. Two DSE with surface areas of 2×2 mm^2 are used to register the total flux of electrons, protons and heavy ions. The DSE with surface area of 3×3 mm^2 are used for registration an low energy protons and electrons with energies more than 3 MeV. Last one DSE, containing two diamond plates with surface area of 4.5×4.5 mm^2 are used for registration an protons with energies more than 30 MeV and heavy ions. The use of DSE ensures that the statistical errors in the measurement of cosmic particles flow in different energy ranges will not exceed 10% for the measurement time from 5 to 10 min.

3. Selection of cosmic rays

Selection of cosmic particles (electrons and protons) is provided by using the four DSEs, that covered by filters with different thicknesses. Figure 4(a, b, c) shows that the signal spectra of 1st, 2nd, 3rd and 4th DSE are shifted in the energy scale in accordance with the thickness of the selective filter. The shift allows observing particles from given energy range in a particular registration channel. The graphs also show that signals from electrons and protons have different amplitudes, and that is why it is possible to separate these signals using amplitude discrimination. The signal range from 0.2 to 0.3 MeV is used to register the electronic component of cosmic rays. The signal range from 2 to 10 MeV is used to register the proton components. To register heavy ions the fifth channel recording is used. Selection of the heavy ions in this channel carried out by selection their amplitudes in the range of amplitudes from 100 to 1000 MeV. The ranges of signals amplitude are given in equivalent energy.

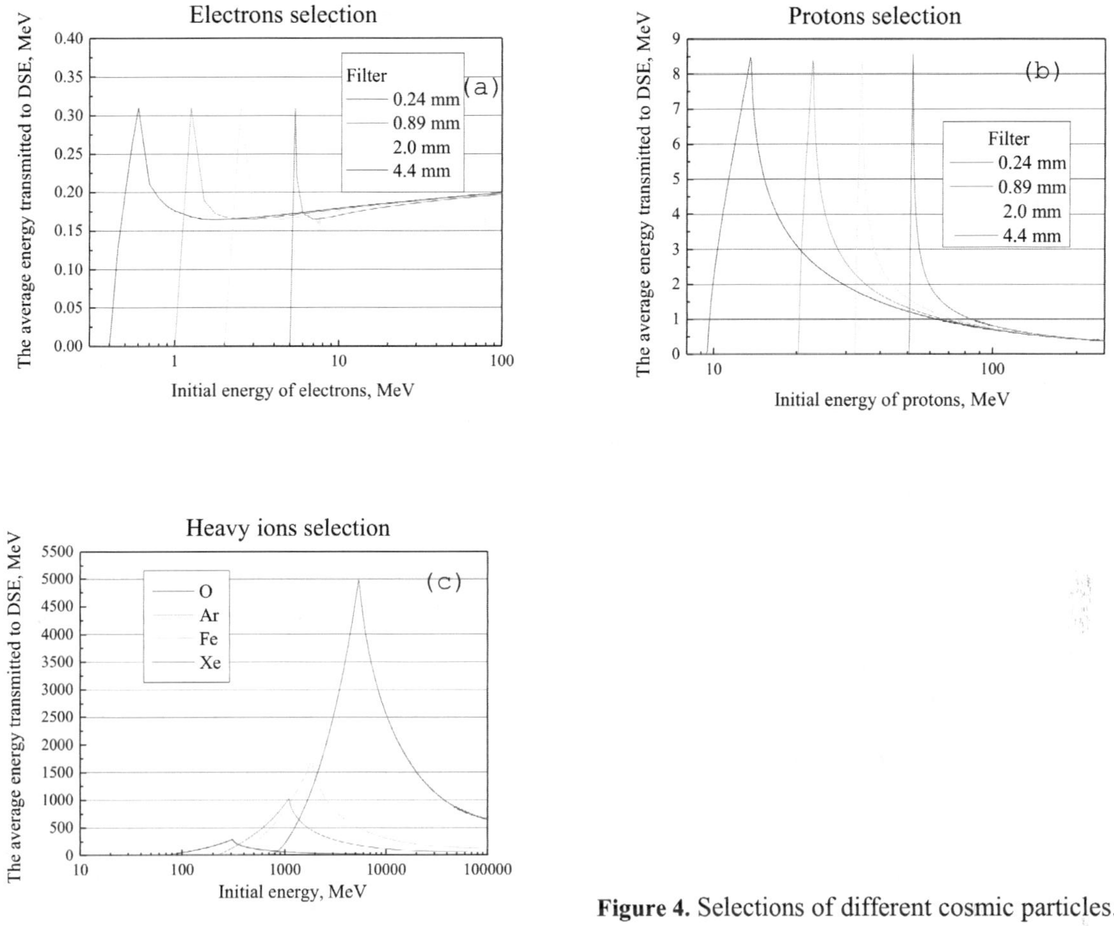

Figure 4. Selections of different cosmic particles.

The absorbed dose from electron and proton components as well as from heavy ions of cosmic rays is determined by calculation based on obtained spectra of cosmic rays.

4. Conclusion

Currently the prototypes of the cosmic particles selective detectors are made and samples are tested. Preparation for the production of flight models is started.

Acknowledgments

This work was partially supported by MEPhI Academic Excellence Project (contract No.02.a03.21.0005, 27.08.2013).

References

[1] Mainwood. Recent developments of diamond detectors for particles and UV radiation 2000 Semicond. Sci. Technol. **15** pp R55 – R63
[2] Zakharchenko K V *et al.* 2015 Spectrometric Diamond Detector of Fluxes of Ionizing Radiation for Space Transportation Systems *Measurement Techniques* **58**(6) 713–8
[3] Altukhov A A *et al.* 2011 (In Russian) *Problems of Atomic Science and Technology* **1** 56 (original Russian title: *Voprosu atomnoi nauki i tehniki)*

The software and hardware for the ground testing of ALFA-ELECTRON space spectrometer

A G Batischev, A M Galper, P Yu Naumov, P P Naumov and A A Solodovnikov

National Research Nuclear University MEPhI (Moscow Engineering Physics Institute), Kashirskoe highway 31, Moscow, 115409, Russia

E-mail: PYNaumov@mephi.ru

Abstract. The complex for ground testing and space detector system calibration has been designed. The fast multilayer scintillation detector of the new telescope-spectrometer for the ALFA-ELECTRON space experiment is in ground testing mode now. The device will planned to install on the outer surface of the Russian Segment of the International Space Station. The basic scheme for the detector amplitude parameters measurement by use of specially designed hardware and software are described and the first prototype testing results are demonstrated.

1. Introduction

In the last 20 years a new class of atmospheric phenomena – the high-altitude electrical discharges (HED) formed above thunderclouds is intensively studied. It is supposed that a variety of observations for atmospheric, ionospheric and magnetospheric phenomena such as «red sprites», «blue jets», «elves», terrestrial gamma flashes and so on are related with them [1, 2].

From a large group of physical objects related to the HED, the acceleration process of high-energy electrons is very important. Direct detection of electron beams will not only confirm the existence of the electrons acceleration in the HED, but also to obtain data about the configuration, size and dynamics of the electric field in the upper atmosphere of the information on the energy spectra and temporal profiles of electron beams. One of the main functions of the ALFA-ELECTRON experiment [3] is electron beams detection from the HED.

2. Physical scheme of spectrometer

The physical arrangement of the detector system is shown in the figure 1. The fast multilayer scintillation detector (MSD) consists of the set of C1 - C10 scintillation detectors. These scintillators have a high transparency and a minimum excitation time. It has been proposed for MSD to use the best (in terms of time and energy resolutions, amplitude characteristics, dimension and mass) Hamamatsu photomultipliers R5611-01A as photodetectors [3].

Structurally, the MSD [4] is composed of the upper detector system (UDS) and the scintillation calorimeter (SC). The UDS includes the detectors C1 and C2, which consists of four identical modules. The C1 detector module group is perpendicular to the C2 detector one. Each module is viewed separately by PMTs.

The SC is located at the distance of 100 mm from the UDS and consists of C3 - C10 detectors, each viewed by two separate PMTs.

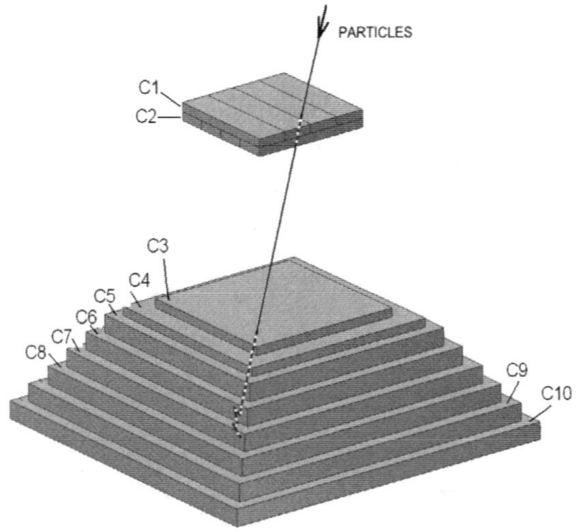

Figure1. The ALFA-ELECTRON multilayer detector system physical scheme (left) and the experimental setup for one-layer detector system testing (right).

The detectors C1 and C2 are used as thin trigger aperture ones, the SC detectors (C3 - C10) are used as total absorption detectors where the particle absorbed will lose all its energy. So, the energy of electrons is measured mainly by the total energy of the particles absorbed in the SC detectors. The C10 detector is a veto detector for passing particles.

Digital controller unit [3] together with the pulse amplitude analysis one provides the separation of particles (electrons and protons) by analyzing the particle energy deposition in scintillation detectors when particles penetrate the layers of the MSD.

3. Description of main possibilities of the hardware and software system

Currently in NRNU MEPhI the hardware and software for testing scintillation detectors system are under final tuning. This complex is based on industrial PC, NIM and VME electronic modules. Instrumentation and program support allowed to get the amplitude spectra in real time from 16 detector channels simultaneously also (with one QDC module – see the text below).

The hardware and software complex includes the following main features:

- provide an amplitude spectra measurement from cosmic radiation sources for a single detector channel (see figure 2) and for 8-16 ones (see figure 3);
- provide the possibility to exposure the thresholds and calculate the spectrum energy resolution (FWHM);
- provide a measurement of the dynamic range of the scintillation detectors;
- provide a verification of prototypes made front-end electronics for compliance with the stated parameters.

Figure 2 shows typical signals from the detector and the diagram for testing of one scintillation detector layer.

The diagram of amplitude characteristic measurement of spectrometer trigger detectors C1 and C2 is shown in the figure 3. Each of the detectors C1 and C2 has 4 channels of registration, because these detectors consist of four identical modules. C3 – C10 detectors has 8 channels. Thus, we have 16 channels in total.

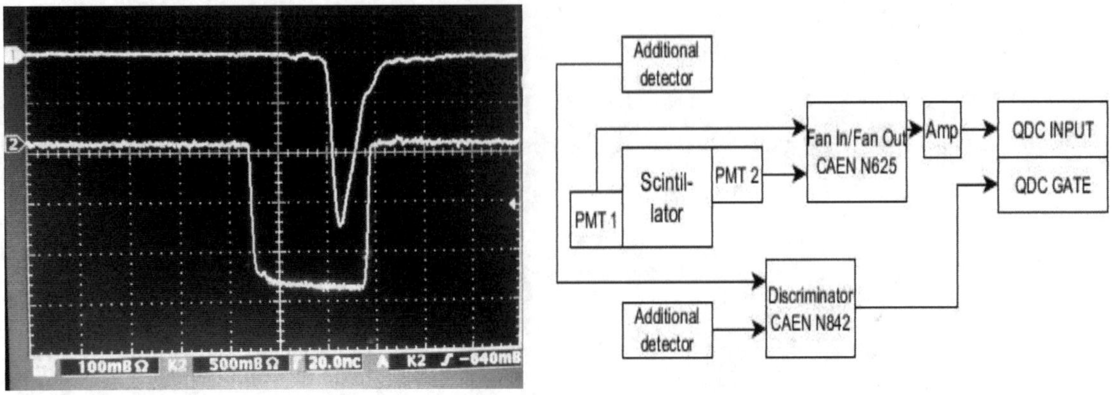

Figure 2. Signals screenshot (left): typical signal from PMT (line#1), QDC Gate signal (line#2) and diagram of one-layer detector calibration (right).

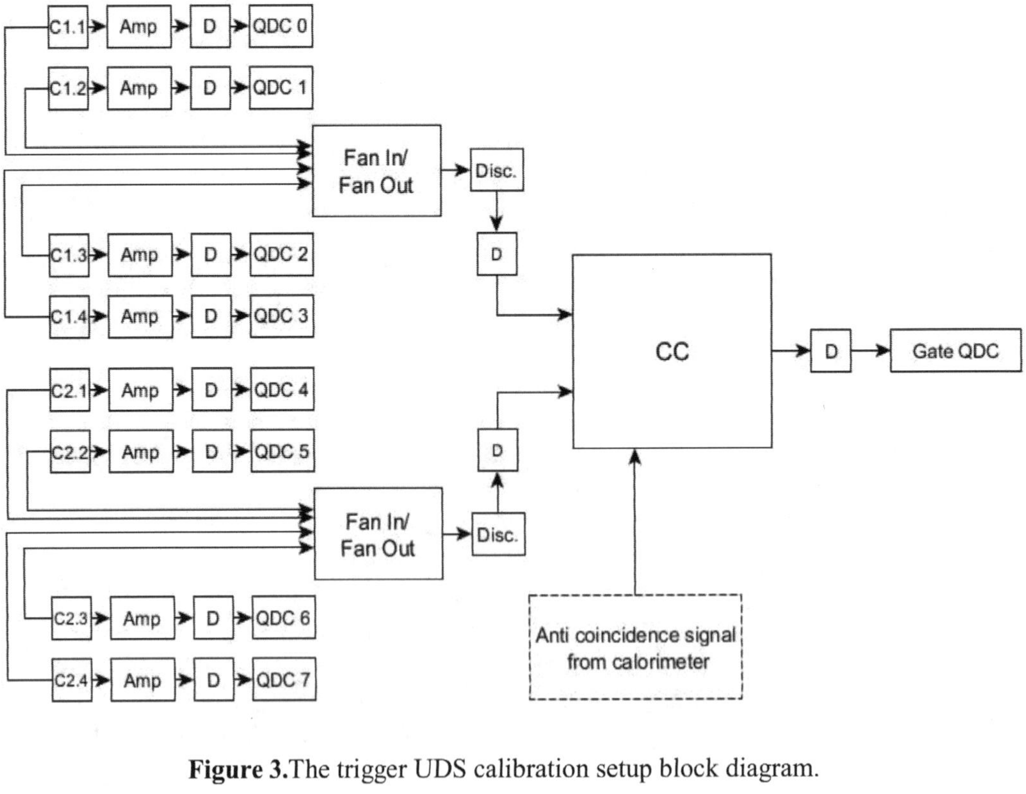

Figure 3. The trigger UDS calibration setup block diagram.

As mentioned above, the C10 detector is a veto detector and close the coincidence scheme (CC) is connected with QDC Gate input.

For organization of measurement control signal (QDC input Gate) the output signals from C1 and C2 detectors are summing from all of the 4 PMTs (C1.1 – C1.4, C2.1 – C2.4) and output signals from C3-C10 detectors are summing from 2 PMTs (for one detector C3.1 – C3.2, for example). The CC allows to detect only those particles that are completely absorbed in the detectors. Calibrated delays (D) are connected to CC inputs for coincidence tuning of pulses.

In this installation we use standard NIM and CAEN analog modules (Fan In/Fan Out, Delay, Discriminators, Coincidence) and CAEN 16-channel QDC V965 controlled by PC program through crate controller V2718KIT VME-PCI Bridge (V27) and PCI Optical Link (A2818) with Optical Fibre5 m duplex (AY2705).

The software for this complex was developed by using C++ (in fact, IDE Qt Creator 5.3) and the main program runs under the Linux OS.

The basic functions of the main program are the following:

-set the parameters of the modules (see below);

-control the data collection process;

-provide a visualization of incoming data in real time;

-work with all of 16 channels in process of registration simultaneously;

-have the opportunity to save and open data files for the further work;

-have the opportunity to restore data if software or hardware will crash;

-provide the data preprocessing and result visualization.

Our control program allows to set the main electronic parameters: threshold, pedestal, sampling time. PC operator can calculate the statistic parameters of spectra (FWHM, maximum of peak, integral background) and paint spectra in different scales (logarithmic and linear) also.

Using CAEN S.p.A. firm hardware and our original software allows to simplify a processes of ground calibration and detector systems testing for current and future space experiments.

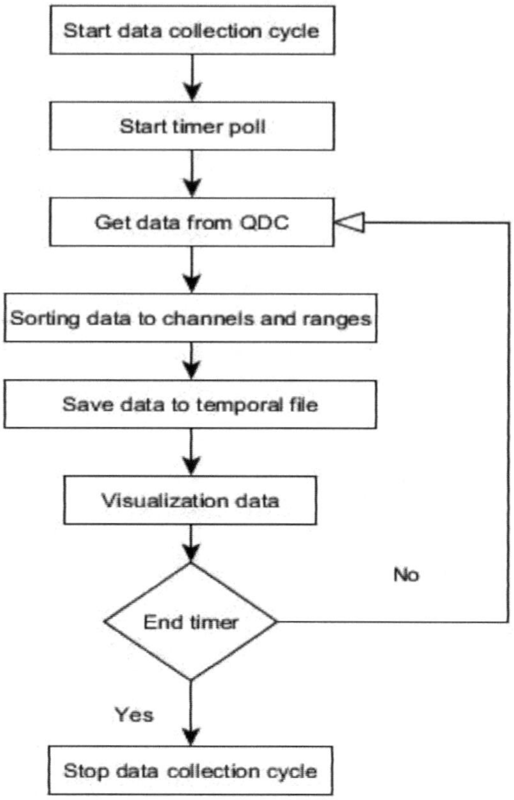

Figure 4. Simplified block diagram of the data acquisition program.

Figure 4 shows the diagram of the data acquisition program implemented in the complex. After the data collection cycle initializing the timer runs and resets the QDC data block located in VME.

Then the data analysis and sorting of the values was obtained for one or more channels (if this option is enabled) and for two affordable QDC dynamic ranges. To prevent data loss (in unforeseen cases), the program provides the preservation of data in a temporary file. Then the visualization process of received data in a bar graph on the PC screen is realized. Figure 5 shows a common view of the ground testing complex and typical experimental spectrum as an example of the program working during detector testing on the cosmic rays fluxes at the ground level. This spectrum has a character form of Landau distribution, which corresponds to the amplitude distribution of cosmic muons at the Earth's surface.

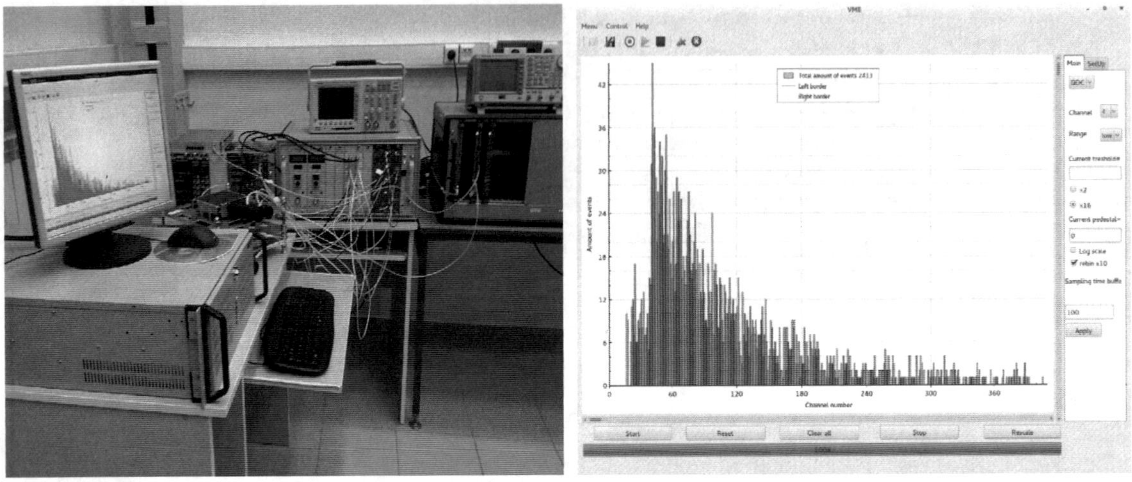

Figure5. The ground testing complex (left) and PC operator window with typical spectrum (right).

4. Conclusion

Currently, in the NRNU MEPhI the new complex for ground testing and calibrations of space detector system have been designed. This complex based on standard NIM and VME electronic modules and is controlled by original PC program. Hardware and software complex was quickly adapted to the new space spectrometer, showed an enormous possibilities and opportunities and can be successfully used in other different ground scientific experiments with fast particle detectors.

Acknowledgements

This work was performed within the framework of the NRNU Fundamental Research and Particle Physics Center supported by NRNU MEPhI Academic Excellence Project (contract N 02.a03.21.0005, 27.08.2013).

References

[1] Bell T F, Pasko V P and Inan U S 1995 *Geophys. Res. Lett.* **22** 2127.
[2] Fishman G J *et al.* 1994 *Science* **264** 1313.
[3] Batischev A G *et al.* 2015 ALFA-ELECTRON Experiment Onboard the International Space Station *Physics Procedia* **74** 139-44.
[4] Batischev A G *et al.* 2015 Multilayer Scintillation Detector for Nuclear Physics Monitoring of Space Weather *Physics Procedia* **74** 132-38.

International Conference on Particle Physics and Astrophysics (ICPPA-2015) IOP Publishing
Journal of Physics: Conference Series **675** (2016) 042029 doi:10.1088/1742-6596/675/4/042029

Asynchronous data readout system for multichannel ASIC

P Y Ivanov* **and E V Atkin**

National Research Nuclear University MEPhI (Moscow Engineering Physics Institute), Kashirskoe highway 31, Moscow, 115409, Russia

E-mail: *xeolit@gmail.com, evatkin@mephi.ru

Abstract. The data readout system of multichannel data-driven ASIC, requiring high-speed (320 Mb/s) output data serialization is described. Its structure, based on a limited number of FIFO blocks, provides a lossless data transfer. The solution has been realized as a separate test IP block in the prototyped 8 channel ASIC, intended for the muon chamber of CBM experiment at FAIR. The block was developed for the UMC 0.18 μm MMRF CMOS process and prototyped via Europractice. Main parameters of the chip are given.

1. Introduction

In modern international accelerator facility experiments, such as ATLAS, ALICE, CMS (at CERN), CBM, PANDA (at FAIR), a multichannel system of data readout and processing is used, where the number of channels ranges from some hundred thousand to millions. The necessity for processing data from a large number of analog sensors sets considerable requirements to the front-end electronics in respect of integration scale, power consumption and chip area. In the work, described herein, there was made a structure analysis and determination of its composing blocks parameters interdependence within the chip, developed earlier [1] for the silicon tracking system of the CBM experiment, and an estimation of its technical characteristics.

The typical structure of the multichannel sensor readout is given in figure 1. It is aimed at an asynchronous data-driven readout with no dead time. Analog signals from sensors arrive at analog inputs A_1 - A_M, and get stored at arrival of a strobe signal into analog memories AM_1 - AM_M. The analog commutator AC allows a single ADC block to accept and digitize signals from m analog memories. By means of commutators AC_i , $i \in [1, M/m]$, the serial scanning of analog memories AM_i and transmission of their signals' values to ADC inputs are implemented.

Being driven by the asynchronous signal START, each ADC digitizes an analog signal and stores its corresponding digital code into a FIFO memory [2]. If signals exist in m analog memories, m words will be written into a FIFO of a single ADC. Therefore, the depth of a FIFO should contain m words.

The data readout unit (DRU) performs scanning the states of all FIFOs, organizes reading and output of digital data into an interface block.

2. Data readout parameters interdependence

The T cycle of processing signals includes the time t_a of an analog signal conversion into a digital form, and its writing into a FIFO, the time of reading data from FIFO and its output

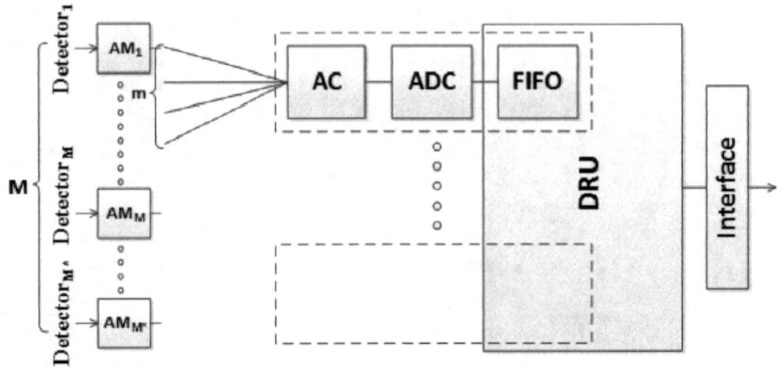

Figure 1. Structure of the multichannel sensor readout.

into an interface, having a bandwidth capacity of P bits/s. In order to reduce the duration of the T cycle, it makes sense to combine the stages of writing and reading into/from a FIFO. The structure contains M/m ADCs which function in parallel. The duration of writing data into all FIFOs does not exceed m$\times$t$_a$. Reading from a FIFO is implemented serially-by-word. The duration of the stage of reading and output depends on the number of words and their bit depth, and also the bandwidth capacity P of the interface. The analysis shows that the duration of the T cycle is defined by the following equation:

$$T = t_a + \frac{M \times n}{P} \tag{1}$$

where n is the bit width of words; M is the number of analog inputs.

In order to reduce the volume of equipment and simplify the circuit, it makes sense to increase the value of m. However, increase of m is limited by the acceptable error of analog signal storage in the memory AM. The requirement for minimization of the T cycle duration places a restriction on the acceptable value of m, for which the time of an analog signal's conversion into a digital form and its writing into a FIFO should not exceed the time of data reading and output

$$m \times t_a \leq \frac{M \times n}{P} \Rightarrow m \leq \frac{M \times n}{t_a \times P} \tag{2}$$

For the specified duration of the T cycle and the output capacity P, the number of inputs M of the device, according to the equation (1), should not exceed the value

$$M \leq \frac{(T - t_a) \times P}{n} \tag{3}$$

In practice, the arrival of signals at all inputs is a single event. An arrival of signals is most probable only at a number of inputs. Then the efficiency of input use is equal to η ($\eta \leq 1$). Therefore, the number of inputs of the device could be increased to $M^* = \frac{M}{\eta}$ for the same output bandwidth capacity P of the interface. To prevent a data loss for a single event, when signals arrive at all M* inputs, the depth of FIFO memory should be increased by $\frac{1}{\eta}$ times.

3. DRU structure

During the cycle T, the DRU block should read data from a FIFO and pass it to the interface at a tempo necessary for its functioning. To satisfy this requirement it makes sense to use a different-depth-FIFO-based pyramid structure of reading data as shown in figure 2.

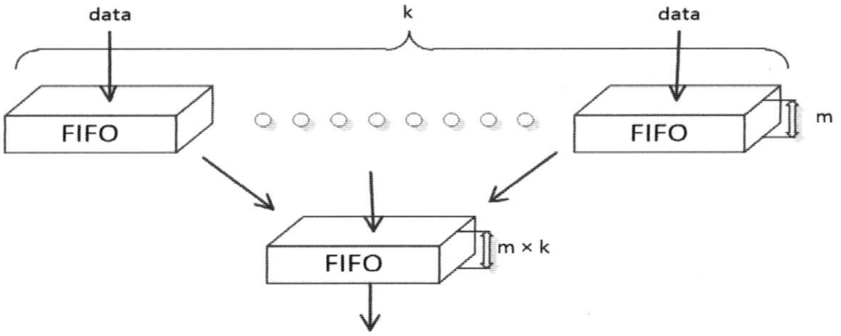

Figure 2. A different-depth-FIFO-based pyramid structure of reading data.

Data scanning and reading from a FIFO are implemented cyclically based on the token ring principle [3].

4. Implementation results

A test version of a DRU block was developed. The following requirements were placed for it: input data bit depth of 5 bits, their arrival rate of 50 MHz, reading results from an output FIFO should be performed at a frequency of 320 MHz, consumption power and area no more than 20 mW 300 × 300 μm correspondingly. The used technology should be 180 nm CMOS of UMC. A full design cycle of the test DRU block was performed. A structure, a Verilog-based behavioral model and functional tests of the DRU block were developed. The block performs scanning, reading data from a FIFO and its output onto a 5-bits-wide bus. The DRU test block topology within the 8-channel chip is highlighted by red outline in figure 3.

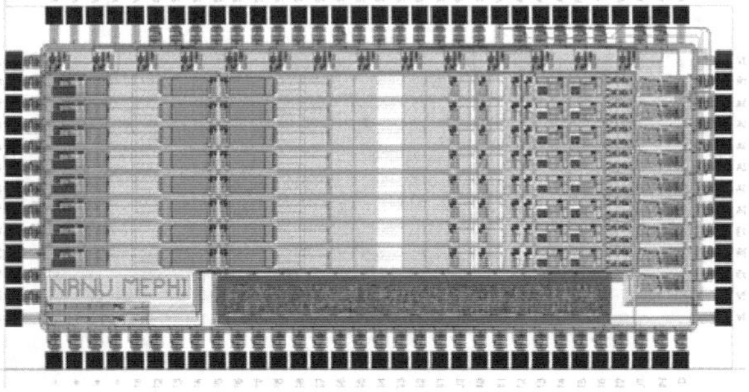

Figure 3. The DRU test block topology within the 8-channels chip.

The obtained characteristics are shown in table 1. As seen in the table the characteristics satisfy the requirements, placed for the multichannel ASIC.

5. Conclusion

- The structure of the multichannel sensor readout is presented. It provides an asynchronous data-driven readout with no dead time.
- There is a tight relationship between the duration of the T cycle, bandwidth capacity of an interface and the number of analog inputs M. For specified T and P values the number of inputs M of the device is limited.

Table 1. Main parameters of the DRU test block

Inside the DRU were used 11 fifo (data width = 5 bit)	
Frequency of reading data, MHz	320
Frequency of writing data, MHz	50
Number of standard cells, gates	2106
Total power, mW	19
Area, μm^2	82212

- The value m is restricted by both analog and digital parts.
- For providing reading and writing data for the duration of the T cycle it is necessary to use a different-depth-FIFO-based pyramid structure of reading data.
- The possibility of the chip creation for the silicon track system of the international experiment CBM with the defined characteristics in the 180 nm CMOS technology is confirmed.

Acknowledgements

This work was supported by the Ministry of Education and Science of the Russian Federation (grant no.14.A12.31.0002) in accordance with the RF government order no. 220. In addition, this work was supported by MEPhI Academic Excellence Project (contract No. 02.a03.21.0005, 27.08.2013).

References

[1] Atkin E *et al.* 2015 *JINST* **10** 04 C04006
[2] Cummings C E 2002 *SNUG* (*http://www.sunburst-design.com/papers*)
[3] Wayne T and Stephen B 1990 *Net. Tech. J.* 64-76

Development of the read-out ASIC for muon chambers

E Atkin[1], I Bulbakov[1], A Gusev[1], E Malankin[1], D Normanov[1], I Sagdiev[1], V Shumikhin[1], O Shumkin[1], P Ivanov[1], S Vinogradov[1], A Voronin[2], V Samsonov[1] and V Ivanov[1]

[1] National Research Nuclear University MEPhI (Moscow Engineering Physics Institute), Kashirskoe highway 31, Moscow, 115409, Russia

[2] Skobeltsyn Institute of Nuclear Physics Lomonosov Moscow State University, 119991, Leninskie gory 1/2, Moscow, Russia

E-mail: evatkin@mephi.ru

Abstract. A front-end prototype ASIC for muon chambers is presented. ASIC was designed and prototyped in the CMOS UMC MMRF 180 nm process via Europractice. The chip includes 8 analog processing channels, each consisting of a preamplifier, two shapers (fast and slow), differential comparator and an area efficient 6 bit SAR ADC with 1.2 mW power consumption at 50 Msps. The chip also includes the threshold DAC and digital serializer. The design has the following features: dynamic range of 100 fC, channel hit rate of 2 MHz, ENC of 1000 e- at 50 pF, power consumption of 10 mW per channel, 6 bit SAR ADC.

1. Introduction

The Compressed Baryonic Matter (CBM) experiment will be one of the major scientific pillars of the future Facility for Antiproton and Ion Research (FAIR) in Darmstadt [1,2]. The muon chamber (MUCH) is part of the equipment. MUCH is based on the gas electron multiplier (GEM) detectors [3]. GEM detectors will be operating at a high gas gain in the range of 1000 to 8000. The MUCH set up has 25000 64-channel chips to handle totally 1.5×10^6 channels. Also, GEM muon chambers will have projective segmentation with the smallest pad size being as small as 4.1×4.1 mm^2 in the inner region and as large as 2.1×2.1 cm^2 in the outer region of the chamber.

This sets up the following ASIC features:

- dynamic range of 1–100 fC;
- wide C_{DET} range of 1–50 pF;
- S/N maximization at the periphery pads;
- hit rate up to 2 MHz at the central pads;
- low (less than 10 mW/ch.) power consumption.

In accordance to the specifications two prototype ASICs were designed.

2. Preamplifier prototype

The preamplifier prototype for CBM MUCH (MUCH ASIC v.1) was designed, fabricated and tested [4]. The prototype includes 10 preamplifier channels. Figures 1 and 2 show the measurement results for the CSA transfer function. MUCH ASIC v.1 has generally met the MUCH front-end requirements and therefore it was taken as a basis for the new prototype chip design described in section 3.

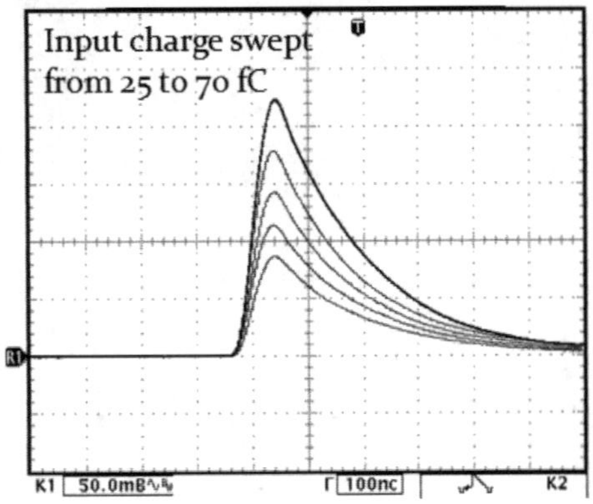

Figure 1. MUCH ASIC v.1 CSA output signal.

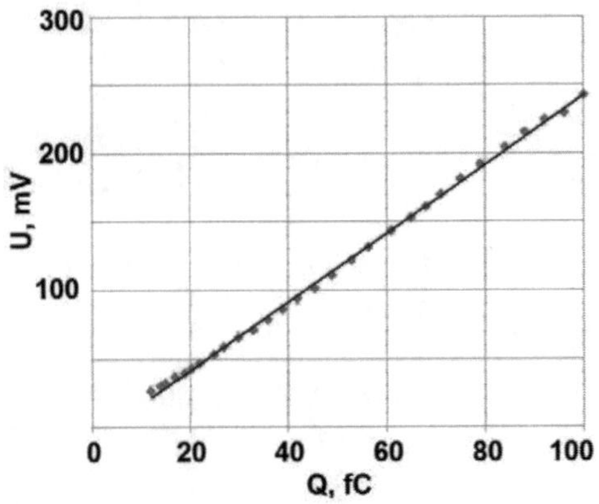

Figure 2. MUCH ASIC v.1 CSA transfer function. Errors lie inside data points. Solid line - linear fit.

3. 8-channels ASIC prototype

The 8-channel ASIC prototype for muon chambers of the CBM experiment (MUCH ASIC v.2) was designed and fabricated in the 0.18 μm CMOS process of UMC (Taiwan).

The chip structure includes 8 analog processing channels, 2 threshold DACs, 8 SAR ADCs and a digital part. Each analog channel contains the preamplifier, followed by two shapers (fast and slow) and differential comparator (see figure 3).

The preamplifier is based on the folded cascode CSA architecture with additional gain boosting. The gain is set to 4 mV/fC and its noise level is estimated by ENC = 1500 el. at 50 pF.

Output CSA response on the test input pulse via the 1.2 pF capacitor has the following parameters: rise time – 30 ns, fall time – 500 ns, signal duration – 600 ns (see figure 4).

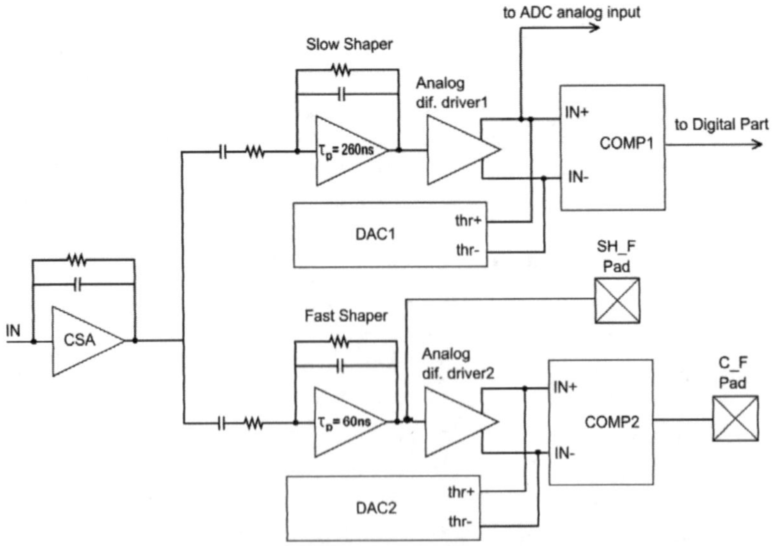

Figure 3. MUCH ASIC v.2 channel structure.

The slow channel is optimized for S/N ratio in order to use it in the periphery, and the fast channel is adapted to the hit rate of the inner detector part, where the occupancy is the highest. Both channels are realized with CR-RC shapers.

Slow and fast shaper output signals were obtained. Shaping times of the slow and fast shapers are equal to 250 ns and 50 ns correspondingly. Due to the double differentiation of the fast channel, the fast shaper output response has a bipolar form. Such a form allows to recognize 2 neighbour hits when the discriminator threshold is low (see figure 5). Channels linearity was studied in the signal dynamic range of 1 – 100 fC. Transfer functions of the CSA, fast and slow channels are shown in figure 6. The ENC of the fast and slow shaper outputs are 2000 el and 1500 el correspondently at 50 pF of the equivalent detector capacitance (see figure 7).

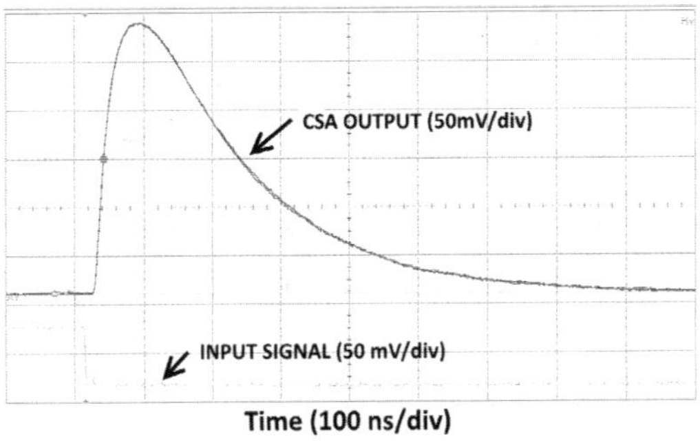

Figure 4. CSA output response.

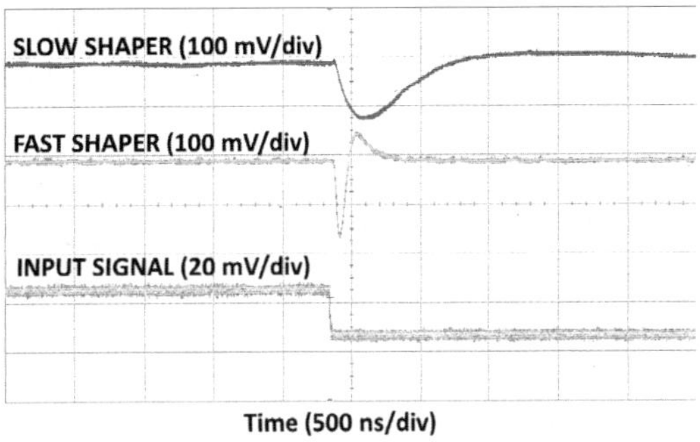

Figure 5. Shapers output response.

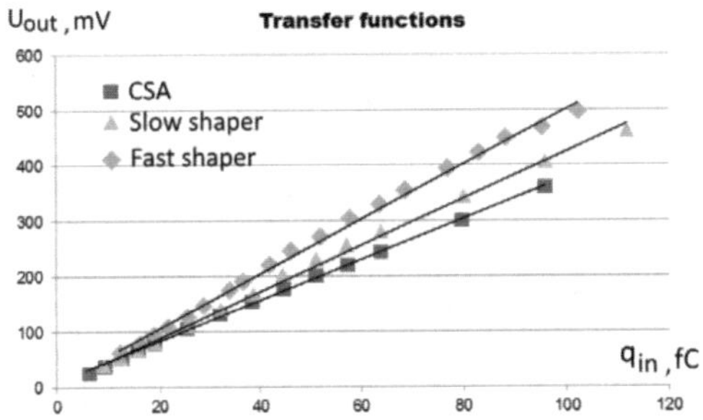

Figure 6. Dynamic range of preamplifier and shapers. Errors lie inside data points. Solid line – linear fit.

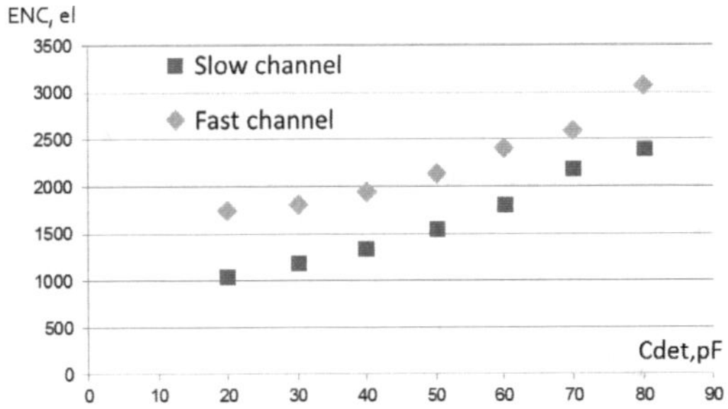

Figure 7. ENC vs C_{det}. Errors lie inside data points.

The schematic of the MUCH ASIC v.2 ADC [5] is shown in figure 8. The single-ended architecture was chosen to save design area. The schematic and layout of comparator were optimized to eliminate errors, arising from the kick-back effect in single-ended architecture. The comparator with the preamplifier stage before latch was used to isolate the capacitor matrix from output signal swing.

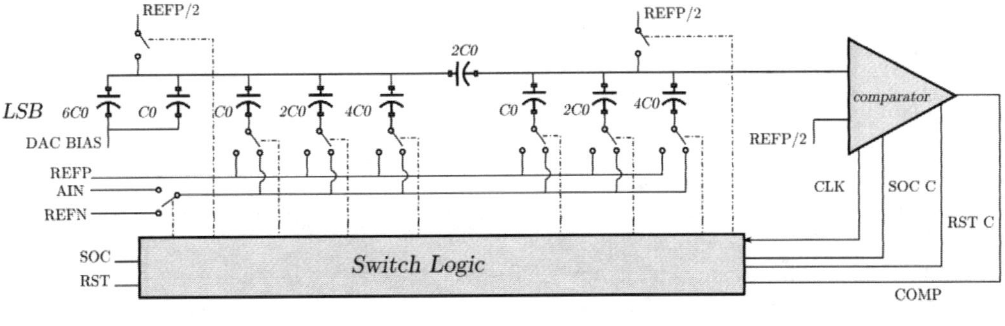

Figure 8. ADC structure.

Prototypes of the gas-filled detectors based on the GEM and TGEM technologies have been studied with the prototype front-end ASIC. A prototype read-out system is being developed jointly by PNPI and MEPhI. ^{55}Fe (5.9 keV) spectrum was obtained (see figure 9). The measured noise of 1000 electrons at a 20pF input capacitance increases up to 2500 electrons at the input capacitance of 80 pF.

Figure 10 shows the MUCH ASIC v.2 die photo.

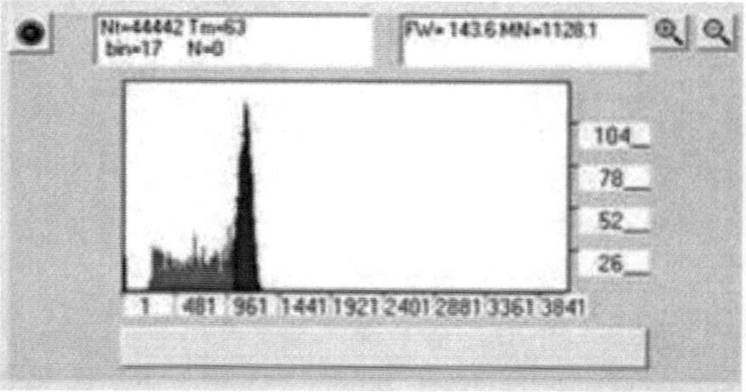

Figure 9. ^{55}Fe spectrum (5.9 keV).

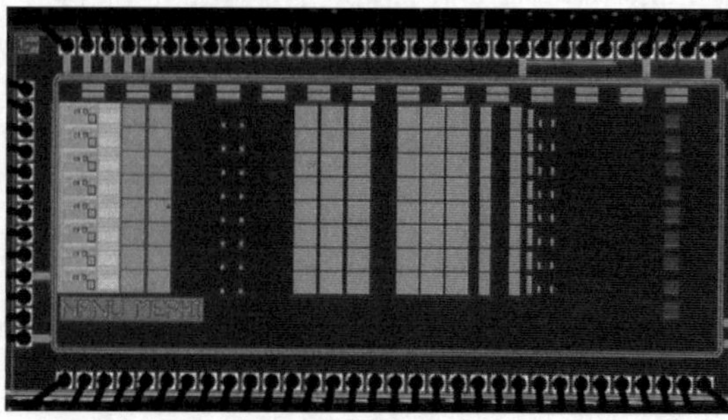

Figure 10. 8-channel prototype chip photo.

Comparison of the simulated and measured parameters is shown in table 1.

Table 1. Comparison of the simulated and measured parameters.

Parameter	Simulation	Measurements
Dynamic range (fC)	1 – 100	1 – 100
CSA rise time (ns)	20	30
Shaping time (fast) (ns)	60	50
Shaping time (slow) (ns)	260	250
Slow channel gain (mV/fC)	5	4.4
Fast channel gain (mV/fC)	5.5	5
Slow channel ENC @ 50 pf C_{det} (el)	800	1500
Fast channel ENC @ 50 pf C_{det} (el)	1200	2000
Power consumption (mW/channel)	10	13

4. Conclusion

The development of the mixed-signal front-end ASIC for muon chambers has been presented. The ASIC was made as a mixed-signal 8 channel chip for reading out the signals of GEM detectors. The next steps towards the 64-channel ASIC version are the development of time stamp with 10 ns resolution, built-in spark protection, digital peak detector, serial interface and slow control.

Acknowledgements

This work was partially supported by MEPhI Academic Excellence Project (contract № 02.a03.21.0005, 27.08.2013)

References

[1] Friman B et al. 2010 *The CBM Physics Book* (Berlin: Springer, New York: Heidelberg)
[2] Johann M 2013 *Nuclear Physics* A **904–905** 941c
[3] Sauli F 1997 GEM: A new concept for electron amplification in gas detectors *Nucl. Instrum. Meth.* A **386** 531
[4] Atkin E, Malankin E and Shumikhin V 2014 A preamplifier for the muon system of the CBM experiment *Instrum. Exp. Tech.* **57** 286
[5] Atkin E *et al.* 2015 Development of the read-out ASIC for muon chambers of the CBM experiment *JINST* **10** C04006

Journal of Physics: Conference Series **675** (2016) 042031 doi:10.1088/1742-6596/675/4/042031

The read-out ASIC for silicon drift detectors

E Atkin[1], P Ivanov[1], A Krivchenko[2], V Levin[2], A Gusev[1], E Malankin[1], D Normanov[1], A Rotin[2], I Sagdiev[1] and V Shumikhin[1]

[1] National Research Nuclear University MEPhI (Moscow Engineering Physics Institute), 115409, Kashirskoe highway 31, Moscow, Russia

[2] Space Research Institute of the Russian Academy of Sciences, 117997, Profsoyuznayas str 84/32, Moscow, Russia

E-mail: vvshumikhin@mephi.ru

Abstract. The paper describes the read-out ASIC for silicon X-ray drift detectors. The ASIC has been designed in CMOS 0.35 μm technology and contains two read-out channels. Each channel includes a preamplifier and shaper. The preamplifier in the first channel has a built-in input transistor, the preamplifier in second channel works with an external JFET, which is built in the detector structure. Preamplifiers have been optimized for operation with detectors with capacitances of 100 fF. The 6-th order shaper has controllable time constants (0.5 – 8 μs).

1. Introduction

The ASIC for operation with the silicon drift detectors (SDD) [1, 2] of PNDetectors company was designed. The detector has a high energy resolution – 130 eV (at -30 °C), leakage current – 10 pA and capacitance – 100 fF. The detector has an optional built-in JFET, which was used at ASIC design. The ASIC has two channels: channel with a built-in input transistor [3] and channel for use with an external JFET built into the detector [4].

The ASIC was designed and prototyped in the 0.35 μm CMOS process of AMS (Austria).

2. The ASIC

2.1. Channel with a built-in input transistor

Channel with a built-in input transistor consists of a CSA, shaper and reset trigger. The CSA has been designed as a cascode circuit with an output source follower. The input preamp stage has been realized on a n-type MOSFET with width – 210 μm, length – 0.35 μm. The input transistor geometry was optimized to achieve minimum noise level at detector capacitance of 100 fF. Figure 1 shows the CSA schematic. The feedback capacitor equals 10 fF and sets a gain of 100 mV/fC. A feedback transistor operates in a switching mode and provides a discharge of the capacitor when the positive pulse is applied to the gate. For generating the reset pulse there was used a Schmidt trigger.

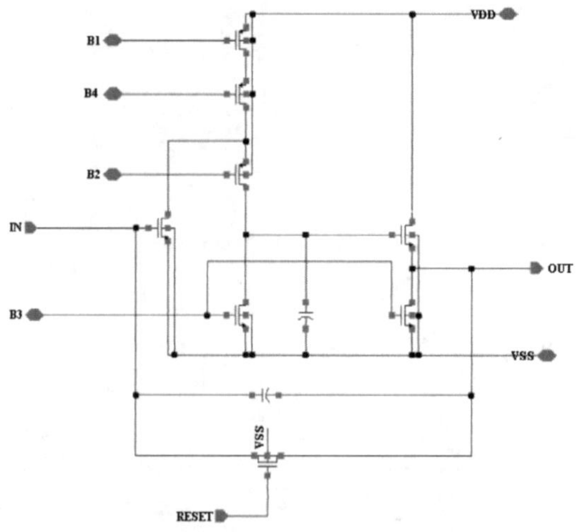

Figure 1. CSA with built-in input transistor.

Figure 2 shows the response of the CSA without leakage current. The rise time of the CSA output signal equals 20 ns. The CSA output signal at the existing leakage current of 10 pA is shown in figure 3. The maximum signal swing at the CSA output is about 1.5 V.

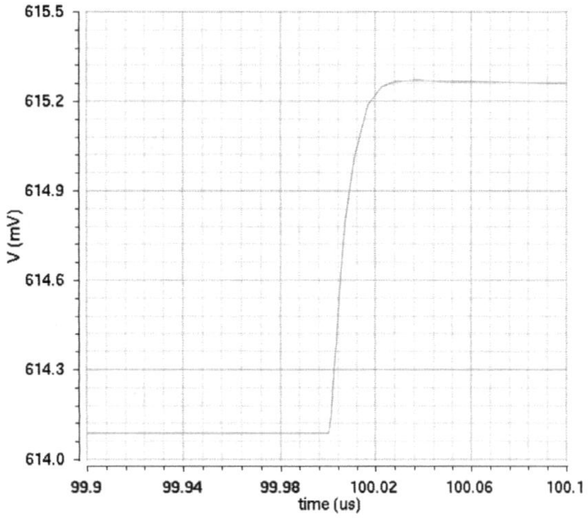

Figure 2. CSA response.

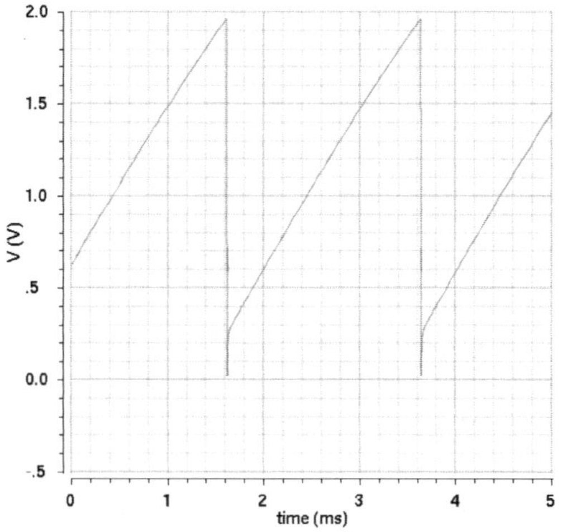

Figure 3. CSA output.

The CSA output is connected to the 6th order shaper. The shaper has a discretely adjustable time constant in the range of 0.5 – 8 μs. Figure 4 shows the structure of the shaper circuit.

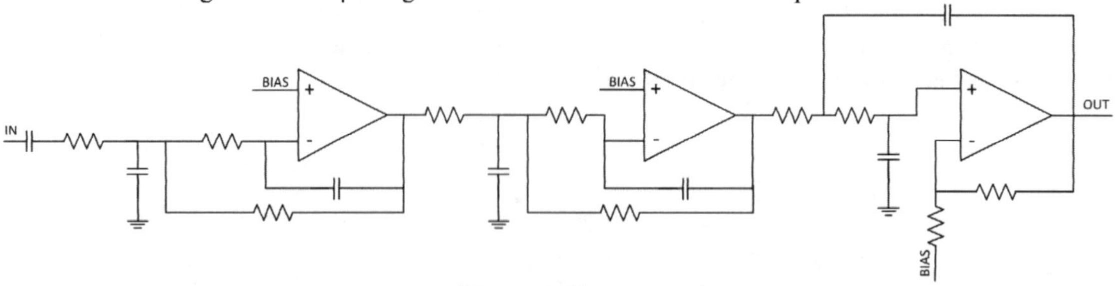

Figure 4. Shaper.

Channel main parameters, obtained from simulations, are summarized below:

- CSA gain: 100 mV/fC;
- CSA rise time: 20 ns;
- ENC: 4 el. at T=-30°C, shaper time constant 8 μs;
- detector capacitance: 100 fF;
- leakage current: up to 10 pA;
- 6th order shaper with adjustable time constants;
- power consumption: 1.3 mW.

2.2. Channel with external transistor

The channel with an external input transistor contains the same blocks as the channel with a built-in input one, namely: preamplifier, shaper and reset trigger.

Figure 5 shows CSA schematic. The CSA was optimized to operate with silicon drift detectors, having integrated JFET. The CSA has pulse reset. The input stage is built as a source follower one. The CSA gain equals 33 mV/fC. It is determined by the feedback capacitance between the inner guard ring and the anode, located on a die of the detector. The feedback capacitance equals 30 fF. In

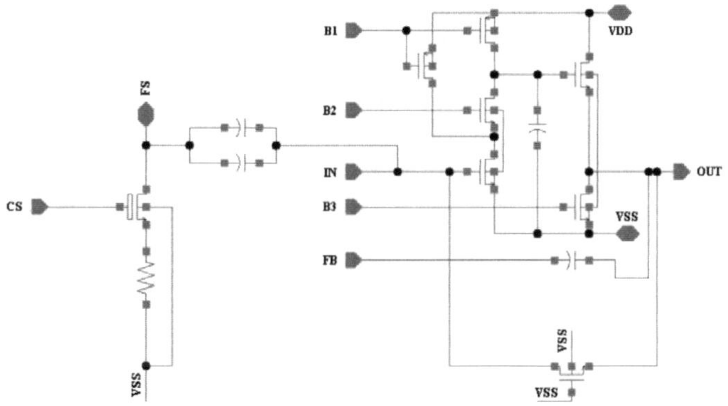

Figure 5. CSA with external input transistor.

the absence of registration of X-ray photons the CSA output voltage increases linearly at a rate determined by the detector leakage current (10 pA) and a value of the feedback capacitance (30 fF). The dynamic range at the CSA output is up to 1.5 V. The rise time of CSA output signal is no more than 100 ns at a load capacitance of 10 pF. Figures 6 and 7 show a typical response of the CSA.

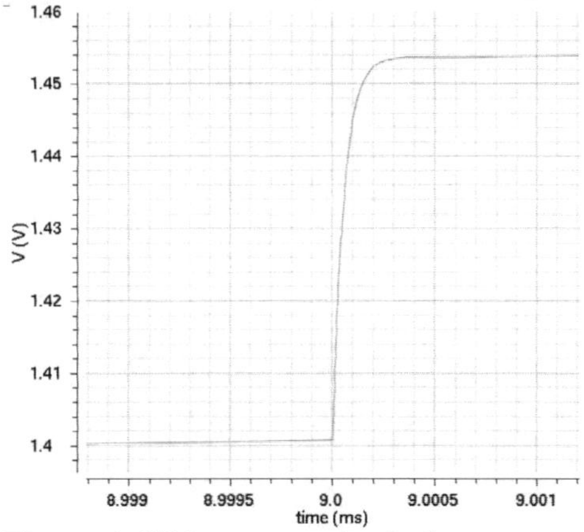

Figure 6. CSA response (no leakage current).

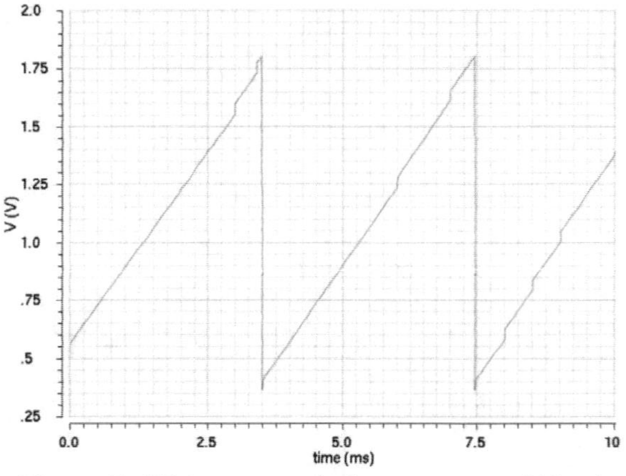

Figure 7. CSA output at leakage current of 10 pA.

The main simulation results of the channel are:

- gain – 33 mV/fC;
- ENC – 12 e;
- power consumption – 0.25 mW (without JFET power consumption);
- dynamic range – 1.3 fC;
- rise time – 100 ns.

3. Layout

The channel was designed and prototyped in the AMS 350 nm CMOS process. The die size is of 3755 x 1330 μm^2. Chip includes 50 pads with ESD protection. Figure 8 shows layout of the ASIC.

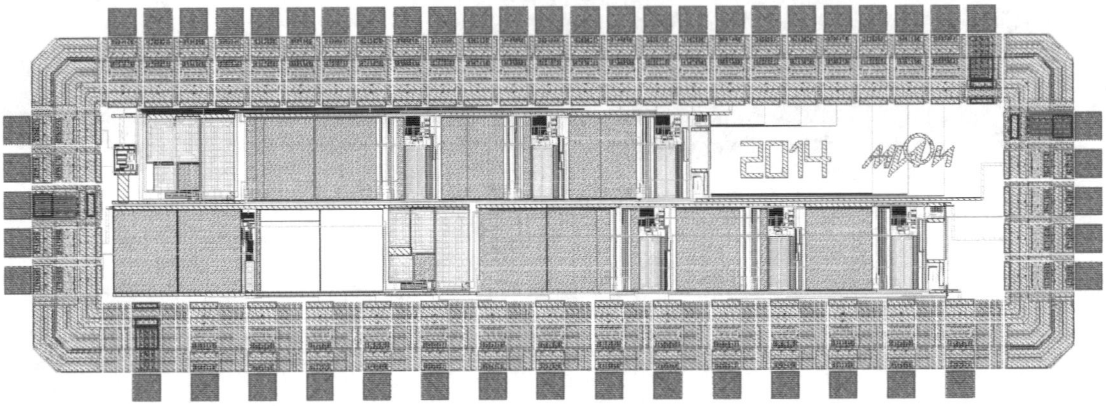

Figure 8. ASIC layout.

4. Conclusion

The read-out ASIC for silicon drift detectors was designed and prototyped. The ASIC has two channels: one with a built-in input n-MOS transistor and another with external JFET. The channel contains a CSA, 6th order shaper and reset trigger. Both preamplifiers were optimized for operation with the PNDetector SDDs. The unpackaged prototype chip was bonded to the test PCB with PNDetector SDD and it is under test now.

Acknowledgements

This work was partially supported by MEPhI Academic Excellence Project (contract № 02.a03.21.0005, 27.08.2013).

References

[1] URL: *http://pndetector.de/*

[2] Gatti E and Rehak P 1984 Semiconductor Drift Chamber - An application of a novel charge transport scheme *Nucl. Instrum. Meth.* A **225** 608-14

[3] Bombelli L *et al.* 2011 "CUBE", a low-noise CMOS preamplifier as alternative to JFET front-end for high-count rate spectroscopy *IEEE Nuclear Science Symposium and Medical Imaging Conf. (NSS/MIC)* pp 1972-5

[4] Tommaso Frizzi *et al.* 2006 The SIDDHARTA chip: a CMOS multi-channel circuit for Silicon Drift Detectors readout in Exotic Atoms Research *IEEE Nuclear Science Symposium Conf. Record* **2** 850-6

Two-phase low-power analogue CMOS peak detector with high dynamic range

E Malankin
National Research Nuclear University MEPhI (Moscow Engineering Physics Institute), Kashirskoe highway 31, Moscow, 115409, Russia

ezmalankin@mephi.ru

Abstract. A low-power two-phase peak detector with wide dynamic range was developed. The PD was designed on the basis ofthe CMOS UMC 180 nm process. This block is considered as a part of the read-out electronics of the CBM experiment at upcoming FAIR accelerator (Germany). Peak detector has the following advantages: wide dynamic range of 5 – 1000 mV, low power consumption of 500 μW. The designed PD meets the requirements to the muon chamber read-out electronics of the CBM experiment. Due to the area efficiency (100x90 μm^2) and low power consumption it can be used in different applications for high-energy physics read-out electronics.

1. Introduction

The conventional analog peak detector (PD) architecture [1] is built basically on the operational amplifier (figure 1). The OpAmp charges the hold capacitance, when the input voltage is higher than the output voltage. However, when the input voltage becomes lower, the hold capacitance is not discharged, because the feedback diode is reverse-biased. In the CMOS peak detectors current source on a MOS transistor is used as a rectifying element instead of the diode in the high gain OpAmp feedback loop a [2].

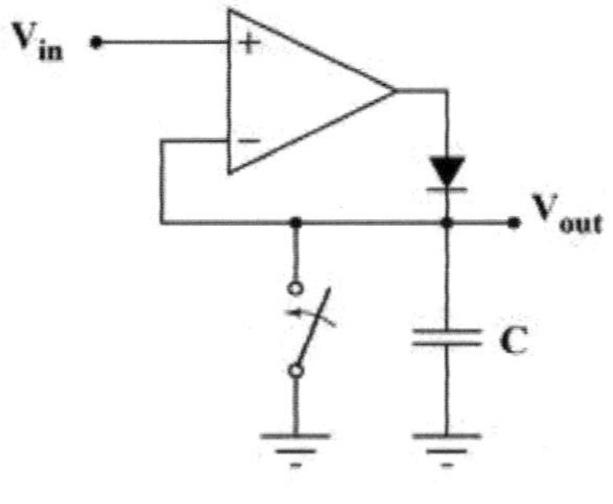

Figure 1. Conventional peak detector.

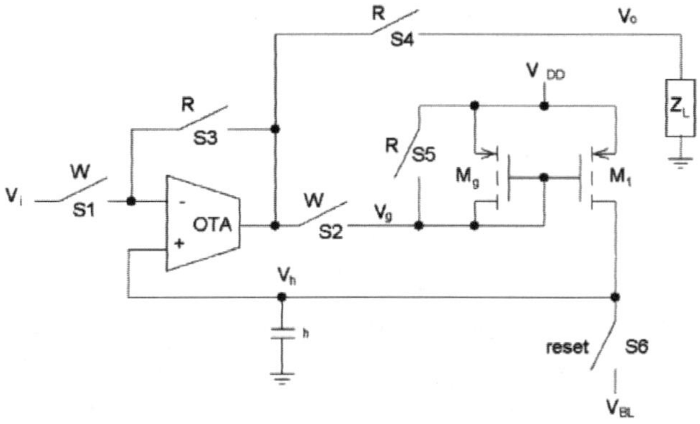

Figure 2. Two-phase peak detector.

In the multi-channel detector read-out systems the two-phase configuration (figure 2) is usually applicable. In this configuration the drawbacks of the classical PD structure are eliminated [3]. In particular, the two-phase configuration has the following advantages:

- absence of the input offset voltage (i.e. the accuracy is not limited by the input offset voltage of the OpAmp);
- common-mode errors have no influence on the accuracy;
- the influence of the input transistors process mismatch is decreased.

An additional requirement set to the peak detector in the muon read-out system in the CBM experiment is the low power consumption (no more than 1 mW), area size of $100x150\mu m^2$, dynamic range of 5 – 500 mV with negative polarity and regular occupancy no less than 0.3 MHz[4].

2. Peak detector schematic and layout

2.1. Peak Detector schematic and simulation

The design was carried out on the basis of the UMC technological libraries with 180 nm design rules. Cadence company software packages were used for schematic development and simulations.

The main block of the two-phase PD is the OTA (figure 3). That amplifier corresponds to the requirements on power consumption and dynamic range. The OTA is optimized for the operation with negative polarity signals with amplitudes in the range of 5 – 500 mV. This range corresponds to the output signals of the shaper in the channel of the muon chambers readout.

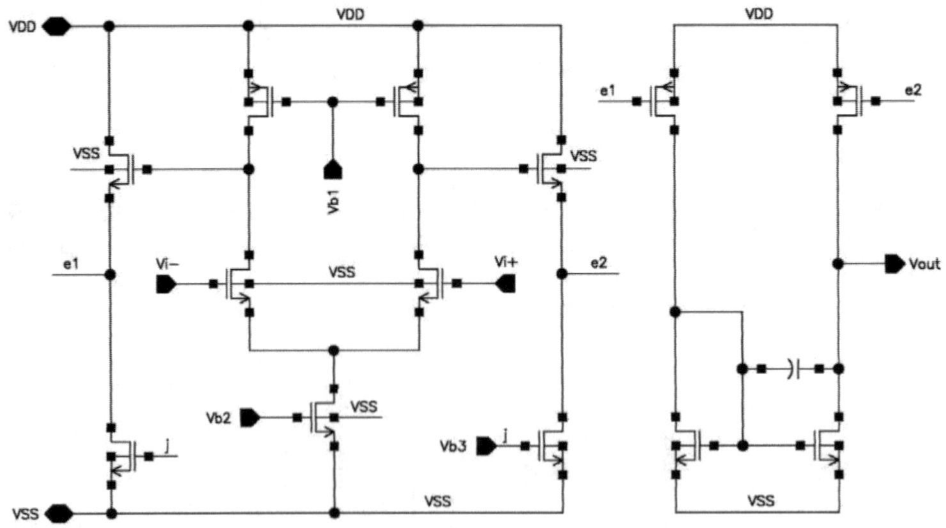

Figure 3. Transconductance operational amplifier.

In figure 4 the schematic of the designed two-phase peak detector is given. This schematic consists of the OTA, rectifying current mirror, hold capacitor, reset scheme, «Peak Find» signal scheme, and analog gates to switch between the "read" and "write" phases.

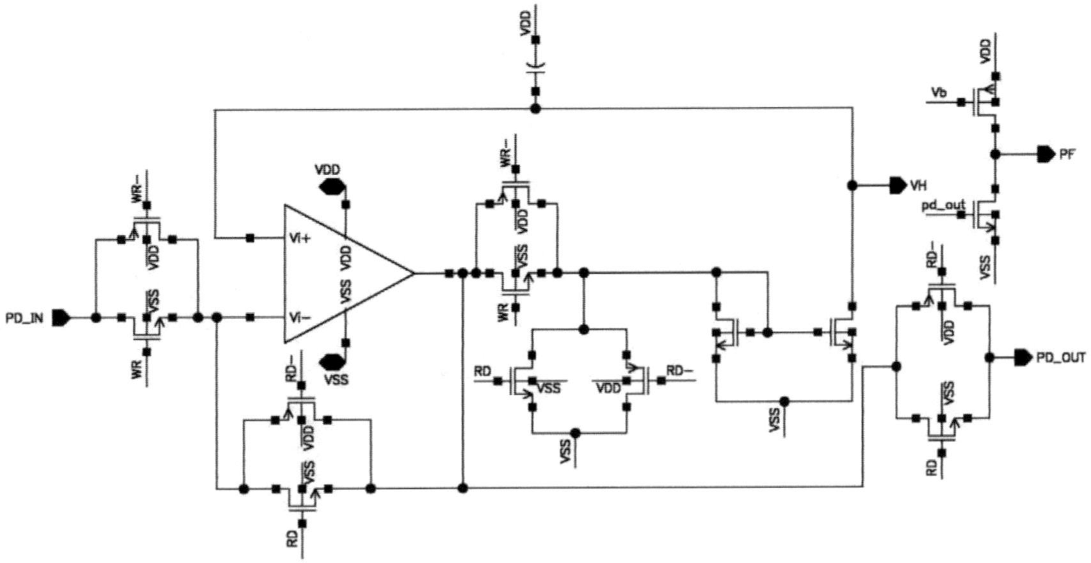

Figure 4. Two-phase peak detector schematic.

During the simulation the main parameters of the two-phase peak-detector were obtained. In figure 5 the functional diagram of the designed peak detector is shown.

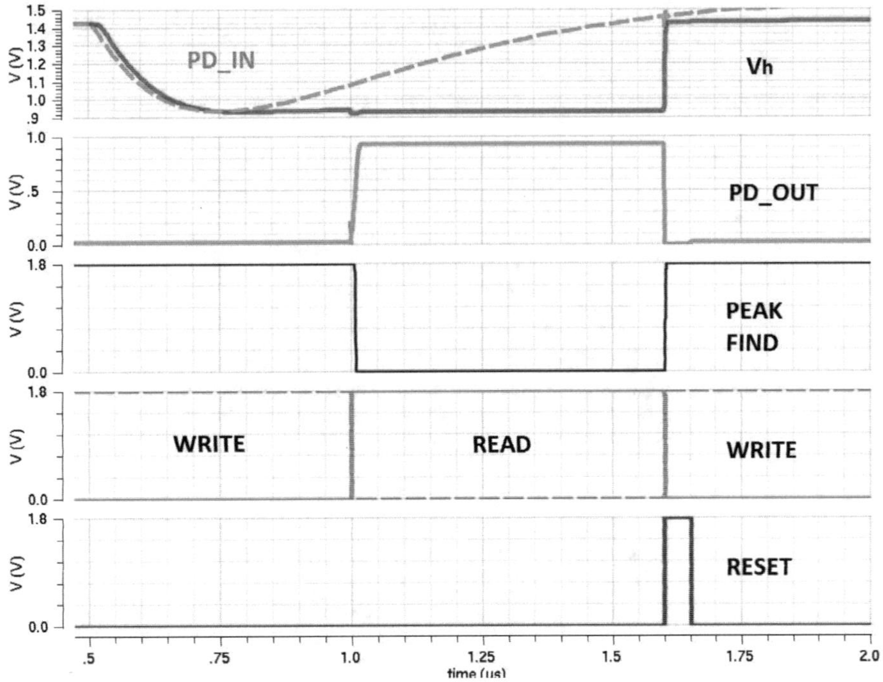

Figure 5. PD functional diagram.

The transfer function is linear all over the input signals range. The whole dynamic range of the peak detector is 5 – 1000 mV (figure 6). As to the Monte-Carlo analysis the distribution of the PD output peak value due to the process variations is no more than 0. 3% at 100mV input signal (figure 7).

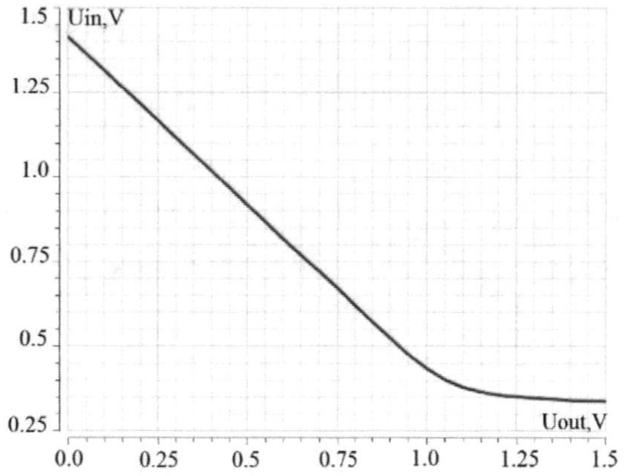

Figure 6. PD transfer function.

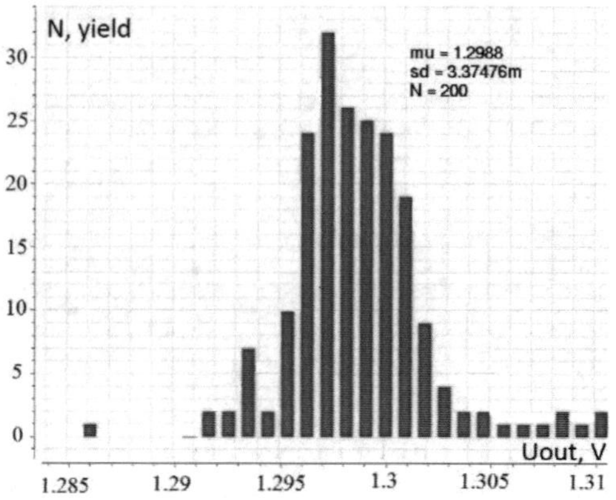

Figure 7. Distribution of the output peak value.

2.2. Peak detector layout

PD layout (figure 8) was designed according to the UMC 180 nm process rules. The area of the Peak Detector is 100x90 μm^2. Here the 100 μm – is the height of the peak detector, which is in correspondence to the analog channel height standard in the read-out electronics for the muon system of the CBM experiment.

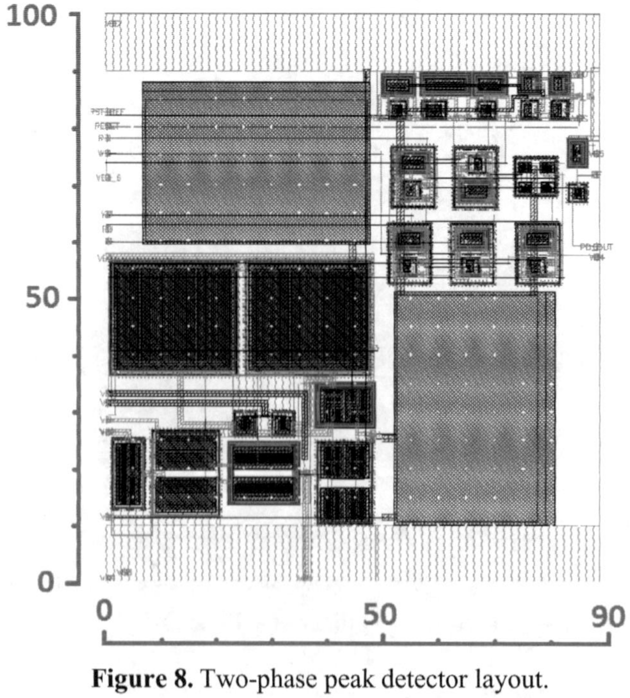

Figure 8. Two-phase peak detector layout.

As a result of the development there was designed the low-power peak detector which could be used as a part of the readout electronics for the muon chambers. The obtained parameters are listed in table 1.

Table 1. Parameters of the peak detector.

Parameter	Specification	Simulation
Input signal dynamic range, mV	5–500	5–1000
Die size, μm^2	100x150	100x90
Regular occupancy, MHz	0.3	0.4
Power consumption, mW	1	0.55

3. Conclusion

The low-power analog peak detector with a wide dynamic range for readout electronics for the muon chambers of the CBM experiment was designed. It was developed on the basis of the UMC 180 nm MMRF CMOS process. The peak detector meets the requirements to the readout electronics for the muon chambers of the CBM experiment.

Acknowledgements

The work was carried out with the financial support ofSAEC “Rosatom” and Helmholtz Association and the Ministry of Education and Science of the Russian Federation (Grant No. 14.A12.31.0002) in accordance with the RF government order No. 220. This work was partially supported by MEPhI Academic Excellence Project (contract № 02.a03.21.0005, 27.08.2013).

References

[1] Kruiskamp M W and Leenaerts D M W1994 *IEEE Trans. Nucl. Sci.* **NS-41**(1) 295
[2] Geronimo G De *et al.* 2002 *Nucl. Instrum. Meth.* A **484** 533–43
[3] Geronimo G De *et al.* 2002 *Nucl. Instrum. Meth.* A **484** 544–56
[4] Atkin E *et al.* 2015 Development of the read-out ASIC for muon chambers of the CBM experiment *JINST* **10** C04006

Trends in the design of microcircuits for front-end electronics (overview)

Yu A Volkov and I I Ilyushchenko
National Research Nuclear University MEPhI (Moscow Engineering Physics Institute), Kashirskoe highway 31, Moscow, 115409, Russia

E-mail: YAVolkov@mephi.ru, IIIlyushchenko@mephi.ru

Abstract. The results of literature analysis devoted to the design of multichannel microcircuits for data acquisition and signal processing from radiation detectors in large physical experiments have been presented. Such parameters of microcircuits as the number of signal processing channels, power consumption per channel, noise level, input charge conversion factor (for the linear part of channel) and so on are considered. We also touched upon the technology of manufacture and the name of physical experiment in the framework of which the design was carried out.

The aim of the paper is to reveal the existing trends in the implementation of microelectronic units, collecting and processing the signals from radiation detectors in large-scale physical experiments. The comparison of present-day microelectronic products is carried out by their most important parameters, particularly: channel number (*Nch*), power consumption per channel (*Pcon*), input noise charge (*Qn*) and conversion factor (*Kcon*). The technology of manufacture is also taken into account.

In our report the parameters of several tens of ICs, presented in scientific papers and conference materials of the last several years were adduced. The summarizing table, given below, contains the data on 26 selected ICs, each having at least 16 channels.

Table. Characteristics of Front-end Electronics ICs.

Reference	Name of ASIC. Purpose and/or name of Project.	N_{ch}	Technology	Q_n	K_{con}	P_{con}
[1]	ASIC for PMT Hamamatsu H8500.	16	BiCMOS, 0.35μm	4fC	-	11 mW
[2]	ASIC for the LAr TOF chamber. MicroBooNE	16	CMOS, 180nm	1200e at 293K, 550e at 77K,	4.7…25 mV/fC	6 mW
[3]	FSDR16. CBM STS.	16	UMC	214e + 13.3e/pF, tp=99 ns	91 mV/fC	10 mW
[4]	SAS. STAR	16	CMOS, 1.2 μm	600e +13e/pF, tp=180 ns	16 mV/fC	60 mW
[4]	ADAM. LHC	32	CMOS	800e +80e/pF, tp= 8 ns	10 mV/fC	-
[5]	APV3. CMS	32	CMOS, 1.2μm HARRIS,	480e +43e/pF	–	2 mW

Journal of Physics: Conference Series **675** (2016) 042033 doi:10.1088/1742-6596/675/4/042033

Source	Name of ASIC. Purpose and/or name of Project.	N_{ch}	Technology	Q_n	K_{con}	P_{con}
[6]	SCT32A(B); silicon strip detector	32	BiCMOS, radiation-hard DMILL	620e+33e/pF	100 mV/fC	–
[7]	ICECAL. calorimeter LHCb, up-dating	32	SiGe BiCMOS, 0.35μm, AMS	1nV/√Hz	-	-
[8]	SPADIC 1.0 for TRD. RICH	32	CMOS, 180 nm UMC,	387e + 11e/pF	-	3.8 mW
[1]	SAMPA. Time-of-flight chamber	32	CMOS, 130 nm TSMC Esup. 1.25V	536e, tp=80 ns	20 mV/fC	8 mW
[1]	ASIC SAMPA**.** Muon tracking chamber	32	CMOS, 130 nm TSMCE=1.25V	950e, Cd=40pF 1600e, Cd=80pF	4 mV/fC	8 mW
[9]	FABRIC. NA50 silicon detector	64	Bipolar Tektronix	476e +63e/pF, tp= 15 ns	100 mV/fC	1.3 mW
[10]	MICROROC. Digital hadron calorimeter	64	SiGe BiCMOS, 0.35μm, AMS	0.24fC, Cd = 80pF	2,38 mV/fC	3.7 mW
[10]	SPACIROC. EUSO-Balloon experiment	64	SiGe BiCMOS, 0.35μm	100% registrat. of charges >50 fC	Kgain =10^6	1 mW
[11]	VMM1. ATLAS small "wheel"	64	CMOS, 130 nm, radiation hard IBM	5000 e, Cd=200pF	9 mV/fC	-
[12]	APVD. CMS silicon tracker	128	CMOS, DMILL	450e+49e/pF, tp=50 ns	108mV/MIP	1 mW
[10]	ALICE128C. ALICE silicon detector	128	CMOS, 1.2 μm, AMS	400e, tp = (1.4…1.8) us	50 mV/MIP	0.34 mW
[10]	SCT128B. HERA-B silicon detector	128	BiCMOS, radiation hard DMILL	559e+35e/pF	107 mV/fC	–
[4]	APV25. CMS tracking system	128	CMOS, 0.25 μm, Harris DMILL	430e+61e/pF, tp= 25ns	–	2 mW
[13]	HELIX128. HERA-B silicon tracker	128	CMOS, 1.2 μm	405e+76e/pF, tp=50нс	63 mV/MIP	2 mW
[14]	HELIX128S-2.1. HERA-B	128	CMOS, 0.8 μm, AMS	33.6 e/pF, tp=(45…100) ns	45 mV/MIP	–
[15]	CAFÉ. LHC silicon tracker	128	Complementary bipolar AT&T	550e +33e/pF, tp=25 ns	–	1.8 mW
[16]	ASIC for the CBM tracking system	128	CMOS, 180 nm, UMC (Taiwan)	2000e, Cd<30pF	50 mV/fC	2 mW
[17]	Binary architecture CMS. SLHC	128	CMOS, 0.13 μm	500e+64 e/pF, tp = 20 ns	_	50 mWE=1,2V
[4]	CBC2. CMS microstrip tracker	128	CMOS, 130 nm, IBM	800e	40 mV/fC	0,3 mW
[18]	Pasta – Front-end electronics ASIC. PANDA	128	CMOS, 130 nm	600e	-	1 mW

This table uses the following denotations and abbreviations: *Cd* – detector output capacitance, e – electron, ADC – analog-to-digital converter, PMT – photoelectron multiplying tube. The abbreviations of well-known physical experiments (ATLAS, CBM, HERA-B and other) and widely used IC manufacturing processes

(CMOS, BiCMOS, SiGeCMOS and other) are also employed. The latter are usually added by design rules, for instance 180nm, 0.35 μm.

We acknowledge, that the IC comparison by some of the parameters is rather conventional. First of all that refers to P_{con} – power consumption per channel, since those channels may differ significantly in their structure and complexity: in some cases a channel consists of only a sequence of analog units, while in other ones an ADC and digital processing units are added.

After analyzing the data of the above presented table one can make the following inferences:

- Typical values of the basic parameters of present-day FEE ICs are as follows: *Pcon* – units of mW per channel, *Qn* – hundreds of electrons, *Kcon* – tens of mV/fC.
- The FEE ICs are most often found with the following numbers of channels: 16, 32, 64 and 128. Present-day technologies allow the manufacturer to create ICs with a greater number of channels, but, from the practical point of view (considering particularly the placement of ICs on a PCB whereon the IC inputs should be connected to the detector outputs), those greater numbers may not appear expedient.
- Among the possible technologies the FEE ASIC designers prefer the one of CMOS, which allows them to solve the problems of minimizing power consumption, cost and chip area in a much simpler way.
- If quite recently the FEE equipment contained only analog electronics, nowadays a distinct trend to include in it some digital processing units is well observed.
- Radiation hardness, that till recently was often demanded for space-intended equipment, is nowadays more and more requested for stationary equipment, especially when the latter is placed near the "beam zone" of accelerator.
- The main way of improving the electrical and mass-dimensional characteristics of FEE units is their implementation in the form of multichannel ASICs.

Acknowledgments
This work was supported by the Ministry of Education and Science of the Russian Federation (grant No.14.A12.31.0002) and by the Center of Fundamental Research and Particle Physics supported by MEPhI Academic Excellence Project (contract No. 02.a03.21.0005, 27.08.2013).

References

[1] Deng S *et al.* 2013 Very fast front end ASIC associated with multi-anode PMTs for a scintillating-fibre beam hodoscope *JINST* **8** C01047 (doi:10.1088/1748-0221/8/01/C01047)
[2] Chen H *et al.* 2012 Readout electronics for the MicroBooNE LAr TPC, with CMOS front end at 89K *JINST* **7** C12004 (doi:10.1088/1748-0221/7/12/C12004)
[3] URL: *https://iri-wiki.uni-frankfurt.de/cms/?q=node/42*
[4] URL: *http://atlas.web.cern.ch/Atlas/internal/tdr.html*
[5] Adloff C *et al.* 2012 MICROROC: MICRO-mesh gaseous structure Read-Out Chip *JINST* **7** C01029 (doi:10.1088/1748 0221/7/01/C01029)
[6] Ahmad S *et al.* 2014 The readout electronic of EUSO-Balloon experiment *JINST* **9**(3) C03050
[7] Metcalfe J *et al.* Design and Performance of the VMM1 ASIC for Micropattern Gas Detectors *TWEPP-2013 23-27 September 2013, Perugia, Italy*
[8] Armbruster T *et al.* 2012 Multi-channel charge pulse amplification, digitization and processing ASIC for detector applications *Nuclear Science Symposium and Medical Imaging Conference (NSS/MIC), IEEE (Oct. 27 2012-Nov. 3 2012, Anaheim, CA)* pp 697–702
[9] Kluge Alexander and Riegler Werner 2013 Upgrade of the ALICE Read-out & Trigger System *CERN-LHCC-2013-019 / LHCC-TDR-015* pp 47–50
[10] The HERA-B Experiment (*http://www-hera-b.desy.de/general/info/*)
[11] The CMS Experiment (*http://cms.web.cern.ch/*)

[12] French M J *et al.* 2001 Design and results from the APV25, a deep sub-micron CMOS front-end chip for the CMS tracker *Nucl. Instrum. Meth.* A **466** 359–65

[13] Raymond M D *et al.* 2012 The CMS binary chip for microstrip tracker readout at the SLHC *JINST* **7** C01033 (DOI:10.1088/1748-0221/7/01/C01033)

[14] Braga D *et al.* 2014 Characterization of the CBC2 readout ASIC for the CMS strip-tracker high-luminosity upgrade *JINST* **9** C03001 (DOI:10.1088/1748-0221/9/03/C03001)

[15] Atkin E *et al.* 2013 A multichannel ASIC for the tracking system of the CBM experiment *Russian Microelectronics* **42**(2) 119-26

[16] 2009 *CMS. The Compact Muon Solenoid Technical Proposal* p 483 (*http://cms.web.cern.ch/cms/index.html*)

[17] Picatoste E *et al.* 2012 Low noise front end ICECAL ASIC for the upgrade of the LHCb calorimeter *JINST* **7** C01080 (doi:10.1088/1748-0221/7/01/C01080)

[18] Goerres A 2014 PASTA status update *PANDA Meeting, MVD Session (10 June 2014, ASR, GSI Darmstadt, Germany (http://drlog.andregoerres.de/talk-pasta-status-update)*

Multichannel readout ASIC design flow for high energy physics and cosmic rays experiments

A Voronin[1,2] and E Malankin[1]
[1] National Research Nuclear University MEPhI (Moscow Engineering Physics Institute), Kashirskoe highway 31, Moscow, 115409, Russia
[2] Skobeltsyn Institute of Nuclear Physics Lomonosov Moscow State University, 119991, Leninskie gory 1/2, Moscow, Russia

E-mail: voronin@silab.sinp.msu.ru, ezmalankin@mephi.ru

Abstract. In the large-scale high energy physics and astrophysics experiments multi-channel readout application specific integrated circuits (ASICs) are widely used. The ASICs for such experiments are complicated systems, which usually include both analog and digital building blocks. The complexity and large number of channels in such ASICs require the proper methodological approach to their design. The paper represents the mixed-signal design flow of the ASICs for high energy physics and cosmic rays experiments. This flow was successfully embedded to the development of the read-out ASIC prototype for the muon chambers of the CBM experiment. The approach was approved in UMC CMOS MMRF 180 nm process. The design flow enable to analyse the mixed-signal system operation on the different levels: functional, behavioural, schematic and post layout including parasitic elements. The proposed design flow allows reducing the simulation period and eliminating the functionality mismatches on the very early stage of the design.

1. ASIC design approach for HEP and Space experiments

General approach for an ASIC mixed-signal design one is able to find in manuscripts [1, 2]. The design of the ASICs for high-energy physics and cosmic ray experiments has a lot of features which require a special approach to the development process. As a peculiarity of the HEP ASIC design it could be concerned the small-batch production and low cost of the design and production. These can't be combined with the industrial scale microelectronics. Moreover, each experiment uses mostly its individual ASIC and sets the requirements to the ASIC parameters (i.e. low noise, dynamic range, high radiation tolerance, occupancy, power consumption, reliability, etc.).

Currently, the ASICs for HEP and astrophysics are multi-channel systems, which include up to 1024 channels. This architecture limits the channel height (25 – 100 μm) and makes preferable the simple schematic solutions with the precisely adjusted parameters. Beside the analog channels the up-to-date readout system includes external control of the ASIC parameters such as regimes, types of processing, thresholds, etc. The internal IC control, electrical calibration, monitoring of the condition, displaying the information on failures, data transfer errors can be embedded in the chip as well. The ASICs for the cosmic ray experiments have requirements which are equal to those in the space equipment.

The last trend in the ASIC for HEP and space research is the implementation of digital signal processing blocks turning the on-chip readout into the SoC. At the same time, analog processing mainly defines the quality of the ASIC full signal processing. The sophisticated architectures of this kind of ASICs require the specific complex approach to their design. The method should be built on the basis of the relevant CAD and EDA tools, by which it is possible to obtain the precise simulation results. It is significant to note, that many ASICs designed for new experiment have an innovation feature [3].

2. ASIC design flow features

Technical specifications arise from the physics simulations (for instance by GEANT) of the expected reactions of particles (figure 1). These set up the requirements to geometry, segmentation, technical realization, power, occupancy, cost and others. Then, using the detector model, it is determined the specifications of electronics (number of channels, power consumption, rate, dynamic range, etc.).

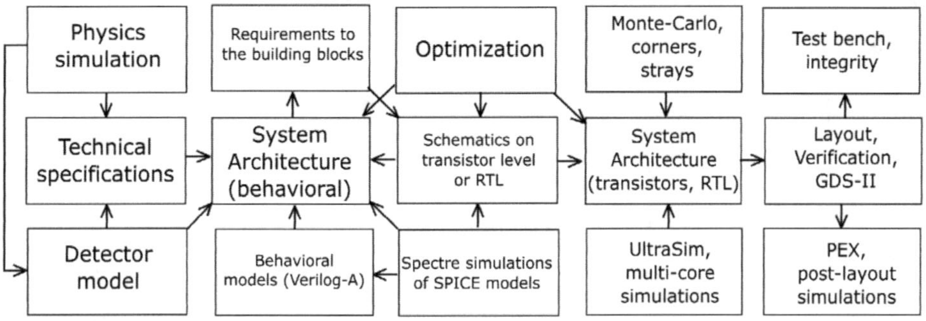

Figure 1. Design flow of mixed-signal ASIC for HEP and cosmic rays experiments.

On the basis of those input data it is developed one or several architecture variants. All the architecture ideas should be simulated on the virtual elements (for example, the behavioral models on Verilog-A) using Cadence tool set. Connection of the detector model and physical simulation of the electronics occupancy gives the correction of the requirements either the compromises to the accuracy, expenses and timelines.

On the platform of the first virtual models the specifications of the building blocks are established in general. The provided studies show that the most adequate approach to the real device are given by the simulations in SPICE environment with SPECTRE simulator. That framework allows carrying out the mixed-signal simulation in the single environment, at the same time giving the number of parameters, taking into account the schematic non-idealities and determining the critical parameters of the schematics. SPECTRE permits to connect the impact generators such as noise, crosstalk at the different points, etc. Monte-Carlo SPICE simulation of the basic elements distribution determines the electronics quality both analog and digital.

The ASIC building blocks designed on the transistor level or as RTL replace the correspondent behavioral part in the system model. When the behavioral block model is changed to the real one, the comparison of the parameters is carried out, as well as the simulation of mismatches in load capacity, signal polarity, time diagrams and etc.

In the process of tuning out the behavioral model into the real (transistor or RTL) one, the simulation period could be significantly increased. In this case the optimization of simulation rate is necessary. There are several approaches which make possible to optimize the speed and accuracy of simulation:

- Utilization of the high-speed simulation of SPICE tools, such as APS, XPS and UltraSim. Those methods use the multi-core algorithms that give the capability to set in motion more processor cores of the calculating server.
- While the digital part volume increases the mixed-signal simulator (AMS) or simplified schematics solution on the high level description languages (Verilog, VHDL, Verilog-A, Verilog-AMS) are used.

Finally, the schematic on the transistor level (or RTL) of the whole chip is assembled. The operation of the overall system could be checked. At this stage it is possible to fix the system functionality errors and make the required corrections of the design.

During the layout design besides the building blocks itself it is necessary to take into account the critical points of the schematic. Such points should have the special pads for experimental laboratory study of the prototype (e.g. with the probe station and picoprobes). It also should be provided the future laboratory tests of the separate blocks.

The verification of the layouts is necessary to fix the design rules errors and the mismatches between the schematic and layout (DRC and LVS).The parasitic extraction (PEX), inclusion of the package and PCB stray elements, post layout simulations are another important step in the ASIC design. Those simulations can give the best approach of the project to the real conditions. This stage is characterized by the following types of the analysis and simulations:

- simulations with the extracted parasitic elements;
- Monte-Carlo simulation of the process variations;
- Simulation of the process corners;
- temperature variations;
- IR-drops on wires in the whole area of the die;
- gate delays simulations;
- integrity (signal transfer either inside the chip and on the PCB).

The flow is finalized by the extraction of the files for manufacturing. The quality of the foundry design kits and some special aspects of the simulation (analog and digital) which can give different results on the speed and accuracy should be taken into account.

3. Approbation of the design approach

The design flow described above was successfully implemented in the development of the prototype readout ASIC for the CBM experiment muon chambers. To show the distinction of the proposed design flow, it is made the comparison with the conventional flow, which was used in another prototype design.

The prototype 8-channel ASIC was developed for the muon chambers of the CBM experiment (figure 2) [3]. The ASIC was designed according to the classical approach of bottom-to-top design flow. At first, the building blocks schematics and layouts were designed. Then, the blocks were connected into the readout channels. The structure of the prototype chip includes 8 analog channels with ADCs. The ASIC was prototyped in the 180 nm CMOS UMC process.

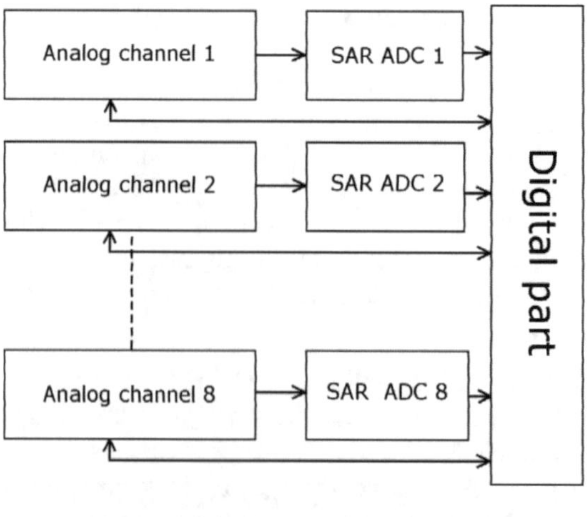

Figure 2. 8-channel prototype ASIC structure.

The experimental studies of the chip showed that the ADC is non-operable. The reason of it was in the mismatch of the signal polarity in the unit starting the ADC conversion. Thus, it becomes impossible to obtain the digitized information from the channels.

This kind of mistake could be eliminated by the application of the complex approach to the design of the multichannel readout ASICs. This method was applied to the development of the next prototype ASIC for muon chambers of the CBM experiment (figure 3).

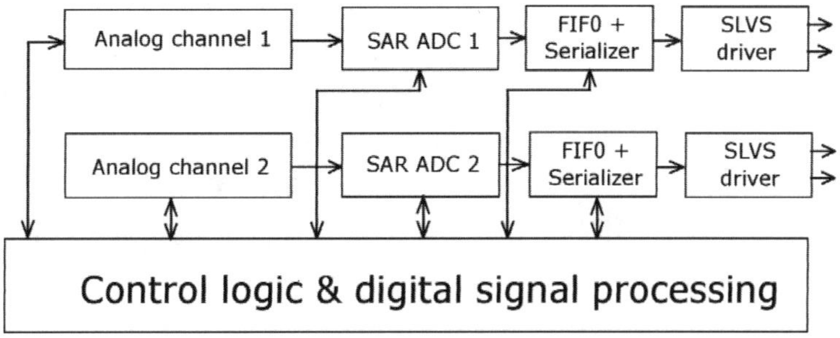

Figure 3. Mixed-signal prototype ASIC structure.

The prototype includes 2 analog channels, SAR ADCs, ADC FIFO in each channel, timestamp block, digital peak detector, e-link drivers and etc. Thus, this prototype is a part of the full readout system of the up-to-date HEP and space experiments.

The design of this chip was provided according to the proposed design flow. The behavioral model of the channel was designed using the Verilog-A models (e.g. amplifiers, ADC, DAC and etc.). The digital part of the channel which controls the ADC operation, as well as the channel memory (FIFO) and serializer were designed with the Verilog-A models. These allow simulating the full mixed-signal system functionality. Then the Verilog-A models were replaced by the real transistor blocks seriously.

The digital blocks were designed, using the standard digital flow. The digital part was converted into the Verilog netlist and then into the schematic block. The digital schematic is based on the special standard cell library of the UMC 180 nm process. The library includes the schematic equivalent of the RTL digital part design. Thus, the mixed-signal system was designed and it was possible to simulate in the analog environment. It became capable to fix the error of simulation and to simulate the different blocks in the complex readout mixed-signal system. The simulations were carried out to check the parameters as the noise, stability, load capacity, parasitic components and etc. Thus, it was fixed all the mismatches between the analog signal polarities and delays. To realize this design flow the most of the Cadence tools as ADE, ADE XL (Monte-Carlo and Corners), simulation accelerating options (APS and XPS), and calculating algorithms were used. It is significant, that the system designer analyzed in parallel with the block design how a block is working with set of other blocks and compared ideal and real schematics functionality. It allows fixing faults during the block and full chip development.

4. Conclusion

The readout ASIC design top-to-bottom-to-top serial flow for high energy physics and cosmic rays experiments was described. The design flow makes possible to analyze the mixed-signal system operation on the different levels: functional, behavioral, schematic and post layout including parasitic elements. The proposed design flow allows reducing the simulation period and eliminating the functionality mismatches on the very early stage of the design. It was successfully embedded to the development of the read-out ASIC prototype for the muon chambers of the CBM experiment. The approach was approved in UMC CMOS MMRF 180 nm process.

Acknowledgements
The work was carried out with the financial support of SAEC "Rosatom" and Helmholtz Association and the Ministry of Education and Science of the Russian Federation (grant No.14.A12.31.0002) in accordance with the RF government order No. 220. This work was partially supported by MEPhI Academic Excellence Project (contract № 02.a03.21.0005, 27.08.2013).

References

[1] Kazyonnov G 2009 (In Russian) *Basics of the integrated circuits and systems design* (Moscow: Binom. Laboratory of knoweledge inc.) p 295 (Original Russian title: *Osnovy proektirovanija integral'nyh shem i sistem* (M.: Binom)
[2] Jess Chen *et al.* 2012 *Mixed-signal methodology guide* (San Jose: Cadence Design Systems Inc)
[3] Ivanov V *et al.* 2015 RU patent 2564956
[4] Atkin E *et al.* Development of the read-out ASIC for muon chambers of the CBM experiment 2015 JINST **10** C0400

High speed SLVS transmitter and receiver

I S Bulbakov[1], E V Atkin[1] and A G Voronin[1,2]

[1] National Research Nuclear University MEPhI (Moscow Engineering Physics Institute), Kashirskoe highway 31, Moscow, 115409, Russia

[2] SINP (Skobeltsyn Institute of Nuclear Physics) Moscow State University, 1(2), Leninskie gory, GSP-1, Moscow 119991, Russia

E-mail: bulbakovis@yandex.ru, evatkin@mephi.ru, voronin@silab.sinp.msu.ru

Abstract. Design of SLVS chip-to-chip communication transmitter/receiver IP block in 180 nm UMC MMRF CMOS process is presented. This block has been developed for study a data transmission over PCBs and/or electrical cables (lines) of few meters length at rates up to 320 Mb/s. Schematic for on-chip tests is also presented. This blocks are used for communication between front-end ASICs and DAQ system.

1. Introduction

Modern experiments in high-energy physics contain a huge number of data processing channels. Therefore, reducing power consumption and increasing data transmission speed of each individual front-end ASIC is an important task. A significant contribution to power consumption is made by circuits of electrical link drivers between FE ASICs and DAQ chips.

The front-end electronics aim to achieve high levels of performance in terms of resolution and accuracy. This performance is limited by the system intrinsic noise, therefore electrical links should be designed to minimize crosstalk and power supply noise. For these reasons, using a low-power low-voltage-swing signaling standards become a very popular in particle physics applications in recent years.

Scalable Low Voltage Signaling (SLVS) standard [1] was chosen for this purpose. It is provide necessary data transmission speed and lower, in compare with Low Voltage Differential Signaling (LVDS) one, power consumption. The protocol is briefly described in section 2. The design of SLVS Transmitter and Receiver IP blocks is presented in section 3. Results of simulation is presented in section 4.

2. SLVS standard

The SLVS standard is defined in [1] and describes a differential current-steering electrical protocol with a voltage swing of 200 mV on a 100 Ω load and a common mode of 200 mV. The differential voltage is therefore 400 mV as depicted in figure 1. The output current is 2 mA, with a power consumption at the load of 0.4 mW.

3. SLVS Transceiver Design

In this Section the ASIC implementation of the SLVS transceiver is presented; this is composed of a pre-driver, transmitter (Tx) and a receiver (Rx) circuits. This IP blocks were designed in a 180 nm UMC MMRF CMOS technology. Typical structure of transceiver is described in [2, 3]

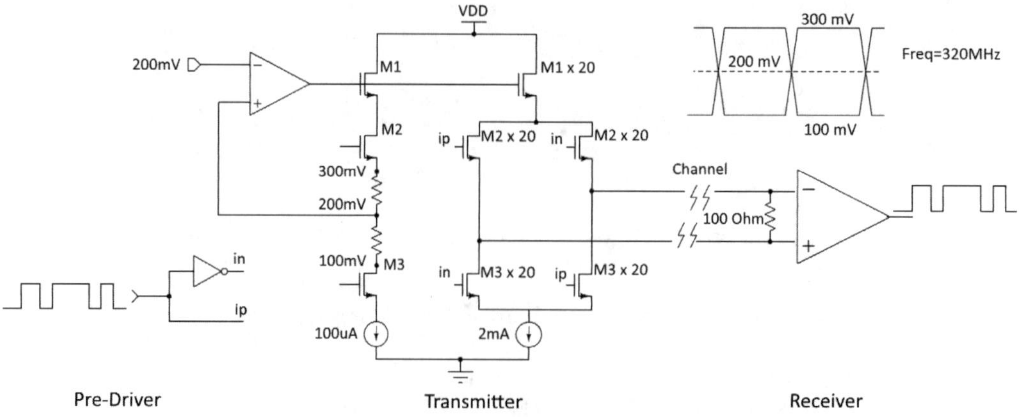

Figure 1. Electrical characteristics and typical transceiver structure of SLVS standard.

and shown in figure 1. The pre-driver circuit splits input digital signal on two signals with reversed phases.

3.1. Transmitter (Tx)

A typical SLVS transmitter operates as a current source with switched polarity. The output current flows through the load resistance, establishing the correct differential output voltage swing.

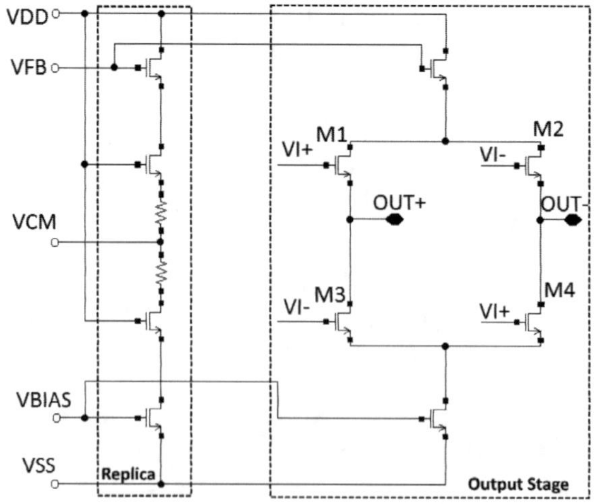

Figure 2. Proposed SLVS transmitter.

The proposed transmitter circuit presented in figure 2 uses the configuration with four MOS switches in H bridge configuration, implemented with the M1-M4 NMOS transistors. The operation of this circuit is described in the following. When M1 and M4 transistors are switched on, the polarity of the output current is positive together with the differential output voltage. On the contrary, when M1 and M4 transistors are switched off, the polarity of the output current and voltage is reversed.

Replica circuit with transistors 20 times smaller than transistors in output stage provides easier control of common mode voltage through negative feedback via an error amplifier. The transmitter consumes 4.9 mW at 1.8 V supply voltage and provides transmission speed up to 320 Mbps and 2 mA output current.

3.2. Receiver (Rx)

The receiver is implemented by a nonlinear differential amplifier, shown in figure 3. It also contains output inverter and buffer. The receiver consumes 500 μW at 1.8 V supply voltage at 320 Mbps.

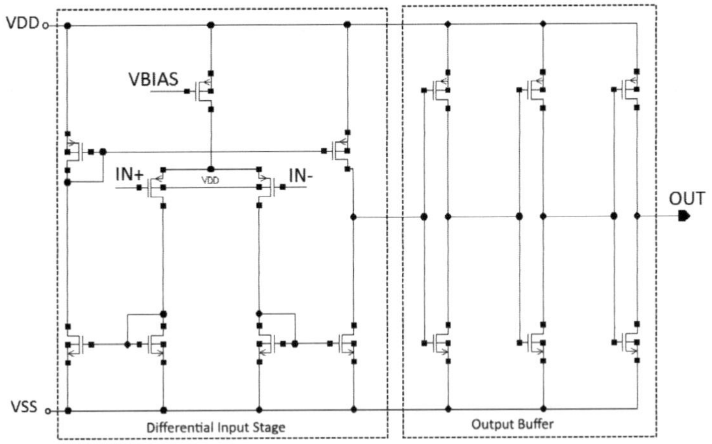

Figure 3. Proposed SLVS receiver.

4. Simulation results

Parasitic elements of chip package was described in [4]. It includes pad and pin capacitance and bond wire inductance. The estimation 1nH/mm for wire bonding was used. Thus, simulation setup contained these parasitics is shown in figure 4.

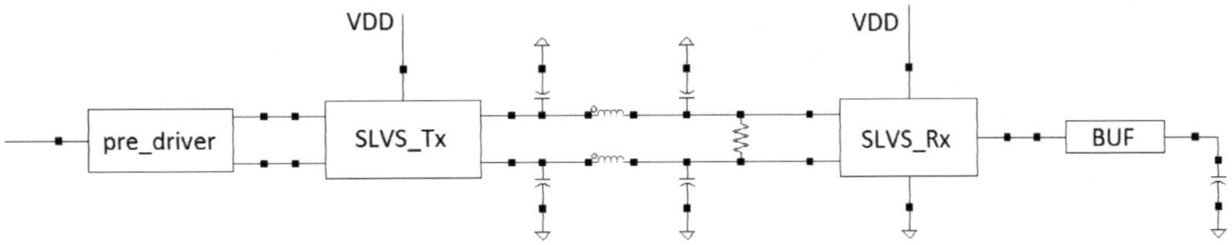

Figure 4. Testbench for SLVS transceiver.

Results of PVT simulation is shown in figure 5 and 6. PVT includes power supply variation from 1.6 V to 2 V, temperature variation from 0 to 60 °C and technological variation taken from design kit models.

Schematics for on-chip tests is shown in figure 7. This allows to get signals for receiver from external pins. Transmitter may works in two modes: 1) uses as input digital signal receiver's output or 2) uses as input signal from external generator.

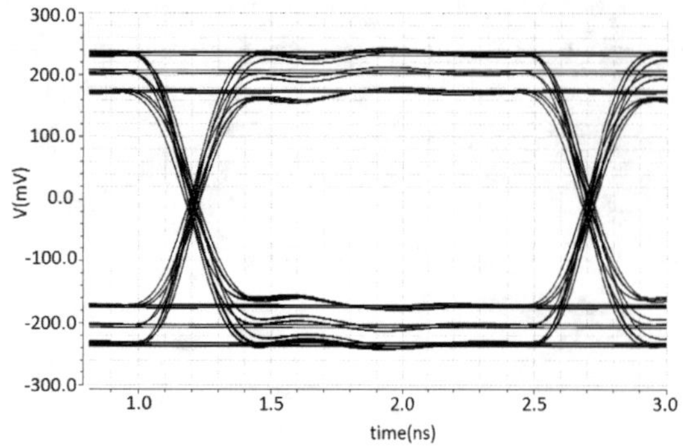

Figure 5. Eye-diagram of the Tx differential output signals.

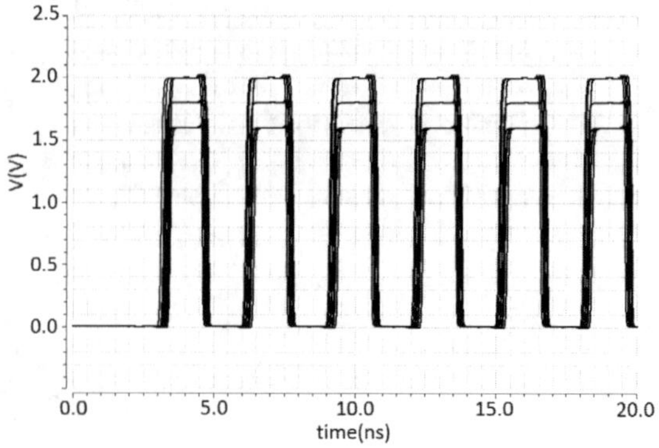

Figure 6. Simulation results of the Rx output signal.

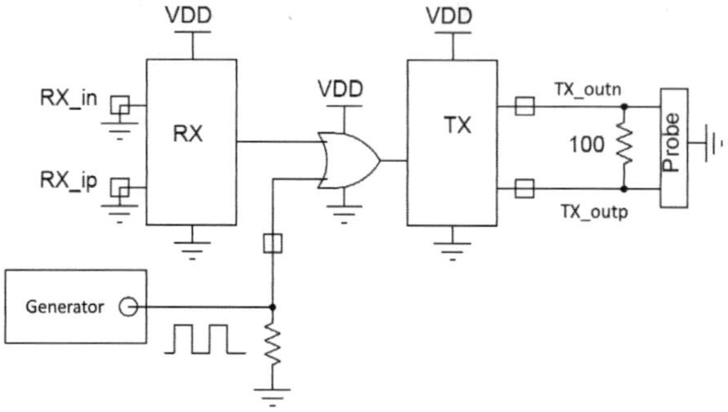

Figure 7. Schematics for on-chip tests.

5. Conclusion

An SLVS transmitter/receiver IP block was designed in 180 nm UMC MMRF CMOS technology and the test chip was submitted for fabrication. Future improvements might include the predetermined output current of transmitter and power-down mode for the receiver and transmitter.

Acknowledgements

This work was partially supported by MEPhI Academic Excellence Project (contract No. 02.a03.21.0005, 27.08.2013).

References

[1] JEDEC Standard JESD8-13 2001 *Scalable Low-Voltage Signaling for 400 mV (SLVS-400)* (Solid State Technology Association, Printed in U.S.A.)

[2] Hugo Daniel Hernandez 2014 *Design review: SLVS Tx and Rx blocks* (SAMPA review) (*indico.cern.ch/event/308952*)

[3] Bonacini S, Kloukinas K and Moreira P 2009 E-link: A radiation-hard low-power electrical link for chip-to-chip communication *Proc. of the TWEPP-09, CERN-2009-06* pp 422–5 (*http://cdsweb.cern.ch/record/1235849*)

[4] Boni A, Pierazzi A and Vecchi D 2001 LVDS I/O Interface for Gb/s-per-Pin Operation in 0.35-μm CMOS *IEEE J. of Solid State Circuits* **36**(4) 706–11

Journal of Physics: Conference Series **675** (2016) 042036 doi:10.1088/1742-6596/675/4/042036

ASIC for calorimetric measurements in the astrophysical experiment NUCLEON

E Atkin[1], A Voronin[1,2], D Karmanov[2], I Kudryashov[2], I Kovalev[2] and V Shumikhin[1]

[1] National Research Nuclear University MEPhI (Moscow Engineering Physics Institute), Kashirskoe highway 31, Moscow, 115409, Russia

[2] Skobeltsyn Institute of Nuclear Physics Lomonosov Moscow State University, 119991, Leninskie gory 1/2, Moscow, Russia

E-mail: voronin@silab.sinp.msu.ru

Abstract. A satellite with the NUCLEON apparatus was launched in Dec. 2014. The space NUCLEON project of ROSCOSMOS is designed to investigate cosmic ray nuclei energy spectra from 100 GeV to 1000 TeV as well as cosmic ray electron spectra from 20 GeV to 3 TeV. The method of energy determination by means of a silicon instrument for measuring the particle charge of cosmic rays and the calorimetric system were developed. The main parameters, that determine the quality of calorimetric systems are linearity of transfer characteristic and the dynamic range of input signals, which should reach 30 000 MIPs (minimum ionizing particles). The ASIC, satisfying these requirements, consisting of 32 channels with a unique dynamic range from 1 to 40000 MIPs, signal to noise ratio not less than 2.5 at a shaper peaking time of 2 μs and a low power consumption of 1.5 mW/channel has been designed. The first results of the ASIC functionality in space are presented.

1. Introduction

The space NUCLEON project of ROSCOSMOS is designed to investigate cosmic ray nuclei energy spectra from 100 GeV to 1000 TeV as well as cosmic ray electron spectra from 20 GeV to 3 TeV. Such investigations are very important for modern astrophysical research. Specific features of the experiment are a limited volume for the placement of electronics equipment and rigid power consumption requirements. This all results in the necessity to design an ASIC for reading the signals from silicon detectors used. The 32-channel ASIC with a unique wide dynamic range from 1 to 40000 mips and a low power consumption of 1.5 mW/channel was designed. The article describes the chip structure and its main parameters.

2. NUCLEON Apparatus

To reach this goal the NUCLEON [1, 2] apparatus consists of several subsystems: the silicon charge measurement system; the carbon target; the scintillator trigger system; the silicon energy measurement system; the silicon minicalorimeter; an auxiliary electronics unit.

The NUCLEON ASIC is used in the 6 planes silicon – tungsten minicalorimeter. Each plane includes a tungsten absorber and four ladders under it. The detector strips in neighbouring planes are oriented in such a way that they are orthogonal to each other. Each detector consists of 64 strips, 1 mm wide and 25 cm long. The detector wafer is 300 μm thick. One ladder contains eight 32-channels

ASICs and four silicon microstrip detectors. Thus the readout of the minicalorimeter is performed from 1536 channels.

3. The ASIC

The ASIC [2, 3] has 32 analog channels and 2 additional testing ones, placed at the die edges in order to reduce edge effects and improve the repeatability of parameters for the working channels. Figure 1 shows the ASIC structure.

The analog channel of ASIC consists of the following blocks: charge sensitive amplifier, shaper and a track and hold (T&H). The outputs of the channels are read out through analog switches, controlled by output logic, based on a shift register. Further the outputs are connected to a voltage-to-current converter. The converter transforms the single ended voltage into a differential current (outp and outn outputs).

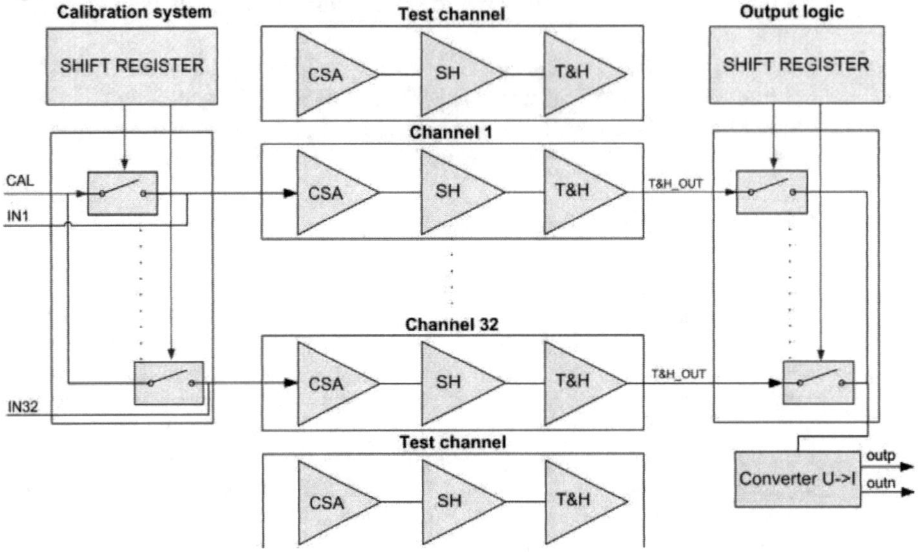

Figure 1. ASIC structure.

Such a structure has been used for the possibility of summing the output signals of multiple ASICs into a single differential current bus.

The calibration system, built in the ASIC, allows supplying the input test pulses subsequently to each working channel via a shift register.

The charge sensitive amplifier (CSA) has a folded cascode structure with a p-MOS input transistor. To achieve a wide dynamic range the CSA has a switchable gain. The division of the full dynamic range (from noise to saturation) into two subranges should be considered as a fundamental special feature of the designed CSA circuit. The inflection point of the transfer characteristic is at the input charge value of about 3 pC and can be adjusted. On the one hand, this allows one to ensure a higher slope of the transfer characteristic in the small amplitude region and thus reach a satisfactory SNR. On the other hand, this allows one to ensure a lower slope in the high amplitude region and thus expand the dynamic range beyond 100 pC. The subranges in the designed scheme are automatically switched.

To attain this aim, an additional circuit is introduced into the CSA structure (see figure 2). At small amplitudes the additional circuit is switched off and exerts no effect on the CSA feedback. In this case, the CSA gain is set by the relatively small feedback capacitor 6.6 pF. It may be calculated by the formula $K_Q = U_{OUT}/Q_{IN} = 1/C1 = 0.15$ V/pC. That provides an optimum signal to noise ratio for a 1 mip signal.

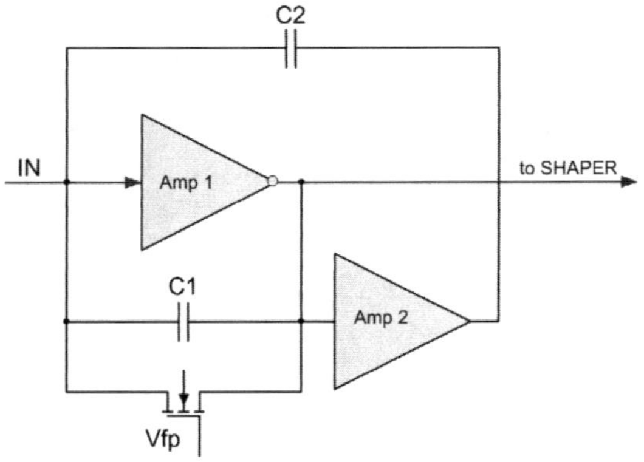

Figure 2. CSA structure.

At higher amplitudes (starting approximately from 3 pC), an additional capacitor C2 is added to the feedback network through an auxiliary amplifier, operating in the class B mode. The gain of this auxiliary amplifier and capacitance of C2 determine the feedback circuit for the CSA at large amplitudes. The equivalent feedback capacitance of the CSA is about 67 pF in this case. To discharge

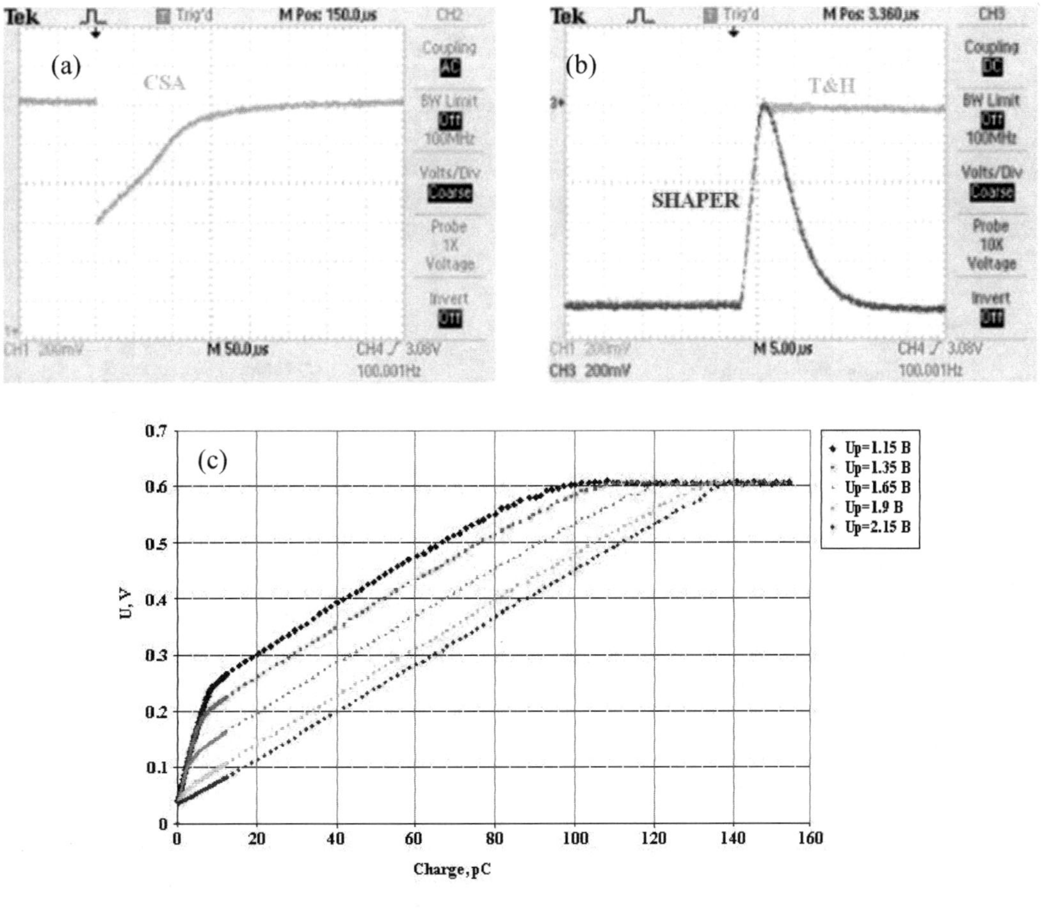

Figure 3. The ASIC output of: a) CSA; b) shaper and T&H. c) Channel transfer function.

the CSA feedback capacitance there was used a long channel (L = 100 μm, W = 1 μm) p-MOS transistor. The shaper has a CR–RC active filter structure. The peaking time of the shaper output is about 2.2 μs. The main purpose of the shaper consists in shortening the pulse duration and improving the signal-to-noise ratio (SNR) in a channel. Maximum swing at the shaper output is 2 V. The track and hold circuit fixes the peak amplitude at the shaper output. To store the T&H signals a capacitor of 1 pF is used. The circuit of a single stage operational amplifier with a 100% feedback is used as a buffer.

The ASIC was designed in the 0.35 μm CMOS technology process of AMIS (OnSemi) [4, 5]. At layout design there were used 5 metal and 2 polysilicon layers. Examples of test are presented on figure 3.

4. First space results

The goals of NUCLEON experiment also methodics, simulations, equipment are detailed described in [1]. In December 2014 the satellite with NUCLEON apparatus was launched on orbit. First experimental minicalorimeter result (for six planes: m0–m5) for the cosmic ray particle energy deposit on the orbit of 475 km is shown in figure 4. It demonstrates the latitudinal energy distribution of a hadron shower. The experimental result confirmed the simulations [1] for the minicalorimeter and are in full correspondence with previous cosmic experiments (ATIC, Pamela, AMS).

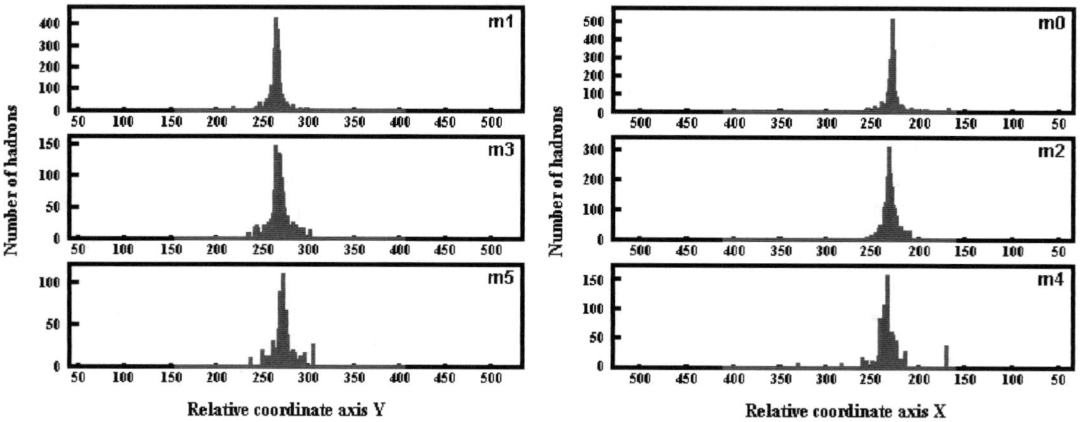

Figure 4. Cosmic ray particle Energy deposit profile of the NUCLEON minicalorimeter (6 silicon plates), obtaining by ASICs. (Y axis – m1, m3, m5; X axis – m0, m2, m4).

5. Conclusion

The 32-channel read-out ASIC for calorimetric measurements in the astrophysical experiment NUCLEON has been developed. It has a unique dynamic range of 120 pC, SNR=2.5 for the minimum ionizing particles (1 MIP) and a low power consumption of 1.5 mW per channel. The experimental results of the laboratory test are presented.

The ASIC was embedded in the hardware of the NUCLEON experiment on board of a satellite, launched to orbit in December 2014. First space results have been given.

Acknowledgements

This work was partially supported by MEPhI Academic Excellence Project (contract № 02.a03.21.0005, 27.08.2013).

References

[1] Atkin E *et al.* 2014 The NUCLEON space experiment for direct high energy cosmic rays investigation in TeV-PeV energy range *Nucl. Instrum. Meth.* A **770** 189-96

[2] Atkin E *et al.* 2015 NUCLEON ASIC and ladder electronics for cosmic ray experiments *JINST* **10**(4) C04042
[3] Atkin E *et al.* 2015 The read-out ASIC for the Space NUCLEON project *JINST* **10**(4) C04005
[4] Atkin E 2009 A high dynamic range ASIC for microstrip detectors *EUROPRACTICE IC Service Activity Report*
[5] Atkin E *et al.* 2011 Readout ASIC for microstrip detectors *Russian Microelectron.* **40** 52-8

Journal of Physics: Conference Series **675** (2016) 042037 doi:10.1088/1742-6596/675/4/042037

Influence of digital processing on the noise levels of spectrometric system for xenon gamma-spectrometer

D Petrenko, Z Uteshev, ANovikov, A Shustov, K Vlasik, I Chernysheva, M Smirnova, K Krivova, V Dmitrenko and S Ulin
National Research Nuclear University MEPhI (Moscow Engineering Physics Institute), Kashirskoe highway 31, Moscow, 115409, Russia

E-mail: denis.petrenko.dp@gmail.com, kfvlasik@mephi.ru, vvdmitrenko@gmail.com, seulin@gmail.com, zmuteshev@mephi.ru

Abstract. Shaping and digital processing of xenon gamma-spectrometer signals were considered. Digital processing influence on the energy resolution of the spectrometric system of xenon gamma-spectrometer was shown. The analysis of contributions to the energy resolution showed that the limit of improvement of the resolution by using of digital processing is reached almost.

1. Introduction

The Xenon gamma ray spectrometer (HPXe spectrometer) is an instrument with a wide range of applications. This device can be applied as well as laboratory equipment and device for industry [1,2]. All these applications need the maximum possible energy resolution. To date, the HPXe spectrometer can achieve energy resolution at the level of 1.7% for the 662 keV gamma source ^{137}Cs. The theoretically possible energy resolution of HPXe detector is 4 keV to 662 keV (i.e. 0.6%) [3]. Accordingly, an important task is to improve the energy resolution of the HPXe spectrometer. One of the ways to improve the energy resolution is the use of digital processing.

2. The influence of digital processing on the energy resolution

The energy resolution of HPXe spectrometer is determined by the self-noise, noise of charge sensitive amplifier, digital processing and different interferences. The self-noise mean that energy resolution of detector is limited by internal processes and properties (fano factor, energy for pair production, inefficiency of shielding grid, recombination and gas purity). Comparative analysis of the energy resolution same detector with the use of digital processing and without it (see figure 1 and figure 2) shows that the digital processing significantly improves the energy resolution.

Accordingly, there are two ways to improve the energy resolution of HPXe spectrometer:

- Reduction of the intrinsic noise of charge sensitive amplifier;
- Improved processing electrical signals from the HPXe spectrometer.

For the case of ideal digital processing of the electrical signals from HPXe spectrometer it is possible to obtain the energy resolution which defined by the self-noise of detector and noise of charge sensitive amplifier only (excluding electrical interferences). Analysis of contributions to energy resolution (see figure 1 and figure 2) shows that the energy resolution in case of ideal digital processing (without contribution from it) is 1.63% for 662 keV (i.e. 11 keV). Therefore it can be argued that the digital processing practically provides the best possible value of the energy resolution for HPXe spectrometer.

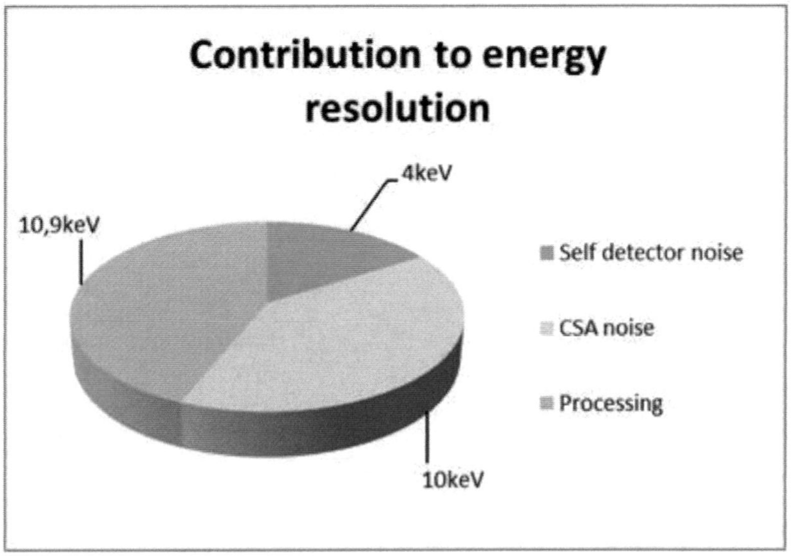

Figure 1. Contribution to energy resolution in case using "Colibri" analog signal processing unit [4]. It shows data for the energy 662 keV.

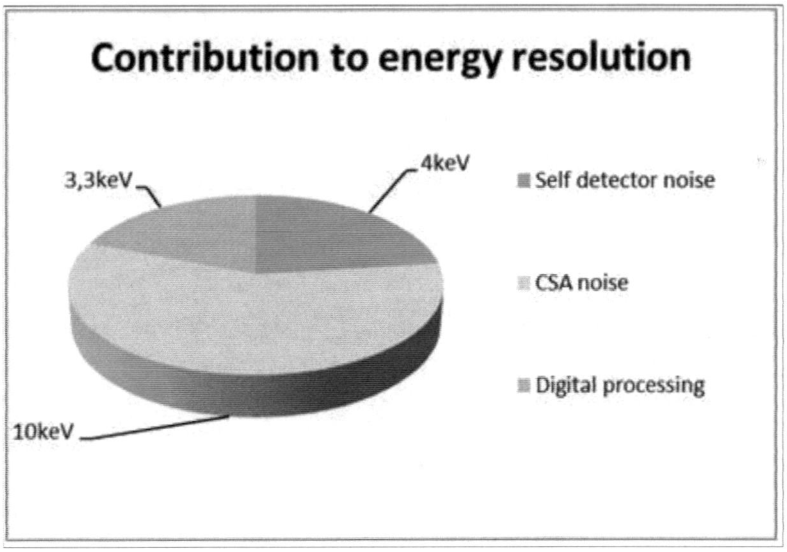

Figure 2. Contribution to energy resolution in case using digital signal processing. It shows data for the energy 662 keV.

3. The shaping and digital processing

The signal shaping must be high-quality for the effective digital processing. In this case, the signal, which input to the digital processing unit, has the shape of exponential discharge. This waveform provides sufficient accuracy of digitizing. The stages of signal conversion are shown in table 1.

The actual digital processing is reduced to the background subtraction of the waveform and its subsequent integration. Figure 3 shows the ideal case where the signal without noise and overlaps enters to the input of the digital processing unit.

Table 1. The stages of signal conversion.

Name	Shape of signal	Kind of signal in the zero approximation
Detector	$i(t)=\frac{\partial q}{\partial t}$	δ-function
Charge sensitive amplifier	$U(t)=\frac{1}{C}\int i(t)dt$	Heaviside step function
Shaper	$U(t)=\theta(t)e^{-t/RC}$	Exponential discharge
Digital processing	$S=\int_{\Delta Tdp} U(t)dt$	Final value

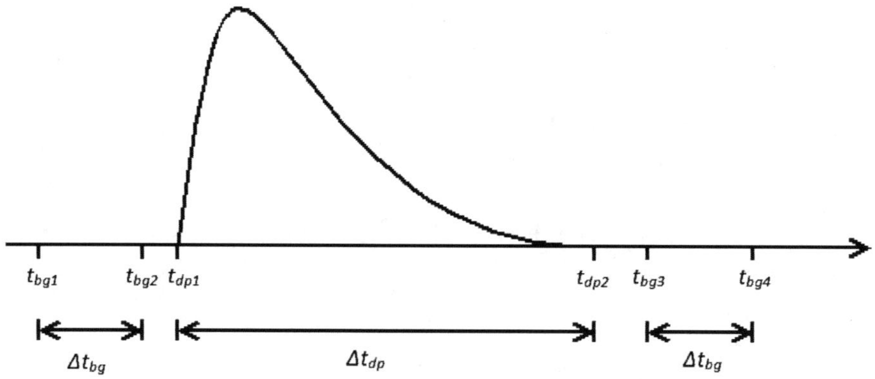

Figure 3. The ideal case where the signal without noise and overlaps enters to the input of the digital processing unit. The intervals Δt_{bg} ($t \in (t_{bg1};t_{bg2})$ and $t \in (t_{bg3};t_{bg4})$) are times when background is calculated. The interval Δt_{dp} ($t \in (t_{dp1};t_{dp2})$) is time when signal is processed.

The noise signal can be considered as a composition of harmonic oscillations of different frequencies. For our system reaction to the impact of oscillations of a single frequency is determined by the following formula:

$$
\begin{aligned}
f(\omega,\varphi) &= -1\frac{1}{\omega}(\cos(\omega t_{dp2}+\varphi)-\cos(\omega t_{dp1}+\varphi)) \\
&-\frac{t_{dp}^2}{2}(\frac{-1}{\omega\Delta t_{bg}}(\cos(\omega t_{bg2}+\varphi)-\cos(\omega t_{bg1}+\varphi)) \\
&-\frac{-1}{\omega\Delta t_{bg}}(\cos(\omega t_{bg4}+\varphi)-\cos(\omega t_{bg3}+\varphi)))
\end{aligned}
\tag{1}
$$

The noise contribution to the energy resolution of each element of the spectrometric channel can be estimated using the Wiener-Khinchin theorem. This necessarily includes knowledge of the system response to a harmonic oscillation and spectral density of the self-noise at each unit of spectrometric channel. Application of this method will simplify the design of new units for spectrometric channel.

4. Conclusion

Modern methods of digital processing are an effective way to improve the energy resolution. The analysis of the energy resolution suggests that the limit of its improvement by using of digital processing is reached almost. The most perspective seems to decrease the self-noise of the electronic units of the spectrometric channel.

References

[1] Novikov A *et al.* 2015 Xenon gamma-ray spectrometer in the experiment Signal on board the spacecraft Interhelioprobe *Proc. SPIE* ***959*** *3*95930L-1
[2] Novikov A *et al.* 2015 Xenon gamma-ray detector for ecological applications. *Journal of Applied Remote Sensing* **9**
[3] Dmitrenko V *et al.* 1983 Compressed-Xenon Ionization Chamber for Gamma Spectroscopy *Sov. Phys.-Tech. Phys* **28** 1440
[4] URL: (*http://greenstar.ru/kolibri.html*)

High-voltage power supply with improved thermostability for Xenon gamma-ray spectrometer

D Petrenko, Z Uteshev, A Novikov, A Shustov, K Vlasik, I Chernysheva, M Smirnova, K Krivova, V Dmitrenko and S Ulin
National Research Nuclear University MEPhI (Moscow Engineering Physics Institute), Kashirskoe highway 31, Moscow, 115409, Russia

E-mail: denis.petrenko.dp@gmail.com, kfvlasik@mephi.ru, vvdmitrenko@gmail.com, seulin@gmail.com, zmuteshev@mephi.ru

Abstract. In this article the high voltage power supply for xenon spectrometer is described. Result of time simulation for output voltage at different temperatures was shown. The experimental data is confirming results of the time simulations. The experimental data showed breadboard model provides a stability of voltage better than 1% of the generated voltage at different temperatures

1. Introduction

The development of a gamma spectrometer "Signal", based on the Xenon gamma ray spectrometer (HPXe spectrometer) is carrying out as part of the experiment "Interheliozond". Since the experiment will be carried out in hard environment it is required to increase the reliability and stability of the HPXe spectrometer [1]. To increase thermostability of HPXe spectrometer is necessary first of all to develop high voltage heat-stable power supply unit. We assume that the stability of 1.5% is sufficient.

2. The original high voltage supply module

The HPXe spectrometer is a pulse ionization chamber, so one of its features is a need to apply the high voltage to the grid and the cathode. The typical high voltage, which has to be applied to these electrodes are 15kV and 20kV. The high voltage supply based on push-pull converter.

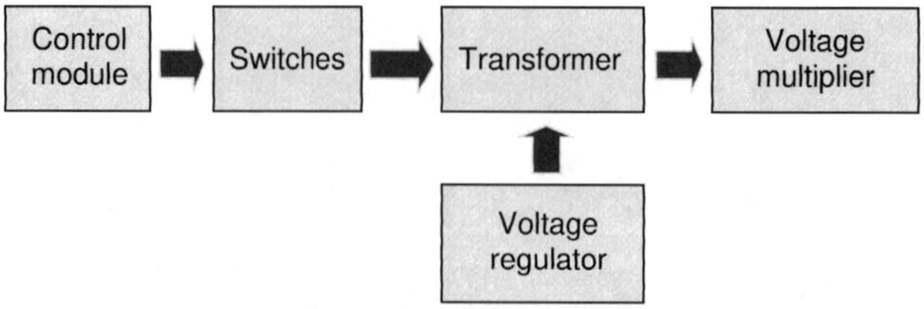

Figure 1. Block diagram of high voltage supply module.

The original circuit is shown on the figure 2. It was analysed using OrCAD software. The work of this circuit has been revealed important shortcomings: voltage spikes of high amplitude which can damage the FETs, voltage spikes on midpoint of transformer after the stack of control pulses and lasting fluctuations arises (see figure 3).

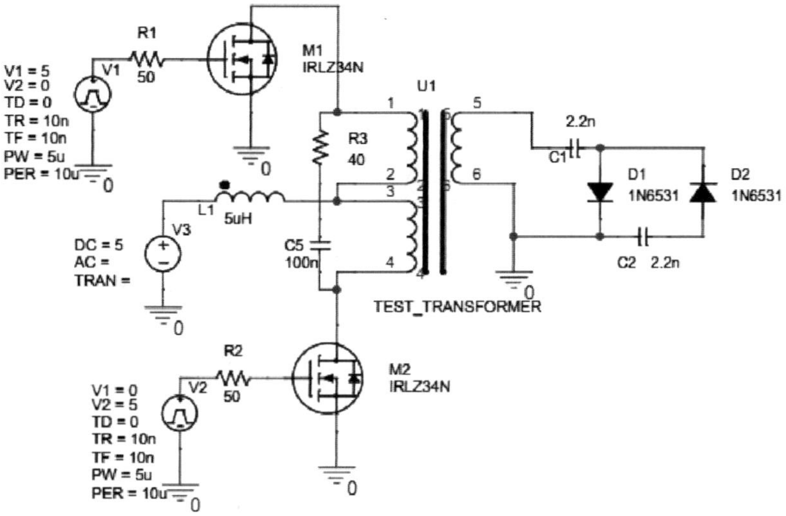

Figure 2. Circuit of high voltage power supply model.

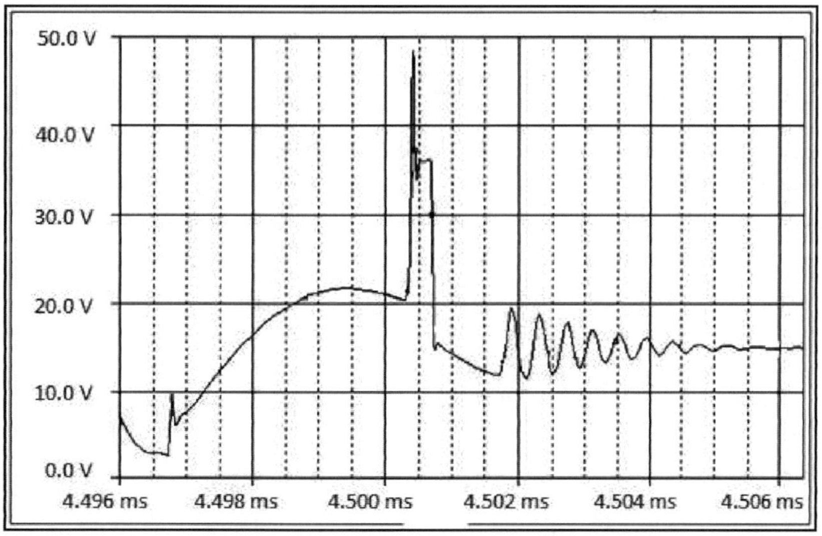

Figure 3. Voltage spikes on midpoint of transformer after the stack of control pulses.

3. The development of optimal circuit of the high voltage supply module

During search of the optimal scheme of high voltage power supply module different variants were investigated, but in this article we will limit ourselves to a description of the final version of thermo-stable high voltage supply module (see figure 4).

The stabilization of the current is provided directly at the FET in the final circuit. Each FET has a feedback using a bipolar transistor. Thus resistors R48 and R49 determine the maximum current flowing through the transformer.

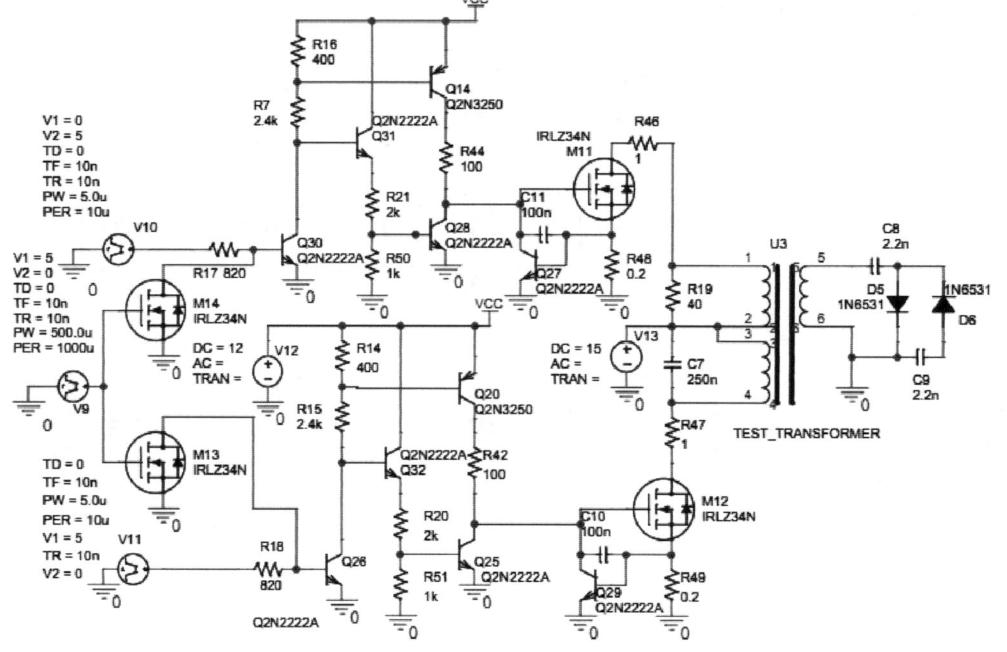

Figure 4. The final version of thermo-stable high voltage supply module.

The optimum waveform observed when the time constant τ of the RC-circuit R42C10 (R44C11) and the period of the control signals are the same. The shape of the input transformer signals will be sinusoidal in case of equality of these values (see figure 5). This waveform is the most preferred for use in the high voltage supply module than rectangular, because it provides a much lower level of interference to nearby electronic components.

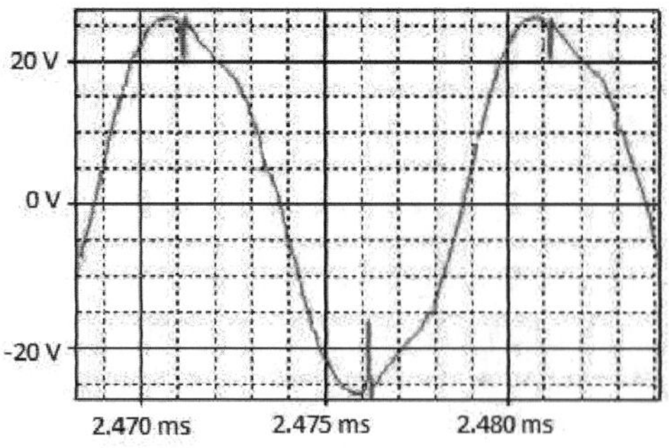

Figure 5. Result of time simulation for transformer differential input voltage. The optimum waveform shown.

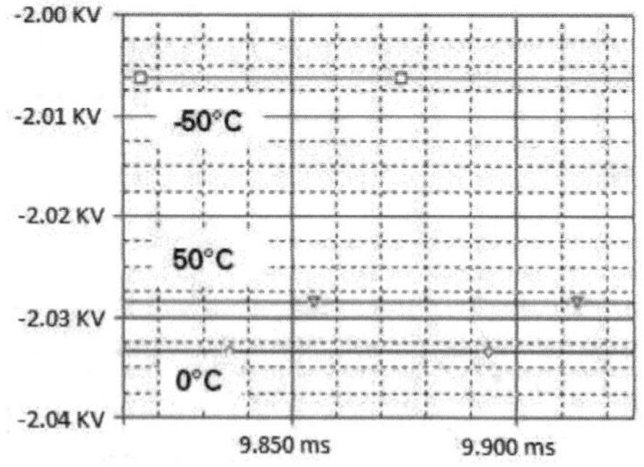

Figure 6. Result of time simulation for module output voltage at different temperatures. The voltage level is constant during simulation.

The voltage level was simulated for different temperatures. According to a simulation (see figure 5) developed circuit provides a change of voltage less than 30V when the temperature was increase from -50°C to 50°C, which is 1.5% of the generated voltage.

4. Experimental results

The developed model was verified with the breadboard. Figure 7 shows the experimental data confirming results of the time simulations. The experimental waveform has a shape close to an optimum waveform (sine wave).

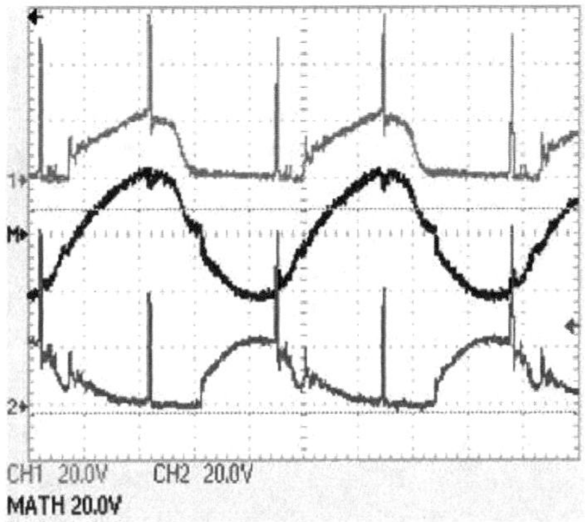

Figure 7. The experimental waveform. The dark line is differential voltage on transformer. The grey lines are input voltages on transformer.

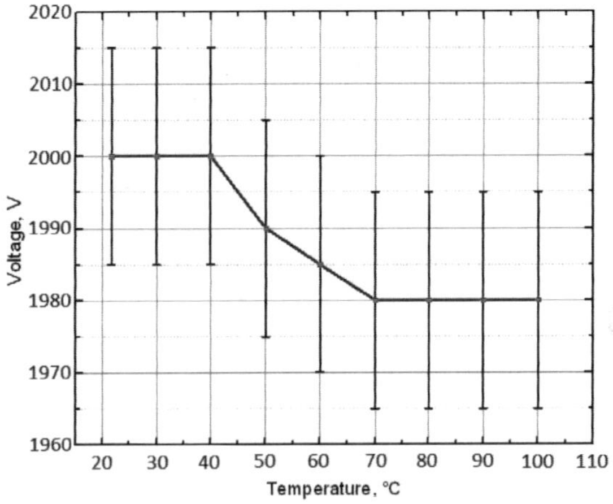

Figure 8. The voltage versus temperature.

According to an experimental data (see figure 8) developed circuit provides a stability of voltage better than 1% of the generated voltage when the temperature was increase from 23°C to 100°C. These results confirm the simulation.

5. Conclusion

The circuit of thermo-stable high voltage supply unit for spectrometer "Signal" was developed using OrCAD software. It is experimentally shown that the breadboard of the new high voltage supply unit provides the expected performance.

Acknowledgments

Authors acknowledge partial support from MEPhI Academic Excellence Project (contract № 02.a03.21.0005, 27.08.2013).

References

[1] Novikov A *et al.* 2015 Xenon gamma-ray spectrometer in the experiment Signal on board the spacecraft Interhelioprobe *Proc. SPIE* ***9593*** 95930L-1

Journal of Physics: Conference Series **675** (2016) 042039 doi:10.1088/1742-6596/675/4/042039

The prospects of creation of well-type xenon gamma-ray spectrometer

D Petrenko, Z Uteshev, A Novikov, A Shustov, K Vlasik, I Chernysheva, M Smirnova, K Krivova, V Dmitrenko and S Ulin
National Research Nuclear University MEPhI (Moscow Engineering Physics Institute), Kashirskoe highway 31, Moscow, 115409, Russia

E-mail: denis.petrenko.dp@gmail.com, kfvlasik@mephi.ru, vvdmitrenko@gmail.com, seulin@gmail.com, zmuteshev@mephi.ru

Abstract. Possibility of creation of a well-type gamma detector based on xenon gamma-ray spectrometer was shown. The results of modelling of xenon well-type detectors are presented.

1. Introduction

The development of modern technologies depends on the methods and equipment. One of the modern equipment, which is used as in scientific research and in industry is well-type detector. In this paper we study the possible characteristics of the well-type xenon detector.

2. Results of modelling of xenon well-type detector

To evaluate the possible characteristics of the xenon well-type detector it was simulated using Geant4. The model used to calculate the characteristics of the detector is shown in figure 1.

Figure 1. Model of xenon well-type detector. Model has the following parameters:

- Density of xenon - 0.3 g/cm^3
- The internal diameter - 30 mm
- The external diameter - 150 mm
- Grid length - 102 mm
- Grid diameter - 60 mm

Based on the presented model of the detector the calculated dependence of the efficiency versus gamma rays energy was obtained. The results are shown in figure 2. For comparison, figure 3 shows the experimental 2 liter high presser xenon detector (HPXe detector) efficiency for a point source at a distance of 20 cm. Experimental points are fitted by function:

$$\eta(E) = \left(a + \frac{b}{E}\right)^{1/2} \tag{1}$$

The comparison shows that theoretically well-type gamma ray xenon detector has higher efficiency than the HPXe detector. This is most pronounced for energies below 100 keV. Comparison of experimental and calculated data by efficiency of HPXe detector was considered in the [1]. The modern 2 liter HPXe detector was described in the [2].

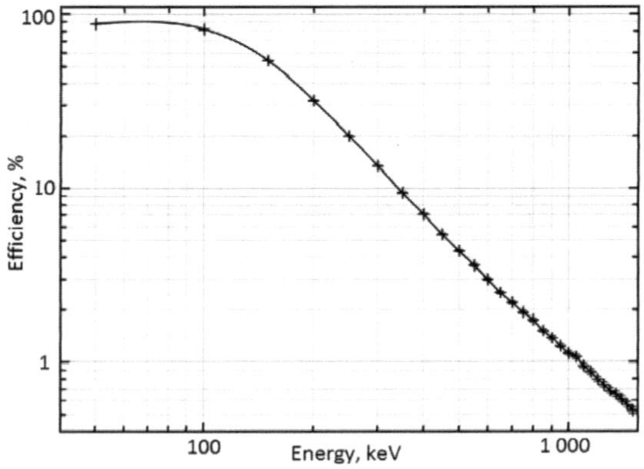

Figure 2. The calculated dependence of the efficiency versus gamma rays energy for xenon well-type detector. Points of calculated dependence are connected by spline curve.

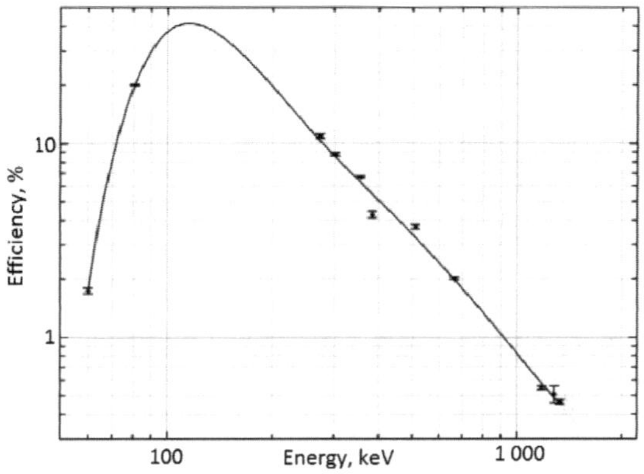

Figure 3. The experimental 2 liter HPXe detector efficiency for a point source at a distance of 20 cm. Solid line is fit of experimental data with equation (1).

Based on the assumption that the energy resolution of the well-type xenon gamma ray detector for energy 662 keV will be 2.5% spectrum for the ^{137}Cs was calculated (see figure 4). For comparison, figure 5 shows the experimental 2 liter HPXe detector spectrum for a point ^{137}Cs source at a distance of 20 cm. The energy resolution of the 2 liter HPXe detector is equal to 2.2% for ^{137}Cs. The spectra of the 2 liter HPXe detector have strongly pronounced peak of the backscatter, which complicates the analysis of the spectra at low energies. The calculated spectra of xenon well-type detector has not strongly pronounced peak of the backscatter.

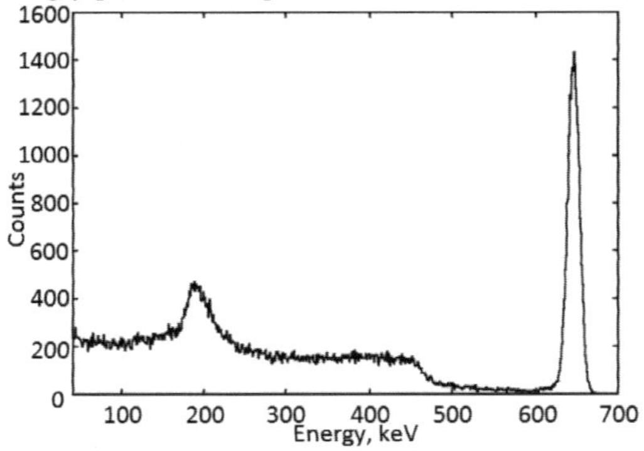

Figure 4. The calculated ^{137}Cs (662 keV) spectrum for well-type xenon detector. Energy resolution (FWHM) is equal to 2.5%.

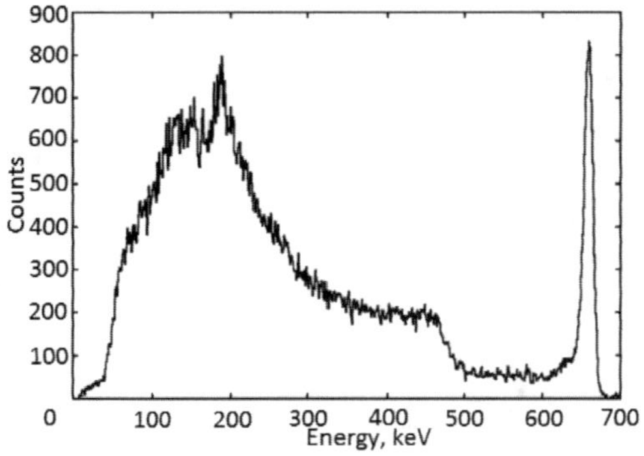

Figure 5. The experimental 2 liter HPXe detector spectrum for a point source ^{137}Cs at a distance of 20 cm. Energy resolution (FWHM) is equal to 2.2%.

3. Conclusion

Modern well-type gamma rays detectors implemented mainly on scintillation crystals and semiconductor devices. Each of these types has fundamental shortcomings. The semiconductor detectors are expensive; at the same time detectors based on scintillation crystals have not good energy resolution for modern application. Today development of low cost well type detector with high energy resolution based on one of these principles of operation is impossible. It can be concluded that it is necessary to develop a detector on different physical principles. Shown results suggest that the well-type xenon gamma ray detector will be able to provide better energy resolution than scintillation detectors at a low manufacturing cost. However, these results need to be verified experimentally.

Acknowledgments

Authors acknowledge partial support from MEPhI Academic Excellence Project (contract № 02.a03.21.0005, 27.08.2013).

References

[1] Veresenko V S A *et al.* 2015 (in Russian) *Proc. Science Session MEPhI-2015* **1** 174 (original Russian title: *Nauchnaya sessiya MEPhI-2015 Sbornik nauchnyh trudov*)

[2] Novikov A *et al.* 2014 New modification of xenon gamma-ray detector with high energy resolution *Opt. Eng.* **53** (2) 021108

Architecture of the multichannel data-driven ASIC

D D Normanov and E V Atkin
National Research Nuclear University MEPhI (Moscow Engineering Physics Institute), Kashirskoe highway 31, Moscow, 115409, Russia

E-mail: dima-norman@yandex.ru, EVAtkin@mephi.ru

Abstract. The development architecture of a multichannel data-driven ASIC is presented. It provides the selection of useful events at an early stage of reading out detector signals. The architecture is based on fast cross-point switches of analog signals, followed by their digitization by a limited set of ADCs and high-speed output data serialization. Such approach reduces the number of subsequent ADCs as well as digital processing channels. That leads to lower power consumption and chip area. The results of a prototype ASIC development, based on this architecture and intended for the CBM experiment at FAIR, are given.

1. Introduction

1.1. CBM experiment

One of the main trends for the modern multichannel experiments is to increase the number of readout channels of information from radiation detectors. For example, one of the key systems in the international experiment CBM [1] (Compressed Baryonic Matter) which is conducted on the Facility Accelerator for Anti-proton and Ion Research (FAIR) in Darmstadt, Germany are the muon chambers. From a hardware point of view the muon chambers, built on avalanche-gas detectors, contain about 1 million channels.

The main objective of the CBM experiment is bombing targets by streams of particles that are accelerated in synchrotrons up to energies of 2 – 45 GeV/nucleon, providing up to 10^7 collisions per second. A stream of charged particles that are produced in these collisions (including secondary) is processed by the muon detector chamber, which should determine the type of particles and restore their tracks. Because the system is multi-channel and the interaction of a single detector with the beam of particles is a random event, the charges generated by the detectors track system will be supplied to the reading – processing chips asynchronously. Accordingly, there is a problem of sequencing (derandomization) them for further processing. The processing should determine the amplitude of the signal, the time it enters the channel and the channel number in which the active signal was detected.
This problem can be solved by a standard digital signal processing, which involves the use of fast multi-bit ADC in each channel, thereby increasing its chip area and power consumption. Since the information received from the detectors is asynchronous, a logical solution is to treat the large number of readout channels by fewer processing channels. To implement this solution the following architecture of data ordering (derandomizer) was proposed.

1.2. Requirements to the read-out ASIC for muon chambers

Proceeding from the need to process a large number of asynchronous data and a very large number of channels, the ASIC reading out the signals from the detectors of muon chambers should meet the following requirements:

- the power consumption of the chips per channel must not exceed 3 – 5 mW (half of this value is traditionally assigned to ADC);
- the maximum height of the channel should not exceed 100 um;
- the IRdrop chip should not exceed 10% of the supply voltage;
- cross-talk between channels must be much less than 5% of the signal amplitude from the detector.

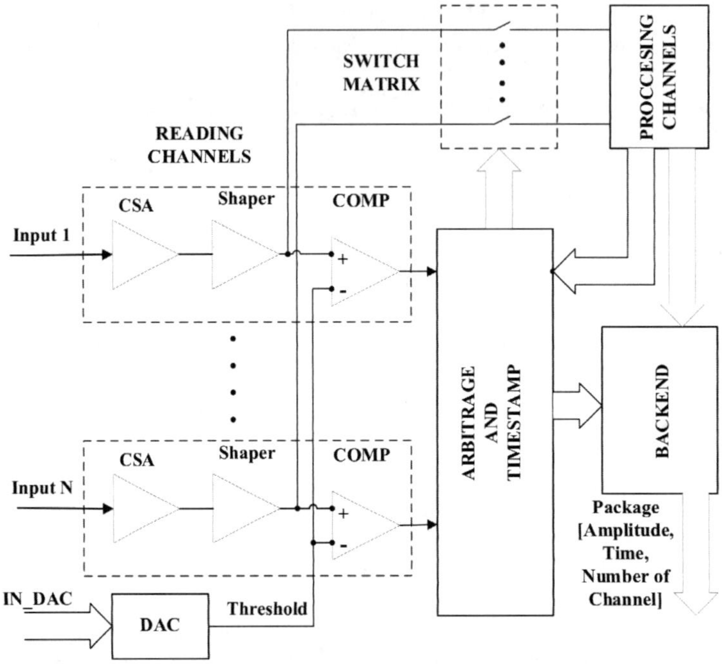

Figure 1. Block Diagram of Architecture Derandomizer.

2. Architecture of derandomizer

2.1. Architecture

Figure 1 shows the architecture of analog derandomizer. At the input of each channel a charge-sensitive amplifier (CSA) is placed, which converts the signal of avalanche-gas detector (in the form of current or charge pulses) in the output voltage. After the CSA in each channel there is an amplifier-shaper (shaper). Its main purpose consists in shortening the duration of the signal and improving the channel's signal/noise ratio. This is followed by comparators (COMP), which determine the presence of the useful signal at the inputs of the matrix keys (SWITCH MATRIX). The threshold for the comparators is set directly from an IC pin or with the aid of threshold DAC.

The arbitration logic (ARBITRAGE) with the comparator signals determines the number of the triggered channels, the availability of free the moment processing channels and connects them with the reading out ones. Thus, the data digitization occurs in an optimal mode – virtually eliminating the dead time in the system at a predetermined statistics at input. The processing channel includes a peak detector and ADC – signals are digitized, whereupon the peak detector selects from them the amplitude. The digital value of the amplitude along with the channel number and the time is transmitted to the backend, which generates from these values the data packet and provides a serial output at a high frequency.

2.2. Timestamp

Determination of the time of arrival (such as a timestamp) consist in the juxtaposition operation of the comparator signal with the external clock frequency of the reading out system. The timestamp calculation unit is an N-bit counter, which is clocked by an external clock signal. When a signal exceeding the comparator threshold appears in the channel, the trigger unit TIMESTAMP toggles, whose signal records the current signal code of the counter in the channel corresponding to each register.

Some problems in determining the time exist since the comparator activation has a propagation delay to the start of the signal, and this delay depends on the amplitude and rise time of the signal, especially if the input signal has a slow front. In this case, a special unit (time walk compensation) corrects the clock pulse number, which determines the time of the event, depending on the signal amplitude.

Synchronously with the operation of the timestamp counter unit, the BACKEND information acquisition unit has its own counter. The latter counts the overflow signals in the timestamp counter and identifies failures in the block.

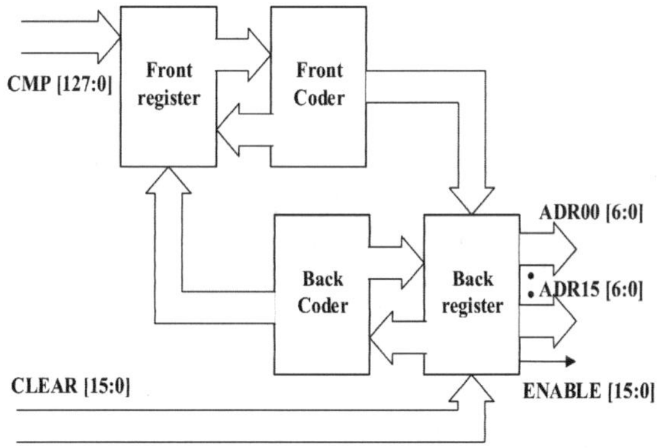

Figure 2. The arbitration logic of derandomizer.

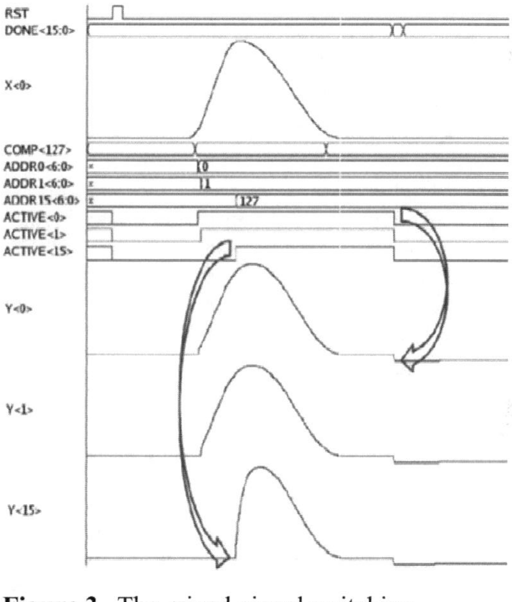

Figure 3. The mixed-signal switching time diagram.

2.3. Arbitrage

Block diagram of the arbitration unit is 128 → 16 (128 channels are readout into 16 channels of processing) is shown in figure 2, and its operation diagram in figure 3. The algorithm of the arbitration logic module is shown below. The input of the Front Register receives a signal from the comparator. If one of the read out channels (or a few) signals exceeds the threshold level of the comparator, then this change will switch one of the outputs of Front Register. If changes have occurred of multiple inputs, the switching will occur for the channel with the lowest number.

For the channel selected by the unit Front Register, (the signal is reset by a reset signal, or by the arrival of a new signal from the output of the comparator), a flag is set, for which the Front Coder unit exposes a signal requesting to allow readout channel switching to the processing channel. The block

Back Coder checks the condition of the availability of a read channel. In the case of finding a free channel for processing the corresponding output is locked in the block Back Register.

On the locked output there is exposed the channel address that arrives on a matrix of keys makes switches the readout channel to the processing channel. Addition to the output of the arbitration, block there is set a signal confirming the switch over (Enable). By front of this signal output at of block Front Register is reset, which corresponds the commutated channel. Front Coder module finds the next readout channel, in which was comparator switching occurred and begins to search for a free readout channel.

Enable signal also resets the peak detector and ADC in the appropriate processing channel. Channel processing remains locked until the arrival of the Peak Find signal of the peak detector.

3. Prototypes

To probate structural solutions and test circuit solutions of the separate derandomizer units there were designed 2 prototypes for the UMC 180 nm MMRF CMOS process (CBM experiment standard). The first prototype [2] has four analog inputs connected to the outputs of four comparators, derandomizer circuit (4 $\rightarrow$ 2), 2 processing channels containing peak detector for measuring signal amplitude. The prototype has proven the feasibility and expedience of creating derandomization systems.

One of the aims for designing the second prototype [3] was to study an analog data-driven architecture. The design was done for an analog derandomizing function of the 24 to 16 structure. That means that the ASIC structure should provide a parallel readout of 24 analog input signals of front-end part and their commutation with 16 outputs. Another aim for the design was to elaborate an appropriate control logic and supplement the ASIC by both an input analog part and an output ADC.

4. Conclusion

The developed architecture of analog data processing on the basis of a multi-channel high-speed analog switch can effectively reduce power consumption and chip area for reading signals from detectors, using 180 nm technology level. The possibility of constructing derandomizers is confirmed by the creation of 2 prototype chips. The results of laboratory tests confirmed the functioning of their inherent principles. This year the run of the 8-channel chip [4] was performed, containing the basic building blocks for the system, with the exception of the derandomizer block. The next step is expected to create a 64 channel chip (64 reading channels 8 channels processing) with the derandomizer architecture.

Acknowledgements

This work was supported by Grant No. 14.A12.31.0002, issued by the Russian Federation Ministry of Education and Science according to the RF Governmental resolution No 220. In addition, this work was supported by MEPhI Academic Excellence Project (contract No. 02.a03.21.0005, 27.08.2013).

References

[1] Friman B *et al.* 2010 *The CBM Physics Book* (Berlin: Springer)
[2] Atkin E V *et al.* 2010 A Method and Integrated Circuit for Processing Random Data Streams in Multichannel Experimental Physical Equipment *Instrum. Exp. Tech.* **53**(4) 524–9
[3] Atkin E V *et al.* 2013 A Multichannel ASIC for the Tracking System of the CBM Experiment Russian Microelectronics **42**(2) 119–26
[4] Atkin E V *et al.* 2015 Development of the read-out ASIC for muon chambers of the CBM experiment *JINST* **10**(4) C04006

Journal of Physics: Conference Series **675** (2016) 042041 doi:10.1088/1742-6596/675/4/042041

The service telemetry and control device for space experiment "GRIS"

A S Glyanenko
National Research Nuclear University MEPhI (Moscow Engineering Physics Institute), Kashirskoe highway 31, Moscow, 115409, Russia

ASGlyanenko@mephi.ru

Abstract. Problems of scientific devices control (for example, fine control of measuring paths), collecting auxiliary (service information about working capacity, conditions of experiment carrying out, etc.) and preliminary data processing are actual for any space device. Modern devices for space research it is impossible to imagine without devices that didn't use digital data processing methods and specialized or standard interfaces and computing facilities. For realization of these functions in "GRIS" experiment onboard ISS for purposes minimization of dimensions, power consumption, the concept "system-on-chip" was chosen and realized. In the programmable logical integrated scheme by Microsemi from ProASIC3 family with maximum capacity up to 3M system gates, the computing kernel and all necessary peripherals are created. In this paper we discuss structure, possibilities and resources the service telemetry and control device for "GRIS" space experiment.

1. Introduction

"GRIS" (Gamma and Roentgen Irradiation of the Sun) is a scientific instrument for detection of hard X-rays and gamma-rays of solar flares and gamma-rays bursts onboard the Service Module "Zvezda" in the Russian Segment of the International Space Station. Information about this experiment is presented in [1]. Not only for this experiment, but for all space experiments the problems of scientific devices control, service data acquisition and preliminary data processing are actual.

Typical control problems are: precision control detector parameters by high-voltage regulation, fine thresholds changing (analog regulation), different switching on/off (in real time), statistical analysis of data flows and change of operation modes of the device, etc.

Typical internal data acquisition problems are: temperature measurements in different equipment parts (in power supplies, on scintillation crystals for energy ranges correction etc.), precision thresholds measurements in discrimination circuits, status different switching on/off (in real time), statistical analysis of data flows and searching for event type solar flare or gamma ray bursts with the change of operation modes of the device, etc.

Next, look at the different approaches to the implementation of the device.

2. Selecting components for the creation of unit

There are 3 main ways to realize onboard specialized control and data acquisition unit:

- the use of standard devices and components;
- the creation of specialized units based on programmable logic integrated schemes (PLIS);
- the development and use of application-specific integrated circuit (ASIC).

Journal of Physics: Conference Series **675** (2016) 042041 doi:10.1088/1742-6596/675/4/042041

Obviously, the use of standard components in comparison with PLIS and ASIC, has major flaws – notably the large size, weight and power consumption. Also look problematic features meet the requirements on reliability (due to excess of functions and/or hardware interfaces) and, probably, radiation tolerance (RT).

The PLIS vs. ASIC:

- there is no time and cost in production of ready software implementation of the node (IP);
- quick adaptation of the design to the requirements of the task;
- cheaper in small and medium batches.

There are two main classes of PLIS: complex programmable logic device (CPLD) and field-programmable gate array (FPGA).

Microsemi and Xilinx are world leaders in PLIS fabrication for space applications: in RAM based technology (multiple-programmable) – Xilinx; Flash based technology (multiple-programmable) and Antifuse technology (once-programmable) – Microsemi.

Using of flash-based interconnects present some unique opportunities and advantages to designers of space-flight electronic hardware:

- The flash cells are reprogrammable. This allows the designer to change the design of the FPGA without removing the FPGA from the board, making prototyping easier. It also allows last-minute design change and code update to provide maximum design flexibility.
- The flash cells are nonvolatile. This means that flash-based FPGAs are standalone devices which do not require the provision of external code-storage devices, unlike SRAM-based FPGAs. This minimizes the board space used, and has an associated saving in mass.
- RT ProASIC3 FPGAs are operating almost at the instant of power-up, which is another advantage of the nonvolatility of the flash programming cells. There is no boot sequence required, as in SRAM-based FPGAs which need to download their configuration code from an external storage device.
- The flash cells do not exhibit single-event upsets in the presence of heavy ion radiation. Therefore no triple-chip redundancy to mitigate configuration upsets is required, unlike SRAM FPGAs.

RT ProASIC3 FPGAs are the first to offer designers of space-flight hardware a Radiation-Tolerent (RT), reprogrammable, nonvolatile logic integration vehicle. They are intended for low-power space applications requiring up to 350 MHz operation and up to 3 million system gates.

RT ProASIC3 devices use the same silicon design and process as the commercial UMC 0.13 μm ProASIC3EL family. RT3PE600L uses the same silicon as the A3PE600L, and RT3PE3000L uses the same silicon as the A3PE3000L.

3. Functional scheme and operation

Functional diagram of the developed device is shown in figure 1. This device is called "controller" on this scheme. All the basic functions of the "system-on-chip" implemented in the FPGA A3PE3000L. Information about this FPGA and any soft-realized modules is presented in [2]. The lack of a significant amount of on-chip Flash-memory necessitates the use of an additional EPROM chip with volume 64KB. As a basic processor in this system is used soft realization of most popular Intel 8051 processor with RISC organization (IP Core8051s).

The Core8051s is a high-performance, eight-bit microcontroller IP Core. It is a fully functional eight-bit embedded controller that executes all ASM51 instructions and has the same instruction set as the 80C31. Core8051s provides software and hardware interrupts. The Core8051s architecture eliminates redundant bus states and implements parallel execution of fetch and execution phases. Since a cycle is aligned with memory fetch when possible, most of the one-byte instructions are performed in a single cycle. Core8051s uses one clock per cycle. This leads to an average performance improvement rate of 8.0 (in terms of MIPS) with respect to the Intel device working with the same clock frequency. Available optional and can be turned off to conserve FPGA resources: multiply and divide instructions (MUL, DIV and DA) in our project is disabled, (but can be implemented as NOP), and the second data pointer (data pointer 1), in our project is used.

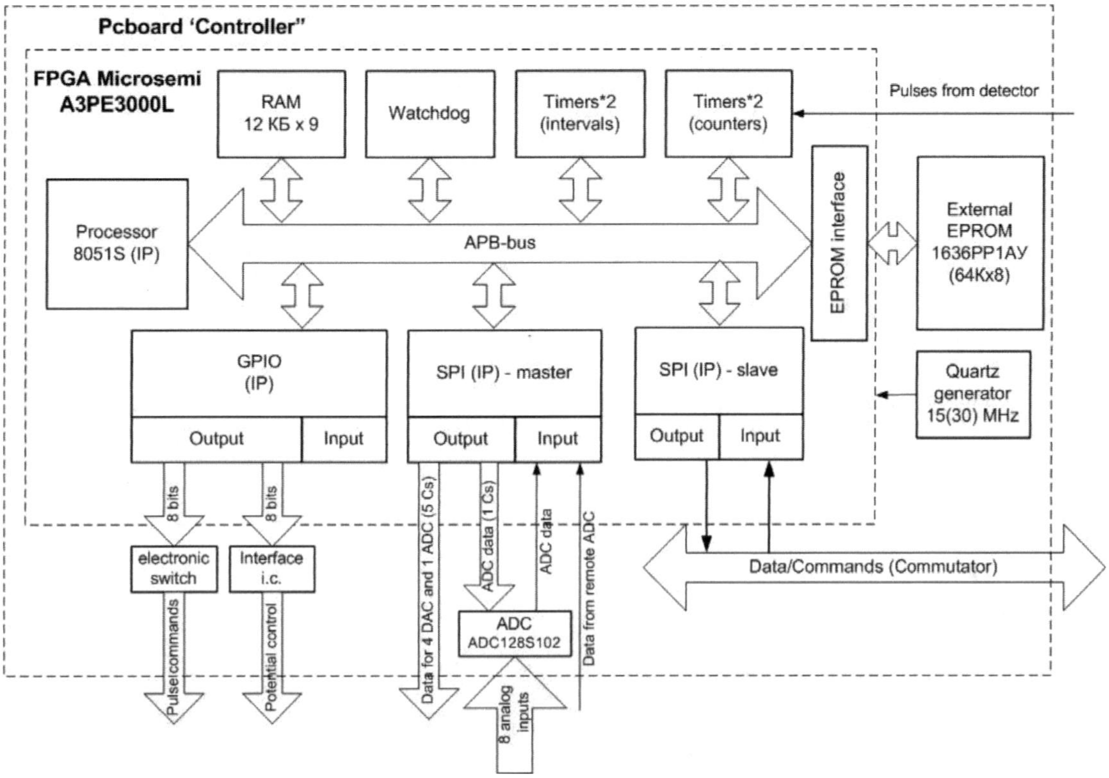

Figure 1. Functional diagram of the developed device.

The advanced peripheral bus (APB) version 3 interface of Core8051s connected to the mirrored master interface of CoreAPB3, with various APB or APB3 slaves connected to the slave interfaces of CoreAPB3 (up to 16 slots). This design implements an 8051-based microcontroller system. In addition to the Core8051s microcontroller core, this system consists of the following peripherals: CoreSPI (two parts), CoreTimer (4 parts), CoreWatchdog, CoreGPIO and Internal RAM (12 KB).

In our design is used 32-bits APB-bus. CoreGPIO is available with the APB slave interface and provide 8 bits-pulse commands and 8-bits potential commands. Also 8 bits can be used as inputs for external status digital signals.

CoreSPI is a serial peripheral interface allowing high-speed synchronous serial data transfers between the 8051 and peripheral devices. This component supports full duplex operation with 8-32 bit serial data transfer and acts as a master or slave. SPI is a point-to-point bus standard used in a variety of embedded applications. One SPI interface is used as slave for data and command interchange with "commutator" device (system for control and data acquisition for all parts of "GRIS"). Other SPI interface is used in master mode to control 4 remote DAC ic's (each 2 channels 12 bit DAC) and 2 8 channels ADC (1 on pcboard of "controller" and one – remote). Thus, this device provides 8 channels DAC and 16 channels ADC.

The CoreTimer module is an APB slave that provides access to an interrupt-generating, programmable decrementing counter. Two counters are used as timers for determination of pulse duration (for pulse commands) and counting interval for statistical analysis of data flows and searching for event type solar flare or gamma ray bursts. Other two timers are used as a counters of photons for event searching.

CoreWatchdog is intended for use with Processor-Based Systems to Protect and Recover from Software Errors. CoreWatchdog is an APB slave that provides a means of recovering from software crashes. When enabled, CoreWatchdog will generate a soft reset for the system if the microprocessor fails to refresh it on a regular basis.

On PCB “controller” are used 8 electronic switches for pulse commands generation and additional interface ic’s for transmitting potential commands to remote units of “GRIS”.

For FPGA power is used +1,5V (for FPGA core) and +3,3V (for FPGA interface part) to ensure compatibility with LVTTL logical levels. Full power dissipation for “controller” PCB is less 100 mW.

4. Conclusion

It was developed service telemetry and control device for internal control, data acquisition and preliminary data processing for “GRIS” experiment onboard ISS based on "system-on-chip" concept.

Acknowledgements

The author would like to thank all colleagues from Astrophysics institute of NRNU “MEPhI” for interest and support in this work.

This work was partially supported by MEPhI Academic Excellence Project (contract № 02.a03.21.0005, 27.08.2013).

References

[1] Kotov Yu D, Yurov V N, Trofimov Yu A, Lupar E E, Glyanenko A S and Faradzhaev R M 2015 *Advances in Space Research* **56** 1797–804

[2] URL: *http://soc.microsemi.com*

Journal of Physics: Conference Series **675** (2016) 042042 doi:10.1088/1742-6596/675/4/042042

A low jitter all – digital phase – locked loop in 180 nm CMOS technology

O V Shumkin[*], V A Butuzov, D D Normanov and P Yu Ivanov
National Research Nuclear University MEPhI (Moscow Engineering Physics Institute), Kashirskoe highway 31, Moscow, 115409, Russia

E-mail: [*]olegbox29@mail.ru

Abstract. An all-digital phase locked loop (ADPLL) was implemented in 180 nm CMOS technology. The proposed ADPLL uses a digitally controlled oscillator to achieve 3 ps resolution. The pure digital phase locked loop is attractive because it is less sensitive to noise and operating conditions than its analog counterpart. The proposed ADPLL can be easily applied to different process as a soft IP block, making it very suitable for system-on-chip applications.

1. Introduction

The PLL is a self-correcting control system in which one signal chases another signal. PLLs can be classified into the four categories shown in Table 1. ADPLL receives only digital signals. The phase and frequency information can be more accurately stored using digital techniques. Digital PLLs allow a faster lock time to be achieved and are attractive for clock generation on high performance microprocessors.

Table 1. Types of PPL.

Types of PPL	Comparison	Evaluation	Storage	Oscillator
Linear PLL (LPLL)	Analog multiplier	LPF	Analog voltage on filter capacitor	VCO
Digital PLL (DPLL)	EXOR PD or JKPD	LPF	Analog voltage on filter capacitor	VCO
Charge-pump PLL (CPPLL)	PFD	Charge pump and LPF	Analog voltage on filter capacitor	VCO
All-digital PLL (ADPLL)	EXOR PD or JKPD or PFD	Digital LPF	Digital word	Digitally controlled oscillator (DCO)

A diagram of an ADPLL is shown in figure 1. An ADPLL consists of the following represented blocks: Phase Frequency Detector Controller (PFD_CTRL), digitally controlled oscillator (DCO) and Frequency Divider (FD). The DCO is the most critical component in ADPLL design.

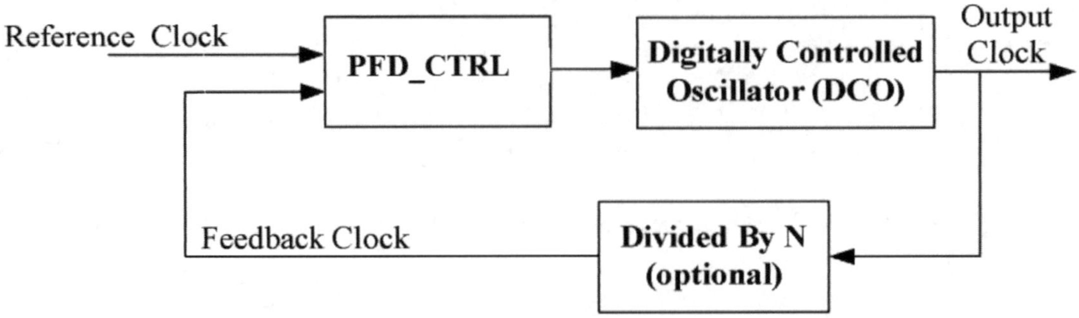

Figure 1. A typical ADPLL Block diagram.

The components of ADPLL are similar to the analog PLL, but the implementation consists of digital components. A voltage controlled oscillator was utilized instead of a digitally controlled oscillator. The PFD_CTRL detects the phase and frequency mismatch of the reference clock and divided DCO clock. Phase Frequency Detector compares the frequency and outputs a signal delay, advance or equivalence. The ADPLL is locked when the PFD_CTRL detects, that the phase and frequency of the two clock inputs match. The output signal of DCO, becomes close to the reference frequency.

The different applications of ADPLL are discussed [1-5]. All the components are created in Verilog, except DCO. DCO contains schematically based components.

2. ADPLL design

2.1. Phase Frequency Detector Controller

A diagram of PFD_CTRL is shown in figure 2. Waveforms of Phase Frequency Detector is shown in figure 3. The phase detector compares the reference frequency (FREF) with a frequency of divider (DCO_CLK). To begin the comparison unit should get a signal started (START). The signal sampling

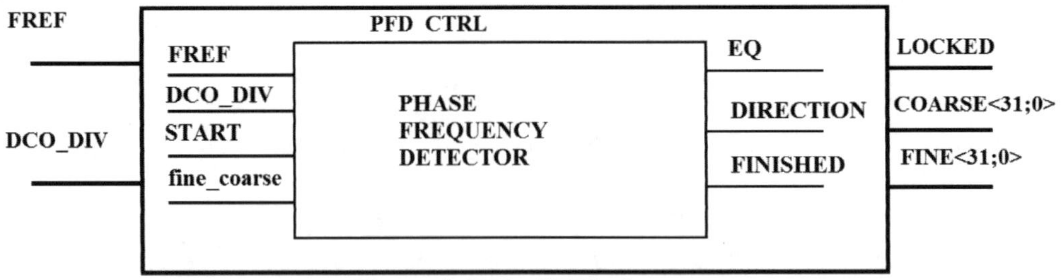

Figure 2. PFD_CTRL diagram.

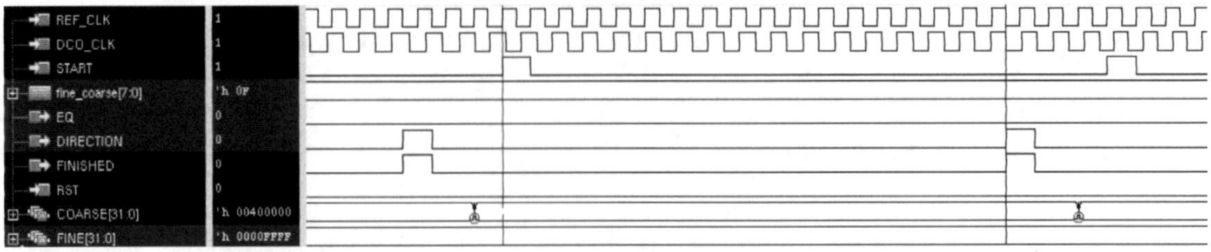

Figure 3. Waveforms of Phase Frequency Detector.

mode comparisons COARSE or FINE (fine_coarse). At the end of phase comparison of the two signals at the end of the iteration signal is generated (FINISHED), the direction (DIREKTION) and equivalence (EQ). FINISHED marks the end of the iteration, the direction indicates that the signal is late or ahead of the reference frequency. The signal EQ marks equity compared frequencies. The length of the account, which can be customized, determines resolution of the phase detector.

A novel frequency detector with variable resolution is used. PFD consists of two counters, which counts clock edges of divided DCO output signal and reference clock signal. When number of reference clock edges achieves maximum, PFD stops and DCO control code changes according to the frequency comparison result. FD accuracy is specified by the maximum counter value, which is different for coarse and fine tuning.

2.2. Digitally Controlled Oscillator

This article describes a digital controlled oscillator for use in an all-digital phase-locked loop. The whole ring contains 32 buffers, which outputs are combined together through three-state invertors and connected to the input of the ring. RST forms the initial step. It also defines the generator is turned on or not: logical 0 – on, 1 – off. The DCO used a two-stage frequency setting – hard tuning (coarse) consisting of the delay circuit and the third state inverters, and fine tuning (fine), constructed from AND-NO elements. The actual ring length is determined by the coarse control one-hot code. In this way, coarse frequency tuning is carried out. To perform fine frequency tuning 32 NAND gates are connected in parallel to the DCO ring. Effective loading capacitance of the NAND gates is determined by the fine control temperature code. The frequency range of the DCO from 170 – 670 MHz With the step of coarse tuning to 15 MHz and 0.5 MHz is fine.

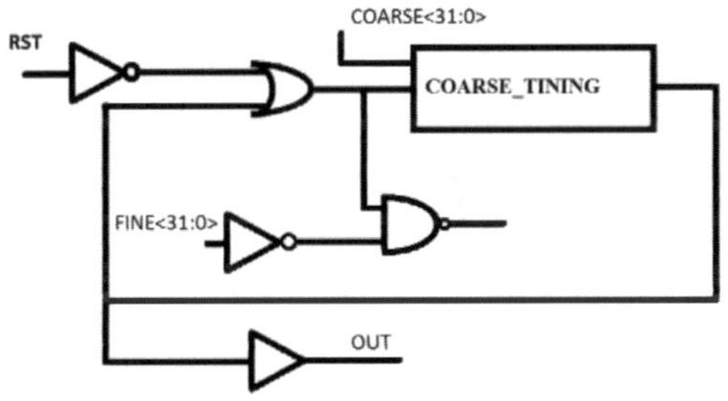

Figure 4. Digitally Controlled Oscillator diagram.

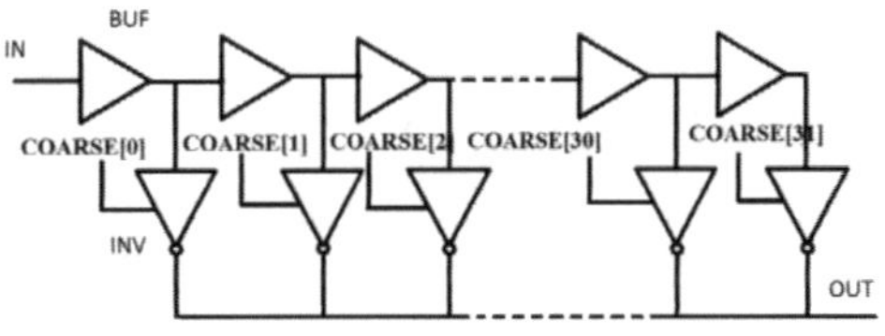

Figure 5. COARSE_TINING of DCO.

2.3 Frequency Divider

The output of the DCO needs to be divided to match the reference frequency. The implementation contains a divide by four frequency state, enabling the ADPLL to have a multiplication factor of 8, 16, 32, 64. The divider is implemented by using two series D Flip-flops by four times. In divider use a multiplexer for selection divider state.

3. Results and comparisons

The ADPLL was found to lock time at 2 μs. The reference frequency was at 80 MHz. For one iteration COARSE setting mode requires 21 clock. For FINE iteration must be 133 cycles. The power of used by the ADPLL system is 6.34 mW. The frequency range of the DCO from 170 - 670 MHz. When the voltage of 0.1 V, the frequency reference DCO falls to 60 MHz. By increasing the temperature by 1 °C rate falls to 1.2 MHz The supply voltage of the system is 1.8 V.

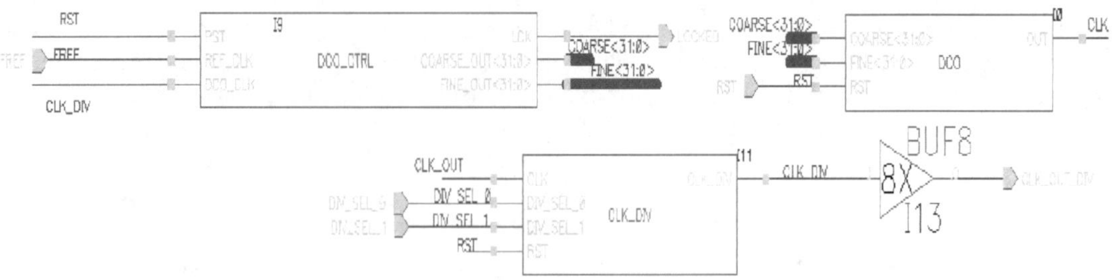

Figure 6. The final system of ADPLL.

4. Conclusion

The paper describes the concept and ADPLL behavioral model of each block. The results can be considered successful - the simulation results are quite good compared to other similar models ADPLL. A good lock time was achieved of 175 cycles and considerations for power were implemented in this project. The work was performed at UMC 180nm technology with a supply voltage 1.8 V. The implementation of the presented ADPLL block focused on the area minimization and design efficiency. The block utilizes only standard digital cells. This significantly reduces design time. The designed ADPLL may be used as IP-block of application-specific SoCs in particular intended for experimental physics' apparatus.

Acknowledgements

This work was supported by Grant No. 14.A12.31.0002, issued by the Russian Federation Ministry of Education and Science according to the RF Governmental resolution No 220. This work was partially supported by MEPhI Academic Excellence Project (contract № 02.a03.21.0005, 27.08.2013).

References

[1] Khalil A H, Ibrahim K T and Salama A E 2002 Digital of ADPLL for good phase and frequency tracking performance *Proc. of the Nineteenth National Radio Science Conf. (Alexandria)* pp 284-90

[2] Olsson T and Nilsson P 2004 A digitally controlled PLL for SoC applications *IEEE J. of Solid-State Circuits* **39** 751-60

[3] Yang S-Y *et al.* 2009 A 7.1 mW 10 GHz all-digital frequency synthesizer with dynamically reconfigurable digital loop filter in 90 nm CMOS *IEEE Int. Solid-State Circuits Conf. - Digest of Technical Papers* pp 90–1

[4] Kumn M and Klingbeil H 2010 An FPGA-Based Linear All- Digital Phase-Locked Loop *IEEE Trans. on circuit and systems* **57**(9)

[5] Yau T Y and Caohuu T 2011 An Efficient All-Digital Phase-Locked Loop with Input Fault Detection *Proc. of IEEE Conference, Information Science and Applications* pp 103-10

Journal of Physics: Conference Series **675** (2016) 042043 doi:10.1088/1742-6596/675/4/042043

Creating a parameterized model of a CMOS transistor with a gate of enclosed layout

S M Vinogradov, E V Atkin and P Y Ivanov

National Research Nuclear University MEPhI (Moscow Engineering Physics Institute), Kashirskoe highway 31, Moscow, 115409, Russia

E-mail: VinogradovSM.work@gmail.com

Abstract. The method of creating a parameterized spice model of an N-channel transistor with a gate of enclosed layout is considered. Formulas and examples of engineering calculations for use of models in the computer-aided Design environment of Cadence Vitruoso are presented. Calculations are made for the CMOS technology with 180 nm design rules of the UMC.

1. Introduction

CMOS integrated circuits nowadays are widely used in many areas with high requirements for radiation resistance, such as high energy physics, nuclear physics and astrophysics. In these experiments, the ionizing radiation may cause various defects in the crystal lattice of the oxide layer of the integrated circuit, which may lead to leakage currents both between the drain and the source of a single transistor, and between the standing side by side transistors. At the same time, the work of a single block, or even the whole integrated circuit can be completely broken. As it is known, N-channel transistors are the most susceptible to this effect[2]. There are different methods to manage the leakage current, but the most common solution is the use of guard rings in combination with an enclosed layout topology of N-channel transistors. Guard rings are used to avoid current leakage between the placed near transistors, and an annular valve, are used to avoid between drain and source. However, it is difficult to fully use these types of transistors in a commercial CMOS technology. Primarily it is caused by poor development or even the lack of parametrized models needed for a full simulation of integrated circuits and the optimization on a range of parameters. This article is dedicated to one of the possible ways to create parametrized models or channel PCELL N-MOSFET ring to complete its technology library of CMOS technology with design rules 180 nm UMC company (Taiwan). The approval of the model is assumed to be done in CAD Virtuoso v.6.1.5 OA of Cadence company, which is used widely by authors in the last years.

2. N-channel enclosed layout transistor(ELT)

2.1. Gate enclosed topology

This model of the transistor [1,3] is able to operate, without an interruption at a total dose of 30 MRad thanks to the enclosed layout structure on the technological library of 240 nm. In Figure 1 there is shown such a ring structure taken however from an other technological library. Where: 1 - P-type guard ring, 2 - ring gate, 3 - drain 4 - source, 5 - symbol layer, 6 - a horizontally stretched transistor line, 7 - a vertical stretched transistor line, 8 - an enclosed gate stretching line.

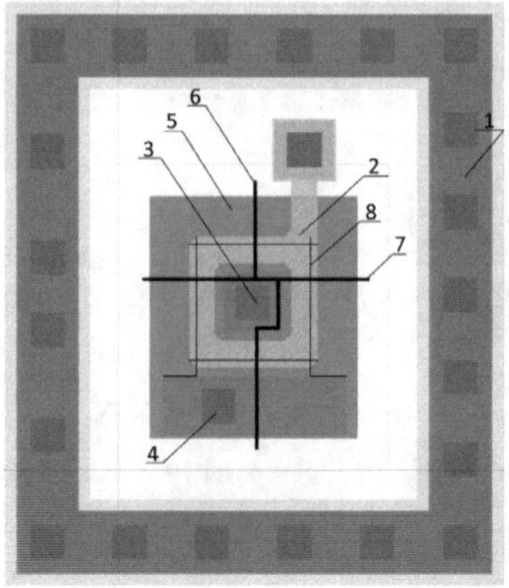

Figure 1. Topological representation of N-channel ELT with minimal geometric dimensions for the 180nm technology of UMC company.

2.2. Model of ELT

In most cases, while describing the transistor with the enclosed topology, one resort to stating its length and width to the length and width of a standard transistor. In contrast to the ring one, standard transistor is well described in various literature and his model comes in standard technology libraries of all companies producing custom integrated circuits. For this reason, the most correct method is to construct a model based on the ELT standard transistor. The formula used to set the annular transistor to standard transistor shown in [2]:

$$\frac{W}{L} = 4\frac{2\alpha}{ln\frac{d'}{d'-2\alpha L}} + 2K\frac{1-\alpha}{\frac{1}{2}ln\frac{1}{\alpha}\sqrt{\alpha^2+2\alpha+5}} + 3\frac{\frac{d-d'}{2}}{L}$$

when $\alpha = 0.05$, K = 3.5 (for channel length < 0.5u), d = inner diameter of the Gate, d= d without the 45 Cut

With this formula, and the standard model of the transistor, we can work to characterize the standard transistor sufficiently.

3. Parametrized Cell

3.1. Parameterized CAD scheme in Virtuoso Cadence

In the CAD Virtuoso Cadence it is possible to use standard tools and specialized programming language SKILL to create Parametrization scheme(PCell). To creation Route of PCell consists of two steps: 1) Creation of a topological representation varies with the basic parameters of the transistor. 2) Creation of a schematic representation for the modeling of the transistor in the electrical circuits with the changes in the major parameters of the transistor.

3.2. Creation of ELT topology

To create a topological representation in addition to the standard software available in CAD, the formulas which will build and change the topology of future transistors are required. These formulas should consider not only the basic parameters of the transistor, and its compatibility with the physical implementation, such as the minimum possible distance between the various layers and their compliance with the minimum step of the grid. The following are the formulas written in SKILL which build the ELT topology:

- Formula of Pcell vertical stretching according to the line 6 (Figure 1)and according to the horizontal line 7 (Figure 1). The formula takes into account the possible stretching speed and corresponds to the grid:

$$fix((\tfrac{0.1(1-(L+1)E}{1-E} + 0.56 + 2L - 2(L - 0.18)) * 100) * 0.01$$

when

$$E = exp(\frac{0.4L}{W - 0.5615509 * KL - 0.15}).$$

- Stretch gate formula. The formula corresponds to the grid:

$$fix(\frac{L}{0.01}) * 0.01$$

- Formula of duplicate source contacts. The formula takes into account the possible stretching gate and corresponds to the grid:
  Step duplication

$$St_s = 0.5,$$

the number of repetitions

$$N_s = fix(2\frac{0.1 * (1 - (L + 1)E)}{1 - E} + 0.72) + 0.5.$$

- Formula of duplicate contacts drain. Formula takes into account the possible gate tension and corresponds to the grid:
  Step duplication

$$St_d = fix(100\frac{\frac{0.1(1-(L+1)*E)}{1-E} - 0.04 - 0.5N_s}{N_s} + 50) * 0.01,$$

the number of repetitions

$$N_d = fix(fix(100\frac{0.1 * (1 - (L + 1) * E)}{1 - E} - 4) * 0.02).$$

3.3. Example of building ELT topology at different preset parameters
When the layout is changed, PCell varies submitted to the parameters. The built topologies are tested with DRC in Calibri software tool. Thus were built topologies for different parameters, they are depicted in Figure 2 and 3.

3.4. Create of ELT schematic
Inherently the ELT model of transistor is a CellView with the possibility of setting the three parameters: W, L, K. According to these parameters, the parameters of the standard N-channel transistor inside CellView and its topology.

The parameters of the transistor pleased inside CellView schematic vary according to the following formulas:

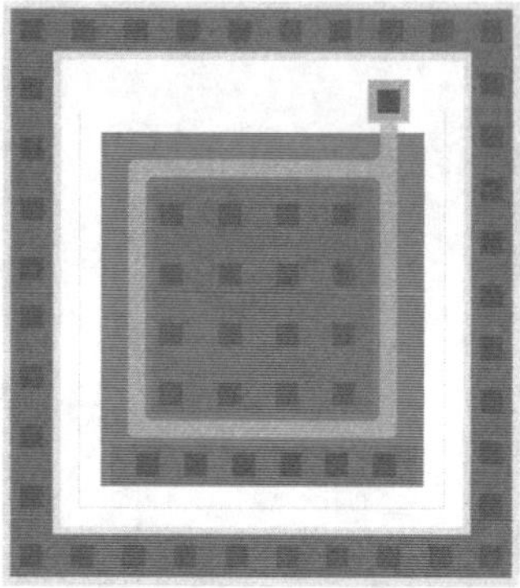

Figure 2. Constructed topology for W = 10 um, L = 0.18 um, K = 3.5.

Figure 3. Constructed topology for W=10 um, L=0.5 um, K=4.

- The formula for calculating the perimeter of the source:

$$P_s = (0.4\frac{1-(pPar("L")+1)Es}{1-Es} + 8pPar("L") + 2.66) * 10^{-7}$$

when

$$Es = exp(\frac{0.4pPar("L")}{pPar("W") - 0.5615508536 * pPar("K") * pPar("L") - 0.15}).$$

- The formula for calculating the area of the source:

$$S_s = (0.139\frac{1-(pPar("L")+1)Es}{1-Es} + 3.01pPar("L") + 0.2303) * 10^{-13}.$$

- The formula for calculating the perimeter drain:

$$P_d = 0.4\frac{1-(pPar("L")+1)Es}{1-Es} * 10^{-7}.$$

- The formula for calculating the area of the drain:

$$S_d = (0.1\frac{1-(pPar("L")+1)Es}{1-Es})^2 * 10^{-13}.$$

4. Conclusion

Using standard Cadence CAD tools and mathematical calculations, there has been built a simple enclosed layout transistor PCell for the CMOS 180 nm UMC technology library.

Acknowledgements

This work was partially supported by MEPhI Academic Excellence Project (contract No. 02.a03.21.0005, 27.08.2013).

References

[1] Kloukinas K *et al.* 1998 Development of a Radiation Tolerant 2.0 V standard cell library using a commercial deep submicron CMOS technology for the LHC experiments *Proc. of the 4th Workshop on electronics for LHC experiments, Rome, 21-25 September 1998* pp 574-80

[2] Bruder M, Fischer P and Harter M 2006 Radiation Hard Design in UMC 018 *Proc. of the CBM Week, 02 March 2006, GSI, Darmstadt*

[3] Anelli G *et al.* 1999 Radiation Tolerant VLSI Circuits in Standard Deep Submicron CMOS Technologies for the LHC Experiments: Practical Design Aspects *IEEE Transactions on Nuclear Science* **46**(6) 1690-6

[4] J. Verbeeck, P. Lerouxb and M. Steyaert, "Tolerant VLSI Circuits in Standard Deep Submicron CMOS Technologies for the LHC Experiments",Proceedings of the IEEE ICMTS 2011.

Optimization of the light yield properties from scintillator tiles read out directly by silicon photomultipliers

B Bobchenko[1,2], M Chadeeva[1,2], M Danilov[1,2], D Mironov[1,2], V Rusinov[1,2] and E Tarkovskiy[1,2]

[1] Institute for Theoretical and Experimental Physics, Moscow, Russia
[2] National Research Nuclear University MEPhI (Moscow Engineering Physics Institute), Kashirskoe highway 31, Moscow, 115409, Russia

E-mail: dima.mironov@cern.ch

Abstract. The research of the light yield from the scintillator tiles with direct readout by silicon photomultipliers (SiPM) has been performed. The tile size is $30 \times 30 \times 3$ mm^3 as planned for the AHCAL of the ILD at the ILC. The different tile geometries were studied. The uniformity of light yield has been optimized and one geometry has been selected. The results are compared to the studies on the same topic, performed by another groups.

1. Introduction

The International Linear Collider is one of the proposed e^+e^- accelerators to further study the properties of the Higgs boson and to search for the dark matter particles. [1] The proposed center of mass energy at the ILC is from 250 GeV to 1 TeV. To achieve a very high energy resolution, which is required for the physics goals, the Particle Flow Approach [2, 3, 4] has been invented. It exploits the possibility to measure the energy of charged particles inside jets in the tracker with the precision much higher, than it is achievable in the calorimeter. The PFA suggests to subtract the energy in the calorimeter, which can be assigned to charged tracks, and consider the rest as an energy of the neutral fraction of the jet. To exploit the possible efficiency of the PFA a very high granularity of the calorimeter is a key requirement. The Analogue Hadron Calorimeter (AHCAL) is on of the projects of such a calorimeter for the ILD detector project at ILC. It is assembled from the scintillator tiles, read out by silicon photomultipliers (SiPM). The transverse tile size has been optimized considering the efficiency of the energy reconstruction and is 30×30 mm^2. The currently proposed tile thickness is 3 mm.

The physics prototype of a highly granular calorimeter was built in 2006 by the CALICE collaboration. It was assembled from 7608 square tiles [5] of different sizes. The big faces were held between two highly reflective foil mirrors and the edges were mated using special chemical treatment. The green wavelength-shifting (WLS) fiber has been used in every tile to collect the light and guide it to the SiPM and increase the efficiency of the detection. The recent development of the SiPMs with increased efficiency and sensitivity to blue light gave a possibility to get rid of the WLS fiber still achieving the required light yield. This vastly reduces manpower needed for the manufacturing and costs of assembly. However, if the SiPM is just placed next

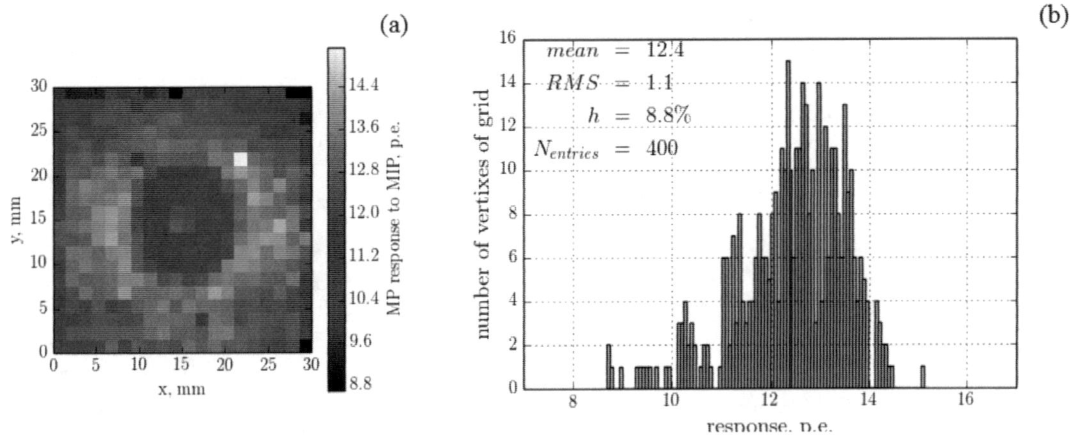

Figure 1. (a) Dependence of the response to MIP in photoelectrons on the beam position in the XY-plane with the optimal geometry[8]. (b) The distribution of the MP responses to MIP in the optimal geometry.

to a flat tile, the response has very big spatial nonuniformity, which affects the calorimeter resolution. To achieve the uniformity of the response, which will allow the required resolution, the dimple design has been introduced in [6]. The goal of this study is the optimization of the geometry of the readout and measurement of the properties of the optimal one.

2. Dimple concept

The minimum ionizing particle (MIP) produces the same amount of light independently of the impact position, when crossing the flat tile. The light which is emited in front of the SiPM has twice higher probability to reach the photodector. To improve the uniformity, one can reduce the mean light emission from MIP in the tile center by reducing the scintillator thickness there by making a dimple in the scintillator in front of the SiPM. In our study the paraboloid dimple is used to optimize the light collection uniformity. The dimple shape is parametrized by its depth d and radius at the tile face r.

The surface mounted SiPM can be plunged into the dimple. This allows using SiPMs soldered directly to the mirrored PCB without introduction of any special electrical connectors.

3. Nonuniformity characterization

To measure the nonuniformity, a rectangular lattice in XY-plane has been used, which is parallel to the big tile face. The SiPM response spectrum to β-particles flying through each vertex of the lattice has been measured. Electrons from ^{90}Sr source were used to simulate a minimum ionising particle (MIP). The response spectrum at each XY-point has been fitted with a Gaussian function around the peak of the distribution. The obtained Gaussian mean, which is close to the most probable (MP) value of the spectrum, has been considered to be the response to a MIP at the given beam position with a correction from [7]. Example of the dependence of this value on the beam position in the XY-plane for the optimal geometry is shown in figure 1a. The distribution of the MP responses to MIP in the optimal geometry is shown in figure 1b. It makes sense to characterize it by the mean M and the spread RMS and take the ratio $h = \frac{RMS}{M}$ as a measure of the response nonuniformity. It allows getting the rough estimate the additional relative error in energy introduced by the nonuniformity as $\frac{h}{\sqrt{N}}$, where N is the number of tiles affected by a shower.

In 2006 WLS-fibered tiles have been tested in a proton beam. When calculating an average using 2 mm lattice, nonuniformity measure h is $\approx 7\%$ with 2 mm scan step and $h \approx 10\%$ with 1 mm step with effective beam size of 1 mm [9].

4. Experimental setup

The dimples have been milled on the big face (30×30 mm^2) using the 3-dimensional engraver with the precision of 10 μm.

The ^{90}Sr source and trigger tile are moved in the XY-plane along the tile big face. The beam direction is along Z axis. The tile is laid down on its big flat face on the fixed mirrored table and covered by a mirror with a hole in the center, in which the SiPM is placed. The surface of the SiPM can be at the mirror level or several millimeters higher or lower. The beam diameter is about 1.3 mm.

In the response conversion to photoelectrons per MIP the 20% SiPM crosstalk and MIP simulation by β-particles have been taken into account.

5. Test of light focusing hypothesis

In the previous note we have used a SiPM from KETEK [10] with 2.2×2.2 mm^2 window and 12100 pixels. It gave us a mean MP response of 12.5 photoelectrons. To test the hypothesis of the focusing of the light we have tried another SiPM from KETEK with 1.2×1.2 mm^2 window and 576 pixels. The surface decrease is 2.8 times, while the observed light yield decrease was only 1.7 times. The nonuniformity increased not significantly from $h = 7.6\%$ to $h = 10.1\%$.

The study of the effect of properties of the surface in front of the SiPM has been performed. No significant differences in light yield have been observed between the initial option (just after milling), polished one and one mated with a hard sandpaper.

6. Test of Mainz geometry

The group from the University of Mainz performed similar studies[11]. They use another scintillator, wrap their tiles with the laser cut 3M ESR foil and use a SiPM from Hamamatsu with 1.3×1.3 mm^2 window and 2668 pixels. Their tiles show the response of 19 photoelectrons per MIP even with a cheaper type of the scintillator. They proposed another measure of nonuniformity H, which is the fraction of the surface, where the response is deviated from the mean response by less than 5%. Below we make the comparison for both the measures h and H. To make a crosscheck with our results the dimples have been milled with a geometry proposed by the University of Mainz in tiles from the scintillator we use. The edges have been mated using special chemical treatment. On our setup with our 1.2×1.2 mm^2 SiPM they show the response of 5.4 photoelectrons per MIP, which is less than the response in our optimal geometry. For Mainz geometry we see the peak of the response in the tile center as in [11]. The nonuniformity value we measure for their tiles is $h = 32\%$, which is significantly higher than any other h value of tiles with dimple and is comparable to $h \approx 40\%$ of flat tiles. We measure H to be $H = 13\%$ for the tiles with their geometry and $H = 28\%$ for our configuration.

7. Conclusion

The direct readout of scintillator tiles with SiPMs is a possible and feasible option for high-granularity calorimetry. The particular geometry to be selected depends on the scintillator material.

Acknowledgments

The work was performed within the framework of the Center of Nanostructured Electronics supported by MEPhI Academic Excellence Project (contract No. 02.a03.21.0005, 27.08.2013) and was partially sponsored by the Russian Ministry of Education and Science contracts 4465.2014.2 and 14.A12.31.0006 and by the RFBR grant 14-02-00873A.

References

[1] Behnke T *et al.* 2013 *arXiv*:1306.6329
[2] Brient J C and Videau H 2001 *arXiv*:0202004
[3] Morgunov V 2002 *CALOR 2002* 70-84
[4] Thomson M 2009 *Nuc. Instrum. Meth.* A **611** 25-40
[5] Adloff C *et al.* (The CALICE collaboration) 2010 *JINST* **5** P05004
[6] Blazey G *et al.* 2009 *Nucl. Instrum. Meth.* A **605** 277-81
[7] Simon F 2011 *CALICE collaboration meeting*
[8] Bobchenko B *et al.* 2015 *Nucl. Instrum. Meth.* A **787** 166-8
[9] Tarkovsky E 2006 *ILC workshop*
[10] URL: *http://www.ketek.net/*
[11] Liu Y 2015 *CALICE collaboration meeting*

Development of event reconstruction algorithm for full-body gamma-camera based on SiPMs

D E Philippov, V N Belyaev, P Zh Buzhan, A L Ilyin, E V Popova[*] and A A Stifutkin
National Research Nuclear University MEPhI (Moscow Engineering Physics Institute), Kashirskoe highway 31, Moscow, 115409, Russia

E-mail: filippovdima@rambler.ru, buzhan@inbox.ru, [*]elenap73@mail.ru

Abstract. The gamma-camera is the detector for nuclear medical imaging where the photomultiplier tubes (PMTs) could be replaced by the silicon photomultipliers (SiPMs). Common systems have the energy resolution about 10% and intrinsic spatial resolution about 3 mm (FWHM). In order to achieve the requirement energy and spatial resolution the classical Anger's logic should be modified. In case of a standard monolithic thallium activated sodium iodide scintillator (500x400x10 mm^3) and SiPM readout it could be done with identification of the clusters. We show that this approach has a good results with the simulated data.

1. Introduction

Most of recent R&D projects in the field of gamma-detectors for medical physics use SiPMs. SiPMs attract a particular interest for developers of detectors for nuclear diagnostics (PET or SPECT), because of its small size, good timing resolution and immunity to magnetic fields. Thus the usage of SiPMs instead of traditional PMTs potentially could improve both the spatial and the timing resolution of the systems and also makes possible the simultaneous use of the nuclear and magnetic-resonance imaging.

The gamma-camera is one of the detectors for nuclear diagnostics where the PMTs could be replaced by the SiPMs. The basic requirements for gamma-cameras are the following: the energy resolution about 10% for 140 keV (the most intensive line of Tc-99m); the operation range of energies from 60 to 360 keV and the intrinsic spatial resolution better than 3 mm in full width on a half of maximum (FWHM).

In this paper we discuss the feasibility to design a gamma-camera with a standard monolithic thallium activated sodium iodide scintillator (NaI(Tl), size of 500x400x10 mm^3), array of 6x6 mm^2 Ketek SiPM PM6660 [1] and 64-channel ASIC MAROC3 [2] as readout electronics. The main problem in such a setup is to achieve the energy resolution at least not worse than those for the standard gamma-cameras. To demonstrate that development of such system is possible, its operation should be first simulated. This simulation has to describe the gamma-quanta interactions, light emission and propagation in scintillator, and the main properties of the SiPM. Also the classical Anger's reconstruction method [3] should be modified to implement it for SiPM's array.

2. Simulation setup

The Monte-Carlo simulation of the full-body gamma camera detection unit was performed using Geant4 (G4) [4] libraries. The simulation has been performed for the following conditions:

- scintillator – NaI(Tl) crystal with size of 500x400x10 mm^3;
- white diffuse reflector was on all sides except the output window;
- crystal was placed in aluminium case with thickness of 2 mm.

For simplicity only one wavelength from real scintillation spectra related to maximum was used, so all optical properties of the components and their surfaces were set only for that mono-line. The light yield of NaI(Tl) was set as 38 photons per keV. The decay time was set as 250 ns. The diffuse reflector was simulated with LUT model of Geant4 [5] and «groundteflonair» surface type with the reflection coefficient of 99.9%. The case output window was 2 mm thick glass coupled with array of SiPMs via optical grease (thickness of 50 um). Each SiPM had size of 7.8x6.8x1.75 mm^3, with sensitive area of 6x6 mm^2. The pitch of the SiPM array was 8 mm for X-axis and 7 mm for Y-axis. All dimensions were taken from existing 64-channel matrix of 6x6 mm^2 Ketek SiPMs (assembled at "Pulsar" in Moscow). This matrix represents one module of the further full-body gamma-camera's detector system. Full detector consists of 62x56 SiPMs which cover almost whole surface of the scintillator. The figure 1 presents schematic model of SiPM array (a), light spread in G4 (b) and SiPM G4 volume (c).

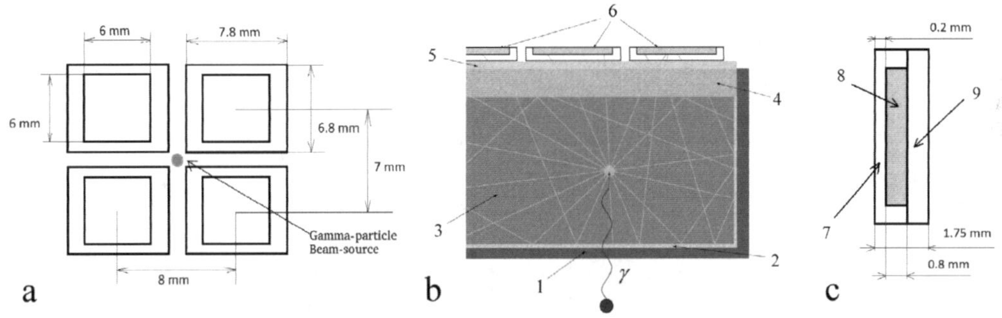

Figure 1. a) four cells from the SiPM array; b) light spread in Geant4; c) SiPM. 1 — aluminium package of the crystal, 2 — diffuse reflector, 3 — NaI(Tl) crystal (n = 1.82), 4 — case output window (n = 1.47), 5 — optical grease (n = 1.51), 6 — photodetector matrix; 7 — epoxy of the SiPM (n = 1.53), 8 — 6x6 mm2 silicon absorbing layer (n = 5.2), 9 — glass-reinforced epoxy laminate FR4 (n = 1.53).

The radionuclide was simulated as a beam point source emitting gamma particles along Z-axis (perpendicular to the detector's surface) with energy of 140 keV at (0;0;50) mm coordinates of world volume in G4 simulation (42 mm from aluminium case).

The output data of G4 simulation was the number of absorbed optical photons in each SiPM for each event. After that G4 data was processed using MATLAB macros to take into account the main properties of SiPMs, such as Dark Counts Rate (DCR) and Photon Detection Efficiency (PDE), and integration time of MAROC3. Thus the number of photons first was converted into number of photoelectrons (or fired microcells, not saturation range) with binomial distribution of PDE of 40% and integration time of 150 ns, and then Poisson distributed dark counts with the rate 14 MHz were added to the signal of each SiPM. Therefore the mean DCR was 2.1 count per event per SiPM and total conversion efficiency of the system (for absorbed light photons) was about 19%.

3. Methods

The final data was processed with three algorithms to reconstruct energy and position information:

- simple Anger's logic with the threshold;
- modified Anger's logic with the threshold and adaptive variable radius;
- modified Anger's logic with the threshold and static cluster's size of N-by-N SiPMs.

3.1. Simple Anger's logic with the threshold

The first algorithm is the standard Anger's logic. The signals above the threshold were used to reconstruct event energy and position, using center-of-gravity formulas (1) for all triggered detectors:

$$\overline{x} = \sum_{i=1}^{N} x_i \cdot A_i \cdot (\sum_{i=1}^{N} A_i)^{-1} \qquad \overline{y} = \sum_{i=1}^{N} y_i \cdot A_i \cdot (\sum_{i=1}^{N} A_i)^{-1} \qquad Energy = \sum_{i=1}^{N} A_i \ , \tag{1}$$

where x_i, y_i - (x;y)-coordinates of i-photodetector, A_i – corresponding amplitude and $\overline{x}$, $\overline{y}$ - center of gravity position.

3.2. Modified Anger's logic with threshold and adaptive variable radius

In the second algorithm the primary cluster was determined as in the first one. All signals from detectors those above the threshold were used to calculate cluster's center coordinates (CCC) using (1) and the standard deviations (σ_x and σ_y) using statistical formulas for discrete distributions (2):

$$\sigma_X = [\sum_{i=1}^{N} (x_i - \overline{x})^2 \cdot A_i \cdot (\sum_{i=1}^{N} A_i)^{-1}]^{1/2} \qquad \sigma_Y = [\sum_{i=1}^{N} (y_i - \overline{y})^2 \cdot A_i \cdot (\sum_{i=1}^{N} A_i)^{-1}]^{1/2} \ . \tag{2}$$

After that the average value of σ_x and σ_y was multiplied by numerical coefficient C and labeled as the cluster's radius. Thus the cluster's radius changes according to width of signals distribution. The signals of the detectors which centers lay inside the cluster were used to get position and energy information using center of gravity formulas (1) again.

3.3. Modified Anger's logic with threshold and static cluster's size of N-by-N SiPMs

The third method differs from second in how the cluster size is defined. Here it was defined as the rectangular arrays of N-by-N SiPMs instead of σ with CCC determined exactly as for the previous method.

4. Results

After the simulated data (from gamma source mentioned above) were processed using all algorithms the following results have been obtained. The energy resolution for the first algorithm was 26.7% without threshold, and 19% and 22.6% with threshold of 5 and 10 photoelectrons (phe), respectively (figure 2a). The plots of the spatial and energy resolution changes versus threshold value for the first algorithm are shown in figure 2b.

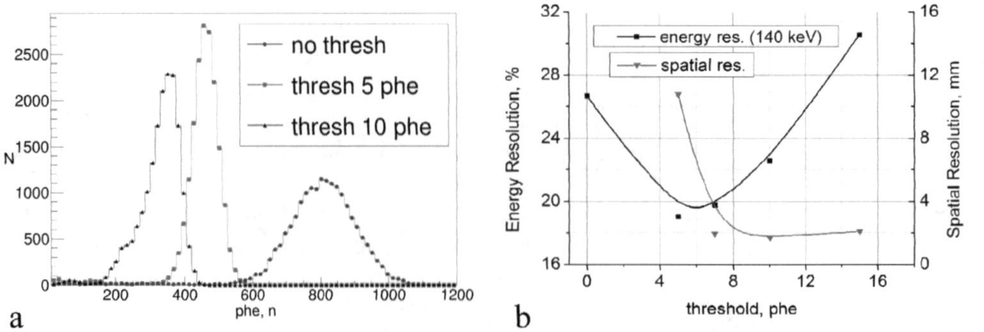

Figure 2. First algorithm, a) energy spectra, thresholds of 0, 5 and 10 phe; b) energy and spatial resolution versus threshold (solid line – b-spline interpolation).

Journal of Physics: Conference Series **675** (2016) 042045 doi:10.1088/1742-6596/675/4/042045

The first algorithm can't be used for the position reconstruction without threshold, but the spatial resolution of 10.8 and 1.7 mm (FWHM) have been obtained for the thresholds of 5 phe and 10 phe, respectively (figure 3).

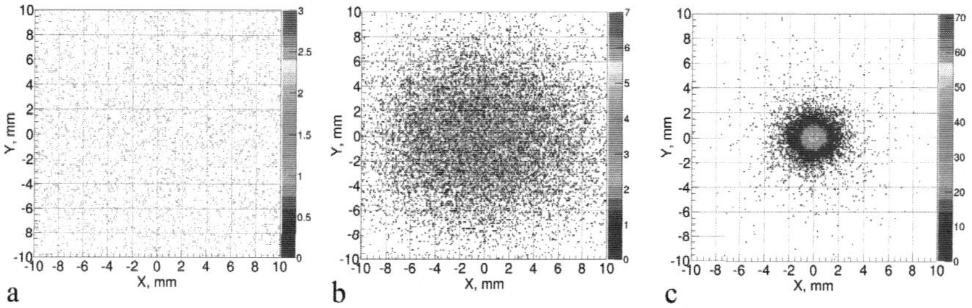

a b c

Figure 3. First algorithm. The distribution of reconstructed positions from point source for different thresholds: a) 0 phe; b) 5 phe; c) 10 phe.

The threshold values of 5 and 10 phe had been also used for reconstruction using the second algorithm. The mean value σ_5 was 35.6 mm for the threshold of 5 phe, and the mean value σ_{10} was 6.6 mm for the threshold of 10 phe (figure 4a).

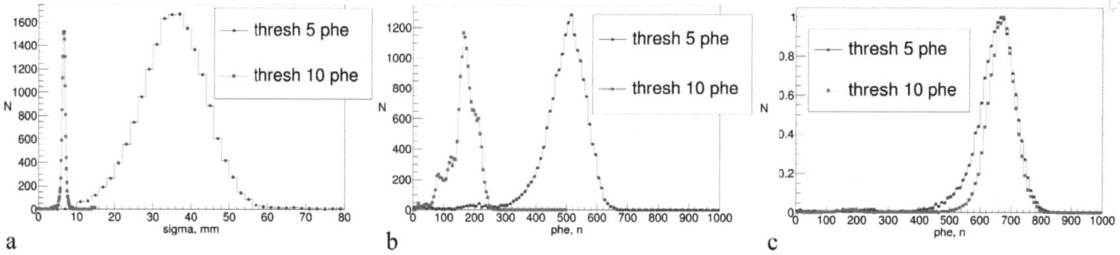

a b c

Figure 4. Second algorithm, a) distribution of averaged by X and Y standard deviation for thresholds of 5 and 10 phe; b) the energy spectra for the thresholds of 5 and 10 phe, radii $1\sigma_5$ and $1\sigma_{10}$; c) radii $2\sigma_5$ and $11\sigma_{10}$ (normalized spectra).

The energy spectra have a distortion of the photopeak (figure 4b) for the clusters, which radius was determined as 1σ (σ_5 and σ_{10}, respectively). Especially, this is clearly seen for the 10 phe threshold. The energy spectra (normalized) are shown in figure 4c with the best energy resolutions of the equal cluster's size for both thresholds (radii $2\sigma_5$ and $11\sigma_{10}$).

Thus for the threshold of 5 phe the optimal radius for the best energy resolution was $2\sigma_5$, and corresponding energy resolution was 18.1%. For the threshold of 10 phe the best energy resolution has been obtained of about 14.9% with the 8-$11\sigma_{10}$ radii. The plots of the spatial and energy resolution as function of the radius (in mm) are given in figure 5 for the thresholds of 5 and 10 phe. The best spatial resolution for the second algorithm was obtained with $1\sigma_5$ and $3\sigma_{10}$ radii and amounted to 3 mm and 1.5 mm, respectively for two thresholds.

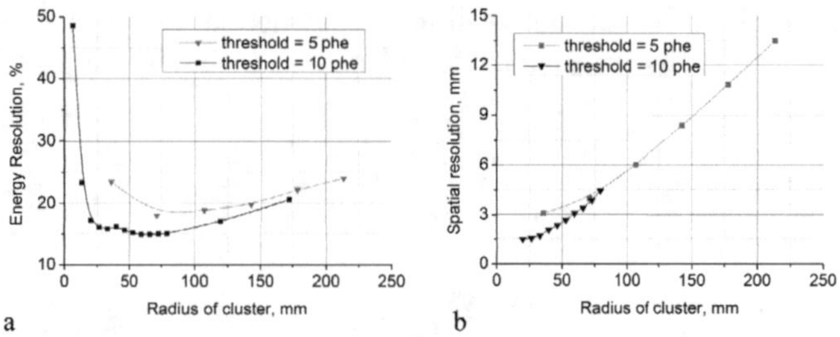

Figure 5. Second algorithm, a) the energy resolution versus the radius; b) the spatial resolution versus the radius (solid line – b-spline interpolation).

For the third algorithm the energy and spatial resolution dependencies on the cluster size was obtained for the threshold of 10 phe, since this threshold in the second algorithm provides the best energy resolution. In figure 6a, the spatial and energy resolution dependencies on cluster size are plotted. As seen on the graph, the best energy resolution of 12.8% of all three algorithms was obtained for the cluster size of 12x12 SiPM. At the same time the best spatial resolution of 1.35 mm was obtained for the 6x6 cluster, while the energy resolution for that cluster was 16.1% (figure 6b).

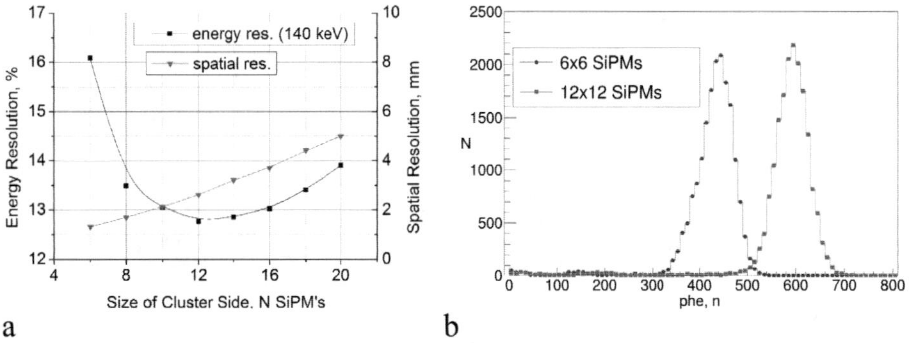

Figure 6. Third algorithm, a) energy and spatial resolution versus cluster's size (N-by-N SiPMs) for the threshold of 10 pe (solid line – b-spline interpolation); b) energy spectra for clusters 6x6 and 12x12 SiPM's.

Additionally, events were generated when the beam source (140 keV) was uniformly distributed over the detector surface. 2D distributions of the reconstructed event's positions for all algorithms are shown in figure 7. The results are shown only for the one detector quadrant in order to reduce the time of the data processing, due to the geometry's symmetry for both X and Y axis.

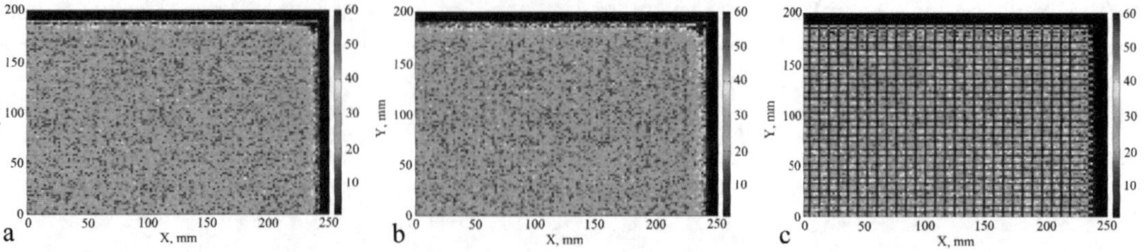

Figure 7. The distribution of reconstructed event's positions from uniformly distributed source (140 keV), threshold of 10 pe. a) first algorithm; b) second algorithm with $3\sigma_{10}$ radius; c) third algorithm with 12x12 SiPM cluster size.

5. Results discussion

5.1. Energy resolution

According to the results, it is not possible to achieve good energy resolution using the standard Anger logic (first method), since the sum of the signals from all the photodetectors increases significantly the noise contribution in the energy resolution. The threshold helps to skip over the detectors whose signal have no useful information, but when the threshold increases, the SiPMs cluster loses useful information and that leads to distortion of the energy spectrum and to the worse energy resolution. Thus, the using of the only one threshold is rather crude approach and insufficient to determine the optimal cluster.

The second and the third methods apply preliminary procedure to define cluster position and size and after that carry on reconstruction using selected devices. It allows to obtain much better results than in the first method. The fixed size rectangular cluster (algorithm 3) providing the highest energy resolution for given detector's geometry. However, such optimization is not sufficient to achieve the required energy resolution of 10%. The improvement of the photodetector parameters is also needed.

5.2. Spatial resolution

The enhancement of the intrinsic resolution is one of the advantages of the SiPM based gamma-camera. According to the results of simulation additional clarification in the cluster selection does not significantly improve the spatial resolution. Thus the spatial resolution unlike the energy one is less affected by influence of the DCR and the only threshold approach is sufficient for the position reconstruction. Also worth noting that the distributions of the reconstructed coordinates from the uniformly distributed source (figure 7) show the uniformity of the first and second methods throughout the plane of the crystal (except the regions close to the edges), unlike the third algorithm, whose response depends on the location of the interaction point relative to the SiPM.

6. Conclusion

The operation of the detection module of the full-body gamma-camera which consists of the monolithic sodium iodide crystal, array of the SiPMs and was irradiated by 140 keV beam point source was simulated. The simulation also includes two main SiPM parameters: the dark count rate and the photon detection efficiency, as well as the integration time of the readout electronics. The results of the position and energy data reconstructions using three algorithms based on the Anger logic are presented.

The major improvement of the suggested reconstruction methods is in the finding of the signal cluster for the further energy and coordinates calculations using centre of gravity formulas (1). The best energy resolution was 12.8% for 140 keV. The best spatial resolution amounted to 1.5 mm (FWHM).

The obtained results show feasibility of the creation of the full-body gamma-camera based on SiPM.

Acknowledgments

This work was supported by Megagrant 2013 program of Russia, agreement №14.A12.31.0006 from 24.06.2013 and partially supported by MEPhI Academic Excellence Project (contract № 02.a03.21.0005, 27.08.2013).

References

[1] PM6660 SiPM Ketek (*http://www.ketek.net/products/sipm/pm6660/*)
[2] ASIC Maroc3 (*http://omega.in2p3.fr/index.php/products/maroc-front-end-chip.html*)
[3] Peterson T and Furenlid L 2011 *Phys. Med. Biol.* **56** R154
[4] Agostinelli S *et al.* 2003 *Nucl. Instr. Meth.* A **506** 252-3
[5] Janecek M and Moses W. 2010 *IEEE Trans. Nucl. Sci.* **57** 967
[6] MATLAB (*www.mathworks.com*)

Journal of Physics: Conference Series **675** (2016) 042046 doi:10.1088/1742-6596/675/4/042046

Simulation of surface radiation defects leakage current SiPM using Synopsys TCAD

P P Parygin*, E V Popova and V M Grachev
National Research Nuclear University MEPhI (Moscow Engineering Physics Institute), Kashirskoe highway 31, Moscow, 1125409, Russia

E-mail:donkarloson@gmail.com

Abstract. Synopsys TCAD is a professional software for the development of the semiconductor technological process and device simulaion. In order to study a radiation damage of the surface of silicon photomultipliers (SiPMs) the simulation of these devices using Synopsys TCAD has been made. Experimental samples were produced by KETEK GmbH and have been irradiated with the different doses of X-rays with an energy of E≈12 keV. The current-voltage characteristics below breakdown measured before and after irradiations have been simulated with TCAD. Obtained curves for experimental and simulation data are presented.

1. Introduction

SiPMs are very promising semiconductor single photon sensitive devices with a high gain [1]. As opposed to conventional vacuum tube photomultipliers (PMTs), SiPMs are not sensitive to the magnetic fields, need much lower bias voltage and they are compact. SiPM is an array of connected in parallel identical single pixels, each of which represents a p-n junction with a serial resistor. Each pixel operates in Geiger mode, when the bias voltage is higher than the breakdown voltage [2].

SiPM has a promising application in such science areas as a high-energy physics, nuclear medicine and astroparticle physics. However, the possibility of using SiPMs in some areas is limited by the radiation tolerance. SiPM radiation hardness is being studied for several years, for example in [3, 4, 5].However, in such works no attempt was made to separate a bulk and surface damage. Recently, the investigation of the pure surface radiation damage has been started. Firstly, it has been done for the Hamamatsu MPPCs [6] and then for the SiPMs produced by KETEK GmbH[7]. Though the detailed analysis were done in these works, the simulation has not been performed.

Synopsys TCAD (Technology Computer-Aided Design) [8] is a powerful tool for the simulation of the semiconductor devices with an environment to analyze the simulation results. In this work, Synopsys TCAD was used to implement a simple design of SiPM to understand the surface radiation damage in SiPM in more detail. Using this model, it is possible to make an analysis of the radiation influence on the surface and bulk regions with respect to surface electric field distribution.

2. Experimental samples and setup

KETEK GbmH (Munich, Germany) [9] manufactured experimental SiPM samples and test structures. Test devices (PNCV)represent a single pixel with p^+-type entrance window (1x1 mm^2) surrounded by n-type highly doped periphery with different gap from 62 μm to 152μm on all four sides. The full structure size is 1.1x1.1 mm^2. In this work the PNCV p-on-n structure with multilayer dielectric on top between p^+-type and n-type (thermal oxide, silicon nitride and deposited oxide) were used.

Keithley 6517B Electrometer [10] was used for the current-voltage characteristic measurements. All measurements were performed under normal conditions. X-ray irradiations were carried out using PHYWE XRE 4.0 Expert Set [11] with anode made of tungsten. Two different doses were applied on the test samples: 200 Gy and 3 kGy, using X-rays with an energy of E≈12 keV. The maximum transmitted energy to silicon atoms for 12 keV photons is 0.011 eV, which is quite below the damage threshold for silicon of 21 eV [12]. Thus, no bulk damage is expected for 12 keV photons. The irradiation of the samples were performed without biasing and the samples were not annealed after the irradiation. The current-voltage characteristics of the irradiated samples were measured immediately after the irradiation.

3. Simulation model

The PNCV SiPM model is presented in figure 1. This structure has a length L = 20 μm and 5.5 μm depth. It consists of the next layers: dielectric layer, which is presented, for simplification, as the single silicon dioxide (0.5 μm depth), boron highly doped p^+ entrance window (0.45 μm depth), phosphorous n-type layer (2.5 μm depth) performed as a retrograde profile and phosphorous n^+ highly doped substrate (from 3 to 5 μm depth). For the device simulation, the bias and the ground contacts have been added. The bias contact extends beyond p^+-layer on silicon dioxide for 7 μm.

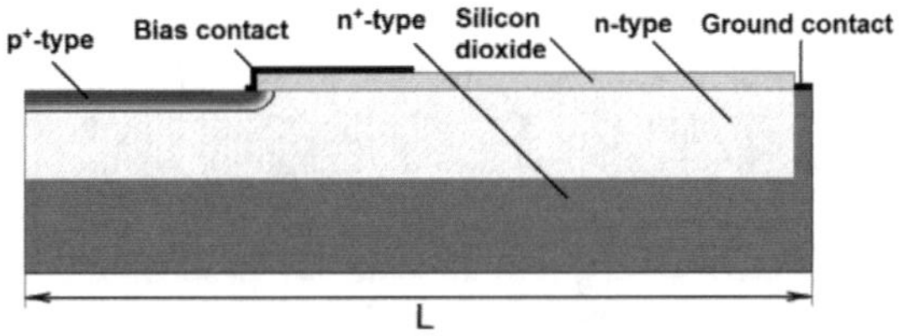

Figure 1. Model structure (sizes are given in text).

The full PNCV structure was implemented as this small model due to the requirement of the fine grid on Si-SiO_2 interface and reduce the simulation time.So, in the depth direction the minimum grid spacing is 1nm that is caused by the thin (up to 10 nm) inversion or accumulation layer on the Si-SiO_2 interface.The minimal grid in the lateral direction is defined by the p^+-n junction and equals to 10nm.

The simulation of the model was performed in Synopsys Sentaurus Device tool for the currents before breakdown.For the current-voltage simulation, the drift-diffusion mode was applied to the system of equations consisting of nonlinear Poisson equation and continuity equations for electrons and holes [13]. Thus, the impact ionization was not used in the simulations. The velocity saturation was taken into account, and for the electric field intensity higher than 10^4 V/cm the saturation velocity is 10^7 cm/s.

For the generation-recombination processes, the Schockley-Read-Hall (SRH) recombination was used. In simulations it appears as SRH recombination for the bulk and surface regions (equations (1)).

$$R_{bulk}^{SRH} = \frac{np - n_i^2}{\tau_p (n + n_1) + \tau_n (p + p_1)} \qquad R_{surf}^{SRH} = \frac{np - n_i^2}{(n + n_1)/S_p + (p + p_1)/S_n} \tag{1}$$

where $n(p)$ – electron (hole) density, n_i –intrinsic concentration, n_1 (p_1) – donor (acceptor) trap concentration (see equations (2)), τ_n (τ_p) – electron (hole) lifetime, S_n (S_p) – electron (hole) surface recombination velocity.

$$n_1 = n_i e^{\frac{E_{trap}}{kT}} \qquad p_1 = n_i e^{\frac{-E_{trap}}{kT}} \tag{2}$$

where E_{trap}– trap energy level, k – Boltzmann constant, T = 293 K – temperature.
E_{trap} value is the deviation of the trap energy level from the middle of the bandgap and it is equal to 0 in our case.

For the bulk SRH recombination we set$\tau_n = \tau_p = 1 \cdot 10^{-2}$ s for the non-irradiated and irradiated sample simulations. For the surface SRH recombination Synopsys TCAD simulates surface recombination velocities S taking into account their doping denedencies according to equation (3):

$$S = S_0 \left[1 + S_{ref} \left(\frac{N_i}{N_{ref}} \right)^{\gamma} \right] \quad (3)$$

whereS_0 – surface recombination parameter, $S_{ref} = 1 \cdot 10^3$– reference value of surface recombination velocity, N_i – ionized dopant concentration, $N_{ref} = 1 \cdot 10^{16}$ cm^{-3}– reference value of ionized dopant concentration, $\gamma = 1$ – degeneracy factor.

Also, we used a simplified model of the $Si–SiO_2$ interface charge. It represents a sum of an oxide charge and interface traps charge. The latter one depends on the type (donor or acceptor), energy level and Fermi energy level [14, 15]. The charge and surface recombination velocity values are different for the non-irradiated and irradiated sample simulations.

4. Results

The test simulations for the "short" (L = 20μm) and "long" (L = 160μm) structures were performed in order to confirm that there is no significant influence on the result currents for different charges and S_0 (figure 2).

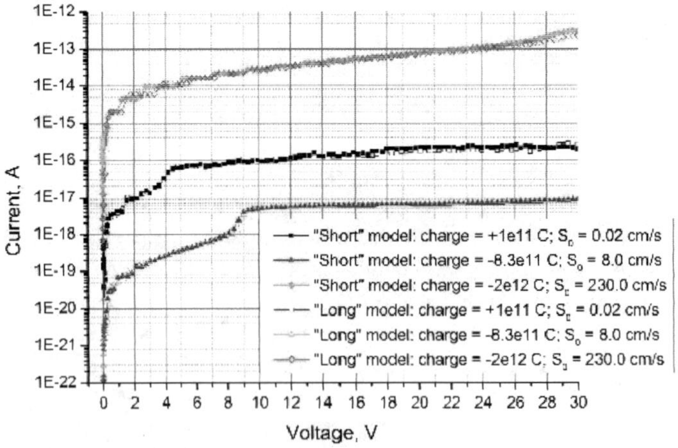

Figure2. Comparison of the short and long model currents.

In order to separate the bulk and surface currents, we performed two simulations. In the first case, the surface generation-recombination processes were deactivated and the only bulk current has been obtained. In the second one, we activate only the surface physical models and deactivate the bulk ones, and as a result, we have the pure surface current. Our model is two-dimensional and, by default, Sentaurus Device assumes a "thickness' (width along z-axis) of 1 μm.

Bulk currents was scaled using coefficient depends on the entrance window area ratio between the model and real PNCV structure. For the surface currents, the periphery perimeter ratio coefficient was used. The full current is equal to the sum of the scaled currents. However, the bulk current is much lower than surface one (figure 3a). It is consistent with data reported in [16].This model of the current calculation was applied to both non-irradiated and irradiated simulations too.

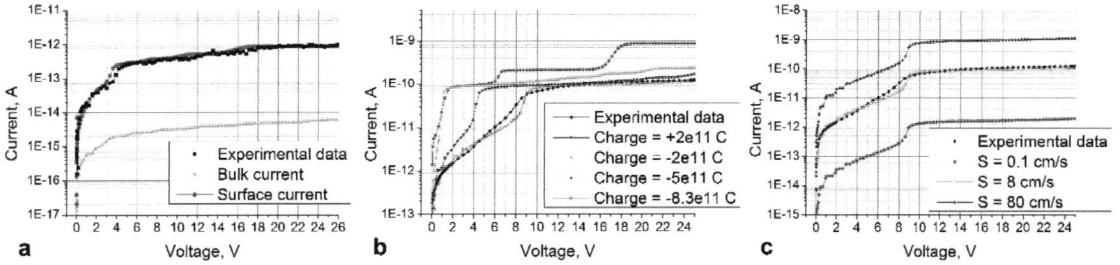

Figure 3. a) bulk and surface currents comparison according to non-irradiated sample; b) influence of the interface charge on I-V curve with constant surface recombination velocity (200 Gy example); c) influence of surface recombination velocity on I–V curve with constant charge (200 Gy example). Experimental errors are within the data points.

The simulation of the irradiated samples were performed with the same bulk lifetime as for non-irradiated samples (since there is no bulk damage), but within creasing the charge in the Si–SiO_2 interface and the surface recombination velocity. Using the different interface charges we are able to adjust the shape of the I–V curve (figure 3b). With surface recombination velocity, it is possible to move up and down the curve so we can reach the required current values as it is shown in figure 3c. For the irradiated samples, the positive charge does not fit and the negative charge [17] was used.

The results of the comparison of the current-voltage characteristics for the simulated and produced sample can be seen in figure 4. The curves are presented for the non-irradiated samples and the samples irradiated with 200 Gy and 3 kGy doses. As one can see, the shape of the simulated and experimental I–V characteristics for the non-irradiated samples is similar, but there is a slope of the simulation curve after some point. It could be caused by the simplicity of the simulation model. As concerns irradiated samples, the shapes and the values of the simulated current-voltage characteristics are in the good agreement with experimental data. The obtained values of the parameters are presented in table 1.

Table 1. The results parameters for the non-irradiated and irradiated samples.

	Interface charge, C	S_0, cm/s
0 Gy	$+1\cdot10^{11}$	0.02
200 Gy	$-8.3\cdot10^{11}$	8.0
3 kGy	$-2\cdot10^{12}$	230.0

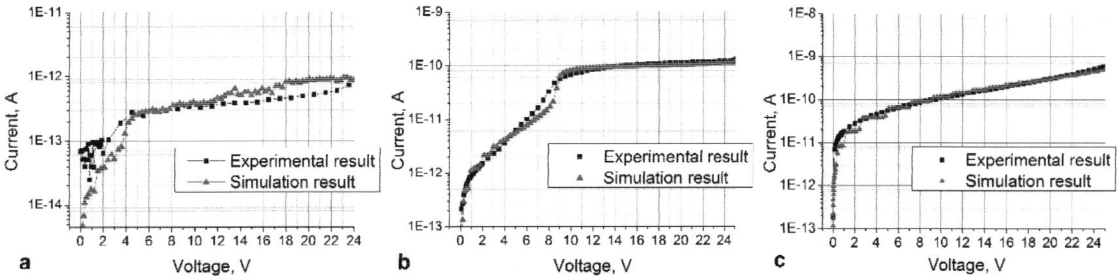

Figure 4. Comparison of experimental and simulation data, a) for non-irradiated samples;b) for 200 Gy dose; c) for 3 kGy dose.

5. Conclusion

In this work, an attempt to model the Silicon photomultipliers with Synopsys TCAD was presented. The simulation results were compared with the PNCV samples manufactured by KETEK GmbH. Using developed model, it is possible to obtain the amount of the charge located on Si-SiO_2 interface and the value of the surface recombination velocity of SiPM. Further work is being started and it concerns the simulation of the complete SiPM.

Acknowledgement

The work has been partially supported by Megagrant 2013 program of Russia, agreement № 14.A12.31.0006 from 24.06.2013 and partially supported by Institute for Experimental Physics, Hamburg University and partially supported by MEPhI Academic Excellence Project (contract № 02.a03.21.0005, 27.08.2013).

Especially, we thank prof. Erika Garutti and Dr. JoernSchwandt from Hamburg University for helpful advices and discussions.

References

[1] Buzhan P *et al.* 2003 Silicon photomultiplier and its possible applications *Nucl. Instrum. Meth.* A **504** 48–52
[2] Dolgoshein B *et al.* 2001 *An Advanced Study of Silicon Photomultiplier* ICFA Instrumentation Bulletin **23** 28–41
[3] Danilov M 2007 *Nucl. Instrum. Meth.* A **581** 451
[4] Musienko Y *et al.* 2007 *Nucl. Instrum. Meth.* A **581** 433
[5] Musienko Y *et al.* 2015 *Nucl. Instrum. Meth.* A **787** 319–22
[6] Xu C*et al.* 2014 Influence of X-ray irradiation on the properties of the Hamamatsu silicon photomultiplier S10362-11-050C *Nucl. Instrum. Meth.* A **762** 149–61
[7] Garutti E *et al.*2014 Silicon photomultiplier characterization and radiation damage investigation for high energy particle physics applications *JINST* **9**(3) C03021
[8] Synopsys TCAD (*http://www.synopsys.com/tools/tcad/*)
[9] KETEK GmbH (*http://www.ketek.net/*)
[10] Keithley 6517B (*http://www.keithley.com/products/dcac/sensitive/highresistance/?mn=6517B/*)
[11] PHYWE XRE 4.0 (*http://www.phywe.com/en/09110-88/*)
[12] Akkerman A *et al.* 2001 Updated NIEL calculations for estimating the damage induced by particles and gamma-rays in Si and GaAs *Radiation Physics and Chemistry* **62** 301
[13] *Synopsys Sentaurus Device manual* 2012
[14] Schwandt J *et al.* 2012 Optimization of the radiation hardness of silicon pixel sensors for high x-ray doses using TCAD simulations *JINST* **7** C01006 (*arXiv*:1111.4901)
[15] Zhang J *et al.* 2012 Investigation of X-ray induced radiation damage at the Si-SiO2 interface of silicon sensors for the European XFEL *JINST* **7** C12012 (*arXiv*:abs/1210.0427)
[16] Engelmann E *et al.* 2015 Extraction of activation energies from temperature dependence of dark currents of SiPM (*in this proceedings*)
[17] Warren W L *et al.*1993 *Journal of Applied Physics* **74** 4034

Experimental study of the possibility of 3D localization of the compact gamma-sources in soft tissues

A K Berdnikova, V N Belyaev, A I Bolozdynya, V A Kantserov and V V Sosnovtsev
National Research Nuclear University MEPhI (Moscow Engineering Physics Institute), Kashirskoe highway 31, Moscow, 115409, Russia

E-mail: akberdnikova@mephi.ru

Abstract. To determine the depth of the area of radiopharmaceutical accumulation a method of simultaneous recording of two lines of gamma rays of different energies and quantitative comparison of the intensity of these lines on the surface of the patient's body is provided. Since the coefficient of linear absorption of gamma radiation in the medium depends not only on the characteristics of the medium, but also on the gamma radiation energy, the intensity of gammas of different energies is attenuated differently after passing through the same absorber layer (soft tissues). Thus, the quantitative comparison of the relative intensities of gamma lines on the surface of the patient's body allows to determine the depth of area of the accumulation of the radiopharmaceutical. The result is achieved by analyzing the energy spectrum of the source, obtained with a semiconductor spectrometer, by measuring the ratio of areas of the absorption peaks of the radioisotope and defining the depth of gamma source using the calibration dependence between the areas ratio and the medium layer thickness. The most widely used medical radioisotope technetium-99m has two gamma-lines - 140 keV and 18.5keV, which allows one to apply the proposed method to search for the sentinel lymph nodes and non-palpable malignant tumors in the soft tissues.

1. Introduction

Radionuclide diagnostics is a branch of nuclear medicine, which currently is one of the most emerging types of the early stage cancer diagnosis [1]. Miniature gamma detectors, or gamma probes, are used to localize small and non-palpable malignant lesions. Intraoperative use of gamma probes for staging and evaluation of the extent of the malignant process is possible due to the structure and physiology of the human lymphatic system. Patient is injected with a radiolabeled nanocolloid, which accumulates in the first lymph node to which cancer cells are most likely to spread from a primary tumor [2]. Gamma probe detects radiation from the radiotracer, thus revealing the affected lymph nodes. One of the devices of such type, gamma-locator, has been developed in NRNU MEPhI [3].

The goal of the present work is to explore the possibility of three-dimensional localization of compact gamma sources with gamma-locator instead of determination of the 2D projection of the point of maximal radiotracer accumulation on the surface if the patient’s body.

2. 3D localization of the compact gamma-sources

If a medical gamma source has two or more emission lines and the energy resolution of detector is high enough, it is possible to measure the intensities of these lines at the surface of the patient’s body after passing through a layer of soft tissues. For example, the most commonly used radionuclide ^{99m}Tc has a 140 keV emission line, and also 18.5 keV x-rays. Quantitative comparison of the intensities of these lines allows to determine the depth of the point of the maximal radiotracer accumulation.

In a narrow beam geometry (without taking into account the scattered radiation), the radiation intensity can be calculated using the following equation:

$$I = I_0 e^{-\mu \cdot x}, \tag{1}$$

where: I – the intensity of gamma rays after passing through the slab of material,
I_0 – the initial intensity,
μ – linear attenuation coefficient of the material,
x – thickness of the layer of the material.

Wide beam geometry is considered to take into account the effect of scattered radiation, which can be denoted as a so-called buildup factor B. Buildup factor depends on the geometry of an experiment, properties of the material and energy of gamma-radiation. In that case the ratio of intensities of two gamma lines of different energies is:

$$I_1 / I_2 = \frac{I_0^1 e^{-\mu_1 x} B_1}{I_0^2 e^{-\mu_2 x} B_2} = I_0^1 / I_0^2 \, e^{-(\mu_1 - \mu_2)x} B_{21}, \tag{2}$$

Where B_{21} is the total buildup factor.

To determine thickness of the soft tissues layer for 3D localization of a malignant formation the calibration of the dependence between the ratio I_1/I_2 and the layer thickness x is needed for a particular gamma detector, primarily because of the buildup factor which can not be determined analytically. Calibration can be performed with an acrylic or a water phantom. After that a gamma probe can be used to obtain a more detailed three dimensional information about placement and structure of the area of an increased radiotracer uptake in the patient's body.

3. Experimental setup

For the feasibility study of the proposed method a ^{137}Cs point-like gamma source, which has a 662 keV emission line (yield 85%) and a 32 keV x-ray line (yield 6%), was used. Activity of the gamma source is approximately 300 kBq.

As a miniature gamma detector a semiconductor probe SDP310 / Z / 60 (S) Standart (manufactured by Baltic Scientific Instruments)was used. This gamma probe contains a quasi-hemispherical detector CdZnTe with a sensitive area of 50÷60 mm^3, the energy resolution is less than 25 keV (~ 3.8%) at 662 keV.

A set of polymethylmethacrylate (PMMA) plates with 1÷5 mm thickness was used as a phantom of human soft tissues. Point-like gamma source was placed between the plates at the increasing distances from the gamma probe. The probe was fixed on the top of the phantom, and the energy spectra were measured at different depths of gamma source.

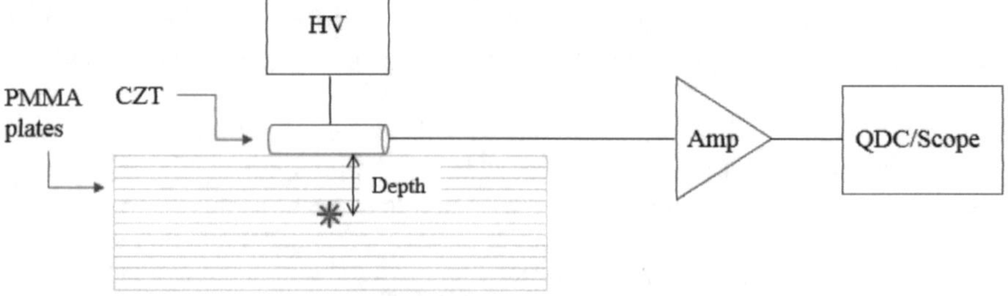

Figure 1. Scheme of an experimental setup.

The scheme of the experiment is presented in the figure 1. Detector (CZT) is biased through a preamplifier with +350V from a high voltage power supply (HV). Signal from the detector preamplifier is shaped and amplified with a spectrometric amplifier (Amp), and then passed to the input of Lecroy WaveRunner oscilloscope, which can be used as a charge-to-digital converter (QDC/Scope).

Experimental energy spectra of ^{137}Cs, obtained at the depths of 5, 15 and 25 mm of the source in PMMA phantom, are presented in the figure 2.

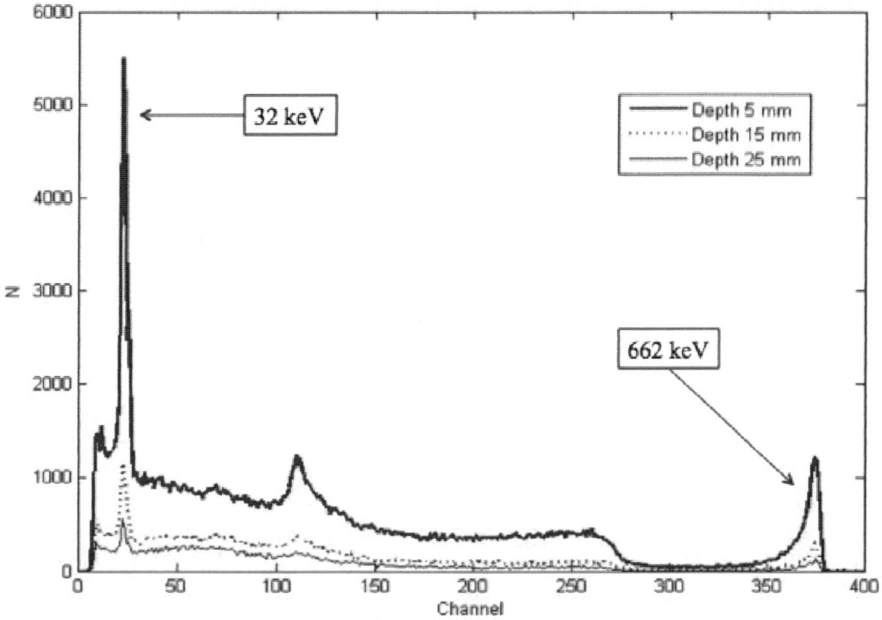

Figure 2. Experimental energy spectra of ^{137}Cs at 5, 15, 25 mm PMMA layer thickness between the source and the CZT gamma probe.

In the presented spectra one can see photopeaks corresponding to emission lines of 32 keV (with a yield of 6%) and 662 keV (85% yield). The peak position in the spectra of these determines the energy of the gamma radiation. Area of the photopeak defines the intensity of radiation corresponding to the energy of 32 keV and 662 keV. Figure 2 shows that the intensity of 32 keV radiation line decreases faster than the intensity of 662 keV gamma line with immersion of gamma-ray source in the material. Dependence between the ratio of the photopeaks areas and the material layer thickness should be exponential according to the equation (2).

4. Experimental results

The ratio of the 32 keV and 662 keV photopeaks areas was measured for energy spectra of ^{137}Cs at the distances of 5, 10, 15 and 20 mm between the source in the phantom and the gamma probe. Area of 32 keV photopeak was calculated with the correction for Compton contrubution, by approximating the Compton continuum with a linear function around the photopeak. The experimental plot of the ratio of the 32 keV photopeak area to the one of 662 keV vs the PMMA layer thickness is presented in the figure 3.

Experimental data can be fitted with the linear equation as a first degree of a Taylor polynomial for an exponential function:

$$y(x) = kx + b, \tag{3}$$

Coefficients k and b were calculated using the least squares method; $k = (-0.6 \pm 0.3) \times 10^{-2}$ mm^{-1}, $b = 3.6 \pm 0.2$.

The resulting linear dependence can serve as a first approximation for calibration of the gamma probe for determination of the depth of the compact gamma source positioning in soft tissues based on the detector's count rate in two energy discrimination windows.

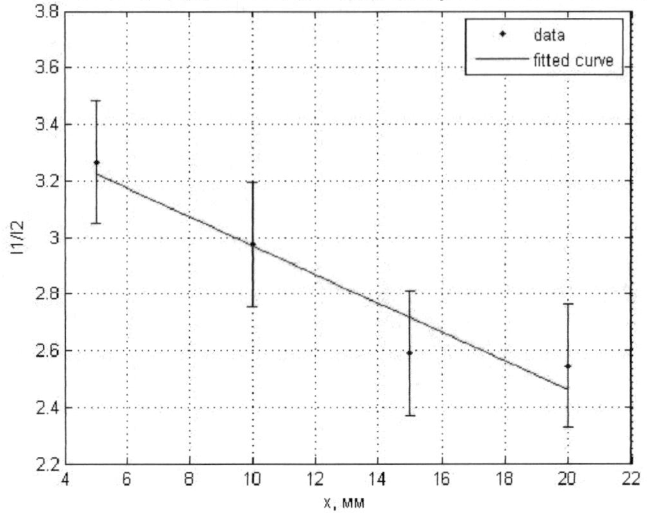

Figure 3. Ratio of 32 keV photopeak area to 662 keV photopeak area vs the thickness of PMMA layer.

5. Conclusion

The feasibility study of the 3D compact gamma source localization method, based on the simultaneous registration of two emission lines of different energies, is presented within this work. Experiments with a point-like ^{137}Cs gamma source placed inside a PMMA phantom and a room-temperature semiconductor gamma probe were carried out to obtain the dependence between the ratio of a 32 keV emission line intensity to the one of 662 keV and the PMMA layer thickness.

Experimental results showed that the dependence at first approximation can be fitted with a linear equation, which can be used for the detecting system calibration. However, a relatively low activity of the gamma source did not allow to determine the calibrating coefficients with the satisfactory accuracy.

Further objective of the study is to provide a set of experiments with a scintillation detector based on a $LaBr_3$:Ce crystal and a silicon photomultiplier to establish a possibility of 3D localization of malignant lesions with the current gamma-locator prototype.

Acknowledgements

This study was supported by RF Government under contracts of NRNU MEPhI with the Ministry of Education and Science of № 11.G34.31.0049 from October 19, 2011 and within the framework of the Centre for Nanostructured Electronics supported by MEPhI Academic Excellence Project (contract № 02.a03.21.0005, 27.08.2013).

References

[1] Wernick M and Arsvold J 2004 *Emission tomography: the fundamentals of PET and SPECT* (Elsevier: Academic Press)

[2] Shneebaum S, Even-Sapir E and Cohen M 1999 Clinical applications of gamma-detection probes—radioguided surgery *Eur. J. Nucl. Med.* **26** 26-35

[3] Yagnyukova A K *et al.* 2015 A γ Probe for Radionuclide Diagnostics of Cancer *Instrum. Exp. Tech.* **58** 153–7

Journal of Physics: Conference Series **675** (2016) 042048 doi:10.1088/1742-6596/675/4/042048

Miniature gamma detector based on inorganic scintillator and SiPM

A K Berdnikova[1], F A Dubinin[2,4,*], V A Kantserov[1,2], A D Orlov[3], D U Pereyma[1], S Z Shmurak[3] and K I Zhukov[2]
[1] National Research Nuclear University MEPhI (Moscow Engineering Physics Institute), Kashirskoe highway 31, Moscow, 115409, Russia
[2] P.N. Lebedev Physical Institute of the Russian Academy of Sciences, Leninskiy Prospekt 53, Moscow,119991, Russia
[3] Institute of Solid State Physics RAS (Russian Academy of Science), Academician Ossipyan Street 2, Chernogolovka, Moscow District, 142432, Russia

E-mail:*ph-and-d@yandex.ru, vadim.kantserov@cern.ch

Abstract. The miniature gamma counter based on a cylindrical $LaBr_3(Ce)$ crystal (5 mm diameter and 10 mm length) and SensL FC30035 silicon photomultiplier (SiPM) is introduced. The main counter characteristics such as relative efficiency, $LaBr_3(Ce)$ self-radioactivity and energy resolution are presented. Capability of using such detector for gamma spectrometry applications is discussed.

1. Introduction

Nuclear technologies have a wide application in many different fields of science and technique, from power industry and extractive industry to medicine. Portable scintillation detectors are widely used in medical diagnostics. NaI(Tl), CsI(Tl), BGO are the most commonly used scintillators. High effective charge and density cause of their widespread use for charged particle and gamma-ray registration. The main characteristics of several inorganic scintillators are presented in the table 1.

Table 1. Basic characteristics of scintillation crystals.

Scintillator	NaI(Tl)	CsI(Tl)	BGO	$LaBr_3(Ce)$
Z_{eff}	51	54	74	45
ρ, g/sm^3	3.67	4.51	7.1	5.08
λ_{max}, nm	415	550	480	365
Hygroscopicity	yes	weak	no	yes
50% attenuation length, mm (662 keV)	25	20	10	18
ΔE/E, % (662 keV)	7	6	10	3
Light yield, phot/keV	38	54	9	63
τ, ns	250	1000	300	16

Vacuum photomultipliers (PMTs) are commonly used for the scintillation light detection. But their size and HV power supply requirements reduce their practical applicability. Silicon photomultipliers (SiPMs) are more preferable to use in miniature portable counters. SiPMs have a relatively small size (thickness is usually less than 3 mm). Bias voltage lies in a range of 20 – 80V which is enough to

[4] To whom any correspondence should be addressed

obtain gain coefficient ~ $10^4 - 10^6$. Photon detection efficiency (PDE) of SiPMs depends on cell size and bias voltage, and achieves 30 – 50% [1].

2. Detector

The detector consist of $LaBr_3(Ce)$ cylindrical scintillator with diameter 5 mm and length 10 mm and SiPM FC30035SMT manufactured by SensL (its characteristics are presented at table 2). Developmental prototype of scintillation crystal was fabricated in the Institute of Solid State Physics RAS. As it is shown in the table 1, light yield and energy resolution of $LaBr_3(Ce)$ are higher in comparison with other common inorganic scintillators. $LaBr_3(Ce)$ also has a relatively short primary decay time. Main negative attributes of $LaBr_3(Ce)$ are its hygroscopicity and self-radiation due to ^{138}La [2].

Table 2. Characteristics of the used SiPMSensL FC30035.

Active area, mm^2	9
Number of cells	4774
Cell size, μm^2	35
Gain*	~ $2 \cdot 10^6$
PDE*, % (λ=365 nm)	25

*at a chosen bias voltage.

The main feature of the detector is that SiPM and scintillator are put together into the single duraluminium housing. On the one hand the housing is watertight, and on the other hand it decreases the light loss between crystal and SiPM boundaries. Scheme of the scintillation package is shown in the figure 1. Amplitude of the SiPM signal is ~ 10 mV at 31 keV gammas (from ^{133}Ba), ~ 400 mV for gammas with energy of 662 keV (^{137}Cs) and the pulse width (FWHM) is around 300 ns. Thus it is not necessary to use a preamplifier and a shaper to adapt the signal for further processing. Waveforms of SiPMs output signals for 31 keV and 662 keV gammas are presented in the figure 2. They were measured by oscilloscope LeCroy 640Zi (50Ohm load) directly without amplifier.

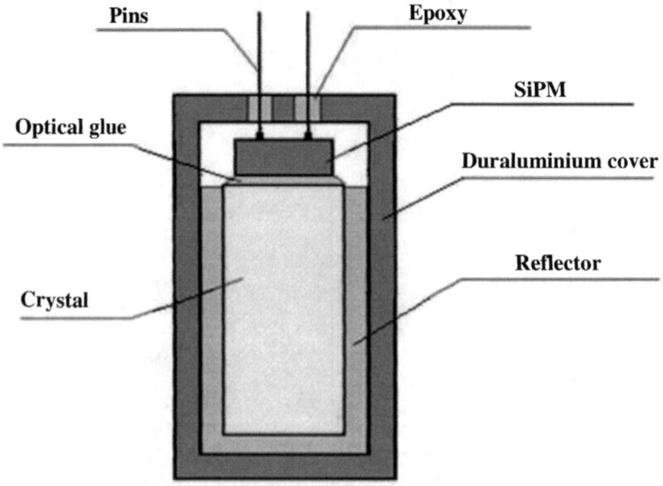

Figure 1. Scintillation unit.

Journal of Physics: Conference Series **675** (2016) 042048 doi:10.1088/1742-6596/675/4/042048

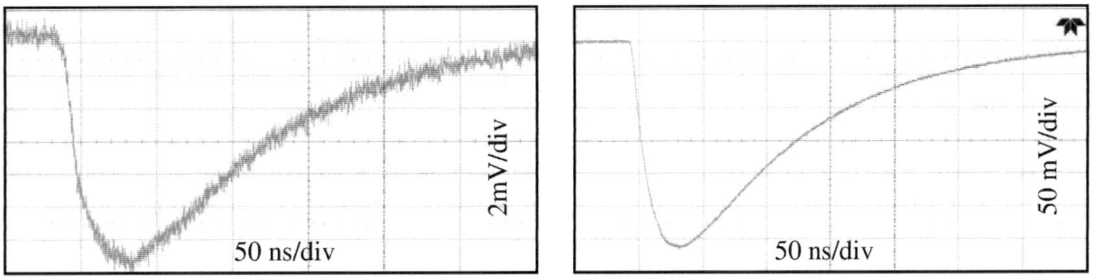

Figure 2. Waveforms of the signal from 32 keV gammas (a) and 662 keV gammas (b). (Oscilloscope LeCroy 640Zi (50Ohm load) without amplifier).

3. Self-radioactivity

Background events caused by self-radioactivity decrease the counter sensitivity. Background count rate determines the minimal activity of a radioactive source that could be registered by the detector.

In order to estimate the level of background radiation of the detector measurements of count rate without sources were provided. Detector was biased with a Polon 1904 power supply, voltage and dark current were controlled by multimeters AM1097 and CHA300. In order to count pulses from the detector discriminator Lecroy 623b and scaler CAEN 145 were used. The threshold of discriminator was set at such level to select signals with an amplitude higher than ones caused by 60 keV gammas from ^{241}Am.

Experimental results showed that the mean value of background count rate is equal 1.13±0.09 pulses/sec.

4. Energy resolution

Due to the high energy resolution and high light yield of the $LaBr_3$(Ce) and the relatively high photon detection efficiency and low noise of the SiPM, the detector under discussion can be used for specrtometry. ^{137}Cs and ^{241}Am charge spectra obtained with the $LaBr_3$(Ce)–SiPM detector are presented in the figures 3 and 4.

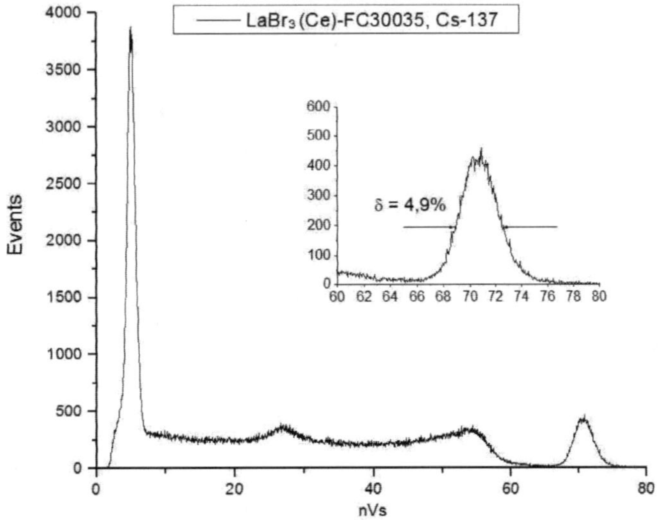

Figure 3. Spectrums of ^{137}Cs obtained with the $LaBr_3$(Ce)–SiPM unit with relative energy resolution δ=4.9% for 662 keV with absolute statistical error 0.005%.

Scheme of power supply consisted of Polon 1904 power supply and multimeters AM1097 and CHA300. Pulses from the detector were directly sent to the oscilloscope Lecroy 640Zi having an inbuilt QDC analyzer. The Gate was around 250 ns for both measurements.

As an energy resolution was used the value $\delta = \frac{FWHM}{N_{max}}$ (relative energy resolution), where FWHM is the Full Width at Half of the Maximum of Gauss distribution fitting the peak in pulse spectra, and N_{max} is the position of Maximum. According to obtained data energy resolution at energy 662 keV (source ^{137}Cs) equal to 4.9% with absolute statistical error 0.005%, at 60 keV (source ^{241}Am) the resolution equal to 16.7% with absolute statistical error 0.05%.

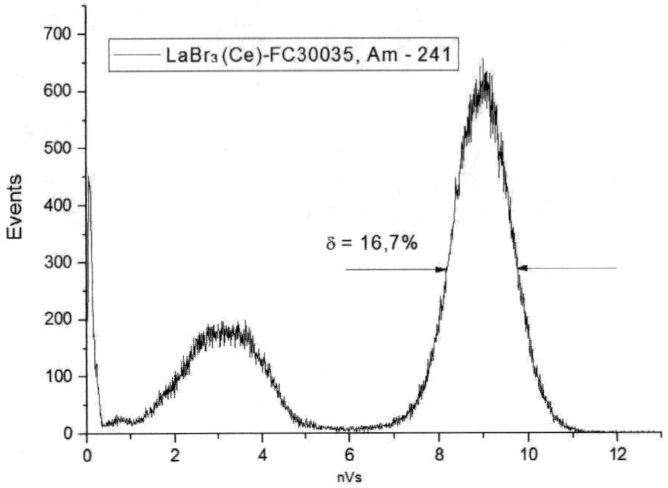

Figure 4. Spectrums of ^{241}Am obtained with the $LaBr_3$(Ce)–SiPM unit with relative energy resolution δ=16.7% for 60keV with absolute statistical error 0.05%.

5. Conclusion

The possibility of the usage of a scintillation detector based on $LaBr_3$(Ce) crystal and SiPM as a miniature gamma counter was considered. Scintillator's self background rate was measured to be 1.13±0.09 pulses/sec. Energy resolution of the detector was measured to be about 4.9% for 662 keV gamma rays. Obtained results show that the detector serve as a good alternative for well-known gamma counters based on NaI(Tl), CsI(Tl), BGO.

Acknowledgments

This work was partially supported by MEPhI Academic Excellence Project (contract № 02.a03.21.0005, 27.08.2013).

References

[1] Klemin S *et al.*2007 (in Russian) *Electronika: STB* **8** 80 (Original Russian title: *Electronika: Nauka, Technologiya, Bisnes*)

[2] Iyudin A F *et al.* 2009 *Instrum. Exp. Tech.* **52** (6) 774

Extraction of activation energies from temperature dependence of dark currents of SiPM

E Engelmann[1], S Vinogradov[2,5], E Popova[3], F Wiest[4], P Iskra[4], W Gebauer[1], S Loebner[1], T Ganka[4], C Dietzinger[4], R Fojt[4] and W Hansch[1]

[1] Universitaet der Bundeswehr Muenchen, Department of Electrical Engineering and Information Technology, Institute for Physics, Werner-Heisenberg-Weg 39, Neubiberg, 85577, Germany

[2] Cockcroft Institute, University of Liverpool, 4 Keckwick Lane, Daresbury, Warrington WA4 4AD, United Kingdom

[3] National Research Nuclear University MEPhI (Moscow Engineering Physics Institute), Kashirskoe highway 31, Moscow, 115409, Russia

[4] KETEK GmbH, Hoferstrasse 3, Munich, 81737, Germany

[5] P. N. Lebedev Physical Institute of the Russian Academy of Sciences, Leninskij Prospekt 53, Moscow, 119991, Russia

E-mail: eugen.engelmann@unibw.de

Abstract. Despite several advantages of Silicon Photomultipliers (SiPM) over Photomultiplier Tubes (PMT) like the increased photon detection efficiency (PDE), the compact design and the insensitivity to magnetic fields, the dark count rate (DCR) of SiPM is still a large drawback. Decreasing of the SiPM dark count rate has become a modern task, which could lead to an enormous enhancement of the application range of this promising photo-detector. The main goal of this work is to gain initial information on the dark generation and identify the dominating contributions to dark currents. The chosen approach to fulfill this task is to extract characteristic activation energies of the contributing mechanisms from temperature dependent investigations of dark currents and DCR. Since conventional methods are not suited for a precise analysis of activation energies, a new method has to be developed. In this paper, first steps towards the development of a reliable method for the analysis of dark currents and dark events are presented.

1. Introduction

The dark count rate (DCR) of the Silicon Photomultiplier (SiPM) is still a limiting factor for the extension of its application range and requires a significant reduction. In order to achieve this requirement, a distinction of contributions determining the DCR is needed to be able to identify the dominating mechanisms. As reported in [1, 3, 4, 5] the extraction of characteristic activation energies from the temperature dependence of the dark current is able to provide information on the dominating mechanisms. Unfortunately these methods are not suited for a precise determination of the needed energy levels, since the investigated dark currents are determined by a mixture of effects depending on voltage on the one hand (e.g. generation current, electric field effects) and overvoltage on the other hand (e.g. gain, crosstalk, afterpulsing). In this paper, we propose a new method to separate the mentioned effects using the responsivity of the detector as an appropriate reference for the measured dark current. This approach allows us to generate a field-independent expression for the initial amount of charge carriers provided to or generated inside the multiplication region I_{ini} (see equation 2). From the Arrhenius plot of I_{ini}, characteristic activation energies can be extracted in an accurate way in order to identify the dominating mechanisms. In this work, the investigations were performed on a PM3350T device provided by KETEK GmbH. The device had an area of 3×3 mm^2 with a micro-cell edge size of $50\mu m$ and implemented optical trenches (see http://www.ketek.net/products/sipm/pm3350/).

2. Conventional method

Using the conventional method for the extraction of activation energies [1], dark currents at a fixed voltage as well as dark currents at a fixed overvoltage were investigated as a function of 1/kT (Arrhenius plot). The breakdown voltage of the investigated SiPM was $27.3V$ at 20°C. The results are shown in figure 1 and 2, respectively. In both plots two clearly distinguishable

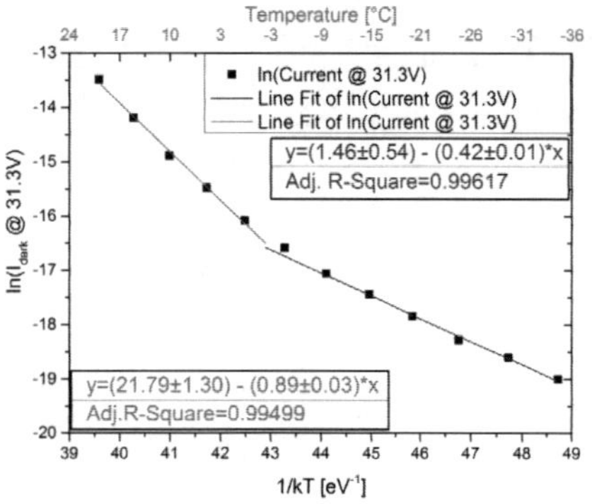

Figure 1. In this figure the Arrhenius plot of dark currents at a fixed voltage of 31.3V is shown.(Error bars are within the data points.)

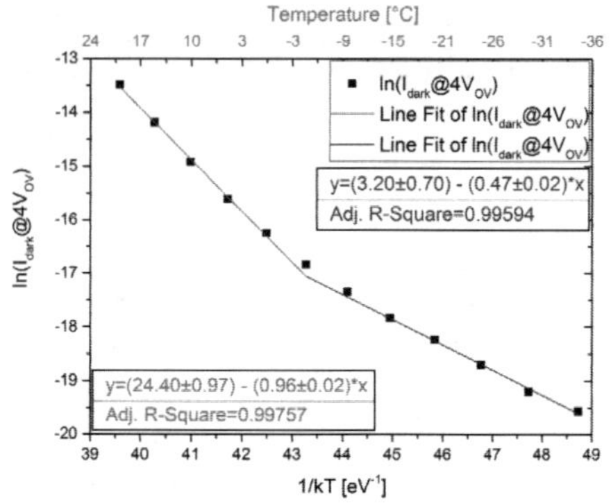

Figure 2. In this figure the Arrhenius plot of dark currents at a fixed overvoltage of 4V is shown.(Error bars are within the data points.)

activation energies can be observed. This result was interpreted as the existence of two mechanisms, dominating in the lower (T<-3°C) and higher (T>-3°C) temperature range, respectively. Despite the qualitative similarities of both plots, the quantitative activation energies differ for both temperature ranges. The extracted energies for the fixed overvoltage case ($V_{OV} = 4V$) are $(0.96 \pm 0.02)eV$ and $(0.47 \pm 0.02)eV$. For the Arrhenius plot at a fixed voltage of 31.3V the energies amount to $(0.89 \pm 0.03)eV$ and $(0.42 \pm 0.01)eV$. Since the underlying effects which lead to dark current or dark count rate depend on the applied electric field as well as on the gain and efficiency of the Geiger discharge development (crosstalk, afterpulsing) of the detector, the SiPM response is dependent on the applied voltage and overvoltage simultaneously. For this reason, the conventional method does not provide reliable results for a sophisticated analysis of activation energies.

3. Proposed method

In order to realize a separation of voltage dependent high-field effects and overvoltage dependent gain and efficiency effects of the detector, we propose an investigation method which is based on two independent measurements of the SiPM response. The first measurement is performed under illuminated conditions (here for $\lambda = 600nm$), the second one under dark conditions.

3.1. First approach

In the first step of the method, the responsivity (R) of the SiPM was defined for every temperature as shown in equation (1). Here I_{ph} is the photo response of the SiPM. The assumption was made, that the responsivity of the device is equal for photon-induced charge carriers (R_{ph}) and charge carriers originating from dark generation (R_{dark}). In figure 3 the responsivity and the dark current of the investigated device are shown at 20°C. Further the dark response of the detector (I_{dark}) can be expressed in terms of the responsivity R_{ph}, the initial dark component before multiplication I_{ini} and a component I_{not_gained}, which is not affected by the responsivity, as shown in equation (2). Here I_{ini} describes the initial amount of generated charge carriers inside or provided to the high-field region before multiplication, I_{not_gained} accounts for charge carriers that do not pass the high-field region and are thus not multiplied.

$$R_{dark}(V_{OV}) \approx R_{ph}(V_{OV}) = \frac{I_{ph}(V_{OV})}{I_{ph}(V_0)} \tag{1}$$

$$I_{dark}(V,T) = I_{ini}(V,T) \cdot R_{ph}(V_{OV}) + I_{not\_gained}(V,T) \tag{2}$$

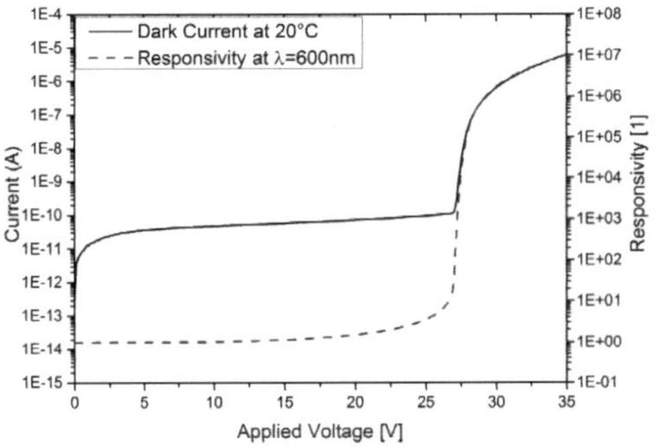

Figure 3. In this figure the responsivity at $\lambda = 600nm$ and I_{dark} at 20°C is shown. (Uncertainties are within the data points.)

Applying equation (2) to dark currents, the initial current component (I_{ini}) can be determined as shown in figure 4 (solid line). The assumption was made, that the dark current consists of two contributing components. The first component was associated with surface currents or peripheral currents which flow at low field conditions (I_{not_gained}). The dominating part of these currents does not reach the high-field region of the device and thus is not multiplied. I_{not_gained} was assumed to increase with increasing depletion width and was modelled with a square root dependence on the applied voltage. The second contribution to the total dark current is a fraction δ of I_{not_gained} which does reach the multiplication region. Modelling the multiplied component I_{ini}, δ was chosen such that the breakdown point (see red circle in figure 4) was matched. The total current is then given by equation (3). The reconstructed currents for three temperatures are shown in figure 5. The experimental data is underestimated by the reconstructed dark currents at higher voltage regions. This result was expected, since no high-field effects were taken into account so far.

$$I_{dark} = (1-\delta)\, I_{not\_gained} + \delta \cdot I_{not\_gained} \cdot R_{ph} \tag{3}$$

$$I_{ini} = \delta \cdot I_{not\_gained} \tag{4}$$

Journal of Physics: Conference Series **675** (2016) 042049 doi:10.1088/1742-6596/675/4/042049

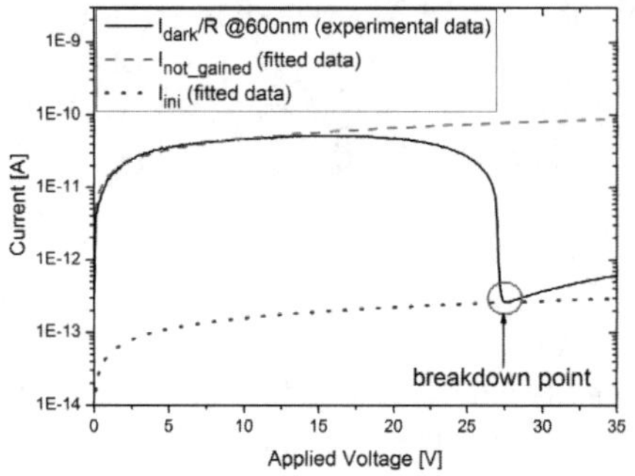

Figure 4. In this figure the determined initial component of the dark current (I_{ini}) as described in equation (2) is shown. (Experimental uncertainties are within the data points.)

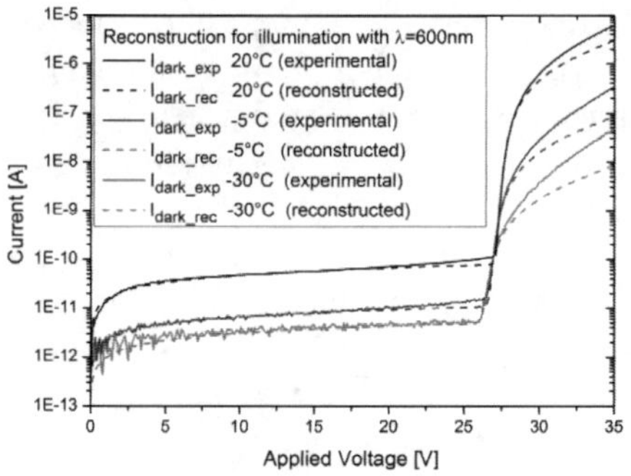

Figure 5. In this figure the reconstructed currents using the model described in equation (3) are shown for three temperatures. (Experimental uncertainties are within the data points.)

3.2. Second approach

In a second approach of the dark current reconstruction an additional parameter (F_{field}) was implemented, which should account for the electric field contributions. The total current was then modelled using equation (5). In order to investigate the multiplied component of the dark current, the difference between the experimental data and the fitted I_{not_gained} was build. The resulting function I_{diff} could be described with a second order polynomial fit in R_{ph}. The applied fit function is shown in equation (6). This approach provides a voltage independent expression for dark generation (I_{ini}) to be investigated in an Arrhenius plot.

$$I_{dark}(T, V, V_{OV}) = (1 - \delta)\, I_{not\_gained}(V, T) + I_{ini}(T) \cdot R_{ph}(V_{OV}) \cdot F_{field}(V) \tag{5}$$

$$I_{diff} = I_{ini} \cdot R_{ph} \cdot F_{field} = I_{ini} \cdot R_{ph} \left(1 + \frac{R_{ph}}{R_{eff}}\right) \tag{6}$$

4. Results

Investigating the parameter I_{ini} as a function of 1/kT, two activation energies could be observed and attributed to the full bandgap energy (E_g) for T>-3°C and the half bandgap energy ($E_g/2$) for T<-3°C (see figure 6). Using equations (7) (here shown for electrons) and (8) as described in [2], a dominating diffusion current for T>-3°C and a dominating generation current for T<-3°C could be identified.

$$I_{diffusion}^{electron} = q\sqrt{\frac{D_n}{\tau_n}}\frac{n_i^2}{N_A} \sim T^{\frac{\gamma}{2}}\left[T^3 \exp\left(-\frac{E_g}{kT}\right)\right] \tag{7}$$

$$I_{generation} = q\frac{n_i W_D}{\tau_g} \sim \exp\left(-\frac{E_g}{2kT}\right) \tag{8}$$

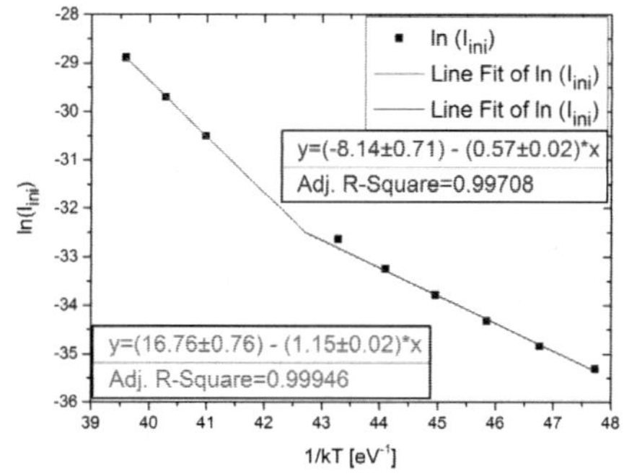

Figure 6. In this figure the Arrhenius plot of I_{ini} is shown. (Error bars are within the data points.)

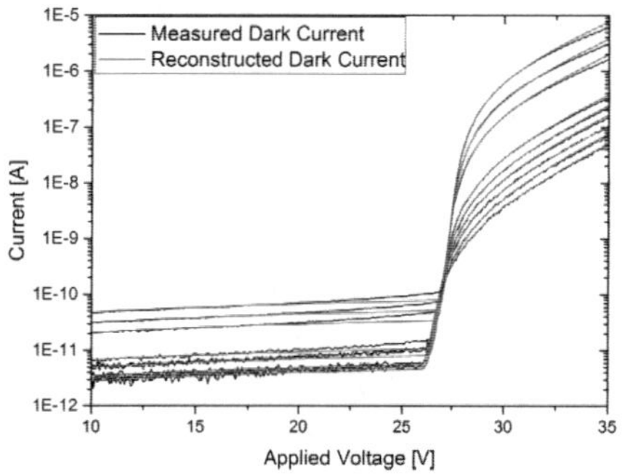

Figure 7. In this figure the reconstructed currents using equation 5 are shown. (Error bars are within the data points.)

Using the second approach, the constructed dark currents reproduce the experimental data to a more precise level (see figure 7). For higher temperatures ($I_{diffusion}$ is dominating), the reconstructed currents overestimate the experimental data in high overvoltage regimes. The observed divergence is owed to the electric field correction F_{field}. The correction must not be applied to the diffusion component of the dark current, since the electric field dependence of $I_{diffusion}$ can be neglected, as reported in [3]. Since the correction was applied to the total multiplied current, it was expected to observe an overestimation in high overvoltage regimes, especially at temperature, for which the diffusion component is dominating.

5. Conclusion

In this work, a first step towards a reliable investigation method of dark current contributions based on the responsivity of the SiPM was presented. The developed model for dark currents could reproduce the experimental data to a sufficiently precise level allowing the separation of field-enhanced effects from dark generation components. In combination with temperature dependent measurements, it was possible to identify a diffusion current to be the dominant contribution to dark events at temperatures T>-3°C, for the investigated SiPM.

Acknowledgements

This work has been partially supported by Megagrant 2013 program of Russia, agreement No 14.A12.31.0006 from 24.06.2013, the EC under Grant Agreement 329100 "SiPM in-depth" and the STFC Cockroft Institute Core Grant No ST/G008248/1.

We would like to express special appreciation to KETEK GmbH, Munich, Germany for the support of the presented investigations.

References

[1] Pagano R *et al.* 2012 Dark Current in Silicon Photomultiplier Pixels: Data and Model *IEEE Trans. Electron. Dev.* **59**(9) 2410-16
[2] Sze S M and NG K K 2007 *Physics of Semiconductor Devices* (John Wiley and Sons, Inc.)
[3] Poyai A *et al.* 2001 Improved extraction of the activation energy of the leakage current in silicon p-n junction diodes *Appl. Phys. Lett.* **78** 1997-9
[4] Auffray E *et al.* 2015 Characterization studies of Silicon Photomultipliers and crystal matrices for a novel time of flight PET detector *arXiv*:1501.04233v1 [physics.ins-det]
[5] Dinu N *et al.* 2010 Temperature and Bias Voltage Dependence of the MPPC Detectors *IEEE Nuclear Science Symposium Conference Record (NSS/MIC)* pp 215-9 (DOI: 10.1109/NSSMIC.2010.5873750)

Journal of Physics: Conference Series **675** (2016) 042050 doi:10.1088/1742-6596/675/4/042050

$LaBr_3(Ce)$ gamma-ray detector for neutron capture therapy

M Smirnova[1,*], E Shmanin[1], A Galavanov[1], A Shustov[1], S Ulin[1], K Vlasik[1], V Dmitrenko[1], A Novikov[1], A Orlov[2], D Petrenko[1], S Shmurak[2] and Z Uteshev[1]
[1]National Research Nuclear University MEPhI (Moscow Engineering Physics Institute), Kashirskoe highway 31, Moscow, 115409, Russia
[2]Institute of Institute of Solid State Physics Russian Academy of Sciences, Chernogolovka, Moscow District, 2 Academician Ossipyan str., 142432 Russia

E-mail: *mariko876@gmail.com, seulin@gmail.com

Abstract. Results of developing of a gamma-ray detector based on $LaBr_3(Ce)$ scintillation crystal for neutron capture therapy are presented. An energy resolution of the detector measured by photomultiplier tube Hamamatsu R6233-100 is showed. It was 2.93% for gamma line 662 keV from a source ^{137}Cs. For radiative capture gamma line of isotope ^{10}B (478 keV) and annihilation line (511 keV) the values were 3.33 and 3.24% respectively. Data analysis of gamma spectra for an estimation of energy resolution threshold required for visual identification gamma lines 478 and 511 keV was made.

1. Introduction

At the present time an increasing of cancer patients is observed. For their treatment different methods of radiotherapy based on the use of radiation sources with high ability are developing dynamically. During treatment session a maximum therapeutic effect is reaching by means of special pharmaceutical product and thermal neutron flux.

Every treatment session requires accurate dosimetry control of absorbed dose. For this purpose a gamma-detector based on scintillation $LaBr_3(Ce)$ crystal was designed for registration of gamma-rays with energy 478 keV. The gamma-rays appear as a result of interaction of thermal neutrons and the pharmaceutical product based on isotope ^{10}B. At the same time, high energy gamma-rays appear due to interaction of thermal neutrons with surrounding materials (for example, material detector and etc.). Positrons, appearing as a result of high energy gamma-rays interaction, annihilate with electrons and produce gamma-rays with energy 511 keV. Gamma-ray detector with high energy resolution is required for visual identification of the gamma lines with 478 and 511 keV.

2. Neutron capture therapy

The special pharmaceutical product based on isotope ^{10}B (the isotope has large thermal neutron capture cross section about 3800 barns [1]) is administered into the patient. After reaching of optimum concentration of preparation in tumour tissue it is exposed by thermal neutron flux irradiation. Result of interaction between boron nucleus and neutron is showed below:

$$^{10}B + n \rightarrow {}^{11}B^* \rightarrow {}^{7}Li^* + {}^{4}He \rightarrow {}^{7}Li + \gamma$$

Journal of Physics: Conference Series **675** (2016) 042050 doi:10.1088/1742-6596/675/4/042050

As result of nuclear interaction ionization radiation injuring cancer cells and gamma-rays with energy of 478 keV are appeared and a number of emitted gamma-rays are proportionate to number of interactions of thermal neutrons with ^{10}B. By means of the gamma-ray detection we can estimate the absorbed dose with high accuracy during the irradiation session.

3. Gamma-ray detector based on scintillation crystal $LaBr_3$(Ce)

Gamma-ray detector consists of scintillation crystal $LaBr_3$(Ce) (25x25 mm) and photomultiplier tube Hamamatsu R6233-100 (diameter of the photocathode is 70 mm) [2]. The prototype of $LaBr_3$(Ce) single crystal with a height of 25 mm and a diameter of 25 mm was grown by the Stockbarger method at Institute of Solid State Physics Russian Academy of Sciences [3]. The crystal was placed into a hermetically sealed container with a quartz window at one of its ends. For processing electrical signals and for forming discrete code was used analog-to-digital converter, mounted in the board SBS-79 by GreenStar Company. For data acquisition and subsequent processing of results of measurements was used special software from the same company.

4. The test results of the gamma-ray detector

Tests of the gamma-ray detector were carried out using different radiation sources from a set of standard calibration gamma-ray sources. The main purposes of this experiment were determination of energy resolution using scintillation crystal $LaBr_3$(Ce) for gamma-rays with 478keV energy and estimation of capability detector to divide gamma lines with energies 478 and 511 keV. As results of experiment gamma spectra were obtained from isotopes: ^{137}Cs (figure 1), ^{60}Co, ^{133}Ba, ^{241}Am. Energy resolution for ^{137}Cs is (2.93±0.03)%.
Based on results of experiment energy resolution as function of gamma-rays energy was fitted (figure 2).

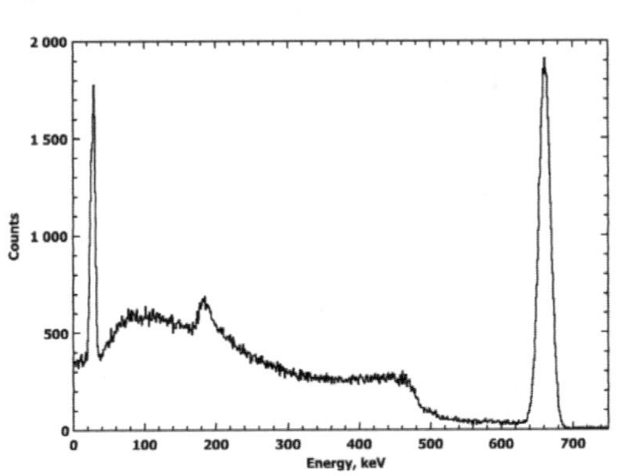

Figure 1. Gamma-ray spectrum from source ^{137}Cs.

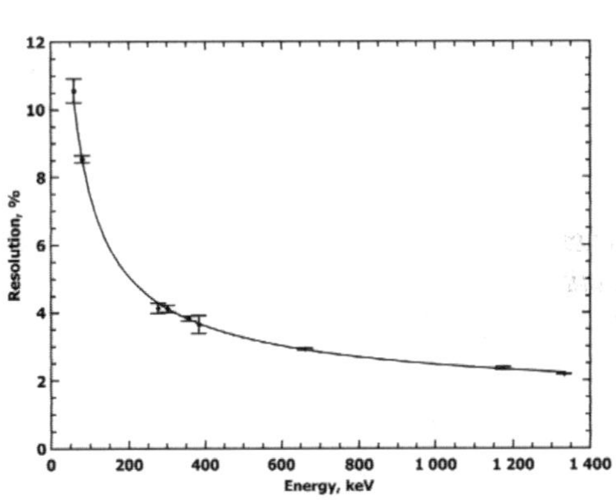

Figure 2. Energy resolution as function of gamma-rays energy.

Experimental data was approximated by the function:

$$\eta(E) = a + \frac{b}{\sqrt{E}} + \frac{c}{E},$$

where $a = 0.87 \pm 0.10$, $b = 43 \pm 3$ and c = 224±25.
Energy resolutions of the gamma-ray detector for gamma-rays with energies 478 and 511 keV are (3.33±0.03)% and (3.24±0.03) % respectively.

5. Energy resolution threshold estimation of gamma-ray detector for gamma lines 478 and 511 keV by Monte-Carlo simulation

Journal of Physics: Conference Series **675** (2016) 042050 doi:10.1088/1742-6596/675/4/042050

Monte-Carlo simulation of energy peaks from gamma-rays with energies 478 and 511 keV carried out to estimate the energy resolution threshold of gamma-ray detector required to identify overlapping gamma peaks [4]. Two Gaussian distributions were created with standard deviations with values obtained during the testing of the detector. Amplitude distributions were equal intensity for both gamma-ray lines with energies 478 and 511 keV. To estimate energy resolution threshold standard deviations is gradually increased with a step of 1.1%. After that new construction of the spectrum with updated standard deviations for Gaussian distributions was obtained (figures 3 and 4).

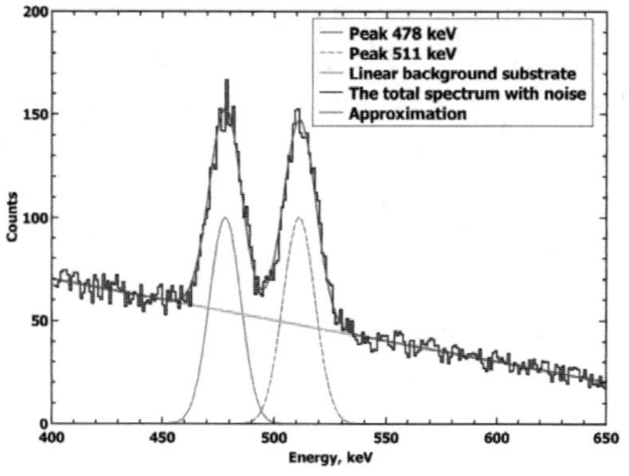

Figure 3. Energy resolutions for gamma lines 478 and 511 keV are 3.33 and 3.24%.

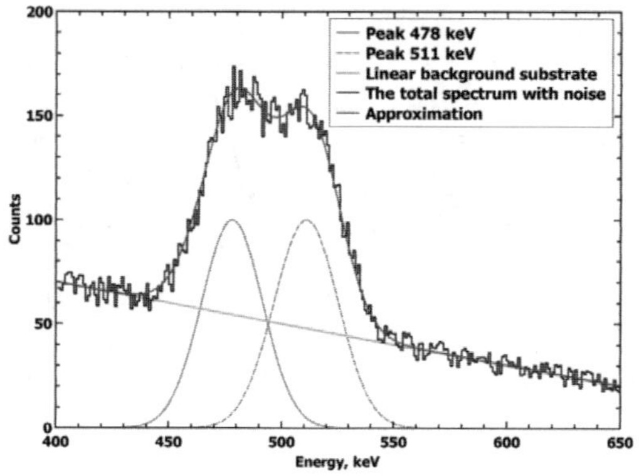

Figure 4. Energy resolutions for gamma lines 478 and 511 keV are 6.68 and 6.58%.

Further increasing of full width at half maximum for the peak (as a consequence of the deterioration of the energy resolution) was allowed to observe (figure 4) overlapping two gamma peaks. Finally visual identification without additional mathematical treatment is not possible. Threshold values of energy resolution for both gamma lines were (6.68±0.06) % and (6.58±0.06) % respectively.

6. Conclusions

According to the experiment the energy resolutions of the gamma-ray detector based on the scintillation crystal $LaBr_3(Ce)$ at energy 662 keV from gamma source ^{137}Cs are 2.93% and for energies 478 and 511 keV 3.33 and 3.24 % respectively. Energy resolution thresholds were obtained by Monte-Carlo simulation and are equal 6.68 and 6.58 % respectively. The gamma-ray detector based on scintillation crystal $LaBr_3(Ce)$ for neutron capture therapy allows to visually identify both gamma lines.

Acknowledgments

Authors wish to acknowledge the Center of Fundamental Researches and Particle Physics. This work was partially supported by MEPhI Academic Excellence Project (contract No. 02.a03.21.0005, 27.08.2013). Authors are also grateful to our colleagues from MEPhI Institute of Astrophysics for assistance with energy resolution measurements.

References

[1] Goorley J *et al.* 2002 *Med. Phys.* **29**(2) 145–56
[2] Hamamatsu Photonics K.K. Editorial Committee 2012 *Photomultipliers tube and assemblies for scintillation counting & high energy physics* (Iwata: Hamamatsu Photonics K. K. Electron Tube Division) pp 22-33 (*http://www.hamamatsu.com/resources/pdf/etd/High_energy_PMT_TPMO0007E.pdf)*
[3] Simutin A *et al.* 2013 *Instr. Exp. Techn.* **56**(5) 531–5
[4] Zamenhof R *et al.* 1996 *Int. J. Radiat. Oncol. Biol. Phys.* **35** 383–97